Encyclopaedia of
Advanced Genetic Analysis

Encyclopaedia of

Advanced Genetic Analysis

Volume 1

Editor:
Veena Kumari

AGROTECH PRESS

JAIPUR ▪ NEW DELHI

AGROTECH PRESS

Regd. Off. "Bony Residency", Gate No. 2
Opp. Tilak Public School, Vishweswariya Nagar,
Gopalpura Road, Jaipur - 302018, India
Ph: 0141 - 2761280, Telefax: 2761381
E-mail : agrotechpress@gmail.com

Delhi Office: 102, 1st Floor, Satyam House,
4327/3, Ansari Road, Darya Ganj,
New Delhi - 110002, India
Ph: 011-45652440
visit us: www.agrotechpress.com

Distributed By
KEYA INTERNATIONAL PRESS LLP
E-mail: keyainternationalpressllp@gmail.com

Published in India, 2024

ISBN : 9789384568207 (5 Vols. Set)

Typeset by Thomson Press

Printed and bound in India by Thomson Press Ltd.,
New Delhi

Contents

Preface

In modern molecular biology and genetics, the genome is the entirety of an organism's hereditary information. It is encoded either in DNA or, for many types of virus, in RNA. The genome includes both the genes and the non-coding sequences of the DNA/RNA. The term was adapted in 1920 by Hans Winkler, professor of botany at the University of Hamburg, Germany. The Oxford English Dictionary suggests the name to be a blend of the words *gene* and *chromosome*. A few related *-ome* words already existed, such as *biome* and *rhizome*, forming a vocabulary into which *genome* fits systematically.

Some organisms have multiple copies of chromosomes: diploid, triploid, tetraploid and so on. In classical genetics, in a sexually reproducing organism the gamete has half the number of chromosomes of the somatic cell and the genome is a full set of chromosomes in a gamete. In haploid organisms, including cells of bacteria, archaea, and in organelles including mitochondria and chloroplasts, or viruses, that similarly contain genes, the single or set of circular and/or linear chains of DNA, likewise constitute the genome. The term *genome* can be applied specifically to mean that stored on a complete set of nuclear DNA but can also be applied to that stored within organelles that contain their own DNA, as with the "mitochondrial genome" or the "chloroplast genome". Additionally, the genome can comprise nonchromosomal genetic elements such as viruses, plasmids, and transposable elements. Genome composition is used to describe the make up of contents of a haploid genome, which should include genome size, proportions of non-repetitive DNA and repetitive DNA in details. By comparing the genome compositions between genomes, scientists can better understand the evolutionary history of a given genome. When talking about genome composition, one should distinguish between prokaryotes and eukaryotes as the big differences on contents structure they have. In prokaryotes, most of the genome (85-90%) is non-repetitive DNA, which means coding DNA mainly forms it, while non-coding regions only take a small part. On the contrary, eukaryotes have the feature of exon-intron organization of protein coding genes; the variation of repetitive DNA content in eukaryotes is also extremely high. When refer to mammalians and plants, the major part of genome is composed by repetitive DNA. Genomes are more than the sum of an organism's genes and have traits that may be measured and studied without reference to the details of any particular genes and their products. Researchers compare traits such as *chromosome number* (karyotype), genome size, gene order, codon usage bias, and GC-content to determine what mechanisms could have produced the great variety of genomes that exist today. Duplications play a major role in shaping the genome. Duplications may range from extension of short tandem repeats, to duplication of a cluster of genes, and all the way to duplications of entire chromosomes or even entire genomes. Such duplications are probably fundamental to the creation of genetic novelty. Horizontal gene transfer is invoked to explain how there is often extreme similarity between small portions of the genomes of two organisms that are otherwise very distantly related. Horizontal gene transfer seems to be common among many microbes. Also, eukaryotic cells seem to have experienced a transfer of some genetic material from their chloroplast and mitochondrial genomes to their nuclear chromosomes.

This book explores current approaches in the study of chromosomes, cytotoxic effects, biotechnological potential of crop residues, plant disease evaluation, biocontrol methods and evaluation by molecular methods, micropropagation, protoplast fusion and molecular modelling uning bioinformatics. The book will be useful for students, teachers and researchers.

—*Editor*

1

Human Genome

The human genome is the genome of *Homo sapiens*, which is stored on 23 chromosome pairs. 22 of these are autosomal chromosome pairs, while the remaining pair is sex-determining. The haploid human genome occupies a total of just over 3 billion DNA base pairs. The Human Genome Project (HGP) produced a reference sequence of the euchromatic human genome, which is used worldwide in biomedical sciences.

The haploid human genome contains ca. 23,000 protein-coding genes, far fewer than had been expected before its sequencing. In fact, only about 1.5% of the genome codes for proteins, while the rest consists of non-coding RNA genes, regulatory sequences, introns, and noncoding DNA (once known as "junk DNA").

Features

Genes

There are estimated to be between 20,000 and 25,000 human protein-coding genes. The estimate of the number of human genes has been repeatedly revised down as genome sequence quality and gene finding methods have improved. Earlier predictions estimated that human cells have as much as 200,000 genes. Surprisingly, the number of human genes seems to be less than a factor of two greater than that of many much simpler organisms, such as the roundworm and the fruit fly. However, human cells make extensive use of alternative splicing to produce several different proteins from a single gene, and the human proteome is thought to be much larger than those of the aforementioned organisms. Besides, most human genes have multiple exons, and human introns are frequently much longer than the flanking exons. Human genes are distributed unevenly across the chromosomes. Each chromosome contains various gene-rich and gene-poor regions, which seem to be correlated with chromosome bands and GC-content. The significance of these nonrandom patterns of gene density is not well understood. In addition to protein coding genes, the human genome contains thousands of RNA genes, including tRNA, ribosomal RNA, microRNA, and other non-coding RNA genes.

Regulatory Sequences

The human genome has many different regulatory sequences which are crucial to controlling gene expression. These are typically short sequences that appear near or within genes. A systematic understanding of these regulatory sequences and how they together act as a gene regulatory network is only beginning to emerge from computational, high-throughput expression and comparative genomics studies.

Some types of non-coding DNA are genetic "switches" that do not encode proteins, but do regulate when and where genes are expressed. Identification of regulatory sequences relies in part on evolutionary conservation. The evolutionary branch between the primates and mouse, for example, occurred 70–90 million years ago. So computer comparisons of gene sequences that identify conserved non-coding sequences will be an indication of their importance in duties such as gene regulation.

Another comparative genomic approach to locating regulatory sequences in humans is the gene sequencing of the puffer fish. These vertebrates have essentially the same genes and regulatory gene sequences as humans, but with only one-eighth the noncoding DNA. The compact DNA sequence of the puffer fish makes it much easier to locate the regulatory genes.

Other DNA

Protein-coding sequences (specifically, coding exons) comprise less than 1.5% of the human genome. Aside from genes and known regulatory sequences, the human genome contains vast regions of DNA the function of which, if any, remains unknown. These regions in fact comprise the vast majority, by some estimates 97%, of the human genome size.

Repeat Elements

- Tandem repeats;
  - Satellite DNA
  - Minisatellite
  - Microsatellite.

- Interspersed repeats;
  - — SINEs
  - — LINEs.

Transposons

- Retrotransposons;
  - — [[Retrotransposon#LTR retrotransposons | LTR]]
    - Ty1-copia
    - Ty3-gypsy.
  - — Non-LTR
    - SINEs
    - LINEs
- DNA Transposons.

Noncoding DNA

There is also a large amount of sequence that does not fall under any known classification. Much of this sequence may be an evolutionary artifact that serves no present-day purpose, and these regions are collectively referred to as noncoding DNA. These regions were once referred to as "junk" DNA; however, there are a variety of emerging indications that many sequences within are likely to function in ways that are not fully understood.

Recent experiments using microarrays have revealed that a substantial fraction of non-genic DNA is in fact transcribed into RNA, which leads to the possibility that the resulting transcripts may have some unknown function.

Also, the evolutionary conservation across the mammalian genomes of much more sequence than can be explained by protein-coding regions indicates that many, and perhaps most, functional elements in the genome remain unknown.

The investigation of the vast quantity of sequence information in the human genome whose function remains unknown is currently a major avenue of scientific inquiry. Meanwhile, considering the global genome DNA information as a whole could provide new ways to understand a possible global level function of non coding DNA.

Information Content

The 2.9 billion base pairs of the haploid human genome correspond to about 691.4 megabytes of data, since every base pair can be coded by 2 bits. The entropy rate of the genome differs significantly between coding and non-coding sequences. It is close to the maximum of 2 bits per base pair for the coding sequences (about 45 million base pairs), but less for the non-coding parts. It ranges between 1.5 and 1.9 bits per base pair for the individual chromosome, except for the Y chromosome, which has an entropy rate below 0.9 bits per base pair.

Sequencing

DNA sequencing determines the order of the nucleotide bases in a genome.

Composite

The Human Genome Project and a parallel project by Celera Genomics each produced and published a haploid human genome sequence, both of which were a composite of the DNA sequence of several individuals.

Personal

A personal genome sequence is a complete sequencing of the chemical base pairs that make up the DNA of a single person. Because medical treatments have different effects on different people because of genetic variations such as single-nucleotide polymorphisms (SNPs), the analysis of personal genomes may lead to personalized medical treatment based on individual genotypes. The completion of the fifth such map was announced in December 2008.

The genome mapped was that of a Korean researcher Seong-Jin Kim. Genome maps had previously been completed for Craig Venter of the U.S. in 2007, James Watson of the U.S. in April 2008, and Yang Huanming of China in November 2008 and Dan Stoicescu in January 2008. Personal genomes had not been sequenced in the Human Genome Project to protect the identity of volunteers who provided DNA samples.

That sequence was derived from the DNA of several volunteers from a diverse population. Another distinction is that the HGP sequence is haploid, however, the sequence maps for Venter and Watson for example are diploid, representing both sets of chromosomes.

Kim's genome had 1.58 million SNPs that had never been reported before and indicates that six out of 10,000 DNA bases are unique to Koreans. Kim's sequence map can be used to assist in building a standard Korean genome, which can then be used to compare the genomes of other Korean individuals for personalized medical treatments.

Mapping

Whereas a genome sequence lists the order of every DNA base in a genome, a genome map identifies the landmarks. A genome map is less detailed than a genome sequence and aids in navigating around the genome.

Variation

An example of a variation map is the HapMap being developed by the International HapMap Project. The HapMap is a haplotype map of the human genome, "which will describe the common patterns of human DNA sequence variation." It catalogs the patterns of small-scale variations in the genome that involve single DNA letters, or bases.

Researchers published the first sequence-based map of large-scale structural variation across the human genome in the journal *Nature* in May 2008. Large-scale structural variations are differences in the genome among people that range from a few thousand to a few million DNA bases; some are gains or losses of stretches of genome sequence and others appear as re-arrangements of stretches of sequence. These variations include differences in the number of copies individuals have of a particular gene, deletions, translocations and inversions.

Variation

Most studies of human genetic variation have focused on single-nucleotide polymorphisms (SNPs), which are substitutions in individual bases along a chromosome. Most analyses estimate that SNPs occur on average somewhere between every 1 in 100 and 1 in 300 base pairs in the euchromatic human genome, although they do not occur at a uniform density.

Thus follows the popular statement that "we are all, regardless of race, genetically 99.9% the same", although this would be somewhat qualified by most geneticists. For example, a much larger fraction of the genome is now thought to be involved in copy number variation. A large-scale collaborative effort to catalogue SNP variations in the human genome is being undertaken by the International HapMap Project.

The genomic loci and length of certain types of small repetitive sequences are highly variable from person to person, which is the basis of DNA fingerprinting and DNA paternity testing technologies. The heterochromatic portions of the human genome, which total several hundred million base pairs, are also thought to be quite variable within the human population (they are so repetitive and so long that they cannot be accurately sequenced with current technology). These regions contain few genes, and it is unclear whether any significant phenotypic effect results from typical variation in repeats or heterochromatin.

Most gross genomic mutations in gamete germ cells probably result in inviable embryos; however, a number of human diseases are related to large-scale genomic abnormalities. Down syndrome, Turner Syndrome, and a number of other diseases result from nondisjunction of entire chromosomes. Cancer cells frequently have aneuploidy of chromosomes and chromosome arms, although a cause and effect relationship between aneuploidy and cancer has not been established.

Genetic Disorders

Most aspects of human biology involve both genetic (inherited) and non-genetic (environmental) factors. Some inherited variation influences aspects of our biology that are not medical in nature (height, eye colour, ability to taste or smell certain compounds, etc). Moreover, some genetic disorders only cause disease in combination with the appropriate environmental factors (such as diet). With these caveats, genetic disorders may be described as clinically defined diseases caused by genomic DNA sequence variation.

In the most straightforward cases, the disorder can be associated with variation in a single gene. For example, cystic fibrosis is caused by mutations in the CFTR gene, and is the most common recessive disorder in caucasian populations with over 1,300 different mutations known. Disease-causing mutations in specific genes are usually severe in terms of gene function, and are fortunately rare, thus genetic disorders are similarly individually rare. However, since there are many genes that can vary to cause genetic disorders, in aggregate they comprise a significant component of known medical conditions, especially in pediatric medicine. Molecularly characterized genetic disorders are those for which the underlying causal gene has been identified, currently there are approximately 2,200 such disorders annotated in the OMIM database.

Studies of genetic disorders are often performed by means of family-based studies. In some instances population based approaches are employed, particularly in the case of so-called founder populations such as those in Finland, French-Canada, Utah, Sardinia, etc. Diagnosis and treatment of genetic disorders are usually performed by a geneticist-physician trained in clinical/medical genetics. The results of the Human Genome Project are likely to provide increased availability of genetic testing for gene-related disorders, and eventually improved treatment. Parents can be screened for hereditary conditions and counselled on the consequences, the probability it will be inherited, and how to avoid or ameliorate it in their offspring. As noted above, there are many different kinds of DNA sequence variation, ranging from complete extra or missing chromosomes down to single nucleotide changes.

It is generally presumed that much naturally occurring genetic variation in human populations is phenotypically neutral, i.e. has little or no detectable effect on the physiology of the individual (although there may be fractional differences in fitness defined over evolutionary time frames). Genetic disorders can be caused by any or all known types of sequence variation. To molecularly characterize a new genetic disorder, it is necessary to establish a causal link between a particular genomic sequence variant and the clinical disease under investigation. Such studies constitute the realm of human molecular genetics. With the advent of the Human Genome and International HapMap Project, it has become feasible to explore subtle genetic influences on many common disease conditions such as diabetes, asthma, migraine, schizophrenia, etc.

Although some causal links have been made between genomic sequence variants in particular genes and some of these diseases, often with much publicity in the general media, these are usually not considered to be genetic disorders *per se* as their causes are complex, involving many different genetic and environmental factors. Thus there may be disagreement in particular cases whether a specific medical condition should be termed a genetic disorder.

Evolution

Comparative genomics studies of mammalian genomes suggest that approximately 5% of the human genome has been conserved by evolution since the divergence of extant lineages approximately 200 million years ago, containing the vast majority of genes. Intriguingly, since genes and known regulatory sequences probably comprise less than 2% of the genome, this suggests that there may be more unknown functional sequence than known functional sequence. A smaller, yet substantial, fraction of human genes seem to be shared among most known vertebrates. The published chimpanzee genome differs from that of the human genome by 1.23% in direct sequence comparisons.

Around 20% of this figure is accounted for by variation within each species, leaving only ~1.06% consistent sequence divergence between humans and chimps at shared genes. This nucleotide by nucleotide difference is dwarfed, however, by the portion of each genome that is not shared, including around 6% of functional genes that are unique to either humans or chimps.

In other words, the considerable observable differences between humans and chimps may be due as much or more to genome level variation in the number, function and expression of genes rather than DNA sequence changes in shared genes. On average, a typical human protein-coding gene differs from its chimpanzee ortholog by only two amino acid substitutions; nearly one third of human genes have exactly the same protein translation as their chimpanzee orthologs. A major difference between the two genomes is human chromosome 2, which is equivalent to a fusion product of chimpanzee chromosomes 12 and 13 (later renamed to chromosomes 2A and 2B, respectively).

Humans have undergone an extraordinary loss of olfactory receptor genes during our recent evolution, which explains our relatively crude sense of smell compared to most other mammals. Evolutionary evidence suggests that the emergence of colour vision in humans and several other primate species has diminished the need for the sense of smell.

Chimpanzee Genome Project

The Chimpanzee Genome Project is an effort to determine the DNA sequence of the Chimpanzee genome. It is expected that by comparing the genomes of humans and other apes, it will be possible to better understand what makes humans distinct from other species.

Chimp-human chromosome differences. The major structural difference is that human chromosome 2 (green colour code) was derived from two smaller chromosomes that are found in other great apes. Parts of human chromosome 2 are scattered among parts of several cat and rat chromosomes in these species that are more distantly related to humans (more ancient common ancestors; about 85 million years since the human/rodent common ancestor: Entrez Pubmed 12552136)

Identification of Human Chromsome

The recognition of chromosome bonds has provided new morphologic features for use in the identification of human chromosomes and chromosomal parts. In addition to the telomere, centromere, and arms, special landmarks (*i.e.*, well-defined bands) were selected to subdivides the arms into regions. The regions are designaled 1,2,3, and 4, moving from the centromere towards the telomere. Using the nomenclater established at the Paris conference is 1971, any particular band or segment of a chromosome can be identified according to the following parameters:

- Chromosome number,
- An arm symbol (p=short arm, q=long),
- The region and band numbers.

For example, 6p23 indicates band number 3 in region 2 of the short arm of chromosome 6. With the use of the banding techniques, particularly the G and Q banding some 320 bands per haploid set can be identified at metaphase. However, if mitosis of followed through late prophase and prometaphase, up to 1256 bands may be recognized.

Determination of Sex

The well-known fact that male and female individuals are statistically found in about equal members suggests that sex is determined by a hereditary mechanism There is physiology and cytologic proof that sex is determined as soon as the egg is fertilized and that it depends on the gametes. Among the physiology evidence is the finding that identical twins—which originate from a single zygote—are always of the same sex. Cytologic evidence was first obtained by McClug, who demonstrated that the karyorpe of a cell is composed not only of common chromosomes (*autosomes*) but also of one or more special chromosomes that are distinguished from the autosomes by morphologic characteristics and behaviour: These were called *accessory chromosomes, allosomes, heterochromsomes, of sex chromosomes.*

The majority of organisms have a pair of sex chromsome, which in the course of evolution, have been specialized for sex determination. One of the sexes has a pair of identical sex chromosomes (XX); the other may have a single sex chromosome, which may be unpaired (XO) or paired with a Y chromosome (XY) The XY pair is also called *hetermorphic* because of the different morphology of the chromosomes.

In the human the garments are not identical with respect to the sex chromosomes. The male is heterozygous (XY) and produces two types of spermatozoa in similar proportions, *i.e.*, one carries the X chromosome, and the other the Y. the female, being homozygous (XX), produces only one kind of gamete (the ovum). Therefore only two combinations of gametes are possible at fertilization and the result is 50 per cent males and 50 per cent females. This type of sex determination is found in all mammals and in certain insects, such as *Drosophila.* In other vertebrates and certain invertebrates, the female is heterogametic, and the male, homogametic. In some animals the sex chromosomes cannot be distinguished from the other. In such cases the sexdetermining genes are confined to a region in a pair of chromosomes.

In the human the Y chromosome contains an essential testis-determining factor and determines the male sex. The genes involved are situated in a region near the centromere. An individual with XO (lacking the Y chromosome) will resemble a female but will have atrophied ovaries (Turner's syndrome).

We have seen earlier that the X and Y chromosomes differ in size and morphology. In the XY pair there is a *homologous region* in which recombination may take place during meiosis and a *differential region* that is unpaired.

Interphase Nuclei

The X chromatin can be bound as a small body in different positions within the nucleus. For example, in nerve cells it may be near the nucleolus in the nucleoplasm, or near the nuclear envelope. In cells of the oral mucosa it is generally attached to the envelope, and in neutrophil leukocytes it may appear as a small rod called the *drumstick.*

The study of sex chromatin has a wide field of medical applications and offers the possibility of relating the origin of certain congenital diseases to chromosome anomalies. Among these application is the diagnosis of sex in intersexual states in postnatal and even in fetal life. The frequency with which sex chromation can be detected in the female varies from tissue to tissue. In nervous tissue the frequency may be 85 per cent, whereas in whole mounts of amniotic or chorionic epithelium it may be as high as 96 per cent. In oral smears the frequency may vary between 20 and 50 per cent in normal females. The differential behaviour of the two X chromosome in the female led Lyon to the so-called *inactive* X *hypothesis* is genetically active; the X undergoing heteropyknosis may be either of maternal or paternal origin, and the decision by which X becomes inactive is taken at random; the inactivation occures early in embryonic life and remains fixed.

It is now admitted that only a part of the X chromosome condenses into a Barr corpuscle; This condensed chromosome is said to contain *facultative heterochromatin,* as opposed to the *constitutive heterochromatin* found in other chromosome takes place even in 3X and 4X individuals, in which only one X remains uncondensed.

The inactivation starts in the human in the late blastocyst stage on about the sixteenth day of embryonic life; the inactivated X chromosome then remains heterochromatic in the somatic cells.

The fact that X inactivation occurs at random has been demonstrated in human diseases linked to the X chromosome. The *Lesch-Nyhansyndrome,* in which a deficiency of one enzyme of the purine metabolism (*i.e.*, hypoxanthine-guanine phosphoribosyl transferase) produces mental retardation and increase uric acid levels, results from a recessive mutation in the X chromsome. When fibroblasts of heterozygous women are cultured in vitro, two types of cell clones are obtained.

Half the clones contain the enzyme, whereas the other half lack the enzyme. During the cell cycle the heterochromatic X chromosome is characterized by its late replication. This can be clearly observed by the use of 5-bromodeoxyuridine, followed by the fluorescent dye. Hoechst 33258 or the Giemsa stain.

The above-mentioned fluorescence method for the study of Q banding led to the demonstration of that a large portion of the Y chromosome is heterochromatic and appears, at interphase, as a strongly fluorescent body, the so- called Y *chromatin*. While the X chromatin is 1.0 to 1.2 μm in diameter, the Y chromatin is only about 0.7-μm in diameter and generally is not attached to the nuclear envelope.

The number of Y chromation bodies is identical to the number of Y chromosome. For example, individuals with XYY have two Y chromatin bodies. The human Y chromosome consists of two segments, one that does not fluoresce and is genetically active. and a second that fluoresces and is genetically inactive. It is this latter segment that may be polymorphic, showing wide variations in size between individuals. Among mammal, only man and the gorilla share this fluorescent region of the Y chromosome, suggesting that it is of recent evolutionalry origin. This region corresponds to a highly repetitive (satellite) DNA, which is specific fore the Y chromosome and probably has no genetic function.

Although the primary determination of sex is made at fertilization the embryo acquires its definite sex characteristics by a more complex mechanism. An epigenetic factor (*i.e.*, hormonal) may assume control of the genetic determination during development, thereby changing the phenotypic direction of sex. Among vertebrates a condition of bisexuality may exist (*e.g.*, the coexistence of structures of the functional sex, together with primordia of the heterologous sex). For example, male amphibians have a redimentary ovary (Bidder's organ) and vestigial oviducts.

In the human embryo until the sixth week the gonads and the primordia of the urogenital tract are identical in males and females. At this time the gonad has already has already been invaded by the primary XX or XY germ cells. At this point, a gene (or set of genes) present in the y chromosome causes the undifferentiated gonad to differentiate into a testis, and the absence of this gene allows the gonad to become an ovary. The development into a testis, and the absence of this gene allows the gonad to become an ovary.

The development into a testis starts as soon as the *gonocytes* (*i.e.*, primordial germ cells) from the yolk sac have finished their migration into the *gonadal ridge*. Gonocytes of the male (XY) migrate deeper into the gonadal blastema, and female gonocytes (XX) remain at the periphery, forming a thick cortical layer.

H-Y Antigen

It is known that the Y chromosome directs the organogenesis of the testis by inducing the production of a protein (or membrane and determines, by cell recognition, the formation of the seminiferous tubule. It has been proposed that a Y-linked determining gene of mammals produces that a Y-linked determining gene of mammals produces this *H-T antigen,* which is found in various cell types in males but is absent in females.

This antigen is also a histocompatibility factor (H) that is found, for example, in the skin of male mice and that determines the rejection of female skin transplants.

Action of Testosterone

The gonad differentiates into a definite testicle at the seventh week, whereas the female gonad differentiates between the eighth and ninth weeks of development, At the time of differentiation an important epigenetic factor is the production of androgens by somatic cells in the embryonic male gonad. In mammals administration of *testosterone* (the male hormone) to the mother produces in the fetus a shift in the differentiation of XX genitalia to a male type, producting what is called *feminine masculine pseudohermaphroditism.* This hormone acting locally accelerates the development of the testis, whereas in the female the absence of the hormone permits the slower development characteristic of the ovary.

Chromosomal Abnormalities

The study of chromosome abnormalities in man, like that of normal cytogenetics, can be divided into periods before and after the development of the banding techniques. In the earlier period it was found that 0.5 per cent of all newborns have gross chromosomal abnormalities, about half of which are in the sex chromosomes and the rest of which are in the autosomes. These abnormalities generally consist of aneuploidy, such as monosomy or trisomy, Structural aberrations, such as translocation, deficiency, duplication, and other more complex alterations, have also been observed;

Diagnosis of chromosomal aberrations can be made even before a child is born. This is done by removing amitotic fluid by puncturing the mother's abdomen, a procedure called *amniocentesis. which* does not entail risks to the fetus. The amniotic fluid contains cells shed from the fetus and which can be cultured.

This technique is useful for karyotyping the babies off mothers with a history of chromosomal abnormality, or children of those who are carriers of a balanced translocation, in whom that risk of conceiving an affected child is high. In addition, this technique permits the *antenatal* determination of the fetal sex, which may be of importance in certain sex-linked diseases.

The improved resolution provided by the banding techniques has not only confirmed the previously observed abnormalities but has also led to the discovery of over 30 new chromosomal syndromes and several malignancies defects.

In aneuploidy the genetic message contained in each chromosome is maintained intact. The alteration is quantitative; it resides in a disequilibrium established by the excess (trisomy) or deficit (monosomy) in the amount of genetic material, Such a dosage effect may be dangerous to the organism and may produce severe anatomical and functional anomalies (*i.e.*, malformations). It will be shown later that trisomy produces changes that are characteristic for each chromosome present in excess; however, in all instances of trisomy the tendency is towards involution of the nervous system, resulting in a more or less severe mental defect.

Another important consequence of aneuploidy is spontaneous abortion. A large proportion of aborted fetuses show trisomy in one of the larger chromosomes in the karyotype. Aneuploidy of one the larger chromosomes is generally lethal, being more severe than an equivalent alteration in a smaller chromosome. Malformation mental retardation, sterility, and spontaneously abortion operate as strong selective mechanisms tending to eliminate from the general population those individuals carrying deleterious genetic imbalances.

Aneuploidy originates by the mechanism called *Non-disjunction.* The immediate cause of non-disjuction is the lagging, during anaphase, of one sister chromatid which during telophase remains in one of the cells together with the other sister chromatid. This change gives rise to a cell line that lacks one chromosome or has one chromosome in excess for that pair (monosomy and trisomy).

Non-disjunction may occur during *meiosis, giving* rise to an aneuploid *ovum. This* when fertilized by a normal spermatozoon, results in a zygote and later an organism in which all cells are aneuploid. In some cases the alteration may be in the male gamete or in both, thus producing more complex types of abnormalities. If non-disjunction occurs in a *mitotic division,* which precedes the formation of germ cells, the effects are similar to those of *meiotic non-disjunction.* If mitotic non-disjunction occurs during embryonic development, a *mosaic* of different cell lines is produced.

We mentioned earlier the various structural aberrations that are found in human chromosomes (*i.e.*, translocation, deficiency, and duplication). The most common of these is the so-called *reciprocal translocation,* in which chromosome segments are exchanged between non-homologous chromosomes. Some of these translocations may result in a chromosome having a length and configuration that may be easily identified. However, in other cases the apparent morphology changes very little. Banding techniques are of fundamental importance in detecting such cases. The most important aberrations or as follows.

Klinefelter's Syndrome

Most affected individuals are practically normal except for minor phenotypic anomalies. They have small testes, frequent gynecomastia (enlarged breasts), tendency to tallness, obesity and underdevelopment of secondary sex characteristics. Spermatogenesis does not occur, thereby resulting in complete infertility. These individuals have a positive sex chromatin and 47 chromosomes (44 autosomes + XXY). Males with 48 chromosome (44 autosome+XXXY) and two Barr corpuscles have also been described.

These individuals have features of klinefelter's syndrome and are mentally retarded. Persons with 49 chromosomes (44 autosomes+XXXXY) have also been reported. They display extensive skeletal anomalies, extreme hypogenitalism, and severe mental deficiency. These persons have three X chromation bodies.

XYY Syndrome

Males having two Y chromosomes have been identified in the past in maximum security institutions. It was proposed that such individuals had a strong tendency to ward anti-social behaviour and aggression. More recently, XYY individuals have been found in the normal population in a proportion of 1 in `650 male infants, suggesting that the correlation with violence may not be a strong as was previously thought.

Gonadal Dysgenesis

Patients with Turner's syndrome usually have a female appearance with short stature, webbed neck (folds of skin extended from the mastoid to the shoulders) and generally infantile internal sexual organs. The ovary does not develop and shows complete absence of germ cells. As result of this ovarian dysgenesis, menstruation does not occur, and secondary sexual characteristics do not develop. The karyotype shows 45 chromosomes (44 autosomes +X), and there is no sexchromation. In this syndrome there is little difference in the gonads up to the third moth of gestation; the ovaries up to the third month of gestation; the ovaries contain approximately the normal number of germ cells (about 2 or 3×10^6 cells) later on there is a rapid atresia of the germ cells, leading to their virtually disappearance after puberty. It is probable that the lack of one X chromosomes determines the progressive ovarian atresia by failure of primordial follicles.

Females with X Polysomy

Triplo-X constitution (47 chromosome; 44 autosomes + XXX) was detected in phenotypically near-normal females. A number of these females are mentally subnormal or psychotic and some of them menstruate. The sex chromatin bodies are found in cells from these women. A few severely retarded patients with three corpuscles of sex chromatin and 48 chromosomes (44 autosomes + XXXX) have been found. These persons are also called metafermales Mixed Chromosomal Aberrations. Kinefelter's syndrome may be found combined with mongolism. Such a person has 48 chromosomes: 45 autosomes (including a trisomic pair 21) +XXY.

Mosaics

Sometimes chromsomal aberrations are produced during development of the embryo. One interesing example is induced by the loss of one Y chromosomes in the first division of the zygote. This may result in twins, of which one has normal male characters and the other has Tuner's syndrome. Sometimes chromosomal aberrations due to non-disjunction are produced during development of the embryo. These individuals possess different chromosomal complements from cell to cell within the same tissue or between tissues, depending on the embryonic stage at which non-disjunction occurred.

Hermaphroditism

A true hermaphrodite is an individual who has both ovarian and testicular tissue. The two types of gonadal tissue may be separated or in close proximity (ovotestis). Frequently, true hermaphrodites haves a 46, XX chromosome complement and are chromation-positive. Other may have a 46, XY mosaicism has also been detected in this condition.

Mongolism

Before the development of the banding techniques about 12 classic syndromes had been described, the majority of which were caused by an extra chromosome or loss of a chromsome segment. Except for the X monosomy (Turner's syndrome), any other monosomy is believed to be inviable (*i.e.*, leads to death of the embryo). This also holds true for all trisomies other than those of chromosomes 13, 18, or 21.

21-Monosomy

Complete deletion of one of the chromosomes in pair 21 is apparently lethal, but there is a syndrome in which a large part of cone is lacking. Children with this condition have a morphologic aspect which is, to some extent, the opposite of mongolism. The nose is prominent, the distance between eyes shorter than normal, and the ears are large and the muscles contracted, It seems that in trisomy and monosomy of the 21st, the phenotype shifts to one or the other side of normal.

18-Trisomy

In this case the child is small and weak, the head is laterally flattened, and the helix of the ear scarcely developed. The hands are short and the digital imprints are rather simple. These children are very retarded mentally and usually die before one year of age.

18-Monosomy

This is the opposite syndrome is which a partial deletion of one chromosome of the pair occures. The ears are voluminous, the fingers long, and the digital imprints are complex and convoluted.

21-Trisomy Mongolism

Among the most important autosomal aberrations is *mongolism,* which is characterized by multiple malformations, mental retardation, and markedly defective development of the central nervous system. It was discovered that the mongoloid has an extra chromosome. Pair 21 is trisomic instead of normal. This aberration probably originates from non-disjunction of pair 21 during meiosis.

The extra chromosome of pair 21 in some cases may become attached to another autosome (translocation), usually to pair 22. The phenotype of a mongoloid is recognizable at birth in most cases. The face of such a patient has a special moon-like aspect, with oblique palpebral fissures, increased separation between the eyes, and a skin fold (epicanthus) at the inner part of the eyes. The nose is flattened the ears and the tongue protrudes.

Mongolism is the most common congenital disease and is present in more than 0.1 per cent of births. Its frequency increases as the mother's age sporadic and in general there is no recurrence in the family. However, in the rarer cases of mongolism by translocation, the disease may affect siblings and may appear in successive generation. Fortunately this "translocation trisomy" represents only 3 to 4 per cent of all call cases of mongolism. In this type there is no change in frequency with the age of the mother, and when the aberration is properly determined by karyotype analysis, the parents should be warned of a repetition of this defect.

13-Trisomy

Multiple and severe body malformations, as well as profound mental deficiency, are characteristic features. The head is small and the eyes are often small, or absent. are-lip, cleft palate, and malformations of the brain are frequent. The internal organs are also severely malformed and in most cases death occurs soon after birth.

One of the most remarkable phenotypes associated with a chromosome structural change is the condition known as *cri du chat* syndrome. The affected baby has a strange mewing cry, multiple malfomations and mental retardation associated with partial deletion of the short arm of chromosome 5. Another type of patient showing facial alteration and skeletal and ophthalmologic abnormalities along with profound mental retadation, is the carrier of a deletion of the long arm of chromosome 18.

The use of the banding techniques has permitted the detection of more than 30 new chromosomal syndromes which result from a partial trisomy or deletion from each of almost all the chromosome. Such alterations have been bound in children with mental retardation or minor congenital defects and also in some with normal intelligence and minor anomalies which were earlier considered as deviations of normality.

The technical advances have permitted the identification of chromosome defects even when there are few clinical symptoms furthermore, these methods have allowed the study of cases of balanced chromosomal translocations, which, as carriers of this type of syndrome, are important to detect. In such cases transmission can be prevented by proper counselling and prenatal diagnosis.

Chromosomal Aberrations

The development of the banding techniques has also resulted in the demonstration of specific chromosomal defects in certain types of tymors. Here only a few examples will be given. The most characteristic of these is the association of *chronic myelogenous leukemia* with the so-called *Philadelphia chromosome* (Ph 1).

This involves a balanced translocation between chromosome 9 and 22. The Ph 1 chromosome is acquired with the disease and is not found in cultured fifbroblasts or in the idential twins of affected patients. In a *retinoblastoma* (*i.e.*, a cancer of the retina) a deletion in the long arm of chromosomes 13 has been identified. Other abnormalities have been observed in lymphomas, meingiomas, and many other tumors. Cancer cells are usually associated with severe chromosomal abnormalities such as polyploidy and aneuploidy.

Starting the Chimpanzee Genome Project

Human and chimpanzee chromosomes are very similar. The primary difference is that humans have one fewer pair of chromosomes than do other great apes. Humans have 23 pairs of chromosomes and other great apes have 24 pairs of chromosomes. In the human evolutionary lineage, two ancestral ape chromosomes fused at their telomeres producing human chromosome 2. There are nine other major chromosomal differences between chimpanzees and humans: chromosome segment inversions on human chromosomes 1, 4, 5, 9, 12, 15, 16, 17, and 18.

After the completion of the Human genome project, a Common Chimpanzee genome project was initiated. In December 2003, a preliminary analysis of 7600 genes shared between the two genomes confirmed that certain genes such as the forkhead-box P2 transcription factor, which is involved in speech development, are different in the human lineage. Several genes involved in hearing were also found to have changed during human evolution, suggesting selection involving human language-related Behaviour. Differences between individual humans and Common Chimpanzees are estimated to be about 10 times the typical difference between pairs of humans.

Draft Genome Sequence of the Common Chimpanzee

Analysis of the genome was published in *Nature* on September 1, 2005, in an article produced by the Chimpanzee Sequencing and Analysis Consortium, a group of scientists which is supported in part by the National Human Genome Research Institute, one of the National Institutes of Health.

The article marked the completion of the draft genome sequence. A database now exists containing the genetic differences between human and chimpanzee genes, with about thirty-five million single-nucleotide changes, five million insertion/deletion events, and various chromosomal rearrangements. Gene duplications account for most of the sequence differences between humans and chimps. Single-base-pair substitutions account for about half as much genetic change as does gene duplication.

Typical human and chimp homologs of proteins differ in only an average of two amino acids. About 30 percent of all human proteins are identical in sequence to the corresponding chimp protein. As mentioned above, gene duplications are a major source of differences between human and chimp genetic material, with about 2.7 percent of the genome now representing differences having been produced by gene duplications or deletions during approximately 6 million years since humans and chimps diverged from their common evolutionary ancestor. The comparable variation within human populations is 0.5 percent.

About 600 genes have been identified that may have been undergoing strong positive selection in the human and chimp lineages; many of these genes are involved in immune system defence against microbial disease or are targeted receptors of pathogenic microorganisms (example: Glycophorin C and *Plasmodium falciparum*). By comparing human and chimp genes to the genes of other mammals, it has been found that genes coding for transcription factors, such as forkhead-box P2 (FOXP2), have often evolved faster in the human relative to chimp; relatively small changes in these genes may account for the morphological differences between humans and chimps. A set of 348 transcription factor genes code for proteins with an average of about 50 percent more amino acid changes in the human lineage than in the chimp lineage.

Six human chromosomal regions were found that may have been under particularly strong and coordinated selection during the past 250,000 years. These regions contain at least one marker allele that seems unique to the human lineage while the entire chromosomal region shows lower than normal genetic variation. This pattern suggests that one or a few strongly selected genes in the chromosome region may have been preventing the random accumulation of neutral changes in other nearby genes.

One such region on chromosome 7 contains the FOXP2 gene (mentioned above) and this region also includes the Cystic fibrosis transmembrane conductance regulator (CFTR) gene, which is important for ion transport in tissues such as the salt-secreting epithelium of sweat glands. Human mutations in the CFTR gene might be selected for as a way to survive cholera.

Another such region on chromosome 4 may contain elements regulating the expression of a nearby protocadherin gene that may be important for brain development and function. Although changes in expression of genes that are expressed in the brain tend to be less than for other organs (such as liver) on average, gene expression changes in the brain have been more dramatic in the human lineage than in the chimp lineage. This is consistent with the dramatic divergence of the unique pattern of human brain development seen in the human lineage compared to the ancestral great ape pattern. The protocadherin-beta gene cluster on chromosome 5 also shows evidence of possible positive selection.

Results from the human and chimp genome analyses should help in understanding some human diseases. Humans appear to have lost a functional caspase-12 gene, which in other primates codes for an enzyme that may protect against Alzheimer's disease. Figures published in *Nature* on September 1, 2005, in an article produced by the Chimpanzee Sequencing and Analysis Consortium, show that 24% of the chimpanzee genome does not align with the human genome. There are 3% further alignment gaps, 1.23% SNP differences, and 2.7% copy number variations totaling at least 30% differences between chimpanzee and *Homo sapiens* genomes.

Genes of the Chromosome 2 Fusion Site

The results of the chimpanzee genome project suggest that when ancestral chromosomes 2A and 2B fused to produce human chromosome 2, no genes were lost from the fused ends of 2A and 2B. At the site of fusion, there are approximately 150,000 base pairs of sequence not found in chimpanzee chromosomes 2A and 2B. Additional linked copies of the PGML/FOXD/CBWD genes exist elsewhere in the human genome, particularly near the p end of chromosome 9.

This suggests that a copy of these genes may have been added to the end of the ancestral 2A or 2B prior to the fusion event. It remains to be determined if these inserted genes confer a selective advantage.

- PGML. The phosphoglucomutase-like gene of human chromosome 2. This gene is incomplete and may not produce a functional transcript.
- FOXD. The forkhead box D4-like gene is an example of an intronless gene. The function of this gene is not known, but it may code for a transcription control protein.
- CBWD. Cobalamin synthetase is a bacterial enzyme that makes vitamin B_{12}. In the distant past, a common ancestor to mice and apes incorporated a copy of a cobalamin synthetase gene. Humans are unusual in that they have several copies of cobalamin synthetase-like genes, including the one on chromosome 2. It remains to be determined what the function of these human cobalamin synthetase-like genes is. If these genes are involved in vitamin B_{12} metabolism, this could be relevant to human evolution. A major change in human development is greater post-natal

brain growth than is observed in other apes. Vitamin B_{12} is important for brain development, and vitamin B_{12} deficiency during brain development results in severe neurological defects in human children.

- CXYorf1-like protein. Several transcripts of unknown function corresponding to this region have been isolated. This region is also present in the closely related chromosome 9p terminal region that contains copies of the PGML/FOXD/CBWD genes.
- Many ribosomal protein L23a pseudogenes are scattered through the human genome.

Mitochondrial Genome

The human mitochondrial genome, while usually not included when referring to the "human genome", is of tremendous interest to geneticists, since it undoubtedly plays a role in mitochondrial disease. It also sheds light on human evolution; for example, analysis of variation in the human mitochondrial genome has led to the postulation of a recent common ancestor for all humans on the maternal line of descent.

Due to the lack of a system for checking for copying errors, Mitochondrial DNA (mtDNA) has a more rapid rate of variation than nuclear DNA. This 20-fold increase in the mutation rate allows mtDNA to be used for more accurate tracing of maternal ancestry.

Studies of mtDNA in populations have allowed ancient migration paths to be traced, such as the migration of Native Americans from Siberia or Polynesians from southeastern Asia. It has also been used to show that there is no trace of Neanderthal DNA in the European gene mixture inherited through purely maternal lineage.

Epigenome

Epigenetics are a variety of features of the human genome that transcend its primary DNA sequence, such as chromatin packaging, histone modifications and DNA methylation, and which are important in regulating gene expression, genome replication and other cellular processes. Epigenetic markers strengthen and weaken transcription of certain genes but do not affect the actual sequence of DNA nucleotides.

Craig Venter # Individual Human Genome Sequenced

John Craig Venter (born October 14, 1946) is an American biologist and entrepreneur, most famous for his role in being one of the first to sequence the human genome and for his role in creating the first cell with a synthetic genome in 2010. Venter founded Celera Genomics, The Institute for Genomic Research and the J. Craig Venter Institute, now working at the latter to create synthetic biological organisms and to document genetic diversity in the world's oceans. He was listed on *Time* magazine's 2007 and 2008 Time 100 list of the most influential people in the world. In 2010, The British Magazine New Statesman listed Craig Venter at 14th in the list of "The World's 50 Most Influential Figures 2010".

Early Life

Venter was born in Salt Lake City, Utah. In his youth, Venter did not take his education seriously, preferring to spend his time on the water in boats or surfing. According to his biography, *A Life Decoded*, he was said to never be a terribly engaged student, having Cs and Ds on his eighth-grade report cards. According to *Time Magazine*, it was not always evident that Venter would become a transformative figure, particularly when he was a boy.

Although he was against the Vietnam War, Venter was drafted and enlisted in the United States Navy where he worked in the intensive-care ward of a field hospital. Being confronted with wounded, maimed, and dying soldiers on a daily basis instilled in him a desire to study medicine — although he later switched to scientific medical research.

Education

Venter graduated from Mills High School and began his college career at a community college, College of San Mateo in California. He received his B.S. degree in biochemistry in 1972, and his Ph.D. degree in physiology and pharmacology in 1975, both from the University of California, San Diego. At UCSD, he studied under biochemist Nathan O. Kaplan, and married former Ph.D. candidate Barbara Rae.

After working as an associate professor, and later as full professor, at the State University of New York at Buffalo, he joined the National Institutes of Health in 1984. In Buffalo, he divorced Dr. Rae-Venter and married his student, Claire M. Fraser, remaining married to her until 2005.

Discovery

While at the NIH, Venter learned of a technique for rapidly identifying all of the mRNAs present in a cell and began to use it to identify human brain genes. The short cDNA sequence fragments discovered by this method are called expressed sequence tags (ESTs) a name coined by Anthony Kerlavage at The Institute for Genomic Research.

The NIH initially led an effort to patent these gene fragments, in which Venter coincidentally and controversially became involved. The NIH later withdrew the patent applications, following public outcry. Subsequent court cases declared that ESTs were not directly patentable.

Human Genome Project

Venter was passionate about the power of genomics to radically transform healthcare. Venter believed that shotgun sequencing was the fastest and most effective way to get useful human genome data. The method was controversial, however, since some geneticists felt it would not be accurate enough for a genome as complicated as that of humans. Frustrated with what Venter viewed as the slow pace of progress in the Human Genome project, and unable to get funds for his ideas, he sought funding from the private sector to fund Celera Genomics.

The goal of the company was to sequence the entire human genome and release it into the public domain for non-commercial use in much less time and for much less cost than the public human genome project. The company planned to monetize their work by creating a value-added database of genomic data to which users could subscribe for a fee.

The goal consequently put pressure on the public genome program and spurred several groups to redouble their efforts to produce the full sequence. DNA from five demographically different individuals was used by Celera to generate the sequence of the human genome; one of the individuals was Venter himself. In 2000, Venter and Francis Collins of the National Institutes of Health and U.S. Public Genome Project jointly made the announcement of the mapping of the human genome in 2000, a full three years ahead of the expected end of the Public Genome Program.

The announcement was made along with US President Bill Clinton, and U.K. Prime Minister Tony Blair. Venter and Collins thus shared an award for "Biography of the Year" from A&E Network. Celera published the first Human Genome in the journal Science, and was soon followed by a Human Genome Project Publication in Nature.

Despite some claims that shotgun sequencing was in some ways less accurate than the clone-by-clone method chosen by the Human Genome Project, the technique became widely accepted by the scientific community and is still the de facto standard used today.

Although Celera was originally set to sequence a composite of DNA samples, partway through the sequencing, Venter switched the samples for his own DNA. After contributing to the Human Genome, and its release it into the public domain, Venter was fired by Celera in early 2002. According to his biography, Venter was ready to leave Celera, and was fired due to conflict with the main investor, Tony White, that had existed since day one of the project. Venter writes that his main goal was always to accelerate science and thereby discovery, and he only sought help from the corporate world when he couldn't find funding in the public sector.

Ocean Sampling

The Global Ocean Sampling Expedition (GOS) is an ocean exploration genome project with the goal of assessing the genetic diversity in marine microbial communities and to understand their role in nature's fundamental processes. Begun as a Sargasso Sea pilot sampling project in August 2003, Craig Venter announced the full Expedition on 4 March 2004. The project, which used Craig Venter's personal yacht, *Sorcerer II*, started in Halifax, Canada, circumnavigated the globe and returned to the U.S. in January 2006.

Current Work

Venter is currently the president of the J. Craig Venter Institute, which conducts research in synthetic biology. In June 2005, he co-founded Synthetic Genomics, a firm dedicated to using modified microorganisms to produce clean fuels and biochemicals. In July 2009, ExxonMobil announced a $600 million collaboration with Synthetic Genomics to research and develop next-generation biofuels. Venter is a member of the USA Science and Engineering Festival's Advisory Board.

Media Coverage

Venter has been the subject of articles in several magazines, including *Wired*, *The Economist*, Australian science magazine *Cosmos*, and *The Atlantic*. Additionally, he was featured on *The Colbert Report* on both February 27, 2007, and October 30, 2007.

Venter appeared in the "Evolution" episode of the documentary television series *Understanding.*

On May 16, 2004, Venter gave the commencement speech at Boston University.

In a 2007 interview with *New Scientist* when asked "Assuming you can make synthetic bacteria, what will you do with them?", Venter replied:

Over the next 20 years, synthetic genomics is going to become the standard for making anything. The chemical industry will depend on it. Hopefully, a large part of the energy industry will depend on it. We really need to find an alternative to taking carbon out of the ground, burning it, and putting it into the atmosphere. That is the single biggest contribution I could make.

Furthermore it suggests that one of the main purposes for creating synthetic bacteria would be to reduce the dependence on fossil fuels through bioremediation.

On May 10, 2007, Venter was awarded an honorary doctorate from Arizona State University., and on October 24 of the same year, he received an honorary doctorate from Imperial College London.

He was on the 2007 Time 100 most influential people in the world list made by Time magazine. In 2007 he also received the Golden Eurydice Award for contributions to Biophilosophy.

On September 4, 2007, a team led by Venter published the first complete (six-billion-letter) genome of an individual human — Venter's own DNA sequence. When on BBC News on October 22, 2007, when asked about his religious view he replied that he thought that a true scientist could not believe in supernatural explanations.

On December 4, 2007, Venter gave the Dimbleby lecture for the BBC in London. He outlined his current work and future developments in genetics.

In February 2008, he gave a speech about his current work at the TED conference.

Venter delivered the 2008 convocation speech for Faculty of Science honours and specialization students at the University of Alberta. A transcription of the speech is available here.

Dr. Venter was featured in Time Magazine's "The Top 10 Everything of 2008" article. Number three in 2008's Top 10 Scientific Discoveries was a piece outlining his work stitching together the 582,000 base pairs necessary to invent the genetic information for a whole new bacterium.

Dr. Venter took part in the inaugural San Diego Science Festival and spoke at its press conference on February 26, 2009.

On April 6, 2009, Venter gave a speech at Arizona State University as part of the Origins Symposium.

For an episode aired on July 27, 2009, Venter was interviewed on his boat by BBC One for the first episode of TV show Bang Goes the Theory.

On May 20, 2010, Venter announced the creation of first self-replicating synthetic bacterial cell.

On November 21, 2010 Steve Kroft profiled J. Craig Venter and his research on 60 minutes.

Individual Human Genome Sequenced

On September 4, 2007, a team led by Sam Levy published the first complete (six-billion-letter) genome of an individual human—Venter's own DNA sequence. Some of the sequences in Venter's genome are associated with wet earwax, increased risk of antisocial Behaviour, Alzheimer's and cardiovascular diseases.

This publication was especially interesting since it contained a diploid instead of a haploid genome and shows promise for personalized medicine via genotyping. This genome, rather immodestly dubbed HuRef by Levy et al., was a landmark accomplishment and as of mid-2010 is probably the highest quality personal genome sequence yet completed. The Human Reference Genome Browser is a web application for the navigation and analysis of Venter's recently published genome.

The HuRef database consists of approximately 32 million DNA reads sequenced using microfluidic Sanger sequencing, assembled into 4,528 scaffolds and 4.1 million DNA variations identified by genome analysis. These variants include single-nucleotide polymorphisms (SNPs), block substitutions, short and large indels, and structural variations like insertions, deletions, inversions and copy number changes.

The browser enables scientists to navigate the HuRef genome assembly and sequence variations, and to compare it with the NCBI human build 36 assembly in the context of the NCBI and Ensembl annotations. The browser provides a comparative view between NCBI and HuRef consensus sequences, the sequence multi-alignment of the HuRef assembly, Ensembl and dbSNP annotations, HuRef variants, and the underlying variant evidence and functional analysis. The interface also represents the haplotype blocks from which diploid genome sequence can be inferred and the relation of variants to gene annotations. The display of variants and gene annotations are linked to external public resources including dbSNP, Ensembl, Online Mendelian Inheritance in Man (OMIM) and Gene Ontology (GO).

Users can search the HuRef genome using HUGO gene names, Ensembl and dbSNP identifiers, HuRef contig or scaffold locations, or NCBI chromosome locations. Users can then easily and quickly browse any genomic region via the simple and intuitive pan and zoom controls; furthermore, data relevant to specific loci can be exported for further analysis.

Mycoplasma Laboratorium

Venter is seeking to patent the first life form created by humanity, possibly to be named *Mycoplasma laboratorium*. There is speculation that this line of research could lead to producing bacteria that have been engineered to perform specific reactions, e.g. produce fuels, make medicines, combat global warming, etc.

In May 2010, a team of scientists led by Venter became the first to successfully create what was described as "synthetic life". This was done by synthesizing a very long DNA molecule containing an entire bacterium genome, and introducing this into another cell, analogous to the accomplishment of Eckard Wimmer's group, who synthesized and ligated an RNA virus genome and "booted" it in cell lysate. The single-celled organism contains four "watermarks" written into its DNA to identify it as synthetic and to help trace its descendants. The watermarks include

1. Code table for entire alphabet with punctuations
2. Names of 46 contributing scientists
3. Three quotations
4. The web address for the cell.

Genetic Distance

Genetic distance refers to the genetic divergence between species or between populations within a species. It considers a variety of parameters used to measure the genetic distance. Smaller genetic distances indicate a close genetic relationship whereas large genetic distances indicate a more distant genetic relationship. Genetic distance can be used to compare the genetic similarity between different species, such as humans and chimpanzees. Within a species genetic distance can be used to measure the divergence between different sub-species.

In its simplest form, the genetic distance between two populations is the difference in frequencies of a trait. For example the frequency of RH negative individuals is 50.4% among Basques, 41.2% in France and 41.1% in England. Thus the genetic difference between the Basques and French is 9.2% and the genetic difference between the French and the English is 0.1% for the RH negative trait. The genetic distance of several individual traits can then be averaged to compute an overall genetic distance.

Measures of Genetic Distance

There are several measures used to indicate genetic distance. These include:

Fixation Index

A commonly used measure of genetic distance is the fixation index which varies between 0 and 1. A value of 0 indicates that two populations are genetically identical whereas a value of 1 indicates that two populations are different species.

Nei's Standard Genetic Distance

This measure assumes that genetic differences arise due to mutations and genetic drift.

Cavalli-Sforza and Edwards 1967

This measure assumes that genetic differences arise due to genetic drift only.

Reynolds, Weir, and Cockerham's 1983

This measure assumes that genetic differences arise due to genetic drift only.

Noncoding DNA

In genetics, noncoding DNA describes components of an organism's DNA sequences that do not encode for protein sequences. In many eukaryotes, a large percentage of an organism's total genome size is noncoding DNA, although the amount of noncoding DNA, and the proportion of coding versus noncoding DNA varies greatly between species.

Much of this DNA has no known biological function and at one time was sometimes referred to as "Junk DNA". However, many types of noncoding DNA sequences do have known biological functions, including the transcriptional and translational regulation of protein-coding sequences. Other noncoding sequences have likely but as-yet undetermined function, an inference from high levels of homology and conservation seen in sequences that do not encode proteins but appear to be under heavy selective pressure.

Fraction of Noncoding Genomic DNA

The amount of total genomic DNA varies widely between organisms, and the proportion of coding and noncoding DNA within these genomes varies greatly as well. More than 98% of the human genome does not encode protein sequences, including most sequences within introns and most intergenic DNA.

While overall genome size, and by extension the amount of noncoding DNA, are correlated to organism complexity, there are many exceptions. For example, the genome of the unicellular *Polychaos dubium* (formerly known as *Amoeba dubia*) has been reported to contain more than 200 times the amount of DNA in humans.

The pufferfish *Takifugu rubripes* genome is only about one eighth the size of the human genome, yet seems to have a comparable number of genes; approximately 90% of the *Takifugu* genome is noncoding DNA and most of the genome size difference appears to lie in the noncoding DNA. The extensive variation in nuclear genome size among eukaryotic species is known as the C-value enigma or C-value paradox.

About 80% of the nucleotide bases in the human genome may be transcribed, but transcription does not necessarily imply function.

Types of Noncoding DNA Sequences

Noncoding Functional RNA: Noncoding RNAs are functional RNA molecules that are not translated into protein. Examples of noncoding RNA include ribosomal RNA, transfer RNA, Piwi-interacting RNA and microRNA. MicroRNAs are predicted to control the translational activity of approximately 30% of all protein-coding genes in mammals and may be vital components in the progression or treatment of various diseases including cancer, cardiovascular disease, and the immune system response to infection.

Cis-regulatory Elements

Cis-regulatory elements are sequences that control the transcription of a gene. Cis-elements may be located in 5' or 3' untranslated regions or within introns. Promoters facilitate the transcription of a particular gene and are typically upstream of the coding region.

Enhancer sequences may exert very distant effects on the transcription levels of genes.

Introns

Introns are non-coding sections of a gene, transcribed into the precursor mRNA sequence, but ultimately removed by RNA splicing during the processing to mature messenger RNA. Many introns appear to be mobile genetic elements.

Studies of group I introns from *Tetrahymena* indicate that some introns appear to be selfish genetic elements, neutral to the host because they remove themselves from flanking exons during RNA processing and do not produce an expression bias between alleles with and without the intron.

Some introns do appear to have significant biological function, possibly through ribozyme functionality that may regulate tRNA and rRNA activity as well as protein-coding gene expression, evident in hosts that have become dependent on such introns over long periods of time; for example, the *trnL-intron* is found in all green plants and appears to have been vertically inherited for several billions of years, including more than a billion years within chloroplasts and an additional 2–3 billion years prior in the cyanobacterial ancestors of chloroplasts.

Pseudogenes

Pseudogenes are DNA sequences, related to known genes, that have lost their protein-coding ability or are otherwise no longer expressed in the cell. Pseudogenes arise from retrotransposition or genomic duplication of functional genes, and become "genomic fossils" that are nonfuctional due to mutations that prevent the transcription of the gene, such as within the gene promoter region, or fatally alter the translation of the gene, such as premature stop codons or frameshifts.

Pseudogenes resulting from the retrotransposition of an RNA intermediate are known as processed pseudogenes; pseudogenes that arise from the genomic remains of duplicated genes or residues of inactivated are nonprocessed pseudogenes.

While Dollo's Law suggests that the loss of function in pseudogenes is likely permanent, silenced genes may actually retain function for several million years and can be "reactivated" into protein-coding sequences and a substantial number of pseudogenes are actively transcribed. Because pseudogenes are presumed to evolve without evolutionary constraint, they can serve as a useful model of the type and frequencies various spontaneous genetic mutations.

Repeat Sequences, Transposons and Viral Elements

Transposons and retrotransposons are mobile genetic elements. Retrotransposon repeated sequences, which include long interspersed nuclear elements (LINEs) and short interspersed nuclear elements (SINEs), account for a large proportion of the genomic sequences in many species. Alu sequences, classified as a short interspersed nuclear element, are the most abundant mobile elements in the human genome. Some examples have been found of SINEs exerting transcriptional control of some protein-encoding genes.

Endogenous retrovirus sequences are the product of reverse transcription of retrovirus genomes into the genomes of germ cells. Mutation within these retro-transcribed sequences can inactivate the viral genome.

Over 8% of the human genome is made up of (mostly decayed) endogenous retrovirus sequences, as part of the over 42% fraction that is recognizably derived of retrotransposons, while another 3% can be identified to be the remains of DNA transposons. Much of the remaining half of the genome that is currently without an explained origin is expected to have found its origin in transposable elements that were active so long ago (> 200 million years) that random mutations have rendered them unrecognizable. Genome size variation in at least two kinds of plants is mostly the result of retrotransposon sequences.

Telomeres

Telomeres are regions of repetitive DNA at the end of a chromosome, which provide protection from chromosomal deterioration during DNA replication.

Functions of Noncoding DNA

Many noncoding DNA sequences have very important biological functions. Comparative genomics reveals that some regions of noncoding DNA are highly conserved, sometimes on time-scales representing hundreds of millions of years, implying that these noncoding regions are under strong evolutionary pressure and positive selection. For example, in the genomes of humans and mice, which diverged from a common ancestor 65–75 million years ago, protein-coding DNA sequences account for only about 20% of conserved DNA, with the remaining majority of conserved DNA represented in noncoding regions. Linkage mapping often identifies chromosomal regions associated with a disease with no evidence of functional coding variants of genes within the region, suggesting that disease-causing genetic variants lie in the noncoding DNA. Some noncoding DNA sequences are genetic "switches" that do not encode proteins, but do regulate when and where genes are expressed.

According to a comparative study of over 300 prokaryotic and over 30 eukaryotic genomes, eukaryotes appear to require a minimum amount of non-coding DNA. This minimum amount can be predicted using a growth model for regulatory genetic networks, implying that it is required for regulatory purposes. In humans the predicted minimum is about 5% of the total genome.

Some noncoding DNA sequences determine the expression levels of various genes. Other sequences of noncoding DNA determine where transcription factors attach. Some specific sequences of noncoding DNA may be features essential to chromosome structure, centromere function and homolog recognition in meiosis.

Noncoding DNA and Evolution

Shared sequences of apparently non-functional DNA are a major line of evidence for common descent.

Pseudogene sequences appear to accumulate mutations more rapidly than coding sequences due to a loss of selective pressure. This allows for the creation of mutant alleles that incorporate new functions that may be favoured by natural selection; thus, pseudogenes can serve as raw material for evolution and can be considered "protogenes".

Junk DNA

Junk DNA, a term that was introduced in 1972 by Susumu Ohno, was a provisional label for the portions of a genome sequence for which no discernible function had been identified. According to a 1980 review in *Nature* by Leslie Orgel and Francis Crick, junk DNA has "little specificity and conveys little or no selective advantage to the organism".

The term is currently, however, an outdated concept, being used mainly in popular science and in a colloquial way in scientific publications, and may have slowed research into the biological functions of noncoding DNA. Several lines of evidence indicate that many "junk DNA" sequences are likely to have unidentified functional activity, and other sequences may have had functions in the past.

Still, a significant amount of the sequence of the genomes of eukaryotic organisms currently appears to fall under no existing classification other than "junk". For example, one experiment removed 0.1% of the mouse genome with no detectable effect on the phenotype. This result suggests that the removed DNA was largely nonfunctional. In addition, these sequences are enriched for the heterochromatic histone modification H3K9me3.

Phylogenetic Footprinting

Phylogenetic footprinting is a technique used to identify transcription factor binding sites (TFBS) within a non-coding region of DNA of interest by comparing it to the orthologous sequence in different species. When this technique is used with a large number of closely related species, this is called phylogenetic shadowing.

Researchers have found that non-coding pieces of DNA contain binding sites for regulatory proteins that govern the spatiotemporal expression of genes. These transcription factor binding sites (TFBS), or regulatory motifs, have proven hard to identify, primarily because they are short in length, and can show sequence variation. The importance of understanding transcriptional regulation to many fields of biology has led researchers to develop strategies for predicting the presence of TFBS, many of which have led to publicly available databases. One such technique is Phylogenetic Footprinting.

Phylogenetic footprinting relies upon two major concepts:

1. The function and DNA binding preferences of transcription factors are well-conserved between diverse species.
2. Important non-coding DNA sequences that are essential for regulating gene expression will show differential selective pressure. A slower rate of change occurs in TFBS than in other, less critical, parts of the non-coding genome.

History

Phylogenetic footprinting was first used and published by Tagle et al. in 1988, which allowed researchers to predict evolutionary conserved cis-regulatory elements responsible for embryonic ε and γ globulin gene expression in primates.

Before phylogenetic footprinting, DNase footprinting was used, where protein would be bound to DNA transcription factor binding sites (TFBS) protecting it from DNase digestion.

One of the problems with this technique was the amount of time and labour it would take. Unlike DNase footprinting, phylogenetic footprinting relies on evolutionary constraints within the genome, with the "important" parts of the sequence being conserved among the different species.

Protocol

It is important when using this technique to decide which genome your sequence should be aligned to. More divergent species will have less sequence homology between orthologous genes. Therefore, the key is to pick species that are related enough to detect homology, but divergent enough to maximize non-alignment "noise". Step wise approach to Phylogenetic footprinting consists of :

1. One should decide on the gene of interest.
2. Carefully choose species with orthologous genes.
3. Decide on the length of the upstream or maybe downstream region to be looked at.
4. Align the sequences.
5. Look for conserved regions and analyse them.

Not all TFBS are Found

Not all transcription binding sites can be found using phylogenetic footprinting due to the statistical nature this technique. Here are several reasons why some TFBS are not found:

Species Specific Binding Sites

Some binding sites seem to have no significant matches in most other species. Therefore, detecting these sites by phylogenetic footprinting is likely impossible unless a large number of closely related species are available.

Very Short Binding Sites

Some binding sites show excellent conservation, but just in a shorter region than the ones we looked for. Such short motifs (e.g., GC-box) are often happened by chance in nonfunctional sequences and detecting these motifs would be challenging.

Less Specific Binding Factors

Some binding sites show some conservation but have had insertions or deletions. It is not obvious if these sequences with insertions or deletions are still functional.

Though they may still be functional if the finding factor is less specific (or less 'picky' if you will). Because deletions and insertions are rare in binding sites, considering insertions and deletions in the sequence would detect a few more true TFBSs, but it could likely include many more false positives.

Not Enough Data

Some motifs are quite well conserved, but they are statistically insignificant a specific dataset. The motif might have appeared in different species by chance. These motifs could be detected if sequences from more organisms are available. So this will be less of a problem in the future.

Compound Binding Regions

Some transcription factors bind as dimers. Therefore, their binding sites may consist of two conserved regions, separated by a few variable nucleotides. Because of the variable internal sequence, the motif cannot be detected.

However, if we could use a program to search for motifs containing a variable sequence in the middle, without counting mutations, these motifs could be discovered.

Accuracy

It is important to keep in mind that not all conserved sequences are under selection pressure. To eliminate false positives statistical analysis must be performed that will show that the motifs reported have a mutation rate meaningfully less than that of the surrounding nonfunctional sequence.

Moreover, results could be more accurate if the prior knowledge about the sequence is considered. For example, some regulatory elements are repeated 15 times in a promoter region (e.g., some metallothionein promoters have up to 15 metal response elements (MREs)).

Thus, to eliminate false motifs with inconsistent order across species, the orientation and order of regulatory elements in a promoter region should be the same in all species. This type of information could help us to identify regulatory elements that are not adequately conserved but occur in several copies in the input sequence.

Transcriptome

The transcriptome is the set of all RNA molecules, including mRNA, rRNA, tRNA, and other non-coding RNA produced in one or a population of cells. The term can be applied to the total set of transcripts in a given organism, or to the specific subset of transcripts present in a particular cell type. Unlike the genome, which is roughly fixed for a given cell line (excluding mutations), the transcriptome can vary with external environmental conditions.

Because it includes all *mRNA* transcripts in the cell, the transcriptome reflects the genes that are being actively expressed at any given time, with the exception of mRNA degradation phenomena such as transcriptional attenuation. The study of *transcriptomics*, also referred to as expression profiling, examines the expression level of mRNAs in a given cell population, often using high-throughput techniques based on DNA microarray technology. The use of next-generation sequencing technology to study the transcriptome at the nucleotide level is known as RNA-Seq.

Transcriptomics is the branch of molecular biology that deals with the study of messenger RNA molecules produced in an individual or population of a particular cell type.

Applications and Analysis

The transcriptomes of stem cells and cancer cells are of particular interest to researchers who seek to understand the processes of cellular differentiation and carcinogenesis. A number of organism-specific transcriptome databases have been constructed and annotated to aid in the identification of genes that are differentially expressed in distinct cell populations or subtypes; however, the analysis of relative mRNA expression levels can be complicated by the fact that relatively small changes in mRNA expression can produce large changes in the total amount of the corresponding protein present in the cell. One analysis method, known as Gene Set Enrichment Analysis, identifies coregulated gene networks rather than individual genes that are up-or down-regulated in different cell populations.

mRNA Regulation

Although microarray studies can reveal the relative amounts of different mRNAs in the cell, levels of mRNA are not directly proportional to the expression level of the proteins they code for. The number of protein molecules synthesized using a given mRNA molecule as a template is highly dependent on translation-initiation features of the mRNA sequence; in particular, the ability of the translation initiation sequence is a key determinant in the recruiting of ribosomes for protein translation. The complete protein complement of a cell or organism is known as the proteome.

A study of 158,807 mouse transcripts revealed that 4520 of these transcripts form antisense partners that are base pair complementary to the exons of genes. These results raise the possibility that significant numbers of "antisense RNA-coding genes" might participate in the regulation of the levels of expression of protein-coding mRNAs.

The Genographic Project

The Genographic Project, launched on April 13, 2005 by the National Geographic Society and IBM, is a multi-year genetic anthropology study that aims to map historical human migration patterns by collecting and analysing DNA samples from hundreds of thousands of people from around the world.

Overview

Field researchers at 11 regional centres around the world collect DNA samples from indigenous populations. The project also sells self-testing kits: for US$100 anyone in the world can order a kit with which a mouth scraping (buccal swab) is obtained, analysed and the DNA information placed on an Internet accessible database.

The genetic markers on mitochondrial DNA (HVR1) and Y-chromosomes (12 microsatellite markers and haplogroup-defining SNPs) are used to trace the participant's distant ancestry, and each customer is provided with their genetic history. As of April 2010 more than 350,000 people had bought a test kit, and the success of the project has spawned a broader interest in direct-to-consumer genetic testing. The Genographic Project is undertaking widespread consultation with indigenous groups from around the world. Genographic Project public participation kits are processed by Family Tree DNA (FTDNA) using the Arizona Research Labs at the University of Arizona.

The project is a privately-funded, not-for-profit collaboration between the National Geographic Society, IBM and the Waitt Family Foundation. Part of the proceeds from the sale of self-testing kits support the Genographic Project's ongoing DNA collection, but the majority are ploughed into a Legacy Fund to be spent on cultural preservation projects nominated by indigenous communities.

Team Members

Team members include:

- Spencer Wells, project director (National Geographic Explorer-in-Residence)
- Jin Li, principal investigator, East Asia
- Theodore Schurr, principal investigator, North America
- Fabricio Santos, principal investigator, South America
- Jaume Bertranpetit, David Comas and Lluis Quintana-Murci, principal investigators, Western Europe and Central Europe
- Pierre Zalloua, principal investigator, Middle East and Northern Africa
- Himla Soodyall, principal investigator, Sub-Saharan Africa
- Elena Balanovska, principal investigator, North Eurasia
- Ramasamy Pitchappan, principal investigator, India
- Alan Cooper, principal investigator, Ancient DNA
- John Mitchell, principal investigator, Australia and New Zealand
- Lisa Matisoo-Smith, principal investigator, Oceania
- Ajay Royyuru, head of computational biology, IBM
- Simon Longstaff, advisory board chair (director of the St James Ethics Centre)
- Meave Leakey, advisory board member
- Merritt Ruhlen, advisory board member
- Colin Renfrew, advisory board member
- Luigi Luca Cavalli-Sforza, advisory board member
- Wade Davis, advisory board member.

Use of Genetic Markers

The Genographic Project relies on the identification of genetic markers. Most human DNA is a shuffled combination of genetic material passed down the generations. There are, however, parts of the human genome that pass unshuffled from parent to child. These segments of DNA are only changed by occasional mutations—random spelling mistakes in the genetic code. When these spelling mistakes are passed down to succeeding generations, they become markers of descent.

Different populations have different genetic markers, and by following them through the generations scientists are able to identify the different branches of the human tree, all the way back to their common African root. Indigenous populations provide geographical and cultural context to the genetic markers in their DNA. These clues can help recreate past migration patterns.

Criticism

Shortly after the announcement of the project in April 2005, the Indigenous Peoples Council on Biocolonialism, (IPCB), released a statement protesting about the project, its connections with the HGDP, and called for a boycott of IBM, Gateway Computers, and National Geographic. Around May 2006, the United Nations Permanent Forum on Indigenous Issues (UNPFII) recommended suspending the project. Concerns were that the knowledge gleaned from the research could clash with long held beliefs leading to the destruction of their culture. They also feared that it could endanger land rights and other benefits.

In May 2006, the representatives of Indigenous went to UNPFII contesting any involvement in the testing. "The Genographic Project is exploitative and unethical because it will use Indigenous peoples as subjects of scientific curiosity in research that provides no benefit to Indigenous peoples, yet subjects them to significant risks. Researchers will take blood or other bodily tissue samples for their own use in order to further their own speculative theories of human history."

UNPFII conducted investigations into the objectives of the Genographic Project, and concluded that since the project was "conceived and has been initiated without appropriate consultation with or regard for the risks to its subjects, the Indigenous peoples, the Council for Responsible Genetics concludes that the Indigenous peoples' representatives are correct and that the Project should be immediately suspended.

As of December 2006 some federally recognized tribes in North America have declined to take part. "What the scientists are trying to prove is that we're the same as the Pilgrims except we came over several thousand years before," said Maurice Foxx, chairman of the Massachusetts Commission on

Indian Affairs and a member of the Mashpee Wampanoag. “Why should we give them that openly?” However, more than 50,000 indigenous participants from the Americas, Africa, Asia, Europe and Oceania had joined the project as of April 2010.

As of 2010, the consensus is that the Genographic Project’s benefits of providing genetic knowledge to humanity have created stronger ties among various cultural groups. While indigenous rights groups’ concerns have been addressed, other cultural interest groups likewise have asserted their rights. Various extended family organizations worldwide support the Genographic Project. They have found connections to their places of heritage. They have discovered kinship relations to other ethnic groups and nationalities, creating stronger and more peaceful ties with other polities. Participants have discovered their “roots,” representing an inalienable right to cultural heritage.

2

Gene Transcription

Cell signalling is part of a complex system of communication that governs basic cellular activities and coordinates cell actions. The ability of cells to perceive and correctly respond to their microenvironment is the basis of development, tissue repair, and immunity as well as normal tissue homeostasis. Errors in cellular information processing are responsible for diseases such as cancer, autoimmunity, and diabetes.

By understanding cell signalling, diseases may be treated effectively and, theoretically, artificial tissues may be created. Traditional work in biology has focused on studying individual parts of cell signalling pathways. Systems biology research helps us to understand the underlying structure of cell signalling networks and how changes in these networks may affect the transmission and flow of information.

Such networks are complex systems in their organization and may exhibit a number of emergent properties including bistability and ultrasensitivity. Analysis of cell signalling networks requires a combination of experimental and theoretical approaches including the development and analysis of simulations and modelling.

Unicellular and Multicellular Organism Cell Signalling

Cell signalling has been most extensively studied in the context of human diseases and signalling between cells of a single organism. However, cell signalling may also occur between the cells of two different organisms. In many mammals, early embryo cells exchange signals with cells of the uterus. In the human gastrointestinal tract, bacteria exchange signals with each other and with human epithelial and immune system cells. For the yeast *Saccharomyces cerevisiae* during mating, some cells send a peptide signal (mating factor *pheromones*) into their environment. The mating factor peptide may bind to a cell surface receptor on other yeast cells and induce them to prepare for mating.

Types of Signals

Cells communicate with each other via direct contact (juxtacrine signalling), over short distances (paracrine signalling), or over large distances and/or scales (endocrine signalling). Some cell-to-cell communication requires direct cell-cell contact. Some cells can form gap junctions that connect their cytoplasm to the cytoplasm of adjacent cells. In cardiac muscle, gap junctions between adjacent cells allows for action potential propagation from the cardiac pacemaker region of the heart to spread and coordinately cause contraction of the heart.

The Notch signalling mechanism is an example of juxtacrine signalling (also known as contact dependent signalling) in which two adjacent cells must make physical contact in order to communicate. This requirement for direct contact allows for very precise control of cell differentiation during embryonic development.

In the worm *Caenorhabditis elegans*, two cells of the developing gonad each have an equal chance of terminally differentiating or becoming a uterine precursor cell that continues to divide. The choice of which cell continues to divide is controlled by competition of cell surface signals. One cell will happen to produce more of a cell surface protein that activates the Notch receptor on the adjacent cell. This activates a feedback loop or system that reduces Notch expression in the cell that will differentiate and increases Notch on the surface of the cell that continues as a stem cell.

Many cell signals are carried by molecules that are released by one cell and move to make contact with another cell. *Endocrine* signals are called hormones. Hormones are produced by endocrine cells and they travel through the blood to reach all parts of the body. Specificity of signalling can be controlled if only some cells can respond to a particular hormone. *Paracrine* signals such as retinoic acid target only cells in the vicinity of the emitting cell. Neurotransmitters represent another example of a paracrine signal.

Some signalling molecules can function as both a hormone and a neurotransmitter. For example, epinephrine and norepinephrine can function as hormones when released from the adrenal gland and are transported to the heart by way of the blood stream. Norepinephrine can also be produced by neurons to function as a neurotransmitter within the brain. Estrogen can be released by the ovary and function as a hormone or act locally via paracrine or autocrine signalling. Active species of oxygen and nitric oxide can also act as cellular messengers. This process is dubbed redox signalling.

Receptors for Cell Signals

Cells receive information from their environment through a class of proteins known as receptors. Notch is a cell surface protein that functions as a receptor. Animals have a small set of genes that code for signalling proteins that interact specifically with Notch receptors and stimulate a response in cells that express Notch on their surface. Molecules that activate (or, in some cases, inhibit) receptors can be classified as hormones, neurotransmitters, cytokines, growth factors but all of these are called receptor ligands. The details of ligand-receptor interactions are fundamental to cell signalling.

Notch acts as a receptor for ligands that are expressed on adjacent cells. While many receptors are cell surface proteins, some are found inside cells. For example, oestrogen is a hydrophobic molecule that can pass through the lipid bilayer of cell surface membranes. Oestrogen receptors inside cells of the uterus can be activated by oestrogen that comes from the ovaries, enters the target cells, and binds to oestrogen receptors.

A number of transmembrane receptors for molecules that include peptide hormones and of intracellular receptors for steroid hormones exist, giving to a cell the ability to respond to a great number of hormonal and pharmacological stimuli. In diseases, often, proteins that interact with receptors are aberrantly activated, resulting in constitutively activated downstream signals.

For several types of intercellular signalling molecules that are unable to permeate the hydrophobic cell membrane due to their hydrophilic nature, the target receptor is expressed on the membrane. When such signalling molecule activates its receptor, the signal is carried into the cell usually by means of a second messenger such as cAMP.

Signalling Pathways

In some cases, receptor activation caused by ligand binding to a receptor is directly coupled to the cell's response to the ligand. For example, the neurotransmitter GABA can activate a cell surface receptor that is part of an ion channel. GABA binding to a GABA A receptor on a neuron opens a chloride-selective ion channel that is part of the receptor. GABA A receptor activation allows negatively-charged chloride ions to move into the neuron, which inhibits the ability of the neuron to produce action potentials.

However, for many cell surface receptors, ligand-receptor interactions are not directly linked to the cell's response. The activated receptor must first interact with other proteins inside the cell before the ultimate physiological effect of the ligand on the cell's behaviour is produced. Often, the behaviour of a chain of several interacting cell proteins is altered following receptor activation. The entire set of cell changes induced by receptor activation is called a signal transduction mechanism or pathway.

In the case of Notch-mediated signalling, the signal transduction mechanism can be relatively simple. Activation of Notch can cause the Notch protein to be altered by a protease. Part of the Notch protein is released from the cell surface membrane and can act to change the pattern of gene transcription in the cell nucleus.

This causes the responding cell to make different proteins, resulting in an altered pattern of cell behaviour. Cell signalling research involves studying the spatial and temporal dynamics of both receptors and the components of signalling pathways that are activated by receptors in various cell types.

This pathway involves changes of protein-protein interactions inside the cell, induced by an external signal. Many growth factors bind to receptors at the cell surface and stimulate cells to progress through the cell cycle and divide. Several of these receptors are kinases that start to phosphorylate themselves and other proteins when binding to a ligand.

This phosphorylation can generate a binding site for a different protein and thus induce protein-protein interaction. The ligand (called epidermal growth factor (EGF)) binds to the receptor (called EGFR). This activates the receptor to phosphorylate itself. The phosphorylated receptor binds to an adaptor protein (GRB2), which couples the signal to further downstream signalling processes.

For example, one of the signal transduction pathways that are activated is called the mitogen-activated protein kinase (MAPK) pathway. The signal transduction component labelled as "MAPK" in the pathway was originally called "ERK," so the pathway is called the MAPK/ERK pathway. The MAPK protein is an enzyme, a protein kinase that can attach phosphate to target proteins such as the transcription factor MYC and, thus, alter gene transcription and, ultimately, cell cycle progression. Many cellular proteins are activated downstream of the growth factor receptors (such as EGFR) that initiate this signal transduction pathway.

Some signalling transduction pathways respond differently depending on the amount of signalling received by the cell. For instance, the hedgehog protein activates different genes, depending on the amount of hedgehog protein present. Complex multi-component signal transduction pathways provide opportunities for feedback, signal amplification, and interactions inside one cell between multiple signals and signalling pathways.

Classification of Intercellular Communication

Within endocrinology (the study of intercellular signalling in animals) and the endocrine system, intercellular signalling is subdivided into the following classifications:

- *Intracrine* signals are produced within the target cell.
- *Autocrine* signals target the cell its self. Sometimes autocrine cells can target cells close by if they are the same type of cell as the emitting cell. An example of this is immune cells.
- *Juxtacrine* signals target adjacent (touching) cells. These signals are transmitted along cell membranes via protein or lipid components integral to the membrane and are capable of affecting either the emitting cell or cells immediately adjacent.
- *Paracrine* signals target cells in the vicinity of the emitting cell. Neurotransmitters represent an example.
- *Endocrine* signals target distant cells. Endocrine cells produce hormones that travel through the blood to reach all parts of the body.

Gene Transcription

Transcription is the process of creating an equivalent RNA copy of a sequence of DNA. Both RNA and DNA are nucleic acids, which use base pairs of nucleotides as a complementary language that can be converted back and forth from DNA to RNA in the presence of the correct enzymes. During transcription, a DNA sequence is read by RNA polymerase, which produces a complementary, antiparallel RNA strand. As opposed to DNA replication, transcription results in an RNA complement that includes uracil (U) in all instances where thymine (T) would have occurred in a DNA complement. Transcription can be explained easily in 4 quick steps. Step 1: DNA unwinds/ "unzips" as the Hydrogen Bonds Break. Step 2: The free nucleotides of the RNA, pair with complementary DNA bases. Step 3: RNA sugar-phosphate backbone forms. (Aided by RNA Polymerase.) Step 4: Hydrogen bonds of the untwisted RNA+DNA "ladder" break, then the RNA leaves the nucleus through the small nucleur pores. This then goes to the cytoplasm to continue on to protein processing.

Transcription is the first step leading to gene expression. The stretch of DNA transcribed into an RNA molecule is called a *transcription unit* and encodes at least one gene. If the gene transcribed encodes for a protein, the result of transcription is messenger RNA (mRNA), which will then be used to create that protein via the process of translation. Alternatively, the transcribed gene may encode for either ribosomal RNA (rRNA) or transfer RNA (tRNA), other components of the protein-assembly process, or other ribozymes. A DNA transcription unit encoding for a protein contains not only the sequence that will eventually be directly translated into the protein (the *coding sequence*) but also *regulatory sequences* that direct and regulate the synthesis of that protein. The regulatory sequence before (upstream from) the coding sequence is called the five prime untranslated region (5'UTR), and the sequence following (downstream from) the coding sequence is called the three prime untranslated region (3'UTR).

Transcription has some proofreading mechanisms, but they are fewer and less effective than the controls for copying DNA; therefore, transcription has a lower copying fidelity than DNA replication.

As in DNA replication, DNA is read from 3' → 5' during transcription. Meanwhile, the complementary RNA is created from the 5' → 3' direction. This means its 5' end is created first in base pairing. Although DNA is arranged as two antiparallel strands in a double helix, only one of the two DNA strands, called the template strand, is used for transcription. This is because RNA is only single-stranded, as opposed to double-stranded DNA. The other DNA strand is called the coding strand, because its sequence is the same as the newly created RNA transcript (except for the substitution of uracil for thymine). The use of only the 3' → 5' strand eliminates the need for the Okazaki fragments seen in DNA replication.

Transcription is divided into 5 stages: *pre-initiation*, *initiation*, *promoter clearance*, *elongation* and *termination.*

Major Steps

Pre-initiation

In eukaryotes, RNA polymerase, and therefore the initiation of transcription, requires the presence of a core promoter sequence in the DNA. Promoters are regions of DNA which promote transcription and in eukaryotes, are found at -30, -75 and -90 base pairs upstream from the start site of transcription. Core promoters are sequences within the promoter which are essential for transcription initiation. RNA polymerase is able to bind to core promoters in the presence of various specific transcription factors. The most common type of core promoter in eukaryotes is a short DNA sequence known as a TATA box, found -30 base pairs from the start site of transcription.

The TATA box, as a core promoter, is the binding site for a transcription factor known as TATA binding protein (TBP), which is itself a subunit of another transcription factor, called Transcription Factor II D (TFIID). After TFIID binds to the TATA box via the TBP, five more transcription factors and RNA polymerase combine around the TATA box in a series of stages to form a preinitiation complex. One transcription factor, DNA helicase, has helicase activity and so is involved in the separating of opposing strands of double-stranded DNA to provide access to a single-stranded DNA template. However, only a low, or basal, rate of transcription is driven by the preinitiation complex alone. Other proteins known as activators and repressors, along with any associated coactivators or corepressors, are responsible for modulating transcription rate.

Thus preinitiation complex contains:- 1.Core Promoter Sequence 2.Transcription Factors 3.DNA Helicase 4.RNA Polymerase 5.Activators and Repressors The transcription preinitiation in archaea is essentially homologous to that of eukaryotes, but is much less complex. The archaeal preinitiation complex assembles at a TATA-box binding site; however, in archaea, this complex is composed of only RNA polymerase II, TBP, and TFB (the archaeal homologue of eukaryotic transcription factor II B (TFIIB)).

Initiation

In bacteria, transcription begins with the binding of RNA polymerase to the promoter in DNA. RNA polymerase is a core enzyme consisting of five subunits: 2 α subunits, 1 β subunit, 1 β' subunit, and 1 ω subunit. At the start of initiation, the core enzyme is associated with a sigma factor (number 70) that aids in finding the appropriate -35 and -10 base pairs downstream of promoter sequences.

Transcription initiation is more complex in eukaryotes. Eukaryotic RNA polymerase does not directly recognize the core promoter sequences. Instead, a collection of proteins called transcription factors mediate the binding of RNA polymerase and the initiation of transcription. Only after certain transcription factors are attached to the promoter does the RNA polymerase bind to it. The completed assembly of transcription factors and RNA polymerase bind to the promoter, forming a transcription initiation complex. Transcription in the archaea domain is similar to transcription in eukaryotes.

Promoter Clearance

After the first bond is synthesized, the RNA polymerase must clear the promoter. During this time there is a tendency to release the RNA transcript and produce truncated transcripts. This is called *abortive initiation* and is common for both eukaryotes and prokaryotes. Abortive initiation continues to occur until the ó factor rearranges, resulting in the transcription elongation complex (which gives a 35 bp moving footprint).

The ó factor is released before 80 nucleotides of mRNA are synthesized. Once the transcript reaches approximately 23 nucleotides, it no longer slips and elongation can occur. This, like most of the remainder of transcription, is an energy-dependent process, consuming adenosine triphosphate (ATP). Promoter clearance coincides with phosphorylation of serine 5 on the carboxy terminal domain of RNA Pol in eukaryotes, which is phosphorylated by TFIIH.

Elongation

One strand of the DNA, the *template strand* (or noncoding strand), is used as a template for RNA synthesis. As transcription proceeds, RNA polymerase traverses the template strand and uses base pairing complementarity with the DNA template to create an RNA copy. Although RNA polymerase traverses the template strand from 3' → 5', the coding (non-template) strand and newly-formed RNA can also be used as reference points, so transcription can be described as occurring 5' → 3'.

This produces an RNA molecule from 5' → 3', an exact copy of the coding strand (except that thymines are replaced with uracils, and the nucleotides are composed of a ribose (5-carbon) sugar where DNA has deoxyribose (one less oxygen atom) in its sugar-phosphate backbone).

Unlike DNA replication, mRNA transcription can involve multiple RNA polymerases on a single DNA template and multiple rounds of transcription (amplification of particular mRNA), so many mRNA molecules can be rapidly produced from a single copy of a gene.

Elongation also involves a proofreading mechanism that can replace incorrectly incorporated bases. In eukaryotes, this may correspond with short pauses during transcription that allow appropriate RNA editing factors to bind. These pauses may be intrinsic to the RNA polymerase or due to chromatin structure.

Termination

Bacteria use two different strategies for transcription termination. In Rho-independent transcription termination, RNA transcription stops when the newly synthesized RNA molecule forms a G-C rich hairpin loop followed by a run of Us. When the hairpin forms, the mechanical stress breaks the weak rU-dA bonds, now filling the DNA-RNA hybrid.

This pulls the poly-U transcript out of the active site of the RNA polymerase, effectively terminating transcription. In the "Rho-dependent" type of termination, a protein factor called "Rho" destabilizes the interaction between the template and the mRNA, thus releasing the newly synthesized mRNA from the elongation complex.

Transcription termination in eukaryotes is less understood but involves cleavage of the new transcript followed by template-independent addition of *A*s at its new 3' end, in a process called polyadenylation.

Measuring and Detecting Transcription

Transcription can be measured and detected in a variety of ways:

- Nuclear Run-on assay: measures the relative abundance of newly formed transcripts
- RNase protection assay and ChIP-Chip of RNAP: detect active transcription sites
- RT-PCR: measures the absolute abundance of total or nuclear RNA levels, which may however differ from transcription rates
- DNA microarrays: measures the relative abundance of the global total or nuclear RNA levels; however, these may differ from transcription rates
- In situ hybridization: detects the presence of a transcript
- MS2 tagging: by incorporating RNA stem loops, such as MS2, into a gene, these become incorporated into newly synthesized RNA. The stem loops can then be detected using a fusion of GFP and the MS2 coat protein, which has a high affinity, sequence specific interaction with the MS2 stem loops. The recruitment of GFP to the site of transcription is visualised as a single fluorescent spot. This remarkable new approach has revealed that transcription occurs in discontinuous bursts, or pulses. With the notable exception of in situ techniques, most other methods provide cell population averages, and are not capable of detecting this fundamental property of genes.
- Northern blot: the traditional method, and until the advent of RNA-Seq, the most quantitative
- RNA-Seq: applies next-generation sequencing techniques to sequence whole transcriptomes, which allows the measurement of relative abundance of RNA, as well as the detection of additional variations such as fusion genes, post-translational edits and novel splice sites.

Transcription Factories

Active transcription units are clustered in the nucleus, in discrete sites called transcription factories or euchromatin. Such sites can be visualized by allowing engaged polymerases to extend their transcripts in tagged precursors (Br-UTP or Br-U) and immuno-labeling the tagged nascent RNA. Transcription factories can also be localized using fluorescence in situ hybridization or marked by antibodies directed against polymerases. There are ~10,000 factories in the nucleoplasm of a HeLa cell, among which are ~8,000 polymerase II factories and ~2,000 polymerase III factories. Each polymerase II factory contains ~8 polymerases. As most active transcription units are associated with only one polymerase, each factory usually contains ~8 different transcription units. These units might be associated through promoters and/ or enhancers, with loops forming a 'cloud' around the factor.

History

A molecule which allows the genetic material to be realized as a protein was first hypothesized by François Jacob and Jacques Monod. RNA synthesis by RNA polymerase was established *in vitro* by several laboratories by 1965; however, the RNA synthesized by these enzymes had properties that suggested the existence of an additional factor needed to terminate transcription correctly. In 1972, Walter Fiers became the first person to actually prove the existence of the terminating enzyme.

Reverse Transcription

Some viruses (such as HIV, the cause of AIDS), have the ability to transcribe RNA into DNA. HIV has an RNA genome that is duplicated into DNA. The resulting DNA can be merged with the DNA genome of the host cell. The main enzyme responsible for synthesis of DNA from an RNA template is called reverse transcriptase. In the case of HIV, reverse transcriptase is responsible for synthesizing a complementary DNA strand (cDNA) to the viral RNA genome.

An associated enzyme, ribonuclease H, digests the RNA strand, and reverse transcriptase synthesises a complementary strand of DNA to form a double helix DNA structure. This cDNA is integrated into the host cell's genome via another enzyme (integrase) causing the host cell to generate viral proteins which reassemble into new viral particles. Subsequently, the host cell undergoes programmed cell death, apoptosis. Some eukaryotic cells contain an enzyme with reverse transcription activity called telomerase. Telomerase is a reverse transcriptase that lengthens the ends of linear chromosomes. Telomerase carries an RNA template from which it synthesizes DNA repeating sequence, or "junk" DNA.

This repeated sequence of DNA is important because every time a linear chromosome is duplicated it is shortened in length. With "junk" DNA at the ends of chromosomes, the shortening eliminates some of the non-essential, repeated sequence rather than the protein-encoding DNA sequence farther away from the chromosome end.

Telomerase is often activated in cancer cells to enable cancer cells to duplicate their genomes indefinitely without losing important protein-coding DNA sequence. Activation of telomerase could be part of the process that allows cancer cells to become technically immortal. However, the true *in vivo* significance of telomerase has still not been empirically proven.

Designing Designs for Research

Reprinted from Trochim, W. and Land, D. (1982). Designing Designs for Research. The Researcher, 1, 1, 1-6.

Much contemporary social research is devoted to examining whether a program, treatment, or manipulation causes some outcome or result. For example, we might wish to know whether a new educational program causes subsequent achievement score gains, whether a special work release program for prisoners causes lower recidivism rates, whether a novel drug causes a reduction in symptoms, and so on. Cook and Campbell (1979) argue that three conditions must be met before we can infer that such a cause-effect relation exists:

1. Covariation. Changes in the presumed cause must be related to changes in the presumed effect. Thus, if we introduce, remove, or change the level of a treatment or program, we should observe some change in the outcome measures.
2. Temporal Precedence. The presumed cause must occur prior to the presumed effect.
3. No Plausible Alternative Explanations. The presumed cause must be the only reasonable explanation for changes in the outcome measures. If there are other factors which could be responsible for changes in the outcome measures we cannot be confident that the presumed cause-effect relationship is correct.

In most social research the third condition is the most difficult to meet. Any number of factors other than the treatment or program could cause changes in outcome measures. Campbell and Stanley (1966) and later, Cook and Campbell (1979) list a number of common plausible alternative explanations (or, threats to internal validity). For example, it may be that some historical event which occurs at the same time that the program or treatment is instituted was responsible for the change in the outcome measures; or, changes in record keeping or measurement systems which occur at the same time as the program might be falsely attributed to the program. The reader is referred to standard research methods texts for more detailed discussions of threats to validity.

This paper is primarily heuristic in purpose. Standard social science methodology textbooks (Cook and Campbell 1979; Judd and Kenny, 1981) typically present an array of research designs and the alternative explanations which these designs rule out or minimize. This tends to foster a "cookbook" approach to research design - an emphasis on the selection of an available design rather than on the construction of an appropriate research strategy. While standard designs may sometimes fit real-life situations, it will often be necessary to "tailor" a research design to minimize specific threats to validity. Furthermore, even if standard textbook designs are used, an understanding of the logic of design construction in general will improve the comprehension of these standard approaches. This paper takes a structural approach to research design. While this is by no means the only strategy for constructing research designs, it helps to clarify some of the basic principles of design logic.

Minimizing Threats to Validity

Good research designs minimize the plausible alternative explanations for the hypothesized cause-effect relationship. But such explanations may be ruled out or minimized in a number of ways other than by design. The discussion which follows outlines five ways to minimize threats to validity, one of which is by research design:

By Argument: The most straightforward way to rule out a potential threat to validity is to simply argue that the threat in question is not a reasonable one. Such an argument may be made either *a priori* or *a posteriori*, although the former will usually be more convincing than the latter. For example, depending on the situation, one might argue that an instrumentation threat is not likely because the same test is used for pre and post test measurements and did not involve observers who might improve, or other such factors. In most cases, ruling out a potential threat to validity by argument alone will be weaker than the other approaches listed below. As a result, the most plausible threats in a study should not, except in unusual cases, be ruled out by argument only.

By Measurement or Observation: In some cases it will be possible to rule out a threat by measuring it and demonstrating that either it does not occur at all or occurs so minimally as to not be a strong alternative explanation for the cause-effect relationship. Consider, for example, a study of the effects of an advertising campaign on subsequent sales of a particular product. In such a study, history (i.e., the occurrence of other events which might lead to an increased desire to purchase the product) would be a plausible alternative explanation. For example, a change in the local economy, the removal of a competing product from the market, or similar events could cause an increase in product sales. One might attempt to minimize such threats by measuring local economic indicators and the availability and sales of competing products. If there is no change in these measures coincident with the onset of the advertising campaign, these threats would be considerably minimized. Similarly, if one is studying the effects of special mathematics training on math achievement scores of children, it might be useful to observe everyday classroom behaviour in order to verify that students were not receiving any additional math training to that provided in the study.

By Design: Here, the major emphasis is on ruling out alternative explanations by adding treatment or control groups, waves of measurement, and the like. This topic will be discussed in more detail below.

By Analysis: There are a number of ways to rule out alternative explanations using statistical analysis. One interesting example is provided by Jurs and Glass (1971). They suggest that one could study the plausibility of an attrition or mortality threat by conducting a two-way analysis of variance. One factor in this study would be the original treatment group designations (i.e., program vs. comparison group), while the other factor would be attrition (i.e., dropout vs. non-dropout group).

The dependent measure could be the pretest or other available pre-program measures. A main effect on the attrition factor would be indicative of a threat to external validity or generalizability, while an interaction between group and attrition factors would point to a possible threat to internal validity. Where both effects occur, it is reasonable to infer that there is a threat to both internal and external validity.

The plausibility of alternative explanations might also be minimized using covariance analysis. For example, in a study of the effects of "workfare" programs on social welfare case loads, one plausible alternative explanation might be the status of local economic conditions. Here, it might be possible to construct a measure of economic conditions and include that measure as a covariate in the statistical analysis.

One must be careful when using covariance adjustments of this type — "perfect" covariates do not exist in most social research and the use of imperfect covariates will not completely adjust for potential alternative explanations. Nevertheless causal assertions are likely to be strengthened by demonstrating that treatment effects occur even after adjusting on a number of good covariates.

By Preventive Action: When potential threats are anticipated they can often be ruled out by some type of preventive action. For example, if the program is a desirable one, it is likely that the comparison group would feel jealous or demoralized. Several actions can be taken to minimize the effects of these attitudes including offering the program to the comparison group upon completion of the study or using program and comparison groups which have little opportunity for contact and communication. In addition, auditing methods and quality control can be used to track potential experimental dropouts or to insure the standardization of measurement.

The five categories listed above should not be considered mutually exclusive. The inclusion of measurements designed to minimize threats to validity will obviously be related to the design structure and is likely to be a factor in the analysis. A good research plan should, where possible. Make use of multiple methods for reducing threats. In general, reducing a particular threat by design or preventive action will probably be stronger than by using one of the other three approaches. The choice of which strategy to use for any particular threat is complex and depends at least on the cost of the strategy and on the potential seriousness of the threat.

Advances in Quasi-Experimentation

Reprinted from Trochim, W. (Ed.), (1986). *Editor's Notes.* Advances in quasi-experimental design and analysis. New Directions for Program Evaluation Series, Number 31, San Francisco, CA: Jossey-Bass.

The intent of this volume is to update, perhaps even to alter, our thinking about quasi-experimentation in applied social research and program evaluation. Since Campbell and Stanley (1963) introduced the term quasi-experiment, we have tended to see this area as involving primarily two interrelated topics: the theory of the validity of casual inferences and a taxonomy of the research designs that enable us to examine causal hypotheses.

We can see this in the leading expositions of quasi-experimentation (Campbell and Stanley, 1963, 1966; Cook and Campbell, 1979) as well as in the standard textbook presentations of the topic (Kidder and Judd, 1986; Rossi and Freeman, 1985), where it is typical to have separate sections or chapters that discuss validity issues first and then proceed to distinguishable quasi-experimental designs (for example, the pretest-posttest non-equivalent group design, the regression-discontinuity design, the interrupted time series design).

My first inclination in editing this volume was to emulate this tradition, beginning the volume with a chapter on validity and following it with a chapter for each of the major quasi-experimental designs that raised the relevant conceptual and analytical issues and discussed recent advances. But, I think, such an approach would have simply contributed to a persistent confusion about the nature of quasi-experimentation and its role in research.

Instead, this volume makes the case that we have moved beyond the traditional thinking on quasi-experiments as a collection of specific designs and threats to validity toward a more integrated, synthetic view of quasi-experimentation as part of a general logical and epistemological framework for research.

To support this view that the notion of quasi-experimentation is evolving toward increasing integration, I will discuss a number of themes that seem to characterize our current thinking and that cut across validity typologies and design taxonomies. This list of themes may also be viewed as a tentative description of the advances in our thinking about quasi-experimentation in social research.

The Role of Judgment

One theme that underlies most of the others and that illustrates our increasing awareness of the tentativeness and frailty of quasi-experimentation concerns the importance of human judgment in research. Evidence bearing on a causal relationship emerges from many sources, and it is not a trivial matter to integrate or resolve conflicts or discrepancies.

In recognition of this problem of evidence, we are beginning to address causal inference as a psychological issue that can be illuminated by cognitive models of the judgmental process. We are also recognizing more clearly the sociological bases of scientific thought (Campbell, 1984) and the fact that science is at root a human enterprise.

Thus, a positivist, mechanistic view is all but gone from quasi-experimental thinking, and what remains is a more judgmental and more scientifically sensible perspective.

The Case for Tailored Designs

Early expositions of quasi-experimentation took a largely taxonomic approach, laying out a collection of relatively discrete research designs and discussing how weak or strong they were for valid causal inference. Almost certainly, early proponents recognized that there was a virtual infinity of design variations and that validity was more complexly related to theory and context than their presentations implied. Nonetheless, what seemed to evolve was a "cookbook" approach to quasi-experimentation that involved "choosing" a design that fit the situation and checking off lists of validity threats.

In an important paper on the coupling of randomized and non-randomized design features, Boruch (1975) explicitly encouraged us to construct research designs as combinations of more elemental units (for example, assignment strategies, measurement occasions) based on the specific contextual needs and plausible alternative explanations for a treatment effect.

This move toward hybrid, tailored, or patched-up designs, which involved suggesting how such designs could be accomplished, is one in which I have been a minor participant (Trochim and Land, 1982; Trochim, 1984).

It is emphasized by Cordray in Chapter One of this volume. The implication for current practice is that we should focus on the advantages of different combinations of design features rather than on a relatively restricted set of prefabricated designs.

In teaching quasi-experimental methods, we need to break away from a taxonomic design mentality and emphasize design principles and issues that cut across the traditional distinctions between true experiments, non-experiments, and quasi-experiments.

The Crucial Role of Theory

Quasi-experimentation and its randomized experimental parent have been criticized for encouraging an atheoretical "black box" mentality of research (see, for instance, Chen and Rossi, 1984; Cronbach, 1982). Persons are assigned to either complex molar program packages or (often) to equally complex comparison conditions.

The machinery of random assignment (or our quasi-experimental attempts to approximate random assignment) are the primary means of defining whether the program has an effect. This *ceteris paribus* mentality is inherently atheoretical and non-contextual: It assumes that the same mechanism works in basically the same way whether we apply it in mental health or criminal justice, income maintenance or education.

There is nothing inherently wrong with this program-group-versus-comparison-group logic. The problem is that it may be a rather crude, uninformative approach. In the two-group case, we are simply creating a dichotomous input into reality. If we observe a posttest difference between groups, it could be explained by this dichotomous program-versus-comparison-group input or by any number of alternative explanations, including differential attrition rates, intergroup rivalry and communication, initial selection differences among groups, or different group histories.

We usually try to deal with these alternative explanations by ruling them out through argument, additional measurement, patched-up design features, and auxiliary analysis. Cook and Campbell (1979), Cronbach (1982), and others strongly favour replication of treatment effects as a standard for judging the validity of a causal assertion, but this advice does little to enhance the validity and informativeness within individual studies or program evaluations.

Chen and Rossi approached this issue by advocating increased attention to social science theory: "not the global conceptual schemes of the grand theorists but much more prosaic theories that are concerned with how human organizations work and how social problems are generated." Evaluators have similarly begun to stress the importance of program theory as the basis for causal assessment (for example, Bickman, in press).

These developments allow increased emphasis to be placed on the role of pattern matching (Trochim, 1985) through the generation of more complex theory-driven predictions that, if corroborated, allow fewer plausible alternative explanations for the effect of a program. Because appropriate theories may not be readily available, especially for the evaluation of contemporary social programs, we are developing methods and processes that facilitate the articulation of the implicit theories which program administrators and stakeholder groups have in mind and which presumably guide the formation and implementation of the program (Trochim, 1985). This theory-driven perspective is consonant with Mark's emphasis in Chapter Three on the study of causal process and with Cordray's discussion in Chapter One on ruling in the program as opposed to ruling out alternative explanations.

Attention to Program Implementation

A theory-driven approach to quasi-experimentation will be futile unless we can demonstrate that the program was in fact carried out or implemented as the theory intended. Consequently, we have seen the development of program implementation theory (for example, McLaughlin, 1984) that directly addresses the process of program execution. One approach emphasizes the development of organizational procedures and training systems that accurately transmit the program and that anticipate likely institutional sources of resistance. Another strategy involves the assessment of program delivery through program audits, management information systems, and the like. This emphasis on program implementation has further obscured the traditional distinction between process and outcome evaluation. At the least, it is certainly clear that good quasi-experimental outcome evaluation cannot be accomplished without attending to program processes, and we are continuing to develop better notions of how to combine these two efforts.

The Importance of Quality Control

Over and over, our experience with quasi-experimentation has shown that even the best-laid research plans often go awry in practice, sometimes with disastrous results. Thus, over the past decade we have begun to pay increasing attention to the integrity and quality of our research methods in real-world settings. One way of achieving this goal is to incorporate techniques used by other professions — accounting, auditing, industrial quality control — that have traditions in data integrity and quality assurance (Trochim and Visco, 1985). For instance, double bookkeeping can be used to keep verifiable records of research participation. Acceptance sampling can be an efficient method for checking accuracy in large data collection efforts, where an exhaustive examination of records is impractical or excessive in cost. These issues are particularly important in quasi-experimentation, where it is incumbent upon the researcher to demonstrate that sampling, measurement, group assignment, and analysis decisions do not interact with program participation in ways that can confound the final interpretation of results.

The Advantages of Multiple Perspectives

We have long recognized the importance of replication and systematic variation in research. In the past few years, Cook (1985) and colleagues Shadish and Houts have articulated a rationale for achieving systematic variation that they term critical multiplism. This perspective rests on the notion that no single realization will ever be sufficient for understanding a phenomenon with validity. Multiple realizations — of research questions, measures, samples, designs, analyses, replications, and so on — are essential for convergence on the truth of a matter. However, such a varied approach can become a methodological and epistemological Pandora's box unless we apply critical judgment in deciding which multiples we will emphasize in a study or set of studies.

Evolution of the Concept of Validity

The history of quasi-experimentation is inseparable from the development of the theory of the validity of causal inference. Much of this history has been played out through the ongoing dialogue between Campbell and Cronbach concerning the definition of validity and the relative importance that should be attributed on the one hand to the establishment of a causal relationship and on the other hand to its generalizability. In the most recent major statement in this area, Cronbach (1982) articulated the UTOS model, which conceptually links the units, treatments, observing operations and settings in a study into a framework that can be used for establishing valid causal inference.

The dialogue continues in Chapter Four of this volume, where Campbell attempts to dispel persistent confusion about the types of validity by tentatively relabeling internal validity as local molar causal validity and external validity as the principle of proximal similarity. It is reasonable to hope that we might achieve a clearer consensus on this issue, as Mark argues in Chapter Three, where he attempts to resolve several different conceptions of validity, including those of Campbell and Cronbach.

Development of Increasingly Complex Realistic Analytic Models

In the past decade, we have made considerable progress toward complicating our statistical analyses to account for increasingly complex contexts and designs. One such advance involves the articulation of causal models of the sort described by Reichardt and Gollob in Chapter Six, especially models that allow for latent variables and that directly model measurement error Joreskog and Sorbom, 1979.

Another important recent development involves analyses that address the problem of selection bias or group nonequivalence — a central issue in quasi-experiments because random assignment is not used and there is no assurance that comparison groups are initially equivalent. At the same time, there is increasing recognition of the implications of not attending to the correct unit of analysis when analyzing the data and of the advantages and implications of conducting analyses at multiple levels. Thus, when we assign classrooms to conditions but analyze individual student data rather than classroom aggregates, we are liable to get a different view of program effects than we are when we analyze at the classroom level, as Shadish, Cook, and Houts argue in Chapter Two.

Other notable advances that are not explicitly addressed in this volume include the development of log linear, probit, and logit models for the analysis of qualitative or nominal level outcome variables (Feinberg, 1980; Forthofer and Lehnen, 1981) and the increasing proliferation of Bayesian statistical approaches to quasi-experimental contexts (Pollard, 1986).

Parallel to the development of these increasingly complex, realistic analytic models, cynicism has deepened about the ability of any single model or analysis to be sufficient. Thus, in Chapter Six Reichardt and Gollob call for multiple analyses to bracket bias, and in Chapter Five Rindskopf recognizes the assumptive notions of any analytic approach to selection bias.

We have virtually abandoned the hope of a single correct analysis, and we have accordingly moved to multiple analyses that are based on systematically distinct assumptional frameworks and that rely in an increasingly direct way on the role of judgment.

All the developments just outlined point to an increasingly realistic and complicated life for quasi-experimentalists. The overall picture that emerges is that all quasi-experimentation is judgmental.

It is based on multiple and varied sources of evidence, it should be multiplistic in realization, it must attend to process as well as to outcome, it is better off when theory driven, and it leads ultimately to multiple analyses that attempt to bracket the program effect within some reasonable range.

In one sense, this is hardly a pretty picture. Our views about quasi-experimentation and its role in causal inference are certainly more tentative and critical than they were in 1965 or perhaps even in 1979.

But, this more integrated and complex view of quasi-experimentation has emerged directly from our experiences in the conduct of such studies. As such, it realistically represents our current thinking about one of the major strands in the evolution of social research methodology in this century.

Analysis

By the time you get to the analysis of your data, most of the really difficult work has been done. It's much more difficult to: define the research problem; develop and implement a sampling plan; conceptualize, operationalize and test your measures; and develop a design structure. If you have done this work well, the analysis of the data is usually a fairly straightforward affair.

In most social research the data analysis involves three major steps, done in roughly this order:

- Cleaning and organizing the data for analysis (Data Preparation)
- Describing the data (Descriptive Statistics)
- Testing Hypotheses and Models (Inferential Statistics)

Data Preparation involves checking or logging the data in; checking the data for accuracy; entering the data into the computer; transforming the data; and developing and documenting a database structure that integrates the various measures. Descriptive Statistics are used to describe the basic features of the data in a study. They provide simple summaries about the sample and the measures. Together with simple graphics analysis, they form the basis of virtually every quantitative analysis of data. With descriptive statistics you are simply describing what is, what the data shows.

Inferential Statistics investigate questions, models and hypotheses. In many cases, the conclusions from inferential statistics extend beyond the immediate data alone. For instance, we use inferential statistics to try to infer from the sample data what the population thinks. Or, we use inferential statistics to make judgments of the probability that an observed difference between groups is a dependable one or one that might have happened by chance in this study. Thus, we use inferential statistics to make inferences from our data to more general conditions; we use descriptive statistics simply to describe what's going on in our data.

In most research studies, the analysis section follows these three phases of analysis. Descriptions of how the data were prepared tend to be brief and to focus on only the more unique aspects to your study, such as specific data transformations that are performed. The descriptive statistics that you actually look at can be voluminous. Usually, the researcher links each of the inferential analyses to specific research questions or hypotheses that were raised in the introduction, or notes any models that were tested that emerged as part of the analysis. In most analysis write-ups it's especially critical to not "miss the forest for the trees." If you present too much detail, the reader may not be able to follow the central line of the results. Often extensive analysis details are appropriately relegated to appendices, reserving only the most critical analysis summaries for the body of the report itself.

DNA Biochemistry

Deoxyribonucleic acid, or DNA, is a nucleic acid that contains the genetic instructions used in the development and functioning of all known living organisms with the exception of some viruses. The main role of DNA molecules is the long-term storage of information. DNA is often compared to a set of blueprints, like a recipe or a code, since it contains the instructions needed to construct other components of cells, such as proteins and RNA molecules. The DNA segments that carry this genetic information are called genes, but other DNA sequences have structural purposes, or are involved in regulating the use of this genetic information.

DNA consists of two long polymers of simple units called nucleotides, with backbones made of sugars and phosphate groups joined by ester bonds. These two strands run in opposite directions to each other and are therefore anti-parallel. Attached to each sugar is one of four types of molecules called bases.

It is the sequence of these four bases along the backbone that encodes information. This information is read using the genetic code, which specifies the sequence of the amino acids within proteins. The code is read by copying stretches of DNA into the related nucleic acid RNA, in a process called transcription.

Within cells, DNA is organized into long structures called chromosomes. These chromosomes are duplicated before cells divide, in a process called DNA replication. Eukaryotic organisms (animals, plants, fungi, and protists) store most of their DNA inside the cell nucleus and some of their DNA in organelles, such as mitochondria or chloroplasts. In contrast, prokaryotes (bacteria and archaea) store their DNA only in the cytoplasm. Within the chromosomes, chromatin proteins such as histones compact and organize DNA. These compact structures guide the interactions between DNA and other proteins, helping control which parts of the DNA are transcribed.

Properties

DNA is a long polymer made from repeating units called nucleotides. As first discovered by James D. Watson and Francis Crick, the structure of DNA of all species comprises two helical chains each coiled round the same axis, and each with a pitch of 34 Ångströms (3.4 nanometres) and a radius of 10 Ångströms (1.0 nanometres). According to another study, when measured in a particular solution, the DNA chain measured 22 to 26 Ångströms wide (2.2 to 2.6 nanometres), and one nucleotide unit measured 3.3 Å (0.33 nm) long. Although each individual repeating unit is very small, DNA polymers can be very large molecules containing millions of nucleotides. For instance, the largest human chromosome, chromosome number 1, is approximately 220 million base pairs long.

In living organisms, DNA does not usually exist as a single molecule, but instead as a pair of molecules that are held tightly together. These two long strands entwine like vines, in the shape of a double helix. The nucleotide repeats contain both the segment of the backbone of the molecule, which holds the chain together, and a base, which interacts with the other DNA strand in the helix. A base linked to a sugar is called a nucleoside and a base linked to a sugar and one or more phosphate groups is called a nucleotide. If multiple nucleotides are linked together, as in DNA, this polymer is called a polynucleotide.

The backbone of the DNA strand is made from alternating phosphate and sugar residues. The sugar in DNA is 2-deoxyribose, which is a pentose (five-carbon) sugar. The sugars are joined together by phosphate groups that form phosphodiester bonds between the third and fifth carbon atoms of adjacent sugar rings.

These asymmetric bonds mean a strand of DNA has a direction. In a double helix the direction of the nucleotides in one strand is opposite to their direction in the other strand: the strands are *antiparallel*. The asymmetric ends of DNA strands are called the 52 (*five prime*) and 32 (*three prime*) ends, with the 5' end having a terminal phosphate group and the 3' end a terminal hydroxyl group. One major difference between DNA and RNA is the sugar, with the 2-deoxyribose in DNA being replaced by the alternative pentose sugar ribose in RNA.

The DNA double helix is stabilized by hydrogen bonds between the bases attached to the two strands. The four bases found in DNA are adenine (abbreviated A), cytosine (C), guanine (G) and thymine (T). These four bases are attached to the sugar/phosphate to form the complete nucleotide.

These bases are classified into two types; adenine and guanine are fused five- and six-membered heterocyclic compounds called purines, while cytosine and thymine are six-membered rings called pyrimidines. A fifth pyrimidine base, called uracil (U), usually takes the place of thymine in RNA and differs from thymine by lacking a methyl group on its ring. Uracil is not usually found in DNA, occurring only as a breakdown product of cytosine. In addition to RNA and DNA, a large number of artificial nucleic acid analogues have also been created to study the proprieties of nucleic acids, or for use in biotechnology.

Grooves

Twin helical strands form the DNA backbone. Another double helix may be found by tracing the spaces, or grooves, between the strands. These voids are adjacent to the base pairs and may provide a binding site. As the strands are not directly opposite each other, the grooves are unequally sized. One groove, the major groove, is 22 Å wide and the other, the minor groove, is 12 Å wide. The narrowness of the minor groove means that the edges of the bases are more accessible in the major groove.

As a result, proteins like transcription factors that can bind to specific sequences in double-stranded DNA usually make contacts to the sides of the bases exposed in the major groove. This situation varies in unusual conformations of DNA within the cell, but the major and minor grooves are always named to reflect the differences in size that would be seen if the DNA is twisted back into the ordinary B form.

Base Pairing

Each type of base on one strand forms a bond with just one type of base on the other strand. This is called complementary base pairing. Here, purines form hydrogen bonds to pyrimidines, with A bonding only to T, and C bonding only to G. This arrangement of two nucleotides binding together across the double helix is called a base pair. As hydrogen bonds are not covalent, they can be broken and rejoined relatively easily.

The two strands of DNA in a double helix can therefore be pulled apart like a zipper, either by a mechanical force or high temperature. As a result of this complementarity, all the information in the double-stranded sequence of a DNA helix is duplicated on each strand, which is vital in DNA replication. Indeed, this reversible and specific interaction between complementary base pairs is critical for all the functions of DNA in living organisms.

The two types of base pairs form different numbers of hydrogen bonds, AT forming two hydrogen bonds, and GC forming three hydrogen bonds. DNA with high GC-content is more stable than DNA with low GC-content, but contrary to popular belief, this is not due to the extra hydrogen bond of a GC base pair but rather the contribution of stacking interactions (hydrogen bonding merely provides specificity of the pairing, not stability). As a result, it is both the percentage of GC base pairs and the overall length of a DNA double helix that determine the strength of the association between the two strands of DNA. Long DNA helices with a high GC content have stronger-interacting strands, while short helices with high AT content have weaker-interacting strands.

In biology, parts of the DNA double helix that need to separate easily, such as the TATAAT Pribnow box in some promoters, tend to have a high AT content, making the strands easier to pull apart. In the laboratory, the strength of this interaction can be measured by finding the temperature required to break the hydrogen bonds, their melting temperature (also called T_m value). When all the base pairs in a DNA double helix melt, the strands separate and exist in solution as two entirely independent molecules. These single-stranded DNA molecules (*ssDNA*) have no single common shape, but some conformations are more stable than others.

Sense and Antisense

A DNA sequence is called "sense" if its sequence is the same as that of a messenger RNA copy that is translated into protein. The sequence on the opposite strand is called the "antisense" sequence. Both sense and antisense sequences can exist on different parts of the same strand of DNA (i.e. both strands contain both sense and antisense sequences). In both prokaryotes and eukaryotes, antisense RNA sequences are produced, but the functions of these RNAs are not entirely clear.

One proposal is that antisense RNAs are involved in regulating gene expression through RNA-RNA base pairing. A few DNA sequences in prokaryotes and eukaryotes, and more in plasmids and viruses, blur the distinction between sense and antisense strands by having overlapping genes. In these cases, some DNA sequences do double duty, encoding one protein when read along one strand, and a second protein when read in the opposite direction along the other strand. In bacteria, this overlap may be involved in the regulation of gene transcription, while in viruses, overlapping genes increase the amount of information that can be encoded within the small viral genome.

Supercoiling

DNA can be twisted like a rope in a process called DNA supercoiling. With DNA in its "relaxed" state, a strand usually circles the axis of the double helix once every 10.4 base pairs, but if the DNA is twisted the strands become more tightly or more loosely wound. If the DNA is twisted in the direction of the helix, this is positive supercoiling, and the bases are held more tightly together. If they are twisted in the opposite direction, this is negative supercoiling, and the bases come apart more easily. In nature, most DNA has slight negative supercoiling that is introduced by enzymes called topoisomerases. These enzymes are also needed to relieve the twisting stresses introduced into DNA strands during processes such as transcription and DNA replication.

Alternate DNA Structures

DNA exists in many possible conformations that include A-DNA, B-DNA, and Z-DNA forms, although, only B-DNA and Z-DNA have been directly observed in functional organisms. The conformation that DNA adopts depends on the hydration level, DNA sequence, the amount and direction of supercoiling, chemical modifications of the bases, the type and concentration of metal ions, as well as the presence of polyamines in solution.

The first published reports of A-DNA X-ray diffraction patterns— and also B-DNA used analyses based on Patterson transforms that provided only a limited amount of structural information for oriented fibers of DNA. An alternate analysis was then proposed by Wilkins *et al.*, in 1953, for the *in vivo* B-DNA X-ray diffraction/scattering patterns of highly hydrated DNA fibers in terms of squares of Bessel functions. In the same journal, James D. Watson and Francis Crick presented their molecular modeling analysis of the DNA

X-ray diffraction patterns to suggest that the structure was a double-helix. Although the 'B-DNA form' is most common under the conditions found in cells, it is not a well-defined conformation but a family of related DNA conformations that occur at the high hydration levels present in living cells. Their corresponding X-ray diffraction and scattering patterns are characteristic of molecular paracrystals with a significant degree of disorder. Compared to B-DNA, the A-DNA form is a wider right-handed spiral, with a shallow, wide minor groove and a narrower, deeper major groove.

The A form occurs under non-physiological conditions in partially dehydrated samples of DNA, while in the cell it may be produced in hybrid pairings of DNA and RNA strands, as well as in enzyme-DNA complexes. Segments of DNA where the bases have been chemically modified by methylation may undergo a larger change in conformation and adopt the Z form.

Here, the strands turn about the helical axis in a left-handed spiral, the opposite of the more common B form. These unusual structures can be recognized by specific Z-DNA binding proteins and may be involved in the regulation of transcription.

Quadruplex Structures

At the ends of the linear chromosomes are specialized regions of DNA called telomeres. The main function of these regions is to allow the cell to replicate chromosome ends using the enzyme telomerase, as the enzymes that normally replicate DNA cannot copy the extreme 32 ends of chromosomes. These specialized chromosome caps also help protect the DNA ends, and stop the DNA repair systems in the cell from treating them as damage to be corrected. In human cells, telomeres are usually lengths of single-stranded DNA containing several thousand repeats of a simple TTAGGG sequence.

These guanine-rich sequences may stabilize chromosome ends by forming structures of stacked sets of four-base units, rather than the usual base pairs found in other DNA molecules. Here, four guanine bases form a flat plate and these flat four-base units then stack on top of each other, to form a stable *G-quadruplex* structure.

These structures are stabilized by hydrogen bonding between the edges of the bases and chelation of a metal ion in the centre of each four-base unit. Other structures can also be formed, with the central set of four bases coming from either a single strand folded around the bases, or several different parallel strands, each contributing one base to the central structure.

In addition to these stacked structures, telomeres also form large loop structures called telomere loops, or T-loops. Here, the single-stranded DNA curls around in a long circle stabilized by telomere-binding proteins. At the very end of the T-loop, the single-stranded telomere DNA is held onto a region of double-stranded DNA by the telomere strand disrupting the double-helical DNA and base pairing to one of the two strands. This triple-stranded structure is called a displacement loop or D-loop.

Branched DNA

In DNA fraying occurs when non-complementary regions exist at the end of an otherwise complementary double-strand of DNA. However, branched DNA can occur if a third strand of DNA is introduced and contains adjoining regions able to hybridize with the frayed regions of the pre-existing double-strand. Although the simplest example of branched DNA involves only three strands of DNA, complexes involving additional strands and multiple branches are also possible. Branched DNA can be used in nanotechnology to construct geometric shapes.

Chemical Modifications

Base Modifications

The expression of genes is influenced by how the DNA is packaged in chromosomes, in a structure called chromatin. Base modifications can be involved in packaging, with regions that have low or no gene expression usually containing high levels of methylation of cytosine bases. For example, cytosine methylation, produces 5-methylcytosine, which is important for X-chromosome inactivation. The average level of methylation varies between organisms - the worm *Caenorhabditis elegans* lacks cytosine methylation, while vertebrates have higher levels, with up to 1% of their DNA containing 5-methylcytosine. Despite the importance of 5-methylcytosine, it can deaminate to leave a thymine base, so methylated cytosines are particularly prone to mutations. Other base modifications include adenine methylation in bacteria, the presence of 5-hydroxymethylcytosine in the brain, and the glycosylation of uracil to produce the "J-base" in kinetoplastids.

Damage

DNA can be damaged by many sorts of mutagens, which change the DNA sequence. Mutagens include oxidizing agents, alkylating agents and also high-energy electromagnetic radiation such as ultraviolet light and X-rays. The type of DNA damage produced depends on the type of mutagen. For example, UV light can damage DNA by producing thymine dimers, which are cross-links between pyrimidine bases. On the other hand, oxidants such as free radicals or hydrogen peroxide produce multiple forms of damage, including base modifications, particularly of guanosine, and double-strand breaks. A typical human cell contains about 150,000 bases that have suffered oxidative damage.

Of these oxidative lesions, the most dangerous are double-strand breaks, as these are difficult to repair and can produce point mutations, insertions and deletions from the DNA sequence, as well as chromosomal translocations.

Many mutagens fit into the space between two adjacent base pairs, this is called *intercalation*. Most intercalators are aromatic and planar molecules; examples include ethidium bromide, daunomycin, and doxorubicin. In order for an intercalator to fit between base pairs, the bases must separate, distorting the DNA strands by unwinding of the double helix. This inhibits both transcription and DNA replication, causing toxicity and mutations. As a result, DNA intercalators are often carcinogens, and benzo[*a*]pyrene diol epoxide, acridines, aflatoxin and ethidium bromide are well-known examples. Nevertheless, due to their ability to inhibit DNA transcription and replication, other similar toxins are also used in chemotherapy to inhibit rapidly growing cancer cells.

Biological Functions

DNA usually occurs as linear chromosomes in eukaryotes, and circular chromosomes in prokaryotes. The set of chromosomes in a cell makes up its genome; the human genome has approximately 3 billion base pairs of DNA arranged into 46 chromosomes. The information carried by DNA is held in the sequence of pieces of DNA called genes. Transmission of genetic information in genes is achieved via complementary base pairing.

For example, in transcription, when a cell uses the information in a gene, the DNA sequence is copied into a complementary RNA sequence through the attraction between the DNA and the correct RNA nucleotides. Usually, this RNA copy is then used to make a matching protein sequence in a process called translation, which depends on the same interaction between RNA nucleotides. In alternative fashion, a cell may simply copy its genetic information in a process called DNA replication. The details of these functions are covered in other articles; here we focus on the interactions between DNA and other molecules that mediate the function of the genome.

Genes and Genomes

Genomic DNA is tightly and orderly packed in the process called DNA condensation to fit the small available volumes of the cell. In eukaryotes, DNA is located in the cell nucleus, as well as small amounts in mitochondria and chloroplasts. In prokaryotes, the DNA is held within an irregularly shaped body in the cytoplasm called the nucleoid. The genetic information in a genome is held within genes, and the complete set of this information in an organism is called its genotype. A gene is a unit of heredity and is a region of DNA that influences a particular characteristic in an organism. Genes contain an open reading frame that can be transcribed, as well as regulatory sequences such as promoters and enhancers, which control the transcription of the open reading frame. In many species, only a small fraction of the total sequence of the genome encodes protein.

For example, only about 1.5% of the human genome consists of protein-coding exons, with over 50% of human DNA consisting of non-coding repetitive sequences. The reasons for the presence of so much non-coding DNA in eukaryotic genomes and the extraordinary differences in genome size, or *C-value*, among species represent a long-standing puzzle known as the "C-value enigma". However, DNA sequences that do not code protein may still encode functional non-coding RNA molecules, which are involved in the regulation of gene expression.

Some non-coding DNA sequences play structural roles in chromosomes. Telomeres and centromeres typically contain few genes, but are important for the function and stability of chromosomes. An abundant form of non-coding DNA in humans are pseudogenes, which are copies of genes that have been disabled by mutation. These sequences are usually just molecular fossils, although they can occasionally serve as raw genetic material for the creation of new genes through the process of gene duplication and divergence.

Transcription and Translation

A gene is a sequence of DNA that contains genetic information and can influence the phenotype of an organism. Within a gene, the sequence of bases along a DNA strand defines a messenger RNA sequence, which then defines one or more protein sequences. The relationship between the nucleotide sequences of genes and the amino-acid sequences of proteins is determined by the rules of translation, known collectively as the genetic code. The genetic code consists of three-letter 'words' called *codons* formed from a sequence of three nucleotides (e.g. ACT, CAG, TTT).

In transcription, the codons of a gene are copied into messenger RNA by RNA polymerase. This RNA copy is then decoded by a ribosome that reads the RNA sequence by base-pairing the messenger RNA to transfer RNA, which carries amino acids. Since there are 4 bases in 3-letter combinations, there are 64 possible codons (4^3 combinations). These encode the twenty standard amino acids, giving most amino acids more than one possible codon. There are also three 'stop' or 'nonsense' codons signifying the end of the coding region; these are the TAA, TGA and TAG codons.

DNA Replication

DNA replication, the basis for biological inheritance, is a fundamental process occurring in all living organisms to copy their DNA. This process is "replication" in that each strand of the original double-stranded DNA molecule serves as template for the reproduction of the complementary strand. Therefore, following DNA replication, two identical DNA molecules have been produced from a single double-stranded DNA molecule.

Cellular proofreading and error toe-checking mechanisms ensure near perfect fidelity for DNA replication. In a cell, DNA replication begins at specific locations in the genome, called "origins". Unwinding of DNA at the origin, and synthesis of new strands, forms a replication fork.

In addition to DNA polymerase, the enzyme that synthesizes the new DNA by adding nucleotides matched to the template strand, a number of other proteins are associated with the fork and assist in the initiation and continuation of DNA synthesis.

DNA replication can also be performed *in vitro* (outside a cell). DNA polymerases, isolated from cells, and artificial DNA primers are used to initiate DNA synthesis at known sequences in a template molecule. The polymerase chain reaction (PCR), a common laboratory technique, employs such artificial synthesis in a cyclic manner to amplify a specific target DNA fragment from a pool of DNA.

DNA usually exists as a double-stranded structure, with both strands coiled together to form the characteristic double-helix. Each single strand of DNA is a chain of four types of nucleotides: adenine, cytosine, guanine, and thymine. A nucleotide is a mono-, di- or triphosphate deoxyribonucleoside; that is, a deoxyribose sugar is attached to one, two or three phosphates.

Chemical interaction of these nucleotides forms phosphodiester linkages, creating the phosphate-deoxribose backbone of the DNA double helix with the bases pointing inward. Nucleotides (bases) are matched between strands through hydrogen bonds to form base pairs. Adenine pairs with thymine and cytosine pairs with guanine.

DNA strands have a directionality, and the different ends of a single strand are called the "3' (three-prime) end" and the "5' (five-prime) end." These terms refer to the carbon atom in deoxyribose to which the next phosphate in the chain attaches. In addition to being complementary, the two strands of DNA are antiparallel: they are orientated in opposite directions. This directionality has consequences in DNA synthesis, because DNA polymerase can only synthesize DNA in one direction by adding nucleotides to the 3' end of a DNA strand.

The pairing of bases in DNA through hydrogen bonding means that the information contained within each strand is redundant. The nucleotides on a single strand can be used to reconstruct nucleotides on a newly synthesized partner strand.

DNA Polymerase

DNA polymerases are a family of enzymes that carry out all forms of DNA replication. A DNA polymerase can only extend an existing DNA strand paired with a template strand; it cannot begin the synthesis of a new strand. To begin synthesis of a new strand, a short fragment of DNA or RNA, called a primer, must be created and paired with the template strand before DNA polymerase can synthesize new DNA. Once a primer pairs with DNA to be replicated, DNA polymerase synthesizes a new strand of DNA by extending the 3' end of an existing nucleotide chain, adding new nucleotides matched to the template strand one at a time via the creation of phosphodiester bonds.

The energy for this process of DNA polymerization comes from two of the three total phosphates attached to each unincorporated base. (Free bases with their attached phosphate groups are called nucleoside triphosphates.) When a nucleotide is being added to a growing DNA strand, two of the phosphates are removed and the energy produced creates a phosphodiester (chemical) bond that attaches the remaining phosphate to the growing chain.

The energetics of this process also help explain the directionality of synthesis - if DNA were synthesized in the 3' to 5' direction, the energy for the process would come from the 5' end of the growing strand rather than from free nucleotides.

DNA polymerases are generally extremely accurate, making less than one error for every 10^7 nucleotides added. Even so, some DNA polymerases also have proofreading ability; they can remove nucleotides from the end of a strand in order to correct mismatched bases.

If the 5' nucleotide needs to be removed during proofreading, the triphosphate end is lost. Hence, the energy source that usually provides energy to add a new nucleotide is also lost.

DNA Replication within the Cell

Origins of Replication

For a cell to divide, it must first replicate its DNA. This process is initiated at particular points within the DNA, known as "origins", which are targeted by proteins that separate the two strands and initiate DNA synthesis. Origins contain DNA sequences recognized by replication initiator proteins (e.g. dnaA in *E coli'* and the Origin Recognition Complex in yeast). These initiator proteins recruit other proteins to separate the two strands and initiate replication forks.

Initiator proteins recruit other proteins to separate the DNA strands at the origin, forming a bubble. Origins tend to be "AT-rich" (rich in adenine and thymine bases) to assist this process, because A-T base pairs have two hydrogen bonds (rather than the three formed in a C-G pair)—strands rich in these nucleotides are generally easier to separate due the positive relationship between the number of hydrogen bonds and the difficulty of breaking these bonds. Once strands are separated, RNA primers are created on the template strands.

More specifically, the leading strand receives one RNA primer per active origin of replication while the lagging strand receives several; these several fragments of RNA primers found on the lagging strand of DNA are called Okazaki fragments, named after their discoverer. DNA polymerase extends the leading strand in one continuous motion and the lagging strand in a discontinuous motion (due to the Okazaki fragments). RNase removes the RNA fragments used to initiate replication by DNA Polymerase, and another DNA Polymerase enters to fill the gaps. When this is complete, a single nick on the leading strand and several nicks on the lagging strand can be found. Ligase works to fill these nicks in, thus completing the newly replicated DNA molecule.

As DNA synthesis continues, the original DNA strands continue to unwind on each side of the bubble, forming 2 replication forks. In bacteria, which have a single origin of replication on their circular chromosome, this process eventually creates a "theta structure" (resembling the Greek letter theta: è). In contrast, eukaryotes have longer linear chromosomes and initiate replication at multiple origins within these.

The Replication Fork

The replication fork is a structure that forms within the nucleus during DNA replication. It is created by helicases, which break the hydrogen bonds holding the two DNA strands together. The resulting structure has two branching "prongs", each one made up of a single strand of DNA. These two strands serve as the template for the leading and lagging strands which will be created as DNA polymerase matches complementary nucleotides to the templates; The templates may be properly referred to as the leading strand template and the lagging strand template.

Leading Strand

The leading strand is the template strand of the DNA double helix so that the replication fork moves along it in the 3' to 5' direction. This allows the new strand synthesized complementary to it to be synthesized 5' to 3' in the same direction as the movement of the replication fork.

On the leading strand, a polymerase "reads" the DNA and adds nucleotides to it continuously. This polymerase is DNA polymerase III (DNA Pol III) in prokaryotes and presumably Pol ε in eukaryotes.

Lagging Strand

The lagging strand is the strand of the template DNA double helix that is oriented so that the replication fork moves along it in a 5' to 3' manner. Because of its orientation, opposite to the working orientation of DNA polymerase III, which moves on a template in a 3' to 5' manner, replication of the lagging strand is more complicated than that of the leading strand.

On the lagging strand, primase "reads" the DNA and adds RNA to it in short, separated segments. In eukaryotes, primase is intrinsic to Pol α. DNA polymerase III or Pol δ lengthens the primed segments, forming Okazaki fragments. Primer removal in eukaryotes is also performed by Pol δ. In prokaryotes, DNA polymerase I "reads" the fragments, removes the RNA using its flap endonuclease domain (RNA primers are removed by 5'-3' exonuclease activity of polymerase I [weaver, 2005], and replaces the RNA nucleotides with DNA nucleotides (this is necessary because RNA and DNA use slightly different kinds of nucleotides). DNA ligase joins the fragments together.

Dynamics at the Replication Fork

As helicase unwinds DNA at the replication fork, the DNA ahead is forced to rotate. This process results in a build-up of twists in the DNA ahead. This build-up would form a resistance that would eventually halt the progress of the replication fork. DNA topoisomerases are enzymes that solve these physical problems in the coiling of DNA. Topoisomerase I cuts a single backbone on the DNA, enabling the strands to swivel around each other to remove the build-up of twists. Topoisomerase II cuts both backbones, enabling one double-stranded DNA to pass through another, thereby removing knots and entanglements that can form within and between DNA molecules. Bare single-stranded DNA has a tendency to fold back upon itself and form secondary structures; these structures can interfere with the movement of DNA polymerase. To prevent this, single-strand binding proteins bind to the DNA until a second strand is synthesized, preventing secondary structure formation.

Clamp proteins form a sliding clamp around DNA, helping the DNA polymerase maintain contact with its template and thereby assisting with processivity. The inner face of the clamp enables DNA to be threaded through it. Once the polymerase reaches the end of the template or detects double stranded DNA, the sliding clamp undergoes a conformational change which releases the DNA polymerase. Clamp-loading proteins are used to initially load the clamp, recognizing the junction between template and RNA primers.

Regulation of Replication

Within eukaryotes, DNA replication is controlled within the context of the cell cycle. As the cell grows and divides, it progresses through stages in the cell cycle; DNA replication occurs during the S phase (Synthesis phase). The progress of the eukaryotic cell through the cycle is controlled by cell cycle checkpoints. Progression through checkpoints is controlled through complex interactions between various proteins, including cyclins and cyclin-dependent kinases. The G1/S checkpoint (or restriction checkpoint) regulates whether eukaryotic cells enter the process of DNA replication and subsequent division. Cells which do not proceed through this checkpoint are quiescent in the "G0" stage and do not replicate their DNA. Replication of chloroplast and mitochondrial genomes occurs independent of the cell cycle, through the process of D-loop replication.

Bacteria

Most bacteria do not go through a well-defined cell cycle and instead continuously copy their DNA; during rapid growth this can result in multiple rounds of replication occurring concurrently. Within *E coli*, the most well-characterized bacteria, regulation of DNA replication can be achieved through several mechanisms, including: the hemimethylation and sequestering of the origin sequence, the ratio of ATP to ADP, and the levels of protein DnaA. These all control the process of initiator proteins binding to the origin sequences.

Because *E coli* methylates GATC DNA sequences, DNA synthesis results in hemimethylated sequences. This hemimethylated DNA is recognized by a protein (SeqA) which binds and sequesters the origin sequence; in addition, dnaA (required for initiation of replication) binds less well to hemimethylated DNA. As a result, newly replicated origins are prevented from immediately initiating another round of DNA replication.

ATP builds up when the cell is in a rich medium, triggering DNA replication once the cell has reached a specific size. ATP competes with ADP to bind to DnaA, and the DnaA-ATP complex is able to initiate replication. A certain number of DnaA proteins are also required for DNA replication — each time the origin is copied the number of binding sites for DnaA doubles, requiring the synthesis of more DnaA to enable another initiation of replication.

Termination of Replication

Because bacteria have circular chromosomes, termination of replication occurs when the two replication forks meet each other on the opposite end of the parental chromosome. *E coli* regulate this process through the use of termination sequences which, when bound by the Tus protein, enable only one direction of replication fork to pass through. As a result, the replication forks are constrained to always meet within the termination region of the chromosome.

Eukaryotes initiate DNA replication at multiple points in the chromosome, so replication forks meet and terminate at many points in the chromosome;

these are not known to be regulated in any particular manner. Because eukaryotes have linear chromosomes, DNA replication often fails to synthesize to the very end of the chromosomes (telomeres), resulting in telomere shortening. This is a normal process in somatic cells — cells are only able to divide a certain number of times before the DNA loss prevents further division. (This is known as the Hayflick limit.) Within the germ cell line, which passes DNA to the next generation, telomerase extends the repetitive sequences of the telomere region to prevent degradation. Telomerase can become mistakenly active in somatic cells, sometimes leading to cancer formation.

Polymerase Chain Reaction

Researchers commonly replicate DNA *in vitro* using the polymerase chain reaction (PCR). PCR uses a pair of primers to span a target region in template DNA, and then polymerizes partner strands in each direction from these primers using a thermostable DNA polymerase. Repeating this process through multiple cycles produces amplification of the targeted DNA region. At the start of each cycle, the mixture of template and primers is heated, separating the newly synthesized molecule and template. Then, as the mixture cools, both of these become templates for annealing of new primers, and the polymerase extends from these. As a result, the number of copies of the target region doubles each round, increasing exponentially.

Interactions with Proteins

All the functions of DNA depend on interactions with proteins. These protein interactions can be non-specific, or the protein can bind specifically to a single DNA sequence. Enzymes can also bind to DNA and of these, the polymerases that copy the DNA base sequence in transcription and DNA replication are particularly important.

DNA-binding Proteins

Structural proteins that bind DNA are well-understood examples of non-specific DNA-protein interactions. Within chromosomes, DNA is held in complexes with structural proteins. These proteins organize the DNA into a compact structure called chromatin. In eukaryotes this structure involves DNA binding to a complex of small basic proteins called histones, while in prokaryotes multiple types of proteins are involved. The histones form a disk-shaped complex called a nucleosome, which contains two complete turns of double-stranded DNA wrapped around its surface.

These non-specific interactions are formed through basic residues in the histones making ionic bonds to the acidic sugar-phosphate backbone of the DNA, and are therefore largely independent of the base sequence. Chemical modifications of these basic amino acid residues include methylation, phosphorylation and acetylation. These chemical changes alter the strength of the interaction between the DNA and the histones, making the DNA more or less accessible to transcription factors and changing the rate of transcription. Other non-specific DNA-binding proteins in chromatin include the high-mobility group proteins, which bind to bent or distorted DNA. These proteins are important in bending arrays of nucleosomes and arranging them into the larger structures that make up chromosomes.

A distinct group of DNA-binding proteins are the DNA-binding proteins that specifically bind single-stranded DNA. In humans, replication protein A is the best-understood member of this family and is used in processes where the double helix is separated, including DNA replication, recombination and DNA repair. These binding proteins seem to stabilize single-stranded DNA and protect it from forming stem-loops or being degraded by nucleases.

In contrast, other proteins have evolved to bind to particular DNA sequences. The most intensively studied of these are the various transcription factors, which are proteins that regulate transcription. Each transcription factor binds to one particular set of DNA sequences and activates or inhibits the transcription of genes that have these sequences close to their promoters. The transcription factors do this in two ways. Firstly, they can bind the RNA polymerase responsible for transcription, either directly or through other mediator proteins; this locates the polymerase at the promoter and allows it to begin transcription. Alternatively, transcription factors can bind enzymes that modify the histones at the promoter; this will change the accessibility of the DNA template to the polymerase.

As these DNA targets can occur throughout an organism's genome, changes in the activity of one type of transcription factor can affect thousands of genes. Consequently, these proteins are often the targets of the signal transduction processes that control responses to environmental changes or cellular differentiation and development. The specificity of these transcription factors' interactions with DNA come from the proteins making multiple contacts to the edges of the DNA bases, allowing them to "read" the DNA sequence. Most of these base-interactions are made in the major groove, where the bases are most accessible.

DNA-modifying Enzymes

Nucleases and Ligases

Nucleases are enzymes that cut DNA strands by catalysing the hydrolysis of the phosphodiester bonds. Nucleases that hydrolyse nucleotides from the ends of DNA strands are called exonucleases, while endonucleases cut within strands. The most frequently used nucleases in molecular biology are the restriction endonucleases, which cut DNA at specific sequences. For instance, the EcoRV enzyme shown to the left recognizes the 6-base sequence 52 - GAT|ATC-32 and makes a cut at the vertical line. In nature, these enzymes protect bacteria against phage infection by digesting the phage DNA when it enters the bacterial cell, acting as part of the restriction modification system. In technology, these sequence-specific nucleases are used in molecular cloning and DNA fingerprinting.

Enzymes called DNA ligases can rejoin cut or broken DNA strands. Ligases are particularly important in lagging strand DNA replication, as they join together the short segments of DNA produced at the replication fork into a complete copy of the DNA template. They are also used in DNA repair and genetic recombination.

Topoisomerases and Helicases

Topoisomerases are enzymes with both nuclease and ligase activity. These proteins change the amount of supercoiling in DNA. Some of these enzymes work by cutting the DNA helix and allowing one section to rotate, thereby reducing its level of supercoiling; the enzyme then seals the DNA break. Other types of these enzymes are capable of cutting one DNA helix and then passing a second strand of DNA through this break, before rejoining the helix. Topoisomerases are required for many processes involving DNA, such as DNA replication and transcription.

Helicases are proteins that are a type of molecular motor. They use the chemical energy in nucleoside triphosphates, predominantly ATP, to break hydrogen bonds between bases and unwind the DNA double helix into single strands. These enzymes are essential for most processes where enzymes need to access the DNA bases.

Polymerases

Polymerases are enzymes that synthesize polynucleotide chains from nucleoside triphosphates. The sequence of their products are copies of existing polynucleotide chains - which are called *templates*. These enzymes function by adding nucleotides onto the 32 hydroxyl group of the previous nucleotide in a DNA strand. As a consequence, all polymerases work in a 52 to 32 direction. In the active site of these enzymes, the incoming nucleoside triphosphate base-pairs to the template: this allows polymerases to accurately synthesize the complementary strand of their template. Polymerases are classified according to the type of template that they use.

In DNA replication, a DNA-dependent DNA polymerase makes a copy of a DNA sequence. Accuracy is vital in this process, so many of these polymerases have a proofreading activity. Here, the polymerase recognizes the occasional mistakes in the synthesis reaction by the lack of base pairing between the mismatched nucleotides. If a mismatch is detected, a 32 to 52 exonuclease activity is activated and the incorrect base removed. In most organisms, DNA polymerases function in a large complex called the replisome that contains multiple accessory subunits, such as the DNA clamp or helicases.

RNA-dependent DNA polymerases are a specialized class of polymerases that copy the sequence of an RNA strand into DNA. They include reverse transcriptase, which is a viral enzyme involved in the infection of cells by retroviruses, and telomerase, which is required for the replication of telomeres. Telomerase is an unusual polymerase because it contains its own RNA template as part of its structure.

Transcription is carried out by a DNA-dependent RNA polymerase that copies the sequence of a DNA strand into RNA. To begin transcribing a gene, the RNA polymerase binds to a sequence of DNA called a promoter and separates the DNA strands. It then copies the gene sequence into a messenger RNA transcript until it reaches a region of DNA called the terminator, where it halts and detaches from the DNA. As with human DNA-dependent DNA polymerases, RNA polymerase II, the enzyme that transcribes most of the genes in the human genome, operates as part of a large protein complex with multiple regulatory and accessory subunits.

Genetic Recombination

A DNA helix usually does not interact with other segments of DNA, and in human cells the different chromosomes even occupy separate areas in the nucleus called "chromosome territories". This physical separation of different chromosomes is important for the ability of DNA to function as a stable repository for information, as one of the few times chromosomes interact is during chromosomal crossover when they recombine. Chromosomal crossover is when two DNA helices break, swap a section and then rejoin.

Recombination allows chromosomes to exchange genetic information and produces new combinations of genes, which increases the efficiency of natural selection and can be important in the rapid evolution of new proteins. Genetic recombination can also be involved in DNA repair, particularly in the cell's response to double-strand breaks.

The most common form of chromosomal crossover is homologous recombination, where the two chromosomes involved share very similar sequences. Non-homologous recombination can be damaging to cells, as it can produce chromosomal translocations and genetic abnormalities. The recombination reaction is catalysed by enzymes known as recombinases, such as RAD51. The first step in recombination is a double-stranded break either caused by an endonuclease or damage to the DNA. A series of steps catalysed in part by the recombinase then leads to joining of the two helices by at least one Holliday junction, in which a segment of a single strand in each helix is annealed to the complementary strand in the other helix. The Holliday junction is a tetrahedral junction structure that can be moved along the pair of chromosomes, swapping one strand for another. The recombination reaction is then halted by cleavage of the junction and re-ligation of the released DNA.

Evolution

DNA contains the genetic information that allows all modern living things to function, grow and reproduce. However, it is unclear how long in the 4-billion-year history of life DNA has performed this function, as it has been proposed that the earliest forms of life may have used RNA as their genetic material. RNA may have acted as the central part of early cell metabolism as it can both transmit genetic information and carry out catalysis as part of ribozymes. This ancient RNA world where nucleic acid would have been used for both catalysis and genetics may have influenced the evolution of the current genetic code based on four nucleotide bases. This would occur, since the number of different bases in such an organism is a trade-off between a small number of bases increasing replication accuracy and a large number of bases increasing the catalytic efficiency of ribozymes.

However, there is no direct evidence of ancient genetic systems, as recovery of DNA from most fossils is impossible. This is because DNA will survive in the environment for less than one million years and slowly degrades into short fragments in solution. Claims for older DNA have been made, most notably a report of the isolation of a viable bacterium from a salt crystal 250 million years old, but these claims are controversial.

DNA Microarray

A DNA microarray is a multiplex technology used in molecular biology. It consists of an arrayed series of thousands of microscopic spots of DNA oligonucleotides, called features, each containing picomoles (10^{-12} moles) of a specific DNA sequence, known as *probes* (or *reporters*). These can be a short section of a gene or other DNA element that are used to hybridize a cDNA or cRNA sample (called *target*) under high-stringency conditions. Probe-target hybridization is usually detected and quantified by detection of fluorophore-, silver-, or chemiluminescence-labelled targets to determine relative abundance of nucleic acid sequences in the target. Since an array can contain tens of thousands of probes, a microarray experiment can accomplish many genetic tests in parallel. Therefore arrays have dramatically accelerated many types of investigation.

In standard microarrays, the probes are attached via surface engineering to a solid surface by a covalent bond to a chemical matrix (via epoxy-silane, amino-silane, lysine, polyacrylamide or others). The solid surface can be glass or a silicon chip, in which case they are colloquially known as an *Affy chip* when an Affymetrix chip is used. Other microarray platforms, such as Illumina, use microscopic beads, instead of the large solid support. DNA arrays are different from other types of microarray only in that they either measure DNA or use DNA as part of its detection system.

DNA microarrays can be used to measure changes in expression levels, to detect single nucleotide polymorphisms (SNPs), or to genotype or resequence mutant genomes. Microarrays also differ in fabrication, workings, accuracy, efficiency, and cost. Additional factors for microarray experiments are the experimental design and the methods of analysing the data.

Microarray technology evolved from Southern blotting, where fragmented DNA is attached to a substrate and then probed with a known gene or fragment. Nucleic Acids Res. 1992 Apr 11;20(7):1679-84. Oligonucleotide hybridizations on glass supports: a novel linker for oligonucleotide synthesis and hybridization properties of oligonucleotides synthesised in situ. Maskos U, Southern EM. The first reported use of this approach was the analysis of 378 arrayed lysed bacterial colonies each harboring a different sequence which were assayed in multiple replicas for expression of the genes in multiple normal and tumour tissue.

This was expanded to analysis of more than 4000 human sequences with computer driven scanning and image processing for quantitative analysis of the sequences in human colonic tumours and normal tissue and then to comparison of colonic tissues at different genetic risk. The use of a collection of distinct DNAs in arrays for expression profiling was also described in 1987, and the arrayed DNAs were used to identify genes whose expression is modulated by interferon. These early gene arrays were made by spotting cDNAs onto filter paper with a pin-spotting device. The use of miniaturized microarrays for gene expression profiling was first reported in 1995, and a complete eukaryotic genome (*Saccharomyces cerevisiae*) on a microarray was published in 1997.

Principle

The core principle behind microarrays is hybridization between two DNA strands, the property of complementary nucleic acid sequences to specifically pair with each other by forming hydrogen bonds between complementary nucleotide base pairs. A high number of complementary base pairs in a nucleotide sequence means tighter non-covalent bonding between the two strands.

After washing off of non-specific bonding sequences, only strongly paired strands will remain hybridized. So fluorescently labelled target sequences that bind to a probe sequence generate a signal that depends on the strength of the hybridization determined by the number of paired bases, the hybridization conditions (such as temperature), and washing after hybridization.

Total strength of the signal, from a spot (feature), depends upon the amount of target sample binding to the probes present on that spot. Microarrays use relative quantitation in which the intensity of a feature is compared to the intensity of the same feature under a different condition, and the identity of the feature is known by its position. An alternative to microarrays is serial analysis of gene expression, where the transcriptome is sequenced allowing an absolute measurement.

Uses and Types

Many types of array exist and the broadest distinction is whether they are spatially arranged on a surface or on coded beads:

- The traditional solid-phase array is a collection of orderly microscopic "spots", called features, each with a specific probe attached to a solid surface, such as glass, plastic or silicon biochip (commonly known as a *genome chip*, *DNA chip* or *gene array*). Thousands of them can be placed in known locations on a single DNA microarray.
- The alternative bead array is a collection of microscopic polystyrene beads, each with a specific probe and a ratio of two or more dyes, which do not interfere with the fluorescent dyes used on the target sequence.

DNA microarrays can be used to detect DNA (as in comparative genomic hybridization), or detect RNA (most commonly as cDNA after reverse transcription) that may or may not be translated into proteins. The process of measuring gene expression via cDNA is called expression analysis or expression profiling.

Table: *Applications include*

Application or technology	***Synopsis***
Gene expression profiling	In an mRNA or gene expression profiling experiment the expression levels of thousands of genes are simultaneously monitored to study the effects of certain treatments, diseases, and developmental stages on gene expression. For example, microarray-based gene expression profiling can be used to identify genes whose expression is changed in response to pathogens or other organisms by comparing gene expression in infected to that in uninfected cells or tissues.

Comparative genomic hybridization	Assessing genome content in different cells or closely related organisms.GeneID Small microarrays to check IDs of organisms in food and feed (like GMO), mycoplasms in cell culture, or pathogens for disease detection, mostly combining PCR and microarray technology.
Chromatin immunoprecipitation on Chip	DNA sequences bound to a particular protein can be isolated by immunoprecipitating that protein (ChIP), these fragments can be then hybridized to a microarray (such as a tiling array) allowing the determination of protein binding site occupancy throughout the genome. Example protein to immunoprecipitate are histone modifications (H3K27me3, H3K4me2, H3K9me3, etc.), Polycomb-group protein (PRC2:Suz12, PRC1:YY1) and trithorax-group protein (Ash1) to study the epigenetic landscape or RNA Polymerase II to study the transcription landscape.
DamID	Analogously to ChIP, genomic regions bound by a protein of interest can be isolated and used to probe a microarray to determine binding site occupancy. Unlike ChIP, DamID does not require antibodies but makes use of adenine methylation near the protein's binding sites to selectively amplify those regions, introduced by expressing minute amounts of protein of interest fused to bacterial DNA adenine methyltransferase.
SNP detection	Identifying single nucleotide polymorphism among alleles within or between populations. Several applications of microarrays make use of SNP detection, including Genotyping, forensic analysis, measuring predisposition to disease, identifying drug-candidates, evaluating germline mutations in individuals or somatic mutations in cancers, assessing loss of heterozygosity, or genetic linkage analysis.
Alternative splicing detection	An '*exon junction array* design uses probes specific to the expected or potential splice sites of predicted exons for a gene. It is of intermediate density, or coverage, to a typical gene expression array (with 1-3 probes per gene) and a genomic tiling array (with hundreds or thousands of probes per gene). It is used to assay the expression of alternative splice forms of a gene. Exon arrays have a different design, employing probes designed to detect each individual exon for known or predicted genes, and can be used for detecting different splicing isoforms.
Fusion genes microarray	A Fusion gene microarray can detect fusion transcripts, *e.g.* from cancer specimens. The principle behind this is building on the alternative splicing microarrays. The oligo design strategy enables combined measurements of chimeric transcript junctions with exon-wise measurements of individual fusion partners.

Tiling array	Genome tiling arrays consist of overlapping probes designed to densely represent a genomic region of interest, sometimes as large as an entire human chromosome. The purpose is to empirically detect expression of transcripts or alternatively splice forms which may not have been previously known or predicted.

Fabrication

Microarrays can be manufactured in different ways, depending on the number of probes under examination, costs, customization requirements, and the type of scientific question being asked. Arrays may have as few as 10 probes or up to 2.1 million micrometre-scale probes from commercial vendors.

Spotted vs. in Situ Synthesised Arrays

Microarrays can be fabricated using a variety of technologies, including printing with fine-pointed pins onto glass slides, photolithography using pre-made masks, photolithography using dynamic micromirror devices, ink-jet printing, or electrochemistry on microelectrode arrays.

In *spotted microarrays*, the probes are oligonucleotides, cDNA or small fragments of PCR products that correspond to mRNAs. The probes are synthesized prior to deposition on the array surface and are then "spotted" onto glass. A common approach utilizes an array of fine pins or needles controlled by a robotic arm that is dipped into wells containing DNA probes and then depositing each probe at designated locations on the array surface. The resulting "grid" of probes represents the nucleic acid profiles of the prepared probes and is ready to receive complementary cDNA or cRNA "targets" derived from experimental or clinical samples.

This technique is used by research scientists around the world to produce "in-house" printed microarrays from their own labs. These arrays may be easily customized for each experiment, because researchers can choose the probes and printing locations on the arrays, synthesize the probes in their own lab (or collaborating facility), and spot the arrays. They can then generate their own labelled samples for hybridization, hybridize the samples to the array, and finally scan the arrays with their own equipment.

This provides a relatively low-cost microarray that may be customized for each study, and avoids the costs of purchasing often more expensive commercial arrays that may represent vast numbers of genes that are not of interest to the investigator. Publications exist which indicate in-house spotted microarrays may not provide the same level of sensitivity compared to commercial oligonucleotide arrays, possibly owing to the small batch sizes and reduced printing efficiencies when compared to industrial manufactures of oligo arrays.

In *oligonucleotide microarrays*, the probes are short sequences designed to match parts of the sequence of known or predicted open reading frames. Although oligonucleotide probes are often used in "spotted" microarrays, the term "oligonucleotide array" most often refers to a specific technique of manufacturing. Oligonucleotide arrays are produced by printing short oligonucleotide sequences designed to represent a single gene or family of gene splice-variants by synthesizing this sequence directly onto the array surface instead of depositing intact sequences. Sequences may be longer (60-mer probes such as the Agilent design) or shorter (25-mer probes produced by Affymetrix) depending on the desired purpose; longer probes are more specific to individual target genes, shorter probes may be spotted in higher density across the array and are cheaper to manufacture.

One technique used to produce oligonucleotide arrays include photolithographic synthesis (Agilent and Affymetrix) on a silica substrate where light and light-sensitive masking agents are used to "build" a sequence one nucleotide at a time across the entire array. Each applicable probe is selectively "unmasked" prior to bathing the array in a solution of a single nucleotide, then a masking reaction takes place and the next set of probes are unmasked in preparation for a different nucleotide exposure. After many repetitions, the sequences of every probe become fully constructed. More recently, Maskless Array Synthesis from NimbleGen Systems has combined flexibility with large numbers of probes.

Two-channel vs. One-channel Detection

Two-colour microarrays or *two-channel microarrays* are typically hybridized with cDNA prepared from two samples to be compared (e.g. diseased tissue versus healthy tissue) and that are labelled with two different fluorophores. Fluorescent dyes commonly used for cDNA labelling include Cy3, which has a fluorescence emission wavelength of 570 nm (corresponding to the green part of the light spectrum), and Cy5 with a fluorescence emission wavelength of 670 nm (corresponding to the red part of the light spectrum). The two Cy-labelled cDNA samples are mixed and hybridized to a single microarray that is then scanned in a microarray scanner to visualize fluorescence of the two fluorophores after excitation with a laser beam of a defined wavelength. Relative intensities of each fluorophore may then be used in ratio-based analysis to identify up-regulated and down-regulated genes.

Oligonucleotide microarrays often carry control probes designed to hybridize with RNA spike-ins. The degree of hybridization between the spike-ins and the control probes is used to normalize the hybridization measurements for the target probes. Although absolute levels of gene expression may be determined in the two-colour array in rare instances, the relative differences in expression among different spots within a sample and between samples is the preferred method of data analysis for the two-colour system. Examples of providers for such microarrays includes Agilent with their Dual-Mode platform, Eppendorf with their DualChip platform for colorimetric Silverquant labelling, and TeleChem International with Arrayit.

In *single-channel microarrays* or *one-colour microarrays*, the arrays provide intensity data for each probe or probe set indicating a relative level of hybridization with the labelled target. However, they do not truly indicate abundance levels of a gene but rather relative abundance when compared to other samples or conditions when processed in the same experiment.

Each RNA molecule encounters protocol and batch-specific bias during amplification, labelling, and hybridization phases of the experiment making comparisons between genes for the same microarray uninformative.

The comparison of two conditions for the same gene requires two separate single-dye hybridizations. Several popular single-channel systems are the Affymetrix "Gene Chip", Illumina "Bead Chip", Agilent single-channel arrays, the Applied Microarrays "CodeLink" arrays, and the Eppendorf "DualChip & Silverquant".

One strength of the single-dye system lies in the fact that an aberrant sample cannot affect the raw data derived from other samples, because each array chip is exposed to only one sample (as opposed to a two-colour system in which a single low-quality sample may drastically impinge on overall data precision even if the other sample was of high quality).

Another benefit is that data are more easily compared to arrays from different experiments so long as batch effects have been accounted for. A drawback to the one-colour system is that, when compared to the two-colour system, twice as many microarrays are needed to compare samples within an experiment.

Microarrays and Bioinformatics

The advent of inexpensive microarray experiments created several specific bioinformatics challenges:

- the multiple levels of replication in experimental design (Experimental design)
- the number of platforms and independent groups and data format (Standardization)
- the treatment of the data (Statistical analysis)
- accuracy and precision (Relation between probe and gene)
- the sheer volume of data and the ability to share it (Data warehousing).

Experimental Design

Due to the biological complexity of gene expression, the considerations of experimental design that are discussed in the expression profiling article are of critical importance if statistically and biologically valid conclusions are to be drawn from the data.

There are three main elements to consider when designing a microarray experiment. First, replication of the biological samples is essential for drawing conclusions from the experiment.

Second, technical replicates (two RNA samples obtained from each experimental unit) help to ensure precision and allow for testing differences within treatment groups. The technical replicates may be two independent RNA extractions or two aliquots of the same extraction.

Third, spots of each cDNA clone or oligonucleotide are present as replicates (at least duplicates) on the microarray slide, to provide a measure of technical precision in each hybridization.

It is critical that information about the sample preparation and handling is discussed, in order to help identify the independent units in the experiment and to avoid inflated estimates of statistical significance.

Standardization

Microarray data is difficult to exchange due to the lack of standardization in platform fabrication, assay protocols, and analysis methods. This presents an interoperability problem in bioinformatics. Various grass-roots open-source projects are trying to ease the exchange and analysis of data produced with non-proprietary chips:

- For example, the "Minimum Information About a Microarray Experiment" (MIAME) checklist helps define the level of detail that should exist and is being adopted by many journals as a requirement for the submission of papers incorporating microarray results. But MIAME does not describe the format for the information, so while many formats can support the MIAME requirements, as of 2007 no format permits verification of complete semantic compliance.
- The "MicroArray Quality Control (MAQC) Project" is being conducted by the US Food and Drug Administration (FDA) to develop standards and quality control metrics which will eventually allow the use of MicroArray data in drug discovery, clinical practice and regulatory decision-making.
- The MGED Society has developed standards for the representation of gene expression experiment results and relevant annotations.

Statistical Analysis

Microarray data sets are commonly very large, and analytical precision is influenced by a number of variables. Statistical challenges include taking into account effects of background noise and appropriate normalization of the data.

Normalization methods may be suited to specific platforms and, in the case of commercial platforms, the analysis may be proprietary. Algorithms that affect statistical analysis include:

- Image analysis: gridding, spot recognition of the scanned image (segmentation algorithm), removal or marking of poor-quality and low-intensity features (called *flagging*).
- Data processing: background subtraction (based on global or local background), determination of spot intensities and intensity ratios, visualisation of data, and log-transformation of ratios, global or local normalization of intensity ratios.
- Identification of statistically significant changes: t-test, ANOVA, Bayesian method or Mann–Whitney test methods tailored to microarray data sets, which take into account multiple comparisons. These methods assess statistical power based on the variation present in the data and the number of experimental replicates, and can help minimize Type I and type II errors in the analyses.
- Network-based methods: Statistical methods that take the underlying structure of gene networks into account, representing either associative or causative interactions or dependencies among gene products.

Microarray data may require further processing aimed at reducing the dimensionality of the data to aid comprehension and more focused analysis. Other methods permit analysis of data consisting of a low number of biological or technical replicates; for example, the Local Pooled Error (LPE) test pools standard deviations of genes with similar expression levels in an effort to compensate for insufficient replication.

Relation between Probe and Gene

The relation between a probe and the mRNA that it is expected to detect is problematic. On the one hand, some mRNAs may cross-hybridize probes in the array that are supposed to detect another mRNA. In addition, mRNAs may experience amplification bias that is sequence or molecule-specific.

On the other hand, probes that are designed to detect the mRNA of a particular gene may be relying on genomic EST information that is incorrectly associated with that gene.

Data Warehousing

Microarray data was found to be more useful when compared to other similar datasets. The sheer volume (in bytes), specialized formats (such as MIAME), and curation efforts associated with the datasets require specialized databases to store the data.

Significance Analysis of Microarrays

Significance analysis of microarrays (SAM) is a statistical technique, established in 2001 by Tusher, Tibshirani and Chu, for determining whether changes in gene expression are statistically significant.

With the advent of DNA microarrays it is now possible to measure the expression of thousands of genes in a single hybridization experiment. The data generated is considerable and a method for sorting out what is significant and what isn't is essential. SAM is distributed by Stanford University in an R-package. SAM identifies statistically significant genes by carrying out gene specific t-tests and computes a statistic d_j for each gene j, which measures the strength of the relationship between gene expression and a response variable. This analysis uses non-parametric statistics, since the data may not follow a normal distribution.

The response variable describes and groups the data based on experimental conditions. In this method, repeated permutations of the data are used to determine if the expression of any gene is significant related to the response.

The use of permutation-based analysis accounts for correlations in genes and avoids parametric assumptions about the distribution of individual genes. This is an advantage over other techniques (for example ANOVA and Bonferroni), which assume equal variance and/or independence of genes.

Basic Protocol

- Perform microarray experiments — DNA microarray with oligo and cDNA primers, SNP arrays, protein arrays, etc.
- Input Expression Analysis in Microsoft Excel
- Run SAM as a Microsoft Excel Add-Ins
- Adjust the Delta tuning parameter to get a significant # of genes along with an acceptable false discovery rate (FDR)) and Assess Sample Size by calculating the mean difference in expression in the SAM Plot Controller
- List Differentially Expressed Genes (Positively and Negatively Expressed Genes)
- Permutations are calculated based on the number of samples

- Block Permutations
- Blocks are batches of microarrays; for example for eight samples split into two groups (control and affected) there are 4!=24 permutations for each block and the total number of permutations is (24)(24)= 196. A minimum of 1000 permutations are recommended; The number of permutations is set by the user when imputing correct values for the data set to run SAM.

Response Formats

Types;

- Quantitative — real-valued (such as heart rate)
- One class — tests whether the mean gene expression differs from zero
- Two class — two sets of measurements
  * Unpaired — measurement units are different in the two groups; e.g. control and treatment groups with samples from different patients
  * Paired — same experimental units are measured in the two groups; eg. samples before and after treatment from the same patients
- Multiclass — more than two groups with each containing different experimental units; generalization of two class unpaired type
- Survival — data of a time until an event (for example death or relapse)
- Time course — each experimental units is measured at more than one time point; experimental units fall into a one or two class design
- Pattern discovery — no explicit response parameter is specified; the user specifies eigengene (principle component) of the expression data and treats it as a quantitative response.

SAM Calculations

SAM calculates a test statistic for relative difference in gene expression based on permutation analysis of expression data and calculates a false discovery rate. Fold changes (t) are specified to guarantee genes called significant change at least a pre-specified amount. This means that the absolute value of the average expression levels of a gene under each of two conditions must be greater than the fold change (t) to be called positive and less than the inverse of the fold change (t) to be called negative.

The SAM Algorithm can be Stated as:

1. Order test statistics according to magnitude
2. For each permutation compute the ordered null (unaffected) scores
3. Plot the ordered test statistic against the expected null scores

4. Call each gene significant if the absolute value of the test statistic for that gene minus the mean test statistic for that gene is greater than a stated threshold
5. Estimate the false discovery rate based on expected versus observed values.

SAM Output

- Significant gene sets;
  - o Positive gene set — higher expression of most genes in the gene set correlates with higher values of the phenotype y.
  - o Negative gene set — lower expression of most genes in the gene set correlates with higher values of the phenotype y.

SAM Features

- Data from Oligo or cDNA arrays, SNP array, protein arrays,etc can be utilized in SAM
- Correlates expression data to clinical parameters
- Correlates expression data with time
- Uses data permutation to estimates False Discovery Rate for multiple testing
- Reports local false discovery rate (the FDR for genes having a similar d_i as that gene) and miss rates
- Can work with blocked design for when treatments are applied within different batches of arrays
- Can adjust threshold determining number of gene called significant.

3

Gene Networks

A gene regulatory network or genetic regulatory network (GRN) is a collection of DNA segments in a cell which interact with each other indirectly (through their RNA and protein expression products) and with other substances in the cell, thereby governing the rates at which genes in the network are transcribed into mRNA. In general, each mRNA molecule goes on to make a specific protein (or set of proteins).

In some cases this protein will be structural, and will accumulate at the cell membrane or within the cell to give it particular structural properties. In other cases the protein will be an enzyme, i.e., a micro-machine that catalyses a certain reaction, such as the breakdown of a food source or toxin. Some proteins though serve only to activate other genes, and these are the transcription factors that are the main players in regulatory networks or cascades. By binding to the promoter region at the start of other genes they turn them on, initiating the production of another protein, and so on. Some transcription factors are inhibitory.

In single-celled organisms, regulatory networks respond to the external environment, optimising the cell at a given time for survival in this environment. Thus a yeast cell, finding itself in a sugar solution, will turn on genes to make enzymes that process the sugar to alcohol. This process, which we associate with wine-making, is how the yeast cell makes its living, gaining energy to multiply, which under normal circumstances would enhance its survival prospects. In multicellular animals the same principle has been put in the service of gene cascades that control body-shape. Each time a cell divides, two cells result which, although they contain the same genome in full, can differ in which genes are turned on and making proteins. Sometimes a 'self-sustaining feedback loop' ensures that a cell maintains its identity and passes it on.

Less understood is the mechanism of epigenetics by which chromatin modification may provide cellular memory by blocking or allowing transcription. A major feature of multicellular animals is the use of morphogen gradients, which in effect provide a positioning system that tells a cell where in the body it is, and hence what sort of cell to become. A gene that is turned on in one cell may make a product that leaves the cell and diffuses through adjacent cells, entering them and turning on genes only when it is present above a certain threshold level.

These cells are thus induced into a new fate, and may even generate other morphogens that signal back to the original cell. Over longer distances morphogens may use the active process of signal transduction. Such signalling controls embryogenesis, the building of a body plan from scratch through a series of sequential steps. They also control maintain adult bodies through feedback processes, and the loss of such feedback because of a mutation can be responsible for the cell proliferation that is seen in cancer. In parallel with this process of building structure, the gene cascade turns on genes that make structural proteins that give each cell the physical properties it needs. It has been suggested that, because biological molecular interactions are intrinsically stochastic, gene networks are the result of cellular processes and not their cause (i.e. Cellular Darwinism). However, recent experimental evidence has favored the attractor view of cell fates.

Genes have sometimes been regarded as nodes in a network, with inputs being proteins such as transcription factors, and outputs being the level of gene expression. The node itself performs a function, and the operation of these functions have been interpreted as performing a kind of information processing within cell and determine cellular behavior.

Gene networks can also be constructed without formulating an explicit causal model. This is often the case when assembling networks from large expression data sets. Covariation and correlation of expression is computed across a large sample of cases and measurements (often transcriptome or proteome data). The source of variation can be either experimental or natural (observational). There are several ways to construct gene expression networks, but one common approach is to compute a matrix of all pair-wise correlations of expression across conditions, time points, or individuals and convert the matrix (after thresholding at some cut-off value) into a graphical representation in which nodes represent genes, transcripts, or proteins and edges connecting these nodes represent the strength of association.

Overview

At one level, biological cells can be thought of as "partially-mixed bags" of biological chemicals – in the discussion of gene regulatory networks, these chemicals are mostly the mRNAs and proteins that arise from gene expression. These mRNA and proteins interact with each other with various degrees of specificity. Some diffuse around the cell. Others are bound to cell membranes, interacting with molecules in the environment. Still others pass through cell membranes and mediate long range signals to other cells in a multi-cellular organism. These molecules and their interactions comprise a gene regulatory network. A typical gene regulatory network looks something like this:

The nodes of this network are proteins, their corresponding mRNAs, and protein/protein complexes. Nodes that are depicted as lying along vertical lines are associated with the cell/environment interfaces, while the others are free-floating and diffusible. Implied are genes, the DNA sequences which are transcribed into the mRNAs that translate into proteins. Edges between nodes represent individual molecular reactions, the protein/protein and protein/ mRNA interactions through which the products of one gene affect those of another, though the lack of experimentally obtained information often implies that some reactions are not modeled at such a fine level of detail.

These interactions can be inductive (the arrowheads), with an increase in the concentration of one leading to an increase in the other, or inhibitory (the filled circles), with an increase in one leading to a decrease in the other. A series of edges indicates a chain of such dependences, with cycles corresponding to feedback loops. The network structure is an abstraction of the system's chemical dynamics, describing the manifold ways in which one substance affects all the others to which it is connected. In practice, such GRNs are inferred from the biological literature on a given system and represent a distillation of the collective knowledge about a set of related biochemical reactions.

Genes can be viewed as nodes in the network, with input being proteins such as transcription factors, and outputs being the level of gene expression. The node itself can also be viewed as a function which can be obtained by combining basic functions upon the inputs (in the Boolean network described below these are Boolean functions, typically AND, OR, and NOT). These functions have been interpreted as performing a kind of information processing within the cell, which determines cellular behavior. The basic drivers within cells are concentrations of some proteins, which determine both spatial (location within the cell or tissue) and temporal (cell cycle or developmental stage) coordinates of the cell, as a kind of "cellular memory". The gene networks are only beginning to be understood, and it is a next step for biology to attempt to deduce the functions for each gene "node", to help understand the behavior of the system in increasing levels of complexity, from gene to signaling pathway, cell or tissue level.

Mathematical models of GRNs have been developed to capture the behavior of the system being modeled, and in some cases generate predictions corresponding with experimental observations. In some other cases, models have proven to make accurate novel predictions, which can be tested experimentally, thus suggesting new approaches to explore in an experiment that sometimes wouldn't be considered in the design of the protocol of an experimental laboratory.

The most common modeling technique involves the use of coupled ordinary differential equations (ODEs). Several other promising modeling techniques have been used, including Boolean networks, Petri nets, Bayesian networks, graphical Gaussian models, Stochastic, and Process Calculi. Conversely, techniques have been proposed for generating models of GRNs that best explain a set of time series observations.

Modelling

Boolean Network

The following example illustrates how a Boolean network can model a GRN together with its gene products (the outputs) and the substances from the environment that affect it (the inputs). Stuart Kauffman was amongst the first biologists to use the metaphor of Boolean networks to model genetic regulatory networks.

1. Each gene, each input, and each output is represented by a node in a directed graph in which there is an arrow from one node to another if and only if there is a causal link between the two nodes.
2. Each node in the graph can be in one of two states: on or off.
3. For a gene, "on" corresponds to the gene being expressed; for inputs and outputs, "on" corresponds to the substance being present.
4. Time is viewed as proceeding in discrete steps. At each step, the new state of a node is a Boolean function of the prior states of the nodes with arrows pointing towards it.

The validity of the model can be tested by comparing simulation results with time series observations.

Continuous Networks

Continuous network models of GRNs are an extension of the boolean networks described above. Nodes still represent genes and connections between them regulatory influences on gene expression. Genes in biological systems display a continuous range of activity levels and it has been argued that using a continuous representation captures several properties of gene regulatory networks not present in the Boolean model.

Formally most of these approaches are similar to an artificial neural network, as inputs to a node are summed up and the result serves as input to a sigmoid function, e.g., but proteins do often control gene expression in a synergistic, i.e. non-linear, way. However there is now a continuous network model that allows grouping of inputs to a node thus realizing another level of regulation. This model is formally closer to a higher order recurrent neural network. The same model has also been used to mimic the evolution of cellular differentiation and even multicellular morphogenesis.

Stochastic Gene Networks

Recent experimental results have demonstrated that gene expression is a stochastic process. Thus, many authors are now using the stochastic formalism, after the work by. Works on single gene expression and small synthetic genetic networks, such as the genetic toggle switch of Tim Gardner and Jim Collins, provided additional experimental data on the phenotypic variability and the stochastic nature of gene expression. The first versions of stochastic models of gene expression involved only instantaneous reactions and were driven by the Gillespie algorithm.

Since some processes, such as gene transcription, involve many reactions and could not be correctly modeled as an instantaneous reaction in a single step, it was proposed to model these reactions as single step multiple delayed reactions in order to account for the time it takes for the entire process to be complete. From here, a set of reactions were proposed that allow generating GRNs. These are then simulated using a modified version of the Gillespie algorithm, that can simulate multiple time delayed reactions (chemical reactions where each of the products is provided a time delay that determines when will it be released in the system as a "finished product").

For example, basic transcription of a gene can be represented by the following single-step reaction (RNAP is the RNA polymerase, RBS is the RNA ribosome binding site, and Pro_i is the promoter region of gene i):

$$RNAP + \Pr o_i \xrightarrow{k_i, bas} \Pr o_i\left(\tau_i^1\right) + RBS_i\left(\tau_i^1\right) + RNAp\left(\tau_I^2\right)$$

A recent work proposed a simulator (SGNSim, Stochastic Gene Networks Simulator), that can model GRNs where transcription and translation are modeled as multiple time delayed events and its dynamics is driven by a stochastic simulation algorithm (SSA) able to deal with multiple time delayed events. The time delays can be drawn from several distributions and the reaction rates from complex functions or from physical parameters. SGNSim can generate ensembles of GRNs within a set of user-defined parameters, such as topology. It can also be used to model specific GRNs and systems of chemical reactions. Genetic perturbations such as gene deletions, gene over-expression, insertions, frame shift mutations can also be modeled as well.

The GRN is created from a graph with the desired topology, imposing in-degree and out-degree distributions. Gene promoter activities are affected by other genes expression products that act as inputs, in the form of monomers or combined into multimers and set as direct or indirect. Next, each direct input is assigned to an operator site and different transcription factors can be allowed, or not, to compete for the same operator site, while indirect inputs are given a target. Finally, a function is assigned to each gene, defining the gene's response to a combination of transcription factors (promoter state). The transfer functions (that is, how genes respond to a combination of inputs) can be assigned to each combination of promoter states as desired.

In other recent work, multiscale models of gene regulatory networks have been developed that focus on synthetic biology applications. Simulations have been used that model all biomolecular interactions in transcription, translation, regulation, and induction of gene regulatory networks, guiding the design of synthetic systems.

Prediction

Other work has focused on predicting the gene expression levels in a gene regulatory network. The approaches used to model gene regulatory networks have been constrained to be interpretable and, as a result, are generally simplified versions of the network. For example, Boolean networks have been used due to their simplicity and ability to handle noisy data but lose data information by having a binary representation of the genes. Also, artificial neural networks omit using a hidden layer so that they can be interpreted, losing the ability to model higher order correlations in the data. Using a model that is not constrained to be interpretable, a more accurate model can be produced. Being able to predict gene expressions more accurately provides a way to explore how drugs affect a system of genes as well as for finding which genes are interrelated in a process. This has been encouraged by the DREAM competition which promotes a competition for the best prediction algorithms. Some other recent work has used artificial neural networks with a hidden layer.

Network Connectivity

Empirical data indicate that biological gene networks are sparsely connected, and that the average number of upstream-regulators per gene is less than two. Theoretical results show that selection for robust gene networks will favor minimally complex, more sparsely connected, networks. These results suggest that a sparse, minimally connected, genetic architecture may be a fundamental design constraint shaping the evolution of gene network complexity.

Techniques and Tools

The following experimental techniques are used to measure gene expression and are listed in roughly chronological order, starting with the older, more established technologies. They are divided into two groups based on their degree of multiplexity.

- Low-to-mid-plex techniques:
  - Reporter gene
  - Northern blot
  - Western blot
  - Fluorescent in situ hybridization
  - Reverse transcription PCR
  - Digital counting of single transcript molecules, see NanoString Technologies
- Higher-plex techniques:
  - SAGE
  - DNA microarray
  - Tiling array
  - RNA-Seq

Transcriptional Noise

A primary cause of the variability (noise) in gene expression occurring between cells in isogenic populations. A major source of transcriptional noise is likely to be transcriptional bursting although other sources of heterogeneity, such as unequal separation of cell contents at mitosis are also likely to contribute considerably. Bursting transcription, as opposed to simple probabilistic models of transcription, reflects multiple states of gene activity, with fluctuations between states separated by irregular intervals, generating uneven protein expression between cells.

Noise in gene expression can have tremendous consequences on cell behaviour, and must be mitigated or integrated. In certain contexts, such as the survival of microbes in rapidly changing stressful environments, or several types of scattered differentiation, the variability may be essential. Variability also impacts upon the effectiveness of clinical treatment, with resistance of bacteria to antibiotics demonstrably caused by non-genetic differences. Variability in gene expression may also contribute to resistance of sub-populations of cancer cells to chemotherapy.

Transcriptional bursting

Transcriptional bursting, also known as transcriptional pulsing, is a fundamental property of genes from bacteria to humans. Transcription of genes, the process which transforms the stable code written in DNA into the mobile RNA message can occur in "bursts" or "pulses". This phenomenon has recently come to light with the advent of new technologies, such as MS2 tagging, to detect RNA production in single cells, allowing precise measurements of RNA number, or RNA appearance at the gene. Other, more widespread techniques, such as Northern Blotting, Microarrays, RT-PCR and RNA-Seq, measure bulk RNA levels from homogenous population extracts.

These techniques lose dynamic information from individual cells, and give the impression transcription is a continuous smooth process. The reality is that transcription is irregular, with strong periods of activity, interspersed by long periods of inactivity. Averaged over millions of cells, this appears continuous. But at the individual cell level, there is considerable variability, and for most genes, very little activity at any one time.

Bursting may result from the stochastic nature of biochemical events superimposed upon a 2 or more step fluctuation. In its most simple form, the gene can exist in 2 states, one where activity is negligible and one where there is a certain probability of activation. Only in the second state does transcription readily occur. Whilst the nuclear and signaling landscapes of complex eukaryotic nuclei are likely to favour more than two simple states- for example, there are over twenty post-translational modifications of nucleosomes known, this simple two step model perhaps provides a reasonable intellectual framework for understanding the changing probabilities affecting transcription.

It seems likely that some rudimentary eukaryotes have genes which do not show bursting. The genes are always in the permissive state, with a simple probability describing the numbers of RNAs generated. What do the repressive and permissive states represent? An attractive idea is that the repressed state is a closed chromatin conformation whilst the permissive state is an open one. Another hypothesis is that the fluctuations reflect transition between bound pre-initiation complexes (permissive) and dissociated ones (restrictive). Bursts may also result from bursty signalling, cell cycle effects or movement of chromatin to and from transcription factories. The bursting phenomenon, as opposed to simple probabilistic models of transcription, can account for the high variability in gene expression occurring between cells in isogenic populations. This variability in turn can have tremendous consequences on cell behaviour, and must be mitigated or integrated. In certain contexts, such as the survival of microbes in rapidly changing stressful environments, or several types of scattered differentiation, the variability may be essential. Variability also impacts upon the effectiveness of clinical treatment, with resistance of bacteria to antibiotics demonstrably caused by non-genetic differences. Variability in gene expression may also contribute to resistance of sub-populations of cancer cells to chemotherapy.

Bookmarking

In genetics and epigenetics, bookmarking is a biological phenomenon believed to function as an epigenetic mechanism for transmitting cellular memory of the pattern of gene expression in a cell, throughout mitosis, to its daughter cells. This is vital for maintaining the phenotype in a lineage of cells so that, for example, liver cells divide into liver cells and not some other cell type. It is characterized by non-compaction of some gene promoters during mitosis. In terms of mechanism, it is believed that:

- at some point prior to the onset of mitosis, the promoters of genes that exist in a transcription-competent state become "marked" in some way,
- this "mark" persists both during and after mitosis, and
- the marking transmits gene expression memory by preventing the mitotic compaction of DNA at this locus, or by facilitating reassembly of transcription complexes on the promoter, or both.

In some cases, bookmarking is mediated by binding of specific factors to the promoter prior to onset of mitosis, but in other cases could be mediated by patterns of histone modification or presence of histone variants that are characteristic of active genes, and which are believed to persist throughout mitosis.

In the case of specific genes, for example, the stress-inducible hsp70 gene, bookmarking may also function as a mechanism for ensuring that the gene can be transcribed early in G1 phase if a stress were to occur at that time. If this gene promoter were compacted it would take time to de-compact in G1, during which time the cell would be unable to transcribe this cytoprotective gene, leaving it vulnerable to stress-induced cell death. In this case, bookmarking appears to be important for cell survival.

Expressed Sequence Tag

An expressed sequence tag or EST is a short sub-sequence of a cDNA sequence. They may be used to identify gene transcripts, and are instrumental in gene discovery and gene sequence determination. The identification of ESTs has proceeded rapidly, with approximately 65.9 million ESTs now available in public databases (e.g. GenBank 18 June 2010, all species). An EST results from one-shot sequencing of a cloned mRNA (i.e. several hundred base pairs of sequence starting from an end of a cDNA). The cDNAs used for EST generation are typically individual clones from a cDNA library.

The resulting sequence is a relatively low quality fragment whose length is limited by current technology to approximately 500 to 800 nucleotides. Because these clones consist of DNA that is complementary to mRNA, the ESTs represent portions of expressed genes. They may be represented in databases as either cDNA/mRNA sequence or as the reverse complement of the mRNA, the template strand.

ESTs can be mapped to specific chromosome locations using physical mapping techniques, such as radiation hybrid mapping, Happy mapping, or FISH. Alternatively, if the genome of the organism that originated the EST has been sequenced, one can align the EST sequence to that genome using a computer.

The current understanding of the human set of genes (as of 2006) includes the existence of thousands of genes based solely on EST evidence. In this respect, ESTs have become a tool to refine the predicted transcripts for those genes, which leads to the prediction of their protein products and ultimately their function. Moreover, the situation in which those ESTs are obtained (tissue, organ, disease state - e.g. cancer) gives information on the conditions in which the corresponding gene is acting. ESTs contain enough information to permit the design of precise probes for DNA microarrays that then can be used to determine the gene expression.

Some authors use the term "EST" to describe genes for which little or no further information exists besides the tag.

The significance of ESTs, their properties, methods to analyze EST dataset and their applications in different areas of biology have been reviewed by Nagaraj et al. (2007).

Sources of Data and Annotations

dbEST

dbEST is a division of Genbank established in 1992. As for GenBank, data in dbEST is directly submitted by laboratories worldwide and is not curated.

EST Contigs

Because of the way ESTs are sequenced, many distinct expressed sequence tags are often partial sequences that correspond to the same mRNA of an organism. In an effort to reduce the number of expressed sequence tags for downstream gene discovery analyses, several groups assembled expressed sequence tags into EST contigs. Example of resources that provide EST contigs include:

- TIGR gene indices
- Unigene
- STACK

Constructing EST contigs is not trivial and may yield artifacts (contigs that contain two distinct gene products). When the complete genome sequence of an organism is available and transcripts are annotated, it is possible to bypass contig assembly and directly match transcripts with ESTs. This approach is used in the TissueInfo system and makes it easy to link annotations in the genomic database to tissue information provided by EST data.

Tissue Information

High-throughput analyses of ESTs often encounter similar data management challenges. A first challenge is that tissue provenance of EST libraries is described in plain English in dbEST. This makes it difficult to write programs that can non ambiguously determine that two EST libraries were sequenced from the same tissue. Similarly, disease conditions for the tissue are not annotated in a computationally friendly manner. For instance, cancer origin of a library is often mixed with the tissue name (e.g., the tissue name "glioblastoma" indicates that the EST library was sequenced from brain tissue and the disease condition is cancer).

With the notable exception of cancer, the disease condition is often not recorded in dbEST entries. The TissueInfo project was started in 2000 to help with these challenges. The project provides curated data (updated daily) to disambiguate tissue origin and disease state (cancer/non cancer), offers a tissue ontology that links tissues and organs by "is part of" relationships (i.e., formalizes knowledge that hypothalamus is part of brain, and that brain is part of the central nervous system) and distributes open-source software for linking transcript annotations from sequenced genomes to tissue expression profiles calculated with data in dbEST.

Paramutation

In epigenetics, paramutation is an interaction between two alleles of a single locus, resulting in a heritable change of one allele that is induced by the other allele. Paramutation violates Mendel's first law, which states that in the process of the formation of the gametes (egg or sperm) the allelic pairs separate, one going to each gamete, and that each allele remains completely uninfluenced by the other. In paramutation an allele in one generation heritably affects the other allele in future generations, even if the allele causing the change is itself not transmitted. What may be transmitted are patterns of DNA methylation or RNAs such as piRNAs, siRNAs, miRNAs or other regulatory RNAs. Through proper breeding, paramutation can result in isogenic sibling plants with drastically different phenotypes.

Paramutation was first discovered and studied in maize (Zea mays) by R.A. Brink at the University of Wisconsin–Madison in the 1950s. Brink noticed that specific weakly expressed alleles of the red1 (r1) locus in maize, which encodes a transcription factor that confers red pigment to corn kernels, can heritably change specific strongly expressed alleles to a weaker expression state. The weaker expression state adopted by the changed allele is heritable and can, in turn, change the expression state of other active alleles in a process termed secondary paramutation. Brink showed that the influence of the paramutagenic allele could persist for many generations.

Interestingly, paramutation can result in a single allele of a gene controlling a spectrum of phenotypes. At r1 in maize, for example, the weaker expression state adopted by an allele following paramutation can range from completely colorless to nearly fully colored kernels. This is an exception to the general observation that continuous variation is controlled by many genes.

Allelic interactions similar to paramutation have since been reported in other organisms, including tomato, pea, and mice.

The molecular basis of paramutation is being unraveled, almost exclusively in maize. Paramutation may share common mechanisms with other epigenetic phenomena, such as gene silencing and genomic imprinting. In maize, paramutation seems to share many traits with the well understood RNA-directed DNA-methylation pathway in Arabidopsis thaliana, even though it has never been observed in the famous model plant. Alleman (2006) reported that, in maize, "paramutation is RNA-directed. Stability of the chromatin states associated with paramutation and transposon silencing requires the mop1 gene, which encodes an RNA-dependent RNA polymerase." Exactly how the RNA produced by this polymerase causes paramutation in maize is not yet understood, but like other epigenetic changes, it involves the covalent modification of DNA and/or the DNA-bound histone proteins without changing the DNA sequence itself.

Sequence Profiling Tool

A sequence profiling tool in bioinformatics is a type of software that presents information related to a genetic sequence, gene name, or keyword input. Such tools generally take a query such as a DNA, RNA, or protein sequence or 'keyword' and search one or more databases for information related to that sequence. Summaries and aggregate results are provided in standardized format describing the information that would otherwise have required visits to many smaller sites or direct literature searches to compile.

Many sequence profiling tools are software portals or gateways that simplify the process of finding information about a query in the large and growing number of bioinformatics databases. The access to these kinds of tools is either web based or locally downloadable executables.

Introduction and Usage

The "post-genomics" era has given rise to a range of web-based tools and software to compile, organize, and deliver large amounts of primary sequence information, as well as protein structures, gene annotations, sequence alignments, and other common bioinformatics tasks.

In general, there exist three types of databases and service providers. The first one includes the popular public-domain or open-access databases supported by funding and grants such as NCBI, ExPASy, Ensembl, and PDB. The second one includes smaller or more specific databases organized and compiled by individual research groups Examples include Yeast Genome Database, RNA database. The third and final one includes private corporate or institutional databases that require payment or institutional affiliation to access. Such examples rare given the globalization of the public databases unless the purported service is 'in-development' or the end point of the analysis is of commercial value.

Typical scenarios of a profiling approach become relevant, particularly, in the cases of the first two groups, where researchers commonly wish to combine information derived from several sources about a single query or target sequence. For example, users might use the sequence alignment and search tool BLAST to identify homologs of their gene of interest in other species, and then use these results to locate a solved protein structure for one of the homologs. Similarly, they might also want to know the likely secondary structure of the mRNA encoding the gene of interest, or whether a company sells a DNA construct containing the gene.

4

Gene Development

Health-care professionals who specialize in genetic disorders use a variety of methods to identify inherited conditions. Analysis of a family medical history, known as pedigree analysis, is used to track the transmission of a condition though generations. Blood tests that identify specific DNA sequences can reveal carriers of a disease-causing gene who have no symptoms of the disease. Geneticists collect a person's medical family history to trace the inheritance of a genetic trait among muliple generations.

The information is placed in a pedigree, which resembles a traditional ultigenerational family tree but includes information about individuals who were diagnosed with a particular disorder or who suffered from certain medical symptoms. A pedigree can help researchers recognize diseases that express themselves in dominant or recessive alleles. Dominant disorders affect eery generation. Recessive disorders may cluster in a sigle generation, reflecting when two parents who both carry a recessive allele for a disease have one or more children who develop the disease.

A pedigree can also identify diseases that show X-linked inheritance. Pedigree analysis can be useful when combined with certain genetic tests. A blood sample taken from a person who is at risk for a genetic disorder can be compared with a DNA sequence known to cause the disorder in question. Other genetic tests can reveal if a person has extra chromosomes, missing chromosomes, or chromosomes that have attached to one another in unusual ways. In some cases, these chromosomal abnormalities may produce genetic disorders in children or they may affect a person's ability to conceive a child.

Genetic testing can also identify disorders in a fetus, enabling parents to learn early in a pregnancy if a fetus will likely be born with health problems or develop them later in life. Presymptomatic testing can identify DNA abnormalities in a person before health problems develop. In the case of certain inherited heart conditions, for example, these tests enable a person to make healthy lifestyle changes or take other preventative measures, such as medications, to lower the risk of illness or death.

Medical genetic testing raises challenging issues because such tests typically provide statistical possibilities rather than a definite prediction of whether a person will develop a given genetic disease. A test result may indicate, for example, that a person has a 75 per cent risk of developing colon cancer by the age of 65. Such results enable a physician to perform appropriate screening tests on the at-risk person in e with many other conditions, including certain cancers, hemophilia, heart disease, and cystic fibrosis. Although the United States Food and Drug Administration (FDA) has approved more than 400 clinical trials in gene therapy, this method of treating disease remains far from an unqualified medical success.

Treatments usually produce some improvement in the underlying condition, but not enough to consider the therapy suitable for large-scale use. The death of a patient involved in a gene therapy experiment in 1999 caused the National Institutes of Health (NIH), a federal agency that monitors gene therapy studies, to reevaluate the safety and effectiveness of gene therapy clinical trials.

Human Genome Project

The Human Genome Project is the most ambitious project in the history of biology. The programme's challenging goal was to identify and sequence all of the DNA in human chromosomes. The project was initiated in 1990 in the United States with government funding, and it rapidly grew into an international consortium of academic centres and drug companies in China, France, Germany, Japan, the United Kingdom, and the United States.

The consortium initially hoped to reach its goal by the year 2005. In 1998 Celera Genomics, a privately funded biotechnology firm, announced that it would sequence the human genome by the year 2000 using different sequencing strategies than those used by the public consortium. This announcement triggered a heated race between Celera Genomics and the public consortium to complete the genome project. In June 2000 both teams declared victory when they jointly announced that they had separately completed a rough draft of the genome. The two teams published their findings simultaneously, although in two different journals, in February 2001. The draft provided a basic outline of 90 per cent of the human genome. Scientists from the public consortium completed the final sequencing of the human genome in April 2003, two years earlier than planned. The completed human genome has provided scientists with a detailed blueprint of our complex genetic code. Large computer databases of genetic information enable scientists to look for patterns and relationships among the actions of different genes.

Among the findings about the human genome was that the number of genes in the human genome is much lower than was predicted—only about 20,000 to 25,000 genes compared to the expected 100,000 genes. This number is a little more than twice the number of genes found in the fruit fly. Scientists are now turning their attention to studying how the relatively low number of genes in the human genome can produce the complex structures found in humans.

Scientists have long known that a single gene produces a single protein and that this single protein subsequently may be processed into several different proteins. In a new science known as proteomics, scientists seek to identify and understand the function of all the proteins in the human body. They theorize that there may be many more proteins than there are genes—that is, more than 25,000. Among other advances, the database of proteins derived from proteomics is expected to help scientists better understand the regulation of gene expression in the body and how it leads to the complexity of cellular structures and functions. In addition, proteomics may lead to the development of breakthrough drugs for a variety of genetic disorders.

Genes and Our Wrld

Breakthroughs in decoding and manipulating the genetic information stored in DNA promise a world of benefits. These scientific advances already help to diagnose and treat disease, develop new medicines, bring criminals to justice, improve our food supply, and clean up the environment. At the same time, however, genetic technologies also present society with the potential for new and serious social or environmental problems. Many of the developments that worry critics of genetic technologies remain on the horizon, but the debate over their inevitable arrival is already in full swing.

Genes and the Environment

Humans have tampered with the genetic composition of other organisms for thousands of years. Most of this manipulation has been decidedly low-tech: domestication of animals and selective breeding of desirable food crops. The development and use of genetic engineering techniques has accelerated the pace at which humans can alter nature, creating some products that have unquestionable benefits and others that raise serious concerns. Some scientists fear that genetic engineering techniques will damage genetic diversity. Domestication and selective breeding, which aim to produce many similar organisms with particular desirable traits, reduce the natural variation of genes within a species. Genetic engineering techniques take this effect a giant step further. Many crop species—including wheat, corn, tomatoes, and strawberries—have been manipulated to maximize yield, appearance, resistance to pests and chemicals, hardiness, and other commercially valuable traits. Once attractive varieties have been developed, agricultural techniques favour wide-scale planting of such genetically similar or identical stocks. The impact of this practic on biodiversity can be ominous because older plant varieties that carry diverse and useful alleles may be lost forever.

The forfeited genetic material could leave a species vulnerable to annihilation from a single factor in the environment, such as an insect pest or an infectious virus. Fortunately, efforts to combat decreaing biodiversity are under way. Farmers, gardeners, government agencies, and other interested parties have collaborated to create seed banks to maintain genetic diversity. These banks catalogue, store, and distribute the seeds of rare or endangered plants, enabling gardeners and farmers to continue cultivating rare plant varieties so that the unique genetic makeup of these plants does not disappear. In the same vein, zoos and other institutions breed endangered species of animals that may no longer be able to survive in their native habitats. Other animal programmes seek to preserve or enhance the genetic variation within certain endangered animal populations.

For example, all cheetahs are almost identical genetically, most likely due to their near extinction about 12,000 years ago. Inbreeding among the few remaining individuals has resulted in a loss of genetic diversity in modern cheetahs that may have affected the cheetah's immune system, leaving the animal vulnerable to disease. Scientists hope to use genetic engineering techniques to introduce new genes into the cheetah population to increase the genetic diversity of the species. The broad use of genetic engineering techniques in agriculture has raised other concerns beyond issue of biodiversity. From a consumer point of view, for instance, new technologies may have compromised a food's taste or nutritional value in exchange for a plant or animal that can be grown faster or at less cost.

In addition, some critics question the safety of genetically engineered foods. They fear that plants or animals that have received new genes will produce proteins that would not be present in nonmanipulated organisms. Such changes could have a serious effect: causing allergies or toxicity in humans who eat these foods, for example, or disrupting a plant's production of key nutrients. Though there has been much speculation about the potential health risks of genetically engineered foods, rigorous scientific investigation into the effects of these foods on humans is just beginning. The potential environmental impact of genetically engineered agriculture is equally controversial. Critics fear that transgenic organisms, which contain DNA from other species, could give rise to populations of genetically altered life forms that could cause disease, displace native species, or otherwise harm delicate ecosystems. Consider the example of a genetically engineered form of oilseed rape, the plant that yields canola oil.

The altered form, which has been grown commercially in the United States since 1993, contains inserted genes that increase its resistance to herbicides (weed-killing chemicals). The altered plants can grow unabated even when a farmer sprays a field with enough herbicide to kill troublesome weeds.

Yet a curious problem has emerged: The transgenic rape can interbreed with weedy relatives that grow nearby. Some scientists fear that this interbreeding could create weed varieties that also have a genetic resistance to known herbicides. The biological and economic impact of such a development could be enormous.

Genes and Society

Advances in genetic technologies allow scientists to take an unprecedented glimpse into the genetic makeup of every person. The information derived from this testing can serve many valuable purposes: It can save lives, assist couples trying to decide whether or not to have children, and help law-enforcement officials solve a crime.Yet breakthroughs in genetic testing also raise some troubling social concerns about privacy and discrimination.

For example, if an individual's genetic information becomes widely available, it could give health insurers cause to deny coverage to people with certain risk factors or encourage employers to reject certain high-risk job applicants. Furthermore, many genetically linked problems are more common among certain racial and ethnic groups—for example, the BRCA1 breast cancer allele is more common in Ashkenazi Jews, and the blood disorder sickle-cell anemia is more prevalent among blacks of African ancestry.

Many minority groups fear that the expansion of genetic testing could create whole new avenues of discrimination. Of particular concern are genetic tests that shed light on traits such as personality, intelligence, and mental health or potential abilities. Genetic tests that indicate a person is unlikely to get along with other people could be used to limit a person's professional advancement. In other cases, tests that identify a genetic risk of heart failure could discourage a person from competing in sports. New technologies that allow the manipulation of genes have raised even more disturbing possibilities. Gene therapy advances, which allow scientists to replace defective genes with normal alleles, give people with typically fatal diseases new hope for healthy lives. To date, gene therapy has focused on manipulating the genetic material in body cells other than gametes, so the changes will not be passed on to fuure generations. However, the application of gene therapy techniques to gametes—the cells involved in reproduction—seems inevitable.

Such manipulation might help prevent the transmission of disease fromone generation to another, but it could also produce unforeseen problems with long-lasting consequences. For instance, many people worry that new genetic techniques could be used to alter or encourage traits now viewed as part of normal human variability, such as shortness or baldness. At various times in the past century, people have advocated efforts to improve the human condition by promoting the perpetuation of certain genes.

This concept, known as eugenics, typically involves encouraging people with "positive" genes to reproduce and discouraging those with "inferior" genes from having offspring. Many people fear that new genetic technologies used to manipulate the human genome could give people previously unattainable methods to resort to extreme forms of eugenics. Advances in genetic technologies have turned some genes into valuable commercial commodities, spawning a host of controversial questions. Who owns a genetically altered organism or the genes it contains? Is it right to patent the use of a naturally occurring gene? Some people feel that genetic material should not be owned or used for profit.

Costa Rica has enacted laws to prevent foreign companies from patenting and then profiting from genes of native Costa Rican plant and animal species. Balancing the need to limit patents on genes are concerns that the profit motive of companies must be protected to maintain incentives to make new discoveries for medical products. The citizens of Iceland, for example, are cooperating with a biotechnology company in a study of the genetic makeup of the Icelandic people. The information compiled will be used to learn about genetic diseases.

Heredity

Humans have had some understanding of heredity since prehistoric times, observing how similar traits pass from parent to offspring and noting that differences arise with ach generation. Most of the mechanisms of heredity, however, were shrouded in mystery until early in the 20th century. Since that time, the rate of discovery has reached a feverish pace, enabling the advancement of modern molecular biology and the current Human Genome Project.

Early Views of Heredity

In ancient times, people understood some basic rules of heredity and used this knowledge to breed domestic animals and crops. By about 5000 BC, for example, people in different parts of the world had begun applying selective breeding techniques to grow new plant varieties, including types of wheat, maize, rice, and date palms, that had never existed in the wild. Ancient people understood that the rules of inheritance also applied to humans. The ancient Greeks were particularly interested in human heredity and evolution. Greek scientists and philosophers hotly debated whether a male or female parent contributed more to an offspring.

In the 4th century BC, Aristotle speculated that acquired characteristics, such as a scar that was incurred during life, could be passed on to offspring. He also believed in a widely held theory known as pangenesis. This theory proposed that particles in the body, called gemmules, reside in the limbs and organs. The gemmules become imprinted with any changes acquired by the body, such as muscle development from exercise. The gemmules then move to the reproductive cells and transfer information about the body's alterations to these cells. The reproductive cells transmit the acquired traits to offspring through particles called pangenes. The theories about the inheritance of acquired characteristics and pangenesis persisted until the middle of the 19th century. French zoologist Jean-Baptiste Lamarck formalized the theory of acquired characteristics in his treaise *Philosophie Zoologique*. Lamarck proposed that organisms evolve by responding to changes in their environment. When organisms undergo a change in order to adjust to their environment, that change acts as a trait that can be passed on to offsprng.

Influences of Darwin and Mendel

A surprising supporter of pangenesis was the British naturalist Charles Robert Darwin, who believed that the theory accounted for the process of heredity and the wide variety of traits seen among offspring. Despite his mistaken belief in pangenesis, Darwin nonetheless had an enormous impact on human understanding of heredity. During his years of extensive worldwide travel, Darwin collected many observations of how related species adapt to their local environments.

Darwin and British naturalist Alfred Wallace independently formulated the theory of natural selection, which holds that members of a given species born with more favorable characteristics to deal with their environment would be most likely to survive to pass on these traits to the ext generation. This important theory was popularized by Darwin's publication *On the Origin of Species*. The book was an immediate sensation, but it raised many questions. Foremost among these was the mystery of how organisms could appear with modified or entirely new traits. At roughly the same time that Darwin published his natural selection theories, the answer to many questions about the mechanisms of heredity were being unraveled by Gregor Mendel, a reclusive

Austrian monk. Mendel conducted a long series of experiments on pea plants during the 1850s and 1860s. Mendel crossbred plants that expressed differing traits, such as height and flower colour. His conclusions from these experiments helped him formulate a comprehensive theory of how such traits pass fromone generation to another. In his studies, Mendel recognized that characteristics were inherited as discrete units, and that each of these was inherited independently of the others. He speculated that each parent has pairs of these units but passes only one to an offspring. He also noted that certain forms of one trait were always dominant over others. Today the units that Mendel described are known as genes.

Emergence of the Science of Genetics

Mendel published his findings in 1866, but they went largely unnoticed for more than three decades. In the year 1900, however, Dutch botanist Hugo Marie de Vries, German botanist Karl Correns, and Austrian botanist Erich Tschermak independently rediscovered the monk's works and verified his conclusions.

Advances in cytology, the science of the structure and function of cells, enabled scientists to more deeply appreciate Mendel's work. In 1902 American biologist Walter S. Sutton and German cell biologist Theodor Boveri separately noted the parallels between Mendel's units and chromosomes. The demonstration of the chromosomal basis of inheritance gave rise to the modern science of genetics. The term *genetics* itself was coined in 1905 by British biologist William Bateson. The terms *gene*and *genotype* were contributed in 1909 by German scientist Wilhelm Johannsen.

In 1905 American biologists Edmund B. Wilson and Nettie Stevens independently discovered and identified the sex chromosomes. Wilson discovered the X chromosome in a butterfly, and Stevens discovered the Y chromosome in a beetle. The discoveries of the X and Y chromosomes helped scientists begin to unravelnew patterns of inheritance. Foremost among this research was the work of American biologist Thomas Hunt Morgan on fruit flies. In 1910 Morgan identified the first proof of a sex-linked trait, an eye-colour characteristic that resides on the X chromosome of fruit flies. With this finding, Morgan became the first scientist to pin down the location of a gene to a specific chromosome. Morgan was also the first to explain the implications of linkage, unuual patterns of inheritance that occur when multiple genes found on the same chromosome are inherited together. A student of Morgan's, American biologist Alfred Sturtevant, found early evidence of the mechanisms of crossing over, the phenomenon in which chromosomes interchange genes.

More definitive proof emerged in the 1930s with work by American geneticists Harriet Creighton and Barbara McClintock. The pair demonstrated gene recombination with experiments on seed colour in corn. McClintock later gained notice for her work on transposable elements, large genetic segments that move within a chromosome or even between chromosomes. Her research into these elements, commonly known as jumping genes, earned McClintock the 1983 Nobel Prize in physiology or mediine.

Breakthroughs in DNA Studies

While cytologists and geneticists were studying the properties and location of genes on chromosomes, other scientists focused their studies on the composition of genes. In 1928 British microbiologist Frederick Griffith ran a series of experiments on two strains of bacteria, one that kills mice and another that is harmless to them. When Griffith injected mice with killed cells of the virulent bacteria, all of the mice survived.

But in a second trial, when Griffith injected a combined cocktail of dead virulent bacteria and live "harmless" bacteria, the mice all died. He concluded that something in the dead virulent cells "transformed" the hereditary material of normally harmless bacteria so that they became killers. Most scientists at the time theorized that the transforming factor was composed of a protein.

The real identity of the transforming factor in this experiment was not identified until 1944, when American geneticists Oswald Avery, Colin MacLeod, and Maclyn McCarty revisited Griffith's research. After isolating different molecular components from dead bacterial cells, Avery and his colleagues determined that DNA was the agent that transformed the live harmless bacteria into killers. Despite a growing body of evidence about the function of DNA, many scientists were not ready to reject proteins as the hereditary material. The debate was largely quieted in 1952 by American geneticists Alfred Hershey and Martha Chase. Hershey and Chase showed that when a type of virus called a bacteriophage infects a bacterium, it is the virus's DNA—not protein—that enters the bacterium to cause infection. Their studies confirmed that DNA contained the virus's genetic information, which triggered vral replication within the bacteria. The experiments of Hershey and Chase convinced most scientists that DNA was the molecule of heredity, but many questions about the structure and mechanisms of DNA remained.

In the early 1950s researchers began to apply techniques of X-ray diffraction to learn about the basic structure of DNA. X-ray diffraction can determine molecular structures by measuring patterns of scattered X rays after they pass through a crystalline substance. British physical chemist Rosalind Franklin and British biophysicist Maurice Wilkins used X-ray diffraction to obtain DNA images of unprecedented clarity. Yet the exact three-dimensional structure of DNA remained unclear. The groundbreakig work of American biochemist James Watson and British biophysicist Francis Crick solved that mystery. In

1953 the two proposed a model of DNA that is still accepted today: A double helix molecule formed by two chains, each composed of alternating sugar and phosphate groups, connected by nitrogenous bases. Watson and Crick were awarded the 1962 Nobel Prize in physiology or medicine for their discoveries.

Watson and Crick speculated that the structure of DNA provided some obvious clues about how the molecule could replicate itself. They proposd a replication model in which each strand of DNA serves as a template for making exact copies. This mode of replication, called semi-conservaive replication, was demonstrated in 1958 by American molecular biologists Matthew Meselson and Franklin Stahl. Their experiments demonstrated the mechanism of replication by tracking DNA containing a heavy nitrogen isotope through a serie of replications. With DNA's structureand replication mechanisms largely solved, scientists turned their attention to identifying the genetic code—learning how a gene's nucleotide sequence determines what type of protein is made. In the late 1950s, South African geneticist Sydney Brenner and other scientists confirmed that RNA acted as an intermediary between DNA and protein production.

Researchers still were uncertain how the sequence of nucleotides in DNA corresponded to the production of specific amino acids. In 1961 Crick and Brenner determined that groups of three nucleotides, now known as codons, code for the 20 amino acids that form the foundation of proteins. The exact relationship between codons and amino acids was clarified after several important discoveries. American biochemists Marshall Nirenberg and J. Heinrich Matthaei synthesized repeated nucleotide sequences that led to the production of repeated single amino acids.

They identified how certain codon combinations code for a specific amino acid. A process developed by American geneticist Har Gobind Khorana helped scientists create a "dictionary" of codons that defined specific amino acids, thus resolving the remaining ambiguities in the genetic code. Only 12 years after the structure of DNA was deduced, the genetic code was solved.

Learning to Manipulate DNA

After scientists had unraveled the structureand replication mechanisms of DNA, many felt that the major discoveries of genetic research were resolved. They predicted that the only task left in genetics was to sort out the molecular details of how genes work. But in the process of studying gene function, researchers developed powerful new molecular techniques, enabling them to analyse and manipulate genes with a speed and precision never before possible. A number of discoveries made during the 1960s and 1970s shed light on how distinct fragments of DNA could be isolated. The work of Swiss molecular biologist Werner Arber focused on specialized enzymes that digest, or "restrict," the DNA of viruses infecting bacteria. These enzymes were subsequently dubed restriction enzymes. In the following decade, scientists learned hat restriction enzymes could also act like molecular scissors to cut DNA.

In 1970 American molecular biologist Hamilton Smith and colleagues determined hat restriction enzymes could cleave DNA molecules at precise and predictable locations. Hamilton concluded that the enzymes were able to recognize specific nucleotide sequences. Scientists quickly realised hat restriction enzymes could be used in the laboratory to manipulate DNA. In 1973 American biochemist Herb Boyer sed restriction enzymes to produce a DNA molecule with genetic material from two different sources. This splicing technique is now know as recombinant DNA. Boyer inserted foreign genes into plasmids and observed that the plasmids could replicate to make many copies of the inserted genes. In subsequent experiments, Boyer, American biochemist Stanley Cohen, and other researchers demonstrated that insertig a recombinant DNA molecule into a host bacteria cell would lead to extremely rpid replication and the production of many identical copies ofthe recombinant DNA.

This process, known as cloning, gave scientists the power to make many copies of desired DNA for molecular study. The speed and efficiency of DNA cloning were vastly improved in the 1980s with the invention of polymerase chain reaction (PCR). Developed by American biochemist Kary Mullis, PCR enables scientists to produce large amounts of DNA sequences in a test tube. In a matter of hours, the process can produce millions of cloned DNA molecules. Yet all of the advances in isolating and replicating DNA would not be possible or be of much use if researchers could not determine the nucleotide sequence of genetic material.

In the late 1970s and early 1980s, British biochemist Frederick Sanger and his associates developed DNA sequencing techniques. Sanger's methods, which used special compounds called dideoxy nucleotides, rapidly yielded the exact nucleotide sequence of a desired sample. With the use of automated equipment, the new techniques transformed genetic sequencing into a speedy, routine laboraory procedure. Many of the new techniques for isolating, sequencing, and replicating DNA have been put to practical use through the field of genetic engineering. The Human Genome Project and the new field of proteomics have both benefited from continuing technical advances and have accelerated the development of new genetic technologies. Modern genetics is poised to radically change the practice of medicine and the biotechnology industry.

Cutting and Measuring DNA

The first step in gene cloning is cutting DNA into appropriately sized pieces. The size of DNA molecules is measured by subjecting the DNA to a technique called "electrophoresis". An electrophoretic apparatus gives DNA an electrical shove (DNA is negatively charged because of its phosphate groups. The electrophoresis apparatus puts the DNA in an electric field, and it moves towards the positive pole), forcing it to move through a porous gel. Two types of gels are commonly u.

Agarose Gels

The most popular variant of electrophoresis is carried out th agarose gels. In the laboratory, suspensions of agarose in buffer are heated to boiling (often in microwave ovens), and the hot solution is poured on plastic or glass plates to a height of a few millimeters or so. Agarose is a derivative of agar — an edible polysaccharide extracted from seaweed. When cooled from heated solutions, solutions of agarose form a nearly transparent gel that looks like a slab of gelatin dessert.

After the gel sets, DNA solutions are placed in little depressions formed by leaving a comb in the gel. The gel is then flooded with a weak salt solution, and a voltage is applied. Since DNA is negatively charged (due to its phosphate groups), it moves towards the positive pole. Large DNA molecules advance slowly in the gel because they are impeded by the gel matrix.

Smaller ones encounter fewer barriers and move more quickly. By plotting the distance migrated against the reciprocal of size (in base pairs) of a group of DNA standards, a fairly straight line can be obtained. From these data, the length of unknowns can be easily calculated. For ease of comparison, the standards are usually run in the same gel beside the unknowns.

Gels of differing porosities can be made by adjusting the concentration of agarose. With 2 and 3 per cent solutions, double-stranded DNA's as small as 50 or 100 base pairs can be resolved. More dilute gels can resolve fragments as large as about 30,000 base pairs (30 kilobase pairs or 30kb).

Polyacrylamide Gels

When very small DNA molecules need to be analysed or when high-resolution separations are required, electrophoresis is carried out in polyacrylamide gels. They, like the gels made from agarose, are water white. Unlike agarose, polyacrylamide gels are polymers of acrylamide, a small synthetic organic compound. Polyacrylaide gel electrophoresis (PAGE) is most often used to separate fragments of DNA that are between 6 and a few thousand base pairs in length. The technique is used mostly in DNA sequencing, a topic that we'll come back to later in the course.

Detecting DNA

DNA doesn't have any colour; it only absorbs light in the ultraviolet range. How is it visualized in a clear gel after electrophoresis? DNA can be detected in both agarose and polyacrylamide gels after staining with various dyes, incuding ethidium bromide; a dye that forms a fluorescent complex upon binding to DNA. Usually, the ethidium bromide is added to the gel before electrophoresis. After the run, the gel is examined and often photographed under an ultraviolet light. Yellow-orange zones (bands) of fluorescence appear (similar to those in the picture at the right), indicating migration of discrete pieces of DNA. Other agents — like methylene blue or silver stains — can also be used to visualize DNA, but they are either less sensitive or less convient, and aren't used as often. By the way, many dyes that bind to DNA are mutagenic; i.e, they cause mutations.

Separating Large Fragments of DNA

Contrary to what we might expect, very large fragments of DNA readily movin agarose gels when an electrical field is applied. But they don't separate according to size. In fact, very big pieces of DNA all migrate at about the same velocity regardless of whether they are 20kb in length or 2000kb.

PFE

Since many techniques, particulary recently derived ones, depend on working with very large DNA fragments, even pieces the size of some chromosomes, several newly developed electrophoretic techniques have been developed. In one of these, called pulsed field electrophoresis (PFE), DNA molecules are analyseon agarose gels but with a modified apparatus. Instead of having only two sets of electrodes situated at opposite ends of the electrophoresis device as in the picture shown above, the pulsed field apparatus bears four sets of electrodes.

Two are designated A^- and A^+ and the other two are called B^- and B^+. The two A's and the two B's are set at an angle of 120° with respect to one another. During electrophoresis, the DNA molecules are subjected to alternating bursts of current from the two A and B pairs, thereby pulling the DNA alternately to the right and to the left as it advances (from south to north in this case) through the gel.

The DNA seems to reorient itself each time the current is switched, and the time that it takes to turn itself around is dependent on its length. The result is that very large pieces of DNA can be separated from another on the basis of their size. The technique also allows one to estimate the length of an unknown when run beside standards of known size. One of the most impresive accomplishments of this technique is the separation of the 16 chromosomes of yeast in a single electrophoretic run.

FIE

Anoher procedure, field inversion electrophoresis, is also effective at separating large molecules of DNA. In it, the electrodes are set up as in ordinary electrophoresis, but the current is switched so that the DNA first moves forward and then backward. Of course, if the switching were to be done in equal time intervals — say, 1 second forward and 1 second back — the DNA wouldn't move at all.

To get it to move at all, electrophoresis is conducted forward for a longer time than backward (or at a greater voltage forward versus backward). One might expect that moving ahead three steps and back two would be equivalent to moving forward one step at a time, but that's not what happens. Apparently the DNA reorients itself during the switching cycles, just as it does in pulsed field electrophoresis. Again, the ability to change directions seems to be dependent on size. This allows different lengths of DNA to be separated from one anoter.

Restriction Endonucleases

One critically important advance that has greatly stimulated the rapid progress in molecular biology and genetic engineering was the discovery of a set of enzymes that are capable of cutting DNA at defined sequences. These enzymes are found in a variety of microorganisms and are caled restriction endonucleases, or more simly, restriction enzymes. The first specific restriction enzyme was discovered by Hamilton Smith in 1970, an accomplishment for which he (and two others) were awarded the Nobel Prize. Since then, some 3,059 similar enzymes have been reported, many of which have different specificites.

Nomenclature

Smith and another Nobel laureate, Daniel Nathans, devisd a nomenclature for these enzymes. In brief, the name of each restriction enzyme derives from the organism from which it is isolated. The first letter of the genus name plus the first two letters of the species name form the first three letters of the restriction enzyme's name. If necessary, a letter indicating strain designation is added, and finally a number is appended that stands for the order in which the enzyme was discovered in each organism.

For example, BamHI is the name of an enzyme that is isolated from the bacteium *Bacillus amyloliquifaciens,* strain H, and it was, presumably, the first restriction endonuclease identified from that source. The first three letters of the name should be italicized, but because italicized letters are often hard to read on a computer.

Restriction Enzymes

The restriction enzymes owe their usefulness to the fact that they bind to DNA at specific DNA sequences, four to eight nucleotides in size, called recognition sites. Once bond, restriction enzymes cut the DNA at or near this site. With a little thought, it should be clear that an enzyme that has a six base pair recognition site will, on the average, produce larger pieces of DNA than one that recognizes a four base site. Expressed quantitatively, the approximate size of the fragments produced by a particular enzyme, given that it is cutting a DNA containing an equal proportion of all four nucleotides, can be calculated from the formula:

$$\text{Average size of fragment} = 4^N$$

where N is the number of bases that the enzyme recognizes. Hence, a four-cutter (the shorthand name for an enzyme that recognizes a site containing four base pairs) is expected to cleave random DNA into fragments of about 4^4 (256) base pairs while an enzyme with a recognition site of six bases will produce pieces (on the average) of about 4^6 (4096) base pairs.

Propertie of Restriction Enzymes

Ends

Another interesting propert of restriction enzymes is that while they often recognize a symmetrical site, they do not always cut at the axis of symmetry. For instance, the enzyme EcoRI. Similarly, the enzyme BglII (from the microorganims, *Bacillus globiggi* and universally and irreverently pronounced BAGEL TWO) recognizes the sequence AGATCT and cuts between the first A and G residues.

In fact, the four nucleotide single-stranded ends are the same for both BglII and BamHI. Moreover, there are at least two other six cutting enzymes that have been discovered that leave the same four nucleotide overhang: BclI and XhoII. These overhanging ends are very useful because — under the proper conditions — they may base pair with each other. In fact, because of their affinity for one another, they are often called cohesive or sticky ends.

Moreover, if molecules with these ends are treated with the appropriate enzyme — DNA ligase — their phosphodiester bonds may be rejoined (ligated). When two ends that originate from digestion by a single enzyme are ligated, the resulting molecule can be cut by the same enzyme again. But if the ends of a DNA molecule that originated with a BamHI cut and a BglII cut are joined together, the new sequence will not be cut with either enzyme.

All restriction endonucleases do not generate 5' single-strand overhangs. In fact, some don't even produce an overhang at all. Several enzymes — like SacI — produce 3' single-stranded sticky ends. And some enzymes — like PvuII — cut at the axis of symmetry, leaving perfectly aligned ends. DNA molecules without overhangs are said to have blunt ends.In addition thereare restriction enzymes that cleave DNA some distance away from the sequence that they recognize.

For example the enzyme HgaI makes staggered cuts that lie 5 and 10 nucleotides away from a 5 base pair sequence, GACGC. This leaves 5' overhanging ends, but, in contrast to the enzymes described above, these will be different almost every time the enzyme cuts.

The Utility ofthe Restriction Enzymes

The discovery of these any restriction endonucleases have allowed genetic engineers to cut pieces of DNA at specific sites and into defined sizes. The result has been that a scientist can work with a collection of molecules all of the same size and with ends of known sequece. Restriction enzymes have proved to be valuable analytical and diagnostic tools as wll.

Restriction Enzmes

Restriction enzymes are DNA-cutting enzymes found in bacteria (and harvested from them for use). Because they cut within the molecule, they are often caled restriction endonucleases. A restriction enzyme recognizes and cuts DNA only at a particular sequence of nucleotides. For example, the bacterium Hemophilus aegypticus produces an enzyme named HaeIII that cuts DNA wherever it encounters the sequence,

5′GGCC3'

3′CCGG5'

The cut is made between the adjacent G and C. This particular sequence occurs at 11 places in the circular DNA molecule of the virus phiX174. Thus treatment of this DNA with the enzyme produces 11 fragments, each with a precise length and nucleotide sequence. These fragments can be separated from one another and the sequence of each determined. HaeIII and AluI cut straight across the double helix producing "blunt" ends. However, any restriction enzymes cut in an offset fashion. The ends of the cut have an overhanging piece of single-stranded DNA. These are called "sticky ends" because they are able to form base pairs with any DNA molecule that contains the complementary sticky end. Any other source of DNA treated with the same enzyme will produce such molecules.

Mixed together, these molecules can join with each other by the base pairing between their sticky ends. The union can be made permanent by another enzyme, DNA ligase, that forms covalent bonds along the backbone of each strand. The result is a molecul of recombinant DNA (rDNA). The ability to prouce recombinant DNA molecules has not only revolutionized the study of genetics, but has laid the foundation for much of the biotechnology industry. The availability of human insulin (for diabetics), human factor VIII (for males with hemophilia A), and other proteins used in human therapy all were made possibl by recombinant NA.

Recombinant DNA and Gene Cloing

Recombinant DNA is DNA that has been created artificially. DNA from two or more sources is incorporated into a sigle recombinant molecule.

Maing Recombinant DNA (rDNA)

- Treat DNA from both sources with the same restriction endonuclease (BamHI in this case).
- BamHI cuts the same site on both molecules

  5' GGATCC 3'

  3' CCTAGG 5'
- The ends of the cut have an overhanging piece of single-stranded DNA.
- These are called "sticky ends" because they are able to base pair with any DNA molecule containing the complementary sticky end.
- In this case, both DNA preparations have complementary sticky ends and thus can pair with each other when mixed.
- DNA ligase covalently links the two into a molecul of recombinant DNA.

To be useful,the recombinant molecule must be replicated many times to provide material for analysis, sequencing, etc. Producing many identical copies of the ame recombinant molecule is called cloning. Cloning can be done in vitro, by a process called the polymerase chain reaction (PCR). Here, however, we shall examine how cloning is done in vivo.

Cloning in vivo can be done in:

- Unicellular microbes like E. coli
- Unicelllar eukaryotes like yeast and
- In mammalian cells grown in tissue culture.

In every case,the recombinant DNA must be taken up by the cell in a form in which it can be replicated and expressed. This is achieved by incorporating the DNA in a vector. A number of viruses (both bacterial and of mammalian cells) can serv as vectors. But here let us examine an example of cloning using E. coli as the host and a plasmid as the vector.

Plasmid

A plasmid is an extra-chromosomal DNA molecule separate from the chromosomal DNA which is capable of replicating independently of the chromosomal DNA. In many cases, it is circular and double-stranded. Plasmids usually occur naturally in bacteria, but are sometimes found in eukaryotic organisms (*e.g.*, the 2-micrometre-ring in Saccharomyces cerevisiae). Plasmid size varies from 1 to over 200 kilobase pairs (kbp). The number of identical plasmids within a single cell can be zero, one, or even thousands under some circumstances. Plasmids can be considered to be part of the mobilome, since they are often associated with conjugation, a mechanism of horizontal gene transfer.The term plasmid was first introduced by the American molecular biologist Joshua Lederberg in 1952.

Plasmids can be considered to be independent life-forms similar to viruses, since both are capable of autonmous replication in suitable (host) environments. However the plasmid-host relationship tends to be more symbiotic than parasitic (although this can also occur for viruses, for example with Endoviruses) since plasmids can endow their hosts with useful packages of DNA to assist mutual survival in times of severe stress.

For example, plasmids can convey antibiotic resistance to host bacteria, who may then survive along with their life-saving guests who are carried along into futurehost generatons.

Vectors

There are two types of plasmid integration into a host bacteria: Non-integrating plasmids replicate as in the top instance; whereas episomes, the lower example, integrate into the host chromosome. Plasmids used in genetic engineering are clled vectors. Plasmids serve as important tools in genetics and biotechnology labs, where they are commonly used to multiply (make many copies of) or express particular genes. Many plasmids are commercially available for such uses. The gene to be replicated is inserted into copies of a plasmid containing genes that make cells resistant to particular antibiotics and a multiple cloning site (MCS, or polylinker), which is a short region containing several commonly used restriction sites allowing the easy insertion of DNA fragments at this location.

Next, the plasmids are inserted into bacteria by a process called transformation. Then, the bacteria are exposed to the particular antibiotics. Only bacteria which take up copies of the plasmid survive the antibiotic, since the plasmid makes them resistant. In particular, the protecting genes are expressed (used to make a protein) and the expressed protein breaks down the antibiotics.

In this way the antibiotics act as a filter to select only the modified bacteria. Now these bacteria can be grown in large amounts, harvested and lysed (often using the alkaline lysis method) to isolate the plasmid of interest. Another major use of plasmids is to make large amounts of proteins.

In this case, researchers grow bacteria containing a plasmid harboring the gene of interest. Just as the bacteria produces proteins to confer its antibiotic resistance, it can also be induced to produce large amounts of proteins from the inserted gene.

This is a cheap and easy way of mass-producing a gene or the protein it then codes for, for example, insulin or even antibiotics. However, a plasmid can only contain inserts of about 1-10 kbp. To clone longer lengths of DNA, lambda phage with lysogeny genes deleted, cosmids, bacterial artificial chromosomes or yeast artificial chromosomes could be used.

Types

One way of grouping plasmids is by their ability to transfer to other bacteria. Conjugative plasmids contain so-called tra-genes, which perform the complex process of conjugation, the transfer of plasmids to another bacterium. Non-conjugative plasmids are incapable of initiating conjugation, hence they can only be transferred with the assistance of conjugative plasmids, by 'accident'. An intermediate class of plasmids are mobilizable, and carry only a subset of the genes required for transfer. They can 'parasitize' a conjugative plasmid, transferring at high frequency only in its presence. Plasmids are now being used to manipulate DNA and may possibly be a tool for curing many diseases. It is possible for plasmids of different types to coexist in a single cell. Seven different plasmids have been found in E. coli. But related plasmids are often incompatible, in the sense that only one of them survives in the cell line, due t the regulation of vital plasmid functions.

Therefore, plasmids can be assigned into compatibility groups. Another way to classify plasmids is by function.

There are five main classes:

1. Fertility-F-plasmids, which contain tra-genes. They are capable of conjugation.
2. Resistance-(R)plasmids, which contain genes that can build a resistance against antibiotics or poisons. Historically known as R-factors, before the nature of plasmids was understood.
3. Col-plasmids, which contain genes that code for (determine the production of) bacteriocins, proteins that can kill other bacteria.
4. Degradative plasmids, which enable the digestion of unusual substances, *e.g.*, toluene or salicylic acid.
5. Virulence plasmids, which turn the bacterium into a pathogen.

Plasmids can belong to more than one of these functional groups. Plasmids that exist only as one or a few copies in each bacterium are, upon cell division, in danger of being lost in one of the segregating bacteria. Such single-copy plasmids have systems which attempt to actively distribute a copy to both daughter cells.

Some plasmids include an addiction system or "postsegregational killing system (PSK)", such as the hok/sok (host killing/suppressor of killing) system of plasmid R in Escherichia coli. They produce both a long-lived poison and a short-lived antidote. Daughter cells that retain a copy of the plasmid survive, while a daughter cell that fails to inherit the plasmid dies or suffers a reduced growth-rate because of the lingering poison from the parent cell.

Plasmid DNA Extraction

As alluded to above, plasmids are often used to purify a specific sequence, since they can easily be purified away from the rest of the genome. For their us as vectors, and for molecular cloning, plasmids often need to be isolated. There are several methods to isolate plasmid DNA from bacteria, the archetypes of which are the miniprep and the maxiprep/bulkprep.

The former can be used to quickly find out whether the plasmid is correct in any of several bacterial clones. The yield is a small amount of impure plasmid DNA, which is sufficient for analysis by restriction digest and for some cloning techniques. In the latter, much larger volumes of bacterial suspension are grown from which a maxi-prep can be performed. Essentially this is a scaled-up miniprep followed by additional purification. This results in relatively large amounts (several micrograms) of very pure plasmid DNA

DNA Ligases

DNA ligases are vital enzymes required for important cellular processes such asDNA replication, repair of damaged DNA and recombination. The enzyme mediates the formation of phosphodiester bonds between adjacent 3'-OH and 5'-phosphate termini, thereby joining the nicks in double stranded DNA. Ligases can be classified into two groups depending on their requirement for ATP or NAD+ as the cofactor. All eukaryotic and virally encoded enzymes are ATP-dependent, whereas most prokaryotic enzymes require NAD^+ for their activity.

ATP-Dependent DNA Ligase

DNA ligase from bacteriophage T7 is a monomer with a molecular weight of 41 kDa. Crystal structures of the apo form of this enzyme and its complex with ATP have been determined at 2.7 Å and 2.6 Å respectively. The structure consists of two distinct domains, a larger N-terminal domain and a C-terminal domain. The ATP-binding site is situated in the N-terminal domain in a pocket beneath one of the beta-sheets. This pocket is lined by a number of motifs that are conserved across a wide family of nucleotidyltransferases.

The structure of the C-terminal domain is remarkably similar to the oligonucleotide binding fold, observed in a number of proteins. The DNA-binding site is proposed to be in a groove running between the two domains. The structure has provided the basis for biochemical experiments. We have made the two domains separately and looked at their biochemical properties. The larger N-terminal domain is an active ligase but with greatly reduced activity. Both domains are able to bind to DNA. The N-terminal domain binds both single and double stranded DNA, while the smaller C-terminal domain is only able to bind double stranded DNA despite having a fold that is similar to single strand DNA binding proteins.

The affinity for nicked DNA comes from a combination of these two DNA-binding affinities at the active site. Interestingly, the N-terminal domain is poor at the adenylation reaction, but this activity is stimulated greatly by addition of the C-terminal domain. Furthermore, we can demonstrate an association of the domains on gel filtration. These data suggest that a conformational change occurs during the adenylation reaction. This would be similar to that observed directly in our crystal structure of an mRNA capping enzyme, a group of enzymes that are closely related to ligases.

NAD-Dependent DNA Ligase

An NAD-dependent DNA ligase from B.stearother-mophilus. The enzyme has been overexpressed and biochemical studies are in progress. Limited proteolysis has been used to investigate the domain structure of the enzyme. There are two domains, with the larger N-terminal domain retaining full self adenylation activity and the smaller C-terminal domain having all of the DNA-binding activity of the full length enzyme. The situation is therefore different to that we found in the ATP-dependent enzyme, where both domains contribute to the activities of the enzyme. Structural studies are underway with these domains. The structure of the large N-terminal domain has been solved and work is in progress to determine the structure of the small domain.

Ribonuclease A

Ribonuclease A (RNase A) is an endonuclease that cleaves single-stranded RNA. Bovine pancreatic RNase A is one of the classic model systems of protein science. The importance of bovine pancreatic RNase A was secured when the Armour purifid a *kilogram* of it, and gave 10 mg samples away free to any interested scientists. The ability to have a single lot of purified enzyme instantly made RNase *the* model system for protein studies.

RNase A was the model protein used to work out many spectroscopic methods for assaying protein structure, including absorbance, circular dichroism/optical rotary dispersion, Raman, EPR and NMR spectroscopy. RNase A was also the first model protein for the development of several chemical structural methods, such as limited proteolysis of disordered segments, chemical modification of exposed side chains, and antigenic recognition. Ribonuclease-S, which is RNase A that has been treated with subtilisin, was the third protein to have its structure solved, in 1967.Studies of the oxidative folding of RNase

A led Chris Anfinsen to enunciate the thermodynamic hypothesis of protein folding, which states that the folded form of a protein represents the minimum of its free energy. RNase A was the first protein for showing the effects of non-native isomers of X-Pro peptide bonds in protein folding.RNase A was the first protein to be studied by multiple sequence alignment and by comparing the properties of evolutionarily related proteins.

Structure and Properties

Labeled ribbon diagram of bovine pancreatic ribonuclease A (PDB accesion code 7RSA). The backbone ribbon is coloured from blue (N-terminus) to red (C-terminus). The side chains of the four disulfide-bonded cysteines are shown in yellow, with their sulpher atoms highlighted as small spheres. Residues important for catalysis are shown in magenta. RNase A is a relatively small protein (124 residues, ~13.7 kDa). It can be characterized as a two-layer á + â protein that is folded in half to resemble a taco, with a deep cleft for binding the RNA substrate.

The first layer is composed of three alpha helices from the N-terminal half of the protein. The second layer consist of three â-hairpins arranged in two a-sheets. The hairpins 61-74 and 105-124 form a four-stranded, antiparallel â-sheet that lies on helix 3 (residues 50-60). The longest â-hairpin 79-104 mates with a short â-strand (residues 42-45) to form a three-stranded, antiparallel â-sheet that lies on helix 2 (residues 24-34).

RNase A has four disulfide bonds in its native state: Cys26-Cys84, Cys58-110, Cys40-95 and Cys65-72. The first two (26-84 and 58-110) are essential for conformational folding; each joins an alpha helix of the first layer to a beta sheet of the second layer, forming a small hyrophobic core in its vicinity. The latter two disulfide bonds (40-95 and 65-72) are less essential for folding; either one can be reduced (but not both) without affecting the native structure under physiological conditions.

These disulfide bonds connect loop segments and are relatively exposed to solvent. Interestingly, the 65-72 disulfide bond has an extraordinarily high propensity to form, significantly more than would be expected from its loop entropy, both as a peptide and in the full-length protein. This suggests that the 61-74 α-hairpin has a high propensity to fold conformationally. RNase A is a basic protein (pI =8.63); its many positive charges are consistent with its binding to RNA (a poly-anion).

More generally, RNase A is unusually polar or, rather, unusually lacking in hydrophobic groups, especially aliphatic ones. This may account for its need of four disulfide bonds to stabilize its structure. The low hydrophobic content may also serve to reuce the physical repulsion between highly charged groups (its own and those of its substrate RNA) and regions of low dielectric constant (the nonpolar residues). The N-terminal a-helix of RNase A (residues 3-13) is connected to the rest of RNase A by a flexible linker (residues 16-23). As shown by F. M. Richards, this linker may be cleaved by subtilisin between residues 20 and 21 without causing the N-terminal helix to dissociate from the rest of RNase A. The peptide-protein complex is called RNase S, the peptide (residues 1-20) is called the S-peptide and the remainder (residues 21-124) is called the S-protein.

The dissociation constant of the S-peptide for the S-protein is roughly 30 pM; this tight binding can be exploited for protein purification by attaching the S-peptide to the protein of interest and passing a mixture over an affinity column with bound S-protein. The RNase S model system has also been used fr studying protein folding by coupling folding and association. The S-peptide was the first peptide from a native protein shown to have (flickering) secondary structure in isolation

Enzymatic Mechanism

The positive charges of RNase A lie mainly in a deep cleft between two lobes. The RNA substrate lies in this cleft and is cleaved by two catalytic histidines, His12 and His119, via a 2'–3' cyclic phosphate intermediate that is stabilized by nearby lysines such as Lys7, Lys41 and Lys66.

Anti-Cancer Effects

RNase A, and to a greater extent its oligomers and some homologs (such as onconase from frogs), have cytotoxic and cytostatic effects, particularly on cancer cells. This has led to the development of onconase as a cancer therapeutic, particularly for external use against skin cancers. As with many protein drugs, the internal use of non-human ribonucleases such as onconase is limited by the patient's immune response.

Heredity

Genetics, study of the function and behaviour of genes. Genes are bits of biochemical instructions found inside the cells of every organism from bacteria to humans. Offspring receive a mixture of genetic information from both parents. This process contributes to the great variation of traits that we see in nature, such as the colour of a flower's petals, the markings on a butterfly's wings, or such human behavioral traits as personality or musical talent.

Geneticists seek to understand how the information encoded in genes is used and controlled by cells and how it is transmitted from on generation to the next. Geneticists also study how tiny variations in genes can disrupt an organism's development or cause disease. Increasingly, modern genetics involves genetic engineering, a technique used by scientists to manipulate genes.

Genetic engineering has produced many advances in medicine and industry, but the potential for abuse of this technique has also presented society with many ethical and legal controversies.

Genetic information is encoded and transmitted fro generation t generation in deoxyribonucleic acid (DNA). DNA is a coiled molecule organized into structures called chromosomes within cells. Segments along the length of a DNA molecule form genes.

Genes direct the synthesis of proteins, the molecular laborers that carry out all life-supporting activities in the cell. Although all humans share the same set of genes, individuals can inherit different forms of a given gene, making each person genetically unique.

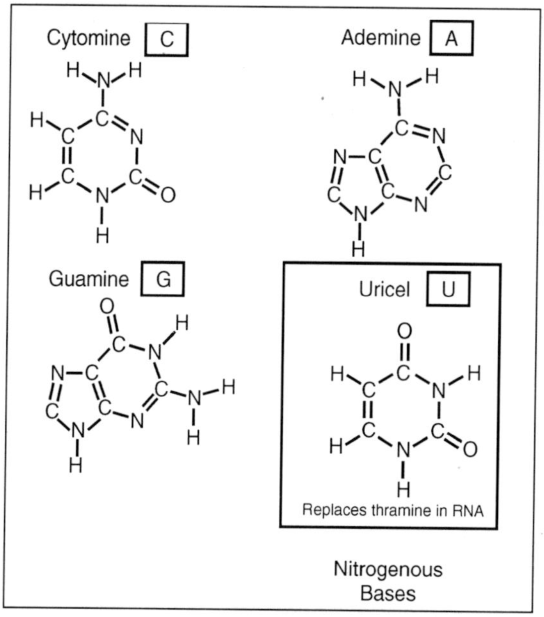

Figure: *Genetics*

Since the earliest days of plant and animal domestication, around 10,000 years ago, humans have understood that characteristic traits of parents could be transmitted to their offspring. The first to speculate about how this process worked were Greek scholars around the 4th century BC, who promoted theories based on conjecture or superstition. Some of these theories remained in favour for several centuries.

The scientific study of genetics did not begin until the late 19th century. In experiments with garden peas, Austrian monk Gregor Mendel described th patterns of inheritance, observing that traits were inherited as separate units. These units are now known as genes. Mendel's work formed the foundation for later scientific achievements that heralded the era of modern genetics.

Principles of Heredity

The site where genes work is the cell. Some organisms, such as paramecia or amoebas, are made up of a single cell. Other organisms are made of many kinds of cells, each having a different function. For instance, a tree contains some cells that form the root system and other cells that form leaves. Each cell's function within an organism is determined by the genetic information encoded in DNA.

In animals, plants, and othe eukaryotes (organisms whose cells contain a nucleus), DNA resides within membrane-bound structures in the cell. These structures include the nucleus, the energy-producing mitochondria, and, in plants, the chloroplasts (structures where photosynthesis takes place). In prokaryotes, one-celled organisms and bacteria that lack internal membrane-bound structures, DNA floats freely within the cell body.

The Importance of Genetics

The modern science of genetics influences many aspects of daily life, from the food we eat to how we identify criminals or treat diseases. In agriculture, genetic advances enable scientists to alter a plant or animal to make it more useful.

For instance, some food crops, such as oranges, potatoes, wheat, and rice, have been genetically altered to withstand insect pests, resulting in a higher crop yield. Tomatoes and apples have been modified so that they resist discoloration or bruising on their way to market, enhancing their appeal on supermarket shelves.

The genetic makeup of cows has been modified to increase their milk production, and cattle raised for beef have been altered so that they grow faster. Genetic technologies have also helped convict criminals.

DNA recovered from semen, blood, skin cells, or hair found at a crime scene can be analysed in a laboratory and compared with the DNA of a suspect. An individual's DNA is as unique as a set of fingerprints, and a DNA match can be used in a courtroom as evidence connecting a person to a crime. Genetics has revolutionized the way industries produce certain substances, many of which formerly required costly and arduous manufacturing methods.

In medicine, scientists can genetically alter bacteria so that they mass-produce specific proteins, such as insulin used by people with diabetes mellitus or human growth hormone used by children who suffer from growth disorders. In other medical applications, genetic technologies have been instrumental in the development of gene therapy. In this still-experimental form of treatment, scientists try to cure disease by replacing malfunctioning genes with healthy ones. Gene therapy has shown promise in treating some devastating conditions, including some forms of cancer and cystic fibrosis. Genetically engineered vaccines are being tested for possible use against the human immunodeficiency virus (HIV), the virus that causes acquired immunodeficiency syndrome (AIDS). The field of human genetics has been energized in recent years by the Human Genome Project, an international collaboration of scientists, governments, and drug companies from around the world. Scientists working on this project have developed detailed maps that identify the chromosomal locations of the estimated 20,000 to 25,000 human genes.

The vast databases emerging from the project help scientists study previously unknown genes as well as many genes all at once to examine how gene activity can cause disease. Scientists expect that the project will lead to the development of new drugs targeted to specific genetic disorders. Despite the benefits derived from genetic advancements, some observers have voiced concerns that genetically engineered organisms could harm people or the environment.

Others fear that new genetic technologies may enable scientists to modify genes that affect characteristics other than those responsible for disease. They warn that determining who has undesirable genetic characteristics may lead to discriminatory practices.

Others are concerned about the common misperception that a person's genes determine all aspects of a person's life, including health and behaviour. This misperception leads people to blame their genetic makeup for problems, leaving no room for the influence of free will, personal responsibility, or hope for change. These and other challenging issues place geneticists at the crossroads of science and social responsibility, where they work to promote understanding of genetic advances and prevent the abuse of them.

Cell Division and Reproduction

Organisms could not grow or function properly if the genetic information encoded in DNA was not passed from cell to cell. DNA is packaged into structures called chromosomes within a cell. Every chromosome in a cell contains many genes, and each gene is located at a particular site, or locus, on the chromosome. Chromosomes vary in size and shape and usually occur in matched pairs called homologues.

The number of homologous chromosomes in a cell depends upon the organism—for example, most cells in the human body contain 23 pairs of chromosomes, while most cells of the fruit fl *Drosophila* contain 4 pairs. Within all organisms, cells divide to produce new cells, each of which requires the genetic information found in DNA. Yet simply splitting the DNA of a dividing cell between two new cells would lead to disaster—the two new cells would have different instructions and each subsequen generation of cells would have less and less genetic information to work with. Imagine how chaotic it would be to rip an architectural blueprint in two, give each half to different contractors, and tell them to construct identical buildings. Just as each contractor would require a full copy of the blueprint to construct a complete building, each new cell needs a complete copy of an organism's genetic information to function properly. Organisms use two types of cell division to ensure that DNA is passed down from cell to cell during reproduction.

Simple one-celled organisms and other organisms that reproduce asexually—that is, without the joining of cells from two different organisms—reproduce by a process called mitosis. During mitosis a cell doubles its DNA before dividing into two cells and distributing the DNA evenly to each resulting cell. Organisms that reproduce sexually use a different type of cell division.

These organisms produce special cells called gametes, or egg and sperm. In the cell division known as meiosis, the chromosomes in a gamete cell are reduced by half. During sexual reproduction, an egg and sperm unite to form a zygote, in which the full number of chromosomes is restored.

Mitosis

Mitosis occurs in five stages: prophase, prometaphase, metaphase, anaphase, and telophase. During prophase, the start of mitosis, the DNA of each chromosome replicates. Each chromosome then reorganizes into paired structures called sister chromatids, with each member of the pair containing a full copy of the DNA sequence.

The sister chromatids condense, thickening until they appear joined at a single site, known as the centromere. Prometaphase is marked by the disintegration of the nuclear membrane. The sister chromatids line up in the middle of the cell halfway between the poles. In metaphase, exactly half of the chromatids face one pole, and half face the other pole. This equilibrium position is called the metaphase plate. In anaphase, the chromatid pairs split apart at the centromere, and each half of the pair then moves towards opposite poles of the cells.

In telophase, the final stage of mitosis, a nuclear membrane forms around the chromosomes at each pole of the cell. Mitosis ends with the formation of two new cells, each with a matching full set of chromosomes as well as an identical complement of cellular structures.

Meiosis

During meiosis, two cell divisions occur to produce four daughter cells from the original parent cell. Each resulting cell has half the chromosomal DNA of the parent cell. A half set of chromosomes in an organism is known as the haploid number. In the first cell division of meiosis the chromosomes of a gamete cell duplicate and join in pairs. The paired chromosomes align at the equator of the cell, and then separate and move to opposite poles in the cell.

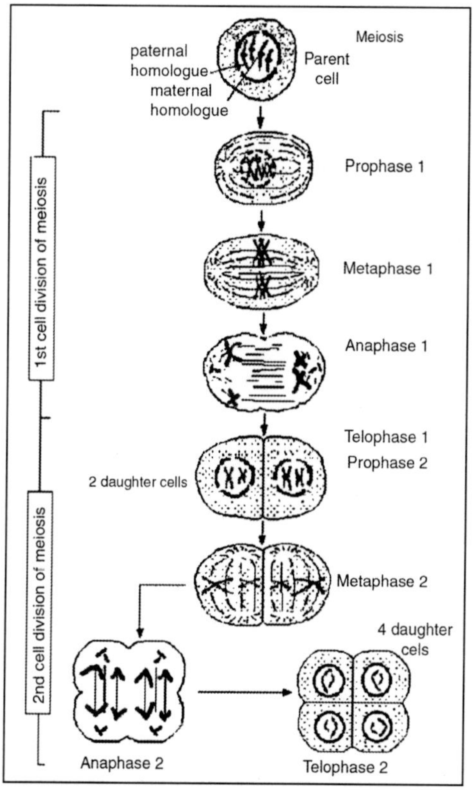

Figure: *Meiosis*

The cell then splits to form two daughter cells. As meiosis proceeds, the two daughter cells undergo another cell division to form four cells, each of which bears half of the number of chromosomes found in the other cells of the organism. Meiosis ensures that reproduction will produce a zygote that has received one set of chromosomes from the male parent and one set of chromosomes from the female parent to form a full set of chromosomes. The entire set of chromosomes in an organism is known as the diploid number. Once formed, the zygote continues to divide and grow through the process of mitosis

Patterns of Inheritance

In life forms that reproduce asexually, such as bacteria and amoebas, all offspring share the exact same genes and are identical to their parents. The genetic transmission that occurs in organisms that reproduce sexually is far more complex. An individual that forms by the union of two gametes inherits its chromosomes from two distinct parents.

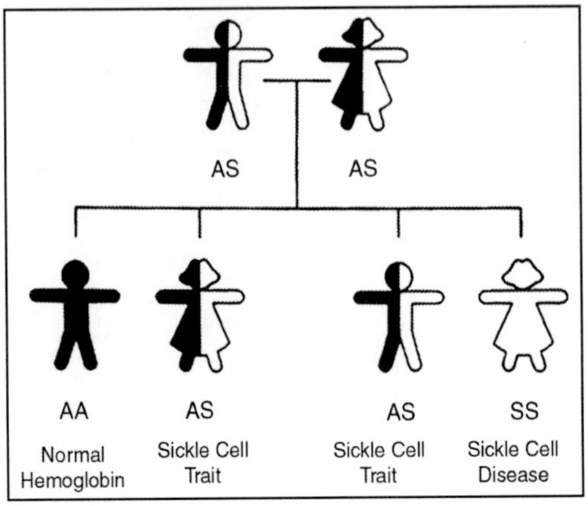

Figure: *Patterns of Inheritance*

Consequently, sexual reproduction guarantees that offspring with new combinations of genes will continually arise. Certai patterns of inheritance were evident long before scientists discovered the molecular structure of DNA and chromosomes. Throughout history, people have recognized that certain traits, whether in humans, animals, or agricultural crops, could be passed fro generation t generation. Yet for centuries, people were unable to reconcile many confusing observations about the mechanisms of inheritance. The first person to make sense of this complex subject was Austrian monk Gregor Mendel, who conducted a series of experiments on pea plants beginning in the 1850s.

Mendel observed the results of crossbreeding plants with different characteristics, such as height, flower colour, and seed shape. His conclusions from these experiments led him to develop explanations for how traits are transmitted fro generation t generation. Mendel's theories form the foundation of modern genetics.

Mendel's Rules

In his research, Mendel observed that characteristics were inherited as separate units, each of which was inherited independently of the others. Mendel suggested that each parent has pairs of these units but contributes only one of each pair to offspring. The units that Mendel described were later given the name *genes*. Mendel recognized that a gene can exist in different forms. Today these alternate forms are known as alleles. For example, pea seeds, the edible part of the plant we call peas, have a texture trait controlled by a single gene.

This gene occurs in two alleles: one corresponding to round (smooth) peas, the other to wrinkled peas. Although an individual can carry only two alleles for a particular gene, each gene may have dozens of different alleles. Mendel's experiments focused on interbreeding different strains of pea plants and then observing the traits that appeared in subsequen generations. When he crossbred plants with round peas and those with wrinkled peas, he discovered that all of the resulting offspring had round peas. Today we know that peas are round and smooth when they contain the right amount of sugar. If peas are missing the gene that produces a protein called starch branching enzyme 1 (SBE1), the peas make too much sugar, causing the peas to swell and then wrinkle and shrivel as they dry.

Mendel concluded that when an organism has two different alleles corresponding to the same genetic trait, one of the two may be dominant. The other allele is said to be recessive, meaning that its presence will be detectable only if an organism has inherited the recessive gene from both parents. For convenience, geneticists designate alleles by a single letter—the dominant allele is represented by a capital letter and the recessive allele by a small letter. In the pea texture example, a plant inherits one allele for pea texture from each parent.

The dominant allele that produces SBE1, resulting in ound, smooth peas, is designated as *R*, while the recessive allele that does not produce SBE1 and produces wrinkled peas is designated as *r*. To determine the set of alleles an organism has for a given trait just by visual observation can often be difficult. In the pea plant example, for instance, plants with smooth peas might be carrying two dominant alleles for that characteristic (RR) or one dominant and one recessive allele (Rr). Geneticists use the termgenotype to refer to the combination of genes that code for a trait, while the termphenotype describes the physical manifestation of that trait.

Therefore, the presence of two dominant alleles for pea texture (*RR*) would reflect thegenotype while a smooth pea indicates thephenotype. Mendel did not limit his experiments to testing the rules of inheritance of single traits. He also studied plant traits involving multiple pairs of genes, breeding plants that have round, yellow seeds with plants that produce wrinkled, green seeds. Such experiments demonstrated that thepatterns of inheritance he observed in his experiments with single traits also apply to cases involving more complex gene combinations.Exceptions to Mendel's Rules

Mendel published his studies in a science journal in 1865, at which time no other scientist commented on his work. Since that time, geneticists have learned that sometimes genes do not easily conform to so-called Mendelianpatterns of inheritance.

Incomplete Dominance

In cases of incomplete dominance, the inheritance of a dominant and a recessive allele results in a blending of traits to produce intermediate characteristics. For example, four-o'clock paint plants may have red, white, or pink flowers. Plants with red flowers have two copies of the dominant allele *R* for red flower colour (*RR*). Plants with white flowers have two copies of the recessive allele *r* for white flower colour (*rr*). Pink flowers result in plants with one copy of each allele (*Rr*), with each allele contributing to a blending of colors.Quantitative Inheritance

Mendel focused his studies on traits determined by a single pair of genes, and the resultingphenotype was easy to distinguish. A tall plant can be markedly different from a short one, and a green pea can easily be distinguished from a yellow one. There are some traits, however, that are not easy to distinguish. Human skin colour, for example, may be any of a wide variety of shades. Traits such as skin colour differ from the ones Mendel studied because they are determined by more than one pair of genes.

In this form of inheritance, known asquantitative inheritance, each pair of genes has only a slight effect on the trait, while the cumulative effect of all the genes determines the physical characteristics of the trait. At least four pairs of genes control human skin colour. Multiple genes also control many traits important in agriculture, such as milk production in cows and ear length in corn.

Multiple Alleles

Another exception to Mendelian genetics involves genes with multiple alleles. Certain traits are controlled by multiple alleles that have complex rules of dominance. In humans, for example, the gene for blood type has three alleles: I_A, I_B, and i. With three alternatives for each member of a gene pair, there are six possible combinations of these genes (I_AI_A, I_BI_B, ii, I_Ai, I_Bi, I_AI_B). Although there are six possible combinations, humans have only four major blood types: A, B, AB, and O.

This results because both I_A and I_B dominate over i, but not over each other, so a person with a gene combination of I_AI_A or I_Ai has blood type A. The gene combinations I_BI_B and I_Bi both produce blood type B. I_AI_B results in a blood type AB, and ii results in blood type O.

Gene Linkage

In his experiments, Mendel was careful to study traits in pea plants where one trait did not appear to influence another, such as the plant's height or the pea's texture. These twophenotypes (height and texture) occur randomly with respect to one another in a manner known as independent assortment. Today scientists understand that independent assortment occurs when the genes affecting thephenotypes are found on different chromosomes.

An exception to independent assortment develops when genes appear near one another on the same chromosome. When genes occur on the same chromosome, they are inherited as a single unit. Genes inherited in this way are said to be linked. For example, in fruit flies the genes affecting eye colour and wing length are inherited together because they appear on the same chromosome. But in many cases, genes on the same chromosome that are inherited together produce offspring with unexpected allele combinations. This results from a process calledcrossing over. Sometimes at the beginning of meiosis, a chromosome pair (made up of a chromosome from the mother and a chromosome from the father) may intertwine and exchange sections of chromosome.

The pair then breaks apart to form two chromosomes with a new combination of genes that differs from the combination supplied by the parents. Through this process of recombining genes, organisms can produce offspring with new combinations of maternal and paternal traits that may contribute to or enhance survival.

Sex-Linked Traits

Most chromosome pairs consist of identical, or homologous, partners. In many species, including humans, there is one pair of chromosomes in which the partners noticeably differ from each other. These are called the sex chromosomes because they determine the differences between males and females.

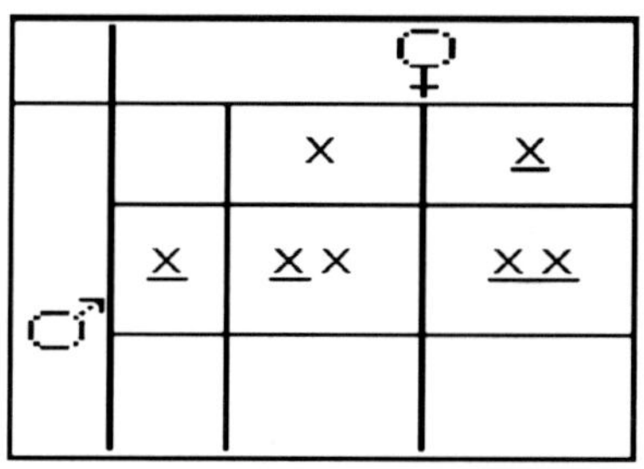

Figure: *Sex-Linked Traits*

Genes located on the sex chromosomes display differentpatterns of inheritance than genes located on other chromosomes. In human females, the sex chromosomes consist of two X chromosomes, while males have an X chromosome and a shorter Y chromosome with many fewer genes. In males the X chromosome contains many genes that have no corresponding gene on the Y chromosome.

A male's X chromosome may contain a recessive allele associated with a genetic disorder, such as hemophilia or Duchenne muscular dystrophy. In this case, males do not have a normal second copy of the gene on the Y chromosome to mask the effects of the recessive gene, and disease typically results. Additional examples of sex-linked traits include red-green colour blindness in humans and eye colour in fruit flies.

The Genetic Code

The structure of DNA encodes all the information every cell needs to function and thrive. In addition, DNA carries hereditary information in a form that can be copied and passed intact fro generation t generation.

A gene is a segment of DNA. The biochemical instructions found within most genes, known as the genetic code, specify the chemical structure of a particular protein. Proteins are composed of long chains of amino acids, and the specific sequence of these amino acids dictates the function of each protein.

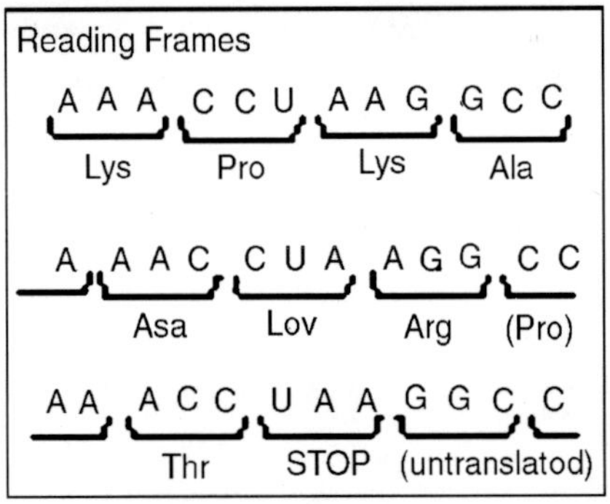

Figure: *Genetic Code*

The DNA structure of a gene determines the arrangement of amino acids in a protein, ultimately determining the type and function of the protein manufactured.

DNA Structure

DNA molecules form from chains of building blocks called nucleotides. Each nucleotide consists of a sugar molecule called deoxyribose that bonds to a phosphate molecule and to a nitrogen-containing compound, known as a base. DNA uses four bases in its structure: adenine (A), cytosine (C), guanine (G), and thymine (T). The order of the bases in a DNA molecule—the genetic code—determines the amino acid sequence of a protein.

In the cells of most organisms, two long strands of DNA join in a single molecule that resembles a spiraling ladder, commonly called a double helix. Alternating phosphate and sugar molecules form each side of this ladder. Bases from one DNA strand join with bases from another strand to form the rungs of the ladder, holding the double helix together.

The pairing of bases in the DNA double helix is highly specific—adenine always joins with thymine, and guanine always links to cytosine. These base combinations, known as complementary base pairing, play a fundamental role in DNA's function by aiding in th replication and storage of genetic information. Complementary base pairing also enables scientists to predict the sequence of bases on one strand of a DNA molecule if they know the order on the corresponding, or complementary, DNA strand. Scientists use complementary base pairing to help identify the genes on a particular chromosome and to develop methods used in genetic engineering. Genes line up in a row along the length of a DNA molecule. In humans a single gene can vary in length from 100 to over 1,000,000 bases.

Genes make up less than 2 per cent of the length of a DNA molecule. The rest of the DNA molecule is made up of long, highly repetitive nucleotide sequences. Once dismissed as "junk" DNA, scientists now believe these nucleotide sequences may play a role in the survival of cells. Identifying the function of these sequences is a thriving field of genetics research

Protein Synthesis

DN replication ensures that the genetic instructions encoded in DNA can be used continuously throug generations to produce the proteins that build and operate the cells of an organism. The process of tapping the genetic code to create proteins, known a protein synthesis, has two crucial steps: transcription and translation.

DN Replication

In order for inherited traits to be transmitted from parent to child, the genetic information encoded in DNA must be copied with great precision during cell division. The accuracy of DN replication depends upon the complementary pairing of bases. Durin replication, the DNA double helix unwinds and bonds joining the base pairs break, separating the DNA molecule into two separate strands.

Each strand of DNA directs the synthesis of another complementary strand. The unpaired bases of each DNA strand attach to bases floating within the cell. But the DNA strand's unpaired bases bond only with specific, complementary bases—for example, an adenine base will bond only with a thymine base and a cytosine bases will pair only with a guanine base. Once all of the bases of a DNA strand bond to complementary bases, the complementary bases then link to each other, forming a new DNA double-helix molecule. Thus the original DNA molecule replicates into two DNA molecules that are exact duplicates.

Transcription

Transcription transfers the genetic code from a molecule of DNA to an intermediary molecule called ribonucleic acid (RNA). The basic nucleotide structure of RNA resembles that of DNA, but the two compounds have three critical differences. First, the structure of RNA incorporates the sugar ribose rather than deoxyribose, the sugar in DNA. Second, RNA uses the base uracil (U) instead of thymine (T). In RNA uracil binds with adenine just as thymine does in DNA. Third, RNA usually exists as a single strand, unlike the double-helix structure that normally characterizes DNA. Transcription involves the production of a special kind of RNA known as messenger RNA (mRNA).

The process begins when the two strands of a DNA molecule separate, a task directed by the enzyme RNA polymerase. After the double helix splits apart, one of the strands serves as a template, or pattern, for the formation of a complementary mRNA molecule. Free-floating individual bases within the cell bind to the bases on the DNA template using complementary base pairing. The individual bases then link together to form a strand of mRNA.

I eukaryotes (organisms whose cells have a nucleus), the mRNA strand undergoes an additional step before the next stage o protein synthesis can occur. The mRNA strand consists of coding regions called exons separated by regions called introns. The introns do not contribute t protein synthesis. Special enzymes in the nucleus remove the introns from the mRNA strand. The remaining exons then link together to form an mRNA strand that contains the entire code for making a protein.

Translation

Once transcription is complete and the genetic code has been copied onto mRNA, the genetic code must be converted into the language of proteins. That is, the information coded in the four bases found in mRNA must be translated into the instructions encoded by the 20 amino acids used in the formation of proteins. This process, called translation, takes place in cellular organelles called ribosomes.

I eukaryotes, mRNA travels out of the nucleus into the cell body to attach to a ribosome. In prokaryotes (organisms without a nucleus), the ribosome clasps mRNA and starts translation before these strands have finished transcription and separated from the DNA. In bot eukaryotes and prokaryotes, the ribosome acts like a workbench and clamp that holds the mRNA strand and coordinates the activity of enzymes and other molecules essential to translation.

Another form of RNA called transfer RNA (tRNA) is found in the cytoplasm of the cell. There are many different types of tRNA, and each type binds with one of the 20 amino acids used in protein formation. One end of a tRNA binds with a specific amino acid. The other end carries three bases, known as an anticodon. The tRNA with an amino acid attached travels to the ribosome where the mRNA is stationed. The anticodon of the tRNA undergoes complementary base pairing with a series of three bases on the mRNA, known as the codon. The mRNA codon codes for the type of amino acid carried by the tRNA. A second tRNA bonds with the next codon on the mRNA. The resident tRNA transfers its amino acid to the amino acid of the incoming tRNA and then leaves the ribosome.

This process continues repeatedly, with new tRNA receiving the growing chain of amino acids, known as a polypeptide chain, from a resident tRNA. The ribosome moves the mRNA strand one codon at a time, making new codons available to bind with tRNAs. The process ends when the entire sequence of mRNA has been translated. The polypeptide chain falls away from the ribosome as a newly formed protein, ready to go to work in the cell.

Mutations

Occasionally mistakes occur during DN replication an protein synthesis. Any alteration in the structure of a gene results in a mutation. Mutations occur during DN replication when the chemical structure of genes undergoes random modifications. Once a change has occurred, the altered genes continue to replicate in their changed form unless another mutation occurs.

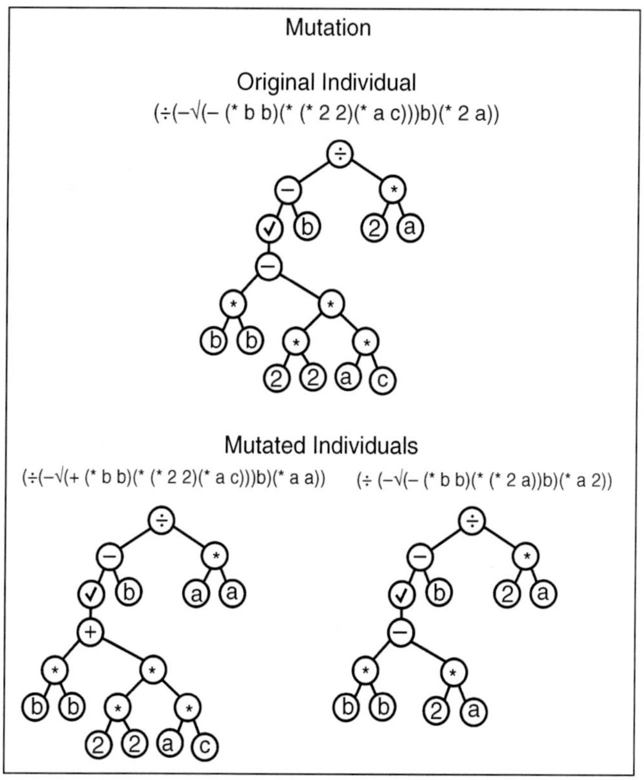

Figure: *Mutations*

Sometimes mutations occur during transcription or translation, causin protein synthesis to go awry. Although mutations may occur in any living cell, they are most important when they occur in gametes because then the change affects the traits of followin generations. Most mutations harm an organism. If a mutation occurs in a gene sequence that codes for a particular protein, the mutation may result in a change in the amino acid sequence directed by the gene. This change, in turn, may affect the function of the protein.

The implications can be significant: The amino acid sequence distinguishing normal hemoglobin from the altered form of hemoglobin responsible for sickle-cell anemia differs by only a single amino acid. Some mutations may be neutral or silent and do not affect the function of a protein. Occasionally a mutation benefits an organism. Over the course of evolutionary time, however, mutations serve the crucial role of providing organisms with previously nonexistent proteins.

In this way, mutations are a driving force behind genetic diversity and the rise of new or more competitive species better able to adapt to changes, such as climate variations, depletion of food sources, or the emergence of new types of disease.

Mutations can produce a change in any region of a DNA molecule. In *point mutation*, for example, a single nucleotide replaces another nucleotide. Although point mutation produces a small change to the DNA sequence, it may cause a change in the amino acid sequence, and thus the function, of a protein. Far more serious are mutations that involve the addition or deletion of one or more bases from a DNA molecule.

Adding or subtracting even a single base from a normal sequence during transcription can disrupt translation by shifting the "reading frame" of every subsequent codon. For example, an mRNA strand may include two codons in the following sequence: AUG UGA. The addition of a cytosine base at the beginning of this sequence shifts the "spelling" of these codons so that they read: CAU GUG.

This may result in an incorrect amino acid sequence during translation, or the protein may be truncated. Known a frameshift mutation, this type of alteration could result in the production of a protein with no real function or one with a harmful effect. Sometimes mutations are caused b transposition, in which long stretches of DNA (containing one or more genes) move from one chromosome to another. These jumping genes, calle transposons, can disrupt transcription and change the type of amino acids inserted into a protein Transposons rearrange and interrupt genes in a way that generally improves the genetic variation of a species.

While mutations can occur spontaneously, some can be caused by exposure to physical or chemical agents in the environment called mutagens. Common environmental mutagens include ultraviolet rays from the sun and various chemicals, such as asbestos, cigarette smoke, and nitrous acid. High-energy radiation, such as medical X rays, can cause DNA strands to break, leading to the deletion of potentially important genetic information. Radiation damage can also affect an entire chromosome, disrupting the function of many genes. In chromosomal translocation, a piece of one chromosome breaks off and merges with another chromosome. In some cases, large sections of chromosomes may break off and be lost.

The cell has highly effective self-repair mechanisms that can correct the harmful changes made by mutations and prevent some mutations from being passed on. Some 50 specialized enzymes locate different types of faulty sequences in the DNA and clip out those flaws. Another repair mechanism scans DNA afte replication and marks mismatched base pairs for repair.

Gene Regulation

The processes that enable information to be copied from genes and then used to synthesize proteins must be regulated if an organism is to survive. Different cells within an organism share the same set of chromosomes. In each cell some genes are active while others are not. For example, in humans only red blood cells manufacture the protein hemoglobin and only pancreas cells make the digestive enzyme known as trypsin, even though both types of cells contain the genes to produce both hemoglobin and trypsin. Each cell produces different proteins according to its needs so that it does not waste energy by producing proteins that will not be used. A variety of mechanisms regulate gene activity in cells. One method involves turning on or off gene transcription, sometimes by blocking the action of RNA polymerase, an enzyme that initiates transcription.

Gene regulation may also involve mechanisms that slow or speed the rate of transcription, using specialized regulatory proteins that bind to DNA. Depending on an organism's particular needs, one regulatory protein may spur transcription for a particular protein, and later, another regulatory protein may slow or halt transcription.

Prokaryotes

The bacteriu *Escherichia coli* (commonly referred to as *E. coli*), found in the intestines of humans and other mammals, provides a good example of gene regulation. *E. coli* uses three enzymes to digest lactose, the primary sugar found in milk. The bacterium produces great quantities of lactose-digesting enzymes when lactose is present and saves energy by not synthesizing the enzymes when the sugar is not available. *E. coli* prefers consuming glucose to lactose, so the bacterium produces these enzymes in the presence of lactose only when no glucose is available. A region of DNA known as an operon controls this gene regulation process. In *E. coli*, the operon includes at least five genes: Three genes, called lac genes, code for the enzymes that digest lactose; one gene encodes for a regulatory protein, called a repressor, that can sense the presence or absence of lactose; and one gene, called an operator, activates transcription, the first step in the synthesis of the lactose-digesting enzymes.

In the absence of lactose, the repressor protein binds with the operator gene to block transcription. This prevents the lac genes from being transcribed and halts production of the lactose-digesting enzymes. If lactose is present, the repressor protein binds to the sugar, leaving the operator gene free to trigger transcription of the lac genes. The transcription of the lac genes produces the mRNA that will direct the production of the three lactose-digesting enzymes

Eukaryotes

Gene regulation i eukaryotes is more complex than in bacteria and other prokaryotes. Even the simples eukaryotes have far more genes than prokaryotes, and these genes must be turned on or off as conditions dictate. Most multicellular organisms contain different types of cells that serve specialized functions. The cells of an animal's heart, blood, skin, liver, and muscles all contain the same genes. But in order to carry out their specific functions within the body, each cell must produce different proteins and respond to changing environmental stimuli, such as glucose levels in the blood or body temperature.

Such specialization is possible only with sophisticated gene regulation Eukaryotes use a variety of mechanisms to ensure that each cell uses the exact proteins it needs at any given moment. In one method, eukaryotic cells use DNA sequences called enhancers to stimulate the transcription of genes located far away from the point on the chromosome where transcription occurs. If a specific protein binds to an enhancer site on the DNA, it causes the DNA to fold so that the enhancer site is brought closer to the site where transcription occurs.

This action can activate or speed up transcription in the genes surrounding the enhancer site, thereby affecting the type and quantity of proteins the cell will produce. Enhancers often exert their effects on large groups of related genes, such as the genes that produce the set of proteins that form a muscle cell. Gene regulation can also take place after transcription has occurred by interfering xwith the steps that modify mRNA before it leaves the nucleus to take part in translation.

This process typically involves removing exons (segments that code for specific proteins) and introns. These sections of the mRNA can be modified in more than one way, enabling a cell to synthesize different proteins depending on its needs.

Genes in Development

Gene regulation helps individual cells within an organism function in a specialized way. Other regulatory mechanisms coordinate the genes that determine how cells develop. All of the specialized cells in an organism, including those of the skin, muscle, bone, liver, and brain, derive from identical copies of a single fertilized egg cell. Each of these cells has the exact same DNA as the original cell, even though they have vastly different appearances and functions. Genes dictate how these cells specialize.

Early in an organism's embryonic development the overall body plan forms. Individual cells commit to a particular layer and region of the embryo, often migrating from one location to another to do so. As the organism grows, cells become part of a particular body organ or tissue, such as skin or muscle. Ultimately, most cells become highly specialized—not only to develop into a neuron rather than a muscle cell, for example, but to become a sensory neuron instead of a motor neuron.

This process of specialization is called differentiation. At each stage of the differentiation process, specific genes known as developmental control genes actively turn on and switch off the genes that differentiate cells. One class of developmental control genes, known as homeotic genes, directs the formation of particular body parts. Activating one set of homeotic genes instructs part of an embryo to develop into a leg, for example, while another set initiates the formation of the head.

If a homeotic gene becomes altered or damaged, an organism's body development can be dramatically disrupted. A change in a single gene in some insects, for instance, can cause a leg to grow where an antenna belongs. Homeotic genes work by regulating the activity of other genes. Homeotic genes code for the production of a regulatory protein that can bind to DNA and thus affect the transcription of one or more genes. This enables homeotic genes to initiate or halt the development and specialization of characteristics in an organism.

Nearly identical homeotic genes have been identified in varied organisms, such as insects, worms, mice, birds, and humans, where they serve similar embryonic development functions. Scientists theorize that homeotic genes first appeared in a single ancestor common to all these organisms. Sometime in evolutionary history, these organisms diverged from their common ancestor, but the homeotic genes continued to be passed down throug generations virtually unchanged during the evolution of these new organisms.

5

Replication of Genetic Molecules

DNA replication, the basis for biological inheritance, is a fundamental process - that occurs in all living organisms - that copies their DNA. This process is replication, in that each strand of the original double-stranded DNA molecule serves as template for the reproduction of the complementary strand. Therefore, following DNA replication, two identical DNA molecules have been produced from a single double-stranded DNA molecule. Cellular proofreading and error toe-checking mechanisms ensure near perfect fidelity for DNA replication

In a cell, DNA replication begins at specific locations in the genome, called "origins". Unwinding of DNA at the origin, and synthesis of new strands, forms a replication fork. In addition to DNA polymerase, the enzyme that synthesizes the new DNA by adding nucleotides matched to the template strand, a number of other proteins are associated with the fork and assist in the initiation and continuation of DNA synthesis.

DNA replication can also be performed in vitro (outside a cell). DNA polymerases, isolated from cells, and artificial DNA primers are used to initiate DNA synthesis at known sequences in a template molecule. The polymerase chain reaction (PCR), a common laboratory technique, employs such artificial synthesis in a cyclic manner to amplify a specific target DNA fragment from a pool of DNA. The process of copying a double-stranded DNA molecule to form two double-stranded molecules DNA replication.

The process of DNA replication is a fundamental process used by all living organisms as it is the basis for biological inheritance. As each DNA strand holds the same genetic information, both strands can serve as templates for the reproduction of the opposite strand. The template strand is preserved in its entirety and the new strand is assembled from nucleotides. This process is called "semiconservative replication". The resulting double-stranded DNA molecules are identical; proofreading and error-checking mechanisms exist to ensure near perfect fidelity. In a cell, DNA replication must happen before cell division can occur. DNA synthesis begins at specific locations in the genome, called "origins", where the two strands of DNA are separated. RNA primers attach to single stranded DNA and the enzyme DNA polymerase extends the primers to form new strands of DNA, adding nucleotides matched to the template strand. The unwinding of DNA and synthesis of new strands forms a replication fork.

In addition to DNA polymerase, a number of other proteins are associated with the fork and assist in the initiation and continuation of DNA synthesis. DNA replication can also be performed artificially, using the same enzymes used within the cell. DNA polymerases and artificial DNA primers are used to initiate DNA synthesis at known sequences in a template molecule. The polymerase chain reaction (PCR), a common laboratory technique, employs artificial synthesis in a cyclic manner to rapidly and specifically amplify a target DNA fragment from a pool of DNA.

Discovery of Replication

The first step in DNA replication is the separation of the two DNA strands that make up the helix that is to be copied. DNA Helicase untwists the helix at locations called replication origins. The replication origin forms a Y shape, and is called a replication fork. The replication fork moves down the DNA strand, usually from an internal location to the strand's end. The result is that every replication fork has a twin replication fork, moving in the opposite direction from that same internal location to the strand's opposite end. Single-stranded binding proteins (SSB) work with helicase to keep the parental DNA helix unwound. It works by coating the unwound strands with rigid subunits of SSB that keep the strands from snapping back together in a helix. The SSB subunits coat the single-strands of DNA in a way as not to cover the bases, allowing the DNA to remain available for base-pairing with the newly synthesized daughter strands.

The two parent strands of DNA are separated to begin replication, one strand is oriented in the 5' to 3' direction while the other strand is oriented in the 3' to 5' direction. DNA replication, however, is inflexible: the enzyme that carries out the replication, DNA polymerase, only functions in the 5' to 3' direction. This characteristic of DNA polymerase means that the daughter strands synthesize through different methods, one adding nucleotides one by one in the direction of the replication fork, the other able to add nucleotides only in chunks. The first strand, which replicates nucleotides one by one is called the leading strand; the other strand, which replicates in chunks, is called the lagging strand.

DNA Replication is Semi-Conservative

DNA replication of one helix of DNA results in two identical helices. If the original DNA helix is called the "parental" DNA, the two resulting helices can be called "daughter" helices. Each of these two daughter helices is a nearly exact copy of the parental helix (it is not 100% the same due to mutations). DNA creates "daughters" by using the parental strands of DNA as a template or guide.

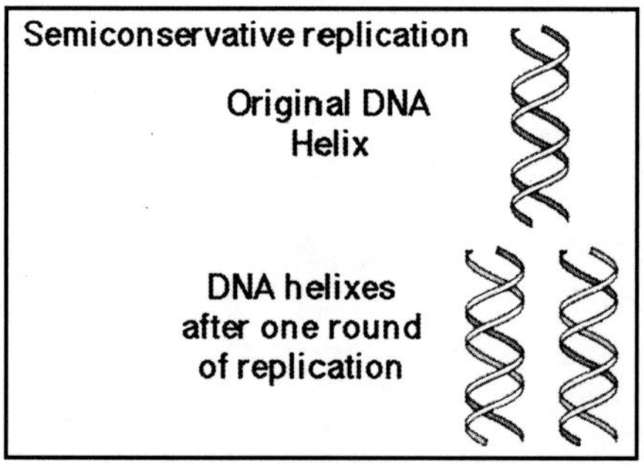

Figure: *Semi-Conservative Nature of DNA Replication*

Each newly synthesized strand of DNA (daughter strand) is made by the addition of a nucleotide that is complementary to the parent strand of DNA. In this way, DNA replication is semi-conservative, meaning that one parent strand is always passed on to the daughter helix of DNA.

The Lagging Strand

Whereas the DNA polymerase on the leading strand can simply follow the replication fork, because DNA polymerase must move in the 5' to 3' direction, on the lagging strand the enzyme must move away from the fork. But if the enzyme moves away from the fork, and the fork is uncovering new DNA that needs to be replicated, then how can the lagging strand be replicated at all? The problem posed by this question is answered through an ingenious method.

The lagging strand replicates in small segments, called Okazaki fragments. These fragments are stretches of 100 to 200 nucleotides in humans (1000 to 2000 in bacteria) that are synthesized in the 5' to 3' direction away from the replication fork. Yet while each individual segment is replicated away from the replication fork, each subsequent Okazaki fragment is replicated more closely to the receding replication fork than the fragment before. These fragments are then stitched together by DNA ligase, creating a continuous strand. This type of replication is called discontinuous. The first synthesized Okazaki fragment on the lagging strand is the furthest away from the replication fork, which is itself receding to the right. Each subsequent Okazaki fragment starts at the replication fork and continues until it meets the previous fragment. The two fragments are then stitched together by DNA ligase.

Because synthesis on the lagging strand takes place in a "backstitching" mechanism, its replication is slightly delayed in relation to synthesis on the leading strand. The lagging strand must wait for a patch of the parent helix to open up a short distance in front of the newly synthesized strand before it can begin its synthesis back to the end of the daughter strand.This "Lag" time does not occur in the leading strand because it synthesizes the new strand by following right behind as the helix unwinds at the replication fork.

Another complication to replication on the lagging strand is the initiation of replication. Whereas the RNA primer on the leading strand only has to trigger the initiation of the strand once, on the lagging strand each individual Okazaki fragment must be triggered. On the lagging strand, then, an enzyme called primase that moves with the replication fork synthesizes numerous RNA primers, each of which triggers the growth of an Okazaki fragment. The RNA primers are eventually removed leaving gaps that are filled by the replication machinery.

Since DNA replication moves along the parent strand in the 5' to 3' direction, replication can occur very easily on the leading strand. The nucleotides are added in the 5' to 3' direction. Triggered by RNA primase, which adds the first nucleotide to the nascent chain, the DNA polymerase simply sits near the replication fork, moving as the fork does, adding nucleotides one after the other, preserving the proper anti-parallel orientation. This sort of replication, since it involves one nucleotide being placed right after another in a series, is called continuous.

DNA Structure

DNA usually exists in a double-stranded structure, with both strands coiled together to form the characteristic double-helix. Each single strand of DNA is a chain of four types of nucleotide: adenine, cytosine, guanine, and thymine. A nucleotide consists of a phosphate and a deoxyribose sugar forming the backbone of the DNA double helix plus a base that points inwards. Nucleotides are matched between strands through hydrogen bonds to form base pairs.

Adenine pairs with thymine and cytosine pairs with guanine. The physical pairing of bases in DNA means that the information contained within each strand is redundant. The nucleotides on a single strand can be used to reconstruct nucleotides on a newly synthesized partner strand. DNA strands have a directionality, and the different ends of a single strand are called the "3' end" and the "5' end" (these refer to the carbon atom in ribose that the next phosphate in the chain attaches to).

In addition to being complementary, the two strands of DNA are antiparallel: they are orientated in opposite directions. This directionality has consequences in DNA synthesis, because DNA polymerase can only synthesize DNA in one direction by adding nucleotides to the 3' end of a DNA strand.

DNA Polymerase

DNA polymerases are a family of enzymes critical for all forms of DNA replication. A DNA polymerase synthesizes a new strand of DNA by extending the 3' end of an existing nucleotide chain, adding new nucleotides matched to the template strand one at a time.

Some DNA polymerases may also have some proofreading ability, removing nucleotides from the end of a strand in order to remove any mismatched bases. DNA polymerases are generally extremely accurate, making less than one error for every million nucleotides added. The energy for the process of DNA polymerization comes from the two additional phosphates attached to each of the unincorporated nucleotides.

These free nucleotides, also known as nucleoside triphosphates, contain a total of three phosphates. When a nucleotide is being added to a growing DNA strand, two of the phosphates are removed and the energy produced is used to attach the remaining phosphate to the growing chain.

The energetics of this process may also explain the directionality of synthesis - if DNA were synthesized in the 3' to 5' direction, the energy for the process would come from the 5' end of the growing strand rather than from free nucleotides. During proofreading, if the 5' nucleotide needed to be removed this triphosphate end would be lost, losing the energy source required to add a new nucleotide to the end. DNA polymerase can only extend an existing DNA strand paired with a template strand, it cannot begin the synthesis of a new strand. To do this a short fragment of DNA or RNA, called a primer, must be created and paired with the template strand before DNA polymerase can synthesize new DNA.

DNA Replication within the Cell

For a cell to divide, it must first replicate its DNA. This process is initiated at particular points within the DNA, known as "origins", which are targeted by proteins that separate the two strands and initiate DNA synthesis. Origins contain DNA sequences recognized by replication initiator proteins (eg. dnaA in *E coli'* and the Origin Recognition Complex in yeast). These initiator proteins recruit other proteins to separate the two strands and initiate replication forks. Initiator proteins recruit other proteins to separate the DNA strands at the origin, forming a bubble. Origins tend to be "AT-rich" (rich in adenine and thymine bases) to assist this process because A-T base pairs have two hydrogen bonds (rather than the three formed in a C-G pair)—strands rich in these nucleotides are generally easier to separate. Once strands are separated, RNA primers are created on the template strands and DNA polymerase extends these to create newly synthesized DNA.

As DNA synthesis continues, the original DNA strands continue to unwind on each side of the bubble, forming replication forks. In bacteria, which have a single origin of replication on their circular chromosome, this process eventually creates a "theta structure". In contrast, eukaryotes have longer linear chromosomes and initiate replication at multiple origins within these.

The Replication Fork

The replication fork is a structure which forms when DNA is being replicated. It is created through the action of helicase, which breaks the hydrogen bonds holding the two DNA strands together. The resulting structure has two branching "prongs", each one made up of a single strand of DNA.

Leading Strand Synthesis

In DNA replication, the leading strand is defined as the new DNA strand at the replication fork that is synthesized in the 5'→ 3' direction in a continuous manner. When the enzyme helicase unwinds DNA, two single stranded regions of DNA (the "replication fork") form. On the leading strand DNA polymerase III is able to synthesize DNA using the free 3' OH group donated by a single RNA primer and continuous synthesis occurs in the direction in which the replication fork is moving.

Lagging Strand Synthesis

The lagging strand is the DNA strand at the opposite side of the replication fork from the leading strand, running in the 3' to 5' direction. Because DNA polymerase cannot synthesize in the 3'→5' direction, the lagging strand is synthesized in short segments known as Okazaki fragments. Along the lagging strand's template, primase builds RNA primers in short bursts. DNA polymerases are then able to use the free 3' OH groups on the RNA primers to synthesize DNA in direction.

The RNA fragments are then removed (different mechanisms are used in eukaryotes and prokaryotes) and new deoxyribonucleotides are added to fill the gaps where the RNA was present. DNA ligase then joins the deoxyribonucleotides together, completing the synthesis of the lagging strand.

Dynamics at the Replication Fork

As helicase unwinds DNA at the replication fork, the DNA ahead is forced to rotate. This process results in a build-up of twists in the DNA ahead. This build-up would form a resistance that would eventually halt the progress of the replication fork. DNA topoisomerases are enzymes that solve these physical problems in the coiling of DNA. Topoisomerase I cuts a single backbone on the DNA, enabling the strands to swivel around each other to remove the build-up of twists. Topoisomerase II cuts both backbones, enabling one double-stranded DNA to pass through another, thereby removing knots and entanglements that can form within and between DNA molecules. Bare single-stranded DNA has a tendency to fold back upon itself and form secondary structures; these structures can interfere with the movement of DNA polymerase. To prevent this, single-strand binding proteins bind to the DNA until a second strand is synthesized, preventing secondary structure formation.

Clamp proteins form a sliding clamp around DNA, helping the DNA polymerase maintain contact with its template and thereby assisting with processivity. The inner face of the clamp enables DNA to be threaded through it. Once the polymerase reaches the end of the template or detects double stranded DNA, the sliding clamp undergoes a conformational change which releases the DNA polymerase. Clamp-loading proteins are used to initially load the clamp, recognizing the junction between template and RNA primers.

Eukaryotes

Within eukaryotes, DNA replication is controlled within the context of the cell cycle. As the cell grows and divides, it progresses through stages in the cell cycle; DNA replication occurs during the S phase (Synthesis phase). The progress of the eukaryotic cell through the cycle is controlled by cell cycle checkpoints. Progression through checkpoints is controlled through complex interactions between various proteins, including cyclins and cyclin-dependent kinases. The G1/S checkpoint (or restriction checkpoint) regulates whether eukaryotic cells enter the process of DNA replication and subsequent division. Cells which do not proceed through this checkpoint are quiescent in the "G0" stage and do not replicate their DNA. Replication of chloroplast and mitochondrial genomes occurs independent of the cell cycle, through the process of D-loop replication.

Bacteria

Most bacteria do not go through a well-defined cell cycle and instead continuously copy their DNA; during rapid growth this can result in multiple rounds of replication occurring concurrently. Within *E coli*, the most well-characterized bacteria, regulation of DNA replication can be achieved through several mechanisms, including: the hemimethylation and sequestering of the origin sequence, the ratio of ATP to ADP, and the levels of protein DnaA. These all control the process of initiator proteins binding to the origin sequences.

Because *E coli* methylates GATC DNA sequences, DNA synthesis results in hemimethylated sequences. This hemimethylated DNA is recognized by a protein (SeqA) which binds and sequesters the origin sequence; in addition, dnaA (required for initiation of replication) binds less well to hemimethylated DNA. As a result, newly replicated origins are prevented from immediately initiating another round of DNA replication.

ATP builds up when the cell is in a rich medium, triggering DNA replication once the cell has reached a specific size. ATP competes with ADP to bind to DnaA, and the DNA-ATP complex is able to initiate replication. A certain number of DnaA proteins are also required for DNA replication — each time the origin is copied the number of binding sites for DnaA doubles, requiring the synthesis of more DnaA to enable another initiation of replication.

Termination of Replication

Because bacteria have circular chromosomes, termination of replication occurs when the two replication forks meet each other on the opposite end of the parental chromosome. *E coli* regulate this process through the use of termination sequences which, when bound by the Tus protein, enable only one direction of replication fork to pass through. As a result, the replication forks are constrained to always meet within the termination region of the chromosome.

Eukaryotes initiate DNA replication at multiple points in the chromosome, so replication forks meet and terminate at many points in the chromosome; these are not known to be regulated in any particular manner. Because eukaryotes have linear chromosomes, DNA replication often fails to synthesize to the very end of the chromosomes (telomeres), resulting in telomere shortening. This is a normal process in somatic cells — cells are only able to divide a certain number of times before the DNA loss prevents further division. (This is known as the Hayflick limit.) Within the germ cell line, which passes DNA to the next generation, the enzyme telomerase extends the repetitive sequences of the telomere region to prevent degradation. Telomerase can become mistakenly active in somatic cells, sometimes leading to cancer formation.

Rolling Circle Replication

Another method of copying DNA, sometimes used *in vivo* by bacteria and viruses, is the process of rolling circle replication. In this form of replication, a single replication fork progresses around a circular molecule to form multiple linear copies of the DNA sequence. In cells, this process can be used to rapidly synthesize multiple copies of plasmids or viral genomes.

In the cell, rolling circle replication is initiated by an initiator protein encoded by the plasmid or virus DNA. This protein is able to nick one strand of the double-stranded, circular DNA molecule at a site called the double-strand origin (DSO) and remains bound to the 5' phosphate end of the nicked strand. The free 3' hydroxyl end is released and can serve as a primer for DNA synthesis. Using the unnicked strand as a template, replication proceeds around the circular DNA molecule, displacing the nicked strand as single-stranded DNA. Continued DNA synthesis produces multiple single-stranded linear copies of the original DNA in a continuous head-to-tail series. *In vivo* these linear copies are subsequently converted to double-stranded circular molecules.

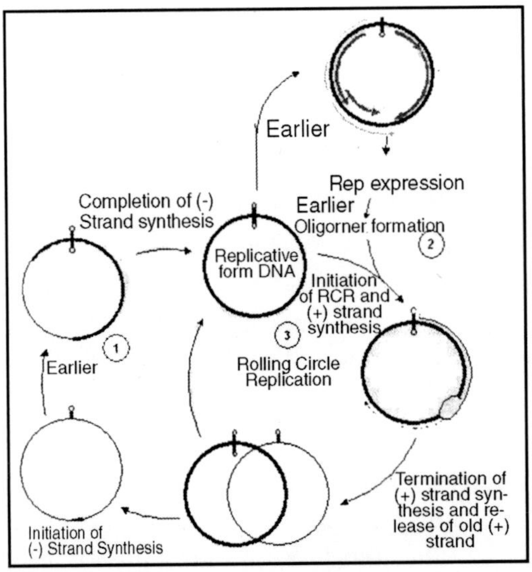

***Figure:** Rolling Circle Replication*

Rolling circle replication can also be performed *in vitro* and has found wide uses in academic research and biotechnology, often used for amplification of DNA from very small amounts of starting material. Replication can be initiated by nicking a double-stranded circular DNA molecule or by hybridizing a primer to a single-stranded circle of DNA. The use of a reverse primer (or random primers) produces hyperbranched rolling circle amplification, resulting in exponential rather than linear growth of the DNA molecule.

Polymerase Chain Reaction

In vitro, researchers commonly replicate DNA using the polymerase chain reaction (PCR). PCR uses a pair of primers to span a target region in template DNA, polymerizing partner strands in each direction. This process can be repeated through multiple cycles through the use of a thermostable polymerase. At the start of each cycle, the mixture of template and primers is heated, separating the newly synthesized molecule and template. Then, as the mixture cools, both of these become templates for new primers to anneal to, and the polymerase extends from these. As a result the number of copies of the target region doubles each round, growing exponentially.

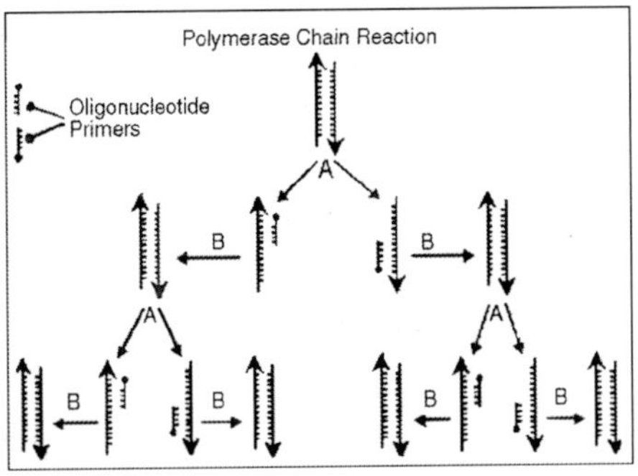

Figure: Polymerase Chain Reaction

Prokaryotic DNA Replication

Duplication of DNA is one of the fundamental properties of "molecule of life". Duplication or replication takes place once in a cell cycle. The duration and initiation point differs from one system to another. When conditions are favorable cell cytoplasmic mass increases and when cytoplasmic mass reaches a ratio to that of the cell size to 2L, bacterial cell division is triggered.

To complete its cell cycle it requires hardly 30 -40 minutes. Several genes involved in different steps of cell division have been characterized. Par excellent system to understand molecular process of prokaryotic cell division is E.coli. The size of the E.coli chromosome is 4.6×10^6 base pairs, but it is compacted by the binding of histone like proteins. First the nucleoid i.e the chromosome frees from the membrane and unwinds. Such DNA initiates replication and completes in a matter of 15-20 minutes.

Then the cytoplasm starts dividing almost in the centre of the cell by central septal ring formation. It is at this time the daughter DNA molecules are partitioned and segregated to the two compartments. The central periseptal ring that started 1-2 minutes after DNA replication now furrows inwards and divides the cytoplasm in the middle region. Then mid-wall splits in the middle and daughter cell are set free, hence the Schizomycetes to bacteria (E.coli) Bacterial cell division requires the participation of the following genes and gene products. Eukaryotic cell division is more complex and highly regulated. The paradigm for it is yeast, for it is a unicellular system, grows fast and requires hardly 60 minutes for one cell cycle. One can obtain different kinds of mutation like conditional and null or auxotrophic kinds. In culture cells, the duration of one cell cycle is about 12 hrs. Onion root tip cells divide once in 12 hrs.

Whereas embryonic cells in its early stages require just 30 minutes for one cell cycle and they go through 6 to 8 such cell divisions very fast. Replication of DNA in cell division is an important phase. The replication is very accurate and it is nucleotide by nucleotide and there is no provision for making any mistakes, even if mistakes are made they are immediately corrected or repaired, otherwise consequences are serious and deleterious.

In general DNA replication is precise, exact and regulated and involves initiation, elongation and termination steps, but mechanisms and rate of replication vary from one system to the other.

Replication is Semi Conservative

Meselson and Stahl showed that DNA replicates in semi conservative mode. There were some doubts about the mode whether it is conservative, dispersive or semi conservative. But Meselson and Stahl used nonradioactive heavy Nitrogen isotope (^{15}N), as the source of Nitrogen. When cells were grown on such source, ^{15}N incorporated nitrogen bases get incorporated into DNA, thus it becomes heavier than the DNA that is grown in normal (^{14}N) nitrogen media.

When two such DNAs, one grown in the presence of ^{15}N source and the other in ^{14}N (normal), are extracted separately and stained with ethidium Bromide, then subjected to equilibrium density centrifugation at ultra speed of 42000rpm for about 24 –36 hrs, the DNA, separates as two distinct bands, the one that is heavier (^{15}N) is the lower band than the normal (^{14}N) upper DNA. If cells grown in heavy isotopes are shifted to normal Nitrogen source and grown for one or two cell generation and if DNA is isolated from it and if one combine this with the first two DNA sources and subject them to equilibrium density ultra centrifugation (using cesium chloride or cesium sulphate), one can observe three bands, the top DNA band is from cells grown in ^{14}N without isotope, the second band is the hybrid band with one strand ^{15}N and the other with normal 14N-Nitrogen and the third band is with heavy nitrogen where both the strands are labeled with ^{15}N.

This experiment virtually and unambiguously proved that DNA replicates in semi conservative mode. Taylor using root tips, labeled with ^{32}P radioisotopes, of Vicia faba demonstrated semi conservative mode of replication at chromosomal level. Semi conservative means that the two daughter molecules produced at the end of replication, each of the molecules retains one parental strand and the other one is the newly made one. So it is semi conservative.

The mode of replication is absolutely semi conservative irrespective the kind of DNA, whether it is ds DNA or ssDNA, whether it is linear or circular, and the basic mechanism is more or less similar; each of them uses their own specialized mechanisms. Experimentally semiconservative mode of replication has been shown by density gradient centrifugation of non-radioactive isotopes such as ^{15}N and ^{14}N sources.

DNA Synthesis

DNA Replication (also known as DNA synthesis) is a process where the double stranded Deoxyribonucleic Acid (DNA) is copied. Replication is an important first step in cell division, as cells must duplicate their entire genetic constitution before they can divide into two daughter cells. DNA replication is also a critical requirement for DNA repair.

Requirements of DNA Replication

Replication requires three important components in order for this process to work.

Template Strand

The DNA serves as a template to guide the incoming nucleotides. The template strand attaches to the RNA primer strand in order to replicate the strand of DNA. The template just guides the nucleotides to the place that they need to go in. This is another component that is needed in order for DNA replication to even take place. If the DNA only had the template strand and none of the rest of the things then the strand could not replicate. The template starnd is very thin. Through an enzyme called primase the DNA template strand is synthesized one nucleotide at a time.

DNA Polymerase

There are many DNA polymerases within a cell, but only one of which is used in the replication of DNA. In order for DNA to replicate, it must have a primer, which is a small strand of RNA. The DNA template strand synthesizes one nucleotide at a time by the use of primase. The adding of nucleotides to the 3' end, which is continued until that section of DNA is replicated. The other DNA polymerases that are within the cell are used in other ways besides replication. DNA polymerase III is one certain polymerase, which aids in replication of what is known as the "lagging strand". DNA ligase is the final enzyme, which combines the lagging strand

Free 3' Hydroxyl

The free 3' hydroxyl is the starter strand called a primer which is required in order for DNA to be replicated. The primer strand is not very big, it is shorter then a single strand of RNA. The primer is the complementary strand to the template strand of DNA. This is another one of the things that are needed in order to make the process of DNA replication even possible. The DNA polymerase is added to the 3' end of the primer, which keeps growing until the section is finally complete. After a while the RNA primer is degraded and disappears.

Types of Replication

During the discovery of DNA replication they also discovered that there are three modes of replication that a DNA strand can take.

The three types of DNA replication are semiconservative, conservative, and dispersive. These three types of replication were tested on in order to see the possible patterns that would result in the complementary base pairing.

Semiconservative Replication

Through this type of replication the parent strand serves as a template for the new strand. The two offsprings would have one of the parent strand and one new strand.

Conservative Replication

Through this type of replication the double helix serves as a template. This although does not contribute to the new double helix.

Dispersive Replication

In this type of replication the fragments from the parent DNA molecule. The DNA serves as a template for the assembly of the new molecule. The new double helix contains old and new parts of the DNA strands.

DNA Repairing

After the DNA is replicated, there are three DNA repair mechanisms.

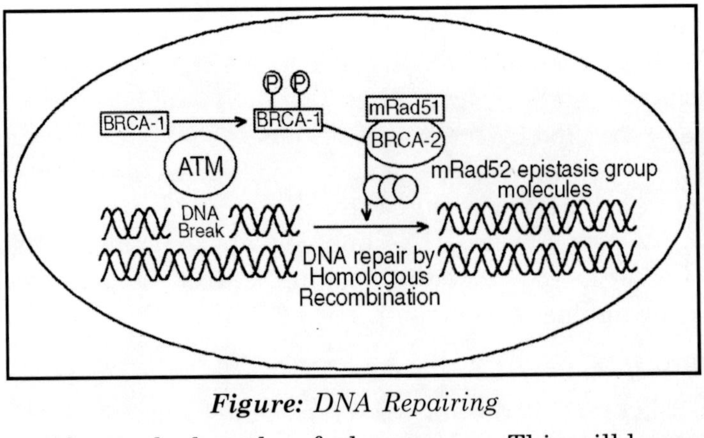

***Figure:** DNA Repairing*

They are to be used when they feel necessary. This will lessen the chance of human mutation. The rate of error in which the DNA polymerases makes a mistake is very high. This is why the DNA has mechanisms to cause fewer mutations. The three mechanisms are:

Proofreading

Every time a nucleotide is introduced to an already growing chain, the DNA polymerase checks the connection. If the pair is mismatched, it will be removed and redone. This lowers the overall rate of mutations.

Mismatch Pair

After the replication takes place, another set of proteins check to make sure that there are no more mismatched base pairs and also which strand is the wrong one.

Excision Pair

There can still be damage while the cell is living. Because of this, there are enzymes constantly checking the cell. If they detect a problem, the enzyme cuts the strand and also cuts away the opposite base. DNA polymerase and ligase then fix this base sequence.

Laboratory Technique

Scientists use two techniques in a laboratory that involve DNA replication. They use these techniques to observe genes and genomes. One of the techniques can make multiple copies of DNA from only a short piece. Another technique allows scientists to determine the base sequence of DNA. The two techniques are:

Polymerase Chain Reaction (PCR)

Through this process, a short strand of DNA is copied repetitively. There are three steps that are repeated in the process. The DNA strands are denatured or heated in order to separate the strand. Next, they add a primer, which was artificially made.

They also add DNA polymerase and the four deoxyribon-ucleotide triphosphates. These three are needed in order to replicate DNA. The final step is then the DNA polymerase catalyzes the production of a complementary strand.

DNA Sequencing

This allows scientists to determines the base sequence of a DNA. The process is the DNA is first denatured and a single strand is placed in a test tube. They then add DNA polymerase, primer, the four dNTP's, and small amounts of ddNTP's. The DNA replicates within the tube and after the DNA fragments are denatured.

The fragments then go through electrophoresis, which sorts the lengths of DNA fragments. The fragments then pass through a laser beam which excites the fluorescent tags. This information is fed into a computer at which tells the sequence.

Mendelson-Stahl Experiment

Through this experiment Meselson- Stahl convinced the scientists that the correct model for DNA replication was the semiconservative replication. They used density labeling in order to distinguish the old and new strands of DNA. In their experiment they used "heavy" isotopes of nitrogen. This nitrogen is a nonradioactive isotope that made the molecule more dense. The chemically identical molecules containing an isotope known as ^{14}N. They needed a way to distinguish the different DNA densities.

Because of this they needed to come up with something to measure the densities of solutions. Meselson, Stahl, and Jerome Vanguard came up with centrifuge, which is a procedure that involves the spinning of solutions at high speed, which causes the particles to separate and form gradient according to the different densities of the solution. The first part of their experiment they grew cultures of Escherichia coli. One of the cultures was grown in ^{15}N, which made the DNA "heavy".

The other culture was placed in a medium using ^{14}N rather than ^{15}N, which made all the DNA "light". In order to see if the use of centrifuge worked they combined the two cultures and when centrifuged it formed two DNA bands. This showed them the different densities with in the two cultures. The next part of their experiment they grew E. Coli in the ^{15}N medium and then transferred the bacteria from that medium to the ^{14}N. Through this they came up with the conclusion that through E. Coli DNA replicates every 20 minutes. They then took DNA samples from each generation through, which they found that the density gradient was different each generation.

Their observations could only be explained with the semiconservative model of DNA replication. Their results showed that when the DNNA first underwent replication, it was in 15N, which caused the DNA to be heavy.

Because in the semiconservative model of DNA replication, the one strand acts as a template for a second strand which the DNA was in a ^{14}N DNA strand and a ^{15}N and they were of intermediate density.

Stages of Gene Expression

Gene Regulation refers to the cellular control of the amount and timing of changes to the appearance of the functional product of a gene. Although a functional gene product may be an RNA or a protein, the majority of known mechanisms regulate protein coding genes. Any step of the gene's expression may be modulated, from DNA-RNA transcription to the post-translational modification of a protein.

Gene regulation is essential for viruses, prokaryotes and eukaryotes as it increases the versatility and adaptability of an organism by allowing the cell to express protein when needed. The first example of gene regulation system was the lac operon, discovered by Jacques Monod, in which protein involved in lactose metabolism are expressed by E.coli only in the presence of lactose and absence of glucose.

Furthermore, gene regulation allows the presence in a multicellular organism of different cells types arranged in a complex pattern, hence different transcriptomes despite them having all the same genome and the generation of patterns by cellular differentiation and morphogenesis.

Regulated Stages of Gene Expression

Any step of gene expression may be modulated, from the DNA-RNA transcription step to post-translational modification of a protein.

The following is a list of stages where gene expression is regulated:

- Chromatin domains
- Transcription
- Post-transcriptional modification
- RNA transport
- Translation
- mRNA degradation
- Post-translational modifications

Modification of DNA

In eukaryotes, the accessibility of large regions of DNA can depend on its chromatin structure which can be altered as a result of histone modifications which are directed by DNA methylation, ncRNA or DNA binding protein.

Structural

Transcription of DNA is dictated by its structure. In general, the density of its packing is indicative of the frequency of transcription. Octameric protein complexes called histones are responsible for the amount of supercoiling of DNA, and these complexes can be temporarily modified by processes such as phosphorylation or more permanently modified by processes such as methylation.

Such modifications are considered to be responsible for more or less permanent changes in gene expression levels. Histne acetylation is also an important process in transcription. Histone acetyltra-nsferase enzymes (HATs) such as CREB-binding protein also dissociate the DNA from the histone complex, allowing transcription to proceed. Often, DNA methylation and histone deacetylation work together in gene silencing. The combination of the two seems to be a signal for DNA to be packed more densely, lowering gene expression.

Chemical

Methylation of DNA is a common method of gene silencing. DNA is typically methylated by methyltransferase enzymes on cytosine nucleotides in a CpG dinucleotide sequence (also called "CpG islands" when densely clustered).

Analysis of the pattern of methylation in a given region of DNA (which can be a promoter) can be achieved through a method called bisulfite mapping. Methylated cytosine residues are unchanged by the treatment, whereas unmethylated ones are changed to uracil.

The differences are analysed by DNA sequencing or by methods developed to quantify SNPs, such as Pyrosequencing (Biotage) or MassArray (Sequenom), measuring the relative amounts of C/T at the CG dinucleotide. Abnormal methylation patterns are thought to be involved in carcinogenesis.

Post transcriptional Regulation

After the DNA is transcribed and mRNA is formed there must be some sort of regulation on how much the mRNA is translated into Proteins. Cells do this by modulating the Capping, Splicing, addition of a Poly(A) Tail, the sequence-specific nuclear export rates and in several contexts sequestration of the RNA transcript.

These processes occur in eukaryotes but not in prokaryotes.

This modulation is a result of a protein or transcript which in turn is regulated and may have an affinity for certain sequences:

- Capping changes the five prime end of the mRNA to a three prime end by 5'-5' linkage, which protects the mRNA from 5' exonuclease, which degrades foreign RNA. The cap also helps in ribosomal binding.
- Splicing removes the introns, noncoding regions that are transcribed into RNA, in order to make the mRNA able to create proteins. Cells do this by spliceosome's binding on either side of an intron, looping the intron into a circle and then cleaving it off. The two ends of the exons are then joined together.
- Addition of poly(A) tail otherwise known as poly-adenylation. Junk RNA is added to the 3' end, and acts as a buffer to the 3' exonuclease in order to increase the half life of mRNA.

In both prokaryotes and eukaryotes a large number of RNA binding protein exist, with often are directed to their target sequence by the secondary structure of the transcript, which may change depending on certain conditions, such as temperature or presence of a ligand (aptamer), some transcripts act as ribozymes and self-regulate their expression.

Regulation of Transcription

Regulation of transcription controls when transcription occurs and how much RNA is created. Transcription of a gene by RNA polymerase can be regulated by at least five mechanisms:

- Specificity factors alter the specificity of RNA polymerase for a given promoter or set of promoters, making it more or less likely to bind to them (*i.e.* sigma factors used in prokaryotic transcription).
- Repressors bind to non-coding sequences on the DNA strand that are close to or overlapping the promoter region, impeding RNA polymerase's progress along the strand, thus impeding the expression of the gene.
- General transcription factors These transcription factors position RNA polymerase at the start of a protein-coding sequence and then release the polymerase to transcribe the mRNA.
- Activators enhance the interaction between RNA polymerase and a particular promoter, encouraging the expression of the gene. Activators do this by increasing the attraction of RNA polymerase for the promoter, through interactions with subunits of the RNA polymerase or indirectly by changing the structure of the DNA.
- Enhancers are sites on the DNA helix that are bound to by activators in order to loop the DNA bringing a specific promoter to the initiation complex.

Examples of Gene Regulation

- Enzyme induction is a process in which a molecule (*e.g.* a drug) induces (*i.e.* initiates or enhances) the expression of an enzyme.
- The induction of heat shock proteins in the fruit fly *Drosophila melanogaster*.
- The Lac operon is an interesting example of how gene expression can be regulated.
- Viruses despite having only a few genes, possess mechanisms to regulate their gene expression, typically into a early and late phase, using collinear systems regulated by anti-terminators (lambda phage) or splicing modulators (HIV).

Up-regulation and Down-regulation

Up-regulation is a process which occurs within a cell triggered by a signal (originating internal or external to the cell) which results in increased expression of one or more genes and as a result the protein(s) encoded by those genes.

Conversely down-regulation is a process resulting in decreased gene and corresponding protein expression:

- *Up:* Regulation occurs for example when a cell is deficient in some kind of receptor. In this case, more receptor protein is synthesized and transported to the membrane of the cell and thus the sensitivity of the cell is brought back to normal reestablishing homeostasis.
- *Down:* Regulation occurs for example when a cell is overly stimulated by a neurotransmitter, hormone, or drug for a prolonged period of time and the expression of the receptor protein is decreased in order to protect the cell.

Gene Regulation can be summarized as how they respond:

- *Inducible Systems:* An inducible system is off unless there is the presence of some molecule (called an inducer) that allows for gene expression. The molecule is said to "induce expression". The manner in which this happens is dependent on the control mechanisms as well as differences between prokaryotic and eukaryotic cells.
- *Repressible Systems:* A repressible system is on except in the presence of some molecule (called a corepressor) that suppresses gene expression. The molecule is said to "repress expression". The manner in which this happens is dependent on the control mechanisms as well as differences between prokaryotic and eukaryotic cells. A large number of studied regulatory systems come from developmental biology. Examples include.

The collinearity of the Hox gene cluster with their nested antero-posterior patterning. It has been speculated that pattern generation of the hand (digits - interdigits) The gradient of Sonic hedgehog (secreted inducing factor) from the zone of polarizing activity in the limb which creates a gradient of active Gli3 which activates Gremlin which inhibits BMPs also secreted in the limb resulting in the formation of an alternating pattern of activity as a result of this reaction-diffusion system.

Somatogenesis is the creation of segmatation (somites) form a uniform tissue (PSM) sequentially from anterior to posterior, this is a achieved in amniotes possibly by means of two opposing gradients, Retinoic acid in the anterior (wavefront) and an oscillating gradient in the posterior (clock) composed of FGF + Notch and Wnt in antiphase.

Sex determination in the soma of a Drosophila requires the sensing of the ratio of autosomal genes to sex chromosome encoded genes, which results in the production of sexless splicing factor in females resulting in the female isoform of doublesex.

Theoretical Circuits

- *Repressor/Inducer*: a activation of a sensor results in the change of exprssion of a gene
- *Negative Feedback*: the gene product downregulates its on production directly or indirectly, which can result in.
  - Keeping transcript levels constant/proportional to a factor
  - Inhibition of run-away reactions when coupled with a positive feedback loop
  - Creating an oscillator by taking advantage in the time delay of transcription and translation, given that the mRNA and protein half-life is shorter
- *Positive Feedback*: the gene product upregulates its own production directly or indirectly, which can result in
  - Signal amplification
  - Bistable switches when two genes inhibit each other and have both positive feedback
  - Pattern generation

Generally, most experiments investigating differential expression used whole cell extracts of RNA, called steady-state levels, to determined which genes changed and by how much they did, these are however not informative of where the regulation has occurred and may actually mask conflicting regulatory processess, it is the most commonly analysed (QPCR and DNA microarray). When studying gene expression there are several methods to look at the various stages. In eukaryotes these include:

The chromatin conformation of the region can be determined by ChIP-chip analysis by pulling down RNA Polymerase II, Histone 3 modifications, Trithorax-group protein,Polycomb-group protein or any other DNA binding element to which a good antibody is available.

Due to post-transcriptional regulation, transcription rates and total RNA levels differ significantly, to measure the transcription rates nuclear run-on assays can be done and newer high-throughput methods are being developed, using thiol labelling instead of radioactivity. Only 5 per cent of the RNA polymerised in the nucleus actually exists and not only introns, abortive products and non-sense transcripts are degradated therefore the differences in nuclear and cytoplasmic levels can be see by separating the two fractions by gentle lysis. Alternative splicing can be analysed with a splicing array or with a tiling array. All in vivo RNA is complexed as RNPs. The quantity of transcripts bound to specific protein can be also analysed by RIP-Chip, for example DCP2 will give an indication of sequestered protein, ribosome bound gives and indication of transcripts active in transcription (although it should be noted that a more dated method, called polysome fractionation, is still popular in some labs).

Protein levels can be analysed by Mass spectrometry, which can only be compare to QPCR data as microarray data is relative and not absolute. RNA and protein degradation rates are measured by means of transcription inhibitors or translation inhibitors (Cycloheximide) respectively. Gene expression is the process by which inheritable information from a gene, such as the DNA sequence, is made into a functional gene product, such as protein or RNA. Several steps in the gene expression process may be modulated, including the transcription step and the post-translational modification of a protein.

Gene regulation gives the cell control over structure and function, and is the basis for cellular differentiation, morphogenesis and the versatility and adaptability of any organism. Gene regulation may also serve as a substrate for evolutionary change, since control of the timing, location, and amount of gene expression can have a profound effect on the functions (actions) of the gene in the organism. Non-protein coding genes (*e.g.* rRNA genes, tRNA genes) are transcribed, but not translated into protein. Genes are expressed by being transcribed into RNA, and this transcript may then be translated into protein. The expression of many genes is regulated afte transcription (*i.e.*, by microRNAs or ubiquitin ligases), so an increase in mRNA concentration need not always increase expression. Nevertheless, mRNA levels can be quantitatively measured by Northern blotting, a process in which a sample of RNA is separated on an agarose gel and hybridized to a radio-labeled RNA probe that is complementary to the target sequence.

Northern blotting requires the use of radioactive reagents and can have lower data quality than more modern methods (due to the fact that quantification is done by measuring band strength in an image of a gel), but it is still often used. It does, for example, offer the benefit of allowing the discrimination of alternately spliced transcripts. A more modern low-throughput approach for measuring mRNA abundance is real-time polymerase chain reaction (The term RT-PCR is used to refer to both reverse transcription PCR as well as real-time PCR, which is also known as quantitative RT-PCR or quantitative PCR (qPCR). With a carefully constructed standard curve qPCR can produce an absolute measurement such as number of copies of mRNA per nanolitre of homogenized tissue.

The lower level of noise in data obtained via qPCR often makes this the method of choice, but the price of the required equipment and reagents can be prohibitive. In addition to low-throughput methods, transcript levels for many genes at once (expression profiling) can be measured with DNA microarray technology or "tag based" technologies like Serial analysis of gene expression (SAGE) or the more advanced version Super SAGE, which can provide a relative measure of the cellular concentration of different messenger RNAs. Recent advances in microarray technology allow for the quantification, on a single array, of transcript levels for every known gene in the human genome. The great advantage of tag-based methods is the "open architecture", allowing for the exact measurement of any transcript, known or unknown. Especially SuperSAGE recommends itself therefore also for studying organisms with unknown genomes.

Protein levels themselves can be estimated by a number of means. The most commonly used method is to perform a Western blot against the protein of interest, whereby cellular lysate is separated on a polyacrylamide gel and then probed with an antibody to the protein of interest. The antibody can either be conjugated to a fluorophore or to horseradish peroxidase for imaging or quantification. Another commonly used method for assaying the amount of a particular protein in a cell is to fuse a copy of the protein to a reporter gene such as Green fluorescent protein, which can be directly imaged using a fluorescent microscope.

Because it is very difficult to clone a GFP-fused protein into its native location in the genome, however, this method often cannot be used to measure endogenous regulatory mechanisms (GFP-fusions are therefore most often expressed on extra-genomic DNA such as an expression vector). Fusing a target protein to a reporter can also change the protein's behaviour, including its cellular localization and expression level. The pattern of detection of a gene or gene product may be described using terms such as facultative, constitutive, circadian, cyclic, housekeeping, or inducible.

Regulation of gene expression is the cellular control of the amount and timing of appearance of the functional product of a gene. Any step of gene expression may be modulated, from the DNA-RNA transcription step to post-translational modification of a protein. Gene regulation gives the cell control over structure and function, and is the basis for cellular differentiation, morphogenesis and the versatility and adaptability of any organism.

Expression System

An expression system consists, minimally, of a source of DNA and the molecular machinery required to transcribe the DNA into mRNA and translate the mRNA into protein using the nutrients and fuel provided. In the broadest sense, this includes every living cell capable of producing protein from DNA. However, an expression system more specifically refers to a laboratory tool, often artificial in some manner, used for assembling the product of a specific gene or genes. It is defined as the "combination of an expression vector, its cloned DNA, and the host for the vector that provide a context to allow foreign gene function in a host cell, that is, produce proteins at a high level". In addition to these biological tools, certain naturally observed configurations of DNA (genes, promoters, enhancers, repressors) and the associated machinery itself are referred to as an expression system, as in the simple repressor 'switch' expression system in Lambda phage.

It is these natural expression systems that inspire artificial expression systems, (such as the Tet-on and Tet-off expression systems). Each expression system has distinct advantages and liabilities, and may be named after the host, the DNA source or the delivery mechanism for the genetic material. For example, common expression systems include bacteria (such as E.coli), yeast (such as S.cerevisiae), plasmid, artificial chromosomes, phage (such as lambda), cell lines, or virus (such as baculovirus, retrovirus, adenovirus).

Over Expression

In the laboratory, the protein encoded by a gene is sometimes expressed in increased quantity. This can come about by increasing the number of copies of the gene or increasing the binding strength of the promoter region. Often, the DNA sequence for a protein of interest will be cloned or subcloned into a plasmid containing the *lac* promoter, which is then transformed into the bacterium *Escherichia coli*. Addition of IPTG (a lactose analog) causes the bacteria to express the protein of interest. However, this strategy does not always yield functional protein, in which case, other organisms or tissue cultures may be more effective. For example, the yeast *Saccharomyces cerevisiae* is often preferred to bacteria for proteins that undergo extensive posttranslational modification. Nonetheless, bacterial expression has the advantage of easily producing large amounts of protein, which is required for X-ray crystallography or nuclear magnetic resonance experiments for structure determination.

Gene Networks

Genes have sometimes been regarded as nodes in a network, with inputs being proteins such as transcription factors, and outputs being the level of gene expression. The node itself performs a function, and the operation of these functions have been interpreted as performing a kind of information processing within cell and determine cellular behaviour.

Control of Gene Expression

The Chromosome of E. Coli

The single chromosome of the common intestinal bacterium *E. coli* is circular and contains some 4.7 million base pairs. It is nearly 1 mm long, but only 2nm wide. The chromosome replicates in a bidirectional method, producing a figure resembling the Greek letter theta. The promoter is the part of the DNA to which the RNA polymerase binds before opening the segment of the DNA to be transcribed.

A segment of the DNA that codes for a specific polypeptide is known as a structural gene. These often occur together on a bacterial chromosome. The location of the polypeptides, which may be enzymes involved in a biochemical pathway, for example, allows for quick, efficient transcription of the mRNAs. Often leader and trailer sequences, which are not translated, occur at the beginning and end of the region. *E. coli* can synthesize 1700 enzymes. Therefore, this small bacterium has the genes for 1700 different mRNAs.

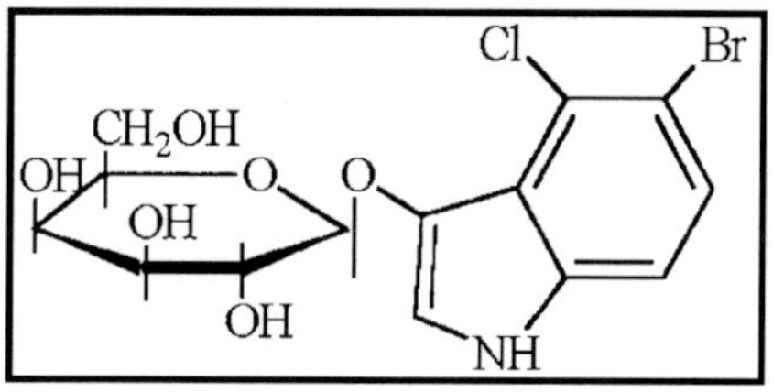

Figure: *Enzyme B-galactosidase*

Lactose, milk sugar, is split by the enzyme b-galactosidase. This enzyme is inducible, since it occurs in large quantities only when lactose, the substrate on which it operates, is present. Conversely, the enzymes for the amino acid tryptophan are produced continuously in growing cells unless tryptophan is present. If tryptophan is present the production of tryptophan-synthesizing enzymes is repressed.

The Operon Model

The operon model of prokaryotic gene regulation was proposed by Fancois Jacob and Jacques Monod. Groups of genes coding for related proteins are arranged in units known as operons. An operon consists of an operator, promoter, regulator, and structural genes. The regulator gene codes for a repressor protein that binds to the operator, obstructing the promoter (thus, transcription) of the structural genes.

The regulator does not have to be adjacent to other genes in the operon. If the repressor protein is removed, transcription may occur. Operons are either inducible or repressible according to the control mechanism. Seventy-five different operons controlling 250 structural genes have been identified for *E. coli*. Both repression and induction are examples of negative control since the repressor proteins turn off transcription.

Plasmids, small DNA fragments, are known from almost all bacterial cells. Plasmids carry between 2 and 30 genes. Some seem to have the ability to move in and out of the bacterial chromosome. An episome is a plasmid incorporated in the bacterial chromosome. Plasmids are self-replicating in a manner like the bacterial chromosome. Many plasmids have been recognized for *E. coli*, including the F plasmids ("sex factors") and R plasmids (drug/antibiotic resistance). The F plasmid contains 25 genes, some of which control the production of F pili (proteins which extend from the surface of F^+, or male, cells to the surface of F^-, or female, cells).

The R plasmid conveys drug resistance on cells having it. As many as 10 resistance genes can be contained on a single R plasmid. The R plasmids can be transferred to other bacteria of the same species, to viruses, and even to bacteria of different species. Drug (antibiotic) resistance has been found among pathogens causing the diseases typhoid fever, gastroenteritus, plague, undulant fever, meningitis, and gonorrhea. In addition to the more common modes of transfer, R plasmids may be passed through the cell membrane. The resistance genes appear to operate by either breaking down the antibiotics or by circumventing the block the antibiotic places on a key bacterial metabolic pathway.

Viruses

Viruses consist of a nucleic acid (DNA or RNA) enclosed in a protein coat (known as a capsid). The capsid may be a single protein repeated over and over, as in tobacco mosaic virus (TMV). It may also be several different proteins, as in the T-even bacteriophages. Once inside the cell, the nucleic acid follows one of two paths: lytic or lysogenic. Retroviruses, such as Human Immunodifficiency Virus (HIV), also include the enzyme reverse transcriptase with the viral RNA. Reverse transcriptase makes a single-stranded viral DNA copy of the single-stranded viral RNA. The single stranded viral DNA is subsequently turned into a double-stranded DNA.

The lytic cycle occurs when the viral DNA immediately takes over the host cell (remember that viruses are obligate intracellular parasites) and begins making new viruses. Eventually the new viruses cause the rupture (or lysis) of the cell, releasing those new viruses to continue the infection cycle. The lysogenic cycle occurs when the viral DNA is incorporated into the host DNA as a prophage.

When the cell replicates the prophage is passed along as if it were host DNA. Sometimes the prophage can emerge from the host chromosome and enter the lytic cycle spontaneously once every 10,000 cell divisions. Ultraviolet light and x-rays may also trigger emergence of the prophage. Transduction is the transfer of host DNA from one cell to another by a virus. Some bacteriophages are temperate since they tend to go lysogenic rather than lytic. These types of viruses are able to transduce fragments of the host DNA.

Transposons are DNA fragments incorporated into the chromosomal DNA. Unlike episomes and prophages, transposons contain a gene producing an enzyme that catalyzes insertion of the transposon at a new site. They also have repeated sequences 20-40 nucleotides in length at each end. Insertion sequences are short (600-1500 base pairs long) simple transposons that do not carry genes beyond those essential for insertion of the transposon into *E. coli*. Complex transposons are much larger and carry additional genes.

Genes incorporated in a complex transposon are known as jumping genes since they can move about on the chromosome (even from chromosome to chromosome). Often the complex transposons are flanked by simple transposons.

The Eukaryotic Chromosome

The eukaryotic chromosome consists of DNA and proteins that appear to play a major role in regulation of eukaryote genes. The DNA of each chromosome is a long single molecule of double stranded DNA. Eukaryotic DNA comes in two forms. Chromatin is the uncoiled form of DNA and is over 50% protein. Chromosomes are coiled DNA/protein that form during the early stages of cell division. The proteins associated with DNA are collectively known as histones. They are relatively short polypeptides which are positively charged (basic) and thus are attracted to the negatively charged (acidic) DNA. Histones are synthesized in quantity during the S-phase of the cell cycle.

One function of theses proteins seems to be the folding and packaging of DNA into chromosome form: the 2 m of DNA in a human cell are packaged into 46 chromosomes with a combined length of 200nm (a nm remember is 10-6m). Some 90 million molecules of histones occur in a single cell, with the majority (30 million) being H1 histones. Five types of histone are known (H1, H2A, H2B, H3, and H4); with the exception of H1 most eukaryote histones are very similar.

A nucleosome is the fundamental packing unit of eukaryotic DNA. The core consists of two molecules each of H2A, H2B, H3, and H4; around which the DNA is wound twice. The H1 histone is outside the core. Between 150-200 nucleotide pairs are associated with the core and linker DNA. This level of packing is known as "beads on a string".

The next level are known as the 30nm strand, whose details of organization are not yet well known. 30 nm strands are further condensed into 300nm wide looped domains. Looped domains are part of condensed sections of chromosomes (the chromosome being 1400nm wide at Metaphase I).

Replication of the Eukaryotic Chromosome

Nucleotide triphosphates (in the forms ATP, GTP, CTP and TTP) are assembled according to the semi-conservative model. Other details of DNA replication are consistent with what we know for prokaryotes. Eukaryotic DNA has many replication forks and also bidirectional synthesis, contrasting to the unidirectional rolling theta prokaryotic method. Eukaryotic DNA is synthesized much slower than prokaryotic DNA, in humans 50 nucleotides per second per replication fork. After replication the new DNA is immediately associated with histones.

Regulation of Eukaryotic Gene Expression

Eukaryotic gene regulation, especially in multicellular organisms, is complicated by the process of development unique to multicellular organisms. Each multicellular organism begins as a single-celled zygote which divides by mitosis. Cells differentiate into functional types by using some genes but ignoring others.

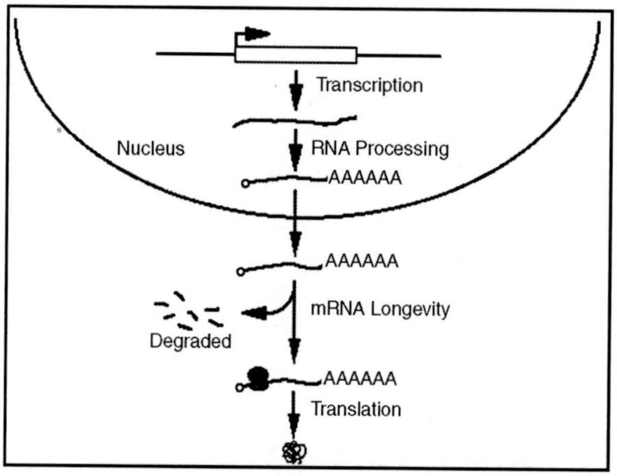

Figure: *Eukaryotic Gene Regulation*

Homeobox genes establish the body plan and position of organs in response to gradients of regulatory molecules. The timing of certain gene expressions seems to follow a sequence, such as the production of different types of fetal hemoglobins by mammalian red blood cells, which switch to adult hemoglobin sometime after birth. Clearly the inactivation of certain genes occurs in every adult cell; therein lies the cure for cancer, old age, etc.

Types of Chromatin

Heterochromatin stains more strongly and is a more condensed chromatin. Euchromatin stains weakly and is more open (less condensed). Euchromatin remains dispersed (uncondensed) during Interphase, when RNA transcription occurs. Some regions of heterochromatin appear to be structural (as in the heterochromatin near the centromere region). Barr bodies, irreversibly inactivated X-chromosomes, are also condensed heterochromatin. Other heterochromatin regions vary from cell to cell. As the cell differentiates, the proportion of heterochromatin to euchromatin increases, reflecting increased specialization of the cell as it matures. Loops (or puffs) in insect chromosomes are areas of active RNA synthesis, suggesting again the functional genes are located in open areas of the chromatin (or, the euchromatin). Eukaryotes also have specific binding proteins working in a similar fashion to prokaryotic mechanisms, however the eukaryotes, as one would expect, have a much more complicated process.

Eukaryotic Genome

We use the term genome to refer to all of the alleles possessed by an organism (or by a population, species, or larger taxonomic group). While the amount of DNA for a diploid cell is constant within a species, the differences can be great between species.

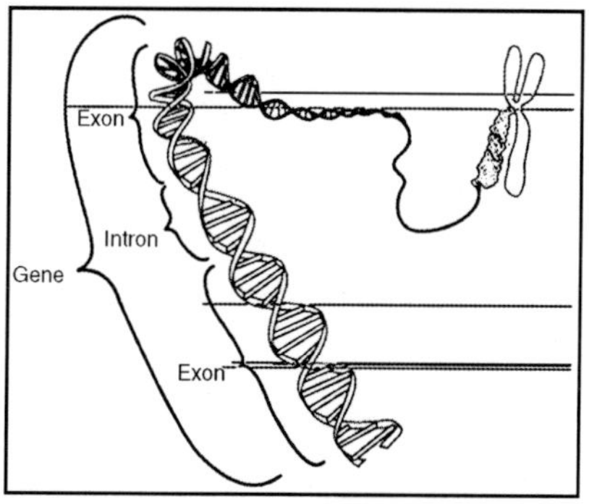

Figure: *Introns and Exons*

Humans have 3.5×10^9 base pairs, *Drosophila* has 1.5×10^8, toads have 3.32×10^9, and salamanders have 8×10^{10} base pairs per haploid genome. Much of the DNA in each cell either has no function or has a function not yet known. Eukaryotes have only 10 per cent of their DNA coding for proteins. Humans may have a little as 1 per cent coding for proteins. Viruses and prokaryotes use a great deal more of their DNA.

Almost half the DNA in eukaryotic cells is repeated nucleotide sequences. Protein-coding sequences are interrupted by non-coding regions. Non-coding interruptions are known as intervening sequences or introns. Coding sequences that are expressed are exons. Most, but not all structural eukaryote genes contain introns. Although transcribed, these introns are excised (cut out) before translation (a seemingly energy inefficient process). The number of introns varies with the particular gene, even occurring in tRNAs, rRNAs and viral genes! Generally the more complex and recently evolved the organism, the more numerous and larger the introns.

Which came first: continuous genes lacking introns or interrupted genes containing introns? Introns have been hypothesized to promote genetic recombination (via crossing-over), thus speeding up the evolution of new proteins. Exons are also thought to code for different functional regions of proteins. Gene families are made up of similar, but not identical, genes. The globin family is the best studied gene family.

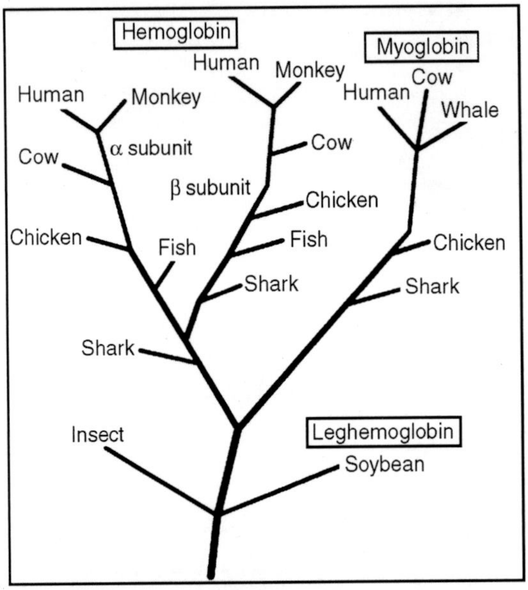

Figure: *Globin Family*

Hemoglobin consists, in humans, of 2 a-chains and 2 b-chains clustered about a common heme. Human beta-globin genes are scattered at five loci on human chromosome 11. These genes are expressed sequentially during development, and are similar with same-length introns in similar positions in each gene. Some of the genes are inactivated copies, others are functional only during certain phases of development. The actin family of genes also exhibits a similar pattern.

Transcription and Processing of mRNA

The process of transcription in eukaryotes is similar to that in prokaryotes, although there are some differences. Eukaryote genes are not grouped in operons as are prokaryote genes. Each eukaryote gene is transcribed separately, with separate transcriptional controls on each gene. Whereas prokaryotes have one type of RNA polymerase for all types of RNA, eukaryotes have a separate RNA polymerase for each type of RNA. One enzyme for mRNA-coding genes such as structural proteins. One enzyme for large rRNAs. A third enzyme for smaller rRNAs and tRNAs.

Prokaryote translation begins even before transcription has finished, while eukaryotes have the two processes separated in time and location (remember the nuclear envelope). After eukaryotes transcribe an RNA, the RNA transcript is extensively modified before export to the cytoplasm. A cap of 7-methylguanine (a series of an unusual base) is added to the 5' end of the mRNA; this cap is essential for binding the mRNA to the ribosome. A string of adenines (as many as 200 nucleotides known as poly-A) is added to the 3' end of the mRNA after transcription.

The function of a poly-A tail is not known, but it can be used to capture mRNAs for study. Introns are cut out of the message and the exons are spliced together before the mRNA leaves the nucleus. There are several examples of identical messages being processed by different methods, often turning introns into exons and vice-versa. Protein molecules are attached to mRNAs that are exported, forming ribonucleo-protein particles (mRNPs) which may help in transport through the nuclear pores and also in attaching to ribosomes.

Antibody-coding Genes

Antibodies are globular complex proteins made by multicellular individuals in response to a specific antigen. The cells making antibodies are lymphocytes, better known as white blood cells. Antibodies immobilize and destroy their specific antigens. A mouse can make 10,000,000 antibodies, more than the total genes of the mouse would suggest. Antibody proteins are composed of two long and two short chains. Each species has a constant region characteristic for the species and type of antibody. The other region is the variable in which antigen-specific amino-acids are located. Susumu Tonegawa tested (and proved) an old hypothesis that the constant and variable regions of antibodies were coded for by different genes. Tonegawa's work further demonstrated that gene fragments in embryos are rearranged to form the variable (functional) genes.

Viruses and Eukaryotes

The viruses of eukaryotes are similar to prokaryote-infecting viruses. Proviruses are viral DNA integrated into the host cell. Some of the DNA viruses can either initiate an infection (lytic in prokaryotes) cycle or can form proviruses. Simian Virus 40 (SV40) causes cancer in hamsters but not in its normal hosts. SV40 can introduce new functional genes into the host DNA, as can a number of other viruses.

Retroviruses can also insert their nucleic acid into host DNA via the reverse transcriptase mode mentioned previously. Reverse transcriptase is carried with the RNA into the host cell. In the process of making the cDNA strand for the viral RNA, the enzyme also makes long terminal repeats (LTRs) sequences at the terminal ends of the cDNA. These LTRs may also make insertion of the viral DNA into the host DNA easier. The inserted viral DNA makes RNA transcripts which are packaged with viral protein coats and reverse transcriptase.

The viral DNA may, depending on its insertion point, cause mutations of the host DNA. Most viral DNA insertions do not damage the host, but rather become part of the host genome and can be passed on if they have managed to infect a germ-line cell. From 0.5 to 1.0 per cent of mouse DNA may be of viral origin.

Eukaryotic Transposons

Eukaryotic transposons resemble prokaryotic transposons in many features. Many eukaryotic transposons are first copied to RNA and then back to DNA before being inserted in a new location.

Genes, Viruses and Cancer

Cancer is a disease in which cells escape the restraints on normal cell growth. Cancer is an inheritable disease (at least from cell to daughter cells). Once a cell has become cancerous, all of its descendant cells are cancerous. Gross chromosomal abnormalities are often visible in cancerous cells. Most carcinogens (cancer-generating factors) are also mutagens (mutation-generating factors). Oncogenes are genes resembling normal genes but in which something has gone wrong, resulting in a cancer. Fifty oncogenes have thus far been discovered. Viruses seem able to cause cancer in three ways. Presence of the viral DNA may disrupt normal host DNA functions.

Viral proteins needed for virus replication may also affect normal host gene regulation. Since most cancer-causing viruses are retroviruses, the virus may serve as a vector for oncogene insertion. Transfers of genes between eukaryotic cells will allow doctors, who have historically been limited to phenotypic cures, to attack disease at the genotypic level. SV40 virus has been used to inject the rabbit beta-globin gene into monkeys. Viruses can thus serve as a possible vector to place healthy (non-mutated) alleles into eggs. Like prokaryotes, eukaryotic organisms do not want to express all of their genes all of the time.

Given the complexity of multicellular eukaryotes, gene regulation in these organisms needs to be very complex. This module provides a brief overview of the various levels of regulation of gene expression in eukaryotes, and takes a look at the basics of transcriptional regulation; a detailed look at eukaryotic transcriptional regulation is beyond the scope of this course.

Gene Regulation in Eukaryotes

Eukaryotes need to regulate their genes for different reasons than prokaryotes. In prokaryotes, gene regulation allowed them to respond to their environment efficiently and economically. While eukaryotes can respond to their environment, the main reason higher eukaryotes need to regulate their genes is cell specialization. Whereas prokaryotes are (relatively speaking) simple, unicellular organisms, multicellular eukaryotes consist of hundreds of different cell types, each differentiated to serve a different specialized function. Each cell type differentiates by activating a different subset of genes.

Because of the multitude of cell types, the regulation of gene expression required to bring about such differentiation is necessarily complex. One way this complexity is demonstrated is in multiple levels of regulation of gene expression.

Levels of Regulation

Before we discuss the specifics of regulation, it is necessary to understand that "gene expression" covers the entire process from transcription through protein synthesis. The final measure of whether or not a gene is "expressed" is if the protein is produced, because it is protein that will ultimately carry out the function specified by the gene. We've seen numerous examples of how eukaryotic cells are more complex than prokaryotic cells. One obvious example of this is the presence of a nucleus in eukaryotic cells, which separates transcription from translation in a way not seen in prokaryotes.

Furthermore, eukaryotic transcripts must be processed before they can be translated. Here is a diagram outlining the steps involved in the production of a protein in eukaryotic cells: Regulation can occur at any point in this pathway; specifically, it occurs at the levels of transcription, RNA processing, mRNA lifetime (longevity), and translation. Each of these types of regulation will be considered in turn.

Regulation of RNA Processing

After transcription, the RNA must be processed before it can be translated. As described elsewhere, RNA processing involves addition of a 5' cap, addition of a 3' poly (A) tail, and removal of introns. This processing represents another level of regulation of gene expression, particularly in regard to splicing out of introns.

Regulation can be of two types:

1. Whether an RNA gets processed;
2. Which exons are retained in the mRNA.

The first type of regulation can determine whether or not an mRNA gets translated. If an RNA is not processed, it will not be transported out of the nucleus, and will not be translated. The second type of regulation can affect the function of the protein produced. Some genes have exons that can be exchanged in a process known as exon shuffling. For example, a gene with four exons might be spliced differently in two different cell types. In each of these cases, the polypeptide produced could have a different function. In mammals, for example, the calcitonin gene produces a hormone in one cell type, and a neurotransmitter in another cell type, due to exon shuffling.

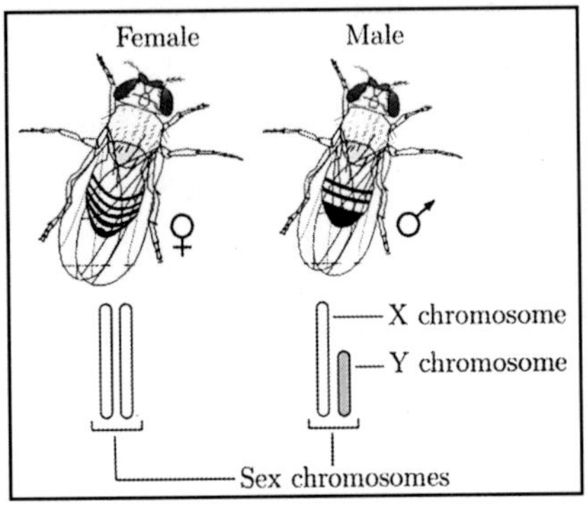

Figure: *Drosophila*

In *Drosophila*, alternate splicing of the *sex-lethal* RNA can produce an mRNA encoding a functional polypeptide, or one with a premature stop codon that encodes a short, nonfunctional polypeptide.

Regulation of RNA Longevity

Imagine two mRNA molecules: one lasts for five minutes in the cytoplasm before being degraded, while the other one manages to linger for an hour before being degraded. If both are translated continually while they exist, it is obvious that more of the second polypeptide will be produced than the first. This is the principle behind regulation of RNA longevity.

mRNAs from different genes have their approximate lifespan encoded in them; this serves to help regulate how much of each polypeptide is produced. The information for lifespan is found in the 3' UTR. The sequence AUUUA, when found in the 3' UTR, is a signal for early degradation (and therefore short lifetime). The more times the sequence is present, the shorter the lifespan of the mRNA. Because it is encoded in the nucleotide sequence, this is a set property of each different mRNA; the longevity of an mRNA can't be varied.

Regulation of Translation

Whether or not an mRNA molecule is translated can be regulated as well. The various mechanisms of translational regulation are incompletely understood, but there are many documented examples (particularly in embryonic develop-ment) of mRNA molecules that are present routinely, but are only translated under certain circumstances. For example, many animals sequester large amounts of mRNA in their eggs, and those mRNA molecules are not translated unless the egg is fertilized. Whether or not a gene is transcribed is the major way that gene expression is regulated in eukaryotes, as it was in prokaryotes. There are some major differences between transcriptional regulation in prokaryotes and eukaryotes. For one thing, because of the complexity of eukaryotic patterns of gene expression, each eukaryotic gene needs its own promoter. In other words, eukaryotic genes are not organized into operons. Another difference is that prokaryotic genes are regulated primarily by repressors. Although repressors occasionally play a role in eukaryotes, eukaryotic genes are primarily regulated by transcriptional activators. These activators are transcription factors.

Regulatory Elements of Eukaryotic Genes

As discussed in the module on transcription, eukaryotic genes have promoters that are recognized by basal transcription factors. In addition to the promoter, eukaryotic genes have one or more enhancers. These are DNA sequences associated with the gene being regulated, and whereas the promoter is responsible for initiating low levels of transcription and determining the transcription start site, enhancers are responsible for increasing ("enhancing") transcription levels, and they are responsible for regulating cell- or tissue-specific transcription (*i.e.* the transcription responsible for differentiation). There are some other basic differences between enhancers and promoters, which are outlined in the module on transcription.

Enhancers function by being recognized and bound to by transcription factors. These are not the basal-type transcription factors (such as TFIID) discussed elsewhere. These are specialized transcription factors, of which there are very many types (all are proteins). A very large number of enhancer elements has been identified and characterized, and each different enhancer has its own transcription factor that it binds to.

Tissue-specific Gene Expression

How is a gene turned on in one cell type, and not in another? It depends primarily on whether the transcription factor for the gene's enhancer is active or not in a cell. To illustrate this, let's consider two different genes - one is regulated by enhancer A, which is recognized by transcription factor A, and the other gene is regulated by enhancer B, which is recognized by transcription factor B.

In one cell type (let's say muscle, for the sake of argument), transcription factor A might be active whereas transcription factor B might not. So how is it determined which transcription factors are active in which cell types? The answer to this question is very complicated, and not completely understood. However, there are a number of mechanisms by which transcription factors can be regulated: Often, the presence or absence of a transcription factor in a cell is the determining step. If the factor is present, the gene is transcribed.

If not, then the gene is not transcribed. The presence of a transcription factor, of course, depends upon the activity of the gene encoding that transcription factor, which forces us to ask how the gene encoding the transcription factor is regulated, which pretty much brings us back to where we began.

Transcription factors can be activated by environmental signals. For example, virtually all organisms have a set of genes called heat shock genes that encode proteins that help the organism survive heat stress. These genes are activated under conditions of heat stress, under the control of a transcription factor called heat shock transcription factor. This factor is always present, but is only activated when greatly increased temperatures are detected. Transcription factors can be activated by signals from other cells in the same organism. Such signals include hormones and growth factors. Hormones must bind to a specific receptor on the target cell, and the receptor mediates the cellular effects of the signal.

There are two basic mechanisms used, one for steroid hormones, and one for peptide hormones:

1. Steroid hormones
2. Peptide hormones.

Steroid Hormones

Steroid hormones are lipid (actually cholesterol) derivatives, such as testosterone and progesterone. These hormones can cross the cytoplasmic membrane into a cell, where they bind to their specific receptor. Steroid receptors are transcription factors, and when they bind to their ligand, they become activated and initiate transcription of a specific set of genes.

Peptide Hormones

Peptide hormones cannot cross the cytoplasmic membrane, and so must bind to a receptor on the cell surface. When bound to its ligand, these receptors initiate a series of biochemical reactions inside the cell, with the ultimate result being the activation of a transcription factor (often by phosphorylation of the transcription factor), which initiates transcription of a specific set of genes.

Developmental Genetics

The primary function of the majority of genes in eukaryotic organisms is to coordinate the development of embryos. This module takes a look at some of the basic principles underlying the role of genes in embryonic development. In most of the other modules of this course, the effects of genes are examined with regard to their effect on the phenotype of the juvenile or adult organism. In reality, however, the vast majority of these phenotypes are established during embryonic development, because most genes function to generate the pattern (the 'shape') of the developing embryo. Loss of function of a particular gene results in an abnormality in development, which manifests itself as a phenotype in the adult. In this sense, virtually all of eukaryotic genetics is developmental genetics, even though we haven't considered it as such.

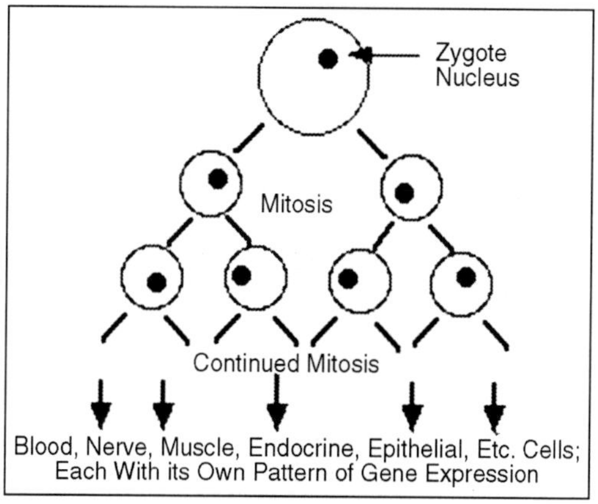

***Figure:** Zygote*

The field of developmental biology is concerned with the function of genes in embryogenesis. How does a single cell, the fertilized egg (or zygote), manage to produce an extremely complex adult organism, which is composed not only of trillions of cells, but of thousands of different types of cells (such as nerve cells, muscle cells, etc.)? There is a simple answer to this question, and that is that cells differentiate. In other words, different cells become specialized to carry out different functions by following different developmental pathways. But how does this occur? How do cells know what pathway to follow?

The Mosaic Theory of Development

One early theory that attempted to explain the process of differentiation was the mosaic theory, as outlined by Wilhelm Roux and August Weismann. This theory proposed that there were determinants that specified the various differentiation pathways. The theory stated that these determinants would all be present in the zygote, but as cell division occurred after fertilization, the determinants would be unequally inherited by the offspring cells.

This is known as qualitative cell division. The set of determinants would thus be divided up until each cell contained only one type of determinant, and that determinant would determine the fate of that cell. When Mendel's work was rediscovered in the early 1900's and the concept of the 'gene' was developed, it was believed that the genes were the determinants that were divided up during cell division. Evidence was accumulated over sixty years that disproved the notion that genes are divided up in development.

The final nail in the coffin of this theory came from nuclear transplantation studies done using amphibians in the early 1960's. In these studies, nuclei were isolated from tadpole intestinal cells (differentiated cells) and injected into eggs that had their own haploid nuclei removed. A small percentage of the injected eg176gs developed into completely normal adult frogs! These frogs had developed using only the genetic information found in a tadpole intestinal nucleus. Therefore, a tadpole intestinal nucleus must contain all of the genetic information necessary to allow the differentiation of every cell type in an adult frog. If the mosaic theory were correct, this would not be the case; a tadpole intestinal nucleus would have only the genetic information necessary to cause the differentiation of an intestinal cell.

Note: The frogs produced by this procedure would be genetically identical to the frog from which the intestinal cell nucleus was obtained. In other words, they would be clones. In fact, these experiments comprised the earliest successful cloning of vertebrate organisms. The recent cloning of Dolly the sheep was done for the same reasons as outlined above: demonstrating that differentiated mammalian nuclei (from mammary gland in this case) have all of the genetic information necessary to drive the development of a normal adult organism.

Differential Gene Expression

If all nuclei in an organism contain the same genetic information, then how does differentiation occur? The explanation for this is found in the Theory of Differential Gene Expression. This theory states that differentiation occurs as a result of expression in a particular cell of only a subset of the total genes present. For example, if a cell expresses only the set of genes that causes muscle differentiation, then that cell will differentiate into a muscle cell. Alternatively, if the cell expresses only the set of genes that causes spleen cell differentiation, then the cell will differentiate into a spleen cell. This is a fairly simple concept, but it hasn't really answered the question.

The question has now become: how do cells activate only a certain set of genes (inactivating all other genes), and how do they know which set of genes to activate? The answer to this question is fairly straightforward as well: the set of genes activated in a cell is dependent on the set of transcription factors found in the cell. Using the muscle cell example from above, the cell activates the muscle-specific genes because it contains transcription factors that specifically activate the muscle-specific genes. If we think about this for a minute, you'll realise that we still haven't answered the question; we've only changed it again.

Now the question becomes: how did the cell come to contain only that specific set of transcription factors? Well, those transcription factors are encoded by genes, and those genes are regulated by other specific transcription factors.

Those transcription factors are in turn encoded by other genes, which are regulated by still other transcription factors, etc. As we can see, there is a hierarchy of genes within each cell. Genes are expressed that encode transcription factors, which activate other genes that encode transcription factors, which activate other genes that bring about differentiation along a specific pathway. How many levels are there to the hierarchy? From what we've seen so far, the hierarchy seems to go on forever. Each set of transcription factors is encoded by genes that are activated by yet another set of transcription factors. It must end somewhere, but where?

Master Control Genes

In one sense, it ends with master control genes. A master control gene is the first gene activated in a hierarchy that leads to differentiation along a particular pathway. Master control genes encode the first transcription factor in a hierarchy; a master control gene product activates the next set of genes that encodes the next set of transcription factors, and the cascade of gene expression has been set in motion.

Let's look again at the muscle cell example considered earlier. The master control 'gene' for muscle development i actually a family of genes called the MyoD gene family. These genes encode HLH-type transcription factors. How do we know these are the master control genes for muscle? Master control genes have a particular property. Because they initiate muscle differentiation, if the MyoD genes are activated in other cell types, they cause those cells to transdifferentiate into muscle.

For example, hepatocytes (liver cells) or adipocytes (fat cells) that are caused to express the MyoD genes (using tricks of molecular biology) will change from their normal phenotype into muscle cells. So how are the master control genes regulated? There are two basic ways that this can happen. One way is through the process of induction. Induction occurs when once cell sends a signal to another cell, telling it to differentiate a certain way. The signal, usually a diffusable protein, causes the recipient cell to activate the appropriate master control gene product. This is how muscle differentiation occurs. Signals from other cells cause the MyoD genes to become active in the cells receiving the signal, and muscle differentiation is initiated.

6

Genome Organization and Evolution

Genomics is a discipline in genetics concerning the study of the genomes of organisms. The field includes intensive efforts to determine the entire DNA sequence of organisms and fine-scale genetic mapping efforts. The field also includes studies of intragenomic phenomena such as heterosis, epistasis, pleiotropy and other interactions between loci and alleles within the genome. In contrast, the investigation of the roles and functions of single genes is a primary focus of molecular biology or genetics and is a common topic of modern medical and biological research. Research of single genes does not fall into the definition of genomics unless the aim of this genetic, pathway, and functional information analysis is to elucidate its effect on, place in, and response to the entire genome's networks.

For the United States Environmental Protection Agency, "the term "genomics" encompasses a broader scope of scientific inquiry associated technologies than when genomics was initially considered. A genome is the sum total of all an individual organism's genes. Thus, genomics is the study of all the genes of a cell, or tissue, at the DNA (genotype), mRNA (transcriptome), or protein (proteome) levels."

History

Genomics was established by Fred Sanger when he first sequenced the complete genomes of a virus and a mitochondrion. His group established techniques of sequencing, genome mapping, data storage, and bioinformatic analyses in the 1970-1980s. A major branch of genomics is still concerned with sequencing the genomes of various organisms, but the knowledge of full genomes has created the possibility for the field of functional genomics, mainly concerned with patterns of gene expression during various conditions. The most important tools here are microarrays and bioinformatics. Study of the full set of proteins in a cell type or tissue, and the changes during various conditions, is called proteomics.

A related concept is materiomics, which is defined as the study of the material properties of biological materials (e.g. hierarchical protein structures and materials, mineralized biological tissues, etc.) and their effect on the macroscopic function and failure in their biological context, linking processes, structure and properties at multiple scales through a materials science approach. The actual term 'genomics' is thought to have been coined by Dr. Tom Roderick, a geneticist at the Jackson Laboratory (Bar Harbor, ME) over beer at a meeting held in Maryland on the mapping of the human genome in 1986.

In 1972, Walter Fiers and his team at the Laboratory of Molecular Biology of the University of Ghent (Ghent, Belgium) were the first to determine the sequence of a gene: the gene for Bacteriophage MS2 coat protein. In 1976, the team determined the complete nucleotide-sequence of bacteriophage MS2-RNA. The first DNA-based genome to be sequenced in its entirety was that of bacteriophage Φ-X174; (5,368 bp), sequenced by Frederick Sanger in 1977.

The first free-living organism to be sequenced was that of *Haemophilus influenzae* (1.8 Mb) in 1995, and since then genomes are being sequenced at a rapid pace. As of September 2007, the complete sequence was known of about 1879 viruses, 577 bacterial species and roughly 23 eukaryote organisms, of which about half are fungi.

Most of the bacteria whose genomes have been completely sequenced are problematic disease-causing agents, such as *Haemophilus influenzae*. Of the other sequenced species, most were chosen because they were well-studied model organisms or promised to become good models. Yeast (*Saccharomyces cerevisiae*) has long been an important model organism for the eukaryotic cell, while the fruit fly *Drosophila melanogaster* has been a very important tool (notably in early pre-molecular genetics). The worm *Caenorhabditis elegans* is an often used simple model for multicellular organisms.

The zebrafish *Brachydanio rerio* is used for many developmental studies on the molecular level and the flower *Arabidopsis thaliana* is a model organism for flowering plants. The Japanese pufferfish (*Takifugu rubripes*) and the spotted green pufferfish (*Tetraodon nigroviridis*) are interesting because of their small and compact genomes, containing very little non-coding DNA compared to most species. The mammals dog (*Canis familiaris*), brown rat (*Rattus norvegicus*), mouse (*Mus musculus*), and chimpanzee (*Pan troglodytes*) are all important model animals in medical research.

Human Genomics

A rough draft of the human genome was completed by the Human Genome Project in early 2001, creating much fanfare. By 2007 the human sequence was declared "finished" (less than one error in 20,000 bases and all chromosomes assembled). Display of the results of the project required significant bioinformatics resources. The sequence of the human reference assembly can be explored using the UCSC Genome Browser.

Bacteriophage Genomics

Bacteriophages have played and continue to play a key role in bacterial genetics and molecular biology. Historically, they were used to define gene structure and gene regulation. Also the first genome to be sequenced was a bacteriophage. However, bacteriophage research did not lead the genomics revolution, which is clearly dominated by bacterial genomics.

Only very recently has the study of bacteriophage genomes become prominent, thereby enabling researchers to understand the mechanisms underlying phage evolution. Bacteriophage genome sequences can be obtained through direct sequencing of isolated bacteriophages, but can also be derived as part of microbial genomes. A detailed database mining of these sequences offers insights into the role of prophages in shaping the bacterial genome.

Cyanobacteria Genomics

At present there are 24 cyanobacteria for which a total genome sequence is available. 15 of these cyanobacteria come from the marine environment. These are six *Prochlorococcus* strains, seven marine *Synechococcus* strains, *Trichodesmium erythraeum* IMS101 and *Crocosphaera watsonii* WH8501. Several studies have demonstrated how these sequences could be used very successfully to infer important ecological and physiological characteristics of marine cyanobacteria.

However, there are many more genome projects currently in progress, amongst those there are further *Prochlorococcus* and marine *Synechococcus* isolates, *Acaryochloris* and *Prochloron*, the N_2-fixing filamentous cyanobacteria *Nodularia spumigena*, *Lyngbya aestuarii* and *Lyngbya majuscula*, as well as bacteriophages infecting marine cyanobaceria.

Thus, the growing body of genome information can also be tapped in a more general way to address global problems by applying a comparative approach.

Some new and exciting examples of progress in this field are the identification of genes for regulatory RNAs, insights into the evolutionary origin of photosynthesis, or estimation of the contribution of horizontal gene transfer to the genomes that have been analysed.

Genomes and Proteomes

The genome of a typical bacterium comes as a single DNA molecule, which, if extended, would be about 2 mm long. (The cell itself has a diameter of about 0.001 mm). The DNA of higher organisms is organized into chromosomes—normal human cells contain 23 chromosome pairs. The total amount of genetic information per cell—the sequence of nucleotides of DNA—is very nearly constant for all members of a species.

Learning Goals

1. To know the basic sizes and organizing principles of simple and complex genomes.
2. To understand ideas of proteomics—the patterns of expression of proteins—and techniques such as microarrays used to study them.
3. To know how genomes are analysed, and the relation of gene sequenceto phenotypic features, including inherited diseases.
4. To absorb a general idea of the contents of particular genomes, and how the genomes of prokaryotes and eukaryotes differ systematically.
5. To recognize both the importance and the difficulty of deriving from a complete genome sequence the amino acid sequences of the proteins encoded, and assigning functions to these proteins.
6. To realize that published genomes record the characteristics of only a single individual, and that there is considerable variation within populations, and great variation between separated populations of organisms belonging to the same species.
7. To appreciate the power of comparative genomics to identify features responsible for differences between species—for instance, what is it that makes us human?

Gene

A gene is a unit of heredity in a living organism. It normally resides on a stretch of DNA that codes for a type of protein or for an RNA chain that has a function in the organism. All living things depend on genes, as they specify all proteins and functional RNA chains. Genes hold the information to build and maintain an organism's cells and pass genetic traits to offspring, although some organelles (e.g. mitochondria) are self-replicating and are not coded for by the organism's DNA.

A modern working definition of a gene is "*a locatable region of genomic sequence, corresponding to a unit of inheritance, which is associated with regulatory regions, transcribed regions, and or other functional sequence regions* Colloquial usage of the term *gene* (e.g. good genes, "hair colour gene") may actually refr to an allele: a *gene* is the basic instruction, a sequence of nucleic acid (DNA or, in the case of certain viruses RNA), while an*allele* is one variant of that gene. Thus, when the mainstream press refers to "having" a "gene" for a specific trait, thismay be incorrect. In many cases all people would have the gene in question, but certain people will have a specific allele of that gene, which results in the trait. In the simplest case, the changes observed may be caused by a single letter of the genetic code - a single nucleotide polymorphism.

The notion of a gene is evolving with the science of genetics, which began when Gregor Mendel noticed that biological variations are inherited from parent organisms as specific, discrete trais. The biological entity responsible for defining traits was later termed a *gene*, but the biological basis for inheritance remained unknown until DNA was identified as the genetic material in the 1940s. All organisms have many genes corresponding to many different biological traits, some of which are immediately visible, such as eye colour or number of limbs, and some of which are not, such as blood type or increased risk for specific diseases, or the thousands of basic biochemical processes that comprise life.

The vast majority of living organisms encode their genes in long strands of DNA. DNA (deoxyribonucleic acid) consists of a chain made from four types of nucleotide subunits, each composed of: a five-carbon sugar (2'-deoxyribose), a phosphate group, and one of the four bases adenine, cytosine, guanine, and thymine. The most common form of DNA in a cell is in a double helix structure, in which two individual DNA strands twist around each other in a right-handed spiral. In this structure, the base pairing rules specify that guanine pairs with cytosine and adenine pairs with thymine. The base pairing between guanine and cytosine forms three hydrogen bonds, whereas the base pairing between adenine and thymine forms two hydrogen bonds. The two strands in a double helix must therefore be *complementary*, that is, their bases must align such that the adenines of one strand are paired with the thymines of the other strand, and so on.

Due to the chemical composition of the pentose residues of the bases, DNA strands have directionality. One end of a DNA polymer contains an exposed hydroxyl group on the deoxyribose; this is known as the 3' end of the molecule. The other end contains an exposed phosphate group; this is the 5' end. The directionality of DNA is vitally important to many cellular processes, since double helices are necessarily directional (a strand running 5'-3' pairs with a complementary strand running 3'-5'), and processes such as DNA replication occur in only one direction. All nucleic acid synthesis in a cell occurs in the 5'-3' direction, because new monomers are added via a dehydration reaction that uses the eposed 3' hydroxyl as a nucleophile. The expression of genes encoded in DNA begins by transcribing the gene into RNA, a second type of nucleic acid that is very similar to DNA, but whose monomers contain the sugar ribose rather than deoxyribose. RNA also contains the base uracil in place of thymie. RNA molecules are less stable than DNA and are typically single-stranded. Genes that encode proteins are composed of a series of three-nucleotide sequences called codons, which serve as the *words* in the genetic *language*. The genetic code specifies the correspondence during protein translation between codons and amino acids. The genetic code is nearly the same for all known organisms.

Physical Definitions

RNA Genes and Genomes

When proteins are manufactured, the gene is first copied into RNA as an intermediate product. In other cases, the RNA molecules are the actual functional products. For example, RNAs known as ribozymes are capable of enzymatic function, and microRNA has a regulatory role. The DNA sequences from which such RNAs are transcribed are known as RNA genes.

Some viruses store their entire genomes in the form of RNA, and contain no DNA at all. Because they use RNA to store genes, their cellular hosts may synthesize their proteins as soon as they are infected and without the delay in waiting for transcription. On the other hand, RNA retroviruses, such as HIV, require the reverse transcription of their genome from RNA into DNA before their proteins can be synthesized.

In 2006, French researchers came across a puzzling example of RNA-mediated inheritance in mouse. Mice with a loss-of-function mutation in the gene Kit have white tails. Offspring of these mutants can have white tails despite having only normal Kit genes. The research team traced this effect back to mutated Kit RNA. While RNA is common as genetic storage material in viruses, in mammals in particular RNA inheritance has been observed very rarely.

Functional Structure of a Gene

All genes have regulatory regions in addition to regions that explicitly code for a protein or RNA product. A regulatory region shared by almost all genes is known as the promoter, which provides a position that is recognized by the transcription machinery when a gene is about to be transcribed and expressed. A gene can have more than one promoter, resulting in RNAs that differ in how far they extend in the 5' end. Although promoter regions have a consensus sequence that is the most common sequence at this position, some genes have "strong" promoters that bind the transcription machinery well, and others have "weak" promoters that bind poorly. These weak promoters usually permit a lower rate of transcription than the strong promoters, because the transcription machinery binds to them and initiates transciption less frequently. Other possible rgulatory regions include enhancers, whih can compensate for a weak promoter. Most regulatory regions are "upstream"—that is, before or toward the 5' end of the tanscription initiation site. Eukaryotic promoter regions are much more complex and difficult to identify than prokaryotic promoters.

Many prokaryotic genes are organized into operons, or groups of genes whose products have related functions and which are transcribed as a unit. By contrast, eukaryotic genes are transcribed only one at a time, but may include long stretches of DNA called introns which are transcribed but never translated into protein (they are spliced out before translation). Splicing can also occur in prokaryotic genes, but is less common than in eukaryotes.

Chromosomes

The total complement of genes in an organism or cell is known as its genome, which may be stored on one or more chromosomes; the region of the chromosome at which a particular gene is located is called its locus. A chromosome consists of a single, very long DNA helix on which thousands of genes are encoded. Prokaryotes—bacteria and archaea—typically store their genomes on a single large, circular chromosome, sometimes supplemented by additional small circles of DNA called plasmids, which usually encode only a few genes and are easily transferable between individuals. For example, the genes for antibiotic resistance are usually encoded on bacterial plasmids and can be passed between individual cells, even those of different species, via horizontal gene transfer. Although some simple eukaryotes also possess plasmids with small numbers of genes, the majority of eukaryotic genes are stored on multiple linear chromosomes, which are packed within the nucleus in complex with storage proteins called histones.

The manner in which DNA is stored on the histone, as well as chemical modifications of the histone itself, are regulatory mechanisms governing whether a particular region of DNA is accessible for gene expression. The ends of eukaryotic chromosomes are capped by long stretches of repetitive sequences called telomeres, which do not code for any gene product but are present to prevent degradation of coding and regulatory regions during DNA replication. The length of the telomeres tends to decrease each time the genome is replicated in preparation for cell division; the loss of telomeres has been proposed as an explanation for cellular senescence, or the loss of the ability to divide, and by extension for the aging process in organisms.

Whereas the chromosomes of prokaryotes are relativel gene-dense, those of eukaryotes often contain so-called "junk DNA", or regions of DNA that serve no obvious function. Smple single-celled eukaryotes have relatively small amounts of such DNA, whereas the genomes of complex multicellular organisms, including humans, contain an absolute majority of DNA without an identified function. However it now appears that, although protein-coding DNA makes up barely 2% of the human genome, about 80% of the bases in the genome may be being expressed, so the term "junk DNA" may be a misnomer.

Gene Expression

In all organisms, there are two major steps separating a protein-coding gene from its protein: First, the DNA on which the gene resides must be *transcribed* from DNA to messenger RNA (mRNA); and, second, it must be *translated* from mRNA to protein. RNA-coding genes must still go through the first step, but are not translated into protein. The process of producing a biologically functional molecule of either RNA or protein is called gene expression, and the resulting molecule itself is called a gene product.

Genetic Code

The genetic code is the set of rules by which a gene is translated into a functional protein. Each gene consists of a specific sequence of nucleotides encoded in a DNA (or sometimes RNA) strand; a correspondence between nucleotides, the basic building blocks of genetic material, and amino acids, the basic building blocks of proteins, must be established for genes to be successfully translated into functional proteins.

Sets of three nucleotides, known as codons, each correspond to a specific amino acid or to a signal; three codons are known as "stop codons" and, instead of specifying a new amino acid, alert the translation machinery that the end of the gene has been reached. There are 64 possible codons (four possible nucleotides at each of three positions, hence 4^3 possible codons) and only 20 standard amino acids; hence the code is redundant and multiple codons can specify the same amino acid. The correspondence between codons and amino acids is nearly universal among all known living organisms.

Transcription

The process of genetic transcription produces a single-stranded RNA molecule known as messenger RNA, whose nucleotide sequence is complementary to the DNA from which it was transcribed. The DNA strand whose sequence matches that of the RNA is known as the coding strand and the strand from which the RNA was synthesized is the template strand. Transcription is performed by an enzyme called an RNA polymerase, which reads the template strand in the 3' to 5' direction and synthesizes the RNA from 5' to 3'. To initiate transcription, the polymerase first recognizes and binds a promoter region of the gene. Thus a major mechanism of gene regulation is the blocking or sequestering of the promoter region, either by tight binding by repressor molecules that physically block the polymerase, or by organizing the DNA so that the promoter region is not accessible.

In prokaryotes, transcription occurs in the cytoplasm; for very long transcripts, translation may begin at the 5' end of the RNA while the 3' end is still being transcribed. In eukaryotes, transcription necessarily occurs in the nucleus, where the cell's DNA is sequestered; the RNA molecule produced by the polymerase is known as the primary transcript and must undergo post-transcriptional modifications before being exported to the cytoplasm for translation. The splicing of inrons present within the transcribed region is a modification unique to eukaryotes; alternative splicing mechanisms can result in mature transcripts from the same gene having different sequences and thus coding for different proteins. This is a major form of regulation in eukaryotic cells.

Translation

Translation is the process by which a mature mRNA molecule is used as a template for synthesizing a new protein. Translation is carried out by ribosomes, large complexes of RNA and protein responsible for carrying out the chemical reactions to add new amino acids to a growing polypeptide chain by the formation of peptide bonds.

The genetic code is read three nucleotides at a time, in units called codons, via interactions with specialized RNA molecules called transfer RNA (tRNA). Each tRNA has three unpaired bases known as the anticodon that are complementary to the codon it reads; the tRNA is also covalently attached to the amino acid specified by the complementary codon. When the tRNA binds to its complementary codon in an mRNA strand, the ribosome ligates its amino acid cargo to the new polypeptide chain, which is synthesized from amino terminus to carboxyl terminus. During and after its synthesis, the new proten must fold to its active three-dimensional structure before it can carry out its cellular function.

DNA Replication and Inheritance

The growth, development, and reproduction of organisms relies on cell division, or the process by which a single cell divides into two usually identical daughter cells. This requires first making a duplicate copy of every gene in the genome in a process called DNA replication. The copies are made by specialized enzymes known as DNA polymerases, which "read" one strand of the double-helical DNA, known as the template strand, and synthesize a new complementary strand. Because the DNA double helix is held together by base pairing, the sequence of one strand completely specifies the sequence of its complement; hence only one strand needs to be read by the enzyme to produce a faithful copy. The process of DNA replication is semiconservative; that is, the copy of the genome inherited by each daughter cell contains one original and one newly synthesized strand of DNA.

After DNA replication is complete, the cell must physically separate the two copies of the genome and divide into two distinct membrane-bound cells. In prokaryotes - bacteria and archaea - this usually occurs via a relatively simple process called binary fission, in which each circular genome attaches to the cell membrane and is separated into the daughter cells as the membrane invaginates to split the cytoplasm into two membrane-bound portions. Binary fission is extremely fast compared to the rates of cell division in eukaryotes. Eukaryotic cell division is a more complex process known as the cell cycle; DNA replication occurs during a phase of this cycle known as S phase, whereas the process of segregating chromosomes and splitting the cytoplasm occurs during M phase. In many single-celled eukaryotes such as yeast, reproduction by budding is common, which results in asymmetrical portions of cytoplasm in the two daughter cells.

Molecular Inheritance

The duplication and transmission of genetic material from one generation of cells to the next is the basis for molecular inheritance, and the link between the classical and molecular pictures of genes. Organisms inherit the characteristics of their parents because the cells of the offspring contain copies of the genes in their parents' cells. In asexually reproducing organisms, the offspring will be a genetic copy or clone of the parent organism. In sexually reproducing organisms, a specialized form of cell division called meiosis produces cells called gametes or germ cells that are haploid, or contain only one copy of each gene.

The gametes produced by females are called eggs or ova, and those produced by males are called sperm. Two gametes fuse to form a fertilized egg, a single cell that once again has a diploid number of genes—each with one copy from the mother and one copy from the father. During the process of meiotic cell division, an event called genetic recombination or *crossing-over* can sometimes occur, in which a length of DNA on one chromatid is swapped with a length of DNA on the corresponding sister chromatid. This has no effect if the alleles on the chromatids are the same, but results in reassortment of otherwise linked alleles if they are different. The Mendelian principle of independent assortment asserts that each of a parent's two genes for each trait will sort independently into gametes; which allele an organism inherits for one trait is unrelated to which allele it inherits for another trait.

This is in fact only true for genes that do not reside on the same chromosome, or are located very far from one another on the same chromosome. The closer two genes lie on the same chromosome, the more closely they will be associated in gametes and the more often they will appear together; genes that are very close are essentially never separated because it is extremely unlikely that a crossover point will occur between them. This is known as genetic linkage.

Prior to Mendel's work, the dominant theory of heredity was one of blending inheritance, which proposes that the traits of the parents blend or mix in a smooth, continuous gradient in the offspring. Although Mendel's work was largely unrecognized after its first publication in 1866, it was rediscovered in 1900 by three European scientists, Hugo de Vries, Carl Correns, and Erich von Tschermak, who had reached similar conclusions from their own research. However, these scientists were not yet aware of the identity of the 'discrete units' on which genetic material resides.

The existence of genes was first suggested by Gregor Mendel (1822–1884), who, in the 1860s, studied inheritance in peaplants (*Pisum sativum*) and hypothesized a factor that conveys traits from parent to offspring. He spent over 10 years of his life on one experiment. Although he did not use the term *gene*, he explained his results in terms of inherited characteristics. Mendel was also the first to hypothesize independent assortment, the distinction between dominant and recessive traits, the distinction between a heterozygote and homozygote, and the difference between what would later be described as genotype (the genetic material of an organism) and phenotype (the visible traits of that organism). Mendel's concept was given a name by Hugo de Vries in 1889, who, at that time probably unaware of Mendel's work, in his book *Intracellular Pangenesis* coined the term "pangen" for "the smallest particle representing] one hereditary characteristic".

Darwin used the term Gemmule to describe a microscopic unit of inheritance, and what would later become known as Chromosomes had been observed separating out during cell division by Wilhelm Hofmeister as early as 1848. The idea that chromosomes are the carriers of inheritance was expressed in 1883 by Wilhelm Roux.

The modern conception of the gene originated with work by Gregor Mendel, a 19th-century Augustinian monk who systematically studied heredity in pea plants. Mendel's work was the first to illustrate particulate inheritance, or the theory that inherited traits are passed from one generation to the next in discrete units that interact in well-defined ways. Danish botanist Wilhelm Johannsen coined the word "gene" ("gen" in Danish and German) in 1909 to describe these fundamental physical and functional units of heredity, while the related word genetics was first used by William Bateson in 1905. The word was derived from Hugo de Vries' 1889 term *pangen* for the same concept, itself a derivative of the word *pangenesis* coined by Darwin (1868). The word pangenesis is made from the Greek word *pan* (a prefix meaning "whole", "encompassing") ad *genesis* ("birth") or *genos* ("origin").

In the early 1900s, Mendel's work rcccivcd renewed attention from scientists. In 190, Thomas Hunt Morgan showed that genes reside on specific chomosomes. He later showed that genes occupy specific locations n the chromosome. With this knowledge, Morgan and his students began the first chromosomal map of the fruit fly *Drosophila*.In 1928, Frederick Griffith showed that genes could be transferred. In what is now known as Griffith's experiment, injections into a mouse of a deadly strain of bacteria that had ben heat-killed transferred genetic information to a safe strain of the same bacteria, killing the mouse.

A series of subsequent discoveries led to the realization decades later that chromosomes within cells are the carriers of genetic material, and that they are made of DNA (deoxyribonucleic acid), a polymeric molecule found in all cells on which the 'discrete units' of Mendelian inheritance are encodd. In 1941, George Wells Beadle and Edward Lawrie Tatum showed that mutations in genes caused errors in specific steps in metabolic pathways. This showed that specific genes code for specific proteins, leading to the "one gene, one enzyme" hypothesis. Oswald Avery, Colin Munro MacLeod, and Maclyn McCarty showed in 1944 that DNA holds the gene's information. In 1953, James D. Watson and Francis Crick demonstrated the molecular structure of DNA. Together, these discoveries established the central dogma of molecular biology, which states that proteins are translated from RNA which is transcribed from DNA.

In 1972, Walter Fiers and his team at the Laboratory of Molecular Biology of the University of Ghent (Ghent Belgium) were the first to determine the sequence of a gene: the gene for Bacteriophage MS2 coat protein. Richard J. Roberts and Phillip Sharp discovered in 1977 that genes can be split into segments. This led to the idea that one gene can make several proteins. Recently (as of 2003–2006), biological results let the notion of gene appear more slippery. In particular, genes do not seem to sit side by side on DNA like discrete beads. Instead, regions of the DNA producing distinct proteins may overlap, so that the idea emerges that "genes are one long continuum".

It was first hypothesized in 1986 by Walter Gilbert that neither DNA nor protein would be required in such a primitive system as that of a very early stage of the earth if RNA could perform as simply a catalyst and genetic information storage processor. The modern study of genetics at the level of DNA is known as molecular genetics and the synthesis of molecular genetics with traditional Darwinian evolution is known as the modern evolutionary synthesis.

Mendelian Inheritance and Classical Genetics

According to the theory of Mendelian inheritance, variations in phenotype—the observable physical and behavioral characteristics of an organism—are due to variations in genotype, or the organism's particular set of genes, each of which specifies a particular trait. Different forms of a gene, which may give rise to different phenotypes, are known as alleles. Organisms such as the pea plants Mendel worked on, along with many plants and animals, have two alleles for each trait, one inherited from each parent. Alleles may be dominant or recessive; dominant alleles give rise to their corresponding phenotypes when paired with any other allele for the same trait, whereas recessive alleles give rise to their corresponding phenotype only when paired with another copy of the same allele.

For example, if the allele specifying tall stems in pea plants is dominant over the allele specifying short stems, then pea plants that inherit one tall allele from one parent and one short allele from the other parent will also have tall stems. Mendel's work found that alleles assort independently in the production of gametes, or germ cells, ensuring variation in the next generation.

Mutation

DNA replication is for the most part extremely accurate, with an error rate per site of around $10^{''6}$ to $10^{''10}$ in eukaryotes. Rare, spontaneous alterations in the base sequence of a particular gene arise from a number of sources, such as errors in DNA replication and the aftermath of DNA damage. These errors are called mutations. The cell contains many DNA repair mechanisms for preventing mutations and maintaining the integrity of the genome; however, in some cases—such as breaks in both DNA strands of a chromosome—repairing the physical damage to the molecule is a higher priority than producing an exact copy. Due to the degeneracy of the genetic code, some mutations in protein-coding genes are *silent*, or produce no change in the amino acid sequence of the protein for which they code; for example, the codons UCU and UUC both code for serine, so the U↔C mutation has no effect on the protein. Mutations that do have phenotypic effects are most often neutral or deleterious to the organism, but sometimes they confer benefits to the organism's fitness.

Mutations propagated to the next generation lead to variations within a species' population. Variants of a single gene are known as alleles, and differences in alleles may give rise to differences in traits. Although it is rare for the variants in a single gene to have clearly distinguishable phenotypic effects, certain well-defined traits are in fact controlled by single genetic loci. A gene's most common allele is called the wild type allele, and rare alleles are called mutants. However, this does not imply that the wild-type allele is the ancestor from which the mutants are descended.

Genome

Chromosomal Organization

The total complement of genes in an organism or cell is known as its genome. In prokaryotes, the vast majority of genes are located on a single chromosome of circular DNA, while eukaryotes usually possess multiple individual linear DNA helices packed into dense DNA-protein complexes called chromosomes. Genes that appear together on one chromosome of one species may appear on separate chromosomes in another species. Many species carry more than one copy of their genome within each of their somatic cells. Cells or organisms with only one copy of each chromosome are called haploid; those with two copies are called diploid; and those with more than two copies are called polyploid. The copies of genes on the chromosomes are not necessaril identical. In sexually reproducing organisms, one copy is normally inherited from each parent.

Number of Genes

Early estimates of the number of human genes that used expressed sequence tag data put it at 50 000–100 000. Following the sequencing of the human genome and other genomes, it has been found that rather few genes (~20 000 in human, mouse and fly, ~13 000 in roundworm, >46 000 in rice) encode all the proteins in an organism. These protein-coding sequences make up 1–2% of the human genome. A large part of the genome is transcribed however, to introns, retrotransposons and seemingly a large array of noncoding RNAs. Total number of proteins (the Earth's proteome) is estimated to be 5 million sequences.

Genetic and Genomic Nomenclature

Gene nomenclature has been established by the HUGO Gene Nomenclature Committee (HGNC) for each known human gene in the form of an approved gene name and symbol (short-form abbreviation). All approved symbols are stored in the HGNC Database. Each symbol is unique and each gene is only given one approved gene symbol. This also facilitates electronic data retrieval from publications. In preference each symbol maintains parallel construction in different members of a gene family and can be used in other species, especially the mouse.

Evolutionary Concept of a Gene

George C. Williams first explicitly advocated the gene-centric view of evolution in his 1966 book *Adaptation and Natural Selection*. He proposed an evolutionary concept of gene to be used when we are talking about natural selection favouring some genes. The definition is: "that which segregates and recombines with appreciable frequency." According to this definition, even an asexual genome could be considered a gene, insofar that it have an appreciable permanency through many generations.

The difference is: the molecular gene *transcribes* as a unit, and the evolutionary gene *inherits* as a unit. Richard Dawkins' books *The Selfish Gene* (1976) and *The Extended Phenotype* (1982) defended the idea that the gene is the only replicator in living systems. This means that only genes transmit their structure largely intact and are potentially immortal in the form of copies. So, genes should be the unit of selection. In *The Selfish Gene* Dawkins attempts to redefine the word 'gene' to mean "an inheritable unit" instead of the generally accepted definition of "a section of DNA coding for a particular protein".

In *River Out of Eden*, Dawkins further refined the idea of gene-centric selection by describing life as a river of compatible genes flowing through geological time. Scoop up a bucket of genes from the river of genes, and we have an organism serving as temporary bodies or survival machines. A river of genes may fork into two branches representing two non-interbreeding species as a result of geographical separation.

Gene Targeting and Implications

Gene targeting is commonly referred to techniques for altering or disrupting mouse genes and provides the mouse models for studying the roles of individual genes in embryonic development, human disorders, aging and diseases. The mouse models, where one or more of its genes are deactivated or made inoperable, are called knockout mice. Since the first reports in which homologous recombination in embryonic stem cells was used to generate gene-targeted mice, gene targeting has proven to be a powerful means of precisely manipulating the mammalian genome, producing at least ten thousand mutant mouse strains and it is now possible to introduce mutations that can be activated at specific time points, or in specific cells or organs, both during development and in the adult animal.

Gene targeting strategies have been expanded to all kinds of modifications, including point mutations, isoform deletions, mutant allele correction, large pieces of chromosomal DNA insertion and deletion, tissue specific disruption combined with spatial and temporal regulation and so on. It is predicted that the ability to generate mouse models with predictable phenotypes will have a major impact on studies of all phases of development, immunology, neurobiology, oncology, physiology, metabolism, and human diseases. Gene targeting is also in theory applicable to species from which totipotent embryonic stem cells can be established, and therefore may offer a potential to the improvement of domestic animals and plants.

Changing Concept

The concept of the gene has changed considerably. From the original definition of a "unit of inheritance", the term evolved to mean a DNA-based unit that can exert its effects on the organism through RNA or protein products. It was also previously believed that one gene makes one protein; this concept was overthrown by the discovery of alternative splicing and trans-splicing.

The definition of a gene is still changing. The first cases of RNA-based inheritance have been discovered in mammals. Evidence is also accumulating that the control regions of a gene do not necessarily have to be close to the coding sequence on the linear molecule or even on the same chromosome. Spilianakis and colleagues discovered that the promoter region of the interferon-gamma gene on chromosome 10 and the regulatory regions of the T(H)2 ytokine locus on chromosome 11 come into close proximity in the nucleus possibly to be jointly regulated. The concept that genes are clearly delimited is also being eroded. There is evidence for fused proteins stemming from two adjacent genes that can produce two separate protein products.

While it is not clear whether these fusion proteins are functional, the phenomenon is more frequent than previously thought. Even more ground-breaking than the discovery of fused genes is the observation that some proteins can be composed of exons from far away regions and even different chromosomes. This new data has led to an updated, and probably tentative, definition of a gene as "a union of genomic sequences encoding a coherent set of potentially overlapping functional products." This new definition categorizes genes by functional products, whether they be proteins or RNA, rather than specific DNA loci; all regulatory elements of DNA are therefore classified as *gene-associated* regions.

A single gene coding for a particular protein corresponds to a sequence of nucleotides along one or more regions of a molecule of DNA. The DNA sequence is collinear with the protein sequence. In species for which the genetic materiais double-stranded DNA, genes may appear on either strand. Bacterial genes are continuous regions of DNA. Therefore the functional unit of genetic sequence information from a bacterium is a string of $3N$ nucleotides encoding a string of N amino acids, or a string of N nucleotides encoding a structural RNA molecule. Such a string, equipped with annotations, would form a typical entry in one of the genetic sequence archives.

Proteomes

The proteome project is a large-scale programme dealing in an integral way with patterns of expression of proteins in biological systems, in ways that complement and extend genome projects.

What kinds of data would we like to measure, and what mature experimental techniques exist to determine them? The basic goal is a spatio-temporal description of the deployment of proteins in the organism. The rates of synthesis of different proteins vary among different tissues and different cell types and states of activity. Methods are available for analysis of transcription patterns of genes. However, because proteins 'turn over' at different rates, it is also necessary to measure proteins directly. High-resolution two-dimension-apolyacylamide gel electrophoresis (2D PAGE) shows the pattern of protein content in a sample. Mass-spectroscpic techniques identify the proteins into hich the sample has been separated.

In principle, a database of amino acid sequences of proteins is inherent in the database of nucleotide sequences of DNA, by virtue of the genetic code. Indeed, new protein sequence data are now being determined almost exclusively by translation of DNA sequences, rather than by direct sequencing of proteins. (Historically the chemical problem of determining amino acid sequences of proteins directly was solved before the genetic code was established and before methods for determination of nucleotide sequences of DNA were developed. F. Sanger's sequencing of insulin in 1955 first proved that proteins had definite amino acid sequences, a proposition that until then was hypothetical.) Should any distinction be made between amino acid sequences determined directly from proteins and those determined by translation from DNA? First, we must assume that it is possible correctly to identify within the DNA data stream the regions that encode proteins.

The pattern-recognition programs that address this question are subject to three types of errors: a genuine protein sequence mabe missed entirely, or an incomplete protein may be reported, or a gene may be incorrectly spliced. Several variations on the theme add to the complexity: Genes for different proteins may overlap, or genes may be assembled from exons in different ways in different tissues. Conversely, some genetic sequences that appear to code for proteins may in fact be defective or not expressed. *A protein inferred* from a genome sequence is a hypothetical object until an experiment verifies its existence.

Second, in many cases the expression of a gene produces a molecule that must be modified within a cell, to make a *mature* protein that differs significantly frothe one suggested by translation of the gene sequence. In many cases the missing details of post-translational modifications—the molecular analogues of body piercing—are quite important. Post-translational modifications include addition of ligands (for instance the covalently-bound haem group of cytochrome *c*), glycosylation, methylation, excision of peptides, and many others. Patterns of disulphide bridges—primary chemical bonds between cysteine residues—cannot be deduced from the amino acid sequence. In some cases, mRNA is edited before translation, creating changes in amino acid sequences that are not inferrable from the genes.

Eavesdropping on the Transmission of Genetic Information

How hereditary information is stored, passed on, and implemented is perhaps *the* fundamental problem of biology. Three types of maps have been essential :

1. Linkage maps of genes
2. Banding patterns of chromosomes
3. DNA sequences.

These represent three very different types of data. Genes, as discovered by Mendel, were entirely abstract entities. Chromosomes are physical objects, the banding patterns their visible landmarks. Only with DNA sequences are we dealing directly with stored hereditary information in its physical form.

It was the very great achievement of the last century of biology to forge connections between these three types of data. The first steps—and giant strides thewere indeed—proved that, for any chromosome, the maps are one-dimensional arrays, and indeed that they are collinear. Any schoolchild now knows that genes are strung out along chromosomes, and that each gene corresponds to a DNA sequence. But the proofs of these statements earned a large number of Nobel prizes.

Splitting a long molecule of DNA—for example, the DNA in an entire chromosome— into fragments of convenient size for cloning and sequencing requires additional maps to report the order of the fragments, so that the entire sequence can be reconstructed from the sequences of the fragments. A restriction endonuclease is an enzyme that cuts DNA at a specific sequence, usually about 6 blong. Cutting DNA with several restriction enzymes with different specificities produces sets of overlapping fragments.

From the sizes of the fragments it is possible to construct a restriction map, stating the order and distance between the restriction enzyme cleavage sites. A mutation in one of these cleavage sites will change the sizes of the fragments produced by the corresponding enzyme, allowing the mutation to be located in the map.

Restriction enzymes can produce fairly large pieces of DNA. Cutting the DNinto smaller pieces, which are cloned and ordered by sequence overlaps (like the example using text on pages 19–20) produces a finer dissection of the DNA called a contig map.

In the past, the connections between chromosomes, genes and DNA sequences have been essential for identifying the molecular deficits underlying inheritediseases, such as Huntington disease or cystic fibrosis. Sequencing of the human genome has changed the situation radically.

Given a disease attributable to a defective protein:

- If we know the protein involved, we can pursue rational approaches to therapy.
- If we know the gene involved, we can devise tests to identify sufferers or carriers.

In many cases, knowledge of the chromosomal location of the gene is unnecessary for either therapy or detection; it is required only for identifying the gene, providing a bridge between the patterns of inheritance and the DNA sequence. (This is not true of diseases arising from chromosome abnormalities.)

For instance, in the case of sickle-cell anaemia, we know the protein involved. The disease arises from a single point mutation in haemoglobin. We can proceed directly to drug design. We need the DNA sequence only for genetic testing and counselling. In contrast, if we know neither the protein nor the gene, we must somehow go from the phenotype back to the gene, a process called positional cloning or reverse genetics. Positional cloning used to involve a kind of 'Tinker to Evers to Chance' cascade from the gene map to the chromosome map to the DNA sequence.

Patterns of inheritance identify the type of genetic defect responsible for a condition. They show, for example, that Huntington disease and cystic fibrosis are caused by single genes. To find the gene associated with cystic fibrosis it was necessary to begin with the gene map, using linkage patterns of heredity in affected families to localize the affected gene to a particular region of a particular chromosome. Knowing the general region of the chromosome, it was then possible to search the DNA of that region to identify candidate genes, and finally to pinpoint the particular gene responsible and sequence it.

In contrast, many diseases do not show simple inheritance, or, even if only a single gene is involved, heredity creates only a predisposition, the clinical consequences of which depend on environmental factors. The full human genome sequence, and measurements of expression patterns, will be essential to identify the genetic components of these more complex cases.

Mappings between the Maps

A gene linkage map can be calibrated to chromosome banding patterns through observation of individuals with deletions or translocations of parts of chromosomes.

The genes responsible for phenotypic changes associated with a deletion must lie within the deletion. Translocations are correlated with altered patterns of linkage and recombination.

There have been several approaches to coordinating chromosome banding patterns with individual DNA sequences of genes:

- In Fluorescent *In Situ* Hybridization (FISH) a probe sequence is labelled with fluorescent dye. The probe is hybridized with the chromosomes, and the chromosomal location where the probe is bound shows up directly in a photograph. Typical resolution is ~105 base pairs, but specialized new techniques can achieve high resolution, down to 1 kbp. Simultaneous FISH with two probes can detect linkage and even estimate genetic distances. This is important in species for which the generation time is long enough to make standard genetic approaches inconvenient. FISH can also detect chromosomal abnormalities.
- Somatic cell hybrids are rodent cells containing few, one, or even partial human chromosomes. (Chromosome fragments are produced by irradiating the human cells prior to fusion. Such lines are called radiation hybrids.) Hybridization of a probe sequence with a panel of somatic cell hybrids, detected by fluorescence, can identify which chromosome contains the probe. This approach has been superseded by use of clones of yeast or bacteria or phage containing fragments of human DNA in artificial chromosomes (YACs and BACs and PAs).

High-resolution Maps

Formerly, genes were the only visible portions of genomes. Now, markers are nlonger limited to genes with phenotypically observable effects, which are anyway too sparse for an adequately high-resolution map of the human genome. Now that we can interrogate DNA sequences directly, any features of DNA that vary among individuals can serve as markers, including:

Variable number tandem repeats (VNTRs), also called minisatellites. VNTRs contain regions 10–100 base pairs long, repeated a variable number of times— ame sequence, different number f repeats. In any individual, VNTRs based on the same repeat motif may apear only once in the genome; or several times, with different lengths on different chromosomes.

The distribution of the sizes of the repeats is the marker. Inheritance of VNTRs can befollowed in a family and mapped to a disease phenotype like any other trait. VNTRs were the first genetic sequence data used for personal identification—genetic fingerprints— in paternity and in criminal cases.

Formerly, VNTRs were observed by producing restriction fragment length polymorphisms (RFLPs) from them. VNTRs are generally flanked by recognition sites for the same restriction enzyme, which will neatly excise them. The results can be spread out on a gel, and detected by Southern blotting. Note the distintion: VNTRs are characteristics of genome sequences; RFLPs are artificial mixtures of short stretches of DNA created in the laboratory in order to identify VNTRs.

It is much easier and more efficient to measure the sizes of VNTRs by amplifying them with PCR, and this method has replaced the use of restriction enzymes. Short tandem reeat polymorphisms (STRPs), also called microsatellites. STRPs are regions of only 2–5 bp but repeated many times; typically 10–30 consecutive copies. They have several advantages as markers over VNTRs, one of which is a more even distribution over the human genome. There is no reason why these markers need lie within expressed genes, and usually they do not. (The CAG repeats in the gene for huntingtin and certain other disease genes are exceptions.)

Panels of microsatellite markers greatly simplify the identificaton of genes. It is interesting to compare a recent project to identify a disease gene now that the human genome sequence is available, with such classic studies as the identification of the gene for cystic fibrosis.

Additional mapping techniques deal more directly with the DNA sequences, and can short-circuit the process of gene identification:

A contig, or contiguous clone map, is a series of overlapping DNA clones of known order along a chromosome from an organism of interest—for instance, human—stored in yeast or bacterial cells as YACs (Yeast Artificial Chromosomes) or BACs (Bacterial Artificial Chromosomes). A contig map can produce a very fine mapping of a genome. In a YAC, human DNA is stably integrated into a small extra chromosome in a yeast cell. A YAC can contain up to 106 base pairs.

In principle, the entire human genome could be represented in 10 000 YAC clones. In a BAC, human DNA is inserted into a plasmid in an *E. coli* cell. (A plasmid is a small piece of double-stranded DNA found in addition to the main genome, usually but not always circular.) A BAC can carry about 250 000 bp. Despite their smaller capacities, BACs are preferred to YACs because of their greater stability and ease of handling.

A sequence tagged site (STS) is a short, sequenced region of DNA, typically 200–600 bp long, that appears in a unique location in the genome. It need not be polymorphic. An STS can be mapped into the genome by using PCR to test for the presence of the sequence in the cells containing a contig map.

One type of STS arises from an expressed sequence tag (EST), a piece of cDNA (complementary DNA; that is, a DNA sequence derived from the messenger RNA of an expressed gene). The sequence contains only the exons of the gene, spliced together to form the sequence that encodes the protein. cDNA sequences can be mapped to chromosomes using FISH, or located within contig maps.

How do contig maps and sequence tagged sites facilitate identifying genes? If you are working with an organism for which the full-genome sequence is not known, but for which full contig maps are available for all chromosomes, you would identify STS markers tightly linked to your gene, then locate these markers in the contig maps.

Picking out Genes in Genomes

Computer programs for genome analysis identify open reading frames or ORFs. An ORF is a region of DNA sequence that begins with an initiation codon (ATG) and ends with a stop codon. An ORF is a potential protein-coding region. Approaches to identifying protein-coding regions choose from or combine two possible approaches:

(1) Detection of regions similar to known coding regions from other organisms. These regions may encode amino acid sequences similar to known proteins, or may be similar to Expressed Sequence Tags (ESTs). Because ESTs are derived from messenger RNA, they correspond to genes known to be transcribed. It is necessary to sequence only a few hundred initial bases of cDNA to give enough information to identify a gene: characterization of genes by ESTs is like indexing poems or songs by their first lines.

(2) *Ab initio* methods, that seek to identify genes from the properties of the DNA sequences themselves. Computer-assisted annotation of genomes is more complete and accurate for bacteria than for eukaryotes. Bacterial genes are relatively easy to identify because they are contiguous—they lack the introns characteristic of eukaryotic genomes—and the intergene spaces are small. In higher organisms, identifying genes is harder. Identification of exons is one problem, assembling them is another. Alternative splicing patterns present a particular difficulty.

A framework for *ab initio* gene identification in eukaryotic genomes includes the following features:

The initial (5_) exon starts with a transcription start point, preceded by a corpromotor site such as the TATA box typically ~30 bp upstream. It is free of in-frame stop codons, and ends immediately before a GT splice signal. (Occasionally a noncoding exon precedes the exon that contains the initiator codon.)

Internal exons, like initial exons, are free of in-frame stop codons. They begin immediately after an AG splice signal and end immediately before a GT splice signal.

The final (3_) exon starts immediately after an AG splice signal and ends with a stop codon, followed by a polyadenylation signal sequence. (Occasionally a noncoding exon follows the exon that contains the stop codon.)

All coding regions have nonrandom sequence characteristics, based partly on codon usage preferences. Empirically, it is found that statistics of hexanucleotides perform best in distinguishing coding from noncoding regions. Starting from a set of known genes from an organism as a training set, pattern recognition programs can be tuned to particular genomes. Accurate gene detection is a crucial component of genome sequence analysis. This problem is an important focus of current research.

Genomes of Prokaryotes

Most prokaryotic cells contain their genetic material in the form of a large single circular piece of double-stranded DNA, usually less than 5 Mbp long. In addition they may contain plasmids. The protein-coding regions of bacterial genomes do not contain introns. In many prokaryotic genomes the protein-coding regions are partially organized into *operons*—tandem genes transcribed into a single messenger RNA molecule under common transcriptional control. In bacteria, the genes of many operons code for proteins with related functions. For instance, successive genes in the trp operon of *E. coli* code for proteins that catalyse successive steps in the biosynthesis of tryptophan. In archaea, a metabolic relationship between genes in operons is less frequently observed.

The Genome of the Bacterium Escherichia Coli

E. coli, strain K-12, has long been the workhorse of molecular biology. The genome of strain MG1655, published in 1997 by the group of F. Blattner at the University of Wisconsin, contains 4 639 221 bp in a single circular DNA molecule, with no plastids. Approximately 89% of the sequence codes for proteins or structural RNAs. An inventory reveals:

- 4284 protein-coding genes
- 122 structural RNA genes
- noncoding repeat sequences
- regulatory elements
- transcription/translation guides
- transposases
- prophage remnants
- insertion sequence elements
- patches of unusual composition, likely to be foreign elements introduced by horizonal transfer.

Analysis of the genome sequence required identification and annotation of protein-coding genes and other functional regions. Many *E. coli* proteins were known before the sequencing was complete, from the many years of intensive investigation: 1853 proteins had been described before publication of the genome sequence. Other genes could be assigned functions from identification of homologues by searching in sequence databanks. The narrower the range of specificity of the function of the homologues, the more precise could be the assignment. Currently, over 80% of proteins can be assigned at least a general function. Other regions of the genome are recognized as regulatory sites, or mobilgenetic elements, also on the basis of similarity to homologous sequences known in other organisms.

The distribution of protein-coding genes over the genome of *E. coli* does not seem to follow any simple rules, either along the DNA or on different strands. Indeed, comparison of strains suggests that genes are mobile.

The *E. coli* genome is relatively gene dense. Genes coding for proteins or structural RNAs occupy ~89% of the sequence. The average size of an open reading frame is 317 amino acids. If the genes were evenly distributed, the average intergenic region would be 130 bp; the observed average distance between genes is 118 bp. However, the sizes of intergenic regions vary considerably. Some intergenic regions are large. These contain sites of regulatory function, and repeated sequences. The longest intergenic region, 1730 bp, contains noncoding repeat sequences.

Approximately three-quarters of the transcribed units contain only one gene; the rest contain several consecutive genes, or operons. It is estimated that the *E. coli* genome contains 630–700 operons. Operons vary in size, although few contain more than five genes. Genes ithin operons tend to have related functions. In some cases, the ame DNA sequence encodes parts of more than one plypeptide chain. One gene codes for both the and _ subunits of DNA polymerase III. Translation of the entire gene forms the subunit. The _ subunit corresponds approximately to the N-terminal two-thirds of the subunit. A frameshift on the ribosome at this point leads to chain termination 50% of the time, causing a 1:1 ratio of expressed and _ subunits.

There do not appear to be any overlapping genes in which different reading frames both code for expressed proteins. In other cases, the same polypeptide chain appears in more than one enzyme. A protein that functions on its own as lipoate dehydrogenase is also an essential subunit of pyruvate dehydrogenase, 2-oxoglutarate dehydrogenase and the glycine cleavage complex.

Having the complete genome, we can examine the protein repertoire of *E. coli*. The largest class of proteins are the enzymes—appoximately 30% of the total genes. Many enzymatic functions are shared by more than one protein. Some of these sets of functionally similar enzymes are very closely related, and appear to have arisen by duplication, either in *E. coli* itself or in an ancestor. Other sets of functionally similar enzymes have very dissimilar sequences, and differ in specificity, regulation, or intracellular location.

Several features of *E. coli's* generous endowment of enzymes give it an extensive and versatile metabolic competence, which allow it to grow and compete under varying conditions:

- It can synthesize all components of proteins and nucleic acids (amino acids and nucleotides), and cofactors.
- It has metabolic flexibility: Both aerobic and anaerobic growth are possible, utilizing different pathways of energy capture. It can grow on many different carbon sources. Not all metabolic pathways are active at any time, the versatility allows response to changes in conditions.
- Even for specific metabolic reactions there are many cases of multiple enzymes. These provide redundancy and contribute to an ability to tune metabolism to varying conditions.
- However, *E. coli* does not possess a complete ange of enzymatic capacity. It cannot fix CO2 or N2.

We have described here some of the *static* features of the *E. coli* genome and itprotein repertoire. Current research is moving into dynamic aspects, to investigations of protein expression patterns in time and space.

The Genome of the Archaeon Methanococcus Jannaschii

S. Luria once suggested that to determine common features of all life one should not try to survey everything, but rather identify the organism most different from us and see what we have in common with it. The assumption was that the way to do this would be to find an organism adapted to the most different environment. Deep-sea exploration has revealed environments as far from the familiar as those portrayed in science fiction. Hydrothermal vents are underwater volcanoes emitting hot lava and gases through cracks in the ocean floor. They create niches for communities of living things disconnected from the surface, depending on the minerals exuded from the vent for their sources of inorganic nutrients.

They support living communities of micro-organisms that are the only known form of life not dependent on sunlight, directly or indirectly, as their enery source.

The micro-organism *Methanococcus jannaschii* was collected from a hydrothermal vent 2600 m deep off the coast of Baja California, Mexico, in 1983. It is a thermophilic organism, surviving at temperatures from 48–94 _C, with aoptimum at 85 _C. It is a strict anaerobe, capable of self-reproduction from inorganic components. Its overall metabolic equation is to synthesize methane from H2 and CO2.

M. jannaschii belongs to the archaea, one of the three major divisions of lfe along with the bacteria and eukarya. The archaea comprise groups of prokaryotes, including organisms adapted to extreme environmental conditions such as high temperature and pressure, or high salt concentration. The genome of *M. jannaschii* was sequenced in 1996 by The Institute for Genomic Research (TIGR). It was the first archaeal genome sequenced. It contains a large chromosome containing a circular double-stranded DNA molecule 1664 976 bp long, and two extrachromosomal elements of 58 407 and 16 550 bp respectively.

There are 1743 predicted coding regions, of which 1682 are on the chromosome, and 44 and 12 are on the large and small extrachromosomal elements respectively. Some RNA genes contain introns. As in other prokaryoti genomes there is little noncoding DNA. *M. jannaschii* would appear to satisfy Luria's goal of finding our most distant extant relative. Comparison of its genome sequence with others shows that it is distantly related to other forms of life. Only 38% of the open reading frames could be assigned a function, on the basis of homology to proteins known from other organisms. However, to everyone's great surprise, archaea are in some ways more closely related to eukaryotes than to bacteria! They are a complex mixture. Iarchaea, proteins involved in transcription, translation and regulation are more similar to those of eukaryotes. Archaeal proteins involved in metabolism are more similar to those of bacteria.

The Genome of One of the Simplest Organisms

Mycoplasma Genitalium

Mycoplasma genitalium is an infectious bacterium, the cause of nongonococcal uretheitis. Its genome was sequenced in 1995 by a collaboration of groups at TIGR, The Johns Hopkins University and The University of North Carolina. The genome is a single DNA molecule containing 580 070 base pairs. This is the smallest cellular genome yet sequenced. So far, *M. genitalium* is the closest we have to a minimal organism, the smallest capable of independent life. (Viruses, in contrast, require the cellular machinery of their hosts.)

The genome is dense in coding regions. 468 genes have been identified as expressed proteins. Some regions of the sequence are gene rich, others gene poor, but overall 85% of the sequence is coding. The average length of a coding region is 1040 bp. As in other bacterial genomes, the coding regions do not contain introns. Further compression of the genome is achieved by overlapping genes. It appears that many of these have arisen through loss of stop codons.

The gene repertoire of *M. genitalium* includes some that encode proteins essential for independent reproduction, such as those involved in DNA replication, transcription, and translation, plus ribosomal and transfer RNAs. Other genes are specific for the infectious activity, including adhesins that mediate binding to infected cells, other molecules for defence against the host's immune system, and many transport proteins. As adaptation to the parasitic lifestyle of the organism, there has been widespread loss of metabolic enzymes, including those responsible for amino acid biosynthesis—indeed, one amino acid is absent from all *M. genitalium* proteins.

Genomes of Eukaryotes

It is rare in science to encounter a completely new world containing phenomena entirely unsuspected. The complexity of the eukaryotic genome is such a world In eukaryotic cells, the majority of DNA is in the nucleus, separated into bundles of nucleoprotein, the chromosomes. Each chromosome contains a single double-stranded DNA molecule. Smaller amounts of DNA appear in organelles—mitochondria and chloroplasts. The organelles originated as intracellular parasites. Organelle genomes usually have the form of circula doublestranded DNA, but are sometimes linear and sometimes appear as multiple circles. The genetic code by which organelle genes are translated differs from that of nuclear genes. Nuclear genomes of different species vary widely in size. The correlation between genome size and complexity of the organism is very rough. It certainly does not support any preconception that humans stand on a pinnacle. In many cases differences in genome size reflect different amounts of simple repetitive sequences.

In addition to variation in DNA content, eukaryotic species vary in the number of chromosomes and distribution of genes among them. Some differences in the distribution of genes among chromosomes involve translocations, or chromosome fragmentations or joinings. For instance, humans have 23 pairs of chromosomes; chimpanzees have 24. Human chromosome 2 is equivalent to a fusion of chimpanzee chromosomes 12 and 13. The difficulty of chromosome pairing during mitosis in a zygote after such an event can contribute to the reproductive isolation required for species separation.

Other differences in chromosome complement reflect duplication or hybridization events. The wheat first used in agriculture, in the Middle East at least 10 000–15 000 years ago, is a diploid called *einkorn* (*Triticum monococcum*), containig 14 pairs of chromosomes. *Emmer* wheat (*T. dicoccum*), also cultivated since palaeolithic times, and *durum* wheat (*T. turgidum*), are merged hybrids f relatives of einkorn with other wild grasses,to form tetraploid species. Additional hybridizatons, to different wild wheats, gave hexaploid forms, including *spelt* (*T. spelta*), and modern common wheat *T. aestivum*. *Triticale*, a robust crop developed in modern agriulture and currently used primarily for animal feed, is an artificial genus arising from crossing durum wheat (*Triticum turgidum*) and rye (*Secale cereale*). Mot triticale varieties are hexaploids.

All these species are still cultivated—some to only minor extents—and have their individual uses in cooking. Spelt, or *farro* in Italian, is the basis of a well-known soup; pasta is made from durum wheat; and bread from *T. aestivum*. Even within single chromosomes, gene families are common in eukaryotes. Some family members are paralogues—related genes that have diverged to provide separate functions in the same species. (Orthologues, in contrast, are homologues that perform the same function in different species. For instance, human and globin are paralogues, and human and horse myoglobin are orthologues.) Other related sequences may be pseudogenes, which may have arisen by duplication, or by retrotransposition from messenger RNA, followed by the accumulation of mutations to the point of loss of function. The human globin gene cluster is a good example.

The Genome of Saccharomyces Cerevisiae (Baker's Yeast)

Yeast is one of the simplest known eukaryotic organisms. Its cells, like our own, contain a nucleus and other specialized intracellular compartments. The sequencing of its genome, by an unusually effective international consortium involving ~100 laboratories, was completed in 1992. The yeast genome contains 12 057 500 bp of nuclear DNA, distributed over 16 chromosomes. The chromosomes range in size over an order of magnitude, from the 1352 kbp chromosome IV to the 230 kbp chromosome I.

The yeast genome contains 5885 predicted protein-coding genes, ~140 genes for ribosomal RNAs, 40 genes for small nuclear RNAs, and 275 transfer RNA genes. In two respects, the yeast genome is denser in coding regions than the known genomes of the more complex eukaryotes *Caenorhabditis elegans*, *Drosophila melanogaster*, and human: (1) Introns are relatively rare, and relatively small. Only 231 genes in yeast contain introns. (2) There are fewer repeat sequences compared with more complex eukaryotes.

A duplication of the entire yeast genome appears to have occurred ~150 million years ago. This was followed by translocations of pieces of the duplicated DNA and loss of one of the copies of most (~92%) of the genes. Of the 5885 potential protein-coding genes, 3408 correspond to known proteins. About 1000 more contain some similarity to known proteins in other species. Another ~800 are similar to ORFs in other genomes that correspond to unknown proteins. Many of these homologues appear in prokaryotes. Only approximately one-third of yeast proteins have identifiable homologues in the human genome.

In taking censuses of genes, it has been useful to classify their functions into broad categories.

- Metabolism
- Energy
- Cell growth, cell division and DNA synthesis
- Transcription
- Protein synthesis
- Protein destination
- Transport facilitation
- Cellular transport and transport mechanisms
- Cellular biogenesis
- Cellular communication/signal transduction
- Cell rescue, defence, cell death and ageing
- Ionic homeostasis
- Cellular organization
- Transposable elements, viral and plasmid proteins
- Unclassified.

Yeast is a testbed for development of methods to assign functions to gene products. The search for homologues has been exhaustive and continues. Collections of mutants exist containing a knockout of every gene. (A unique sequence 'bar code' introduced into each mutant facilitates identification of the ones that grow under selected conditions.) A surprisingly-large fraction of these single-gene knockout strains are viable—80%. Cellular localization and expression patterns are being investigated. Several types of measurements, including those based on activation of transcription by pairs of proteins that can form dimers—the two-hybrid method, are producing catalogues of interprotein interactions.

The Genome of Caenorhabditis Elegans

The nematode worm *Caenorhabditis elegans* entered biological research in the 1960s, at the express invitation of S. Brenner. He recognized its potential as a sufficiently complex organism to be interesting but simple enough to permit complete analysis, at the cellular level, of its development and neural circuitry. The *C. elegans* genome, completed in 1998, provided the first full DNA sequence of a multicellular organism. The *C. elegans* genome contains ~97 Mbp of DNA distributed on paired chromosomes I, II, III, IV, V and X. There is no Y chromosome. Different genders in *C. elegans* appear in the XX genotype, a selffertilizing hermaphrodite; and the XO genotype, a male.

Several kinds of RNA genes have been identified. The *C. elegans* genome contains 659 genes for tRNA, almost half of them (44%) on the X chromosome. Spliceosomal RNAs appear in dispersed copies, often identical. (Spliceosomes are the organelles that convert pre-mRNA transcripts to mature mRNA by excising introns.) Ribosomal RNAs appear in a long tandem array at the end of chromosome I. 5S RNAs appear in a tandem array on chromosome V. Some RNA genes appear in introns of protein-coding genes. The *C. elegans* genome contains many repeat sequences. Approximately 2.6% of the genoe consists of tandem repeats. Approximately 3.6% of the genome contains inverted repeats; these appear prefrentially within introns, rather than between genes. Repeats of the hexaer sequence TTAGGC appear in many places. There are also simple duplications, involving hundreds to tens of thousands of kbp.

The Genome of Drosophila Melanogaster

Drosophila melanogaster, the fruit fly, has been the subject of detailed studies of genetics and development for a century. Its genome sequence, the product of a collaboration between Celera Genomics and the Berkeley Drosophila Genome Project, was announced in 1999.

The chromosomes of *D. melanogaster* are nucleoprotein complexes. Approximately one-third of the genome is contained in heterochromatn, highly coiled and compact (and therefore densely staining) regions flanking the centromeres. The other two-thirds is euchromatin, a relatively uncoiled, less-compact form. Most of the active genes are in the euchromatin. The heterochromatin in *D. melanogaster* contains many tandem repeats of the sequence AATAACATAG, and relatively few genes.

The total chromosomal DNA of *Drosophila melanogaster* contains about ~180 Mbp. The released sequence consists of the euchromatic portion, about 120 Mbp. The genome is distributed over five chromosomes: three large autosomes, a Y chromosome, and a fifth tiny chromosome containing only ~1 Mbp of euchromatin. The fly's 13 601 genes are approximately double the number in yeast, but are fewer than in *C. elegans*, perhaps a surprise. The average density of genes in the euchromatin sequence is 1 gene/9 kbp; much lower than the typical 1gene/kbp densities of prokaryotic genomes.

Despite the fact that insects are not very closely related to mammals, the fly genome is useful in the study of human disease. It contains homologues of 289 human genes implicated in various diseases, including cancer, and cardiovascular, neurological, endocrinological, renal, metabolic, and haematological diseases. Some of these homologues have different functions in humans and flies. Other human disease-associated genes can be introduced into, and studied in, the fly. For instance, the gene for human spinocerebellar ataxia type 3, when expressed in the fly, produces similar neuronal cell degeneration. There are now fly models for Parkinson disease and malaria.

The noncoding regions of the *D. melanogaster* genome must contain regions controlling spatio-temporal patterns of development. The developmental biology of the fly has been studied very intensively. It is therefore an organism in which the study of the genomics of development will prove extremely informative.

The Genome of Arabidopsis Thaliana

As a flowering plant *Arabidopsis thaliana* is a very distant relative of the other higher eukaryotic organisms for which genome sequences are available. It invites comparative analysis to identify common and specialised features. The *Arabidopsis thaliana* genome contains ~125 Mbp of DNA, of which 115.4 Mbp was reported in 2000 by The *Arabidopsis* Genome Initiative. There are five pairs of chromosomes, containing 25 498 predicted genes. The genome is relatively compact, with 1 gene/4.6 kbp on average. The genes of *Arabidopsis* are relatively small. Exons are typically 250 bp long, and introns relatively small, with mean length 170 bp. Typical of plant genes is an enrichment of coding regions in GC content.

Most *Arabidopsis* proteins have homologues in animals, but some systems are unique, among higher organisms, to plants, including cell wall production and photosynthesis. Many proteins shared with animals have diverged widely since the last common ancestor. Typical of another difference between plants and animals, 25% of the nuclear genes have signal sequences governing their transport into organelles—mitochondria and chloroplasts—compared to 5% of mitochon-drialtargeted nuclear genes in animals. Traces of large- and small-scale duplication events appear in the *Arabidopsis* genome. 58% of the genome contains 24 duplicated segments, _ 100 kbp in length. The genes in the duplicated segments are usually *not* identifiable homologues. However, 4140 genes (17% of all genes) appear in 1528 tandem arrays of gene families, containing up to 23 members. The pattern of duplications suggests that an ancestor of *Arabidopsis* underwent a whole-genome duplication, to form a tetraploid species.

The Genome of Homo Sapiens (The Human Genome)

In February 2001, the International Human Genome Sequencing Consortium and Celera Genomics published, separately, drafts of the human genome. The sequence amounted to ~3.2 _ 109 bp, thirty times larger than the genomes of *C. elegans* or *D. melanogaster*. One reason for this disparity in size is that coding sequences form less than 5% of the human genome; repeat sequences over 50%. Perhaps the most surprising feature was the small number of genes identified.

The finding of only about 20 000–25 000 genes suggests that alternative splicing patterns make a very significant contribution to our protein repertoire. It is estimated that ~50% of genes have alternative splicing patterns. The published human genome is distributed over 22 chromosome pairs plus the X and Y chromosomes. The DNA contents of the autosomes range from 279 Mbp down to 48 Mbp. The X chromosome contains 163 Mbp and the Y chromosome only 51 Mbp. The exons of human proein-coding genes are relatively small compared to those in other known eukaryotic genomes. The introns are relatively long. As a result many proteincoding genes span long stretches of DNA. For instance, the dystrophin gene, coding for a 3685 amino acid protein, is _2.4 Mbp long.

Protein Coding Genes

Analysis of the human protein repertoire implied by the genome sequence has proved difficult because of the problems in reliably detecting genes, and alternative splicing patterns.

Repeat Sequences

Repeat sequences compriseover 50% of the genome:

- Transposable elements, or interspersed repeats—almost half the entire genome! These include the LINEs and SINEs.
- etroposed pseudogenes
- Simple 'stutters'—repeats of short oligomers. These include the minisatellites anmicrosatellites. Trinucleotide repeats such as CAG, corresponding to glutamine repeats in the corrsponding protein, are implicated in numerous diseases.
- Segmental duplications, of blocks of ~10–300 kbp. Interchromosomal duplications appear on nonhomologous chromosomes, sometimes at multiple sites. Some intrachromosomal duplications include closely spaced duplicated regions many kbp long of very similar sequence implicated in genetic diseases; for example Charcot-Marie-Tooth syndrome type 1A, a progressive peripheral neuropathy resulting from duplication of a region containing the gene for peripheral myelin protein 22.
- Blocks of tandem repeats, including gene families

RNA

RNA genes in the human genome include:

1. 497 transfer RNA genes. One large cluster contains 140 tRNA genes within a 4Mbp region on chromosome 6.
2. Genes for 28S and 5.8S ribosomal RNAs appear in a 44 kbp tandem

repeat unit of 150–200 copies. 5S RNA genes also appear in tandem arrays containing 200–300 genes, the largest of which is on chromosome 1.

3. Small nucleolar RNAs include two families of molecules that cleave and process ribosomal RNAs.

4.Splicesomal snRNAs, including the U1, U2, U4, U5, and U6 snRNAs, many owhich appear in clusters of tandem repeats of nearly-identical sequences, or inverted repeats.

Single-nucleotide Polymorphisms (SNPs)

A single-nucleotide polymorphism or SNP (pronounced 'snip') is a genetic variation between individuals, limited to a single base pair which can be substituted, inserted or deleted. Sickle-cell anaemia is an example of a disease caused by a specific SNP: an A → T mutation in the _-globin gene changes a Glu→ Val, creating a sticky surface on the haemoglobin molecule that leads to polymerization of the deoxy form.

SNPs are distributed throughout the genome, occurring on average every 5000 base pairs. Although they arose by mutation, mny positions containing SNPs have low mutation rates, and provide stable markers for mapping genes. A consortium of four major academic genome centres and eleven private companies is creating a high-quality, high-density human SNP map, with a commitment to open public access. A recently published version contains 1.8 million SNPs. Not all SNPs are linked to diseases. Many are not within functional regions (although the density of SNPs is higher than the average in regions containing genes).

Some SNPs that occur within exons are mutations to synonymos codons, or cause substitutions that do not significantly affect protein function. Other types of SNPs can cause more than local perturbation to a protein: (1) A mutation from a sense codon to a stop codon, or vice versa, will cause either premature truncation of protein synthesis or 'readthrough'. (2) A deletion or insertion will cause a phase shift in translation.

The A, B and O allelles of the genes for blood groups illustrate these ossibilities. A and B alleles differ by four SP substitutions. They code for related enzyme that add different saccharide units to an antigen on the surface of red blood cells.

The O allelle has undergone a mutation causing a phase shift, and produces no active enzyme. The red blood cells of type O individuals contain neither the A nor the B antigen. This is why people with type O blood are universal donors in blood transfusions. The loss of activity of the protein does not seem to carry any adverse consequences. Indeed, individuals of blood types B and

O have greater resistance to smallpox.

Stong correlation of a disease with a specific NP is advantageous in clinical work, because it is relatively easy to test for affected people or carriers. ut if a disease arises from dysfunction of a specific protein, there ought to be many sites of mutations that could cause inactivation. A particular site may predominate if (1) all bearers of the gene are descendants of a single individual in whom he mutation occurred, and/or (2) the disease results from a *gain* rather than loss of a specific property, such as in the ability of sickle cell haemoglobin to polymerize, and/or (3) the mutation rateat a particular site is unusually high, as in the Gly380→Arg mutation in the fibroblast growth receptor gene FGFR3, associated with achondroplasia (a syndrome including short stature).

In contrast, many independent mutations have been detected in the BRCA1 and BRCA2 genes, loci associated with increased disposition to early-onset breast and ovarian cancer. The normal gene products function as tumour suppressors. Insertion or deletion mutants causing phase shifts generally produce a missing or inactive protein. But it cannot be deduced a priori whether a noel substitutio mutant in BRCA1 or BRCA2 confers increased risk or not.

Treatments of diseases caused by defective or absent proteins include:

1. Providing normal protein We have mentioned insulin for diabetes, and Factor VIII in the most common type of hemophilia. Another example is the administration of human growth hormone in patients with an absence or severe reduction in normal levels. Use of recombinant prteins eliminates the risk of transmission of AIDS through blood transfusions or of Creutzfeld-Jakob disease from growth hormone isolated from crude pituitary extracts.
2. Lifestyle adjustments that make the function unnecessary Phenylketoneuria (PKU) is a genetic disease caused by deficiency in phenylalanine hydroxylase, the enzyme that converts phenylalanine to tyrosine. Accumulation of high levels of phenylalaninc causes developmental defects, including mental retardation. The symptoms can be avoided by a phenylalanine-free diet. Screening of newborns for high blood phenylalanine levels is legally required in the United States and many other countries.
3. Gene therapy to replace absent protcins is an active field of research. Other clinical applictions of SNPs reflect correlations between genotype and reaction to therapy (pharmacogenomics). For example, a SNP in the gene for N-acetyl transferase (NAT–2) is correlated with peripheral neuropathy—weakness, numbness and pain in the arms, legs, hands

or feet—as a side effect of treatment with isoniazid (isonicotinic acid hydrazide), a common treatment for tuberculosis. Patients who test positive for this SNP are given alternative treatment.

Genetic Diversity in Anthropology

SNP data are of great utility in anthropology, giving clues to historical variations in population size, and migration patterns. Degrees of genetic diversity are interpretable in terms of the size of the founding population. Founders are the original set of individuals from whom an entire population is descended.

Founders can be either original colonists, such as the Polynesians who first settled New Zealand, or merely the survivors of a nearextinction. Cheetahs show the effects of a population bottleneck, estimated to have occurred 10 000 years ago. All living cheetahs are as closely related to onanother as typical siblings in other species. Extrapolations of mitochondrial DNA variation in contemporary humans suggest a single maternal ancestor who lived 14 000–200 000 years ago. Calling her Eve suggests that she was the first woman. But Mitochondrial Eve was in fact the founder of a surviving population following extinction of descendants of all her contemporaries.

Population-specific SNPs are informative about migrations. Mitochondrial sequences provide information about female ancestors and Y chromosome sequences provide information about male ancestors. For example, it has been suggested that the population of Iceland—first inhabited about 1100 years ago— is descended from Scandinavian males, and from females from both Scandinavia and the British Isles. Mediaeval Icelandic writings refer to raids on settlemets in the British Isles.

A fascinating relationship between human DNA sequences and language families has been investigatd by L. L. Cavalli-Sforza and colleagues. Tese studies have proved useful in working out interrelationships among American Indian languages. They confirm that the Basques, knon to be a linguistically isolated population, have been genetically isolated also.

In the discovery of isolated populations, anthropological genetics provides data useful in medicine, for mapping disease genes is easier if the background variation is low. Genetically isolated populations in Europe include, in addition to the Basques, the Finns, Icelanders, Welsh, and Lapps.

Genetic Diversity and Personal Identification

Variations in our DNA sequences give us individual fingerprints, useful for identification and for establishment of relationships, including but not limited to questions of paternity. The use of DNA analysis as evidence in criminal trials is now well-established.

Genetic fingerprinting techniques were originally based on patterns of variable number tandem repeats (VNTRs), but have been extended to include analysis of other features including mitochondrial DNA sequences.

For most of us, all our mitochondria are genetically identical, a condition called homoplasmy. However, some individuals contain mitochondria with different DNA sequences; this is heteroplasmy. Such sequence variaton in a disease gene can complicate the observed inheritance pattern of the disease.

The most famous case of heteroplasmy involved Tsar Nicholas II of Russia. After the revolution in 1917 the Tsar and his family were taken to exile in Yekaterinburg in Central Russia. uring the night of 16–17 July 1918, the Tsar, Tsarina Alexandra, at least three of their five children, plus their physician and three servants who had accompanied the family, were killed, and their bodies buried in a secret grave. When the remains were rediscovered, assembly of the bones and examination of the dental work suggested—and sequence analysis confirmed—that the remains included an expected family group. The identity of the remains of the Tsarina were proved by matching the mitochondrial DNA sequence with that of a maternal relative, Prince Philip, Chancellor of the University of Cambridge, Duke of Edinburgh, and grandnephew of the Tsarina.

However, comparisons of mitochondrial DNA sequences of the putativremains of Nicholas II with those of two maternal relatives revealed a difference at base 16169: the Tsar had a C and the relatives a T. The extreme political and even religious sensitivities mandated that no doubts were tolerable. To confirm the identity beyond any reasonable question, the body of Grand Duke Georgij, brother of the Tsar, was exhumed.

Genetic Analysis of Cattle Domestication

Animal resources are an integral and essential aspect of human culture. Analysis of DNA sequences sheds light on their historical development and on the genetic variety characterizing modern breeding populations.

Contemporary domestic cattle include those familiar in Western Europe and North America, *Bos taurus*; and the zebu of Africa and India, *Bos indicus*. The most obvious difference in external appearance is the humpback of the zebu. It has been widely believed that the domestication of cattle occurred once, about 8000–10 000 years ago, and that the two species subsequently diverged.

Analysis of mitochondrial DNA sequences from European, African, and Asian cattle suggest, however, that (1) all European and African breeds are more closely related to each other than either is to Indian breeds, and (2) the two groups diverged about 200 000 years ago, implying recent independent domestications of different species. The similarity in physical appearance of

African and Indian zebu (and other similarities at the molecular level; for instance, VNTR markers in nuclear DNA) must then be attributable to importation of male cattle from India to East Africa.

7

Genetic Engineering

Genetic engineering, also called genetic modification, is the direct human manipulation of an organism's genome using modern DNA technology. It involves the introduction offoreign DNA or synthetic genes into the organism of interest. The introduction of new DNA does not require the use of classical genetic methods, however traditional breedingmethods are typically used for the propagation of recombinant organisms.

An organism that is generaed through the introduction of recombinant DNA is considered to be a genetically modified organism. The first organisms genetically engineered were bacteria in 1973 and then mice in 1974. Insulin-producing bacteria were commercialized in 1982 and genetically modified food has been sold since 1994.

The most common form of genetic engineering involves the insertion of new genetic material at an unspecified location in the host genome. This is accomplished by isolating and copying the genetic material of interest using molecular cloning methods to generate a DNA sequence containing the required genetic elements for expression, and then inserting this construct into the host organism. Other forms of genetic engineering include gene targeting and knocking out specific genes via engineered nucleases such aszinc finger nuclease or engineered homing endonucleases.

Genetic engineering techniques have been applied in numerous fields including research, biotechnology, and medicine. Medicines such as insulin and human growth hormone are now produced in bacteria, experimental mice such as the oncomouse and the knockout mouse are being used for research purposes and insect resistant and/or herbicide tolerant crops have been commercialized. Genetically engineered plants and animals capable of producing biotechnology drugs more cheaply than current methods (calledpharming) are also being developed and in 2009 the FDA approved the sale of the pharmaceutical protein antithrombin produced in the milk of genetically engineered goats.

Definition

Genetic engineering alters the genetic makeup of an organism using techniques that introduce heritable material prepared outside the organism either directly into the host or into a cell that is thenfused or hybridized with the host. This involves using recombinant nucleic acid (DNA or RNA) techniques to form new combinations of heritable genetic material followed by the incorporation of that material either indirectly through a vector system or directly through micro-injection, macro-injection and micro-encapsulation techniques. Genetic engineering does not include traditional animal andplant breeding, in vitro fertilisation, induction of polyploidy, mutagenesis and cell fusion techniques that do not use recombinant nucleic acids or a genetically modified organism in the process. Cloning and stem cell research, although not considered genetic engineering, are closely related and genetic engineering can be used within them. Synthetic biology is an emerging discipline that takes genetic engineering a step further by introducing artificially synthesized genetic material from raw materials into an organism.

If genetic material from another species is added to the host, the resulting organism is called transgenic. If genetic material from the same species or a species that can naturally breed with the host is used the resulting organism is called cisgenic. Genetic engineering can also be used to remove genetic material from the target organism, creating a knock out organism. In Europe genetic modification is synonymous with genetic engineering while within the United States of America it can also refer to conventional breeding methods. Within the scientific community, the term genetic engineering is not commonly used; more specific terms such as transgenic are preferred.

History

Humans have altered the genomes of species for thousands of years through artificial selection and more recently mutagenesis. Genetic engineering as the direct manipulation of DNA by humans outside breeding and mutations has only existed since the 1970s. The term "genetic engineering" was first coined by Jack Williamson in his science fiction novel Dragon's Island, published in 1951,one year before DNA's role in heredity was confirmed by Alfred Hershey and Martha Chase, and two years before James Watson and Francis Crick showed that the DNA molecule has a double-helix structure.

In 1972 Paul Berg created the first recombinant DNA molecules by combining DNA from the monkey virus SV40 wih that of the lambda virus. In 1973 Herbert Boyer and Stanley Cohen created the first transgenic organism by inserting antibiotic resistance genes into the plasmid of an E. coli bacterium. A year later Rudolf Jaenisch created a transgenic mouse by introducing foreign DNA into its embryo, making it the world's first transgenic animal.

In 1976 Genentech, the first genetic engineering company was founded by Herbert Boyer and Robert Swanson and a year later and the company produced a human protein (somatostatin) in E.coli. Genentech announced the production of genetically engineered human insulin in 1978. In 1980, the U.S. Supreme Court in the Diamond v. Chakrabarty case ruled that genetically altered life could be patented. The insulin produced by bacteria, branded humulin, was approved for release by the Food and Drug Administration in 1982.

The first field trials of genetically engineered plants occurred in France and the USA in 1986, tobacco plants were engineered to be resistant to herbicides. The People's Republic of China was the first country to commercialize transgenic plants, introducing a virus-resistant tobacco in 1992.

In 1994 Calgene attained approval to commercially release the Flavr Savr tomato, a tomato engineered to have a longer shelf life. In 1994, the European Union approved tobacco engineered to be resistant to the herbicide bromoxynil, making it the first genetically engineered crop commercialized in Europe.

In 1995, Bt Potato was approved safe by the Environmental Protection Agency, making it the first pesticide producing crop to be approved in the USA. In 2009 11 transgenic crops were grown commercially in 25 countries, the largest of which by area grown were the USA, Brazil, Argentina, India, Canada, China, Paraguay and South Africa.

In 2010, scientists at the J. Craig Venter Institute, announced that they had created the first synthetic bacterial genome, and added it to a cell containing no DNA. The resulting bacterium, namedSynthia, was the world's first synthetic life form.

Process

Isolating the Gene

First, the gene to be inserted into the genetically modified organism must be chosen and isolated. Presently, most genes transferred into plants provide protection against insects or tolerance to herbicides. In animals the majority of genes used are growth hormone genes. Once chosen the genes must be isolated. This typically involves multiplying the gene using polymerase chain reaction (PCR). If the chosen gene or the donor organism's genome has been well studied it may be present in a genetic library. If the DNA sequence is known, but no copies of the gene are available, it can be artificially synthesized. Once isolated, the gene is inserted into a bacterial plasmid.

Figure: *Elements of genetic engineering*

Constructs

The gene to be inserted into the genetically modified organism must be combined with other genetic elements in order for it to work properly. The gene can also be modified at this stage for better expression or effectiveness. As well as the gene to be inserted most constructs contain a promoter and terminatorregion as well as a selectable marker gene. The promoter region initiates transcription of the gene and can be usd to control the location and level of gene expression, while the terminator region ends transcription. The selectable marker, which in most cases confers antibiotic resistance to the organism it is expressed in, is needed to determine which cells are transformed with the new gene. The constucts are made using recombinant DNA techniques, such asrestriction digests, ligations and molecular cloning.

Gene Targeting

The most common form of genetic engineering involves inserting new genetic material randomly within the host genome. Other techniques allow new genetic material to be inserted at a specific location in the host genome or generate mutations at desired genomic loci capable of knocking out endogenous genes. The technique of gene targeting uses homologous recombination to target desired changes to a specific endogenous gene. This tends to occur at a relatively low frequency in plants and animals and generally requires the use of selectable markers. he frequency of gene targeting can be greatly enhanced with the use of enginered nucleases such as zinc finger nucleases, engineered homing endoucleases, or nucleases created from TAL effectors. In addition to enhancing gene targeting, engineered nuclases can also be used to introduce mutations at endogenous genes that generate a gene knockout.

Transformation

About 1% of bacteria are naturally ble to take up foreign DNA but it can also be induced in other bacteria. Stressing the bacteriafor example, with a heat shockor an electric shock, can make the cell membrane permeable to DNA that may then incorporate into their genome or exist as extrachromosomal DNA. DNA is generally inserted into animal cells using microinjection, where it can be injected through the cells nuclear envelope directly into the nucleus or through the use ofviral vectors. In plants the DNA is generally inserted using Agrobacterium-mediated recombination or biolistics.

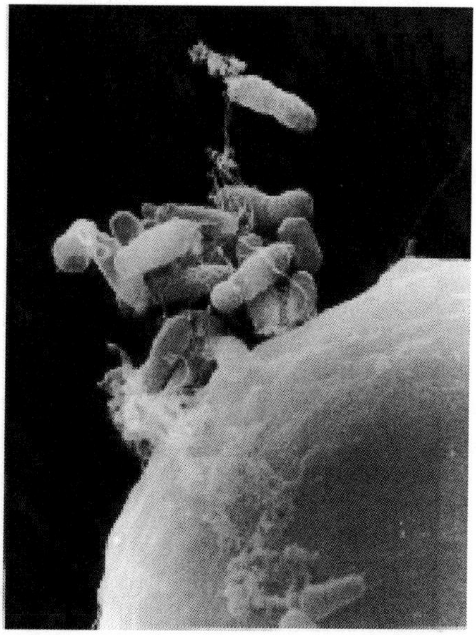

Figure: *A. tumefaciens attaching itself to a carrot cell*

In Agrobacterium-mediated recombination the plasmid construct must also contain T-DNA. Agrobacterium naturally inserts DNA from a tumor inducing plasmid into any susceptible plant's genome it infects, causing crown gall disease. The T-DNA region of this plasmid is responsible for insertion of the DNA. The genes to be inserted are cloned into a binary vector, which contains T-DNA and can be grown in both E. Coli and Agrobacterium. Once the binary vector is constructed the plasmid is transformed into Agrobacterium containing no plasmids and plant cells are infected.

The Agrobacterium will then naturally insert the genetic material into the plant cells. In biolistics particles of gold or tungsten are coated with DNA and then shot into young plant cells or plant embryos. Some genetic material will enter the cells and transform them. This method can be used on plants that are not susceptible to Agrobacterium infection and also allows transformation of plant plastids. Another transformation method for plant and animal cells is electroporation. Electroporation involves subjecting the plant or animal cell to an electric shock, which can make the cell membrane permeable to plasmid DNA. In some cases the electroporated cells will incorporate the DNA into their genome. Due to the damage caused to the cells and DNA the transformation efficiency of biolistics and electroporation is lower than agrobacterial mediated transformation and microinjection.

Selection

Not all the organism's cells will be transformed with the new genetic material; in most cases a selectable marker is used to differentiate transformed from untransformed cells. If a cell has been successfully transformed with the DNA it will also contain the marker gene. By growing the cells in the presence of an antibiotic or chemical that selects or marks the cells expressing that gene it is possible to separate the transgenic events from the non-transgenic. Another method of screening involves using a DNA probe that will only stick to the inserted gene. A number of strategies have been developed that can remove the selectable marker from the mature transgenic plant.

Regeneration

As often only a single cell is transformed with genetic material the organism must be regrown from that single cell. As bacteria consist of a single cell and reproduce clonally regeneration is not necessary. In plants this is accomplished through the use of tissue culture. Each plant species has different requirements for successful regeneration through tissue culture. If successful an adult plant is produced that contains the transgene in every cell. In animals it is necessary to ensure that the inserted DNA is present in the embryonic stem cells. When the offspring is produced they can be screened for the presence of the gene. All offspring from the first generation will be heterozygous for the inserted gene and must be mated together to produce a homozygous animal.

Confirmation

The finding that a recombinant organism contains the inserted genes is not usually sufficient to ensure that the genes will be expressed in an appropriate manner in the intended tissues of the recombinant organism. To examine the presence of the gene, further analysis frequently uses PCR, Southern hybridization, and DNA sequencing, which serve to determine the chromosomal location and copy number of the inserted gene. To examine expression of the trans-gene, an extensive analysis of transcription, RNA processing patterns, and the expression and localization of the protein product(s) is usually necessary, using methods including northern hybridization, quantitative RT-PCR, Western blot, immunofluorescence and phenotypic analysis. When appropriate, the organism's offspring are studied to confirm that the transgene and associated phenotype are stably inherited.

Applications

Genetic engineering has applications in medicine, research, industry and agriculture and can be used on a wide range of plants, animals and micro organism.

Medicine

In medicine genetic engineering has been used to mass-produce insulin, human growth hormones, follistim (for treating infertility), human albumin, monoclonal antibodies, antihemophilic factors,vaccines and many other drugs. Vaccination generally involves injecting weak live, killed or inactivated forms of viruses or their toxins into the person being immunized.

Genetically engineered viruses are being developed that can still confer immunity, but lack the infectious sequences. Mouse hybridomas, cells fused together to create monoclonal antibodies, have been humanised through genetic engineering to create human monoclonal antibodies.

Genetic engineering is used to create animal models of human diseases. Genetically modified mice are the most common genetically engineered animal model. They have been used to study and model cancer (the oncomouse), obesity, heart disease, diabetes, arthritis, substance abuse, anxiety, aging and Parkinson disease. Potential cures can be tested against these mouse models. Also genetically modified pigs have been bred with the aim of increasing the success of pig to human organ transplantation.

Gene therapy is the genetic engineering of humans by replacing defective human genes with functional copies. This can occur in somatic tissue or germline tissue. If the gene is inserted into the germline tissue it can be passed down to that person's descendants. Gene therapy has been used to treat patients suffering from immune deficiencies (notably Severe combined immunodeficiency) and trials have been carried out on other genetic disorders.

The success of gene therapy so far has been limited and a patient (Jesse Gelsinger) has died during a clinical trial testing a new treatment. There are also ethical concerns should the technology be used not just for treatment, but for enhancement, modification or alteration of a human beings' appearance, adaptability, intelligence, character or behavior. The distinction between cure and enhancement can also be difficult to establish. Transhumanists consider the enhancement of humans desirable.

Research

Genetic engineering is an important tool for natural scientists. Genes and other genetic information from a wide range of organisms are transformed into bacteria for storage and modification, creating genetically modified bacteria in the process. Bacteria are cheap, easy to grow, clonal, multiply quickly, relatively easy to transform and can be stored at -80 °C almost indefinitely. Once a gene is isolated it can be stored inside the bacteria providing an unlimited supply for research.

Figure: *Knockout mice*

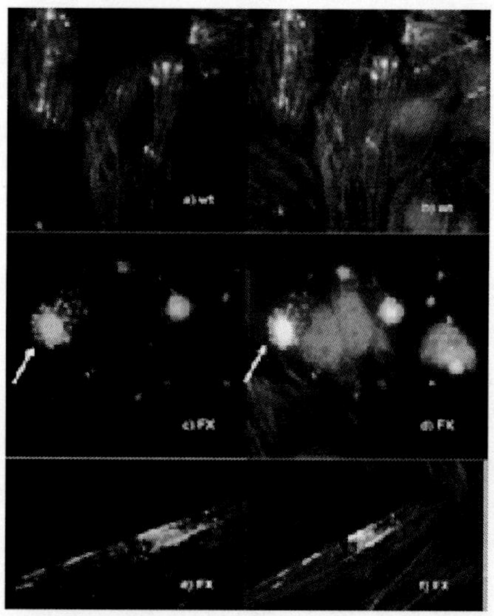

Figure: *Human cells in which some proteins are fused with green fluorescent protein to allow them to be visualised*

Organisms are genetically engineered to discover the functions of certain genes. This could be the effect on the phenotype of the organism, where the gene is expressed or what other genes it interacts with. These experiments generally involve loss of function, gain of function, tracking and expression.

- Loss of function experiments, such as in a gene knockout experiment, in which an organism is engineered to lack the activity of one or more genes. A knockout experiment involves the creation and manipulation of a DNA construct in vitro, which, in a simple knockout, consists of a copy of the desired gene, which has been altered such that it is non-functional. Embryonic stem cells incorporate the altered gene, which replaces the already present functional copy. These stem cells are injected into blastocysts, which are implanted into surrogate mothers. This allows the experimenter to analyze the defects caused by this mutation and thereby determine the role of particular genes. It is used especially frequently in developmental biology. Another method, useful in organisms such as Drosophila (fruit fly), is to induce mutations in a large population and then screen the progeny for the desired mutation. A similar process can be used in both plants and prokaryotes.

- Gain of function experiments, the logical counterpart of knockouts. These are sometimes performed in conjunction with knockout experiments to more finely establish the function of the desired gene. The process is much the same as that in knockout engineering, except that the construct is designed to increase the function of the gene, usually by providing extra copies of the gene or inducing synthesis of the protein more frequently.
- Tracking experiments, which seek to gain information about the localization and interaction of the desired protein. One way to do this is to replace the wild-type gene with a 'fusion' gene, which is a juxtaposition of the wild-type gene with a reporting element such as green fluorescent protein (GFP) that will allow easy visualization of the products of the genetic modification. While this is a useful technique, the manipulation can destroy the function of the gene, creating secondary effects and possibly calling into question the results of the experiment. More sophisticated techniques are now in development that can track protein products without mitigating their function, such as the addition of small sequences that will serve as binding motifs to monoclonal antibodies.
- Expression studies aim to discover where and when specific proteins are produced. In these experiments, the DNA sequence before the DNA that codes for a protein, known as a gene's promoter, is reintroduced into an organism with the protein coding region replaced by a reporter gene such as GFP or an enzyme that catalyzes the production of a dye. Thus the time and place where a particular protein is produced can be observed. Expression studies can be taken a step further by altering the promoter to find which pieces are crucial for the proper expression of the gene and are actually bound by transcription factor proteins; this process is known as promoter bashing.

Industrial

By engineering genes into bacterial plasmids it is possible to create a biological factory that can produce proteins and enzymes. Some genes do not work well in bacteria, so yeast, a eukaryote, can also be used. Bacteria and yeast factories have been used to produce medicines such as insulin, human growth hormone, and vaccines, supplements such as tryptophan, aid in the production of food (chymosin in cheese making) and fuels. Other applications involving genetically engineered bacteria being investigated involve making the bacteria perform tasks outside their natural cycle, such as cleaning up oil spills, carbon and other toxic waste.

Agriculture

One of the best-known and controversial applications of genetic engineering is the creation of genetically modified food. There are three generations of genetically modified crops. First generation crops have been commercialized and most provide protection from insects and/or resistance to herbicides. There are also fungal and virus resistant crops developed or in development. They have been developed to make the insect and weed management of crops easier and can indirectly increase crop yield.

Figure: *Bt-toxins present in peanut leaves (bottom image) protect it from extensive damage caused by European corn borerlarvae (top image).*

The second generation of genetically modified crops being developed aim to directly improve yield by improving salt, cold or drought tolerance and to increase the nutritional value of the crops.

The third generation consists of pharmaceutical crops, crops that contain edible vaccines and other drugs. Some agriculturally important animals have been genetically modified with growth hormones to increase their size while others have been engineered to express drugs and other proteins in their milk.

The genetic engineering of agricultural crops can increase the growth rates and resistance to different diseases caused by pathogens and parasites. This is beneficial as it can greatly increase the production of food sources with the usage of fewer resources that would be required to host the world's growing populations. These modified crops would also reduce the usage of chemicals, such as fertilizers and pesticides, and therefore decrease the severity and frequency of the damages produced by these chemical pollution.

Ethical and safety concerns have been raised around the use of genetically modified food. A major safety concern relates to the human health implications of eating genetically modified food, in particular whether toxic or allergic reactions could occur. Gene flow into related non-transgenic crops, off target effects on beneficial organisms and the impact on biodiversity are important environmental issues. Ethical concerns involve religious issues, corporatecontrol of the food supply, intellectual property rights and the level of labeling needed on genetically modified products.

Other Uses

In materials science, a genetically modified virus has been used to construct a more environmentally friendly lithium-ion battery. Some bacteria have been genetically engineered to create black and white photographs while others have potential to be used as sensors by expressing a fluorescent protein under certain environmental conditions. Genetic engineering is also being used to create BioArt and novelty items such as blue roses, and glowing fish.

Opposition and Criticism

A 2010 study of Canola found transgenes in 80% of wild (uncultivated or "feral") varieties in North Dakota, meaning 80% of the plants which had established themselves in the area were genetically engineered varieties. The researchers stated that "we found the highest densities of [such transgene-containing] plants near agricultural fields and along major freeways, but we were also finding plants in the middle of nowhere" adding that "over time,..the build-up of different types of herbicide resistance in feral [natural] canola and closely related weeds, like field mustard, could make it more difficult to manage these plants using herbicides."

Chromosome

A chromosome is an organized structure of DNA and protein found in cells. It is a single piece of coiled DNA containing many genes, regulatory elements and other nucleotide sequences. Chromosomes also contain DNA-bound proteins, which serve to package the DNA and control its functions.

Chromosomes vary widely between different organisms. The DNA molecule may be circular or linear, and can be composed of 100,000 to 10,000,000,000 nucleotides in a long chain. Typically, eukaryotic cells (cells with nuclei) have large linear chromosomes and prokaryotic cells (cells without defined nuclei) have smaller circular chromosomes, although there are many exceptions to this rule. Also, cells may contain more than one type of chromosome; for example, mitochondria in most eukaryotes and chloroplasts in plants have their own small chromosomes.

In eukaryotes, nuclear chromosomes are packaged by proteins into a condensed structure called chromatin. This allows the very long DNA molecules to fit into the cell nucleus. The structure of chromosomes and chromatin varies through the cell cycle. Chromosomes are the essential unit for cellular division and must be replicated, divided, and passed successfully to their daughter cells so as to ensure the genetic diversity and survival of theirprogeny. Chromosomes may exist as either duplicated or unduplicated. Unduplicated chromosomes are single linear strands, whereas duplicated chromosomes (copied during synthesis phase) contain two copies joined by a centromere.

Compaction of the duplicated chromosomes during mitosis and meiosis results in the classic four-arm structure (pictured to the right). Chromosomal recombination plays a vital role in genetic diversity. If these structures are manipulated incorrectly, through processes known as chromosomal instability and translocation, the cell may undergo mitotic catastrophe and die, or it may unexpectedly evade apoptosis leading to the progression of cancer.

In practice "chromosome" is a rather loosely defined term. In prokaryotes and viruses, the term genophore is more appropriate when no chromatin is present. However, a large body of work uses the term chromosome regardless of chromatin content. In prokaryotes, DNA is usually arranged as a circle, which is tightly coiled in on itself, sometimes accompanied by one or more smaller, circular DNA molecules called plasmids.

These small circular genomes are also found in mitochondria and chloroplasts, reflecting their bacterial origins. The simplest genophores are found in viruses: these DNA orRNA molecules are short linear or circular genophores that often lack structural proteins. The word chromosome comes from the (chroma, colour) and (soma, body) due to their property of being very strongly stained by particular dyes.

History

In a series of experiments beginning in the mid-1880s, Theodor Boveri gave the definitive demonstration that chromosomes are the vectors of heredity. His two principles were the continuity of chromosomes and the individuality of chromosomes. It is the second of these principles that was so original. Wilhelm Roux suggested that each chromosome carries a different genetic load. Boveri was able to test and confirm this hypothesis.

Aided by the rediscovery at the start of the 1900s of Gregor Mendel's earlier work, Boveri was able to point out the connection between the rules of inheritance and the behaviour of the chromosomes. Boveri influenced two generations of American cytologists: Edmund Beecher Wilson, Walter Sutton and Theophilus Painter were all influenced by Boveri (Wilson and Painter actually worked with him). In his famous textbook The Cell in Development and Heredity, Wilson linked together the independent work of Boveri and Sutton (both around 1902) by naming the chromosome theory of inheritance the "Sutton-Boveri Theory" (the names are sometimes reversed). Ernst Mayr remarks that the theory was hotly contested by some famous geneticists: William Bateson, Wilhelm Johannsen, Richard Goldschmidt and T.H. Morgan, all of a rather dogmatic turn of mind. Eventually, complete proof came from chromosome maps in Morgan's own lab.

In Eukaryotes

Eukaryotes (cells with nuclei such as those found in plants, yeast, and animals) possess multiple large linear chromosomes contained in the cell's nucleus. Each chromosome has one centromere, with one or two arms projecting from the centromere, although, under most circumstances, these arms are not visible as such. In addition, most eukaryotes have a small circular mitochondrial genome, and some eukaryotes may have additional small circular or linear cytoplasmic chromosomes. In the nuclear chromosomes of eukaryotes, the uncondensed DNA exists in a semi-ordered structure, where it is wrapped around histones (structural proteins), forming a composite material calledchromatin.

Chromatin

Chromatin is the complex of DNA and protein found in the eukaryotic nucleus, which packages chromosomes. The structure of chromatin varies significantly between different stages of the cell cycle, according to the requirements of the DNA.

Interphase Chromatin

During interphase (the period of the cell cycle where the cell is not dividing), two types of chromatin can be distinguished:

- Euchromatin, which consists of DNA that is active, e.g., being expressed as protein.
- Heterochromatin, which consists of mostly inactive DNA. It seems to serve structural purposes during the chromosomal stages. Heterochromatin can be further distinguished into two types:
- Constitutive heterochromatin, which is never expressed. It is located around the centromere and usually contains repetitive sequences.
- Facultative heterochromatin, which is sometimes expressed.

Individual chromosomes cannot be distinguished at this stage – they appear in the nucleus as a homogeneous tangled mix of DNA and protein.

Metaphase Chromatin and Division

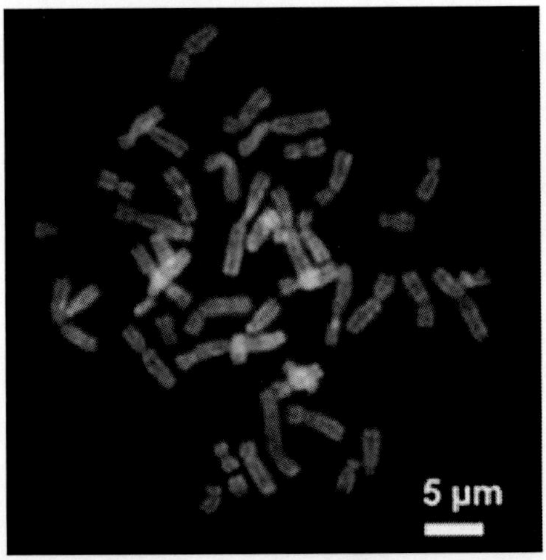

***Figure:** Human chromosomes duringmetaphase.*

In the early stages of mitosis or meiosis (cell division), the chromatin strands become more and more condensed. They cease to function as accessible genetic material (transcription stops) and become a compact transportable form.

This compact form makes the individual chromosomes visible, and they form the classic four arm structure, a pair of sister chromatids attached to each other at the centromere. The shorter arms are called p arms (from the French petit, small) and the longer arms are called q arms (q follows p in the Latin alphabet; q-g "grande"). This is the only natural context in which individual chromosomes are visible with an optical microscope.

During mitosis, microtubules grow from centrosomes located at opposite ends of the cell and also attach to the centromere at specialized structures calledkinetochores, one of which is present on each sister chromatid. A special DNA base sequence in the region of the kinetochores provides, along with special proteins, longer-lasting attachment in this region. The microtubules then pull the chromatids apart toward the centrosomes, so that each daughter cell inherits one set of chromatids. Once the cells have divided, the chromatids are uncoiled and DNA can again be transcribed. In spite of their appearance, chromosomes are structurally highly condensed, which enables these giant DNA structures to be contained within a cell nucleus.

In Humans

Chromosomes can be divided into two types—autosomes, and sex chromosomes. Certain genetic traits are linked to a person's sex and are passed on through the sex chromosomes. The autosomes contain the rest of the genetic hereditary information. All act in the same way during cell division. Human cells have 23 pairs of large linear nuclear chromosomes (22 pairs of autosomes and one pair of sex chromosomes), giving a total of 46 per cell.

In addition to these, human cells have many hundreds of copies of themitochondrial genome. Sequencing of the human genome has provided a great deal of information about each of the chromosomes. Below is a table compiling statistics for the chromosomes, based on the Sanger Institute's human genome information in the Vertebrate Genome Annotation (VEGA) database. Number of genes is an estimate as it is in part based on gene predictions. Total chromosome length is an estimate as well, based on the estimated size of unsequenced heterochromatin regions.

Estimated number of genes and base pairs on each human chromosome. The right-hand legend of this chart is mislabeled "Total bases". It should say "Total base pairs".

Chromosome	*Genes*	*Totalbase pairs*	*Sequenced base pairs[5]*
1	4,220	247,199,719	224,999,719
2	1,491	242,751,149	237,712,649
3	1,550	199,446,827	194,704,827
4	446	191,263,063	187,297,063
5	609	180,837,866	177,702,766
6	2,281	170,896,993	167,273,993
7	2,135	158,821,424	154,952,424
8	1,106	146,274,826	142,612,826
9	1,920	140,442,298	120,312,298
10	1,793	135,374,737	131,624,737
11	379	134,452,384	131,130,853

12	1,430	132,289,534	130,303,534
13	924	114,127,980	95,559,980
14	1,347	106,360,585	88,290,585
15	921	100,338,915	81,341,915
16	909	88,822,254	78,884,754
17	1,672	78,654,742	77,800,220
18	519	76,117,153	74,656,155
19	1,555	63,806,651	55,785,651
20	1,008	62,435,965	59,505,254
21	578	46,944,323	34,171,998
22	1,092	49,528,953	34,893,953
X (sex chromosome)			
1,846	154,913,754	151,058,754	
Y (sex chromosome)	454	57,741,652	25,121,652
Total	32,185	3,079,843,747	2,857,698,560

In Prokaryotes

The prokaryotes – bacteria and archaea – typically have a single circular chromosome, but many variations exist. Most bacteria have a single circular chromosome that can range in size from only 160,000 base pairs in the endosymbiotic bacterium Candidatus Carsonella ruddii, to 12,200,000 base pairs in the soil-dwelling bacterium Sorangium cellulosum. Spirochaetes of the genus Borreliaare a notable exception to this arrangement, with bacteria such as Borrelia burgdorferi, the cause of Lyme disease, containing a single linear chromosome.

Structure in Sequences

Prokaryotic chromosomes have less sequence-based structure than eukaryotes. Bacteria typically have a single point (the origin of replication) from which replication starts, whereas some archaea contain multiple replication origins. The genes in prokaryotes are often organized in operons, and do not usually contain introns, unlike eukaryotes.

DNA Packaging

Prokaryotes do not possess nuclei. Instead, their DNA is organized into a structure called the nucleoid. The nucleoid is a distinct structure and occupies a defined region of the bacterial cell. This structure is, however, dynamic and is maintained and remodeled by the actions of a range of histone-like proteins, which associate with the bacterial chromosome. In archaea, the DNA in chromosomes is even more organized, with the DNA packaged within structures similar to eukaryotic nucleosomes. Bacterial chromosomes tend to be tethered to the plasma membrane of the bacteria. In molecular biology application, this allows for its isolation from plasmid DNA by centrifugation of lysed bacteria and pelleting of the membranes (and the attached DNA). Prokaryotic chromosomes and plasmids are, like eukaryotic DNA, generally supercoiled. The DNA must first be released into its relaxed state for access for transcription, regulation, and replication.

Number of Chromosomes in Various Organisms

Eukaryotes

These tables give the total number of chromosomes (including sex chromosomes) in a cell nucleus. For example, human cells are diploid and have 22 different types of autosome, each present as two copies, and two sex chromosomes. This gives 46 chromosomes in total. Other organisms have more than two copies of their chromosomes, such as bread wheat, which is hexaploid and has six copies of seven different chromosomes – 42 chromosomes in total.

Aparticular eukaryotic species all have the same number of nuclear chromosomes. Other eukaryotic chromosomes, i.e., mitochondrial and plasmid-like small chromosomes, are much more variable in number, and there may be thousands of copies per cell.

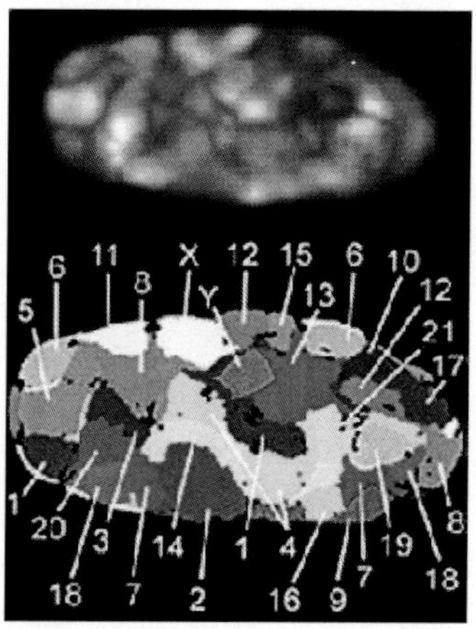

***Figure:** The 23 human chromosome territories during prometaphase infibroblast cells.*

Asexually reproducing species have one set of chromosomes, which are the same in all body cells. However, asexual species can be either haploid or diploid. Sexually reproducing species have somatic cells (body cells), which are diploid [2n] having two sets of chromosomes, one from the mother and one from the father. Gametes, reproductive cells, are haploid [n]: They have one set of chromosomes. Gametes are produced by meiosis of a diploid germ line cell. During meiosis, the matching chromosomes of father and mother can exchange small parts of themselves (crossover), and thus create new chromosomes that are not inherited solely from either parent. When a male and a female gamete merge (fertilization), a new diploid organism is formed.

Some animal and plant species are polyploid [Xn]: They have more than two sets of homologous chromosomes. Plants important in agriculture such as tobaccoor wheat are often polyploid, compared to their ancestral species.

Wheat has a haploid number of seven chromosomes, still seen in some cultivars as well as the wild progenitors. The more-common pasta and bread wheats are polyploid, having 28 (tetraploid) and 42 (hexaploid) chromosomes, compared to the 14 (diploid) chromosomes in the wild wheat.

Prokaryotes

Prokaryote species generally have one copy of each major chromosome, but most cells can easily survive with multiple copies. For example, Buchnera, asymbiont of aphids has multiple copies of its chromosome, ranging from 10–400 copies per cell.

However, in some large bacteria, such as Epulopiscium fishelsoni up to 100,000 copies of the chromosome can be present. Plasmids and plasmid-like small chromosomes are, as in eukaryotes, very variable in copy number. The number of plasmids in the cell is almost entirely determined by the rate of division of the plasmid – fast division causes high copy number.

Karyotype

In general, the karyotype is the characteristic chromosome complement of a eukaryote species. The preparation and study of karyotypes is part ofcytogenetics.

Although the replication and transcription of DNA is highly standardized in eukaryotes, the same cannot be said for their karyotypes, which are often highly variable. There may be variation between species in chromosome number and in detailed organization. In some cases, there is significant variation within species. Often there is:

1. variation between the two sexes
2. variation between the germ-line and soma (between gametes and the rest of the body)
3. variation between members of a population, due to balanced genetic polymorphism
4. geographical variation between races
5. mosaics or otherwise abnormal individuals.

Also, variation in karyotype may occur during development from the fertilised egg. The technique of determining the karyotype is usually called karyotyping. Cells can be locked part-way through division (in metaphase) in vitro (in a reaction vial) with colchicine. These cells are then stained, photographed, and arranged into a karyogram, with the set of chromosomes arranged, autosomes in order of length, and sex chromosomes (here X/Y) at the end: Like many sexually reproducing species, humans have special gonosomes (sex chromosomes, in contrast to autosomes). These are XX in females and XY in males.

Historical Note

Investigation into the human karyotype took many years to settle the most basic question: How many chromosomes does a normal diploid human cell contain? In 1912, Hans von Winiwarter reported 47 chromosomes in spermatogonia and 48 in oogonia, concluding an XX/XO sex determination mechanism.

Painter in 1922 was not certain whether the diploid number of man is 46 or 48, at first favouring 46. He revised his opinion later from 46 to 48, and he correctly insisted on humans having an XX/XY system.

New techniques were needed to definitively solve the problem:

1. Using cells in culture
2. Pretreating cells in a hypotonic solution, which swells them and spreads the chromosomes
3. Arresting mitosis in metaphase by a solution of colchicine
4. Squashing the preparation on the slide forcing the chromosomes into a single plane
5. Cutting up a photomicrograph and arranging the result into an indisputable karyogram.

It took until 1954 before the human diploid number was confirmed as 46. Considering the techniques of Winiwarter and Painter, their results were quite remarkable. Chimpanzees (the closest living relatives to modern humans) have 48 chromosomes.

Aberrations

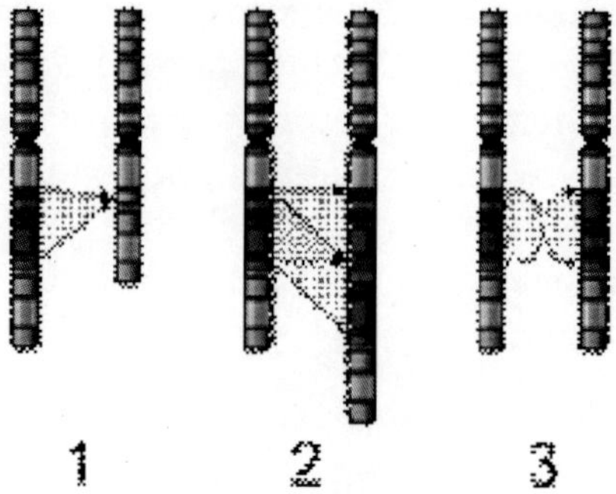

The three major single chromosome mutations; deletion (1), duplication (2) and inversion (3).

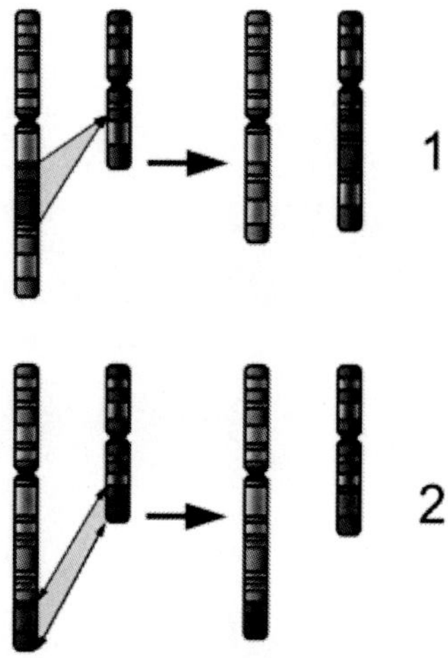

The two major two-chromosome mutations; insertion (1) and translocation (2).

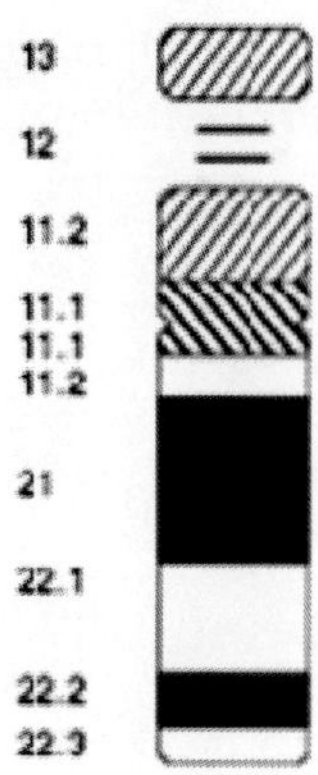

Figure: *In Down syndrome, there are three copies of chromosome 21*

Chromosomal aberrations are disruptions in the normal chromosomal content of a cell and are a major cause of genetic conditions in humans, such as Down syndrome, although most aberrations have little to no effect. Some chromosome abnormalities do not cause disease in carriers, such as translocations, orchromosomal inversions, although they may lead to a higher chance of birthing a child with a chromosome disorder. Abnormal numbers of chromosomes or chromosome sets, called aneuploidy, may be lethal or may give rise to genetic disorders. Genetic counseling is offered for families that may carry a chromosome rearrangement.

The gain or loss of DNA from chromosomes can lead to a variety of genetic disorders. Human examples include:

- Cri du chat, which is caused by the deletion of part of the short arm of chromosome 5. "Cri du chat" means "cry of the cat" in French; the condition was so-named because affected babies make high-pitched cries that sound like those of a cat. Affected individuals have wide-set eyes, a small head and jaw, moderate to severe mental health issues, and are very short.
- Down syndrome, the most common trisomy, usually caused by an extra copy of chromosome 21 (trisomy 21). Characteristics include decreased muscle tone, stockier build, asymmetrical skull, slanting eyes and mild to moderate developmental disability.
- Edwards syndrome, or trisomy-18, the second-most-common trisomy. Symptoms include motor retardation, developmental disability and numerous congenital anomalies causing serious health problems. Ninety percent of those affected die in infancy; however, those that live past their first birthday are usually quite healthy thereafter. They have characteristic clenched hands and overlapping fingers.
- Isodicentric 15, also called idic(15), partial tetrasomy 15q, or inverted duplication 15 (inv dup 15).
- Jacobsen syndrome, which is very rare. It is also called the terminal 11q deletion disorder. Those affected have normal intelligence or mild developmental disability, with poor expressive language skills. Most have a bleeding disorder called Paris-Trousseau syndrome.
- Klinefelter's syndrome (XXY). Men with Klinefelter syndrome are usually sterile, and tend to be taller and have longer arms and legs than their peers. Boys with the syndrome are often shy and quiet, and have a higher incidence of speech delay and dyslexia. Without testosterone treatment, some may developgynecomastia during puberty.
- Patau Syndrome, also called D-Syndrome or trisomy-13. Symptoms are somewhat similar to those of trisomy-18, without the characteristic

folded hand.

- Small supernumerary marker chromosome. This means there is an extra, abnormal chromosome. Features depend on the origin of the extra genetic material. Cat-eye syndrome and isodicentric chromosome 15 syndrome (or Idic15) are both caused by a supernumerary marker chromosome, as isPallister-Killian syndrome.
- Triple-X syndrome (XXX). XXX girls tend to be tall and thin and have a higher incidence of dyslexia.
- Turner syndrome (X instead of XX or XY). In Turner syndrome, female sexual characteristics are present but underdeveloped. Females with Turner syndrome often have a short stature, low hairline, abnormal eye features and bone development and a "caved-in" appearance to the chest.
- XYY syndrome. XYY boys are usually taller than their siblings. Like XXY boys and XXX girls, they are more likely to have learning difficulties.
- Wolf-Hirschhorn syndrome, which is caused by partial deletion of the short arm of chromosome 4. It is characterized by severe growth retardation and severe to profound mental health issues.

DNA

The structure of the DNA double helix. The atoms in the structure are colour coded by element and the detailed structure of two base pairs is shown in the bottom right.

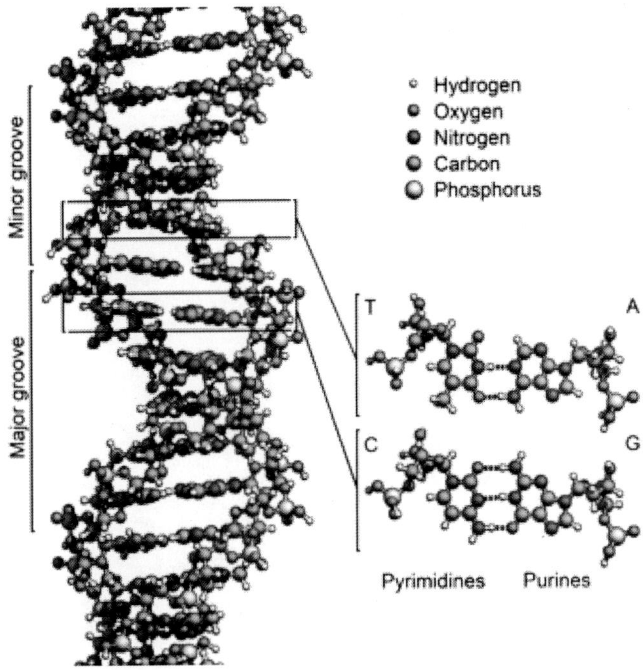

Deoxyribonucleic acid DNA is a nucleic acid that contains the genetic instructions used in the development and functioning of all known living organisms (with the exception of RNA viruses). The DNA segments that carry this genetic information are called genes, but other DNA sequences have structural purposes, or are involved in regulating the use of this genetic information. Along with RNA and proteins, DNA is one of the three major macromolecules that are essential for all known forms of life. DNA consists of two long polymers of simple units called nucleotides, with backbones made of sugars and phosphate groups joined byester bonds. These two strands run in opposite directions to each other and are therefore anti-parallel. Attached to each sugar is one of four types of molecules called nucleobases (informally, bases). It is the sequence of these four nucleobases along the backbone that encodes information. This information is read using the genetic code, which specifies the sequence of the amino acids within proteins. The code is read by copying stretches of DNA into the related nucleic acid RNA in a process called transcription.

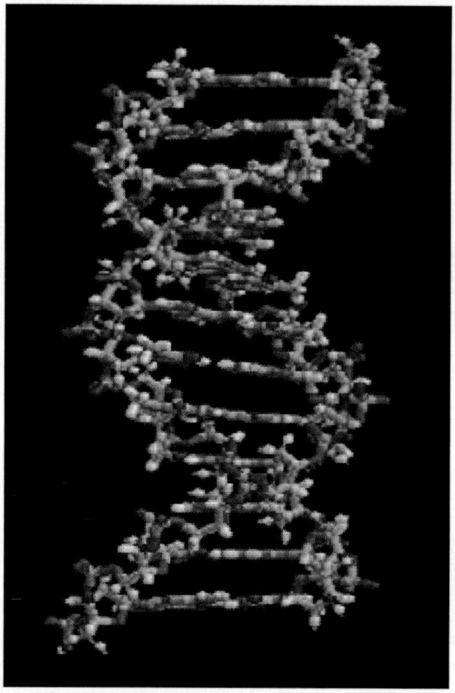

Figure: *The structure of part of a DNAdouble helix*

Within cells DNA is organized into long structures called chromosomes. During cell division these chromosomes are duplicated in the process of DNA replication, providing each cell its own complete set of chromosomes. Eukaryotic organisms (animals, plants, fungi, andprotists) store most of their DNA inside the cell nucleus and some of their DNA in organelles, such as mitochondria or chloroplasts. In contrast, prokaryotes (bacteria and archaea) store their DNA only in the cytoplasm. Within the chromosomes, chromatin proteins such as histones compact and organize DNA. These compact structures guide the interactions between DNA and other proteins, helping control which parts of the DNA are transcribed.

DNA is a long polymer made from repeating units called nucleotides. As first discovered by James D. Watson and Francis Crick, the structure of DNA of all species comprises two helical chains each coiled round the same axis, and each with a pitch of 34 Ångströms(3.4 nanometres) and a radius of 10 Ångströms (1.0 nanometres). According to another study, when measured in a particular solution, the DNA chain measured 22 to 26 Ångströms wide (2.2 to 2.6 nanometres), and one nucleotide unit measured 3.3 Å (0.33 nm) long. Although each individual repeating unit is very small, DNA polymers can be very large molecules containing millions of nucleotides. For instance, the largest human chromosome, chromosome number 1, is approximately

220 million base pairs long.

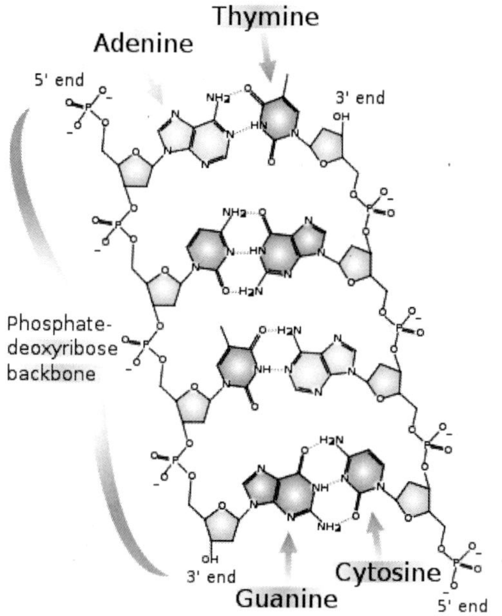

Figure: *Chemical structure of DNA. Hydrogen bonds shown as dotted lines.*

In living organisms DNA does not usually exist as a single molecule, but instead as a pair of molecules that are held tightly together. These two long strands entwine like vines, in the shape of a double helix. The nucleotide repeats contain both the segment of the backbone of the molecule, which holds the chain together, and a nucleobase, which interacts with the other DNA strand in the helix. A nucleobase linked to a sugar is called a nucleoside and a base linked to a sugar and one or more phosphate groups is called a nucleotide. Polymers comprising multiple linked nucleotides (as in DNA) are called a polynucleotide.

The backbone of the DNA strand is made from alternating phosphate and sugar residues. The sugar in DNA is 2-deoxyribose, which is apentose (five-carbon) sugar. The sugars are joined together by phosphate groups that form phosphodiester bonds between the third and fifth carbon atoms of adjacent sugar rings. These asymmetric bonds mean a strand of DNA has a direction. In a double helix the direction of the nucleotides in one strand is opposite to their direction in the other strand: the strands are antiparallel. The asymmetric ends of DNA strands are called the 52 (five prime) and 32 (three prime) ends, with the 5' end having a terminal phosphate group and the 3' end a terminal hydroxyl group. One major difference between DNA and RNA is the sugar, with the 2-deoxyribose in DNA being replaced by the alternative pentose sugar ribosein RNA.

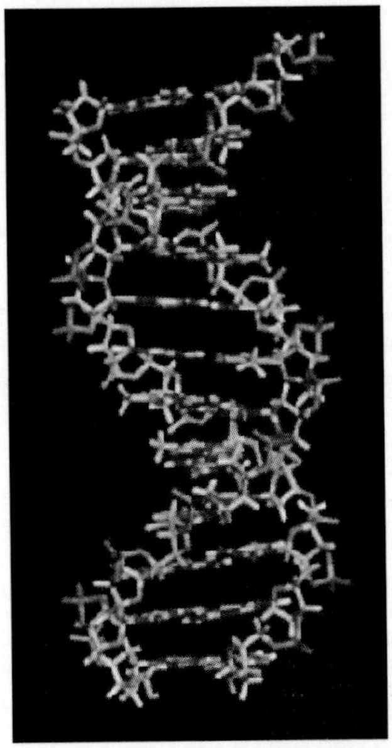

Figure: *A section of DNA. The bases lie horizontally between the two spiraling strands. Animated version at File:DNA orbit animated.gif.*

The DNA double helix is stabilized primarily by two forces: hydrogen bonds between nucleotides and base-stacking interactions among thearomatic nucleobases. In the aqueous environment of the cell, the conjugated ð bonds of nucleotide bases align perpendicular to the axis of the DNA molecule, minimizing their interaction with the solvation shell and therefore, the Gibbs free energy. The four bases found in DNA areadenine (abbreviated A), cytosine (C), guanine (G) and thymine (T). These four bases are attached to the sugar/ phosphate to form the complete nucleotide, as shown for adenosine monophosphate.

The nucleobases are classified into two types: the purines, A and G, being fused five- and six-membered heterocyclic compounds, and the pyrimidines, the six-membered rings C and T. A fifth pyrimidine nucleobase, uracil (U), usually takes the place of thymine in RNA and differs from thymine by lacking a methyl group on its ring. Uracil is not usually found in DNA, occurring only as a breakdown product of cytosine. In addition to RNA and DNA a large number of artificial nucleic acid analogues have also been created to study the proprieties of nucleic acids, or for use in biotechnology.

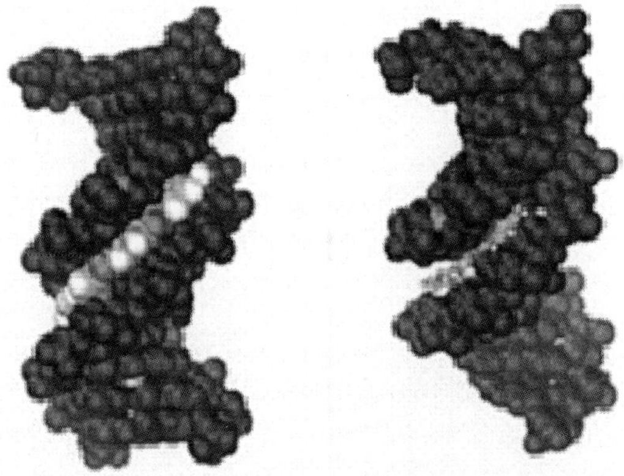

Figure: *Major and minor grooves of DNA. Minor groove is a binding site for the dye Hoechst 33258.*

Grooves

Twin helical strands form the DNA backbone. Another double helix may be found by tracing the spaces, or grooves, between the strands. These voids are adjacent to the base pairs and may provide a binding site. As the strands are not directly opposite each other, the grooves are unequally sized. One groove, the major groove, is 22 Å wide and the other, the minor groove, is 12 Å wide. The narrowness of the minor groove means that the edges of the bases are more accessible in the major groove. As a result, proteins like transcription factors that can bind to specific sequences in double-stranded DNA usually make contacts to the sides of the bases exposed in the major groove. This situation varies in unusual conformations of DNA within the cell, but the major and minor grooves are always named to reflect the differences in size that would be seen if the DNA is twisted back into the ordinary B form.

Base Pairing

In a DNA double helix, each type of nucleobase on one strand normally interacts with just one type of nucleobase on the other strand. This is called complementarybase pairing. Here, purines form hydrogen bonds to pyrimidines, with A bonding only to T, and C bonding only to G. This arrangement of two nucleotides binding together across the double helix is called a base pair. As hydrogen bonds are not covalent, they can be broken and rejoined relatively easily. The two strands of DNA in a double helix can therefore be pulled apart like a zipper, either by a mechanical force or high temperature. As a result of this complementarity, all the information in the double-stranded sequence of a DNA helix is duplicated on each strand, which is vital in DNA replication. Indeed, this reversible and specific interaction between complementary base pairs is critical for all the functions of DNA in living organisms.

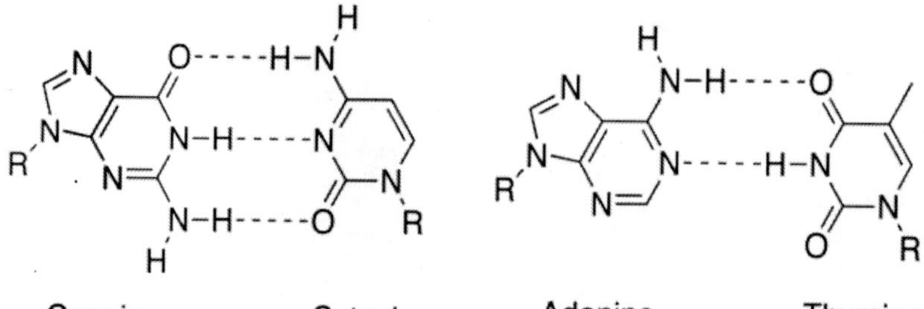

Top, a GC base pair with three hydrogen bonds. Bottom, an AT base pair with two hydrogen bonds. Non-covalent hydrogen bonds between the pairs are shown as dashed lines.

The two types of base pairs form different numbers of hydrogen bonds, AT forming two hydrogen bonds, and GC forming three hydrogen bonds. DNA with high GC-content is more stable than DNA with low GC-content. Although it is often stated that this is due to the added stability of an additional hydrogen bond, this is incorrect. DNA with high GC-content is more stable due to intra-strand base stacking interactions.

As noted above, most DNA molecules are actually two polymer strands, bound together in a helical fashion by noncovalent bonds; this double stranded structure (dsDNA) is maintained largely by the intrastrand base stacking interactions, which are strongest for G,C stacks. The two strands can come apart – a process known as melting – to form two ss DNA molecules.

Melting occurs when conditions favor ssDNA; such conditions are high temperature, low salt and high pH (low pH also melts DNA, but since DNA is unstable due to acid depurination, low pH is rarely used). The stability of the dsDNA form depends not only on the GC-content (% G,C basepairs) but also on sequence (since stacking is sequence specific) and also length (longer molecules are more stable).

The stability can be measured in various ways; a common way is the "melting temperature", which is the temperature at which 50% of the ds molecules are converted to ss molecules; melting temperature is dependent on ionic strength and the concentration of DNA.

As a result, it is both the percentage of GC base pairs and the overall length of a DNA double helix that determine the strength of the association between the two strands of DNA. Long DNA helices with a high GC-content have stronger-interacting strands, while short helices with high AT content have weaker-interacting strands. In biology, parts of the DNA double helix that need to separate easily, such as the TATAAT Pribnow box in some promoters, tend to have a high AT content, making the strands easier to pull apart.

In the laboratory, the strength of this interaction can be measured by finding the temperature required to break the hydrogen bonds, their melting temperature (also called T_m value). When all the base pairs in a DNA double helix melt, the strands separate and exist in solution as two entirely independent molecules. These single-stranded DNA molecules (ssDNA) have no single common shape, but some conformations are more stable than others.

Sense and Antisense

A DNA sequence is called "sense" if its sequence is the same as that of a messenger RNA copy that is translated into protein. The sequence on the opposite strand is called the "antisense" sequence. Both sense and antisense sequences can exist on different parts of the same strand of DNA (i.e. both strands contain both sense and antisense sequences). In both prokaryotes and eukaryotes, antisense RNA sequences are produced, but the functions of these RNAs are not entirely clear. One proposal is that antisense RNAs are involved in regulating gene expression through RNA-RNA base pairing.

A few DNA sequences in prokaryotes and eukaryotes, and more in plasmids and viruses, blur the distinction between sense and antisense strands by having overlapping genes. In these cases, some DNA sequences do double duty, encoding one protein when read along one strand, and a second protein when read in the opposite direction along the other strand. In bacteria, this overlap may be involved in the regulation of gene transcription, while in viruses, overlapping genes increase the amount of information that can be encoded within the small viral genome.

Supercoiling

DNA can be twisted like a rope in a process called DNA supercoiling. With DNA in its "relaxed" state, a strand usually circles the axis of the double helix once every 10.4 base pairs, but if the DNA is twisted the strands become more tightly or more loosely wound.

If the DNA is twisted in the direction of the helix, this is positive supercoiling, and the bases are held more tightly together. If they are twisted in the opposite direction, this is negative supercoiling, and the bases come apart more easily. In nature, most DNA has slight negative supercoiling that is introduced by enzymes calledtopoisomerases. These enzymes are also needed to relieve the twisting stresses introduced into DNA strands during processes such as transcription and DNA replication.

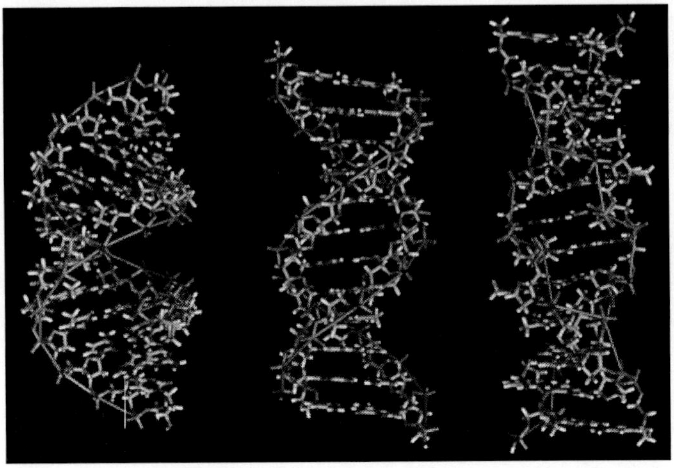

***Figure:** From left to right, the structures of A, B and Z DNA*

Alternate DNA Structures

DNA exists in many possible conformations that include A-DNA, B-DNA, and Z-DNA forms, although, only B-DNA and Z-DNA have been directly observed in functional organisms. The conformation that DNA adopts depends on the hydration level, DNA sequence, the amount and direction of supercoiling, chemical modifications of the bases, the type and concentration of metal ions, as well as the presence of polyamines in solution.

The first published reports of A-DNA X-ray diffraction patterns— and also B-DNA used analyses based on Patterson transforms that provided only a limited amount of structural information for oriented fibers of DNA. An alternate analysis was then proposed by Wilkins et al., in 1953, for the in vivo B-DNA X-ray diffraction/scattering patterns of highly hydrated DNA fibers in terms of squares of Bessel functions. In the same journal, James D. Watson and Francis Crick presented their molecular modeling analysis of the DNA X-ray diffraction patterns to suggest that the structure was a double-helix.

Although the 'B-DNA form' is most common under the conditions found in cells, it is not a well-defined conformation but a family of related DNA conformations that occur at the high hydration levels present in living cells. Their corresponding X-ray diffraction and scattering patterns are characteristic of molecular paracrystals with a significant degree of disorder.

Compared to B-DNA, the A-DNA form is a wider right-handed spiral, with a shallow, wide minor groove and a narrower, deeper major groove. The A form occurs under non-physiological conditions in partially dehydrated samples of DNA, while in the cell it may be produced in hybrid pairings of DNA and RNA strands, as well as in enzyme-DNA complexes. Segments of DNA where the bases have been chemically modified by methylation may undergo a larger change in conformation and adopt the Z form. Here, the strands turn about the helical axis in a left-handed spiral, the opposite of the more common B form. These unusual structures can be recognized by specific Z-DNA binding proteins and may be involved in the regulation of transcription.

Alternate DNA Chemistry

For a number of years exobiologists have proposed the existence of a shadow biosphere, a postulated microbial biosphere of Earth that uses radically diferent biochemical and molecular processes than currently known life. One of theproposals was the existence of lifeforms that use arsenic insteadof phosphorus in DNA.

A December 2010 NASA press conference stated that the bacterium GFAJ-1, which has evolved in an arsenic-rich environment, is the irst terrestrial lifeform found which may have this ability. The bacterium was found in Mono Lake, east of Yosemite National Park. GFAJ-1 is a rod-shaped extremophile bacterium in th family Halomonadaceae that, when starved of phosphorus, may be capable of incorporating the usually posonous element arsenic in its DNA. This discovery may lend weight to the long-standing idea that extraterrestrial life could have a different chemical makeup from life on Earth.

The research was carried out by a team led by Felisa Wolfe-Simn, a geomicrobiologist and geobiochemist, a Postdoctoral Fellow f the NASA Astrobiolog Institute with Arizona State University. This finding has, however, faced strong criticism from the scientfic community; scientists have argued that there is no evidence that arsenic is actually incorporated int biomolecules.Independent conformatin of this finding has also not yet been possible.

Quadruplex Structures

At the ends of the linear chromosomes are specialized regions of DNA called telomeres. The main unction of these regions is to allow the cell to replicat chromosome ends using the enzymetelomerase, as the enzymes that normally replicate DNA cannot copy the extreme 32 ends of chromosomes. These specialized chromosome caps also help protect the DNA ends, ad stop theDNA repair systems in the cell from treating them as damage to e corrected. In human cells, telomeres are usually lengthsof single-stranded DNA containing several thousand repeats of a simple TTAGGG sequence.

These guanine-rich sequences may stabilize chromosome ends by forming structures of stacked sets of four-base units, rather than the usual base pairs found in other DNA molecules. Here, four guanine bases form a flat plate and these flat four-base units then stack on top of each other, to form a stable G-quadruplex tructure.

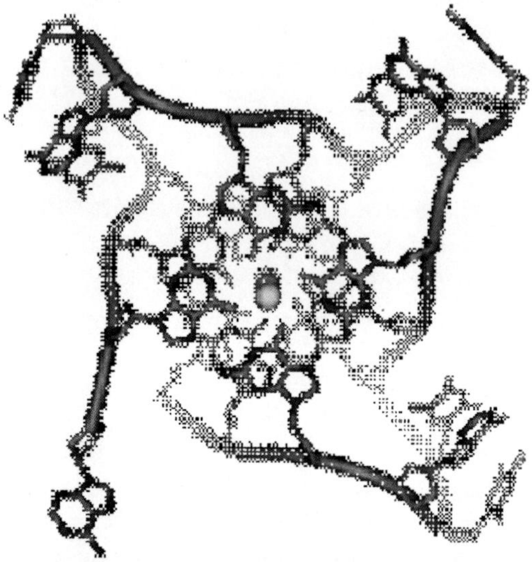

Figure: *DNA quadruplex fomed by telomererepeats. The looped conformation of the DNA backbone is very different from the typical DNA helix.*

These structures are stabilized by hydrogen bonding between the edges of the bases and chelation of a metal ion in the centre of each four-base unit. Other structures can also e formed, with the central set of four bases coming from either a single strand folded around the bases, or several different parallel strands, each contributing one base to the central structure.

In addition to these stacked structures, telomeres also form large loop structures called telomere loops, or T-loops. Here, the single-stranded DNA curls around in a long circle stabilized by telomere-binding proteins.

At the very end of the T-loop, the single-stranded telomere DNA is held onto region of double-stranded DNA by the telomere strand disrupting the double-helical DNA and base pairing to one of the two strands. This triple-stranded structure is called a displacement loop or D-loop.

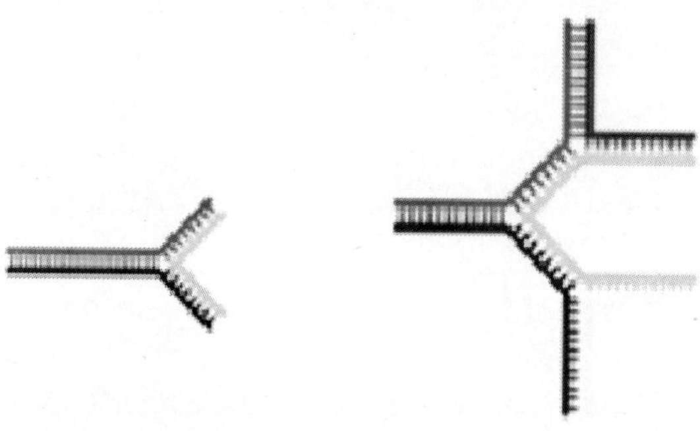

***Figure:** Branched DNA can form networks containing multipe branches.*

Branched DNA

In DNA fraying occurs when non-complementary regions existat the end of an otherwise complementry double-strand of DNA. However, branched DNA can occur if a third strand of DNA is introduced and contans adjoining regions able to hybridize with the frayed regions of the pre-existing double-strand. Although the simplest example of branched DNA involves only three strands of DNA, complexes involving additional strands and multiple branches are also possible.Branched DNA can be used i nanotechnology to construct geometric shapes.

Vibration

DNA may carry out low-frequency collective motion as observed by the Raman spectroscopy and analyzed with a quasi-continuum model.

Base Modifications

The expression ofgenes is influenced by how the DNA is packaged in chromosomes, in a structure called chromatin. Base modifications can be involved in packaging, with regions that have low or no gene expression usually containing high levels of methylation of cytosine bases. For example, cytosine methylation, produces 5-methylcytosine, which s important for X-chromosome inactivation. The average level of methylation varies between organisms – the worm Caenorhabditis elegans lacks cytosine methylation while vertebrates have higher levels, with up to % of their DNA containing 5-methylcytosine. Despite the importance of 5-methylcytosine, it can deaminate to leave a thymine base, so methylated cytosines are particularly prone to mutations. Other base modifications include adenine methylation in bacteria, the presence of 5-hydroxymethylcytosine in the brin, and the glycosylation ofuracil to produce the "J-base" in kinetoplastids.

Damage

DNA can be damaged by many sorts of mutagens, which change the DNA sequence. Mutagens include oxidizing aents, alkylating agents and also high-energy electromagnetic radiation such as ultraviolet light and X-rays.

The type of DNA damage produced depends on the type of mutagen. For example, UV light can damage DNA by producing thymine dimers, which are cross-links between pyrimidine bases.

On the other hand, oxidants such asfree radicals or hydrogen peroxide produce multiple forms of damage, including base modifications, particularly of guanosine, and double-strand breaks. A typical human cell contains about

150,000 bases that have suffered oxidative damage. Of these oxidative lesions, the most dangerous are double-strand breaks, as these are difficult to repair and can produce point mutations, insertions and deletions from the DNA sequence, as well as chromosomal transloctions. Many mutagens fit into the space between two adjacent base pairs, this is called intercalation. Most intercalators are aromatic and planar molecules; examples include ethdium bromide, acridines, daunomycin, and doxorubicin. In order for an intercalator to fit between base pairs, the bases must separate, distorting the DNA strands by unwinding of the double helix.

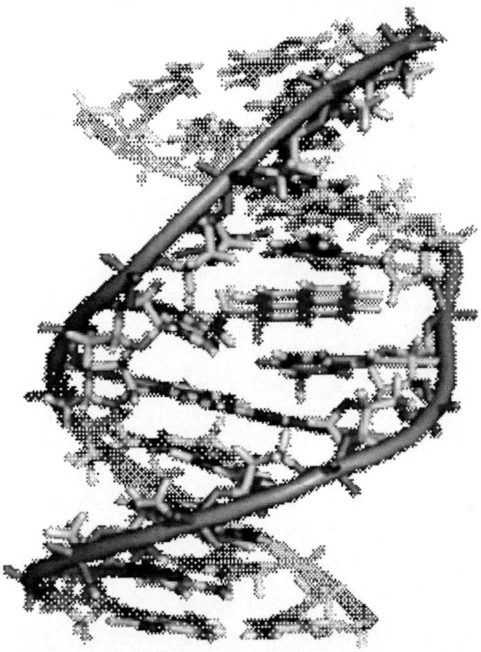

Figure: *A covalent adduct between ametabolically activated form ofbenzo[a]pyrene, the major mutagen intobacco smoke, and DNA*

This inhibits both transcription and DNA replication, causing toxicity and mutations. As a result, DNAintercalators may be carcinogens, and in the case of thalidomide, a teratogen. Others such as benzo[a]pyrene diol epoxide and aflatoxin form DNA adducts which induce errors in replication. Nevertheless, due to their ability to inhibit DNA transcription and replication, other similar toxins are also used in chemotherapy to inhibit rapidly growig cancer cells.

Biological Functions

DNA usually occurs as linear chromosomes in eukaryotes, and circular chromosomes in prokaryotes. The set of chromosomes in a cell makes up itsgenome; the human genome has approximately 3 billion base pairs of DNA arranged into 46 chromosomes. The information carried by DNA is held in thesequence of pieces of DNA called genes. Transmission of genetic information in genes is achieved via complementary base pairing.

For example, in transcription, when a cell uses the information in a gene, the DNA sequence is copied into a complementary RNA sequence through the attraction between the DNA and the correct RNA nucleotide. Usually, this RNA copy is then used to make a matching protein sequence in a process called translation, which depends on the same inteaction between RNA nucleotides. In alternative fashion, a cel may simply copy its genetic information in a process called DNA replication.

The details of thesefunctions are covered in other articles; here we focus on the interactions between DNA and other molecules that mediate the function of the genome.

Genes and Genomes

Genomic DNA i tightly and orderly packed in the process called DNA cndensation to fit the small available volumes of the cell. In eukaryotes, DNA is located in the cell nucleus, as well as small amounts in mitochondria and chloroplasts. In prokaryotes, the DNA is held within an irregularly shapd body in the cytoplasm called the nucleoid.

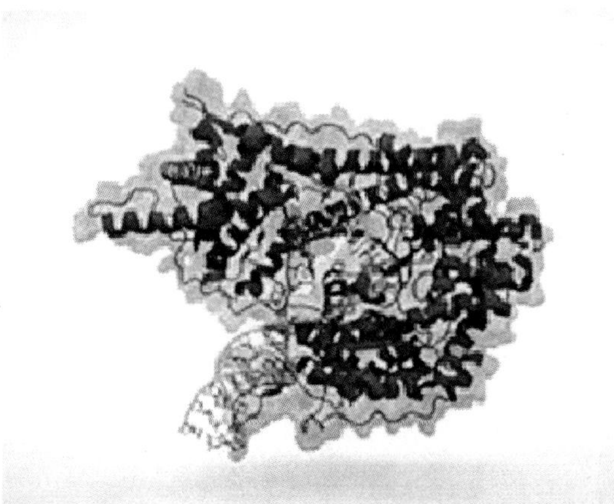

Figure: *T7 RNA polymerase (blue) producing a mRNA (green) from a DNA template (orange).*

The genetic iformation in a genome is held within genes, and the complete set of this information in an organism is called its genotype. A gene is a unit of heredity and is a region of DNA that influences a particularcharacteristic in an organism.

Genes contain an open reading frame that can be transcribed, as well as regulatory sequences such as promoters and enhancers, which control thetranscription of the open reading frame.

In many species, only a small fraction of the total sequence of the genome encods protein.

For example, only about 1.5% of the human genome consists of protein-coding exons, with over 50% of human DNA consisting of non-coding repetitive sequences.

The reasons for the presence of so much noncoding DNA in eukaryotic genomes and the extraordinary differences in genome size, or C-value, among species represent a long-standing puzzle known as the "C-value enigma".

However, DNA sequences that do not code protein may still encode functional non-coding RNA molecules, which are involved in the regulation of gene expression.

Some noncoding DNA sequences play structural roles in chromosomes. Telomeres and centromeres typically contain few genes, but are important for the function and stability of chromosomes. An abundant form of noncoding DNA in humans are pseudogenes, which are copies of genes that have been disabled by mutation. These sequences areusually just molecular fossils, although they can occasionally serve as raw enetic material for the creation of new genes through the process of gene duplication and divergence.

Transcription and Translation

A gene is a sequence of DNA that contains genetic information and can influence the phenotype of an organism. Within a gene, the sequence of bases along a DNA strand defines a messenger RNA sequence, which then defines one or more protein sequence. The relationship between the nucleotide sequeces of genes and the amino-acid sequences of proteins is determined by the rules of translation, known collectively as the genetic code. The genetic code consists of three-letter 'words' called codons formed from a sequence of three nucleotides (e.g. ACT, CAG, TTT).

In transcription, the codons of a gene are copied into messenger RNA by RNA polymerase. This RNA copy is then decoded by a ribosome that reads the RNA sequence by base-pairing the messenger RNA to transfer RNA, which carries amino acids. Since there are 4 bases in 3-letter combinations, there are 64 possible codons (4 combinations). These encode the twenty standard amino acids, giving most amino acids more than one possible coon. There are also three 'stop' or nonsense' codons signifying the end of the coding region;

these are the TAA, TGA and TAG codons.

DNA replicatin. The double helix is unwound by a helicase and topoisomerase. Next, one DNA polymerase produces the leading strand copy. Another DNA polymerase binds to the lagging strand. This enzyme makes discontinuous segments (called Okazaki fragments) before DNA liase joins them together.

Replication

Cell division is essential for an organism to grow, but, when a cell divides, it must replicate the DNA in its genome so that the two daughter cells have the same genetic information as their parent. The double-stranded structure of DNA provides a simple mechanism for DNA replication. Here the two strands are sparated and then each strand'scomplementary DNA sequence is recreated by an enzymecalled DNA polymerase. Thisenzyme makes the complementary strand by finding the correct base through complementary base pairing, and bonding it onto the original strand. As DNA olymerases can only extend DNA strand in a 52 to 32 direction, different mechanisms are used to cop the antiparallel strands of the double helix. In this way, the base on the old strand dicttes which base apears on the new strand, and the cell ends up with a perfect copy of its DNA.

Interactionswith Proteins

All the function of DNA depend on interactions with proteins. These protin interactions can be non-specific, or the protein can bind specifically to a single DNA sequence. Enzymes can also bnd to DNA and of these, the polymerases that copy the DNA base sequence in transcription and DNA replication re particularly important.

DNA-binding Proteins

Structuralproteins that bind DNA are well-understood examples of non-specific DNA-protein interactions. Within chromosomes, DNA is hel in complexes with structural proteins. These proteins organize the DNA into a compact structure called hromatin. In eukaryotes this structure involves DNA binding to a complex of small basic proteins called hstones, while in prokaryotes multiple types of proteins are involved.

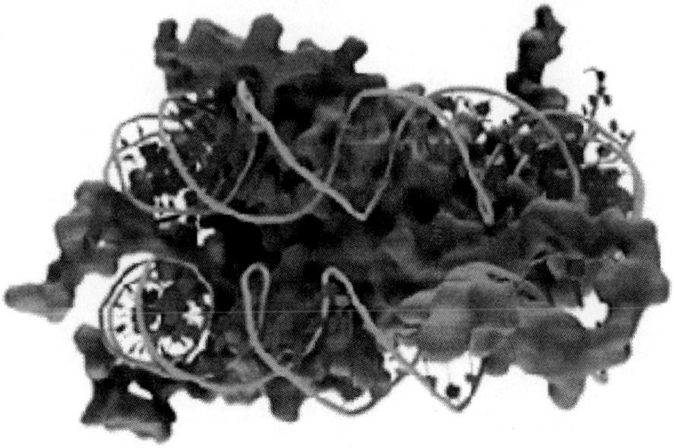

***Figure:** Interaction of DNA (shown in orange) withhistones (shown in blue). These proteins' basic amino acids bind to the acidic phosphate groups on DNA.*

The histones form a disk-shaped complex called a nucleosome, which contains two complete turns of double-stranded DNA wrapped around its surface. These non-specific interactons are formed through basic residues in the histones making ionic bonds to the acidic sugar-phosphate backbone of the DNA, and are therefore largely independent ofthe base sequence. Chemical modifications of these basic amino acid residues include methylation, phosphorylationand acetylation. These chemical changes alter the strength of the interaction between the DNA and the histones, making the DNA more or less accessible to transcription factors and changin the rate of transcription. Other non-specfic DNA-binding proteins in chromatin include the high-mobility group proteins, which bind to bent o distorted DNA. These proteins are important in bending arrays of nucleosomes and arranging them into the larger structures that make up hromosomes.

A distinct group of DNA-binding proteins are the DNA-binding proteins that specifically bind single-stranded DNA. In humans, repliction protein A is the best-understood member of this familyand is used in processes where the double helix is separated, including DNA replication, recombination and DNA repair. These binding proteins seem to stabilize single-stranded DNA and protect it from forming stem-loops or being degraded by nucleases.

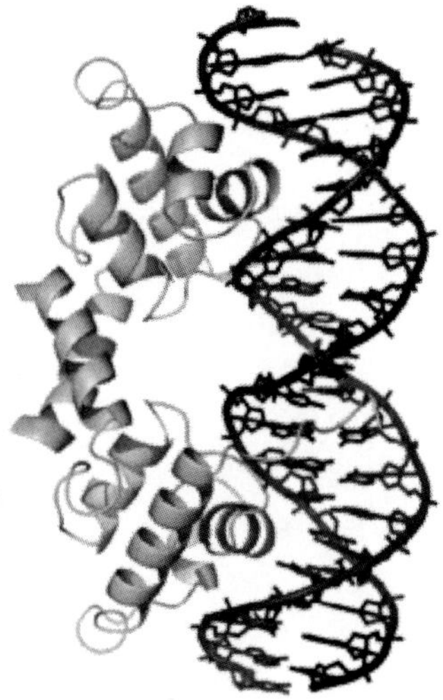

Figure: *The labda repressor helix-turn-helix transcription factor bound to its DNA target*

In contrast, other proteins have evolved to bind to particular DNA sequences. The most intensively studied of these are the various transcription factors, which are proteins that regulate transcription. Each transcription factor binds to one particular set of DNA sequences and acivates or inhibits the transcription of genes that have these sequences close to their promoters. The transcription factors do this in two ways. Firstly, they can bind the RNA polymerase responsible for transcription, either directly or through other mediator proteins; this locates the polymerase at the promoter and allows it to begin transcription. Alternatively, transcription factors can bind enzymes that modify the histones at the promoter; this will change the accessibility of the DNA template to the polymerase.

As these DNA targets can occur throughout an organism's genome, changes in the activity of one type of transcription factor can affect thousands of genes.Consequently, these proteins are often the targets of the signal transduction processes that control resposes to environmental changes or cellular differentiation anddevelopment. The specificity of these transcription factors' interactions with DNA come from the proteins making multipe contacts to the edges of the DNA bases, allowing them to "read" the DNA sequence. Most of these base-interactions are made in the major groove, where the bases are most accessible.

DNA-modifying Enzymes

Nuceases and Ligases

Nucleases are enzymes that cut DNA strands by catalyzing the hydrolysis of the phosphodiester bonds. Nucleases that hydrolyse

nucleotides from the ends of DNA strands are called exonucleases, while endonucleases cut within strands. The most frequently used nucleases in molecular biology are the restriction endonucleases, which cut DNA at specific sequences. For instance, the EcoRV enzyme shown to the left recognizes the 6-base sequence 52 -GAT|ATC-32 andmakes a cut at the vertical line. In nature, these enzymes protect bacteria against phage infection by digesting the phage DNA when t enters the bacterial cell, acting as part of the restriction modification system. In technology, these sequence-specific nucleases are used in molecular cloning and DNA fingerprinting.

Enzymes called DNA ligases can rejoin cut or broken DNA strands. Ligases are particularly important in lagging strand DNA replication, as they join together the short egments of DNA produced at the replication fork into a compete copy of the DNA template. They are also used in DNA repair and genetic recombination.

Topoisomerses and Helicases

Topoisomerases are enzymes with both nuclease and ligase activity. These proteins change the amont of supercoiling in DNA. Some of these enzymes work by cutting the DNA helix and allowing one section to rotate, thereby reducing its level of supercoiling; the enzyme then seals the DNA break. Other types of these enzymes are capale of cutting one DNA helix and then passing a second strand of DNA through this break, before rejoining the helix. Topoisomerases are required for many processes involving DNA, uch as DNA replication and transcription.Helicases are poteins that are a type of molecular motor. They use the chemical energy in nucleoside triphosphates, predominantly ATP, to break hydrogen bonds betwen bases and unwind the DNA double helix into single strnds. These enzymes are essential for most processes where enzymes need to access the DNA bases.

Polymerases

Polymerases are enzymes that synthesize polynuclotide chains from nucleoside triphosphates. The sequence of their products are copies of existing polynucleotide chains – which are calledtemplates. These enzymes function by adding nucleotides onto the 32 hydroxyl group of the previous nucleotide in a DNA strand. As a consequence, all polymerases work in a 52 to 32 direction. In the active site of these enzymes, the ncoming nucleoside triphosphate base-pairs to the template: this allows polymerases to accurately synthesize the complementary strand of their template. Polymerases are classified according to the type of template that they use.

In DNA replication, a DNA-dependent DNA polymerase makes a copy of a DNA sequence. Accuracy is vital in this process, so many of these polymerases have a proofreading activity. Here, the polymerase recognizes the occasional mistakes in the synthesis reaction by the lack of base pairing between the mismatched nucleotides. If a mismatch is detected, a 32 to 52 exonuclease activity is activated and the incorrect base removed. In most organisms, DNA polymerases function in a large complex called the replisome that contains multiple accessory subunits, such as the DNA clampor helicases.

RNA-dependent DNA polymerases are a specialized class of polymerases that copy the sequence of an RNA strand into DNA. They include reverse transcriptase, which is a viral enzyme involved in the infection of cells by retroviruses, and telomerase, which is required for the replication of telomeres. Telomerase is an unusual polymerase because it contains its own RNA template as part of its structure.

Transcription is carried out by a DNA-dependent RNA polymerase that copies the sequence of a DNA strand into RNA. To begin transcribing a gene, the RNA polymerase binds to a sequence of DNA called a promoter and separates the DNA strands. It the copies the gene sequence into a messenger RNA trancript until it reaches a region of DNA called the terminator, where it halts and detaches from the DNA. As ith human DNA-dependent DNA polymerases, RNA polymerase II, the enzyme that tanscribes most of the genes in the human genome, operates as part of a large protein complexwith multiple regulatory and accessory subunits.

Genetic Recombination

Structure of the Holliday junctionintermediate in genetic recombination. The four separate DNA strands are coloured red, blue, green and yellow.

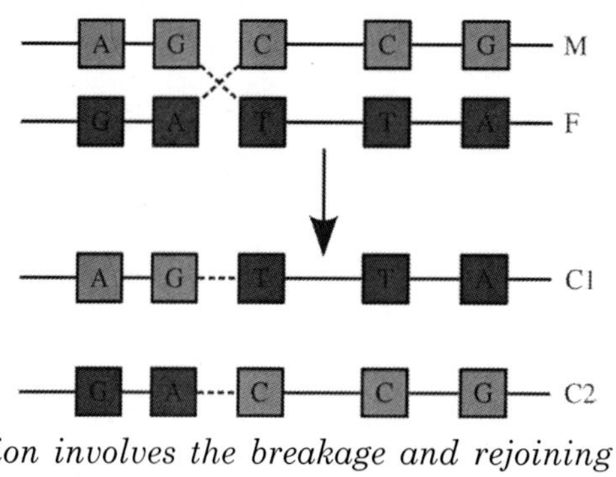

Figure: *Recombination involves the breakage and rejoining of two chromosomes (M and F) to produce two re-arranged chromosomes (C1 and C2).*

A DNA helix usually does not interact with other segments of DNA, and in human cells the different chromosomes even occupy separate areas in the nucleus called "chromosome territories".

This physcal separation of different chromosomes is important for the ability of DNA to function as a stable repository for information, as one of the

8

Genetic Assembly

Genome assembly refers to the process of taking a large number of short DNA sequences and putting them back together to create a representation of the original chromosomes from which the DNA originated.

In a shotgun sequencing project, all the DNA from a source (usually a single organism, anything from a bacterium to a mammal) is first fractured into millions of small pieces.

These pieces are then "read" by automated sequencing machines, which can read up to 900 nucleotides or bases at a time. (The four bases are adenine, guanine, cytosine, and thymine, represented as AGCT.) A genome assembly algorithm works by taking all the pieces and aligning them to one another, and detecting all places where two of the short sequences, or reads, overlap. These overlapping reads can be merged together, and the process continues.

Genome assembly is a very difficult computational problem, made more difficult because many genomes contain large numbers of identical sequences, known as repeats.

These repeats can be thousands of nucleotides long, and some occur in thousands of different locations, especially in the large genomes of plants and animals.

The resulting (draft) genome sequence is produced by combining the information sequenced contigs and then employing linking information to create scaffolds. Scaffolds are positioned along thephysical map of the chromosomes creating a "golden path".

Assembly Software

Originally, most large-scale DNA sequencing centers developed their own software for assembling the sequences that they produced. However, this has changed as the software has grown more complex and as the number of sequencing centers has increased. An example of such assembler Short Oligonucleotide Analysis Package developed by BGI for de novo assembly of human-sized genomes, alignment, SNP detection, resequencing, indel finding, and structural variation analysis.

Genome Annotation

Genome annotation is the process of attaching biological information to sequences. It consists of two main steps:

1. identifying elements on the genome, a process called gene prediction, and
2. attaching biological information to these elements.

Automatic annotation tools try to perform all this by computer analysis, as opposed to manual annotation (a.k.a. curation) which involves human expertise. Ideally, these approaches co-exist and complement each other in the same annotation pipeline.

The basic level of annotation is using BLAST for finding similarities, and then annotating genomes based on that. However, nowadays more and more additional information is added to the annotation platform. The additional information allows manual annotators to deconvolute discrepancies between genes that are given the same annotation.

Some databases use genome context information, similarity scores, experimental data, and integrations of other resources to provide genome annotations through their Subsystems approach. Other databases (e.g. Ensembl) rely on both curated data sources as well as a range of different software tools in their automated genome annotation pipeline.

Structural annotation consists of the identification of genomic elements.

- ORFs and their localisation
- gene structure
- coding regions
- location of regulatory motifs

Functional annotation consists of attaching biological information to genomic elements.

- biochemical function
- biological function

- involved regulation and interactions
- expression

These steps may involve both biological experiments and in silico analysis. A variety of software tools have been developed to permit scientists to view and share genome annotations.

Genome annotation is the next major challenge for the Human Genome Project, now that the genome sequences of human and several model organisms are largely complete.

Identifying the locations of genes and other genetic control elements is often described as defining the biological "parts list" for the assembly and normal operation of an organism. Scientists are still at an early stage in the process of delineating this parts list and in understanding how all the parts "fit together".

Genome annotation is an active area of investigation and involves a number of different organizations in the life science community which publish the results of their efforts in publicly available biological databases accessible via the web and other electronic means. Here is an alphabetical listing of on-going projects relevant to genome annotation:

- ENCyclopedia Of DNA Elements (ENCODE)
- Entrez Gene
- Ensembl
- GENCODE
- Gene Ontology Consortium
- GeneRIF
- RefSeq
- Uniprot
- Vertebrate and Genome Annotation Project (Vega)

At Wikipedia, genome annotation has started to become automated under the auspices of the Gene Wiki portal which operates a bot that harvests gene data from research databases and creates gene stubs on that basis.

Genome Project Finished

A genome, there are usually regions that are difficult to sequence (often regions with highly repetitive DNA). Thus, 'completed' genome sequences are rarely ever complete, and terms such as 'working draft' or 'essentially complete' have been used to more accurately describe the status of such genome projects. Even when every base pair of a genome sequence has been determined, there are still likely to be errors present because DNA sequencing is not a completely accurate process. It could also be argued that a complete genome project should include the sequences of mitochondria and (for plants) chloroplasts as these organelles have their own genomes.

It is often reported that the goal of sequencing a genome is to obtain information about the complete set of genes in that particular genome sequence. The proportion of a genome that encodes for genes may be very small (particularly in eukaryotes such as humans, where coding DNA may only account for a few percent of the entire sequence).

However, it is not always possible (or desirable) to only sequence the coding regions separately. Also, as scientists understand more about the role of this noncoding DNA (often referred to as junk DNA), it will become more important to have a complete genome sequence as a background to understanding the genetics and biology of any given organism.

In many ways genome projects do not confine themselves to only determining a DNA sequence of an organism. Such projects may also include gene prediction to find out where the genes are in a genome, and what those genes do. There may also be related projects to sequence ESTs or mRNAs to help find out where the genes actually are.

Historical and Technological Perspectives

Historically, when sequencing eukaryotic genomes (such as the worm Caenorhabditis elegans) it was common to first map the genome to provide a series of landmarks across the genome. Rather than sequence a chromosome in one go, it would be sequenced piece by piece (with the prior knowledge of approximately where that piece is located on the larger chromosome).

Changes in technology and in particular improvements to the processing power of omputers, means that genomes can now be 'shotgun sequenced' in one go (there are caveats to this approach though when compared to the traditiona approach).

Improvements in DNA sequencing technology has meant that the cost of sequencing a new genome sequence has steadily fallen (in terms of cost per base pair) and newer technology has also meant that genomes can be sequenced far more quickly.

When research agencies decide what new genomes to sequence, the emphasis has been on species which are either high importance as model organism or have a relevance to human health (e.g. pathogenic bacteria or vectors of disease such as mosquitos) or species which have commercial importance (e.g. livestock and crop plants). Secondary emphasis is placed on species whose genomes will help answer important questions in molecular evolution (e.g. the common chimpanzee).

In the future, it is likely that it will become even cheaper and quicker to sequence a genome. This will allow for complete genome sequences to be determined from many different individuals of the same species. For humans, this will allow us to better understand aspects of human genetic diversity.

Genome Projects

***Figure:** L1 Dominette 01449, the Hereford who serves as the subject of the Bovine Genome Project*

Many organisms have genome projects that have either been completed or will be completed shortly, including:

- Humans, Homo sapiens;
- Palaeo-Eskimo, an ancient-human
- Neanderthal, "Homo neanderthalensis".
- Common Chimpanzee Pan troglodytes; s
- Domestic Cow
- Bovine Genome
- Honey Bee Genome Sequencing Consortium
- Human microbiome project
- International Grape Genome Program
- International HapMap Project

Relative Ploidy

An important use of MLPA is to determine relative ploidy. For example, probes may be designed to target various regions of chromosome 21 of a human cell. The signal strengths of the probes are compared with those obtained from a reference DNA sample known to have two copies of the chromosome. If an extra copy is present in the test sample, the signals are exected to be 1.5 times the intensities of the respective probes from the reference. If only one copy is present the proportion is expected to be 0.5. If the sampe has two copies, the relative probe strengths are expected to be equal.

Dosage Quotient Analysis

Dosage quotient analysis is the usual method of interpreting MLPA data. If a and b are the signals from two amplicons in the patient sample, and A and B are the corresponding amplicons in the experimental control, then the dosage quotient DQ = (a/b) / (A/B). Although dosage quotients may be calculated for any pair of amplicons, it is usually the case that one of the pair is an internal reference probe.

Advantages of MLPA

MLPA facilitates the amplification and detection of multiple targets with a single primer pair. In a standard multiplex PCR reaction, each fragment needs a unique amplifying primer pair. These primers being present in a large quantity result in various problems such as dimerization and false priming. With MLPA, amplification of probes can be achieved. Thus, many sequences (up to 40) can be amplified and quantified using just a single primer pair. MLPA reaction is fast, cheap and very simple to perform.

Applications of MLPA

MLPA has a variety of applications including detection of mutations and single nucleotide polymorphisms, analysis of DNA methylation, relative mRNA quantification, chromosomal characterisation of cell lines and tissue samples, detection of gene copy number,detection of duplications and deletions in human cancer predisposition genes such as BRCA1, BRCA2, hMLH1and hMSH2 and aneuploidy determination. MLPA has potential application in prenatal diagnosis both invasive and noninvasive.

Sequence Profiling tool

A sequence profiling tool in bioinformatics is a type of software that presents information related to a genetic sequence, gene name, or keyword input. Such tools generally take a query such as aDNA, RNA, or protein sequence or 'keyword' and search one or more databases for information related to that sequence. Summaries and aggregate results are provided in standardized format describing the information that would otherwise have required visits to many smaller sites or direct literature searches to compile.

Many sequence profiling tools are software portals or gateways that simplify the process of finding information about a query in the large and growing number of bioinformatics databases. The access to these kinds of tools is either web based or locally downloadable executables.

Introduction and Usage

The "post-genomics" era has given rise to a range of web-based tools and software to compile, organize, and deliver large amounts of primary sequence information, as well as protein structures, gene annotations, sequence alignments, and other common bioinformatics tasks. In general, there exist three types of databases and service providers. The first one includes the popular public-domain or open-access databases supported by funding and grants such as NCBI,ExPASy, Ensembl, and PDB. The second one includes smaller or more specific databases organized and compiled by individual research groups Examples include Yeast Genome Database, RNA database.

The third and final one includes private corporate or institutional databases that require payment or institutional affiliation to access. Such examples rare given the globalization of the public databases unless the purported service is 'in-development' or the end point of the analysis is of commercial value.

Typical scenarios of a profiling approach become relevant, particularly, in the cases of the first two groups, where researchers commonly wish to combine information derived from several sources about a single query or target sequence. For example, users might use the sequence alignment and search tool BLAST to identify homologs of their gene of interest in other species, and then use these results to locate a solved protein structure for one of the homologs.

Similarly, they might also want to know the likely secondary structure of the mRNA encoding the gene of interest, or whether a company sells a DNA construct containing the gene. Sequence profiling tools serve to automate and integrate the process of seeking such disparate information by rendering the process of searching several different external databases transparent to the user. Many public databases are already extensively linked so that complementary information in another database is easily accessible; for example, Genbank and the PDB are closely intertwined.

However, specialized tools organized and hosted by specific research groups can be difficult to integrate into this linkage effort because they are narrowly focused, are frequently modified, or use custom versions of common file formats. Advantages of sequence profiling tools include the ability to use multiple of these specialized tools in a single query and present the output with a common interface, the ability to direct the output of one set of tools or database searches into the input of another, and the capacity to disseminate hosting and compilation obligations to a network of research groups and institutions rather than a single centralized repository.

Keyword Based Profilers

Most of the profiling tools available on the web today fall into this category. The user, upon visiting the site/tool, enters any relevant information like a keyword e.g. dystrophy, diabetes etc., or GenBank accession numbers, PDB ID. All the relevant hits by the search are presented in a format unique to each tool's main focus. Profiling tools based on keyword searches are essentially search enginesthat are highly specialized for bioinformatics work, thereby eliminating a clutter of irrelevant or non-scholarly hits that might occur with a traditional search engine like Google. Most keyword-based profiling tools allow flexible types of keyword input, accession numbers from indexed databases as well as traditional keyword descriptors.

Each profiling tool has its own focus and area of interest. For example, the NCBI search engine Entrez segregates its hits by category, so that users looking for protein structure information can screen out sequences with no corresponding structure, while users interested in perusing the literature on a subject can view abstracts of papers published in scholarly journals without distraction from gene or sequence results. The Pubmed biosciences literature database is a popular tool for literature searches, though this service is nearly equaled with the more general Google Scholar.

Keyword-based data aggregation services like the Bioinformatic Harvester performs provide reports from a variety of third-party servers in an as-is format so that users need not visit the website or install the software for each individual component service. This is particularly invaluable given the rapid emergence of various sites providing different sequence analysis and manipulation tools. Another aggregative web portal, the Human Protein Reference Database (Hprd), contains manually annotated and curated entries for human proteins.

The information provided is thus both selective and comprehensive, and the query format is flexible and intuitive. The pros of developing manually curated databases include presentation of proofread material and the concept of 'molecule authorities' to undertake the responsibility of specific proteins. However, the cons are that they are typically slower to update and may not contain very new or disputed data.

Sequence Data Based Profilers

A typical sequence profiling tool carries this further by using an actual DNA, RNA, or protein sequence as an input and allows the user to visit different web-based analysis tools to obtain the information desired. Such tools are also commonly supplied with commercial laboratory equipment like gene sequencers or sometimes sold as software applications for molecular biology.

In another public-database example, the BLAST sequence search report from NCBI provides a link from its alignment report to other relevant information in its own databases, if such specific information exists.

For example, a retrieved record that contains a human sequence will carry a separate link that connects to its location on a human genome map; a record that contains a sequence for which a 3-D structure has been solved would carry a link that connects it to its structure database. Sequerome, a public service tool, links the entire BLAST report to many third party servers/sites that provide highly specific services in sequence manipulations such as restriction enzyme maps, open reading frame analyses for nucleotide sequences, and secondary structure prediction.

The tool provides added advantage of maintaining a research log of the operations performed by the user, which can be then conveniently archived using 'mail', 'print' or 'save' functionality. Thus an entire operation of researching on a sequence using different research tools and thus carrying a project to its completion can be completed within one browser interface.

Consequently, future generation of sequence profiling tools would include ability to collaborate online with researchers to share project logs and research tools, annotate results of sequence analysis or lab work, customize and automate the processing of sets of sequence data etc.

InstaSeq is a Google powered search tool that allows the user to directly enter a sequence and search the entire World Wide Web. This unique search engine, which is the only one of its kind, is in contrast to searching specific databases e.g.GenBank.

As a result the user can end up with a privately hosted document or a page from a lesser known database from just about anywhere in the world. Though the presence of sequence based profilers are far and few in the present scenario, their key role will become evident when huge amounts of sequence data need to be cross processed across portals and domains.

Future Growth and Directions

The proliferation of bioinformatics tools for genetic analysis aids researchers in identifying and categorizing genes and gene sets of interest in their work; however, the large variety of tools that perform substantially similar aggregative and analytical functions can also confuse and frustrate new users. The decentralization encouraged by aggregative tools allows individual research groups to maintain specialized servers dedicated to specific types of data analysis in the expectation that their output will be collected into a larger report on a gene or protein of interest to other researchers.

Data produced by microarray experiments, two-hybrid screening, and other high-throughput biological experiments is voluminous and difficult to analyze by hand; the efforts of structural genomic-scollaborations that are aimed at quickly solving large numbers of highly varied protein structures also increase the need for integration between sequence and structure databases and portals. This impetus toward developing more comprehensive and more user-friendly methods of sequence profiling makes this an active area of research among current genomics researchers.

DNA Sequencing

DNA sequencing includes several methods and technologies that are used for determining the order of the nucleotide bases—adenine, guanine, cytosine, and thymine—in a molecule of DNA.

Knowledge of DNA sequences has become indispensable for basic biological research, other research branches utilizing DNA sequencing, and in numerous applied fields such as diagnostic, biotechnology, forensic biology and biological systematics. The advent of DNA sequencing has significantly accelerated biological research and discovery.

The rapid speed of sequencing attained with modern DNA sequencing technology has been instrumental in the sequencing of the human genome, in the Human Genome Project. Related projects, often by scientific collaboration across continents, have generated the complete DNA sequences of many animal, plant, and microbial genomes.

The first DNA sequences were obtained in the early 1970s by academic researchers using laborious methods based on two-dimensional chromatography. Following the development of dye-based sequencing methods with automated analysis, DNA sequencing has become easier and orders of magnitude faster.

History

RNA sequencing was one of the earliest forms of nucleotide sequencing. The major landmark of RNA sequencing is the sequence of the first complete gene and the complete genome of Bacteriophage MS2, identified and published by Walter Fiers and his coworkers at the University of Ghent (Ghent, Belgium), between 1972 and 1976. Prior to the development of rapid DNA sequencing methods in the early 1970s by Frederick Sanger at the University of Cambridge, in England and Walter Gilbert and Allan Maxam at Harvard, a number of laborious methods were used. For instance, in 1973, Gilbert and Maxam reported the sequence of 24 basepairs using a method known as wandering-spot analysis.

The chain-termination method developed by Sanger and coworkers in 1977 soon became the method of choice, owing to its relative ease and reliability. It involves separating DNA bases from different DNA fragments.

Maxam–Gilbert Sequencing

In 1976–1977, Allan Maxam and Walter Gilbert developed a DNA sequencing method based on chemical modification of DNA and subsequent cleavage at specific bases. Although Maxam and Gilbert published their chemical sequencing method two years after the ground-breaking paper of Sanger and Coulson on plus-minus sequencing, Maxam–Gilbert sequencing rapidly became more popular, since purified DNA could be used directly, while the initial

Sanger method required that each read start be cloned for production of single-stranded DNA. However, with the improvement of the chain-termination method Maxam-Gilbert sequencing has fallen out of favour due to its technical complexity prohibiting its use in standard molecular biology kits, extensive use of hazardous chemicals, and difficulties with scale-up.

The method requires radioactive labeling at one 5' end of the DNA (typically by a kinase reaction using gamma-P ATP) and purification of the DNA fragment to be sequenced. Chemical treatment generates breaks at a small proportion of one or two of the four nucleotide bases in each of four reactions (G, A+G, C, C+T). For example, the purines (A+G) are depurinated using formic acid, the guanines (and to some extent the adenines) are methylated by dimethyl sulfate, and the pyrimidines (C+T) are methylated using hydrazine. The addition of salt (sodium chloride) to the hydrazine reaction inhibits the methylation of thymine for the C-only reaction.

The modified DNAs are then cleaved by hot piperidine at the position of the modified base. The concentration of the modifying chemicals is controlled to introduce on average one modification per DNA molecule. Thus a series of labeled fragments is generated, from the radiolabeled end to the first "cut" site in each molecule.

The fragments in the four reactions are electrophoresed side by side in denaturing acrylamide gels for size separation. To visualize the fragments, the gel is exposed to X-ray film for autoradiography, yielding a series of dark bands each corresponding to a radiolabeled DNA fragment, from which the sequence may be inferred.

Also sometimes known as "chemical sequencing", this method led to the Methylation Interference Assay used to map DNA-binding sites for DNA-binding proteins.

Chain-termination Methods

Because the chain-terminator method (or Sanger method after its developer Frederick Sanger) is more efficient and uses fewer toxic chemicals and lower amounts of radioactivity than the method of Maxam and Gilbert, it rapidly became the method of choice. The key principle of the Sanger method was the use of dideoxy-nucleotidetriphosphates (ddNTPs) as DNA chain terminators.

The classical chain-termination method requires a single-stranded DNA template, a DNA primer, a DNA polymerase, normal deoxynucleotidetriphosphates (dNTPs), and modified nucleotides (dideoxyNTPs) that terminate DNA strand elongation. These ddNTPs will also be radioactively or fluorescently labelled for detection in automated sequencing machines. The DNA sample is divided into four separate sequencing reactions, containing all four of the standard deoxynucleotides (dATP, dGTP, dCTP and dTTP) and the DNA polymerase.

To each reaction is added only one of the four dideoxynucleotides (ddATP, ddGTP, ddCTP, or ddTTP) which are the chain-terminating nucleotides, lacking a 3'-OH group required for the formation of a phosphodiester bond between two nucleotides, thus terminating DNA strand extension and resulting in DNA fragments of varying length.

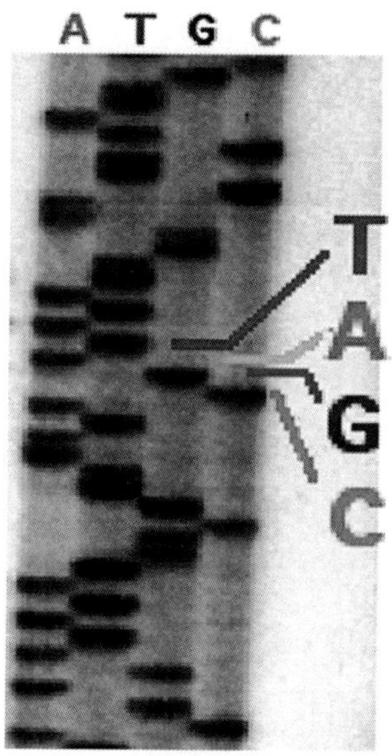

Figure: *Part of a radioactively labelled sequencing gel*

The newly synthesized and labelled DNA fragments are heat denatured, and separated by size (with a resolution of just one nucleotide) by gel electrophoresis on a denaturing polyacrylamide-urea gel with each of the four reactions run in one of four individual lanes (lanes A, T, G, C); the DNA bands are then visualized byautoradiography or UV light, and the DNA sequence can be directly read off the X-ray film or gel image.

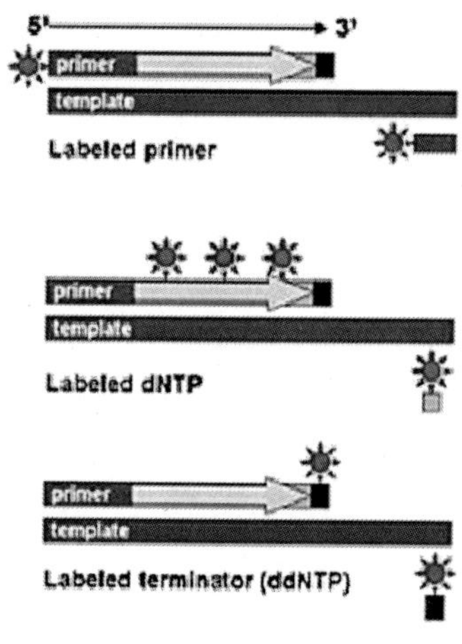

In the image on the right, X-ray film was exposed to the gel, and the dark bands correspond to DNA fragments of different lengths. A dark band in a lane indicates a DNA fragment that is the result of chain termination after incorporation of a dideoxynucleotide (ddATP, ddGTP, ddCTP, or ddTTP). The relative positions of the different bands among the four lanes are then used to read (from bottom to top) the DNA sequence.

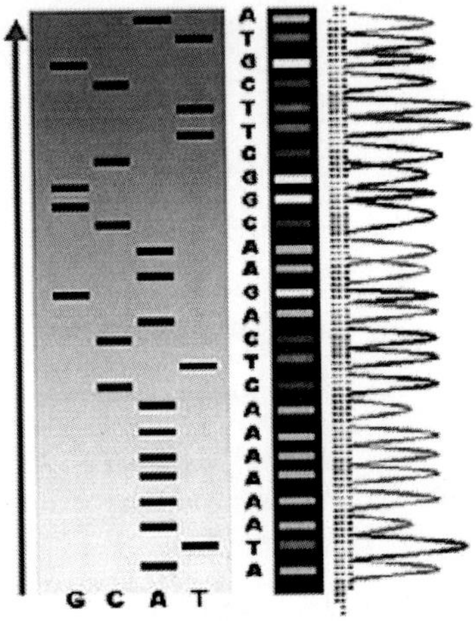

Fig. 5: *Sequence ladder by radioactive sequencing compared to fluorescent peaks (click to expand)*

DNA fragments are labelled with a radioactive or fluorescent tag on the primer, in the new DNA strand with a labeled dNTP, or with a labeled ddNTP. Technical variations of chain-termination sequencing include tagging with nucleotides containing radioactive phosphorus forradiolabelling, or using a primer labeled at the 5' end with a fluorescent dye. Dye-primer sequencing facilitates reading in an optical system for faster and more economical analysis and automation. The later development by Leroy Hood and coworkers of fluorescently labeled ddNTPs and primers set the stage for automated, high-throughput DNA sequencing.

Chain-termination methods have greatly simplified DNA sequencing. For example, chain-termination-based kits are commercially available that contain the reagents needed for sequencing, pre-aliquoted and ready to use. Limitations include non-specific binding of the primer to the DNA, affecting accurate read-out of the DNA sequence, and DNA secondary structures affecting the fidelity of the sequence.

Dye-terminator Sequencing

Dye-terminator sequencing utilizes labelling of the chain terminator ddNTPs, which permits sequencing in a single reaction, rather than four reactions as in the labelled-primer method. In dye-terminator sequencing, each of the four dideoxynucleotide chain terminators is labelled with fluorescent dyes, each of which emit light at differentwavelengths. Owing to its greater expediency and speed, dye-terminator sequencing is now the mainstay in automated sequencing. Its limitations include dye effects due to differences in the incorporation of the dye-labelled chain terminators into the DNA fragment, resulting in unequal peak heights and shapes in the electronic DNA sequence trace chromatogram after capillary electrophoresis.

This problem has been addressed with the use of modified DNA polymerase enzyme systems and dyes that minimize incorporation variability, as well as methods for eliminating "dye blobs". The dye-terminator sequencing method, along with automated high-throughput DNA sequence analyzers, is now being used for the vast majority of sequencing projects.

Challenges

Common challenges of DNA sequencing include poor quality in the first 15–40 bases of the sequence and deteriorating quality of sequencing traces after 700–900 bases. Base calling software typically gives an estimate of quality to aid in quality trimming.

In cases where DNA fragments are cloned before sequencing, the resulting sequence may contain parts of the cloning vector. In contrast, PCR-based cloning and emerging sequencing technologies based on pyrosequencing often avoid using cloning vectors. Recently, one-step Sanger sequencing (combined amplification and sequencing) methods such as Ampliseq and SeqSharp have been developed that allow rapid sequencing of target genes without cloning or prior amplification.

Current methods can directly sequence only relatively short (300–1000 nucleotides long) DNA fragments in a single reaction. The main obstacle to sequencing DNA fragments above this size limit is insufficient power of separation for resolving large DNA fragments that differ in length by only one nucleotide. In all cases the use of a primer with a free 3' end is essential.

Automation and Sample Preparation

Automated DNA-sequencing instruments (DNA sequencers) can sequence up to 384 DNA samples in a single batch (run) in up to 24 runs a day.

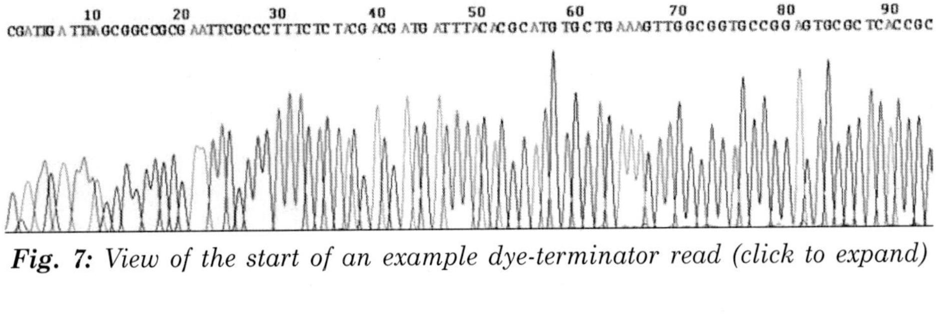

Fig. 7: *View of the start of an example dye-terminator read (click to expand)*

DNA sequencers carry out capillary electrophoresis for size separation, detection and recording of dye fluorescence, and data output as fluorescent peak tracechromatograms. Sequencing reactions by thermocycling, cleanup and re-suspension in a buffer solution before loading onto the sequencer are performed separately. A number of commercial and non-commercial software packages can trim low-quality DNA traces automatically. These programs score the quality of each peak and remove low-quality base peaks (generally located at the ends of the sequence). The accuracy of such algorithms is below visual examination by a human operator, but sufficient for automated processing of large sequence data sets.

High-throughput Sequencing

The high demand for low-cost sequencing has driven the development of high-throughput sequencing technologies that parallelize the sequencing process, producing thousands or millions of sequences at once. High-throughput sequencing technologies are intended to lower the cost of DNA sequencing beyond what is possible with standard dye-terminator methods.

The first of the "next-generation" sequencing technologies, MPSS was developed in the 1990s at Lynx Therapeutics, a company founded in 1992 by Sydney Brenner and Sam Eletr. MPSS was a bead-based method that used a complex approach of adapter ligation followed by adapter decoding, reading the technology was so complex, MPSS was only performed 'in-house' by Lynx Therapeutics and no machines were sold; when the merger with Solexa later led to the development of sequencing-by-synthesis, a more simple approach with numerous advantages, MPSS became obsolete.

However, the essential properties of the MPSS output were typical of later "next-gen" data types, including hundreds of thousands of short DNA sequences. In the case of MPSS, these were typically used for sequencing cDNA for measurements of gene expression levels. Lynx Therapeutics merged with Solexa in 2004, and this company was later purchased by Illumina.

Polony Sequencing

Polony sequencing, developed in the laboratory of George Church at Harvard, was among the first next-generation sequencing systems used to sequence a full genome in 2005. It combined an in vitro paired-tag library with emulsion PCR, an automated microscope, and ligation-based sequencing chemistry to sequence an E. coli genome at an accuracy of > 99.9999% and a cost approximately 1/10 that of Sanger sequencing. The technology was licensed to Agencourt Biosciences, subsequently spun out into Agencourt Personal Genomics, and ultimately incorporated into the Applied Biosystems SOLiD platform.

Pyrosequencing

A parallelized version of pyrosequencing was developed by 454 Life Sciences, which has since been acquired by Roche Diagnostics. The method amplifies DNA inside water droplets in an oil solution (emulsion PCR), with each droplet containing a single DNA template attached to a single primer-coated bead that then forms a clonal colony. The sequencing machine contains many picolitre-volume wells each containing a single bead and sequencing enzymes. Pyrosequencing uses luciferase to generate light for detection of the individual nucleotides added to the nascent DNA, and the combined data are used to generate sequence read-outs. This technology provides intermediate read length and price per base compared to Sanger sequencing on one end and Solexa and SOLiD on the other.

Illumina (Solexa) Sequencing

Solexa, now part of Illumina, developed a sequencing technology based on reversible dye-terminators. DNA molecules are first attached to primers on a slide and amplified so that local clonal colonies are formed (bridge amplification). Four types of ddNTPs are added, and non-incorporated nucleotides are washed away. Unlike pyrosequencing, the DNA can only be extended one nucleotide at a time. A camera takes images of the fluorescently labeled nucleotides, then the dye along with the terminal 3' blocker is chemically removed from the DNA, allowing the next cycle.

SOLiD Sequencing

Applied Biosystems' SOLiD technology employs sequencing by ligation. Here, a pool of all possible oligonucleotides of a fixed length are labeled according to the sequenced position. Oligonucleotides are annealed and ligated; the preferential ligation by DNA ligase for matching sequences results in a signal informative of the nucleotide at that position. Before sequencing, the DNA is amplified by emulsion PCR. The resulting bead, each containing only copies of the same DNA molecule, are deposited on a glass slide. The result is sequences of quantities and lengths comparable to Illumina sequencing.

Ion Semiconductor Sequencing

Ion Torrent Systems Inc. developed a system based on using standard sequencing chemistry, but with a novel, semiconductor based detection system. This method of sequencing is based on the detection of hydrogen ions that are released during the polymerisation of DNA, as opposed to the optical methods used in other sequencing systems. A microwell containing a template DNA strand to be sequenced is flooded with a single type of nucleotide. If the introduced nucleotide is complementary to the leading template nucleotide it is incorporated into the growing complementary strand. This causes the release of a hydrogen ion that triggers a hypersensitive ion sensor, which indicates that a reaction has occurred. If homopolymer repeats are present in the template sequence multiple nucleotides will be incorporated in a single cycle. This leads to a corresponding number of released hydrogens and a proportionally higher electronic signal.

DNA Nanoball Sequencing

DNA nanoball sequencing is a type of high throughput sequencing technology used to determine the entire genomic sequence of an organism. The company Complete Genomics uses this technology to sequence samples that researchers submit from several projects. The method uses rolling circle replication to amplify small fragments of genomic DNA into DNA nanoballs. Unchained sequencing by ligation is then used to determine the nucleotide sequence. This method of DNA sequencing allows large numbers of DNA nanoballs to be sequenced per run and at low reagent costs compared to other next generation sequencing platforms. However, only short sequences of DNA are determined from each DNA nanoball which makes mapping the short reads to a reference genome difficult. This technology has been used for multiple genome sequencing projects and is scheduled to be used for more.

Helioscope(TM) Single Molecule Sequencing

Based on "true single molecule sequencing" technology, Helioscope sequencing uses DNA fragments with added polyA tail adapters, which are attached to the flow cell surface. The next steps involve extension-based sequencing with cyclic washes of the flow cell with fluorescently labeled nucleotides (one nucleotide type at a time, as with the Sanger method). The reads are performed by the Helioscope sequencer. The reads are short, up to 55 bases per run, but recent improvemend of the methodology allowes more accurate reads of homopolymers (stretches of one type of nucleotides) and RNA sequencing.

Single Molecule SMRT(TM) sequencing

SMRT sequencing is based on the sequencing by synthesis approach. The DNA is synthesisd in so calles zero-mode wave-guides (ZMWs) - small well-like containers with the capturing tools located at the bottom of the well. The sequencing is performed with use of unmodified polymerase (attached to the ZMW bottom) and fluorescently labelled nucleotides flowing freely in the solution. The wells are constructed in a way that only the fluorescence occurring by the bottom of the well is detected. The fluorescent label is detached from the nucleotide at its incorporation into the DNA strand, leaving an unmodified DNA strand. According to Pacific Biosciences, the SMTR technology developer, this methodology allows detection of nucleotide modifications (such ad cytosine methylation). This happens through the observation of polymerase kinetics. This approach allows reads of 1000 nucleotides.

Single Molecule Real Time (RNAP) Sequencing

This method is based on RNA polymerase (RNAP), which is attached to a polystyrene bead, with distal end of sequenced DNA is attached to another bead, with both beads being placed in optical traps. RNAP motion during transcription brings the beads in closer and their relative distance changes, which can then be recorded at a single nucleotide resolution. The sequence is deduced based on the four readouts with lowered concentrations of each of the four nucleotide types (similarly to Sangers method).

Nanopore DNA Sequencing

This method is based on the readout of electrical signal occurring at nucleotides passing by alpha-hemolysin pores covalently bound with cyclodextrin. The DNA passing through the nanopore changes its ion current. This change is dependent on the shape, size and length of the DNA sequence. Each type of the nucleotide blocks the ion flow through the pore for a different period of time. The method has a potential of development as it does not require modified nucleotides, however single nucleotide resolution is not yet available.

VisiGen Biotechnologies Approach

VisiGen Biotechnologies introduced a specially engineered DNA polymerase for use in their sequencing. This polymerase acts as a sensor - having incorporated a donor fluorescent dye by its active centre. This donor dye acts by FRET (fluorescent resonant energy transfer), inducing fluorescence of differently labeled nucleotides.

This approach allows reads performed at the spead at which polymerase incorporates nucleotides into the sequence (several hundred per second). The nucleotide fluorochrome is released after the incorporation into the DNA strand. The expected read lengths in this approach should reach 1000 nucleotides, however this will have to be confirmed.

Future Methods

Sequencing by hybridization is a non-enzymatic method that uses a DNA microarray. A single pool of DNA whose sequence is to be determined is fluorescently labeled and hybridized to an array containing known sequences. Strong hybridization signals from a given spot on the array identifies its sequence in the DNA being sequenced. Mass spectrometry may be used to determine mass differences between DNA fragments produced in chain-termination reactions.

DNA sequencing methods currently under development include labeling the DNA polymerase, reading the sequence as a DNA strand transits through nanopores, and microscopy-based techniques, such as AFM or transmission electron microscopy that are used to identify the positions of individual nucleotides within long DNA fragments (>5,000 bp) by nucleotide labeling with heavier elements (e.g., halogens) for visual detection and recording.

Third generation technologies aim to increase throughput and decrease the time to result and cost by eliminating the need for excessive reagents and harnessing the processivity of DNA polymerase.

In microfluidic Sanger sequencing the entire thermocycling amplification of DNA fragments as well as their separation by electrophoresis is done on a single glass wafer (approximately 10 cm in diameter) thus reducing the reagent usage as well as cost. In some instances researchers have shown that they can increase the throughput of conventional sequencing through the use of microchips. Research will still need to be done in order to make this use of technology effective.

In October 2006, the X Prize Foundation established an initiative to promote the development of full genome sequencing technologies, called the Archon X Prize, intending to award $10 million to "the first Team that can build a device and use it to sequence 100 human genomes within 10 days or less, with an accuracy of no more than one error in every 100,000 bases sequenced, with sequences accurately covering at least 98% of the genome, and at a recurring cost of no more than $10,000 (US) per genome."

Each year NHGRI promotes grants for new research and developments in genomics. 2010 grants and 2011 candidates include continuing work in microfluidic, polony and base-heavy sequencing methodologies

Major Landmarks in DNA Sequencing

- 1953 Discovery of the structure of the DNA double helix.
- 1972 Development of recombinant DNA technology, which permits isolation of defined fragments of DNA; prior to this, the only accessible samples for sequencing were from bacteriophage or virus DNA.
- 1977 The first complete DNA genome to be sequenced is that of bacteriophage öX174.
- 1977 Allan Maxam and Walter Gilbert publish "DNA sequencing by chemical degradation". Frederick Sanger, independently, publishes "DNA sequencing with chain-terminating inhibitors".
- 1984 Medical Research Council scientists decipher the complete DNA sequence of the Epstein-Barr virus, 170 kb.
- 1986 Leroy E. Hood's laboratory at the California Institute of Technology and Smith announce the first semi-automated DNA sequencing machine.
- 1987 Applied Biosystems markets first automated sequencing machine, the model ABI 370.

- 1990 The U.S. National Institutes of Health (NIH) begins large-scale sequencing trials on Mycoplasma capricolum, Escherichia coli, Caenorhabditis elegans, and Saccharomyces cerevisiae (at US$0.75/base).
- 1991 Sequencing of human expressed sequence tags begins in Craig Venter's lab, an attempt to capture the coding fraction of the human genome.
- 1995 Craig Venter, Hamilton Smith, and colleagues at The Institute for Genomic Research (TIGR) publish the first complete genome of a free-living organism, the bacterium Haemophilus influenzae. The circular chromosome contains 1,830,137 bases and its publication in the journal Science marks the first use of whole-genome shotgun sequencing, eliminating the need for initial mapping efforts.
- 1996 Pål Nyrén and his student Mostafa Ronaghi at te Royal Institute of Technology in Stockholm publish their method of pyrosequencing
- 1998 Phil Green and Brent Ewing of the University of Washington publish "phred" for sequencer data analysis.
- 2000 Lynx Therapeutics publishes and markets "MPSS" - a parallelized, adapter/ligation-mediated, bead-based sequencing technology, launching "next-generation" sequencing.
- 2001 A draft sequence of the human genome is published.
- 2004 454 Life Sciences markets a parallelized version of pyrosequencing. The first version of their machine reduced sequencing costs 6-fold compared to automated Sanger sequencing, and was the second of a new generation of sequencing technologies, after MPSS.

Genome

In modern molecular biology and genetics, the genome is the entirety of an organism's hereditary information. It is encoded either in DN or, for many types of virus, in RNA. The genome includes both the genes andthe non-coding sequences of the DNA/RNA.

Origin of Term

The term was adapted in 1920 by Hans Winkler, Professor of Botany at the University of Hamburg, Germany. In Greek, the word genome (ãßíïìáé) means "I become, I am born, to come into being". The Oxford English Dictionary suggests the name to be a blend of the words gene and chromosome. A few related -ome words already existed, such asbiome and rhizome, forming a vocabulary into which genome fits systematically.

Overview

Some organisms have multiple copies of chromosomes, diploid, triploid, tetraploid and soon. In classical genetics, in a sexually reproducing organism (typically eukarya) thegamete has half the number of chromosomes of the somatic cell and the genome is a full set of chromosmes in a gamete. In haploid organisms, including cells of bacteria,archaea, and in organelles including mitochondria and chloroplasts, or viruses, that similarly contin genes, the single or set of circular and/or linear chains of NA (or RNA forsome viruses), likewise constitute the genome.

The term genome can be applied specifically to mean that stored on a complete set of nuclear DNA (i.e., the "nuclear genome") but can also be applied to that stored within organelles that contain their own DNA, as with the "mitochondrial genome" or the "chloroplast genome". Additionally, the genome can comprise nonchromosomal genetic elements such as viruses, plasmids, and transposable elements. When people say that the genome of a sexually reproducing species has been "sequenced", typically they are referring to a determination of the sequences of one set ofautosomes and one of each type of sex chromosome, which together represent both of the possible sexes.

Even in species that exist in only one sex, what is described as "a genome sequence" may be a composite read from the chromosomes of various individuals. In general use, thephrase "genetic makeup" is sometimes used conversationally to mean the genome of a particular individual or organism. The study of the global properties of genomes of related organism is usually referred to asgenomics, which distinguishes it from genetics which generally studies the properties of single genes or groups of genes.

Both the number of base pairs and the number of genes vary widely from one species to another, and there is only a rough correlation between the two (an observation known as the C-value paradox). At present, the highest known number of genes is around 60,000, for the protozoan causingtrichomoniasis.

Types

Most biological entities that are more complex than a virus sometimes or always carry additional genetic material besides that which resides in their chromosomes. In some contexts, such as sequencing the genome of a pathogenic microbe, "genome" is meant to include information stored on this auxiliary material, which is carried in plasmids. In such circumstances then, "genome" describes ll of the genes and information on non-coding DNA that have the potential to be present. In eukaryotes such as plants, protozoa and animals, however, "genome" carries the typical connotation of only information on chromosomal DNA. So although these organisms contain chloroplastsand/or mitochondria that have their own DNA, the genetic information contained by DNA within these organelles is not considered part of the genome. In fact, mitochondria are sometimes said to have their own genome often referred to as the "mitochondrial genome". The DNA found within the chloroplast may be referred to as the "plastome".

Bibliography

Ayala, F. J.: *The Genetic Structure of Populations*, W.H. Freeman & Co., San Francisco, California, 1977.

Beckner, M.: *The Biological Way of Thought*, University of California Press, Berkeley and Los Angeles, 1968.

Berg, Paul, and Singer, Maxine: *Dealing with Genes: The Language of Heredity. Mill Valley*, University Science Books, CA, 1992.

Brookes, Martin: *Get a Grip on Genetics, Time Life Books*, East Sussex, England, 1998.

Dando, M.: *Biological Warfare in the 21st Century: Biotechnology and the Proliferation of Biological Weapons*. London: Brassey's, 1994.

Darlington, C. D., and Mather, K.: *The Elements of Genetics*, London, Allen and Unwin, 1949.

Dawkins, R.: *The Selfish Gene*. Oxford: Oxford University Press, 1976.

DeVita, V.T.: *Cancer: Principles and Practice of Oncology*, Lippincott, Philadelphia, 2005.

Diaz, E.: *Microbial Biodegradation: Genomics and Molecular Biology,* Caister Academic Press, UK, 2008.

Fisher, R. A.: *The Genetical Theory of Natural Selection*, Clarendon Press, Oxford, 1930.

Fritz, A.: *International Classification of Diseases for Oncology*, World Health Organization, Geneva, 2000.

Gillis, Justin: *Drug Firms, Gene Labs to Map Genetic Code*, The Daily News, Longview, WA, 1999.

Glover, D. M., and Hames, B. D.: *Genes and Embryos*: New York, Oxford University Press, 1989.

Griffiths A.J.F.: *Introduction to Genetic Analysis*, W.H. Freeman and Company, New York, USA, 2005.

Hamerton, J. L.: *Human Cytogenetics*, Academic Press, New York, 1971.

Hartl D, Jones E: *Genetics: Analysis of Genes and Genomes,* Jones & Bartlett, 2005.

Hartl, Daniel L.: *Basic Genetics*, Jones and Bartlett Publishers, Boston, 1991.

Jonoska, N.: *Self-Assembling DNA Graphs, DNA-Based Computers VIII*, Springer-Verlag, Berlin, 2003.

Klotz, J. W.: *Genes, Genesis, and Evolution*: St. Louis, Mo., Concordia Publishing Co., 1970.

Lumsden C., and E. Wilson: *Genes, Mind and Culture: The Coevolutionary Process.* Cambridge, MA: Harvard University Press, 1981.

Mayr, E.: *Change of Genetic Environment and Evolution*, Allen and Unwin, London, 1954.

Nottingham, S.F.: *Eat Your Genes: How Genetically Modified Food is Entering Our Diet*, Zed Books, London, 1998.

Prakash, Chandra Prakash : *Cell Physiology*, Discovery, Delhi, 2010.

Robin Whyman: *Applied Organometallic Chemistry and Catalysis*, Oxford University Press, Delhi, 2005.

Roobeek, A: *The Biotechnology Revolution?,* Blackwell, Oxford, 1995.

Sturtevant, Alfred: *History of Genetics*, Harper and Row, New York, 1965.

Suzuki, D., and Knudtson, P.: *Genethics: The Clash Between the New Genetics and Human Values*, Harvard University Press, Cambridge, Mass, 1989.

Wilson, E. O., and Lumden, C.: *Genes, Mind, and Culture: The Evolutionary Process*, Harvard University Press, Cambridge, Mass, 1981.

Wilson, E. O., and Lumden, C.: *Genes, Mind, and Culture: The Evolutionary Process*: Cambridge, Mass., Harvard University Press, 1981.

Index

H

L

M

N

P

R

S

T

V

❑❑❑

Encyclopaedia of

Advanced Genetic Analysis

Encyclopaedia of

Advanced Genetic Analysis

Volume 2

Editor:
Veena Kumari

AGROTECH PRESS

JAIPUR ▪ NEW DELHI

AGROTECH PRESS

Regd. Off. "Bony Residency", Gate No. 2
Opp. Tilak Public School, Vishweswariya Nagar,
Gopalpura Road, Jaipur - 302018, India
Ph: 0141 - 2761280, Telefax: 2761381
E-mail : agrotechpress@gmail.com

Delhi Office: 102, 1st Floor, Satyam House,
4327/3, Ansari Road, Darya Ganj,
New Delhi - 110002, India
Ph: 011-45652440
visit us: www.agrotechpress.com

Distributed By
KEYA INTERNATIONAL PRESS LLP
E-mail: keyainternationalpressllp@gmail.com

Published in India, 2024

ISBN : 9789384568207 (5 Vols. Set)

Typeset by Thomson Press

Printed and bound in India by Thomson Press Ltd.,
New Delhi

Contents

Preface

Linkage analysis allows geneticists to determine which genes are on the same chromosome. Geneticist would now like to extend that analysis to the developement of detailed genetic and molecular maps of different species. These maps would describe the order of genes and molecular markers on each chromosome of that species. Collectively, the maps of all of the chromosomes would describe the genetic organization of that species. The long term goal is to apply this information. For example, once a human gene is mapped to a specific chromosome, the map information can be used to locate molecular markers linked to that gene.

The molecular marker can then be used as a tool to begin efforts to clone the gene itself. Once the gene is cloned and characterized that information might give clues to possible clinical treatments of the disease. These maps can be developed by several methods. These include in situ hybridization, somatic cell hybrids, meiotic recombination maps, and fine-structure mapping. We will look at each of these different types of techniques. He technique of in situ hybridization is used to locate the chromosomal location of a specific DNA probe. The theory is the same as for Southern hybridization, except that the DNA to which the probe will hybridize is the actual chromosome. The probe is labelled with a flourescent probe. It is then added to a chromosomal preparation from the species of interest. After a sufficient time for annealing to occur, the chromosomes are viewed using a floruescent microscope. The probe will hybridize to the chromosome carrying the sequence of interest. If the species has been characterized cytogenetically, the marker can be assigned to the appropriate chromosome. Because this technique uses a flourescent probe it is called flourescence in situ hybridization or FISH. It has also be called chromosome painting. Somatic cell hybrids are culture lines that contain the entire complement of the mouse genome and a few human chromosomes. These culture lines are developed by mixing human and mouse cells in the presence of the Sendai virus. The virus facilitates the fusing of the two cell types to form a hybrid cell. For a reason that is not entirely known, most, but not all, human chromosomes are lost from the hybrid cell lines. Usually a few human chromosomes are retained. Because the human and mouse chromosomes can be distinguished by chromosome staining techniques, it can be determined which human cells are retained with a specific cell line. Typically, the mouse cell line is mutant for a specific function. Because the hybrids are selected on cells in which the mutant cells could not grow, any surviving hybrid lines would contain the human chromosome that complements the mouse mutation and several other human chromosomes. The linkage relationship between two or three genes or markers can be determined by analysing the number of recombination events that occur in a specific mating of individuals polymorphic for the genes in question. In general, the term for this procedure is called meitotic recombination mapping. The same scientific principles apply when developing a detailed map of any chromosome. The only difference is that the segregating population is not scored for just two or three genes but for a large number of genes. Molecular maps can be used to identify a marker for a specific gene. These markers are quite useful For a specific gene that is difficult to score or is expressed late in the life cycle. Maps can also be used as a starting point for cloning a gene. A fine structure map of the species is quite useful for this purpose. Yeast artifical chromosome clones and recently, bacterial artificial chromosome clones are key tools for developing a fine structure map. If you have identified markers that is flank your gene and identified a clone to which the markers reside, you are on your way to determining where that gene resides. What you are going to do is a procedure termed chromosomal walking. This involves identifying the clone to which your markers reside and mapping the STS sites on that clone in your populaton. If the linkage distance does not improve, you next go to STS sites located on adjacent clones. Your repeat this process until you have a markers associated with a clone that co-segregate with your gene.

The book will serve as a text book to the students and would also be a useful reference book to the postgraduate students and research scientists working in the areas of life sciences, genomics, biotechnology and molecular biology as well as masters in computer applications who are interested in bioinformatics.

—Editor

1

Genomes and Genetic Variation

A genome does not capture the genetic diversity or the genetic polymorphism of a species. For example, the human genome sequence in principle could be determined from just half the information on the DNA of one cell from one individual. To learn what variations in genetic information underlie particular traits or diseases requires comparisons across individuals.

This point explains the common usage of "genome" (which parallels a common usage of "gene") to refer not to the information in any particular DNA sequence, but to a whole family of sequences that share a biological context.

Although this concept may seem counter intuitive, it is the same concept that says there is no particular shape that is the shape of a cheetah. Cheetahs vary, and so do the sequences of their genomes. Yet both the individual animals and their sequences share commonalities, so one can learn something about cheetahs and "cheetah-ness" from a single example of either.

Sequencing and Mapping

The Human Genome Project was organized to map and to sequence the human genome. Other genome projects include mouse, rice, the plant Arabidopsis thaliana, the puffer fish, and bacteria like E. coli. In 1976, Walter Fiers at the University of Ghent (Belgium) was the first to establish the complete nucleotide sequence of a viral RNA-genome (bacteriophage MS2).

The first DNA-genome project to be completed was the Phage Ö-X174, with only 5386 base pairs, which was sequenced by Fred Sanger in 1977. The first bacterial genome to be completed was that of Haemophilus influenzae, completed by a team at The Institute for Genomic Research in 1995. A few months later, the first eukaryotic genome was completed, with the 16 chromosomes of budding yeast Saccharomyces cerevisiae being released as a result of a European-led effort begun in the mid-1980s.

The development of new technologies has made it dramatically easier and cheaper to do sequencing, and the numer of complete genome sequences is growing rapidly. Among many genome databases, the one maintained by the US National Institutes of Health is inclusive.

These new technologies open up the prospect of personal genome sequencing as an important diagnostic tool. A major step toward that goal was the completion of the decipherment of the full genomeof DNA pioneer James D. Watson in 2007.

Whereas a genome sequence lists the order of every DNA base in a genome, a genome map identifies the landmarks. A genome map is less detailed than a genome sequence and aids in navigating around the genome. A fundamental step in the Human genome project was the release of a detailed genomic map by Jean Weissenbach and his team at the Genoscope in Paris.

Genome Evolution

Genomes are more than the sum of an organism's genes and have traits that may be measured and studied without reference to the details of any particular genes and their products. Researchers compare traits such as chromosome number (karyotype), genome size, gene order, codon usage bias, and GC-content to determine what mechanisms could have produced the great variety of genomes that exist today.

Duplications play a major role in shaping the genome. uplications may range from extension of short tandem repeats, to duplication of a cluster of genes, and all the way to duplications of entire chromosomes or even entire genomes. Such duplications are probably fundamental to the creation of genetic novelty.

Horizontal gene transfer is invoked to explain how there is often extreme similarity between small portions of the genomes o two organisms that are otherwise very distantly related. Horizontal gene transfer seems to be common among many microbes. Also, eukaryotic cells seem to have experienced a transfer of some genetic material from their chloroplast and mitochondrial genomes to their nuclear chromosomes.

Heredity

Heredity is the passing of traits to offspring (from its parent or ancestors). This is the process by which an offspring cell or organism acquires or becomes predisposed to the characteristics of its parent cell or oganism. Through heredity, variations exhibited by individuals can accumulate and cause some species to evolve. The study of heredity inbiology is called genetics, which includes the field of epigenetics.

Overview

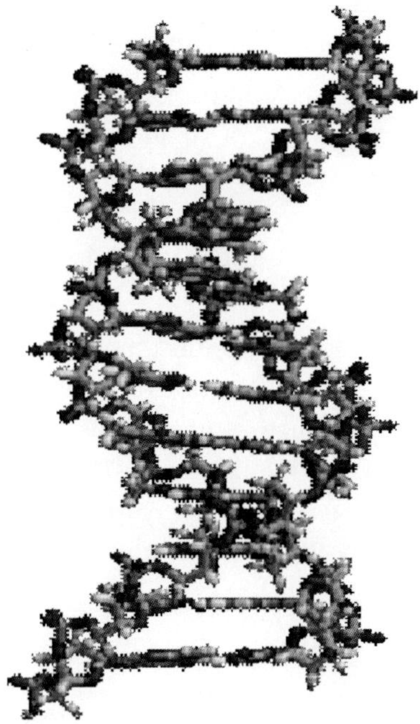

***FIgure:** DNA structure. Bases are in the centre, surrounded by phosphate–sugar chains in adouble helix.*

In humans, eye color is an inherited characteristic and an individual might inherit the "brown-eye trait" from one of the parents. Inherited traits are controlled by genes and the complete set of genes within an organism's genome is called its genotype.

The complete set of observable traits that make up the structure and behaviour of an organism is called its phenotype. These traits come from the interaction of its genotype with the environment. As a result, many aspects of an organism's phenotype are not inherited. For example,suntanned skin comes from the interaction between a person's genotype and sunlight; thus, suntans are not passed on to people's children. However, some people tan more easily than others, due to differences in their genotype; a striking example are people with the inherited trait ofalbinism, who do not tan at all and are very sensitive to sunburn.

Heritable traits are known to be passed from one generation to the next via DNA, a molecule that encodes genetic information. DNA is a longpolymer composed of four types of bases. The sequence of bases along a particular DNA molecule specify the genetic information, in a manner similar to a sequence of letters spelling out a word. Before a cell divides, the DNA is copied, so that each of the resulting two cells will inherit the DNA sequence.

Portions of a DNA molecule that specify a single functional unit are called genes; different genes have different sequences of bases. Within cells, the long strands of DNA form condensed structures called chromosomes.

The specific location of a DNA sequence within a chromosome is known as a locus. If the DNA sequence at a locus varies between individuals, the different forms of this sequence are called alleles. DNA sequences can change thrugh mutations, producing new alleles. If a mutation occurs witin a gene, the new allele may affect the trait that the gene controls, altering the phenotype of the organism.

However, while this simple correspondence between an allele and a trait works in some cases, most traits are more complex and are controlled by multiple interacting genes within and among organisms. Developmental biologists suggest that complex interactions in genetic networks and communication among cells can lead to heritable variations that may underlay some of the mechanics in developmental plasticity and canalization.

Recent findings have confirmed important examples of heritable changes that cannot be explained by direct agency of the DNA molecule. These phenomena are classed as epigenetic inheritance systems that are causally or independently evolving over genes. Research into modes and mechanisms of epigenetic inheritance is still in its scientific infancy, however, this area of research has attracted much recent activity as it broadens the scope of heritability and evolutionary biology in general.

DNA methylation marking chromatin, self-sustaining metabolic loops, gene silencing by RNA interference, and the three dimensional conformation of proteins (such as prions) are areas where epigenetic inheritance systems have been discovered at the organismic level.

Heritability may also occur at even larger scales. For example, ecological inheritance through the process of niche construction is defined by the regular and repeated activities of organisms in their environment. This generates a legacy of effect that modifies and feeds back into the selection regime of subsequent generations. Descendants inherit genes plus environmental characteristics generated by the ecological actions of ancestors.

Other examples of heritability in evolution that are not under the direct control of genes include the inheritance of cultural traits, group heritability, and symbiogenesis. These examples of heritability that operate above the gene are covered broadly under the title of multilevel or hierarchical selection, which has been a subject of intense debate in the history of evolutionary science.

Relation to Theory of Evolution

When Charles Darwin proposed his theory of evolution in 1859, one of its major problems was the lack of an underlying mechanism for heredity. Darwin believed in a mix of blending inheritance and the inheritance of acquired traits (pangenesis). Blending inheritance would lead to uniformity across populations in only a few generations and thus would remove variation from a population on which natural selection could act.

This led to Darwin adopting some Lamarckian ideas in later editions of On the Origin of Species and his later biological works. Darwin's primary approach to heredity was to outline how it appeared to work (noticing that traits could be inherited which were not expressed explicitly in the parent at the time of reproduction, that certain traits could be sex-linked, etc.) rather than suggesting mechanisms.

Darwin's initial model of heredity was adopted by, and then heavily modified by, his cousin Francis Galton, who laid the framework for the biometric school of heredity. Galton rejected the aspects of Darwin's pangenesis model which relied on acquired traits.

The inheritance of acquired traits was shown to have little basis in the 1880s when August Weismann cut the tails off many generations of mice and found that their offspring continued to develop tails.

History

The ancients had a variety of ideas about heredity: Theophrastus proposed that male flowers caused female flowers to ripen; Hippocrates speculated that "seeds" were produced by various body parts and transmitted to offspring at the time of conception, and Aristotle thought that male and female semen mixed at conception. Aeschylus, in 458 BC, proposed the male as the parent, with the female as a "nurse for the young life sown within her." Various hereditary mechanisms were envisaged without being properly tested or quantified. These included blending inheritance and the inheritance of acquired traits. Nevertheless, people were able to develop domestic breeds of animals as well as crops through artificial selection. The inheritance of acquired traits also formed a part of early Lamarckian ideas on evolution.

During the 18th century, Dutch microscopist Antonie van Leeuwenhoek (1632–1723) discovered "animalcules" in the sperm of humans and other animals. Some scientists speculated they saw a "little man" (homunculus) inside each sperm. These scientists formed a school of thought known as the "spermists." They contended the only contributions of the female to the next generation were the womb in which the homunculus grew, and prenatal influences of the womb. An opposing school of thought, the ovists, believed that the future human was in the egg, and that sperm merely stimulated the growth of the egg. Ovists thought women carried eggs containing boy and girl children, and that the gender of the offspring was determined well before conception.

Gregor Mendel: Father of Genetics

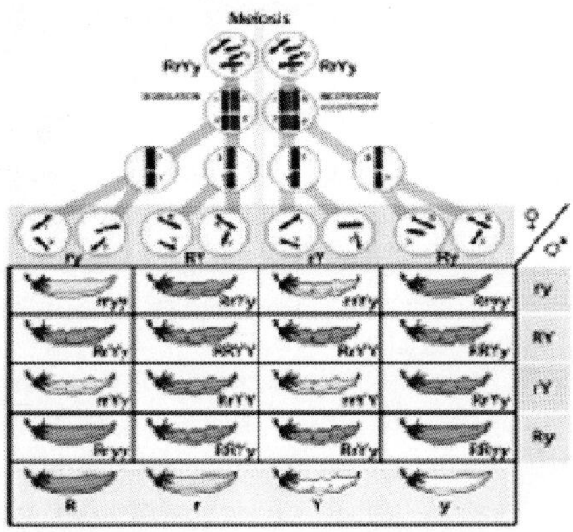

FIgure: *Table showing how the genes exchange according to segregation or independent assortment during meiosis and how this translates into Mendel's laws*

The idea of particulate inheritance of genes can be attributed to the Moravian monk Gregor Mendel who published his work on pea plants in 1865. However, his work was not widely known and was rediscovered in 1901. It was initially assumed the Mendelian inheritance only accounted for large (qualitative) differences, such as those seen by Mendel in his pea plants—and the idea of additive effect of (quantitative) genes was not realised until R.A. Fisher's (1918) paper, "The Correlation Between Relatives on the Supposition of Mendelian Inheritance" Mendel's overall contribution gave scientists an useful overview that traits were inheritable. As of today, his pea plant demonstration became the foundation of the study of Mendelian Traits. These traits can be traced on a single loci.

Modern Development of Genetics and Heredity

In the 1930s, work by Fisher and others resulted in a combination of Mendelian and biometric schools into the modern evolutionary synthesis. The modern synthesis bridged the gap between experimental geneticists and naturalists; and between both and palaeontologists, stating that:

1. All evolutionary phenomena can be explained in a way consistent with known genetic mechanisms and the observational evidence of naturalists.

2. Evolution is gradual: small genetic changes, recombination ordered by natural selection. Discontinuities amongst species (or other taxa) are explained as originating gradually through geographical separation and extinction (not saltation).
3. Selection is overwhelmingly the main mechanism of change; even slight advantages are important when continued. The object of selection is the phenotype in its surrounding environment. The role of genetic drift is equivocal; though strongly supported initially by Dobzhansky, it was downgraded later as results from ecological genetics were obtained.
4. The primacy of population thinking: the genetic diversity carried in natural populations is a key factor in evolution. The strength of natural selection in the wild was greater than expected; the effect of ecological factors such as niche occupation and the significance of barriers to gene flow are all important.
5. In palaeontology, the ability to explain historical observations by extrapolation from micro to macro-evolution is proposed. Historical contingency means explanations at different levels may exist. Gradualism does not mean constant rate of change.

The idea that speciation occurs after populations are reproductively isolated has been much debated. In plants, polyploidy must be included in any view of speciation. Formulations such as 'evolution consists primarily of changes in the frequencies of alleles between one generation and another' were proposed rather later.

The traditional view is that developmental biology ('evo-devo') played little part in the synthesis, but an account of Gavin de Beer's work by Stephen Jay Gould suggests he may be an exception.

Almost all aspects of the synthesis have been challenged at times, with varying degrees of success. There is no doubt, however, that the synthesis was a great landmark in evolutionary biology. It cleared up many confusions, and was directly responsible for stimulating a great deal of research in the post-World War II era. Trofim Lysenko however caused a backlash of what is now called Lysenkoism in the Soviet Union when he emphasised Lamarckian ideas on the inheritance of acquired traits. This movement affected agricultural research and led to food shortages in the 1960s and seriously affected the USSR.

Common Genetic Disorders

- Down syndrome
- Huntington's disease
- Phenylketonuria (PKU)

- FOXP2
- Hemophilia

Types of Heredity

Dominant and Recessive

An allele is said to be dominant if it is always expressed in the appearance of an organism (phenotype). For example, in peas the allele for green pods, G, is dominant to that for yellow pods, g. Since the allele for green pods is dominant, pea plants with the pair of alleles GG (homozygote) or Gg (heterozygote) will have green pods. The allele for yellow pods is recessive. The effects of this allele are only seen when it is present in both chromosomes, gg (homozygote).

The description of a mode of biological inheritance consists of three main categories:

1. Number of Involved loci

- Monogenetic (also called "simple") – one locus
- Oligogenetic – few loci
- Polygenetic – mny loci

2. Involved Chromosomes

- Autosomal – loci are not situated on a sex chromosome
- Gonosomal – loci are situated on a sex chromosome
- X-chromosomal – loci are situated on the X-chromosome (the more common case)
- Y-chromosomal – loci are situated on the Y-chromosome
- Mitochondrial – loci are situated on the mitochondrial DNA

3. Correlation Genotype–phenotype

- Dominant
- Intermediate (also called "codominant")
- Recessive

These three categories are part of every exact description of a mode of inheritance in the above order. Additionally, more specifications may be added as follows:

4. Coincidental and Environmental Interactions

- Penetrance
- Complete

- Incomplete (percentual number)
- Expressivity
- Invariable
- Variable
- Heritability (in polygenetic and sometimes also in oligogenetic modes of inheritance)
- Maternal or paternal imprinting phenomena.

5. Sex-linked Interactions

- Sex-linked inheritance (gonosomal loci)
- Sex-limited phenotype expression (e.g., cryptorchism)
- Inheritance through the maternal line (in case of mitochondrial DNA loci)
- Inheritance through the paternal line (in case of Y-chromosomal loci)

6. Locus–locus Interactions

- Epistasis with other loci (e.g., overdominance)
- Gene coupling with other loci
- Homozygotous lethal factors
- Semi-lethal factors

Determination and description of a mode of inheritance is primarily achieved through statistical analysis of pedigree data. In case the involved loci are known, methods of molecular genetics can also be employed.

Mutation

In molecular biology and genetics, mutations are changes in a genomic sequence: the DNA sequence of a cell's genome or the DNA or RNAsequence of a virus.

They can be defined as sudden and spontaneous changes in the cell. Mutations are caused by radiation, viruses,transposons and mutagenic chemicals, as well as errors that occur during meiosis or DNA replication. They can also be induced by the organism itself, by cellular processes such as hypermutation.

A mutation is passed to the offspring stably, unless it is a dynamic mutation.

Mutation can result in several different types of change in sequences;(DNA) these can either have no effect, alter the product of a gene, or prevent the gene from functioning properly or completely. Studies in the fly Drosophila melanogaster suggest that if a mutation changes a proteinproduced by a gene, this will probably be harmful, with about 70 percent of these mutations having damaging effects, and the remainder being either neutral or weakly beneficial. Due to the damaging effects that mutations can have on genes, organisms have mechanisms such as DNA repair to remove mutations.

Viruses that use RNA as their genetic material have rapid mutation rates, which can be an advantage since these viruses will evolve constantly and rapidly, and thus evade the defensive responses of e.g. the human immune system.

Description

Mutations can involve large sections of DNA becoming duplicated, usually through genetic recombination. These duplications are a major source of raw material for evolving new genes, with tens to hundreds of genes duplicated in animal genomes every million years. Most genes belong to larger families of genes of shared ancestry.

Novel genes are produced by several methods, commonly through the duplication and mutation of an ancestral gene, o by recombining parts of different genes to form new combiations with new functions.

Here, domains act as moduls, each with a particular and independent function, that an be mixed together to produce genes encoding new proteins with novel propertie.

For example, the human eye uses four genes to make structures that sense light: three for color vision and one for night vision; all four arose from a single ancestral gene.

Another advantage of duplicating a gene (or even anentire genome) is that this increases redundancy; this allows one gene in the pair to acquire a new function while the other copy performs the original function. Other types of mutation occasionally create new genes from previously noncoding DNA.

Changes in chromosome number may involve even larger mutations, where segments of the DNA within chromosomes break and then rearrange. For example, in the Homininae, two chromosomes fused to produce human chromosome 2; this fusion did not occur in the lineage of the other apes, and they retain these separate chromosomes.

In evolution, the most important role of such chromosomal rearrangements may be to accelerate the divergence of a population into new species by making populations less likely to interbreed, and thereby preserving genetic differences between these populations.

Sequences of DNA that can move about the genome, such as transposons, make up a major fraction of the genetic material of plants and animals, and may have been important in the evolution of genomes.

For example, more than a million copies of the Alu sequence are present in the human genome, and these sequences have now been recruited to perform functions such as regulatinggene expression. Another effect of these mobile DNA sequences is that when they move within a genome, they can mutate or delete existing genes and thereby produce genetic diversity.

Nonlethal mutations accumulate within the gene pool and increase the amount of genetic variation. The abundance of some genetic changes within the gene pool can be reduced by natural selection, while other "more favorable" mutations may accumulate and result in adaptive changes.

For example, a butterfly may produce offspring with new mutations. The majority of these mutations will have no effect; but one might change the color of one of the butterfly's offspring, making it harder (or easier) for predators to see. If this color change is advantageous, the chance of this butterfly surviving and producing its own offspring are a little better, and over time the number of butterflies with this mutation may form a larger percentage of the population.

Neutral mutations are defined as mutations whose effects do not influence the fitness of an individual. These can accumulate over time due to genetic drift. It is believed that the overwhelming majority of mutations have no significant effect on an organism's fitness.

Also, DNA repair mechanisms are able to mend most changes before they become permanent mutations, and many organisms have mechanisms for eliminating otherwise permanently mutated somatic cells. Mutation is generally accepted by biologists as the mechanism by which natural selection acts, generating advantageous new traits that survive and multiply in offspring as well as disadvantageous traits, in less fit offspring, that tend to die out.

Causes

Two classes of mutations are spontaneous mutations (molecular decay) and induced mutations caused by mutagens.

Spontaneous Mutation

Spontaneous mutations on the molecular level can be caused by:

- Tautomerism – A base is changed by the repositioning of a hydrogen atom, altering the hydrogen bonding pattern of that base resulting in incorrect base pairing during replication.

- Depurination – Loss of a purine base (A or G) to form an apurinic site (AP site).
- Deamination – Hydrolysis changes a normal base to an atypical base containing a keto group in place of the original amine group.

  Examples include C '! U and A '! HX (hypoxanthine), which can be corrected by DNA repair mechanisms; and 5MeC (5-methylcytosine) '! T, which is less lkely to be detected as a mutation because thymine is a normal DNA base.
- Slipped strnd mispairing - Denaturation of the new strand from the template during rplication, followed by renaturationin a different spot ("slipping"). This can lead to insertions or deletions.

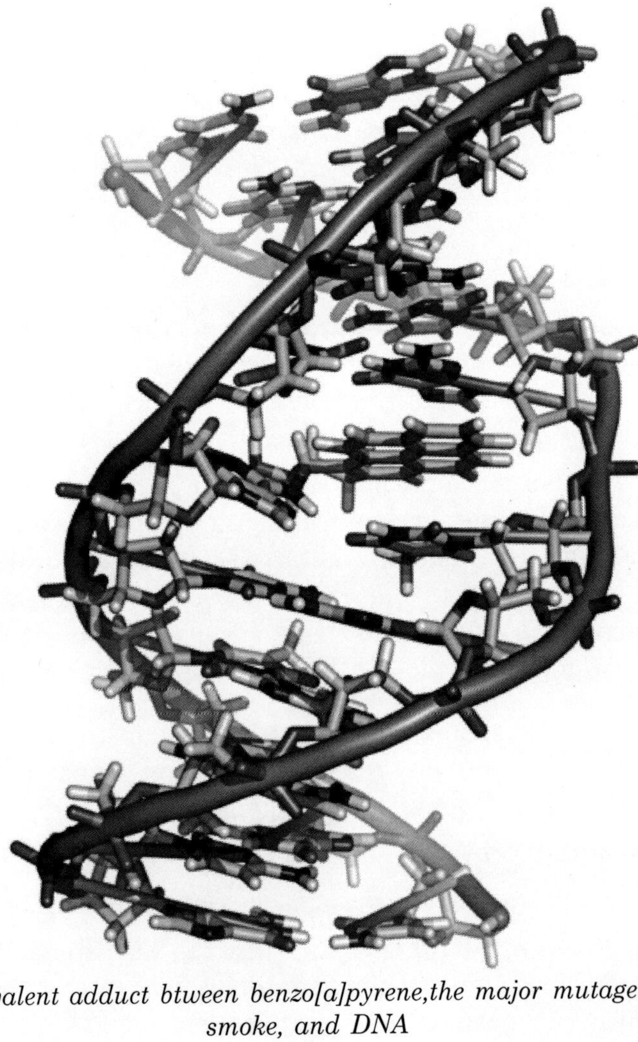

FIgure: *A covalent adduct btween benzo[a]pyrene,the major mutagen in tobacco smoke, and DNA*

Induced Mutation

Induced mutations on the molecular level can be caused by:

- Chemicals
  - Hydroxylamine NH_2OH
  - Base analogs (e.g. BrdU)
  - Alkylating agents (e.g. N-ethyl-N-nitrosourea) These agents can mutate both replicating and non-replicating DNA. In contrast, a base analog can only mutate the DNA when the analog is incorporated in replicating the DNA. Each of these classes of chemical mutagens has certain effects that then lead to transitions, transversions, or deletions.
  - Agents that form DNA adducts (e.g. ochratoxin A metabolites)
  - DNA intercalating agents (e.g. ethidium bromide)
  - DNA crosslinkers
  - Oxidative damage
  - Nitrous acid converts amine groups on A and C to diazo groups, altering their hydrogen bonding patterns which leads to incorrect base pairing during replication.
- Radiation
  - Ultraviolet radiation (nonionizing radiation). Two nucleotide bases in DNA – cytosine and thymine – are most vulnerable to radiation that can change their properties.

    UV light can induce adjacent pyrimidine bases in a DNA strand to become covalently joined as a pyrimidine dimer.

    UV radiation, particularly longer-wave UVA, can also cause oxidative damage to DNA.
  - Ionizing radiation
  - Radioactive decay, such as C in DNA
- Viral infections

DNA has so-called hotspots, where mutations occur up to 100 times more frequently than the normal mutation rate. A hotspot can be at an unusual base, e.g., 5-methylcytosine.

Mutation rates also vary across species. Evolutionary biologists have theorized that higher mutation rates are beneficial in some situations, because they allow organisms to evolve and therefore adapt more quickly to their environments.

For example, repeated exposure of bacteria to antibiotics, and selection of resistant mutants, can result in the selection of bacteria that have a much higher mutation rate than the original population (mutator strains).

Classification of Mutation Types

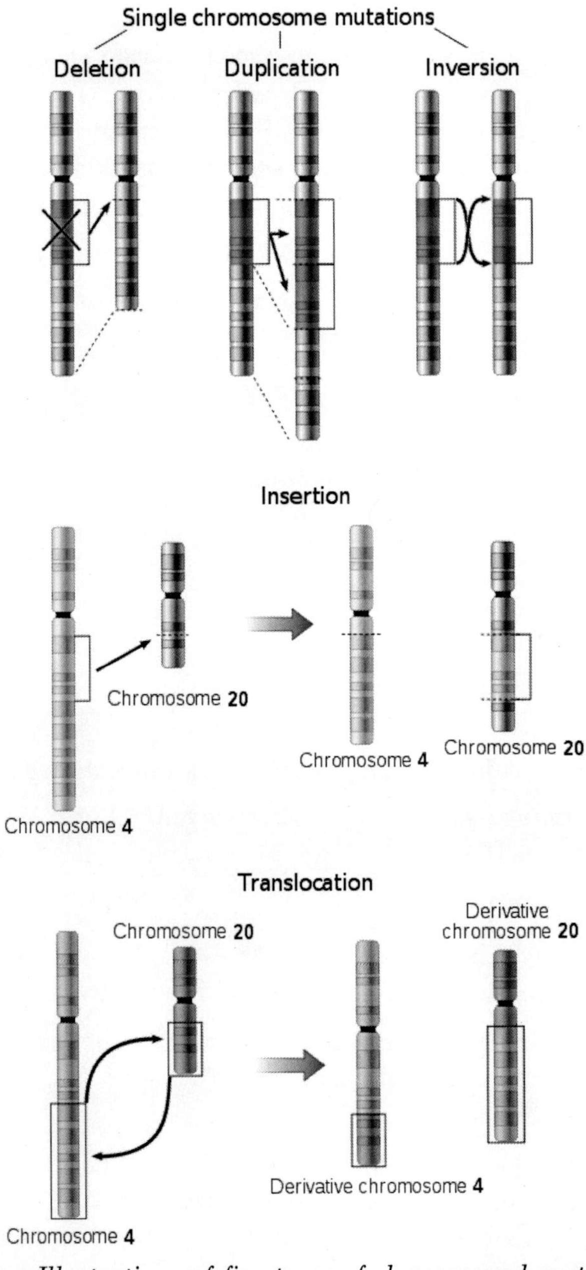

***FIgure:** Illustrations of five types of chromosomal mutations.*

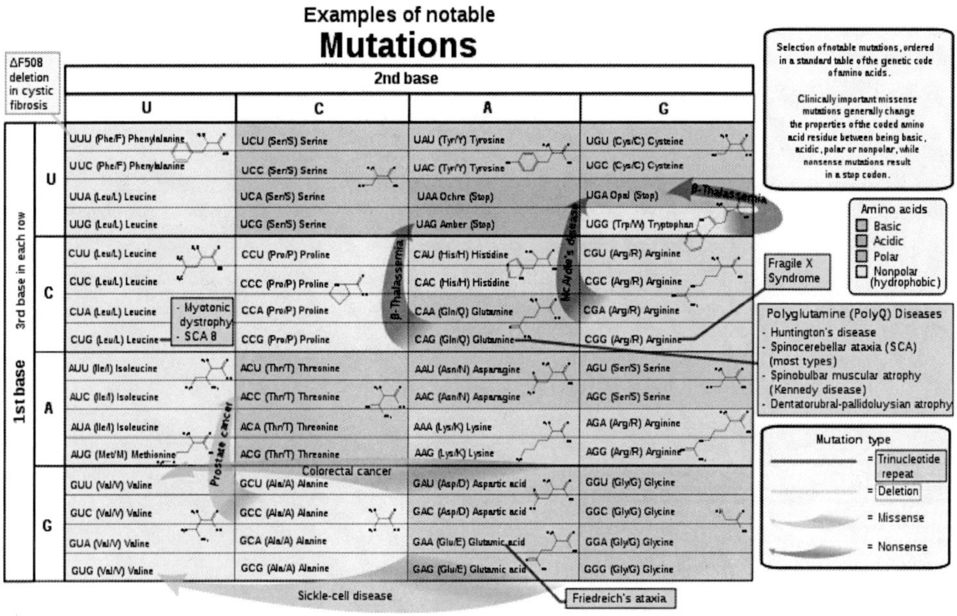

FIgure:*Selection of disease-causing mutations, in a standard table of the genetic code of amino acids.*

By Effect on Structure

The sequence of a gene can be altered in a number of ways. Gene mutations have varying effects on health depending on where they occur and whether they alter the function of essential proteins. Mutations in the structure of genes can be classified as:

- Small-scale mutations, such as those affecting a small gene in one or a few nucleotides, including:
- Point mutations, often caused by chemicals or malfunction of DNA replication, exchange a single nucleotide for another. These changes are classified as transitions or transversions. Most common is the transition that exchanges a purine for a purine (A "! G) or a pyrimidine for a pyrimidine, (C "! T).

  A transition can be caused by nitrous acid, base mis-pairing, or mutagenic base analogs such as5-bromo-2-deoxyuridine (BrdU). Less common is a transversion, which exchanges a purine for a pyrimidine or a pyrimidine for a purine (C/T "! A/G).

  An example of a transversion is adenine (A) being converted into a cytosine (C). A point mutation can be reversed by another point mutation, in which the nucleotide is changed back to its original state (true reversion) or by second-site reversion (a complementary mutation elsewhere that results in regained gene functionality). Point mutations that occur within the protein coding region of a gene may be classified into three kinds, depending upon what the erroneous codon codes for:

- Silent mutations: which code for the same amino acid.
- Missense mutations: which code for a different amino acid.
- Nonsense mutations: which code for a stop and can truncate the protein.
- Insertions add one or more extra nucleotides into the DNA. They are usually caused by transposable elements, or errors during replication of repeating elements (e.g. AT repeats). Insertions in the coding region of a gene may alter splicing of the mRNA(splice site mutation), or cause a shift in the reading frame (frameshift), both of which can significantly alter the gene product. Insertions can be reverted by excision of the transposable element.
- Deletions remove one or more nucleotides from the DNA. Like insertions, these mutations can alter the reading frame of the gene. They are generally irreversible: though exactly the same sequence might theoretically be restored by an insertion, transposable elements able to revert a very short deletion (say 1–2 bases) in any location are either highly unlikely to exist or do not exist at all. Note that a deletion is not the exact opposite of an insertion: the former is quite random while the latter consists of a specific sequence inserting at locations that are not entirely random or even quite narrowly defined.
- Large-scale mutations in chromosomal structure, including:
- Amplifications (or gene duplications) leading to multiple copies of all chromosomal regions, increasing the dosage of the genes located within them.
- Deletions of large chromosomal regions, leading to loss of the genes within those regions.
- Mutations whose effect is to juxtapose previously separate pieces of DNA, potentially bringing together separate genes to form functionally distinct fusion genes (e.g. bcr-abl). These include:
- Chromosomal translocations: interchange of genetic parts from nonhomologous chromosomes.
- Interstitial deletions: an intra-chromosomal deletion that removes a segment of DNA from a single chromosome, thereby apposing previously distant genes. For example, cells isolated from a human astrocytoma, a type of brain tumor, were found to have a chromosomal deletion removing sequences between the "fused in glioblastoma"gene and the

receptor tyrosine kinase "ros", producing a fusion protein (FIG-ROS). The abnormal FIG-ROS fusion protein has constitutively active kinase activity that causes oncogenic transformation (a transformation from normal cells to cancer cells).

- Chromosomal inversions: reversing the orientation of a chromosomal segment.
- Loss of heterozygosity: loss of one allele, either by a deletion or recombination event, in an organism that previously had two different alleles.

By Effect on Function

- Loss-of-function mutations are the result of gene product having less or no function. When the allele has a complete loss of function (null allele) it is often called an amorphic mutation. Phenotypes associated with such mutations are most often recessive. Exceptions are when the organism is haploid, or when the reduced dosage of a normal gene product is not enough for a normal phenotype (this is calledhaploinsufficiency).
- Gain-of-function mutations change the gene product such that it gains a new and abnormal function. These mutations usually have dominant phenotypes. Often called a neomorphic mutation.
- Dominant negative mutations (also called antimorphic mutations) have an altered gene product that acts antagonistically to the wild-type allele. These mutations usually result in an altered molecular function (often inactive) and are characterised by a dominant or semi-dominant phenotype. In humans, Marfan syndrome is an example of a dominant negative mutation occurring in anautosomal dominant disease. In this condition, the defective glycoprotein product of the fibrillin gene (FBN1) antagonizes the product of the normal allele.
- Lethal mutations are mutations that lead to the death of the organisms which carry the mutations.
- A back mutation or reversion is a point mutation that restores the original sequence and hence the original phenotype.

By Effect on Fitness

In applied genetics it is usual to speak of mutations as either harmful or beneficial.

- A harmful mutation is a mutation that decreases the fitness of the organism.

- A beneficial mutation is a mutation that increases fitness of the organism, or which promotes traits that are desirable.

In theoretical population genetics, it is more usual to speak of such mutations as deleterious or advantageous. In the neutral theory of molecular evolution, genetic drift is the basis for most variation at the molecular level.

- A neutral mutation has no harmful or beneficial effect on the organism. Such mutations occur at a steady rate, forming the basis for the molecular clock.
- A deleterious mutation has a negative effect on the phenotype, and thus decreases the fitness of the organism.
- An advantageous mutation has a positive effect on the phenotype, and thus increases the fitness of the organism.
- A nearly neutral mutation is a mutation that may be slightly deleterious or advantageous, although most nearly neutral mutations are slightly deleterious.

In reality, viewing the fitness effects ofmutations in these discrete categories is an oversimplification. Attempts have been made to infer the distributionof fitness effects using mutagenesis experiments or theoretical models applied to molecular sequence data. However, the current distribution is still uncertain, and some aspects of the distribution likely vary between species.

By Inheritance

- inheritable generic in pro-generic tissue or cells on path to be changed to gametes.
- non inheritable somatic (e.g., carcinogenic mutation)
- non inheritable post mortem aDNA mutation in decaying remains.

By pattern of inheritance The human genome contains two copies of each gene – a paternal and a maternal allele.

- A heterozygous mutation is a mutation of only one allele.
- A homozygous mutation is an identical mutation of both the paternal and maternal alleles.
- Compound heterozygous mutations or a genetic compound comprises two different mutations in the paternal and maternal alleles.
- A wildtype or homozygous non-mutated organism is one in which neither allele is mutated. (Just not a mutation)

By Impact on Protein Sequence

- A frameshift mutation is a mutation caused by insertion or deletion of a number of nucleotides that is not evenly divisible by three from a DNA sequence. Due to the triplet nature of gene expression by codons, the insertion or deletion can disrupt the reading frame, or the grouping of the codons, resulting in a completely different translation from the original. The earlier in the sequence the deletion or insertion occurs, the more altered the protein produced is.

In contrast, any insertion or deletion that is evenly divisible by three is termed an in-frame mutation

- A nonsense mutation is a point mutation in a sequence of DNA that results in a premature stop codon, or a nonsense codon in the transcribed mRNA, and possibly a truncated, and often nonfunctional protein product.
- Missense mutations or nonsynonymous mutations are types of point mutations where a single nucleotide is changed to cause substitution of a different amino acid. This in turn can render the resulting protein nonfunctional. Such mutations are responsible for diseases such as Epidermolysis bullosa, sickle-cell disease, and SOD1 mediated ALS (Boillée 2006, p. 39).
- A neutral mutation is a mutation that occurs in an amino acid codon which results in the use of a different, but chemically similar, amino acid. The similarity between the two is enough that little or no change is often rendered in the protein. For example, a change from AAA to AGA will encode arginine, a chemically similar molecule to the intended lysine. Neutral mutations occur because of the degenerate nature of the genetic code.
- Silent mutations are mutations that do not result in a change to the amino acid sequence of a protein. They may occur in a region that does not code for a protein, or they may occur within a codon in a manner that does not alter the final amino acid sequence. The phrase silent mutation is often used interchangeably with the phrase synonymous mutation; however, synonymous mutations are a subcategory of the former, occurring only within exons. The name silent could be a misnomer. For example, a silent mutation in the exon/intron border may lead to alternative splicing by changing the splice site.

By Inheritance Ability

A mutation has caused this garden moss rose to produce flowers of different colors. This is a somatic mutation that may also be passed on in the germ line. In multicellular organisms with dedicated reproductive cells, mutations can be subdivided into germ line mutations, which can be passed on to descendants through their reproductive cells, and somatic mutations (also called acquired mutations), which involve cells outside the dedicated reproductive group and which are not usually transmitted to descendants.

Figure: *A germline mutation gives rise to a constitutional mutation in the offspring, that is, a mutation that is present in every cell. A constitutional mutation can also occur very soon after fertilisation, or continue from a previous constitutional mutation in a parent.*

If the organism can reproduce asexually through mechanisms such as cuttings or budding the distinction can become blurred. For example, plants can sometimes transmit somatic mutations to their descendants asexually or sexually where flower buds develop in somatically mutated parts of plants. A new mutation that was not inherited from either parent is called a de novo mutation. The source of the mutation is unrelated to the consequence, although the consequences are related to which cells were mutated.

Special Classes

- Conditional mutation is a mutation that has wild-type (or less severe) phenotype under certain "permissive" environmental conditions and a mutant phenotype under certain "restrictive" conditions. For example, a temperature-sensitive mutation can cause cell death at high temperature (restrictive condition), but might have no deleterious consequences at a lower temperature (permissive condition).

Nomenclature

A committee of the Human Genome Variation Society (HGVS) has developed the standard human sequence variant nomenclature, which should be used by researchers and DNA diagnostic centers to generate unambiguous mutation descriptions. In principle, this nomenclature can also be used to describe mutations in other organisms. The nomenclature specifies the type of mutation and base or amino acid changes.

- Nucleotide substitution (e.g. 76A>T) - The number is the position of the nucleotide from the 5' end, the first letter represents the wild type nucleotide, and the second letter represents the nucleotide which replaced the wild type. In the given example, the adenine at the 76th position was replaced by a thymine.
- If it becomes necessary to differentiate between mutations in genomic DNA, mitochondrial DNA, and RNA, a simple convention is used. For example, if the 100th base of a nucleotide sequence mutated from G to C, then it would be written as g.100G>C if the mutation occurred in genomic DNA, m.100G>C if the mutation occurred in mitochondrial DNA, or r.100g>c if the mutation occurred in RNA. Note that for mutations in RNA, the nucleotide code is written in lower case.
- Amino acid substitution (e.g. D111E) – The first letter is the one letter code of the wild type amino acid, the number is the position of the amino acid from the N-terminus, and the second letter is the one letter code of the amino acid present in the mutation. Nonsense mutations are represented with an X for the second amino acid (e.g. D111X).
- Amino acid deletion (e.g. ÄF508) – The Greek letter Ä (delta) indicates a deletion. The letter refers to the amino acid present in the wild type and the number is the position from the N terminus of the amino acid were it to be present as in the wild type.

Harmful Mutations

Changes in DNA caused by mutation can cause errors in protein sequence, creating partially or completely non-functional proteins. To function correctly, each cell depends on thousands of proteins to function in the right places at the right times.

When a mutation alters a protein that plays a critical role in the body, a medical condition can result. A condition caused by mutations in one or more genes is called a genetic disorder. Some mutations alter a gene's DNA base sequence but do not change the function of the protein made by the gene.

Studies of the fly Drosophila melanogastersuggest that if a mutation does change a protein, this will probably be harmful, with about 70 percent of these mutations having damaging effects, and the remainder being either neutral or weakly beneficial. However, studies in yeast have shown that only 7% of mutations that are not in genes are harmful.

If a mutation is present in a germ cell, it can give rise to offspring that carries the mutation in all of its cells. This is the case in hereditary diseases. On the other hand, a mutation may occur in asomatic cell of an organism. Such mutations will be present in all descendants of this cell within the same organism, and certain mutations can cause the cell to become malignant, and thus causecancer.

Often, gene mutations that could cause a genetic disorder are repaired by the DNA repair system of the cell. Each cell has a number of pathways through which enzymes recognize and repair mistakes in DNA. Because DNA can be damaged or mutated in many ways, the process of DNA repair is an important way in which the body protects itself from disease.

Beneficial Mutations

Although mutations that change protein sequences are predominantly harmful to an organism; on occasion, the effect can be neutral or positive in a given environment.

In this case, the mutation may enable the mutant organism to withstand particular environmental stresses better than wild-type organisms, or reproduce more quickly. In these cases a mutation will tend to become more common in a population through natural selection.

For example, a specific 32 base pair deletion in human CCR5 (CCR5-Ä32) confers HIV resistance to homozygotes and delays AIDS onset in heterozygotes. The CCR5 mutation is more common in those of European descent.

One possible explanation of the etiology of the relatively high frequency of CCR5-Ä32 in the European population is that it conferred resistance to the bubonic plague in mid-14th century Europe. People with this mutation were more likely to survive infection; thus its frequency in the population increased.

This theory could explain why this mutation is not found in southern Africa, where the bubonic plague never reached. A newer theory suggests that the selective pressure on the CCR5 Delta 32 mutation was caused by smallpox instead of the bubonic plague.

Another example is Sickle cell disease, a blood disorder in which the body produces an abnormal type of the oxygen-carrying substance hemoglobin in the red blood cells. One-third of all indigenousinhabitants of Sub-Saharan Africa carry the gene, because in areas where malaria is common, there is a survival value in carrying only a single sickle-cell gene (sickle cell trait). Those with only one of the two alleles of the sickle-cell disease are more resistant to malaria, since the infestation of the malaria plasmodium is halted by the sickling of the cells which it infests.

Prion Mutation

Prions are proteins and do not contain genetic material. However, prion replication has been shown to be subject to mutation and natural selection just like other forms of replication.

Nucleotide

Nucleotides are molecules that, when joined together, make up the structural units of RNA and DNA. In addition, nucleotides participate in cellular signaling (cyclic guanosine monophosphate andcyclic adenosine monophosphate), and are incorporated into important cofactors of enzymatic reactions (coenzyme A, flavin adenine dinucleotide, flavin mononucleotide, and nicotinamide adenine dinucleotide phosphate). Nucleotide derivatives such as the nucleoside triphosphates play central roles in metabolism, in which capacity they serve as sources of chemical energy (adenosine triphosphate and guanosine triphosphate).

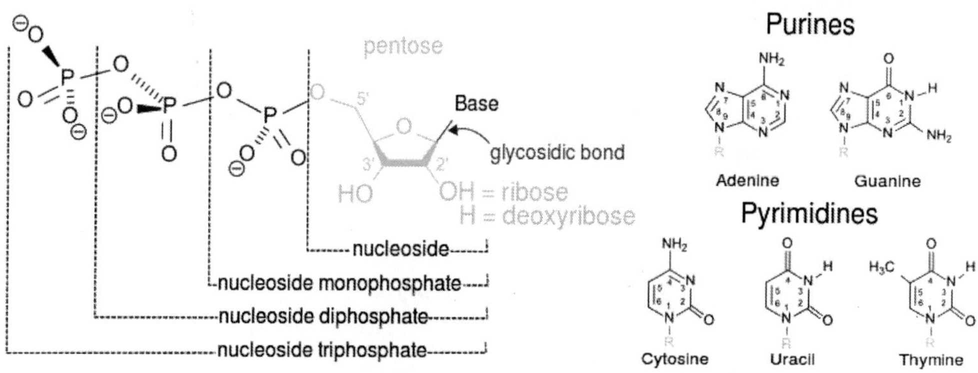

Figure: Structural elements of the most common nucleotides

Nucleotide Structure

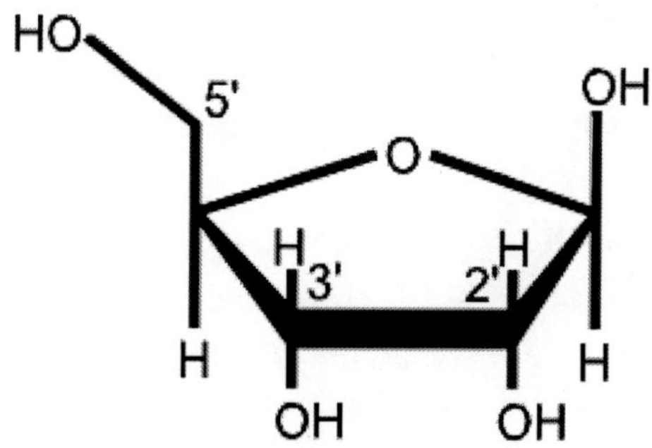

Figure: Ribose structure indicating numbering of carbon atoms

A nucleotide is composed of a nucleobase (nitrogenous base), a five-carbon sugar (either ribose or 2'-deoxyribose), and one phosphate group. Together, the nucleobase and sugar compose a nucleoside. The phosphate groups form bonds with either the 2, 3, or 5-carbon of the sugar, with the 5-carbon site most common. Cyclic nucleotides form when the phosphate group is bound to two of the sugar's hydroxyl groups. Ribonucleotides are nucleotides where the sugar is ribose, and deoxyribonucleotides contain the sugar deoxyribose. Nucleotides can contain either a purine or a pyrimidine base.

Nucleic acids are polymeric macromolecules made from nucleotide monomers. In DNA, the purine bases are adenine and guanine, while the pyrimidines arethymine and cytosine. RNA uses uracil in place of thymine. Adenine always pairs with thymine by 2 hydrogen bonds, while guanine pairs with cytosine through 3 hydrogen bonds, each due to their unique structures.

Synthesis

Nucleotides can be synthesized by a variety of means both in vitro and in vivo. In vivo, nucleotides can be synthesized de novo or recycled through salvage pathways. The components used in de novo nucleotide synthesis are derived from biosynthetic precursors of carbohydrate and amino acid metabolism, and from ammonia and carbon dioxide. The liver is the major organ of de novo synthesis of all four nucleotides. De novo synthesis of pyrimidines and purines follows two different pathways. Pyrimidines are synthesized first from aspartate and carbamoyl-phosphate in the cytoplasm to the common precursor ring structure orotic acid, onto which a phosphorylated ribosyl unit is covalently linked. Purines, however, are first synthesized from the sugar template onto which the ring synthesis occurs. For reference, the syntheses of the purine and pyrimidine nucleotides are carried out by several enzymes in the cytoplasm of the cell, not within a specific organelle. Nucleotides undergo breakdown such that useful parts can be reused in synthesis reactions to create new nucleotides.

In vitro, protecting groups may be used during laboratory production of nucleotides. A purified nucleoside is protected to create a phosphoramidite, which can then be used to obtain analogues not found in nature and/or to synthesize an oligonucleotide.

Pyrimidine Ribonucleotide Synthesis

The color scheme is as follows: enzymes, coenzymes, substrate names,inorganic molecules The synthesis of the pyrimidines CTP and UTP occurs in the cytoplasm and starts with the formation of carbamoyl phosphate from glutamine and CO_2. Next, aspartate undergoes a condensation reaction with carbamoyl-phosphate to form orotic acid.

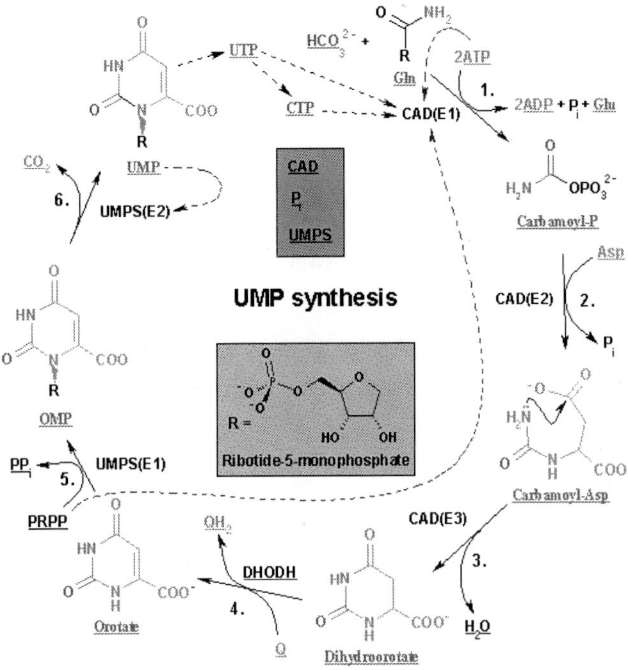

Figure: *The synthesis of UMP.*

In a subsequent cyclization reaction, the enzyme Aspartate carbamoyltransferase forms N-carbamoyl-aspartate which is converted into dihydroorotic acid by Dihydroorotase. The latter is converted to orotate by Dihydroorotate oxidase. The net reaction is:

(S)-Dihydroorotate + O_2 = Orotate + H_2O_2

Orotate is covalently linked with a phosphorylated ribosyl unit. The covalent linkage between the ribose and pyrimidine occurs at position C_1 of the ribose unit, which contains a pyrophosphate, and N_1 of the pyrimidine ring. Orotate phosphoribosyltransferase (aka "PRPP transferase") catalyzes the net reaction yielding orotidine monophosphate (OMP):

Orotate + 5-Phospho-á-D-ribose 1-diphosphate (aka. "PRPP") = Orotidine 5'-phosphate + Pyrophosphate

Orotidine-5-phosphate is decarboxylated by Orotidine-5'-phosphate decarboxylase to form uridine monophosphate (UMP). PRPP transferase catalyzes both the ribosylation and decarboxylation reactions, forming UMP from orotic acid in the presence of PRPP.

It is from UMP that other pyrimidine nucleotides are derived. UMP is phosphorylated by two kinases to uridine triphosphate (UTP) via two sequential reactions with ATP. First the diphosphate form UDP is produced, which in turn is phosphorylated to UTP. Both steps are fueled by ATP hydrolysis:

ATP + UMP = ADP + UDP UDP + ATP = UTP + ADP

CTP is subsequently formed by amination of UTP by the catalytic activity of CTP synthetase. Glutamine is the NH_3 donor and the reaction is fueled by ATP hydrolysis, too:

UTP + Glutamine + ATP + H_2O = CTP + ADP + P_i

Cytidine monophosphate (CMP) is derived from cytidine triphosphate (CTP) with subsequent loss of two phosphates.

Purine Ribonucleotide Synthesis

The atoms which are used to build the purine nucleotides come from a variety of sources:

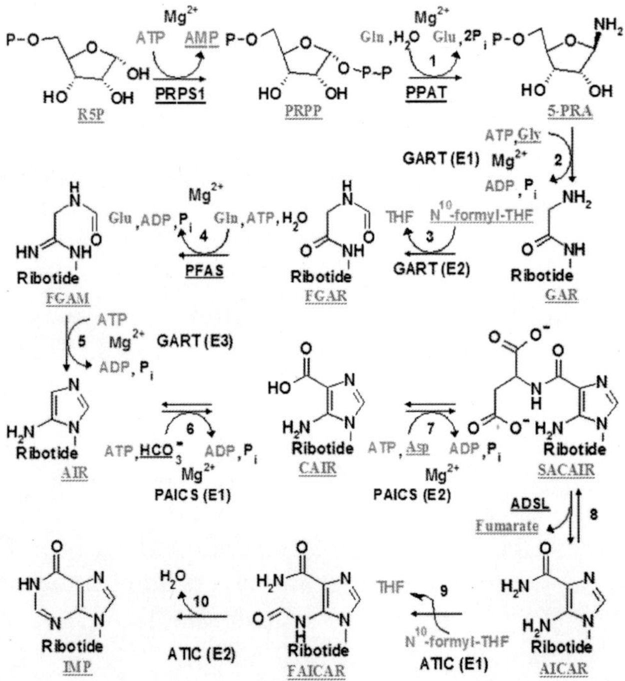

The synthesis of IMP. The color scheme is as follows: enzymes,coenzymes, substrate names, metal ions, inorganic molecules

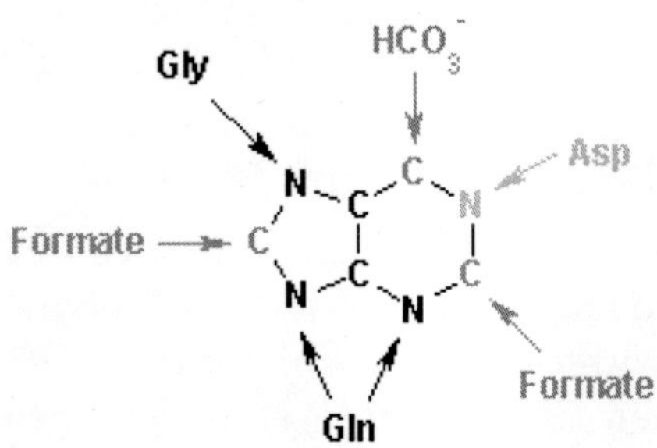

The biosynthetic origins of purine ring atoms

N1 arises from the amine group of Asp

C2 and C8 originate from formate

N3 and N9 are contributed by the amide group of Gln

C4, C5 and N7 are derived from Gly

C6 comes from HCO3- (CO2)

The de novo synthesis of purine nucleotides by which these precursors are incorporated into the purine ring proceeds by a 10-step pathway to the branch-point intermediate IMP, the nucleotide of the base hypoxanthine. AMP and GMP are subsequently synthesized from this intermediate via separate, two-step pathways. Thus, purine moieties are initially formed as part of theribonucleotides rather than as free bases.

Six enzymes take part in IMP synthesis. Three of them are multifunctional:

- GART (reactions 2, 3, and 5)
- PAICS (reactions 6, and 7)
- ATIC (reactions 9, and 10)

The pathway starts with the formation of PRPP. PRPS1 is the enzyme that activates R5P, which is formed primarily by thepentose phosphate pathway, to PRPP by reacting it with ATP. The reaction is unusual in that a pyrophosphoryl group is directly transferred from ATP to C_1 of R5P and that the product has the á configuration about C1. This reaction is also shared with the pathways for the synthesis of Trp, His, and the pyrimidine nucleotides. Being on a major metabolic crossroad and requiring much energy, this reaction is highly regulated.

In the first reaction unique to purine nucleotide biosynthesis, PPAT catalyzes the displacement of PRPP's pyrophosphategroup (PP_i) by an amide nitrogen donated from either glutamine (N), glycine (N&C), aspartate (N), folic acid (C_1), or CO_2. This is the committed step in purine synthesis. The reaction occurs with the inversion of configuration about ribose C_1, thereby forming â-5-phosphorybosylamine (5-PRA) and establishing the anomeric form of the future nucleotide.

Next, a glycine is incorporated fueled by ATP hydrolysis and the carboxyl group forms an amine bond to the NH_2 previously introduced. A one-carbon unit from folic acid coenzyme N_{10}-formyl-THF is then added to the amino group of the substituted glycine followed by the closure of the imidazole ring. Next, a second NH_2 group is transferred from a glutamine to the first carbon of the glycine unit. A carboxylation of the second carbon of the glycin unit is concomittantly added. This new carbon is modified by the additional of a third

NH_2 unit, this time transferred from an aspartate residue. Finally, a second one-carbon unit from formyl-THF is added to the nitrogen group and the ring covalently closed to form the common purine precursor inosine monophosphate (IMP).

Inosine monophosphate is converted to adenosine monophosphate in two steps. First, GTP hydrolysis fuels the addition of aspartate to IMP by adenylosuccinate synthase, substituting the carbonyl oxygen for a nitrogen and forming the intermediate adenylosuccinate. Fumarate is then cleaved off forming adenosine monophosphate. This step is catalyzed by adenylosuccinate lyase.

Inosine monophosphate is converted to guanosine monophosphate by the oxidation of IMP forming xanthylate, followed by the insertion of an amino group at C_2. NAD is the electron acceptor in the oxidation reaction. The amide group transfer from glutamine is fueled by ATP hydrolysis.

Pyramidine and Purine Degradation

In humans, pyrimidine rings (C, T, U) can be degraded completely to CO_2 and NH_3 (urea excretion). That having been said, purine rings (G, A) cannot. Instead they are degraded to the metabolically inert uric acid which is then excreted from the body. Uric acid is formed when GMP is split into the base guanine and ribose. Guanine is deaminated to xanthine which in turn is oxidized to uric acid. This last reaction is irreversible. Similarly, uric acid can be formed when AMP is deaminated to IMP from which the ribose unit is removed to form hypoxanthine. Hypoxanthine is oxidized to xanthine and finally to uric acid. Instead of uric acid secretion, guanine and IMP can be used for recycling purposes and nucleic acid synthesis in the presence of PRPP and aspartate (NH_3 donor).

Length Unit

Nucleotide (abbreviated nt) is a common length unit for single-stranded RNA, similar to how base pair is a length unit for double-stranded DNA.

Abbreviation Codes for Degenerate Bases

The IUPAC has designated the symbols for nucleotides. Apart from the five (A, G, C, T/U) bases, often degenerate bases are used especially for designing PCR primers. These nucleotide codes are listed here.

IUPAC nucleotide code	Base
A	*Adenine*
C	*Cytosine*
G	*Guanine*
T (or U)	*Thymine (or Uracil)*

R	*A or G*
Y	*C or T (U)*
S	*G or C*
W	*A or T (U)*
K	*G or T (U)*
M	*A or C*
B	*C or G or T (U)*
D	*A or G or T (U)*
H	*A or C or T (U)*
V	*A or C or G*
N	*any base*
. or -	*gap*

Genetic Variation

Genetic variation, variation in alleles of genes, occurs both within and among populations. Genetic variation is important because it provides the "raw material" for natural selection. Genetic variation is brought about by mutation, a change in a chemical structure of a gene. Polyploidy is an example of chromosomal mutation. Polyploidy is a condition wherein organisms have three or more sets of genetic variation (3n or more).

Among Individuals within a Population

Genetic variation among individuals within a population can be identified at a variety of levels. It is possible to identify genetic variation from observations of phenotypic variation in either quantitative traits (traits that vary continuously and are coded for by many genes, e.g., leg length in dogs) or discrete traits (traits that fall into discrete categories and are coded for by one or a few genes (e.g., white, pink, red petal color in certain flowers).

Genetic variation can also be identified by examining variation at the level of enzymes using the process of protein electrophoresis. Polymorphic genes have more than one allele at each locus. Half of the genes that code for enzymes in insects and plants may be polymorphic, whereas polymorphisms are less common in vertebrates.

Ultimately, genetic variation is caused by variation in the order of bases in the nucleotides in genes. New technology now allows scientists to directly sequence DNA which has identified even more genetic variation than was previously detected by protein electrophoresis. Examination of DNA has shown genetic variation in both coding regions and in the non-coding intron region of genes. Genetic variation will result in phenotypic variation if variation in the order of nucleotides in the DNA sequence results in a difference in the order of amino acids in proteins coded by that DNA sequence, and if the resultant differences in amino acid sequence influence the shape, and thus the function of the enzyme.

Between Populations

Geographic variation in genes often occurs among populations living in different locations. Geographic variation may be due to differences in selective pressures or to genetic drift.

Measurement

Genetic variation within a population is commonly measured as the percentage of gene loci that are polymorphic or the percentage of gene loci in individuals that are heterozygous.

Sources

Mutations are the ultimate source of genetic variation because they alter the order of bases in the nucleotides of DNA. Mutations are likely to be rare and most mutations are neutral or deleterious, but in some instances the new alleles can be favored by natural selection.

Genetic variation can also be produced by the recombination of chromosomes that occurs during sexual reproduction, called independent assortment.

Crossing over and random segregation during meiosis can result in the production of new alleles or new combinations of alleles.

Variation and recombination can be facilitated by transposable and transposed genetic elements, commonly known as endogenous retroviruses, LINEs, SINEs, etc.

Maintenance in Populations

A variety of factors maintain genetic variation in populations. Potentially harmful recessive alleles can be hidden from selection in the heterozygous individuals in populations of diploid organisms (recessive alleles are only expressed in the less common homozygous individuals). Natural selection can also maintain genetic variation in balanced polymorphisms. Balanced polymorphisms may occur when heterozygotes are favored or when selection is frequency dependent.

Evolution

Evolution is any change across successive generations in the heritable characteristics of biological populations. Evolutionary processes give rise to diversity at every level of biological organisation, including species, individual organisms and molecules such as DNA and proteins. Life on Earth originated and then evolved from a universal common ancestor approximately 3.7 billion years ago. Repeated speciation and the divergence of life can be traced through shared sets of biochemical and morphological traits, or by shared DNA sequences.

These homologous traits and sequences are more similar among species that share a more recent common ancestor, and can be used to reconstruct evolutionary histories, using both existing species and the fossil record. Existing patterns of biodiversity have been shaped both by speciation and by extinction.

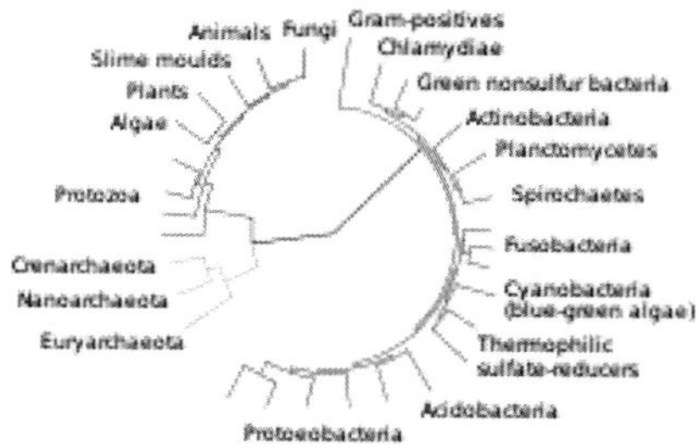

The theory of evolution by natural selection was formulated by Charles Darwin, who was the first to recognise natural selection as an important cause of evolution. Evolution by natural selection is a consequence of three widely accepted premises: 1) more offspring are produced than can possibly survive, 2) traits vary among individuals, leading to differential rates of survival and reproduction, and 3) trait differences are heritable.

Thus, when members of a population die, they are replaced by individuals that are not born from random parents. Instead, these new members are born from parents that are betteradapted to the environment in which natural selection took place.

This can cause the evolution of traits that are seemingly fitted for the functional roles they perform. Natural selection is the only known cause of adaptation, but not the only known cause of evolution. Other, nonadaptive causes of evolutioninclude mutation and genetic drift.

In the early 20th century, genetics was integrated with Darwin's theory of evolution by natural selection.

All other branches of biology were made compatible with the resulting synthesis while previously held notions about evolution, such as orthogenesis and "progress", were made obsolete. Scientists continue to study evolution using both observational data and experiments in both the field and the laboratory.

Biologists agree that descent with modification is one of the most reliably established facts in science. Discoveries in evolutionary biology have made a significant impact not just within the traditional branches of biology, but also in other cademic disciplines (e.g., anthropology and psychology) and on society at large.

History of Evolutionary Thought

The proposal that one type of animal could descend from an animal of another type goes back to some of the first pre-Socratic Greek philosophers, such as Anaximander and Empedocles. In contrast to these materialistic views, Aristotle understood all natural things, not only living things, as being imperfect actualizations of different fixed natural possibilities, known as "forms", "ideas", or (in Latin translations) "species".

This was part of his teleological understanding of nature in which all things have an intended role to play in a divine cosmic order. Variations of this idea became the standard understanding of the Middle Ages, and were integrated into Christian learning, but Aristotle did not demand that real types of animals corresponded one-for-one with exact metaphysical forms, and specifically gave examples of how new types of living things could come to be.

In the 17th century the new method of modern science rejected Aristotle's approach, and sought explanations of natural phenomena in terms of laws of nature which were the same for all visible things, and did not need to assume any fixed natural categories, nor any divine cosmic order. But this new approach was slow to take root in the biological sciences, which became the last bastion of the concept of fixed natural types.

John Ray used one of the previously more general terms for fixed natural types, "species", to apply to animal and plant types, but unlike Aristotle he strictly identified each type of living thing as a species, and proposed that each species can be defined by the features that perpetuate themselves each generation.

These species were designed by God, but showing differences caused by local conditions. The biological classification introduced by Carolus Linnaeus in 1735 also viewed species as fixed according to a divine plan.

Other naturalists of this time speculated on evolutionary change of species over time according to natural laws. Maupertuis wrote in 1751 of natural modifications occurring during reproduction and accumulating over many generations to produce new species.

Buffon suggested that species could degenerate into different organisms, and Erasmus Darwin proposed that all warm-blooded animals could have descended from a single micro-organism (or "filament"). The first fully-fledged evolutionary scheme was Lamarck's "transmutation" theory of 1809 which envisaged spontaneous generation continually producing simple forms of life developed greater complexity in parallel lineages with an inherent progressive tendency, and that on a local level these lineages adapted to the environment by inheriting changes caused by use or disuse in parents. (The latter process was later called Lamarckism.)

These ideas were condemned by establishment naturalists as speculation lacking empirical support. In particular Georges Cuvier insisted that species were unrelated and fixed, their similarities reflecting divine design for functional needs.

In the meantime, Ray's ideas of benevolent design had been developed by William Paley into a natural theology which proposed complex adaptations as evidence of divine design, and was admired by Charles Darwin.

In 1842 Charles Darwin penned his first sketch of what became On the Origin of Species.

The critical break of biology from the concept of fixed species is generally seen as the theory of evolution of Charles Darwin. Partly influenced by An Essay on the Principle of Population by Thomas Robert Malthus, Darwin noted that population growth would lead to a "struggle for existence" where favorable variations could prevail as others perished. Each generation, many offspring fail to survive to an age of reproduction because of limited resources.

This could explain the diversity of animals and plants from a common ancestry through the working of natural laws working the same for all types of thing. Darwin was developing his theory of "natural selection" from 1838 onwards until Alfred Russel Wallace sent him a similar theory in 1858. Both men presented theirseparate papers to the Linnean Society of London.

At the end of 1859, Darwin's publication of On the Origin of Species explained natural selection in detail and in a way that lead to an increasingly wide acceptance of Darwinian evolution.

Thomas Henry Huxley applied Darwin's ideas to humans, usingpaleontology and comparative anatomy to provide strong evidence that humans and apes shared a common ancestry. Some were disturbed by this since it implied that humans did not have a special place in the universe.

Precise mechanisms of reproductive heritability and the origin of new traits remained a mystery. Towards this end, Darwin developed his provisional theory ofpangenesis.

In 1865 Gregor Mendel reported that traits were inherited in a predictable manner through the independent assortment and segregation of elements (later known as genes).

Mendel's laws of inheritance eventually supplanted most of Darwin's pangenesis theory. August Weismann made the important distinction between germ cells (sperm and eggs) and somatic cells of the body, demonstrating that heredity passes through the germ line only.

Hugo de Vries connected Darwin's pangenesis theory to Wiesman's germ/ soma cell distinction and proposed that Darwin's pangenes were concentrated in the cell nucleus and when expressed they could move into the cytoplasm to change the cells structure.

De Vries was also one of the researchers who made Mendel's work well-known, believing that Mendelian traits corresponded to the transfer of heritable variations along the germline. To explain how new variants originate, De Vries developed a mutation theory that led to a temporary rift between those who accepted Darwinian evolution and biometricians who allied with de Vries.

At the turn of the 20th century, pioneers in the field of population genetics, such as J.B.S. Haldane, Sewall Wright, and Ronald Fisher, set the foundations of evolution onto a robust statistical philosophy. The false contradiction between Darwin's theory, genetic mutations, and Mendelian inheritance was thus reconciled.

In the 1920s and 1930s a modern evolutionary synthesis connected natural selection, mutation theory, and Mendelian inheritance into a unified theory that applied generally to any branch of biology. The modern synthesis was able to explain patterns observed across species in populations, through fossil transitions in palaeontology, and even complex cellular mechanisms in developmental biology.

The publication of the structure of DNA by James Watson and Francis Crick in 1953 demonstrated a physical basis for inheritance. Molecular biology improved our understanding of the relationship between genotype and phenotype. Advancements were also made in phylogenetic systematics, mapping the transition of traits into a comparative and testable framework through the publication and use of evolutionary trees.

In 1973, evolutionary biologist Theodosius Dobzhansky penned that "nothing in biology makes sense except in the light of evolution", because it has brought to light the relations of what first seemed disjointed facts in natural history into a coherent explanatory body of knowledge that describes and predicts many observable facts about life on this planet.

Since then, the modern synthesis has been further extended to explain biological phenomena across the full and integrative scale of the biological hierarchy, from genes to species. This extension has been dubbed "eco-evo-devo".

Heredity

Evolution in organisms occurs through changes in heritable traits – particular characteristics of an organism. In humans, for example, eye colour is an inherited characteristic and an individual might inherit the "brown-eye trait" from one of their parents.

Inherited traits are controlled by genes and the complete set of genes within an organism's genome is called its genotype.

The complete set of observable traits that make up the structure and behaviour of an organism is called its phenotype. These traits come from the interaction of its genotype with the environment. As a result, many aspects of an organism's phenotype are not inherited.

For example, suntanned skin comes from the interaction between a person's genotype and sunlight; thus, suntans are not passed on to people's children. However, some people tan more easily than others, due to differences in their genotype; a striking example are people with the inherited trait of albinism, who do not tan at all and are very sensitive to sunburn.

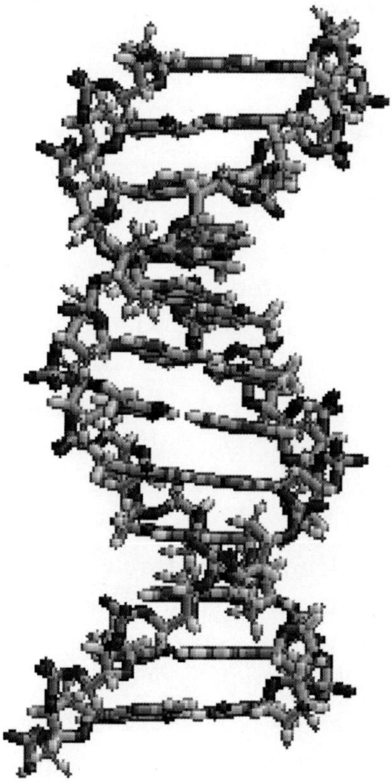

***Figure:** DNA structure. Bases are in the centre, surrounded by phosphate–sugar chains in adouble helix.*

Heritable traits are known to be passed from one generation to the next via DNA, a molecule that encodes genetic information. DNA is a long polymer composed of four types of bases.

The sequence of bases along a particular DNA molecule specify the genetic information, in a manner similar to a sequence of letters spelling out a sentence. Before a cell divides, the DNA is copied, so that each of the resulting two cells will inherit the DNA sequence. Portions of a DNA molecule that specify a single functional unit are called genes; different genes have different sequences of bases. Within cells, the long strands of DNA form condensed structures called chromosomes.

The specific location of a DNA sequence within a chromosome is known as a locus. If the DNA sequence at a locus varies between individuals, the different forms of this sequence are called alleles. DNA sequences can change through mutations, producing new alleles. If a mutation occurs within a gene, the new allele may affect the trait that the gene controls, altering the phenotype of the organism.

However, while this simple correspondence between an allele and a trait works in some cases, most traits are more complex and are controlled by multiple interacting genes.

Recent findings have confirmed important examples of heritable changes that cannot be explained by changes to the sequence of nucleotides in the DNA. These phenomena are classed as epigenetic inheritance systems.

DNA methylation marking chromatin, self-sustaining metabolic loops, gene silencing by RNA interferenceand the three dimensional conformation of proteins (such as prions) are areas where epigenetic inheritance systems have been discovered at the organismic level.

Developmental biologists suggest that complex interactions in genetic networks and communication among cells can lead to heritable variations that may underlay some of the mechanics in developmental plasticity and canalization. Heritability may also occur at even larger scales.

For example, ecological inheritance through the process of niche construction is defined by the regular and repeated activities of organisms in their environment. This generates a legacy of effects that modify and feed back into the selection regime of subsequent generations.

Descendants inherit genes plus environmental characteristics generated by the ecological actions of ancestors. Other examples of heritability in evolution that are not under the direct control of genes include the inheritance of cultural traits and symbiogenesis.

Variation

Figure: *White peppered moth*

An individual organism's phenotype results from both its genotype and the influence from the environment it has lived in. A substantial part of the variation in phenotypes in a population is caused by the differences between their genotypes.

Figure: *Black morph in peppered moth evolution*

The modern evolutionary synthesis defines evolution as the change over time in this genetic variation. The frequency of one particular allele will become more or less prevalent relative to other forms of that gene. Variation disappears when a new allele reaches the point of fixation — when it either disappears from the population or replaces the ancestral allele entirely.

Natural selection will only cause evolution if there is enough genetic variation in a population. Before the discovery of Mendelian genetics, one common hypothesis was blending inheritance. But with blending inheritance, genetic variance would be rapidly lost, making evolution by natural selection implausible.

The Hardy-Weinberg principle provides the solution to how variation is maintained in a population with Mendelian inheritance. The frequencies of alleles (variations in a gene) will remain constant in the absence of selection, mutation, migration and genetic drift. Variation comes from mutations in genetic material, reshuffling of genes through sexual reproduction and migration between populations (gene flow). Despite the constant introduction of ne variation through mutation and gene flow, most of the genome f a species is identical in all individuals of that species. However, even relatively small differences in genotype can lead to dramatic differences in phenotype: for example, chimpanzees and humans differ in only about 5% of their genomes.

Mutation

Mutations are changes in the DNA sequence of a cell's genome. When mutations occur, they can either have no effect, alter the product of a gene, or prevent the gene from functioning. Based on studies in the fly Drosophila melanogaster, it has been suggested that if a mutation changes a protein produced by a gene, this will probably be harmful, with about 70% of these mutations having damaging effects, and the remainder being either neutral or weakly beneficial.

Mutations can involve large sections of a chromosome becoming duplicated (usually by genetic recombination), which can introduce extra copies of a gene into a genome.Extra copies of genes are a major source of the raw material needed for new genes to evolve.

This is important because most new genes evolve within gene families from pre-existing genes that share common ancestors. For example, the human eye uses four genes to make structures that sense light: three for colour vision and one for night vision; all four are descended from a single ancestral gene.

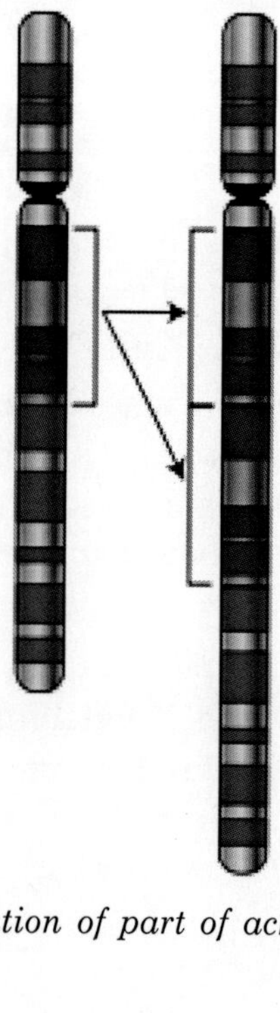

Figure: *Duplication of part of achromosome.*

New genes can be generated from an ancestral gene when a duplicate copy mutates and acquires a new function. This process is easier once a gene has been duplicated because it increases the redundancy of the system; one gene in the pair can acquire a new function while the other copy continues to perform its original function.

Other types of mutations can even generate entirely new genes from previously noncoding DNA.

The generation of new genes can also involve small parts of several genes being duplicated, with these fragments then recombining to form new combinations with new functions.

When new genes are assembled from shuffling pre-existing parts, domains act as modules with simple independent functions, which can be mixed together to produce new combinations with new and complex functions.

For example, polyketide synthases are large enzymes that make antibiotics; they contain up to one hundred independent domains that each catalyze one step in the overall process, like a step in an assembly line.

2

Genome Evolution

Genomic evolution is a set of phenomena involved in the changing of the structure of a genome through evolution. The study of genome evolution involves multiple fields such as structural analysis of the genome, the study of genomic parasites, gene and ancient genome duplications, polyploidy, and comparative genomics. Evolutionary biologists are interested in five specific questions in regards to evolution of the genome, these are:

1. How did the genome evolve into its current size?
2. What is the content within the genome, is it mostly junk or not?
3. What is the distribution of genes within a genome?
4. What is the composition of the nucleotides within the genome?
5. How does translation of the genetic code evolve?

Genome Size

Genome size is all the DNA that makes the genome. A genome can consist of genetic regions and noncoding regions. Genetic regions are those that encode proteins while noncoding regions refer to promoters and junk DNA. The C-value is another term for the genome size. Within a species the C-value does not show much variation, but there is a significant difference in the C-value between species.

Prokaryotic Genome

Prokaryotes are unicellular organisms that do not have membrae-bound organelles and lack a structurally distinct nucleus. Research on prokaryotic genomes shows that there is a significant positive correlation between the C-value of prokaryotes and the amount of genes that compose the genome. This indicates that gene size is the main factor influencing the size of the genome.

Eukaryotic Genome

In eukaryotic organisms, there is a paradox observed, namely that the number of genes that make up the genome does not correlate with genome size. In other ords, the genome size is much larger than would be expected given the total number of protein coding genes.

Related Fields

An important area within the study of molecular evolution is the use of molecular data to determine the correct biological classification of organisms. This is called molecular systematics or molecular phylogenetics.

Tools and concepts developed in the study of molecular evolution are now commonly used for comparative genomics and molecular genetics, while the influx of new data from these fields has been spurring advancement in molecular evolution.

Key Researchers in Molecular Evolution

Some researchers who have made key contributions to the develoment of the field:

- Motoo Kimura — Neutral theory
- Masatoshi Nei — Adaptive evolution
- Walter M. Fitch — Phylogenetic reconstruction
- Walter Gilbert — RNA world
- Joe Felsenstein — Phylogenetic methods
- Susumu Ohno — Gene duplication
- John H. Gillespie — Mathematics of adaptation
- Dan Graur - Neutral models of molecular evolution
- Wen-Hsiung Li - Neutral models of molecular evolution

Journals and Societies

Journals dedicated to molecular evolution include Molecular Biology and Evolution, Journal of Molecular Evolution, and Molecular Phylogenetics and Evolution.

Research in molecular evolution is also published in journals of genetics, molecular biology, genomics, systematics, or evolutionary biology. The Society for Molecular Biology and Evolution publishes the journal "Molecular Biology and Evolution" and holds an annual international meeting.

Population Genetics

Population genetics is the study of allele frequency distribution and change under the influence of the four main evolutionary processes: natural selection,genetic drift, mutation and gene flow. It also takes into account the factors of recombination, population subdivision and population structure. It attempts to explain such phenomena as adaptation and speciation.

Population genetics was a vital ingredient in the emergence of the modern evolutionary synthesis. Its primary founders were Sewall Wright, J. B. S. Haldaneand R. A. Fisher, who also laid the foundations for the related discipline of quantitative genetics.

Fundamentals

Population genetics is the study of the frequency and interaction of alleles and genes in populations. A population is a set of organisms in which any pair of members can breed together. This implies that all members belong to the same species and live near each other.

For example, all of the moths of the same species living in an isolated forest are a population. A gene in this population may have several alternate forms, which account for variations between thephenotypes of the organisms. An example might be a gene for coloration in moths that has two alleles: black and white.

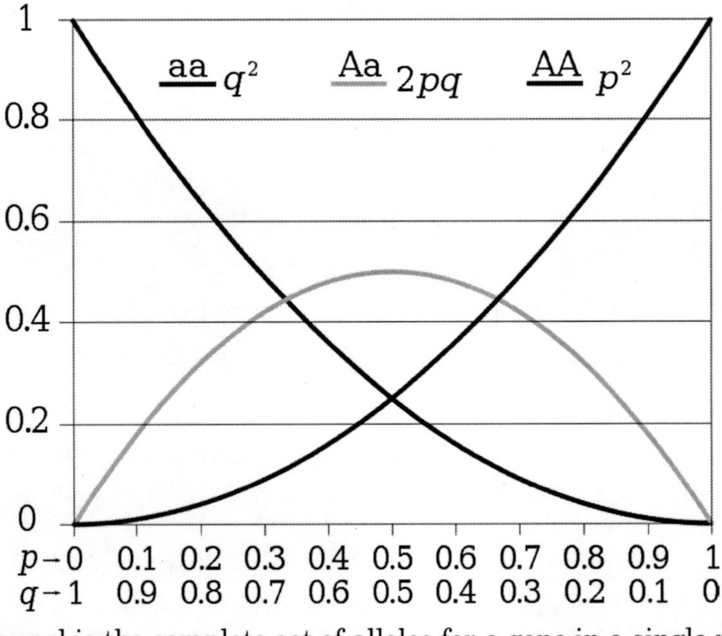

A gene pool is the complete set of alleles for a gene in a single population; theallele fequency for an allele is the fraction of the genes in the pool that is composed of that allele (for example, what fraction of moth coloration genes are the black allele). Evolution occurs when there are changes in the frequencies of alleles within a populatio; for example, the allele for black color in a population of moths becoming more common.

Hary–Weinberg principle for two alleles: the horizontal axis shows the two allele frequencies p and q and the vertical axis shows the genotype frequencies. Each graph shows one of the three possible genotypes.

Hardy–Weinberg Principle

Natural selection will only cause evolution if there is enough genetic variation in a population. Before the discovery of Mendelian genetics, one common hypothesis was blending inheritance. But with blending inheritance, genetic variance would be rapidly lost, making evolution by natural selection implausible.

The Hardy-Weinberg principle provides the solution to how variation is maintained in a population withMendelian inheritance. According to this principle, the frequencies of alleles (variations in a gene) will remain constant in the absence of selection, mutation, migration and genetic drift. The Hardy-Weinberg "equilibrium" refers to this stability of allele frequencies over time.

A second component of the Hardy-Weinberg principle concerns the effects of a single generation of random mating. In this case, the genotype frequencies can be predicted from the allele frequencies.

For example, in the simplest case of a single locus with two alleles: thedominant allele is denoted A and the recessive a and their frequencies are denoted by p and q; freq(A) = p; freq(a) = q; p + q = 1.

If the genotype frequencies are in Hardy-Weinberg proportions resulting from random mating, then we will have freq(AA) = p for the AAhomozygotes in the population, freq(aa) = q for the aa homozygotes, and freq(Aa) = 2pq for the heterozygotes.

Genetic Genealogy

Concepts

" Population genetics

" Haplogroup/ Haplotype

" Most recent common ancestor

" Human mitochondrial DNA haplogroups

" Human Y-chromosome DNA haplogroups

" Genomics

Other

- " Y-chromosome haplogroups by populations
- " Genealogical DNA test
- " Personal genomics
- " Genographic Project

This box: view · talk · edit

Scope and Theoretical Considerations

The mathematics of population genetics were originally developed as part of the modern evolutionary synthesis. According to Beatty (1986), it defines the core of the modern synthesis.

According to Lewontin (1974), the theoretical task for population genetics is a process in two spaces: a "genotypic space" and a "phenotypic space". The challenge of a complete theory of population genetics is to provide a set of laws that predictably map a population of genotypes (G_1) to aphenotype space (P_1), where selection takes place, and another set of laws that map the resulting population (P_2) back to genotype space (G_2) where Mendeliangenetics can predict the next generation of genotypes, thus completing the cycle. Even leaving aside for the moment the non-Mendelian aspects of molecular genetics, this is clearly a gargantuan task. Visualizing this transformation schematically:

$$G_1 \overset{T_1}{\rightarrow} P_1 \overset{T_2}{\rightarrow} P_2 \overset{T_3}{\rightarrow} G_2 \overset{T_4}{\rightarrow} G'_1 \rightarrow \cdots$$

(adapted from Lewontin 1974, p. 12). XD

T_1 represents the genetic and epigenetic laws, the aspects of functional biology, or development, that transform a genotype into phenotype. We will refer to this as the "genotype-phenotype map". T_2 is the transformation due to natural selection, T_3 are epigenetic relations that predict genotypes based on the selected phenotypes and finally T_4 the rules of Mendelian genetics.

In practice, there are two bodies of evolutionary theory that exist in parallel, traditional population genetics operating in the genotype space and the biometric theory used in plant and animal breeding, operating in phenotype space. The missing part is the mapping between the genotype and phenotype space.

This leads to a "sleight of hand" (as Lewontin terms it) whereby variables in the equations of one domin, re considered parameters or constants, where, in a full-treatment they would be transformed themselves by the evolutionary process and are in reality functions of the state variables in the other domain. The "sleight of hand" is assuming that we know this mapping. Proceeding as if we do understand it is enough to analyze many cases of interest. For example, if the phenotype is almost one-to-one with genotype (sickle-cell disease) or the time-scale is sufficiently short, the "constants" can be treated as such; however, there are many situations where it is inaccurate.

The Four Processes

Natural Selection

Natural selection is the fact that some traits make it more likely for an organism to survive and reproduce. Population genetics describes natural selection by defining fitness as a propensity or probability of survival and reprduction in a particular environment.

The fitness is normally given by the symbol w=1+s where s is the selection coefficient. Natural selection acts on phenotypes, or the observable characteistics of organisms, but the genetically heritable basis of any phenotype which gives a reproductive advantage will become more common in a population. In this way, natural selection converts differences in fitness into changes in allele frequency in a population over successive generations.

Before the advent of population genetics, many biologists doubted that small difference n fitness were sufficient to make a large differene to evolution. Population geneticists addressed this concern in part by comparing selection to genetic drift. Selection can overcome genetic drift when s is greater than 1 divided by the effective population size. When this criterion is met, the probability that a new advantageous mutant becomes fixed is approximately equal to s. The time until fixation of such an allele depends little on genetic drift, and is approximately proportional to log(sN)/s.

Genetic Drift

Genetic drift is a change in allele frequencies caused by random sampling. That is, the alleles in the offspring are a random sample of those in the parents. Genetic drift may cause gene variants to disappear completely, and thereby reduce genetic variability. In contrast to natural selection, which makes gene variants more common or less common depending on their reproductive success, the changes due to genetic drift are not driven by environmental or adaptive pressures, and may be beneficial, neutral, or detrimental to reproductive success.

The effect of genetic drift is larger for alleles present in a smaller number of copies, and smaller when an allele is present in many copies. Vigorous debates wage among scientists over the relative importance of genetic drift compared with natural selection. Ronald Fisher held the view that genetic drift plays at the most a minor role in evolution, and this remained the dominant view for several decades. In 1968 Motoo Kimura rekindled the debate with his neutral theory of molecular evolution which claims that most of the changes in the genetic material are caused by neutral mutations and genetic drift. The role of genetic drift by means of sampling error in evolution has been criticized by John H Gillespie and Will Provine, who argue that selection on linked sites is a more important stochastic force.

The population genetics of genetic drift are described using either branching processes or a diffusion equation describing changes in allele frequency. These approaches are usually applied to the Wright-Fisher and Moran models of population genetics. Assuming genetic drift is the only evolutionary force acting on an allele, after t generations in many replicated populations, starting with allele frequencies of p and q, the variance in allele frequency across those populations is

$$V_t \approx pq\left(1 - \exp\left\{-\frac{t}{2N_e}\right\}\right)$$

Mutation

Mutation is the ultimate source of genetic variation in the form of new alleles. Mutation can result in several different types of change in DNA sequences; these can either have no effect, alter the product of a gene, or prevent the gene from functioning. Studies in the fly Drosophila melanogaster suggest that if a mutation changes a protein produced by a gene, this will probably be harmful, with about 70 percent of these mutations having damaging effects, and the remainder being either neutral or weakly beneficial.

Mutations can involve large sections of DNA becoming duplicated, usually through genetic recombination. These duplications are a major source of raw material for evolving new genes, with tens to hundreds of genes duplicated in animal genomes every million years.

Most genes belong to larger families of genes of shared ancestry. Novel genes are produced by several methods, commonly through the duplication and mutation of an ancestral gene, or by recombining parts of different genes to form new combinations with new functions.

Here, domains act as modules, each with a particular and independent function, that can be mixed together to produce genes encoding new proteins with novel properties. For example, the human eye uses four genes to make structures that sense light: three for color vision and one for night vision; all four arose from a single ancestral gene.

Another advantage of duplicating a gene (or even an entire genome) is that this increasesredundancy; this allows one gene in the pair to acquire a new function while the other copy performs the original function. Other types of mutation occasionally create new genes from previously noncoding DNA.

In addition to being a major source of variation, mutation may also function as a mechanism of evolution when there are different probabilities at the molecular level for different mutations to occur, a process known as mutation bias.

If two genotypes, for example one with the nucleotide G and another with the nucleotide A in the same position, have the same fitness, but mutation from G to A happens more often than mutation from A to G, then genotypes with A will tend to evolve. Different insertion vs. deletion mutation biases in different taxa can lead to the evolution of different genome sizes. Developmental or mutational biases have also been observed in morphological evolution. For example, according to the phenotype-first theory of evolution, mutations can eventually cause the genetic assimilation of traits that were previously induced by the environment.

Mutation bias effects are superimposed on other processes. If selection would favor either one out of two mutations, but there is no extra advantage to having both, then the mutation that occurs the most frequently is the one that is most likely to become fixed in a population. Mutations leading to the loss of function of a gene are much more common than mutations that produce a new, fully functional gene. Most loss of function mutations are selected against. But when selection is weak, muation bias towards loss of function canaffect evolution. For example, pigments are no longer useful when animals live in the darknessof caves, and tend to be lost.

This kind of loss of function can occur because of mutation bias, and/or because the function had a cost, and once the benefit of the function disappeared, natural selection leads to the loss. Loss of sporulation ability in a bacterium during laboratory evolution appears to have been caused by mutation bias, rather than natural selection against the cost of maintaining sporulation ability. When there is no selection for loss of function, the speed at which loss evolves depends more on the mutation rate than it does on theeffective population size, indicating that it is driven more by mutation bias than by genetic drift.

Evolution of Mutation Rate

Due to the damaging effects that mutations can have on cells, organisms have evolved mechanisms such as DNA repair to remove mutations. Therefore, the optimal mutation rate for a species is a trade-off between costs of a high mutation rate, such as deleterious mutations, and the metabolic costs of maintaining systems to reduce the mutation rate, such as DNA repair enzymes.

Viruses that use RNA as their genetic material have rapid mutation rates, which can be an advantage since these viruses will evolve constantly and rapidly, and thus evade the defensive responses of e.g. the human immune system.

Gene Flow & Transfer

Gene flow is the exchange of genes between populations, which are usually of the same species. Examples of gene flow within a species include the migration and then breeding of organisms, or the exchange of pollen. Gene transfer between species includes the formation of hybrid organisms and horizontal gene transfer.

Gene Flow

Migration into or out of a population can change allele frequencies, as well as introducing genetic variation into a population. Immigration may add new genetic material to the established gene pool of a population. Conversely, emigration may remove genetic material. As barriers to reproduction between two diverging populations are required for the populations to become new species, gene flow may slow this process by spreading genetic differences between the populations. Gene flow is hindered by mountain ranges, oceans and deserts or even man-made structures such as the Great Wall of China, which has hindered the flow of plant genes.

Depending on how far two species have diverged since their most recent common ancestor, it may still be possible for them to produce offspring, as with horses and donkeys mating to producemules. Such hybrids are generally infertile, due to the two different sets of chromosomes being unable to pair up during meiosis.

In this case, closely related species may regularly interbreed, but hybrids will be selected against and the species will remain distinct. However, viable hybrids are occasionally formed and these new species can either have properties intermediate between their parent species, or possess a totally new phenotype. The importance of hybridization in creating new species of animals is unclear, although cases have been seen in many types of animals, with thegray tree frog being a particularly well-studied example.

Hybridization is, however, an important means of speciation in plants, since polyploidy (having more than two copies of each chromosome) is tolerated in plants more readily than in animals.Polyploidy is important in hybrids as it allows reproduction, with the two different sets of chromosomes each being able to pair with an identical partner during meiosis. Polyploids also have more genetic diversity, which allows them to avoid inbreeding depression in small populations.

Horizontal Gene Transfer

Horizontal gene transfer is the transfer of genetic material from one organism to another organism that is not its offspring; this is most common among bacteria. In medicine, this contributes to the spread of antibiotic resistance, as when one bacteria acquires resistance genes it can rapidly transfer them to other species. Horizontal transfer of genes from bacteria to eukaryotes such as the yeast Saccharomyces cerevisiae and the adzuki bean beetle Callosobruchus chinensis may also have occurred.

An example of larger-scale transfers are the eukaryotic bdelloid rotifers, which appear to have received a range of genes from bacteria, fungi, and plants. Viruses can also carry DNA between organisms, allowing transfer of genes even across biological domains. Large-scale gene transfer has also occurred between the ancestors of eukaryotic cells and prokaryotes, during the acquisition of chloroplasts and mitochondria.

Recombination

Basic models of population genetics consider only one gene locus at a time. Because of epistasis, the phenotypic effect of an allele at one locus may depend on which alleles are present at many other loci. If all genes are in linkage equilibrium, the effect of an allele at one locus can be averaged across the gene pool at other loci. In reality, one allele is frequently found in linkage disequilibriumwith genes at other loci, especially with genes located nearby on the same chromosome.

Recombination breaks up this linkage disequilibrium too slowly to avoid genetic hitchhiking, where an allele at one locus rises to high frequency because it is linked to an allele under selection at a nearby locus. This is a problem for population genetic models that treat one gene locus at a time. It can, however, be exploited as a method for detecting the action of natural selection via selective sweeps. In the extreme case of primarily asexual populations, different population genetic equations can be derived and solved, and these behave quite differently to the sexual case.

Genetic Structure

Because of physical barriers to migration, along with limited tendency for individuals to move or spread (vaglity), and tendency to remain or come back to natal place (philopatry), natural populations rarely all interbreed as convenient in thoretical random models (panmixy) (Buston et al., 2007). There is usually a geographic range within which individuals are more closely related to one another than those randomly selected from the general population. This is described as the extent to which a population is genetically strctured (Repaci et al., 2007). Genetic structuring can be caused by migration due to historical climate change, species range expansion or current availability of habitat.

Microbial Population Genetics

Microbial population genetics is a rapidly advancing field of investigation with relevance to many other theoretical and applied areas of scientific investigations. The population genetics of microorganismsays the foundations for tracking the origin and evolution of antibiotic resistance and deadly infectious pathogens. Population genetics of microorganisms is lso an essential factor for devising strategies for the conservation and better utilization of beneficial microbes (Xu, 2010).

Population Genetics

Population genetics was developed as a reconciliation of the Mendelian and biometrician models. A key step was the work of the British biologist and statisticianR.A. Fisher.

In a series of papers starting in 1918 and culminating in hs 1930 book The Genetical Theory of Natural Selection, Fisher showed that the continuous variation measured by the biometricians could be produced by the combined action of many discrete genes, and that natural selection could changegene frequencies in a population, resulting in evolution (though lacking the knowledge of what an actual gene was at this time, it should be said in this sense he understood phenotypic trait frequency, rather than specifically identifiable gene frequency).

In a series of papers beginning in 1924, another British geneticist,J.B.S. Haldane, applied statistical analysis to real-world examples of natural selection, such as the evolution of industrial melanism in peppered moths, and showed that natural selection worked at an even faster rate than Fisher assumed.

The American biologist Sewall Wright, who had a background in animal breeding experiments, focused on combinations of interacting genes, and the effects of inbreeding on small, relatively isolated populations that exhibited genetic drift. In 1932, Wright introduced the conceptof an adaptive landscape and argued that genetic drift and inbreeding could drive a small, isolated sub-population away from an adaptive peak, allowing natural selection to drive it towards different adaptive peaks.

Fisher and Wright had some fundamental disagreements and a controversy about the relative roles of selection and drift continued for much of the century between the Americans and the British. The Frenchman Gustave Malécot was also important early in the development of the discipline.

The work of Fisher, Haldane and Wright founded the discipline of population genetics. This integrated natural selection with Mendelian genetics, which was the critical first step in developing a unified theory of how evolution worked. John Maynard Smith was Haldane's pupil, whilst W.D. Hamilton was heavily influenced by the writings of Fisher. The American George R. Price worked with both Hamilton and Maynard Smith. American Richard Lewontin and Japanese Motoo Kimura were heavily influenced by Wright.

Modern Evolutionary Synthesis

In the first few decades of the 20th century, most field naturalists continued to believe that Lamarckian and orthogenic mechanisms of evolution provided the best explanation for the comlexity they observed in the living world. However, as the field of genetics continued to develop, those views became less tenable.

Theodosius Dobzhansky, a postdoctoral worker in T. H. Morgan's lab, had been influenced by the work on genetic diversity y Russian geneticists such as Sergei Chetverikov. He helped to bridge the divide between the foundations of microevolution developed by the population geneticists and the patterns of macroevolution observed by field biologists, with his 1937 book Genetics and the Origin of Species.

Dobzhansky examined the genetic diversity of wild populations and showed that, contrary to the assumptions of the population geneticists, these populations had large amounts of genetic diversity, with marked differences between sub-populations. The book also took the highly mathematical work of the population geneticists and put it into a more accessible form.

In Great Britain E.B. Ford, the pioneer of ecological genetics, continued throughout the 1930s and 1940s to demonstrate the power of selection due to ecological factors including the ability to maintain genetic diversity throughgenetic polymorphisms such as human blood types. Ford's work would contribute to a shift in emphasis during the course of the modern synthesis towards natural selection over genetic drift.

Nevertheless, ongoing studies of eukaryotic transposable elements, and of their impact on speciation, point again to a major role of nonadaptive processes such as mutation andgenetic drift. Mutation and genetic drift are also viewed as major factors in the evolution of genome complexity

Mendelian Inheritance

Mendelian inheritance (or Mendelian genetics or Mendelism) is a scientific description of how hereditary characteristics are passed from parent organisms to their offspring; it underlies much of genetics. This theoretical framework was initially derived from the work of Gregor Johann Mendel published in 1865 and 1866 which was re-discovered in 1900; it was initially very controversial. When Mendel's theories were integrated with the chromosome theory of inheritance by Thomas Hunt Morgan in 1915, they became the core of classical genetics.

History

The laws of inheritance were derived by Gregor Mendel, a 19th century Austrian Priest/monk conducting hybridization experiments in garden peas

(Pisum sativum). Between 1856 and 1863, he cultivated and tested some 29,000 pea plants. From these experiments he deduced two generalizations which later became known as Mendel's Principles of Heredity or Mendelian inheritance. He described these principles in a two part paper, Experiments on Plant Hybridization that he read to the Natural History Society of Brnoon February 8 and March 8, 1865, and which was published in 1866.

Mendel's conclusions were largely ignored. Although they were not completely unknown to biologists of the time, they were not seen as generally applicable, even by Mendel himself, who thought they only applied to certain categories of species or traits. A major block to understanding their significance was the importance attached by 19th century biologists to the apparent blending of inherited traits in the overall appearance of the progeny, now known to be due to multigene interactions, in contrast to the organ-specific binary characters studied by Mendel.

In 1900, however, his work was "re-discovered" by three European scientists, Hugo de Vries, Carl Correns, and Erich von Tschermak. The exact nature of the "re-discovery" has been somewhat debate: De Vries published first on the subject, mentioning Mendel in a footnote, while Correns pointed out Mendel's priority after having read De Vries's paper and realizing that he himself did not have priority. De Vries may not have acknowledged truthfully how much of his knowledge of the laws came from his own work, or came only after reading Mendel's paper. Later scholars have accused Von Tschermak of not truly understanding the results at all. Regardless, the "re-discovery" made Mendelism an important but controversial theory. Its most vigorous promoter in Europe was William Bateson, who coined the term "genetics" and "allele" to describe many of its tenets.

The model of heredity was highly contested by other biologists because it implied that heredity was discontinuous, inopposition to the apparently continuous variation observable for many traits. Many biologists also dismissed the theory because they were not sure it would apply to all species. However later work by biologists and statisticians such as R.A. Fisher showed that if multiple Mendelian factors were involved in the expression of an individual trait, they could produce the diverse results observed.

Thomas Hunt Morgan and his assistants later integrated the theoretical model of Mendel with the chromosome theory of inheritance, in which the chromosomes of cells were thought to hold the actual hereditary material, and create what is now known as classical genetics, which was extremely successful and cemented Mendel's place in history. Mendel's findings allowed other scientists to predict the expression of traits on the basis of mathematical probabilities.

A large contribution to Mendel's success can be traced to his decision to start his crosses only with plants he demonstrated were true-breeding. He also only measured absolute (binary) characteristics, such as color, shape, and position of the offspring, ather than quantitative characteristics. He expressed his results numerically and subjected them to statistical analysis. His method of data analysis and his large sample size gave credibility to his data. He also had the foresight to follow several successive generations (f2, f3) of his pea plants and record their variations.

Finally, he performed "test crosses" (back-crossing descendants of the initial hybridization to the initial true-breeding lines) to reveal the presence and proportion of recessive characters. Without his hard work and careful attention to procedure and detail, Mendel's work could not have had the impact it made on the world of genetics.

Mendel's Laws

Mendel discovered that when crossing white flower and purple flower plants, the result is not a blend. Rather than being a mix of the two, the offspring was purple flowered. He then conceived the idea of heredity units, which he called "factors", one of which is a recessive characteristic and the other dominant. Mendel said that factors, later called genes, normally occur in pairs in ordinary body cells, yet segregate during the formation of sex cells. Each member of the pair becomes part of the separate sex cell.

The dominant gene, such as the purple flower in Mendel's plants, will hide the recessive gene, the white flower. After Mendel self-fertilized the F1 generation and obtained the 3:1 ratio, he correctly theorized that genes can be paired in three different ways for each trait: AA, aa, and Aa. The capital "A" represents the dominant factor and lowercase "a" represents the recessive. (The last combination listed above, Aa, will occur roughly twice as often as each of the other two, as it can be made in two different ways, Aa or aA.)

Mendel stated that each individual has two factors for each trait, one from each parent. The two factors may or may not contain the same information. If the two factors are identical, the individual is called homozygous for the trait. If the two factors have different information, the individual is called heterozygous. The alternative forms of a factor are called alleles.

The genotype of an individual is made up of the many alleles it possesses. An individual's physical appearance, or phenotype, is determined by its alleles as well as by its environment. An individual possesses two alleles for each trait; one allele is given by the female parent and the other by the male parent. They are passed on when an individual matures and produces gametes: egg and sperm. When gametes form, the paired alleles separate randomly so that each gamete receives a copy of one of the two alleles. The presence of an allele doesn't promise that the trait will be expressed in the individual that posseses it. In heterozygous individuals the only alele that is expressed is the dominant. The recessive allele is present but its expression is hidden.

Mendel summarized his findings in two laws; the Law of Segregation and the Law of Independent Assortment.

Law of Segregation (The "First Law")

The Law of Segregation states that every individual possesses a pair of alleles for any particular trait and that each parent passes a randomly selected copy (allele) of only one of these to its offspring. The offspring then receives its own pair of alleles for that trait. Whichever of the two alleles in the offspring is dominant determines how the offspring expresses that trait (e.g. the color of a plant, the color of an animal's fur, the color of a person's eyes).

More precisely the law states that when any individual produces gametes, the copies of a gene separate so that each gamete receives only one copy (allele). A gamete will receive one allele or the other. The direct proof of this was later found following the observation of meiosis by two independent scientists, the German botanist, Oscar Hertwig in 1876, and the Belgian zoologist, Edouard Van Beneden in 1883. In meiosis, the paternal and maternal chromosomes get separated and the alleles with the traits of a character are segregated into two different gametes.

The two coexisting alleles of an individual for each trait segregate (separate) during gamete formation so that each gamete gets only one of the two alleles. Alleles again unite at random fertilization of gametes.

Law of Independent Assortment (The "Second Law")

The Law of Independent Assortment, also known as "Inheritance Law" states that separate genes for separate traits are passed independently of one another from parents to offspring. That is, the biological selection of a particular gene in the gene pair for one trait to be passed to the offspring has nothing to do with the selection of the gene for any other trait. More precisely the law states that alleles of different genes assort independently of one another during gamete formation.

While Mendel's experiments with mixing one trait always resulted in a 3:1 ratio between dominant and recessive phenotypes, his experiments with mixing two traits (dihybrid cross) showed 9:3:3:1 ratios. But the 9:3:3:1 table shows that each of the two genes are independently inherited with a 3:1 phenotypic ratio. Mendel concluded that different traits are inherited independently of each other, so that there is no relation, for example, between a cat's color and tail length. This is actually only true for genes that are not linked to each other. Independent assortment occurs during meiosis I in eukaryotic organisms, specifically anaphase I of meiosis, to produce a gamete with a mixture of the organism's maternal and paternal chromosomes. Along with chromosomal crossover, this process aids in increasing genetic diversity by producing novel genetic combinations.

Of the 46 chromosomes in a normal diploid human cell, half are maternally-derived (from the mother's egg) and half are paternally-derived (from the father's sperm). This occurs as sexual reproductioninvolves the fusion of two haploid gametes (the egg and sperm) to produce a new organism having the full complement of chromosomes. During gametogenesis—the production of new gametes by an adult—the normal complement of 46 chromosomes needs to be halved to 23 to ensure that the resulting haploid gamete can join with another gamete to produce a diploid organism. An error in the number of chromosomes, such as those caused by a diploid gamete joining with a haploid gamete, is termed aneuploidy.

In independent assortment the chromosomes that end randomly sorted from all possible combinations of maternal and paternal chromosomes. Because gametes end up with a random mix instead of a pre-defined "set" from either parent, gametes are therefore considered assorted independently. As such, the gamete can end up with any combination of paternal or maternal chromosomes. Any of the possible combinations of gametes formed from maternal and paternal chromosomes will occur with equal frequency. For human gametes, with 23 pairs of chromosomes, the number of possibilities is 2 or 8,388,608 possible combinations. The gametes will normally end up with 23 chromosomes, but the origin of any particular one will be randomly selected from paternal or maternal chromosomes. This contributes to the genetic variability of progeny.

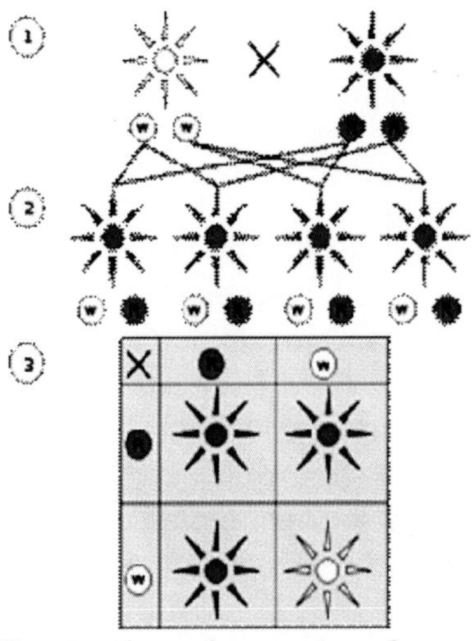

***Figure:** Dominant and recessive phenotypes.*

(1) Parental generation. (2) F_1 geeration. (3) F_2 generation. Dominant (red) and recessive (white) phenotype look alike in the F_1 (first) generation and show a 3:1 ratio in the F_2(second) generation

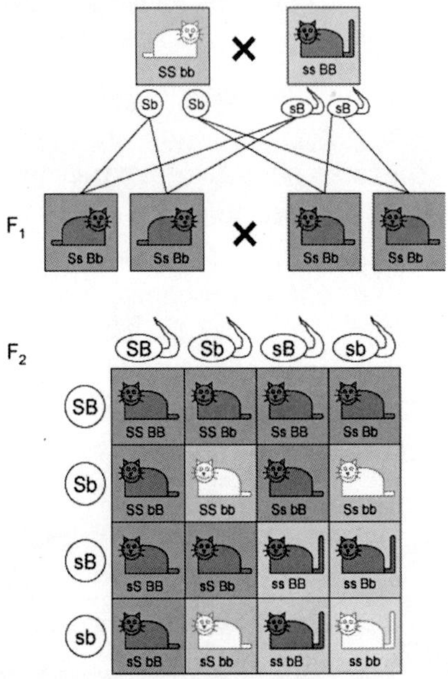

Figure: The phenotypes of two independent traits show a 9:3:3:1 ratio in the F_2 generation. In this example, coat color is indicated by B (brown, dominant) or b(white) while tail length is indicated by S (short, dominant) or s (long). When parents are homozygous for each trait ('SSbb and ssBB), their children in the F_1generation are heterozygous at both loci and only show the dominant phenotypes. If the children mate with each other, in the F_2 generation all combination of coat color and tail length occur: 9 are brown/short (purple boxes), 3 are white/short (pink boxes), 3 are brown/long (blue boxes) and 1 is white/long (green box).

Background

The reason for these laws is found in the nature of the cell nucleus. It is made up of several chromosomes carrying the genetic traits. In a normal cell, each of these chromosomes has two parts, the chromatids. A reproductive cell, which is created in a process called meiosis, usually contains only one of those chromatids of each chromosome. By merging two of these cells (usually one male and one female), the full set is restored and the genes are mixed. The resulting cell becomes a new embryo. The fact that this new life has half the genes of each parent (23 from mother, 23 fro father for total of 46 in the case of humans) is one reason for the Mendelian laws. The second most important reason is the varying dominance of diferent genes, causing some traits to appear unevenly instead of averaging out (whereby dominant doesn't mean more likely to reproduce—recessive genes can become the most common, too).

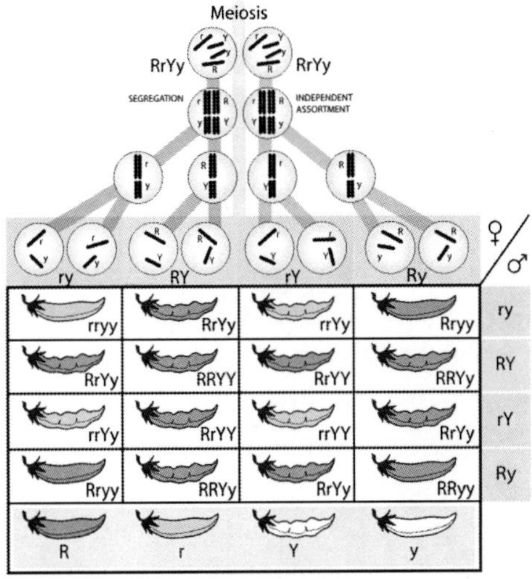

Figure: *Table showing how the genes exchange according to segregation or independent assortment during meiosis and how this translates into Mendel's laws*

Figure: *Mendel's law: Forget-me-not*

There are several advantages of this method (sexual reproduction) over reproduction without genetic exchange:

1. Instead ofnearly identical copies of an organism, a broad range of offspring develops, allowing more different abiities and evolutionary strategies.
2. There are usually some errors in every cell nucleus. Copying the genes usually adds more of them. By distributing them randomly ver different

chromosomes and mixing the genes, such errors will be distributed unevenly over the different children. Some of them will herefore have only very few such problems. This helps reduce problems with copying errors somwhat.

3. Genes can spread faster from one part of a population to another. This is for instance useful if there's a temporary isolation of two groups. New genes developing in each of the populations don't get reduced to half when one sid replaces the other, they mix and form a population with the advantages of both sides.
4. ometimes, a mutation can have positive side effects. For example, sickle cell aneia is a mutation that can cause the benefit of malaria resistance. The mechanism behind the Mendelian laws can make it possble for some offspring to carry the advantages without the disadvantages until furthermutations solve he problems.

Mendelian Trait

A Mendelian trait is one that is controlled by a single locus and shows a simple Mendelian inheritance pattern. In such cases, a mutation in a single gene can cause a disease that is inherited according to Mendel's laws. Examples include sickle-cell anemia, Tay-Sachs disease, cystic fibrosis and xeroderma pigmentosa. A disease controlled by a single gene contrasts with a multi-factorial disease, like arthritis, which is affected by several loci (and the environment) as well as those diseases inherited in a non-Mendelian fashion. The Mendelian Inheritance in Man database is a catalog of, among other things, genes in which Mendelian traits cause disease.

3

Genetic Disorders

A genetic disorder is an illness caused by abnormalities in genes or chromosomes. While some diseases, such as cancer, are due in part to genetic disorders, they can also be caused by environmental factors. Most disorders are quite rare and affect one person in every several thousands or millions. Some types of recessive gene disorders confer an advantage in the heterozygous state in certain environments.

Single Gene Disorder

A single gene disorder is the result of a single mutated gene. There are estimated to be over 4000 human diseases caused by single gene defects. Single gene disorders can be passed on to subsequent generations in several ways. Genomic imprinting and uniparental disomy, however, may affect inheritance patterns. The divisions between recessive and dominant types are not "hard and fast" although the divisions between autosomal and X-linked types are (since the latter types are distinguished purely based on the chromosomal location of the gene).

For example, achondroplasia is typically considered a dominant disorder, but children with two genes for achondroplasia have a severe skeletal disorder that achondroplasics could be viewed as carriers of. Sickle-cell anemia is also considered a recessive condition, but heterozygous carriers have increased resistance to malaria in early childhood, which could be described as a related dominant condition. When a couple where one partner or both are sufferers or carriers of a single gene disorder and wish to have a child they can do so through IVF whichs means they can then have PGD (pre-implantation genetic diagnosis) to check whether the fertilised egg has had the genetic disorder passed on. Autosomal Dominant.

Only one mutated copy of the gene will be necessary for a person to be affected by an autosomal dominant disorder. Each affected person usually has one affected parent. There is a 50% chance that a child will inherit the mutated gene. Conditions that are autosomal dominant often have low penetrance, which means that although only one mutated copy is needed, a relatively small proportion of those who inherit that mutation go on to develop the disease. Examples of this type of disorder are Huntington's disease, neurofibromatosis type 1, Marfan syndrome, hereditary nonpolyposis colorectal cancer, and hereditary multiple exostoses, which is a highly penetrant autosomal dominant disorder. Birth defects are also called congenital anomalies.

Autosomal Recessive

Two copies of the gene must be mutated for a person to be affected by an autosomal recessive disorder. An affected person usually has unaffected parents who each carry a single copy of the mutated gene (and are referred to as carriers). Two unaffected people who each carry one copy of the mutated gene have a 25% chance with each pregnancy of having a child affected by the disorder. Examples of this type of disorder are cystic fibrosis, sickle-cell disease (also partial sickle-cell disease), Tay-Sachs disease, Niemann-Pick disease, spinal muscular atrophy, Roberts sydrome, and Dry (otherwise known as "rice-brand") earwax.

X-linked Dominant

X-linked dominant disorders are caused by mutations in genes on the X chromosome. Only a few disorders have this inheritance pattern, with a prime example being X-linked hypophosphatemic rickets. Males and females are both affected in these disorders, with males typically being more severely affected than females. Some X-linked dominant conditions such as Rett syndrome, incontinentia pigmenti type 2 and Aicardi syndrome are usually fatal in males either in utero or shortly after birth, and are therefore predominantly seen in females.

Exceptions to this finding are extremely rare cases in which boys with Klinefelter syndrome (47,XXY) also inherit an X-linked dominant condition and exhibit symptoms more similar to those of a female in terms of disease severity. The chance of passing on an X-linked dominant disorder differs between men and women. The sons of a man with an X-linked dominant disorder will all be unaffected (since they receive their father's Y chromosome), and his daughters will all inherit the condition. A woman with an X-linked dominant disorder has a 50% chance of having an affected fetus with each pregnancy, although it should be noted that in cases such as incontinentia pigmenti only female offspring are generally viable. In addition, although these conditions do not alter fertility per se, individuals with Rett syndrome or Aicardi syndrome rarely reproduce.

X-linked Recessive

X-linked recessive conditions are also caused by mutations in genes on the X chromosome. Males are more frequently affected than females, and the chance of passing on the disorder differs between men and women. The sons of a man with an X-linked recessive disorder will not be affected, and his daughters will carry one copy of the mutated gene. A woman who is a carrier of an X-linked recessive disorder (XX) has a 50% chance of having sons who are affected and a 50% chance of having daughters who carry one copy of the mutated gene and are therefore carriers. X-linked recessive conditions include the serious diseases Hemophilia A, Duchenne muscular dystrophy, and Lesch-Nyhan syndrome as well as common and less serious conditions such as male pattern baldness and red-green color blindness. X-linked recessive conditions can sometimes manifest in females due to skewed X-inactivation or monosomy X (Turner syndrome).

Y-linked

Y-linked disorders are caused by mutations on the Y chromosome. Because males inherit a Y chromosome from their fathers, *every* son of an affected father will be affected. Because females inherit an X chromosome from their fathers, female offspring of affected fathers are *never* affected.

Since the Y chromosome is relatively small and contains very few genes, there are relatively few Y-linked disorders. Often the symptoms include infertility, which may be circumvented with the help of some fertility treatments. Examples are male infertility and hypertrichosis pinnae.

Mitochondrial

This type of inheritance, also known as maternal inheritance, applies to genes in mitochondrial DNA. Because only egg cells contribute mitochondria to the developing embryo, only mothers can pass on mitochondrial conditions to their children. An example of this type of disorder is Leber's hereditary optic neuropathy. underwear

Multifactorial and Polygenic (Complex) Disorders

Genetic disorders may also be complex, multifactorial, or polygenic, meaning that they are likely associated with the effects of multiple genes in combination with lifestyle and environmental factors. Multifactorial disorders include heart disease and diabetes. Although complex disorders often cluster in families, they do not have a clear-cut pattern of inheritance. This makes it difficult to determine a person's risk of inheriting or passing on these disorders. Complex disorders are also difficult to study and treat because the specific factors that cause most of these disorders have not yet been identified.

On a pedigree, polygenic diseases do tend to "run in families", but the inheritance does not fit simple patterns as with Mendelian diseases. But this does not mean that the genes cannot eventually be located and studied. There is also a strong environmental component to many of them (e.g., blood pressure).

- asthma
- autoimmune diseases such as multiple sclerosis
- cancers
- ciliopathies
- cleft palate
- diabetes
- heart disease
- hypertension
- inflammatory bowel disease
- mental retardation
- mood disorder
- obesity
- refractive error.

Prognosis and Treatment of Genetic Disorders

Genetic disorders rarely have effective treatments, though gene therapy is being tested as a possible treatment for some genetic diseases, including some forms of retinitis pigmentosa

- Gauchers disease is a genetic disease affecting metabolism. It is more treatable then most other genetic diseases, and can be treated with enzyme replacement therapy, medication miglustat, and bone marrow transplantion.

Hypothyroidism

What is Hypothyroidism?

The thyroid is the largest endocrine gland in the body. It sits just below the larynx (voice box) and wraps around the trachea (windpipe). The thyroid gland produces thyroid hormone, which helps the body grow and develop. It also plays an important role in the body's metabolism (the processes in the body that use energy, such as eating, breathing, and regulating heat).

Hypothyroidism (or underactive thyroid) is a common condition in which the thyroid gland produces too little thyroid hormone. About 1 in 5,000 babies is born with congenital hypothyroidism, in which the thyroid fails to grow normally and cannot produce enough of its hormone.

There is no known cause for most cases of congenital hypothyroidism. But about 10 to 20 percent of the time, the condition is caused by an inherited defect that alters the production of thyroid hormone.

The most common inherited form of hypothyroidism is a defect of the TPO (thyroid peroxidase) gene on chromosome 2. This gene plays an important role in thyroid hormone production.

How do People get Hypothyroidism?

Hypothyroidism may be caused by 1) an autoimmune disease that attacks the thyroid gland, 2) surgery or radiation to treat thyroid cancer and other conditions, or 3) rare and random genetic events in which a mutation is acquired during early development.

What are the Symptoms of Hypothyroidism?

In babies with the inherited form of hypothyroidism, the condition affects growth and cognitive development. It may cause mental retardation, delayed puberty, stunted growth, and ataxia (inability to coordinate muscle movements).

In adults, hypothyroidism slows the body's metabolism, making the patient feel mentally and physically sluggish. Symptoms may include weakness, fatigue, muscle aches, mood swings, hair loss, memory loss, or slow speech.

A person's symptoms will depend upon how little thyroid hormone they produce, and for how long they have had the disorder.

When the body is deprived of thyroid hormone, the pituitary gland works overtime, producing extra thyroid-stimulating hormone (TSH). This glut of TSH may enlarge the thyroid into a condition called a goiter.

How do Doctors Diagnose Hypothyroidism?

Babies are normally screened for hypothyroidism 24 hours after birth. A tiny sample of blood taken from the baby's heel is tested for low thyroid hormone levels or high thyroid-stimulating hormone (TSH) levels.

How is Hypothyroidism Treated?

Hormone replacement therapy: people with hypothyroidism must take a synthetic form of thyroid hormone every day to reduce their symptoms.

Interesting Facts about Hypothyroidism

One in 5,000 people has hypothyroidism.

Colon Cancer

What is Colon Cancer?

Cells normally grow and divide only when they are needed to keep our bodies functioning properly. But sometimes, the mechanisms that regulate cell growth stop working and cells divide out of control to form tumors. This is called cancer. When cancer develops in the cells lining the colon (the first part of the large intestine), it is called colon cancer.

People who have a history of colon cancer in their family are at greater risk of getting the disease themselves. The risk increases when a relative has had the disease before age 50. These families are considered high-risk, because they may have inherited one of two rare genetic conditions: FAP (familial adenomatous polyposis) or HNPCC (hereditary non-polyposis colon cancer).

FAP is caused by mutations of the APC (adenomatous polyposis coli) gene on chromosome 5. APC is a tumour suppressor gene, which means that it prevents uncontrolled cell growth. People who inherit a mutated form of this gene develop growths called polyps in their colon.

By age 15, they may have hundreds of these polyps. Polyps are not cancerous at first, but if they aren't treated, they will develop into colon cancer.

HNPCC (also called Lynch syndrome) is caused by mutations in one of several genes that fix damaged DNA. People who inherit one of these mutations have a much greater risk of accumulating mutations that will lead to uncontrolled cel growth and cancer.

How do People get Colon Cancer?

FAP and HNPCC are both inherited in an autosomal dominant pattern. If a parent has FAP or HNPCC, his or her children run a 50 percent risk of inheriting the mutated gene.

Usually when a person inherits a defective gene it does not necessarily mean he or she will develop a malignant cancer.

However, the APC gene strikingly predisposes one to colon cancer. People who inherit one bad copy of the APC gene are practically guaranteed to develop colon cancer by age 40.

Similarly, people who inherit one bad copy of a gene associated with HNPCC have an 80 percent chance of getting colon cancer before. HNPCC also increases a person's risk of developing other cancers, including ovarian, stomach, brain, and liver.

What are the Symptoms of Colon Cancer?

Colon cancer affects the stomach and bowels. Common symptoms include: diarrhea or constipation,blood in the stool, vomiting, bloating, cramps, and unexplained weight loss.

How do Doctors Diagnose Colon Cancer?

When a patient shows symptoms of colon cancer, his or her doctor can screen for the disease using one of several tests:

- Fecal Occult Blood Test (FOBT) - Colon cancer can sometimes cause tiny dots of blood, too small for the eye to see, in the feces The FOBT test uses a special chemical to check the patient's stool sample for these traces of blood.
- Flexible-Sigmoidoscopy - Using a thin flexible tube alled a simoidoscope, the doctor looks inside the patient's colon for growths called polyps.
- Double Contrast Barium Enema (DCBA) - A silvery-white metallic substance called barium is inserted up the patient's colon through the rectum. The barium outlines the patient's colon on an x-ray screen.
- Colonoscopy - Using a thin instrument called a colonoscope, the doctor looks inside the patient's colon. During the procedure, the doctor removes pieces of tissue (called a biopsy) to test them for cancer. If the doctor finds any polyps, he or she can also remove them. A newer method, called virtual colonoscopy, looks at the colon without going into the body, with an MRI or CT scan.
- DNA-Based stool test - This test examines DNA taken from a patient's stool sample to look for genetic defects associated with colon cancer.

How is Colon Cancer Treated?

Colon cancer is very treatable. In fact, about90 percent of patients survive the disase after treatment. First, doctors stage the disease to see how far it has progressed. If the cancer has not pread to other tissues of the body, it can be treated with: special chemicals (chemotherapy) or radition (powerful x-rays) that kill all rapidly dividing cells in the body, including cancer cells surgery to remove the polyps and/or cancerous part of the colon.

Interesting Facts about Colon Cancer

People who have FAP can develop hundreds and even thousands of polyps in their colon, whereas people with HNPCC develop relatively few.

The progression from a benign to a malignant cancer typically requires multiple mutations that allow cells to acquire new and abnormal characteristics, such as an increased growth rate, inability to adhere or stick to neighbouring cells, propensity to migrate to other places in the body, etc. For example, at least seven mutations are required to produce a malignant colon tumour.

Inherited cancers often provide clues about the genes mutated in noninherited (sporadic) cancers. For example, mutations in the APC gene are found not only in FAP tumors but in 85% of all sporadic colon tumors as well.

Breast Cancer and Ovarian Cancer

What is Breast Cancer and Ovarian Cancer?

Cells normally grow and divide only when they are needed to keep our bodies functioning properly. But sometimes, the mechanisms that regulate cell growth stop working and cells divide out of control to form tumors. This is called cancer. When cancer develops in the cells of breast or ovarian tissue it is called breast or ovarian cancer, respectively.

Most people who develop breast or ovarian cancer have no history of the disease in their family. In fact, only 5 to 10 percent of all breast and ovarian cancers are caused by inherited genetic factors. These rare cases typically result from inherited mutations in either the BRCA1 or BRCA2 gene.

The BRCA1 and BRCA2 genes are called tumour suppressor genes because they prevent uncontrolled cell growth. BRCA1 is located on chromosome 17, and BRCA2 is located on chromosome 13. Scientists believe BRCA1 and BRCA2 work by fixing damaged or broken DNA.

Women who inherit a mutated copy of the BRCA1 or BRCA2 gene accumulate broken and deformed chromosomes, and therefore have a greater chance of accumulating mutations that will lead to uncontrolled cell growth and cancer. Men who inherit the defective genes are also more likely to develop breast and/or prostrate cancer. (Yes, men can get breast cancer.)

How do you get Breast or Ovarian Cancers?

High-risk families include those who have inherited a mutation in either the BRCA1 or BRCA2 gene. The mutated BRCA1 and BRCA2 genes are inherited in an autosomal dominant pattern.

A child only needs to inherit one copy of the mutated gene to have an increased cancer risk. Children who have a parent with the BRCA1 or BRCA2 mutation have a 50 percent chance of inheriting the mutation.

Just because a person inherits the defective gene does not mean he or she will develop cancer, but inheritance greatly increases the risk. Out of every 100 women who inherit a mutated BRCA1 or BRCA2 gene, as many as 60 will develop breast cancer by age 50; by age 70, approximately 80 will develop breast cancer.

How do Doctors Test for BRCA1 and BRCA2 Mutations?

A person with a strong family history of breast or ovarian cancer is a likely candidate for genetic screening. By analyzing a sample of the patient's blood, doctors can identify whether the person has inherited the RCA1 or BRCA2 mutation. The test cannot tell if or when the person will develop breast or ovarian cancer; it can only tell if he or she is at risk because of the faulty gene.

What can a Woman do if she Inherits these Mutations?

Some women who discover that they've inherited the mutated BRCA1 and BRCA2 genes undergo special treatments to protect themselves from breast and ovarian cancer. They may have one or both breasts removed (called a mastectomy or double mastectomy), and their ovaries removed.

They may take the medicine tamoxifen, which is believed to protect against breast cancer.

Interesting Facts about Breast and Ovarian Cancer

The progression from a benign to a malignant cancer typically requires multiple mutations that allow cells to acquire new and abnormal characteristics, such as an increased growth rate, inability to adhere or stick to neighbouring cells, propensity to migrate to other places in the body, etc. Genes involved in the repair of DNA damage (such as BRCA1 and BRCA2) are often associated with cancer. This is because they allow mutations to accumulate at a much faster rate.

Usually, people inherit one of several hundred different mutations of the BRCA1 and BRCA2 genes. But for some reason, Eastern European (Ashkenazi) Jews seem to inherit only three of theses different mutations. Ashkenazi Jews are also 10 times more likely to have mutations in the BRCA1 and BRCA2 genes than any other ethnic group.

Alzheimer's Disease

What is Alzheimer's Disease?

Alzheimer's is a disease that causes dementia, or loss of brain function. It affects the parts of the brain that deal with memory, thought, and language.

The brain of a person with Alzheimer's contains abnormal clumps of cellular debris and protein(plaques) and collapsed microtubules (support structures of the cell). Microtubule disintegration is caused by a malfunctioning protein called tau, which normally stabalizes the microtubules.

In Alzheimer's patients, tau proteins instead cluster together to form disabling tangles. These plaques and tangles damage the healthy cells around them. The brain also produces smaller amounts of neurotransmitters (acetylcholine, serotonin, and norepinephrine), chemicals that allow nerve cells to talk to one another.

The most common form of the disease, which strikes after age 65, is linked to the apolipoprotein E (apoE) gene on chromosome 19. Scientists don't know how apoE4 increases the risk of developing Alzheimer's. They do know that everyone has apoE, which comes in three forms.

One of the forms (apoE4) increases a person's risk of developing Alzheimer's. The other two forms seem to protect against the disease. While people who inherit the apoE4 form of the gene are at increased risk for the disease, they will not necessarily develop it.

Mutations in genes found on chromosomes 1, 14, and 21 are linked to rarer forms of the disease, which strike earlier in life.

How do People get Alzheimer's Disease?

Scientists don't know exactly how people develop Alzheimer's, but they believe it is caused by a combination of genes and environmental factors (multifactorial disorder). The early-onset forms of Alzheimer's are inherited in an autosomal dominant pattern, which means that only one parent has to pass down a defective copy of the gene for their child to develop the disorder.

What are the Symptoms of Alzheimer's Disease?

Because Alzheimer's destroys brain cells, people who have the disorder slowly lose their ability to think clearly. At first, they may forget words or names, or have trouble finding things. As the disorder worsens, they may forget how to do simple tasks (such as walking to a friend's house or brushing their hair). Some people with Alzheimer's also feel nervous or sad.

How do Doctors Diagnose Alzheimer's Disease?

There is no one test for Alzheimer's. Doctors use several different tests to check a patient's memory, language skills, and problem solving abilities. These tests don't diagnose Alzheimer's, but they can rule out other disorders that have similar symptoms.

How is Alzheimer's Disease Treated?

There is no cure for Alzheimer's, but a few medicines can slow its symptoms. A drug called Aricept increases the amount of the neurotransmitter acetylcholine in the brain. A new medicine, Namenda, protects brain cells from a chemical called glutamate, which can damage nerve cells. Doctors may also give their Alzheimer's patients antidepressants or anti-anxiety medicines to ease some of their symptoms.

People with Alzheimer's often need a caregiver - someone to help them get around and do the things they were once able to do themselves.

Interesting Facts about Alzheimer's Disease

Alzheimer's was named after the German doctor, Alois Alzheimer, who first named the disorder in 1906.

The older a person gets, the higher his or her risk of getting Alzheimer's. Only about 1 or 2 people out of 100 have Alzheimer's at age 65; whereas, one out of every five people has the disorder by age 80.

Down Syndrome

What is Down Syndrome?

Down syndrome is a developmental disorder caused by an extra copy of chromosome 21 (which is why the disorder is also called "trisomy 21"). Having an extra copy of this chromosome means that each gene may be producing more protein product than normal. Cells seem to tolerate this better than having not enough protein, or having altered protein due to a mutation in the DNA sequence. However, producing too much protein can also have serious consequences, as seen in Down syndrome. Genes on chromosome 21 that specifically contribute to the various symptoms of Down syndrome are now being identified.

How do People get Down Syndrome?

Down syndrome is typically caused by what is called nondisjunction. If a pair of number 21 chromosomes fails to separate during the formation of an egg (or sperm), this is referred to as nondisjunction. When that egg unites with a normal sperm to form an embryo, that embryo ends up with three copies of chromosome 21 instead of the normal two. The extra chromosome is then copied in every cell of the baby's body.

Interestingly, nondisjunction events seem to occur more frequently in older women. This may explain why the risk of having a baby with Down syndrome is greater among mothers age 35 and older.

In rare cases Down syndrome is caused by a Robertsonian translocation, which occurs when the long arm of chromosome 21 breaks off and attaches to another chromosome at the centromere. The carrier of such a translocation will not have Down syndrome, but can produce children with Down syndrome. Javascript is required to view this content.

What are the Symptoms of Down Syndrome?

People with Down syndrome have very distinct facial features: a flat face, a small broad nose, abnormally shaped ears, a large tongue, and upward slanting eyes with small folds of skin in the corners. People with Down syndrome have an increased risk of developing a number of medically significant problems:

respiratory infections, gastrointestinal tract obstruction (blocked digestive tract), leukemia, heart defects, hearing loss, hypothyroidism, and various eye abnormalities. They also exhibit moderate to severe mental retardation; children with Down syndrome usually develop more slowly than their peers, and have trouble learning to walk, talk, and take care of themselves.

Because of these medical problems most people with Down syndrome have a decreased life expectancy. About half live to be 50 years of age.

How do Docors Diagnose Down Syndrome?

Two types o tests check for Down syndrome during a woman's pregnancy: screening and diagnostic tests.

Screening tests identify a mother who is likely carrying a baby with Down syndrome. Themost common screening tests are the Triple Screen and the Alpha-Fetoprotein Plus. These tests measure levels of certain substances in the blood.

Alternatively, ultrasounds (which use sound waves to look inside the mother's uterus) allow the doctor to examine the fetus in the womb for the physical signs of Down syndrome.

To confirm a positive result identified in a screening test, one of the following diagnostic tests can be performed: chorionic villus sampling (CVS), amniocentesis, and percutaneous umbilical blood sampling (PUBS).

Each takes a sample from the placenta, amniotic fluid, or umbilical cord, respectively, to examine the baby's chromosomes and determine if he or she has an extra chromosome 21.

If Down syndrome is not diagnosed in the womb, doctors can usually recognize it after the baby is born by the distinctive facial features. The diagnosis is confirmed with a karyotype - an examination of the baby's chromosomes.

How is Down Syndrome Treated?

No cure exists for Down syndrome. But physical therapy and/or speech therapy can help people with the disorder develop more normally. Screening for common medical problems associated with the disorder, followed by corrective surgery, can often improve quality of life.

Moreover, enriched environments significantly increase their capacity to learn and lead a meaningful life.

Interesting Facts about Down Syndrome

Down syndrome is really the only trisomy compatible with life. Only two other trisomies have been observed in babies born alive (trisomies 13 and 18), but babies born with these trisomies have only a 5% chance of surviving longer than one year. In 90% of Trisomy 21 cases, the additional chromosome comes from the mother's egg rather than the father's sperm.

Down syndrome is the most common genetic disorder caused by a chromosomal abnormality. It affects 1 out of every 800 to 1,000 babies. Down syndrome was originally described in 1866 by John Langdon Down. It wasn't until 1959 that a French doctor, named Jerome Lejeune, discovered it was caused by the inheritance of an extra chromosome 21.

Severe Combined Immunodeficiency

What is Severe Combined Immunodeficiency (SCID)?

SCID is a group of very rare-and potentially fatal-inherited disorders related to the immune system. The immune system normally fights off attacks from dangerous bacteria and viruses. People with SCID have a defect in their immune system that leaves them vulnerable to potentially deadly infections.

There are several types of SCID. The most common form is caused by a mutation in the SCIDX1 gene located on the X chromosome. This gene encodes a protein that is used to construct a receptor called IL2RG (interleukin-2 receptor). These receptors reside in the plasma membrane of immune cells. Their job is to allow two types of immune cells - T cells and B cells - to communicate. When the gene is mutated, the receptors cannot form and are absent from immune cells. As a result, the immune cells can't communicate with one another about invaders in the environment. Not enough T and B cells are produced to fight off the infection, and the body is left defenseless.

Another form of SCID is caused by a mutation on chromosome 20 and is characterized by a deficiency of the enzyme adenosine deaminase.

How do People Get SCID?

The most common form of SCID exhibits an X-linked recessive pattern of inheritance, and is therefore referred to as X-linked SCID. When a gene is located on the X chromosome, males are more often affected than females. Males do not have a second X chromosome to compensate for the defective one. They only need to inherit one bad copy of he gene to have the symptoms of the disorder. Females, on the other hand, have two X chromosomes. If they inherit one defective X chromosome, they still have its healthy pair. They don't develop symptoms of the disorder, but they still carry the faulty gene and can pass it on to their children.

What are the Symptoms of SCID?

Symptoms usually appear in the first few months of life. Because the immune system cannot protect the baby's body, babieswith the disorder tend to get one infection after another. Some of these bacterial infections may be life-threatening, including pneumonia (lung infection), meningitis (brain infection), and sepsis (blood infection). To make matters worse, SCID patients often don't respond to the antibiotics used to treat bacterial infections. They may suffer more frequently from ear infections, sinus infections, a chronic cough, and rashes on the skin.

Early diagnosis of SCID is very important, because without quick treatment, children with the disease aren't likely to live past age 2.

How do Doctors Diagnose SCID?

SCID can be identified before the baby is born by removing and testing cells from the placenta (chorionic villus sampling or CVS), or by removing and testing a sample of the fluid surrounding the baby (amniocentesis). Most babies are diagnosed with SCID in the first 6 months of life. The most common screening methods are an immune function test, and a blood test that detects low white blood cell counts, as well as low levels of immune cells (T cells and B cells).

How is SCID Treated?

Children with SCID must be careful to stay away from germ-rich environments (such as day care centres and crowded shopping malls) where they could pick up a potentially life-threatening infection. The most effective treatment is a bone marrow transplant. Unspecialized stem cells (that will form blood and immune cells) are taken from the bone marrow of a healthy donor and injected into the SCID patient. Ideally, these new cells will stimulate the production of the needed immune cells. Transplants done within the first few months of life are most successful. The tissue must be "matched" to the patient, however, which can limit the usefulness of this therapy. Siblings make the best donors as their cells likely contain a similar genetic make-up.

Gene therapy for this disorder may soon be possible. This therapy would compensate for the faulty gene by injecting healthy copies of the gene into a patient's bone marrow stem cells.

Interesting Facts about SCID

About 1 out of every 100,000 babies is born with SCID.

SCID is sometimes called Bubble Boy disease. In the 1970s, a boy named David Vetter had to live in a plastic bubble for 12 years because of SCID.

4

Genetic Code

The genetic code is the set of rules by which information encoded in genetic material (DNA or mRNA sequences) is translated into proteins (amino acid sequences) by living cells. The code defines a mapping between tri-nucleotide sequences, called codons, and amino acids. With some exceptions, a triplet codon in a nucleic acid sequence specifies a single amino acid. Because the vast majority of genes are encoded with exactly the same code, this particular code is often referred to as the canonical or standard genetic code, or simply *the* genetic code, though in fact there are many variant codes. For example, protein synthesis in human mitochondria relies on a genetic code that differs from the standard genetic code.

Not all genetic information is stored using the genetic code. All organisms' DNA contains regulatory sequences, intergenic segments, and chromosomal structural areas that can contribute greatly to phenotype. Those elements operate under sets of rules that are distinct from the codon-to-amino acid paradigm underlying the genetic code.

Discovery

After the structure of DNA was deciphered by James Watson, Francis Crick, Maurice Wilkins and Rosalind Franklin, serious efforts to understand the nature of the encoding of proteins began. George Gamow postulated that a three-letter code must be employed to encode the 20 standard amino acids used by living cells to encode proteins, because 3 is the smallest integer n such that 4 is at least 20.

The fact that codons consist of three DNA bases was first demonstrated in the Crick, Brenner et al. experiment. The first elucidation of a codon was done by Marshall Nirenberg and Heinrich J. Matthaei in 1961 at the National Institutes of Health. They used a cell-free system to translate a poly-uracil RNA sequence (i.e., UUUUU...) and discovered that the polypeptide that they had synthesized consisted of only the amino acid phenylalanine. They thereby deduced that the codon UUU specified the amino acid phenylalanine. This was followed by experiments in the laboratory of Severo Ochoa demonstrating that the poly-adenine RNA sequence (AAAAA...) coded for the polypeptide, poly-lysine. and the poly-cytosine RNA sequence (CCCCC...) coded for the polypeptide, poly-proline.

Therefore the codon AAA specified the amino acid lysine, and the codon CCC specified the amino acid proline. Using different copolymers most of the remaining codons were then determined. Extending this work, Nirenberg and Philip Leder revealed the triplet nature of the genetic code and allowed the codons of the standard genetic code to be deciphered. In these experiments various combinations of mRNA were passed through a filter which contained ribosomes, the components of cells that translate RNA into protein. Unique triplets promoted the binding of specific tRNAs to the ribosome. Leder and Nirenberg were able to determine the sequences of 54 out of 64 codons in their experiments.

Subsequent work by Har Gobind Khorana identified the rest of the genetic code. Shortly thereafter, Robert W. Holley determined the structure of transfer RNA (tRNA), the adapter molecule that facilitates the process of translating RNA into protein. This work was based upon earlier studies by Severo Ochoa, who received the Nobel prize in 1959 for his work on the enzymology of RNA synthesis. In 1968, Khorana, Holley and Nirenberg received the Nobel Prize in Physiology or Medicine for their work.

Transfer of Information Via the Genetic Code

The genome of an organism is inscribed in DNA, or in the case of some viruses, RNA. The portion of the genome that codes for a protein or an RNA is referred to as a gene. Those genes that code for proteins are composed of tri-nucleotide units called codons, each coding for a single amino acid. Each nucleotide sub-unit consists of a phosphate, deoxyribose sugar and one of the 4 nitrogenous nucleobases.

The purine bases adenine (A) and guanine (G) are larger and consist of two aromatic rings. The pyrimidine bases cytosine (C) and thymine (T) are smaller and consist of only one aromatic ring. In the double-helix configuration, two strands of DNA are joined to each other by hydrogen bonds in an arrangement known as base pairing. These bonds almost always form between an adenine base on one strand and a thymine on the other strand and between a cytosine base on one strand and a guanine base on the other. This means that the number of A and T residues will be the same in a given double helix, as will the number of G and C residues. In RNA, thymine (T) is replaced by uracil (U), and the deoxyribose is substituted by ribose.

Each protein-coding gene is transcribed into a template molecule of the related polymer RNA, known as messenger RNA or mRNA. This, in turn, is translated on the ribosome into an amino acid chain or polypeptide. The process of translation requires transfer RNAs specific for individual amino acids with the amino acids covalently attached to them, guanosine triphosphate as an energy source, and a number of translation factors. tRNAs have anticodons complementary to the codons in mRNA and can be "charged" covalently with amino acids at their 3' terminal CCA ends. Individual tRNAs are charged with specific amino acids by enzymes known as aminoacyl tRNA synthetases, which have high specificity for both their cognate amino acids and tRNAs. The high specificity of these enzymes is a major reason why the fidelity of protein translation is maintained.

There are $4^3 = 64$ different codon combinations possible with a triplet codon of three nucleotides; all 64 codons are assigned for either amino acids or stop signals during translation.

If, for example, an RNA sequence, UUUAAACCC is considered and the reading frame starts with the first U (by convention, 5' to 3'), there are three codons, namely, UUU, AAA and CCC, each of which specifies one amino acid. This RNA sequence will be translated into an amino acid sequence, three amino acids long. A comparison may be made with computer science, where the codon is similar to a word, which is the standard "chunk" for handling data (like one amino acid of a protein), and a nucleotide is similar to a bit, in that it is the smallest unit.

Salient Features

Sequence Reading Frame

A codon is defined by the initial nucleotide from which translation starts. For example, the string GGGAAACCC, if read from the first position, contains the codons GGG, AAA and CCC; and, if read from the second position, it contains the codons GGA and AAC; if read starting from the third position, GAA and ACC. Every sequence can thus be read in three reading frames, each of which will produce a different amino acid sequence (in the given example, Gly-Lys-Pro, Gly-Asn, or Glu-Thr, respectively). With double-stranded DNA there are six possible reading frames, three in the forward orientation on one strand and three reverse on the opposite strand.

The actual frame in which a protein sequence is translated is defined by a start codon, usually the first AUG codon in the mRNA sequence.

Start/Stop Codons

Translation starts with a chain initiation codon (start codon). Unlike stop codons, the codon alone is not sufficient to begin the process. Nearby sequences (such as the Shine-Dalgarno sequence in *E. coli*) and initiation factors are also required to start translation.

The most common start codon is AUG which is read as methionine or, in bacteria, as formylmethionine. Alternative start codons (depending on the organism), include "GUG" or "UUG", which normally code for valine or leucine, respectively. However, when used as a start codon, these alternative start codons are translated as methionine or formylmethionine.

The three stop codons have been given names: UAG is *amber*, UGA is *opal* (sometimes also called *umber*), and UAA is *ochre*. "Amber" was named by discoverers Richard Epstein and Charles Steinberg after their friend Harris Bernstein, whose last name means "amber" in German.

The other two stop codons were named "ochre" and "opal" in order to keep the "color names" theme. Stop codons are also called "termination" or "nonsense" codons and they signal release of the nascent polypeptide from the ribosome due to binding of release factors in the absence of cognate tRNAs with anticodons complementary to these stop signals.

Effect of Mutations

During the process of DNA replication, errors occasionally occur in the polymerization of the second strand. These errors, called mutations, can have an impact on the phenotype of an organism, especially if they occur within the protein coding sequence of a gene. Error rates are usually very low—1 error in every 10–100 million bases—due to the "proofreading" ability of DNA polymerases.

Missense mutations and nonsense mutations are examples of point mutations, which can cause genetic diseases such as sickle-cell disease and thalassemia respectively. Clinically important missense mutations generally change the properties of the coded amino acid residue between being basic, acidic polar or non-polar, whereas nonsense mutations result in a stop codon.

Mutations that disrupt the reading frame sequence by indels (insertions or deletions) of a non-multiple of 3 nucleotide bases are known as frameshift mutations. These mutations usually result in a completely different translation from the original, and are also very likely to cause a stop codon to be read, which truncates the creation of the protein.

These mutations may impair the function of the resulting protein, and are thus rare in *in vivo* protein-coding sequences. One reason inheritance of frameshift mutations is rare is that if the protein being translated is essential for growth under the selective pressures the organism faces, absence of a functional protein may cause death before the organism is viable. Frameshift mutations may result in severe genetic diseases such as Tay-Sachs disease.

Although most mutations that change protein sequences are harmful or neutral, some mutations have a positive effect on an organism. These mutations may enable the mutant organism to withstand particular environmental stresses better than wild-type organisms, or reproduce more quickly.

In these cases a mutation will tend to become more common in a population through natural selection. Viruses that use RNA as their genetic material have rapid mutation rates, which can be an advantage since these viruses will evolve constantly and rapidly, and thus evade the defensive responses of e.g. the human immune system. In large populations of asexually reproducing organisms, for example, E. coli, multiple beneficial mutations may co-occur, causing competition among them, this phenomenon is called clonal interference.

Degeneracy of the Genetic Code

The genetic code has redundancy but no ambiguity. For example, although codons GAA and GAG both specify glutamic acid (redundancy), neither of them specifies any other amino acid (no ambiguity). The codons encoding one amino acid may differ in any of their three positions. For example the amino acid glutamic acid is specified by GAA and GAG codons (difference in the third position), the amino acid leucine is specified by UUA, UUG, CUU, CUC, CUA, CUG codons (difference in the first or third position), while the amino acid serine is specified by UCA, UCG, UCC, UCU, AGU, AGC (difference in the first, second or third position).

A position of a codon is said to be a fourfold degenerate site if any nucleotide at this position specifies the same amino acid. For example, the third position of the glycine codons (GGA, GGG, GGC, GGU) is a fourfold degenerate site, because all nucleotide substitutions at this site are synonymous; i.e., they do not change the amino acid. Only the third positions of some codons may be fourfold degenerate.

A position of a codon is said to be a twofold degenerate site if only two of four possible nucleotides at this position specify the same amino acid. For example, the third position of the glutamic acid codons (GAA, GAG) is a twofold degenerate site. In twofold degenerate sites, the equivalent nucleotides are always either two purines (A/G) or two pyrimidines (C/U), so only transversional substitutions (purine to pyrimidine or pyrimidine to purine) in twofold degenerate sites are nonsynonymous. A position of a codon is said to be a non-degenerate site if any mutation at this position results in amino acid substitution. There is only one threefold degenerate site where changing to three of the four nucleotides may have no effect on the amino acid (depending on what it is changed to), while changing to the fourth possible nucleotide always results in an amino acid substitution. This is the third position of an isoleucine codon: AUU, AUC, or AUA all encode isoleucine, but AUG encodes methionine. In computation this position is often treated as a twofold degenerate site.

There are three amino acids encoded by six different codons: serine, leucine, and arginine. Only two amino acids are specified by a single codon. One of these is the amino-acid methionine, specified by the codon AUG, which also specifies the start of translation; the other is tryptophan, specified by the codon UGG. The degeneracy of the genetic code is what accounts for the existence of synonymous mutations.

Degeneracy results because a triplet code designates 20 amino acids and a stop codon. Because there are four bases, triplet codons are required to produce at least 21 different codes. For example, if there were two bases per codon, then only 16 amino acids could be coded for ($4^2=16$). Because at least 21 codes are required, then 4^3 gives 64 possible codons, meaning that some degeneracy must exist.

These properties of the genetic code make it more fault-tolerant for point mutations. For example, in theory, fourfold degenerate codons can tolerate any point mutation at the third position, although codon usage bias restricts this in practice in many organisms; twofold degenerate codons can tolerate one out of the three possible point mutations at the third position. Since transition mutations (purine to purine or pyrimidine to pyrimidine mutations) are more likely than transversion (purine to pyrimidine or vice-versa) mutations, the equivalence of purines or that of pyrimidines at twofold degenerate sites adds a further fault-tolerance.

A practical consequence of redundancy is that some errors in the genetic code only cause a silent mutation or an error that would not affect the protein because the hydrophilicity or hydrophobicity is maintained by equivalent substitution of amino acids; for example, a codon of NUN (where N = any nucleotide) tends to code for hydrophobic amino acids. NCN yields amino acid residues that are small in size and moderate in hydropathy; NAN encodes average size hydrophilic residues. These tendencies may result from the shared ancestry of the aminoacyl tRNA synthetases related to these codons.

Even so, single point mutations can still cause dysfunctional proteins. For example, a mutated hemoglobin gene causes sickle-cell disease. In the mutant hemoglobin a hydrophilic glutamate (Glu) is substituted by the hydrophobic valine (Val), that is, GAA or GAG becomes GUA or GUG. The substitution of glutamate by valine reduces the solubility of â-globin which causes hemoglobin to form linear polymers linked by the hydrophobic interaction between the valine groups causing sickle-cell deformation of erythrocytes. Sickle-cell disease is generally not caused by a *de novo* mutation. Rather it is selected for in malarial regions (in a way similar to thalassemia), as heterozygous people have some resistance to the malarial *Plasmodium* parasite (heterozygote advantage).

These variable codes for amino acids are allowed because of modified bases in the first base of the anticodon of the tRNA, and the base-pair formed is called a wobble base pair. The modified bases include inosine and the Non-Watson-Crick U-G basepair.

Variations to the Standard Genetic Code

While slight variations on the standard code had been predicted earlier, none were discovered until 1979, when researchers studying human mitochondrial genes discovered they used an alternative code. Many slight variants have been discovered since, including various alternative mitochondrial codes, as well as small variants such as *Mycoplasma* translating the codon UGA as tryptophan and *Candida* species translating CUG as a serine rather than a leucine. In bacteria and archaea, GUG and UUG are common start codons. However, in rare cases, certain specific proteins may use alternative initiation (start) codons not normally used by that species.

In certain proteins, non-standard amino acids are substituted for standard stop codons, depending upon associated signal sequences in the messenger RNA: UGA can code for selenocysteine and UAG can code for pyrrolysine as discussed in the relevant articles. Selenocysteine is now viewed as the 21st amino acid, and pyrrolysine is viewed as the 22nd.

Notwithstanding these differences, all known codes have strong similarities to each other, and the coding mechanism is the same for all organisms: three-base codons, tRNA, ribosomes, reading the code in the same direction and translating the code three letters at a time into sequences of amino acids.

Expanded Genetic Code

An expanded genetic code refers to an artificially modified genetic code in which one or more specific codons have been allocated to encode an amino acid which is not among the twenty/twentytwo found in nature.

In order to understand how expansion of the genetic code is achieved, the translational mechanism catalysed by ribosomes should be considered: the central parts are the transfer RNAs (tRNA) which are used as keys to decode a three nucleotide code on the RNA being translated into its equivalent amino acid. To do this the tRNA recognises a specific codon thanks to a complementary sequence called the anticodon on one of its loops.

As a result generally there are one (or more) tRNAs for each codon (or group of degenerate codons). The encoding of a codon to its amino acid is a result of the aminoacyl tRNA synthetase which adds the aminoacyl group to its allocated tRNA. However, the aminoacyl tRNA synthetase often recognises a specific motif irrespective of the anticodon, meaning that if the anticodon were to be mutated the encoding of that amino acid would change to a new codon.

For successful translation of a novel amino acid, the codon to which the amino acid is reassigned must be free or unfavoured and the novel tRNA and synthetase set (called the orthogonal set when including the codon) must not crosstalk with the endogenous tRNA and synthetase sets, while still being functionally compatible with the ribosome and other components of the translation apparatus. The tRNA synthetase pair is taken from a distant organism, generally from a different domain, and the active site of the synthetase is modified to accept the non-natural amino acid.

The possibility of reassigning codons was realized by Normanly et al. in 1990 when a viable mutant strain of E. coli read through the amber codon. As a result the amber codon became the choice codon to be assigned a novel amino acid. Later, in the Schultz lab the tRNATyr/tyrosyl-tRNA synthetase (TyrRS) from Methanococcus jannaschii was used to introduce a tyrosine instead of STOP, the default value of the amber codon. As mentioned, this was possible because of the differences between the endogenous bacterial synthases and the orthologous archeal synthase which do not recognise each other.

Directed Evolution

Several methods for selecting the synthetase which accepts only the non natural amino acid have been developed. One of which is by using a combination of positive and negative selection

Some amino acids which have been added in order to label protein.

This orthologous set can then be mutated and screened through directed evolution to accept a different, even novel, amino acid. Mutations to the plasmid containing the pair can be introduced by error-prone PCR or through degenerate primers for the synthetase's active site.

Selection involves multiple rounds of a two-step process, where the plasmid is transferred into cells expressing chloramphenicol acetyl transferase with a premature amber codon.

In the presence of toxic chloramphenicol and the non-natural amino acid, the surviving cells will have overridden the amber codon using the orthogonal tRNA aminoacylated with either the standard amino acids or the non-natural one. To remove the former, the plasmid is inserted into cells with a barnase gene (toxic) with a premature amber codon but without the non-natural amino acid, removing all the orthogonal synthases which do not specifically recognize the non-natural amino acid.

In addition to the recoding of the tRNA to a different codon, they can be mutated to recognize a four base codon, allowing additional free coding options. The non natural amino acid, as a result, introduces diverse physicochemical and biological properties in order to be used as a tool to explore protein structure and function or to create novel or enhanced protein for practical purposes.

Diversity

The orthogonal pairs of synthase and tRNA which work for one organism may not work for another as the synthase may mis-aminoacylate endogenous tRNAs or the tRNA be mis-aminoacylated itself by an endogenous synthase. As a result the sets created to date different between organisms.

Orthogonal Sets in E. Coli

- tRNA-TyrRS pair from the archaeon *Methanococcus jannaschii*
- tRNA–LysRS pair from the archaeon *Pyrococcus horikoshii*
- tRNA–GluRS pair from *Methanosarcina mazei*
- leucyl-tRNA synthetase from *Methanobacterium thermoautotrophicum* and a mutant leucyl tRNA derived from *Halobacterium* sp.

Orthogonal Sets in Yeast

- tRNA-TyrRS pair from *Escherichia coli*
- tRNA–LeuRS pair from *Escherichia coli*
- tRNA from human and GlnRS from *Escherichia coli.*

Orthogonal Sets in Mammalian Cells

- tRNA-TyrRS pair from *Bacillus stearothermophilus*
- modified tRNA-TrpRS pair from *Bacillus subtilis* trp
- tRNA–LeuRS pair from *Escherichia coli.*

Protein Studies

With an expanded genetic code, the unnatural amino acid can be genetically directed to any chosen site in the protein of interest.

The high efficiency and fidelity allows a better control of the placement of the modification compared to modifying the protein post-translationally, which generally will target all amino acids of the same type, such as the thiol group of cysteine and the -amino group of lysine. Also, an expanded genetic code allows modifications to be carried out *in vivo*.

The ability to site-specifically direct lab-synthesized chemical moieties into proteins allows many types of studies which would otherwise be extremely difficult.

- Probing Protein Structure and Function: by using amino acids with slightly different size such as o-Methyltyrosine or dansylalanine instead of tyrosine, and by inserting genetically-coded reporter moieties (color-changing and/or spin-active) into selected protein sites, chemical information about the protein's structure and function can be measured.
- Identifying and Regulating Protein Activity: by using photocaged aminoacids, protein function can be "switched" on or off by illuminating the organism.
- Changing the mode of action of a protein: one can start with the gene for a protein which binds a certain sequence of DNA, and, by inserting a chemically active amino acid into the binding site, convert it to a protein which cuts the DNA, rather than binding it.
- Improving immunogenicity and overcoming self-tolerance: by replacing strategically-chosen tyrosines with *p*-nitro phenylalanine, a tolerated self-protein can be made immunogenic.

An example of the possible application for this method is the biomedical where "chemical warheads" can be added to protein which target specific cellular components.

Since 2001, 40 non-natural amino acids have been added into protein by creating a unique codon (recoding) and a corresponding transfer-RNA:aminoacyl – tRNA-synthetase pair to encode it with diverse physicochemical and biological properties in order to be used as a tool to exploring protein structure and function or to create novel or enhanced proteins.

Theories on the Origin of the Genetic Code

Despite the minor variations that exist, the genetic code used by all known forms of life is nearly universal. However, there are a huge number of possible genetic codes. If amino acids are randomly associated with triplet codons, there will be 1.5 x 10 possible genetic codes. The question arises: why this code? How did it originate?

Phylogenetic analysis of transfer RNA suggests that tRNA molecules evolved before the present set of aminoacyl-tRNA synthetases.

Theoretically the genetic code could be completely random (a "frozen accident"), completely non-random (optimal) or a combination of random and nonrandom. There are sufficient data to refute the first possibility. For a start, a quick view on the table of the genetic code already shows a clustering of amino acid assignments.

Furthermore, amino acids that share the same biosynthetic pathway tend to have the same first base in their codons, and amino acids with similar physical properties tend to have similar codons. There are four themes running through the many theories that seek to explain the evolution of the genetic code (and hence the origin of these patterns):

- Chemical principles govern specific RNA interaction with amino acids. Aptamer experiments showed that some amino acids have a selective chemical affinity for the base triplets that code for them. Recent experiments show that of the 8 amino acids tested, 6 show some RNA triplet-amino acid association. This has been called the stereochemical code. The stereochemical code could have created an ancient core of assignments. The current complex translation mechanism involving tRNA and associated enzymes may be a later development, and that originally, protein sequences were directly templated on base sequences.
- Biosynthetic expansion. The standard modern genetic code grew from a simpler earlier code through a process of "biosynthetic expansion". Here the idea is that primordial life "discovered" new amino acids (e.g., as by-products of metabolism) and later back-incorporated some of these into the machinery of genetic coding. Although much circumstantial evidence has been found to suggest that fewer different amino acids were used in the past than today, precise and detailed hypotheses about exactly which amino acids entered the code in exactly what order have proved far more controversial.
- Natural selection has led to codon assignments of the genetic code that minimize the effects of mutations. A recent hypothesis suggests that the triplet code was derived from codes that used longer than triplet codons. Longer than triplet decoding has higher degree of codon redundancy and is more error resistant than the triplet decoding. This feature could allow accurate decoding in the absence of highly complex translational machinery such as the ribosome.
- The biocommunicative approach investigates nucleotide sequences as code, i.e. language-like text, which follows in parallel three (3) kinds of rules: combinatorial (syntactic), context-sensitive (pragmatic) and content-specific (semantic). Natural genome editing from this perspective is competent viral-driven generation and integration of meaningful

nucleotide sequences into pre-existing genomic content arrangements and the ability to (re-)combine and (re-)regulate them according to context-dependent (i.e. adaptational) purposes of the host organism.

- Information-theoretic approaches treat the genetic code in terms of an error-prone information channel. The inherent noise in the channel poses the organism with a fundamental question: how to construct a genetic code that can withstand the impact of noise while accurately and efficiently translating information?

  These "rate-distortion" models suggest that the genetic code originated as a result of the interplay of the three conflicting evolutionary forces: the needs for diverse amino-acids, for error-tolerance and for minimal cost of resources.

  The code emerges at a coding phase transition when the mapping of codons to amino-acids becomes nonrandom. The emergence of the code is governed by the topology defined by the most probable errors and is related to the map colouring problem.

- Genetic code and Codon populations: In other way, considering the global information level of Universal Genetic Code map, studying the whole Human genome codon populations show that the position of each codon within this table precisely dictates its population. So the Universal Genetic Code Table not only maps codons to amino acids, but serves perhaps as a global checksum matrix at the balance level of the whole genome: frequencies of the 64 codons in the whole human genome scale are a self-similar fractal expansion of the universal genetic code map.

  This global matrix-like function of Genetic Code table was published in September 2010 by the journal *Interdisciplinary Science* reporting that *Codon populations in single-stranded whole human genome DNA Are fractal and fine-tuned by the Golden Ratio 1.618*. In this paper, Jean-Claude Perez shows that Universal Genetic Code Table kernel governs not only the micro scale (codons i.e amino acids) but the macro scale as well. Particularly, the 6 folding steps of codon populations modeled by the binary divisions of the *Dragon fractal paper folding curve* extend the Chargaff's second rule from the domain of single TCAG nucleotides to the larger domain of codon triplets in the case of the whole Human genome.

Control of Gene Expression in Prokaryotes

Control of gene expression in prokaryotes involves control of "operons".

Operon consists of:

1. promoter - DNA sequence where RNA polymerase (transcription enzyme) binds.
2. operator - DNA sequence where an active repressor protein binds.
3. structural genes - DNA coding for protein (usually several that are functionally related, e.g. enzymes of a metabolic pathway).

Regulatory components (NOT part of the operon):

1. regulatory gene - DNA sequence coding for production of a repressor protein (often located away from operon it regulates).
2. repressor protein - in active form, it binds to operator to prevent transcription of structural genes.

Repressible Operon - ex. tryptophan operon : Repressible operons are usually in the "on" condition, i.e. their gene products are generally required most of the time for cell maintenance, etc. Repressor protein is made in an inactive form (it needs to bind to a corepressor to make it an active repressor which can then bind to operator).

Corepressor is usually an end-product of the metabolic pathway coded by the operon, e.g. tryptophan in our example here. Corepressor must accumulate to high levels before it can bind to inactive repressor and make it active.

Inducible Operon - ex. Lactose Operon

Inducible operons are usually in the "off" condition, i.e. their gene products are only occasionally (or never) needed by the cell. Repressor protein is made in an active form (capable of binding to operator by itself).

Repressor protein can be inactivated by binding to an inducer (lactose in our example here) which then allows the operon to be activated.

Negative vs. Positive Controls

Above examples simply show operons being turned "on" or "off" at the operator, which represents a form of negative control.

Some operons, when turned "on", do not get transcribed at a rate sufficient to see a noticable effect because the RNA polymerase does not bind to promoter effectively.

These operons can be transcribed at a greatly accelerated rate by making the RNA polymerase binding much more efficient. This requires an activated "helper" for the RNA polymerase binding.

Gene Regulation in Eukaryotes

Developmental Biologists have long been interested in how a complex, multicellular organism is able to arise from a single fertilized egg (zygote), i.e. how cells arising from a common lineage (all arising from the zygote) are able to differentiate into the numerous, complex cell types that make up a multicellular organism.

Since the zygote contains ALL the genes necessary for the development of an organism, how is it that only a limited number of these genes are expressed in certain cells so that a multitude of cell types arise from the zygote, e.g. expression of genes for nerve cells, bone cells, skin cells, etc.

At one time, it was generally believed that progressive determination was caused by groups of cells actually losing most of the zygotes original set of genes until a particular cell type contained only genes for forming that cell type, e.g. a cell that eventually becomes a nerve cell has gradually lost all of the genes of the original zygote except those for nerve cell formation.

Today we know that this is generally not the case, although there are exception.

Mitosis faithfully reproduces the full chromosome complement (and all of their genes) to be equally distributed to the new daughter cells, i.e. the daughter cells retain the total genetic potential of the original zygote, "totipotent cells".

Much experimental evidence now supports this concept of the genetic equivalence of nuclei in a multicellular organism.

Nuclear transplantation experiments performed by Robert Briggs and Thomas King, and John Gurdon showed that cells of early embryos retain full genetic potential of zygote.

F.C. Steward was the first to show that individual cells taken from a mature carrot plant could grow into new carrot plants (clones).

More recently the cloning of "Dolly" the sheep, which was not all that important scientifically except that it showed that it could be done in mammals (humans?).

Regeneration, in both plants and animals, have long supported the concept of totipotent cells or the genetic equivalence of somatic cell nuclei.

Therefore, the fact that a differentiating cell is using only a limited number of the total available genes, the problem becomes a differential activation/inactivation of available genes.

For example, cells that eventually become nerve cells do so by activating only the genes necessary for nerve cell formation and likewise inactivating all other genes.

The ways in which cells control gene expression has become a major focus of biological research in the last half of the 20th Century and with the more recent genetic engineering techniques will continue to be a major focus well into this century.

Most of what we first learned about differential gene activation/inactivation came from work on bacteria, e.g. *Escherichia coli.*

Although these cells do not undergo the complex developmental phenomena associated with multicellular organism, they do require activation and inactivation of genes to maintain their existence, and their relative simplicity lent them very well to early studies of gene control.

Many of the models for gene control in prokaryotes have been used as model systems to initially investigate eukaryotic gene control.

However, as one might suspect, the controls used to regulate euakyotic genes appear to be much more complex and must take into account certain aspects of the eukaryote genome that are non-existent in prokaryotes.

1. Very large number of genes present (30,000+) human genes; although only about 3-5% is being used at any one time; recall that *E. coli* has only about 2500 genes.

2. Most eukaryote DNA does not code for RNA or protein (does not function as "genes").

3. Presence of chromatin and the structural complexity of the chromosome.

4. Compartmentalization of most DNA into a nucleus.

5. Complexity in variety of genes being activated at any one time (recall functionally linked genes into operons in bacteria), etc.

One way of looking at "gene expression" is to consider that a gene is expressed (or "activated") only when the product that the gene coes for is produced and functional, e.g. an "active" enzyme.

Considering this very broad description for "gene activation", the eukaryotic cell possesses numerous places where control could take place. Some of these are, at present, only possibilities while others have good experimental support for their existence.

Potential Levels of Eukaryotic Gene Control

Pretranscriptional Control

The physical form of eukaryotic chromatin may be able to block the availability of genes for transcription by forming chromatin structures that "hide" the DNA from trancription factors and enzymes.

"DNA packing" involved basic proteins called HISTONES (H1, H2A, H2B, H3, H4) and various other chromosomal proteins.

An extreme form of DNA packing occurs when chromosomes condense during Prophase of cell division. Very little, if any, gene activity occurs while mitotic/meiotic chromosomes are formed.

Heterochromatin is chromatin in a highly compacted form that is very low or completely devoid of transcriptional activity. It can occur in chromatin found in the Interphase cell.

The tight ciling might be due to high methylation of the DNA in these areas.

Regardless of the cause of heterochromatin, experimenal evidence has shown very little transcriptional activity associciated with these areas.

DNA methylation, in general, is a means of inactivating genes over extended time.

An extreme example of gene inactivation by heterochromatin formation is the Barr body typically found in human females. In this example, an entire chromosome's genes are prevented from being transcribed (activated).

A chromosome may have heterochromatin areas interspersed with less compacted areas called euchromatin, which are thought to be areas where the DNA is more available for transcription.

Even in euchromatin, the DNA is tightly wrapped around nucleosomes formed by hisones. To make this DNA available for transcription, the histones of thenucleosome appear to undergo acetylation (addition of acetyl groups) which causes a "relaxation" of the DNA/nucleosome association.

This relaxing of the DNA/nucleosome assoication makes the DNA more available for transciption.

Transcriptional Control

Regulatory proteins have also been shown to exist in eukaryotes and may be either activators or repressors of gene activity.

Since the majority of the eukaryotic genes are inactive at any one time, it is believed that most regulators are activators.

Enhancers have also been shown to exist in eukaryotes. These are small stretches of DNA that have the ability to enhance transcriptional activity after binding activator proteins and then interacting at the promoter region to enhance RNA polymerase binding and activity.

Recall that transcription factors are also needed for RNA polymerase binding and activity.

Enhancers are found some distance away from the promoter and may be "upstream" or "downstream" from the promoter.

Posttranscriptional Control

The initial transcript (RNA) in eukaryotes is not functional but must undergo modification to become functional.

rRNA, tRNA, and mRNA must all undergo varying degrees of posttranscriptional modification before they can become functional.

In general, these modifications may include:

1. the cutting out of functional pieces of RNA from a much larger nonfunctional transcript.
2. some of the above pieces may be functional but some require the bonding together of excised pieces to form a single functional piece (exons/introns).
3. mRNA requires the most extensive posttranscriptional modifications, which in addition to the above, require chemical modifications at both ends of the molecule. These modifications may help the transport out of the nucleus, protect the molecule from nuclease attack in the cytoplasm, or somehow be required for initiation of translation.

mRNA is generally consider a "short-lived" molecule in both prokaryotes and eukaryotes, and was once used as a prime criterion for proving that a RNA was in fact a mRNA.

However, exceptions to this generality have been shown in eukaryotes, particularly in certain highly specialized cells (e.g. maturing vertebrate erythrocytes) and developing embryos.

In the case of the maturing vertebrate red blood cell (called erythropoiesis), the erythroblast is a nucleated cell that gradually shuts down most of it's genes except for mRNA for globin production. Huge amounts of globin mRNA accumulate and eventually the cell becomes anucleate, however, the globin mRNA remains for a considerable time afterwards while the cell makes and accumulates very large amounts of hemoglobin.

Unfertilized eggs have been shown to manufacture and accumulate large amounts of mRNA coding for proteins necessary for the rapid, early development of the embryo; these mRNAs are stored in an inactive form in the egg. Activation of the egg, by sperm or artificial means, activates these mRNAs for embryo protein production.

Translational Control

The observation that eggs may contain "long-lived" mRNA but that this RNA is not translated until egg activation has been shown to be a block at translation. A certain initiation factor(s) necessary for translation rapidly becomes available after fertilization (activation).

Also, the amount of heme remaining in maturing red blood cells helps determine how much more globin will be made (translation of globin mRNA). The poliovirus is thought to select the translation of its "uncapped" mRNAs by somehow destroying the host cell's cap recognition system on the ribosome complex.

Posttranslational Control

Even after the protein is first made, it is not necessarily functional.

Insulin is initially made as a single protein (inactive proinsulin) which must fold back upon itself, form three disulfide bonds, and finally have a small stretch of amino acids removed by a peptidase before it becomes active insulin consisting of two polypeptides held together by two of the disulfide bonds.

Many of the digestive enzymes of the small intestine are synthesized as zymogens (inactive forms) that must undergo enzymatic cleavage to become active, e.g. trypsinogen/trypsin, chymotrypsinogen/chymotrypsin, pepsinogen/ pepsin.

Some proteins (malformed, no longer needed, etc.) are "marked" for degradation by proteosomes.

Eukaryotic Gene Organization and its Evolution

Unlike the prokaryotes, eukaryotes do not possesss an "operon type" arrangement of functionally related genes. (recall prokaryote operon).

Instead, it has been shown that genes coding for functionally related products, e.g. a series of enzymes in a metabolic pathway, are scattered throughout the genome and each controlled by a separate promoter, etc.

This gene arrangement may at first appear disorganized but the reality is that it allows much greater variety in the use of these "scattered" genes. Genes can be activated in numerous combinations. A multigene family is a collection of genes within a genome that are similar or identical in sequence.

The 28S, 18S and 5.8S rRNA genes occur in eukaryotes as tandem arrays of several hundred to several thousand located in association with a nucleosome found inside the nucleus.

The family of globin genes are sequence related but not identical and function in making different hemoglobins at different developmental stages. Both types of multigene families most likely arose by gene duplication and mutation.

Molecular Biology of Cancer

Cancer is caused by inapproriate gene activity associated with the cell cycle.

Some chararcteristics of malignant cancers (solid types):

1. Loss of cell cycle control
2. Cell surface alterations
3. Secretion of distinct products, e.g. proteolytic enzymes, embryonic antigens, angiogenin, etc.
4. Morphological alterations, enlarged nuclei, aberrant nucleoli, numerous microvilli
5. Highly vascularized
6. Metastasis (travel to distant tissues and establish secondary or "metastatic" cancers).

The genetic basis of cancer was first discovered through the work with retroviruses that were shown to cause cancer in monkeys, chickens, mice, etc.

The viral genes were identified that could be directly linked to the cause of cancer in experimental animals.

These genes were called oncogenes ("cancer genes").

Upon closer examination, researchers found that in fact the oncogenes were not unique but were in fact mutated forms of normal genes (proto-oncogene) found in eukaryotic organisms; including humans.

These normal genes, for the most part, are involved with some aspect of regulating the cell cycle.

Three ways for conversion of a proto-oncogene into an oncogene:

1. translocation or transposition to new promoter
2. gene amplification
3. point mutation within gene.

In addition to proto-oncogenes that can become oncogenes, there are normal, tumour-suppressor genes that are mutated during the formation of cancers.

Tumour-suppressor genes produce products that normally block the cell cycle; which would normally prevent cancers.

The cancerous state results from alteration of several genes, both proto-oncogenes and tumour suppressor genes.

5

Microbial Genetics

Introduction to Microbial Genetics

Microorganisms have the ability to acquire genes and thereby undergo the process of recombination. In recombination, a new chromosome with a genotype different from that of the parent results from the combination of genetic material from two organisms. This new arrangement of genes is usuall accompanied by new chemical or physical properties.

In microorganisms, several kinds of recombination are known to occur. The most common form is general recombination, which sually involves a recirocal exchange of DNA between a pair of DNA sequences. It occurs anywhere on the microbial chromosome and is typified by the exchanges occurring in bacterial transformation, bacterial recombination, and bacterial transduction. A second type of recombination, called site-specific recombination, involves the integration of a viral genome into the bacteial chromosome.

A third type is replicative rcombination, which is due to the movement of genetic elements as they switch position from one place on the chromosome to another. The principles of recombination apply to prokaryotic microorganisms but not to eukaryotic microorganism. Eukaryotes exhibit a complete sexual life cycle, including miosis. In this process, new combinations of a particular gene frm during the process of crossing over. This process occurs between homologous chromosomes and is not seen in bacteria, whre only a single chromosome exists. Much of the wok in microbial genetics has been performed with bacteria, and the unique features of microbial genetics are usually those associaed with prokaryotes such as bacteria.

The Bacterial Chromosome and Plasmid

While eukaryotes have two or more chromosomes, prokaryotes such as bacteria possess a single hromosome composed f double-stranded DNA in a loop. The DNA is located in the nucleoid of the cell and is not asociated with protein. In *Escherichia coli*, the length of the chromosome, when open, is many times the length of the cel. Many bacteria (and some yeasts or othr fungi) also possess looped bits of DA known as plasmids, which exist and replicate independently of the chromosome.

Plasmids have relatively few genes (fewer than 30). The genetic information of the plasmid is usually not essential to survival of the host bacteria. Plasmids can be removed from the host cell in the process of curing. Curing may occur spontaneously or may be induced by treatments such as ultraviolet light. Certain plasmids, called episomes, may be integrated into the bacterial chromosome. Others contain genes for certain types of pili and are able to transfer copies of themselves to other bacteria. Such plasmids are referred to as conjugative plasmds.

A special plasmid called a fertility (F) factor plays an important role in conjugation. The F factor contains genes that encourage cellular attachment during conjugation and accelerate plasmid transfer between conjugating baterial cells. Those cells contributing DNA are called $^+$ (onor) cells, while those receiving DNA are the F^- (recipient) cells. The F factor can exist outside the bacterial chromosome or ay be integrated into the hromosome.

Plasmids contain genes that impart antibiotic resistance. Up to eight genes for resisting eight different antibiotics may be found on a single plasmid. Genes that enode a series of bacteriocins are also found on plasmids. Bacteriocins are bacterial proteins capable of destroying other bacteria. Still other plasmids increase the pathogenicity of their host bacteria because the plasmid contains genes for toxin synthesis. Transposable elements.

Transposable elements, als known as transposons, are segments of DNA that move about within the chromosome and estblish new genetic sequences. First discovered by Barbara McClintock in the 1940s, transposons behave somewhat like lysogenic viruses except that they cannot exist apart from the chromosome or reproduce themselves.

The simplest transposons, inserion sequences, are short sequences of DNA bounded at both ends by identical sequences of nucleotides in reverse orentation (inverted repeats). Insertion sequences can insert within a gene and cause a rearrangement mutation of the genetic material. If the sequence carries a stop coon, it may block transcription of the DNA during protein synthesis. Insertion sequences may also encourage the movement of drug-resistance genes between plasmids and chromosomes.

Mutation

In molecular biology and genetics, mutationsare changes in a genomic sequence: the DNA sequence of a cell's genome or the DNA or RNA sequence of a virus. Mutations are caused by radiation, viruses, transposons and mutagenic chemicals, as wellas errors that occur during meiosis or DNA replication. They can also be induced by the organism itself, by cellular processes such as hypermutation.

Mutation can result n several different types of change in DNA sequences; these can either have no effect, alter the product of a gene, or prevent the gene from functioning properly or completely. Studies in the fly Drosophila melanogaster suggest that if a mutation changes a protein produced by a gene, this will probably be harmful, with about 70 per cent of these mutations having damaging effects, and the remainder being either neutral or weakly beneficial.

Due to the damaging effects that mutations can have on enes, organisms have mechanisms such as DNA repair to remove mutations. Therefore, the optimal mutation rate for a species is a trade-off between costs of a high mutation rate, such as deleterious mutations, and the metabolic costs of maintaining systms to reduce the mutation rate, such as DNA repair enzymes. Viruses that use RNA as their genetic material have rapid mutation rates, which can be an advantage since these viruses will evolve constantly and rapidly, and thus evade the defensive responses of *e.g.* the human immune system.

Description

Mutations can involve large sections of DNA becoming duplicated, usually through genetic recombination. These duplications are a major source of raw material for evolving new genes, with tens o hundreds of genes duplicated in animal genomes every million years. Most genes belong to larger families of genes of shared ancestry. Novel genes are produced by several methods, commonly through the duplication and mutation of an ancestral gene, or by recombining parts of different genes to form new combinations with new functions. Here, domains act as modules, each with a particular and independent function, that can be mixed together to roduce genes encoding new proteins with novel properties. For example, the human eye uses four genes to make structures that sense light: three for color vision and one for night vision; all four arose from a single ancestral gene.

Another advantage of duplicating a gene (or even an entire genome) is that this increases redundancy; this allows one gene in the pair to acquire a new function while the other copy performs the original function. Other types of mutation occasioally create new genes from previously noncoding DNA.

Changes in chromosome number may invlve even larger mutations, where segments of the DNA within chromosomes break and then rearrange.

For example, two chromosomes in the Homo genus fused to produce human chromosome 2; this fusion did not occur in the lineage of the other apes, and they retain these separate chromosomes. In evolution, the most important role of such chromosomal rearrangements may be to accelerate the divergence of a population into new species by making populations less likely to interbreed, and thereby preserving genetic diferences between these populations.

Sequences of DNA that can move about the genome, such as transposons, make up a major fraction of the genetic material of plants and animals and may have been important in the evolution of genomes. For example, more than a million copies of the Alu sequence re present in the human genome, and these sequences have now been recruited to perform functions such as regulatinggene expression.

Another effect of these mobile DNA sequences is that when they move within a genome, they can mutate or delete existing genes and thereby produce genetic diversity. In multicellular organisms with dedicated reproductive cells, mutations can be subdivided into germ line mutations, which can be passed on to descendants through their reproductive cells, and somatic mutations (also called acquired mutations), which involve cells outside the dedicated reproductive group and which are not usually tansmitted to descendants.

If the organism can reproduce asexually through mechanisms such as cuttings or budding the distinction can become blurred. For example, plants can sometimes transmit somatic mutations to their descendants asexually or sexually where flower buds develop in somatically mutated parts of plants. A new mutation that was not inherited from either parent is called a de novo mutation.

The source of the mutation is unrelated to the consequence, although the consequences are related to which cells were mutated. Nonlethal mutations accumulate within the gene pool and increase the amount of genetic variation. The abundance of some genetic changes within the gene pool can be reduced by natural selection, while other "more favourable" mutations may accumulate and result in adaptive changes.

For example, a butterfly may produce offspring with new mutations. The majority of these mutations will have no effect; but one might change the color of one of the butterfly's offspring, making it harder (or easier) for predators to see. If this color change is advantageous, the chance of this butterfly surviving and producing its own offspring are a little better, and over time the number of butterflies with this mutation may form a larger percentage of the population. Neutral mutations are defined as mutations whose effects do not influence the fitness of an individual. These can accumulate over time due to genetic drift. It is believed that the overwhelming majority of mutations have no sinificant effect on an organism's fitness.

Also, DNA repair mechanisms are able to mend most changes before they become permanent mutations, and many organisms have mechanisms for eliminating otherwise permanently mutated somatic cells.

Mutation is generally accepted by biologists as the mechanism by which natural selection acts, generating advantageous new traits that survive and multiply in offspring as well as disadvantageous traits, in less fit offspring, that tend to die out.

Causes

Two classes of mutations are spontaneous mutations (molecular decay) and induced mutations caused by mutagens.

Spontaneous mutations on the molecular level can be caused by:

- *Tautomerism*: A base is changed by the repositioning of a hydrogen atom, altering the hydrogen boning pattern of that base resulting in incorrect base pairing during replication.
- *Depurination*: Loss of a purine base (A or G) to form an apurinic site (AP site).
- *Deamination*: Hydrolysis changes a normal base to an atypical base containing a keto group in place of the original amine group. Examples include C → U and → HX (hypoxanthine), which can be corrected by DNA repair mechanisms; and 5MeC (5-methylcytosine) → T, which is less likely to be deected as a mutation because thymine is a normal DNA base.
- *Slipped strand mispairing*: Denaturation of the new strand from the template during replication, followed by renaturation in a different spot ("slipping"). This can lead to insertions or deletions.

Induced mutations on the molecular level can be caused by:

- Chemicals
  - Hydroxylamine NH_2OH
  - Base analogs (*e.g.* BrdU)
  - Alkylating agents (*e.g.* N-ethyl-N-nitrosourea) These agentscan mutate both replicating and non-replicating DNA In contrast, a base analog can only mutate the DNA whn the analog is incorporated in replicating the DNA. Each of these classes of chemical mutagens has certain effects that then lead to transitions, ransversions, or deletions.

- Agents that formDNA adducts (*e.g.* ochratoxin A metabolites)
- DNAintercalating agents (*e.g.* ethidium bromide)
- DNA crosslinkers
- Oxidative damage
- Nitrous acid converts amine groups on A and C to diazo groups, altering their hyrogen bonding patterns which leads to incorrect base pairing during replication.

• Radiation
  - Ultraviolet radiation nonionizing radiation). Two nucleotide bases in DNA–cytosine and thymine–are mot vulnerable to radiation that can change their properties. UVlight can induce adjacent pyrimidine bases in a DNA strand to become covalently joined as a pyrimidine dimer. UV radiation, particularly loner-wave UVA, can also cause oxidative damage toDNA.
  - Ionizing radiation

• Viral infections

DNA has so-called hotspots, where mutations occur up to 100 times more frequently than the normal mutation rate. A hotspot can be at an unusual base, *e.g.*, 5-methylcytosine. Mutation rates also var across species. Evolutionary biologists have theorized that higher mutation rates are beneficial in some situations, because they allow organisms to evolve and therefore adapt more quickly to their environments. For example, repeated exposure of bacteria to antibiotics, and selection of resistnt mutants, can result in the selection of bacteria that have a much higher mutation rate than the original population (mutator strains).

Classification of Mutation Types

By Effect on Structure

The sequence of a gene can be altered in a number of ways. Gene mutations have varying effects on health depending on where they occur and whether they alter the function of essential proteins.

Mutations in the structure of genes can be classified as:

• *Small-scale mutations, such as those affecting a small gene in one or a few nucleotides, including:*
  - Point mutatins, often caused by chemicals or malfunction of DNA replication, exchange a single nucleotide for another. These changes are classified as transitions or transversions. Most common is the ransition that exchanges a purine for a purine (A ↔ G) or a pyrimidine

for a pyrimidine, (C ↔ T). A transition can be caused by nitrous acid, base mis-pairing, or mutagenic base analogs such as 5-bromo-2-deoxyuridine (BrdU). Less common is a transversion, which exchanges a purine for a pyrimidine or a pyrimidine for a purine (C/T ↔ A/G). An example of a transversion is adenine (A) being converted into a cytosine (C).

A point mutation can be reversed by another point mutation, in which the nucleotide is changed back to its original state (true reversion) or by second-site reversion (a complementary mutation elsewhere that results in regained gene functionality). Point mutations that occur within the protein coding region of a gene may be classified into three kinds, depending upon what the erroneous codon codes for:

1. *Silent mutations*: Which code for the same amino acid.
2. *Missense mutations*: Which code for a different amino acid.
3. *Nonsense mutations*: Which code for a stop and can truncate the protein.
   - Insertons add one or more extra nucleotides into the DNA. They are usually caused by transposable elements, or errors during replication of repeating elements (*e.g.* AT repeats). Insertions in the coding region of a gene may alter splicing of the mRNA (splice site mutation), or cause a shift in the reading frame (frameshift), both of which can significantly alter the gene product. Insertions can be reverted by excision of the transposable element.
   - Deetions remove one or more nucleotides from the DNA. Like insertions, these mutations can alter the reading frame of the gene. They are generally irreversible: though exactly the same sequence might theoretically be restored by an insertion, transposable elements able to revert a very short deletion (say 1-2 bases) in any location are either highly unlikely to exist or do not exist at all. Note that a deletion is not the exact opposite of an insertion: the former is quite random while the latter consists of a specific sequence inserting at locations that are not entirely random or even quite narrowly defined.

- *Large-scale mutations in chromosomal structure, including*:
  - Amplifications (or gene duplications) leading to multiple copies of all chromosomal regions, increasing the dosage of the genes located within them.
  - Deletions of large chromosomal regions, leading to loss of the genes within those regions.

- Mutations whose effct is to juxtapose previously separate pieces of DNA, potentially bringing together separate genes to form functionally distinct fusion genes (*e.g.* bcr-abl).

  These include:

1. *Chromosomal translocations*: Interchange of genetic parts from nonhomologous chromo-somes.
2. *Interstitial deletions*: An itra-chromosomal deletion that removes a segment of DNA from a single chromosome, thereby apposing previously distant genes. Fo example, cells isolated from a human astrocytoma, a type of brain tumor, were found to have a chromosomal deletion removing sequences between the "fused in glioblastoma" gene and the receptor tyrosine kinase "ros", producing a fusion protein (FIG-ROS). The abnormal FIG-ROS fusion protein has constitutively active kinase activity that causes oncogenic transformation (a transfor-mation from normal cells to cancer cells).
3. *Chromosomal inversions*: Reversing the orientation of a chromosomal segment.
   - *Loss of heterozygosity*: Loss of one allele, either by a deletion or recombination event, in an organism that previously had two different alleles.

By Effect on Function

- Loss-of-function mutations are the result of gene product having less or no function. When the allele has a complete loss of function (null allele) it is often called an amorphic mutation. Phenotypes associated with such mutations are most often recessive. Exceptions are when the organism is haploid, or when the reduced dosage of a normal gene product is not enough for a normal phenotype (this is called haploinsufficiency).
- Gain-of-function mutations change the gene product such that it gains a new and abnormal function. These mutations usually have dominant phenotypes. Often called a neomorphic mutation.
- Dominant negative mutations (also called antimorphic mutations) have an altered gene product that acts antagonistically to the wild-type allele. These mutations usually result in an altered molecular function (often inactive) and are characterised by a dominant or semi-dominant phenotype.
- In humans, Marfan syndrome is an example of a dominant negative mutation occurring in an autosomal dominant disease.

In this condition, the defective glycoprotein product of the fibrillin gene (FBN1) antagonizes the product of the normal allele.

- Lethal mutations are mutations that lead to the death of the organisms which carry the mutations.
- A back mutation or reversion is a point mutation that restores the original sequence and hence the original phenotype.

By Effect on Fitness

In applied genetics it is usual to speak of mutations as either harmful or beneficial.

- A harmful mutation is a mutation that decreases the fitness of the organism.
- A beneficial mutation is a mutation that increases fitness of the organism, or which promotes traits that are desirable.

In theoretical population genetics, it is more usual to speak of such mutations as deleterious or advantageous.

In the neutral theory of molecular evolution, genetic drift is the basis for most variation at the molecular level.

- A neutral mutation has no harmful or beneficial effect on the organism. Such mutations occur at a steady rate, forming the basis for the molecular clock.
- A deleterious mutation has a negative effect on the phenotype, and thus decreases the fitness of the organism.
- An advantageous mutation has a positive effect on the phenotype, and thus increases the fitness of the organism.
- A nearly neutral mutation is a mutation that may be slightly deleterious or advantageous, although most nearly neutral mutations are slightly deleterious.

In reality, viewing the fitness effects of mutations in these discrete categories is an oversimplification. Attempts have been made to infer the distribution of fitness effects using mutagenesis experiments or theoretical models applied to molecular sequence data. However, the current distribution is still uncertain, and some aspects of the distribution likely vary between species.

By inheritance:

- Inheritable generic in pro-generic tissue or cells on path to be changed to gametes.
- Non inheritable somatic (*e.g.*, cacinogenic mutation

- Non inheritable post mortem aDNA mutation in decaying remains.

By pattern of inheritance The human genome contains two copies of each gene–a paterna and a maternal allele:

- A heterozygous mutation is a mutation of only one allele.
- A homozygous mutation is an identical mutation of both the paternal and maternal alleles.
- Compound heterozygous mutations or a genetic compound comprises two diffeent mutations in the paternal and maternal alleles.
- A wildtype or homozygous non-mutated organism is one in which neither allele is mutated. (Just not a mutation)

By Impact on Protein Sequence

- A frameshift mutation is a mutation caused by insertion or deletion of a number of nucleoides that is not evenly divisible by three from a DNA sequence. Due to the triplet nature of gene expression by codons, the insertion or deletion can disrupt the reading frame, or the grouping of the codons, resulting in a completely different translation from the original. The earlier in the sequence the deletion or insertion occurs, the more altered the protein produced is.
- A nonsnse mutation is a point mutation in a sequence of DNA that results in a premature stop codon, or a nonsense codon in the transcribed mRNA, and possibly a truncated, and often nonfunctional protein product.
- Missense mutations or nonsynonymous mutations are types of point mutations where a single nucleotide is changed to cause substitution of a different amino acid. This in turn can render the resulting protein nonfunctional. Such mutations are responsible for diseases such as Epidermolysis bullosa, sickle-cell disease, and SOD1 mediated ALS.
- A neutral mutation is a mutation that occurs in an amino acid codon which results in the use of a different, but chemically similar, amino acid. The similarity between the two is enough that little or no change is often rendered in the protein. For example, a change from AAA to AGA will encode arginine, a chemically similar molecule to the intended lysine.
- Silent mutations are mutations that do not result in a change to the amino acid sequence of a protein. They may occur in a region that does not code for a protein, or they may occur within a codon in a manner that does not alter the final amino acid sequence. The phrase silent mutation is often used interchangeably with the phrase synonymous

mutation; however, synonymous mutations are a subcategory of the former, occurring only within exons. The name silent could be a misnomer. For example, a silent mutation in the exon/intron border may lead to alternative splicing by changing the splice site, thereby leading to a changed protein.

Special classes:

- Conditional mutation is a mutation that has wild-type (or less severe) phenotype under certain "permissive" environmental conditions and a mutant phenotype under certain "restrictive" conditions. For example, a temperature-sensitive mutation can cause cell death at high temperture (restrictive condition), but might have no deleterious consequences at a lower temperature (pemissive condition).

Harmful Mutations

Changes in DNA caused by mutation can cause errors in protein sequence, creating partially or completely non-functional proteins. To function correctly, each cell depends on thousands of proteins to function in the right places at the right times. When a mutation alters a protein that plays a critical role in the body, a medical condition can result. A condition caused by mutations in one or more genes is calle a genetic disorder. Some mutations alter a gene's DNA base sequence but do not change the function of the protein made by the gene. Studies of the fly Drosophila melanogaster suggest that if a mutation does change a protein, this will probably be harmful, with about 70 per cent of these mutations having damaging effects, and the remainder being either neutral or weakly beneficial.

However, studies in yeast have shown that only 7% of mutations that are not in genes are harmful. If a mutation is present in a germ cell, it can give rise to offspring that carries the mutation in all of its cells. This is the case in hereditary diseases. On the other hand, a mutation may occur in a somatic cell of an organism. Such mutations will be present in all descendants of this cell within the same organism, and certain mutations can cause the cell to become malignant, and thus cause cancer. Often, gene mutations tha could cause a genetic disorder are repaired by the DNA repair system of the cell. Each cell has a number of pathways though which enymes recognize and repair mistakes in DNA. Because DNA ca be damaged or mutated in many ways, the process of DNA repair is an important way in which the body protects itself from disease.

Beneficial Mutations

Although most mutations that change protein sequences are neutral or harmful, some mutations have a positive effect on an organism. In this case, the mutation may enable the mutant organism to withstand particular environmental stresses better than wild-type organisms, or reproduce more quickly. In these cases a mutation will tend to become more common in a population through natural selection. For example, a specific 32 base pair deletion in human CCR5 (CCR5-Δ32) confers HIV resistance to homozygotes and delays AIDS onset in heterozygotes.

The CCR5 mutation is more common in those of European descent. One possible explanation of the etiology of the relatively high frequency of CCR5-Δ32 in the European population is that it conferred resistance to the bubonic plague in mid-14th century Europe. People with this mutation were more likely to survive infection; thus its frequency in the population increased.

This theory could explain why this mutation is not found in southern Africa, where the bubonic plague never reached. A newer theory suggests that the selective pressure on the CCR5 Delta 32 mutation was caused by smallpox instead of the bubonic plague. Another example, is Sickle cell disease which is a blood disorder in which the body produces an abnormal type of the oxygen-carrying substance hemoglobin in the red blood cells.

One-third of all indigenous inhabitants of Sub-Saharan Africa carry the gene, because in areas where malaria is common, there is a survival value in carrying only a single sickle-cell gene (sickle cell trait). Those with only one of the two alleles of the sickle-cell disease are more resistant to malari

6

Gene Transcription

Cell signalling is part of a complex system of communication that governs basic cellular activities and coordinates cell actions. The ability of cells to perceive and correctly respond to their microenvironment is the basis of development, tissue repair, and immunity as well as normal tissue homeostasis. Errors in cellular information processing are responsible for diseases such as cancer, autoimmunity, and diabetes. By understanding cell signalling, diseases may be treated effectively and, theoretically, artificial tissues may be created.

Traditional work in biology has focused on studying individual parts of cell signalling pathways. Systems biology research helps us to understand the underlying structure of cell signalling networks and how changes in these networks may affect the transmission and flow of information. Such networks are complex systems in their organization and may exhibit a number of emergent properties including bistability and ultrasensitivity. Analysis of cell signalling networks requires a combination of experimental and theoretical approaches including the development and analysis of simulations and modelling.

Unicellular and Multicellular Organism Cell Signalling

Cell signalling has been most extensively studied in the context of human diseases and signalling between cells of a single organism. However, cell signalling may also occur between the cells of two different organisms. In many mammals, early embryo cells exchange signals with cells of the uterus. In the human gastrointestinal tract, bacteria exchange signals with each other and with human epithelial and immune system cells. For the yeast *Saccharomyces cerevisiae* during mating, some cells send a peptide signal (mating factor *pheromones*) into their environment. The mating factor peptide may bind to a cell surface receptor on other yeast cells and induce them to prepare for mating.

Types of Signals

Cells communicate with each other via direct contact (juxtacrine signalling), over short distances (paracrine signalling), or over large distances and/or scales (endocrine signalling). Some cell-to-cell communication requires direct cell-cell contact. Some cells can form gap junctions that connect their cytoplasm to the cytoplasm of adjacent cells. In cardiac muscle, gap junctions between adjacent cells allows for action potential propagation from the cardiac pacemaker region of the heart to spread and coordinately cause contraction of the heart.

The Notch signalling mechanism is an example of juxtacrine signalling (also known as contact dependent signalling) in which two adjacent cells must make physical contact in order to communicate. This requirement for direct contact allows for very precise control of cell differentiation during embryonic development.

In the worm *Caenorhabditis elegans*, two cells of the developing gonad each have an equal chance of terminally differentiating or becoming a uterine precursor cell that continues to divide. The choice of which cell continues to divide is controlled by competition of cell surface signals. One cell will happen to produce more of a cell surface protein that activates the Notch receptor on the adjacent cell. This activates a feedback loop or system that reduces Notch expression in the cell that will differentiate and increases Notch on the surface of the cell that continues as a stem cell.

Many cell signals are carried by molecules that are released by one cell and move to make contact with another cell. *Endocrine* signals are called hormones. Hormones are produced by endocrine cells and they travel through the blood to reach all parts of the body. Specificity of signalling can be controlled if only some cells can respond to a particular hormone.

Paracrine signals such as retinoic acid target only cells in the vicinity of the emitting cell. Neurotransmitters represent another example of a paracrine signal. Some signalling molecules can function as both a hormone and a neurotransmitter.

For example, epinephrine and norepinephrine can function as hormones when released from the adrenal gland and are transported to the heart by way of the blood stream. Norepinephrine can also be produced by neurons to function as a neurotransmitter within the brain. Estrogen can be released by the ovary and function as a hormone or act locally via paracrine or autocrine signalling. Active species of oxygen and nitric oxide can also act as cellular messengers. This process is dubbed redox signalling.

Receptors for Cell Signals

Cells receive information from their environment through a class of proteins known as receptors. Notch is a cell surface protein that functions as a receptor. Animals have a small set of genes that code for signalling proteins that interact specifically with Notch receptors and stimulate a response in cells that express Notch on their surface. Molecules that activate (or, in some cases, inhibit) receptors can be classified as hormones, neurotransmitters, cytokines, growth factors but all of these are called receptor ligands. The details of ligand-receptor interactions are fundamental to cell signalling.

Notch acts as a receptor for ligands that are expressed on adjacent cells. While many receptors are cell surface proteins, some are found inside cells. For example, oestrogen is a hydrophobic molecule that can pass through the lipid bilayer of cell surface membranes. Oestrogen receptors inside cells of the uterus can be activated by oestrogen that comes from the ovaries, enters the target cells, and binds to oestrogen receptors.

A number of transmembrane receptors for molecules that include peptide hormones and of intracellular receptors for steroid hormones exist, giving to a cell the ability to respond to a great number of hormonal and pharmacological stimuli. In diseases, often, proteins that interact with receptors are aberrantly activated, resulting in constitutively activated downstream signals.

For several types of intercellular signalling molecules that are unable to permeate the hydrophobic cell membrane due to their hydrophilic nature, the target receptor is expressed on the membrane. When such signalling molecule activates its receptor, the signal is carried into the cell usually by means of a second messenger such as cAMP.

Signalling Pathways

In some cases, receptor activation caused by ligand binding to a receptor is directly coupled to the cell's response to the ligand. For example, the neurotransmitter GABA can activate a cell surface receptor that is part of an ion channel. GABA binding to a GABA A receptor on a neuron opens a chloride-selective ion channel that is part of the receptor. GABA A receptor activation allows negatively-charged chloride ions to move into the neuron, which inhibits the ability of the neuron to produce action potentials.

However, for many cell surface receptors, ligand-receptor interactions are not directly linked to the cell's response. The activated receptor must first interact with other proteins inside the cell before the ultimate physiological effect of the ligand on the cell's behaviour is produced. Often, the behaviour of a chain of several interacting cell proteins is altered following receptor activation. The entire set of cell changes induced by receptor activation is called a signal transduction mechanism or pathway.

In the case of Notch-mediated signalling, the signal transduction mechanism can be relatively simple. Activation of Notch can cause the Notch protein to be altered by a protease. Part of the Notch protein is released from the cell surface membrane and can act to change the pattern of gene transcription in the cell nucleus.

This causes the responding cell to make different proteins, resulting in an altered pattern of cell behaviour. Cell signalling research involves studying the spatial and temporal dynamics of both receptors and the components of signalling pathways that are activated by receptors in various cell types.

This pathway involves changes of protein-protein interactions inside the cell, induced by an external signal. Many growth factors bind to receptors at the cell surface and stimulate cells to progress through the cell cycle and divide. Several of these receptors are kinases that start to phosphorylate themselves and other proteins when binding to a ligand. This phosphorylation can generate a binding site for a different protein and thus induce protein-protein interaction.

The ligand (called epidermal growth factor (EGF)) binds to the receptor (called EGFR). This activates the receptor to phosphorylate itself. The phosphorylated receptor binds to an adaptor protein (GRB2), which couples the signal to further downstream signalling processes. For example, one of the signal transduction pathways that are activated is called the mitogen-activated protein kinase (MAPK) pathway. The signal transduction component labelled as "MAPK" in the pathway was originally called "ERK," so the pathway is called the MAPK/ERK pathway.

The MAPK protein is an enzyme, a protein kinase that can attach phosphate to target proteins such as the transcription factor MYC and, thus, alter gene transcription and, ultimately, cell cycle progression. Many cellular proteins are activated downstream of the growth factor receptors (such as EGFR) that initiate this signal transduction pathway.

Some signalling transduction pathways respond differently depending on the amount of signalling received by the cell. For instance, the hedgehog protein activates different genes, depending on the amount of hedgehog protein present.

Complex multi-component signal transduction pathways provide opportunities for feedback, signal amplification, and interactions inside one cell between multiple signals and signalling pathways.

Classification of Intercellular Communication

Within endocrinology (the study of intercellular signalling in animals) and the endocrine system, intercellular signalling is subdivided into the following classifications:

- *Intracrine* signals are produced within the target cell.
- *Autocrine* signals target the cell its self. Sometimes autocrine cells can target cells close by if they are the same type of cell as the emitting cell. An example of this is immune cells.
- *Juxtacrine* signals target adjacent (touching) cells. These signals are transmitted along cell membranes via protein or lipid components integral to the membrane and are capable of affecting either the emitting cell or cells immediately adjacent.
- *Paracrine* signals target cells in the vicinity of the emitting cell. Neurotransmitters represent an example.
- *Endocrine* signals target distant cells. Endocrine cells produce hormones that travel through the blood to reach all parts of the body.

Gene Transcription

Transcription is the process of creating an equivalent RNA copy of a sequence of DNA. Both RNA and DNA are nucleic acids, which use base pairs of nucleotides as a complementary language that can be converted back and forth from DNA to RNA in the presence of the correct enzymes. During transcription, a DNA sequence is read by RNA polymerase, which produces a complementary, antiparallel RNA strand. As opposed to DNA replication, transcription results in an RNA complement that includes uracil (U) in all instances where thymine (T) would have occurred in a DNA complement. Transcription can be explained easily in 4 quick steps. Step 1: DNA unwinds/ "unzips" as the Hydrogen Bonds Break. Step 2: The free nucleotides of the RNA, pair with complementary DNA bases. Step 3: RNA sugar-phosphate backbone forms. (Aided by RNA Polymerase.) Step 4: Hydrogen bonds of the untwisted RNA+DNA "ladder" break, then the RNA leaves the nucleus through the small nucleur pores. This then goes to the cytoplasm to continue on to protein processing.

Transcription is the first step leading to gene expression. The stretch of DNA transcribed into an RNA molecule is called a *transcription unit* and encodes at least one gene. If the gene transcribed encodes for a protein, the result of transcription is messenger RNA (mRNA), which will then be used to create that protein via the process of translation. Alternatively, the transcribed gene may encode for either ribosomal RNA (rRNA) or transfer RNA (tRNA), other components of the protein-assembly process, or other ribozymes. A DNA transcription unit encoding for a protein contains not only the sequence that will eventually be directly translated into the protein (the *coding sequence*) but also *regulatory sequences* that direct and regulate the synthesis of that protein. The regulatory sequence before (upstream from) the coding sequence is called the five prime untranslated region (5'UTR), and the sequence following (downstream from) the coding sequence is called the three prime untranslated region (3'UTR).

Transcription has some proofreading mechanisms, but they are fewer and less effective than the controls for copying DNA; therefore, transcription has a lower copying fidelity than DNA replication.

As in DNA replication, DNA is read from 3' → 5' during transcription. Meanwhile, the complementary RNA is created from the 5' → 3' direction. This means its 5' end is created first in base pairing. Although DNA is arranged as two antiparallel strands in a double helix, only one of the two DNA strands, called the template strand, is used for transcription. This is because RNA is only single-stranded, as opposed to double-stranded DNA. The other DNA strand is called the coding strand, because its sequence is the same as the newly created RNA transcript (except for the substitution of uracil for thymine). The use of only the 3' → 5' strand eliminates the need for the Okazaki fragments seen in DNA replication.

Transcription is divided into 5 stages: *pre-initiation*, *initiation*, *promoter clearance*, *elongation* and *termination*.

Major Steps

Pre-initiation

In eukaryotes, RNA polymerase, and therefore the initiation of transcription, requires the presence of a core promoter sequence in the DNA. Promoters are regions of DNA which promote transcription and in eukaryotes, are found at -30, -75 and -90 base pairs upstream from the start site of transcription. Core promoters are sequences within the promoter which are essential for transcription initiation. RNA polymerase is able to bind to core promoters in the presence of various specific transcription factors. The most common type of core promoter in eukaryotes is a short DNA sequence known as a TATA box, found -30 base pairs from the start site of transcription.

The TATA box, as a core promoter, is the binding site for a transcription factor known as TATA binding protein (TBP), which is itself a subunit of another transcription factor, called Transcription Factor II D (TFIID). After TFIID binds to the TATA box via the TBP, five more transcription factors and RNA polymerase combine around the TATA box in a series of stages to form a preinitiation complex. One transcription factor, DNA helicase, has helicase activity and so is involved in the separating of opposing strands of double-stranded DNA to provide access to a single-stranded DNA template.

However, only a low, or basal, rate of transcription is driven by the preinitiation complex alone. Other proteins known as activators and repressors, along with any associated coactivators or corepressors, are responsible for modulating transcription rate.

Thus preinitiation complex contains:- 1.Core Promoter Sequence 2.Transcription Factors 3.DNA Helicase 4.RNA Polymerase 5.Activators and Repressors The transcription preinitiation in archaea is essentially homologous to that of eukaryotes, but is much less complex. The archaeal preinitiation complex assembles at a TATA-box binding site; however, in archaea, this complex is composed of only RNA polymerase II, TBP, and TFB (the archaeal homologue of eukaryotic transcription factor II B (TFIIB)).

Initiation

In bacteria, transcription begins with the binding of RNA polymerase to the promoter in DNA. RNA polymerase is a core enzyme consisting of five subunits: 2 α subunits, 1 β subunit, 1 β' subunit, and 1 ω subunit. At the start of initiation, the core enzyme is associated with a sigma factor (number 70) that aids in finding the appropriate -35 and -10 base pairs downstream of promoter sequences.

Transcription initiation is more complex in eukaryotes. Eukaryotic RNA polymerase does not directly recognize the core promoter sequences. Instead, a collection of proteins called transcription factors mediate the binding of RNA polymerase and the initiation of transcription. Only after certain transcription factors are attached to the promoter does the RNA polymerase bind to it. The completed assembly of transcription factors and RNA polymerase bind to the promoter, forming a transcription initiation complex. Transcription in the archaea domain is similar to transcription in eukaryotes.

Promoter Clearance

After the first bond is synthesized, the RNA polymerase must clear the promoter. During this time there is a tendency to release the RNA transcript and produce truncated transcripts. This is called *abortive initiation* and is common for both eukaryotes and prokaryotes. Abortive initiation continues to occur until the ó factor rearranges, resulting in the transcription elongation complex (which gives a 35 bp moving footprint).

The ó factor is released before 80 nucleotides of mRNA are synthesized. Once the transcript reaches approximately 23 nucleotides, it no longer slips and elongation can occur. This, like most of the remainder of transcription, is an energy-dependent process, consuming adenosine triphosphate (ATP). Promoter clearance coincides with phosphorylation of serine 5 on the carboxy terminal domain of RNA Pol in eukaryotes, which is phosphorylated by TFIIH.

Elongation

One strand of the DNA, the *template strand* (or noncoding strand), is used as a template for RNA synthesis. As transcription proceeds, RNA polymerase traverses the template strand and uses base pairing complementarity with the DNA template to create an RNA copy.

Although RNA polymerase traverses the template strand from 3' → 5', the coding (non-template) strand and newly-formed RNA can also be used as reference points, so transcription can be described as occurring 5' → 3'.

This produces an RNA molecule from 5' → 3', an exact copy of the coding strand (except that thymines are replaced with uracils, and the nucleotides are composed of a ribose (5-carbon) sugar where DNA has deoxyribose (one less oxygen atom) in its sugar-phosphate backbone).

Unlike DNA replication, mRNA transcription can involve multiple RNA polymerases on a single DNA template and multiple rounds of transcription (amplification of particular mRNA), so many mRNA molecules can be rapidly produced from a single copy of a gene.

Elongation also involves a proofreading mechanism that can replace incorrectly incorporated bases. In eukaryotes, this may correspond with short pauses during transcription that allow appropriate RNA editing factors to bind. These pauses may be intrinsic to the RNA polymerase or due to chromatin structure.

Termination

Bacteria use two different strategies for transcription termination. In Rho-independent transcription termination, RNA transcription stops when the newly synthesized RNA molecule forms a G-C rich hairpin loop followed by a run of Us. When the hairpin forms, the mechanical stress breaks the weak rU-dA bonds, now filling the DNA-RNA hybrid. This pulls the poly-U transcript out of the active site of the RNA polymerase, effectively terminating transcription.

In the "Rho-dependent" type of termination, a protein factor called "Rho" destabilizes the interaction between the template and the mRNA, thus releasing the newly synthesized mRNA from the elongation complex.

Transcription termination in eukaryotes is less understood but involves cleavage of the new transcript followed by template-independent addition of *A*s at its new 3' end, in a process called polyadenylation.

Measuring and Detecting Transcription

Transcription can be measured and detected in a variety of ways:

- Nuclear Run-on assay: measures the relative abundance of newly formed transcripts
- RNase protection assay and ChIP-Chip of RNAP: detect active transcription sites
- RT-PCR: measures the absolute abundance of total or nuclear RNA levels, which may however differ from transcription rates
- DNA microarrays: measures the relative abundance of the global total or nuclear RNA levels; however, these may differ from transcription rates
- In situ hybridization: detects the presence of a transcript
- MS2 tagging: by incorporating RNA stem loops, such as MS2, into a gene, these become incorporated into newly synthesized RNA. The stem loops can then be detected using a fusion of GFP and the MS2 coat protein, which has a high affinity, sequence specific interaction with the MS2 stem loops. The recruitment of GFP to the site of transcription is visualised as a single fluorescent spot. This remarkable new approach has revealed that transcription occurs in discontinuous bursts, or pulses. With the notable exception of in situ techniques, most other methods provide cell population averages, and are not capable of detecting this fundamental property of genes.
- Northern blot: the traditional method, and until the advent of RNA-Seq, the most quantitative
- RNA-Seq: applies next-generation sequencing techniques to sequence whole transcriptomes, which allows the measurement of relative abundance of RNA, as well as the detection of additional variations such as fusion genes, post-translational edits and novel splice sites.

Transcription Factories

Active transcription units are clustered in the nucleus, in discrete sites called transcription factories or euchromatin. Such sites can be visualized by allowing engaged polymerases to extend their transcripts in tagged precursors (Br-UTP or Br-U) and immuno-labeling the tagged nascent RNA. Transcription factories can also be localized using fluorescence in situ hybridization or marked by antibodies directed against polymerases. There are ~10,000 factories in the nucleoplasm of a HeLa cell, among which are ~8,000 polymerase II factories and ~2,000 polymerase III factories. Each polymerase II factory contains ~8 polymerases. As most active transcription units are associated with only one polymerase, each factory usually contains ~8 different transcription units. These units might be associated through promoters and/ or enhancers, with loops forming a 'cloud' around the factor.

History

A molecule which allows the genetic material to be realized as a protein was first hypothesized by François Jacob and Jacques Monod. RNA synthesis by RNA polymerase was established *in vitro* by several laboratories by 1965; however, the RNA synthesized by these enzymes had properties that suggested the existence of an additional factor needed to terminate transcription correctly. In 1972, Walter Fiers became the first person to actually prove the existence of the terminating enzyme.

Reverse Transcription

Some viruses (such as HIV, the cause of AIDS), have the ability to transcribe RNA into DNA. HIV has an RNA genome that is duplicated into DNA. The resulting DNA can be merged with the DNA genome of the host cell. The main enzyme responsible for synthesis of DNA from an RNA template is called reverse transcriptase. In the case of HIV, reverse transcriptase is responsible for synthesizing a complementary DNA strand (cDNA) to the viral RNA genome.

An associated enzyme, ribonuclease H, digests the RNA strand, and reverse transcriptase synthesises a complementary strand of DNA to form a double helix DNA structure. This cDNA is integrated into the host cell's genome via another enzyme (integrase) causing the host cell to generate viral proteins which reassemble into new viral particles. Subsequently, the host cell undergoes programmed cell death, apoptosis. Some eukaryotic cells contain an enzyme with reverse transcription activity called telomerase. Telomerase is a reverse transcriptase that lengthens the ends of linear chromosomes. Telomerase carries an RNA template from which it synthesizes DNA repeating sequence, or "junk" DNA.

This repeated sequence of DNA is important because every time a linear chromosome is duplicated it is shortened in length. With "junk" DNA at the ends of chromosomes, the shortening eliminates some of the non-essential, repeated sequence rather than the protein-encoding DNA sequence farther away from the chromosome end.

Telomerase is often activated in cancer cells to enable cancer cells to duplicate their genomes indefinitely without losing important protein-coding DNA sequence. Activation of telomerase could be part of the process that allows cancer cells to become technically immortal. However, the true *in vivo* significance of telomerase has still not been empirically proven.

Designing Designs for Research

Reprinted from Trochim, W. and Land, D. (1982). Designing Designs for Research. The Researcher, 1, 1, 1-6.

Much contemporary social research is devoted to examining whether a program, treatment, or manipulation causes some outcome or result. For example, we might wish to know whether a new educational program causes subsequent achievement score gains, whether a special work release program for prisoners causes lower recidivism rates, whether a novel drug causes a reduction in symptoms, and so on. Cook and Campbell (1979) argue that three conditions must be met before we can infer that such a cause-effect relation exists:

1. Covariation. Changes in the presumed cause must be related to changes in the presumed effect. Thus, if we introduce, remove, or change the level of a treatment or program, we should observe some change in the outcome measures.
2. Temporal Precedence. The presumed cause must occur prior to the presumed effect.
3. No Plausible Alternative Explanations. The presumed cause must be the only reasonable explanation for changes in the outcome measures. If there are other factors which could be responsible for changes in the outcome measures we cannot be confident that the presumed cause-effect relationship is correct.

In most social research the third condition is the most difficult to meet. Any number of factors other than the treatment or program could cause changes in outcome measures. Campbell and Stanley (1966) and later, Cook and Campbell (1979) list a number of common plausible alternative explanations (or, threats to internal validity).

For example, it may be that some historical event which occurs at the same time that the program or treatment is instituted was responsible for the change in the outcome measures; or, changes in record keeping or measurement systems which occur at the same time as the program might be falsely attributed to the program. The reader is referred to standard research methods texts for more detailed discussions of threats to validity.

This paper is primarily heuristic in purpose. Standard social science methodology textbooks (Cook and Campbell 1979; Judd and Kenny, 1981) typically present an array of research designs and the alternative explanations which these designs rule out or minimize. This tends to foster a "cookbook" approach to research design - an emphasis on the selection of an available design rather than on the construction of an appropriate research strategy.

While standard designs may sometimes fit real-life situations, it will often be necessary to "tailor" a research design to minimize specific threats to validity. Furthermore, even if standard textbook designs are used, an understanding of the logic of design construction in general will improve the comprehension of these standard approaches. This paper takes a structural approach to research design. While this is by no means the only strategy for constructing research designs, it helps to clarify some of the basic principles of design logic.

Minimizing Threats to Validity

Good research designs minimize the plausible alternative explanations for the hypothesized cause-effect relationship. But such explanations may be ruled out or minimized in a number of ways other than by design. The discussion which follows outlines five ways to minimize threats to validity, one of which is by research design:

By Argument: The most straightforward way to rule out a potential threat to validity is to simply argue that the threat in question is not a reasonable one. Such an argument may be made either *a priori* or *a posteriori*, although the former will usually be more convincing than the latter.

For example, depending on the situation, one might argue that an instrumentation threat is not likely because the same test is used for pre and post test measurements and did not involve observers who might improve, or other such factors. In most cases, ruling out a potential threat to validity by argument alone will be weaker than the other approaches listed below. As a result, the most plausible threats in a study should not, except in unusual cases, be ruled out by argument only.

By Measurement or Observation: In some cases it will be possible to rule out a threat by measuring it and demonstrating that either it does not occur at all or occurs so minimally as to not be a strong alternative explanation for the cause-effect relationship. Consider, for example, a study of the effects of an advertising campaign on subsequent sales of a particular product. In such a study, history (i.e., the occurrence of other events which might lead to an increased desire to purchase the product) would be a plausible alternative explanation.

For example, a change in the local economy, the removal of a competing product from the market, or similar events could cause an increase in product sales. One might attempt to minimize such threats by measuring local economic indicators and the availability and sales of competing products. If there is no change in these measures coincident with the onset of the advertising campaign, these threats would be considerably minimized. Similarly, if one is studying the effects of special mathematics training on math achievement scores of children, it might be useful to observe everyday classroom behaviour in order to verify that students were not receiving any additional math training to that provided in the study.

By Design: Here, the major emphasis is on ruling out alternative explanations by adding treatment or control groups, waves of measurement, and the like. This topic will be discussed in more detail below.

By Analysis: There are a number of ways to rule out alternative explanations using statistical analysis. One interesting example is provided by Jurs and Glass (1971). They suggest that one could study the plausibility of an attrition or mortality threat by conducting a two-way analysis of variance. One factor in this study would be the original treatment group designations (i.e., program vs. comparison group), while the other factor would be attrition (i.e., dropout vs. non-dropout group).

The dependent measure could be the pretest or other available pre-program measures. A main effect on the attrition factor would be indicative of a threat to external validity or generalizability, while an interaction between group and attrition factors would point to a possible threat to internal validity. Where both effects occur, it is reasonable to infer that there is a threat to both internal and external validity.

The plausibility of alternative explanations might also be minimized using covariance analysis. For example, in a study of the effects of "workfare" programs on social welfare case loads, one plausible alternative explanation might be the status of local economic conditions.

Here, it might be possible to construct a measure of economic conditions and include that measure as a covariate in the statistical analysis. One must be careful when using covariance adjustments of this type — "perfect" covariates do not exist in most social research and the use of imperfect covariates will not completely adjust for potential alternative explanations. Nevertheless causal assertions are likely to be strengthened by demonstrating that treatment effects occur even after adjusting on a number of good covariates.

By Preventive Action: When potential threats are anticipated they can often be ruled out by some type of preventive action. For example, if the program is a desirable one, it is likely that the comparison group would feel jealous or demoralized. Several actions can be taken to minimize the effects of these attitudes including offering the program to the comparison group upon completion of the study or using program and comparison groups which have little opportunity for contact and communication.

In addition, auditing methods and quality control can be used to track potential experimental dropouts or to insure the standardization of measurement.

The five categories listed above should not be considered mutually exclusive. The inclusion of measurements designed to minimize threats to validity will obviously be related to the design structure and is likely to be a factor in the analysis. A good research plan should, where possible. Make use of multiple methods for reducing threats. In general, reducing a particular threat by design or preventive action will probably be stronger than by using one of the other three approaches. The choice of which strategy to use for any particular threat is complex and depends at least on the cost of the strategy and on the potential seriousness of the threat.

Advances in Quasi-Experimentation

Reprinted from Trochim, W. (Ed.), (1986). *Editor's Notes*. Advances in quasi-experimental design and analysis. New Directions for Program Evaluation Series, Number 31, San Francisco, CA: Jossey-Bass.

The intent of this volume is to update, perhaps even to alter, our thinking about quasi-experimentation in applied social research and program evaluation. Since Campbell and Stanley (1963) introduced the term quasi-experiment, we have tended to see this area as involving primarily two interrelated topics: the theory of the validity of casual inferences and a taxonomy of the research designs that enable us to examine causal hypotheses.

We can see this in the leading expositions of quasi-experimentation (Campbell and Stanley, 1963, 1966; Cook and Campbell, 1979) as well as in the standard textbook presentations of the topic (Kidder and Judd, 1986; Rossi and Freeman, 1985), where it is typical to have separate sections or chapters that discuss validity issues first and then proceed to distinguishable quasi-experimental designs (for example, the pretest-posttest non-equivalent group design, the regression-discontinuity design, the interrupted time series design).

My first inclination in editing this volume was to emulate this tradition, beginning the volume with a chapter on validity and following it with a chapter for each of the major quasi-experimental designs that raised the relevant conceptual and analytical issues and discussed recent advances. But, I think, such an approach would have simply contributed to a persistent confusion about the nature of quasi-experimentation and its role in research.

Instead, this volume makes the case that we have moved beyond the traditional thinking on quasi-experiments as a collection of specific designs and threats to validity toward a more integrated, synthetic view of quasi-experimentation as part of a general logical and epistemological framework for research.

To support this view that the notion of quasi-experimentation is evolving toward increasing integration, I will discuss a number of themes that seem to characterize our current thinking and that cut across validity typologies and design taxonomies. This list of themes may also be viewed as a tentative description of the advances in our thinking about quasi-experimentation in social research.

The Role of Judgment

One theme that underlies most of the others and that illustrates our increasing awareness of the tentativeness and frailty of quasi-experimentation concerns the importance of human judgment in research. Evidence bearing on a causal relationship emerges from many sources, and it is not a trivial matter to integrate or resolve conflicts or discrepancies.

In recognition of this problem of evidence, we are beginning to address causal inference as a psychological issue that can be illuminated by cognitive models of the judgmental process. We are also recognizing more clearly the sociological bases of scientific thought (Campbell, 1984) and the fact that science is at root a human enterprise. Thus, a positivist, mechanistic view is all but gone from quasi-experimental thinking, and what remains is a more judgmental and more scientifically sensible perspective.

The Case for Tailored Designs

Early expositions of quasi-experimentation took a largely taxonomic approach, laying out a collection of relatively discrete research designs and discussing how weak or strong they were for valid causal inference. Almost certainly, early proponents recognized that there was a virtual infinity of design variations and that validity was more complexly related to theory and context than their presentations implied. Nonetheless, what seemed to evolve was a "cookbook" approach to quasi-experimentation that involved "choosing" a design that fit the situation and checking off lists of validity threats.

In an important paper on the coupling of randomized and non-randomized design features, Boruch (1975) explicitly encouraged us to construct research designs as combinations of more elemental units (for example, assignment strategies, measurement occasions) based on the specific contextual needs and plausible alternative explanations for a treatment effect. This move toward hybrid, tailored, or patched-up designs, which involved suggesting how such designs could be accomplished, is one in which I have been a minor participant (Trochim and Land, 1982; Trochim, 1984). It is emphasized by Cordray in Chapter One of this volume.

The implication for current practice is that we should focus on the advantages of different combinations of design features rather than on a relatively restricted set of prefabricated designs. In teaching quasi-experimental methods, we need to break away from a taxonomic design mentality and emphasize design principles and issues that cut across the traditional distinctions between true experiments, non-experiments, and quasi-experiments.

The Crucial Role of Theory

Quasi-experimentation and its randomized experimental parent have been criticized for encouraging an atheoretical "black box" mentality of research (see, for instance, Chen and Rossi, 1984; Cronbach, 1982). Persons are assigned to either complex molar program packages or (often) to equally complex comparison conditions.

The machinery of random assignment (or our quasi-experimental attempts to approximate random assignment) are the primary means of defining whether the program has an effect. This *ceteris paribus* mentality is inherently atheoretical and non-contextual: It assumes that the same mechanism works in basically the same way whether we apply it in mental health or criminal justice, income maintenance or education.

There is nothing inherently wrong with this program-group-versus-comparison-group logic. The problem is that it may be a rather crude, uninformative approach. In the two-group case, we are simply creating a dichotomous input into reality. If we observe a posttest difference between groups, it could be explained by this dichotomous program-versus-comparison-group input or by any number of alternative explanations, including differential attrition rates, intergroup rivalry and communication, initial selection differences among groups, or different group histories. We usually try to deal with these alternative explanations by ruling them out through argument, additional measurement, patched-up design features, and auxiliary analysis. Cook and Campbell (1979), Cronbach (1982), and others strongly favour replication of treatment effects as a standard for judging the validity of a causal assertion, but this advice does little to enhance the validity and informativeness within individual studies or program evaluations.

Chen and Rossi approached this issue by advocating increased attention to social science theory: "not the global conceptual schemes of the grand theorists but much more prosaic theories that are concerned with how human organizations work and how social problems are generated." Evaluators have similarly begun to stress the importance of program theory as the basis for causal assessment (for example, Bickman, in press). These developments allow increased emphasis to be placed on the role of pattern matching (Trochim, 1985) through the generation of more complex theory-driven predictions that, if corroborated, allow fewer plausible alternative explanations for the effect of a program. Because appropriate theories may not be readily available, especially for the evaluation of contemporary social programs, we are developing methods and processes that facilitate the articulation of the implicit theories which program administrators and stakeholder groups have in mind and which presumably guide the formation and implementation of the program (Trochim, 1985). This theory-driven perspective is consonant with Mark's emphasis in Chapter Three on the study of causal process and with Cordray's discussion in Chapter One on ruling in the program as opposed to ruling out alternative explanations.

Attention to Program Implementation

A theory-driven approach to quasi-experimentation will be futile unless we can demonstrate that the program was in fact carried out or implemented as the theory intended. Consequently, we have seen the development of program implementation theory (for example, McLaughlin, 1984) that directly addresses the process of program execution. One approach emphasizes the development of organizational procedures and training systems that accurately transmit the program and that anticipate likely institutional sources of resistance. Another strategy involves the assessment of program delivery through program audits, management information systems, and the like.

This emphasis on program implementation has further obscured the traditional distinction between process and outcome evaluation. At the least, it is certainly clear that good quasi-experimental outcome evaluation cannot be accomplished without attending to program processes, and we are continuing to develop better notions of how to combine these two efforts.

The Importance of Quality Control

Over and over, our experience with quasi-experimentation has shown that even the best-laid research plans often go awry in practice, sometimes with disastrous results.

Thus, over the past decade we have begun to pay increasing attention to the integrity and quality of our research methods in real-world settings. One way of achieving this goal is to incorporate techniques used by other professions — accounting, auditing, industrial quality control — that have traditions in data integrity and quality assurance (Trochim and Visco, 1985). For instance, double bookkeeping can be used to keep verifiable records of research participation. Acceptance sampling can be an efficient method for checking accuracy in large data collection efforts, where an exhaustive examination of records is impractical or excessive in cost. These issues are particularly important in quasi-experimentation, where it is incumbent upon the researcher to demonstrate that sampling, measurement, group assignment, and analysis decisions do not interact with program participation in ways that can confound the final interpretation of results.

The Advantages of Multiple Perspectives

We have long recognized the importance of replication and systematic variation in research. In the past few years, Cook (1985) and colleagues Shadish and Houts have articulated a rationale for achieving systematic variation that they term critical multiplism. This perspective rests on the notion that no single realization will ever be sufficient for understanding a phenomenon with validity. Multiple realizations — of research questions, measures, samples, designs, analyses, replications, and so on — are essential for convergence on the truth of a matter. However, such a varied approach can become a methodological and epistemological Pandora's box unless we apply critical judgment in deciding which multiples we will emphasize in a study or set of studies.

Evolution of the Concept of Validity

The history of quasi-experimentation is inseparable from the development of the theory of the validity of causal inference. Much of this history has been played out through the ongoing dialogue between Campbell and Cronbach concerning the definition of validity and the relative importance that should be attributed on the one hand to the establishment of a causal relationship and on the other hand to its generalizability. In the most recent major statement in this area, Cronbach (1982) articulated the UTOS model, which conceptually links the units, treatments, observing operations and settings in a study into a framework that can be used for establishing valid causal inference. The dialogue continues in Chapter Four of this volume, where Campbell attempts to dispel persistent confusion about the types of validity by tentatively relabeling internal validity as local molar causal validity and external validity as the principle of proximal similarity. It is reasonable to hope that we might achieve a clearer consensus on this issue, as Mark argues in Chapter Three, where he attempts to resolve several different conceptions of validity, including those of Campbell and Cronbach.

Development of Increasingly Complex Realistic Analytic Models

In the past decade, we have made considerable progress toward complicating our statistical analyses to account for increasingly complex contexts and designs. One such advance involves the articulation of causal models of the sort described by Reichardt and Gollob in Chapter Six, especially models that allow for latent variables and that directly model measurement error Joreskog and Sorbom, 1979.

Another important recent development involves analyses that address the problem of selection bias or group nonequivalence — a central issue in quasi-experiments because random assignment is not used and there is no assurance that comparison groups are initially equivalent. At the same time, there is increasing recognition of the implications of not attending to the correct unit of analysis when analyzing the data and of the advantages and implications of conducting analyses at multiple levels.

Thus, when we assign classrooms to conditions but analyze individual student data rather than classroom aggregates, we are liable to get a different view of program effects than we are when we analyze at the classroom level, as Shadish, Cook, and Houts argue in Chapter Two. Other notable advances that are not explicitly addressed in this volume include the development of log linear, probit, and logit models for the analysis of qualitative or nominal level outcome variables (Feinberg, 1980; Forthofer and Lehnen, 1981) and the increasing proliferation of Bayesian statistical approaches to quasi-experimental contexts (Pollard, 1986).

Parallel to the development of these increasingly complex, realistic analytic models, cynicism has deepened about the ability of any single model or analysis to be sufficient. Thus, in Chapter Six Reichardt and Gollob call for multiple analyses to bracket bias, and in Chapter Five Rindskopf recognizes the assumptive notions of any analytic approach to selection bias. We have virtually abandoned the hope of a single correct analysis, and we have accordingly moved to multiple analyses that are based on systematically distinct assumptional frameworks and that rely in an increasingly direct way on the role of judgment.

All the developments just outlined point to an increasingly realistic and complicated life for quasi-experimentalists. The overall picture that emerges is that all quasi-experimentation is judgmental. It is based on multiple and varied sources of evidence, it should be multiplistic in realization, it must attend to process as well as to outcome, it is better off when theory driven, and it leads ultimately to multiple analyses that attempt to bracket the program effect within some reasonable range.

In one sense, this is hardly a pretty picture. Our views about quasi-experimentation and its role in causal inference are certainly more tentative and critical than they were in 1965 or perhaps even in 1979. But, this more integrated and complex view of quasi-experimentation has emerged directly from our experiences in the conduct of such studies. As such, it realistically represents our current thinking about one of the major strands in the evolution of social research methodology in this century.

Analysis

By the time you get to the analysis of your data, most of the really difficult work has been done. It's much more difficult to: define the research problem; develop and implement a sampling plan; conceptualize, operationalize and test your measures; and develop a design structure. If you have done this work well, the analysis of the data is usually a fairly straightforward affair.

In most social research the data analysis involves three major steps, done in roughly this order:

- Cleaning and organizing the data for analysis (Data Preparation)
- Describing the data (Descriptive Statistics)
- Testing Hypotheses and Models (Inferential Statistics)

Data Preparation involves checking or logging the data in; checking the data for accuracy; entering the data into the computer; transforming the data; and developing and documenting a database structure that integrates the various measures. Descriptive Statistics are used to describe the basic features of the data in a study. They provide simple summaries about the sample and the measures. Together with simple graphics analysis, they form the basis of virtually every quantitative analysis of data. With descriptive statistics you are simply describing what is, what the data shows.

Inferential Statistics investigate questions, models and hypotheses. In many cases, the conclusions from inferential statistics extend beyond the immediate data alone. For instance, we use inferential statistics to try to infer from the sample data what the population thinks. Or, we use inferential statistics to make judgments of the probability that an observed difference between groups is a dependable one or one that might have happened by chance in this study. Thus, we use inferential statistics to make inferences from our data to more general conditions; we use descriptive statistics simply to describe what's going on in our data.

In most research studies, the analysis section follows these three phases of analysis. Descriptions of how the data were prepared tend to be brief and to focus on only the more unique aspects to your study, such as specific data transformations that are performed. The descriptive statistics that you actually look at can be voluminous. Usually, the researcher links each of the inferential analyses to specific research questions or hypotheses that were raised in the introduction, or notes any models that were tested that emerged as part of the analysis. In most analysis write-ups it's especially critical to not "miss the forest for the trees." If you present too much detail, the reader may not be able to follow the central line of the results. Often extensive analysis details are appropriately relegated to appendices, reserving only the most critical analysis summaries for the body of the report itself.

DNA Biochemistry

Deoxyribonucleic acid, or DNA, is a nucleic acid that contains the genetic instructions used in the development and functioning of all known living organisms with the exception of some viruses. The main role of DNA molecules is the long-term storage of information. DNA is often compared to a set of blueprints, like a recipe or a code, since it contains the instructions needed to construct other components of cells, such as proteins and RNA molecules. The DNA segments that carry this genetic information are called genes, but other DNA sequences have structural purposes, or are involved in regulating the use of this genetic information.

DNA consists of two long polymers of simple units called nucleotides, with backbones made of sugars and phosphate groups joined by ester bonds. These two strands run in opposite directions to each other and are therefore anti-parallel. Attached to each sugar is one of four types of molecules called bases.

It is the sequence of these four bases along the backbone that encodes information. This information is read using the genetic code, which specifies the sequence of the amino acids within proteins. The code is read by copying stretches of DNA into the related nucleic acid RNA, in a process called transcription.

Within cells, DNA is organized into long structures called chromosomes. These chromosomes are duplicated before cells divide, in a process called DNA replication. Eukaryotic organisms (animals, plants, fungi, and protists) store most of their DNA inside the cell nucleus and some of their DNA in organelles, such as mitochondria or chloroplasts. In contrast, prokaryotes (bacteria and archaea) store their DNA only in the cytoplasm. Within the chromosomes, chromatin proteins such as histones compact and organize DNA. These compact structures guide the interactions between DNA and other proteins, helping control which parts of the DNA are transcribed.

Properties

DNA is a long polymer made from repeating units called nucleotides. As first discovered by James D. Watson and Francis Crick, the structure of DNA of all species comprises two helical chains each coiled round the same axis, and each with a pitch of 34 Ångströms (3.4 nanometres) and a radius of 10 Ångströms (1.0 nanometres). According to another study, when measured in a particular solution, the DNA chain measured 22 to 26 Ångströms wide (2.2 to 2.6 nanometres), and one nucleotide unit measured 3.3 Å (0.33 nm) long. Although each individual repeating unit is very small, DNA polymers can be very large molecules containing millions of nucleotides. For instance, the largest human chromosome, chromosome number 1, is approximately 220 million base pairs long.

In living organisms, DNA does not usually exist as a single molecule, but instead as a pair of molecules that are held tightly together. These two long strands entwine like vines, in the shape of a double helix. The nucleotide repeats contain both the segment of the backbone of the molecule, which holds the chain together, and a base, which interacts with the other DNA strand in the helix. A base linked to a sugar is called a nucleoside and a base linked to a sugar and one or more phosphate groups is called a nucleotide. If multiple nucleotides are linked together, as in DNA, this polymer is called a polynucleotide.

The backbone of the DNA strand is made from alternating phosphate and sugar residues. The sugar in DNA is 2-deoxyribose, which is a pentose (five-carbon) sugar. The sugars are joined together by phosphate groups that form phosphodiester bonds between the third and fifth carbon atoms of adjacent sugar rings.

These asymmetric bonds mean a strand of DNA has a direction. In a double helix the direction of the nucleotides in one strand is opposite to their direction in the other strand: the strands are *antiparallel*. The asymmetric ends of DNA strands are called the 52 (*five prime*) and 32 (*three prime*) ends, with the 5' end having a terminal phosphate group and the 3' end a terminal hydroxyl group. One major difference between DNA and RNA is the sugar, with the 2-deoxyribose in DNA being replaced by the alternative pentose sugar ribose in RNA.

The DNA double helix is stabilized by hydrogen bonds between the bases attached to the two strands. The four bases found in DNA are adenine (abbreviated A), cytosine (C), guanine (G) and thymine (T). These four bases are attached to the sugar/phosphate to form the complete nucleotide.

These bases are classified into two types; adenine and guanine are fused five- and six-membered heterocyclic compounds called purines, while cytosine and thymine are six-membered rings called pyrimidines. A fifth pyrimidine base, called uracil (U), usually takes the place of thymine in RNA and differs from thymine by lacking a methyl group on its ring. Uracil is not usually found in DNA, occurring only as a breakdown product of cytosine. In addition to RNA and DNA, a large number of artificial nucleic acid analogues have also been created to study the proprieties of nucleic acids, or for use in biotechnology.

Grooves

Twin helical strands form the DNA backbone. Another double helix may be found by tracing the spaces, or grooves, between the strands. These voids are adjacent to the base pairs and may provide a binding site. As the strands are not directly opposite each other, the grooves are unequally sized. One groove, the major groove, is 22 Å wide and the other, the minor groove, is 12 Å wide. The narrowness of the minor groove means that the edges of the bases are more accessible in the major groove. As a result, proteins like transcription factors that can bind to specific sequences in double-stranded DNA usually make contacts to the sides of the bases exposed in the major groove. This situation varies in unusual conformations of DNA within the cell, but the major and minor grooves are always named to reflect the differences in size that would be seen if the DNA is twisted back into the ordinary B form.

Base Pairing

Each type of base on one strand forms a bond with just one type of base on the other strand. This is called complementary base pairing. Here, purines form hydrogen bonds to pyrimidines, with A bonding only to T, and C bonding only to G. This arrangement of two nucleotides binding together across the double helix is called a base pair. As hydrogen bonds are not covalent, they can be broken and rejoined relatively easily.

The two strands of DNA in a double helix can therefore be pulled apart like a zipper, either by a mechanical force or high temperature. As a result of this complementarity, all the information in the double-stranded sequence of a DNA helix is duplicated on each strand, which is vital in DNA replication. Indeed, this reversible and specific interaction between complementary base pairs is critical for all the functions of DNA in living organisms.

The two types of base pairs form different numbers of hydrogen bonds, AT forming two hydrogen bonds, and GC forming three hydrogen bonds. DNA with high GC-content is more stable than DNA with low GC-content, but contrary to popular belief, this is not due to the extra hydrogen bond of a GC base pair but rather the contribution of stacking interactions (hydrogen bonding merely provides specificity of the pairing, not stability).

As a result, it is both the percentage of GC base pairs and the overall length of a DNA double helix that determine the strength of the association between the two strands of DNA. Long DNA helices with a high GC content have stronger-interacting strands, while short helices with high AT content have weaker-interacting strands.

In biology, parts of the DNA double helix that need to separate easily, such as the TATAAT Pribnow box in some promoters, tend to have a high AT content, making the strands easier to pull apart. In the laboratory, the strength of this interaction can be measured by finding the temperature required to break the hydrogen bonds, their melting temperature (also called T_m value). When all the base pairs in a DNA double helix melt, the strands separate and exist in solution as two entirely independent molecules. These single-stranded DNA molecules (*ssDNA*) have no single common shape, but some conformations are more stable than others.

Sense and Antisense

A DNA sequence is called "sense" if its sequence is the same as that of a messenger RNA copy that is translated into protein. The sequence on the opposite strand is called the "antisense" sequence. Both sense and antisense sequences can exist on different parts of the same strand of DNA (i.e. both strands contain both sense and antisense sequences). In both prokaryotes and eukaryotes, antisense RNA sequences are produced, but the functions of these RNAs are not entirely clear. One proposal is that antisense RNAs are involved in regulating gene expression through RNA-RNA base pairing.

A few DNA sequences in prokaryotes and eukaryotes, and more in plasmids and viruses, blur the distinction between sense and antisense strands by having overlapping genes. In these cases, some DNA sequences do double duty, encoding one protein when read along one strand, and a second protein when read in the opposite direction along the other strand. In bacteria, this overlap may be involved in the regulation of gene transcription, while in viruses, overlapping genes increase the amount of information that can be encoded within the small viral genome.

Supercoiling

DNA can be twisted like a rope in a process called DNA supercoiling. With DNA in its "relaxed" state, a strand usually circles the axis of the double helix once every 10.4 base pairs, but if the DNA is twisted the strands become more tightly or more loosely wound. If the DNA is twisted in the direction of the helix, this is positive supercoiling, and the bases are held more tightly together. If they are twisted in the opposite direction, this is negative supercoiling, and the bases come apart more easily. In nature, most DNA has slight negative supercoiling that is introduced by enzymes called topoisomerases. These enzymes are also needed to relieve the twisting stresses introduced into DNA strands during processes such as transcription and DNA replication.

Alternate DNA Structures

DNA exists in many possible conformations that include A-DNA, B-DNA, and Z-DNA forms, although, only B-DNA and Z-DNA have been directly observed in functional organisms. The conformation that DNA adopts depends on the hydration level, DNA sequence, the amount and direction of supercoiling, chemical modifications of the bases, the type and concentration of metal ions, as well as the presence of polyamines in solution.

The first published reports of A-DNA X-ray diffraction patterns— and also B-DNA used analyses based on Patterson transforms that provided only a limited amount of structural information for oriented fibers of DNA. An alternate analysis was then proposed by Wilkins *et al.*, in 1953, for the *in vivo*

B-DNA X-ray diffraction/scattering patterns of highly hydrated DNA fibers in terms of squares of Bessel functions. In the same journal, James D. Watson and Francis Crick presented their molecular modeling analysis of the DNA X-ray diffraction patterns to suggest that the structure was a double-helix.

Although the 'B-DNA form' is most common under the conditions found in cells, it is not a well-defined conformation but a family of related DNA conformations that occur at the high hydration levels present in living cells. Their corresponding X-ray diffraction and scattering patterns are characteristic of molecular paracrystals with a significant degree of disorder. Compared to B-DNA, the A-DNA form is a wider right-handed spiral, with a shallow, wide minor groove and a narrower, deeper major groove.

The A form occurs under non-physiological conditions in partially dehydrated samples of DNA, while in the cell it may be produced in hybrid pairings of DNA and RNA strands, as well as in enzyme-DNA complexes. Segments of DNA where the bases have been chemically modified by methylation may undergo a larger change in conformation and adopt the Z form. Here, the strands turn about the helical axis in a left-handed spiral, the opposite of the more common B form. These unusual structures can be recognized by specific Z-DNA binding proteins and may be involved in the regulation of transcription.

Quadruplex Structures

At the ends of the linear chromosomes are specialized regions of DNA called telomeres. The main function of these regions is to allow the cell to replicate chromosome ends using the enzyme telomerase, as the enzymes that normally replicate DNA cannot copy the extreme 32 ends of chromosomes. These specialized chromosome caps also help protect the DNA ends, and stop the DNA repair systems in the cell from treating them as damage to be corrected. In human cells, telomeres are usually lengths of single-stranded DNA containing several thousand repeats of a simple TTAGGG sequence.

These guanine-rich sequences may stabilize chromosome ends by forming structures of stacked sets of four-base units, rather than the usual base pairs found in other DNA molecules. Here, four guanine bases form a flat plate and these flat four-base units then stack on top of each other, to form a stable *G-quadruplex* structure. These structures are stabilized by hydrogen bonding between the edges of the bases and chelation of a metal ion in the centre of each four-base unit. Other structures can also be formed, with the central set of four bases coming from either a single strand folded around the bases, or several different parallel strands, each contributing one base to the central structure.

In addition to these stacked structures, telomeres also form large loop structures called telomere loops, or T-loops. Here, the single-stranded DNA curls around in a long circle stabilized by telomere-binding proteins. At the very end of the T-loop, the single-stranded telomere DNA is held onto a region of double-stranded DNA by the telomere strand disrupting the double-helical DNA and base pairing to one of the two strands. This triple-stranded structure is called a displacement loop or D-loop.

Branched DNA

In DNA fraying occurs when non-complementary regions exist at the end of an otherwise complementary double-strand of DNA. However, branched DNA can occur if a third strand of DNA is introduced and contains adjoining regions able to hybridize with the frayed regions of the pre-existing double-strand. Although the simplest example of branched DNA involves only three strands of DNA, complexes involving additional strands and multiple branches are also possible. Branched DNA can be used in nanotechnology to construct geometric shapes.

Chemical Modifications

Base Modifications

The expression of genes is influenced by how the DNA is packaged in chromosomes, in a structure called chromatin. Base modifications can be involved in packaging, with regions that have low or no gene expression usually containing high levels of methylation of cytosine bases. For example, cytosine methylation, produces 5-methylcytosine, which is important for X-chromosome inactivation. The average level of methylation varies between organisms - the worm *Caenorhabditis elegans* lacks cytosine methylation, while vertebrates have higher levels, with up to 1% of their DNA containing 5-methylcytosine. Despite the importance of 5-methylcytosine, it can deaminate to leave a thymine base, so methylated cytosines are particularly prone to mutations. Other base modifications include adenine methylation in bacteria, the presence of 5-hydroxymethylcytosine in the brain, and the glycosylation of uracil to produce the "J-base" in kinetoplastids.

Damage

DNA can be damaged by many sorts of mutagens, which change the DNA sequence. Mutagens include oxidizing agents, alkylating agents and also high-energy electromagnetic radiation such as ultraviolet light and X-rays. The type of DNA damage produced depends on the type of mutagen. For example, UV light can damage DNA by producing thymine dimers, which are cross-links between pyrimidine bases. On the other hand, oxidants such as free radicals or hydrogen peroxide produce multiple forms of damage, including base modifications, particularly of guanosine, and double-strand breaks. A typical human cell contains about 150,000 bases that have suffered oxidative damage.

Of these oxidative lesions, the most dangerous are double-strand breaks, as these are difficult to repair and can produce point mutations, insertions and deletions from the DNA sequence, as well as chromosomal translocations.

Many mutagens fit into the space between two adjacent base pairs, this is called *intercalation*. Most intercalators are aromatic and planar molecules; examples include ethidium bromide, daunomycin, and doxorubicin. In order for an intercalator to fit between base pairs, the bases must separate, distorting the DNA strands by unwinding of the double helix.

This inhibits both transcription and DNA replication, causing toxicity and mutations. As a result, DNA intercalators are often carcinogens, and benzo[*a*]pyrene diol epoxide, acridines, aflatoxin and ethidium bromide are well-known examples. Nevertheless, due to their ability to inhibit DNA transcription and replication, other similar toxins are also used in chemotherapy to inhibit rapidly growing cancer cells.

Biological Functions

DNA usually occurs as linear chromosomes in eukaryotes, and circular chromosomes in prokaryotes. The set of chromosomes in a cell makes up its genome; the human genome has approximately 3 billion base pairs of DNA arranged into 46 chromosomes. The information carried by DNA is held in the sequence of pieces of DNA called genes. Transmission of genetic information in genes is achieved via complementary base pairing.

For example, in transcription, when a cell uses the information in a gene, the DNA sequence is copied into a complementary RNA sequence through the attraction between the DNA and the correct RNA nucleotides. Usually, this RNA copy is then used to make a matching protein sequence in a process called translation, which depends on the same interaction between RNA nucleotides. In alternative fashion, a cell may simply copy its genetic information in a process called DNA replication. The details of these functions are covered in other articles; here we focus on the interactions between DNA and other molecules that mediate the function of the genome.

Genes and Genomes

Genomic DNA is tightly and orderly packed in the process called DNA condensation to fit the small available volumes of the cell. In eukaryotes, DNA is located in the cell nucleus, as well as small amounts in mitochondria and chloroplasts. In prokaryotes, the DNA is held within an irregularly shaped body in the cytoplasm called the nucleoid. The genetic information in a genome is held within genes, and the complete set of this information in an organism is called its genotype. A gene is a unit of heredity and is a region of DNA that influences a particular characteristic in an organism. Genes contain an open reading frame that can be transcribed, as well as regulatory sequences such as promoters and enhancers, which control the transcription of the open reading frame. In many species, only a small fraction of the total sequence of the genome encodes protein. For example, only about 1.5% of the human genome consists of protein-coding exons, with over 50% of human DNA consisting of non-coding repetitive sequences. The reasons for the presence of so much non-coding DNA in eukaryotic genomes and the extraordinary differences in genome size, or *C-value*, among species represent a long-standing puzzle known as the "C-value enigma". However, DNA sequences that do not code protein may still encode functional non-coding RNA molecules, which are involved in the regulation of gene expression.

Some non-coding DNA sequences play structural roles in chromosomes. Telomeres and centromeres typically contain few genes, but are important for the function and stability of chromosomes. An abundant form of non-coding DNA in humans are pseudogenes, which are copies of genes that have been disabled by mutation. These sequences are usually just molecular fossils, although they can occasionally serve as raw genetic material for the creation of new genes through the process of gene duplication and divergence.

Transcription and Translation

A gene is a sequence of DNA that contains genetic information and can influence the phenotype of an organism. Within a gene, the sequence of bases along a DNA strand defines a messenger RNA sequence, which then defines one or more protein sequences. The relationship between the nucleotide sequences of genes and the amino-acid sequences of proteins is determined by the rules of translation, known collectively as the genetic code. The genetic code consists of three-letter 'words' called *codons* formed from a sequence of three nucleotides (e.g. ACT, CAG, TTT).

In transcription, the codons of a gene are copied into messenger RNA by RNA polymerase. This RNA copy is then decoded by a ribosome that reads the RNA sequence by base-pairing the messenger RNA to transfer RNA, which carries amino acids. Since there are 4 bases in 3-letter combinations, there are 64 possible codons (4^3 combinations). These encode the twenty standard amino acids, giving most amino acids more than one possible codon. There are also three 'stop' or 'nonsense' codons signifying the end of the coding region; these are the TAA, TGA and TAG codons.

DNA Replication

DNA replication, the basis for biological inheritance, is a fundamental process occurring in all living organisms to copy their DNA. This process is "replication" in that each strand of the original double-stranded DNA molecule serves as template for the reproduction of the complementary strand. Therefore, following DNA replication, two identical DNA molecules have been produced from a single double-stranded DNA molecule.

Cellular proofreading and error toe-checking mechanisms ensure near perfect fidelity for DNA replication. In a cell, DNA replication begins at specific locations in the genome, called "origins".

Unwinding of DNA at the origin, and synthesis of new strands, forms a replication fork. In addition to DNA polymerase, the enzyme that synthesizes the new DNA by adding nucleotides matched to the template strand, a number of other proteins are associated with the fork and assist in the initiation and continuation of DNA synthesis.

DNA replication can also be performed *in vitro* (outside a cell). DNA polymerases, isolated from cells, and artificial DNA primers are used to initiate DNA synthesis at known sequences in a template molecule.

The polymerase chain reaction (PCR), a common laboratory technique, employs such artificial synthesis in a cyclic manner to amplify a specific target DNA fragment from a pool of DNA.

DNA usually exists as a double-stranded structure, with both strands coiled together to form the characteristic double-helix. Each single strand of DNA is a chain of four types of nucleotides: adenine, cytosine, guanine, and thymine. A nucleotide is a mono-, di- or triphosphate deoxyribonucleoside; that is, a deoxyribose sugar is attached to one, two or three phosphates. Chemical interaction of these nucleotides forms phosphodiester linkages, creating the phosphate-deoxribose backbone of the DNA double helix with the bases pointing inward. Nucleotides (bases) are matched between strands through hydrogen bonds to form base pairs. Adenine pairs with thymine and cytosine pairs with guanine.

DNA strands have a directionality, and the different ends of a single strand are called the "3' (three-prime) end" and the "5' (five-prime) end." These terms refer to the carbon atom in deoxyribose to which the next phosphate in the chain attaches. In addition to being complementary, the two strands of DNA are antiparallel: they are orientated in opposite directions. This directionality has consequences in DNA synthesis, because DNA polymerase can only synthesize DNA in one direction by adding nucleotides to the 3' end of a DNA strand.

The pairing of bases in DNA through hydrogen bonding means that the information contained within each strand is redundant. The nucleotides on a single strand can be used to reconstruct nucleotides on a newly synthesized partner strand.

DNA Polymerase

DNA polymerases are a family of enzymes that carry out all forms of DNA replication. A DNA polymerase can only extend an existing DNA strand paired with a template strand; it cannot begin the synthesis of a new strand. To begin synthesis of a new strand, a short fragment of DNA or RNA, called a primer, must be created and paired with the template strand before DNA polymerase can synthesize new DNA. Once a primer pairs with DNA to be replicated, DNA polymerase synthesizes a new strand of DNA by extending the 3' end of an existing nucleotide chain, adding new nucleotides matched to the template strand one at a time via the creation of phosphodiester bonds.

The energy for this process of DNA polymerization comes from two of the three total phosphates attached to each unincorporated base. (Free bases with their attached phosphate groups are called nucleoside triphosphates.) When a nucleotide is being added to a growing DNA strand, two of the phosphates are removed and the energy produced creates a phosphodiester (chemical) bond that attaches the remaining phosphate to the growing chain.

The energetics of this process also help explain the directionality of synthesis - if DNA were synthesized in the 3' to 5' direction, the energy for the process would come from the 5' end of the growing strand rather than from free nucleotides.

DNA polymerases are generally extremely accurate, making less than one error for every 10^7 nucleotides added. Even so, some DNA polymerases also have proofreading ability; they can remove nucleotides from the end of a strand in order to correct mismatched bases. If the 5' nucleotide needs to be removed during proofreading, the triphosphate end is lost. Hence, the energy source that usually provides energy to add a new nucleotide is also lost.

DNA Replication within the Cell

Origins of Replication

For a cell to divide, it must first replicate its DNA. This process is initiated at particular points within the DNA, known as "origins", which are targeted by proteins that separate the two strands and initiate DNA synthesis. Origins contain DNA sequences recognized by replication initiator proteins (e.g. dnaA in *E coli'* and the Origin Recognition Complex in yeast). These initiator proteins recruit other proteins to separate the two strands and initiate replication forks.

Initiator proteins recruit other proteins to separate the DNA strands at the origin, forming a bubble. Origins tend to be "AT-rich" (rich in adenine and thymine bases) to assist this process, because A-T base pairs have two hydrogen bonds (rather than the three formed in a C-G pair)—strands rich in these nucleotides are generally easier to separate due the positive relationship between the number of hydrogen bonds and the difficulty of breaking these bonds. Once strands are separated, RNA primers are created on the template strands.

More specifically, the leading strand receives one RNA primer per active origin of replication while the lagging strand receives several; these several fragments of RNA primers found on the lagging strand of DNA are called Okazaki fragments, named after their discoverer. DNA polymerase extends the leading strand in one continuous motion and the lagging strand in a discontinuous motion (due to the Okazaki fragments). RNase removes the RNA fragments used to initiate replication by DNA Polymerase, and another DNA Polymerase enters to fill the gaps. When this is complete, a single nick on the leading strand and several nicks on the lagging strand can be found. Ligase works to fill these nicks in, thus completing the newly replicated DNA molecule.

As DNA synthesis continues, the original DNA strands continue to unwind on each side of the bubble, forming 2 replication forks. In bacteria, which have a single origin of replication on their circular chromosome, this process eventually creates a "theta structure" (resembling the Greek letter theta: è). In contrast, eukaryotes have longer linear chromosomes and initiate replication at multiple origins within these.

The Replication Fork

The replication fork is a structure that forms within the nucleus during DNA replication. It is created by helicases, which break the hydrogen bonds holding the two DNA strands together. The resulting structure has two branching "prongs", each one made up of a single strand of DNA. These two strands serve as the template for the leading and lagging strands which will be created as DNA polymerase matches complementary nucleotides to the templates; The templates may be properly referred to as the leading strand template and the lagging strand template.

Leading Strand

The leading strand is the template strand of the DNA double helix so that the replication fork moves along it in the 3' to 5' direction. This allows the new strand synthesized complementary to it to be synthesized 5' to 3' in the same direction as the movement of the replication fork. On the leading strand, a polymerase "reads" the DNA and adds nucleotides to it continuously. This polymerase is DNA polymerase III (DNA Pol III) in prokaryotes and presumably Pol ε in eukaryotes.

Lagging Strand

The lagging strand is the strand of the template DNA double helix that is oriented so that the replication fork moves along it in a 5' to 3' manner. Because of its orientation, opposite to the working orientation of DNA polymerase III, which moves on a template in a 3' to 5' manner, replication of the lagging strand is more complicated than that of the leading strand.

On the lagging strand, primase "reads" the DNA and adds RNA to it in short, separated segments. In eukaryotes, primase is intrinsic to Pol α. DNA polymerase III or Pol δ lengthens the primed segments, forming Okazaki fragments. Primer removal in eukaryotes is also performed by Pol δ. In prokaryotes, DNA polymerase I "reads" the fragments, removes the RNA using its flap endonuclease domain (RNA primers are removed by 5'-3' exonuclease activity of polymerase I [weaver, 2005], and replaces the RNA nucleotides with DNA nucleotides (this is necessary because RNA and DNA use slightly different kinds of nucleotides). DNA ligase joins the fragments together.

Dynamics at the Replication Fork

As helicase unwinds DNA at the replication fork, the DNA ahead is forced to rotate. This process results in a build-up of twists in the DNA ahead. This build-up would form a resistance that would eventually halt the progress of the replication fork. DNA topoisomerases are enzymes that solve these physical problems in the coiling of DNA. Topoisomerase I cuts a single backbone on the DNA, enabling the strands to swivel around each other to remove the build-up of twists. Topoisomerase II cuts both backbones, enabling one double-stranded DNA to pass through another, thereby removing knots and entanglements that can form within and between DNA molecules. Bare single-stranded DNA has a tendency to fold back upon itself and form secondary structures; these structures can interfere with the movement of DNA polymerase. To prevent this, single-strand binding proteins bind to the DNA until a second strand is synthesized, preventing secondary structure formation.

Clamp proteins form a sliding clamp around DNA, helping the DNA polymerase maintain contact with its template and thereby assisting with processivity. The inner face of the clamp enables DNA to be threaded through it. Once the polymerase reaches the end of the template or detects double stranded DNA, the sliding clamp undergoes a conformational change which releases the DNA polymerase. Clamp-loading proteins are used to initially load the clamp, recognizing the junction between template and RNA primers.

Regulation of Replication

Within eukaryotes, DNA replication is controlled within the context of the cell cycle. As the cell grows and divides, it progresses through stages in the cell cycle; DNA replication occurs during the S phase (Synthesis phase). The progress of the eukaryotic cell through the cycle is controlled by cell cycle checkpoints. Progression through checkpoints is controlled through complex interactions between various proteins, including cyclins and cyclin-dependent kinases. The G1/S checkpoint (or restriction checkpoint) regulates whether eukaryotic cells enter the process of DNA replication and subsequent division. Cells which do not proceed through this checkpoint are quiescent in the "G0" stage and do not replicate their DNA. Replication of chloroplast and mitochondrial genomes occurs independent of the cell cycle, through the process of D-loop replication.

Bacteria

Most bacteria do not go through a well-defined cell cycle and instead continuously copy their DNA; during rapid growth this can result in multiple rounds of replication occurring concurrently. Within *E coli*, the most well-characterized bacteria, regulation of DNA replication can be achieved through several mechanisms, including: the hemimethylation and sequestering of the origin sequence, the ratio of ATP to ADP, and the levels of protein DnaA. These all control the process of initiator proteins binding to the origin sequences.

Because *E coli* methylates GATC DNA sequences, DNA synthesis results in hemimethylated sequences. This hemimethylated DNA is recognized by a protein (SeqA) which binds and sequesters the origin sequence; in addition, dnaA (required for initiation of replication) binds less well to hemimethylated DNA. As a result, newly replicated origins are prevented from immediately initiating another round of DNA replication.

ATP builds up when the cell is in a rich medium, triggering DNA replication once the cell has reached a specific size. ATP competes with ADP to bind to DnaA, and the DnaA-ATP complex is able to initiate replication. A certain number of DnaA proteins are also required for DNA replication — each time the origin is copied the number of binding sites for DnaA doubles, requiring the synthesis of more DnaA to enable another initiation of replication.

Termination of Replication

Because bacteria have circular chromosomes, termination of replication occurs when the two replication forks meet each other on the opposite end of the parental chromosome. *E coli* regulate this process through the use of termination sequences which, when bound by the Tus protein, enable only one direction of replication fork to pass through. As a result, the replication forks are constrained to always meet within the termination region of the chromosome.

Eukaryotes initiate DNA replication at multiple points in the chromosome, so replication forks meet and terminate at many points in the chromosome;

these are not known to be regulated in any particular manner. Because eukaryotes have linear chromosomes, DNA replication often fails to synthesize to the very end of the chromosomes (telomeres), resulting in telomere shortening. This is a normal process in somatic cells — cells are only able to divide a certain number of times before the DNA loss prevents further division. (This is known as the Hayflick limit.) Within the germ cell line, which passes DNA to the next generation, telomerase extends the repetitive sequences of the telomere region to prevent degradation. Telomerase can become mistakenly active in somatic cells, sometimes leading to cancer formation.

Polymerase Chain Reaction

Researchers commonly replicate DNA *in vitro* using the polymerase chain reaction (PCR). PCR uses a pair of primers to span a target region in template DNA, and then polymerizes partner strands in each direction from these primers using a thermostable DNA polymerase. Repeating this process through multiple cycles produces amplification of the targeted DNA region. At the start of each cycle, the mixture of template and primers is heated, separating the newly synthesized molecule and template. Then, as the mixture cools, both of these become templates for annealing of new primers, and the polymerase extends from these. As a result, the number of copies of the target region doubles each round, increasing exponentially.

Interactions with Proteins

All the functions of DNA depend on interactions with proteins. These protein interactions can be non-specific, or the protein can bind specifically to a single DNA sequence. Enzymes can also bind to DNA and of these, the polymerases that copy the DNA base sequence in transcription and DNA replication are particularly important.

DNA-binding Proteins

Structural proteins that bind DNA are well-understood examples of non-specific DNA-protein interactions. Within chromosomes, DNA is held in complexes with structural proteins. These proteins organize the DNA into a compact structure called chromatin. In eukaryotes this structure involves DNA binding to a complex of small basic proteins called histones, while in prokaryotes multiple types of proteins are involved.

The histones form a disk-shaped complex called a nucleosome, which contains two complete turns of double-stranded DNA wrapped around its surface. These non-specific interactions are formed through basic residues in the histones making ionic bonds to the acidic sugar-phosphate backbone of the DNA, and are therefore largely independent of the base sequence. Chemical modifications of these basic amino acid residues include methylation, phosphorylation and acetylation. These chemical changes alter the strength of the interaction between the DNA and the histones, making the DNA more or less accessible to transcription factors and changing the rate of transcription. Other non-specific DNA-binding proteins in chromatin include the high-mobility group proteins, which bind to bent or distorted DNA. These proteins are important in bending arrays of nucleosomes and arranging them into the larger structures that make up chromosomes.

A distinct group of DNA-binding proteins are the DNA-binding proteins that specifically bind single-stranded DNA. In humans, replication protein A is the best-understood member of this family and is used in processes where the double helix is separated, including DNA replication, recombination and DNA repair. These binding proteins seem to stabilize single-stranded DNA and protect it from forming stem-loops or being degraded by nucleases.

In contrast, other proteins have evolved to bind to particular DNA sequences. The most intensively studied of these are the various transcription factors, which are proteins that regulate transcription. Each transcription factor binds to one particular set of DNA sequences and activates or inhibits the transcription of genes that have these sequences close to their promoters. The transcription factors do this in two ways. Firstly, they can bind the RNA polymerase responsible for transcription, either directly or through other mediator proteins; this locates the polymerase at the promoter and allows it to begin transcription. Alternatively, transcription factors can bind enzymes that modify the histones at the promoter; this will change the accessibility of the DNA template to the polymerase.

As these DNA targets can occur throughout an organism's genome, changes in the activity of one type of transcription factor can affect thousands of genes. Consequently, these proteins are often the targets of the signal transduction processes that control responses to environmental changes or cellular differentiation and development. The specificity of these transcription factors' interactions with DNA come from the proteins making multiple contacts to the edges of the DNA bases, allowing them to "read" the DNA sequence. Most of these base-interactions are made in the major groove, where the bases are most accessible.

DNA-modifying Enzymes

Nucleases and Ligases

Nucleases are enzymes that cut DNA strands by catalysing the hydrolysis of the phosphodiester bonds. Nucleases that hydrolyse nucleotides from the ends of DNA strands are called exonucleases, while endonucleases cut within strands. The most frequently used nucleases in molecular biology are the restriction endonucleases, which cut DNA at specific sequences. For instance, the EcoRV enzyme shown to the left recognizes the 6-base sequence 52 - GAT | ATC-32 and makes a cut at the vertical line. In nature, these enzymes protect bacteria against phage infection by digesting the phage DNA when it enters the bacterial cell, acting as part of the restriction modification system. In technology, these sequence-specific nucleases are used in molecular cloning and DNA fingerprinting.

Enzymes called DNA ligases can rejoin cut or broken DNA strands. Ligases are particularly important in lagging strand DNA replication, as they join together the short segments of DNA produced at the replication fork into a complete copy of the DNA template. They are also used in DNA repair and genetic recombination.

Topoisomerases and Helicases

Topoisomerases are enzymes with both nuclease and ligase activity. These proteins change the amount of supercoiling in DNA. Some of these enzymes work by cutting the DNA helix and allowing one section to rotate, thereby reducing its level of supercoiling; the enzyme then seals the DNA break. Other types of these enzymes are capable of cutting one DNA helix and then passing a second strand of DNA through this break, before rejoining the helix. Topoisomerases are required for many processes involving DNA, such as DNA replication and transcription.

Helicases are proteins that are a type of molecular motor. They use the chemical energy in nucleoside triphosphates, predominantly ATP, to break hydrogen bonds between bases and unwind the DNA double helix into single strands. These enzymes are essential for most processes where enzymes need to access the DNA bases.

Polymerases

Polymerases are enzymes that synthesize polynucleotide chains from nucleoside triphosphates. The sequence of their products are copies of existing polynucleotide chains - which are called *templates*. These enzymes function by adding nucleotides onto the 32 hydroxyl group of the previous nucleotide in a DNA strand. As a consequence, all polymerases work in a 52 to 32 direction. In the active site of these enzymes, the incoming nucleoside triphosphate base-pairs to the template: this allows polymerases to accurately synthesize the complementary strand of their template. Polymerases are classified according to the type of template that they use.

In DNA replication, a DNA-dependent DNA polymerase makes a copy of a DNA sequence. Accuracy is vital in this process, so many of these polymerases have a proofreading activity. Here, the polymerase recognizes the occasional mistakes in the synthesis reaction by the lack of base pairing between the mismatched nucleotides. If a mismatch is detected, a 32 to 52 exonuclease activity is activated and the incorrect base removed. In most organisms, DNA polymerases function in a large complex called the replisome that contains multiple accessory subunits, such as the DNA clamp or helicases.

RNA-dependent DNA polymerases are a specialized class of polymerases that copy the sequence of an RNA strand into DNA. They include reverse transcriptase, which is a viral enzyme involved in the infection of cells by retroviruses, and telomerase, which is required for the replication of telomeres. Telomerase is an unusual polymerase because it contains its own RNA template as part of its structure.

Transcription is carried out by a DNA-dependent RNA polymerase that copies the sequence of a DNA strand into RNA. To begin transcribing a gene, the RNA polymerase binds to a sequence of DNA called a promoter and separates the DNA strands. It then copies the gene sequence into a messenger RNA transcript until it reaches a region of DNA called the terminator, where it halts and detaches from the DNA. As with human DNA-dependent DNA polymerases, RNA polymerase II, the enzyme that transcribes most of the genes in the human genome, operates as part of a large protein complex with multiple regulatory and accessory subunits.

Genetic Recombination

A DNA helix usually does not interact with other segments of DNA, and in human cells the different chromosomes even occupy separate areas in the nucleus called "chromosome territories". This physical separation of different chromosomes is important for the ability of DNA to function as a stable repository for information, as one of the few times chromosomes interact is during chromosomal crossover when they recombine. Chromosomal crossover is when two DNA helices break, swap a section and then rejoin.

Recombination allows chromosomes to exchange genetic information and produces new combinations of genes, which increases the efficiency of natural selection and can be important in the rapid evolution of new proteins. Genetic recombination can also be involved in DNA repair, particularly in the cell's response to double-strand breaks.

The most common form of chromosomal crossover is homologous recombination, where the two chromosomes involved share very similar sequences. Non-homologous recombination can be damaging to cells, as it can produce chromosomal translocations and genetic abnormalities. The recombination reaction is catalysed by enzymes known as recombinases, such as RAD51. The first step in recombination is a double-stranded break either caused by an endonuclease or damage to the DNA. A series of steps catalysed in part by the recombinase then leads to joining of the two helices by at least one Holliday junction, in which a segment of a single strand in each helix is annealed to the complementary strand in the other helix. The Holliday junction is a tetrahedral junction structure that can be moved along the pair of chromosomes, swapping one strand for another. The recombination reaction is then halted by cleavage of the junction and re-ligation of the released DNA.

Evolution

DNA contains the genetic information that allows all modern living things to function, grow and reproduce. However, it is unclear how long in the 4-billion-year history of life DNA has performed this function, as it has been proposed that the earliest forms of life may have used RNA as their genetic material. RNA may have acted as the central part of early cell metabolism as it can both transmit genetic information and carry out catalysis as part of ribozymes. This ancient RNA world where nucleic acid would have been used for both catalysis and genetics may have influenced the evolution of the current genetic code based on four nucleotide bases. This would occur, since the number of different bases in such an organism is a trade-off between a small number of bases increasing replication accuracy and a large number of bases increasing the catalytic efficiency of ribozymes.

However, there is no direct evidence of ancient genetic systems, as recovery of DNA from most fossils is impossible. This is because DNA will survive in the environment for less than one million years and slowly degrades into short fragments in solution. Claims for older DNA have been made, most notably a report of the isolation of a viable bacterium from a salt crystal 250 million years old, but these claims are controversial.

DNA Microarray

A DNA microarray is a multiplex technology used in molecular biology. It consists of an arrayed series of thousands of microscopic spots of DNA oligonucleotides, called features, each containing picomoles (10^{-12} moles) of a specific DNA sequence, known as *probes* (or *reporters*). These can be a short section of a gene or other DNA element that are used to hybridize a cDNA or cRNA sample (called *target*) under high-stringency conditions. Probe-target hybridization is usually detected and quantified by detection of fluorophore-, silver-, or chemiluminescence-labelled targets to determine relative abundance of nucleic acid sequences in the target. Since an array can contain tens of thousands of probes, a microarray experiment can accomplish many genetic tests in parallel. Therefore arrays have dramatically accelerated many types of investigation.

In standard microarrays, the probes are attached via surface engineering to a solid surface by a covalent bond to a chemical matrix (via epoxy-silane, amino-silane, lysine, polyacrylamide or others). The solid surface can be glass or a silicon chip, in which case they are colloquially known as an *Affy chip* when an Affymetrix chip is used. Other microarray platforms, such as Illumina, use microscopic beads, instead of the large solid support. DNA arrays are different from other types of microarray only in that they either measure DNA or use DNA as part of its detection system.

DNA microarrays can be used to measure changes in expression levels, to detect single nucleotide polymorphisms (SNPs), or to genotype or resequence mutant genomes. Microarrays also differ in fabrication, workings, accuracy, efficiency, and cost. Additional factors for microarray experiments are the experimental design and the methods of analysing the data.

Microarray technology evolved from Southern blotting, where fragmented DNA is attached to a substrate and then probed with a known gene or fragment. Nucleic Acids Res. 1992 Apr 11;20(7):1679-84. Oligonucleotide hybridizations on glass supports: a novel linker for oligonucleotide synthesis and hybridization properties of oligonucleotides synthesised in situ. Maskos U, Southern EM. The first reported use of this approach was the analysis of 378 arrayed lysed bacterial colonies each harboring a different sequence which were assayed in multiple replicas for expression of the genes in multiple normal and tumour tissue.

This was expanded to analysis of more than 4000 human sequences with computer driven scanning and image processing for quantitative analysis of the sequences in human colonic tumours and normal tissue and then to comparison of colonic tissues at different genetic risk. The use of a collection of distinct DNAs in arrays for expression profiling was also described in 1987, and the arrayed DNAs were used to identify genes whose expression is modulated by interferon. These early gene arrays were made by spotting cDNAs onto filter paper with a pin-spotting device. The use of miniaturized microarrays for gene expression profiling was first reported in 1995, and a complete eukaryotic genome (*Saccharomyces cerevisiae*) on a microarray was published in 1997.

Principle

The core principle behind microarrays is hybridization between two DNA strands, the property of complementary nucleic acid sequences to specifically pair with each other by forming hydrogen bonds between complementary nucleotide base pairs. A high number of complementary base pairs in a nucleotide sequence means tighter non-covalent bonding between the two strands.

After washing off of non-specific bonding sequences, only strongly paired strands will remain hybridized. So fluorescently labelled target sequences that bind to a probe sequence generate a signal that depends on the strength of the hybridization determined by the number of paired bases, the hybridization conditions (such as temperature), and washing after hybridization. Total strength of the signal, from a spot (feature), depends upon the amount of target sample binding to the probes present on that spot. Microarrays use relative quantitation in which the intensity of a feature is compared to the intensity of the same feature under a different condition, and the identity of the feature is known by its position. An alternative to microarrays is serial analysis of gene expression, where the transcriptome is sequenced allowing an absolute measurement.

Uses and Types

Many types of array exist and the broadest distinction is whether they are spatially arranged on a surface or on coded beads:

- The traditional solid-phase array is a collection of orderly microscopic "spots", called features, each with a specific probe attached to a solid surface, such as glass, plastic or silicon biochip (commonly known as a *genome chip*, *DNA chip* or *gene array*). Thousands of them can be placed in known locations on a single DNA microarray.
- The alternative bead array is a collection of microscopic polystyrene beads, each with a specific probe and a ratio of two or more dyes, which do not interfere with the fluorescent dyes used on the target sequence.

DNA microarrays can be used to detect DNA (as in comparative genomic hybridization), or detect RNA (most commonly as cDNA after reverse transcription) that may or may not be translated into proteins. The process of measuring gene expression via cDNA is called expression analysis or expression profiling.

Applications include:

Application or Technology	*Synopsis*
Gene expression profiling	In an mRNA or gene expression profiling experiment the expression levels of thousands of genes are simultaneously monitored to study the effects of certain treatments, diseases, and developmental stages on gene expression. For example, microarray-based gene expression profiling can be used to identify genes whose expression is changed in response to pathogens or other organisms by comparing gene expression in infected to that in uninfected cells or tissues.

Comparative genomic hybridization	Assessing genome content in different cells or closely related organisms. GeneID Small microarrays to check IDs of organisms in food and feed (like GMO), mycoplasms in cell culture, or pathogens for disease detection, mostly combining PCR and microarray technology.
Chromatin immuno-precipitation on Chip	DNA sequences bound to a particular protein can be isolated by immunoprecipitating that protein (ChIP), these fragments can be then hybridized to a microarray (such as a tiling array) allowing the determination of protein binding site occupancy throughout the genome. Example protein to immunoprecipitate are histone modifications (H3K27me3, H3K4me2, H3K9me3, etc.), Polycomb-group protein (PRC2:Suz12, PRC1:YY1) and trithorax-group protein (Ash1) to study the epigenetic landscape or RNA Polymerase II to study the transcription landscape.
DamID	Analogously to ChIP, genomic regions bound by a protein of interest can be isolated and used to probe a microarray to determine binding site occupancy. Unlike ChIP, DamID does not require antibodies but makes use of adenine methylation near the protein's binding sites to selectively amplify those regions, introduced by expressing minute amounts of protein of interest fused to bacterial DNA adenine methyltransferase.
SNP detection	Identifying single nucleotide polymorphism among alleles within or between populations. Several applications of microarrays make use of SNP detection, including Genotyping, forensic analysis, measuring predisposition to disease, identifying drug-candidates, evaluating germline mutations in individuals or somatic mutations in cancers, assessing loss of heterozygosity, or genetic linkage analysis.
Alternative splicing detection	An '*exon junction array* design uses probes specific to the expected or potential splice sites of predicted exons for a gene. It is of intermediate density, or coverage, to a typical gene expression array (with 1-3 probes per gene) and a genomic tiling array (with hundreds or thousands of probes per gene). It is used to assay the expression of alternative splice forms of a gene. Exon arrays have a different design, employing probes designed to detect each individual exon for known or predicted genes, and can be used for detecting different splicing isoforms.
Fusion genes microarray	A Fusion gene microarray can detect fusion transcripts, *e.g.* from cancer specimens. The principle behind this is building on the alternative splicing microarrays. The oligo design strategy enables combined measurements of chimeric transcript junctions with exon-wise measurements of individual fusion partners.

Tiling array	Genome tiling arrays consist of overlapping probes designed to densely represent a genomic region of interest, sometimes as large as an entire human chromosome. The purpose is to empirically detect expression of transcripts or alternatively splice forms which may not have been previously known or predicted.

Fabrication

Microarrays can be manufactured in different ways, depending on the number of probes under examination, costs, customization requirements, and the type of scientific question being asked. Arrays may have as few as 10 probes or up to 2.1 million micrometre-scale probes from commercial vendors.

Spotted vs. in Situ Synthesised Arrays

Microarrays can be fabricated using a variety of technologies, including printing with fine-pointed pins onto glass slides, photolithography using pre-made masks, photolithography using dynamic micromirror devices, ink-jet printing, or electrochemistry on microelectrode arrays.

In *spotted microarrays*, the probes are oligonucleotides, cDNA or small fragments of PCR products that correspond to mRNAs. The probes are synthesized prior to deposition on the array surface and are then "spotted" onto glass. A common approach utilizes an array of fine pins or needles controlled by a robotic arm that is dipped into wells containing DNA probes and then depositing each probe at designated locations on the array surface.

The resulting "grid" of probes represents the nucleic acid profiles of the prepared probes and is ready to receive complementary cDNA or cRNA "targets" derived from experimental or clinical samples. This technique is used by research scientists around the world to produce "in-house" printed microarrays from their own labs. These arrays may be easily customized for each experiment, because researchers can choose the probes and printing locations on the arrays, synthesize the probes in their own lab (or collaborating facility), and spot the arrays. They can then generate their own labelled samples for hybridization, hybridize the samples to the array, and finally scan the arrays with their own equipment.

This provides a relatively low-cost microarray that may be customized for each study, and avoids the costs of purchasing often more expensive commercial arrays that may represent vast numbers of genes that are not of interest to the investigator. Publications exist which indicate in-house spotted microarrays may not provide the same level of sensitivity compared to commercial oligonucleotide arrays, possibly owing to the small batch sizes and reduced printing efficiencies when compared to industrial manufactures of oligo arrays.

In *oligonucleotide microarrays*, the probes are short sequences designed to match parts of the sequence of known or predicted open reading frames. Although oligonucleotide probes are often used in "spotted" microarrays, the term "oligonucleotide array" most often refers to a specific technique of manufacturing. Oligonucleotide arrays are produced by printing short oligonucleotide sequences designed to represent a single gene or family of gene splice-variants by synthesizing this sequence directly onto the array surface instead of depositing intact sequences. Sequences may be longer (60-mer probes such as the Agilent design) or shorter (25-mer probes produced by Affymetrix) depending on the desired purpose; longer probes are more specific to individual target genes, shorter probes may be spotted in higher density across the array and are cheaper to manufacture. One technique used to produce oligonucleotide arrays include photolithographic synthesis (Agilent and Affymetrix) on a silica substrate where light and light-sensitive masking agents are used to "build" a sequence one nucleotide at a time across the entire array. Each applicable probe is selectively "unmasked" prior to bathing the array in a solution of a single nucleotide, then a masking reaction takes place and the next set of probes are unmasked in preparation for a different nucleotide exposure. After many repetitions, the sequences of every probe become fully constructed. More recently, Maskless Array Synthesis from NimbleGen Systems has combined flexibility with large numbers of probes.

Two-channel vs. One-channel Detection

Two-colour microarrays or *two-channel microarrays* are typically hybridized with cDNA prepared from two samples to be compared (e.g. diseased tissue versus healthy tissue) and that are labelled with two different fluorophores. Fluorescent dyes commonly used for cDNA labelling include Cy3, which has a fluorescence emission wavelength of 570 nm (corresponding to the green part of the light spectrum), and Cy5 with a fluorescence emission wavelength of 670 nm (corresponding to the red part of the light spectrum). The two Cy-labelled cDNA samples are mixed and hybridized to a single microarray that is then scanned in a microarray scanner to visualize fluorescence of the two fluorophores after excitation with a laser beam of a defined wavelength. Relative intensities of each fluorophore may then be used in ratio-based analysis to identify up-regulated and down-regulated genes.

Oligonucleotide microarrays often carry control probes designed to hybridize with RNA spike-ins. The degree of hybridization between the spike-ins and the control probes is used to normalize the hybridization measurements for the target probes. Although absolute levels of gene expression may be determined in the two-colour array in rare instances, the relative differences in expression among different spots within a sample and between samples is the preferred method of data analysis for the two-colour system. Examples of providers for such microarrays includes Agilent with their Dual-Mode platform, Eppendorf with their DualChip platform for colorimetric Silverquant labelling, and TeleChem International with Arrayit.

In *single-channel microarrays* or *one-colour microarrays*, the arrays provide intensity data for each probe or probe set indicating a relative level of hybridization with the labelled target. However, they do not truly indicate abundance levels of a gene but rather relative abundance when compared to other samples or conditions when processed in the same experiment. Each RNA molecule encounters protocol and batch-specific bias during amplification, labelling, and hybridization phases of the experiment making comparisons between genes for the same microarray uninformative.

The comparison of two conditions for the same gene requires two separate single-dye hybridizations. Several popular single-channel systems are the Affymetrix "Gene Chip", Illumina "Bead Chip", Agilent single-channel arrays, the Applied Microarrays "CodeLink" arrays, and the Eppendorf "DualChip & Silverquant". One strength of the single-dye system lies in the fact that an aberrant sample cannot affect the raw data derived from other samples, because each array chip is exposed to only one sample (as opposed to a two-colour system in which a single low-quality sample may drastically impinge on overall data precision even if the other sample was of high quality). Another benefit is that data are more easily compared to arrays from different experiments so long as batch effects have been accounted for. A drawback to the one-colour system is that, when compared to the two-colour system, twice as many microarrays are needed to compare samples within an experiment.

Microarrays and Bioinformatics

The advent of inexpensive microarray experiments created several specific bioinformatics challenges:

- the multiple levels of replication in experimental design (Experimental design)
- the number of platforms and independent groups and data format (Standardization)
- the treatment of the data (Statistical analysis)
- accuracy and precision (Relation between probe and gene)
- the sheer volume of data and the ability to share it (Data warehousing).

Experimental Design

Due to the biological complexity of gene expression, the considerations of experimental design that are discussed in the expression profiling article are of critical importance if statistically and biologically valid conclusions are to be drawn from the data.

There are three main elements to consider when designing a microarray experiment. First, replication of the biological samples is essential for drawing conclusions from the experiment.

Second, technical replicates (two RNA samples obtained from each experimental unit) help to ensure precision and allow for testing differences within treatment groups. The technical replicates may be two independent RNA extractions or two aliquots of the same extraction.

Third, spots of each cDNA clone or oligonucleotide are present as replicates (at least duplicates) on the microarray slide, to provide a measure of technical precision in each hybridization. It is critical that information about the sample preparation and handling is discussed, in order to help identify the independent units in the experiment and to avoid inflated estimates of statistical significance.

Standardization

Microarray data is difficult to exchange due to the lack of standardization in platform fabrication, assay protocols, and analysis methods. This presents an interoperability problem in bioinformatics. Various grass-roots open-source projects are trying to ease the exchange and analysis of data produced with non-proprietary chips:

- For example, the "Minimum Information About a Microarray Experiment" (MIAME) checklist helps define the level of detail that should exist and is being adopted by many journals as a requirement for the submission of papers incorporating microarray results. But MIAME does not describe the format for the information, so while many formats can support the MIAME requirements, as of 2007 no format permits verification of complete semantic compliance.
- The "MicroArray Quality Control (MAQC) Project" is being conducted by the US Food and Drug Administration (FDA) to develop standards and quality control metrics which will eventually allow the use of MicroArray data in drug discovery, clinical practice and regulatory decision-making.
- The MGED Society has developed standards for the representation of gene expression experiment results and relevant annotations.

Statistical Analysis

Microarray data sets are commonly very large, and analytical precision is influenced by a number of variables. Statistical challenges include taking into account effects of background noise and appropriate normalization of the data. Normalization methods may be suited to specific platforms and, in the case of commercial platforms, the analysis may be proprietary. Algorithms that affect statistical analysis include:

- Image analysis: gridding, spot recognition of the scanned image (segmentation algorithm), removal or marking of poor-quality and low-intensity features (called *flagging*).
- Data processing: background subtraction (based on global or local background), determination of spot intensities and intensity ratios, visualisation of data, and log-transformation of ratios, global or local normalization of intensity ratios.
- Identification of statistically significant changes: t-test, ANOVA, Bayesian method or Mann–Whitney test methods tailored to microarray data sets, which take into account multiple comparisons. These methods assess statistical power based on the variation present in the data and the number of experimental replicates, and can help minimize Type I and type II errors in the analyses.
- Network-based methods: Statistical methods that take the underlying structure of gene networks into account, representing either associative or causative interactions or dependencies among gene products.

Microarray data may require further processing aimed at reducing the dimensionality of the data to aid comprehension and more focused analysis.

Other methods permit analysis of data consisting of a low number of biological or technical replicates; for example, the Local Pooled Error (LPE) test pools standard deviations of genes with similar expression levels in an effort to compensate for insufficient replication.

Relation between Probe and Gene

The relation between a probe and the mRNA that it is expected to detect is problematic. On the one hand, some mRNAs may cross-hybridize probes in the array that are supposed to detect another mRNA. In addition, mRNAs may experience amplification bias that is sequence or molecule-specific.

On the other hand, probes that are designed to detect the mRNA of a particular gene may be relying on genomic EST information that is incorrectly associated with that gene.

Data Warehousing

Microarray data was found to be more useful when compared to other similar datasets. The sheer volume (in bytes), specialized formats (such as MIAME), and curation efforts associated with the datasets require specialized databases to store the data.

Significance Analysis of Microarrays

Significance analysis of microarrays (SAM) is a statistical technique, established in 2001 by Tusher, Tibshirani and Chu, for determining whether changes in gene expression are statistically significant.

With the advent of DNA microarrays it is now possible to measure the expression of thousands of genes in a single hybridization experiment. The data generated is considerable and a method for sorting out what is significant and what isn't is essential. SAM is distributed by Stanford University in an R-package.

SAM identifies statistically significant genes by carrying out gene specific t-tests and computes a statistic d_j for each gene j, which measures the strength of the relationship between gene expression and a response variable. This analysis uses non-parametric statistics, since the data may not follow a normal distribution.

The response variable describes and groups the data based on experimental conditions. In this method, repeated permutations of the data are used to determine if the expression of any gene is significant related to the response.

The use of permutation-based analysis accounts for correlations in genes and avoids parametric assumptions about the distribution of individual genes. This is an advantage over other techniques (for example ANOVA and Bonferroni), which assume equal variance and/or independence of genes.

Basic Protocol

- Perform microarray experiments — DNA microarray with oligo and cDNA primers, SNP arrays, protein arrays, etc.
- Input Expression Analysis in Microsoft Excel
- Run SAM as a Microsoft Excel Add-Ins
- Adjust the Delta tuning parameter to get a significant # of genes along with an acceptable false discovery rate (FDR)) and Assess Sample Size by calculating the mean difference in expression in the SAM Plot Controller
- List Differentially Expressed Genes (Positively and Negatively Expressed Genes)
- Permutations are calculated based on the number of samples
- Block Permutations

o Blocks are batches of microarrays; for example for eight samples split into two groups (control and affected) there are 4!=24 permutations for each block and the total number of permutations is (24)(24)= 196. A minimum of 1000 permutations are recommended; The number of permutations is set by the user when imputing correct values for the data set to run SAM.

Response Formats

Types;

- o Quantitative — real-valued (such as heart rate)
- o One class — tests whether the mean gene expression differs from zero
- o Two class — two sets of measurements
  - * Unpaired — measurement units are different in the two groups; e.g. control and treatment groups with samples from different patients
  - * Paired — same experimental units are measured in the two groups; eg. samples before and after treatment from the same patients
- o Multiclass — more than two groups with each containing different experimental units; generalization of two class unpaired type
- o Survival — data of a time until an event (for example death or relapse)
- o Time course — each experimental units is measured at more than one time point; experimental units fall into a one or two class design
- o Pattern discovery — no explicit response parameter is specified; the user specifies eigengene (principle component) of the expression data and treats it as a quantitative response.

SAM Calculations

SAM calculates a test statistic for relative difference in gene expression based on permutation analysis of expression data and calculates a false discovery rate. Fold changes (t) are specified to guarantee genes called significant change at least a pre-specified amount. This means that the absolute value of the average expression levels of a gene under each of two conditions must be greater than the fold change (t) to be called positive and less than the inverse of the fold change (t) to be called negative.

The SAM Algorithm can be Stated as:

1. Order test statistics according to magnitude
2. For each permutation compute the ordered null (unaffected) scores
3. Plot the ordered test statistic against the expected null scores
4. Call each gene significant if the absolute value of the test statistic for that gene minus the mean test statistic for that gene is greater than a stated threshold
5. Estimate the false discovery rate based on expected versus observed values.

SAM Output

- Significant gene sets;
    - o Positive gene set — higher expression of most genes in the gene set correlates with higher values of the phenotype y.
    - o Negative gene set — lower expression of most genes in the gene set correlates with higher values of the phenotype y.

SAM Features

- Data from Oligo or cDNA arrays, SNP array, protein arrays,etc can be utilized in SAM
- Correlates expression data to clinical parameters
- Correlates expression data with time
- Uses data permutation to estimates False Discovery Rate for multiple testing
- Reports local false discovery rate (the FDR for genes having a similar d_i as that gene) and miss rates
- Can work with blocked design for when treatments are applied within different batches of arrays
- Can adjust threshold determining number of gene called significant.

7

Transformation and Transfection of Gene

Many methods of transfection and transformation - two ways of expressing a foreign or modified gene in an organism - are effective in only a small percentage of a population subjected to the techniques. Thus, a method for identifying those few successful gene uptake events is necessary. Reporter genes used in this way are normally expressed under their own promoter independent from that of the introduced gene of interest; the reporter gene can be expressed constitutively (that is, it is "always on") or inducibly with an external intervention such as the introduction of IPTG in the â-galactosidase system.

As a result, the reporter gene's expression is independent of the gene of interest's expression, which is an advantage when the gene of interest is only expressed under certain specific conditions or in tissues that are difficult to access. In the case of selectable-marker reporters such as CAT, the transfected population of bacteria can be grown on a substrate that contains chloramphenicol. Only those cells that have successfully taken up the construct containing the CAT gene will survive and multiply under these conditions.

Gene Expression Assays

Reporter genes can also be used to assay for the expression of the gene of interest, which may produce a protein that has little obvious or immediate effect on the cell culture or organism. In these cases the reporter is directly attached to the gene of interest to create a gene fusion. The two genes are under the same promoter elements and are transcribed into a single messenger RNA molecule.

The mRNA is then translated into protein. In these cases it is important that both proteins be able to properly fold into their active conformations and interact with their substrates despite being fused. In building the DNA construct, a segment of DNA coding for a flexible polypeptide linker region is usually included so that the reporter and the gene product will only minimally interfere with one another.

Promoter Assays

Reporter genes can be used to assay for the activity of a particular promoter in a cell or organism. In this case there is no separate "gene of interest"; the reporter gene is simply placed under the control of the target promoter and the reporter gene product's activity is quantitatively measured. The results are normally reported relative to the activity under a "consensus" promoter known to induce strong gene expression.

Further Uses

A more complex usage of reporter genes on a large scale is in two-hybrid screening, which aims to identify proteins that natively interact with one another in vivo.

GUS Reporter System

The GUS reporter system (GUS: beta-glucuronidase) is a reporter gene system, particularly useful in plant molecular biology. Several kinds of GUS reporter gene assay are actually available, depending on the substrate used. The term GUS staining refers to the most common of these, a histochemical technique.

Purpose

The purpose of this technique is to analyze the activity of a promoter (in terms of expression of a gene under that promoter) either in a quantitative way or through visualization of its activity in different tissues. The technique is based on beta-glucuronidase, an enzyme from the bacterium Escherichia coli; this enzyme, when incubated with some specific colorless or non-fluorescent substrates, can transform them into coloured or fluorescent products.

Substrates

There are actually different possible glucuronides that can be used as substrates for the beta-glucuronidase, depending on the type of detection needed (histochemical, spectrophotometrical, fluorimetrical).

The most common substrate for GUS histochemical staining is 5-bromo-4-chloro-3-indolyl glucuronide (X-Gluc): the product of the reaction is in this case a clear blue color. Other common substrates are p-nitrophenyl â-D-glucuronide for the spectrophotometrical assay and 4-methylumbelliferyl-beta-D-glucuronide (MUG) for the fluorimetrical assay.

History

The system was originally developed by Richard Anthony Jefferson during his Ph.D. at the University of Colorado at Boulder. He adapted the technique for the use with plants as he worked in the Plant Breeding Institute of Cambridge, between 1985 and 1987 .

Since then thousands of labs have used the system, making it probably the most widely used tool in plant molecular biology, as underlined by over 6000 citations in scientific literature.

Target Organisms

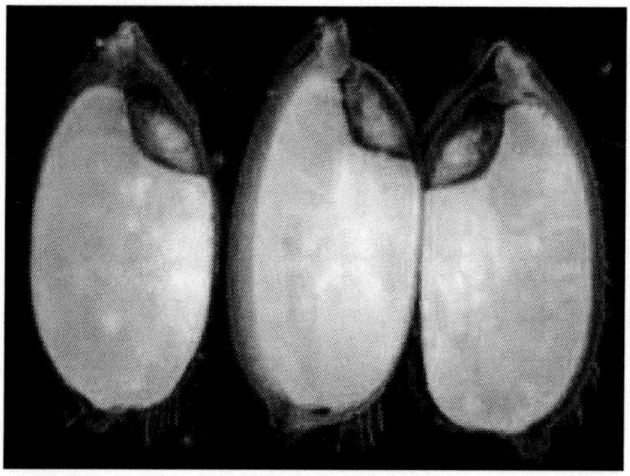

Rice Embryo Showing GUS Expression

An organism is suitable for a GUS assay if it has no beta-glucuronidase or if the activity is very low (background activity). For this reason the assay is not useful in most vertebrates and many molluscs.

Since there is no detectable GUS activity in higher plants, mosses, algae, ferns, fungi and most bacteria, the assay is perfectly suited for these organisms. Thus it is used widely in plant science.

Other Reporter Systems

The GUS system is not the only available gene reporter system for the analysis of promoter activity. Other competing systems are based on e.g. luciferase, GFP, beta-galactosidase, chloramphenicol acetyltransferase (CAT), alkaline phosphatase. The use of one or the other system is mainly dependent on the organism of interest.

Other Uses

The GUS assay, as well as other reporter gene systems, can be used for other kinds of studies other than the classical promoter activity assay. Reporter systems have been used for the determination of the efficiency of gene delivery systems, the intracellular localization of a gene product, the detection of protein-protein or protein-DNA interactions, the efficiency of translation initiation signals and the success of molecular cloning efforts.

Northern Blot: The northern blot is a technique used in molecular biology research to study gene expression by detection of RNA (or isolated mRNA) in a sample.

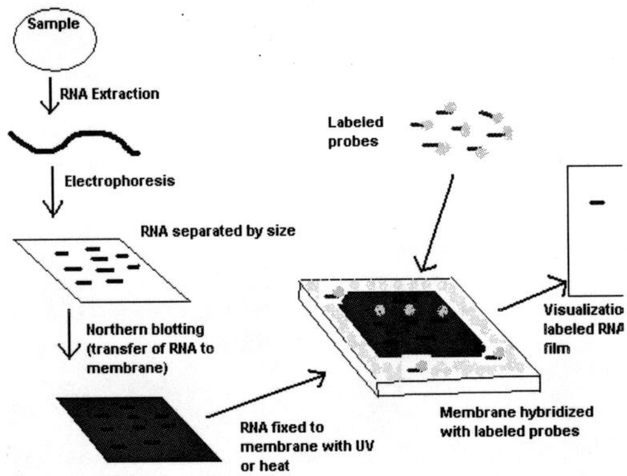

Figure : *Flow diagram outlining the general procedure for RNA detection by northern blotting.*

With northern blotting it is possible to observe cellular control over structure and function by determining the particular gene expression levels during differentiation, morphogenesis, as well as abnormal or diseased conditions. Northern blotting involves the use of electrophoresis to separate RNA samples by size and detection with a hybridization probe complementary to part of or the entire target sequence.

The term 'northern blot' actually refers specifically to the capillary transfer of RNA from the electrophoresis gel to the blotting membrane. However, the entire process is commonly referred to as northern blotting. The northern blot technique was developed in 1977 by James Alwine, David Kemp, and George Stark at Stanford University. Northern blotting takes its name from its similarity to the first blotting technique, the Southern blot, named for biologist Edwin Southern. The major difference is that RNA, rather than DNA, is analyzed in the northern blot.

Procedure

A general blotting procedure starts with extraction of total RNA from a homogenized tissue sample or from cells. Eukaryotic mRNA can then be isolated through the use of oligo (dT) cellulose chromatography to isolate only those RNAs with a poly(A) tail. RNA samples are then separated by gel electrophoresis. Since the gels are fragile and the probes are unable to enter the matrix, the RNA samples, now separated by size, are transferred to a nylon membrane through a capillary or vacuum blotting system.

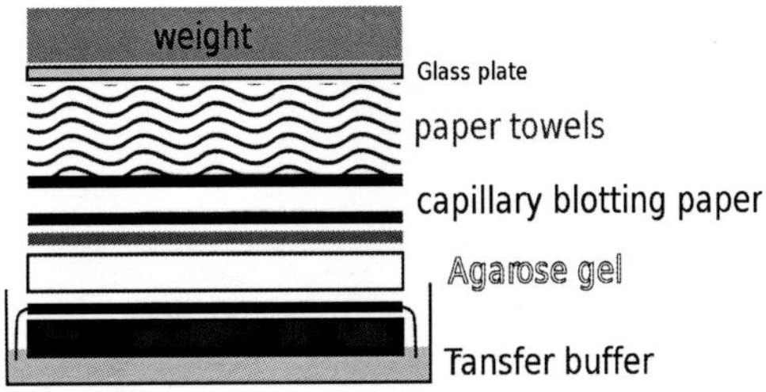

Figure : *Capillary blotting system setup for the transfer of RNA from an electrophoresis gel to a blotting membrane.*

A nylon membrane with a positive charge is the most effective for use in northern blotting since the negatively charged nucleic acids have a high affinity for them. The transfer buffer used for the blotting usually contains formamide because it lowers the annealing temperature of the probe-RNA interaction, thus preventing RNA degradation by high temperatures. Once the RNA has been transferred to the membrane, it is immobilized through covalent linkage to the membrane by UV light or heat.

After a probe has been labeled, it is hybridized to the RNA on the membrane. Experimental conditions that can affect the efficiency and specificity of hybridization include ionic strength, viscosity, duplex length, mismatched base pairs, and base composition. The membrane is washed to ensure that the probe has bound specifically and to avoid background signals from arising. The hybrid signals are then detected by X-ray film and can be quantified by densitometry. To create controls for comparison in a northern blot, samples not displaying the gene product of interest can be used after determination by microarrays or RT-PCR.

Gels

The RNA samples are most commonly separated on agarose gels containing formaldehyde as a denaturing agent for the RNA to limit secondary structure. The gels can be stained with ethidium bromide (EtBr) and viewed under UV light to observe the quality and quantity of RNA before blotting. Polyacrylamide gel electrophoeresis with urea can also be used in RNA separation but it is most commonly used for fragmented RNA or microRNAs. An RNA ladder is often run alongside the samples on an electrophoresis gel to observe the size of fragments obtained but in total RNA samples the ribosomal subunits can act as size markers. Since the large ribosomal subunit is 28S (approximately 5kb) and the small ribosomal subunit is 18S (approximately 2kb) two prominent bands will appear on the gel, the larger at close to twice the intensity of the smaller.

Probes

Probes for northern blotting are composed of nucleic acids with a complementary sequence to all or part of the RNA of interest, they can be DNA, RNA, or oligonucleotides with a minimum of 25 complementary bases to the target sequence. RNA probes (riboprobes) that are transcribed in vitro are able to withstand more rigorous washing steps preventing some of the background noise. Commonly cDNA is created with labelled primers for the RNA sequence of interest to act as the probe in the northern blot.

The probes need to be labelled either with radioactive isotopes (P) or with chemiluminescence in which alkaline phosphatase or horseradish peroxidase breakdown chemiluminescent substrates producing a detectable emission of light. The chemiluminescent labelling can occur in two ways: either the probe is attached to the enzyme, or the probe is labelled with a ligand (e.g. biotin) for mwhich the antibody (e.g. avidin or streptavidin) is attached to the enzyme. X-ray film can detect both the radioactive and chemiluminescent signals and many researchers prefer the chemiluminescent signals because they are faster, more sensitive, and reduce the health hazards that go along with radioactive labels. The same membrane can be probed up to five times without a significant loss of the target RNA.

Applications

Northern blotting allows one to observe a particular gene's expression pattern between tissues, organs, developmental stages, environmental stress levels, pathogen infection, and over the course of treatment. The technique has been used to show overexpression of oncogenes and downregulation of tumor-suppressor genes in cancerous cells when compared to 'normal' tissue, as well as the gene expression in the rejection of transplanted organs. If an upregulated gene is observed by an abundance of mRNA on the northern blot the sample can then be sequenced to determine if the gene is known to researchers or if it is a novel finding.

The expression patterns obtained under given conditions can provide insight into the function of that gene. Since the RNA is first separated by size, if only one probe type is used variance in the level of each band on the membrane can provide insight into the size of the product, suggesting alternative splice products of the same gene or repetitive sequence motifs. The variance in size of a gene product can also indicate deletions or errors in transcript processing, by altering the probe target used along the known sequence it is possible to determine which region of the RNA is missing.

BlotBase is an online database publishing northern blots. BlotBase has over 700 published northern blots of human and mouse samples, in over 650 genes across more than 25 different tissue types. Northern blots can be searched by a blot ID, paper reference, gene identifier, or by tissue. The results of a search provide the blot ID, species, tissue, gene, expression level, blot image (if available), and links to the publication that the work originated from. This new database provides sharing of information between members of the science community that was not previously seen in northern blotting as it was in sequence analysis, genome determination, protein structure, etc.

Advantages and Disadvantages

Analysis of gene expression can be done by several different methods including RT-PCR, RNase protection assays, microarrays, serial analysis of gene expression (SAGE), as well as northern blotting. Microarrays are quite commonly used and are usually consistent with data obtained from northern blots; however, at times northern blotting is able to detect small changes in gene expression that microarrays cannot. The advantage that microarrays have over northern blots is that thousands of genes can be visualized at a time, while northern blotting is usually looking at one or a small number of genes.

A problem in northern blotting is often sample degradation by RNases (both endogenous to the sample and through environmental contamination), which can be avoided by proper sterilization of glassware and the use of RNase inhibitors such as DEPC (diethylpyrocarbonate).

The chemicals used in most northern blots can be a risk to the researcher, since formaldehyde, radioactive material, ethidium bromide, DEPC, and UV light are all harmful under certain exposures. Compared to RT-PCR, northern blotting has a low sensitivity, but it also has a high specificity which is important to reduce false positive results.

The advantages of using northern blotting include the detection of RNA size, the observation of alternate splice products, the use of probes with partial homology, the quality and quantity of RNA can be measured on the gel prior to blotting, and the membranes can be stored and reprobed for years after blotting.

Reverse Northern Blot

A variant of the procedure known as the reverse northern blot is occasionally used. In this procedure, the substrate nucleic acid (that is affixed to the membrane) is a collection of isolated DNA fragments, and the probe is RNA extracted from a tissue and radioactively labelled.

The use of DNA microarrays that have come into widespread use in the late 1990s and early 2000s is more akin to the reverse procedure, in that they involve the use of isolated DNA fragments affixed to a substrate, and hybridization with a probe made from cellular RNA. Thus the reverse procedure, though originally uncommon, enabled northern analysis to evolve into gene expression profiling, in which many (possibly all) of the genes in an organism may have their expression monitored.

Fluorescent in Situ Hybridization

FISH (fluorescence in situ hybridization) is a cytogenetic technique developed by Christoph Lengauer that is used to detect and localize the presence or absence of specific DNA sequences on chromosomes. FISH uses fluorescent probes that bind to only those parts of the chromosome with which they show a high degree of sequence complementarity. Fluorescence microscopy can be used to find out where the fluorescent probe bound to the chromosomes. FISH is often used for finding specific features in DNA for use in genetic counselling, medicine, and species identification. FISH can also be used to detect and localize specific mRNAs within tissue samples. In this context, it can help define the spatial-temporal patterns of gene expression within cells and tissues.

Probes

Probes are often derived from fragments of DNA that were isolated, purified, and amplified for use in the Human Genome Project. The size of the human genome is so large, compared to the length that could be sequenced directly, that it was necessary to divide the genome into fragments. (In the eventual analysis, these fragments were put into order by digesting a copy of each fragment into still smaller fragments using sequence-specific endonucleases, measuring the size of each small fragment using size-exclusion chromatography, and using that information to determine where the large fragments overlapped one another.)

To preserve the fragments with their individual DNA sequences, the fragments were added into a system of continually replicating bacteria populations. Clonal populations of bacteria, each population maintaining a single artificial chromosome, are stored in various laboratories around the world. The artificial chromosomes (BAC) can be grown, extracted, and labeled, in any lab. These fragments are on the order of 100 thousand base-pairs, and are the basis for most FISH probes.

Preparation and Hybridization Process

First, a probe is constructed. The probe must be large enough to hybridize specifically with its target but not so large as to impede the hybridization process. The probe is tagged directly with fluorophores, with targets for antibodies or with biotin. Tagging can be done in various ways, such as nick translation, or PCR using tagged nucleotides.

Then, an interphase or metaphase chromosome preparation is produced. The chromosomes are firmly attached to a substrate, usually glass. Repetitive DNA sequences must be blocked by adding short fragments of DNA to the sample.

The probe is then applied to the chromosome DNA and incubated for approximately 12 hours while hybridizing. Several wash steps remove all unhybridized or partially-hybridized probes. The results are then visualized and quantified using a microscope that is capable of exciting the dye and recording images. If the fluorescent signal is weak, amplification of the signal may be necessary in order to exceed the detection threshold of the microscope.

Fluorescent signal strength depends on many factors such as probe labeling efficiency, the type of probe, and the type of dye. The dye incorporation rate can be quantified photometric with specialized nano-volume photometer (starting with 0.3 μl probe volume). Fluorescently-tagged antibodies or streptavidin are bound to the dye molecule. These secondary components are selected so that they have a strong signal. FISH experiments designed to detect or localize gene expression within cells and tissues rely on the use of a reporter gene, such as one expressing green fluorescent protein, to provide the fluorescence signal.

Variations on Probes and Analysis

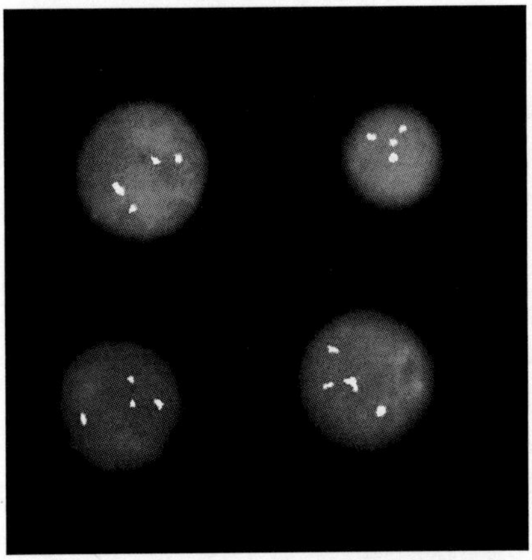

Figure : *Interphase cells positive for a chromosomal t(9;22) rearrangement.*

Fish is a very general technique. The differences between the various FISH techniques are usually due to variations in the sequence and labeling of the probes; and how they are used in combination. These few modifications make possible all FISH techniques.

Probe size is important because longer probes hybridize less specifically than shorter probes. The overlap defines the resolution of detectable features. For example, if the goal of an experiment is to detect the breakpoint of a translocation, then the overlap of the probes — the degree to which one DNA sequence is contained in the adjacent probes — defines the minimum window in which the breakpoint may be detected.

The mixture of probe sequences determines the type of feature the probe can detect. Probes that hybridize along an entire chromosome are used to count the number of a certain chromosome, show translocations, or identify extra-chromosomal fragments of chromatin. This is often called "whole-chromosome painting." If every possible probe is used, every chromosome, (the whole genome) would be marked fluorescently, which would not be particularly useful for determining features of individual sequences. However, a mixture of smaller probes can be created that is specific to a particular region (locus) of DNA; these mixtures are used to detect deletion mutations. When combined with a specific colour, a locus-specific probe mixture is used to detect very specific translocations. Special locus-specific probe mixtures are often used to count chromosomes, by binding to the centromeric regions of chromosomes, which are unique enough to identify each chromosome (with the exception of Chromosome 13, 14, 21, 22.)

A variety of other techniques use mixtures of differently-colored probes. A range of colors in mixtures of fluorescent dyes can be detected, so each human chromosome can be identified by a characteristic color using whole-chromosome probe mixtures and a variety of ratios of colors. Although there are more chromosomes than easily-distinguishable fluorescent dye colors, ratios of probe mixtures can be used to create secondary colors. Similar to comparative genomic hybridization, the probe mixture for the secondary colors is created by mixing the correct ratio of two sets of differently-colored probes for the same chromosome.

This technique is sometimes called M-FISH. The same physics that make a variety of colors possible for M-FISH can be used for the detection of translocations. That is, colors that are adjacent appear to overlap; a secondary color is observed. Some assays are designed so that the secondary color will be present or absent in cases of interest.

An example is the detection of BCR/ABL translocations, where the secondary color indicates disease. This variation is often called double-fusion FISH or D-FISH. In the opposite situation—where the absence of the secondary color is pathological—is illustrated by an assay used to investigate translocations where only one of the breakpoints is known or constant. Locus-specific probes are made for one side of the breakpoint and the other intact chromosome. In normal cells, the secondary colour is observed, but only the primary colour is observed when the translocation occurs. This technique is sometimes called "break-apart FISH".

Stellaris FISH Probes

Stellaris FISH, formerly known as Single Molecule RNA FISH, is a method of detecting and quantifying mRNA and other long RNA molecules in a thin layer of tissue sample. Targets can be reliably imaged through the application of multiple short singly labeled oligonucleotide probes. The binding of up to 48 fluorescent labeled oligos to a single molecule of mRNA provides sufficient fluorescence to accurately detect and localize each target mRNA in a wide-field fluorescent microscopy image. Probes not binding to the intended sequence do not achieve sufficient localized fluorescence to be distinguished from background. Single molecule RNA FISH assays can be performed in simplex or multiplex, and can be used as a follow-up experiment to qPCR, or imaged simultaneously with a fluorescent antibody assay. The technology has potential applications in cancer diagnosis, neuroscience, gene expression analysis, and companion diagnostics.

Fiber FISH

In an alternative technique to interphase or metaphase preparations, fiber FISH, interphase chromosomes are attached to a slide in such a way that they are stretched out in a straight line, rather than being tightly coiled, as in conventional FISH, or adopting a random conformation, as in interphase FISH. This is accomplished by applying mechanical shear along the length of the slide, either to cells that have been fixed to the slide and then lysed, or to a solution of purified DNA. A technique known as chromosome combing is increasingly used for this purpose. The extended conformation of the chromosomes allows dramatically higher resolution - even down to a few kilobases. The preparation of fiber FISH samples, although conceptually simple, is a rather skilled art, and only specialized laboratories use the technique routinely.

Q-FISH

Q-FISH combines FISH with PNAs and computer software to quantify fluorescence intensity. This technique is used routinely in telomere length research.

Flow-FISH

Flow-FISH uses flow cytometry to perform FISH automatically using per-cell fluorescence measurements.

Medical Applications

Often parents of children with a developmental delay want to know more about their child's conditions before choosing to have another child. These concerns can be addressed by analysis of the parents' and child's DNA. In cases where the child's developmental delay is not understood, the cause of it can potentially be determined using FISH and cytogenetic techniques.

Examples of diseases that are diagnosed using FISH include Prader-Willi syndrome, Angelman syndrome, 22q13 deletion syndrome, chronic myelogenous leukemia, acute lymphoblastic leukemia, Cri-du-chat, Velocardiofacial syndrome, and Down syndrome. FISH on sperm cells is indicated for men with an abnormal somatic or meiotic karyotype as well as those with oligozoospermia, since approximately 50% of oligozoospermic men have an increased rate of sperm chromosome abnormalities. The analysis of chromosomes 21, X, and Y is enough to identify oligozoospermic individuals at risk.

In medicine, FISH can be used to form a diagnosis, to evaluate prognosis, or to evaluate remission of a disease, such as cancer. Treatment can then be specifically tailored.

A traditional exam involving metaphase chromosome analysis is often unable to identify features that distinguish one disease from another, due to subtle chromosomal features; FISH can elucidate these differences.

FISH can also be used to detect diseased cells more easily than standard Cytogenetic methods, which require dividing cells and requires labor and time-intensive manual preparation and analysis of the slides by a technologist. FISH, on the other hand, does not require living cells and can be quantified automatically, a computer counts the fluorescent dots present. However, a trained technologist is required to distinguish subtle differences in banding patterns on bent and twisted metaphase chromosomes.

Species Identification

FISH is often used in clinical studies. If a patient is infected with a suspected pathogen, bacteria, from the patient's tissues or fluids, are typically grown on agar to determine the identity of the pathogen. Many bacteria, however, even well-known species, do not grow well under laboratory conditions.

FISH can be used to detect directly the presence of the suspect on small samples of patient's tissue.

FISH can also be used to compare the genomes of two biological species, to deduce evolutionary relationships. A similar hybridization technique is called a zoo blot. Bacterial FISH probes are often primers for the 16s rRNA region. FISH is widely used in the field of microbial ecology, to identify microorganisms. Biofilms, for example, are composed of complex (often) multi-species bacterial organizations.

Preparing DNA probes for one species and performing FISH with this probe allows one to visualize the distribution of this specific species within the biofilm. Preparing probes (in two different colors) for two species allows to visualize/study co-localization of these two species in the biofilm, and can be useful in determining the fine architecture of the biofilm.

Lab-on-a-chip and FISH

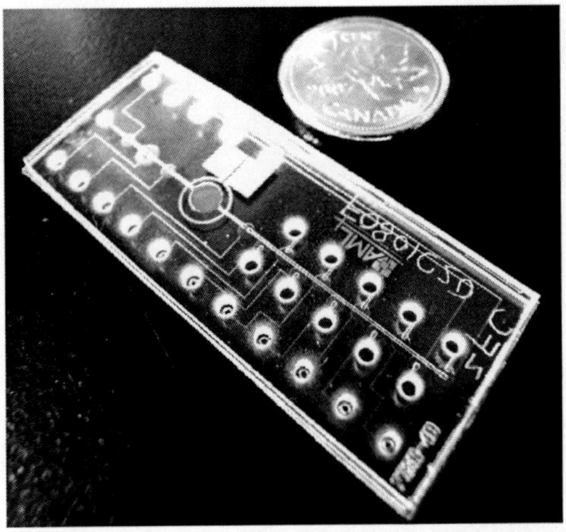

Microfluidic chip that automates the interphase FISH procedure. The microchip shown requires only minutes of setup time by technician, as opposed to the hours or days of labour needed to perform FISH with conventional equipment.

Although interphase fluorescence in situ hybridization (FISH) is a sensitive diagnostic tool used for the detection of chromosomal abnormalities on cell-by-cell basis, the cost-per-test and the technical complexity of current FISH protocols has inhibited its widespread utilization. Lab-on-a-chip or microfluidic devices, incorporate networks of microchannels that can miniaturize, integrate and automate conventional analytical techniques onto chip-style platforms. Since microchannels permit sophisticated levels of fluid control (down to picolitres), these devices can reduce analysis times, lower reagent consumption, and minimize human intervention. Currently, FISH has been performed on glass microfluidic platforms that standardize much of the protocol offering repeatable results that are accurate, cost-effective and easier to obtain in a clinical setting.

Compared to conventional FISH methods, these first implementations of on-chip FISH provide a 10-fold higher throughput and a 10-fold reduction in the cost of testing, enabling the simultaneous assessment of several chromosomal abnormalities or patients. It is increasingly essential that diagnostic tests determine the type and extent of chromosomal abnormalities for more informed diagnosis and for appropriate choice of treatment strategies.

Since the on-chip FISH technique is 10-20 times more cost-effective than conventional methods, and can be fully integrated and automated, this technology will make widespread genetic testing of patients more accessible in a clinical setting. Recently, the first demonstration of Metaphase FISH on chip has led to renewed efforts towards automating the metaphase FISH protocol. Metaphase FISH had continued to be difficult to integrate owing to the complex sample preparation protocol often spanning over 3 weeks. New reports confirm that a research group in Denmark have tested successfully a novel lab on chip device to integrate the entire sample preparation protocol for Metaphase FISH called FISHprep.

Virtual Karyotype

Virtual karyotyping is another cost-effective, clinically available alternative to FISH panels uses thousands to millions of probes on a single array to detect copy number changes, genome-wide, at unprecedented resolution. Currently, this type of analysis will only detect gains and losses of chromosomal material and will not detect balanced rearrangements, such as translocations and inversions which are hallmark aberrations seen in many types of leukemia and lymphoma.

Molecular Cytogenetics

Molecular cytogenetics involves the combination of molecular biology and cytogenetics. In general this involves the use of a series of techniques referred to as fluorescence in situ hybridization, or FISH, in which DNA probes are labeled with different colored fluorescent tags to visualize one or more specific regions of the genome. FISH can either be performed as a direct approach to metaphase chromosomes or interphase nuclei. Alternatively, an indirect approach can be taken in which the entire genome can be assessed for copy number changes using virtual karyotyping. Virtual karyotypes are generated from arrays made of thousands to millions of probes, and computational tools are used to recreate the genome in silico.

Virtual Karyotype

Virtual karyotype detects genomic copy number variations at a higher resolution level than conventional karyotyping or chromosome-based comparative genomic hybridization (CGH).

Background

A karyotype is the characteristic chromosome complement of a eukaryote species. A karyotype is typically presented as an image of the chromosomes from a single cell arranged from largest (chromosome 1) to smallest (chromosome 22), with the sex chromosomes (X and/or Y) shown last. Historically, karyotypes have been obtained by staining cells after they have been chemically arrested during cell division. Karyotypes have been used for several decades to identify chromosomal abnormalities in both germline and cancer cells. Conventional karyotypes can assess the entire genome for changes in chromosome structure and number, but the resolution is relatively coarse, with a detection limit of 5-10Mb.

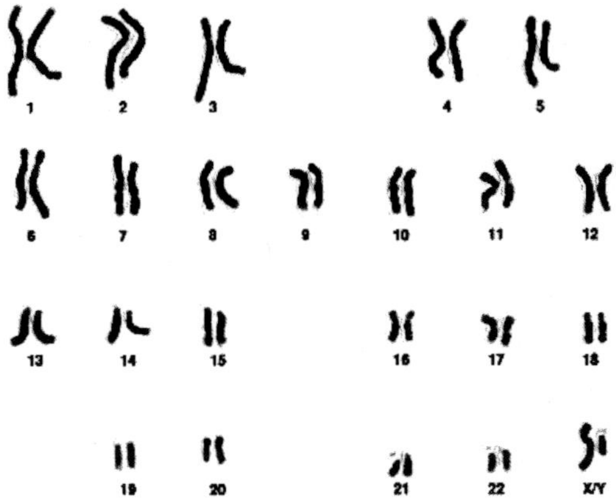

***Figure :** Karyotype of human male using Giemsa staining*

Virtual Karyotype

Recently, platforms for generating high-resolution karyotypes in silico from disrupted DNA have emerged, such as array comparative genomic hybridization (arrayCGH) and SNP arrays. Conceptually, the arrays are composed of hundreds to millions of probes which are complementary to a region of interest in the genome. The disrupted DNA from the test sample is fragmented, labeled, and hybridized to the array. The hybridization signal intensities for each probe are used by specialized software to generate a log2ratio of test/normal for each probe on the array. Knowing the address of each probe on the array and the address of each probe in the genome, the software lines up the probes in chromosomal order and reconstructs the genome in silico.

Virtual karyotypes have dramatically higher resolution than conventional cytogenetics. The actual resolution will depend on the density of probes on the array. Currently, the Affymetrix SNP6.0 is the highest density commercially available array for virtual karyotyping applications. It contains 1.8 million polymorphic and non-polymorphic markers for a practical resolution of 10-20kb—about the size of a gene. This is approximately 1000-fold greater resolution than karyotypes obtained from conventional cytogenetics.

How to Get a Virtual Karyotype

Virtual karyotypes can be performed on germline samples for constitutional disorders, and clinical testing is available from dozens of CLIA certified laboratories (GeneTests.org). Virtual karyotyping can also be done on fresh or formalin-fixed paraffin-embedded tumors. CLIA-certified laboratories offering testing on tumors include Creighton Medical Laboratories (fresh and paraffin embedded tumor samples) and CombiMatrix Molecular Diagnostics (fresh tumor samples).

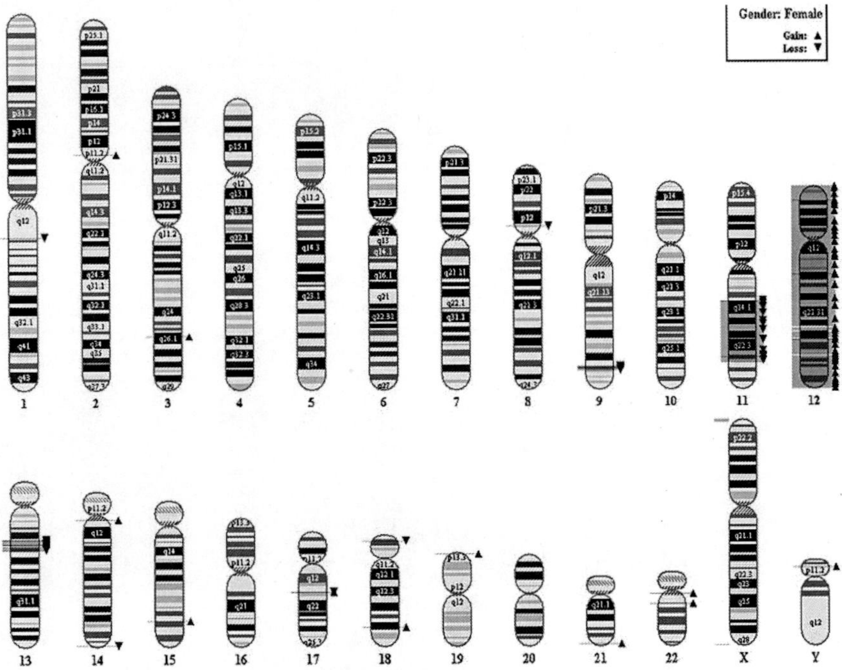

Figure : *Virtual karyotype of a chronic lymphocytic leukemia sample using a SNP array.*

Different Platforms for Virtual Karyotyping

Array-based karyotyping can be done with several different platforms, both laboratory-developed and commercial. The arrays themselves can be genome-wide (probes distributed over the entire genome) or targeted (probes for genomic regions known to be involved in a specific disease) or a combination of both.

Further, arrays used for karyotyping may use non-polymorphic probes, polymorphic probes (i.e., SNP-containing), or a combination of both. Non-polymorphic probes can provide only copy number information, while SNP arrays can provide both copy number and loss-of-heterozygosity (LOH) status in one assay.

The probe types used for non-polymorphic arrays include cDNA, BAC clones (e.g., BlueGnome), and oligonucleotides (e.g., Agilent, Santa Clara, CA, USA or Nimblegen, Madison, WI, USA). Commercially available oligonucleotide

SNP arrays can be solid phase (Affymetrix, Santa Clara, CA, USA) or bead-based (Illumina, SanDiego, CA, USA). Despite the diversity of platforms, ultimately they all use genomic DNA from disrupted cells to recreate a high resolution karyotype in silico.

The end product does not yet have a consistent name, and has been called virtual karyotyping, digital karyotyping, molecular allelokaryotyping, and molecular karyotyping. Other terms used to describe the arrays used for karyotyping include SOMA (SNP oligonucleotide microarrays) and CMA (chromosome microarray).

Some consider all platforms to be a type of array comparative genomic hybridization (arrayCGH), while others reserve that term for two-dye methods, and still others segregate SNP arrays because they generate more and different information than two-dye arrayCGH methods.

Applications

Detecting Copy-number Changes

Copy number changes can be seen in both germline and tumor samples. Copy number changes can be detected by arrays with non-polymorphic probes, such as arrayCGH, and by SNP-based arrays. Human beings are diploid, so a normal copy number is always two for the non-sex chromosomes.

Deletions: A deletion is the loss of genetic material. The deletion can be heterozygous (copy number of 1) or homozygous (copy number of 0, nullisomy). Microdeletion syndromes are examples of constitutional disorders due to small deletions in germline DNA. Deletions in tumor cells may represent the inactivation of a tumor suppressor gene, and may have diagnostic, prognostic, or therapeutic implications.

Gains: A copy number gain represents the gain of genetic material. If the gain is of just one additional copy of a segment of DNA, it may be called a duplication.

If there is one extra copy of an entire chromosome, it may be called a trisomy. Copy number gains in germline samples may be disease-associated or may be a benign copy number variant. When seen in tumor cells, they may have diagnostic, prognostic, or therapeutic implications.

Amplifications: Technically, an amplification is a type of copy number gain in which there is a copy number >10. In the context of cancer biology, amplifications are often seen in oncogenes. This could indicate a worse prognosis, help categorize the tumor, or indicate drug eligibility. An example of drug eligibility is Her2Neu amplification and Herceptin, and an image of Her2Neu amplification detected by SNP array virtual karyotyping is provided.

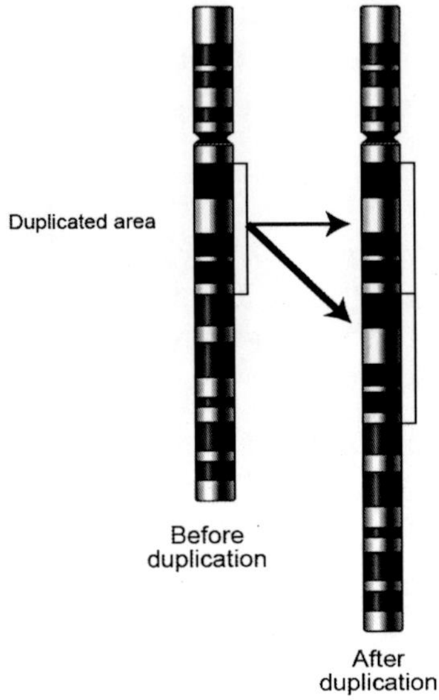

Figure : *Schematic of a region of a chromosome before & after a duplication event*

Loss of heterozygosity (LOH), Autozygous segments, and uniparental disomy Autozygous segments and uniparental disomy (UPD) are diploid/'copy neutral' genetic findings and therefore are only detectable by SNP-based arrays. Both autozygous segments and UPD will show loss of heterozygosity (LOH) with a copy number of two by SNP array karyotyping.

The term Runs of Homozgygosity (ROH), is a generic term that can be used for either autozygous segments or UPD. Autozygous segment: An autozygous segment is bi-parental and seen only in the germline.

They are extended runs of homozygous markers in the genome, and they occur when an identical haplotype block is inherited from both parents. They are also called "identical by descent" (IBD) segments, and they can be used for homozygosity mapping.

Uniparental Disomy: UPD occurs when both copies of a gene or genomic region are inherited from the same parent. This is uniparental, in contrast to autozygous segments which are bi-parental.

When present in the germline, they can be harmless or associated with disease, such as Prader-Willi or Angelman syndromes. Also in contrast to autozygosity, UPD can develop in tumor cells, and this is referred to as acquired UPD or copy neutral LOH in the literature.

Acquired UPD is quite common in both hematologic and solid tumors, and is reported to constitute 20 to 80% of the LOH seen in human tumors. Acquired UPD can serve as the 2nd hit in the Knudson Two Hit Hypothesis of Tumorigenesis, and thus can be the biological equivalent of a deletion. Because this type of lesion cannot be detected by arrayCGH, FISH, or conventional cytogenetics, SNP-based arrays are preferred for virtual karyotyping of tumors.

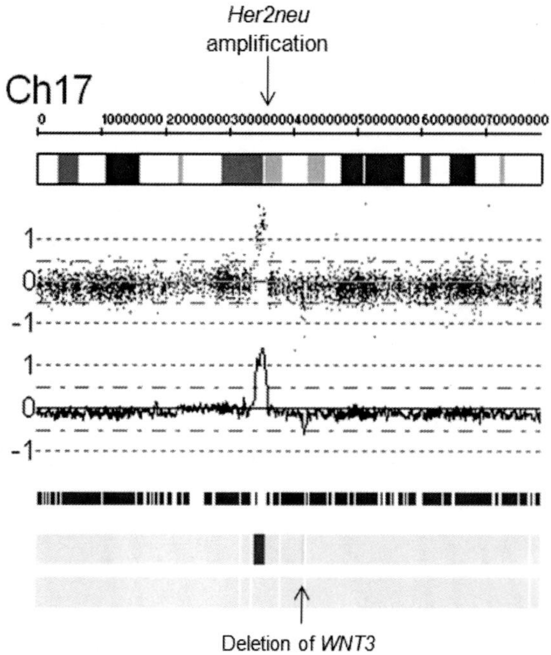

Figure : *Her2 Amplification by SNP array virtual karyotype.*

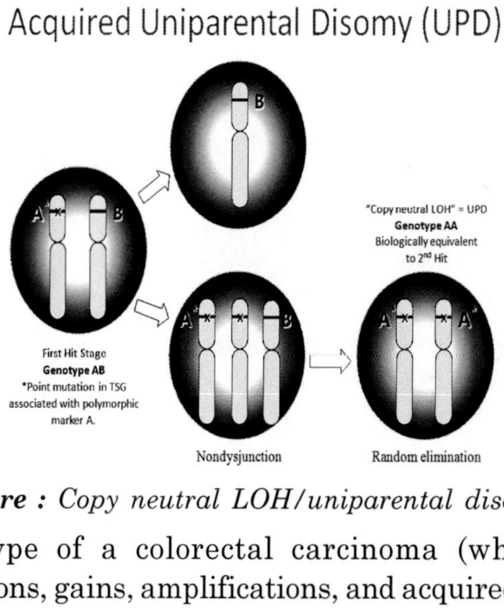

Figure : *Copy neutral LOH/uniparental disomy*

Virtual karyotype of a colorectal carcinoma (whole genome view) demonstrating deletions, gains, amplifications, and acquired UPD (copy neutral LOH). is a SNP array virtual karyotype from a colorectal carcinoma demonstrating deletions, gains, amplifications, and acquired UPD (copy neutral LOH).

Examples of Clinical Cancer Applications

A virtual karyotype can be generated from nearly any tumor, but the clinical meaning of the genomic aberrations identified are different for each tumor type. Clinical utility varies and appropriateness is best determined by an oncologist or pathologist in consultation with the laboratory director of the lab performing the virtual karyotype.

Below are examples of types of cancers where the clinical implications of specific genomic aberrations are well established. This list is representative, not exhaustive.

The web site for the Cytogenetics Laboratory at Wisconsin State Laboratory of Hygiene has additional examples of clinically relevant genetic changes that are readily detectable by virtual karyotyping.

Neuroblastoma

Based on a series of 493 neuroblastoma samples, it has been reported that overall genomic pattern, as tested by array-based karyotyping, is a predictor of outcome in neuroblastoma:

- Tumors presenting exclusively with whole chromosome copy number changes were associated with excellent survival.
- Tumors presenting with any kind of segmental chromosome copy number changes were associated with a high risk of relapse.
- Within tumors showing segmental alterations, additional independent predictors of decreased overall survival were MYCN amplification, 1p and 11q deletions, and 1q gain.

Earlier publications categorized neuroblastomas into three major subtypes based on cytogenetic profiles:

- Subtype 1: favorable neuroblastoma with near triploidy and a predominance of numerical gains and losses, mostly representing non-metastatic NB stages 1, 2 and 4S.
- Subtypes 2A and 2B: found in unfavorable widespread neuroblastoma, stages 3 and 4, with 11q loss and 17q gain without MYCN amplification (subtype 2A) or with MYCN amplification often together with 1p deletions and 17q gain (subtype 2B).

Wilms' Tumor

Tumor-specific loss-of-heterozygosity (LOH) for chromosomes 1p and 16q identifies a subset of Wilms' tumor patients who have a significantly increased risk of relapse and death. LOH for these chromosomal regions can now be used as an independent prognostic factor together with disease stage to target intensity of treatment to risk of treatment failure.

Renal-cell Carcinoma

Renal epithelial neoplasms have characteristic cytogenetic aberrations that can aid in classification. See also Atlas of Genetics and Cytogenetics in Oncology and Haematology.

- Clear cell carcinoma: loss of 3p
- Papillary carcinoma: trisomy 7 and 17
- Chromophobe carcinoma: hypodiploid with loss of chromosomes 1, 2, 6, 10, 13, 17, 21

Array-based karyotyping can be used to identify characteristic chromosomal aberrations in renal tumors with challenging morphology. Array-based karyotyping performs well on paraffin embedded tumors and is amenable to routine clinical use.

In addition, recent literature indicates that certain chromosomal aberrations are associated with outcome in specific subtypes of renal epithelial tumors.

Clear cell renal carcinoma: del 9p and del 14q are poor prognostic indicators. Papillary renal cell carcinoma: duplication of 1q marks fatal progression.

Chronic Lymphocytic Leukemia

Array-based karyotyping is a cost-effective alternative to FISH for detecting chromosomal abnormalities in chronic lymphocytic leukemia (CLL). Several clinical validation studies have shown >95% concordance with the standard CLL FISH panel.

In addition, many studies using array-based karyotyping have identified 'atypical deletions' missed by the standard FISH probes and acquired uniparental disomy at key loci for prognostic risk in CLL.,

Four main genetic aberrations are recognized in CLL cells that have a major impact on disease behavior.

1. Deletions of part of the short arm of chromosome 17 (del 17p) which target p53 are particularly deleterious. Patients with this abnormality have significantly short interval before they require therapy and a shorter survival. This abnormality is found in 5–10% of patients with CLL.
2. Deletions of the long arm on chromosome 11 (del 11q) are also unfavorable although not to the degree seen with del 17p. The abnormality targets the ATM gene and occurs infrequently in CLL (5–10%).
3. Trisomy 12, an additional chromosome 12, is a relatively frequent finding occurring in 20–25% of patients and imparts an intermediate prognosis.

4. Deletion of 13q14 (del 13q14) is the most common abnormality in CLL with roughly 50% of patients with cells containing this defect. When del 13q14 is seen in isolation, patients have the best prognosis and most will live many years, even decades, without the need for therapy.

Multiple Myeloma

Avet-Loiseau, et al. in the Journal of Clinical Oncology, used SNP array karyotyping of 192 multiple myeloma (MM) samples to identify genetic lesions associated with prognosis, which were then validated in a separate cohort (n = 273). In MM, lack of a proliferative clone makes conventional cytogenetics informative in only ~30% of cases. FISH panels are useful in MM, but standard panels would not detect several key genetic abnormalities reported in this study.

1. Virtual karyotyping identified chromosomal abnormalities in 98% of MM cases
2. del(12p13.31)is an independent adverse marker
3. amp(5q31.1) is a favorable marker
4. The prognostic impact of amp(5q31.1) over-rides that of hyperdiploidy and also identifies patients who greatly benefit from high-dose therapy.

Array-based karytyping cannot detect balanced translocations, such as t(4;14) seen in ~15% of MM. Therefore, FISH for this translocation should also be performed if using SNP arrays to detect genome-wide copy number alterations of prognostic significance in MM.

Medulloblastoma

Array-based karyotyping of 260 medulloblastomas by Pfister S, et al. resulted in the following clinical subgroups based on cytogenetic profiles:

- Poor prognosis: gain of 6q or amplification of MYC or MYCN
- Intermediate: gain of 17q or an i(17q) without gain of 6q or amplfication of MYC or MYCN
- Excellent prognosis: 6q and 17q balanced or 6q deletion

Oligodendroglioma

The 1p/19q co-deletion is considered a “genetic signature” of oligodendroglioma. Allelic losses on 1p and 19q, either separately or combined, are more common in classic oligodendrogliomas than in either astrocytomas or oligoastrocytomas. In one study, classic oligodendrogliomas showed 1p loss in 35 of 42 (83%) cases, 19q loss in 28 of 39 (72%), and these were combined in 27 of 39 (69%) cases; there was no significant difference in 1p/19q loss of heterozygosity status between low-grade and anaplastic oligodendrogliomas.

1p/19q co-deletion has been correlated with both chemosensitivity and improved prognosis in oligodendrogliomas. Most larger cancer treatment centers routinely check for the deletion of 1p/19q as part of the pathology report for oligodendrogliomas. The status of the 1p/19q loci can be detected by FISH or virtual karyotyping. Virtual karyotyping has the advantage of assessing the entire genome in one assay, as well as the 1p/19q loci. This allows assessment of other key loci in glial tumors, such as EGFR and TP53 copy number status.

Whereas the prognostic relevance of 1p and 19q deletions is well established for anaplastic oligodendrogliomas and mixed oligoastrocytomas, the prognostic relevance of the deletions for low-grade gliomas is more controversial. In terms of low-grade gliomas, a recent study also suggests that 1p/19q co-deletion may be associated with a (1;19)(q10;p10) translocation which, like the combined 1p/19q deletion, is associated with superior overall survival and progression-free survival in low-grade glioma patients.

Oligodendrogliomas show only rarely mutations in the p53 gene, which is in contrast to other gliomas. Epidermal growth factor receptor amplification and whole 1p/19q codeletion are mutually exclusive and predictive of completely different outcomes, with EGFR amplification predicting poor prognosis.

Glioblastoma

Yin et al studied 55 glioblastoma and 6 GBM cell lines using SNP array karyotyping. Acquired UPD was identified at 17p in 13/61 cases. A significantly shortened survival time was found in patients with 13q14 (RB) deletion or 17p13.1 (p53) deletion/acquired UPD. Taken together,these results suggest that this technique is a rapid,robust,and inexpensive method to profile genome-wide abnormalities in GBM.Because SNP array karyotyping can be performed on paraffin embedded tumors, it is an attractive option when tumor cells fail to grow in culture for metaphase cytogenetics or when the desire for karyotyping arises after the specimen has been formalin fixed.

The importance of detecting acquired UPD (copy neutral LOH) in glioblastoma:

- Of patients with 17p abnormality, ~50% were deletions and ~50% were aUPD
- Both 17p del and 17p UPD were associated with worse outcome.
- 9/13 had homozygous TP53 mutations underlying the 17p UPD.

In addition, in cases with uncertain grade by morphology, genomic profiling can assist in diagnosis.

- Concomitant gain of 7 and loss of 10 is essentially pathognomonic for GBM

- EGFR amplification, loss of PTEN (on 10q), and loss of p16 (on 9p) occur almost exclusively in glioblastoma and can provide means to distinguish anaplastic astrocytoma from glioblastoma.

Acute Lymphoblastic Leukemia

Cytogenetics, the study of characteristic large changes in the chromosomes of cancer cells, has been increasingly recognized as an important predictor of outcome in acute lymphoblastic leukemia (ALL).

NB: Balanced translocations cannot be detected by array-based karyotyping.

Some cytogenetic subtypes have a worse prognosis than others. These include:

- A translocation between chromosomes 9 and 22, known as the Philadelphia chromosome, occurs in about 20% of adult and 5% in pediatric cases of ALL.
- A translocation between chromosomes 4 and 11 occurs in about 4% of cases and is most common in infants under 12 months.
- Not all translocations of chromosomes carry a poorer prognosis. Some translocations are relatively favorable. For example, Hyperdiploidy (>50 chromosomes) is a good prognostic factor.
- Genome-wide assessment of copy number changes can be done by conventional cytogenetics or virtual karyotyping. SNP array virtual karyotyping can detect copy number changes and LOH status, while arrayCGH can detect only copy number changes. Copy neutral LOH (acquired uniparental disomy) has been reported at key loci in ALL, such as CDKN2A gene at 9p, which have prognostic significance. SNP array virtual karyotyping can readily detect copy neutral LOH. Array CGH, FISH, and conventional cytogenetics cannot detect copy neutral LOH.

Cytogenetic change	Risk category
Philadelphia chromosome	*Poor prognosis*
t(4;11)(q21;q23)	*Poor prognosis*
t(8;14)(q24.1;q32)	*Poor prognosis*
Complex karyotype (more than four abnormalities)	*Poor prognosis*
Low hypodiploidy or near triploidy	*Poor prognosis*
High hyperdiploidy	*Good prognosis*
del(9p)	*Good prognosis*

- Correlation of prognosis with bone marrow cytogenetic finding in acute lymphoblastic leukemia

Prognosis	Cytogenetic findings
Favorable	*Hyperdiploidy > 50 ; t (12;21)*
Intermediate	*Hyperdioloidy 47 -50; Normal(diploidy); del (6q); Rearrangements of 8q24*
Unfavorable	*Hypodiploidy-near haploidy; Near tetraploidy; del (17p); t (9;22); t (11q23)*

- Unclassified ALL is considered to have an intermediate prognosis.

Myelodysplastic Syndrome

Myelodysplastic syndrome (MDS) has remarkable clinical, morphological, and genetic heterogeneity. Cytogenetics play a decisive role in the World Health Organization's classification-based International Prognostic Scoring System (IPSS) for MDS.

- Good Prognosis: normal karyotype, isolated del(5q), isolated del(20q), -Y
- Poor Prognosis: complex abnormalities (ie, >=3 abnormalities), –7 or del(7q)
- Intermediate Prognosis: all other abnormalities, including trisomy 8 and del(11q)

In a comparison of metaphase cytogenetics, FISH panel, and SNP array karyotyping for MDS, it was found that each technique provided a similar diagnostic yield. No single method detected all defects, and detection rates improved by ~5% when all three methods were used.

Acquired UPD, which is not detectable by FISH or cytogenetics, has been reported at several key loci in MDS using SNP array karyotyping, including deletion of 7/7q.

Myeloproliferative Neoplasms/Myeloproliferative Disorders

Philadelphia chromosome–negative myeloproliferative neoplasms (MPNs) including polycythemia vera,essential thrombocythemia, and primary myelofibrosis show an inherent tendency for transformation into leukemia (MPN-blast phase),which is accompanied by acquisition of additional genomic lesions. In a study of 159 cases, SNP-array analysis was able to capture practically all cytogenetic abnormalities and to uncover additional lesions with potentially important clinical implications.

- The number of genomic alterations was more than 2 to 3 times greater in the blast phase as in the chronic phase of the disease.
- Deletion of 17p (TP53)was significantly associated with prior exposure to hydroxyurea as well as a complex karyotype in samples with MPN-blast crisis. Interestingly, not only deletion, but also 17p copy neutral LOH, was associated with a complex karyotype, a poor prognostic

marker in myeloid malignancies. Copy neutral LOH (acquired UPD)is readily detectably by SNP array karyotype, but not by cytogenetics, FISH, or arrayCGH.

- Blast phase patients with loss of chromosomal material on 7q showed poor survival. Loss of 7q is known to be predictive for rapid progression and poor response in AML therapy.MPN-blast phase patients with cytogenetically undetectable 7q copy neutral-LOH had comparable survival rates to those with 7/7q in their leukemic cells.
- 9p copy neutral-LOH with homozygous JAK2 mutation was also linked to an inferior outcome in MPN-blast crisis in comparison with patients with either heterozygous JAK2V617F or wild-type JAK2. In contrast to LOH on 17p, the prognostic impact of 9pCNN-LOH was independent of established risk factors such as 7/7q, 5q, or complex karyotype.

Colorectal Cancer

Identification of biomarkers in colorectal cancer is particularly important for patients with stage II disease, where less than 20% have tumor recurrence. 18q LOH is an established biomarker associated with high risk of tumor recurrence in stage II colon cancer.. Figure 7 shows a SNP array karyotype of a colorectal carcinoma (whole genome view).

Colorectal cancers are classified into specific tumor phenotypes based on molecular profiles which can be integrated with the results of other ancillary tests, such as microsatellite instability testing, IHC, and KRAS mutation status:

- Chromosomal instability (CIN) which have allelic imbalance at a number of chromosomal loci, including 5q, 8p, 17p, and 18q.
- Microsatellite instability (MSI) which tend to have diploid karyotypes.

Malignant Rhabdoid Tumors

Malignant rhabdoid tumors are rare, highly aggressive neoplasms found most commonly in infants and young children. Due to their heterogenous histologic features, diagnosis can often be difficult and misclassifications can occur. In these tumors, the INI1 gene (SMARCB1)on chromosome 22q functions as a classic tumor suppressor gene. Inactivation of INI1 can occur via deletion, mutation, or acquired UPD. In a recent study, SNP array karyotyping identified deletions or LOH of 22q in 49/51 rhabdoid tumors. Of these, 14 were copy neutral LOH (or acquired UPD), which is detectable by SNP array karyotyping, but not by FISH, cytogenetics, or arrayCGH. MLPA detected a single exon homozygous deletion in one sample that was below the resolution of the SNP array.

SNP array karyotyping can be used to distinguish, for example, a medulloblastoma with an isochromosome 17q from a primary rhabdoid tumor with loss of 22q11.2. When indicated, molecular analysis of INI1 using MLPA and direct sequencing may then be employed. Once the tumor-associated changes are found, an analysis of germline DNA from the patient and the parents can be done to rule out an inherited or de novo germline mutation or deletion of INI1, so that appropriate recurrence risk assessments can be made.

Uveal Melanoma

The most important genetic alteration associated with poor prognosis in uveal melanoma is loss of an entire copy of Chromosome 3 (Monosomy 3), which is strongly correlated with metastatic spread. Gains on chromosomes 6 and 8 are often used to refine the predictive value of the Monosomy 3 screen, with gain of 6p indicating a better prognosis and gain of 8q indicating a worse prognosis in disomy 3 tumors. In rare instances, monosomy 3 tumors may duplicate the remaining copy of the chromosome to return to a disomic state referred to as isodisomy. Isodisomy 3 is prognostically equivalent to monosomy 3, and both can be detected by tests for chromosome 3 loss of heterozygosity.

Limitations

Unlike karyotypes obtained from conventional cytogenetics, virtual karyotypes are reconstructed by computer programs using signals obtained from disrupted DNA. In essence, the computer program will correct translocations when it lines up the signals in chromosomal order. Therefore, virtual karyotypes cannot detect balanced translocations and inversions. They also can only detect genetic aberrations in regions of the genome that are represented by probes on the array. In addition, virtual karyotypes generate a relative copy number normalized against a diploid genome, so tetraploid genomes will be condensed into a diploid space unless renormalization is performed.

Renormalization requires an ancillary cell-based assay, such as FISH, if one is using arrayCGH. For karyotypes obtained from SNP-based arrays, tetraploidy can often be inferred from the maintenance of heterozygosity within a region of apparent copy number loss. Low-level mosaicism or small subclones may not be detected by virtual karyotypes because the presence of normal cells in the sample will dampen the signal from the abnormal clone.

The exact point of failure, in terms of the minimal percentage of neoplastic cells, will depend on the particular platform and algorithms used. Many copy number analysis software programs used to generate array-based karyotypes will falter with less than 25–30% tumor/abnormal cells in the sample. However, in oncology applications this limitation can be minimized by tumor enrichment strategies and/or software optimized for use with oncology samples. The analysis algorithms are evolving rapidly, and some are even designed to thrive on 'normal clone contamination', so it is anticipated that this limitation will continue to dissipate.

Happy Mapping

HAPPY Mapping, first proposed by Paul H. Dear and Peter R. Cook in 1989, is a method used in molecular biology to study the linkage between two or more DNA sequences. According to the Single Molecule Genomics Group, it is "Mapping based on the analysis of approximately HAPloid DNA samples using the PolYmerase chain reaction".

In genomics, HAPPY mapping can be applied to assess the synteny and orientation of various DNA sequences across a particular genome - the generation of a "genomic" map.

As with linkage mapping, HAPPY mapping relies on the differential probability of two or more DNA sequences being separated. In genetic mapping, the probability of a recombination event between two genetic loci on the same chromosome is directly proportional to the distance between them.

HAPPY mapping replaces recombination with fragmentation - instead of relying on recombination to separate genetic loci, the entire genome is fragmented, for example, by radiation or mechanical shearing. If the DNA is broken on a random basis, the longer the distance between two DNA sequences, the higher the chances of it to break between the two, and vice versa.

HAPPY mapping retains the benefits of genetic mapping while removing some of the problems associated with recombination. I.e., the need for polymorphism, and breeding. Also, recombination can be locale specific whereas breakage of genomic DNA by radiation or mechanical shearing seems to be more random.

Reverse Transcription PCR

Reverse transcription polymerase chain reaction (RT-PCR) is a variant of polymerase chain reaction (PCR), a laboratory technique commonly used in molecular biology to generate many copies of a DNA sequence, a process termed "amplification".

In RT-PCR, however, an RNA strand is first reverse transcribed into its DNA complement (complementary DNA, or cDNA) using the enzyme reverse transcriptase, and the resulting cDNA is amplified using traditional PCR or real-time PCR. Reverse transcription PCR is not to be confused with real-time polymerase chain reaction (Q-PCR/qRT-PCR), which is also sometimes abbreviated as RT-PCR.

RT-PCR Principles and Procedure

RT-PCR utilizes a pair of primers, which are complementary to a defined sequence on each of the two strands of the cDNA. These primers are then extended by a DNA polymerase and a copy of the strand is made after each cycle, leading to exponential amplification.

RT-PCR includes three major steps. The first step is reverse transcription (RT), in which RNA is reverse transcribed to cDNA using reverse transcriptase. This step is very important in order to perform PCR since DNA polymerase can act only on DNA templates. The RT step can be performed either in the same tube with PCR (one-step PCR) or in a separate one (two-step PCR) using a temperature between 40°C and 50°C, depending on the properties of the reverse transcriptase used.

The next step involves the denaturation of the dsDNA at 95°C, so that the two strands separate and the primers can bind again at lower temperatures and begin a new chain reaction. Then, the temperature is decreased until it reaches the annealing temperature which can vary depending on the set of primers used, their concentration, the probe and its concentration (if used), and the cations concentration. The main consideration when choosing the optimal annealing temperature is the melting temperature (Tm) of the primers and probes (if used). The annealing temperature chosen for a PCR depends directly on length and composition of the primers. This is the result of the difference of hydrogen bonds between A-T (2 bonds) and G-C (3 bonds). An annealing temperature about 5 degrees below the lowest Tm of the pair of primers is usually used.

The final step of PCR amplification is DNA extension from the primers. This is done with thermostable Taq DNA polymerase, usually at 72°C, the temperature at which the enzyme works optimally. The length of the incubation at each temperature, the temperature alterations, and the number of cycles are controlled by a programmable thermal cycler. The analysis of the PCR products depends on the type of PCR applied. If a conventional PCR is used, the PCR product is detected using agarose gel electrophoresis and ethidium bromide (or other nucleic acid staining).

Conventional RT-PCR is a time-consuming technique with important limitations when compared to real-time PCR techniques. This, combined with the fact that ethidium bromide has low sensitivity, yields results that are not always reliable. Moreover, there is an increased cross-contamination risk of the samples since detection of the PCR product requires the post-amplification processing of the samples. Furthermore, the specificity of the assay is mainly determined by the primers, which can give false-positive results. However, the most important issue concerning conventional RT-PCR is the fact that it is a semi- or even a low-quantitative technique, whereas the amplicon can be visualized only after the amplification ends.

Real-time RT-PCR provides a method in which the amplicons can be visualized as the amplification progresses using a fluorescent reporter molecule. There are three major kinds of fluorescent reporters used in real time RT-PCR, which are general non-specific DNA Binding Dyes such as SYBR Green I, TaqMan Probes and Molecular Beacons (including Scorpions). The real-time PCR thermal cycler has a fluorescence detection threshold, below which it cannot discriminate the difference between an amplification generated signal and background noise. On the other hand, the fluorescence increases as the amplification progresses and the instrument performs data acquisition during the annealing step of each cycle.

The number of amplicons will reach the detection baseline after a specific cycle, which depends on the initial concentration of the target DNA sequence. The cycle at which the instrument can discriminate the amplification generated fluorescence from the background noise is called the threshold cycle (Ct). The higher the initial DNA concentration, the lower its Ct will be.

Use of Reverse Transcription Polymerase Chain Reaction

The exponential amplification via reverse transcription polymerase chain reaction provides for a highly sensitive technique in which a very low copy number of RNA molecules can be detected. RT-PCR is widely used in the diagnosis of genetic diseases and, semiquantitatively, in the determination of the abundance of specific different RNA molecules within a cell or tissue as a measure of gene expression. Northern blot analysis is used to study the RNA's gene expression further. RT-PCR can also be very useful in the insertion of eukaryotic genes into prokaryotes.

Because most eukaryotic genes contain introns which are present in the genome but not in the mature mRNA, the cDNA generated from a RT-PCR reaction is the exact (without regard to the error prone nature of reverse transcriptases) DNA sequence which would be directly translated into protein after transcription. When these genes are expressed in prokaryotic cells for the sake of protein production or purification, the RNA produced directly from transcription need not undergo splicing as the transcript contains only exons (prokaryotes, such as E.coli, lack the mRNA splicing mechanism of eukaryote RT-PCR is commonly used in studying the genomes of viruses whose genomes are composed of RNA, such as Influenzavirus A and retroviruses like HIV.

Polymerase Chain Reaction

The polymerase chain reaction (PCR) is a scientific technique in molecular biology to amplify a single or a few copies of a piece of DNA across several orders of magnitude, generating thousands to millions of copies of a particular DNA sequence.

Developed in 1983 by Kary Mullis, PCR is now a common and often indispensable technique used in medical and biological research labs for a variety of applications. These include DNA cloning for sequencing, DNA-based phylogeny, or functional analysis of genes; the diagnosis of hereditary diseases; the identification of genetic fingerprints (used in forensic sciences and paternity testing); and the detection and diagnosis of infectious diseases. In 1993, Mullis was awarded the Nobel Prize in Chemistry along with Michael Smith for his work on PCR.

The method relies on thermal cycling, consisting of cycles of repeated heating and cooling of the reaction for DNA melting and enzymatic replication of the DNA. Primers (short DNA fragments) containing sequences complementary to the target region along with a DNA polymerase (after which the method is named) are key components to enable selective and repeated amplification.

As PCR progresses, the DNA generated is itself used as a template for replication, setting in motion a chain reaction in which the DNA template is exponentially amplified. PCR can be extensively modified to perform a wide array of genetic manipulations.

Almost all PCR applications employ a heat-stable DNA polymerase, such as Taq polymerase, an enzyme originally isolated from the bacterium Thermus aquaticus. This DNA polymerase enzymatically assembles a new DNA strand from DNA building-blocks, the nucleotides, by using single-stranded DNA as a template and DNA oligonucleotides (also called DNA primers), which are required for initiation of DNA synthesis.

The vast majority of PCR methods use thermal cycling, i.e., alternately heating and cooling the PCR sample to a defined series of temperature steps. These thermal cycling steps are necessary first to physically separate the two strands in a DNA double helix at a high temperature in a process called DNA melting.

At a lower temperature, each strand is then used as the template in DNA synthesis by the DNA polymerase to selectively amplify the target DNA. The selectivity of PCR results from the use of primers that are complementary to the DNA region targeted for amplification under specific thermal cycling conditions.

PCR Principles and Procedure

PCR is used to amplify a specific region of a DNA strand (the DNA target). Most PCR methods typically amplify DNA fragments of up to ~10 kilo base pairs (kb), although some techniques allow for amplification of fragments up to 40 kb in size.

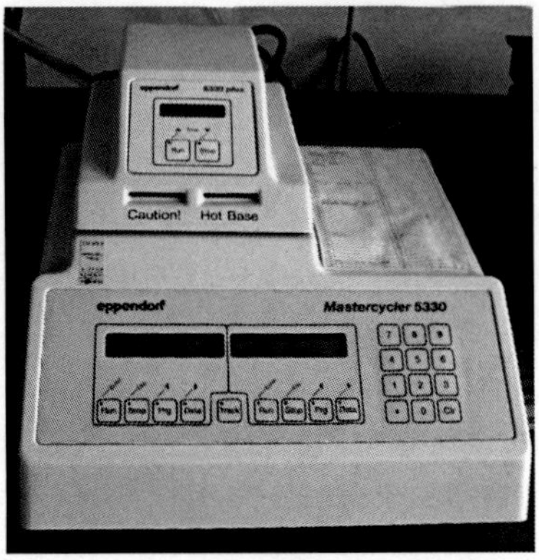

Figure : *A thermal cycler for PCR*

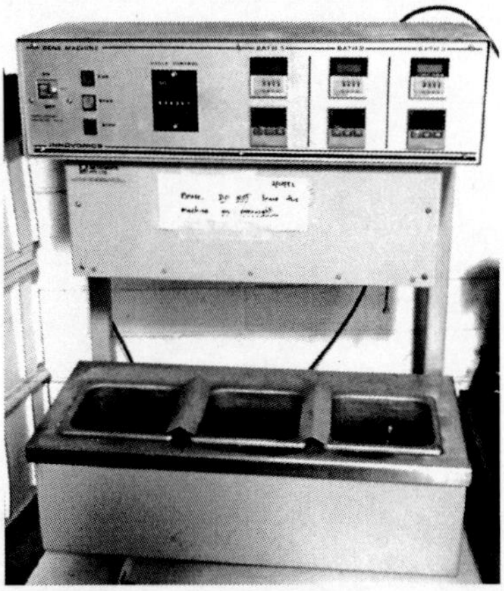

Figure : *An older model three-temperature thermal cycler for PCR*

A basic PCR set up requires several components and reagents. These components include:

- DNA template that contains the DNA region (target) to be amplified.
- Two primers that are complementary to the 3' (three prime) ends of each of the sense and anti-sense strand of the DNA target.
- Taq polymerase or another DNA polymerase with a temperature optimum at around 70 °C.
- Deoxynucleoside triphosphates (dNTPs; nucleotides containing triphosphate groups), the building-blocks from which the DNA polymerase synthesizes a new DNA strand.
- Buffer solution, providing a suitable chemical environment for optimum activity and stability of the DNA polymerase.

- Divalent cations, magnesium or manganese ions; generally Mg is used, but Mn can be utilized for PCR-mediated DNA mutagenesis, as higher Mn concentration increases the error rate during DNA synthesis
- Monovalent cation potassium ions.

The PCR is commonly carried out in a reaction volume of 10–200 ìl in small reaction tubes (0.2–0.5 ml volumes) in a thermal cycler. The thermal cycler heats and cools the reaction tubes to achieve the temperatures required at each step of the reaction . Many modern thermal cyclers make use of the Peltier effect, which permits both heating and cooling of the block holding the PCR tubes simply by reversing the electric current. Thin-walled reaction tubes permit favorable thermal conductivity to allow for rapid thermal equilibration. Most thermal cyclers have heated lids to prevent condensation at the top of the reaction tube. Older thermocyclers lacking a heated lid require a layer of oil on top of the reaction mixture or a ball of wax inside the tube.

Procedure

Typically, PCR consists of a series of 20-40 repeated temperature changes, called cycles, with each cycle commonly consisting of 2-3 discrete temperature steps, usually three (Fig. 2). The cycling is often preceded by a single temperature step (called hold) at a high temperature (>90°C), and followed by one hold at the end for final product extension or brief storage. The temperatures used and the length of time they are applied in each cycle depend on a variety of parameters. These include the enzyme used for DNA synthesis, the concentration of divalent ions and dNTPs in the reaction, and the melting temperature (Tm) of the primers.

- Initialization step: This step consists of heating the reaction to a temperature of 94–96 °C (or 98 °C if extremely thermostable polymerases are used), which is held for 1–9 minutes. It is only required for DNA polymerases that require heat activation by hot-start PCR.
- Denaturation step: This step is the first regular cycling event and consists of heating the reaction to 94–98 °C for 20–30 seconds. It causes DNA melting of the DNA template by disrupting the hydrogen bonds between complementary bases, yielding single-stranded DNA molecules.
- Annealing step: The reaction temperature is lowered to 50–65 °C for 20–40 seconds allowing annealing of the primers to the single-stranded DNA template. Typically the annealing temperature is about 3-5 degrees Celsius below the Tm of the primers used. Stable DNA-DNA hydrogen bonds are only formed when the primer sequence very closely matches the template sequence. The polymerase binds to the primer-template hybrid and begins DNA synthesis.

- Extension/elongation step: The temperature at this step depends on the DNA polymerase used; Taq polymerase has its optimum activity temperature at 75–80 °C, and commonly a temperature of 72 °C is used with this enzyme. At this step the DNA polymerase synthesizes a new DNA strand complementary to the DNA template strand by adding dNTPs that are complementary to the template in 5' to 3' direction, condensing the 5'-phosphate group of the dNTPs with the 3'-hydroxyl group at the end of the nascent (extending) DNA strand. The extension time depends both on the DNA polymerase used and on the length of the DNA fragment to be amplified. As a rule-of-thumb, at its optimum temperature, the DNA polymerase will polymerize a thousand bases per minute. Under optimum conditions, i.e., if there are no limitations due to limiting substrates or reagents, at each extension step, the amount of DNA target is doubled, leading to exponential (geometric) amplification of the specific DNA fragment.

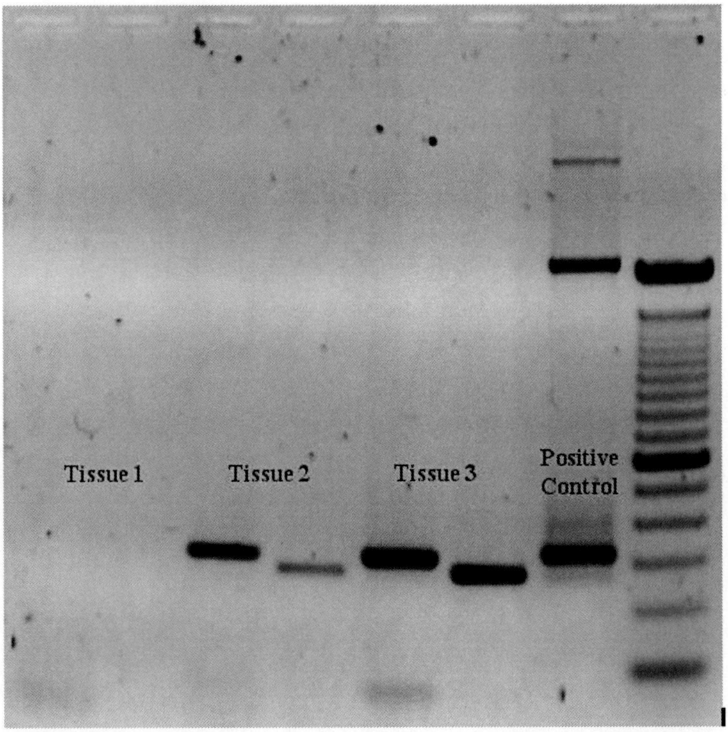

Figure : *Ethidium bromide-stained PCR products after gel electrophoresis. Two sets of primers were used to amplify a target sequence from three different tissue samples. No amplification is present in sample #1; DNA bands in sample #2 and #3 indicate successful amplification of the target sequence. The gel also shows a positive control, and a DNA ladder containing DNA fragments of defined length for sizing the bands in the experimental PCRs.*

- Final elongation: This single step is occasionally performed at a temperature of 70–74 °C for 5–15 minutes after the last PCR cycle to ensure that any remaining single-stranded DNA is fully extended.
- Final hold: This step at 4–15 °C for an indefinite time may be employed for short-term storage of the reaction.

To check whether the PCR generated the anticipated DNA fragment (also sometimes referred to as the amplimer or amplicon), agarose gel electrophoresis is employed for size separation of the PCR products. The size(s) of PCR products is determined by comparison with a DNA ladder (a molecular weight marker), which contains DNA fragments of known size, run on the gel alongside the PCR products.

PCR Stages

The PCR process can be divided into three stages: Exponential amplification: At every cycle, the amount of product is doubled (assuming 100% reaction efficiency). The reaction is very sensitive: only minute quantities of DNA need to be present. Leveling off stage: The reaction slows as the DNA polymerase loses activity and as consumption of reagents such as dNTPs and primers causes them to become limiting. Plateau: No more product accumulates due to exhaustion of reagents and enzyme.

PCR Optimization

In practice, PCR can fail for various reasons, in part due to its sensitivity to contamination causing amplification of spurious DNA products. Because of this, a number of techniques and procedures have been developed for optimizing PCR conditions. Contamination with extraneous DNA is addressed with lab protocols and procedures that separate pre-PCR mixtures from potential DNA contaminants. This usually involves spatial separation of PCR-setup areas from areas for analysis or purification of PCR products, use of disposable plasticware, and thoroughly cleaning the work surface between reaction setups.

Primer-design techniques are important in improving PCR product yield and in avoiding the formation of spurious products, and the usage of alternate buffer components or polymerase enzymes can help with amplification of long or otherwise problematic regions of DNA. Addition of reagents, such as formamide, in buffer systems may increase the specificity and yield of PCR.

Application of PCR

Selective DNA Isolation

PCR allows isolation of DNA fragments from genomic DNA by selective amplification of a specific region of DNA. This use of PCR augments many methods, such as generating hybridization probes for Southern or northern hybridization and DNA cloning, which require larger amounts of DNA, representing a specific DNA region. PCR supplies these techniques with high amounts of pure DNA, enabling analysis of DNA samples even from very small amounts of starting material.

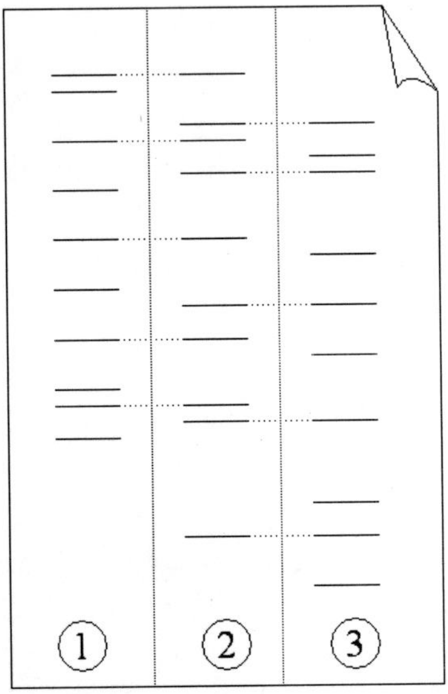

Figure : *Electrophoresis of PCR-amplified DNA fragments. (1) Father. (2) Child. (3) Mother. The child has inherited some, but not all of the fingerprint of each of its parents, giving it a new, unique fingerprint.*

Other applications of PCR include DNA sequencing to determine unknown PCR-amplified sequences in which one of the amplification primers may be used in Sanger sequencing, isolation of a DNA sequence to expedite recombinant DNA technologies involving the insertion of a DNA sequence into a plasmid or the genetic material of another organism. Bacterial colonies (E. coli) can be rapidly screened by PCR for correct DNA vector constructs. PCR may also be used for genetic fingerprinting; a forensic technique used to identify a person or organism by comparing experimental DNAs through different PCR-based methods.

Some PCR 'fingerprints' methods have high discriminative power and can be used to identify genetic relationships between individuals, such as parent-child or between siblings, and are used in paternity testing. This technique may also be used to determine evolutionary relationships among organisms.

Amplification and Quantification of DNA

Because PCR amplifies the regions of DNA that it targets, PCR can be used to analyze extremely small amounts of sample. This is often critical for forensic analysis, when only a trace amount of DNA is available as evidence.

PCR may also be used in the analysis of ancient DNA that is tens of thousands of years old. These PCR-based techniques have been successfully used on animals, such as a forty-thousand-year-old mammoth, and also on human DNA, in applications ranging from the analysis of Egyptian mummies to the identification of a Russian tsar.

Quantitative PCR methods allow the estimation of the amount of a given sequence present in a sample—a technique often applied to quantitatively determine levels of gene expression. Real-time PCR is an established tool for DNA quantification that measures the accumulation of DNA product after each round of PCR amplification.

DNA in Forensic Entomology

PCR in Diagnosis of Diseases

PCR permits early diagnosis of malignant diseases such as leukemia and lymphomas, which is currently the highest-developed in cancer research and is already being used routinely. PCR assays can be performed directly on genomic DNA samples to detect translocation-specific malignant cells at a sensitivity that is at least 10,000-fold higher than that of other methods.

PCR also permits identification of non-cultivatable or slow-growing microorganisms such as mycobacteria, anaerobic bacteria, or viruses from tissue culture assays and animal models. The basis for PCR diagnostic applications in microbiology is the detection of infectious agents and the discrimination of non-pathogenic from pathogenic strains by virtue of specific genes. Viral DNA can likewise be detected by PCR. The primers used need to be specific to the targeted sequences in the DNA of a virus, and the PCR can be used for diagnostic analyses or DNA sequencing of the viral genome. The high sensitivity of PCR permits virus detection soon after infection and even before the onset of disease. Such early detection may give physicians a significant lead in treatment. The amount of virus ("viral load") in a patient can also be quantified by PCR-based DNA quantitation techniques .

Variations on the Basic PCR Technique

- Allele-specific PCR: a diagnostic or cloning technique based on single-nucleotide polymorphisms (SNPs) (single-base differences in DNA). It requires prior knowledge of a DNA sequence, including differences between alleles, and uses primers whose 3' ends encompass the SNP. PCR amplification under stringent conditions is much less efficient in the presence of a mismatch between template and primer, so successful amplification with an SNP-specific primer signals presence of the specific SNP in a sequence.

- Assembly PCR or Polymerase Cycling Assembly (PCA): artificial synthesis of long DNA sequences by performing PCR on a pool of long oligonucleotides with short overlapping segments. The oligonucleotides alternate between sense and antisense directions, and the overlapping segments determine the order of the PCR fragments, thereby selectively producing the final long DNA product.
- Asymmetric PCR: preferentially amplifies one DNA strand in a double-stranded DNA template. It is used in sequencing and hybridization probing where amplification of only one of the two complementary strands is required. PCR is carried out as usual, but with a great excess of the primer for the strand targeted for amplification. Because of the slow (arithmetic) amplification later in the reaction after the limiting primer has been used up, extra cycles of PCR are required.

  A recent modification on this process, known as Linear-After-The-Exponential-PCR (LATE-PCR), uses a limiting primer with a higher melting temperature (Tm) than the excess primer to maintain reaction efficiency as the limiting primer concentration decreases mid-reaction.
- Helicase-dependent amplification: similar to traditional PCR, but uses a constant temperature rather than cycling through denaturation and annealing/extension cycles. DNA helicase, an enzyme that unwinds DNA, is used in place of thermal denaturation.
- Hot start PCR: a technique that reduces non-specific amplification during the initial set up stages of the PCR. It may be performed manually by heating the reaction components to the denaturation temperature (e.g., 95°C) before adding the polymerase.

  Specialized enzyme systems have been developed that inhibit the polymerase's activity at ambient temperature, either by the binding of an antibody or by the presence of covalently bound inhibitors that dissociate only after a high-temperature activation step. Hot-start/cold-finish PCR is achieved with new hybrid polymerases that are inactive at ambient temperature and are instantly activated at elongation temperature.
- Intersequence-specific PCR (ISSR): a PCR method for DNA fingerprinting that amplifies regions between simple sequence repeats to produce a unique fingerprint of amplified fragment lengths.
- Inverse PCR: is commonly used to identify the flanking sequences around genomic inserts. It involves a series of DNA digestions and self ligation, resulting in known sequences at either end of the unknown sequence.

- Ligation-mediated PCR: uses small DNA linkers ligated to the DNA of interest and multiple primers annealing to the DNA linkers; it has been used for DNA sequencing, genome walking, and DNA footprinting.
- Methylation-specific PCR (MSP): developed by Stephen Baylin and Jim Herman at the Johns Hopkins School of Medicine, and is used to detect methylation of CpG islands in genomic DNA. DNA is first treated with sodium bisulfite, which converts unmethylated cytosine bases to uracil, which is recognized by PCR primers as thymine. Two PCRs are then carried out on the modified DNA, using primer sets identical except at any CpG islands within the primer sequences. At these points, one primer set recognizes DNA with cytosines to amplify methylated DNA, and one set recognizes DNA with uracil or thymine to amplify unmethylated DNA. MSP using qPCR can also be performed to obtain quantitative rather than qualitative information about methylation.
- Miniprimer PCR: uses a thermostable polymerase (S-Tbr) that can extend from short primers ("smalligos") as short as 9 or 10 nucleotides. This method permits PCR targeting to smaller primer binding regions, and is used to amplify conserved DNA sequences, such as the 16S (or eukaryotic 18S) rRNA gene.
- Multiplex Ligation-dependent Probe Amplification (MLPA): permits multiple targets to be amplified with only a single primer pair, thus avoiding the resolution limitations of multiplex PCR.
- Multiplex-PCR: consists of multiple primer sets within a single PCR mixture to produce amplicons of varying sizes that are specific to different DNA sequences. By targeting multiple genes at once, additional information may be gained from a single test-run that otherwise would require several times the reagents and more time to perform. Annealing temperatures for each of the primer sets must be optimized to work correctly within a single reaction, and amplicon sizes. That is, their base pair length should be different enough to form distinct bands when visualized by gel electrophoresis.
- Nested PCR: increases the specificity of DNA amplification, by reducing background due to non-specific amplification of DNA. Two sets of primers are used in two successive PCRs. In the first reaction, one pair of primers is used to generate DNA products, which besides the intended target, may still consist of non-specifically amplified DNA fragments. The product(s) are then used in a second PCR with a set of primers whose binding sites are completely or partially different from and located 3' of each of the primers used in the first reaction. Nested PCR is often more successful in specifically amplifying long DNA fragments

than conventional PCR, but it requires more detailed knowledge of the target sequences.

- Overlap-extension PCR or Splicing by overlap extension (SOE) : a genetic engineering technique that is used to splice together two or more DNA fragments that contain complementary sequences. It is used to join DNA pieces containing genes, regulatory sequences, or mutations; the technique enables creation of specific and long DNA constructs.
- Quantitative PCR (Q-PCR): used to measure the quantity of a PCR product (commonly in real-time). It quantitatively measures starting amounts of DNA, cDNA, or RNA. Q-PCR is commonly used to determine whether a DNA sequence is present in a sample and the number of its copies in the sample. Quantitative real-time PCR has a very high degree of precision. QRT-PCR (or QF-PCR) methods use fluorescent dyes, such as Sybr Green, EvaGreen or fluorophore-containing DNA probes, such as TaqMan, to measure the amount of amplified product in real time. It is also sometimes abbreviated to RT-PCR (Real Time PCR) or RQ-PCR. QRT-PCR or RTQ-PCR are more appropriate contractions, since RT-PCR commonly refers to reverse transcription PCR , often used in conjunction with Q-PCR.
- Reverse Transcription PCR (RT-PCR): for amplifying DNA from RNA. Reverse transcriptase reverse transcribes RNA into cDNA, which is then amplified by PCR. RT-PCR is widely used in expression profiling, to determine the expression of a gene or to identify the sequence of an RNA transcript, including transcription start and termination sites. If the genomic DNA sequence of a gene is known, RT-PCR can be used to map the location of exons and introns in the gene. The 5' end of a gene (corresponding to the transcription start site) is typically identified by RACE-PCR (Rapid Amplification of cDNA Ends).
- Solid Phase PCR: encompasses multiple meanings, including Polony Amplification (where PCR colonies are derived in a gel matrix, for example), Bridge PCR (primers are covalently linked to a solid-support surface), conventional Solid Phase PCR (where Asymmetric PCR is applied in the presence of solid support bearing primer with sequence matching one of the aqueous primers) and Enhanced Solid Phase PCR (where conventional Solid Phase PCR can be improved by employing high Tm and nested solid support primer with optional application of a thermal 'step' to favour solid support priming).
- Thermal asymmetric interlaced PCR (TAIL-PCR): for isolation of an unknown sequence flanking a known sequence. Within the known sequence, TAIL-PCR uses a nested pair of primers with differing

annealing temperatures; a degenerate primer is used to amplify in the other direction from the unknown sequence.

- Touchdown PCR (Step-down PCR): a variant of PCR that aims to reduce nonspecific background by gradually lowering the annealing temperature as PCR cycling progresses. The annealing temperature at the initial cycles is usually a few degrees (3-5°C) above the T_m of the primers used, while at the later cycles, it is a few degrees (3-5°C) below the primer T_m. The higher temperatures give greater specificity for primer binding, and the lower temperatures permit more efficient amplification from the specific products formed during the initial cycles.
- PAN-AC: uses isothermal conditions for amplification, and may be used in living cells.
- Universal Fast Walking: for genome walking and genetic fingerprinting using a more specific 'two-sided' PCR than conventional 'one-sided' approaches (using only one gene-specific primer and one general primer — which can lead to artefactual 'noise') by virtue of a mechanism involving lariat structure formation. Streamlined derivatives of UFW are LaNe RAGE (lariat-dependent nested PCR for rapid amplification of genomic DNA ends), 5'RACE LaNe and 3'RACE LaNe.

History

A 1971 paper in the Journal of Molecular Biology by Kleppe and co-workers first described a method using an enzymatic assay to replicate a short DNA template with primers in vitro. However, this early manifestation of the basic PCR principle did not receive much attention, and the invention of the polymerase chain reaction in 1983 is generally credited to Kary Mullis.

When Mullis developed the PCR in 1983, he was working in Emeryville, California for Cetus Corporation, one of the first biotechnology companies. There, he was responsible for synthesizing short chains of DNA. Mullis has written that he conceived of PCR while cruising along the Pacific Coast Highway one night in his car. He was playing in his mind with a new way of analyzing changes (mutations) in DNA when he realized that he had instead invented a method of amplifying any DNA region through repeated cycles of duplication driven by DNA polymerase. In Scientific American, Mullis summarized the procedure:

"Beginning with a single molecule of the genetic material DNA, the PCR can generate 100 billion similar molecules in an afternoon. The reaction is easy to execute. It requires no more than a test tube, a few simple reagents, and a source of heat." He was awarded the Nobel Prize in Chemistry in 1993 for his invention, seven years after he and his colleagues at Cetus first put his proposal to practice. However, some controversies have remained about the intellectual and practical contributions of other scientists to Mullis' work, and whether he had been the sole inventor of the PCR principle.

At the core of the PCR method is the use of a suitable DNA polymerase able to withstand the high temperatures of >90 °C (194 °F) required for separation of the two DNA strands in the DNA double helix after each replication cycle.

The DNA polymerases initially employed for in vitro experiments presaging PCR were unable to withstand these high temperatures. So the early procedures for DNA replication were very inefficient and time consuming, and required large amounts of DNA polymerase and continuous handling throughout the process.

The discovery in 1976 of Taq polymerase — a DNA polymerase purified from the thermophilic bacterium, Thermus aquaticus, which naturally lives in hot (50 to 80 °C (122 to 176 °F)) environments such as hot springs — paved the way for dramatic improvements of the PCR method. The DNA polymerase isolated from T. aquaticus is stable at high temperatures remaining active even after DNA denaturation, thus obviating the need to add new DNA polymerase after each cycle. This allowed an automated thermocycler-based process for DNA amplification.

Patent Wars

The PCR technique was patented by Kary Mullis and assigned to Cetus Corporation, where Mullis worked when he invented the technique in 1983. The Taq polymerase enzyme was also covered by patents. There have been several high-profile lawsuits related to the technique, including an unsuccessful lawsuit brought by DuPont. The pharmaceutical company Hoffmann-La Roche purchased the rights to the patents in 1992 and currently holds those that are still protected.

A related patent battle over the Taq polymerase enzyme is still ongoing in several jurisdictions around the world between Roche and Promega. The legal arguments have extended beyond the lives of the original PCR and Taq polymerase patents, which expired on March 28, 2005.

NanoString Technologies

NanoString Technologies is a privately held life sciences company that develops and manufactures solutions for detecting and counting large sets of target molecules in biological samples. The company was founded by Dr. Krassen Dimitrov in 2003 and is based in Seattle, WA. The company's products are based on a digital molecular barcoding technology invented by Dr. Dimitrov at the Institute for Systems Biology (ISB) in Seattle in the lab of Dr. Leroy

Hood. This technology underlies the company's nCounter™ Analysis System which became commercially available worldwide in 2008. NanoString's products are used for a variety of discovery and translational research applications including gene expression, miRNA analysis, and copy number variation analysis. NanoString is also developing the technology for use in molecular diagnostics. The company's investors include Clarus Ventures, OVP Venture Partners and Draper Fisher Jurvetson.

The Technology

The Nanostring technology was invented by Dr. Krassen Dimitrov, and is protected by U.S. Patent 7,473,767. nCounter Analysis System utilizes a digital technology that is based on direct multiplexed measurement of gene expression that is capable of high level precision and sensitivity at less than 1 transcript copy per cell. The technology uses molecular "barcodes" and single-molecule imaging to detect and count hundreds of unique transcripts in a single reaction.

Each color-coded barcode is attached to a single target-specific probe corresponding to a gene of interest. Mixed together with controls, they form a multiplexed assay. The degree of multiplexing is in the hundreds, which is less than that of microarrays.

However, the nanostring technology has higher throughput, accuracy and sensitivity than microarrays, which makes it preferable for low-multiplex applications, such as biomarker validation or molecular diagnostics.

The nanostring process includes the following steps:

- Hybridization: NanoString's Technology employs two ~50 base probes per mRNA that hybridize in solution. The Reporter Probe carries the signal; the Capture Probe allows the complex to be immobilized for data collection.
- Purification and Immobilization: After hybridization, the excess probes are removed and the probe/target complexes are aligned and immobilized in the nCounter Cartridge.
- Data Collection: Sample Cartridges are placed in the Digital Analyzer instrument for data collection. Color codes on the surface of the cartridge are counted and tabulated for each target molecule.

The nCounter Analysis System primarily addresses the target validation and routine testing segments of the genomics research market in both academia and the pharmaceutical industry. Cancer research and biomarker validation are two of the areas where NanoString has seen the most rapid adoption of its nCounter Analysis System.

Products

NanoString's products include:

The nCounter Analysis System: The system consists of two instruments; the Prep Station which is an automated fluidic instrument that immobilizes CodeSet complexes for data collection and the Digital Analyzer which derives data by counting of fluorescent barcodes.

CodeSets: CodeSets are custom-made or pre-designed sets of color-coded probes pre-mixed with a set of system controls for use in a variety of basic research and translational medicine applications including gene expression, miRNA analysis, and copy number variation.

- Higher-plex techniques:
    - o SAGE

Serial analysis of gene expression (SAGE) is a technique used by molecular biologists to produce a snapshot of the messenger RNA population in a sample of interest in the form of small tags that correspond to fragments of those transcripts. The original technique was developed by Dr. Victor Velculescu at the Oncology Center of Johns Hopkins University and published in 1995.

Several variants have been developed since, most notably a more robust version, LongSAGE, RL-SAGE and the most recent SuperSAGE.

Many of these have improved the technique with the capture of longer tags, enabling more confident identification of a source gene.

Overview

Briefly, SAGE experiments proceed as follows:

1. Isolate the mRNA of an input sample (e.g. a tumour).
2. Extract a small chunk of sequence from a defined position of each mRNA molecule.
3. Link these small pieces of sequence together to form a long chain (or concatemer).
4. Clone these chains into a vector which can be taken up by bacteria.
5. Sequence these chains using modern high-throughput DNA sequencers.
6. Process this data with a computer to count the small sequence tags.

A more in-depth, technical explanation of the technique is available here.

Analysis

The output of SAGE is a list of short sequence tags and the number of times it is observed. Using sequence databases a researcher can usually determine, with some confidence, the original mRNA (and therefore which gene) the tag was extracted from. Statistical methods can be applied to tag and count lists from different samples in order to determine which genes are more highly expressed. For example, a normal tissue sample can be compared against a corresponding tumour to determine which genes tend to be more (or less) active.

Applications

Although SAGE was originally conceived for use in cancer studies, it has been successfully used to describe the transcriptome of other diseases and in a wide variety of organisms.

Comparison to DNA Microarrays

The general goal of the technique is similar to the DNA microarray. However, SAGE sampling is based on sequencing mRNA output, not on hybridization of mRNA output to probes, so transcription levels are measured more quantitativly than by microarray.

In addition, the mRNA sequences do not need to be known a priori, so genes or gene variants which are not known can be discovered. Microarray experiments are much cheaper to perform, so large-scale studies do not typically use SAGE. Quantifying gene expressions is more exact in SAGE because it involves directly counting the number of transcripts whereas spot intensities in microarrays fall in non-discrete gradients and are prone to background noise.

Variant Protocols: miRNA Cloning

MicroRNAs, or miRNAs for short, are small (~22nt) segments of RNA which have been found to play a crucial role in gene regulation. One of the most commonly used methods for cloning and identifying miRNAs within a cell or tissue was developed in the Bartel Lab and published in a paper by Lau et al. (2001). Since then, several variant protocols have arisen, but most have the same basic format. The procedure is quite similar to SAGE:

The small RNA are isolated, then linkers are added to each, and the RNA is converted to cDNA by RT-PCR. Following this, the linkers, containing internal restriction sites, are digested with the appropriate restriction enzyme and the sticky ends are ligated together into concatamers. Following concatenation, the fragments are ligated into plasmids and are used to transform bacteria to generate many copies of the plasmid containing the inserts. Those may then be sequenced to identify the miRNA present, as well as analysing expression levels of a given miRNA by counting the number of times it is present, similar to SAGE.

DNA Microarray

A DNA microarray (also commonly known as gene chip, DNA chip, or biochip) is a collection of microscopic DNA spots attached to a solid surface. Scientists use DNA microarrays to measure the expression levels of large numbers of genes simultaneously or to genotype multiple regions of a genome.

Each DNA spot contains picomoles (10 moles) of a specific DNA sequence, known as probes (or reporters).

These can be a short section of a gene or other DNA element that are used to hybridize a cDNA or cRNA sample (called target) under high-stringency conditions.

Probe-target hybridization is usually detected and quantified by detection of fluorophore-, silver-, or chemiluminescence-labeled targets to determine relative abundance of nucleic acid sequences in the target.

8

Genetic Recombination

Structure of the Holliday junction intermediate in genetic recombination. The four separate DNA strands are coloured red, blue, green and yellow.

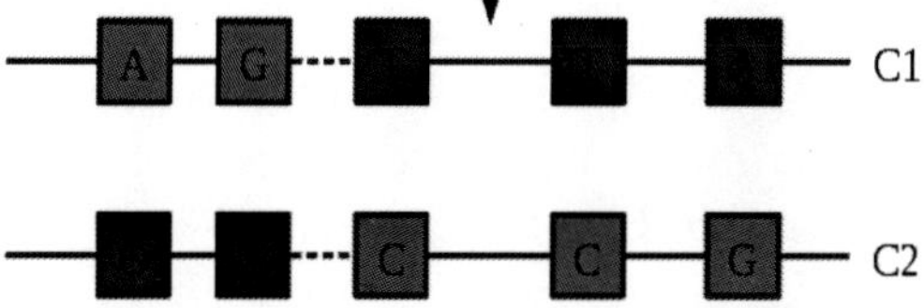

Figure: *Recombination involves the breakage and rejoining of two chromosomes (M and F) to produce two re-arranged chromosomes (C1 and C2).*

A DNA helix usually does not interact with other segments of DNA, and in human cells the different chromosomes even occupy separate areas in the nucleus called "chromosome territories". This physical separation of different chromosomes is important for the ability of DNA to function as a stable repository for information, as one of the few times chromosomes interact is during chromosomal crossover when they recombine. Chromosomal crossover is when two DNA helices break, swap a section and then rejoin.

Recombination allows chromosomes to exchange genetic information and produces new combinations of genes, which increases the efficiency of natural selection and can be important in the rapid evolution of new proteins. Genetic recombination can also be involved in DNA repair, particularly in the cell's response to double-strand breaks.

The most common form of chromosomal crossover is homologous recombination, where the two chromosomes involved share very similar sequences. Non-homologous recombination can be damaging to cells, as it can produce chromosomal translocations and genetic abnormalities. The recombination reaction is catalyzed by enzymes known as recombinases, such as RAD51.

The first step in recombination is a double-stranded break either caused by an endonuclease or damage to the DNA. A series of steps catalyzed in part by the recombinase then leads to joining of the two helices by at least one Holliday junction, in which a segment of a single strand in each helix is annealed to the complementary strand in the other helix. The Holliday junction is a tetrahedral junction structure that can be moved along the pair of chromosomes, swapping one strand for another. The recombination reaction is then halted by cleavage of the junction and re-ligation of the released DNA.

Evolution

DNA contains the genetic information that allows all modern living things to function, grow and reproduce. However, it is unclear how long in the 4-billion-year history of life DNA has performed this function, as it has been proposed that the earliest forms of life may have used RNA as their genetic material. RNA may have acted as the central part of early cell metabolism as it can both transmit genetic information and carry out catalysis as part of ribozymes.

This ancient RNA world where nucleic acid would have been used for both catalysis and genetics may have influenced the evolution of the current genetic code based on four nucleotide bases. This would occur, since the number of different bases in such an organism is a trade-off between a small number of bases increasing replication accuracy and a large number of bases increasing the catalytic efficiency of ribozymes.

However, there is no direct evidence of ancient genetic systems, as recovery of DNA from most fossils is impossible. This is because DNA will survive in the environment for less than one million years and slowly degrades into short fragments in solution. Claims for older DNA have been made, most notably a report of the isolation of a viable bacterium from a salt crystal 250 million years old, but these claims are controversial. On August 8, 2011, a report, based on NASA studies with meteorites found on Earth, was published suggesting building blocks of DNA (adenine, guanine and related organic molecules) may have been formed extraterrestrially in outer space.

Uses in Technology

Genetic Engineering

Methods have been developed to purify DNA from organisms, such as phenol-chloroform extraction, and to manipulate it in the laboratory, such as restriction digests and the polymerase chain reaction. Modern biology and biochemistry make intensive use of these techniques in recombinant DNA technology. Recombinant DNA is a man-made DNA sequence that has been assembled from other DNA sequences. They can be transformed into organisms in the form of plasmids or in the appropriate format, by using a viral vector. The genetically modified organisms produced can be used to produce products such as recombinant proteins, used in medical research, or be grown in agriculture.

Forensics

Forensic scientists can use DNA in blood, semen, skin, saliva or hair found at a crime scene to identify a matching DNA of an individual, such as a perpetrator. This process is formally termed DNA profiling, but may also be called "genetic fingerprinting". In DNA profiling, the lengths of variable sections of repetitive DNA, such as short tandem repeats and minisatellites, are compared between people. This method is usually an extremely reliable technique for identifying a matching DNA. However, identification can be complicated if the scene is contaminated with DNA from several people. DNA profiling was developed in 1984 by British geneticist Sir Alec Jeffreys, and first used in forensic science to convict Colin Pitchfork in the 1988 Enderby murders case.

People convicted of certain types of crimes may be required to provide a sample of DNA for a database. This has helped investigators solve old cases where only a DNA sample was obtained from the scene. DNA profiling can also be used to identify victims of mass casualty incidents. On the other hand, many convicted people have been released from prison on the basis of DNA techniques, which were not available when a crime had originally been committed.

Bioinformatics

Bioinformatics involves the manipulation, searching, and data mining of biological data, and this includes DNA sequence data. The development of techniques to store and search DNA sequences have led to widely applied advances in computer science, especially string searching algorithms, machine learning and database theory. String searching or matching algorithms, which find an occurrence of a sequence of letters inside a larger sequence of letters, were developed to search for specific sequences of nucleotides. The DNA sequenced may be aligned with other DNA sequences to identify homologous sequences and locate the specific mutations that make them distinct. These techniques, especially multiple sequence alignment, are used in studying phylogenetic relationships and protein function. Data sets representing entire genomes' worth of DNA sequences, such as those produced by the Human Genome Project, are difficult to use without the annotations that identify the locations of genes and regulatory elements on each chromosome. Regions of DNA sequence that have the characteristic patterns associated with protein- or RNA-coding genes can be identified by gene finding algorithms, which allow researchers to predict the presence of particular gene products and their possible functions in an organism even before they have been isolated experimentally. Entire genomes may also be compared which can shed light on the evolutionary history of particular organism and permit the examination of complex evolutionary events.

DNA Nanotechnology

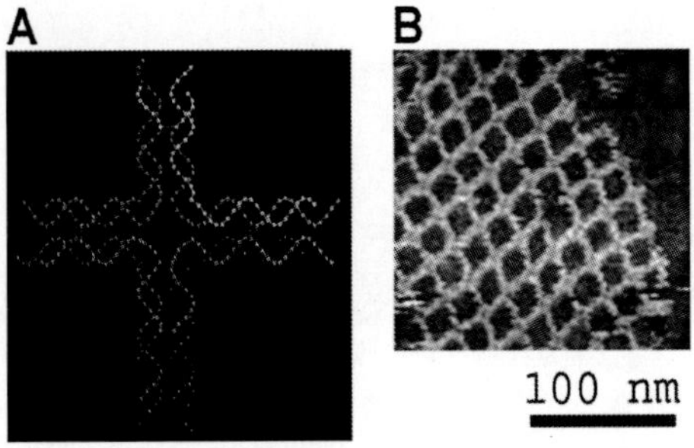

The DNA structure at left (schematic shown) will self-assemble into the structure visualized by atomic force microscopy at right. DNA nanotechnology is the field that seeks to design nanoscale structures using the molecular recognition properties of DNA molecules.

DNA nanotechnology uses the unique molecular recognition properties of DNA and other nucleic acids to create self-assembling branched DNA complexes with useful properties. DNA is thus used as a structural material rather than as a carrier of biological information. This has led to the creation of two-dimensional periodic lattices (both tile-based as well as using the "DNA origami" method) as well as three-dimensional structures in the shapes of polyhedra. Nanomechanical devices and algorithmic self-assembly have also been demonstrated, and these DNA structures have been used to template the arrangement of other molecules such as gold nanoparticles and streptavidin proteins. DNA nanotechnology

DNA nanotechnology is a branch of nanotechnology which uses the molecular recognition properties of DNA and other nucleic acids to create designed, artificial structures out of DNA for technological purposes. In this field, DNA is used as a structural material rather than as a carrier of genetic information, making it an example of bionanotechnology. DNA nanotechnology has applications in molecular self-assembly and in DNA computing.

Although DNA is usually considered in the context of molecular biology as the carrier of genetic information in living cells, DNA nanotechnology considers DNA solely as a chemical and as a material, and is usually pursued outside of any biological context. DNA nanotechnology makes use of the fact that, due to the specificity of Watson-Crick base pairing, only portions of the strands which are complementary to each other will bind to each other to form duplex DNA. DNA nanotechnology attempts to rationally design sets of DNA strands so that desired portions of each strand will assemble in the correct positions for some desired target structure, a process called nucleic acid design.

Although the field is usually called DNA nanotechnology, its principles apply equally well to other nucleic acids such as RNA and PNA, and structures incorporating these have been made. For this reason the field is occasionally referred to as nucleic acid nanotechnology.

Fundamental Concepts

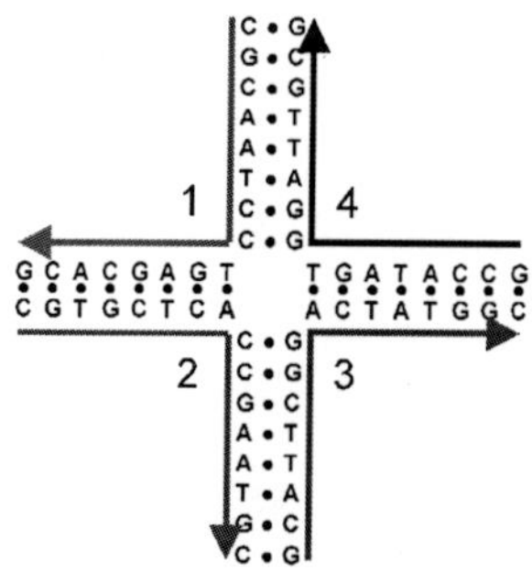

These four strands associate into a DNA four-arm junction because this structure maximizes the number of correct base pairs, with A's matched to T's and C's matched to G's. See this image for a more realistic model of the four-arm junction showing its tertiary structure.

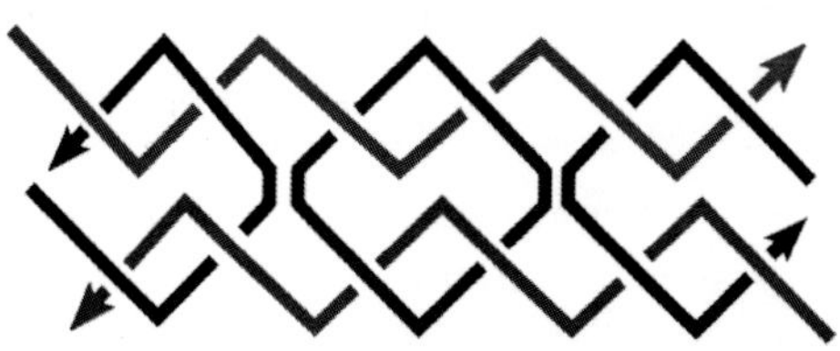

A double-crossover (DX) molecule. This molecule consists of five DNA single strands which form two double-helical domains, on the left and the right in this image. There are two crossover points where the strands cross from one domain into the other.

DNA nanotechnology creates complex structures out of nucleic acids by making use of the specificity of base pairing in nucleic acid molecules. The structure of a nucleic acid molecule consists of a sequence of nucleotides, distinguished by which nucleobase they contain. In DNA, the four bases used are adenine (A), cytosine (C), guanine (G), and thymine (T).

Nucleic acids have the property that two molecules will bind to each other to form a double helix only if the two sequences are complementary, meaning that they form matching sequences of base pairs, with A's only binding to T's, and C's only to G's. Because the formation of correctly matched base pairs is energetically favorable, nucleic acid strands are expected in most cases to bind to each other in the conformation that maximizes the number of correctly paired bases. This property, that the sequence determines the pattern of binding and the overall structure, is used by the field of DNA nanotechnology in that sequences are artificially designed so that a desired structure is favored to form.

Nearly all structures in DNA nanotechnology make use of branched DNA structures containing junctions, as opposed to most biological DNA which exists in a linear double helix form. One of the simplest branched structures, and the first made, is a four-arm junction which can be made using four individual DNA strands which are complementary to each other in the correct pattern. Unlike in natural Holliday junctions, in the artificial immobile four-arm junction shown below, the base sequence of each arm is different, meaning that the junction point is fixed in a certain position.

Junctions can be used in more complex molecules. One of the more widely used of these is the "double-crossover" or DX motif. A DX molecule can be thought of as two DNA duplexes positioned parallel to each other, with two crossover points where strands cross from one duplex into the other. Each junction point is itself topologically a four-arm junction. This molecule has the advantage that the junction points are now constrained to a single orientation as opposed to being flexible as in the four-arm junction. This makes the DX motif suitable as a structural building block for larger DNA complexes.

Design

DNA nanostructures must be rationally designed so that the individual nucleic acid strands will assemble into the desired structures. The design process of such nanostructures usually begins with the specification of a desired target structure and/or functionality. Then, the overall secondary structure of the target molecule is designed, meaning the arrangement of nucleic acid strands within the structure, and which portions of those strands should be bound to each other. The last step is the primary structure design, the specification of the actual base sequences of each nucleic acid strand.

Structural Design

The first step in designing a nucleic acid nanostructure is to decide how a given structure should be represented by a specific arrangement of nucleic acid strands. This design step thus determines the secondary structure, or the series of base pairs which hold the individual strands together in the desired shape. There are several approaches which have been demonstrated:

- Tile-based structures. This approach breaks the target structure into smaller units with strong binding between the strands contained in each unit, and relatively weaker interactions between the units. It is often used to make periodic lattices, but can also be used to implement algorithmic self-assembly, making them one platform for DNA computing.
- Folding structures. An alternative to the tile-based approach, folding approaches make the nanostructure out of a single long strand. This long strand can either have a designed sequence which folds due to its interactions with itself, or it can be folded into the desired shape by using shorter, "staple" strands. This latter method is called DNA origami, which allows the creation of two- and three-dimensional shapes at the nanoscale using DNA.
- Kinetic assembly. Recently, there has been interest in controlling the kinetics of DNA self-assembly, so that transient dynamics can also be programmed into the assembly. Such a method also has the advantage of proceeding isothermally and thus not requiring a thermal annealing step required by solely thermodynamic approaches.

Sequence Design

After any of the above approaches are used to design the secondary structure of a target molecule, an actual sequence of nucleotides must be devised which will form into the desired structure. Nucleic acid design is the process of assigning a specific nucleic acid base sequence to each strand so that they will associate into a desired conformation. Nucleic acid design is central to the field of DNA nanotechnology. Most methods seek to designing sequences so that the target structure is a thermodynamic minimum, and mis-assembled structures have higher energies and are thus disfavored.

This is done either through heuristic methods such as sequence symmetry minimization and coding theory based approaches, or by explicitly using a full nearest-neighbor thermodynamic model. Geometric models are also used to examine tertiary structure of the nanostructures and ensure that the complexes are not overly strained. Nucleic acid design has similar goals to protein design: in both, the sequence of monomers is designed to favor the desired folded or associated structure and to disfavor alternate structures. Nucleic acid design has the advantage of being a much computationally simpler problem, since the simplicity of Watson-Crick base pairing rules leads to simple heuristic methods which yield experimentally robust designs. However, nucleic acid structures are less versatile than proteins in their functionality.

Structural DNA Nanotechnology

Structural DNA nanotechnology, sometimes abbreviated as SDN, focuses on synthesizing and characterizing nucleic acid complexes and materials with various nanoscale structures. Structural DNA nanotechnology is largely based on the fact that the three-dimensional structure of DNA—the nucleic acid double helix— has a robust, defined geometry which makes it possible to predict and design the structures of more complex DNA molecules. Many such structures have been created, including two- and three-dimensional structures; and periodic, aperiodic, and discrete structures.

Periodic Lattices

Smaller nucleic acid assemblies can be equipped with sticky ends in order to combine them into a two-dimensional periodic lattice. The earliest example of this was the array of DX, or double-crossover, molecules. Each DX molecule can be designed with four sticky ends, one at each end of the two double-helical domains, and these sticky ends can be designed with sequences that cause the DX units to combine into a specific tessellated pattern. They thus form extended flat sheets which are essentially rigid two-dimensional crystals of DNA. Two-dimensional arrays have been made out of other motifs as well, including the Holliday junction rhombus array as well as various DX-based arrays making use of a double-cohesion scheme.

Creating three-dimensional lattices out of DNA was the earliest goal of DNA nanotechnology, but proved to be one of the most difficult to realize. Success in constructing three-dimensional DNA lattices was finally reported in 2009 using a motif based on the concept of tensegrity, a balance between tension and compression forces.

Nanotubes

In addition to flat sheets, DX arrays have been made to form hollow nanotubes of 4–20 nm diameter. These DNA nanotubes are somewhat similar in size and shape to carbon nanotubes, but the carbon nanotubes are stronger and better conductors, whereas the DNA nanotubes are more easily modified and connected to other structures. There have been multiple schemes for constructing DNA nanotubes, one of which uses the inherent curvature of DX tiles to form a DX lattice to curl around itself and close into a tube. An alternative design uses single-stranded “tiles” for which the rigidity of the tube is an emergent property. This method also has the benefit of being able to determine the circumference of the nanotube in a simple, modular fashion.

Polyhedra

A number of three-dimensional DNA molecules have been made which have the connectivity of a polyhedron such as an octahedron or cube. In other words, the DNA duplexes trace the edges of a polyhedron with a DNA junction at each vertex. The earliest demonstrations of DNA polyhedra involved multiple ligations and solid-phase synthesis steps to create catenated polyhedra. More recent work has yielded polyhedra whose synthesis is much easier. These include a DNA octahedron made from a long single strand designed to fold into the correct conformation, as well as a tetrahedron which can be produced from four DNA strands in a single step, pictured at the top of this article.

Arbitrary Shapes

Nanostructures of arbitrary shapes are usually made using the DNA origami method. DNA origami makes use of a long natural virus strand as a "scaffold" strand, and computationally designs shorter "staple" strands which bind to portions of the scaffold strand and force it to fold into the desired shape. This method has the advantage of being easy to design, as the base sequence is predetermined by the scaffold strand sequence, and it also does not require high strand purity and accurate stoichiometry, as most other DNA nanotechnology methods do. DNA origami was first demonstrated for two-dimensional shapes; demonstrated designs included the smiley face and a coarse map of North America. This was later extended to solid three-dimensional shapes.

In addition, structures have been constructed with two-dimensional faces which fold into an overall three-dimensional shape, akin to a cardboard box. These can be programmed to open and release their cargo in response to a stimulus, making them potentially useful as programmable molecular cages.

Functional Nucleic Acid Nanostructures

DNA nanotechnology focuses on creating molecules with designed functionalities as well as structures. These include both dynamic functionality within the nucleic acid structure itself, for example with computation and mechanical motion, as well as by including other components such as small molecules or nanoparticles which have their own functionalities. Many classes of functional systems have been demonstrated.

Nanoarchitecture

The idea of using DNA arrays to template the assembly of other functional molecules was first suggested by Nadrian Seeman in 1987, but only recently has progress been made in reducing these kinds of schemes to practice. In 2006, researchers covalently attached gold nanoparticles to a DX-based tile and showed that self-assembly of the DNA structures also assembled the nanoparticles hosted on them. Also that year, Dwyer and LaBean demonstrated the letters "D" "N" and "A" created on a 4x4 DX array using streptavidin, and a hierarchical assembly based on this approach was also demonstrated that scales to larger arrays (8X8 and 8.96 MD). A non-covalent hosting scheme was shown in 2007, using Dervan polyamides on a DX array to arrange streptavidin proteins on specific kinds of tiles on the DNA array. There has also been interest in using DNA nanotechnology to assemble molecular electronics devices. To this end, DNA has been used to assemble single walled carbon nanotubes into field-effect transistors.

Algorithmic Self-assembly

Figure: *The Sierpinski gasket fractal.*

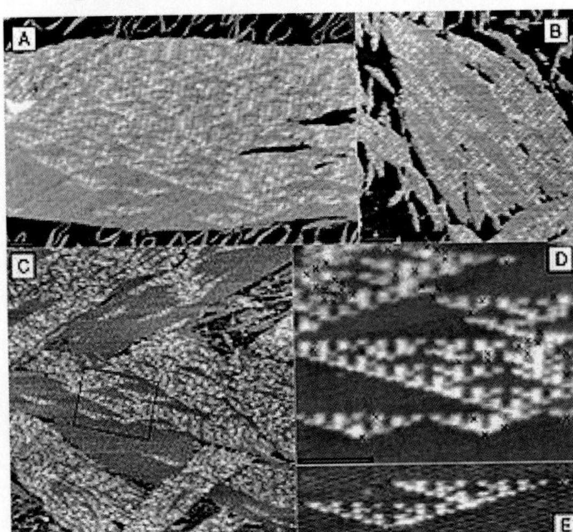

Figure: *DNA arrays that display a representation of the Sierpinski gasket on their surfaces.*

DNA nanotechnology has been applied to the related field of DNA computing. The DX tiles can have their sticky end sequences chosen so that they act as Wang tiles, allowing them to perform computation. A DX array has been demonstrated whose assembly encodes an XOR operation; this allows the DNA array to implement a cellular automaton which generates a fractal called the Sierpinski gasket. Another system has the function of a binary counter, displaying a representation of increasing binary numbers as it grows. These results show that computation can be incorporated into the assembly of DNA arrays, increasing its scope beyond simple periodic arrays.

Note that DNA computing overlaps with, but is distinct from, DNA nanotechnology. The latter uses the specificity of Watson-Crick basepairing to make novel structures out of DNA. These structures can be used for DNA computing, but they do not have to be. Additionally, DNA computing can be realized without using the types of molecules made possible by DNA nanotechnology.

Nanomechanical Devices

DNA complexes have been made which change their conformation upon some stimulus. These are intended to have applications in nanorobotics. DNA machines have also been made which show a twisting motion. The first such device made use of the transition between the B-DNA and Z-DNA forms to respond to a change in buffer conditions. This reliance on buffer conditions, however, caused all devices to change state at the same time. A subsequent system, called "molecular tweezers," changes from an open to a closed state based upon the presence of control strands, allowing multiple devices to be individually operated in solution. This was followed up by another system which relies on the presence of control strands to switch from a paranemic-crossover (PX) conformation to a double-junction (JX2) conformation.

Nucleic acid nanomachines have been made which exhibit directional motion along a linear track, called DNA walkers. A large number of schemes have been demonstrated. One strategy is to control the motion of the walker along the track using control strands which need to be manually added in sequence. Another approach is to make use of restriction enzymes or deoxyribozymes to cleave the strands and cause the walker to move forward, which has the advantage of running autonomously. A later system extended the concept of DNA walkers to walk upon a two-dimensional surface rather than a linear track, and demonstrated the ability to selectively pick up and move molecular cargo. Additionally, a linear walker has been demonstrated which performs DNA-templated synthesis as the walker advances along the track, allowing autonomous multistep chemical synthesis directed by the walker.

Applications

DNA nanotechnology provides one of the only ways to form designed, complex structures with precise control over nanoscale features. The field is beginning to see application to solve basic science problems in structural biology and biophysics. One such application, the earliest envisaged for the field, is in crystallography, where molecules which are hard to crystallize by themselves could be arranged and oriented within a three-dimensional nucleic acid lattice, thus allowing determination of their structure. DNA origami rods have also been used to replace liquid crystals in residual dipolar coupling experiments in protein NMR spectroscopy; using DNA origami is advantageous because, unlike liquid crystals, they are tolerant of the detergents needed to suspend membrane proteins in solution. It has been demonstrated that DNA walkers can be used as nanoscale assembly lines to move nanoparticles and direct chemical synthesis. Furthermore, DNA origami structures have aided in the biophysical studies of enzyme function and protein folding.

DNA nanotechnology is also moving towards potential real-world applications. It has been suggested that the ability of nucleic acid arrays to arrange other molecules has potential applications in molecular scale electronics, with the assembly of a nucleic acid lattice templating the assembly of molecular electronic elements such as molecular wires. DNA nanotechnology has also been called a form of programmable matter because of the coupling of computation to its material properties.

There are also potential applications for DNA nanotechnology in nanomedicine, making use of its ability to perform computation in a biocompatible format to make "smart drugs" for targeted drug delivery. One such system being investigated uses a hollow DNA box containing proteins which induce apoptosis, or cell death, which will only open when in proximity to a cancer cell.

Materials and Methods

The sequences of the individual DNA strands which make up the target structure are designed computationally, using molecular modeling and thermodynamic modeling software. Once the sequences have been designed, the nucleic acid molecules themselves are synthesized through standard oligonucleotide synthesis methods. This process is usually automated by using a machine called an oligonucleotide synthesizer, and nucleic acids of custom sequence are commercially available from many vendors. For methods which require pure strands of known concentration, the nucleic acid strands can be purified using denaturing gel electrophoresis, and concentrations are determined by one of several nucleic acid quantitation methods using ultraviolet absorbance spectroscopy.

The fully formed target structures are usually characterized by native gel electrophoresis, which provides information about the size and shape of DNA molecules. An electrophoretic mobility shift assay can indicate whether a stucture incorporates all the individual desired strands. Fluorescent labeling and Förster resonance energy transfer (FRET) are also used to characterize the structure of the molecules.

Nucleic acid structures can be directly imaged by atomic force microscopy, which is well-suited to extended two-dimensional structures, but is less useful for discrete three-dimensional structures due to the interaction of the microscope tip with the fragile nucleic acid structure. For these latter structures transmission electron microscopy and cryo-electron microscopy are important methods. Extended three-dimensional lattices are analyzed by X-ray crystallography.

History

The concept of DNA nanotechnology was invented by Nadrian Seeman in the early 1980s. Seeman was originally concerned with using a three-dimensional DNA lattice to orient target molecules, which would simplify their crystallographic study by eliminating the difficult process of obtaining pure crystals. This idea had reportedly come to him in fall 1980, after realizing the similarity between the M. C. Escher woodcut Depth and an array of DNA six-arm junctions.

To this end, Seeman's laboratory in 1991 published the synthesis of a cube made of DNA, the first three-dimensional nanoscale object, for which he received the 1995 Feynman Prize in Nanotechnology, which was followed by a DNA truncated octahedron. However, it soon became clear that these molecules, polygonal shapes with flexible junctions as their vertices, were not rigid enough to form extended three-dimensional lattices. Seeman developed the more rigid double-crossover (DX) motif, and in collaboration with Erik Winfree, in 1998 published the creation of two-dimensional lattices of DX tiles. The synthesis of a three-dimensional lattice was finally published by Seeman in 2009, nearly thirty years after he had set out to do so.

These tile-based structures had the advantage that they provided the capability to implement DNA computing, which was demonstrated by Winfree and Paul Rothemund in their 2004 paper on the algorithmic self-assembly of a Sierpinski gasket structure, and for which they shared the 2006 Feynman Prize in Nanotechnology. Winfree's key insight was that the DX tiles could be used as Wang tiles, meaning that their assembly was capable of performing computation. The first DNA nanomachine—a motif which changes its structure in response to an input—was demonstrated in 1999 by Seeman. An improved system was demonstrated by Bernard Yurke the following year, which was also the first nucleic acid device to make use of toehold-mediated strand displacement. The next advance was to translate this into mechanical motion, and in 2004 and 2005, a number of DNA walker systems were demonstrated by the groups of Seeman, Niles Pierce, Andrew Turberfield, and Chengde Mao.

In 2006, Rothemund first demonstrated the new DNA origami technique for easily and robustly creating folded DNA molecules of any shape. Rothemund had conceived of this method as being conceptually intermediate between Seeman's DX lattices, which used many short strands, and William Shih's DNA octahedron, which consisted mostly of one very long strand; Rothemund's DNA origami contains a long strand whose folding is assisted by a number of short strands. This method allowed the creation of much larger structures than were previously possible, which are less technically demanding to design and synthesize. DNA origami was the cover story of Nature on March 15, 2006. Rothemund's research demonstrating two-dimensional DNA origami structures was followed by the demonstration of solid three-dimensional DNA origami by Shih in 2009, while the labs of Jørgen Kjems and Hao Yan demonstrated hollow three-dimensional structures made out of two-dimensional faces.

DNA nanotechnology as a field was initially met with some skepticism due to the unusual non-biological use of nucleic acids as a material for building structures and doing computation, as well as the preponderance of proof of principle experiments which extended the capabilities of the field but were far from actual applications. Seeman's seminal 1991 paper on the synthesis of the DNA cube was rejected by the journal Science after one reviewer praised its originality while another criticized it for its lack of biological relevance. By the early 2010s, however, the field was considered to have increased its capabilities to the point that applications for basic science research were beginning to be realized, and practical applications in medicine and other fields were beginning to be considered feasible. In addition, the field had grown from very few active laboratories in 2001, to at least 60 in 2010, which increased the talent pool and thus the number of advancements in the field during that decade.

History and Anthropology

Because DNA collects mutations over time, which are then inherited, it contains historical information, and, by comparing DNA sequences, geneticists can infer the evolutionary history of organisms, their phylogeny. This field of phylogenetics is a powerful tool in evolutionary biology. If DNA sequences within a species are compared, population geneticists can learn the history of particular populations. This can be used in studies ranging from ecological genetics to anthropology; For example, DNA evidence is being used to try to identify the Ten Lost Tribes of Israel.

DNA has also been used to look at modern family relationships, such as establishing family relationships between the descendants of Sally Hemings and Thomas Jefferson. This usage is closely related to the use of DNA in criminal investigations detailed above. Indeed, some criminal investigations have been solved when DNA from crime scenes has matched relatives of the guilty individual.

History of DNA Research

DNA was first isolated by the Swiss physician Friedrich Miescher who, in 1869, discovered a microscopic substance in the pus of discarded surgical bandages. As it resided in the nuclei of cells, he called it "nuclein". In 1878, Albrecht Kossel isolated the non-protein component of "nuclein", nucleic acid, and later isolated its five primary nucleobases. In 1919, Phoebus Levene identified the base, sugar and phosphate nucleotide unit. Levene suggested that DNA consisted of a string of nucleotide units linked together through the phosphate groups. However, Levene thought the chain was short and the bases repeated in a fixed order. In 1937 William Astbury produced the first X-ray diffraction patterns that showed that DNA had a regular structure.

In 1927 Nikolai Koltsov proposed that inherited traits would be inherited via a "giant hereditary molecule" which would be made up of "two mirror strands that would replicate in a semi-conservative fashion using each strand as a template". In 1928, Frederick Griffith discovered that traits of the "smooth" form of the Pneumococcus could be transferred to the "rough" form of the same bacteria by mixing killed "smooth" bacteria with the live "rough" form. This system provided the first clear suggestion that DNA carries genetic information—the Avery–MacLeod–McCarty experiment—when Oswald Avery, along with coworkers Colin MacLeod and Maclyn McCarty, identified DNA as the transforming principle in 1943. DNA's role in heredity was confirmed in 1952, when Alfred Hershey and Martha Chase in the Hershey–Chase experiment showed that DNA is the genetic material of the T2 phage.

In 1953, James D. Watson and Francis Crick suggested what is now accepted as the first correct double-helix model of DNA structure in the journal Nature. Their double-helix, molecular model of DNA was then based on a single X-ray diffraction image (labeled as "Photo 51") taken by Rosalind Franklin and Raymond Gosling in May 1952, as well as the information that the DNA bases are paired — also obtained through private communications from Erwin Chargaff in the previous years. Chargaff's rules played a very important role in establishing double-helix configurations for B-DNA as well as A-DNA.

Experimental evidence supporting the Watson and Crick model were published in a series of five articles in the same issue of Nature. Of these,

Franklin and Gosling's paper was the first publication of their own X-ray diffraction data and original analysis method that partially supported the Watson and Crick model; this issue also contained an article on DNA structure by Maurice Wilkins and two of his colleagues, whose analysis and in vivo B-DNA X-ray patterns also supported the presence in vivo of the double-helical DNA configurations as proposed by Crick and Watson for their double-helix molecular model of DNA in the previous two pages of Nature. In 1962, after Franklin's death, Watson, Crick, and Wilkins jointly received the Nobel Prize in Physiology or Medicine. However, Nobel rules of the time allowed only living recipients, but a vigorous debate continues on who should receive credit for the discovery.

In an influential presentation in 1957, Crick laid out the central dogma of molecular biology, which foretold the relationship between DNA, RNA, and proteins, and articulated the "adaptor hypothesis". Final confirmation of the replication mechanism that was implied by the double-helical structure followed in 1958 through the Meselson–Stahl experiment. Further work by Crick and coworkers showed that the genetic code was based on non-overlapping triplets of bases, called codons, allowing Har Gobind Khorana, Robert W. Holley and Marshall Warren Nirenberg to decipher the genetic code. These findings represent the birth of molecular biology.

Crystallography

Crystallography is the experimental science of the arrangement of atoms in solids. The word "crystallography" derives from the Greek words crystallon = cold drop / frozen drop, with its meaning extending to all solids with some degree of transparency, and grapho = write.

Before the development of X-ray diffraction crystallography, the study of crystals was based on their geometry. This involves measuring the angles of crystal faces relative to theoretical reference axes (crystallographic axes), and establishing the symmetry of the crystal in question. The former is carried out using a goniometer. The position in 3D space of each crystal face is plotted on a stereographic net, e.g. Wulff net or Lambert net. In fact, the pole to each face is plotted on the net. Each point is labelled with its Miller index. The final plot allows the symmetry of the crystal to be established.

Crystallographic methods now depend on the analysis of the diffraction patterns of a sample targeted by a beam of some type. Although X-rays are most commonly used, the beam is not always electromagnetic radiation. For some purposes electrons or neutrons are used. This is facilitated by the wave properties of the particles. Crystallographers often explicitly state the type of illumination used when referring to a method, as with the terms X-ray diffraction, neutron diffraction and electron diffraction.

These three types of radiation interact with the specimen in different ways. X-rays interact with the spatial distribution of the valence electrons, while electrons are charged particles and therefore feel the total charge distribution of both the atomic nuclei and the surrounding electrons. Neutrons are scattered by the atomic nuclei through the strong nuclear forces, but in addition, the magnetic moment of neutrons is non-zero. They are therefore also scattered by magnetic fields. When neutrons are scattered from hydrogen-containing materials, they produce diffraction patterns with high noise levels. However, the material can sometimes be treated to substitute hydrogen for deuterium. Because of these different forms of interaction, the three types of radiation are suitable for different crystallographic studies.

Theory

Generally, an image of a small object is made using a lens to focus the illuminating radiation, as is done with the rays of the visible spectrum in light microscopy. However, the wavelength of visible light (about 4000 to 7000 angstroms) is three orders of magnitude longer than the length of typical atomic bonds and atoms themselves (about 1 to 2 angstroms).

Therefore, obtaining information about the spatial arrangement of atoms requires the use of radiation with shorter wavelengths, such as X-rays or neutrons beam. Employing shorter wavelengths implied abandoning microscopy and true imaging, however, because there exists no material from which a lens capable of focusing this type of radiation can be created. (That said, scientists have had some success focusing X-rays with microscopic Fresnel zone plates made from gold, and by critical-angle reflection inside long tapered capillaries.) Diffracted x-ray or neutrons beams cannot be focused to produce images, so the sample structure must be reconstructed from the diffraction pattern.

Sharp features in the diffraction pattern arise from periodic, repeating structure in the sample, which are often very strong due to coherent reflection of many photons from many regularly spaced instances of similar structure, while non-periodic components of the structure result in diffuse (and usually weak) diffraction features. Said more simply, areas with a higher density and repetition of atom order tend to reflect more light toward one point in space when compared to those areas with fewer atoms and less repetition.

Because of their highly ordered and repetitive structure, crystals give diffraction patterns of sharp Bragg reflection spots, and are ideal for analyzing the structure of solids.

Notation

- Coordinates in square brackets such as [100] denote a direction vector (in real space).

- Coordinates in angle brackets or chevrons such as <100> denote a family of directions which are related by symmetry operations. In the cubic crystal system for example, <100> would mean [100], [010], [001] or the negative of any of those directions.
- Miller indices in parentheses such as (100) denote a plane of the crystal structure, and regular repetitions of that plane with a particular spacing. In the cubic system, the normal to the (hkl) plane is the direction [hkl], but in lower-symmetry cases, the normal to (hkl) is not parallel to [hkl].
- Indices in curly brackets or braces such as {100} denote a family of planes and their normals which are equivalent in cubic materials due to symmetry operations, much the way angle brackets denote a family of directions. In non-cubic materials, <hkl> is not necessarily perpendicular to {hkl}.

Technique

Some materials studied using crystallography, proteins for example, do not occur naturally as crystals. Typically, such molecules are placed in solution and allowed to crystallize over days, weeks, or months through vapor diffusion. A drop of solution containing the molecule, buffer, and precipitants is sealed in a container with a reservoir containing a hygroscopic solution. Water in the drop diffuses to the reservoir, slowly increasing the concentration and allowing a crystal to form. If the concentration were to rise more quickly, the molecule would simply precipitate out of solution, resulting in disorderly granules rather than an orderly and hence usable crystal.

Once a crystal is obtained, data can be collected using a beam of radiation. Although many universities that engage in crystallographic research have their own X-ray producing equipment, synchrotrons are often used as X-ray sources, because of the purer and more complete patterns such sources can generate. Synchrotron sources also have a much higher intensity of X-ray beams, so data collection takes a fraction of the time normally necessary at weaker sources. Complementary neutron crystallography techniques are used to enhance hydrogen atoms positions. Such techniques are available in Neutron facilities.

Producing an image from a diffraction pattern requires sophisticated mathematics and often an iterative process of modelling and refinement. In this process, the mathematically predicted diffraction patterns of an hypothesized or "model" structure are compared to the actual pattern generated by the crystalline sample. Ideally, researchers make several initial guesses, which through refinement all converge on the same answer. Models are refined until their predicted patterns match to as great a degree as can be achieved without radical revision of the model. This is a painstaking process, made much easier today by computers.

The mathematical methods for the analysis of diffraction data only apply to patterns, which in turn result only when waves diffract from orderly arrays. Hence crystallography applies for the most part only to crystals, or to molecules which can be coaxed to crystallize for the sake of measurement. In spite of this, a certain amount of molecular information can be deduced from the patterns that are generated by fibers and powders, which while not as perfect as a solid crystal, may exhibit a degree of order.

This level of order can be sufficient to deduce the structure of simple molecules, or to determine the coarse features of more complicated molecules. For example, the double-helical structure of DNA was deduced from an X-ray diffraction pattern that had been generated by a fibrous sample.

Crystallography in Materials Engineering

Crystallography is a tool that is often employed by materials scientists. In single crystals, the effects of the crystalline arrangement of atoms is often easy to see macroscopically, because the natural shapes of crystals reflect the atomic structure. In addition, physical properties are often controlled by crystalline defects. The understanding of crystal structures is an important prerequisite for understanding crystallographic defects. Mostly, materials do not occur in a single crystalline, but poly-crystalline form, such that the powder diffraction method plays a most important role in structural determination.

A number of other physical properties are linked to crystallography. For example, the minerals in clay form small, flat, platelike structures. Clay can be easily deformed because the platelike particles can slip along each other in the plane of the plates, yet remain strongly connected in the direction perpendicular to the plates. Such mechanisms can be studied by crystallographic texture measurements.

In another example, iron transforms from a body-centered cubic (bcc) structure to a face-centered cubic (fcc) structure called austenite when it is heated. The fcc structure is a close-packed structure, and the bcc structure is not, which explains why the volume of the iron decreases when this transformation occurs.

Crystallography is useful in phase identification. When performing any process on a material, it may be desired to find out what compounds and what phases are present in the material. Each phase has a characteristic arrangement of atoms. Techniques like X-ray or neutron diffraction can be used to identify which patterns are present in the material, and thus which compounds are present. Crystallography covers the enumeration of the symmetry patterns which can be formed by atoms in a crystal and for this reason has a relation to group theory and geometry.

Biology

X-ray crystallography is the primary method for determining the molecular conformations of biological macromolecules, particularly protein and nucleic acids such as DNA and RNA. In fact, the double-helical structure of DNA was deduced from crystallographic data. The first crystal structure of a macromolecule was solved in 1958 A three-dimensional model of the myoglobin molecule obtained by X-ray analysis. The Protein Data Bank (PDB) is a freely accessible repository for the structures of proteins and other biological macromolecules. Computer programs like RasMol or Pymol can be used to visualize biological molecular structures. Neutron crystallography is often used to help refine structures obtained by x-ray methods or to solve a specific bond; the methods are often viewed as complementary, as x-rays are sensitive to electron positions and scatter most strongly off heavy atoms, while neutrons are sensitive to nucleus positions and scatter strongly off many light isotopes, including hydrogen and deuterium. Electron crystallography has been used to determine some protein structures, most notably membrane proteins and viral capsids.

DNA Error

The Singularity is Near

The Singularity Is Near: When Humans Transcend Biology is a 2005 update of Raymond Kurzweil's 1999 book, The Age of Spiritual Machines and his 1990 book The Age of Intelligent Machines. In it, as in the two previous versions, Kurzweil attempts to give a glimpse of what awaits us in the near future. He proposes a coming technological singularity, and how we would thus be able to augment our bodies and minds with technology. He describes the singularity as resulting from a combination of three important technologies of the 21st century: genetics, nanotechnology, and robotics (including artificial intelligence).

Four central postulates of the book are as follows:

1. A technological-evolutionary point known as "the singularity" exists as an achievable goal for humanity.
2. Through a law of accelerating returns, technology is progressing toward the singularity at an exponential rate.
3. The functionality of the human brain is quantifiable in terms of technology that we can build in the near future.

4. Medical advancements make it possible for a significant number of his generation (Baby Boomers) to live long enough for the exponential growth of technology to intersect and surpass the processing of the human brain.

Kurzweil's speculative reasoning and selective use of growth indicators has been heavily debated and challenged. In response to this, in the last chapter he gives responses to some of the criticisms he has received.

Postulates

Acceptance and Striving for the Idea of Living Forever

Kurzweil asserts that a technological-evolutionary jump known as "the singularity" will be recognized as an achievable goal by humanity. Kurzweil acknowledges that belief in the singularity promotes the paradigm shift necessary for its advancement. By promoting the "truth" of its coming through predictions that seem remarkable at the time but inevitable after the fact (a global computer network, a computer beating the chess champion, etc.), Kurzweil's popular series of books reinforces the belief that a singularity is unavoidable.

The Law of Accelerating Returns

Kurzweil asserts in his Law of Accelerating Returns that technology is progressing toward the Singularity at an exponential rate, relying almost entirely on empirical data. He expands on Moore's Law with models showing that not only the return, but the rate of return is increasing exponentially.

An Objective Measurement of Cerebral Processing Power

Kurzweil asserts that the functionality of the brain is quantifiable in terms of technology that we can build in the near future. Kurzweil's earlier books showed cerebral processing power as primarily the number of computations in a square inch multiplied by the area of the brain. In this update, however, he acknowledges the possibility of Penrose-Hameroff Microtubule quantum processing (Orch-OR) and states that if his calculations of the processing capability of the brain are off by a factor of a billion, the double-exponential growth of technology will still catch up to it twenty-four years after his original projections. The Orch-OR theory is generally discredited among neuroscientists.

In a rebuttal paper, Hameroff asserts that the quantum processing power required for consciousness is at an order of magnitude greater than can be expressed through conventional systems of processing measurement. This argument seems to ignore Kurzweil's premise that accelerating returns in development of present technologies could, in a nominal period, overcome such a barrier. Additionally, other technologies could emerge which greatly lower the time required to reach the Singularity. A notable example would be more advanced forms of quantum computing capable of full neural emulation, which on a large enough scale, could function equivalently to the biological human brain.

Sufficient Medical Advancements

Kurzweil asserts that medical advancements will keep his generation alive long enough for the exponential growth of technology to intersect and surpass the processing of the human brain. Kurzweil explains how nanobots will eventually be able to repair and replace any part of the body that wears out, but relies on other methods of medical technology to prolong our lives long enough to reach the singularity. The usefulness of this medical postulate then becomes a function of how long it will take to reach the singularity, something that has been thrown into question due to the possibility of quantum brain processing explored in many recent books by scientists, such as Roger Penrose's The Emperor's New Mind.

DNA Errors

According to Ray Kurzweil, DNA errors (in the form of either pathogens, viruses, and/or cancer cells) are the causes of most major illnesses. These illnesses range from cancer to Alzheimer's disease and even Down syndrome. 130 different nanotechnology-related devices were invented in 2006 that could possibly eliminate the DNA errors that cause cancer someday. A majority of the statements about nanobots eliminating DNA errors and pathogens were emphasized frequently in Kurzweil's futurist book The Singularity Is Near; published in 2005 by an American publishing company.

The only permanent way to get rid of the DNA errors (that are a part of cancer cells and/or the various pathogens found in other illnesses) is by injecting nanobots (ideally multifunctional nanoparticles) into the patient's body. The high levels of SP2 protein that are found in cancer cells would be neutralized. The same nanobots that would lower the levels of SP2 protein inside the patient's stem cells could also repair the long-term damage to the person's DNA strand so that his/her descendants are less likely to acquire cancer and/other diseases in the first place.

All methods of death that are not caused by accidents are caused by a long line of DNA errors that are never corrected over a very long period of time. However, this method may not be possible or even affordable until the year 2025 (the Kanzius RF Therapy is scheduled to begin testing in 2012 but will be extremely difficult to acquire for many years). All research about curing cancer with nanorobotics has been currently restricted to small animals like mice. Thanks to the cost-performance ratio, nanomedicine is becoming more affordable for the average person year by year. It may become possible to have extremely effective and extremely inexpensive treatments for AIDS that will eliminate the DNA errors that are causing the disease through nanomedicine.

Even with the developed countries of the world competing against each other to be the first to create a sustainable nanorobotics program that can cure cancers and other diseases caused by DNA errors, government oversight (in addition to further oversight by the United Nations) will allow the DNA error-eliminating procedures to become increasingly affordable to everyone.

Organization

- Prologue: The Power of Ideas
- Chapter 1: The Six Epochs
- Chapter 2: A Theory of Technology Evolution
- Chapter 3: Achieving the Computational Capacity of the Human Brain
- Chapter 4: Achieving the Software of Human Intelligence
- Chapter 5: GNR (genetics, nanotech, robots)
- Chapter 6: The Impact
- Chapter 7: Ich bin ein Singularitarian
- Chapter 8: The Deeply Intertwined Promise and Peril of GNR
- Chapter 9: Response to critics
- Epilogue

Kurzweil first defines the Singularity as a point in the future when technological advances begin to happen so rapidly that normal humans cannot keep pace, and are "cut out of the loop." Kurzweil emphasizes that this will have a profound, disruptive effect on human societies and on everyday life, and will mark the end of human history as we know it. In place of normal humans, Strong Artificial Intelligences and cybernetically augmented humans will become the dominant forms of sentient life on the Earth. The Singularity will be initiated once self-improving Artificial Intelligences (sentient, highly intelligent computer programs capable of quickly redesigning themselves and their offspring to improve performance) come into existence.

Underlying all of Kurzweil's ideas regarding the progress of technology and the Singularity is the Law of Accelerating Returns. The Law states that technological progress occurs exponentially instead of linearly, meaning that each new advancement enables several higher advancements instead of just one higher advancement, and concordantly, every year, more useful inventions and discoveries are made than were made in the last.

The Law of Accelerating Returns has a very important consequence in that extrapolation of exponentially improving technology trends into the future suggests, by Kurzweil's analysis, that highly advanced technologies will arrive far sooner than linear-thinking people assume. The creation of the modern Internet and the completion of the Human Genome Project are prominent examples illustrative of this point. Both were multi-year projects that relied on computer technology to reach completion.

In both cases, critics derided them as hopeless since, in the beginning, both relied on computers that would have taken decades to process all of the necessary data. However, these critics had failed to take into account the exponentially improving nature of computer processing speeds and price-performance, and thus failed to see that, within a few years, the two projects would have access to vastly superior computers that would drastically shorten their timelines for completion.

Kurzweil lays forth his idea that the fate of the universe is to progress through six different epochs, characterized by the major paradigm shift that takes place after each one. The universe has already passed through four of these epochs and we are entering epoch five. Each step indirectly produces the next through chemical-, biological-, and technological Evolution (respectively).

The Six Epochs are subject to the Law of Accelerating Returns, which states that each transition occurs more rapidly than the last. Kurzweil supports this final postulate with logarithmic graphs of the chronology of important events in the history of the Universe (i.e. - the Big Bang, the origin of life, the birth of the human race, the creation of the first computer).

Epoch 1. Physics and Chemistry: This epoch starts at the beginning of the universe. In this epoch, information is mostly held in subatomic structures such as particles and atoms. That is, the most complicated stable objects in the universe do not exceed the molecular scale in size or complexity.

Epoch 2. Biology and DNA: This epoch starts with the beginning of life on Earth, suddenly giving rise to more complicated, yet stable, organisms that are capable of growth and self-sustainment. However, in this epoch organisms do not change within their lifetimes as evolution takes thousands of generations. Moreoever, in this stage, genetic information is stored in DNA molecules.

Epoch 3. Brains: The evolution of life gradually produced more and more complex organisms, necessitating the need for fast central control and thus giving rise to the evolution of brains. With brains, organisms can now change their behavior dynamically to suit changes in the environment and can also learn from past experiences. Evolutionary information is, in this stage, stored in neural patterns.

Epoch 4. Technology: Evolution of brains culminates with the evolution of humans, which possess the ability to create technology. In this stage, technological designs are also subject to evolution and information is held in hardware and software designs.

Epoch 5. The Merger of Human Technology with Human Intelligence

This epoch, which Kurzweil argues we are in the process of entering, is where technology reaches a level of sophistication and fine-structuring comparable with that of biology, allowing the two to merge to create higher forms of life and intelligence.

Epoch 6. The Universe Wakes Up: After mastering the methods of technology and biology, Kurzweil predicts that human/machine civilization will expand its frontiers into the universe, gradually (or perhaps explosively) consuming the contents of the cosmos until the universe reaches a 'saturated' state where all inanimate matter has been converted to substrates for computation and intelligence, and a truly universal super-intelligence takes form.

Next, Kurzweil briefly lays out a number of postulates and consequences in the first chapter as they relate to the Singularity. As stated, the Law of Accelerating Returns implies exponential- and non-linear growth in technological capabilities and human knowledge. Information technologies are experiencing superexponential growth since human economic investment in them increases as they become more powerful.

Human brain-scanning techniques are reliant upon computers and advanced machines, which are all subject to the Law of Accelerating Returns. Therefore, our ability to understand the way the human brain works is improving tremendously every year. Based on the current trends, brain scans should give us a sound understanding of how the human brain works by the mid-2020s. Using that information, scientists should be able to create simulated human brains inside of computers, leading to the first Artificial Intelligence (a thinking computer capable of passing the Turing Test) by 2029. The computer hardware of that era should also be powerful enough to meet the cognitive needs of a simulated human mind.

Artificial Intelligence would inevitably prove superior to human intelligence since the former would combine the fast speeds, memory capacity and recall, and instant downloading/learning abilities of computers with the creativity and pattern-recognition abilities of the human mind, effectively constituting a being with the strengths of both and the weaknesses of neither. Artificial Intelligences would also have the ability to edit their own software at will to instantly improve themselves, which is an ability humans naturally lack and can only approximate through laborious genetic engineering techniques.

Biology presents inherent and insurmountable limitations that originate at the molecular level. These limitations handicap both potential mental abilities and physical abilities.

This reality will inevitably lead to the creation of synthetic replacement parts that are far more durable and capable than natural human organs, limbs and cells. Some humans will choose to incorporate these engineered prostheses into their bodies, becoming cybernetically enhanced.

Nanotechnology will play a major role in human brain augmentation with the advent of brain nanomachines designed specifically for interacting with human neurons. Brain cybernetics of this type would allow human users to vastly expand their cognitive abilities, to experience full-immersion virtual reality, and to directly interface with computers and other mentally augmented humans. Being naturally more powerful, the cybernetic portion of augmented human minds will come to predominate.

A very broad range of advanced nanomachines will also be created, allowing the humans and machines of the future to alter their world with incredible power. Aside from the aforementioned brain augmentation role, nanotechnology could be used for a variety of tasks including a total repair of the Earth's environment and the near instantaneous manipulation of physical objects via foglets so as to blur the distinction between "real" and "virtual" reality.

Kurzweil foresees the computers of the future using three-dimensional computer chips composed of nanotubes. This architecture will allow them to operate at terahertz speeds.

Computer advances like this will eventually make A.I.'s so powerful that they will completely take over the cutting edge of all scientific research and development, and will generate new advancements (including improvements to their own programming) so quickly that normal humans will not be able to follow what is going on. By this point, the Singularity will definitely have been reached.

In the far future, A.I.'s and cybernetic humans will live almost exclusively in full-immersion virtual reality worlds, which will themselves be contained in advanced computers.

Computers will grow ever more advanced, but at some point, the bottom limit to transistor size as defined by the laws of Physics will be reached, and performance improvements will only be possible through the creation of new computers or the expansion of existing ones.

More and more inanimate matter on the Earth will be restructured to form useful computer substrate, but it will eventually prove insufficient, and the future machine race will radiate out from the Earth in all directions, "saturating" first the solar system, then the Milky Way galaxy, and eventually the entire cosmos with "intelligence" by converting unstructured, inanimate "dumb" matter (e.g., Moon rocks, dead gas giant planets, meteoroids) into structured "smart" matter that lives in the sense that it supports thinking, feeling A.I.'s. Entire celestial bodies of countless number across the universe will be totally converted into computer substrate.

The length of time it will take to finish the task of "waking up" the universe depends heavily on whether the speed of light can be surpassed or circumvented—the maximum rate of travel of course limiting the speed at which the advanced Earth civilization can spread across space to new locations.

Kurzweil concludes the chapter by stating his belief that, while A.I.'s will inevitably prove vastly superior to humans in every way, he expects them to respect human life and to embody human values.

9

Classical Genetics

Mutation

DNA replication is for the most part extremely accurate, with an error rate per site of around 10 to 10 in eukaryotes. Rare, spontaneous alterations in the base sequence of a particular gene arise from a number of sources, such as errors in DNA replication and the aftermath of DNA damage.

These errors are called mutations. The cell contains many DNA repair mechanisms for preventing mutations and maintaining the integrity of the genome; however, in some cases—such as breaks in both DNA strands of a chromosome—repairing the physical damage to the molecule is a higher priority than producing an exact copy.

Due to the degeneracy of the genetic code, some mutations in protein-coding genes are silent, or produce no change in the amino acid sequence of the protein for which they code; for example, the codons UCU and UUC both code for serine, so the $U \rightarrow C$ mutation has no effect on the protein.

Mutations that do have phenotypic effects are most often neutral or deleterious to the organism, but sometimes they confer benefits to the organism's fitness. Mutations propagated to the next generation lead to variations within a species' population.

Variants of a single gene are known as alleles, and differences in alleles may give rise to differences in traits. Although it is rare for the variants in a single gene to have clearly distinguishable phenotypic effects, certain well-defined traits are in fact controlled by single genetic loci. A gene's most common allele is called the wild type allele, and rare alleles are called mutants. However, this does not imply that the wild-type allele is the ancestor from which the mutants are descended.

Genome

Chromosomal Organization

The total complement of genes in an organism or cell is known as its genome. In prokaryotes, the vMast majority of genes are located on a single chromosome of circular DNA, while eukaryotes usually possess multiple individual linear DNA helices packed into dense DNA-protein complexes called chromosomes.

Genes that appear together on one chromosome of one species may appear on separate chromosomes in another species. Many species carry more than one copy of their genome within each of their somatic cells. Cells or organisms with only one copy of each chromosome are called haploid; those with two copies are called diploid; and those with more than two copies are called polyploid. The copies of genes on the chromosomes are not necessarily identical. In sexually reproducing organisms, one copy is normally inherited from each parent.

Number of Genes

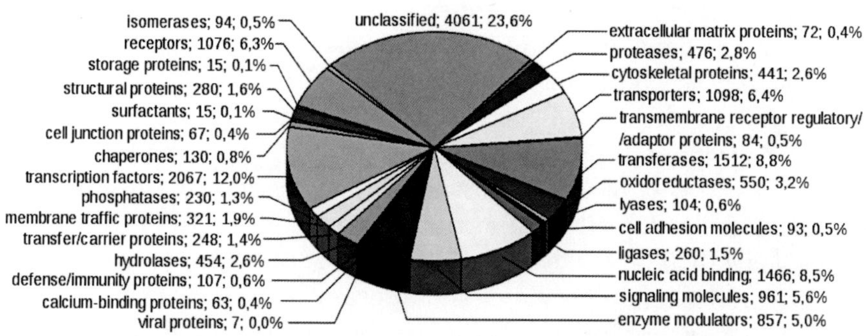

The human genome, categorized by function of each gene product, given both as number of genes and as percentage of all genes. Early estimates of the number of human genes that used expressed sequence tag data put it at 50 000–100 000. Following the sequencing of the human genome and other genomes, it has been found that rather few genes (~20 000 in human, mouse and fly, ~13 000 in roundworm, >46 000 in rice) encode all the proteins in an organism. These protein-coding sequences make up 1–2% of the human genome. A large part of the genome is transcribed however, to introns, retrotransposons and seemingly a large array of noncoding RNAs. Total number of proteins (the Earth's proteome) is estimated to be 5 million sequences.

Genetic and Genomic Nomenclature

Gene nomenclature has been established by the HUGO Gene Nomenclature Committee (HGNC) for each known human gene in the form of an approved gene name and symbol (short-form abbreviation). All approved symbols are stored in the HGNC Database. Each symbol is unique and each gene is only given one approved gene symbol. This also facilitates electronic data retrieval from publications. In preference each symbol maintains parallel construction in different members of a gene family and can be used in other species, especially the mouse.

Evolutionary Concept of a Gene

George C. Williams first explicitly advocated the gene-centric view of evolution in his 1966 book Adaptation and Natural Selection. He proposed an evolutionary concept of gene to be used when we are talking about natural selection favoring some genes. The definition is: "that which segregates and recombines with appreciable frequency." According to this definition, even an asexual genome could be considered a gene, insofar that it have an appreciable permanency through many generations.

The difference is: the molecular gene transcribes as a unit, and the evolutionary gene inherits as a unit. Richard Dawkins' books The Selfish Gene (1976) and The Extended Phenotype (1982) defended the idea that the gene is the only replicator in living systems. This means that only genes transmit their structure largely intact and are potentially immortal in the form of copies. So, genes should be the unit of selection. In The Selfish Gene Dawkins attempts to redefine the word 'gene' to mean "an inheritable unit" instead of the generally accepted definition of "a section of DNA coding for a particular protein". In River Out of Eden, Dawkins further refined the idea of gene-centric selection by describing life as a river of compatible genes flowing through geological time. Scoop up a bucket of genes from the river of genes, and we have an organism serving as temporary bodies or survival machines. A river of genes may fork into two branches representing two non-interbreeding species as a result of geographical separation.

Gene Targeting and Implications

Gene targeting is commonly referred to techniques for altering or disrupting mouse genes and provides the mouse models for studying the roles of individual genes in embryonic development, human disorders, aging and diseases. The mouse models, where one or more of its genes are deactivated or made inoperable, are called knockout mice. Since the first reports in which homologous recombination in embryonic stem cells was used to generate gene-targeted mice, gene targeting has proven to be a powerful means of precisely manipulating the mammalian genome, producing at least ten thousand mutant mouse strains and it is now possible to introduce mutations that can be activated at specific time points, or in specific cells or organs, both during development and in the adult animal.

Gene targeting strategies have been expanded to all kinds of modifications, including point mutations, isoform deletions, mutant allele correction, large pieces of chromosomal DNA insertion and deletion, tissue specific disruption combined with spatial and temporal regulation and so on. It is predicted that the ability to generate mouse models with predictable phenotypes will have a major impact on studies of all phases of development, immunology, neurobiology, oncology, physiology, metabolism, and human diseases. Gene targeting is also in theory applicable to species from which totipotent embryonic stem cells can be established, and therefore may offer a potential to the improvement of domestic animals and plants.

Changing Concept

The concept of the gene has changed considerably. From the original definition of a "unit of inheritance", the term evolved to mean a DNA-based unit that can exert its effects on the oMrganism through RNA or protein products. It was also previously believed that one gene makes one protein; this concept was overthrown by the discovery of alternative splicing and trans-splicing. The definition of a gene is still changing. The first cases of RNA-based inheritance have been discovered in mammals. Evidence is also accumulating that the control regions of a gene do not necessarily have to be close to the coding sequence on the linear molecule or even on the same chromosome. Spilianakis and colleagues discovered that the promoter region of the interferon-gamma gene on chromosome 10 and the regulatory regions of the T(H)2 cytokine locus on chromosome 11 come into close proximity in the nucleus possibly to be jointly regulated.

The concept that genes are clearly delimited is also being eroded. There is evidence for fused proteins stemming from two adjacent genes that can produce two separate protein products. While it is not clear whether these fusion proteins are functional, the phenomenon is more frequent than previously thought. Even more ground-breaking than the discovery of fused genes is the observation that some proteins can be composed of exons from far away regions and even different chromosomes. This new data has led to an updated, and probably tentative, definition of a gene as "a union of genomic sequences encoding a coherent set of potentially overlapping functional products". This new definition categorizes genes by functional products, whether they be proteins or RNA, rather than specific DNA loci; all regulatory elements of DNA are therefore classified as gene-associated regions.

Big Gene

The 'big genes' are a class of genes. A big gene is a single gene whose nuclear transcript spans 500 kb or more of chromosomal DNA.

Character

The largest of the big genes is the gene for dystrophin, which spans 2.3 Mb. Many big genes have modestly sized mRNAs; the exons encoding these RNAs typically encompass about 1% of the total chromosomal gene region in which they occur.

Synthesis

Full length cDNA sequences are often essential for the delineation of exon/intron structures in big genes. But because big genes are synthesized very slowly, their synthesis cannot be completed in rapidly dividing cells. It is in neurons, which do not divide, that a large fraction of the big genes are expressed.

Copy Number Variation

Copy-number variations (CNVs)—a form of structural variation—are alterations of the DNA of a genome that results in the cell having an abnormal number of copies of one or more sections of the DNA. CNVs correspond to relatively large regions of the genome that have been deleted (fewer than the normal number) or duplicated (more than the normal number) on certain chromosomes.

For example, the chromosome that normally has sections in order as A-B-C-D might instead have sections A-B-C-C-D (a duplication of "C") or A-B-D (a deletion of "C"). This variation accounts for roughly 12% of human genomic DNA and each variation may range from about one kilobase (1,000 nucleotide bases) to several megabases in size. CNVs contrast with single-nucleotide polymorphisms (SNPs), which affect only one single nucleotide base.

Sources

CNVs may either be inherited or caused by de novo mutation. A recently proposed mechanism for the cause of some CNVs was fork stalling and template switching, a replication misstep. However, this model was subsequently superseded by microhomology-mediated break-induced replication (MMBIR). CNVs can be caused by structural rearrangements of the genome such as deletions, duplications, inversions, and translocations. Low copy repeats (LCRs), which are region-specific repeat sequences, are susceptible to such genomic rearrangements resulting in CNVs. Factors such as size, orientation, percentage similarity and the distance between the copies influence the susceptibility of

LCRs to genomic rearrangement. Segmental Duplications (SDs) map near ancestral duplication sites in a phenomenon called duplication shadowing which describes the observation of a ~10 fold increased probability of duplication in regions flanking duplications versus other random regions.

Identification

Copy number variation can be discovered by cytogenetic techniques such as fluorescent in situ hybridization, comparative genomic hybridization, array comparative genomic hybridization, and by virtual karyotyping with SNP arrays. Recent advances in DNA sequencing technology have further enabled the identification of CNVs by next-generation sequencing. CNVs can be limited to a single gene or include a contiguous set of genes. CNVs can result in having either too many or too few of the dosage-sensitive genes, which may be responsible for a substantial amount of human phenotypic variability, complex behavioral traits and disease susceptibility.

In certain cases, such as rapidly growing Escherichia coli cells, the gene copy number can be 4-fold greater for genes located near the origin of DNA replication, rather than at the terminus of DNA replication. Elevating the gene copy number of a particular gene can increase the expression of the protein that it encodes.

Prevalence in Humans

The fact that DNA copy number variation is a widespread and common phenomenon among humans was first uncovered following the completion of the human genome project. It is estimated that approximately 0.4% of the genomes of unrelated people typically differ with respect to copy number. De novo CNVs have been observed between identical twins who otherwise have identical genomes.

Role in Disease

Like other types of genetic variation, some CNVs have been associated with susceptibility or resistance to disease. Gene copy number can be elevated in cancer cells. For instance, the EGFR copy number can be higher than normal in non-small cell lung cancer. In addition, a higher copy number of CCL3L1 has been associated with lower susceptibility to HIV infection, and a low copy number of FCGR3B (the CD16 cell surface immunoglobulin receptor) can increase susceptibility to systemic lupus erythematosus and similar inflammatory autoimmune disorders. Copy number variation has also been associated with autism, schizophrenia, and idiopathic learning disability.

However, although once touted as the explanation for the elusive hereditary causes of complex diseases like rheumatoid arthritis, the most common CNVs have little or no role in causing disease.

Among common functional CNVs, gene gains outnumber losses, suggesting that many of them are favored in evolution and, therefore, beneficial in some way. One example of CNV is the human salivary amylase gene (AMY1). This gene is typically present as two diploid copies in chimpanzees. Humans average over 6 copies and may have as many as 15. This is thought to be an adaptation to a high-starch diet that improves the ability to digest starchy foods.

DNA

Deoxyribonucleic acid is a nucleic acid that contains the genetic instructions used in the development and functioning of all known living organisms (with the exception of RNA viruses). The DNA segments that carry this genetic information are called genes, but other DNA sequences have structural purposes, or are involved in regulating the use of this genetic information. Along with RNA and proteins, DNA is one of the three major macromolecules that are essential for all known forms of life.

DNA consists of two long polymers of simple units called nucleotides, with backbones made of sugars and phosphate groups joined by ester bonds. These two strands run in opposite directions to each other and are therefore anti-parallel. Attached to each sugar is one of four types of molecules called nucleobases (informally, bases). It is the sequence of these four nucleobases along the backbone that encodes information. This information is read using the genetic code, which specifies the sequence of the amino acids within proteins. The code is read by copying stretches of DNA into the related nucleic acid RNA in a process called transcription.

Within cells DNA is organized into long structures called chromosomes. During cell division these chromosomes are duplicated in the process of DNA replication, providing each cell its own complete set of chromosomes. Eukaryotic organisms (animals, plants, fungi, and protists) store most of their DNA inside the cell nucleus and some of their DNA in organelles, such as mitochondria or chloroplasts. In contrast, prokaryotes (bacteria and archaea) store their DNA only in the cytoplasm. Within the chromosomes, chromatin proteins such as histones compact and organize DNA. These compact structures guide the interactions between DNA and other proteins, helping control which parts of the DNA are transcribed.

Properties

DNA is a long polymer made from repeating units called nucleotides. As first discovered by James D. Watson and Francis Crick, the structure of DNA of all species comprises two helical chains each coiled round the same axis, and each with a pitch of 34 Ångströms (3.4 nanometres) and a radius of 10 Ångströms (1.0 nanometres). According to another study, when measured in a particular solution, the DNA chain measured 22 to 26 Ångströms wide (2.2 to 2.6 nanometres), and one nucleotide unit measured 3.3 Å (0.33 nm) long. Although each individual repeating unit is very small, DNA polymers can be very large molecules containing millions of nucleotides. For instance, the largest human chromosome, chromosome number 1, is approximately 220 million base pairs long.

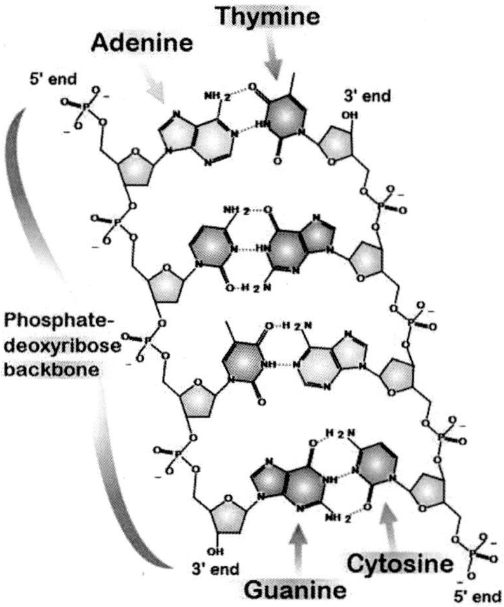

Figure: *Chemical structure of DNA. Hydrogen bonds shown as dotted lines.*

In living organisms DNA does not usually exist as a single molecule, but instead as a pair of molecules that are held tightly together. These two long strands entwine like vines, in the shape of a double helix. The nucleotide repeats contain both the segment of the backbone of the molecule, which holds the chain together, and a nucleobase, which interacts with the other DNA strand in the helix. A nucleobase linked to a sugar is called a nucleoside and a base linked to a sugar and one or more phosphate groups is called a nucleotide. Polymers comprising multiple linked nucleotides (as in DNA) are called a polynucleotide.

The backbone of the DNA strand is made from alternating phosphate and sugar residues. The sugar in DNA is 2-deoxyribose, which is a pentose (five-carbon) sugar. The sugars are joined together by phosphate groups that form phosphodiester bonds between the third and fifth carbon atoms of adjacent sugar rings. These asymmetric bonds mean a strand of DNA has a direction. In a double helix the direction of the nucleotides in one strand is opposite to their direction in the other strand: the strands are antiparallel. The asymmetric ends of DNA strands are called the 52 (five prime) and 32 (three prime) ends, with the 5' end having a terminal phosphate group and the 3' end a terminal hydroxyl group. One major difference between DNA and RNA is the sugar, with the 2-deoxyribose in DNA being replaced by the alternative pentose sugar ribose in RNA.

The DNA double helix is stabilized primarily by two forces: hydrogen bonds between nucleotides and base-stacking interactions among the aromatic nucleobases. In the aqueous environment of the cell, the conjugated ð bonds of nucleotide bases align perpendicular to the axis of the DNA molecule, minimizing their interaction with the solvation shell and therefore, the Gibbs free energy. The four bases found in DNA are adenine (abbreviated A), cytosine (C), guanine (G) and thymine (T). These four bases are attached to the sugar/ phosphate to form the complete nucleotide, as shown for adenosine monophosphate.

The nucleobases are classified into two types: the purines, A and G, being fused five- and six-membered heterocyclic compounds, and the pyrimidines, the six-membered rings C and T. A fifth pyrimidine nucleobase, uracil (U), usually takes the place of thymine in RNA and differs from thymine by lacking a methyl group on its ring. Uracil is not usually found in DNA, occurring only as a breakdown product of cytosine. In addition to RNA and DNA a large number of artificial nucleic acid analogues have also been created to study the proprieties of nucleic acids, or for use in biotechnology.

Grooves

Twin helical strands form the DNA backbone. Another double helix may be found by tracing the spaces, or grooves, between the strands. These voids are adjacent to the base pairs and may provide a binding site. As the strands are not directly opposite each other, the grooves are unequally sized. One groove, the major groove, is 22 Å wide and the other, the minor groove, is 12 Å wide.

The narrowness of the minor groove means that the edges of the bases are more accessible in the major groove. As a result, proteins like transcription factors that can bind to specific sequences in double-stranded DNA usually make contacts to the sides of the bases exposed in the major groove. This situation varies in unusual conformations of DNA within the cell, but the major and minor grooves are always named to reflect the differences in size that would be seen if the DNA is twisted back into the ordinary B form.

Base Pairing

In a DNA double helix, each type of nucleobase on one strand normally interacts with just one type of nucleobase on the other strand. This is called complementary base pairing. Here, purines form hydrogen bonds to pyrimidines, with A bonding only to T, and C bonding only to G. This arrangement of two nucleotides binding together across the double helix is called a base pair. As hydrogen bonds are not covalent, they can be broken and rejoined relatively easily.

The two strands of DNA in a double helix can therefore be pulled apart like a zipper, either by a mechanical force or high temperature. As a result of this complementarity, all the information in the double-stranded sequence of a DNA helix is duplicated on each strand, which is vital in DNA replication. Indeed, this reversible and specific interaction between complementary base pairs is critical for all the functions of DNA in living organisms.

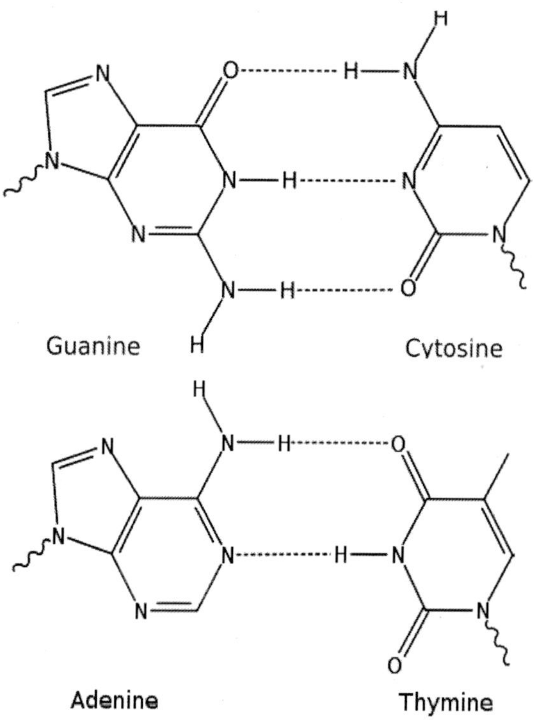

Figure: *Top, a GC base pair with three hydrogen bonds. Bottom, an AT base pair with two hydrogen bonds. Non-covalent hydrogen bonds between the pairs are shown as dashed lines.*

The two types of base pairs form different numbers of hydrogen bonds, AT forming two hydrogen bonds, and GC forming three hydrogen bonds. DNA with high GC-content is more stable than DNA with low GC-content. Although it is often stated that this is due to the added stability of an additional hydrogen bond, this is incorrect. DNA with high GC-content is more stable due to intra-strand base stacking interactions.

As noted above, most DNA molecules are actually two polymer strands, bound together in a helical fashion by noncovalent bonds; this double stranded structure (dsDNA) is maintained largely by the intrastrand base stacking interactions, which are strongest for G,C stacks.

The two strands can come apart – a process known as melting – to form two ss DNA molecules. Melting occurs when conditions favor ssDNA; such conditions are high temperature, low salt and high pH (low pH also melts DNA, but since DNA is unstable due to acid depurination, low pH is rarely used). The stability of the dsDNA form depends not only on the GC-content (% G,C basepairs) but also on sequence (since stacking is sequence specific) and also length (longer molecules are more stable).

The stability can be measured in various ways; a common way is the "melting temperature", which is the temperature at which 50% of the ds molecules are converted to ss molecules; melting temperature is dependent on ionic strength and the concentration of DNA. As a result, it is both the percentage of GC base pairs and the overall length of a DNA double helix that determine the strength of the association between the two strands of DNA. Long DNA helices with a high GC-content have stronger-interacting strands, while short helices with high AT content have weaker-interacting strands. In biology, parts of the DNA double helix that need to separate easily, such as the TATAAT Pribnow box in some promoters, tend to have a high AT content, making the strands easier to pull apart. In the laboratory, the strength of this interaction can be measured by finding the temperature required to break the hydrogen bonds, their melting temperature (also called T_m value). When all the base pairs in a DNA double helix melt, the strands separate and exist in solution as two entirely independent molecules. These single-stranded DNA molecules (ssDNA) have no single common shape, but some conformations are more stable than others.

Base Pair

In molecular biology and genetics, the linking between two nitrogenous bases on opposite complementary DNA or certain types of RNA strands that are connected via hydrogen bonds is called a base pair (often abbreviated bp). In the canonical Watson-Crick DNA base pairing, adenine (A) forms a base pair with thymine (T) and guanine (G) forms a base pair with cytosine (C). In RNA, thymine is replaced by uracil (U). Alternate hydrogen bonding patterns, such as the wobble base pair and Hoogsteen base pair, also occur—in particular, in RNA—giving rise to complex and functional tertiary structures. Pairing is the mechanism by which codons on messenger RNA molecules are recognized by anticodons on transfer RNA during protein translation. Some DNA- or RNA-binding enzymes can recognize specific base pairing patterns that identify particular regulatory regions of genes. The size of an individual gene or an organism's entire genome is often measured in base pairs because DNA is usually double-stranded. Hence, the number of total base pairs is equal to the number of nucleotides in one of the strands (with the exception of non-coding single-stranded regions of telomeres). The haploid human genome (23 chromosomes) is estimated to be about 3 billion base pairs long and to contain 20,000–25,000 distinct genes. A kilobase (kb) is a unit of measurement in molecular biology equal to 1000 base pairs of DNA or RNA.

Hydrogen Bonding and Stability

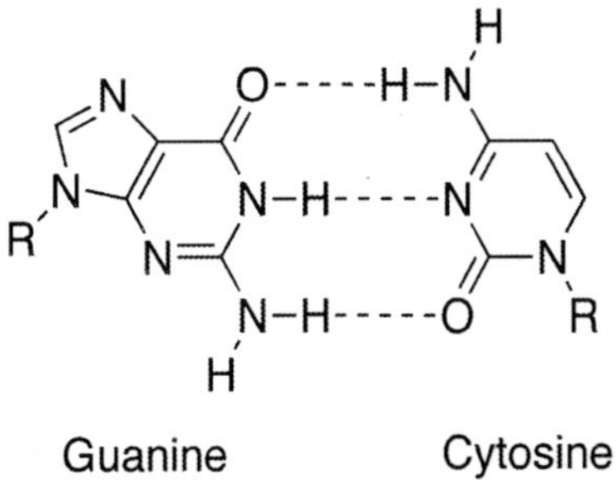

Figure: *A GC base pair demonstrating three intermolecular hydrogen bonds*

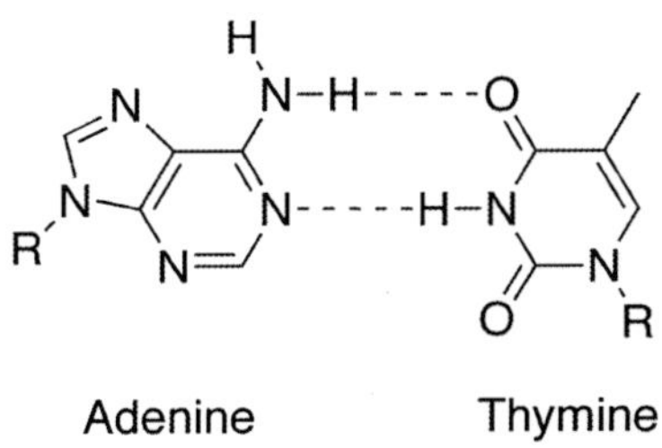

Figure: *An AT base pair demonstrating two intermolecular hydrogen bonds*

Hydrogen bonding is the chemical interaction that underlies the base-pairing rules described above. Appropriate geometrical correspondence of hydrogen bond donors and acceptors allows only the "right" pairs to form stably. DNA with high GC-content is more stable than DNA with low GC-content, but, contrary to popular belief, the hydrogen bonds do not stabilize the DNA significantly, and stabilization is mainly due to stacking interactions.

The larger nucleobases, adenine and guanine, are members of a class of double-ringed chemical structures called purines; the bigger or smaller nucleobases, cytosine and thymine (and uracil), are members of a class of single-ringed chemical structures called pyrimidines. Purines are complementary only with pyrimidines: pyrimidine-pyrimidine pairings are energetically unfavorable because the molecules are too far apart for hydrogen bonding to be established; purine-purine pairings are energetically unfavorable because the molecules are too close, leading to overlap repulsion. The only other possible pairings are GT and AC; these pairings are mismatches because the pattern of hydrogen donors and acceptors do not correspond. The GU pairing, with two hydrogen bonds, does occur fairly often in RNA.

Paired DNA and RNA molecules are comparatively stable at room temperature but the two nucleotide strands will separate above a melting point that is determined by the length of the molecules, the extent of mispairing (if any), and the GC content. Higher GC content results in higher melting temperatures; it is, therefore, unsurprising that the genomes of extremophile organisms such as Thermus thermophilus are particularly GC-rich. On the converse, regions of a genome that need to separate frequently — for example, the promoter regions for often-transcribed genes — are comparatively GC-poor (for example, see TATA box). GC content and melting temperature must also be taken into account when designing primers for PCR reactions.

Base Stacking

Base stacking interactions in DNA and RNA are due to dispersion attraction, short-range exchange repulsion, and electrostatic interactions, which also contribute to stability. Again, GC stacking interactions with adjacent bases tend to be more favorable. (Note, however, that a GC stacking interaction with the next base pair is geometrically different from a CG interaction.) Base stacking effects are especially important in the secondary structure and tertiary structure of RNA; for example, RNA stem-loop structures are stabilized by base stacking in the loop region.

Base Analogs and Intercalators

Chemical analogs of nucleotides can take the place of proper nucleotides and establish non-canonical base-pairing, leading to errors (mostly point mutations) in DNA replication and DNA transcription. This is due to their isosteric chemistry. One common mutagenic base analog is 5-bromouracil, which resembles thymine but can base-pair to guanine in its enol form. Other chemicals, known as DNA intercalators, fit into the gap between adjacent bases on a single strand and induce frameshift mutations by "masquerading" as a base, causing the DNA replication machinery to skip or insert additional nucleotides at the intercalated site. Most intercalators are large polyaromatic compounds and are known or suspected carcinogens. Examples include ethidium bromide and acridine.

Examples

The following DNA sequences illustrate pair double-stranded patterns. By convention, the top strand is written from the 5' end to the 3' end; thus, the bottom strand is written 3' to 5'.

A base-paired DNA sequence:

ATCGATTGAGCTCTAGCG

TAGCTAACTCGAGATCGC

The corresponding RNA sequence, in which uracil is substituted for thymine where uracil takes its place in the RNA strand:

AUCGAUUGAGCUCUAGCG

UAGCUAACUCGAGAUCGC

Length Measurements

The following abbreviations are commonly used to describe the length of a D/RNA molecule:

- bp = base pair(s)—one bp corresponds to circa 3.4 Å of length along the strand
- kb (= kbp) = kilo base pairs = 1,000 bp
- Mb = mega base pairs = 1,000,000 bp
- Gb = giga base pairs = 1,000,000,000 bp.

In case of single stranded DNA/RNA units of nucleotides are used, abbreviated nt (or knt, Mnt, Gnt), as they are not paired. For distinction between units of computer storage and bases kbp, Mbp, Gbp, etc. may be used for basepairs. The length of 16S rDNA for bacteria is 1542 base-pairs in length. The Centimorgan is also often used to imply distance along a chromosome, but the number of base-pairs it corresponds to varies widely. In the Human genome, the centimorgan is about 1 million base pairs.

Sense and Antisense

A DNA sequence is called “sense” if its sequence is the same as that of a messenger RNA copy that is translated into protein. The sequence on the opposite strand is called the “antisense” sequence. Both sense and antisense sequences can exist on different parts of the same strand of DNA (i.e. both strands contain both sense and antisense sequences). In both prokaryotes and eukaryotes, antisense RNA sequences are produced, but the functions of these RNAs are not entirely clear. One proposal is that antisense RNAs are involved in regulating gene expression through RNA-RNA base pairing.

A few DNA sequences in prokaryotes and eukaryotes, and more in plasmids and viruses, blur the distinction between sense and antisense strands by having overlapping genes. In these cases, some DNA sequences do double duty, encoding one protein when read along one strand, and a second protein when read in the opposite direction along the other strand. In bacteria, this overlap may be involved in the regulation of gene transcription, while in viruses, overlapping genes increase the amount of information that can be encoded within the small viral genome.

Supercoiling

DNA can be twisted like a rope in a process called DNA supercoiling. With DNA in its "relaxed" state, a strand usually circles the axis of the double helix once every 10.4 base pairs, but if the DNA is twisted the strands become more tightly or more loosely wound. If the DNA is twisted in the direction of the helix, this is positive supercoiling, and the bases are held more tightly together. If they are twisted in the opposite direction, this is negative supercoiling, and the bases come apart more easily. In nature, most DNA has slight negative supercoiling that is introduced by enzymes called topoisomerases. These enzymes are also needed to relieve the twisting stresses introduced into DNA strands during processes such as transcription and DNA replication.

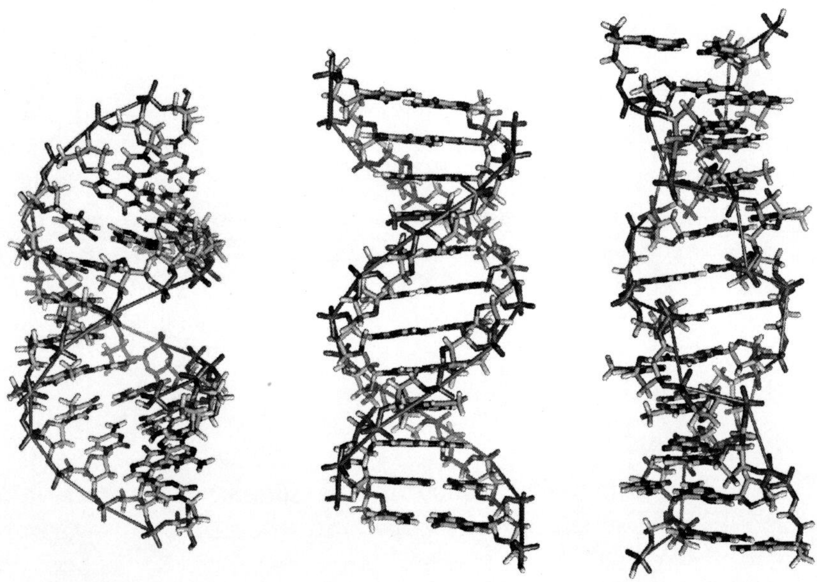

Figure: *From left to right, the structures of A, B and Z DNA*

Alternate DNA Structures

DNA exists in many possible conformations that include A-DNA, B-DNA, and Z-DNA forms, although, only B-DNA and Z-DNA have been directly observed in functional organisms. The conformation that DNA adopts depends on the hydration level, DNA sequence, the amount and direction of supercoiling, chemical modifications of the bases, the type and concentration of metal ions, as well as the presence of polyamines in solution.

The first published reports of A-DNA X-ray diffraction patterns— and also B-DNA used analyses based on Patterson transforms that provided only a limited amount of structural information for oriented fibers of DNA. An alternate analysis was then proposed by Wilkins et al., in 1953, for the in vivo B-DNA X-ray diffraction/scattering patterns of highly hydrated DNA fibers in terms of squares of Bessel functions. In the same journal, James D. Watson and Francis Crick presented their molecular modeling analysis of the DNA X-ray diffraction patterns to suggest that the structure was a double-helix.

Although the 'B-DNA form' is most common under the conditions found in cells, it is not a well-defined conformation but a family of related DNA conformations that occur at the high hydration levels present in living cells. Their corresponding X-ray diffraction and scattering patterns are characteristic of molecular paracrystals with a significant degree of disorder.

Compared to B-DNA, the A-DNA form is a wider right-handed spiral, with a shallow, wide minor groove and a narrower, deeper major groove. The A form occurs under non-physiological conditions in partially dehydrated samples of DNA, while in the cell it may be produced in hybrid pairings of DNA and RNA strands, as well as in enzyme-DNA complexes. Segments of DNA where the bases have been chemically modified by methylation may undergo a larger change in conformation and adopt the Z form. Here, the strands turn about the helical axis in a left-handed spiral, the opposite of the more common B form. These unusual structures can be recognized by specific Z-DNA binding proteins and may be involved in the regulation of transcription.

Alternate DNA Chemistry

For a number of years exobiologists have proposed the existence of a shadow biosphere, a postulated microbial biosphere of Earth that uses radically different biochemical and molecular processes than currently known life. One of the proposals was the existence of lifeforms that use arsenic instead of phosphorus in DNA.

A December 2010 NASA press conference stated that the bacterium GFAJ-1, which has evolved in an arsenic-rich environment, is the first terrestrial lifeform found which may have this ability. The bacterium was found in Mono Lake, east of Yosemite National Park. GFAJ-1 is a rod-shaped extremophile bacterium in the family Halomonadaceae that, when starved of phosphorus, may be capable of incorporating the usually poisonous element arsenic in its DNA. This discovery may lend weight to the long-standing idea that extraterrestrial life could have a different chemical makeup from life on Earth.

The research was carried out by a team led by Felisa Wolfe-Simon, a geomicrobiologist and geobiochemist, a Postdoctoral Fellow of the NASA Astrobiology Institute with Arizona State University. This finding has, however, faced strong criticism from the scientific community; scientists have argued that there is no evidence that arsenic is actually incorporated into biomolecules. Independent conformation of this finding has also not yet been possible.

Quadruplex Structures

At the ends of the linear chromosomes are specialized regions of DNA called telomeres. The main function of these regions is to allow the cell to replicate chromosome ends using the enzyme telomerase, as the enzymes that normally replicate DNA cannot copy the extreme 32 ends of chromosomes. These specialized chromosome caps also help protect the DNA ends, and stop the DNA repair systems in the cell from treating them as damage to be corrected. In human cells, telomeres are usually lengths of single-stranded DNA containing several thousand repeats of a simple TTAGGG sequence.

These guanine-rich sequences may stabilize chromosome ends by forming structures of stacked sets of four-base units, rather than the usual base pairs found in other DNA molecules. Here, four guanine bases form a flat plate and these flat four-base units then stack on top of each other, to form a stable G-quadruplex structure. These structures are stabilized by hydrogen bonding between the edges of the bases and chelation of a metal ion in the centre of each four-base unit. Other structures can also be formed, with the central set of four bases coming from either a single strand folded around the bases, or several different parallel strands, each contributing one base to the central structure.

In addition to these stacked structures, telomeres also form large loop structures called telomere loops, or T-loops. Here, the single-stranded DNA curls around in a long circle stabilized by telomere-binding proteins. At the very end of the T-loop, the single-stranded telomere DNA is held onto a region of double-stranded DNA by the telomere strand disrupting the double-helical DNA and base pairing to one of the two strands. This triple-stranded structure is called a displacement loop or D-loop.

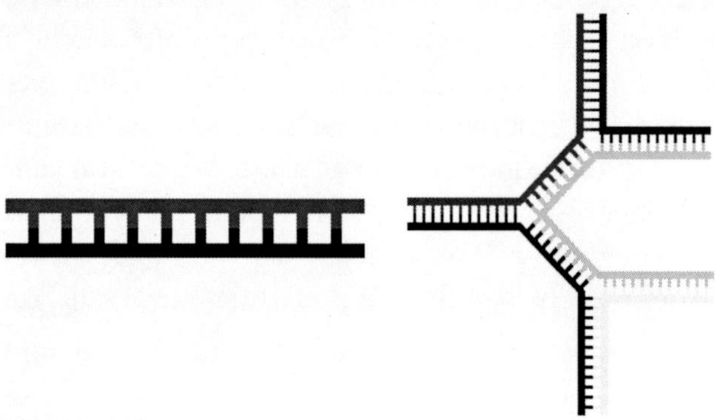

Single branch *Multiple branches*

Figure: *Branched DNA can form networks containing multiple branches.*

Branched DNA

In DNA fraying occurs when non-complementary regions exist at the end of an otherwise complementary double-strand of DNA. However, branched DNA can occur if a third strand of DNA is introduced and contains adjoining regions able to hybridize with the frayed regions of the pre-existing double-strand. Although the simplest example of branched DNA involves only three strands of DNA, complexes involving additional strands and multiple branches are also possible. Branched DNA can be used in nanotechnology to construct geometric shapes.

Vibration

DNA may carry out low-frequency collective motion as observed by the Raman spectroscopy and analyzed with a quasi-continuum model.

Chemical Modifications

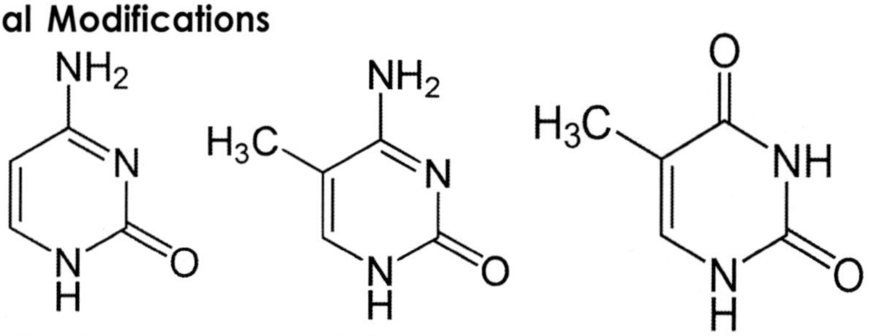

Figure: *Structure of cytosine with and without the 5-methyl group. Deamination converts 5-methylcytosine into thymine.*

Base Modifications

The expression of genes is influenced by how the DNA is packaged in chromosomes, in a structure called chromatin. Base modifications can be involved in packaging, with regions that have low or no gene expression usually containing high levels of methylation of cytosine bases. For example, cytosine methylation, produces 5-methylcytosine, which is important for X-chromosome inactivation.

The average level of methylation varies between organisms – the worm Caenorhabditis elegans lacks cytosine methylation, while vertebrates have higher levels, with up to 1% of their DNA containing 5-methylcytosine. Despite the importance of 5-methylcytosine, it can deaminate to leave a thymine base, so methylated cytosines are particularly prone to mutations.

Other base modifications include adenine methylation in bacteria, the presence of 5-hydroxymethylcytosine in the brain, and the glycosylation of uracil to produce the "J-base" in kinetoplastids.

Damage

DNA can be damaged by many sorts of mutagens, which change the DNA sequence. Mutagens include oxidizing agents, alkylating agents and also high-energy electromagnetic radiation such as ultraviolet light and X-rays. The type of DNA damage produced depends on the type of mutagen. For example, UV light can damage DNA by producing thymine dimers, which are cross-links between pyrimidine bases. On the other hand, oxidants such as free radicals or hydrogen peroxide produce multiple forms of damage, including base modifications, particularly of guanosine, and double-strand breaks. A typical human cell contains about 150,000 bases that have suffered oxidative damage. Of these oxidative lesions, the most dangerous are double-strand breaks, as these are difficult to repair and can produce point mutations, insertions and deletions from the DNA sequence, as well as chromosomal translocations.

Many mutagens fit into the space between two adjacent base pairs, this is called intercalation. Most intercalators are aromatic and planar molecules; examples include ethidium bromide, acridines, daunomycin, and doxorubicin. In order for an intercalator to fit between base pairs, the bases must separate, distorting the DNA strands by unwinding of the double helix. This inhibits both transcription and DNA replication, causing toxicity and mutations. As a result, DNA intercalators may be carcinogens, and in the case of thalidomide, a teratogen. Others such as benzo[a]pyrene diol epoxide and aflatoxin form DNA adducts which induce errors in replication. Nevertheless, due to their ability to inhibit DNA transcription and replication, other similar toxins are also used in chemotherapy to inhibit rapidly growing cancer cells.

Biological Functions

DNA usually occurs as linear chromosomes in eukaryotes, and circular chromosomes in prokaryotes. The set of chromosomes in a cell makes up its genome; the human genome has approximately 3 billion base pairs of DNA arranged into 46 chromosomes. The information carried by DNA is held in the sequence of pieces of DNA called genes. Transmission of genetic information in genes is achieved via complementary base pairing.

For example, in transcription, when a cell uses the information in a gene, the DNA sequence is copied into a complementary RNA sequence through the attraction between the DNA and the correct RNA nucleotides. Usually, this RNA copy is then used to make a matching protein sequence in a process called translation, which depends on the same interaction between RNA nucleotides. In alternative fashion, a cell may simply copy its genetic information in a process called DNA replication. The details of these functions are covered in other articles; here we focus on the interactions between DNA and other molecules that mediate the function of the genome.

Genes and Genomes

Genomic DNA is tightly and orderly packed in the process called DNA condensation to fit the small available volumes of the cell. In eukaryotes, DNA is located in the cell nucleus, as well as small amounts in mitochondria and chloroplasts. In prokaryotes, the DNA is held within an irregularly shaped body in the cytoplasm called the nucleoid. The genetic information in a genome is held within genes, and the complete set of this information in an organism is called its genotype. A gene is a unit of heredity and is a region of DNA that influences a particular characteristic in an organism. Genes contain an open reading frame that can be transcribed, as well as regulatory sequences such as promoters and enhancers, which control the transcription of the open reading frame.

In many species, only a small fraction of the total sequence of the genome encodes protein. For example, only about 1.5% of the human genome consists of protein-coding exons, with over 50% of human DNA consisting of non-coding repetitive sequences. The reasons for the presence of so much noncoding DNA in eukaryotic genomes and the extraordinary differences in genome size, or C-value, among species represent a long-standing puzzle known as the "C-value enigma". However, DNA sequences that do not code protein may still encode functional non-coding RNA molecules, which are involved in the regulation of gene expression.

Some noncoding DNA sequences play structural roles in chromosomes. Telomeres and centromeres typically contain few genes, but are important for the function and stability of chromosomes. An abundant form of noncoding DNA in humans are pseudogenes, which are copies of genes that have been disabled by mutation. These sequences are usually just molecular fossils, although they can occasionally serve as raw genetic material for the creation of new genes through the process of gene duplication and divergence.

Transcription and Translation

A gene is a sequence of DNA that contains genetic information and can influence the phenotype of an organism. Within a gene, the sequence of bases along a DNA strand defines a messenger RNA sequence, which then defines one or more protein sequences. The relationship between the nucleotide sequences of genes and the amino-acid sequences of proteins is determined by the rules of translation, known collectively as the genetic code. The genetic code consists of three-letter 'words' called codons formed from a sequence of three nucleotides (e.g. ACT, CAG, TTT).

In transcription, the codons of a gene are copied into messenger RNA by RNA polymerase. This RNA copy is then decoded by a ribosome that reads the RNA sequence by base-pairing the messenger RNA to transfer RNA, which carries amino acids. Since there are 4 bases in 3-letter combinations, there are 64 possible codons (4 combinations). These encode the twenty standard amino acids, giving most amino acids more than one possible codon. There are also three 'stop' or 'nonsense' codons signifying the end of the coding region; these are the TAA, TGA and TAG codons.

Replication

Cell division is essential for an organism to grow, but, when a cell divides, it must replicate the DNA in its genome so that the two daughter cells have the same genetic information as their parent. The double-stranded structure of DNA provides a simple mechanism for DNA replication. Here, the two strands are separated and then each strand's complementary DNA sequence is recreated by an enzyme called DNA polymerase. This enzyme makes the complementary strand by finding the correct base through complementary base pairing, and bonding it onto the original strand. As DNA polymerases can only extend a DNA strand in a 52 to 32 direction, different mechanisms are used to copy the antiparallel strands of the double helix. In this way, the base on the old strand dictates which base appears on the new strand, and the cell ends up with a perfect copy of its DNA.

Interactions with Proteins

All the functions of DNA depend on interactions with proteins. These protein interactions can be non-specific, or the protein can bind specifically to a single DNA sequence. Enzymes can also bind to DNA and of these, the polymerases that copy the DNA base sequence in transcription and DNA replication are particularly important.

DNA-binding Proteins

Structural proteins that bind DNA are well-understood examples of non-specific DNA-protein interactions. Within chromosomes, DNA is held in complexes with structural proteins. These proteins organize the DNA into a compact structure called chromatin. In eukaryotes this structure involves DNA binding to a complex of small basic proteins called histones, while in prokaryotes multiple types of proteins are involved. The histones form a disk-shaped complex called a nucleosome, which contains two complete turns of double-stranded DNA wrapped around its surface.

These non-specific interactions are formed through basic residues in the histones making ionic bonds to the acidic sugar-phosphate backbone of the DNA, and are therefore largely independent of the base sequence. Chemical modifications of these basic amino acid residues include methylation, phosphorylation and acetylation. These chemical changes alter the strength of the interaction between the DNA and the histones, making the DNA more or less accessible to transcription factors and changing the rate of transcription. Other non-specific DNA-binding proteins in chromatin include the high-mobility group proteins, which bind to bent or distorted DNA. These proteins are important in bending arrays of nucleosomes and arranging them into the larger structures that make up chromosomes.

A distinct group of DNA-binding proteins are the DNA-binding proteins that specifically bind single-stranded DNA. In humans, replication protein A is the best-understood member of this family and is used in processes where the double helix is separated, including DNA replication, recombination and DNA repair. These binding proteins seem to stabilize single-stranded DNA and protect it from forming stem-loops or being degraded by nucleases.

In contrast, other proteins have evolved to bind to particular DNA sequences. The most intensively studied of these are the various transcription factors, which are proteins that regulate transcription. Each transcription factor binds to one particular set of DNA sequences and activates or inhibits the transcription of genes that have these sequences close to their promoters. The transcription factors do this in two ways.

Firstly, they can bind the RNA polymerase responsible for transcription, either directly or through other mediator proteins; this locates the polymerase at the promoter and allows it to begin transcription. Alternatively, transcription factors can bind enzymes that modify the histones at the promoter; this will change the accessibility of the DNA template to the polymerase.

As these DNA targets can occur throughout an organism's genome, changes in the activity of one type of transcription factor can affect thousands of genes. Consequently, these proteins are often the targets of the signal transduction processes that control responses to environmental changes or cellular differentiation and development. The specificity of these transcription factors' interactions with DNA come from the proteins making multiple contacts to the edges of the DNA bases, allowing them to "read" the DNA sequence. Most of these base-interactions are made in the major groove, where the bases are most accessible.

DNA-modifying Enzymes

Nucleases and Ligases

Nucleases are enzymes that cut DNA strands by catalyzing the hydrolysis of the phosphodiester bonds. Nucleases that hydrolyse nucleotides from the ends of DNA strands are called exonucleases, while endonucleases cut within strands. The most frequently used nucleases in molecular biology are the restriction endonucleases, which cut DNA at specific sequences. For instance, the EcoRV enzyme shown to the left recognizes the 6-base sequence 52 -

GAT|ATC-32 and makes a cut at the vertical line. In nature, these enzymes protect bacteria against phage infection by digesting the phage DNA when it enters the bacterial cell, acting as part of the restriction modification system. In technology, these sequence-specific nucleases are used in molecular cloning and DNA fingerprinting.

Enzymes called DNA ligases can rejoin cut or broken DNA strands. Ligases are particularly important in lagging strand DNA replication, as they join together the short segments of DNA produced at the replication fork into a complete copy of the DNA template. They are also used in DNA repair and genetic recombination.

Topoisomerases and Helicases

Topoisomerases are enzymes with both nuclease and ligase activity. These proteins change the amount of supercoiling in DNA. Some of these enzymes work by cutting the DNA helix and allowing one section to rotate, thereby reducing its level of supercoiling; the enzyme then seals the DNA break. Other types of these enzymes are capable of cutting one DNA helix and then passing a second strand of DNA through this break, before rejoining the helix. Topoisomerases are required for many processes involving DNA, such as DNA replication and transcription.

Helicases are proteins that are a type of molecular motor. They use the chemical energy in nucleoside triphosphates, predominantly ATP, to break hydrogen bonds between bases and unwind the DNA double helix into single strands. These enzymes are essential for most processes where enzymes need to access the DNA bases.

Polymerases

Polymerases are enzymes that synthesize polynucleotide chains from nucleoside triphosphates. The sequence of their products are copies of existing polynucleotide chains – which are called templates. These enzymes function by adding nucleotides onto the 32 hydroxyl group of the previous nucleotide in a DNA strand. As a consequence, all polymerases work in a 52 to 32 direction. In the active site of these enzymes, the incoming nucleoside triphosphate base-pairs to the template: this allows polymerases to accurately synthesize the complementary strand of their template. Polymerases are classified according to the type of template that they use.

In DNA replication, a DNA-dependent DNA polymerase makes a copy of a DNA sequence. Accuracy is vital in this process, so many of these polymerases have a proofreading activity. Here, the polymerase recognizes the occasional mistakes in the synthesis reaction by the lack of base pairing between the mismatched nucleotides. If a mismatch is detected, a 32 to 52 exonuclease activity is activated and the incorrect base removed. In most organisms, DNA polymerases function in a large complex called the replisome that contains multiple accessory subunits, such as the DNA clamp or helicases.

RNA-dependent DNA polymerases are a specialized class of polymerases that copy the sequence of an RNA strand into DNA. They include reverse transcriptase, which is a viral enzyme involved in the infection of cells by retroviruses, and telomerase, which is required for the replication of telomeres. Telomerase is an unusual polymerase because it contains its own RNA template as part of its structure.

Transcription is carried out by a DNA-dependent RNA polymerase that copies the sequence of a DNA strand into RNA. To begin transcribing a gene, the RNA polymerase binds to a sequence of DNA called a promoter and separates the DNA strands. It then copies the gene sequence into a messenger RNA transcript until it reaches a region of DNA called the terminator, where it halts and detaches from the DNA. As with human DNA-dependent DNA polymerases, RNA polymerase II, the enzyme that transcribes most of the genes in the human genome, operates as part of a large protein complex with multiple regulatory and accessory subunits.

10

Prognosis and Treatment of Genetic Disorders

Genetic disorders rarely have effective treatments, though gene therapy is being tested as a possible treatment for some genetic diseases, including some forms of retinitis pigmentosa Huntington's disease clinical research

There are different therapies under investigation for Huntington's disease. Huntington's disease is a genetic neurological disorder characterized after onset by uncoordinated, jerky body movements and a decline in some mental abilities for which there is no cure or effective treatments. Animal models help in the ongoing investigations using intrabody therapy, gene silencing, or stem cell implants, but all are at their first steps. Several clinical trials of various compounds are also in development. Many of these trials use the unified Huntington's disease rating scale (UHDRS) developed by the Huntington's Study Group. This provides an overall rating system based on motor, behavioral, cognitive, and functional assessments, backed up by a reference database of previous results provided by HSG for reference. Another avenue of research is understanding the physical mechanics of protein folding, HD is one of several targets of the folding@home project, one of the largest distributed computing systems on the World Wide Web. Animal models

Appropriate animal models are critical for understanding the fundamental mechanisms causing the disease and for supporting the early stages of drug development. Neurochemically-induced mice or monkeys were first available, but they did not mimic the progressive features of the disease. After the HD gene was discovered, transgenic animals exhibiting HD were generated by inserting a CAG repeat expansion into the genome, of mice (strain R6/2),), Drosophila fruit flies, and more recently monkeys. Expression without insertion of a DNA repeat in nematode worms also produced a valuable model.

Intrabody Therapy

Genetically-engineered intracellular antibody fragments called intrabodies have shown therapeutic results, by inhibiting mHtt aggregation, in Drosophila models. This was achieved using an intrabody called C4 sFv, a single chain variable fragment which binds to the end of mHtt within a cell. This therapy prevented larval and pupal mortality (without therapy 77% died) and delayed neurodegeneration in the adult, significantly increasing their lifespan. Intrabodies have shown promise as a useful tool for drug discovery, and may potentially be used as a therapy for HD and other neurodegenerative disorders caused by protein mis-folding or abnormal protein interactions.

Gene Silencing

As HD has been conclusively linked to a single gene, gene silencing is potentially possible. Researchers have investigated using gene knockdown of mHTT as a potential treatment. Using a mouse model, siRNA therapy achieved a 60% reduction in expression of mHTT and progression of the disease was stalled. However, this study used the human form of the mHTT protein in mice, and was therefore only able to directly target the mHTT, leaving endogenous, wild-type mouse Htt gene expression unaffected. The widely-held hypothesis that silencing the wild type mouse Htt in middle aged mice, the age of onset of HD, renders undesirable, gene silencing to treat HD, is now largely discredited by experiment. These data indicate that the coincident silencing of the wild-type and mutant htt may be considered as a therapeutic tool for HD. This is despite the unchallenged experimental evidence that wild-type Htt is critical for development in mouse models. Another study showed that mouse models already in late stages of the disease recovered motor function after expression of mHTT was stopped.

Stem Cell Treatments

Stem cell treatments are a type of intervention strategy that introduces new cells into damaged tissue in order to treat disease or injury. Many medical researchers believe that stem cell treatments have the potential to change the face of human disease and alleviate suffering. The ability of stem cells to self-renew and give rise to subsequent generations with variable degrees of differentiation capacities, offers significant potential for generation of tissues that can potentially replace diseased and damaged areas in the body, with minimal risk of rejection and side effects. A number of stem cell therapies exist, but most are at experimental stages or costly, with the notable exception of bone marrow transplantation. Medical researchers anticipate that adult and embryonic stem cells will soon be able to treat cancer, Type 1 diabetes mellitus, Parkinson's disease, Huntington's disease, Celiac Disease, cardiac failure, muscle damage and neurological disorders, and many others. Nevertheless, before stem cell therapeutics can be applied in the clinical setting, more research is necessary to understand stem cell behavior upon transplantation as well as the mechanisms of stem cell interaction with the diseased/injured microenvironment.

Stem cell implants are a stem cell treatment where damaged neurons are replaced by transplantation of stem cells (or possibly neural stem cells—a type of somatic [adult] stem cell) into affected regions of the brain. Hypothetically, embryonic stem cells can be differentiated into neuronal precursors in vitro, and transplanted into damaged areas of the brain to generate replacement neurons; if enough damaged neurons can be replaced and develop the correct synaptic connectivity, symptoms could be alleviated. This treatment would not prevent further neuronal damage, so it would have to be an ongoing treatment. Experiments have yielded some positive results in animal models, but stem cell treatment for HD remains highly speculative and has not been tested in clinical trials. Current treatments

For over 30 years, bone marrow, and more recently, umbilical cord blood stem cells, have been used to treat cancer patients with conditions such as leukemia and lymphoma. During chemotherapy, most growing cells are killed by the cytotoxic agents. These agents, however, cannot discriminate between the leukemia or neoplastic cells, and the hematopoietic stem cells within the bone marrow. It is this side effect of conventional chemotherapy strategies that the stem cell transplant attempts to reverse; a donor's healthy bone marrow reintroduces functional stem cells to replace the cells lost in the host's body during treatment.

Hematopoietic Stem Cell Transplantation

Hematopoietic stem cell transplantation (HSCT) is the transplantation of multipotent hematopoietic stem cell or blood, usually derived from bone marrow, peripheral blood stem cells, or umbilical cord blood. Stem cell transplantation is a medical procedure in the fields of hematology and oncology, most often performed for people with diseases of the blood, bone marrow, or certain cancers. Hematopoietic stem cell transplantation remains a risky procedure with many possible complications; it has traditionally been reserved for patients with life-threatening diseases. While occasionally used experimentally in nonmalignant and nonhematologic indications such as severe disabling auto-immune disease and cardiovascular disease, the risk of fatal complications appears too high to gain wider acceptance.

A total of 50,417 first hematopoietic stem cell transplants were reported as taking place worldwide in 2006, according to a global survey of 1327 centers in 71 countries conducted by the Worldwide Network for Blood and Marrow Transplantation. Of these, 28,901 (57%) were autologous and 21,516 (43%) were allogeneic (11,928 from family donors and 9,588 from unrelated donors). The main indications for transplant were lymphoproliferative disorders (54.5%) and leukemias (33.8%), and the majority took place in either Europe (48%) or the Americas (36%). In 2009, according to the world marrow donor association, stem cell products provided for unrelated transplantation worldwide had increased to 15,399 (3,445 bone marrow donations, 8,162 peripheral blood stem cell donations, and 3,792 cord blood units).

While it remains a procedure involving a high degree of risk, one must keep in mind that many disorders (mainly acute leukemias and neoplastic lymphoproliferative disorders) treated by this procedure—which is usually not attempted unless necessary because of prior treatment failures or the patient's emergent condition (unresponsive or severe acute leukemia, relapsed acute leukemia, or progressing, formerly chronic leukemias that have not responded to other regimens and have entered blast crisis)- themselves will usually carry a mortality and morbidity rate that is as high and often higher than that for the procedure. Further, less severe complications from this procedure, and even those that carry a worse prognosis such as many cases of acute graft versus host disease, can be dealt with at least to a point, if the patient was in acceptable health beforehand and the condition was noticed and aggressively treated early, with modern standards of tertiary medical care.

History

Georges Mathé, a French oncologist, performed the first bone marrow transplant in 1959 on five Yugoslavian nuclear workers whose own marrow had been damaged by irradiation caused by a Criticality accident at the Vinèa Nuclear Institute, but all of these transplants were rejected. Mathé later pioneered the use of bone marrow transplants in the treatment of leukemia.

Stem cell transplantation was pioneered using bone-marrow-derived stem cells by a team at the Fred Hutchinson Cancer Research Center from the 1950s through the 1970s led by E. Donnall Thomas, whose work was later recognized with a Nobel Prize in Physiology or Medicine. Thomas' work showed that bone marrow cells infused intravenously could repopulate the bone marrow and produce new blood cells. His work also reduced the likelihood of developing a life-threatening complication called graft-versus-host disease.

The first physician to perform a successful human bone marrow transplant on a disease other than cancer was Robert A. Good at the University of Minnesota in 1968.

Indications

Many recipients of HSCTs are multiple myeloma or leukemia patients who would not benefit from prolonged treatment with, or are already resistant to, chemotherapy. Candidates for HSCTs include pediatric cases where the patient has an inborn defect such as severe combined immunodeficiency or congenital neutropenia with defective stem cells, and also children or adults with aplastic anemia who have lost their stem cells after birth. Other conditions treated with stem cell transplants include sickle-cell disease, myelodysplastic syndrome, neuroblastoma, lymphoma, Ewing's Sarcoma, Desmoplastic small round cell tumor, chronic granulomatous disease and Hodgkin's disease. More recently non-myeloablative, or so-called "mini transplant," procedures have been developed that require smaller doses of preparative chemo and radiation. This has allowed HSCT to be conducted in the elderly and other patients who would otherwise be considered too weak to withstand a conventional treatment regimen.

List of conditions treated with hematopoietic stem cell transplantation. Hematopoietic stem cell transplantation may be used to treat a number of conditions both congenital and acquired.

Acquired

- Malignancies
  - o Hematological
    - § Leukemias
      - § Acute lymphoblastic leukemia (ALL)
      - § Acute myeloid leukemia (AML)
      - § Chronic lymphocytic leukemia (CLL)
      - § Chronic myelogenous leukemia (CML), accelerated phase or blast crisis
    - § Lymphomas
      - § Hodgkin's disease
      - § Non-Hodgkin's lymphoma
    - § Myelomas
      - § Multiple myeloma (Kahler's disease)
  - o Solid tumor cancers
    - § Neuroblastoma
    - § Desmoplastic small round cell tumor
    - § Ewing's sarcoma
    - § Choriocarcinoma

- Hematologic disease
  - Phagocyte disorders
    - Myelodysplasia
  - Anemias
    - Paroxysmal nocturnal hemoglobinuria (PNH; severe aplasia)
    - Aplastic anemia
    - Acquired pure red cell aplasia
  - Myeloproliferative disorders
    - Polycythemia vera
    - Essential thrombocytosis
    - Myelofibrosis
- Metabolic disorders
  - Amyloidoses
    - Amyloid light chain (AL) amyloidosis
- Environmentally-induced diseases
  - Radiation poisoning
- Viral diseases
  - HIV

Congenital

- Lysosomal storage disorders
  - Lipidoses (disorders of lipid storage)
    - Neuronal ceroid lipofuscinoses
      - Infantile neuronal ceroid lipofuscinosis (INCL, Santavuori disease,)
      - Jansky-Bielschowsky disease (late infantile neuronal ceroid lipofuscinosis)
    - Sphingolipidoses
      - Niemann-Pick disease
      - Gaucher disease
    - Leukodystrophies
      - Adrenoleukodystrophy
      - Metachromatic leukodystrophy
      - Krabbe disease (globoid cell leukodystrophy)

- Mucopolysaccharidoses
  - Hurler syndrome (MPS I H, á-L-iduronidase deficiency)
  - Scheie syndrome (MPS I S)
  - Hurler-Scheie syndrome (MPS I H-S)
  - Hunter syndrome (MPS II, iduronidase sulfate deficiency)
  - Sanfilippo syndrome (MPS III)
  - Morquio syndrome (MPS IV)
  - Maroteaux-Lamy syndrome (MPS VI)
  - Sly syndrome (MPS VII)
- Glycoproteinoses
  - Mucolipidosis II (I-cell disease)
  - Fucosidosis
  - Aspartylglucosaminuria
  - Alpha-mannosidosis
- Other
  - Wolman disease (acid lipase deficiency)

- Immunodeficiencies
  - T-cell deficiencies
    - Ataxia telangiectasia
    - DiGeorge syndrome
  - Combined T- and B-cell deficiencies
    - Severe combined immunodeficiency (SCID), all types
  - Well-defined syndromes
    - Wiskott-Aldrich syndrome
  - Phagocyte disorders
    - Kostmann syndrome
    - Shwachman-Diamond syndrome
  - Immune dysregulation diseases
    - Griscelli syndrome, type II
  - Innate immune deficiencies
    - NF-Kappa-B Essential Modulator (NEMO) deficiency (Inhibitor of Kappa Light Polypeptide Gene Enhancer in B Cells Gamma Kinase deficiency)

- Hematologic diseases
    - § Hemoglobinopathies
    - § Sickle cell disease
    - § â thalassemia major (Cooley's anemia)
  - o Anemias
    - § Aplastic anemia
      - § Diamond-Blackfan anemia
      - § Fanconi anemia
  - o Cytopenias
    - § Amegakaryocytic thrombocytopenia
  - o Hemophagocytic syndromes
    - § Hemophagocytic lymphohistiocytosis (HLH)

Graft Types

Autologous

Autologous HSCT requires the extraction (apheresis) of haematopoietic stem cells (HSC) from the patient and storage of the harvested cells in a freezer. The patient is then treated with high-dose chemotherapy with or without radiotherapy with the intention of eradicating the patient's malignant cell population at the cost of partial or complete bone marrow ablation (destruction of patient's bone marrow function to grow new blood cells). The patient's own stored stem cells are then returned to his/her body, where they replace destroyed tissue and resume the patient's normal blood cell production. Autologous transplants have the advantage of lower risk of infection during the immune-compromised portion of the treatment since the recovery of immune function is rapid. Also, the incidence of patients experiencing rejection (graft-versus-host disease) is very rare due to the donor and recipient being the same individual. These advantages have established autologous HSCT as one of the standard second-line treatments for such diseases as lymphoma. However, for others such as Acute Myeloid Leukemia, the reduced mortality of the autogenous relative to allogeneic HSCT may be outweighed by an increased likelihood of cancer relapse and related mortality, and therefore the allogeneic treatment may be preferred for those conditions. Researchers have conducted small studies using non-myeloablative hematopoietic stem cell transplantation as a possible treatment for type I (insulin dependent) diabetes in children and adults. Results have been promising; however, as of 2009 it was premature to speculate whether these experiments will lead to effective treatments for diabetes.

Allogeneic

Allogeneic HSCT involves two people: the (healthy) donor and the (patient) recipient. Allogeneic HSC donors must have a tissue (HLA) type that matches the recipient. Matching is performed on the basis of variability at three or more loci of the HLA gene, and a perfect match at these loci is preferred. Even if there is a good match at these critical alleles, the recipient will require immunosuppressive medications to mitigate graft-versus-host disease. Allogeneic transplant donors may be related (usually a closely HLA matched sibling), syngeneic (a monozygotic or 'identical' twin of the patient - necessarily extremely rare since few patients have an identical twin, but offering a source of perfectly HLA matched stem cells) or unrelated (donor who is not related and found to have very close degree of HLA matching). Unrelated donors may be found through a registry of bone marrow donors such as the National Marrow Donor Program. People who would like to be tested for a specific family member or friend without joining any of the bone marrow registry data banks may contact a private HLA testing laboratory and be tested with a mouth swab to see if they are a potential match. A "savior sibling" may be intentionally selected by preimplantation genetic diagnosis in order to match a child both regarding HLA type and being free of any obvious inheritable disorder. Allogeneic transplants are also performed using umbilical cord blood as the source of stem cells. In general, by transplanting healthy stem cells to the recipient's immune system, allogeneic HSCTs appear to improve chances for cure or long-term remission once the immediate transplant-related complications are resolved.

A compatible donor is found by doing additional HLA-testing from the blood of potential donors. The HLA genes fall in two categories (Type I and Type II). In general, mismatches of the Type-I genes (i.e. HLA-A, HLA-B, or HLA-C) increase the risk of graft rejection. A mismatch of an HLA Type II gene (i.e. HLA-DR, or HLA-DQB1) increases the risk of graft-versus-host disease. In addition a genetic mismatch as small as a single DNA base pair is significant so perfect matches require knowledge of the exact DNA sequence of these genes for both donor and recipient. Leading transplant centers currently perform testing for all five of these HLA genes before declaring that a donor and recipient are HLA-identical. Race and ethnicity are known to play a major role in donor recruitment drives, as members of the same ethnic group are more likely to have matching genes, including the genes for HLA.

Sources and Storage of Cells

To limit the risks of transplanted stem cell rejection or of severe graft-versus-host disease in allogeneic HSCT, the donor should preferably have the same human leukocyte antigens (HLA) as the recipient. About 25 to 30 percent of allogeneic HSCT recipients have an HLA-identical sibling. Even so-called "perfect matches" may have mismatched minor alleles that contribute to graft-versus-host disease.

Bone Marrow

In the case of a bone marrow transplant, the HSC are removed from a large bone of the donor, typically the pelvis, through a large needle that reaches the center of the bone. The technique is referred to as a bone marrow harvest and is performed under general anesthesia.

Peripheral Blood Stem Cells

Peripheral stem cell transplantation is a method of replacing blood-forming cells destroyed by cancer treatment. Immature blood cells (stem cells) in the circulating blood that are similar to those in the bone marrow are given to the patient after treatment. This helps the bone marrow recover and continue producing healthy blood cells. Transplantation may be autologous (an individual's own blood cells saved earlier), allogeneic (blood cells donated by someone else), or syngeneic (blood cells donated by an identical twin). Also called peripheral stem cell support.

Peripheral blood stem cells are now the most common source of stem cells for allogeneic HSCT. They are collected from the blood through a process known as apheresis. The donor's blood is withdrawn through a sterile needle in one arm and passed through a machine that removes white blood cells. The red blood cells are returned to the donor. The peripheral stem cell yield is boosted with daily subcutaneous injections of Granulocyte-colony stimulating factor, serving to mobilize stem cells from the donor's bone marrow into the peripheral circulation.

Amniotic Fluid

It is also possible to extract hematopoietic stem cells from amniotic fluid for both autologous or heterologous use at the time of childbirth.

Umbilical Cord Blood

Umbilical cord blood is obtained when a mother donates her infant's umbilical cord and placenta after birth. Cord blood has a higher concentration of HSC than is normally found in adult blood. However, the small quantity of blood obtained from an umbilical cord (typically about 50 mL) makes it more suitable for transplantation into small children than into adults. Newer techniques using ex-vivo expansion of cord blood units or the use of two cord blood units from different donors are being explored to allow cord blood transplants to be used in adults. Cord blood is typically harvested from the umbilical cord of a child being born after preimplantation genetic diagnosis

(PGD) for human leucocyte antigen (HLA) matching in order to donate to an ill sibling requiring HSCT.

Storage of HSC

Unlike other organs, bone marrow cells can be frozen (cryopreserved) for prolonged periods, without damaging too many cells. This is a necessity with autologous HSC because the cells must be harvested from the recipient months in advance of the transplant treatment. In the case of allogeneic transplants, fresh HSC are preferred, in order to avoid cell loss that might occur during the freezing and thawing process. Allogeneic cord blood is stored frozen at a cord blood bank because it is only obtainable at the time of childbirth. To cryopreserve HSC, a preservative, DMSO, must be added, and the cells must be cooled very slowly in a controlled-rate freezer to prevent osmotic cellular injury during ice crystal formation. HSC may be stored for years in a cryofreezer, which typically utilizes liquid nitrogen.

Conditioning Regimens

Myeloablative Transplants

The chemotherapy or irradiation given immediately prior to a transplant is called the conditioning or preparative regimen, the purpose of which is to help eradicate the patient's disease prior to the infusion of HSC and to suppress immune reactions. The bone marrow can be ablated with dose-levels that cause minimal injury to other tissues. In allogeneic transplants a combination of cyclophosphamide with busulfan or total body irradiation is commonly employed. This treatment also has an immunosuppressive effect that prevents rejection of the HSC by the recipient's immune system. The post-transplant prognosis often includes acute and chronic graft-versus-host disease that may be life-threatening. However in certain leukemias this can coincide with protection against cancer relapse owing to the graft versus tumor effect. Autologous transplants may also use similar conditioning regimens, but many other chemotherapy combinations can be used depending on the type of disease.

Non-myeloablative Allogeneic Transplants

This is a newer treatment approach using lower doses of chemotherapy and radiation, which are too low to eradicate all the bone marrow cells of a recipient. Instead, non-myeloablative transplants run lower risks of serious infections and transplant-related mortality while relying upon the graft versus tumor effect to resist the inherent increased risk of cancer relapse. Also significantly, while requiring high doses of immunosuppressive agents in the early stages of treatment, these doses are less than for conventional transplants. This leads to a state of mixed chimerism early after transplant where both recipient and donor HSC coexist in the bone marrow space.

Decreasing doses of immunosuppressive therapy then allows donor T-cells to eradicate the remaining recipient HSC and to induce the graft versus tumor effect. This effect is often accompanied by mild graft-versus-host disease, the appearance of which is often a surrogate for the emergence of the desirable graft versus tumor effect, and also serves as a signal to establish an appropriate dosage level for sustained treatment with low levels of immunosuppressive agents. Because of their gentler conditioning regimens, these transplants are associated with a lower risk of transplant-related mortality and therefore allow patients who are considered too high-risk for conventional allogeneic HSCT to undergo potentially curative therapy for their disease. These new transplant strategies are still somewhat experimental, but are being used more widely on elderly patients unfit for myeloablative regimens and for whom the higher risk of cancer relapse may be acceptable.

Engraftment

After several weeks of growth in the bone marrow, expansion of HSC and their progeny is sufficient to normalize the blood cell counts and reinitiate the immune system. The offspring of donor-derived hematopoietic stem cells have been documented to populate many different organs of the recipient, including the heart, liver, and muscle, and these cells had been suggested to have the abilities of regenerating injured tissue in these organs. However, recent research has shown that such lineage infidelities does not occur as a normal phenomenon.

Complications

HSCT is associated with a high treatment-related mortality in the recipient (10% or higher) , which limits its use to conditions that are themselves life-threatening. Major complications are veno-occlusive disease, mucositis, infections (sepsis), graft-versus-host disease and the development of new malignancies.

Infection

Bone marrow transplantation usually requires that the recipient's own bone marrow be destroyed ("myeloablation"). Prior to "engraftment" patients may go for several weeks without appreciable numbers of white blood cells to help fight infection. This puts a patient at high risk of infections, sepsis and septic shock, despite prophylactic antibiotics. However, antiviral medications, such as acyclovir and valacyclovir, are quite effective in prevention of HSCT-related outbreak of herpetic infection in seropositive patients. The immunosuppressive agents employed in allogeneic transplants for the prevention or treatment of graft-versus-host disease further increase the risk of opportunistic infection. Immunosuppressive drugs are given for a minimum of 6-months after a transplantation, or much longer if required for the treatment of graft-versus-host disease. Transplant patients lose their acquired immunity, for example immunity to childhood diseases such as measles or polio. For this reason transplant patients must be re-vaccinated with childhood vaccines once they are off immunosuppressive medications.

Veno-occlusive Disease

Severe liver injury can result from hepatic veno-occlusive disease (VOD). Elevated levels of bilirubin, hepatomegaly and fluid retention are clinical hallmarks of this condition. There is now a greater appreciation of the generalized cellular injury and obstruction in hepatic vein sinuses, and hepatic VOD has lately been referred to as sinusoidal obstruction syndrome (SOS). Severe casesof SOS are associated with a high mortality rate. Anticoagulants or defibrotide may be effective in reducing the severity of VOD but may also increase bleeding complications. Ursodiol has been shown to help prevent VOD, presumably by facilitating the flow of bile.

Mucositis

The injury of the mucosal lining of the mouth and throat is a common regimen-related toxicity following ablative HSCT regimens. It is usually not life-threatening but is very painful, and prevents eating and drinking. Mucositis is treated with pain medications plus intravenous infusions to prevent dehydration and malnutrition.

Graft-versus-host Disease

Graft-versus-host disease (GVHD) is a common complication after a stem cell transplant or bone marrow transplant from another person (an allogeneic transplant). Immune cells (white blood cells) in the donated marrow or stem cells (the graft) recognize the recipient (the host) as "foreign". The transplanted immune cells then attack the host's body cells. Graft versus host disease can also take place during a blood transfusion under certain circumstances.

Graft-versus-host disease (GVHD) is an inflammatory disease that is unique to allogeneic transplantation. It is an attack of the "new" bone marrow's immune cells against the recipient's tissues. This can occur even if the donor and recipient are HLA-identical because the immune system can still recognize other differences between their tissues. It is aptly named graft-versus-host disease because bone marrow transplantation is the only transplant procedure in which the transplanted cells must accept the body rather than the body accepting the new cells. Acute graft-versus-host disease typically occurs in the first 3 months after transplantation and may involve the skin, intestine, or the liver.. High-dose corticosteroids such as prednisone are a standard treatment; however this immuno-suppressive treatment often leads to deadly infections. Chronic graft-versus-host disease may also develop after allogeneic transplant.

It is the major source of late treatment-related complications, although it less often results in death. In addition to inflammation, chronic graft-versus-host disease may lead to the development of fibrosis, or scar tissue, similar to scleroderma; it may cause functional disability and require prolonged immunosuppressive therapy. Graft-versus-host disease is usually mediated by T cells, which react to foreign peptides presented on the MHC of the host .

Causes

According to the 1966 Billingham Criteria, 3 criteria must be met in order for GVHD to occur.

- An immuno-competent graft is administered, with viable and functional immune cells.
- The recipient is immunologically disparate - histo-incompatible.
- The recipient is immuno-compromised and therefore cannot destroy or inactivate the transplanted cells.

After bone marrow transplantation, T cells present in the graft, either as contaminants or intentionally introduced into the host, attack the tissues of the transplant recipient after perceiving host tissues as antigenically foreign. The T cells produce an excess of cytokines, including TNF-á and interferon-gamma (IFNã). A wide range of host antigens can initiate graft-versus-host-disease, among them the human leukocyte antigens (HLAs). However, graft-versus-host disease can occur even when HLA-identical siblings are the donors. HLA-identical siblings or HLA-identical unrelated donors often have genetically different proteins (called minor histocompatibility antigens) that can be presented by Major histocompatibility complex (MHC) molecules to the donor's T-cells, which see these antigens as foreign and so mount an immune response. While donor T-cells are undesirable as effector cells of graft-versus-host-disease, they are valuable for engraftment by preventing the recipient's residual immune system from rejecting the bone marrow graft (host-versus-graft). In addition, as bone marrow transplantation is frequently used to treat cancer, mainly leukemias, donor T-cells have proven to have a valuable graft-versus-tumor effect. A great deal of current research on allogeneic bone marrow transplantation involves attempts to separate the undesirable graft-vs-host-disease aspects of T-cell physiology from the desirable graft-versus-tumor effect.

Types

In the clinical setting, graft-versus-host-disease is divided into acute and chronic forms.

- The acute or fulminant form of the disease (aGVHD) is normally observed within the first 100 days post-transplant, and is a major challenge to transplants owing to associated morbidity and mortality.

- The chronic form of graft-versus-host-disease (cGVHD) normally occurs after 100 days. The appearance of moderate to severe cases of cGVHD adversely influences long-term survival.

This distinction is not arbitrary: Acute and chronic graft-versus-host-disease appear to involve different immune cell subsets, different cytokine profiles, somewhat different host targets, and respond differently to treatment.

Clinical Manifestation

In the classical sense, acute graft-versus-host-disease is characterized by selective damage to the liver, skin (rash), and mucosa, and the gastrointestinal tract. Newer research indicates that other graft-versus-host-disease target organs include the immune system (the hematopoietic system, e.g., the bone marrow and the thymus) itself, and the lungs in the form of idiopathic pneumonitis. Chronic graft-versus-host-disease also attacks the above organs, but over its long-term course can also cause damage to the connective tissue and exocrine glands. Acute GVHD of the GI tract can result in severe intestinal inflammation, sloughing of the mucosal membrane, severe diarrhea, abdominal pain, nausea, and vomiting. This is typically diagnosed via intestinal biopsy. Liver GVHD is measured by the bilirubin level in acute patients. Skin GVHD results in a diffuse maculopapular rash, sometimes in a lacy pattern.

It appears in both acute and chronic GVHD, mucosal damage to the vagina can result in severe pain and scarring. This can result in an inability to have sexual intercourse. Acute GVHD is staged as follows: overall grade (skin-liver-gut) with each organ staged individually from a low of 1 to a high of 4. Patients with grade IV GVHD usually have a poor prognosis. If the GVHD is severe and requires intense immunosuppression involving steroids and additional agents to get under control, the patient may develop severe infections as a result of the immunosuppression and may die of infection.

In the oral cavity, chronic graft-versus-host-disease manifests as lichen planus with a higher risk of malignant transformation to oral squamous cell carcinoma in comparison to the classical oral lichen planus. Graft-versus-host-disease-associated oral cancer may have more aggressive behavior with poorer prognosis, when compared to oral cancer in non-hematopoietic stem cell transplantation patients.

Transfusion-associated GVHD

Transfusion-associated graft versus host disease (TA-GvHD) is a rare complication of blood transfusion, in which the donor T lymphocytes mount an immune response against the recipient's lymphoid tissue. Donor lymphocytes are usually identified as foreign and destroyed by the recipient's immune system.

However, in situations where the recipient is immunocompromised (inborn immunodeficiency, acquired immunodeficiency, malignancy), or when the donor is homozygous and the recipient is heterozygous for an HLA haplotype (as can occur in directed donations from first-degree relatives), the recipient's immune system is not able to destroy the donor lymphocytes. This can result in graft versus host disease.

This type of GVHD is associated with transfusion of un-irradiated blood to immunocompromised recipients. It can also occur in situations in which the blood donor is homozygous and the recipient is heterozygous for an HLA haplotype.

It is associated with higher mortality (80-90%) due to involvement of bone marrow lymphoid tissue, however the clinical manifestations are similar to GVHD resulting from bone marrow transplantation. Transfusion-associated GVHD is rare in modern medicine. It is almost entirely preventable by controlled irradiation of blood products to inactivate the white blood cells (including lymphocytes) within. Epidemiology and pathogenesis

The incidence in immunocompromised patients receiving blood transfusions is estimated to be 0.1 - 1.0%, mortality around 80 - 90%. Mortality is higher in TA-GvHD than in GvHD associated with bone marrow transplantation, where the engrafted lymphoid cells in the bone marrow are of donor origin; therefore, the immune reaction is not directed against them.

The most common cause of death in TA-GvHD is infections and hemorrages, secondary to pancytopenia and liver dysfunction.

Presentation and Diagnosis

Clinical Manifestations

The clinical presentation is the same as GvHD occurring in other settings, such as bone marrow transplantation. TA-GvHD can develop four to thirty days after the transfusion. Typical symptoms include:

- fever
- erythematous maculopapular rash, which can progress to generalised erythroderma
- toxic epidermal necrolysis in extreme cases

Other symptoms can include cough, abdominal pain, vomiting, and profuse diarrhea (up to 8 liters/day).

Laboratory Manifestations

Laboratory findings include pancytopenia, abnormal liver enzymes, and electrolyte imbalance (when diarrhea is present).

Diagnosis

TA-GvHD can be suspected from a biopsy of the affected skin, and established by HLA analysis of the circulating lymphocytes. This testing can identify circulating lymphocytes with a different HLA type than the tissue cells of the host.

Treatment and Prevention

Treatment is only supportive, as no available form of therapy has proven effective in treating TA-GvHD. Prevention includes gamma irradiation of the lymphocyte-containing blood products. This procedure should be performed in transfusions when:

- the recipient is immunocompromised
- the blood components are from a family donor
- HLA-matched platelets are transferred.

Another means of prevention is the use of third- or fourth-generation leukoreduction filters, although the efficacy of this procedure has not yet been documented.

In Thymus Transplantation

Thymus transplantation may be said to be able to cause a special type of GVHD because the recipients thymocytes would use the donor thymus cells as models when going through the negative selection to recognize self-antigens, and could therefore still mistake own structures in the rest of the body for being non-self. This is a rather indirect GVHD because it is not directly cells in the graft itself that causes it but cells in the graft that make the recipient's T cells act like donor T cells. It can be seen as a multiple-organ autoimmunity in xenotransplantation experiments of the thymus between different species. Autoimmune disease is a frequent complication after human allogeneic thymus transplantation, found in 42% of subjects over 1 year post transplantation. However, this is partially explained by the fact that the indication itself, that is, complete DiGeorge syndrome, increases the risk of autoimmune disease.

Prevention

- DNA-based tissue typing allows for more precise HLA matching between donors and transplant patients, which has been proven to reduce the incidence and severity of GVHD and to increase long-term survival.
- The T-cells of umbilical cord blood (UCB) have an inherent immunological immaturity, and the use of UCB stem cells in unrelated donor transplants has a reduced incidence and severity of GVHD.
- Methotrexate, ciclosporin and tacrolimus are common drugs used for GVHD prophylaxis.

- Graft-versus-host-disease can largely be avoided by performing a T-cell-depleted bone marrow transplant. However, these types of transplants come at a cost of diminished graft-versus-tumor effect, greater risk of engraftment failure, or cancer relapse, and general immunodeficiency, resulting in a patient more susceptible to viral, bacterial, and fungal infection. In a multi-center study, disease-free survival at 3 years was not different between T cell-depleted and T cell-replete transplants.

Treatment of GVHD

Intravenously administered corticosteroids, such as prednisone, are the standard of care in acute GVHD and chronic GVHD. The use of these corticosteroids is designed to suppress the T-cell-mediated immune onslaught on the host tissues; however, in high doses, this immune-suppression raises the risk of infections and cancer relapse. Therefore, it is desirable to taper off the post-transplant high-level steroid doses to lower levels, at which point the appearance of mild GVHD may be welcome, especially in HLA mis-matched patients, as it is typically associated with a graft-versus-tumor effect.

The use of irradiated blood components prevent GVHD. The irradiation of blood eliminates the proliferative capacity of lymphocytes present in red cell, platelet and freshly collected plasma components. The penetrated photons of radiation beam into blood components cause the formation of electrically charged particles or secondary electrons. These electrons damage the DNA of lymphocytes either by direct interaction or by reacting initicially with cell water to form free radicals. The damaged lymphocytes are unable to proliferate in the host and therefore cannot mediate transfusion-associated GVHD.

Investigational Therapies for Graft-versus-host Disease

There are a large number of clinical trials either ongoing or recently completed in the investigation of graft-versus-host disease treatment and prevention.

Graft-versus-tumor Effect

Graft-versus-tumor effect (GVT) or “graft versus leukemia” effect is the beneficial aspect of the Graft-versus-Host phenomenon. For example, HSCT patients with either acute and in particular chronic graft-versus-host disease after an allogeneic transplant tend to have a lower risk of cancer relapse. This is due to a therapeutic immune reaction of the grafted donor T lymphocytes against the diseased bone marrow of the recipient. This lower rate of relapse accounts for the increased success rate of allogeneic transplants, compared to transplants from identical twins, and indicates that allogeneic HSCT is a form of immunotherapy. GVT is the major benefit of transplants that do not employ the highest immuno-suppressive regimens. Graft versus tumor is mainly beneficial in diseases with slow progress, e.g. chronic leukemia, low-grade lymphoma, and some cases multiple myeloma. However, it is less effective in rapidly growing acute leukemias. If cancer relapses after HSCT, another transplant can be performed, infusing the patient with a greater quantity of donor white blood cells.

Oral Carcinoma

Patients after HSCT are at a higher risk for oral carcinoma. Post-HSCT oral cancer may have more aggressive behavior with poorer prognosis, when compared to oral cancer in non-HSCT patients.

Prognosis

Prognosis in HSCT varies widely dependent upon disease type, stage, stem cell source, HLA-matched status (for allogeneic HCST) and conditioning regimen. A transplant offers a chance for cure or long-term remission if the inherent complications of graft versus host disease, immuno-suppressive treatments and the spectrum of opportunistic infections can be survived. In recent years, survival rates have been gradually improving across almost all populations and sub-populations receiving transplants.

Mortality for allogeneic stem cell transplantation can be estimated using the prediction model created by Sorror et al., using the Hematopoietic Cell Transplantation-Specific Comorbidity Index (HCT-CI). The HCT-CI was derived and validated by investigators at the Fred Hutchinson Cancer Research Center (Seattle, WA). The HCT-CI modifies and adds to a well-validated comorbidity index, the Charlson Comorbidity Index (CCI) (Charlson et al.) The CCI was previously applied to patients undergoing allogeneic HCT but appears to provide less survival prediction and discrimination than the HCT-CI scoring system.

Risks to Donor

The risks of a complication depend on patient characteristics, health care providers and the apheresis procedure, and the colony-stimulating factor used (G-CSF, GM-CSF). G-CSF drugs include Filgrastim (Neupogen, Neulasta), and lenograstim (Graslopin).

Drug Risks

Filgrastim is typically dosed in the 10 microgram/kg level for 4–5 days during the harvesting of stem cells. The documented adverse effects of filgrastim include splenic rupture (indicated by left upper abdominal or shoulder pain, risk 1 in 40000), Adult respiratory distress syndrome (ARDS), alveolar hemorrage, and allergic reactions (usually expressed in first 30 minutes, risk

1 in 300). In addition, platelet and hemoglobin levels dip post-procedure, not returning to normal until one month.

The question of whether patients over 65 react the same as patients under 65 has not been sufficiently examined. Coagulation issues and inflammation of atherosclerotic plaques are known to occur as a result of G-CSF injection. G-CSF has also been described to induce genetic changes in mononuclear cells of normal donors. There is evidence that myelodysplasia (MDS) or acute myeloid leukaemia (AML) can be induced by GCSF in susceptible individuals.

Access Risks

Blood was drawn peripherally in a majority of patients, but a central line to jugular/subclavian/femoral veins may be used in 16% of women and 4% of men. Adverse reactions during apheresis were experienced in 20% of women and 8% of men, these adverse events primarily consisted of numbness/tingling, multiple line attempts, and nausea.

Clinical Observations

A study involving 2408 donors (18–60 years) indicated that bone pain (primarily back and hips) as a result of filgrastim treatment is observed in 80% of donors by day 4 post-injection. This pain responded to acetaminophen or ibuprofen in 65% of donors and was characterized as mild to moderate in 80% of donors and severe in 10%. Bone pain receded post-donation to 26% of patients 2 days post-donation, 6% of patients one week post-donation, and <2% 1 year post-donation. Donation is not recommended for those with a history of back pain. Other symptoms observed in more than 40% of donors include myalgia, headache, fatigue, and insomnia. These symptoms all returned to baseline 1 month post-donation, except for some cases of persistent fatigue in 3% of donors. . In one metastudy that incorporated data from 377 donors, 44% of patients reported having adverse side effects after peripheral blood HSCT. Side effects included pain prior to the collection procedure as a result of GCSF injections, post-procedural generalized skeletal pain, fatigue and reduced energy.

Severe Reactions

A study that surveyed 2408 donors found that serious adverse events (requiring prolonged hospitalization) occurred in 15 donors (at a rate of 0.6%), although none of these events were fatal. Donors were not observed to have higher than normal rates of cancer with up to 4–8 years of follow up. One study based on a survey of medical teams covered approximately 24,000 peripheral blood HSCT cases between 1993 and 2005, and found a serious cardiovascular adverse reaction rate of about 1 in 1500. This study reported a cardiovascular-related fatality risk within the first 30 days HSCT of about 2 in 10000. For this same group, severe cardiovascular events were observed with a rate of about 1 in 1500. The most common severe adverse reactions were pulmonary edema/deep vein thrombosis, splenic rupture, and myocardial infarction. Haematological malignancy induction was comparable to that observed in the general population with only 15 reported cases within 4 years. ..

Donor Registration and Recruitment

At the end of 2010, 14.9 million people had registered their willingness to be a bone marrow donor with one of the 64 registries from 45 countries participating in Bone Marrow Donors Worldwide. 12.2 million of these registered donors had been ABDR typed, allowing easy matching. A further 453,000 cord blood units had been received by one of 44 cord blood units from 26 countries participating. The highest total number of bone marrow donors registered were those from the USA (6.4 million), and the highest number per capita were those from Cyprus (10.6% of the population).

Experimental HIV treatment

In 2007, a team of doctors in Berlin, Germany, including Gero Hütter, performed a stem cell transplant for a leukaemia patient who was also HIV-positive. From 60 matching donors, they selected a [CCR5]-Ä32 homozygous individual with two genetic copies of a rare variant of a cell surface receptor. This genetic trait confers resistance to HIV infection by blocking attachment of HIV to the cell. Roughly one in 1000 people of European ancestry have this inherited mutation, but it is rarer in other populations. The transplant was repeated a year later after a relapse. Over three years after the initial transplant and despite discontinuing antiretroviral therapy, researchers cannot detect HIV in the transplant recipient's blood or in various biopsies. Levels of HIV-specific antibodies have also declined, leading to speculation that the patient may have been functionally cured of HIV. However, scientists emphasise that this is an unusual case. Potentially fatal transplant complications (the "Berlin patient" suffered from graft-versus-host disease and leukoencephalopathy) mean that the procedure could not be performed in others with HIV, even if sufficient numbers of suitable donors were found.

Potential Treatments

Brain Damage

Stroke and traumatic brain injury lead to cell death, characterized by a loss of neurons and oligodendrocytes within the brain. Healthy adult brains contain neural stem cells which divide to maintain general stem cell numbers, or become progenitor cells. In healthy adult animals, progenitor cells migrate within the brain and function primarily to maintain neuron populations for olfaction (the sense of smell). Interestingly, in pregnancy and after injury, this system appears to be regulated by growth factors and can increase the rate at which new brain matter is formed. Although the reparative process appears to initiate following trauma to the brain, substantial recovery is rarely observed in adults, suggesting a lack of robustness. Stem cells may also be used to treat brain degeneration, such as in Parkinson's and Alzheimer's disease.

Cancer

The development of gene therapy strategies for treatment of intra-cranial tumours offers much promise, and has shown to be successful in the treatment of some dogs; although research in this area is still at an early stage. Using conventional techniques, brain cancer is difficult to treat because it spreads so rapidly. Researchers at the Harvard Medical School transplanted human neural stem cells into the brain of rodents that received intracranial tumours. Within days, the cells migrated into the cancerous area and produced cytosine deaminase, an enzyme that converts a non-toxic pro-drug into a chemotheraputic agent. As a result, the injected substance was able to reduce the tumor mass by 81 percent. The stem cells neither differentiated nor turned tumorigenic. Some researchers believe that the key to finding a cure for cancer is to inhibit proliferation of cancer stem cells. Accordingly, current cancer treatments are designed to kill cancer cells. However, conventional chemotherapy treatments cannot discriminate between cancerous cells and others. Stem cell therapies may serve as potential treatments for cancer. Research on treating Lymphoma using adult stem cells is underway and has had human trials. Essentially, chemotherapy is used to completely destroy the patients own lymphocytes, and stem cells injected, eventually replacing the immune system of the patient with that of the healthy donor.

Spinal cord injury

A team of Korean researchers reported on November 25, 2003, that they had transplanted multipotent adult stem cells from umbilical cord blood to a patient suffering from a spinal cord injury and that following the procedure, she could walk on her own, without difficulty. The patient had not been able to stand up for roughly 19 years. For the unprecedented clinical test, the scientists isolated adult stem cells from umbilical cord blood and then injected them into the damaged part of the spinal cord. According to the October 7, 2005 issue of The Week, University of California, Irvine researchers transplanted multipotent human fetal-derived neural stem cells into paralyzed mice, resulting in locomotor improvements four months later. The observed recovery was associated with differentiation of transplanted cells into new neurons and oligodendrocytes- the latter of which forms the myelin sheath around axons of the central nervous system, thus insulating neural impulses and facilitating communication with the brain. In January 2005, researchers at the University of Wisconsin–Madison differentiated human blastocyst stem cells into neural stem cells, then into pre-mature motor neurons, and finally into spinal motor neurons, the cell type that, in the human body, transmits messages from the brain to the spinal cord and subsequently mediates motor function in the periphery.

The newly generated motor neurons exhibited electrical activity, the signature action of neurons. Lead researcher Su-Chun Zhang described the process as "[teaching] the blastocyst stem cells to change step by step, where each step has different conditions and a strict window of time." Transformation of blastocyst stem cells into motor neurons had eluded researchers for decades. While Zhang's findings were a significant contribution to the field, the ability of transplanted neural cells to establish communication with neighboring cells remains unclear. Accordingly, studies using chicken embryos as a model organism can be an effective proof-of-concept experiment. If functional, the new cells could be used to treat diseases like Lou Gehrig's disease, muscular dystrophy, and spinal cord injuries.

Heart Damage

Several clinical trials targeting heart disease have shown that adult stem cell therapy is safe, effective, and equally efficient in treating old and recent infarcts. Adult stem cell therapy for treating heart disease was commercially available in at least five continents at the last count (2007).

Possible mechanisms of recovery include:

- Generation of heart muscle cells
- Stimulation of growth of new blood vessels to repopulate damaged heart tissue
- Secretion of growth factors
- Assistance via some other mechanism

It may be possible to have adult bone marrow cells differentiate into heart muscle cells.

Haematopoiesis (Blood Cell Formation)

The specificity of the human immune cell repertoire is what allows the human body to defend itself from rapidly adapting antigens. However, the immune system is vulnerable to degradation upon the pathogenesis of disease, and because of the critical role that it plays in overall defense, its degradation is often fatal to the organism as a whole. Diseases of hematopoietic cells are called hematopathology. The specificity of the immune cells is what allows recognition of foreign antigens, causing further challenges in the treatment of immune disease. Identical matches between donor and recipient must be made for successful transplantation treatments, but matches are uncommon, even between first-degree relatives. Research using both hematopoietic adult stem cells and embryonic stem cells has provided insight into the possible mechanisms and methods of treatment for many of these ailments.

Fully mature human red blood cells may be generated ex vivo by hematopoietic stem cells (HSCs), which are precursors of red blood cells. In this process, HSCs are grown together with stromal cells, creating an environment that mimics the conditions of bone marrow, the natural site of red blood cell growth.

Erythropoietin, a growth factor, is added, coaxing the stem cells to complete terminal differentiation into red blood cells. Further research into this technique should have potential benefits to gene therapy, blood transfusion, and topical medicine.

Baldness

Hair follicles also contain stem cells, and some researchers predict research on these follicle stem cells may lead to successes in treating baldness through an activation of the stem cells progenitor cells. This treatment is expected to work by activating already existing stem cells on the scalp. Later treatments may be able to simply signal follicle stem cells to give off chemical signals to nearby follicle cells which have shrunk during the aging process, which in turn respond to these signals by regenerating and once again making healthy hair. Most recently, Dr. Aeron Potter of the University of California has claimed that stem cell therapy led to a significant and visible improvement in follicular hair growth . Results from his experiments are under review by the journal Science (journal).

Missing Teeth

In 2004, scientists at King's College London discovered a way to cultivate a complete tooth in mice and were able to grow them stand-alone in the laboratory. Researchers are confident that this technology can be used to grow live teeth in human patients.

In theory, stem cells taken from the patient could be coaxed in the lab into turning into a tooth bud which, when implanted in the gums, will give rise to a new tooth, and would be expected to grow within two months. It will fuse with the jawbone and release chemicals that encourage nerves and blood vessels to connect with it. The process is similar to what happens when humans grow their original adult teeth.

Many challenges remain, however, before stem cells could be a choice for the replacement of missing teeth in the future.

Deafness

Heller has reported success in re-growing cochlea hair cells with the use of embryonic stem cells.

Blindness and Vision Impairment

Since 2003, researchers have successfully transplanted corneal stem cells into damaged eyes to restore vision. "Sheets of retinal cells used by the team are harvested from aborted fetuses, which some people find objectionable." When these sheets are transplanted over the damaged cornea, the stem cells stimulate renewed repair, eventually restore vision. The latest such development was in June 2005, when researchers at the Queen Victoria Hospital of Sussex, England were able to restore the sight of forty patients using the same technique. The group, led by Dr. Sheraz Daya, was able to successfully use adult stem cells obtained from the patient, a relative, or even a cadaver. Further rounds of trials are ongoing. In April 2005, doctors in the UK transplanted corneal stem cells from an organ donor to the cornea of Deborah Catlyn, a woman who was blinded in one eye when acid was thrown in her eye at a nightclub. The cornea, which is the transparent window of the eye, is a particularly suitable site for transplants. In fact, the first successful human transplant was a cornea transplant. The absence of blood vessels within the cornea makes this area a relatively easy target for transplantation. The majority of corneal transplants carried out today are due to a degenerative disease called keratoconus.

The University Hospital of New Jersey reports that the success rate for growth of new cells from transplanted stem cells varies from 25 percent to 70 percent. In 2009, researchers at the University of Pittsburgh Medical center demonstrated that stem cells collected from human corneas can restore transparency without provoking a rejection response in mice with corneal damage.

Amyotrophic Lateral Sclerosis

Stem cells have resulted in significant locomotor improvements in rats with an Amyotrophic lateral sclerosis-like disease. In a rodent model that closely mimics the human form of ALS, animals were injected with a virus to kill the spinal cord motor nerves which mediate movement. Animals subsequently received stem cells in the spinal cord. Transplanted cells migrated to the sites of injury, contributed to regeneration of the ablated nerve cells, and restored locomotor function.

Graft vs. Host Disease and Crohn's Disease

Phase III clinical trials expected to end in second-quarter 2008 were conducted by Osiris Therapeutics using their in-development product Prochymal, derived from adult bone marrow. The target disorders of this therapeutic are graft-versus-host disease and Crohn's disease.

Neural and Behavioral Birth Defects

A team of researchers led by Prof. Joseph Yanai were able to reverse learning deficits in the offspring of pregnant mice who were exposed to heroin and the pesticide organophosphate. This was done by direct neural stem cell transplantation into the brains of the offspring. The recovery was almost 100 percent, as shown both in behavioral tests and objective brain chemistry tests. Behavioral tests and learning scores of the treated mice showed rapid improvement after treatment, providing results that rivaled non-treated mice. On the molecular level, brain chemistry of the treated animals was also restored to normal. Through the work, which was supported by the US National Institutes of Health, the US-Israel Binational Science Foundation and the Israel anti-drug authorities, the researchers discovered that the stem cells worked even in cases where most of the cells died out in the host brain.

The scientists found that before they die the neural stem cells succeed in inducing the host brain to produce large numbers of stem cells which repair the damage. These findings, which answered a major question in the stem cell research community, were published in January 2008 in the leading journal, Molecular Psychiatry. Scientists are now developing procedures to administer the neural stem cells in the least invasive way possible - probably via blood vessels, making therapy practical and clinically feasible. Researchers also plan to work on developing methods to take cells from the patient's own body, turn them into stem cells, and then transplant them back into the patient's blood via the blood stream. Aside from decreasing the chances of immunological rejection, the approach will also eliminate the controversial ethical issues involved in the use of stem cells from human embryos.

Diabetes

Diabetes patients lose the function of insulin-producing beta cells within the pancreas. Human embryonic stem cells may be grown in cell culture and stimulated to form insulin-producing cells that can be transplanted into the patient. However, clinical success is highly dependent on the development of the following procedures:

- Transplanted cells should proliferate
- Transplanted cells should differentiate in a site-specific manner
- Transplanted cells should survive in the recipient (prevention of transplant rejection)
- Transplanted cells should integrate within the targeted tissue

- Transplanted cells should integrate into the host circuitry and restore function

Orthopaedics

Clinical case reports in the treatment of orthopaedic conditions have been reported. To date, the focus in the literature for musculoskeletal care appears to be on mesenchymal stem cells. Centeno et al. have published MRI evidence of increased cartilage and meniscus volume in individual human subjects. The results of trials that include a large number of subjects, are yet to be published. However, a published safety study conducted in a group of 227 patients over a 3-4 year period shows adequate safety and minimal complications associated with mesenchymal cell transplantation.

Wakitani has also published a small case series of nine defects in five knees involving surgical transplantation of mesenchymal stem cells with coverage of the treated chondral defects.

Wound Healing

Stem cells can also be used to stimulate the growth of human tissues. In an adult, wounded tissue is most often replaced by scar tissue, which is characterized in the skin by disorganized collagen structure, loss of hair follicles and irregular vascular structure. In the case of wounded fetal tissue, however, wounded tissue is replaced with normal tissue through the activity of stem cells. A possible method for tissue regeneration in adults is to place adult stem cell "seeds" inside a tissue bed "soil" in a wound bed and allow the stem cells to stimulate differentiation in the tissue bed cells. This method elicits a regenerative response more similar to fetal wound-healing than adult scar tissue formation. Researchers are still investigating different aspects of the "soil" tissue that are conducive to regeneration.

Infertility

Culture of human embryonic stem cells in mitotically inactivated porcine ovarian fibroblasts (POF) causes differentiation into germ cells (precursor cells of oocytes and spermatozoa), as evidenced by gene expression analysis. Human embryonic stem cells have been stimulated to form Spermatozoon-like cells, yet still slightly damaged or malformed. It could potentially treat azoospermia.

Bibliography

Ayala, F. J.: *The Genetic Structure of Populations*, W.H. Freeman & Co., San Francisco, California, 1977.

Beckner, M.: *The Biological Way of Thought*, University of California Press, Berkeley and Los Angeles, 1968.

Berg, Paul, and Singer, Maxine: *Dealing with Genes: The Language of Heredity. Mill Valley*, University Science Books, CA, 1992.

Brookes, Martin: *Get a Grip on Genetics, Time Life Books*, East Sussex, England, 1998.

Dando, M.: *Biological Warfare in the 21st Century: Biotechnology and the Proliferation of Biological Weapons*. London: Brassey's, 1994.

Darlington, C. D., and Mather, K.: *The Elements of Genetics*, London, Allen and Unwin, 1949.

Dawkins, R.: *The Selfish Gene*. Oxford: Oxford University Press, 1976.

DeVita, V.T.: *Cancer: Principles and Practice of Oncology*, Lippincott, Philadelphia, 2005.

Diaz, E.: *Microbial Biodegradation: Genomics and Molecular Biology,* Caister Academic Press, UK, 2008.

Fisher, R. A.: *The Genetical Theory of Natural Selection*, Clarendon Press, Oxford, 1930.

Fritz, A.: *International Classification of Diseases for Oncology*, World Health Organization, Geneva, 2000.

Gillis, Justin: *Drug Firms, Gene Labs to Map Genetic Code*, The Daily News, Longview, WA, 1999.

Glover, D. M., and Hames, B. D.: *Genes and Embryos*: New York, Oxford University Press, 1989.

Griffiths A.J.F.: *Introduction to Genetic Analysis*, W.H. Freeman and Company, New York, USA, 2005.

Hamerton, J. L.: *Human Cytogenetics*, Academic Press, New York, 1971.

Hartl D, Jones E: *Genetics: Analysis of Genes and Genomes,* Jones & Bartlett, 2005.

Hartl, Daniel L.: *Basic Genetics*, Jones and Bartlett Publishers, Boston, 1991.

Jonoska, N.: *Self-Assembling DNA Graphs, DNA-Based Computers VIII*, Springer-Verlag, Berlin, 2003.

Klotz, J. W.: *Genes, Genesis, and Evolution*: St. Louis, Mo., Concordia Publishing Co., 1970.

Lumsden C., and E. Wilson: *Genes, Mind and Culture: The Coevolutionary Process.* Cambridge, MA: Harvard University Press, 1981.

Mayr, E.: *Change of Genetic Environment and Evolution*, Allen and Unwin, London, 1954.

Nottingham, S.F.: *Eat Your Genes: How Genetically Modified Food is Entering Our Diet*, Zed Books, London, 1998.

Prakash, Chandra Prakash : *Cell Physiology*, Discovery, Delhi, 2010.

Robin Whyman: *Applied Organometallic Chemistry and Catalysis*, Oxford University Press, Delhi, 2005.

Roobeek, A: *The Biotechnology Revolution?,* Blackwell, Oxford, 1995.

Sturtevant, Alfred: *History of Genetics*, Harper and Row, New York, 1965.

Suzuki, D., and Knudtson, P.: *Genethics: The Clash Between the New Genetics and Human Values*, Harvard University Press, Cambridge, Mass, 1989.

Wilson, E. O., and Lumden, C.: *Genes, Mind, and Culture: The Evolutionary Process*, Harvard University Press, Cambridge, Mass, 1981.

Wilson, E. O., and Lumden, C.: *Genes, Mind, and Culture: The Evolutionary Process*: Cambridge, Mass., Harvard University Press, 1981.

Index

H

I

M

N

O

P

R

T

U

V

W

❑❑❑

Encyclopaedia of

Advanced Genetic Analysis

Encyclopaedia of

Advanced Genetic Analysis

Volume 3

Editor:
Veena Kumari

AGROTECH PRESS
JAIPUR ▪ NEW DELHI

AGROTECH PRESS

Regd. Off. "Bony Residency", Gate No. 2
Opp. Tilak Public School, Vishweswariya Nagar,
Gopalpura Road, Jaipur - 302018, India
Ph: 0141 - 2761280, Telefax: 2761381
E-mail : agrotechpress@gmail.com

Delhi Office: 102, 1st Floor, Satyam House,
4327/3, Ansari Road, Darya Ganj,
New Delhi - 110002, India
Ph: 011-45652440
visit us: www.agrotechpress.com

Distributed By
KEYA INTERNATIONAL PRESS LLP
E-mail: keyainternationalpressllp@gmail.com

Published in India, 2024

ISBN : 9789384568207 (5 Vols. Set)

Typeset by Thomson Press

Printed and bound in India by Thomson Press Ltd.,
New Delhi

Contents

Preface

Genomics is a discipline in genetics that applies recombinant DNA, DNA sequencing methods, and bioinformatics to sequence, assemble, and analyze the function and structure of genomes (the *complete* set of DNA within a single cell of an organism). The field includes efforts to determine the entire DNA sequence of organisms and fine-scale genetic mapping. The field also includes studies of intragenomic phenomena such as heterosis, epistasis, pleiotropy and other interactions between loci and alleles within the genome. In contrast, the investigation of the roles and functions of single genes is a primary focus of molecular biology or genetics and is a common topic of modern medical and biological research. Research of single genes does not fall into the definition of genomics unless the aim of this genetic, pathway, and functional information analysis is to elucidate its effect on, place in, and response to the entire genome's networks. Following James D. Watson and Francis Crick's discovery of the structure of DNA in 1953 and Fred Sanger's publication of the Amino acid sequence of insulin in 1955, nucleic acid sequencing became a major target of early molecular biologists. In 1964, Robert W. Holley and colleagues published the first nucleic acid sequence ever determined, the ribonucleotide sequence of alanine transfer RNA. Extending this work, Marshall Nirenberg and Philip Leder revealed the triplet nature of the genetic code and were able to determine the sequences of 54 out of 64 codons in their experiments. In 1972, Walter Fiers and his team at the Laboratory of Molecular Biology of the University of Ghent (Ghent, Belgium) were the first to determine the sequence of a gene: the gene for Bacteriophage MS2 coat protein. Fiers' group expanded on their MS2 coat protein work, determining the complete nucleotide-sequence of bacteriophage MS2-RNA (whose genome encodes just four genes in 3569 base pairs and Simian virus 40 in 1976 and 1978, respectively.

Genetic Engineering (GE) is a highly complicated and advanced branch of science which involves a wide range of techniques used in changing the genetic material in the DNA code in a living organism. 'Genetic Engineering' means the deliberate modification of the characters of an organism by the manipulation of its genetic material. Genetic engineering comes under the broad heading of Biotechnology. There is a great scope in this field as the demand for genetic engineers are growing in India as well as abroad. A cell is the smallest living unit, the basic structural and functional unit of all living matter, whether a plant, an animal, humans or a fungus. While some organisms are single celled, others like plants, animals, humans etc. are made up of a lot more cells. For e.g., humans have approximately 3 million cells. A cell is composed of a 'cell membrane' enclosing the whole cell, many 'organelles' equivalent to the organs in the body and a 'nucleus' which is the command centre of the cell. Inside the nucleus are the chromosomes which is the storage place for all genetic information which determines the nature and characteristics of an organism. This information is written along the thin thread, called DNA, a nucleic acid which constitutes the genes (units of heredity). The DNA governs cell growth and is responsible for the transmission of genetic information from one generation to the next. Genetic engineering aims to re-arrange the sequence of DNA in gene using artificial methods. The work of a genetic engineer involves extracting the DNA out of one organism, changing it using chemicals or radiation and subsequently putting it back into the same or a different organism. For eg: genes and segments of DNA from one species is taken and put into another species. They also study how traits and characteristics are transmitted through the generations, and how genetic disorders are caused. Their research involves researching the causes and discovering potential cures if any. Genetic engineering have specialisations related to plants, animals and human beings. Genetic engineering in plants and animals may be to improve certain natural characteristics of value, to increase resistance to disease or damage and to develop new characteristics etc. It is used to change the colour, size, texture etc. of plants otherwise known as GM (Genetically Modified) foods. GE in humans can be to correct severe hereditary defects by introducing normal genes into cells in place of missing or defective ones.

There are three key features of the book. First of all, the subject matter has been discussed in a lucid way with important points highlighted to aid students for better comprehension. Efforts have been made for a wide coverage along with in depth analysis where needed.

—*Editor*

1

Myriad Genetics Case

In 2009, a lawsuit, Association For Molecular Pathology, et al., vs. United States Patent and Trademark Office, et al. was filed against Myriad Genetics and the U.S. Patent and Trademark Office by professional medical organizations, doctors, and patients, represented by the American Civil Liberties Union and the Public Patent Foundation of the Benjamin N. Cardozo School of Law, New York, in the U.S. District Court for the Southern District of New York

The complaint challenged specific claims on isolated genes and diagnostic methods in seven of Myriad's 23 patents on BRCA1 and BRCA2. Some of these patents are owned by Myriad and some of them were licensed from University of Utah. More specifically, the plaintiffs wanted certain claims declared invalid on the grounds that they are not patentable subject matter- that the isolated genes are unpatentable products of nature and that the diagnostic method claims are mere thought processes that do not yield any real world transformations.

Many patent lawyers had predicted that the courts would throw out this case. However, United States District Court Judge Robert W. Sweet accepted the case, and on March 29, 2010 ruled that the claims were invalid. Other claims, which were not challenged in the lawsuit, still stand.

Judge Sweet's 152–page decision ruled that the challenged claims to the isolated gene sequences had been "improperly granted" because they claimed unpatentable subject matter; the claims to the diagnostic methods were found invalid under the recent In re Bilski decision. Because the case could be decided with patent law, Judge Sweet did not look at the challenge on First

Amendment grounds and dismissed them without prejudice. However, on March 30, 2010, Myriad announced that it will appeal the decision. On June 16, 2010, Myriad filed its Notice of Appeal. Oral arguments were heard at the Federal Circuit on April 4, 2011.

On July 29, 2011 the Federal Circuit overturned the District court in part, and upheld the decision in part. The Federal Circuit overturned the District Court's finding that the claims covering isolated gene sequences are invalid and also overturned the invalidity of some of the diagnostic claims; the Federal Circuit upheld the finding that the claims for the diagnostic methods that only compare or analyze sequences - that have no transformative step - are invalid.

Further appeals are likely; on October 12, 2011 the ACLU and the Public Patent Foundation said they intended to petition the Supreme Court to overturn the decision.

Gene Pool

In population genetics, a gene pool is the complete set of unique alleles in a species or population.

Description

A large gene pool indicates extensive genetic diversity, which is associated with robust populations that can survive bouts of intense selection. Meanwhile, low genetic diversity (see inbreeding and population bottlenecks) can cause reduced biological fitness and an increased chance of extinction, although as explained by genetic drift new genetic variants, that may cause an increase in the fitness of organisms, are more likely to fix in the population if it is rather small.

When all individuals in a population are identical with regard to a particular phenotypic trait, the population is said to be monomorphic. When the individuals show several variants of a particular trait they are said to be polymorphic.

Gene Pool Concept in Crop Breeding

Harlan and de Wet (1971) proposed classifying each crop and its related species by gene pools rather than by formal taxonomy.

1. Primary gene pool (GP-1): Members of this gene pool are probably in the same "species" (in conventional biological usage) and can intermate freely. Harlan and de Wet wrote, "Among forms of this gene pool, crossing is easy; hybrids are generally fertile with good chromosome pairing; gene segregation is approximately normal and gene transfer is generally easy.". They also advised subdividing each crop gene pool in two:

   * Subspecies A: Cultivated races

   * Subspecies B: Spontaneous races (wild or weedy)

2. Secondary gene pool (GP-2): Members of this pool are probably normally classified as different species than the crop species under consideration (the primary gene pool). However, these species are closely related and can cross and produce at least some fertile hybrids (F1s). As would be expected by members of different species, there are some reproductive barriers between members of the primary and secondary gene pools:
   * hybrids may be weak
   * hybrids may be partially sterile
   * chromosomes may pair poorly or not at all
   * recovery of desired phenotypes may be difficult in subsequent generations
   * However, "The gene pool is available to be utilized, however, if the plant breeder or geneticist is willing to put out the effort required."
3. Tertiary gene pool (GP-3): Members of this gene pool are more distantly related to the members of the primary gene pool. The primary and tertiary gene pools can be intermated, but gene transfer between them is impossible without the use of "rather extreme or radical measures" such as:
   * embryo rescue (or embryo culture, a form of plant organ culture)
   * induced polyploidy (chromosome doubling)
   * bridging crosses (e.g., with members of the secondary gene pool).

Gene Flow

In population genetics, gene flow (also known as gene migration) is the transfer of alleles of genes from one population to another.

Migration into or out of a population may be responsible for a marked change in allele frequencies (the proportion of members carrying a particular variant of a gene). Immigration may also result in the addition of new genetic variants to the established gene pool of a particular species or population.

There are a number of factors that affect the rate of gene flow between different populations. One of the most significant factors is mobility, as greater mobility of an individual tends to give it greater migratory potential. Animals tend to be more mobile than plants, although pollen and seeds may be carried great distances by animals or wind. Maintained gene flow between two populations can also lead to a combination of the two gene pools, reducing the genetic variation between the two groups. It is for this reason that gene flow strongly acts against speciation, by recombining the gene pools of the groups, and thus, repairing the developing differences in genetic variation that would have led to full speciation and creation of daughter species. For example, if a species of grass grows on both sides of a highway, pollen is likely to be transported from one side to the other and vice versa. If this pollen is able to fertilize the plant where it ends up and produce viable offspring, then the alleles in the pollen have effectively been able to move from the population on one side of the highway to the other.

Barriers to Gene Flow

Physical barriers to gene flow are usually, but not always, natural. They may include impassable mountain ranges, oceans, or vast deserts. In some cases, they can be artificial, man-made barriers, such as the Great Wall of China, which has hindered the gene flow of native plant populations. One of these native plants, Ulmus pumila, demonstrated a lower prevalence of genetic differentiation than the plants Vitex negundo, Ziziphus jujuba, Heteropappus hispidus, and Prunus armeniaca whose habitat is located on the opposite side of the Great Wall of China where Ulmus pumila grows. This is because Ulmus pumila has wind-pollination as its primary means of propagation and the latter-plants carry out pollination through insects. Samples of the same species which grow on either side have been shown to have developed genetic differences, because there is little to no gene flow to provide recombination of the gene pools. Barriers to gene flow need not always be physical. Species can live in the same environment, yet show very limited gene flow due to limited hybridization or hybridization yielding unfit hybrids.

Gene Flow in Humans

Gene flow has been observed in humans. For example, in the United States, gene flow was observed between a white European population and a black West African population, which were recently brought together. In West Africa, where malaria is prevalent, the Duffy antigen provides some resistance to the disease, and this allele is thus present in nearly all of the West African population. In contrast, Europeans have either the allele Fy or Fy , because malaria is almost non-existent. By measuring the frequencies of the West African and European groups, scientists found that the allele frequencies became mixed in each population because of movement of individuals. It was also found that this gene flow between European and West African groups is much greater in the Northern U.S. than in the South.

Gene Flow Between Species

Gene flow can occur between species, either through hybridization or gene transfer from bacteria or virus to new hosts. Gene transfer, defined as the movement of genetic material across species boundaries, which includes horizontal gene transfer, antigenic shift, and reassortment is sometimes an important source of genetic variation. Viruses can transfer genes between species. Bacteria can incorporate genes from other dead bacteria, exchange genes with living bacteria, and can exchange plasmids across species boundaries. "Sequence comparisons suggest recent horizontal transfer of many genes among diverse species including across the boundaries of phylogenetic "domains". Thus determining the phylogenetic history of a species can not be done conclusively by determining evolutionary trees for single genes." Biologist Gogarten suggests "the original metaphor of a tree no longer fits the data from recent genome research". Biologists [should] instead use the metaphor of a mosaic to describe the different histories combined in individual genomes and use the metaphor of an intertwined net to visualize the rich exchange and cooperative effects of horizontal gene transfer.

"Using single genes as phylogenetic markers, it is difficult to trace organismal phylogeny in the presence of HGT [horizontal gene transfer]. Combining the simple coalescence model of cladogenesis with rare HGT [horizontal gene transfer] events suggest there was no single last common ancestor that contained all of the genes ancestral to those shared among the three domains of life.

Each contemporary molecule has its own history and traces back to an individual molecule cenancestor. However, these molecular ancestors were likely to be present in different organisms at different times."

Genetic Pollution

Purebred, naturally-evolved, region-specific, wild species can be threatened with extinction through the process of genetic pollution, potentially causing uncontrolled hybridization, introgression and genetic swamping. These processes can lead to homogenization or replacement of local genotypes as a result of either a numerical and/or fitness advantage of introduced plant or animal. Nonnative species can bring about a form of extinction of native plants and animals by hybridization and introgression either through purposeful introduction by humans or through habitat modification, bringing previously isolated species into contact. These phenomena can be especially detrimental for rare species coming into contact with more abundant ones. Interbreeding between the species can cause a 'swamping' of the rarer species' gene pool, creating hybrids that drive the originally purebred native stock to complete extinction.

The extent of this facet of gene flow is not always apparent from morphological (outward appearance) observations alone. Some degree of gene flow may be due to normal, evolutionarily constructive processes, and all constellations of genes and genotypes cannot be preserved. That being said, hybridization with or without introgression may threaten a rare species' existence nonetheless.

Models of Gene Flow

Models of gene flow can be derived from population genetics, e.g. Sewall Wright's neighborhood model, Wright's island model and the stepping stone model.

Gene Flow Mitigation

This section does not cite any references or sources. Please help improve this section by adding citations to reliable sources. Unsourced material may be challenged and removed. (March 2010)

When cultivating genetically modified (GM) plants or livestock, it becomes necessary to prevent "genetic pollution" i.e. their genetic modification from reaching other conventionally hybridized or wild native plant and animal populations by using gene flow mitigation usually through unintentional cross pollination and crossbreeding. Reasons to limit gene flow may include biosafety or agricultural co-existence, in which GM and non-GM cropping systems work side by side. Scientists in several large research programmes are investigating methods of limiting gene flow in plants. Among these programmes are Tran-scontainer, which investigates methods for biocontainment, SIGMEA, which focuses on the biosafety of genetically modified plants, and Co-Extra, which studies the co-existence of GM and non-GM product chains. Generally, there are three approaches to gene flow mitigation: keeping the genetic modification out of the pollen, preventing the formation of pollen, and keeping the pollen inside the flower.

- The first approach requires transplastomic plants. In transplastomic plants, the modified DNA is not situated in the cell's nucleus but is present in plastids, which are cellular compartments outside the nucleus. An example for plastids are chloroplasts, in which photosynthesis occurs. In some plants, the pollen does not contain plastids and, consequently, any modification located in plastids cannot be transmitted by the pollen.
- The second approach relies on male sterile plants. Male sterile plants are unable to produce functioning flowers and therefore cannot release viable pollen. Cytoplasmic male sterile plants are known to produce higher yields. Therefore, researchers are trying to introduce this trait to genetically modified crops.
- The third approach works by preventing the flowers from opening. This trait is called cleistogamy and occurs naturally in some plants. Cleistogamous plants produce flowers which either open only partly or not at all. However, it remains unclear how reliable cleistogamy is for gene flow mitigation: a Co-Extra research project on rapeseed investigating the matter has published preliminary results which cast doubt on the attainment of a high degree of reliability.

Biological Dispersal

Biological dispersal refers to species movement away from an existing population or away from the parent organism. Through simply moving from one habitat patch to another, the dispersal of an individual has consequences not only for individual fitness, but also for population dynamics, population genetics, and species distribution. Understanding dispersal and the consequences both for evolutionary strategies at a species level, and for processes at an ecosystem level, requires understanding on the type of dispersal, the dispersal range of a given species, and the dispersal mechanisms involved.

Biological dispersal may be contrasted with geodispersal, which is the mixing of previously isolated populations (or whole biotas) following the erosion of geographic barriers to dispersal or gene flow (Lieberman, 2005;Albert and Reis, 2011).

Types of Dispersal

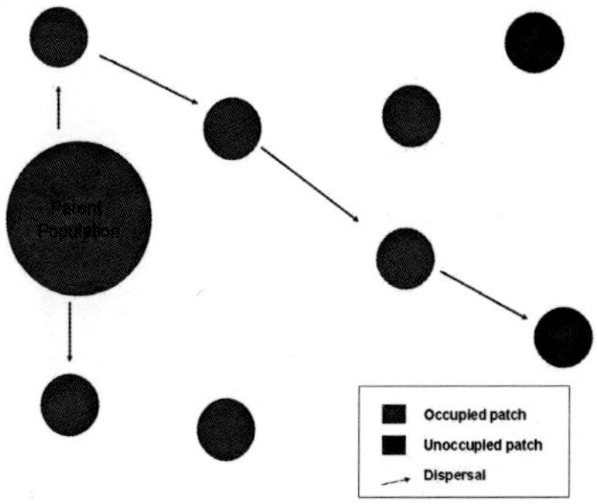

Figure: *Dispersal from parent population*

At some time during its life, an organism, whether animal or plant, moves, or is moved, so that it or its offspring do not die exactly where they were born. Such movement is called dispersal. Some organisms are motile throughout their lives, but others are adapted to move or be moved at precise, limited phases of their life cycles. This is commonly called the dispersive phase of the life cycle. The strategies of organisms' entire life cycles often are predicated on the nature and circumstances of their dispersive phases.

Density Independent Dispersal

Organisms have evolved adaptations for dispersal that take advantage of various forms of kinetic energy occurring naturally in the environment. This is referred to as density independent or passive dispersal and operates on many groups of organisms (some invertebrates, fish, insects and sessile organisms such as plants) that depend on animal vectors, wind, gravity or current for dispersal.

Density Dependent Dispersal

Density dependent or active dispersal for many animals largely depends on factors such as local population size, resource competition, habitat quality, and habitat size.

Due to population density, dispersal may relieve pressure for resources in an ecosystem, and competition for these resources may be a selection factor for dispersal mechanisms. Dispersal of organisms is a critical process for understanding both geographic isolation in evolution through gene flow and the broad patterns of current geographic distributions (biogeography).

Dispersal Range

"Dispersal range" refers to the distance a species can move from an existing population or the parent organism. An ecosystem depends critically on the ability of individuals and populations to disperse from one habitat patch to another. Therefore, biological dispersal is critical to the stability of ecosystems.

Environmental Constraints

Few species are ever evenly or randomly distributed within or across landscapes. In general, species significantly vary across the landscape in association with environmental features that influence their reproductive success and population persistence. Spatial patterns in environmental features (e.g. resources) permit individuals to escape unfavorable conditions and seek out new locations.

This allows the organism to "test" new environments for their suitability, provided they are within animal's geographic range. In addition, the ability of a species to disperse over a gradually changing environment could enable a population to survive extreme conditions. (i.e. climate change).

Dispersal Barriers

A dispersal barrier may mean that the dispersal range of a species is much smaller than the species distribution. An artificial example is habitat fragmentation due to human land use. Natural barriers to dispersal that limit species distribution include mountain ranges and rivers. An example is the separation of the ranges of the two species of chimpanzee by the Congo River.

On the other hand, human activities may also expand the dispersal range of a species by providing new dispersal methods (e.g., ships).

Dispersal Mechanisms

Most animals are capable of locomotion and the basic mechanism of dispersal is movement from one place to another. Locomotion allows the organism to "test" new environments for their suitability, provided they are within the animal's range. Movements are usually guided by inherited behaviors.

The formation of barriers to dispersal or gene flow between adjacent areas can isolate populations on either side of the emerging divide. The geographic separation and subsequent genetic isolation of portions of an ancestral population can result in speciation.

Plant Dispersal Mechanisms

Figure: *Burs are an example of a seed dispersion mechanism which uses a biotic vector, in this case animals with fur.*

Seed dispersal is the movement or transport of seeds away from the parent plant. Plants have limited mobility and consequently rely upon a variety of dispersal vectors to transport their propagules, including both abiotic and biotic vectors. Seeds can be dispersed away from the parent plant individually or collectively, as well as dispersed in both space and time. The patterns of seed dispersal are determined in large part by the dispersal mechanism and this has important implications for the demographic and genetic structure of plant populations, as well as migration patterns and species interactions. There are five main modes of seed dispersal: gravity, wind, ballistic, water and by animals.

Animal Dispersal Mechanisms

Non-motile Animals

There are numerous animal forms that are non—motile, such as sponges, bryozoans, tunicates, sea anemones, corals, and oysters. In common, they are all either marine or aquatic. It may seem curious that plants have been so successful at stationary life on land, while animals have not, but the answer lies in the food supply. Plants produce their own food from sunlight and carbon dioxide—both generally more abundant on land than in water. Animals fixed in place must rely on the surrounding medium to bring food at least close enough to grab, and this occurs in the three-dimensional water environment, but with much less abundance in the atmosphere.

All of the marine and aquatic invertebrates whose lives are spent fixed to the bottom (more or less; anemones are capable of getting up and moving to a new location if conditions warrant) produce dispersal units. These may be specialized "buds", or motile sexual reproduction products, or even a sort of alteration of generations as in certain cnidaria.

Corals provide a good example of how sedentary species achieve dispersion. Corals reproduce by releasing sperm and eggs directly into the water. These release events are coordinated by lunar phase in certain warm months, such that all corals of one or many species on a given reef will release on the same single or several consecutive nights. The released eggs are fertilized, and the resulting zygote develops quickly into a multicellular planula. This motile stage then attempts to find a suitable substratum for settlement. Most are unsuccessful and die or are fed upon by zooplankton and bottom dwelling predators such as anemones and other corals. However, untold millions are produced, and a few do succeed in locating spots of bare limestone, where they settle and transform by growth into a polyp. All things being favorable, the single polyp grows into a coral head by budding off new polyps to form a colony.

Motile Animals

The majority of all animals are motile. Although motile animals can, in theory, disperse themselves by their spontaneous and independent locomotive powers, a great many species utilize the existing kinetic energies in the environment, resulting in passive movement. Dispersal by water currents is especially associated with the physically small inhabitants of marine waters known as zooplankton. The term plankton comes from the Greek, ðëáãêôïí, meaning "wanderer" or "drifter".

Dispersal by Dormant Stages

Many animal species, especially freshwater invertebrates, are able to disperse by wind or by transfer with an aid of larger animals (birds, mammals or fishes) as dormant eggs, dormant embryos or, in some cases, dormant adult stages. Tardigrades, some rotifers and some copepods are able to withstand desiccation as adult dormant stages. Many other taxa (Cladocera, Bryozoa, Hydra, Copepoda and so on) can disperse as dormant eggs or embryos. Freshwater sponges usually have special dormant propagules called gemmulae for such a dispersal. Many kinds of dispersal dormant stages are able to withstand not only desiccation and low and high temperature, but also action of digestive enzymes during their transfer through digestive tracts of birds and other animals, high concentration of salts and many kinds of toxicants. Such dormant-resistant stages made possible the long-distance dispersal from one water body to another and broad distribution ranges of many freshwater animals.

Genetic Erosion

Genetic erosion is a process whereby an already limited gene pool of an endangered species of plant or animal diminishes even more when individuals from the surviving population die off without getting a chance to meet and breed with others in their endangered low population.

Genetic erosion occurs because each individual organism has many unique genes which get lost when it dies without getting a chance to breed. Low genetic diversity in a population of wild animals and plants leads to a further diminishing gene pool, inbreeding and a weakening immune system and fast tracks that species towards eventual extinction.

All the world's endangered species are plagued by varying degrees of genetic erosion and most need a human assisted breeding program to keep their population viable and to keep them from going extinct in the long run.

The smaller the population is, the more magnified the effect of genetic erosion gets when each surviving individual of the species is lost without getting a chance to breed. Genetic erosion gets compounded and accelerated by habitat fragmentation, today most endangered species live in smaller and smaller chunks of fragmented habitat interspersed with human settlements and farmland making it impossible for them to naturally meet and breed with others of their kind, many die off without getting a chance to breed and pass on their genes in the living population. The gene pool of a species or a population is the complete set of unique alleles that would be found by inspecting the genetic material of every living member of that species or population. A large gene pool indicates extensive genetic diversity, which is associated with robust populations that can survive bouts of intense selection. Meanwhile, low genetic diversity (see inbreeding and population bottlenecks) can cause reduced biological fitness and an increased chance of extinction.

Processes and Consequences

"A population bottleneck creates a shrinking gene pool that leaves fewer and fewer mating partners. What are the genetic implications? The animals become part of a high stakes poker game — with a crooked dealer. After beginning with a 52-card deck, the players wind up with, say, five cards that they are dealt over and over. As they begin to inbreed, congenital effects appear, both physical and reproductive. Often abnormal sperm increase; infertility rises; the birthrate falls. Most perilous in the long run, each animal's immune defense system is weakened. Thus, even if an endangered species in a bottleneck can withstand whatever human development may be eating away at its habitat, it still faces the threat of an epidemic that could well be fatal to the entire population."

Genetic Erosion in Agricultural and Livestock Biodiversity

Genetic erosion in agricultural and livestock biodiversity is the loss of genetic diversity, including the loss of individual genes, and the loss of particular combinants of genes (or gene complexes) such as those manifested in locally adapted landraces of domesticated animals or plants adapted to the natural environment in which they originated. The term genetic erosion is sometimes used in a narrow sense, such as for the loss of alleles or genes, as well as more broadly, referring to the loss of varieties or even species. The major driving forces behind genetic erosion in crops are: variety replacement, land clearing, overexploitation of species, population pressure, environmental degradation, overgrazing, policy and changing agricultural systems. The main factor, however, is the replacement of local varieties of domestic plants and animals by other varieties or species.

A large number of varieties can also often be dramatically reduced when commercial varieties are introduced into traditional farming systems. Many researchers believe that the main problem related to agro-ecosystem management is the general tendency towards genetic and ecological uniformity imposed by the development of modern agriculture. Human intervention, modern science and safeguards to guard against genetic erosion

In-situ Conservation

With advances in modern science several techniques and safeguards have emerged to check the relentless advance of genetic erosion and the resulting acceleration of endangered species towards extinction.

However many of these techniques and safeguards are too expensive yet to be practical, the best way to protect species is to protect their habitat and to let them live in it naturally. Wildlife sanctuaries and national parks have been created to preserve entire ecosystems with all the web of species which call them home.

Wildlife corridors are created to join fragmented habitats (Habitat fragmentation) to enable endangered species to travel, meet and breed with others of their kind. Scientific conservation and modern wildlife management techniques with the help of scientifically trained staff help manage these protected ecosystems and the wildlife found in them.

Wild animals are also translocated and reintroduced to other locations physically when fragmented wildlife habitat is too far and isolated to be able to link it with a wildlife corridor or when local extinction has already occurred.

Ex-situ Conservation

Modern policies of the zoo associations and zoos around the world have changed to putting extreme importance on keeping and breeding wild sourced species and subspecies of animals in their registered endangered species breeding programs which will have a chance to be reintroduced and survive in the wild. The main objectives of zoos today have changed to breed species and subspecies to assist conservation efforts in the wild. Zoos do this by maintaining extremely detailed scientific breeding records (i.e., studbooks) and loaning their wild animals to other zoos around the country and indeed globally for breeding to safeguard against inbreeding.Costly and sometimes controversial ultra modern ex-situ conservation techniques have emerged for saving the genetic biodiversity on our planet and the diversity in their gene pool by guarding against genetic erosion through modern concepts like seedbanks, sperm banks and tissue banks. Sperms, eggs and embryos can now be frozen and kept in these banks which are sometimes called Modern Noah's Ark or Frozen Zoos.

Cryopreservation techniques are used to freeze these living materials and keep them alive by storing them submerged in liquid nitrogen tanks. Thus, preserved material can then be used for artificial insemination, in vitro fertilization, embryo transfer and cloning to protect diversity in the gene pool of critically endangered species.

It is today possible to save endangered species from extinction by preserving parts like tissue, sperms, eggs etc. even after the death of a critically endangered animal or collected from one found freshly dead in captivity or from the wild and resurrect it with the help of cloning and give it another chance to breed its genes into the living population of the respective species which is threatened with extinction. Resurrection of dead critically endangered wildlife with the help of cloning is still being perfected and is still too expensive to be practical but with time and advancement is science it may well become a routine procedure. Recently strategies have been made to find an integrated approach to in situ and ex situ conservation.

Genetic Admixture

Genetic admixture occurs when individuals from two or more previously separated populations begin interbreeding. Admixture results in the introduction of new genetic lineages into a population. It has been known to slow local adaptation by introducing foreign, unadapted genotypes (known as gene swamping). It also prevents speciation by homogenizeing populations. Genetic admixture often occurs when a geographic barrier separating populations ,such as a river or isthmus, is removed or when anthropogenic activities result in movement of populations (for example invasive species).

One example of genetic admixture resulting from the introduction of an invasive species is provided by the Cuban brown anole. Several isolated populations of this species exist in the native range of Cuba. However in the introduced range of Florida, these populations freely interbreed, forming an admixed population.

Miscegenation

Miscegenation (Latin miscere "to mix" + genus "kind") is the mixing of different racial groups through marriage, cohabitation, sexual relations, and procreation.

The term miscegenation has been used since the 19th century to refer to interracial marriage and interracial sex, and more generally to the process of racial admixture, which has taken place since ancient history. The term entered historical records during European colonialism and the Age of Discovery, but societies such as China and Japan also had restrictions on marrying with peoples they considered different. Historically the term has been used in the context of laws banning interracial marriage and sex, so-called anti-miscegenation laws. It is a potentially offensive word.

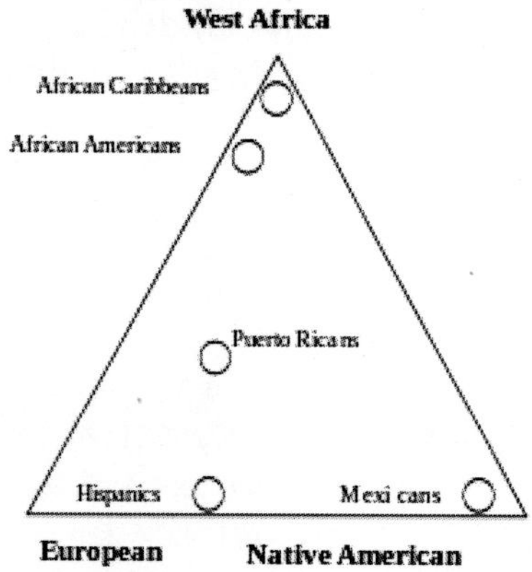

Figure: Triangle plot shows average admixture of five North American ethnic groups. Individuals that self-identify with each group can be found at many locations on the map, but on groups tend to cluster differently.

Usage

Today, the word miscegenation is avoided by many scholars, because the term suggests an actual biological phenomenon, rather than its nature as a categorization imposed on certain relationships. The word is considered offensive by many, and other terms such as "interracial", "interethnic" or "cross-cultural" are more common in contemporary usage. The term remains in use among scholars when referring to past practices concerning multiraciality, such as anti-miscegenation laws that banned interracial marriages.

In Spanish, Portuguese and French, the words used to describe the mixing of races are mestizaje, mestiçagem and métissage. These words, much older than the term miscegenation, are derived from the Late Latin mixticius for "mixed", which is also the root of the Spanish word mestizo. Portuguese also uses miscigenação, derived from the same Latin root as the English word. These non-English terms for "race-mixing" are not considered as offensive as "miscegenation", although they have historically been tied to the caste system (Casta) that was established during the colonial era in Spanish-speaking Latin America. Some groups in South America, however, consider the use of the word mestizo offensive because it was used during the times of the colony to refer specifically to the mixing between the conquistadores and the indigenous people. Today, the mixes among races and ethnicities are diverse, so it is considered preferable to use the term "mixed-race" or simply "mixed" (mezcla). In Portuguese-speaking Latin America (i.e., Brazil), a milder form of caste system existed, although it also provided for legal and social discrimination among individuals belonging to different races, since slavery for blacks existed until the late 19th century. Miscegenation occurred significantly from the very first settlements, with their descendants achieving high rank in government and society. To this day, the Brazilian class system is drawn mostly around socio-economic lines, not racial ones (in a manner similar to other former Portuguese colonies).

The concept of miscegenation is tied to concepts of racial difference. As the different connotations and etymologies of miscegenation and mestizaje suggest, definitions of race, "race mixing" and multiraciality have diverged globally as well as historically, depending on changing social circumstances and cultural perceptions. Mestizo are people of mixed white and indigenous, usually Amerindian ancestry, who do not self-identify as indigenous peoples or Native Americans. In Canada, however, the Métis, who also have partly Amerindian and partly white, often French-Canadian, ancestry, have identified as an ethnic group and are a constitutionally recognized aboriginal people. (In Canada these are referred to as métis. A very famous one was Louis Riel.)

The differences between related terms and words which encompass aspects of racial admixture show the impact of different historical and cultural factors leading to changing social interpretations of race and ethnicity. Thus the Comte de Montlosier, in exile during the French Revolution, equated class difference in 18th-century France with racial difference. Borrowing Boulainvilliers' discourse on the "Nordic race" as being the French aristocracy that invaded the plebeian "Gauls", he showed his contempt for the lowest social class, the Third Estate, calling it "this new people born of slaves ... mixture of all races and of all times".

Etymological History

Miscegenation comes from the Latin miscere, "to mix" and genus, "kind". The word was coined in the U.S. in 1863, and the etymology of the word is tied up with political conflicts during the American Civil War over the abolition of slavery and over the racial segregation of African-Americans. The reference to genus was made to emphasize the supposedly distinct biological differences between whites and non-whites, though all humans belong to the same genus, Homo, same species, Homo sapiens and same subspecies, Homo sapiens sapiens. The word was coined in an anonymous propaganda pamphlet published in New York City in December 1863, during the American Civil War. The pamphlet was entitled Miscegenation: The Theory of the Blending of the Races, Applied to the American White Man and Negro. It purported to advocate the intermarriage of whites and blacks until they were indistinguishably mixed, as a desirable goal, and further asserted that this was the goal of the Republican Party. The pamphlet was a hoax, concocted by Democrats, to discredit the Republicans by imputing to them what were then radical views that offended against the attitudes of the vast majority of whites, including those who opposed slavery. There was already much opposition to the war effort; in New York in particular the opposition reached heights of the Draft Riots, that included numerous lynchings.

The pamphlet and variations on it were reprinted widely in both the north and south by Democrats and Confederates. Only in November 1864 was the pamphlet exposed as a hoax. The hoax pamphlet was written by David Goodman Croly, managing editor of the New York World, a Democratic Party paper, and George Wakeman, a World reporter.

By then, the word miscegenation had entered the common language of the day as a popular buzzword in political and social discourse. The issue of miscegenation, raised by the opponents of Abraham Lincoln, featured prominently in the election campaign of 1864. In the United States, miscegenation has referred primarily to the intermarriage between whites and non-whites, especially blacks. Before the publication of Miscegenation, the word amalgamation, borrowed from metallurgy, had been in use as a general term for ethnic and racial intermixing. A contemporary usage of this metaphor was Ralph Waldo Emerson's private vision in 1845 of America as an ethnic and racial smelting-pot, a variation on the concept of the melting pot. Opinions in the U.S on the desirability of such intermixing, including that between white Protestants and Irish Catholic immigrants, were divided. The term miscegenation was coined to refer specifically to the intermarriage of blacks and whites, with the intent of galvanising opposition to the war.

The Concept of Miscegenation

Franz Boas (considered to be the 'father' of American cultural anthropology) as well as many of his students, such as Ashley Montagu, considered race to be an invalid concept. From this point of view, if the concept of race is invalid, then miscegenation as the 'crossing' of races is equally invalid. These ideas are pursued in greater depth in Ashley Montagu's books on the subject.

Laws Banning Miscegenation

Laws banning "race-mixing" were enforced in Nazi Germany (the Nuremberg Laws) from 1935 until 1945, in certain U.S. states from the Colonial era until 1967 and in South Africa during the early part of the Apartheid era. All these laws primarily banned marriage between spouses of different racially or ethnically defined groups, which was termed "amalgamation" or "miscegenation" in the U.S. The laws in Nazi Germany and many of the U.S. states, as well as South Africa, also banned sexual relations between such individuals. In the United States, the various state laws prohibited the marriage of whites and blacks, and in many states also the intermarriage of whites with Native Americans or Asians. In the U.S., such laws were known as anti-miscegenation laws. From 1913 until 1948, 30 out of the then 48 states enforced such laws.

Although an "Anti-Miscegenation Amendment" to the United States Constitution was proposed in 1871, in 1912–1913, and in 1928, no nation-wide law against racially mixed marriages was ever enacted. In 1967, the United States Supreme Court unanimously ruled in Loving v. Virginia that anti-miscegenation laws are unconstitutional. With this ruling, these laws were no longer in effect in the remaining 16 states that still had them.

The laws in U.S. states were established to maintain "racial purity" and white supremacy. Such laws were passed in South Africa because of fears that the white minority would be "bred-out" by the black majority.

The Nazi ban on interracial marriage and interracial sex was enacted in September 1935 as part of the Nuremberg Laws, the Gesetz zum Schutze des deutschen Blutes und der deutschen Ehre (The Law for the Protection of German Blood and German Honour).

The Nuremberg Laws classified Jews as a race, and forbade marriage and extramarital sexual relations between persons of Jewish origin and persons of "German or related blood". Such intercourse was condemned as Rassenschande (lit. "race-disgrace") and could be punished by imprisonment (usually followed by deportation to a concentration camp) and even by death.

The Prohibition of Mixed Marriages Act in South Africa, enacted in 1949, banned intermarriage between different racial groups, including between whites and non-whites. The Immorality Act, enacted in 1950, also made it a criminal offense for a white person to have any sexual relations with a person of a different race. Both laws were repealed in 1985.

History of ethnoracial admixture and attitudes towards miscegenation

Africa

Africa has a long history of interracial mixing with Arabs and later European men having sexual relations with black African women as well as taking them as wives. Arabs played a big role in the African slave trade and unlike the trans-atlantic trade most of the black African slaves in the Arab slave trade were women. Most of them were used as sexual slaves by the Arab men and some were even taken as wives.

Sir Richard Francis Burton writes during his expedition to Africa about relationships between black women and white men. He writes, "The women are well disposed toward strangers of fair complexion, apparently with the permission of their husbands." There are several mulatto populations throughout Africa mostly the results of interracial relationships between Arab and European men and black women.

In South Africa there are big mulatto communities like the Coloureds and Griqua formed by White colonists taking native African wives. In Namibia there is a community called the Rehoboth Basters formed by the interracial marriage of Dutch/German men and black African women.

In the former Portuguese Africa (now known as Angola, Mozambique and Cape Verde) racial mixing between white Portuguese and black Africans was fairly common, especially in Cape Verde where the majority of the population is of mixed descent.

There have been some recorded cases of Chinese merchants and labourers taking African wives throughout Africa as many Chinese workers were employed to build railways and other infrastructural projects in Africa. These labour groups were made up completely of men with very few Chinese women coming to Africa.

In West Africa, especially Nigeria there are many cases of Lebanese men taking African women. Many such mixed people have gained prominent positions in Africa. Flight Lieutenant Jerry John Rawlings, who has a Scottish father and a black Ghanaian mother became the President and Head of State of Ghana. Jean Ping, the son of a Chinese trader and a black Gabonese mother, became the Deputy Prime Minister as well as the Foreign minister of Gabon and is currently the Chairperson of the Commission of the African Union in 2009. Ian Khama son of a Botswanian leader and a white mother is the President of Botswana and the paramount chief of the Bamangwato people.

Indian men, who have long been traders in East Africa, at times married among local African women. The British Empire brought many Indian workers into East Africa to build the Uganda Railway. Indians eventually populated South Africa, Kenya, Uganda, Tanzania, Malawi, Rwanda, Zambia, Zimbabwe and Zaire in small numbers. These interracial unions were mostly unilateral marriages between Indian men and East African women.

Réunion

The majority of the population of Réunion is defined as mixed race. In the last 350 years, various ethnic groups (Africans, Chinese, English, French, Gujuratis, Tamil Indians) have arrived and settled on the island. There have been mixed race people on the island since its first permanent inhabitation in 1665.

Americas and Canada

The 2006 Census counted 289,400 mixed unions (marriages and common-law unions) involving a visible minority person with a non-visible minority person or a person from a different visible minority group. This was a 33.1% increase from 2001. Among all mixed unions in 2006, 247,600 couples were in unions involving a visible minority person and someone who was not a visible minority. The groups most likely to be involved in a mixed union were Japanese (74.4%), Latin Americans (47%) and blacks (40.6%).

Jamaica and Haiti

By some estimates, 80,000 North American and European women (most of them over the age of 40) visit Jamaica and Haiti every year for sex with young men (mostly in their 20s). They're called "milk bottles". HIV/AIDS infection rates in the Caribbean are much higher than in Canada or the U.S. Even so, female sex tourists in the Caribbean are not especially preoccupied by the risk.

United States

Historically, "race mixing" between black and white people was taboo in the United States. So-called anti-miscegenation laws, barring blacks and whites from marrying or having sex, were established in colonial America as early as 1691.

The 1691 Virginia law was amended in 1705 to remove Indian-white intermarriage from the prohibition. Thomas Jefferson's policy proposal for dealing with Native Americans was "to let our settlements and theirs meet and blend together, to intermix, and become one people." He regretted in 1813 that white-Indian war had prevented this: "They would have mixed their blood with ours, and been amalgamated and identified with us within no distant period of time."

The taboo among American whites surrounding white-black intermarriage can be seen as a historical consequence of the oppression and racial segregation of African-Americans. In many U.S. states interracial marriage was already illegal when the term miscegenation was invented in 1863. The first laws banning interracial marriage were introduced in the late 17th century in the slave-holding colonies of Virginia (1691) and Maryland (1692). Later these laws also spread to colonies and states where slavery did not exist.

It has also been argued that the first laws banning interracial marriage were a response by the planter elite to the problems they were facing due to the socio-economic dynamics of the plantation system in the Southern colonies. The bans in Virginia and Maryland were established at a time when slavery was not yet fully institutionalized. At the time, most forced laborers on the plantations were indentured servants, and they were mostly white. Some historians have suggested that the at-the-time unprecedented laws banning interracial marriage were originally invented by planters as a divide and rule tactic after the uprising of servants in Bacon's Rebellion. According to this theory, the ban on interracial marriage was issued to split up the racially mixed, increasingly mixed-race labour force into whites, who were given their freedom, and blacks, who were later treated as slaves rather than as indentured servants. By forbidding interracial marriage, it became possible to keep these two new groups separated and prevent a new rebellion.

In 1918, there was considerable controversy in Arizona when an Indian farmer B. K. Singh married the sixteen year-old daughter of one of his white tenants. During and after slavery, most American whites regarded interracial marriage between whites and blacks as taboo. However, during slavery many white American men and women did conceive children with black partners. These children automatically became slaves if the mother was a slave or were born free if the mother was free, as slavery was matrilineal. Some children were freed by their slave-holding fathers or bought to be emancipated if the father was not the owner. Many children of these unions formed enclaves under names such as Colored and Gens de couleur, etc. Most mixed-raced descendants merged into the African-American ethnic group during Jim Crow, while over the centuries a minority of mixed-raced Americans passed and became white, and others exist to this day in small mixed enclaves of Mestees such as the Melungeons and Lumbee.

Genetic research suggests that a considerable minority of white Americans (estimated at 1/3 of the population by some geneticists such as Mark Shriver) has some distant African-American ancestry, and that the majority of black Americans have some European ancestry. After the Civil War and the abolition of slavery in 1865, the intermarriage of white and black Americans continued to be taboo, especially but not only in the former slave states.

The Motion Picture Production Code of 1930, also known as Hays Code, explicitly stated that the depiction of "miscegenation... is forbidden." One important strategy intended to discourage the marriage of white Americans and Americans of partly African descent was the promulgation of the one-drop theory, which held that any person with any known African ancestry, however remote, must be regarded as "black". This definition of blackness was encoded in the anti-miscegenation laws of various U.S. states, such as Virginia's Racial Integrity Act of 1924. The plaintiffs in Loving v. Virginia, Mildred Jeter and Richard Loving became the historically most prominent interracial couple in the US through their legal struggle against this act. Accusations of support for miscegenation were commonly made by slavery defenders against Abolitionists before the Civil War. After the War, similar charges were used by white segregationists against advocates of equal rights for African Americans. They were said to be secretly plotting the destruction of the white race through miscegenation. In the 1950s, segregationists alleged a Communist plot funded by the Soviet Union with that goal. In 1957, segregationists cite the anti-semitic hoax A Racial Program for the Twentieth Century as evidence for these claims.

In 1958, the Christian fundamentalist preacher Jerry Falwell, at the time a defender of segregation, in a sermon railed against integration, warning that it would lead to miscegenation, which would "destroy our [white] race eventually.

Asians were also specifically included in some state laws. California continued to ban Asian/white marriages until the Perez v. Sharp decision in 1948.

In the United States, segregationists and Christian identity groups have claimed that several passages in the Bible, for example the stories of Phinehas and of the so-called "curse of Ham", should be understood as referring to miscegenation and that certain verses expressly forbid it. Most theologians read these verses and references as forbidding inter-religious marriage, rather than inter-racial marriage.

Miscegenation has become increasingly accepted in the United States since the Civil Rights movement and up to the present day. The most notable American of mixed race is the current President of the United States, Barack Obama, who is the product of a "mixed" marriage between a black father and white mother. Nevertheless, as late as 2009, a Louisiana justice of the peace refused to issue a marriage license to an interracial couple, justifying the decision on grounds of concern for any children the couple might have.

In recent years, interracial pornographic films, which most commonly refers to black male/white female films, have increased in popularity, becoming one of the fastest-growing and biggest-selling genres. Interracial films that include black men and white women together have a majority audience of white male viewers.

According to Renée Christine Romano, "Alvin Poussaint, a black psychiatrist who studied Freedom Summer volunteers, theorized that black men in the movement saw white women more as symbolic conquests than as individuals. The black men he studied viewed "sexual intimacy with the white girls as a weapon of revenge against white society," Poussaint argued."

Asia

Inter-ethnic marriage in Southeast Asia dates back to the spread of Indian culture, Hinduism and Buddhism to the region. From the 1st century onwards, mostly male traders and merchants from the Indian subcontinent frequently intermarried with the local female populations in Cambodia, Burma, Champa, central Siam, the Malay Peninsula, and Malay Archipelago. Many Indianized kingdoms arose in Southeast Asia during the Middle Ages. From the 9th century onwards, a large number of mostly male Arab traders from the Middle East settled down in the Malay Peninsula and Malay Archipelago, and they intermarried with the local Malay, Indonesian and Filipina female populations. This contributed to the spread of Islam in Southeast Asia. From the 14th to the 17th centuries, many Chinese, Indian and Arab traders settled down within the maritime kingdoms of Southeast Asia and intermarried with the local female populations. This tradition continued among Portuguese traders who also intermarried with the local populations. In the 16th and 17th centuries, thousands of Japanese people also travelled to Southeast Asia and intermarried with the local women there.

The white men who sailed to Asia in the 16th century left a generation of mixed-race offspring that, at the high point of empire building, was more than one-million strong. Some 100,000 Amerasians stayed in Vietnam after the fall of Saigon. During and after the Indonesian National Revolution (1945–1965) around 300,000 people, pre-dominantly Eurasians, left Indonesia to go to the Netherlands.

In the 19th century and early 20th century, there was a network of small numbers of Chinese prostitutes being trafficked across Asia, in countries such as China, Japan, Korea, Singapore and British India, in what was then known as the 'Yellow Slave Traffic'. There was also a network of prostitutes from continental Europe being trafficked to India, Ceylon, Singapore, China and Japan at around the same time, in what was then known as the 'White Slave Traffic'.

During World War II, Japanese soldiers engaged in war rape during their invasions across East Asia and Southeast Asia. The term "comfort women" is a euphemism for the estimated 200,000, mostly Korean and Chinese, women who were forced into prostitution in Japanese military brothels during World War II. Some Dutch women, captured in Dutch colonies in Asia, were also forced into sexual slavery.

Sex tourism has emerged in the late 20th century as a controversial aspect of Western tourism and globalization. Sex tourism is typically undertaken internationally by tourists from wealthier countries. Author Nils Ringdal alleged that three out of four men between the ages of 20 and 50 who have visited Asia or Africa have paid for sex.

Central Asia

Today central asians are a mixed race of various peoples such as Mongols, Turkics, Iranics. The Mongol invasion of Central Asia in 13th century resulted in the massacre of mostly Iranic population and other Indo-European people as well as large numbers of intermarriage and assimilation. Modern Genetic shows that Central Asian Turkic people and Hazara are an mixture of Northeast asians and Indo-european people. Caucasian ancestry is prevelant in almost all central asian turkic people. Kazakhs, Hazara, Karakalpaks have more European mtdna than European y-dna, Kyrgyz have mostly European y-dna with substanial European mtdna. Other Turkic people like Uyghurs, Uzbeks, have mostly European y-dna but also an significant high percentages of European mtdna. Turkmen have predominately European y-dna and mtdna.

China

There have been various periods in the history of China where large numbers of Arabs, Persians and Turks from the "Western Regions" (Central Asia and West Asia) migrated to China, beginning with the arrival of Islam during the Tang Dynasty in the 7th century. Due to the majority of these immigrants being male, they often intermarried with local Chinese females. While intermarriage was initially discouraged by the Tang Dynasty, it was later encouraged during the Song Dynasty, which allowed third-generation immigrants with official titles to intermarry with Chinese imperial princesses. Immigration to China increased under the Mongol Empire, when large numbers of West and Central Asians were brought over to help govern Yuan China in the 13th century.

Iranian, Arab, and Turkic women also migrated to China and mixed with Chinese. Iranian women as dancers were in demand in China during this period. During the Sui dynasty, ten young dancing girls were sent from Persia to China. During the Tang dynasty bars were often attended by Iranian or Sogdian waitresses who performed dances for clients.

During the Five Dynasties and Ten Kingdoms Period(Wudai) (907-960), there are examples of Persian women marrying Chinese emperors. Some Chinese officials from the Song Dynasty era also married women from Dashi (Arabia).

Of the Han Chinese Li family in Quanzhou, Li Nu, the son of Li Lu, visited Hormuz in Persia in 1376, married a Persian or an Arab girl, and brought her back to Quanzhou. He then converted to Islam, Li Nu was the ancestor of the Ming Dynasty reformer Li Chih.

By the 14th century, the total population of Muslims in China had grown to 4 million. After Mongol rule had been overthrown by the Ming Dynasty in 1368, this led to a violent Chinese backlash against West and Central Asians. In order to contain the violence, the Ming administration instituted a policy where all West and Central Asian males were required to intermarry with native Chinese females, hence assimilating them into the local population. Their descendants are today known as the Hui people. Caucasian ancestry is also prevalent among the Hui people, 6.7% Hui people's maternal genetics have an Caucasian origin, while slightly over 30% paternal genetics also have an Caucasian origin. Han women who married Hui men became Hui, and Han men who married Hui women also became Hui.

In the frontier districts of Sichuan, numerous half Chinese-Tibetans were found. Tibetan women were glad to marry Chinese traders and soldiers. Some Chinese traders married Tibetan girls. Traders and officials in ancient times were often forbidden to bring Chinese women with them to Tibet, so they had to marry Tibetan women, the male offspring where considered Chinese and female offspring as Tibetan. Special names were used for these children of Chinese fathers and Tibetan mothers. They are often assimilated into the Tibetan population. Chinese and Nepalese in Tibet married Tibetan women.

In Qinghai, premarital sex between Tibetan girls and Han Chinese was common, some Tibetan girls boasted of their sexual conquests of Han Chinese boys. Dolkar, a Tibetan girl working as a guard at Drapchi Prison married a Chinese.

Chinese men also married Turkic Uyghur women in Xinjiang from 1880–1949. Sometimes poverty influenced Uyghur women to marry Chinese. These marriages were not recognized by local mullahs since Muslims women were not allowed to marry non Muslim men under Islamic law.

This did not stop the women because they enjoyed advantages, not being subject to Islamic law and they were not subjected to certain taxes. Uyghur women married to Chinese also did not have to wear a veil and they received their husband's property upon his death. These women were forbidden from having burial in Muslim graves. The children of Chinese men and Uyghur women were considered as Uyghur. Some Chinese soldiers had Uyghur women as temporary wives, and after the man's military service was up, the wife was left behind or sold, and if it was possible, sons were taken, and daughters were sold.

European travellers noted that many Han Chinese in Xinjiang married Uyghur (who were called turki) women and had children with them. A Chinese was spotted with a "young" and "good looking" Uyghur wife and another Chinese left behind his Uyghur wife and child in Khotan.

After 1950, some intermarriage between Han and Uyghur continued. A Han married a Uyghur woman in 1966 and had three daughters with her, and other cases on intermarriage also continued.

Hong Kong

South Asians have been living in Hong Kong throughout the colonial period, before the partition of India into the nations of India and Pakistan. They migrated to Hong Kong and worked as police officers as well as army officers during colonial rule. 25,000 of the Muslims in Hong Kong trace their roots back to what is now Pakistan. Around half of them belong to 'local boy' families, Muslims of mixed Chinese and South Asian ancestry, descended from early Indian/Pakistani Muslim immigrants who took local Chinese wives and brought their children up as Muslims.

Indian Subcontinent

The Indian subcontinent has a long history of inter-ethnic marriage dating back to ancient history. Various groups of people have been intermarrying for millennia in South Asia, including groups as diverse as the Dravidian, Indo-Aryan, Austro-Asiatic and Tibeto-Burman peoples. Invading Macedonians, Greeks, Scythians, Huns, Persians, Mongols (known as Mughals), and Europeans took Indian wives.

3,000–8,000 years ago, Indo-European-speaking nomadic groups from Europe, the Near East, Anatolia, and the Caucasus migrated to India. According to 19th-century British historians, it was these "Aryans" who established the caste system, an elitist form of social organization that separated the "light-skinned" Indo-Aryan conquerors from the "conquered dark-skinned" indigenous Dravidian population through enforcement of "racial endogamy". Much of this was simply conjecture, fueled by British imperialism; British policies of divide and rule as well as enumeration of the population into rigid categories during the tenure of British rule in India contributed towards the hardening of these segregated caste identities. Since the independence of India from British rule, the British fantasy of an "Aryan Invasion and subjugation of the dark skinned Dravidians in India" has become a staple polemic in South Asian geopolitics, including the propaganda of Indophobia in Pakistan. There is no decisive theory as to the origins of the caste system in India, and globally renowned historians and archaeologists like Jim Shaffer, J.P. Mallory, Edwin Bryant, and others, have disputed the claim of "Aryan Invasion".

Some researchers claim that genetic similarities to Europeans were more common in members of the higher ranks. Their findings, published in Genome Research, supported the idea that members of higher castes are more closely related to Europeans than are the lower castes. According to the research, invading European populations were predominantly male who intermarried with local females and formed the upper castes i.e. the local females had upward mobility in caste which was denied to local males. However, other researchers have criticized and contradicted this claim. A study by Joanna L. Mountain et al. of Stanford University concluded that there was "no clear separation into three genetically distinct groups along caste lines", although "an inferred tree revealed some clustering according to caste affiliation". A 2006 study by Ismail Thanseem et al. of Centre for Cellular and Molecular Biology (India) concluded that the "lower caste groups might have originated with the hierarchical divisions that arose within the tribal groups with the spread of Neolithic agriculturalists, much earlier than the arrival of Aryan speakers", and "the Indo-Europeans established themselves as upper castes among this already developed caste-like class structure within the tribes." A 2006 genetic study by the National Institute of Biologicals in India, testing a sample of men from 32 tribal and 45 caste groups, concluded that the Indians have acquired very few genes from Indo-European speakers. More recent studies have also debunked the claims that so-called "Aryans" and "Dravidians" have a "racial divide". A study conducted by the Centre for Cellular and Molecular Biology in 2009 (in collaboration with Harvard Medical School, Harvard School of Public Health and the Broad Institute of Harvard and MIT) analyzed half a million genetic markers across the genomes of 132 individuals from 25 ethnic groups from 13 states in India across multiple caste groups. The study establishes, based on the impossibility of identifying any genetic indicators across caste lines, that castes in South Asia grew out of traditional tribal organizations during the formation of Indian society, and were not the product of any Aryan invasion and subjugation of Dravidian people.

Rape and enslavement of Hindu women by invading Islamic armies was very common. For example, according to the Chachnama by Qazi Ismail, during the Arab invasion of Sindh (712 CE), Muhammad bin Qasim first attacked and conquered Debal. It was situated on the sea-coast (not far from modern Karachi).

All males of the age of seventeen and upwards were put to the sword and their women and children were enslaved. 700 Hindu women, who were under the protection of Budh (that is, had taken shelter in the temple), were all captured with their valuable ornaments, and adorned with jewels. Muhammad despatched one-fifth of the legal spoil to Hajjaj which included seventy-five damsels, while the other four-fifths were distributed among the soldiers. Thereafter whichever places he attacked like Rawar, Sehwan, Dhalila, Brahmanabad and Multan, Hindu soldiers and men with arms were slain, many women of the higher classes immolated themselves in Jauhar. Most others became prize of the victors. These women were enslaved and converted, and batches of them were despatched to the Caliph in regular installments. For example, after Rawar was taken, many women and children were enslaved, of whom 30 were young ladies of royal blood. Muhammad Qasim sent all these to Hajjaj, who forwarded them to Walid the Khalifa. He sold some of these female slaves of royal birth, and some he presented to others. In Northern India, female slaves captured after every campaign of the marching army, were converted and married to Muslim soldiers who settled down in colonies established in places like Mansura, Kuzdar, Mahfuza and Multan. During the Islamic involvement in India, it was normal for kings to possess harems filled with native Hindu women won as booties of war. The most famous one was of Akbar's harem, which had over 5000 women. Most of these Muslim soldiers were Turks, Afghans, Persians, Mongols (Mughals), other Central Asians and Arabs. There are even cases of Indian princesses marrying kings abroad. For example, the Korean text Samguk Yusa about the Gaya kingdom (it was absorbed by the kingdom of Silla later), indicates that in 48 AD, King Kim Suro of Gaya (the progenitor of the Gimhae Kim clan) took a princess (Princess Heo) from the "Ayuta nation" (which is the Korean name for the city of Ayodhya in North India) as his bride and queen.

Princess Heo belonged to the Mishra royal family of Ayodhya. According to the Samguk Yusa, the princess had a dream about a heavenly fair handsome king from a faraway land who was awaiting heaven's anointed ride. After Princess Heo had the dream, she asked her parents, the king and queen of Ayodhya, for permission to set out and seek the foreign prince, which the king and queen urged with the belief that god orchestrated the whole fate. That king was no other than King Kim Suro of the Korean Gaya kingdom.

In Goa, a Portuguese colony in India, during the late 16th century and 17th century, there was a community of over thousand Japanese slaves and traders, who were either Japanese Christians fleeing persecution in Japan, or young Japanese women and girls brought or captured as sexual slaves by Portuguese traders and their South Asian lascar crew members from Japan. In both cases, they often intermarried with the local population in Goa.

Interracial marriages between European men and Indian women were very common during colonial times. Most of these Indian women usually were Muslim belonging to aristocratic families and families with royal ancestry. According to the historian William Dalrymple, about one in three European men had Indian wives in colonial India. This was primarily because the Europeans (mostly Dutch, British, French and Portuguese and up to a lesser extent Swedes and Danes) came to India when they were young and there were very few white women available in India. The most famous of such interracial liaisons was between the Hyderabadi noblewoman Khair-un-Nissa and the Scottish resident James Achilles Kirkpatrick.

The 600,000-strong Anglo-Indian community was formed by British soldiers taking Indian women as wives. Such relationships have had a strong influence on the arts. Lakmé, an opera by the Frenchman Léo Delibes, deals with the romantic relationship between the British officer Gérald and the daughter of a Hindu high priest Lakmé (Laxmi in Sanskrit).

In Ceylon (present day Sri Lanka), interracial relationships between Dutch, British and Portuguese men and local women were common. The 65,000-strong Burgher community was formed by the interracial marriages of Dutch and Portuguese men with local Sinhalese and Tamil women. In addition to intermarriage, inter-ethnic prostitution in India was also fairly common at the time, when

British officers would frequently visit Indian nautch dancers. In the mid-19th century, there were around 40,000 British soldiers but fewer than 2,000 British officials present in India. Many British and other European officers had their own harems made up of Indian women similar to those the Nawabs and kings of India had. In the 19th century and early 20th century, thousands of women and girls from continental Europe were also trafficked into British India (and Ceylon), where they worked as prostitutes servicing both British soldiers and local Indian (and Ceylonese) men.

As British females began arriving in British India in large numbers from the early-to-mid-19th century, miscegenation became increasingly uncommon in India as British women increasingly disapproved of relationships between local woman and their men. Interracial relationships were also despised after the events of the Sepoy Mutiny of 1857, also known as "India's First War of Independence", where some Indian sepoys rebelled against the British East India Company.

The idea of protecting British female chastity from the "lustful Indian male" had a significant influence on the policies of the British Raj in order to prevent racial miscegenation between the British females and the native Indian male population. While some restrictive policies were imposed on British females in order to protect them from miscegenation, most of these policies were directed against native Indian males.

For example, the 1883 Ilbert Bill, which would have granted Indian judges the right to judge British offenders, was opposed by many British colonialists on the grounds that Indian judges cannot be trusted in dealing with cases involving British females. In the aftermath of the 1919 Amritsar Massacre, the long-held stereotype of Indian males as dark-skinned rapists lusting after white British females was challenged by several novels such as A Passage to India (1924) and The Jewel in the Crown (1966), both of which involve an Indian male being wrongly accused of raping a British female.

When Burma was ruled under the administration of British India, millions of Indians, mostly Muslim, migrated there. The small population mixed descendants of Indian males and local Burmese females are called "Zerbadees", often in a pejorative sense implying mixed race.

Japan

Inter-ethnic marriage in Japan dates back to the 7th century, when Chinese and Korean immigrants began intermarrying with the local Japanese population. In the 1590s, over 50,000 Koreans were forcibly brought to Japan during Hideyoshi's invasions of Korea, where they intermarried with the local population. In the 16th and 17th centuries, around 58,000 Japanese travelled abroad, many of whom intermarried with the local women in Southeast Asia. During the anti-Christian persecutions in 1596, many Japanese Christians fled to Macau and other Portuguese colonies such as Goa, where there was a community of Japanese slaves and traders by the early 17th century. Intermarriage with the local populations in these Portuguese colonies also took place. Portuguese traders in Japan also intermarried with the local Christian women.

From the 15th century, Chinese, Korean and other Far Eastern visitors frequented brothels in Japan. This practice later continued among visitors from the "Western Regions", mainly European traders. This began with the arrival of Portuguese ships to Japan in the 16th century. Portuguese visitors and their South Asian (and sometimes African) crewmembers often engaged in slavery in Japan, where they brought Japanese slaves who were taken to Macau and other Portuguese colonies in Southeast Asia, the Americas, and India. Later European East India companies, including those of the Dutch and British, also engaged in prostitution in Japan.

In the early part of the Shôwa era, Japanese governments executed a eugenic policy to limit the birth of children with inferior traits, as well as aiming to protect the life and health of mothers. Family Center staff also attempted to discourage marriage between Japanese women and Korean men who had been recruited from the peninsula as laborers following its annexation by Japan in 1910. In 1942, a survey report argued that "the Korean laborers brought to Japan, where they have established permanent residency, are of the lower classes and therefore of inferior constitution...By fathering children with Japanese women, these men could lower the caliber of the Yamato minzoku."

In 1928, journalist Shigenori Ikeda promoted the 21 December as the blood-purity day (junketsu de) and sponsored free blood-test at the Tokyo Hygiene laboratory. By the early 1930s, detailed "eugenic marriage" questionnaires were printed or inserted in popular magazines for public consumption. Promoters like Ikeda were convinced that these marriage surveys would not only insure the eugenic fitness of spouses but also help avoid class differences that could disrupt and even destroy marriage. The goal was to create a database of individuals and their entire households which would enable eugenicists to conduct in-depth surveys of any given family's genealogy.

To prevent venereal diseases and rape by Japanese soldiers and to provide comfort to soldiers and head off espionage, the Imperial Japanese Army established "comfort stations" in the Greater East Asia Co-Prosperity Sphere where around 200,000 women, mostly from Korea and China, were recruited or kidnapped by the Kempeitai or the Tokeitai as comfort women.

One of the last eugenic measures of the Shôwa regime was taken by the Higashikuni government. On 19 August 1945, the Home Ministry ordered local government offices to establish a prostitution service for Allied soldiers to preserve the "purity" of the "Japanese race". The official declaration stated that : "Through the sacrifice of thousands of "Okichis" of the Shôwa era, we shall construct a dike to hold back the mad frenzy of the occupation troops and cultivate and preserve the purity of our race long into the future...."

According to Peter Schrijvers in "The GI War against Japan: American Soldiers in Asia and the Pacific during World War II", rape "reflects a burning need to establish total dominance of the other" the enemy.

According to Xavier Guillaume, US soldiers' rape of Japanese women was "general practice". Schrijvers states regarding rapes on Okinawa that "The estimate of one Okinawan historian for the entire three-month period of the campaign exceeds 10,000. A figure that does not seem unlikely when one realizes that during the first 10 days of the occupation of Japan there were 1,336 reported cases of rape of Japanese women by American soldiers in Kanagawa prefecture alone".

However, despite being told by the Japanese military that they would suffer rape, torture and murder at the hands of the Americans, Japanese civilians "were often surprised at the comparatively humane treatment they received from the American enemy." According to Islands of Discontent: Okinawan Responses to Japanese and American Power by Mark Selden, the Americans "did not pursue a policy of torture, rape, and murder of civilians as Japanese military officials had warned." Japanese society, with its ideology of homogeneity, has traditionally been intolerant of ethnic and other differences.

Men or women of mixed ancestry, foreigners, and members of minority groups faced discrimination in a variety of forms. In 2005, a United Nations report expressed concerns about racism in Japan and that government recognition of the depth of the problem was not total. In 2005, Japanese Minister Taro Aso called Japan a "one race" nation.

Korea

Inter-ethnic marriage in Korea dates back to the arrival of Muslims in Korea during the Middle Ages, when Persian and Turkic navigators, traders and slaves settled in Korea and married local Korean people. Some assimilation into Buddhism and Shamanism eventually took place, owing to Korea's geographical isolation from the Muslim world. There are several Korean clans that are descended from such intermarriages. For example, the Deoksu Jang clan, claiming some 30,000 Korean members, view Jang Sunnyong, a Central Asian who married a Korean female, as their ancestor. Another clan, Gyeongju Seol, claiming at least 2,000 members in Korea, view a Central Asian (probably an Uyghur) named Seol Son as their ancestor. There are even cases of Korean kings marrying princesses from abroad. For example, the Korean text Samguk

Yusa about the Gaya kingdom (it was absorbed by the kingdom of Silla later), indicate that in 48 AD, King Kim Suro of Gaya (the progenitor of the Gimhae Kim clan) took a princess (Princess Heo) from the "Ayuta nation" (which is the Korean name for the city of Ayodhya in North India) as his bride and queen. Princess Heo belonged to the Mishra royal family of Ayodhya. According to the Samguk Yusa, the princess had a dream about a heavenly fair handsome king from a far away land who was awaiting heaven's anointed ride.

After Princess Heo had the dream, she asked her parents, the king and queen of Ayodhya, for permission to set out and seek the foreign prince, which the king and queen urged with the belief that god orchestrated the whole fate. That king was no other than King Kim Suro of the Korean Gaya kingdom.

6,423 Korean women married US military personnel as war brides during and immediately after the Korean War. The average number of Korean women marrying US military personnel each year was about 1,500 per year in the 1960s and 2,300 per year in the 1970s.

International marriages now make up 13% of all marriages in South Korea. Most of these marriages are unions between a Korean male and a foreign female usually from China, Japan, Vietnam, the Philippines, United States, Mongolia, Thailand, and Russia. On the other hand, Korean females have married foreign males from Japan, China, the United States, Bangladesh, Pakistan, Philippines, and Nepal. Between 1990 and 2005, there have been 159,942 Korean males and 80,813 Korean females married to foreigners.

South Korea is among the world's most ethnically homogeneous nations. Koreans have traditionally valued an unmixed blood as the most important feature of Korean identity. The term "Kosian", referring to someone who has a Korean father and a non-Korean mother, is considered offensive by some who prefer to identify themselves or their children as Korean. Moreover, the Korean office of Amnesty International has claimed that the word "Kosian" represents racial discrimination. Kosian children, like those of other mixed-race backgrounds in Korea, often face discrimination. There are an estimated 35,000 mixed-raced South Koreans, most of them half Caucasian, according to the Pearl Buck Foundation. Discrimination is far worse against those who have African American fathers.

Malaysia and Singapore

In West Malaysia and Singapore, the majority of inter-ethnic marriages are between Chinese and Indians. The offspring of such marriages are informally known as "Chindian", though the Malaysian government only classifies them by their father's ethnicity. As the majority of these intermarriages usually involve an Indian groom and Chinese bride, the majority of Chindians in Malaysia are usually classified as "Indian" by the Malaysian government. As for the Malays, who are predominantly Muslim, legal restrictions in Malaysia make it uncommon for them to intermarry with either the Indians, who are predominantly Hindu, or the Chinese, who are predominantly Buddhist and Taoist. Non-Muslims are required to convert to Islam in order to marry Muslims. However, this has not entirely stopped intermarriage between the Malays and the Chinese and Indians. There are more Indian-Malay marriages than there are Chinese-Malay ones; this has led to the emergence of a large Muslim Indian society that is most famously known for their 24-hour coffee shops and active involvement in the textile industry. The Muslim Chinese community, on the other hand, is small and has only a negligible impact on the socio-economy and demography of the region.

It is common for Arabs in Singapore and Malaysia to take local Malay wives, due to a common Islamic faith. The Chitty people, in Singapore and the Malacca state of Malaysia, are a Tamil people with considerable Malay descent, which was due to the first Tamil settlers taking local wives, since they did not bring along any of their own women with them. According to government statistics, the population of Singapore as of September 2007 was 4.68 million, of whom multiracial people, including Chindians and Eurasians, formed 2.4%.

In the East Malaysian states of Sabah and Sarawak, there have been many incidents of intermarriage between Chinese and native tribes such as the Murut and Dusun in Sabah, and the Iban and Bisaya in Sarawak. This phenomenon has resulted in a potpourri of cultures in both states where many people claiming to be of native descent have some Chinese blood in them, and many Chinese have native blood in them. The offspring of these mixed marriages are called 'Sino-(name of tribe)', e.g. Sino-Dusun. Normally, if the father is Chinese, the offspring will adopt Chinese culture and if the father is native then native culture will be adopted, but this is not always the case. These Sino-natives are usually fluent in Malay and English. A smaller number are able to speak Chinese dialects and Mandarin, especially those who have received education in vernacular Chinese schools.

Burma

Burmese Muslims are the descendants of Bengalis, Indian Muslims, Arabs, Persians, Turks, Pathans, Chinese Muslims and Malays who settled and intermarried with the local Burmese population and other Burmese ethnic groups such as the Rakhine, Shan, Karen, and Mon.

The oldest Muslim group in Burma (Myanmar) are the Rohingya people, who some believed are descended from Bengalis who intermarried with the native females in the Rakhine State after the 7th century, but this is just an theory. When Burma was ruled by the British India administration, millions of Indians, mostly Muslim, migrated there. The small population of mixed descendants of Indian males and local Burmese females are called "Zerbadees", often in a pejorative sense implying mixed race. The Panthays, a group of Chinese Muslims descended from West Asians and Central Asians, migrated from China and also intermarried with local Burmese females.

In addition, Burma has an estimated 52,000 Anglo-Burmese people, descended from British and Burmese people. Anglo-Burmese people frequently intermarried with Anglo-Indian immigrants, who eventually assimilated into the Anglo-Burmese community.

Philippines

Historically, admixture has been an ever present and pervading phenomenon in the Philippines. The Philippines were originally settled by Australoid peoples called Negritos (different from other australoid groups) which now form the country's aboriginal community. Some Admixture may have occurred between this earlier group and the mainstream Malayo-Polynesian population.

A considerable number of the population in the town of Cainta, Rizal, are descended from Indian soldiers who mutinied against the British Indian Army when the British briefly occupied the Philippines in 1762 to 1763.

These Indian soldiers called Sepoy settled in town and intermarried with native females. The Sepoy ancestry of Cainta is very visible today, particularly in Barrio Dayap near Brgy. Sto Nino. Their unique physical characteristics make them distinct from the average Filipinos.

There has been a Chinese presence in the Philippines since the 9th century. However, large-scale migrations of Chinese to the Philippines only started during the Spanish colonial era, when the world market was opened to the Philippines. It is estimated that among Filipinos, 10%–20% have some Chinese ancestry and 1.5% are "full-blooded" Chinese.

According to the American anthropologist Dr. H. Otley Beyer, the ancestry of Filipinos is 2% Arab. This dates back to when Arab traders intermarried with the local Malay Filipina female populations during the pre-Spanish history of the Philippines. Major Arab migration to the Philippines coincided with the spread of Islam in the region.

Filipino-Muslim royal families from the Sultanate of Sulu and the Sultanate of Maguindanao claim Arab descent even going as far as claiming direct lineage from the prophet Mohammad. Such intermarriage mostly took place around the Mindanao island area, but the arrival of Spanish Conquistadors to the Philippines abruptly halted the spread of Islam further north into the Philippines. Intermarriage with Spanish people later became more prevalent after the Philippines was colonized by the Spanish Empire.

When the Spanish colonized the Philippines, a significant portion of the Filipino population mixed with the Spanish. When the United States took the Philippines from Spain during the Spanish-American War, much intermixing of Americans, both white and black, took place on the island of Luzon where the USA had a Naval Base and Air Force Base even after the USA gave the Philippines independence after World War II.

First children and descendants of male Filipino population with Spanish surnames who intermarried with white American female population may be considered Spanish mestizos. The descendants of Filipinos and Europeans are today known as mestizos, following the term used in other former Spanish colonies.

Much mixing with the Japanese also took place due to the war rapes of Filipina women during World War II. Today there is an increasing number of Japanese men marrying Filipina woman and fathering children by them whose family remain behind in the Philippines and are financially supported by their Japanese fathers who make regular visits to the Philippines. Today mixed-race marriages have a mixed reaction in the Philippines, most urban centers like Manila and Cebu are more willing to accept interracial marriages than rural areas, plus there is more approval if the Filipina marries out than a Filipino male.

Europe and Germany

During the tempestuous years following World War I, the French Army occupied the Rhineland, utilising African soldiers amongst their forces. Their children were known as "Rhineland Bastards".

Beginning in 1933, the mainstream Nazi anti-Semitism considered the Jews as being a group of people bound by close, so-called genetic (blood) ties, to form a unit, which one could not join or secede from. The influence of Jews had been declared to have a detrimental impact on Germany, in order to justify the discriminations and persecutions of Jews. To be spared from those, one had to prove one's affiliation with the group of the Aryan race, as conceived by the Nazis.

It was paradoxical that neither genetic tests nor allegedly racial outward features in one's physiognomy determined one's affiliation, although the Nazis talked a lot about physiognomy, but only the records of the religious affiliations of one's grandparents decided it. However, while earlier the grandparents had still been able to choose their religion, their grandchildren in the Nazi era were compulsorily categorised as Jews, thus non-Aryans, if three or four grandparents had been enrolled as members of a Jewish congregation, regardless of whether the persecuted themselves were Jews according to the Halachah (roughly meaning: Jewish by birth from a Jewish mother or by conversion), apostates, irreligionists or Christians. The Nuremberg Laws of 1935 forbade persons racially regarded as so-called Aryans and non-Aryans to marry; this included all marriages where at least one partner was a German citizen. Non-Aryans comprised mostly Jewish Germans and Gentile Germans of Jewish descent. However, Germans of extra-European and especially of African descent and Germans regarded as belonging to the minority group of Sinti and Roma were also considered as non-Aryans.

Eventually children – whenever born – within a mixed marriage, as well as children from extramarital mixed relationships born until July 31, 1936, were discriminated against as Mischlinge. However, children later born to mixed parents, not yet married as at the passing of the Nuremberg Laws, were to be discriminated against as Geltungsjuden, regardless of whether the parents had meanwhile married abroad or remained unmarried. Eventually children who were enrolled in a Jewish congregation were also subject to discrimination as Geltungsjuden.

Geltungsjuden were subjected to varying degrees of forced labour in 1940, partly ordered for all Jewish-classified spouses, either only for Jewish-classified husbands or only exempting Jewish-classified wives taking care of minor children. No documents indicate the exemption of a mixed marriage and especially of its Jewish-classified spouse from some persecutions.

Systematic deportations of Jewish Germans and Gentile Germans of Jewish descent started on October 18, 1941. German Jews and German Gentiles of Jewish descent living in mixed marriages were in fact mostly spared from deportation. In the event that a mixed marriage ended by the death of the so-called Aryan spouse or the divorce of the Jewish-classified spouse, the Jewish-classified spouse residing within Germany was usually deported soon after unless the couple still had minor children not counted as Geltungsjuden.

In March 1943 an attempt to deport the Berlin-based Jews and Gentiles of Jewish descent, living in non-privileged mixed marriages, failed due to public protest by their in-laws of so-called Aryan kinship (Rosenstraße protest). Also the Aryan-classified husbands and Mischling-classified children (starting at the age of 16) from mixed marriages were taken by the Organisation Todt for forced labour, starting in autumn 1944.

A last attempt, undertaken in February/March 1945, ended because the extermination camps were already liberated. However, 2,600 from all over the Reich were deported to Theresienstadt, of whom most survived the last months until their liberation.

With the defeat of Nazi Germany in 1945 the laws banning so-called mixed marriages were lifted again. If couples who had already lived together during the Nazi era had remained unmarried due to the legal restrictions then got married after the war, their date of marriage was legally retroactively backdated if they wished it to the date they formed a couple. Even if one spouse was already dead, the marriage could be retroactively recognised. In the West German Federal Republic of Germany 1,823 couples applied for recognition, which was granted in 1,255 cases.

It is estimated that up to 7,000 postwar black German children with black GI fathers and white German mothers were adopted by Americans.

Hungary

The Avars, Asiatic nomads who during the late 6th and 7th centuries had formed an extensive empire largely inhabited by conquered Slavs, made the agricultural Slavs pay taxes, and used their wives and daughters as concubines.

The Hungarians are thought to have originated in an ancient Finno-Ugric population that originally inhabited the forested area between the Volga River and the Ural Mountains. At the time of the Magyar migration in the 10th century, the present-day Hungary was inhabited by Slavs, numbering about 200,000, who were either assimilated or enslaved by the Magyars.

During the Russian campaign in the 13th century, the Mongols drove some 40,000 Cuman families, a nomadic tribe, west of the Carpathian Mountains. The Iranian Jassic people came to Hungary together with the Cumans after they were defeated by the Mongols. Over the centuries they were fully assimilated into the Hungarian population. Rogerius, a monk who survived the Mongol invasion of Hungary, pointed out not only the genocidal element of the occupation, but also that the Mongols especially "found pleasure" in humiliating women.

Iberian Peninsula

In ancient history, the Iberian Peninsula was frequently invaded by foreigners who intermarried with the native population. One of the earliest foreign groups to arrive in the region were the Indo-European Celts who intermarried with the pre-Indo-European Iberians in prehistoric Iberia. They were later followed by the Semitic Phoenicians and Carthaginians and the Indo-European Romans who intermarried with the pre-Roman peoples of the Iberian Peninsula during Classical Antiquity. They were in turn followed by the Germanic Visigoths, Suebi and Vandals and the Iranic Sarmatians and Alans who also intermarried with the local population in Hispania during late Antiquity.

In the 6th century, the region was reconquered by the Byzantine Empire (Eastern Roman Empire), when Byzantine Greeks also settled there, before the region was lost again to the Visigothic Kingdom less than a century later. After the Umayyad conquest of Hispania in the 8th century, the Islamic state of Al-Andalus was established in the Iberian Peninsula. Islamic marital law is by itself an anti-miscegenation law, since it allows a Muslim male to marry Christian and Jewish females but not a Muslim female to marry a Christian or Jewish male. The offspring of such marriages were known as Muladi or Muwallad, an Arabic term still used in the modern Arab world to refer to people with Arab fathers and non-Arab mothers. Some sources consider this term the origin for the Spanish word Mulato. However, the Real Academia Española does not endorse such etymology. In addition, many Muladi were also descended from Saqaliba (Slavic) slaves taken from Eastern Europe via the Arab slave trade.

After the Reconquista, which was completed in 1492, most of the Moors were forced to either flee to Morocco or convert to Christianity. The ones who converted to Christianity were known as Moriscoes, and they were often persecuted by the Spanish Inquisition as suspects of heresy on the basis of the Limpieza de sangre ("Cleanliness of blood") doctrine, under which anti-miscegenation laws were implemented in the Iberian Peninsula.

Anyone whose ancestors had miscegenated with the Moors or Jews were suspicious of secretly practicing Islam or Judaism, so were often particularly monitored by the Inquisition. The claim to universal hidalguía (lowest nobility) of the Basques was justified by erudites like Manuel de Larramendi (1690–1766) because the Arab invasion had not reached the Basque territories, so it was believed that Basques had maintained their original purity, while the rest of Spain was suspect of miscegenation. In fact, the Arab invasion also reached the Basque country and there had been a significant Jewish minority in Navarre, but the hidalguía helped many Basques to official positions in the administration. In December 2008, an important genetic study revealed that the religious conversions of Jews and Muslims have had a profound impact on the population of the Iberian Peninsula. This study indicated a Sephardic Jewish mean admixture of about 20% and a North African admixture of about 11%.

Italian Peninsula

As was the case in other regions conquered by Muslims, it was acceptable in Islamic marital law for a Muslim male to marry Christian and Jewish females in southern Italy when under Islamic rule between the 8th and 11th centuries. In this case, most intermarriages were between Arab and Berber males from North Africa and the local Greek, Roman and Italian females of Sicily and southern Italy.

Such intermarriages were particularly common in the Emirate of Sicily, where one writer visiting the place in the 970s expressed shock at how common it was in rural areas. After the Norman conquest of southern Italy, all Muslim citizens (whether foreign, native or mixed) of the Kingdom of Sicily were known as "Moors". After a brief period of Arab-Norman culture had flourished under the reign of Roger II of Sicily, later rulers had forced the Moors to either convert to Christianity or be expelled from the kingdom.

In Malta, Arabs and Italians from neighbouring Sicily and Calabria intermarried with the local inhabitants, who were descended from Phoenicians, Greeks, Romans and Vandals. The Maltese people are descended from such unions, and the Maltese language is descended from Siculo-Arabic.

In the Republic of Venice in northern Italy, it was common for foreign Arab and Berber traders, known to Europeans as the "Moors", to take local Italian wives. This became a subject matter in several William Shakespeare plays, most notably Othello, involving an inter-ethnic relationship between a Moorish Othello and his Venetian wife Desdemona, based on Giovanni Battista Giraldi's "Un Capitano Moro" which was itself inspired by an actual incident that occurred in Venice around 1508. At times, the Italian city-states also played an active role in the Arab slave trade, where Moorish and Italian traders occasionally exchanged slaves. Leonardo da Vinci's mother Caterina, for example, was most likely a slave from the Middle East.

During World War II, France's Moroccan troops known as Goumiers committed war rapes in Italy after the Battle of Monte Cassino and in Germany. In Italy, victims of the mass rape committed after the Battle of Monte Cassino by Goumiers are known as Marocchinate. According to Italian sources, more than 7,000 Italian civilians, including women and children, were raped by Goumiers.

Southeastern and Eastern Europe

Vikings explored and eventually settled in territories in Slavic-dominated areas of Eastern Europe. By 950 AD these settlements were largely Slavicized through intermarriage with the local population. Eastern Europe was also an important source for the Arab slave trade at the time, when Saqaliba (Slavic) slaves were taken to the Arab World, where the women and girls often served in harems, some of whom married their Arab masters. When the Mongol Empire annexed much of Eastern Europe in the 13th century, the Mongols also intermarried with the local population and often engaged in war rape during the Mongol invasion of Europe.

In the 11th century, the Byzantine territory of Anatolia was conquered by the Seljuq Turks, who came from Turkestan in Central Asia. Their Ottoman Turkish descendants went on to annex the Balkans and much of Eastern Europe in the 15th and 16th centuries. Due to Islamic marital law allowing a Muslim male to marry Christian and Jewish females, it was common in the Ottoman Empire for Turkish males to intermarry with European females. For example, various sultans of the Ottoman Dynasty often had Greek (Rûm), Slavic (Saqaliba), Venetian, Northcaucasian and French wives.

Some of these European wives exerted great influence upon the empire as Valide Sultan ("Mother-Sultan"), some famous examples including Roxelana, a Slavic harem slave who later became Suleiman the Magnificent's favourite wife, and Aimée du Buc de Rivéry, wife of Abdul Hamid I and cousin of French Empress Josephine. Due to the common occurrence of such intermarriages in the Ottoman Empire, they have had a significant impact on the ethnic makeup of the modern Turkish population in Turkey, which now differs from that of the Turkic population in Central Asia. In addition to intermarriage, the large harems of Ottoman sultans often consisted almost entirely of female concubines who were of Christian European origin.

The degree of miscegenation is very high in the former Soviet Union. Interethnic marriages made up about 22% of all marriages in Moscow, according to the figures from 1995. The majority of such marriages consist among Russian, Ukrainians, Byelorussians and other Slavic.

Annually, many Russian women end up as prostitutes in India, China, Japan, or South Korea. Bars in major Chinese cities now offer blond, blue-eyed Russian "hostesses", while in Tokyo, Russian girls are the latest addition to the menu in fancy "hostess" bars. Russian prostitutes are found everywhere in Macau. Their customers are almost exclusively wealthy Chinese, Japanese and Korean businessmen. Eastern European women also end up in Turkey and the United Arab Emirates.

Low levels of mixed ancestry are, in some areas (especially urban), almost universal, and generally go entirely ignored and unnoticed unless persons wish to identify themselves with ethnic minorities. Highly visible divergence from the local ethnic majority is also treated differently, depending on whether the individual identifies with the local culture or not. In modern times, attitudes towards miscegenation in the former Soviet Union vary greatly, depending on the race and gender of each partner. For example, unions between Slavic males and Oriental (Asian) or Turkic, Northcaucasian, Southcaucasian and Finno-Ugric women are almost universally tolerated, and their children are generally identified and treated as members of the local ethnic majority.

However, unions between Slavic women and visibly non-Slavic men may meet varying degrees of discrimination, from light to none for Asian men (depending also on origin, whether they are immigrants or were born in the Soviet Union, and where in the Soviet Union they were born), to some hostility for Finno-Ugric, Turkic and Northcaucasian men (although much of this is due to the assumption of their faith as Muslim) and Jews, and quite high intolerance towards those who marry blacks or have children by them (young African-Russians in Moscow are often scornfully called 'Children of the Olympics', under the assumption that they were conceived by visiting tourists during the 1980 Moscow Olympic Games). The situation is also highly affected by self-identification, since many people of Asian or Turkic blood have assimilated to the point where they identify themselves as Russian/Ukrainian/etc. and are socially accepted as such.

United Kingdom

Even though 80 percent of the genetic characteristics of most indigenous Britons today have been passed down from a few thousand Ice Age hunters and the genetic makeup of today's indigenous population is much the same as it was 12,000 years ago, Britain has a fairly long history of inter-ethnic marriage among the various European populations that inhabited the island, including the Celtic, Roman, Anglo-Saxon and Anglo-Norman peoples. Although it is debateable whether these can be described as ethnic groups with distinct cultural identities. The Celts cannot be described with a single ethnic identity and various tribal names were used and a 'Celtic' culture was shared in Europe despite the variety of languages. Norman and Vikings people rarely held onto an identity and allegiances were more to leaders and family rather than an ethnic group, there is no evidence the Vikings for example acted as an distinct ethnic group in Britain and so called Scandinavian markers may have been in Britain thousands of years in the Neolithic. Also there were Scandinavians and pre-Viking Scandinavian influence amongst the Anglo-Saxons as shown in Beowulf and Sutton Hoo.

There is no evidence the Anglo-Saxons used that term or indeed forced their languages or culture on local Britons, indeed there is no evidence the majority of local Britons used the term Celt or even spoke a Celtic language, indeed a Germanic language which evolved into English may have been spoke in large parts of Britain as Oppenheimer suggests . The numbers of Vikings, Anglo-Saxons, Romans may be widely exaggerated, Viking, Anglo-Saxons armies may have been small . Intermarriage with non-European populations began in the late 15th century, with the arrival of the Romani people, who have Indian origins. The Romani in Britain intermarried with the local population and became known to the Romani as the Romanichal. In India, the British East India Company and other European soldiers intermarried with Indian women. The offspring of these mixed marriages between the British and Indians were known as Anglo-Indians. Indian wives sometimes accompanied their husbands back to Britain.

Inter-ethnic relationships have become increasingly accepted over the last several decades. As of 2001, 2% of all marriages in Britain are inter-ethnic. Despite having a much lower non-white population (9%), mixed marriages in the United Kingdom are as common as in the United States, although America has many fewer specific definitions of race (four racial definitions as opposed to the United Kingdom's 86). As of 2005, it is estimated that nearly half of British-born African-Caribbean males, a third of British-born African-Caribbean females, and a fifth of Indian and African males, have white partners. As of 2009, one in 10 children in the UK lives in a mixed-race family and two out of five Chinese women have partners of a different race.

Middle East

A Stanford team found the greatest diversity outside Africa among people living in the wide crescent of land stretching from the eastern shore of the Mediterranean to northern India. Not only was the region among the first colonized by the African migrants, they theorize, but the large number of European and East Asian genes among the population indicates that it has long been a human highway, with large numbers of migrants from both directions conquering, trading and generally reproducing along its entire length. The same team also found out that the Bedouin nomads of the Middle East actually have some European and South Asian blood.

In ancient times a Celtic people known as Galatians settled in what is present day Turkey and thus interbred with the people there. During the empire of Alexander the Great, many Greek/Macedonian soldiers had inter-ethnic relationships with women throughout the Middle East all the way to northwestern India. Later, coastal North Africa and parts of the Middle East were part of the Roman Empire and many Roman men were posted as soldiers there. Many of them had inter-ethnic relationships with local Middle-Eastern women. The Germanic people known as the Vandals also conquered parts of coastal North Africa during the great migration period which led to opportunities of inter-ethnic relationships between Germanic men and local North African women.

Inter-ethnic sexual slavery was common during the Arab slave trade throughout the Middle Ages and early modern period, when women and girls captured from non-Arab lands often ended up as sexual slaves in the harems of the Arab World. Most of these slaves came from places such as Sub-Saharan Africa (mainly Zanj), South Asia (Hindus), the North Caucasus (mainly Circassians), Central Asia (mainly Tartars), and Central and Eastern Europe (mainly Saqaliba). The Barbary pirates also captured 1,250,000 slaves from Western Europe and North America between the 16th and 19th centuries. It was also common for Arab conquerors, traders and explorers to marry local females in the lands they conquered or traded with, in various different parts of Africa, Asia.

Inter-ethnic relationships were generally accepted in Arabic society and formed a fairly common theme in medieval Arabic literature and Persian literature. For example, the Kurdish poet Nizami, who had himself married his Kipchak slave girl, wrote The Seven Beauties (1196). Its frame story involves a Persian prince marrying seven foreign princesses, including Byzantine, Chinese, Indian, Khwarezmian, Maghrebian, Slavic and Tartar princesses. Hadith Bayad wa Riyad, a 12th-century Arabic tale from Al-Andalus, was a love story involving an Iberian girl and a Damascene man. The One Thousand and One Nights tale of "The Man of Al-Yaman and His Six Slave-Girls" involves a Yemeni man's relationship with foreign slave girls, four of which are white, black, brown and yellow. Another One Thousand and One Nights tale, "The Ebony Horse", involves the Prince of Persia, Qamar al-Aqmar, rescuing his lover, the Princess of Sana'a, from the Byzantine Emperor who also wishes to marry her. One study found that some Arabic-speaking populations—Palestinians, Jordanians, Syrians, Iraqis, and Bedouins—have what appears to be substantial mtDNA gene flow from sub-Saharan Africa, amounting to 10–15% of lineages within the past three millennia. In the case of Yemenites, the average is higher at 35%. Of particular historic interest might be the finding that with almost no exceptions the sub-Saharan gene flow was exclusively female, at least in part as a result of the Arab slave trade.

Between the 11th and 13th centuries, medieval Western Asia was repeatedly invaded by Europeans (Crusades) and Mongols (Mongol Empire) which led to opportunities for inter-ethnic relationships between European, Mongol and Central Asian soldiers with local Arab women. As well as many Europeans, there were North Africans, South Asians and Central Asians who worked as mercenaries and traders in the area, most of them converting to Islam and taking local women as wives. From 839 AD, Viking Varangian mercenaries who were in the service of the Byzantine Empire, notably Harald Sigurdsson, campaigned in North Africa, Jerusalem and other places in the Middle East during the Byzantine-Arab Wars, and interbred with the local population as spoils of warfare or through eventual settling with many Scandinavian and Slavic Viking men taking Arab or Anatolian women as wives. There is archaeological evidence these Vikings had established contact with the city of Baghdad, at the time the center of the Islamic Empire, and connected with the populace there.

A genetic anthropological study known as The Genographic Project has found what is believed to be faint genetic traces left by medieval Crusaders in the Middle East. The team has uncovered a specific DNA signature in Lebanon that is probably linked to the Christian crusades of the 7th and 8th centuries. It is believed the Crusaders were welcomed by Christian Arabs long suffering under Islamic rule and many offered their daughters in marriage to the European Crusaders who originated from European kingdoms mostly France, England and the Holy Roman Empire.

In the Ottoman Empire, in addition to the Ottoman elites often taking large numbers of European wives and concubines (see Southeastern and Eastern Europe section), there were also opportunities for the reverse, when the empire recruited young Christian boys (mostly Europeans in addition to Christian Arabs) to become the elite troop of the Turkish Empire, the Janissaries. These Janissaries were stationed throughout the Turkish empire including the Middle-East and North Africa leading to inter-ethnic relationships between European men and women from the Middle East and North Africa. Further inter-ethnic marriages took place during the European colonial age when Great Britain, France and Italy ruled various parts of this region. The majority of these marriages were between European men and local Arab and Berber women. The concubines of the Ottoman Sultan consisted chiefly of purchased slaves. Because Islamic law forbade Muslims to enslave fellow Muslims, the Sultan's concubines were generally of Christian origin. The mother of a Sultan, though technically a slave, received the extremely powerful title of Valide Sultan, and at times became effective ruler of the Empire. One notable example was Kösem Sultan, daughter of a Greek Christian priest, who dominated the Ottoman Empire during the early decades of the 17th century. Another notable example was Roxelana, the favourite wife of Suleiman the Magnificent.

Inter-ethnic sexual slavery still continues today in a smaller form in the Arab states of the Persian Gulf, where women and children are trafficked from the post-Soviet states, Eastern Europe, Far East, Africa, South Asia and other parts of the Middle East.

Israel

The modern State of Israel was established as a nation-state for the Jewish people. The Jewish identity contains elements of religion (Judaism), ethnicity, and a sense of a common lineage.

In this sense, Jewish miscegenation could be viewed on two levels; one based on belonging to the Jewish ethnic group or Jewish people, and the other based on the race of a given Jew. Jewish miscegenation based on Jewishness (belonging to the Jewish ethnic group or Jewish people) would be defined on whether one parent is not Jewish, independent of whether either the Jewish or non-Jewish parent are of the same or different races. Racial miscegenation would be defined as the union between a Jew of a given race with a person of a different race, be the other person a Jew or not. Two Jewish people may still be considered "interracial" if those two Jews are of different races, although it would not be considered exogamous in the context of Jewish ethnicity, as both are still Jews.

In Israel, all marriages must be approved by religious celebrants, while civil marriages are legally recognized if performed abroad. Rules governing marriage are based on strict religious guidelines of each religion. By Israeli law, authority over all issues related to Judaism in Israel, including marriage, falls under the Orthodox Chief Rabbinate of Israel. Orthodox Judaism is the only form of Judaism recognized by the state, and marriages performed in Israel by non-Orthodox Rabbis are not recognized.

The Rabbinate prohibits marriage in Israel of halakhic Jews (i.e. people born to a Jewish mother or Jewish by conversion), whether they are Orthodox Jews or not, to partners who are non-Jewish or who are of Jewish descent that runs through the paternal line (i.e. not Jewish according to halakha). As a result, in the state of Israel, people of differing religious traditions cannot legally marry someone in another religion and multi-faith couples must leave the country to get married, most often to Central and Northern Greece . The only other option in Israel for the marriage of a halakhic Jew (Orthodox or not) to a non-Jew, or for that matter, a Christian to a non-Christian or Muslim to a non-Muslim, is for one partner to formally convert to the other's religion, be it to Judaism (Orthodox only), a Christian denomination (such as Eastern Orthodox or Maronite) or a denomination of Islam (such as Sunni or Shia).

As for persons with patrilineal Jewish descent (i.e. not recognized as Jewish according to halakha) who wish to marry a halakhic Jew (i.e. born to a Jewish mother or is Jewish by Orthodox conversion) who is Orthodox or otherwise, is also required to formally convert to Judaism (Orthodox only) or they cannot legally marry.

According to a Haaretz article "Justice Ministry drafts civil marriage law for 'refuseniks'" 300,000 people, or 150,000 couples, are affected by marriage restrictions based on the partners' disparate religious traditions or non-halakhic Jewish status.

Israeli law concerns itself with miscegenation based on Jewish ethnicity, not miscegenation based on race. Therefore, there are no restrictions on interracial marriages between Jews of different Jewish ethnic divisions, or between other co-religionists of different races, although social stigma may still exist.

Many Israeli Jews oppose mixed relationships between Jewish women and Arab men. A 2007 opinion survey found that more than half of Israeli Jews believed intermarriage is equivalent to "national treason". A group of 35 Jewish men, known as "Fire for Judaism", in Pisgat Ze'ev have started patrolling the town in an effort to stop Jewish women from dating Arab men.

The municipality of Petah Tikva has also announced an initiative to prevent interracial relationships, providing a telephone hotline for friends and family to "inform" on Jewish girls who date Arab men as well as psychologists to provide counselling. The town of Kiryat Gat launched a school programme in schools to warn Jewish girls against dating local Bedouin men.

In February 2010 Maariv has reported that the Tel Aviv municipality has instituted an official, government-sponsored "counselling program" to discourage Jewish girls from dating and marrying Arab boys. The Times has also reported on a vigilante parents' group policing the Jerusalem neighborhood of Pisgat Ze'ev to intimidate and discourage local Arab-Jewish couples.

The Jewish anti-missionary group Yad L'achim has also performed paramilitary "rescue operations" of Jewish women from non-Jewish husbands and celebrates the "rescued women" on their website. However, according to Halakha, descendants from a Jewish mother are always defined to be Jewish.

Oceania

Australia

Analysis of the 2006 census reveals that 52% of Aboriginal men and 55% of Aboriginal women were married to non-Aboriginal Australians.

New Zealand

Most people who are entitled to call themselves Maori under the Maori Affairs Amendment Act 1974 are less than half Maori. Two-thirds of Maori babies, half of Pacific babies, and a third of white and Asian babies belonged to more than one ethnic group.

Portuguese Colonies

According to Gilberto Freyre, a Brazilian sociologist, miscegenation was commonplace in the Portuguese colonies, and was even supported by the court as a way to boost low populations and guarantee a successful and cohesive settlement. Thus, settlers often released African slaves to become their wives. The children were guaranteed full Portuguese citizenship, provided the parents were married. Some former Portuguese colonies have large mixed-race populations, for instance, Brazil, Cape Verde, Mozambique, Timor Leste, Macau and São Tomé and Príncipe. In the case of Brazil, the influential "Indianist" novels of José de Alencar (O Guarany, Iracema, and Ubirajara) perhaps went farther than in the other colonies, advocating miscegenation in order to create a truly Brazilian race. Mixed marriages between Portuguese and locals in former colonies were very common in all Portuguese colonies. Miscegenation was still common in Africa until the independence of the former Portuguese colonies in the mid-1970s.

Demographics of Ethnoracial Admixture

U.S.

According to the U.S. Census, in 2000 there were 504,119 Asian-white marriages, 287,576 black-white marriages, and 31,271 Asian-black marriages. The black-white marriages increased from 65,000 in 1970 to 558,000 in 2010, according to Census Bureau figures.

In the United States, rates of interracial cohabitation are significantly higher than those of marriage. Although only 7% of married African American men have Caucasian American wives, 13% of cohabitating African American men have Caucasian American partners. 25% of married Asian American women have Caucasian spouses, but 45% of cohabitating Asian American women are with Caucasian American men. Of cohabiting Asian men, slightly over 37 percent of Asian men have white female partners. In 2006, 41% of Asian American-born women were registered as having White husbands. Almost 30%

of Asians and Latinos outmarry, with 86.8 and 90% of these intermarriages, respectively, being to a white person. According to Karyn Langhorne Folan, "...although the most recent census available reported that 70% of African American women are single, African American women have the greatest resistance to marrying 'out' of the race."

One study on college students noted that, while 90% of the 80 black men reported experiences in interracial dating, only 12% of the 140 white men reported such experiences. Some studies have shown as many as 45% of white women have dated men outside their race. The massive sexual survey revealed that 19% of black males had engaged in sexual activity with white women. A Gallup poll on interracial dating in June 2006 found that 95% of 18- to 29-year-olds approve of blacks' and whites' dating each other. About 60% of that age group said they have dated someone of another race. In 1980, just 17% said they had dated someone from a different racial background.

According to the 2010 Census, the number of interracial marriages in the U.S. is up 20% since 2000, to around 4.5 million. "A record 14.6% of all new marriages in the United States in 2008 were between spouses of a different race or ethnicity from one another. ... Rates more than doubled among whites and nearly tripled among blacks. But for both Hispanics and Asians, rates were nearly identical in 2008 and 1980.", according to a Pew Research Center analysis of demographic data from the U.S. Census Bureau.

According to studies by Jenifer L. Bratter and Rosalind B. King made publicly available on the Education Resources Information Center, White female-Black male and White female-Asian male marriages are more prone to divorce than White-White pairings. Conversely, unions between White males and non-White females (and between Hispanics and non-Hispanic persons) have similar or lower risks of divorce than White-White marriages, unions between white-black female last longer than white-white pairings or white-Asian pairings.

Brazil

Multiracial Brazilians make up 42.6% of Brazil's population, 79.782 million people, and they live in all regions of Brazil. Multiracial Brazilians are mainly people of mixed European, African, East Asian (mostly Japanese) and Amerindian ancestry.

Genetic Studies of Racial Admixture

Miscegenation between two populations reduces the genetic distance between the populations. During the Age of Discovery which began in the early 15th century, European explorers sailed all across the globe reaching all the major continents. In the process they came into contact with many populations that had been isolated for thousands of years. The Tasmanian aboriginals were one of the most isolated groups on the planet. They were driven to extinction by European explorers, however a number of their descendants survive today as a result of admixture with Europeans. This is an example of how modern migrations have begun to reduce the genetic divergence of the human race.

The demographic composition of the old world has not changed significantly since the age of discovery. However, the new world demographics were radically changed within a short time following the voyage of Columbus. The colonization of Americas brought Native Americans into contact with the distant populations of Europe, Africa and Asia. As a result many countries in the Americas have significant and complex multiracial populations. Furthermore many who identify themselves by only one race still have multiracial ancestry.

Admixture in the United States

Table: *Admixture in European-American population*

European admixture	*Frequency*
90-100	68%
80-89.9	22%
70-79.9	8%
60-69.9	< 1%
50-59.9	< 1%
40-49.9	< 1%
0-39.9	0

Some claim the vast majority of African-Americans possess varying degrees of European admixture (the average Black American is 20% European) although studies suggest the Native American admixture in Black Americans is highly exaggerated; some estimates put average African-American possession of European admixture at 25% with figures as high as 50% in the Northeast and less than 10% in the south. A recent study by Mark D. Shriver of a European-American sample found that the average admixture in the white population is 0.7% African and 3.2% Native American. However, 70% of the sample had no African admixture. The other 30% had African admixture ranging from 2% to 20% with an average of 2.3%. By extrapolating these figures to the whole population some scholars suggest that up to 74 million European-Americans may have African admixture in the same range (2–20%).

Dr Mark Shriver, the team leader of the study, found that he had 11% West African ancestry though he identifies as white. Studies based on skin reflectance have shown the color line in the US applied selective pressure on genes that code for skin color but did not apply any selective pressure on other invisible African genes. Since there are an estimated 6 genetic loci involved in skin color determination it is possible for someone to have 15–20% African admixture and not possess any of alleles that code for dark skin. This is the basis of the passing phenomenon. Thus African admixture amongst white Americans can increase without any significant change in skin tone. Conversely amongst African-Americans, an amount of African Admixture is directly correlated with darker skin since no selective pressure is applied; as a result, African-Americans may have a much wider range of African admixture (>0–100%), whereas European-Americans have a lower range (2–20%). A small overlap exists so that it is possible that someone who identifies himself as white may have more African admixture than a person who identifies himself as black.

A statistical analysis done in 1958 using historical census data and historical data on immigration and birth rates, concluded that 21 percent of the white population had black ancestors. The growth in the white population could not be attributed to births in the white population and immigration from Europe alone, but had received significant contribution from the African American population as well. The author states in 1958:

The data presented in this study indicate that the popular belief in the non-African background of white persons is invalid. Over twenty-eight million white persons are descendants of persons of African origin. Furthermore, the majority of the persons with African ancestry are classified as white.

In the United States intermarriage among Filipinos with other races is common. They have the largest number of interracial marriages among Asian immigrant groups, as documented in California. It is also noted that 21.8% of Filipino Americans are of mixed blood, second among Asian Americans, and is the fastest growing.

Admixture in Latin America

Prior to the European conquest of the Americas the demographics of Latin America was naturally 100% American Indian. Today those who identify themselves as Native Americans are small minorities in many countries. For example the CIA lists Argentina's native population at 0.9%, Brazil's at 0.4%, and Uruguay's at 0%.

The early conquest of Latin America was primarily carried out by male soldiers and sailors from Spain and Portugal. Since they carried very few European women on their journeys the new settlers married and fathered children with Amerindian women and also with women imported from Africa. This process of miscegenation was even encouraged by the Spanish monarchy and it led to the system of stratification known as the Casta. This system had Europeans (Spaniards and Portuguese) at the top of the hierarchy followed by those of mixed race. Unmixed Blacks and Native Americans were at the bottom. A philosophy of whitening emerged in which Amerindian and African culture was stigmatized in favor of European values. Many Amerindian languages were lost as mixed race offspring adopted Spanish and Portuguese as their first languages. Only towards the end of the 19th Century and beginning of the 20th century did large numbers of Europeans begin to migrate to South America and consequently altering its demographics.

In addition many Africans were shipped to regions all over the Americas and were present in many of the early voyages of the conquistadors. Brazil has the largest population of African descendants outside Africa. Other countries such as Jamaica, Cuba, Puerto Rico, Dominican Republic, Haiti, Venezuela, Colombia, and Ecuador still have sizeable populations identified as Black.

However countries such as Argentina and Chile do not have a visible African presence today. Census information from the early 19th century shows that people categorized as Black made up to 30% of the population, or around 400,000 people.

Though almost completely absent today, their contribution to Argentine culture is significant include the tango, the milonga and the zamba, words of Bantu origin.

Table: *Demographics of Brazil in 1835, 1940, 2000 and 2008*

Year	*White*	*Brown*	*Black*
1835	24.4%	18.2%	51.4%
1940	64%	21%	14%
2000	53.7%	38.5%	6.2%
2008	48.8%	43.8%	6.5%

The ideology of whitening encouraged non-whites to seek white or lighter skinned partners. This dilution of non-white admixture would be beneficial to their offspring as they would face less stigmatization and find it easier to assimilate into mainstream society. After successive generations of European gene flow, non-white admixture levels would drop below levels at which skin color or physical appearance is not affected thus allowing individuals to identify as white. In many regions, the native and black populations were simply overwhelmed by a succession of waves of European immigration.

Historians and scientists are thus interested in tracing the fate of Native Americans and

Africans from the past to the future. The questions remain about what proportion of these populations simply died out and what proportion still has descendants alive today including those who do not racially identify themselves as their ancestors would have. Admixture testing has thus become a useful objective tool in shedding light on the demographic history of Latin America.

Recent Studies

Unlike in the United States, there were no anti-miscegenation policies in Latin America. Though still a racially stratified society there were no significant barriers to gene flow between the three populations. As a result admixture profiles are a reflection of the colonial populations of Africans, Europeans and Amerindians. The pattern is also sex biased in that the African and Amerindian maternal lines are found in significantly higher proportions than African or Amerindian Y chromosomal lines. This is an indication that the primary mating pattern was that of European males with Amerindian or African females. According to the study more than half the white populations of the Latin American countries studied have some degree of either native American or African admixture (MtDNA or Y Chromosome). In countries such as Chile and Colombia almost the entire white population was shown to have some non-white admixture

Following the dispersal of Humans from Africa 50,000 – 70,000 years ago South America was the last continent to be occupied by humans. Thus the largest geographic distance between continents is between Africa and South America. Since genetic distance increases with geographic distance the two most genetically divergent groups are Africans and Native Americans based on distance. The arrival of Africans in Brazil and subsequent mixing with native South Americans entails the creation of intermediate populations, such as the Zambo or Garifuna between the two divergent groups.

Frank Moya Pons, a Dominican historian documented that Spanish colonists intermarried with Taíno women, and, over time, these mestizo descendants intermarried with Africans, creating a tri-racial Creole culture. 1514 census records reveal that 40% of Spanish men in the colony of Santo Domingo had Taíno wives. A recent study conducted in Puerto Rico suggests that over 61% of the population possess Amerindian mtDNA.

Admixture in the Philippines

Historically, admixture has been an ever present and pervading phenomenon in the Philippines. The Philippines was originally settled by Australoid peoples called Negritos which now form the country's aboriginal community. Admixture occurred between this earlier group and the mainstream Malayo-Polynesian population.

There has been Indian migration to and influence in the Philippines since the precolonial era. About 25% of the words in the Tagalog language are Sanskrit terms and about 5% of the country's population possess Indian ancestry from antiquity. There has been a Chinese presence in the Philippines since the 9th century. However, large-scale migrations of Chinese to the Philippines only started during the Spanish colonial era, when the world market was opened to the Philippines. It is estimated that among Filipinos, 10%–20% have some

Chinese ancestry and 1.5% are "full-blooded" Chinese. According to the American anthropologist Dr. H. Otley Beyer, the ancestry of Filipinos is 2% Arab. This dates back to when Arab traders intermarried with the local Malay Filipina female populations during the pre-Spanish history of the Philippines. A recent genetic study by Stanford University indicates that at least 3.6% of the population are European or of part European descent from both Spanish and United States colonization.

Admixture Among the Romani People

Genetic evidence has shown that the Romani people ("Gypsies") originated from the Indian subcontinent and mixed with the local populations in Central Asia, the Middle East, and Europe. In the 1990s, it was discovered that Romani populations carried large frequencies of particular Y chromosomes (inherited paternally) that otherwise exist only in populations from South Asia, in addition to fairly significant frequencies of particular mitochondrial DNA (inherited maternally) that is rare outside South Asia. 47.3% of Romani males carry Y chromosomes of haplogroup H-M82 which is rare outside of the Indian subcontinent. Mitochondrial haplogroup M, most common in Indian subjects and rare outside Southern Asia, accounts for nearly 30% of Romani people.

A more detailed study of Polish Roma shows this to be of the M5 lineage, which is specific to India. Moreover, a form of the inherited disorder congenital myasthenia is found in Romani subjects. This form of the disorder, caused by the 1267delG mutation, is otherwise only known in subjects of Indian ancestry. This is considered to be the best evidence of the Indian ancestry of the Romanies. The Romanis have been described as "a conglomerate of genetically isolated founder populations", while a number of common Mendelian disorders among Romanies from all over Europe indicates "a common origin and founder effect". A study from 2001 by Gresham et al. suggests "a limited number of related founders, compatible with a small group of migrants splitting from a distinct caste or tribal group".

Also the study pointed out that "genetic drift and different levels and sources of admixture, appear to have played a role in the subsequent differentiation of populations". The same study found that "a single lineage ... found across Romani populations, accounts for almost one-third of Romani males. A similar preservation of a highly resolved male lineage has been reported elsewhere only for Jewish priests". A 2004 study by Morar et al. concluded that the Romani are "a founder population of common origins that has subsequently split into multiple socially divergent and geographically dispersed Gypsy groups". The same study revealed that this population "was founded approximately 32–40 generations ago, with secondary and tertiary founder events occurring approximately 16–25 generations ago".

Hybrid

In biology and specifically, genetics, the term hybrid has several meanings, all referring to the offspring of sexual reproduction.

1. In general usage, hybrid is synonymous with heterozygous: any offspring resulting from the mating of two distinctly homozygous individuals
2. a genetic hybrid carries two different alleles of the same gene
3. a structural hybrid results from the fusion of gametes that have differing structure in at least one chromosome, as a result of structural abnormalities
4. a numerical hybrid results from the fusion of gametes having different haploid numbers of chromosomes
5. a permanent hybrid is a situation where only the heterozygous genotype occurs, because all homozygous combinations are lethal.

From a taxonomic perspective, hybrid refers to offspring resulting from the interbreeding between two animals or plants of different taxa.

1. Hybrids between different subspecies within a species (such as between the Bengal tiger and Siberian tiger) are known as intra-specific hybrids. Hybrids between different species within the same genus (such as between lions and tigers) are sometimes known as interspecific hybrids or crosses. Hybrids between different genera (such as between sheep and goats) are known as intergeneric hybrids. Extremely rare interfamilial hybrids have been known to occur (such as the guineafowl hybrids). No interordinal (between different orders) animal hybrids are known.
2. The second type of hybrid consists of crosses between populations, breeds or cultivars within a single species. This meaning is often used in plant and animal breeding, where hybrids are commonly produced and selected because they have desirable characteristics not found or inconsistently present in the parent individuals or populations. This flow of genetic material between populations or races is often called hybridization.

Types of Hybrids

Depending on the parents, there are a number of different types of hybrids;

- Single cross hybrids — result from the cross between two true breeding organisms and produces an F1 generation called an F1 hybrid (F1 is short for Filial 1, meaning "first offspring"). The cross between two different homozygous lines produces an F1 hybrid that is heterozygous; having two alleles, one contributed by each parent and typically one is dominant and the other recessive. The F1 generation is also phenotypically homogeneous, producing offspring that are all similar to each other.

- Double cross hybrids — result from the cross between two different F1 hybrids.
- Three-way cross hybrids — result from the cross between one parent that is an F1 hybrid and the other is from an inbred line.
- Triple cross hybrids — result from the crossing of two different three-way cross hybrids.
- Population hybrids — result from the crossing of plants or animals in a population with another population. These include crosses between organisms such as interspecific hybrids or crosses between different races.

Etymology

According to the Oxford English Dictionary, the word is derived from Latin hybrida , meaning the "offspring of a tame sow and a wild boar", "child of a freeman and slave", etc. The term entered into popular use in English in the 19th century, though examples of its use have been found from the early 17th century.

Interspecific Hybrids

Interspecific hybrids are bred by mating two species, normally from within the same genus. The offspring display traits and characteristics of both parents. The offspring of an interspecific cross are very often sterile; thus, hybrid sterility prevents the movement of genes from one species to the other, keeping both species distinct.

Sterility is often attributed to the different number of chromosomes the two species have, for example donkeys have 62 chromosomes, while horses have 64 chromosomes, and mules and hinnies have 63 chromosomes.

Mules, hinnies, and other normally sterile interspecific hybrids cannot produce viable gametes, because, the extra chromosome cannot make a homologous pair at meiosis, meiosis is disrupted, and viable sperm and eggs are not formed. However, fertility in female mules has been reported with a donkey as the father.

Most often other processes occurring in plants and animals keep gametic isolation and species distinction. Species often have different mating or courtship patterns or behaviors, the breeding seasons may be distinct and even if mating does occur antigenic reactions to the sperm of other species prevent fertilization or embryo development.

The Lonicera fly is the first known animal species that resulted from natural hybridization. Until the discovery of the Lonicera fly, this process was known to occur in nature only among plants. While it is possible to predict the genetic composition of a backcross on average, it is not possible to accurately predict the composition of a particular backcrossed individual, due to random segregation of chromosomes. In a species with two pairs of chromosomes, a twice backcrossed individual would be predicted to contain 12.5% of one species' genome (say, species A).

However, it may, in fact, still be a 50% hybrid if the chromosomes from species A were lucky in two successive segregations, and meiotic crossovers happened near the telomeres. The chance of this is fairly high:

$$\left(\frac{1}{2}\right)^{(2\times2)} = \frac{1}{16}$$

(where the "two times two" comes about from two rounds of meiosis with two chromosomes); however, this probability declines markedly with chromosome number and so the actual composition of a hybrid will be increasingly closer to the predicted composition.

Hybrids are often named by the portmanteau method, combining the names of the two parent species. For example, a zeedonk is a cross between a zebra and a donkey. Since the traits of hybrid offspring often vary depending on which species was mother and which was father, it is traditional to use the father's species as the first half of the portmanteau. For example, a liger is a cross between a male lion and a female tiger, while a tiglon is a cross between a male tiger and a female lion.

2

Habitat Fragmentation

Humans have been introducing species world wide to environments for a long time both directly such as establishing a population to be used as a biological control and indirectly such as accidental escapes of individuals out of agriculture. '

This causes drastic global affects on various populations with hybridization being one of the reasons introduced species can be so detrimental. When habitats become broken apart, one of two things can occur, genetically speaking. The first is that populations that were once connected can be cut off from one another, preventing their genes from interacting. Occasionally, this will result in a population of one species breeding with a population of another species as a means of surviving such as the case with the red wolves.

Their population numbers being so small, they needed another means of survival. Habitat fragmentation also led to the influx of generalist species into areas where they would not have been, leading to competition and in some cases interbreeding/incorporation of a population into another. In this way, habitat fragmentation is essentially an indirect method of introducing species to an area.

The Hybridization Continuum

There is a kind of continuum with three semi-distinct categories dealing with anthropogenic hybridization: hybridization without Introgression, hybridization with widespread introgression, and essentially a Hybrid swarm. Depending on where a population falls along this continuum, the management plans for that population will change. Hybridization is currently an area of great discussion within Wildlife management and habitat management fields. Global climate change, which is also anthropogenically caused, is creating other changes such as difference in population distributions which are indirect causes for an increase in anthropogenic hybridization.

Consequences

Hybridization can be a less discussed way toward Extinction then within detection of where a population lies along the hybrid continuum. The dispute of hybridization is how to manage the resulting hybrids. When a population experiences hybridization with substantial introgression, there still exists parent types of each set of individuals. When a complete hybrid swarm is created, all the individuals are hybrids.

Management of Hybrids

Conservationists disagree on when is the proper time to give up on a population that is becoming a hybrid swarm or to try and save the still existing pure individuals. Once it becomes a complete mixture, we should look to conserve those hybrids to avoid their loss. Most leave it as a case-by-case basis, depending on detecting of hybrids within the group. It is nearly impossible to regulate hybridization via policy because hybridization can occur beneficially when it occurs "naturally" and there is the matter of protecting those previously mentioned hybrid swarms because if they are the only remaining evidence of prior species, they need to be conserved as well.

This section does not cite any references or sources. Please help improve this section by adding citations to reliable sources. Unsourced material may be challenged and removed. (August 2009)

In some species, hybridisation plays an important role in evolutionary biology. While most hybrids are disadvantaged as a result of genetic incompatibility, the fittest survive, regardless of species boundaries.

They may have a beneficial combination of traits allowing them to exploit new habitats or to succeed in a marginal habitat where the two parent species are disadvantaged.

This has been seen in experiments on sunflower species. Unlike mutation, which affects only one gene, hybridisation creates multiple variations across genes or gene combinations simultaneously. Successful hybrids could evolve into new species within 50-60 generations. This leads some scientists to speculate that life is a genetic continuum rather than a series of self-contained species.

Where there are two closely related species living in the same area, less than 1 in 1000 individuals are likely to be hybrids because animals rarely choose a mate from a different species (otherwise species boundaries would completely break down). In some closely related species there are recognized "hybrid zones".

Some species of Heliconius butterflies exhibit dramatic geographical polymorphism of their wing patterns, which act as aposematic signals advertising their unpalatability to potential predators. Where different-looking geographical races abut, inter-racial hybrids are common, healthy and fertile. Heliconius hybrids can breed with other hybrid individuals and with individuals of either parental race. These hybrid backcrosses are disadvantaged by natural selection because they lack the parental form's warning coloration, and are therefore not avoided by predators.

A similar case in mammals is hybrid White-Tail/Mule Deer. The hybrids don't inherit either parent's escape strategy. White-tail Deer dash while Mule Deer bound. The hybrids are easier prey than the parent species.

In birds, healthy Galapagos Finch hybrids are relatively common, but their beaks are intermediate in shape and less efficient feeding tools than the specialised beaks of the parental species so they lose out in the competition for food. Following a major storm in 1983, the local habitat changed so that new types of plants began to flourish, and in this changed habitat, the hybrids had an advantage over the birds with specialised beaks - demonstrating the role of hybridization in exploiting new ecological niches. If the change in environmental conditions is permanent or is radical enough that the parental species cannot survive, the hybrids become the dominant form. Otherwise, the parental species will re-establish themselves when the environmental change is reversed, and hybrids will remain in the minority.

Natural hybrids may occur when a species is introduced into a new habitat. In Britain, there is hybridisation of native European Red Deer and introduced Chinese Sika Deer. Conservationists want to protect the Red Deer, but the environment favors the Sika Deer genes. There is a similar situation with White-headed Ducks and Ruddy Ducks.

Expression of Parental Traits in Hybrids

When two distinct types of organisms breed with each other, the resulting hybrids typically have intermediate traits (e.g., one parent has red flowers, the other has white, and the hybrid, pink flowers). Commonly, hybrids also combine traits seen only separately in one parent or the other (e.g., a bird hybrid might combine the yellow head of one parent with the orange belly of the other). Most characteristics of the typical hybrid are of one of these two types, and so, in a strict sense, are not really new. However, an intermediate trait does differ from those seen in the parents (e.g., the pink flowers of the intermediate hybrid just mentioned are not seen in either of its parents). Likewise, combined traits are new when viewed as a combination. In a hybrid, any trait that falls outside the range of parental variation is termed heterotic. Heterotic hybrids do have new traits, that is, they are not intermediate. Positive heterosis produces more robust hybrids, they might be stronger or bigger; while the term negative heterosis refers to weaker or smaller hybrids. Heterosis is common in both animal and plant hybrids. For example, hybrids between a lion and a tigress ("ligers") are much larger than either of the two progenitors, while a tigon (lioness × tiger) is smaller.

Also the hybrids between the Common Pheasant (Phasianus colchicus) and domestic fowl (Gallus gallus) are larger than either of their parents, as are those produced between the Common Pheasant and hen Golden Pheasant (Chrysolophus pictus). Spurs are absent in hybrids of the former type, although present in both parents.

When populations hybridize, often the first generation (F_1) hybrids are very uniform. Typically, however, the individual members of subsequent hybrid generations are quite variable. High levels of variability in a natural population, then, are indicative of hybridity. Researchers use this fact to ascertain whether a population is of hybrid origin. Since such variability generally occurs only in later hybrid generations, the existence of variable hybrids is also an indication that the hybrids in question are fertile.

Genetic Mixing and Extinction

Regionally developed ecotypes can be threatened with extinction when new alleles or genes are introduced that alter that ecotype. This is sometimes called genetic mixing. Hybridization and introgression of new genetic material can lead to the replacement of local genotypes if the hybrids are more fit and have breeding advantages over the indigenous ecotype or species.

These hybridization events can result from the introduction of non native genotypes by humans or through habitat modification, bringing previously isolated species into contact. Genetic mixing can be especially detrimental for rare species in isolated habitats, ultimately affecting the population to such a degree that none of the originally genetically distinct population remains.

Effect on Biodiversity and Food Security

In agriculture and animal husbandry, the Green Revolution's use of conventional hybridization increased yields by breeding "high-yielding varieties". The replacement of locally indigenous breeds, compounded with unintentional cross-pollination and crossbreeding (genetic mixing), has reduced the gene pools of various wild and indigenous breeds resulting in the loss of genetic diversity.

Since the indigenous breeds are often well-adapted to local extremes in climate and have immunity to local pathogens this can be a significant genetic erosion of the gene pool for future breeding. Therefore, commercial plant geneticists strive to breed "widely adapted" cultivars to counteract this tendency.

Limiting Factors

A number of conditions exist that limit the success of hybridization, the most obvious is great genetic diversity between most species. But in animals and plants that are more closely related hybridization barriers can include morphological differences, differing times of fertility, mating behaviors and cues, physiological rejection of sperm cells or the developing embryo. In plants, barriers to hybridization include blooming period differences, different pollinator vectors, inhibition of pollen tube growth, somatoplastic sterility, cytoplasmic-genic male sterility and structural differences of the chromosomes.

Mythical and Legendary Hybrids

Ancient folktales often contain mythological creatures, sometimes these are described as hybrids (e.g., Hippogriff as the offspring of a griffin and a horse, and the Minotaur which is the offspring of Pasiphaë and a white bull). More often they are kind of chimera, i.e., a composite of the physical attributes of two or more kinds of animals, mythical beasts, and often humans, with no suggestion that they are the result of interbreeding, e.g., Harpies, mermaids, and centaurs.

Historical Development

Artificial selection was practiced by the Romans. Treatises as much as 2,000 years old give advice on selecting animals for different purposes, and these ancient works cite still older authorities, such as Mago the Carthaginian. The notion of artificial selection was later expressed by the Persian Muslim polymath Abu Rayhan Biruni in the 11th century.

He noted the idea in his book titled India, and gave various examples. The agriculturist selects his corn, letting grow as much as he requires, and tearing out the remainder. The forester leaves those branches which he perceives to be excellent, whilst he cuts away all others. The bees kill those of their kind who only eat, but do not work in their beehive.

Charles Darwin coined the term as an illustration of his proposed wider process of natural selection. Darwin noted that many domesticated animals and plants had special properties that were developed by intentional animal and plant breeding from individuals that showed desirable characteristics, and discouraging the breeding of individuals with less desirable characteristics.

Slow though the process of selection may be, if feeble man can do much by his powers of artificial selection, I can see no limit to the amount of change, to the beauty and infinite complexity of the co-adaptations between all organic beings, one with another and with their physical conditions of life, which may be effected in the long course of time by nature's power of selection. We are profoundly ignorant of the causes producing slight and unimportant variations;

and we are immediately made conscious of this by reflecting on the differences in the breeds of our domesticated animals in different countries,—more especially in the less civilized countries where there has been but little artificial selection.

Contrast to Natural Selection

There is no real difference in the genetic processes underlying artificial and natural selection, and the concept of artificial selection was used by Charles Darwin as an illustration of the wider process of natural selection. The selection process is termed "artificial" when human preferences or influences have a significant effect on the evolution of a particular population or species.

Indeed, many evolutionary biologists view domestication as a type of natural selection and adaptive change that occurs as organisms are brought under the control of human beings.

However, it is useful to distinguish between artificial selection that is unintentional or involves manipulating the environment only, and artificial selection that alter internal DNA sequences in the laboratory. Genetic manipulation in the labs have little in common to processes that occur in nature.

Admixture Mapping

Admixture mapping is a method of gene mapping that makes use of a population of mixed ancestry (an admixed population) to find the genetic loci that contribute to differences in diseases or other phenotypes found between the different ancestral populations.

The method is best applied to populations with recent admixture from two populations that were previously genetically isolated for tens of thousands of years, such as African Americans (admixture of African and European populations). The method attempts to correlate the degree of ancestry near a genetic locus with the phenotype or disease of interest. Genetic markers which differ in frequency between the ancestral populations are needed across the genome.

Admixture mapping is based on the assumption that differences in disease rates or phenotypes are due in part to differences in the frequencies of disease-causing or phenotype-causing genetic variants between populations. In an admixed population, these causal variants will occur more frequently on chromosomal segments inherited from one or another ancestral population.

The first admixture scans were published in 2005 and since then genetic contributors to a variety of disease and trait differences have been mapped. These include hypertension, multiple sclerosis, BMI, and prostate cancer in African Americans. By 2010, high-density mapping panels had been constructed for African Americans, Latino/Hispanics, and Uyghurs.

Founder Effect

In population genetics, the founder effect is the loss of genetic variation that occurs when a new population is established by a very small number of individuals from a larger population. It was first fully outlined by Ernst Mayr in 1952, using existing theoretical work by those such as Sewall Wright. As a result of the loss of genetic variation, the new population may be distinctively different, both genetically and phenotypically, from the parent population from which it is derived. In extreme cases, the founder effect is thought to lead to the speciation and subsequent evolution of new species.

The original population has nearly equal numbers of blue and red individuals. The three smaller founder populations show that one or the other color may predominate (founder effect), due to random sampling of the original population. A population bottleneck may also cause a founder effect even though it is not strictly a new population.

The founder effect is a special case of genetic drift. In addition to founder effects, the new population is often a very small population and so shows increased sensitivity to genetic drift, an increase in inbreeding, and relatively low genetic variation. This can be observed in the limited gene pool of Iceland, Faroe Islands, Easter Islanders and those native to Pitcairn Island. Another example is the legendarily high deaf population of Martha's Vineyard, which resulted in the development of Martha's Vineyard Sign Language.

Serial Founder Effect

Serial founder effects have occurred when populations migrate over long distances. Such long distance migrations typically involve relatively rapid movements followed by periods of settlement. The populations in each migration carry only a subset of the genetic diversity carried from previous migrations. As a result, genetic differentiation tends to increase with geographic distance as described by the "Isolation by distance" model. The migration of humans out of Africa is characterized by serial founder effects. Africa has the highest degree of genetic diversity, which is consistent with an African origin of modern humans. After the initial migration from Africa, the Indian subcontinent was the first major settling point for modern humans. Consequently, India has the second highest genetic diversity in the world. In general, the genetic diversity of the Indian continent is a subset of Africa, and the genetic diversity outside Africa is a subset of India.

Founder Effects in Island Ecology

Founder populations are essential to the study of island biogeography and island ecology. A natural "blank slate" is not easily found, but a classic series of studies on founder population effects were done following the catastrophic

1883 eruption of Krakatoa, which erased all life on the island. Another continuing study has been following the biocolonization of Surtsey, Iceland, a new volcanic island that erupted offshore between 1963 and 1967. An earlier event, the Toba eruption in Sumatra of about 73,000 YBP, covered some parts of India with 3–6 metres (9.8–20 ft) of ash, and must have coated the Nicobar Islands and Andaman Islands, much nearer in the ash fallout cone, with life-smothering layers, forcing them to restart their biodiversity from zero.

Founder Effects in Human Populations

Due to various migrations throughout human history, founder effects are somewhat common among humans in different times and places. The effective founder population of Quebec was only 2,600. After twelve to sixteen generations, with an eightyfold growth but only minimal gene dilution from intermarriage, Quebec has what geneticists call optimal linkage disequilibrium (genetic sharing). The result: far fewer genetic variations, including those that have been well-studied because they are connected with inheritable diseases.

Founder effects can also occur naturally as competing genetic lines die out. This means that an effective founder population consists only of those whose genetic print is identifiable in subsequent populations. Because in sexual reproduction, genetic recombination ensures that with each generation, only half the genetic material of a parent is represented in the offspring, some genetic lines may die out entirely, even though there are numerous progeny.

A recent study concluded that of the people migrating across the Bering land bridge at the close of the ice age, only 70 left their genetic print in modern descendants, a minute effective founder population— this can be easily misread as though implying that only 70 people crossed to North America. The misinterpretations of "Mitochondrial Eve" are a case in point: it may be hard to explain that a "mitochondrial Eve" was not the only woman of her time.

In humans, founder effects can arise from cultural isolation, and inevitably, endogamy. For example, the Amish populations in the United States exhibit founder effects. This is because they have grown from a very few founders, have not recruited newcomers, and tend to marry within the community. Though still rare, phenomena such as polydactyly (extra fingers and toes, a symptom of Ellis-van Creveld syndrome) are more common in Amish communities than in the American population at large.

Similarly, there is a high frequency of fumarase deficiency among the 10,000 members of the Fundamentalist Church of Jesus Christ of Latter Day Saints, a community which practices both endogamy and polygyny, where it is estimated 75 to 80 percent of the community are blood relatives of just two men - founders John Y. Barlow and Joseph Smith Jessop.

An 1814, 15 British colonists founded a settlement on Tristan da Cunha, a group of small islands in the Atlantic Ocean, midway between Africa and South America. One of the early colonists apparently carried a recessive allele for retinitis pigmentosa, a progressive form of blindness that afflicts homozygous individuals. Of the founding colonists' 240 descendants on the island in the late 1960s, 4 had retinitis pigmentosa. The frequency of the allele that causes this disease is ten times higher on Tristan da Cunha than in the populations from which the founders came.

Cleidocranial dysostosis is found among the South African Muslim population due to a Chinese man who was affected by this disease. More severe illnesses exist among certain Jewish groups. Ashkenazi Jews, for example, have a particularly high chance of suffering from Tay-Sachs disease, a fatal condition in young children.

Genetic Drift

Genetic drift or allelic drift is the change in the frequency of a gene variant (allele) in a population due to random sampling. The alleles in the offspring are a sample of those in the parents, and chance has a role in determining whether a given individual survives and reproduces. A population's allele frequency is the fraction of the copies of one gene that share a particular form. Genetic drift may cause gene variants to disappear completely and thereby reduce genetic variation.

The effect of genetic drift is larger for alleles present in a smaller number of copies, and smaller when an allele is present in many copies. Vigorous debates occurred over the relative importance of natural selection versus neutral processes, including genetic drift. Ronald Fisher held the view that genetic drift plays at the most a minor role in evolution, and this remained the dominant view for several decades. In 1968 Motoo Kimura rekindled the debate with his neutral theory of molecular evolution, which claims that most instances where a genetic change spreads across a population (although not necessarily changes in phenotypes) are caused by genetic drift.

Analogy with Cans of Paint

A painter P_1 has 20 cans of paint, 10 red and 10 blue, in the back of his truck. Another painter (P_2, representing the following generation) comes up with 20 empty cans that he wants filled. The first painter, obliging, fills an empty can from one of his own randomly selected cans, and returns his can to the back of his truck. He continues this way, filling empty cans from his (which we will pretend never run out of paint), until all 20 are filled. P_2 takes his 20 cans, now full of paint, off to the back of his own truck. P_3 awaits him with 20 empty cans. Continue ad infinitum. Eventually, a painter will walk away with 20 cans all of the same color.

It is obvious that this can happen at any generation - if, for example, P_1 had by chance always chosen one of his red cans when filling P_2's empties, P_2 would have walked away with 20 cans of red paint - and, therefore, would not have any blue paint to give to P_3, who would have to be content with all red. Slightly less obvious is that the odds of this happening become smaller if the number of cans to be filled (that is, the size of the population) is larger. With 20 cans, the odds of P_1 always picking a red can are (1/2) - about one in a million. Slim, but with 200 cans the odds become 1 in 1.6x10 - less probable than picking out a specific atom from amongst all the atoms in the sun.

More cans gives us a better statistical sample, meaning that the number of red and blue cans is likely to remain much closer to what it was in the previous generation. Genetic drift is weaker in large populations - the frequency of an allele (i.e., the fraction of cans that are red, in our example) will change more slowly between generations when the population size is large.

Conversely, drift is fast in small populations - with only 4 cans, it might only take a few iterations before they are all the same color. Genetic drift thus tends to eliminate variation more quickly in small populations; large populations will tend to have greater genetic diversity.

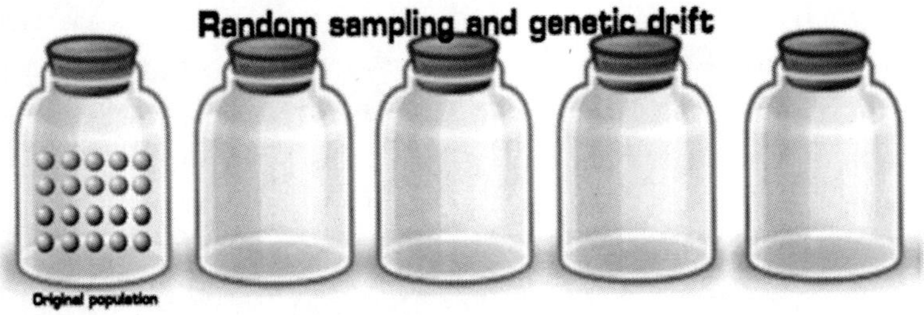

Figure: *In this simulation, there is fixation in the blue "allele" within five generations.*

Probability and Allele Frequency

In probability theory, the law of large numbers predicts little change taking place over time when the population is large. When the reproductive population is small, however, the effects of sampling error can alter the allele frequencies significantly. Genetic drift is therefore considered to be a consequential mechanism of evolutionary change primarily within small, isolated populations.

This effect can be illustrated with a simplified example. Consider a very large colony of bacteria isolated in a drop of solution. The bacteria are genetically identical except for a single gene with two alleles labeled A and B. Half the bacteria have allele A and the other half have allele B. Thus both A and B have allele frequency 1/2. A and B are neutral alleles—meaning they do not affect the bacteria's ability to survive and reproduce.

This being the case, all bacteria in this colony are equally likely to survive and reproduce. The drop of solution then shrinks until it has only enough food to sustain four bacteria. All the others die without reproducing. Among the four who survive, there are sixteen possible combinations for the A and B alleles:

(A-A-A-A), (B-A-A-A), (A-B-A-A), (B-B-A-A),

(A-A-B-A), (B-A-B-A), (A-B-B-A), (B-B-B-A),

(A-A-A-B), (B-A-A-B), (A-B-A-B), (B-B-A-B),

(A-A-B-B), (B-A-B-B), (A-B-B-B), (B-B-B-B).

If each of the combinations with the same number of A and B respectively are counted, we get the following table. The probabilities are calculated with the slightly faulty premise that the peak population size was infinite.

A	*B*	*Combinations*	*Probability*
4	0	1	1/16
3	1	4	4/16
2	2	6	6/16
1	3	4	4/16
0	4	1	1/16

The probability of any one possible combination is:

$$\frac{1}{2}\cdot\frac{1}{2}\cdot\frac{1}{2}\cdot\frac{1}{2}=\frac{1}{16}$$

where 1/2 (the probability of the A or B allele for each surviving bacterium) is multiplied four times (the total sample size, which in this example is the total number of surviving bacteria).

As seen in the table, the total number of possible combinations to have an equal (conserved) number of A and B alleles is six, and its probability is 6/16. The total number of possible alternative combinations is ten, and the probability of unequal number of A and B alleles is 10/16.

The total number of possible combinations can be represented as binomial coefficients and they can be derived from Pascal's triangle. The probability for any one of the possible combinations can be calculated with the formula

$$\binom{N}{k}(1/2)^N$$

where N is the number of bacteria and k is the number of A (or B) alleles in the combination. The function '()' signifies the binomial coefficient and can be expressed as "N choose k". Using the formula to calculate the probability that between them the surviving four bacteria have two A alleles and two B alleles.

$$\binom{4}{2}\left(\frac{1}{2}\right)^4 = 6 \cdot \frac{1}{16} = \frac{6}{16}$$

Genetic drift occurs when a population's allele frequencies change due to random events. In this example the population contracted to just four random survivors, a phenomenon known as population bottleneck.

The original colony began with an equal distribution of A and B alleles but chances are that the remaining population of four members has an unequal distribution. The probability that this surviving population will undergo drift (10/16) is higher than the probability that it will remain the same (6/16).

Mathematical Models of Genetic Drift

Mathematical models of genetic drift can be solved using either branching processes or a diffusion equation describing changes in allele frequency.

Wright-Fisher Model

Consider a gene with two alleles, A or B. In diploid populations consisting of N individuals there are 2N copies of each gene. An individual can have two copies of the same allele or two different alleles. We can call the frequency of one allele p and the frequency of the other q.

The Wright-Fisher model assumes that generations do not overlap. For example, annual plants have exactly one generation per year. Each copy of the gene found in the new generation is drawn independently at random from all copies of the gene in the old generation. The formula to calculate the probability of obtaining k copies of an allele that had frequency p in the last generation is then

$$\frac{(2N)!}{k!(2N-k)!} p^k q^{2n-k}$$

where the symbol "!" signifies the factorial function. This expression can also be formulated using the binomial coefficient,

$$\binom{2N}{k} p^k q^{2N-k}$$

Moran Model

The Moran model assumes overlapping generations. At each time step, one individual is chosen to reproduce and one individual is chosen to die. So in each timestep, the number of copies of a given allele can go up by one, go down by one, or can stay the same. This means that the transition matrix is tridiagonal, which means that mathematical solutions are easier for the Moran model than for the Wright-Fisher model.On the other hand, computer simulations are usually easier to perform using the Wright-Fisher model, because fewer time steps need to be calculated.

In the Moran model, it takes N timesteps to get through one generation, where N is the effective population size. In the Wright-Fisher model, it takes just one. In practice, the Moran model and Wright-Fisher model give qualitatively similar results, but genetic drift runs twice as fast in the Moran model.

Other Models of Drift

If the variance in the number of offspring is much greater than that given by the binomial distribution assumed by the Wright-Fisher model, then given the same overall speed of genetic drift (the variance effective population size), genetic drift is a less powerful force compared to selection.

Random Effects Other Than Sampling Error

Random changes in allele frequencies can also be caused by effects other than sampling error, for example random changes in selection pressure. One important alternative source of stochasticity, perhaps more important than genetic drift, is genetic draft. Genetic draft is the effect on a locus by selection on linked loci. The mathematical properties of genetic draft are different from those of genetic drift. The direction of the random change in allele frequency is autocorrelated across generations.

Drift and Fixation

The Hardy–Weinberg principle states that within sufficiently large populations, the allele frequencies remain constant from one generation to the next unless the equilibrium is disturbed by migration, genetic mutation, or selection. Populations do not gain new alleles from the random sampling of alleles passed to the next generation, but the sampling can cause an existing allele to disappear.

Because random sampling can remove, but not replace, an allele, and because random declines or increases in allele frequency influence expected allele distributions for the next generation, genetic drift drives a population towards genetic uniformity over time. When an allele reaches a frequency of 1 (100%) it is said to be "fixed" in the population and when an allele reaches a frequency of 0 (0%) it is lost. Once an allele becomes fixed, genetic drift comes to a halt, and the allele frequency cannot change unless a new allele is introduced in the population via mutation or gene flow.

Thus even while genetic drift is a random, directionless process, it acts to eliminate genetic variation over time.

Time to Fixation or Loss

Assuming genetic drift is the only evolutionary force acting on an allele, at any given time the probability that an allele will eventually become fixed in the population is simply its frequency in the population at that time. For example, if the frequency p for allele A is 75% and the frequency q for allele B is 25%, then given unlimited time the probability A will ultimately become fixed in the population is 75% and the probability that B will become fixed is 25%. The expected number of generations for fixation to occur is proportional to the population size, such that fixation is predicted to occur much more rapidly in smaller populations. Normally the effective population size, which is smaller than the total population, is used to determine these probabilities. The effective population (N_e) takes into account factors such as the level of inbreeding, the stage of the lifecycle in which the population is the smallest, and the fact that some neutral genes are genetically linked to others that are under selection.

The effective population size may not be the same for every gene in the same population. One forward-looking formula used for approximating the expected time before a neutral allele becomes fixed through genetic drift, according to the Wright-Fisher model, is

$$\bar{T}_{fixed} = \frac{-4N_e(1-p)In(1-p)}{p}$$

where T is the number of generations, N_e is the effective population size, and p is the initial frequency for the given allele. The result is the number of generations expected to pass before fixation occurs for a given allele in a population with given size (N_e) and allele frequency (p).

The expected time for the neutral allele to be lost through genetic drift can be calculated as

$$\bar{T}_{lost} = \frac{-4N_e p}{1-p} In\, p.$$

When a mutation appears only once in a population large enough for the initial frequency to be negligible, the formulas can be simplified to

$$\bar{T}_{fixed} = 4N_e$$

for average number of generations expected before fixation of a neutral mutation, and

$$\bar{T}_{lost} = 2\left(\frac{N_e}{N}\right) In(2N)$$

for the average number of generations expected before the loss of a neutral mutation.

Time to Loss with Both Drift and Mutation

The formulae above apply to an allele that is already present in a population, and which is subject to neither mutation nor selection. If an allele is lost by mutation much more often than it is gained by mutation, then mutation, as well as drift, may influence the time to loss. If the allele prone to mutational loss begins as fixed in the population, and is lost by mutation at rate m per replication, then the expected time in generations until its loss in a haploid population.

Genetic Drift Versus Natural Selection

Although both processes affect evolution, genetic drift operates randomly while natural selection functions non-randomly. While natural selection has a direction, guiding evolution towards heritable adaptations to the current environment, genetic drift has no direction and is guided only by the mathematics of chance. As a result, drift acts upon the genotypic frequencies within a population without regard to their phenotypic effects. In contrast, selection favors the spread of alleles whose phenotypic effects increase survival and/or reproduction of their carriers. Selection lowers the frequencies of alleles that cause unfavorable traits, and ignores those that are neutral.

In natural populations, genetic drift and natural selection do not act in isolation; both forces are always at play. However, the degree to which alleles are affected by drift or selection varies according to population size. The magnitude of drift on allele frequencies per generation is larger in small populations. The magnitude of drift is large enough to overwhelm selection when the selection coefficient is less than 1 divided by the effective population size. As a result, drift affects the frequency of more alleles in small populations than in large ones. When the allele frequency is very small, drift can also overpower selection—even in large populations. For example, while disadvantageous mutations are usually eliminated quickly in large populations, new advantageous mutations are almost as vulnerable to loss through genetic drift as are neutral mutations. Not until the allele frequency for the advantageous mutation reaches a certain threshold will genetic drift have an effect.

The mathematics of genetic drift depend on the effective population size, but it is not clear how this is related to the actual number of individuals in a population. Genetic linkage to other genes that are under selection can reduce the effective population size experienced by a neutral allele. With a higher recombination rate, linkage decreases and with it this local effect on effective population size. This effect is visible in molecular data as a correlation between local recombination rate and genetic diversity, and negative correlation between gene density and diversity at noncoding sites. Stochasticity associated with linkage to other genes that are under selection is not the same as sampling error, and is sometimes known as genetic draft in order to distinguish it from genetic drift.

Allopatric Speciation

Allopatric speciation (from the ancient Greek allos, "other" + Greek patrida, "fatherland") or geographic speciation is speciation that occurs when biological populations of the same species become isolated due to geographical changes such as mountain building or social changes such as emigration. The isolated populations then undergo genotypic and/or phenotypic divergence as: (a) they become subjected to different selective pressures, (b) they independently undergo genetic drift, and (c) different mutations arise in the populations' gene pools.

The separate populations over time may evolve distinctly different characteristics. If the geographical barriers are later removed, members of the two populations may be unable to successfully mate with each other, at which point, the genetically isolated groups have emerged as different species. Allopatric isolation is a key factor in speciation and a common process by which new species arise. Adaptive radiation, as observed by Charles Darwin in Galapagos finches, is a consequence of allopatric speciation among island populations.

Isolating Mechanisms

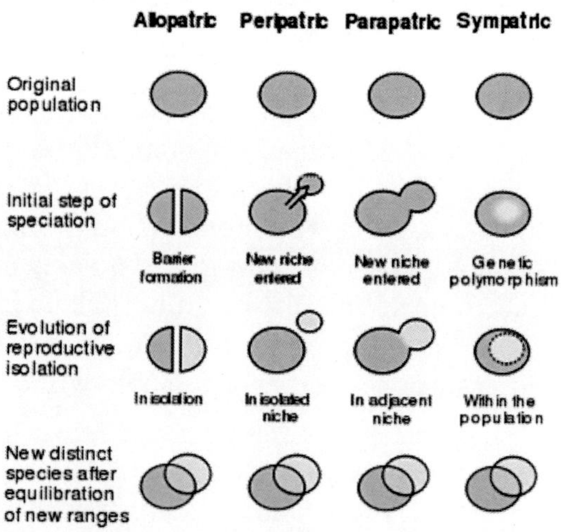

Figure: *Comparison of allopatric, peripatric, parapatric and sympatric speciation*

Allopatric speciation may occur when a species is subdivided into two genetically isolated populations. Allopatric and allopatry are terms from biogeography, referring to organisms whose ranges are entirely separate such that they do not occur in any one place together. If these organisms are closely related (e.g. sister species), such a distribution is usually the result of allopatric speciation. Separation may be attributed to either geological processes or population dispersal.

Geographical Isolation

Geological processes can fragment a population through such events as emergence of mountain ranges, canyon formation, glacial processes, the formation or destruction of land bridges, or the subsidence of large bodies of water. On a global scale, plate tectonics are major geological factors leading to separation of populations and the resulting distribution of species.

Approximately 50,000 years ago, the Death Valley region of the western United States had a rainy climate which produced an interconnecting system of freshwater rivers and lakes. Climatic changes resulted in a drying trend that has continued for the last 10,000 years. As the lakes and rivers shrank, fish populations became geologically isolated. The few remaining (separated) springs are currently home to a variety of fish, many sharing a close common ancestor; yet each has uniquely adapted to its own particular pool.

The extent to which a geological barrier can effectively isolate a population correlates to the mobility of the organism or its offspring. For example physical barriers such as canyons may effectively block migration and dispersal of small mammals; however, have little impact on flying birds or wind-borne seeds.

Population Dispersal

Population dispersal is used to describe migratory events, either in the form of range expansion (natural movement away from parents) or jump dispersal (crossing of barriers), which may lead to genetic isolation. If the smaller population fragment becomes genetically isolated from the parental group, it may be subjected to its own unique mutations, selection forces, and genetic drift effects; thus, it will follow its own evolutionary pathway. Migrations or accidental relocations (such as birds being blown off course) may lead to population fragments; whereby groups merely become separated by distance. Once gene flow between the two groups is disrupted, speciation becomes a possibility.

Allopatric Speciation in Peripheral Populations

When populations become genetically isolated, heritable variations may accumulate so that they become different from the parental population. Given sufficient time, these variations may lead to reproductive isolation. Portions of a populations that exist along the edges of the parent population's geographic territory have higher likelihood of developing reproductive isolation. Such peripheral populations are likely to possess genes that are different from the parental population. After isolation, the founding population is less likely to represent the gene pool of the parent population. In addition, peripheral isolates are likely to represent a small number of individuals, meaning their gene pool is more susceptible to the effects of genetic drift (random chance). Furthermore, it is likely that the peripheral population will inhabit an environment different from its ancestral gene pool, likely causing it to be subjected to different selective pressures as it colonizes new areas. The outer periphery of a population's habitat tends to be extreme; hence, the reason range expansion is kept in check. For most peripheral isolates, it is more likely that they die off rather than survive and speciate.

Genesis of Reproductive Barriers

Adaptive divergence may occur when a population becomes geographically divided: followed by an accumulation of genetic differences as they adapt to their own unique environments. Reproductive barriers do not evolve as a consequence of external forces that drive populations toward speciation. Rather, the evolution of reproductive isolation, leading to speciation, is generally thought to be an incidental by-product of genetic divergence, particularly adaptive changes that evolve through natural selection in response to different environmental conditions in separate geographic areas.

The Biological Species Concept, proposed by Ernst Mayr, in the 1942, emphasizes reproductive isolation as the basis of defining a species. The definition states: "A species is defined as a population or group of populations whose members have the potential to interbreed with one another in nature and to produce viable offspring, but cannot produce viable, fertile offspring with members of other species." Mayr, a proponent of allopatric speciation, hypothesized that adaptive genetic changes that accumulate between allopatric populations cause negative epistasis in hybrids, resulting in sterility of the offspring. If there is considerable genetic and phenotypic change without the loss of the capacity for interbreeding, then such hybridization is simply prevented by the geographical separation of populations. In this case the populations are normally regarded as subspecies.

The frequency of other types of speciation, such as sympatric speciation, parapatric speciation, and heteropatric speciation, is debated. Proponents of peripatric speciation contend that small population size in the peripheral isolate (sometimes referred to as a "splinter population") increases genetic drift, which can be a more powerful force than natural selection in small populations. It deconstructs complex genotypes, allowing the creation of novel gene combinations. Both forms need not be mutually exclusive. In practice, passive isolation or fragmentation as well as active dispersal seem to play a role in many cases of speciation.

Alternative Modes of Speciation

Sympatric speciation represents an alternative method of speciation that does not require physical separation; instead speciation occurs within a population sharing the same geographic boundaries. For example, the development of polyploidy in plant species can lead to a new species arising within the geographic range of its parent population.

In parapatric speciation there is no physical barrier to gene exchange within the population. Instead, the population is continuous; however, mating is not random. Individuals mate with their closest neighbors rather than with individuals in a more distant location. Divergence may occur as a consequence of both reduced gene flow and natural selection, imposed by the large distance between individuals within a population's habitat. Allopatric speciation is thought to be the dominant mode of speciation.

The African Elephant has always been regarded as a single species but, because of morphological and DNA differences, some scientists classify them into three subspecies. Researchers at the University of California, San Diego have argued that divergence due to geographical isolation has gone further, and the elephants of West Africa should be regarded as a separate species from both the savanna elephants of Central, Eastern and Southern Africa, or the forest elephants of Central Africa. A similar situation exists with the Asian Elephant, which has four distinct living sub-species.

Other cases arise where two populations that are quite distinct morphologically, and are native to different continents, have been classified as different species; but when members of one species are introduced into the other's range, they are found to interbreed freely, showing that they were in fact only geographically isolated subspecies.

This was found to be the case when the Mallard was introduced into New Zealand and interbred freely with the native Grey Duck, which had been classified as a separate species. It is controversial whether its specific status can now be retained.

Small Population Size

Small populations behave differently from larger populations. They often result in population bottlenecks, which have harmful consequences for the survival of that population.

Demographic Effects

The influence of stochastic variation in demographic (reproductive and mortality) rates is much higher for small populations than large ones. Stochastic variation in demographic rates causes small populations to fluctuate randomly in size. The smaller the population the greater the probability that fluctuations will lead to extinction. They are subject to a higher chance of extinction because they are more vulnerable to genetic drift, resulting in stochastic variation in their gene pool, their demography and their environment.

One demographic consequence of a small population size, the probability that all offspring in a generation are of the same sex, and where males and females are equally likely to be produced (see sex ratio), is easy to calculate: it is given by 1 / 2 (The chance of all animals being females is 1 / 2 ; the same holds for all males, thus this result).

This can be a problem in very small populations. In 1977, the last 18 Kakapo on a Fiordland island in New Zealand were all male, though the probability of this would only be 0.0000076 if determined by chance (however, females are generally more predated than males and kakapo may be subject to sex allocation). With a population of just three individuals the probability of them all being the same sex is 0.25. Put another way, for every four species reduced to three individuals (or more precisely three individuals in the effective population), one will go extinct within one generation just because they are all the same sex. If the population remains at this size for several generations, such an event becomes almost inevitable.

Environmental Effects

Stochastic variation in the environment (year to year variation in rainfall, temperature) can produce temporally correlated birth and death rates (i.e. 'good' years when birth rates are high and death rates are low and 'bad' years when birth rates are low and death rates are high) that lead to fluctuations in the population size. Again, smaller populations are more likely to go extinct due to these environmentally generated population fluctuations than are large populations.

Genetic Consequences

Conservationists are often worried about a loss of genetic variation in small populations. There are two types of genetic variation that are important when dealing with small populations.

- The degree of homozygosity within individuals in a population; i.e. the proportion of an individual's loci that contain homozygous rather than heterozygous alleles. Many deleterious alleles are only harmful in the homozygous form.
- The degree of monomorphism/polymorphism within a population; i.e. how many different alleles of the same gene exist in the gene pool of a population. Polymorphism may be particularly important at loci involved in the immune response. There are two mechanisms operating in small populations that influence these two types of genetic variation.

- Genetic drift — Genetic variation is determined by the joint action of natural selection and genetic drift (chance). In small populations the relative importance of genetic drift (chance) is higher; deleterious alleles can become more frequent and 'fixed' in a population due to chance. Any allele, deleterious, beneficial or neutral is more likely to be lost from a small population (gene pool) than a large one. This results in a reduction in the number of forms of alleles in a small population and in extreme cases to monomorphism where there is only one form of the allele. Continued fixation of deleterious alleles in small populations is called Muller's ratchet, and can lead to mutational meltdown.
- Inbreeding — In a small population, related individuals are more likely to breed together. The offspring of related parents have a far higher number of homozygous loci than the offspring of unrelated parents.

  There are two types of potential consequence of loss of genetic variation:
- Inbreeding depression — Inbreeding depression is usually taken to mean any immediate harmful effect, on individuals or the population, of a decrease in either type of genetic variation. Inbreeding depression can almost never be found in declining populations that were not very large to begin with; it is somewhat common in large populations becoming small though. The reason is purging selection, most efficient in populations that are strongly but not dangerously inbred.
- The ability of the population to adapt/evolve to changing conditions, "without variability evolution is impossible" . It is obvious that the absolute size of a population limits the absolute degree of allelic diversity. On the other hand, should an advantageous mutation arise, it is likely to show its effect sooner and more thoroughly.

The effective population size is commonly lower than the actual population size.

Genetic Pollution

Genetic pollution is a controversial term for uncontrolled gene flow into wild populations. This gene flow is undesirable according to some environmentalists and conservationists, including groups such as Greenpeace, TRAFFIC, and GeneWatch UK.

Usage

- Some conservation biologists and conservationists have used genetic pollution for a number of years as a term to describe gene flow (which they disapprove of) from a domestic, feral, non-native or invasive species to a wild indigenous population.

- The term is of late being associated with the gene flow from a genetically engineered (GE) organism to a non GE organism, frequently by those disapproving of such gene flow.

Invasive Species

Conservation biologists and conservationists have for a number of years used the term to describe gene flow from domestic, feral, non-native and invasive species into wild indigenous species, which they consider undesirable. For example, TRAFFIC is the international wildlife trade monitoring network that works to limit trade in wild plants and animals so that it is not a threat to conservationist goals.

They promote awareness of the effects of introduced invasive species that may "hybridize with native species, causing genetic pollution". The Joint Nature Conservation Committee (JNCC) is the statutory adviser to the Government of United Kingdom and international nature conservation. Its work contributes to maintaining and enriching biological diversity and educating about the effects of the introduction of invasive/non-native species. In this context they have advised that invasive species: "will alter the genetic pool (a process called genetic pollution), which is an irreversible change." Effect upon endangered species. Compromise of certain wild plant or animal endangered species may have particular significance in the case of these relatively small populations of threatened taxa.

Genetic Engineering

In the field of agriculture, agroforestry and animal husbandry genetic pollution is being used to describe gene flows between GE species and wild relatives. An early use of the term genetic pollution in this later sense appears in a wide-ranging review of the potential ecological effects of genetic engineering in The Ecologist magazine in July 1989. It was also popularized by environmentalist Jeremy Rifkin in his 1998 book The Biotech Century. While intentional crossbreeding between two genetically distinct varieties is described as hybridization with the subsequent introgression of genes, Rifkin, who had played a leading role in the ethical debate for over a decade before, used genetic pollution to describe what he considered to be problems that might occur due the unintentional process of (modernly) genetically modified organisms (GMOs) dispersing their genes into the natural environment by breeding with wild plants or animals.

The usage of genetic pollution by the Food and Agriculture Organization of the United Nations (FAO) is currently defined as: "Uncontrolled spread of genetic information (frequently referring to transgenes) into the genomes of organisms in which such genes are not present in nature." Since 2005 there has existed a GM Contamination Register, launched for GeneWatch UK and

Greenpeace International that records all incidents of intentional or accidental release of organisms genetically modified using modern techniques. In a 10 year study of four different crops, none of the genetically engineered plants were found to be more invasive or more persistent than their conventional counterparts. An often cited claimed example of genetic pollution is the reputed discovery of transgenes from GE maize in landraces of maize in Oaxaca, Mexico. The report from Quist and Chapela, has since been discredited on methodological grounds.

The scientific journal that originally published the study concluded that "the evidence available is not sufficient to justify the publication of the original paper." More recent attempts to replicate the original studies have concluded that genetically modified corn is absent from southern Mexico in 2003 and 2004. The goal of genetic engineering crop plants to help advance tenability and condition of the world food supply has at times had difficulties with public health concerns raised about the safety of the food from the end product.

A 2004 study performed near an Oregon field trial for a genetically modified variety of creeping bentgrass (Agrostis stolonifera) revealed that the transgene and its associate trait (resistance to the glyphosate herbicide) could be transmitted by wind pollination to resident plants of different Agrostis species, up to 14 km from the test field. In 2007, the Scotts Company, producer of the genetically modified bentgrass, agreed to pay a civil penalty of $500,000 to the United States Department of Agriculture (USDA). The USDA alleged that Scotts "failed to conduct a 2003 Oregon field trial in a manner which ensured that neither glyphosate-tolerant creeping bentgrass nor its offspring would persist in the environment".

Controversial Term

Whether genetic pollution or similar terms, such as "genetic deterioration", "genetic swamping", "genetic takeover" and "genetic aggression", are an appropriate scientific description of the biology of invasive species is debated. Hymer and Simberloff argue that these types of terms: "...imply either that hybrids are less fit than the parentals, which need not be the case, or that there is an inherent value in "pure" gene pools". They recommend that gene flow from invasive species be termed genetic mixing since: " "Mixing" need not be value-laden, and we use it here to denote mixing of gene pools whether or not associated with a decline in fitness". Environmentalists such as Patrick Moore, an ex-member and cofounder of Greenpeace, questions if the term genetic pollution is more political than scientific. The term is considered to arouse emotional feelings towards the subject matter. In an interview he comments: "If you take a term used quite frequently these days, the term "genetic pollution," otherwise referred to as genetic contamination, it is a propaganda term, not a technical or scientific term. Pollution and contamination are both value judgments. By using the word “genetic” it gives the public the impression that they are talking about something scientific or technical—as if there were such a thing as genes that amount to pollution.”

Breeding Back

Breeding back can be described as either a natural or a human attempt to assemble or re-assemble the genes of an extinct species, subspecies or domesticated breed, which may still be present in the larger gene pool of the overall species or those of multiple interbreedable species. Breeding back is controversial, especially claims that an extinct animal has been recreated. Similar appearance (phenotypical reconstruction) does not assure similar behavior or biology. For some of the animals that are being bred back, questions remain about the ecological niche, hardiness, and disease resistance of the original creatures. For instance, the aurochs died out almost 400 years ago and the records kept cannot definitively answer some of these questions.

Dogs

The American Alsatian dog breed has been created to resemble the bone and body structure of the Dire Wolf, although not the temperament. Other dog breeds that attempt to recreate a ‘wolf-look’, without actually cross breeding with wolves, are the Tamaskan Dog, Utonagan, and the Northern Inuit dog breeds. Some speculate that it may be possible to breed back the now extinct Japanese Wolf, since Asian wolves possibly crossed with Japanese wolfdogs.

Grazing Animals

Other selectively bred examples of breeding back include that of the aurochs, the extinct forerunner of domestic cattle. The products of these attempts are called the Heck cattle. The Heck horse, a phenotypic copy of the tarpan has also been produced, although it lacks the upright manes. Another prominent breeding back effort is the Quagga Project to bring back the extinct subspecies of the Plains Zebra called Quagga. Back breeding is also being used to recreate domestic breeds that have disappeared, such as the Cumberland Pig.

3

Technological Applications

Gene Knockdown

The RNA interference pathway is often exploited in experimental biology to study the function of genes in cell culture and in vivo in model organisms. Double-stranded RNA is synthesized with a sequence complementary to a gene of interest and introduced into a cell or organism, where it is recognized as exogenous genetic material and activates the RNAi pathway. Using this mechanism, researchers can cause a drastic decrease in the expression of a targeted gene. Studying the effects of this decrease can show the physiological role of the gene product. Since RNAi may not totally abolish expression of the gene, this technique is sometimes referred as a "knockdown", to distinguish it from "knockout" procedures in which expression of a gene is entirely eliminated.

Extensive efforts in computational biology have been directed toward the design of successful dsRNA reagents that maximize gene knockdown but minimize "off-target" effects. Off-target effects arise when an introduced RNA has a base sequence that can pair with and thus reduce the expression of multiple genes at a time. Such problems occur more frequently when the dsRNA contains repetitive sequences. It has been estimated from studying the genomes of H. sapiens, C. elegans, and S. pombe that about 10% of possible siRNAs will have substantial off-target effects. A multitude of software tools have been developed implementing algorithms for the design of general, mammal-specific, and virus-specific siRNAs that are automatically checked for possible cross-reactivity.

Depending on the organism and experimental system, the exogenous RNA may be a long strand designed to be cleaved by dicer, or short RNAs designed to serve as siRNA substrates. In most mammalian cells, shorter RNAs are used because long double-stranded RNA molecules induce the mammalian interferon response, a form of innate immunity that reacts nonspecifically to foreign genetic material.

Mouse oocytes and cells from early mouse embryos lack this reaction to exogenous dsRNA and are therefore a common model system for studying gene-knockdown effects in mammals. Specialized laboratory techniques have also been developed to improve the utility of RNAi in mammalian systems by avoiding the direct introduction of siRNA, for example, by stable transfection with a plasmid encoding the appropriate sequence from which siRNAs can be transcribed, or by more elaborate lentiviral vector systems allowing the inducible activation or deactivation of transcription, known as conditional RNAi.

Functional Genomics

Most functional genomics applications of RNAi in animals have used C. elegans and Drosophila, as these are the common model organisms in which RNAi is most effective. C. elegans is particularly useful for RNAi research for two reasons: firstly, the effects of the gene silencing are generally heritable, and secondly because delivery of the dsRNA is extremely simple. Through a mechanism whose details are poorly understood, bacteria such as E. coli that carry the desired dsRNA can be fed to the worms and will transfer their RNA payload to the worm via the intestinal tract.

This "delivery by feeding" is just as effective at inducing gene silencing as more costly and time-consuming delivery methods, such as soaking the worms in dsRNA solution and injecting dsRNA into the gonads. Although delivery is more difficult in most other organisms, efforts are also underway to undertake large-scale genomic screening applications in cell culture with mammalian cells.

Approaches to the design of genome-wide RNAi libraries can require more sophistication than the design of a single siRNA for a defined set of experimental conditions. Artificial neural networks are frequently used to design siRNA libraries and to predict their likely efficiency at gene knockdown. Mass genomic screening is widely seen as a promising method for genome annotation and has triggered the development of high-throughput screening methods based on microarrays.

However, the utility of these screens and the ability of techniques developed on model organisms to generalize to even closely related species has been questioned, for example from C. elegans to related parasitic nematodes. Functional genomics using RNAi is a particularly attractive technique for genomic mapping and annotation in plants because many plants are polyploid, which presents substantial challenges for more traditional genetic engineering methods. For example, RNAi has been successfully used for functional genomics studies in bread wheat (which is hexaploid) as well as more common plant model systems Arabidopsis and maize.

Figure: *A normal adult Drosophila fly, a common model organism used in RNAi experiments.*

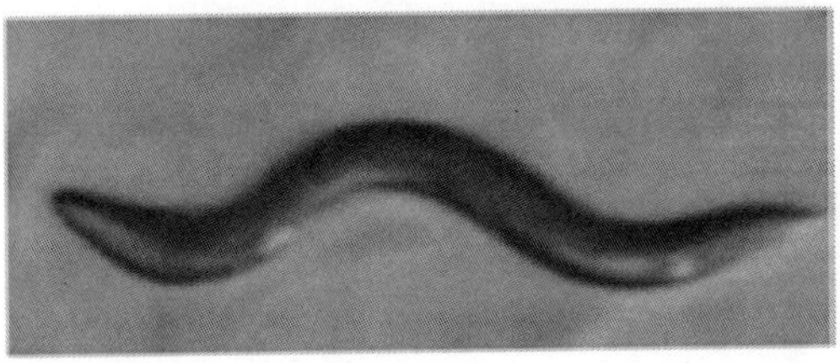

Figure: *An adult C. elegans worm, grown under RNAi suppression of a nuclear hormone receptor involved in desaturase regulation. These worms have abnormal fatty acid metabolism but are viable and fertile.*

Medicine

It may be possible to exploit RNA interference in therapy. Although it is difficult to introduce long dsRNA strands into mammalian cells due to the interferon response, the use of short interfering RNA mimics has been more successful. Among the first applications to reach clinical trials were in the treatment of macular degeneration and respiratory syncytial virus, RNAi has also been shown to be effective in the reversal of induced liver failure in mouse models.

Other proposed clinical uses center on antiviral therapies, including topical microbicide treatments that use RNAi to treat infection (at Harvard University Medical School; in mice, so far) by herpes simplex virus type 2 and the inhibition of viral gene expression in cancerous cells, knockdown of host receptors and coreceptors for HIV, the silencing of hepatitis A and hepatitis B genes, silencing of influenza gene expression, and inhibition of measles viral replication. Potential treatments for neurodegenerative diseases have also been proposed, with particular attention being paid to the polyglutamine diseases such as Huntington's disease. RNA interference is also often seen as a promising way to treat cancer by silencing genes differentially upregulated in tumor cells or genes involved in cell division. A key area of research in the use of RNAi for clinical applications is the development of a safe delivery method, which to date has involved mainly viral vector systems similar to those suggested for gene therapy.

Despite the proliferation of promising cell culture studies for RNAi-based drugs, some concern has been raised regarding the safety of RNA interference, especially the potential for "off-target" effects in which a gene with a coincidentally similar sequence to the targeted gene is also repressed. A computational genomics study estimated that the error rate of off-target interactions is about 10%. One major study of liver disease in mice led to high death rates in the experimental animals, suggested by researchers to be the result of "oversaturation" of the dsRNA pathway, due to the use of shRNAs that have to be processed in the nucleus and exported to the cytoplasm using an active mechanism.

All these are considerations that are under active investigation, to reduce their impact in the potential therapeutic applications for RNAi. RNA interference-based applications are being developed to target persistent HIV-1 infection. Viruses like HIV-1 are particularly difficult targets for RNAi-attack because they are escape-prone, which requires combinatorial RNAi strategies to prevent viral escape. The future of antiviral RNAi therapeutics is very promising, but it remains of critical importance to include many controls in pre-clinical test models to unequivocally demonstrate sequence-specific action of the RNAi inducers.

Biotechnology

RNA interference has been used for applications in biotechnology, particularly in the engineering of food plants that produce lower levels of natural plant toxins. Such techniques take advantage of the stable and heritable RNAi phenotype in plant stocks. For example, cotton seeds are rich in dietary protein but naturally contain the toxic terpenoid product gossypol, making them unsuitable for human consumption. RNAi has been used to produce cotton stocks whose seeds contain reduced levels of delta-cadinene synthase, a key enzyme in gossypol production, without affecting the enzyme's production in other parts of the plant, where gossypol is important in preventing damage from plant pests. Similar efforts have been directed toward the reduction of the cyanogenic natural product linamarin in cassava plants. Although no plant products that use RNAi-based genetic engineering have yet passed the experimental stage, development efforts have successfully reduced the levels of allergens in tomato plants and decreased the precursors of likely carcinogens in tobacco plants. Other plant traits that have been engineered in the laboratory include the production of non-narcotic natural products by the opium poppy, resistance to common plant viruses, and fortification of plants such as tomatoes with dietary antioxidants. Previous commercial products, including the Flavr Savr tomato and two cultivars of ringspot-resistant papaya, were originally developed using antisense technology but likely exploited the RNAi pathway.

Genome-scale RNAi Screening

The genome-scale RNAi research relies on the high-throughput screening (HTS) technology. The RNAi HTS technology allows genome-wide loss-of-function screening and is broadly used in the identification of genes associated with specific biological phenotypes. This technology has been hailed as the second genomics wave, following the first genomics wave of gene expression microarray and single nucleotide polymorphism discovery platforms . One of the major advantages of the genome-scale RNAi screening is its ability to simultaneously interrogate thousands of genes. With the ability of generating a large amount of data per experiment, the genome-scale RNAi screening has led to an explosion in the rate of generating data.

Consequently, one of the most fundamental challenges in the genome-scale RNAi researches is to glean biological significance from mounds of data, which requires the adoption of suitable statistics/bioinformatics methods such as those presented in a recently published book for analyzing RNAi screens. The basic process of cell-based RNAi screening includes (i) the choice of an RNAi library, (ii) the selection of a robust and stable type of cells, (iii) the transfection of the selected cells with RNAi agents from the chosen RNAi library, (iv) necessary treatment or incubation, (v) signal detection, (vi) statistical and bioinformatics analysis, and (vii) the determination of important genes or therapeutical targets .

History and Discovery

Example petunia plants in which genes for pigmentation are silenced by RNAi. The left plant is wild-type; the right plants contain transgenes that induce suppression of both transgene and endogenous gene expression, giving rise to the unpigmented white areas of the flower.

The discovery of RNAi was preceded first by observations of transcriptional inhibition by antisense RNA expressed in transgenic plants, and more directly by reports of unexpected outcomes in experiments performed by plant scientists in the United States and the Netherlands in the early 1990s. In an attempt to alter flower colors in petunias, researchers introduced additional copies of a gene encoding chalcone synthase, a key enzyme for flower pigmentation into petunia plants of normally pink or violet flower color.

The overexpressed gene was expected to result in darker flowers, but instead produced less pigmented, fully or partially white flowers, indicating that the activity of chalcone synthase had been substantially decreased; in fact, both the endogenous genes and the transgenes were downregulated in the white flowers. Soon after, a related event termed quelling was noted in the fungus Neurospora crassa, although it was not immediately recognized as related. Further investigation of the phenomenon in plants indicated that the downregulation was due to post-transcriptional inhibition of gene expression via an increased rate of mRNA degradation.

This phenomenon was called co-suppression of gene expression, but the molecular mechanism remained unknown. Not long after, plant virologists working on improving plant resistance to viral diseases observed a similar unexpected phenomenon. While it was known that plants expressing virus-specific proteins showed enhanced tolerance or resistance to viral infection, it was not expected that plants carrying only short, non-coding regions of viral RNA sequences would show similar levels of protection. Researchers believed that viral RNA produced by transgenes could also inhibit viral replication.

The reverse experiment, in which short sequences of plant genes were introduced into viruses, showed that the targeted gene was suppressed in an infected plant. This phenomenon was labeled "virus-induced gene silencing" (VIGS), and the set of such phenomena were collectively called post transcriptional gene silencing.

After these initial observations in plants, many laboratories around the world searched for the occurrence of this phenomenon in other organisms. Craig C. Mello and Andrew Fire's 1998 Nature paper reported a potent gene silencing effect after injecting double stranded RNA into C. elegans. In investigating the regulation of muscle protein production, they observed that neither mRNA nor antisense RNA injections had an effect on protein production, but double-stranded RNA successfully silenced the targeted gene. As a result of this work, they coined the term RNAi. Fire and Mello's discovery was particularly notable because it represented the first identification of the causative agent for the phenomenon. Fire and Mello were awarded the Nobel Prize in Physiology or Medicine in 2006 for their work.

Gene Therapy for Color Blindness

Gene therapy for color blindness is an experimental gene therapy aiming to convert congenitally colorblind individuals to trichromats by introducing a photopigment gene that they lack. Though partial color blindness is considered only a mild disability and is controversial whether it is even a disorder, it is a condition that affects many people, particularly males.

Complete color blindness, or achromatopsia, is very rare but more severe. While never demonstrated in humans, animal studies have shown that it is possible to confer color vision by injecting a gene of the missing photopigment using gene therapy. As of 2011 there is no medical entity offering this treatment, and no clinical trials available for volunteers.

Color Blindness

The retina of the human eye contains photoreceptive cells called cones that allow color vision. A normal trichromat individual possesses three different types of cones to distinguish different colors within the visible spectrum from 380 nm to 740 nm. The three types of cones are designated L, M, and S cones, and each type is sensitive to a certain range of wavelength of light depending on what photopigment it contains.

More specifically, the L cone absorbs around 560 nm, the M cone absorbs near 530 nm, and the S cone absorbs near 420 nm. Contrary to popular belief, the peak absorption frequency for L, M, and S cones do not exactly correspond to red, green, and blue wavelength. Rather, the peak frequency for the L cone is orange, yellowish green in M cones, and blue-violet in S cones. These cones transduce the absorbed light into electrical information to be relayed to neurons in the retina such as retinal bipolar cells and retinal ganglion cells, before reaching the brain.

The signals from different cones are added or subtracted from each other to process the color of incoming light. For instance, the color red stimulate L cones more than M cones, whereas the color green stimulates the L and M cones more than the S cones. The colors are perceived in an opponent process, such that red and green are perceived in opposition, as are blue and yellow, black and white.

The gene loci coding for the photopigments: M-opsin and L-opsin are located in close proximity within the X chromosome and are highly polymorphic. Among the population, some have a deleted gene for the M photopigment in the X chromosome (such as in deuteranopia), whereas others have a mutated form of the gene (such as in deuteranomaly). Individuals who can express only two types of opsins in the cones are called dichromats. Because males have only one copy of the X chromosome, dichromatism is much more prevalent among men. With only two types of cones, dichromats are less capable of distinguishing between two colors. In the most common form of color blindness, deuteranopes have difficulty discriminating between red and green color. This is shown by their poor performance in Ishihara test. Although dichromatism poses little problem for daily life, dichromats may find some color-coded diagrams and maps difficult to read.

Less common forms of dichromacy include protoanopia (lack of L-cones), and tritanopia (lack of S-cones). If a person lacks two types of photopigments, they are considered monochromats. People lacking the three types of photopigments are said to have complete color blindness or achromatopsia. Color blindness can also result from damages to the visual cortex in the brain.

Theory

Experiments using a variety of mammals (including primates) demonstrated that it is possible to confer color vision to animals by introducing an opsin gene that the animal previously lacked. Using a replication-defective recombinant adeno-associated virus (rAAV) as a vector, the cDNA of the opsin gene found in the L or M cones can be delivered to some fraction of the cones within the retina via subretinal injection. Upon gaining the gene, the cone begins to express the new photopigment. The effect of therapy lasts until the cones die or the inserted DNA is lost within the cones.

While gene therapy for humans has been ongoing with some success, a gene therapy for humans to gain color vision has not been attempted to date. However, demonstrations using several mammals (including primates such as a squirrel monkey) suggest that the therapy should be feasible for humans as well. It is also theoretically possible for trichromats to be “upgraded” to tetrachromats by introducing new opsin genes.

Motivation

The goal of the gene therapy is to make some of the cones in the retina of a dichromat individual to express the missing photopigment. Although partial color blindness is considered to be a mild disability and even an advantage under certain circumstances (such as spotting camouflaged objects), it can pose challenges for certain occupations such as driving or piloting an aircraft. More generally, color codes in maps and figures may be difficult to read for individuals with color blindness.

Because only a single gene codes for a photopigment and the gene is only expressed in the retina, it is a relatively easy condition to treat using gene therapy compared to other genetic diseases. However, there remains the question of whether the therapy is worthwhile, for an individual to undergo an invasive subretinal injection to temporarily treat a condition that is more of an inconvenience than a disorder.

However, complete color blindness, or achromatopsia, is very rare but more severe. Indeed, achromats cannot see any color, have a strong photophobia (blindness in full sun), and a reduced visual acuity (generally 20/200 after correction). Moreover, the research may have strong implications toward genetic therapy of other cone diseases. Other cone diseases such as Leber's congenital amaurosis, cone-rod dystrophy, and certain types of maculopathies may be treatable using the same techniques as the gene therapy used for color blindness.

Research so Far

There has been ongoing research for gene therapy to treat Leber's congenital amaurosis, a genetic disorder in photoreceptors that can lead to vision loss and blindness. These treatments use AAV vector and is delivered in much the same way as the gene therapy for color blindness.

Jacobs et al. published their research in the journal: Science in 2007, on their work introducing a human L-cone photopigment in mice. Since the mice possess only S cones and M cones, they are dichromats. The researchers replaced M-opsin with a cDNA of L-opsin in the X chromosome of some mice. By breeding these "knock-in" transgenic mice, they generated heterozygous females with both an M cone and an L cone. These mice had improved range of color vision and have gained trichromacy, as tested by electroretinogram and behavioral tests. However, this is more difficult to apply in the form of gene therapy.

In a paper published in the journal: Visual Neuroscience by Mauck et al., researchers used recombinant AAV vector to introduce the green fluorescent protein (GFP) gene in the cones of gerbils. The genetic insert was designed to only be expressed in S or M cones, and the study observed the expression of GFP in vivo over time. The study characterized the time of onset of expression, and also demonstrated that gene expression could stabilize if a sufficiently high dose of the viral vector is given.

Mancuso et al. published their research in the journal: Nature in 2009, on converting adult dichromat squirrel monkeys into trichromats using gene therapy. New world monkeys such as squirrel monkeys lack the L-opsin gene and are incapable of discriminating between certain shades of red and green. The researchers used recombinant AAV vector to deliver a human L-opsin gene into the monkey's retina. Cones that gained the missing genes began expressing the new photopigment.

The researchers raised two possibilities if the therapy worked—that the monkeys would either remain dichromatic with greater sensitivity for longer wavelength of light, or they would become trichromats. Electroretinogram recordings demonstrated that they are able to discriminate blue-green from red-violet, and have indeed gained trichromacy. The treated monkeys were also more successful when their color vision was tested with a modified Ishihara test. In 2007, Alexander JJ et al. used gene therapy to restore some of the sight of mice with achromatopsia.

The results were positive for 80% of the mice treated. Moreover, a paper by Komáromy et al., published in 2010, deals with gene therapy for a form of achromatopsia in dogs. Cone function and day vision have been restored for at least 33 months in two young dogs with achromatopsia. However, this therapy was less efficient for older dogs.

Theoretical Questions

According to research by David H. Hubel and Torsten Wiesel, suturing shut one eye of monkeys at an early age resulted in an irreversible loss of vision in that eye, even after the suture was removed. The study concluded that the neural circuitry for vision is wired during a "critical period" in childhood, after which the visual circuitry can no longer be rewired to process new sensory input. Contrary to this finding, Mancuso et al.'s success in conferring trichromacy to adult squirrel monkeys suggests that it is possible to adapt the preexisting circuit to allow greater acuity in color vision. The researchers concluded that integrating the stimulus from the new photopigment as an adult was not analogous to vision loss following visual deprivation.

It is yet unknown how the animals that gain a new photopigment are perceiving the new color. While the article by Mancuso et al. states that the monkey has indeed gained trichromacy and gained the ability to discriminate between red and green, they claim no knowledge of how the animal internally perceives the sensation.

While red/green color blindness among deuteranopes can be treated by introducing M-opsin genes, rarer forms of color blindness such as tritanopia can in principle be treated as well. For tritanopia, the S-opsin gene must be introduced instead of M-opsin gene.

Challenges

Despite the success in animals, there still remain challenges to conducting gene therapy on humans for treating color blindness.

Safety

How to deliver the viral vector into the retina is probably the main obstacle to making gene therapy a practical treatment for color blindness. Because the virus has to be injected directly by using a needle to penetrate the sclera of the eye, the treatment may be highly unpleasant and is a risk for eye infection. Without a way to deliver the virus noninvasively, the treatment is rather risky for the benefit gained.

As of now there is no knowledge of how frequently the gene needs to be injected to maintain trichromacy among congenitally colorblind individuals. At the time of publication, Mancuso et al. reports that the treated squirrel monkeys have maintained 2 years of color vision after the treatment.

If repeat injections are needed, there is also the concern of the body developing an immune reaction to the virus. If a body develops sensitivity to the viral vector, the success of the therapy could be jeopardized and/or the body may respond unfavorably. An editorial by J. Bennett points to Mancuso et al.'s use of an "unspecified postinjection corticosteroid therapy". Bennett suggests that the monkeys may have experienced inflammation due to the injection.

However, the AAV virus that is commonly used for this study is non-pathogenic, and the body is less likely to develop an immune reaction. Needless to say, an extensive review of the safety of the treatment must precede any human trials. Obviously, the subject should first be evaluated to identify which photopigment they need to gain trichromacy. Also, while gene therapy may treat congenital color blindness (such as dichromacy), it is not intended to treat non-retinal forms of color blindness such as damage to the visual cortex of the brain.

Other Challenges

As a way to introduce new genetic information to change a person's phenotype, a gene therapy for color blindness is open to the same ethical questions and criticisms as gene therapy in general. Given the large number of people with color blindness, there is also the question of whether color blindness is a disorder or a phenotypic variant in the human population.

In fact, dichromacy can even be advantageous under certain circumstances such as when identifying a camouflaged object from the background. Furthermore, even if gene therapy succeeds in converting incomplete colorblind individuals to trichromats, the degree of satisfaction among the subjects is unknown, though who watches black and white television by choice now? The following sentence reflects bias of those not afflicted with color vision deficiency- There probably is little increase in performance on day-to-day tasks, and it is uncertain how the quality of life will improve (or worsen) after the therapy.

Life Extension

Life extension science, also known as anti-aging medicine, experimental gerontology, and biomedical gerontology, is the study of slowing down or reversing the processes of aging to extend both the maximum and average lifespan. Some researchers in this area, and "life extensionists" or "longevists" (who wish to achieve longer lives for themselves), believe that future breakthroughs in tissue rejuvenation with stem cells, molecular repair, and organ replacement (such as with artificial organs or xenotransplantations) will eventually enable humans to have indefinite lifespans (agerasia) through complete rejuvenation to a healthy youthful condition.

The sale of putative anti-aging products such as nutrition, physical fitness, skin care, hormone replacements, vitamins, supplements and herbs is a lucrative global industry, with the US market generating about $50 billion of revenue each year. Medical experts state that the use of such products has not been shown to affect the aging process, and many claims of anti-aging medicine advocates have been roundly criticized by medical experts, including the American Medical Association. Bioethicists question whether and how the human lifespan should be extended.

Average and Maximum Lifespans

During the process of aging, an organism accumulates damage to macromolecules, its cells, its tissues and its organs. This accumulated damage is the result of oxidation damage to the cell contents caused by free radicals. The maximum life span for humans is currently maximized at approximately 120 years, whereas the maximum lifespan of a mouse, commonly used as a model in research on aging, is about four years.

Genetic differences between humans and mice that may account for these different aging rates include efficiency of DNA repair, types and quantities of antioxidant enzymes, and different rates of free radical production. The most important and challenging factor remains the telomere limitation of each individual species.

Average lifespan in a population is lowered by infant and child mortality, which are frequently linked to infectious diseases or nutrition problems. Later in life, vulnerability to accidents and age-related chronic disease such as cancer or cardiovascular disease play an increasing role in mortality. Extension of expected lifespan can often be achieved by access to improved medical care, vaccinations, good diet, exercise and avoidance of hazards such as smoking.

Maximum lifespan is determined by the rate of aging for a species inherent in its genes and by environmental factors. One widely recognized method of extending maximum lifespan in organisms such as nematodes is calorie restriction. Another technique used evolutionary pressure such as breeding from only older members.

Theoretically, extension of maximum lifespan could be achieved by reducing the rate of aging damage, by periodic replacement of damaged tissues, or by molecular repair or rejuvenation of deteriorated cells and tissues and the enhancement of telomerase enzyme activity. Future research will be geared towards telomere repair strategies.

Current Anti-aging Strategies and Issues

Diets and Supplements

Much life extension research focuses on nutrition—diets or supplements—as a means to extend lifespan, although few of these have been systematically tested for significant longevity effects. The many diets promoted by anti-aging advocates are often contradictory. Two diets with different approaches and some support from scientific research are the Paleolithic diet and Caloric restriction.

The free-radical theory of aging suggests that antioxidant supplements, such as Vitamin C, Vitamin E, Q_{10}, lipoic acid, carnosine, and N-acetylcysteine, might extend human life. However, combined evidence from several clinical trials suggest that â-Carotene supplements and high doses of Vitamin E increase mortality rates. Other substances proposed to extend lifespan include oxytocin, insulin, human chorionic gonadotropin (hCG), and erythropoietin (EPO). Resveratrol is a sirtuin stimulant that appears to extend lifespan in simple organisms such as nematodes and short-lived fish.

Some supplements, including the minerals selenium or zinc have been reported to extend the lifespan of rats and mice, though none has been proven to do so in humans, and significant toxic effects were observed. Metformin may also extend life span in mice.

Hormone Treatments

The anti-aging industry offers several hormone therapies. Some of these have been criticized for possible dangers to the patient and a lack of proven effect. For example, the American Medical Association has been critical of some anti-aging hormone therapies.

Even if some recent clinical studies have shown that low-dose GH treatment for adults with GH deficiency changes the body composition by increasing muscle mass, decreasing fat mass, increasing bone density and muscle strength, improves cardiovascular parameters (i.e. decrease of LDL cholesterol), and affects the quality of life without significant side effects.

The evidence for use of growth hormone as an anti-aging therapy is mixed and based on animal studies. An early study suggested that supplementation of mice with growth hormone increased average life expectancy. Additional animal experiments have suggested that growth hormone may generally act to shorten maximum lifespan; knockout mice lacking the receptor for growth hormone live especially long. Furthermore, mouse models lacking the insulin-like growth factor also live especially long and have low levels of growth hormone. Scientific controversy regarding anti-aging nutritional supplementation and medicine

Some critics dispute the portrayal of aging as a disease. For example, Leonard Hayflick, who determined that fibroblasts are limited to around 50 cell divisions, reasons that aging is an unavoidable consequence of entropy. Hayflick and fellow biogerontologists Jay Olshansky and Bruce Carnes have strongly criticized the anti-aging industry in response to what they see as unscrupulous profiteering from the sale of unproven anti-aging supplements.

Ethics and politics of anti-aging nutritional supplementation and medicine

Politics relevant to the substances of life extension pertain mostly to communications and availability. In the United States, product claims on food and drug labels are strictly regulated. The First Amendment (freedom of speech) protects third-party publishers' rights to distribute fact, opinion and speculation on life extension practices. Manufacturers and suppliers also provide informational publications, but because they market the substances, they are subject to monitoring and enforcement by the Federal Trade Commission (FTC), which polices claims by marketers. What constitutes the difference between truthful and false claims is hotly debated and is a central controversy in this arena.

Proposed Strategies of Life Extension

Nanotechnology

Future advances in nanomedicine could give rise to life extension through the repair of many processes thought to be responsible for aging. Raymond Kurzweil, a futurist and transhumanist, believes that advanced medical nanorobotics could completely remedy the effects of aging by 2030.

Cloning and Body Part Replacement

Some life extensionists suggest that therapeutic cloning and stem cell research could one day provide a way to generate cells, body parts, or even entire bodies (generally referred to as reproductive cloning) that would be genetically identical to a prospective patient. Recently, the US Department of Defense initiated a program to research the possibility of growing human body parts on mice. Complex biological structures, such as mammalian joints and limbs, have not yet been replicated. Dog and primate brain transplantation experiments were conducted in the mid-20th century but failed due to rejection and the inability to restore nerve connections. As of 2006, the implantation of bio-engineered bladders grown from patients' own cells has proven to be a viable treatment for bladder disease. Proponents of body part replacement and cloning contend that the required biotechnologies are likely to appear earlier than other life-extension technologies.

The use of human stem cells, particularly embryonic stem cells, is controversial. Opponents' objections generally are based on interpretations of religious teachings or ethical considerations. Proponents of stem cell research point out that cells are routinely formed and destroyed in a variety of contexts. Use of stem cells taken from the umbilical cord or parts of the adult body may not provoke controversy.

The controversies over cloning are similar, except general public opinion in most countries stands in opposition to reproductive cloning. Some proponents of therapeutic cloning predict the production of whole bodies, lacking consciousness, for eventual brain transplantation.

Cryonics

For cryonicists (advocates of cryopreservation), storing the body at low temperatures after death may provide an "ambulance" into a future in which advanced medical technologies may allow resuscitation and repair. They speculate cryogenic temperatures will minimize changes in biological tissue for many years, giving the medical community ample time to cure all disease, rejuvenate the aged and repair any damage that is caused by the cryopreservation process. Cryonicists do not believe that legal death is "real death" because stoppage of heartbeat and breathing—the usual medical criteria for legal death—occur before biological death of cells and tissues of the body.

Even at room temperature, cells may take hours to die and days to decompose. Although neurological damage occurs within 4–6 minutes of cardiac arrest, the irreversible neurodegenerative processes do not manifest for hours. Cryonicists state that rapid cooling and cardio-pulmonary support applied immediately after certification of death can preserve cells and tissues for long-term preservation at cryogenic temperatures. People, particularly children, have survived up to an hour without heartbeat after submersion in ice water. In one case, full recovery was reported after 45 minutes underwater.

To facilitate rapid preservation of cells and tissue, cryonics "standby teams" are available to wait by the bedside of patients who are to be cryopreserved to apply cooling and cardio-pulmonary support as soon as possible after declaration of death.

No mammal has been successfully cryopreserved and brought back to life, and resuscitation from cryonics is not possible with current science. Some scientists still support the idea based on their expectations of the capabilities of future science.

Strategies for Engineered Negligible Senescence (SENS)

Another proposed life extension technology would combine existing and predicted future biochemical and genetic techniques. One such theoretical strategy proposes a cure for cancer, stem cell treatments, addition of new enzymes to the human body and moving mitochondrial DNA to the cellular nucleus. There is no scientific evidence that supports this strategy, and has been called pseudoscientific because its proposed techniques are speculative. Strategies for Engineered Negligible Senescence (SENS) is the name Aubrey de Grey gives to his proposal to research regenerative medical procedures to periodically repair all the age-related damage in the human body, thereby maintaining a youthful state indefinitely. The term first appeared in print in de Grey's 1999 book The Mitochondrial Free Radical Theory of Aging, and was later prefaced with the term "strategies" in the article Time to Talk SENS: Critiquing the Immutability of Human Aging De Grey argues for a "goal-directed rather than curiosity-driven" approach to the science of aging, and to this purpose he tentatively identifies seven "damages of aging" and their potential methods of treatments.

SENS has received media attention but has been criticized by some scientists. While many biogerontologists (scientists who study aging) find it "worthy of discussion" and SENS conferences feature important research in the field, some contend that de Grey's programme is too speculative given current scientific research, referring to it as "fantasy rather than science".

Framework

The goal of the Strategies for Engineered Negligible Senescence (SENS) is the complete reversal of all age-related illnesses and indefinite extension of the healthy human lifespan. The SENS project consists in implementing a series of periodic medical interventions designed to repair, prevent or render irrelevant all the types of molecular and cellular damage that cause age-related pathology and degeneration, in order to avoid debilitation and death from age-related causes.

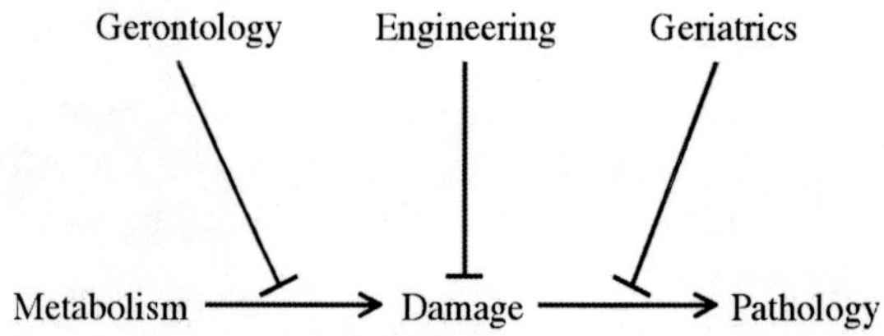

Figure: *The arrows with flat heads are a notation meaning "inhibits," used in the literature of gene expression and gene regulation.*

De Grey defines aging as "the set of accumulated side effects from metabolism that eventually kills us", and, more specifically, as follows: "a collection of cumulative changes to the molecular and cellular structure of an adult organism, which result in essential metabolic processes, but which also, once they progress far enough, increasingly disrupt metabolism, resulting in pathology and death." As de Grey states, "geriatrics is the attempt to stop damage from causing pathology; traditional gerontology is the attempt to stop metabolism from causing damage; and the SENS (engineering) approach is to eliminate the damage periodically, so keeping its abundance below the level that causes any pathology." His approach to biomedical gerontology ("anti-aging medicine") is thus distinctive because of its emphasis on rejuvenation rather than attempting to slow the aging process.

He identifies what he believes to be the seven biological causes of senescence. According to de Grey, the causes of aging in humans are:

1. cell loss or atrophy (without replacement)
2. oncogenic nuclear mutations and epimutations,
3. cell senescence (Death-resistant cells),
4. mitochondrial mutations,
5. Intracellular junk or junk inside cells (lysosomal aggregates),
6. extracellular junk or junk outside cells (extracellular aggregates),
7. random extracellular cross-linking.

For each of these problems, de Grey outlines possible solutions, with a research and a clinical component. The clinical component is required because in some of the proposed therapies, feasibility has already been proven, but not completely applied and approved for use by human beings. He believes we will be able to apply these solutions before we completely understand the targeted aging mechanisms, which will take longer. He states that the goals work together to eliminate known causes of human senescence, are concrete, seem achievable, and are considered feasible by experts in the applicable fields. The goals were said to be taken from classical literature describing the biological causes of senescence.

De Grey proposes that engineered negligible senescence therapies could extend the human lifetime by many centuries or millennia, as early therapies give them enough time to seek more effective therapies later on. He describes an actuarial escape velocity of life extension, when advances in senescence treatment come rapidly enough to save the lives of the oldest beneficiaries of the previous treatments.

Types of Aging Damage and Treatment Schemes

Cancer-causing Nuclear Mutations/Epimutations—OncoSENS

These are changes to the nuclear DNA (nDNA), the molecule that contains our genetic information, or to proteins which bind to the nDNA. Certain mutations can lead to cancer, and, according to de Grey, non-cancerous mutations and epimutations do not contribute to aging within a normal lifespan, so cancer is the only endpoint of these types of damage that must be addressed. A mutation in a functional gene of a cell can cause that cell to malfunction or to produce a malfunctioning product, because of the sheer number of cells, de Grey believes that redundancy takes care of this problem, although cells that have mutated to produce toxic products might have to be disabled.

In de Grey's opinion, the effect of mutations and epimutations that really matters is cancer, this is because if even one cell turns into a cancer cell it might spread and become deadly. This would need to be corrected by a cure for cancer, if any is ever found.

The SENS program focuses on a strategy called "whole-body interdiction of lengthening telomeres" (WILT), which would be made possible by periodic regenerative medicine treatments. Restoring telomeres would protect the ends of DNA from being cut off after successive divisions since each one normally removes some which is at first the telomere.

Mitochondrial Mutations—MitoSENS

Mitochondria are components in our cells that are important for energy production. They contain their own genetic material, and mutations to their DNA can affect a cell's ability to function properly. Indirectly, these mutations may accelerate many aspects of aging. Because of the highly oxidative environment in mitochondria and their lack of the sophisticated repair systems found in cell nucleus, mitochondrial mutations are believed to a be a major cause of progressive cellular degeneration.

This would be corrected by allotopic expression—moving the DNA for mitochondria completely within the cellular nucleus, where it is better protected. In humans, all but 13 proteins are already protected in this way. De Grey argues that experimental evidence demonstrates that the operation is feasible. However, a 2003 study showed that some mitochondrial proteins are too hydrophobic to survive the transport from the cytoplasm to the mitochondria, and this is perhaps one of the reasons that not all of the mitochondrial genes have migrated to the nucleus during the course of evolution.

Intracellular Junk—LysoSENS

Our cells are constantly breaking down proteins and other molecules that are no longer useful or which can be harmful. Those molecules which can't be digested simply accumulate as junk inside our cells, which is readily detected in the form of lipofuscin granules.

Atherosclerosis, macular degeneration, liver spots on the skin and all kinds of neurodegenerative diseases (such as Alzheimer's disease) are associated with this problem. Junk inside cells might be removed by adding new enzymes to the cell's natural digestion organ, the lysosome. These enzymes would be taken from bacteria, molds and other organisms that are known to completely digest animal bodies.

Extracellular Junk—AmyloSENS

Harmful junk protein can also accumulate outside of our cells. Junk outside cells might be removed by enhanced phagocytosis (the normal process used by the immune system), and small drugs able to break chemical beta-bonds. The large junk in this class can be removed surgically. Junk here means useless things accumulated by a body, but which cannot be digested or removed by its processes, such as the amyloid plaques characteristic of Alzheimer's disease and other amyloidoses. The oft-mentioned "toxins" that are identified as causes of many diseases most likely fit under this class.

Cell Loss and Atrophy—RepleniSENS

Some of the cells in our bodies cannot be replaced, or can be only replaced very slowly—more slowly than they die. This decrease in cell number affects some of the most important tissues of the body. Muscle cells are lost in skeletal muscles and the heart, causing them to become frailer with age. Loss of neurons in the substantia nigra causes Parkinson's disease, while loss of immune cells impairs the immune system. Cell depletion can be partly corrected by therapies involving exercise and growth factors. But stem cell therapy, regenerative medicine and tissue engineering are almost certainly required for any more than just partial replacement of lost cells. De Grey points out that this research of stem cell treatments is playing an increasingly important role in the international scientific community and progress is already occurring on many fronts. However, a large number of details are involved, and most such treatments are still experimental.

Cell Senescence—ApoptoSENS

This is a phenomenon where the cells are no longer able to divide, but also do not die and let others divide. They may also do other things that they are not supposed to do, like secreting proteins that could be harmful. Degeneration of joints, immune senescence, accumulation of visceral fat and type 2 diabetes are caused by this. Cells sometimes enter a state of resistance to signals sent, as part of a process called apoptosis, to instruct cells to destroy themselves. An example is the state known as cellular senescence. Cells in this state could be eliminated by forcing them to apoptose, and healthy cells would multiply to replace them. Cell killing with suicide genes or vaccines is suggested for making the cells undertake apoptosis.

Extracellular Crosslinks—GlycoSENS

Cells are held together by special linking proteins. When too many cross-links form between cells in a tissue, the tissue can lose its elasticity and cause problems including arteriosclerosis, presbyopia and weakened skin texture. These are chemical bonds between structures that are part of the body, but not within a cell. In senescent people many of these become brittle and weak.

De Grey proposes to further develop small-molecular drugs and enzymes to break links caused by sugar-bonding, known as advanced glycation endproducts, and other common forms of chemical linking.

Scientific Controversy

While some fields mentioned as branches of SENS are broadly supported by the medical research community, i.e. stem cell research (RepleniSENS), anti-Alzheimers research (AmyloSENS) and oncogenomics (OncoSENS), the SENS programme as a whole has been a highly controversial proposal, with many critics arguing that the SENS agenda is fanciful and the highly complicated biomedical phenomena involved in the aging process contain too many unknowns for SENS to be scientific or implementable in the foreseeable future. Cancer may well deserve special attention as an aging-associated disease (OncoSENS), but the SENS claim that nuclear DNA damage only matters for aging because of cancer has been challenged.

In November 2005, 28 biogerontologists published a statement of criticism in EMBO reports, "Science fact and the SENS agenda: what can we reasonably expect from ageing research?," arguing "each one of the specific proposals that comprise the SENS agenda is, at our present stage of ignorance, exceptionally optimistic," and that some of the specific proposals "will take decades of hard work [to be medically integrated], if [they] ever prove to be useful." The researchers argue that while there is "a rationale for thinking that we might eventually learn how to postpone human illnesses to an important degree," increased basic research, rather than the goal-directed approach of SENS, is presently the scientifically appropriate goal. This article was written in response to a July 2005 EMBO reports article previously published by de Grey and a response from de Grey was published in the same November issue. De Grey summarizes these events in "The biogerontology research community's evolving view of SENS," published on the Methuselah Foundation website.

De Grey Technology Review Controversy

In February 2005, Technology Review, which is owned by the Massachusetts Institute of Technology, published an article by Sherwin Nuland, a Clinical Professor of Surgery at Yale University and the author of "How We Die", that drew a skeptical portrait of Aubrey de Grey, at the time a computer associate in the Flybase Facility of the Department of Genetics at the University of Cambridge. While admiring de Grey's intelligence, Nuland concluded that he "would surely destroy us in attempting to preserve us" because living for such long periods would undermine what it means to be human. In the same issue of the magazine, Jason Pontin, the editor in chief and publisher of Technology Review, criticised de Grey. Writing an editor's letter, titled Against Transcendence, he questioned the usefulness and appropriateness of introducing transcendentalism, or transhumanism, into science. The April 2005 issue of Technology Review contained a reply by Aubrey de Grey and numerous comments from readers.

During June 2005 David Gobel, CEO and Co-founder of Methuselah Foundation offered Technology Review $20,000 to fund a prize competition to publicly clarify the viability of the SENS approach. In July 2005, Pontin announced a $20,000 prize, funded 50/50 by Methuselah Foundation and MIT Technology Review, open to any molecular biologist, with a record of publication in biogerontology, who could prove that SENS was "so wrong that it is unworthy of learned debate."

Technology Review received five submissions to its Challenge. In March, of 2006, Technology Review announced that it had chosen a panel of judges for the Challenge. Three met the terms of the prize competition. They were published by Technology Review on June 9, 2006. Accompanying the three submissions were rebuttals by de Grey, and counter-responses to de Grey's rebuttals. On July 11, 2006, Technology Review published the results of the SENS Challenge.

In the end, no one won the $20,000 prize. The judges felt that no submission met the criterion of the challenge and disproved SENS, although they unanimously agreed that one submission, by Preston Estep and his colleagues, was the most eloquent. Craig Venter succinctly expressed the prevailing opinion: "Estep et al. ... have not demonstrated that SENS is unworthy of discussion, but the proponents of SENS have not made a compelling case for it." Summarizing the judges' deliberations, Pontin wrote, "SENS is highly speculative. Many of its proposals have not been reproduced, nor could they be reproduced with today's scientific knowledge and technology. Echoing Myhrvold, we might charitably say that de Grey's proposals exist in a kind of antechamber of science, where they wait (possibly in vain) for independent verification.

SENS does not compel the assent of many knowledgeable scientists; but neither is it demonstrably wrong." In a letter of dissent dated July 11, 2006 in Technology Review, Estep et al. criticized the ruling of the judges.

Social and Economic Implications

According to de Grey, of the roughly 150,000 people who die each day across the globe, about two thirds — 100,000 per day — die of age-related causes. In industrialized nations, the proportion is much higher, reaching 90%. De Grey and other scientists in the general field have argued that the costs of a rapidly growing aging population will increase to the degree that the costs of an accelerated pace of aging research are easy to justify in terms of future costs avoided. Olshansky et al. 2006 argue, for example, that the total economic cost of Alzheimer's disease in the US alone will increase from $80–100 billion today to more than $1 trillion in 2050. "Consider what is likely to happen if we don't [invest further in aging research].

Take, for instance, the impact of just one age-related disorder, Alzheimer disease (AD). For no other reason than the inevitable shifting demographics, the number of Americans stricken with AD will rise from 4 million today to as many as 16 million by midcentury. This means that more people in the United States will have AD by 2050 than the entire current population of the Netherlands. Globally, AD prevalence is expected to rise to 45 million by 2050, with three of every four patients with AD living in a developing nation. The US economic toll is currently $80–$100 billion, but by 2050 more than $1 trillion will be spent annually on AD and related dementias. The impact of this single disease will be catastrophic, and this is just one example."

Genetic Modification

Gene therapy, in which artificial genes are integrated with an organism to replace mutated or otherwise deficient genes, has been proposed as a future strategy to prevent aging. Targeting catalase to the mitochondria resulted in a 20% lifespan increase in transgenic mice, and improved performance in AAV therapeutically infected mice.

Fooling Genes

In The Selfish Gene, Richard Dawkins describes an approach to life-extension that involves "fooling genes" into thinking the body is young. Dawkins attributes inspiration for this idea to Peter Medawar. The basic idea is that our bodies are composed of genes that activate throughout our lifetimes, some when we are young, and others when we are older. Presumably, these genes are activated by environmental factors (this is known as epigenetics), and the changes caused by these genes activating can be lethal.

It is a statistical certainty that we possess more lethal genes that activate in later life than in early life. Therefore, to extend life we should be able to prevent these genes from switching on, and we should be able to do so by "identifying changes in the internal chemical environment of a body that take place during aging... and by simulating the superficial chemical properties of a young body".

History of Life Extension and the Life Extension Movement

In 1970, the American Aging Association was formed under the impetus of Denham Harman, originator of the free radical theory of aging. Harman wanted an organization of biogerontologists that was devoted to research and to the sharing of information among scientists interested in extending human lifespan.

In 1976, futurists Joel Kurtzman and Philip Gordon wrote No More Dying. The Conquest Of Aging And The Extension Of Human Life, (ISBN 0-440-36247-4) the first popular book on research to extend human lifespan. Subsequently, Kurtzman was invited to testify before the House Select Committee on Aging, chaired by Claude Pepper of Florida, to discuss the impact of life extension on the Social Security system. Saul Kent published The Life Extension Revolution (ISBN 0-688-03580-9) in 1980 and created a nutraceutical firm called the Life Extension Foundation, a non-profit organization that promotes dietary supplements. The Life Extension Foundation publishes a periodical called Life Extension Magazine. The 1982 bestselling book Life Extension: A Practical Scientific Approach (ISBN 0-446-51229-X) by Durk Pearson and Sandy Shaw further popularized the phrase "life extension".

In 1983, Roy Walford, a life-extensionist and gerontologist, published a popular book called Maximum Lifespan. In 1988, Walford and his student Richard Weindruch summarized their research into the ability of calorie restriction to extend the lifespan of rodents in The Retardation of Aging and Disease by Dietary Restriction (ISBN 0-398-05496-7). It had been known since the work of Clive McCay in the 1930s that calorie restriction can extend the maximum lifespan of rodents. But it was the work of Walford and Weindruch that gave detailed scientific grounding to that knowledge. Walford's personal interest in life extension motivated his scientific work and he practiced calorie restriction himself. Walford died at the age of 80 from complications caused by amyotrophic lateral sclerosis.

Money generated by the non-profit Life Extension Foundation allowed Saul Kent to finance the Alcor Life Extension Foundation, the world's largest cryonics organization. The cryonics movement had been launched in 1962 by Robert Ettinger's book, The Prospect of Immortality. In the 1960s, Saul Kent had been a co-founder of the Cryonics Society of New York. Alcor gained national prominence when baseball star Ted Williams was cryonically preserved by Alcor in 2002 and a family dispute arose as to whether Williams had really wanted to be cryopreserved.

Regulatory and legal struggles between the Food and Drug Administration (FDA) and the Life Extension Foundation included seizure of merchandise and court action. In 1991, Saul Kent and Bill Faloon, the principals of the Foundation, were jailed. The LEF accused the FDA of perpetrating a "Holocaust" and "seeking gestapo-like power" through its regulation of drugs and marketing claims. In 1992, the American Academy of Anti-Aging Medicine (A4M) was formed to create what it considered an anti-aging medical specialty distinct from geriatrics, and to hold trade shows for physicians interested in anti-aging medicine. The American Board of Medical Specialties recognizes neither anti-aging medicine nor the A4M's professional standing.

Ethics and Politics of Life Extension

Leon Kass (chairman of the US President's Council on Bioethics from 2001 to 2005) has questioned whether potential exacerbation of overpopulation problems would make life extension unethical. He states his opposition to life extension with the words:

"simply to covet a prolonged life span for ourselves is both a sign and a cause of our failure to open ourselves to procreation and to any higher purpose ... [The] desire to prolong youthfulness is not only a childish desire to eat one's life and keep it; it is also an expression of a childish and narcissistic wish incompatible with devotion to posterity."

John Harris, former editor-in-chief of the Journal of Medical Ethics, argues that as long as life is worth living, according to the person himself, we have a powerful moral imperative to save the life and thus to develop and offer life extension therapies to those who want them.

Transhumanist philosopher Nick Bostrom has argued that any technological advances in life extension must be equitably distributed and not restricted to a privileged few. In an extended metaphor entitled "The Fable of the Dragon-Tyrant", Bostrom envisions death as a monstrous dragon who demands human sacrifices. In the fable, after a lengthy debate between those who believe the dragon is a fact of life and those who believe the dragon can and should be destroyed, the dragon is finally killed. Bostrom argues that political inaction allowed many preventable human deaths to occur.

Aging as a Disease

Most mainstream medical organizations and practitioners do not consider aging to be a disease. David Sinclair says: "I don't see aging as a disease, but as a collection of quite predictable diseases caused by the deterioration of the body". The two main arguments used are that aging is both inevitable and universal while diseases are not.

However not everyone agrees. Harry R. Moody, Director of Academic Affairs for AARP, notes that what is normal and what is disease strongly depends on a historical context. David Gems, Assistant Director of the Institute of Healthy Ageing, strongly argues that aging should be viewed as a disease. In response to the universality of aging, David Gems notes that it is as misleading as arguing that Basenji are not dogs because they do not bark. Because of the universality of aging he calls it a 'special sort of disease'. Robert M. Perlman, coined the terms 'aging syndrome' and 'disease complex' in 1954 to describe aging.

The discussion whether aging should be viewed as a disease or not has important implications. It would stimulate pharmaceutical companies to develop life extension therapies and it would also increase the regulation of the anti-aging market by the FDA. Anti-aging now falls under the regulations for cosmetic medicine which are less tight than those for drugs.

Pharmacological Gene Therapy

Pharmacological gene therapy is a new field that combines pharmacological therapy and gene therapy. It is used either to prevent a defective gene from producing its protein or to increase the concentration of normal protein produced in the body by insertion of DNA or RNA fragments. It can also be using to generate immunity from contagious disease, such as TB, via the process of DNA vaccination.

Gene Therapeutics can be used to treat such conditions as Cystic Fibrosis (via the addition of a normal CFTR gene), Haemophilia A or even some of the complications of AIDS.

It uses a number of methods including: Gene augmentation Targeted Inhibition (using Antisense or antigene technology) Cell Killing (either direct or assisted) and DNA vaccination.

Predictive Medicine

Predictive medicine is a rapidly emerging field of medicine that entails predicting disease and instituting preventive measures in order to either prevent the disease altogether or significantly decrease its impact upon the patient (such as by preventing mortality or limiting morbidity). While different prediction methodologies exist, such as genomics, proteomics, and cytomics, the most fundamental way to predict future disease is based on genetics. Although proteomics and cytomics allow for the early detection of disease, much of the time those detect biological markers that exist because a disease process has already started.

However, comprehensive genetic testing (such as through the use of DNA arrays or full genome sequencing) allows for the estimation of disease risk years to decades before any disease even exists, or even whether a healthy fetus is at higher risk for developing a disease in adolescence or adulthood. Individuals who are more susceptible to disease in the future can be offered lifestyle advice or medication with the aim of preventing the predicted illness.

Current genetic testing guidelines supported by the health care professionals discourage purely predictive genetic testing of minors until they are competent to understand the relevancy of genetic screening so as to allow them to participate in the decision about whether or not it is appropriate for them. Genetic screening of newborns and children in the field of predictive medicine is deemed appropriate if there is a compelling clinical reason to do so, such as the availability of prevention or treatment as a child that would prevent future disease.

The Goal

The goal of predictive medicine is to predict future disease so that health care professionals and the patient themselves can be proactive in instituting lifestyle modifications and increased physician surveillance, such as bi-annual full body skin exams by a dermatologist or internist if their patient is found to have an increased risk of melanoma or an EKG and cardiology examination by a cardiologist if a patient is found to be at increased risk for a cardiac arrhythmia. Predictive medicine is intended not for patients but more so for healthy individuals, its purpose being to determine whether susceptibility to a particular disease is increased or not.

A tremendous number of association studies have been published in scientific literature that show associations between specific genetic variants in a person's genetic code and a specific disease. Association and correlation studies have found that a female individual with a mutation in the BRCA1 gene has a 65% cumulative risk of breast cancer.

Genetic variants in the Factor V gene is associated with an increased tendency to form blood clots, such as deep vein thrombosis (DVTs). Genetics tests are expected to reach the market more quickly than new medicines. Myriad Genetics is already generating revenue from genetic tests for BRCA1 and BRCA2.

Predictive medicine changes the fundamental paradigm of medicine from being reactive to being proactive and has the potential to significantly decrease the incidence and prevalence of both common and rare diseases.

Predictive Medicine

Available types of predictive medicine through health care professionals include:

- *Newborn screening:* Newborn screening is conducted just after birth to identify genetic disorders that can be treated early in life. This testing of infants for certain disorders is one of the most widespread uses of genetic screening - all US states currently test infants for phenylketonuria and congenital hypothyroidism. US state law mandates collecting a sample by pricking the heel of a newborn baby to obtain enough blood to fill a few circles on filter paper labeled with names of infant, parent, hospital, and primary physician.
- *Diagnostic testing:* Diagnostic testing is conducted to aid in the specificity diagnosis or detection of a disease. It is often used to confirm a particular diagnosis when a certain condition is suspected based on the subject's mutations and physical symptoms. The diversity in diagnostic testing ranges from common consulting room tests such as measuring blood pressure and urine tests to more invasive protocols such as biopsies.

- *Medical bioinformatics:* Medical bioinformatics involves determining individual cell molecular parameters by cytomics and single cell-based microarrays. A challenge remains to efficiently extract relevant predictive medicine parameters. This information is frequently extracted by computer-assisted identification and characterization of a few cell populations or gene clusters of interest.
- *Prenatal testing:* Prenatal testing is used to look for diseases and conditions in a fetus or embryo before it is born. This type of testing is offered for couples who have an increased risk of having a baby with a genetic or chromosomal disorder. Screening can determine the sex of the fetus. Prenatal testing can help a couple decide whether to abort the pregnancy. Like diagnostic testing, prenatal testing can be noninvasive or invasive. Non-invasive techniques include examinations of the woman's womb through ultrasonography or maternal serum screens. These non-invasive techniques can evaluate risk of a condition, but cannot determine with certainty if the fetus has a condition. More invasive prenatal methods are slightly more risky for the fetus and involve needles or probes being inserted into the placenta or chorionic villus sampling.
- *Carrier testing:* Carrier testing is done to identify people who carry one copy of a gene mutation that, when present in both copies, causes a genetic disorder. This type of testing is offered to individuals who have genetic disorder in their family history or to people in ethnic groups with increased risk of certain genetic diseases. If both parents are tested, carrier testing can provide information about a couple's risk of having a child with a genetic disorder.
- *Preconception testing:* Preconception testing is conducted on a person or two potential parents before a child is even conceived. This allows prospective parents to understand the risk of diseases and likelihood of traits in their future offspring. Companies have already started to apply preconception testing through sperm banks and egg donation services. A company called Existence Genetics announced in 2010 that it has invented a new patent-pending preconception testing technology that they call the Pythia Approach (it's scientific name is Offspring Projections Through the Combined Analysis of Different Individuals, or OP-CADI) that uses a computer system to combine the genetic makeup of two potential parents in-order to predict the risk of a large number of common and rare diseases, as well as the likelihood of specific traits, in future offspring if those two potential parents were to have a child.

Health Benefits

The future of medicine's focus may potentially shift from treating existing diseases, typically late in their progression, to preventing disease before it sets in. Predictive medicine is based on probabilities: while it evaluates susceptibility to diseases, it is not able to predict with 100% certainty that a specific disease will occur. Unlike many preventive interventions that are directed at groups (e.g., immunization programs), predictive medicine is conducted on an individualized basis.

For example, glaucoma is a monogenic disease whose early detection can allow to prevent permanent loss of vision. Predictive medicine is expected to be most effective when applied to polygenic multifactorial disease that are prevalent in industrialized countries, such as diabetes mellitus, hypertension, and myocardial infarction. With careful usage, predictive medicine methods such as genetic screens can help diagnose inherited genetic disease caused by problems with a single gene (such as cystic fibrosis) and help early treatment. Some forms of cancer and heart disease are inherited as single-gene diseases and some people in these high-risk families may also benefit from access to genetic tests. As more and more genes associated with increased susceptibility to certain diseases are reported, predictive medicine becomes more useful.

Direct-to-Consumer Genetic Testing

Direct-to-Consumer (DTC) genetic testing enables a consumer to screen his or her own genes without having to go though a health care professional. They can be ordered without the permission of a physician. Variety in DTC tests range from those testing for mutations associated with cystic fibrosis to breast cancer alleles. DTC tests make the applicability of predictive medicine very real and accessible to consumers. Benefits of DTC testing include this accessibility, privacy of genetic information, and promotion of proactive health care. Risks of obtaining DTC testing are the lack of governmental regulation and the interpreting of genetic information without professional counseling.

Limitations of Predictive Medicine

On a protein level, structure is more conserved than sequence. Therefore, in many diseases, having the faulty gene still does not necessarily mean someone will get the disease. Common, complex diseases in the wider population are affected not only by heredity, but also by external causes such as lifestyle and environment. Therefore, genes are not perfect predictors of future health; individuals with both the high risk form of the gene and those without are all candidates to get the disease. Multiple factors in the environment, particular smoking, diet and exercise, infection, and pollution; play important roles and can be more important than genetic make-up. This makes the results and risks determined by predictive medicine more difficult to quantify. Furthermore, the potential false positives or false negatives that may arise from a predictive genetic screen can cause substantial unnecessary strain on the individual. Targeting medication to people who are genetically susceptible to a disease but do not yet show the symptoms of it can be a questionable measure.

In large populations, there is concern that likely most of the people taking preventative medications would never have developed the disease anyway. Many medications carry undesirable side effects that high risk individuals must then cope with. In contrast, several populations-based prevention measures (such as encouraging healthy diets or banning tobacco advertising) carry a far lower likelihood of adverse effects and are also less expensive.

Another potential downfall of commercially available genetic testing lies within the psychological impacts of accessibility to such data. For single-gene inherited diseases, counseling and the right to refuse a test (the right "not to know") have been found to be important.

However, adequate individual counseling can be difficult to employ to the potentially large proportion of the population likely to be identified as at high risk of common complex disease. Some people are vulnerable to adverse psychological reactions to genetic predictions of stigmatized or feared conditions, such as cancer or mental illness.

4

Technologies Used for Immunomic Research

Immunomic Microarrays

Several types of microarrays have been created to specifically observe immune system response and interactions. Antibody microarrays use antibodies as probes and antigens as targets. They can be used to directly measure the antigen concentrations for which the antibody probes are specific.

Peptide microarrays use antigen peptides as probes and serum antibodies as targets. These can be used for functional immunomic applications to the understanding of autoimmune diseases and allergies, definition of B-cell epitopes, vaccine studies, detection assays, and analysis of antibody specificity. MHC microarrays are the most recent development in immunomic arrays and use peptide-MHC complexes and their co-stimulatory molecules as probes and T-cell populations as targets. Bound T-cells are activated and secrete cytokines, which are captured by specific detection antibodies. This microarray can map MHC-restricted T cell epitopes.

Lymphochip

The Lymphochip is a specialized human cDNA microarray enriched for genes related to immune function. 17,853 cDNA clones were taken from three sources. The first set of clones were selected if identified expressed sequence tags (ESTs) were unique or enriched specifically in lymphoid cDNA libraries; these represent ~80% of the Lymphochip clones. The second set of clones was identified during first-generation microarray analysis of immune responses.

Finally, 3,183 genes that are known or suspected to have roles in immune function, oncogenesis, apoptosis, cell proliferation, or being open reading frames from pathogenic human viruses were used on the Lymphochip. New genes are frequently being added.

T-cell-and-B-cell-epitope Mapping Tools

Epitope mapping identifies the sites of antibodies to which their target antigens bind. In the past, scientists would have to isolate antigens, digest them into smaller fragments, and determine which of these fragments stimulated T- and B- cell responses to define an antibody's epitope. Immunomics harnesses the power of bioinformatics and offers mapping algorithms that accelerate the discovery of epitope sequences. These algorithms are relevant to vaccine design and for characterizing and modifying immune responses in the context of autoimmunity, endocrinology, allergy, transplantation, diagnostics and engineering of therapeutic proteins.

T-cell and B-cell epitope mapping algorithms can computationally predict epitopes based on the genomic sequence of pathogens, without prior knowledge of a protein's structure or function. A series of steps are used to identify epitopes:

1. Comparison between virulent and avirulent organisms identify candidate genes that code for epitopes that solicit T-cell responses by looking for sequences that are unique to virulent strains. Additionally, differential microarray technologies can discover pathogen-specific genes that are upregulated during host-interaction and may be relevant for analysis because they are critical to the function of the pathogen.
2. Immunoinformatics tools predict regions of these candidate genes that interact with T cells by scanning genome-derived protein sequences of a pathogen.
3. These predicted peptides are synthesized and used in in vitro screening against T cells. Recognizing a positive immune response can suggest that this peptide contains an epitope that stimulates immune response in the course of natural infection or disease.

Epitope Mapping Tools Available

- EpiMatrix
- TEPITOPE
- Multipred
- MHC Thread
- MHCPred
- NetMHC

- LpPep
- BIMAS

Tetramer Staining by Flow Cytometry

The guiding principle behind flow cytometry is that cells or subcellular particles are tagged with fluorescent probes are passed through a laser beam and sorted by the strength of fluorescence emitted by cells contained in the droplets. MHC tetramer staining by flow cytometry identifies and isolates specific T cells based on the binding specificity of their cell surface receptors with fluorescently-tagged MHC-peptide complexes.

ELISPOT

ELISPOT is a modified version of the ELISA immunoassay and is a common method of monitoring immune responses.

Immunomic Contributions to Understanding the Immune System

Immunomics has made a considerable impact on the understanding of the immune system by uncovering differences in gene expression profiles of cell types, characterizing immune response, illuminating immune cell lineages and relationship, and establishing gene regulatory networks. Whereas the following list of contributions is not complete, it is meant to demonstrate the broad application of immunomic research and powerful consequences on immunology.

States of Immune Cell Activation or Differentiation

Blymphocyte Activation or Anergy

Microarrays have discovered gene expression patterns that correlate with antigen-induced activation or anergy in B lymphocytes. Lymphocyte anergy pathways involve induction of some, but not all of the signaling pathways used during lymphocyte activation. For example, NFAT and MAPK/ERK kinase pathways are expressed in anergic (or "tolerant) cell lines, whereas NF-kB and c-Jun N-terminal kinases pathways are not.

Of the 300 genes that were altered in expression after antigen-stimulated naïve B cells, only 8 of these genes were regulated in tolerant B cells. Understanding these "tolerance" pathways have important implications for designing immunosuppressive drugs. These gene expression signatures of tolerant B cells could be used during drug screens to probe for compounds that mimic the functional effects of natural tolerance.

Lymphocyte Differentiation

Gene expression profiles during human lymphocyte differentiation has followed mature, naïve B cells from their resting state through germinal center reactions and into terminal differentiation. These studies have shown that germinal center B cells represent a distinct stage in differentiation because the gene expression profile is different than activated peripheral B cells.

Although no in vitro culture system has been able to induce resting peripheral B cells to adopt a full germinal center phenotype, these gene expression profiles can be used to measure the success of in vitro cultures in mimicking the germinal center state as they are being developed.

Lymphoid Malignancies

About 9 of every 10 human lymphoid cancers derive from B cells. Distinct immunome-wide expression patterns in a large number of diffuse large cell lymphoma (DLCL)– the most common form of non-Hodgkin's lymphoma – have identified at least two different subtypes in what was previously thought to be a single disease.

One subset of these DLCLs shows a similar gene expression pattern to that of normal germinal center B cells and implies that the tumor cell originated from a germinal center B cell. Other surveys of B cell malignancies show that follicular lymphomas share expression features with germinal center B cells, whereas chronic lymphocytic leukemia cells resemble resting peripheral blood lymphocytes. Furthermore, heterogeneity in each of these cell lines also suggest that different subtypes exist within each type of lymphoma, just as it has been shown in DLCL. Such knowledge can be used to direct patients to the most appropriate therapy.

Immune Response

Macrophage Responses to Bacteria

Microarrays have analyzed global responses of macrophages to different microorganisms and have confirmed that these responses sustain and control inflammatory processes, and also kill microorganisms. These independent studies have been able to better describe how macrophages mount attacks against different microorganisms. A "core transcriptional response" was observed to induce 132 genes and repress 59 genes. Induced genes include pro-inflammatory chemokines and cytokines, and their respective receptors. A "pathogen-specific response" was also observed.

Dendritic Response to Pathogen

Dendritic cells (DCs) help macrophages sustain inflammatory processes and participate in the innate immune system response, but can also prime adaptive immunity. Gene expression analyses have shown that DCs can "multi-task" by temporally segregating their different functions. Soon after recognizing an infectious agent, immature DCs transition to a state of early activation via a core response characterized by rapid downregulation of genes involved with pathogen recognition and phagocytosis, upregulation of cytokine and chemokine genes to recruit other immune cells to the side of inflammations; and expression of genes that control migratory capacity. Early activated DCs are enabled to migrate from non-lymphoid tissues to lymph nodes, where they can prime T-cell responses. These early DCs responses are related to innate immunity and comprise of the "core transcriptional response" of DCs. Pathogen-specific responses have a stronger influence on the DC's ability to regulate adaptive immunity.

Distinguishing Immune Cell Types, Lineages, and Relationships

Comparing distinctions between immune cells' overall transcriptional program can generate plots that position each cell type to best reflect it's expression profile relative to all other cells and can reveal interesting relationships between cell types. For example, the transcriptional profiles from thymic medullary epithelial immune cells mapped closer to lymphocytes than to other epithelia. This can suggest that a functional interaction exists between these two cells type and requires the sharing of particular transcripts and proteins. When comparing gene expression profiles from cells of the blood system, T-cell and B-cell subsets tightly group with their respective cell types.

By looking at the transcriptional profile of different T-cells, scientists have shown that natural killer T-cells are a close variant of conventional CD4+ T cells, rather than an intermediary cell-type between T cells and natural killer cells. Additionally, DCs, natural killer cells, and B cells are tightly grouped based of their global expression profiles. It may have been expected that B lymphocytes and T lymphocytes would cluster separately from each other, or that natural killer cells would be more closely related to T cells because they share common precursors, cytolytic activity, and similar activation markers. Therefore, immunomics has established relationship between cell lineages that depart from classical views. Additionally, it may better explain the observed plasticity in lymphoid and myeloid cell differentiation because of the considerable overlap between global expression profiles of these different lineages.

Establishing Regulatory Networks in Immune Cells

Networks represent the broadest level of genetic interactions and aim to link all genes and transcripts in the immunological genome. Cellular phenotypes and differentiation states are ultimately established by the activity of these networks of co-regulated genes. One of the most complete networks in immunology has deciphered regulatory connections among normal and transformed human B cells. This analysis suggests a hierarchical network where a small number of highly connected genes (called "hubs") regulated most interactions. Proto-oncogene MYC emerged as a major hub and highly influential regulator for B cells. Notably, MYC was found to directly control BYSL, a highly conserved, but poorly characterized gene, and is the largest hub in the whole B cell network. This suggests that BYSL encodes an important cellular molecule and a critical effecter of MYC function, and motivates additional studies to elucidate its function. Therefore, using gene expression data to create networks can reveal genes highly influential in immune cell differentiation that pre-genomic technologies had not yet identified.

Applications to Medicine

Vaccine Development

As quoted by Stefania Bambini and Rino Rappuoli, "New powerful genomics technologies have increased the number of disease that can be addressed by vaccination, and decreased the time for discover research and vaccine development." The availability of complete genome sequences of pathogens in combination with high-throughput genomics technologies have helped to accelerate vaccine development. Reverse vaccinology uses genomic sequences of viral, bacterial, or parasitic pathogens to identify genes potentially encoding genes that promote pathogenesis. The first application of reverse vaccinology identified vaccine candidates against "'Neisseria meningitidis"' serogroup B. Computational tools identified 600 putative surface-exposed or secreted proteins from the complete genome sequence of a MenB pathogenic strain, on the basis of sequence features.

These putative proteins were expressed in E. coli, purified, and used to immunize mice. Tests using mice immune sera estimated the ability of antibodies to protect against these proteins. The proteins able to solicit a robust immune response were checked for sequence conservation across a panel of meningitides strains and allowed for further selection of antigen able to elicit an immune response against most strains in the panel. On the basis of these antigen sequences, scientists have been able to develop a universal "cocktail" vaccine against Neisseria meninitidis that uses five antigens to promote immunity. Similar approaches have been used for a variety of other human pathogens, such as Streptococcus pneumoniae, Chlamydia pneumoniae, Bacillus anthracis, Porphyromonas gingivalis, Mycobacterium tuberculosis, Helicobacter pylori, amongst others. Additionally, studies have started for the development of vaccines against viruses.

Diagnosing and Predicting Disease

The inventory of receptors and signal transduction pathways that immune cells use to monitor and defend the body gives rise to signature patterns of altered gene expression in peripheral blood cells that reflect the character of the infection or injury. Therefore, recognizing characteristic expression profiles of peripheral blood cells may be a powerful diagnostic tool by recruiting these cells as "spies" to detect occult diseases or agents that cannot be readily cultured from the host.

For example, cytomegalovirus (CMV) infection of fibroblasts and HTLV-I infection of T lymphocytes revealed distinct gene expression profiles. CMV infection provoked a unique interferon response whereas HTLV-1 infection induced NF-kB target genes. A type of white blood cells have also been tested again bacterial exposures and immunome expression varied based on the type of bacterial strain used.

Monitoring the change of peripheral blood gene expression can also help determine the course of infection and help treat patients with a therapy tailored to their disease stage. This approach has already been used against sepsis – a disease that progresses through a predictable line of events. Changes gene expression signatures may precede clinical exacerbation of symptoms, like in multiple sclerosis, and allow physicians to nip these "flare-ups" in the bud.

The Immunological Genome Project

The immune system is a network of genetic and signaling pathways connected by a network of interacting cells. The Immunological Genome Project seeks to generate a complete compendium of protein-coding gene expression for all cell populations in the mouse immune system. It analyzes both steady-state conditions within different cell populations, and in response to genetic and/or environmental perturbations created by natural genetic polymorphism, gene knock-out, gene knock-down by RNAi, or drug treatment. Computational tools to reverse-engineer or predict immune cell regulatory networks use these expression profiles.

By 2008, the ImmGen project involved seven immunology and three computational biology laboratories across the United States and over 200 cell populations involved in the immune system had been identified and described. This consortium has created a data browser to explore the expression patterns of particular genes, networks of co-regulated genes, and genes that can reliably distinguish cell types. Raw data is also accessible from the NCBI's Gene Expression Omnibus.,

Immunomic Databases

- Immune Response in silico (IRIS)
- Reference Database of Immune Cells
- Immunological Genome Project
- Immune Epitope Database and Analysis Resource (IEDB)
- IMGT

- SYFPEiTHi
- AniJen
- MHCBN
- IPD
- Epitome
- Allergome
- Nitrogenomics

Nitrogenomics as the branch of the study of genomics pertaining to nitrogen utilisation and assimilation in organisms. Nitrogen is a primary nutrient essential for sustaining the life of every organism. Nitrogen is freely available in the atmosphere and Earth's crust and is generally assimilated by plants and microorganisms, then moved to higher organisms through the food chain. Genomics of nitrogen assimilation lie at different levels of organisms, from microbes to higher organisms, where different genetic controls regulate the actual assimilation.

This term nitrogenomics was not available before April 2004, when I made the posting of this term for the first time in Wikipedia. The growing awareness of genomic sciences and their enormous application potential, led me to think of a specialised term for the molecular genetics and genomics of the nitrogen utilisation in the organic world. It rather amused me that the words 'nitrogen' and 'genomics' fused well. Though I coined the term without much thinking of its potential as a branch of science, I now feel that the science which is described within this term can be the science of life itself.

Nitrogen is as essential as carbon, hydrogen and oxygen in the organic world, as it forms the primary component of amino acids which makes proteins, the building and guiding blocks of organic life. But the most intriguing part of the story is that nitrogen is not freely available as in the case of other primary elements of life. In fact it is mostly available abundantly in the unavailable forms. So how then nitrogen comes into life? Its an intricate cycle that involves enormous microorganisms, entire plant kingdom and the whole animal kingdom or every living organisms in interaction with the forces of nature! The science behind this is immense.

Metagenomics

Metagenomics is the study of metagenomes, genetic material recovered directly from environmental samples. The broad field may also be referred to as environmental genomics, ecogenomics or community genomics. Traditional microbiology and microbial genome sequencing rely upon cultivated clonal cultures. Metagenomics offers a powerful lens for viewing the microbial world that has the potential to revolutionize understanding of the entire living world.

Early environmental gene sequencing cloned specific genes (often the 16S rRNA gene) to produce a profile of diversity in a natural sample. Such work revealed that the vast majority of microbial biodiversity had been missed by cultivation-based methods. Recent studies use "shotgun" Sanger sequencing or massively parallel pyrosequencing to get largely unbiased samples of all genes from all the members of the sampled communities.

History

Origin of the Term: The term "metagenomics" was first used by Jo Handelsman, Jon Clardy, Robert M. Goodman, and others, and first appeared in publication in 1998. The term metagenome referenced the idea that a collection of genes sequenced from the environment could be analyzed in a way analogous to the study of a single genome. The exploding interest in environmental genetics, along with the buzzword-like nature of the term, has resulted in the broader use of metagenomics to describe any sequencing of genetic material from environmental (i.e. uncultured) samples, even work that focuses on one organism or gene. Recently, Kevin Chen and Lior Pachter (researchers at the University of California, Berkeley) defined metagenomics as "the application of modern genomics techniques to the study of communities of microbial organisms directly in their natural environments, bypassing the need for isolation and lab cultivation of individual species."

Environmental Gene Surveys

Conventional sequencing begins with a culture of identical cells as a source of DNA. However, early metagenomic studies revealed that there are probably large groups of microorganisms in many environments that cannot be cultured and thus cannot be sequenced. These early studies focused on 16S ribosomal RNA sequences which are relatively short, often conserved within a species, and generally different between species. Many 16S rRNA sequences have been found which do not belong to any known cultured species, indicating that there are numerous non-isolated organisms out there.

Early molecular work in the field was conducted by Norman R. Pace and colleagues, who used PCR to explore the diversity of ribosomal RNA sequences. The insights gained from these breakthrough studies led Pace to propose the idea of cloning DNA directly from environmental samples as early as 1985.

This led to the first report of isolating and cloning bulk DNA from an environmental sample, published by Pace and colleagues in 1991 while Pace was in the Department of Biology at Indiana University. Considerable efforts ensured that these were not PCR false positives and supported the existence of a complex community of unexplored species. Although this methodology was limited to exploring highly conserved, non-protein coding genes, it did support early microbial morphology-based observations that diversity was far more complex than was known by culturing methods. Soon after that, Healy reported the metagenomic isolation of functional genes from "zoolibraries" constructed from a complex culture of environmental organisms grown in the laboratory on dried grasses in 1995. After leaving the Pace laboratory, Ed DeLong continued in the field and has published work that has largely laid the groundwork for environmental phylogenies based on signature 16S sequences, beginning with his group's construction of libraries from marine samples.

Longer Sequences from Environmental Samples

Recovery of DNA sequences longer than a few thousand base pairs from environmental samples was very difficult until recent advances in molecular biological techniques, particularly related to constructing libraries in bacterial artificial chromosomes (BACs), provided better vectors for molecular cloning.

Shotgun Metagenomics

Advances in bioinformatics, refinements of DNA amplification, and proliferation of computational power have greatly aided the analysis of DNA sequences recovered from environmental samples. These advances have enabled the adaptation of shotgun sequencing to metagenomic samples. The approach, used to sequence many cultured microorganisms as well as the human genome, randomly shears DNA, sequences many short sequences, and reconstructs them into a consensus sequence.

In 2002, Mya Breitbart, Forest Rohwer, and colleagues used environmental shotgun sequencing to show that 200 liters of seawater contains over 5000 different viruses. Subsequent studies showed that there are >1000 viral species in human stool and possibly a million different viruses per kilogram of marine sediment, including many bacteriophages. Essentially all of the viruses in these studies were new species. In 2004, Gene Tyson, Jill Banfield, and colleagues at the University of California, Berkeley and the Joint Genome Institute sequenced DNA extracted from an acid mine drainage system. This effort resulted in the complete, or nearly complete, genomes for a handful of bacteria and archaea that had previously resisted attempts to culture them. It was now possible to study entire genomes without the biases associated with laboratory cultures.

Global Ocean Sampling Expedition

Beginning in 2003, Craig Venter, leader of the privately-funded parallel of the Human Genome Project, has led the Global Ocean Sampling Expedition, circumnavigating the globe and collecting metagenomic samples throughout. All of these samples are sequenced using shotgun sequencing, in hopes that new genomes (and therefore new organisms) would be identified. The pilot project, conducted in the Sargasso Sea, found DNA from nearly 2000 different species, including 148 types of bacteria never before seen. As of 2009, Venter has circumnavigated the globe and thoroughly explored the West Coast of the United States, and is currently in the midst of a two-year expedition to explore the Baltic, Mediterranean and Black Seas.

Pyrosequencing

In 2006 Robert Edwards, Forest Rohwer, and colleagues at San Diego State University published the first sequences of environmental samples generated with so-called next generation sequencing, in this case chip-based pyrosequencing developed by 454 Life Sciences. This technique for sequencing DNA generates shorter fragments than conventional techniques, however this limitation is compensated for by the very large number of sequences generated. In addition, this technique does not require cloning the DNA before sequencing, removing one of the main biases in metagenomics.

Software

A major problem with metagenomes is binning. Binning is the process of identifying from what organism a particular sequence has originated. Traditionally, BLAST is a method used to rapidly search for similar sequences in existing public databases. More advanced methods have been employed to bin sequences. Big successes have been achieved for a family of methods using intrinsic features of the sequence, such as oligonucleotide frequencies.

These methods include TETRA (Teeling et al., 2004), Phylopythia (McHardy et al., 2007), TACOA (Diaz et al., 2009), PCAHIER (Zheng and Wu, 2010), DiScRIBinATE (Ghosh et al., 2010), SPHINX (Mohammed et al., 2011), and Parallel-META (Su et al., 2011). In 2007, Daniel Huson and Stephan Schuster developed and published the first stand-alone metagenome analysis tool, MEGAN, which can be used to perform a first analysis of a metagenomic shotgun dataset.

This tool was originally developed to analyse the metagenome of a mammoth sample. However in a recent study by Monzoorul et al. 2009, it was shown that adopting the LCA approach (of MEGAN) solely based on bit-score of the alignment leads to a number of false positive assignments especially in the context of metagenomic sequences originating from new organisms. This study proposed a new approach called SOrt-ITEMS which used several alignment parameters to increase the accuracy of assignments.

MG-RAST

In 2007, Folker Meyer and Robert Edwards and a team at Argonne National Laboratory and the University of Chicago released the Metagenomics RAST server (MG-RAST) a community resource for metagenome data set analysis. As of October 2011 3.7 Terabases (10^12 bases) of DNA have been analyzed by MG-RAST, more than 4300 public data sets are freely available for comparison within MG-RAST. Over 7000 users now have submitted a total of 38,000 metagenomes to MG-RAST. The server also acts as the de-fact repository for metagenomics data.

Applications

Metagenomics can improve strategies for monitoring the impact of pollutants on ecosystems and for cleaning up contaminated environments. Increased understanding of how microbial communities cope with pollutants is helping assess the potential of contaminated sites to recover from pollution and increase the chances of bioaugmentation or biostimulation trials to succeed.

Recent progress in mining the rich genetic resource of non-culturable microbes has led to the discovery of new genes, enzymes, and natural products. The impact of metagenomics is witnessed in the development of commodity and fine chemicals, agrochemicals and pharmaceuticals where the benefit of enzyme-catalyzed chiral synthesis is increasingly recognized.

Metagenomic sequencing is being used to characterize the microbial communities from 15-18 body sites from at least 250 individuals. This is part of the Human Microbiome initiative with primary goals to determine if there is a core human microbiome, to understand the changes in the human microbiome that can be correlated with human health, and to develop new technological and bioinformatics tools to support these goals.

It is well known that the vast majority of microbes have not been cultivated. Functional metagenomics strategies are being used to explore the interactions between plants and microbes through cultivation-independent study of the microbial communities.

Finally, metagenomic sequencing is particularly useful in the study of viral communities. As viruses lack a shared universal phylogenetic marker (as are 16S RNA for bacteria and archaea, and 18S RNA for eukarya), the only way to access the genetic diversity of the viral community from an environmental sample is through metagenomics. Viral metagenomes (also called viromes) should thus provide more and more information about viral diversity and evolution.

Microbial Diversity

Much of the interest in metagenomics comes from the discovery that the vast majority of microorganisms had previously gone unnoticed. Traditional microbiological methods relied upon laboratory cultures of organisms. Surveys of ribosomal RNA (rRNA) genes taken directly from the environment revealed that cultivation based methods find less than 1% of the bacteria and archaea species in a sample.

Gene Surveys

Shotgun sequencing and screens of clone libraries reveal genes present in environmental samples. This provides information both on which organisms are present and what metabolic processes are possible in the community. This can be helpful in understanding the ecology of a community, particularly if multiple samples are compared to each other.

Environmental Genomes

Shotgun metagenomics also is capable of sequencing nearly complete microbial genomes directly from the environment. Because the collection of DNA from an environment is largely uncontrolled, the most abundant organisms in an environmental sample are most highly represented in the resulting sequence data.

To achieve the high coverage needed to fully resolve the genomes of underrepresented community members, large samples, often prohibitively so, are needed. On the other hand, the random nature of shotgun sequencing ensures that many of these organisms will be represented by at least some small sequence segments. Due to the limitations of microbial isolation methods, the vast majority of these organisms would go unnoticed using traditional culturing techniques.

Community Metabolism

This section does not cite any references or sources. Please help improve this section by adding citations to reliable sources. Unsourced material may be challenged and removed. (October 2009)

Many bacterial communities show significant division of labor in metabolism. Waste products of some organisms are metabolites for others. Working together they turn raw resources into fully metabolized waste. Using comparative gene studies and expression experiments with microarrays or proteomics researchers can piece together a metabolic network that goes beyond species boundaries.

Such studies require detailed knowledge about which versions of which proteins are coded by which species and even by which strains of which species. Therefore, community genomic information is another fundamental tool (with metabolomics and proteomics) in the quest to determine how metabolites are transferred and transformed by a community.

Predictive Medicine

Predictive medicine is a rapidly emerging field of medicine that entails predicting disease and instituting preventive measures in order to either prevent the disease altogether or significantly decrease its impact upon the patient (such as by preventing mortality or limiting morbidity). While different prediction methodologies exist, such as genomics, proteomics, and cytomics, the most fundamental way to predict future disease is based on genetics. Although proteomics and cytomics allow for the early detection of disease, much of the time those detect biological markers that exist because a disease process has already started.

However, comprehensive genetic testing (such as through the use of DNA arrays or full genome sequencing) allows for the estimation of disease risk years to decades before any disease even exists, or even whether a healthy fetus is at higher risk for developing a disease in adolescence or adulthood. Individuals who are more susceptible to disease in the future can be offered lifestyle advice or medication with the aim of preventing the predicted illness.

Current genetic testing guidelines supported by the health care professionals discourage purely predictive genetic testing of minors until they are competent to understand the relevancy of genetic screening so as to allow them to participate in the decision about whether or not it is appropriate for them. Genetic screening of newborns and children in the field of predictive medicine is deemed appropriate if there is a compelling clinical reason to do so, such as the availability of prevention or treatment as a child that would prevent future disease.

Examples of Predictive Medicine

Available types of predictive medicine through health care professionals include:

- *Newborn screening:* Newborn screening is conducted just after birth to identify genetic disorders that can be treated early in life. This testing of infants for certain disorders is one of the most widespread uses of genetic screening - all US states currently test infants for phenylketonuria and congenital hypothyroidism. US state law mandates collecting a sample by pricking the heel of a newborn baby to obtain enough blood to fill a few circles on filter paper labeled with names of infant, parent, hospital, and primary physician.
- *Diagnostic testing:* Diagnostic testing is conducted to aid in the specificity diagnosis or detection of a disease. It is often used to confirm a particular diagnosis when a certain condition is suspected based on the subject's mutations and physical symptoms. The diversity in diagnostic testing ranges from common consulting room tests such as measuring blood pressure and urine tests to more invasive protocols such as biopsies.
- *Medical bioinformatics:* Medical bioinformatics involves determining individual cell molecular parameters by cytomics and single cell-based microarrays. A challenge remains to efficiently extract relevant predictive

medicine parameters. This information is frequently extracted by computer-assisted identification and characterization of a few cell populations or gene clusters of interest.

- *Prenatal testing:* Prenatal testing is used to look for diseases and conditions in a fetus or embryo before it is born. This type of testing is offered for couples who have an increased risk of having a baby with a genetic or chromosomal disorder. Screening can determine the sex of the fetus. Prenatal testing can help a couple decide whether to abort the pregnancy. Like diagnostic testing, prenatal testing can be noninvasive or invasive. Non-invasive techniques include examinations of the woman's womb through ultrasonography or maternal serum screens. These non-invasive techniques can evaluate risk of a condition, but cannot determine with certainty if the fetus has a condition. More invasive prenatal methods are slightly more risky for the fetus and involve needles or probes being inserted into the placenta or chorionic villus sampling.
- *Carrier testing:* Carrier testing is done to identify people who carry one copy of a gene mutation that, when present in both copies, causes a genetic disorder. This type of testing is offered to individuals who have genetic disorder in their family history or to people in ethnic groups with increased risk of certain genetic diseases. If both parents are tested, carrier testing can provide information about a couple's risk of having a child with a genetic disorder.
- *Preconception testing:* Preconception testing is conducted on a person or two potential parents before a child is even conceived. This allows prospective parents to understand the risk of diseases and likelihood of traits in their future offspring. Companies have already started to apply preconception testing through sperm banks and egg donation services. A company called Existence Genetics announced in 2010 that it has invented a new patent-pending preconception testing technology that they call the Pythia Approach (it's scientific name is Offspring Projections Through the Combined Analysis of Different Individuals, or OP-CADI) that uses a computer system to combine the genetic makeup of two potential parents in-order to predict the risk of a large number of common and rare diseases, as well as the likelihood of specific traits, in future offspring if those two potential parents were to have a child.

Health Benefits

The future of medicine's focus may potentially shift from treating existing diseases, typically late in their progression, to preventing disease before it sets in. Predictive medicine is based on probabilities: while it evaluates susceptibility to diseases, it is not able to predict with 100% certainty that a specific disease will occur. Unlike many preventive interventions that are directed at groups (e.g., immunization programs), predictive medicine is conducted on an individualized basis.

For example, glaucoma is a monogenic disease whose early detection can allow to prevent permanent loss of vision. Predictive medicine is expected to be most effective when applied to polygenic multifactorial disease that are prevalent in industrialized countries, such as diabetes mellitus, hypertension, and myocardial infarction. With careful usage, predictive medicine methods such as genetic screens can help diagnose inherited genetic disease caused by problems with a single gene (such as cystic fibrosis) and help early treatment.

Some forms of cancer and heart disease are inherited as single-gene diseases and some people in these high-risk families may also benefit from access to genetic tests. As more and more genes associated with increased susceptibility to certain diseases are reported, predictive medicine becomes more useful.

Direct-to-Consumer Genetic Testing

Direct-to-Consumer (DTC) genetic testing enables a consumer to screen his or her own genes without having to go though a health care professional. They can be ordered without the permission of a physician. Variety in DTC tests range from those testing for mutations associated with cystic fibrosis to breast cancer alleles. DTC tests make the applicability of predictive medicine very real and accessible to consumers. Benefits of DTC testing include this accessibility, privacy of genetic information, and promotion of proactive health care. Risks of obtaining DTC testing are the lack of governmental regulation and the interpreting of genetic information without professional counseling.

Limitations of Predictive Medicine

On a protein level, structure is more conserved than sequence. Therefore, in many diseases, having the faulty gene still does not necessarily mean someone will get the disease. Common, complex diseases in the wider population are affected not only by heredity, but also by external causes such as lifestyle and environment. Therefore, genes are not perfect predictors of future health; individuals with both the high risk form of the gene and those without are all candidates to get the disease. Multiple factors in the environment, particular smoking, diet and exercise, infection, and pollution; play important roles and can be more important than genetic make-up. This makes the results and risks determined by predictive medicine more difficult to quantify. Furthermore, the potential false positives or false negatives that may arise from a predictive genetic screen can cause substantial unnecessary strain on the individual.

Targeting medication to people who are genetically susceptible to a disease but do not yet show the symptoms of it can be a questionable measure. In large populations, there is concern that likely most of the people taking preventative medications would never have developed the disease anyway. Many medications carry undesirable side effects that high risk individuals must then cope with. In contrast, several populations-based prevention measures (such as encouraging healthy diets or banning tobacco advertising) carry a far lower likelihood of adverse effects and are also less expensive.

Another potential downfall of commercially available genetic testing lies within the psychological impacts of accessibility to such data. For single-gene inherited diseases, counseling and the right to refuse a test (the right "not to know") have been found to be important. However, adequate individual counseling can be difficult to employ to the potentially large proportion of the population likely to be identified as at high risk of common complex disease. Some people are vulnerable to adverse psychological reactions to genetic predictions of stigmatized or feared conditions, such as cancer or mental illness.

Ethics and Law

Predictive medicine ushers in a number of sensitive legal and ethical issues. There is a delicate balance that presides over predictive medicine and occupational health: if an employee were dismissed because he was found to be at risk of a certain chemical agent used in his workplace, would his termination be considered discrimination or an act of prevention? Several organizations believe that legislation is needed to prevent insurers and employers from using predictive genetic test results to decide who gets insurance or a job: "Ethical considerations, and legal, are fundamental to the whole issue of genetic testing.

The consequences for individuals with regard to insurance and employment are also of the greatest importance, together with the implications for stigma and discrimination." In the future, people may be required to reveal genetic predictions about their health to their employers or insurers. The grim prospect of discrimination based on a person's genetic make-up can lead to a "genetic underclass" which does not receive equal opportunity for insurance and employment.

Currently in the United States, health insurers do not require applicants for coverage to undergo genetic testing. Genetic information is under the same protection of confidentiality as other sensitive health information under the Health Insurance Portability and Accountability Act (HIPAA) when health insurers come across it. In the United States, the Genetic Information Nondiscrimination Act, signed into law by President Bush on May 21, 2008;

prohibits health insurers from denying coverage or charging differentials in premiums, and bars employers from making job placement or hiring/firing decisions based on individuals' genetic predispositions.

Literature Resources

In 2010, Penguin Publisher's published Outsmart Your Genes, by Brandon Colby, MD. This book is written for the lay reader and discusses the new field of predictive medicine and genetic testing from the perspective of a physician, and includes a number of pop-culture references such as to Sergei Grinkov, David Bloom, and the NFL. In the second half of Outsmart Your Genes, Dr. Colby provides specific examples of how genetic testing can be utilized to lower a person's risk of a large number of diseases, including Alzheimer's, cancer, heart disease, stroke, multiple sclerosis, asthma, and many others.

In 2010, Harper published The Language of Life: DNA and the Revolution in Personalized Medicine, by Francis S. Collins. The book is also written for the lay reader and discusses how genetic testing may be utilized to usher in personalized medicine. Dr. Collins describes general concepts of how genetics may be utilized to combat a large number of diseases, including Alzheimer's and heart disease.

Personal Genomics

Personal genomics is the branch of genomics concerned with the sequencing and analysis of the genome of an individual. The genotyping stage employs different techniques, including single-nucleotide polymorphism (SNP) analysis chips (typically 0.02% of the genome), or partial or full genome sequencing. Once the genotypes are known, the individual's genotype can be compared with the published literature to determine likelihood of trait expression and disease risk.

Automated sequencers have increased the speed and reduced the cost of sequencing, making it possible to offer genetic testing to consumers.

Use of Personal Genomics in Predictive medicine

Predictive medicine is the use of the information produced by personal genomics techniques when deciding what medical treatments are appropriate for a particular individual.

An example of the use of predictive medicine is pharmacogenomics, in which genetic information can be used to select the most appropriate drug to prescribe to a patient. The drug should be chosen to maximize the probability of obtaining the desired result in the patient and minimize the probability that the patient will experience side effects. Genetic information may allow physicians to tailor therapy to a given patient, in order to increase drug efficacy and minimize side effects. There are only a few examples in which this information is currently useful in clinical practice. Disease risk may be calculated based on genetic markers and genome-wide association studies, though most common medical conditions are multifactorial and the actual risk to the individual depends on both genetic and environmental components.

Cost of Sequencing an Individual's Genome

The cost of sequencing a human genome is dropping rapidly, due to the continual development of new, faster, cheaper DNA sequencing technologies such as "next generation DNA sequencing".

The National Human Genome Research Institute, part of the U.S. National Institutes of Health, has set a target to be able to sequence a human-sized genome for US$100,000 by 2009 and US$1,000 by 2014.

There are 6 billion base pairs in the diploid human genome. Statistical analysis reveals that a coverage of approximately ten times is required to get coverage of both alleles in 90% human genome from 25 base-pair reads with shotgun sequencing.

This means a total of 60 billion base pairs that must be sequenced. An Applied Biosystems SOLiD, Illumina or Helicos sequencing machine can sequence 2 to 10 billion base pairs in each $8,000 to $18,000 run. The purchase cost, personnel costs and data processing costs must also be taken into account. Sequencing a human genome cost approximately $300,000 in 2008.

In 2009, Complete Genomics of Mountain View announced that it would provide full genome sequencing for $5,000, from June 2009. This will only be available to institutions, not individuals. Approximately 1 million babies are born in Canada each year; to sequence all of their genomes would cost approximately $1 billion per year, or just 1% of Canada's total healthcare budget. Given the ethical concerns about presymptomatic genetic testing of minors, it is likely that personal genomics will first be applied to adults who can provide consent to undergo such testing.

In June 2009, Illumina announced the launch of its own Personal Full Genome Sequencing Service at a depth of 30X for $48,000 per genome. Only one year later, in 2010, they cut the price 60% to $19,500. Prices are expected to drop further over the next few years through economies of scale and increased competition.

Knome's whole genome sequencing approach aims, instead, to read every site in the whole euchromatic portion of a person's genome (roughly 3 billion sites). While significantly more expensive than SNP chip-based genotyping, this approach yields significantly more data, identifying both novel and known sequence variants, some of which may be particularly relevant in efforts to understand personal health, as well as ancestry.

Table: *Timeline of personal genomes sequenced*

Year	*Cost Source*	*Personal genomes sequenced*	*Company*
2003	$3,000,000,000	1	Various
2009	$48,000	100	Illumina
2010	$19,500	1000+	Illumina
2011	$19,500 to $4,000	?	Illumina

Comparative Genomics

Comparative genomics analysis characterizes the differences and similarities between whole genomes. It may be applied to both genomes from individuals from different species or individuals from the same species, generally at lower cost than sequencing from scratch.

Services Already Available

Companies which offer genome-wide personal genomics services have already gone to market and are selling their services direct to the consumer.

- deCODEme.com charges $1100 to carry out genotyping of approximately 1 million SNPs and provides risk estimates for 47 diseases as well as ancestry analyses.
- Existence Genetics provides genetic testing services through healthcare providers and health and wellness organizations, with prices starting at $350. This company provides testing for over 700 different diseases and traits, including preventable sudden cardiac death, malignant hyperthermia.
- Navigenics, began offering SNP-based genomic risk assessments as of April 2008. Navigenics is medically focused and emphasizes a clinician's and genetic counselor's role in interpreting results. Affymetrix Genome-Wide Human SNP Array 6.0 , which genotypes 900,000 SNPs.
- Pathway Genomics analyzes over 100 genetic markers to identify genetic risk for common health conditions such as melanoma, prostate cancer and rheumatoid arthritis.
- 23andMe sells mail order kits for SNP genotyping. The $99 kit, with $9.00/month subscription, or $399 without a subscription contains a saliva sample collection container and instructions. The consumer mails the sample to 23andMe for microarray analysis. The information is stored in a user profile and used to estimate the genetic risk of the consumer for 178 diseases and conditions, as well as ancestry analysis. 23andMe utilizes a DNA array manufactured by Illumina.

- SNPedia is a wiki that collects and shares information about the consequences of DNA variations, and through the associated program Promethease, anyone who has obtained DNA data about themselves (from any company) can get a free, independent report containing risk assessments and related information.
- Bioresolve describes a similar service to that of 23andMe; the Ottawa Better Business Bureau gave the company an "F" reliability rating.
- Knome provides full genome (98% genome) sequencing services for $4,998 for whole genome sequencing and interpretation for consumers, or $29,500 for whole genome sequencing and analysis for researchers, depending on requirements.
- HelloGene and HelloGenome personal genome information services describe genotyping and full genome sequencing launched by Theragen in Korea. HelloGenome is the first commercial whole genome sequencing service in Asia while HelloGene is the first in Korea. HelloGene uses Affymetrix SNP chips while HelloGenome uses Solexa machines.
- Illumina, Oxford Nanopore Technologies, Sequenom, Pacific Biosciences, Complete Genomics and 454 Life Sciences are commercializing full genome sequencing but do not provide genetic analysis or counselling.

Ethical Issues

Genetic discrimination is discriminating on the basis of information obtained from an individual's genome. Genetic non-discrimination laws have been enacted in some US states and, at the federal level, by the Genetic Information Nondiscrimination Act (GINA). The GINA legislation prevents discrimination by health insurers and employers, but does not apply to life insurance or long-term care insurance.

Other Issues

Full sequencing of the genome can identify polymorphisms that are so rare that no conclusions may be drawn about their impact, creating uncertainty in the analysis of individual genomes, particularly in the context of clinical care. Czech medical geneticist Eva Machácková writes: "In some cases it is difficult to distinguish if the detected sequence variant is a causal mutation or a neutral (polymorphic) variation without any effect on phenotype. The interpretation of rare sequence variants of unknown significance detected in disease-causing genes becomes an increasingly important problem."

Human Genome

The human genome is the genome of Homo sapiens, which is stored on 23 chromosome pairs plus the small mitochondrial DNA. 22 of the 23 chromosomes are autosomal chromosome pairs, while the remaining pair is sex-determining. The haploid human genome occupies a total of just over 3 billion DNA base pairs. The Human Genome Project (HGP) produced a reference sequence of the euchromatic human genome, which is used worldwide in biomedical sciences.

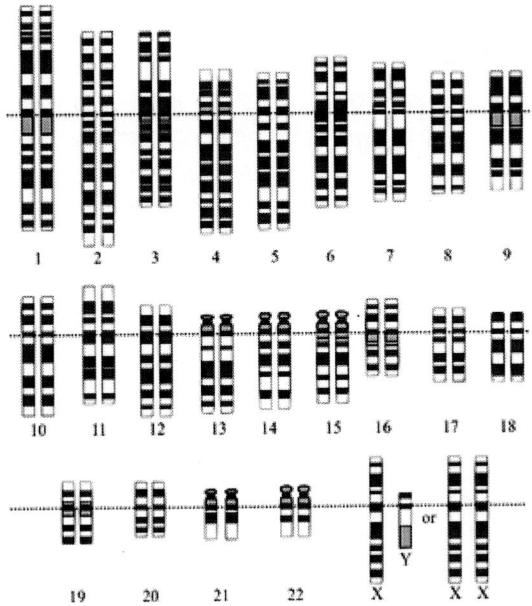

***Figure:** A graphical representation of the normal human karyotype*

The haploid human genome contains ca. 23,000 protein-coding genes, far fewer than had been expected before its sequencing. In fact, only about 1.5% of the genome codes for proteins, while the rest consists of non-coding RNA genes, regulatory sequences, introns, and noncoding DNA (once known as "junk DNA").

Features

Genes

There are estimated to be between 20,000 and 25,000 human protein-coding genes. The estimate of the number of human genes has been repeatedly revised down as genome sequence quality and gene finding methods have improved. Earlier predictions estimated that human cells had as many as 2,000,000 genes.

Surprisingly, the number of human genes seems to be less than a factor of two greater than that of many much simpler organisms, such as the roundworm and the fruit fly. However, a larger proportion of human genes are related to central nervous system and especially brain development.

Human genes are distributed unevenly across the chromosomes. Each chromosome contains various gene-rich and gene-poor regions, which seem to be correlated with chromosome bands and GC-content. The significance of these nonrandom patterns of gene density is not well understood. In addition to protein coding genes, the human genome contains thousands of RNA genes, including tRNA, ribosomal RNA, microRNA, and other non-coding RNA genes.

Regulatory Sequences

The human genome has many different regulatory sequences which are crucial to controlling gene expression. These are typically short sequences that appear near or within genes. A systematic understanding of these regulatory sequences and how they together act as a gene regulatory network is only beginning to emerge from computational, high-throughput expression and comparative genomics studies. Some types of non-coding DNA are genetic "switches" that do not encode proteins, but do regulate when and where genes are expressed. Identification of regulatory sequences relies in part on evolutionary conservation. The evolutionary branch between the primates and mouse, for example, occurred 70–90 million years ago. So computer comparisons of gene sequences that identify conserved non-coding sequences will be an indication of their importance in duties such as gene regulation. Another comparative genomic approach to locating regulatory sequences in humans is the gene sequencing of the puffer fish. These vertebrates have essentially the same genes and regulatory gene sequences as humans, but with only one-eighth the noncoding DNA. The compact DNA sequence of the puffer fish makes it much easier to locate the regulatory genes.

Other DNA

Protein-coding sequences (specifically, coding exons) comprise less than 1.5% of the human genome. Aside from genes and known regulatory sequences, the human genome contains vast regions of DNA the function of which, if any, remains unknown. These regions in fact comprise the vast majority, by some estimates 97%, of the human genome size. Much of this is composed of:

Repeat Elements

- Tandem repeats
  - Satellite DNA
  - Minisatellite
  - Microsatellite
- Interspersed repeats
  - SINEs
  - LINEs

Transposons

- Retrotransposons
  - o LTR
    - — Ty1-copia
    - — Ty3-gypsy
  - o Non-LTR
    - — SINEs
    - — LINEs
- DNA Transposons

Noncoding DNA

Many DNA sequences that do not code for gene expression have important biological functions as indicated by comparative genomics studies that report some sequences of noncoding DNA that are highly conserved, sometimes on time-scales representing hundreds of millions of years, implying that these noncoding regions are under strong evolutionary pressure and positive selection. These noncoding sequences were once referred to as "junk" DNA and there are many sequences that are likely to function, but in ways that are not fully understood.

Recent experiments using microarrays have revealed that a substantial fraction of non-genic DNA is in fact transcribed into RNA, which leads to the possibility that the resulting transcripts may have some unknown function. Also, the evolutionary conservation across the mammalian genomes of much more sequence than can be explained by protein-coding regions indicates that many, and perhaps most, functional elements in the genome remain unknown.

The investigation of the vast quantity of sequence information in the human genome whose function remains unknown is currently a major avenue of scientific inquiry. Meanwhile, considering the global genome DNA information as a whole could provide new ways to understand a possible global level function of non coding DNA.

Information content

The 2.9 billion base pairs of the haploid human genome correspond to a maximum of about 725 megabytes of data, since every base pair can be coded by 2 bits. Since individual genomes vary by less than 1% from each other, they can be losslessly compressed to roughly 4 megabytes.

The entropy rate of the genome differs significantly between coding and non-coding sequences. It is close to the maximum of 2 bits per base pair for the coding sequences (about 45 million base pairs), but less for the non-coding parts. It ranges between 1.5 and 1.9 bits per base pair for the individual chromosome, except for the Y-chromosome, which has an entropy rate below 0.9 bits per base pair.

Sequencing

DNA sequencing determines the order of the nucleotide bases in a genome.

Composite

The Human Genome Project and a parallel project by Celera Genomics each produced and published a haploid human genome sequence, both of which were a composite of the DNA sequence of several individuals.

Personal

A personal genome sequence is a complete sequencing of the chemical base pairs that make up the DNA of a single person. Because medical treatments have different effects on different people because of genetic variations such as single-nucleotide polymorphisms (SNPs), the analysis of personal genomes may lead to personalized medical treatment based on individual genotypes.

The completion of the fifth such map was announced in December 2008. The genome mapped was that of a Korean researcher Seong-Jin Kim. Genome maps had previously been completed for Craig Venter of the U.S. in 2007, James Watson of the U.S. in April 2008, and Yang Huanming of China in November 2008 and Dan Stoicescu in January 2008. Personal genomes had not been sequenced in the Human Genome Project to protect the identity of volunteers who provided DNA samples. That sequence was derived from the DNA of several volunteers from a diverse population. Another distinction is that the HGP sequence is haploid, however, the sequence maps for Venter and Watson for example are diploid, representing both sets of chromosomes.

Kim's genome had 1.58 million SNPs that had never been reported before and indicates that six out of 10,000 DNA bases are unique to Koreans. Kim's sequence map can be used to assist in building a standard Korean genome, which can then be used to compare the genomes of other Korean individuals for personalized medical treatments.

Mapping

Whereas a genome sequence lists the order of every DNA base in a genome, a genome map identifies the landmarks. A genome map is less detailed than a genome sequence and aids in navigating around the genome.

Variation

An example of a variation map is the HapMap being developed by the International HapMap Project. The HapMap is a haplotype map of the human genome, "which will describe the common patterns of human DNA sequence variation." It catalogs the patterns of small-scale variations in the genome that involve single DNA letters, or bases. Researchers published the first sequence-based map of large-scale structural variation across the human genome in the journal Nature in May 2008. Large-scale structural variations are differences in the genome among people that range from a few thousand to a few million DNA bases; some are gains or losses of stretches of genome sequence and others appear as re-arrangements of stretches of sequence. These variations include differences in the number of copies individuals have of a particular gene, deletions, translocations and inversions.

Variation

Most studies of human genetic variation have focused on single-nucleotide polymorphisms (SNPs), which are substitutions in individual bases along a chromosome. Most analyses estimate that SNPs occur 1 in 1000 base pairs, on average, in the euchromatic human genome, although they do not occur at a uniform density. Thus follows the popular statement that "we are all, regardless of race, genetically 99.9% the same", although this would be somewhat qualified by most geneticists. For example, a much larger fraction of the genome is now thought to be involved in copy number variation. A large-scale collaborative effort to catalog SNP variations in the human genome is being undertaken by the International HapMap Project.

The genomic loci and length of certain types of small repetitive sequences are highly variable from person to person, which is the basis of DNA fingerprinting and DNA paternity testing technologies. The heterochromatic portions of the human genome, which total several hundred million base pairs, are also thought to be quite variable within the human population (they are so repetitive and so long that they cannot be accurately sequenced with current technology). These regions contain few genes, and it is unclear whether any significant phenotypic effect results from typical variation in repeats or heterochromatin.

Most gross genomic mutations in gamete germ cells probably result in inviable embryos; however, a number of human diseases are related to large-scale genomic abnormalities. Down syndrome, Turner Syndrome, and a number of other diseases result from nondisjunction of entire chromosomes. Cancer cells frequently have aneuploidy of chromosomes and chromosome arms, although a cause and effect relationship between aneuploidy and cancer has not been established.

Genetic Disorders

Most aspects of human biology involve both genetic (inherited) and non-genetic (environmental) factors. Some inherited variation influences aspects of our biology that are not medical in nature (height, eye color, ability to taste or smell certain compounds, etc.). Moreover, some genetic disorders only cause disease in combination with the appropriate environmental factors (such as diet). With these caveats, genetic disorders may be described as clinically defined diseases caused by genomic DNA sequence variation.

In the most straightforward cases, the disorder can be associated with variation in a single gene. For example, cystic fibrosis is caused by mutations in the CFTR gene, and is the most common recessive disorder in caucasian populations with over 1,300 different mutations known. Disease-causing mutations in specific genes are usually severe in terms of gene function, and are fortunately rare, thus genetic disorders are similarly individually rare. However, since there are many genes that can vary to cause genetic disorders, in aggregate they comprise a significant component of known medical conditions, especially in pediatric medicine. Molecularly characterized genetic disorders are those for which the underlying causal gene has been identified, currently there are approximately 2,200 such disorders annotated in the OMIM database.

Studies of genetic disorders are often performed by means of family-based studies. In some instances population based approaches are employed, particularly in the case of so-called founder populations such as those in Finland, French-Canada, Utah, Sardinia, etc. Diagnosis and treatment of genetic disorders are usually performed by a geneticist-physician trained in clinical/medical genetics. The results of the Human Genome Project are likely to provide increased availability of genetic testing for gene-related disorders, and eventually improved treatment.

Parents can be screened for hereditary conditions and counselled on the consequences, the probability it will be inherited, and how to avoid or ameliorate it in their offspring. As noted above, there are many different kinds of DNA sequence variation, ranging from complete extra or missing chromosomes down to single nucleotide changes. It is generally presumed that much naturally occurring genetic variation in human populations is phenotypically neutral, i.e. has little or no detectable effect on the physiology of the individual (although there may be fractional differences in fitness defined over evolutionary time frames). Genetic disorders can be caused by any or all known types of sequence variation.

To molecularly characterize a new genetic disorder, it is necessary to establish a causal link between a particular genomic sequence variant and the clinical disease under investigation. Such studies constitute the realm of human molecular genetics. With the advent of the Human Genome and International HapMap Project, it has become feasible to explore subtle genetic influences on many common disease conditions such as diabetes, asthma, migraine, schizophrenia, etc.

Although some causal links have been made between genomic sequence variants in particular genes and some of these diseases, often with much publicity in the general media, these are usually not considered to be genetic disorders per se as their causes are complex, involving many different genetic and environmental factors. Thus there may be disagreement in particular cases whether a specific medical condition should be termed a genetic disorder.

Evolution

Comparative genomics studies of mammalian genomes suggest that approximately 5% of the human genome has been conserved by evolution since the divergence of extant lineages approximately 200 million years ago, containing the vast majority of genes. Intriguingly, since genes and known regulatory sequences probably comprise less than 2% of the genome, this suggests that there may be more unknown functional sequence than known functional sequence. A smaller, yet substantial, fraction of human genes seem to be shared among most known vertebrates.

The published chimpanzee genome differs from that of the human genome by 1.23% in direct sequence comparisons. Around 20% of this figure is accounted for by variation within each species, leaving only 1.06% consistent sequence divergence between humans and chimps at shared genes. This nucleotide by nucleotide difference is dwarfed, however, by the portion of each genome that is not shared, including around 6% of functional genes that are unique to either humans or chimps. In other words, the considerable observable differences between humans and chimps may be due as much or more to genome level variation in the number, function and expression of genes rather than DNA sequence changes in shared genes.

Indeed, even within humans, there has been found to be a previously unappreciated amount of copy number variation (CNV) which can make up as much as 5 - 15% of the human genome. In other words, between humans, there could be +/- 500,000,000 base pairs of DNA, some being active genes, others inactivated, or active at different levels. The full significance of this finding remains to be seen. On average, a typical human protein-coding gene differs from its chimpanzee ortholog by only two amino acid substitutions; nearly one third of human genes have exactly the same protein translation as their chimpanzee orthologs. A major difference between the two genomes is human chromosome 2, which is equivalent to a fusion product of chimpanzee chromosomes 12 and 13 (later renamed to chromosomes 2A and 2B, respectively).

Humans have undergone an extraordinary loss of olfactory receptor genes during our recent evolution, which explains our relatively crude sense of smell compared to most other mammals. Evolutionary evidence suggests that the emergence of color vision in humans and several other primate species has diminished the need for the sense of smell.

Mitochondrial Genome

The human mitochondrial genome, while usually not included when referring to the "human genome", is of tremendous interest to geneticists, since it undoubtedly plays a role in mitochondrial disease. It also sheds light on human evolution; for example, analysis of variation in the human mitochondrial genome has led to the postulation of a recent common ancestor for all humans on the maternal line of descent.

Due to the lack of a system for checking for copying errors, Mitochondrial DNA (mtDNA) has a more rapid rate of variation than nuclear DNA. This 20-fold increase in the mutation rate allows mtDNA to be used for more accurate tracing of maternal ancestry.

Studies of mtDNA in populations have allowed ancient migration paths to be traced, such as the migration of Native Americans from Siberia or Polynesians from southeastern Asia. It has also been used to show that there is no trace of Neanderthal DNA in the European gene mixture inherited through purely maternal lineage.

Epigenome

Epigenetics are a variety of features of the human genome that transcend its primary DNA sequence, such as chromatin packaging, histone modifications and DNA methylation, and which are important in regulating gene expression, genome replication and other cellular processes. Epigenetic markers strengthen and weaken transcription of certain genes but do not affect the actual sequence of DNA nucleotides.

Omics

The English-language neologism omics informally refers to a field of study in biology ending in -omics, such as genomics or proteomics. The related suffix -ome is used to address the objects of study of such fields, such as the genome or proteome, respectively. Functional genomics aims at identifying the functions of as many genes as possible of a given organism.

It combines different -omics techniques such as transcriptomics and proteomics with saturated mutant collections. The suffix -ome as used in molecular biology refers to a totality of some sort; it is an example of a "neo-suffix" formed by abstraction from various Greek terms in -ùìá, a sequence that does not form an identifiable suffix in Greek.

Origin

The Oxford English Dictionary (OED) distinguishes three different fields of application for the -ome suffix:

1. in medicine, forming nouns with the sense "swelling, tumour"

2. in botany or zoology, forming nouns in the sense "a part of an animal or plant with a specified structure"
3. in cellular and molecular biology, forming nouns with the sense "all constituents considered collectively".

The -ome suffix originated as a variant of -oma, and became productive in the last quarter of the 19th century. It was originally found in terms like sclerome or rhizome. All of these are terms derived from Greek words in -ùìá, a sequence that is not a single suffix, but analyzable as -ù-ìá, the -ù- belonging to the word stem, usually a verb, and the -ìá being a genuine Greek suffix forming abstract nouns.

OED suggests that the third definition originated as a back-formation from mitome, Early attestations include biome (1916) and genome (first coined as German Genom in 1920).

The association with chromosome in molecular biology is by false etymology. The word chromosome derives from the genuinely contains the preceding is not a stem-forming suffix but part of the word's root. Because genome refers to the complete genetic makeup of an organism, a neo-suffix -ome suggested itself as referring to "wholeness" or "completion".

Bioinformaticians and molecular biologists figured amongst the first scientists to start to apply the "-ome" suffix widely. Some early advocates were bioinformaticians in Cambridge, UK, where there were many early bioinformatics labs such as the MRC centre, Sanger centre, and EBI (European Bioinformatics Institute). For example, the MRC centre is where the first genome and proteome projects were carried out.

Unrelated Words in Omics

The word "comic" does not use the "omics" suffix; it derives from Greek (merriment) (an adjectival suffix), rather than presenting a truncation of similarly, the word "economy" is assembled from Greek (household) (law or custom), and "economic(s)" from suffix -omics is sometimes used to create portmanteau words to refer to schools of economics such as Reaganomics.

Current Usage

Many "omes" beyond the original "genome" have become useful and have been widely adopted by research scientists. "Proteomics" has become well-established as a term for studying proteins at a large scale. "Omes" can provide an easy short-hand to encapsulate a field; for example, an interactomics study is clearly recognisable as relating to large-scale analyses of gene-gene, protein-protein, or protein-ligand interactions. Researchers are rapidly taking up omes and omics, as shown by the explosion of the use of these terms in PubMed since the mid '90s.

Proteomics

Proteomics is the large-scale study of proteins, particularly their structures and functions. Proteins are vital parts of living organisms, as they are the main components of the physiological metabolic pathways of cells. The term "proteomics" was first coined in 1997 to make an analogy with genomics, the study of the genes. The word "proteome" is a blend of "protein" and "genome", and was coined by Marc Wilkins in 1994 while working on the concept as a PhD student. The proteome is the entire complement of proteins, including the modifications made to a particular set of proteins, produced by an organism or system. This will vary with time and distinct requirements, or stresses, that a cell or organism undergoes.

Complexity of the Problem

After genomics, proteomics is considered the next step in the study of biological systems. It is much more complicated than genomics mostly because while an organism's genome is more or less constant, the proteome differs from cell to cell and from time to time. This is because distinct genes are expressed in distinct cell types. This means that even the basic set of proteins which are produced in a cell needs to be determined.

In the past this was done by mRNA analysis, but this was found not to correlate with protein content. It is now known that mRNA is not always translated into protein, and the amount of protein produced for a given amount of mRNA depends on the gene it is transcribed from and on the current physiological state of the cell. Proteomics confirms the presence of the protein and provides a direct measure of the quantity present.

Post-translational Modifications

Not only does the translation from mRNA cause differences, many proteins are also subjected to a wide variety of chemical modifications after translation. Many of these post-translational modifications are critical to the protein's function.

Phosphorylation

One such modification is phosphorylation, which happens to many enzymes and structural proteins in the process of cell signaling. The addition of a phosphate to particular amino acids—most commonly serine and threonine mediated by serine/threonine kinases, or more rarely tyrosine mediated by tyrosine kinases—causes a protein to become a target for binding or interacting with a distinct set of other proteins that recognize the phosphorylated domain.

Because protein phosphorylation is one of the most-studied protein modifications, many "proteomic" efforts are geared to determining the set of phosphorylated proteins in a particular cell or tissue-type under particular circumstances. This alerts the scientist to the signaling pathways that may be active in that instance.

Ubiquitination

Ubiquitin is a small protein that can be affixed to certain protein substrates by enzymes called E3 ubiquitin ligases. Determining which proteins are poly-ubiquitinated can be helpful in understanding how protein pathways are regulated. This is therefore an additional legitimate "proteomic" study. Similarly, once it is determined which substrates are ubiquitinated by each ligase, determining the set of ligases expressed in a particular cell type will be helpful.

Additional Modifications

Listing all the protein modifications that might be studied in a "Proteomics" project would require a discussion of most of biochemistry; therefore, a short list will serve here to illustrate the complexity of the problem. In addition to phosphorylation and ubiquitination, proteins can be subjected to (among others) methylation, acetylation, glycosylation, oxidation and nitrosylation. Some proteins undergo ALL of these modifications, often in time-dependent combinations, aptly illustrating the potential complexity one has to deal with when studying protein structure and function.

Distinct Proteins are Made Under Distinct Settings

Even if one is studying a particular cell type, that cell may make different sets of proteins at different times, or under different conditions. Furthermore, as mentioned, any one protein can undergo a wide range of post-translational modifications.

Therefore a "proteomics" study can become quite complex very quickly, even if the object of the study is very restricted. In more ambitious settings, such as when a biomarker for a tumor is sought – when the proteomics scientist is obliged to study sera samples from multiple cancer patients – the amount of complexity that must be dealt with is as great as in any modern biological project.

Limitations to Genomic Study

Scientists are very interested in proteomics because it gives a much better understanding of an organism than genomics. First, the level of transcription of a gene gives only a rough estimate of its level of expression into a protein. An mRNA produced in abundance may be degraded rapidly or translated inefficiently, resulting in a small amount of protein. Second, as mentioned above many proteins experience post-translational modifications that profoundly affect their activities; for example some proteins are not active until they become phosphorylated. Methods such as phosphoproteomics and glycoproteomics are used to study post-translational modifications. Third, many transcripts give rise to more than one protein, through alternative splicing or alternative post-translational modifications. Fourth, many proteins form complexes with other proteins or RNA molecules, and only function in the presence of these other molecules. Finally, protein degradation rate plays an important role in protein content.

Methods of Studying Proteins

One way in which a particular protein can be studied is to develop an antibody which is specific to that modification. For example, there are antibodies which only recognize certain proteins when they are tyrosine-phosphorylated, known as phospho-specific antibodies; also, there are antibodies specific to other modifications. These can be used to determine the set of proteins that have undergone the modification of interest.

For sugar modifications, such as glycosylation of proteins, certain lectins have been discovered which bind sugars. These too can be used.

A more common way to determine post-translational modification of interest is to subject a complex mixture of proteins to electrophoresis in "two-dimensions", which simply means that the proteins are electrophoresed first in one direction, and then in another, which allows small differences in a protein to be visualized by separating a modified protein from its unmodified form. This methodology is known as "two-dimensional gel electrophoresis".

Recently, another approach has been developed called PROTOMAP which combines SDS-PAGE with shotgun proteomics to enable detection of changes in gel-migration such as those caused by proteolysis or post translational modification. Determining the existence of proteins in complex mixtures Classically, antibodies to particular proteins or to their modified forms have been used in biochemistry and cell biology studies. These are among the most common tools used by practicing biologists today.

For more quantitative determinations of protein amounts, techniques such as ELISAs can be used. For proteomic study, more recent techniques such as matrix-assisted laser desorption/ionization (MALDI) have been employed for rapid determination of proteins in particular mixtures and increasingly electrospray ionization (ESI).

Computational Methods in Studying Protein Biomarkers

Computational predictive models have shown that extensive and diverse feto-maternal protein trafficking occurs during pregnancy and can be readily detected non-invasively in maternal whole blood. This computational approach circumvented a major limitation, the abundance of maternal proteins interfering with the detection of fetal proteins, to fetal proteomic analysis of maternal blood. Computational models can use fetal gene transcripts previously identified in maternal whole blood to create a comprehensive proteomic network of the term neonate. Such work shows that the fetal proteins detected in pregnant woman's blood originate from a diverse group of tissues and organs from the developing fetus. The proteomic networks contain many biomarkers that are proxies for development and illustrate the potential clinical application of this technology as a way to monitor normal and abnormal fetal development.

An information theoretic framework has also been introduced for biomarker discovery, integrating biofluid and tissue information. This new approach takes advantage of functional synergy between certain biofluids and tissues with the potential for clinically significant findings not possible if tissues and biofluids were considered individually. By conceptualizing tissue-biofluid as information channels, significant biofluid proxies can be identified and then used for guided development of clinical diagnostics. Candidate biomarkers are then predicted based on information transfer criteria across the tissue-biofluid channels. Significant biofluid-tissue relationships can be used to prioritize clinical validation of biomarkers.

Establishing Protein–protein Interactions

Most proteins function in collaboration with other proteins, and one goal of proteomics is to identify which proteins interact. This is especially useful in determining potential partners in cell signaling cascades. Several methods are available to probe protein–protein interactions. The traditional method is yeast two-hybrid analysis. New methods include protein microarrays, immunoaffinity chromatography followed by mass spectrometry, dual polarisation interferometry, Microscale Thermophoresis and experimental methods such as phage display and computational methods

Practical Applications of Proteomics

One of the most promising developments to come from the study of human genes and proteins has been the identification of potential new drugs for the treatment of disease. This relies on genome and proteome information to identify proteins associated with a disease, which computer software can then use as targets for new drugs. For example, if a certain protein is implicated in a disease, its 3D structure provides the information to design drugs to interfere with the action of the protein. A molecule that fits the active site of an enzyme, but cannot be released by the enzyme, will inactivate the enzyme. This is the basis of new drug-discovery tools, which aim to find new drugs to inactivate proteins involved in disease. As genetic differences among individuals are found, researchers expect to use these techniques to develop personalized drugs that are more effective for the individual.

Biomarkers

The FDA defines a biomarker as, "A characteristic that is objectively measured and evaluated as an indicator of normal biologic processes, pathogenic processes, or pharmacologic responses to a therapeutic intervention".

Understanding the proteome, the structure and function of each protein and the complexities of protein–protein interactions will be critical for developing the most effective diagnostic techniques and disease treatments in the future.

An interesting use of proteomics is using specific protein biomarkers to diagnose disease. A number of techniques allow to test for proteins produced during a particular disease, which helps to diagnose the disease quickly. Techniques include western blot, immunohistochemical staining, enzyme linked immunosorbent assay (ELISA) or mass spectrometry.

Proteogenomics

In what is now commonly referred to as proteogenomics, proteomic technologies such as mass spectrometry are used for improving gene annotations. Parallel analysis of the genome and the proteome facilitates discovery of post-translational modifications and proteolytic events , especially when comparing multiple species (comparative proteogenomics).

Current Research Methodologies

Fluorescence two-dimensional differential gel electrophoresis (2-D DIGE) can be used to quantify variation in the 2-D DIGE process and establish statistically valid thresholds for assigning quantitative changes between samples. Comparative proteomic analysis can reveal the role of proteins in complex biological systems, including reproduction. For example, treatment with the insecticide triazophos causes an increase in the content of brown planthopper (Nilaparvata lugens (Stål)) male accessory gland proteins (Acps) that can be transferred to females via mating, causing an increase in fecundity (i.e. birth rate) of females.

To identify changes in the types of accessory gland proteins (Acps) and reproductive proteins that mated female planthoppers received from male planthoppers, researchers conducted a comparative proteomic analysis of mated N. lugens females. The results indicated that these proteins participate in the reproductive process of N. lugens adult females and males. Proteome analysis of Arabidopsis peroxisomes has been established as the major unbiased approach for identifying new peroxisomal proteins on a large scale.

There are many approaches to characterizing the human proteome, which is estimated to contain between 20,000 and 25,000 non-redundant proteins. The number of unique protein species likely increase by between 50,000 and

500,000 due to RNA splicing and proteolysis events, and when post-translational modification are also considered, the total number of unique human proteins is estimated to range in the low millions.

In addition, first promising attempts to decipher the proteome of animal tumors have recently been reported. Bioinformatics

Bioinformatics /Ìbaj.oŠÌjnfYrÈmætjks/ is the application of computer science and information technology to the field of biology and medicine. Bioinformatics deals with algorithms, databases and information systems, web technologies, artificial intelligence and soft computing, information and computation theory, software engineering, data mining, image processing, modeling and simulation, signal processing, discrete mathematics, control and system theory, circuit theory, and statistics, for generating new knowledge of biology and medicine, and improving & discovering new models of computation (e.g.

DNA computing, neural computing, evolutionary computing, immuno-computing, swarm-computing, cellular-computing). Java, XML, Perl, C, C++, Python, R, SQL, CUDA and MATLAB are the some of the more prominent software technologies used in this field.

Introduction

Bioinformatics was applied in the creation and maintenance of a database to store biological information at the beginning of the "genomic revolution", such as nucleotide and amino acid sequences. Development of this type of database involved not only design issues but the development of complex interfaces whereby researchers could both access existing data as well as submit new or revised data.

In order to study how normal cellular activities are altered in different disease states, the biological data must be combined to form a comprehensive picture of these activities. Therefore, the field of bioinformatics has evolved such that the most pressing task now involves the analysis and interpretation of various types of data, including nucleotide and amino acid sequences, protein domains, and protein structures. The actual process of analyzing and interpreting data is referred to as computational biology. Important sub-disciplines within bioinformatics and computational biology include:

- the development and implementation of tools that enable efficient access to, and use and management of, various types of information.
- the development of new algorithms (mathematical formulas) and statistics with which to assess relationships among members of large data sets, such as methods to locate a gene within a sequence, predict protein structure and/or function, and cluster protein sequences into families of related sequences.

The primary goal of bioinformatics is to increase the understanding of biological processes. What sets it apart from other approaches, however, is its focus on developing and applying computationally intensive techniques (e.g., pattern recognition, data mining, machine learning algorithms, and visualization) to achieve this goal. Major research efforts in the field include sequence alignment, gene finding, genome assembly, drug design, drug discovery, protein structure alignment, protein structure prediction, prediction of gene expression and protein–protein interactions, genome-wide association studies and the modeling of evolution.

Interestingly, the term bioinformatics was coined before the "genomic revolution". Paulien Hogeweg and Ben Hesper introduced the term in 1978 to refer to "the study of information processes in biotic systems"., a definition that placed bioinformatics as a field parallel to biophysics or biochemistry (e.g. biochemistry, the study of chemical processes in biological systems) . However, its primary use since at least the late 1980s has been to describe the application of computer science and information sciences to the analysis of biological data, particularly in those areas of genomics involving large-scale DNA sequencing.

Bioinformatics now entails the creation and advancement of databases, algorithms, computational and statistical techniques and theory to solve formal and practical problems arising from the management and analysis of biological data. Over the past few decades rapid developments in genomic and other molecular research technologies and developments in information technologies have combined to produce a tremendous amount of information related to molecular biology. It is the name given to these mathematical and computing approaches used to glean understanding of biological processes.

Common activities in bioinformatics include mapping and analyzing DNA and protein sequences, aligning different DNA and protein sequences to compare them and creating and viewing 3-D models of protein structures.

There are two fundamental ways of modelling a Biological system (e.g. living cell) both coming under Bioinformatic approaches.

- Static
  - o Sequences – Proteins, Nucleic acids and Peptides
  - o Structures – Proteins, Nucleic acids, Ligands (including metabolites and drugs) and Peptides
  - o Interaction data among the above entities including microarray data and Networks of proteins, metabolites
- Dynamic
  - o Systems Biology comes under this category including reaction fluxes and variable concentrations of metabolites

- o Multi-Agent Based modelling approaches capturing cellular events such as signalling, transcription and reaction dynamics

A broad sub-category under bioinformatics is structural bioinformatics.

Major Research Areas

Sequence Analysis

Since the Phage Φ-X174 was sequenced in 1977, the DNA sequences of thousands of organisms have been decoded and stored in databases. This sequence information is analyzed to determine genes that encode polypeptides (proteins), RNA genes, regulatory sequences, structural motifs, and repetitive sequences. A comparison of genes within a species or between different species can show similarities between protein functions, or relations between species (the use of molecular systematics to construct phylogenetic trees). With the growing amount of data, it long ago became impractical to analyze DNA sequences manually. Today, computer programs such as BLAST are used daily to search sequences from more than 260 000 organisms, containing over 190 billion nucleotides.

These programs can compensate for mutations (exchanged, deleted or inserted bases) in the DNA sequence, to identify sequences that are related, but not identical. A variant of this sequence alignment is used in the sequencing process itself. The so-called shotgun sequencing technique (which was used, for example, by The Institute for Genomic Research to sequence the first bacterial genome, Haemophilus influenzae) does not produce entire chromosomes, but instead generates the sequences of many thousands of small DNA fragments (ranging from 35 to 900 nucleotides long, depending on the sequencing technology).

The ends of these fragments overlap and, when aligned properly by a genome assembly program, can be used to reconstruct the complete genome. Shotgun sequencing yields sequence data quickly, but the task of assembling the fragments can be quite complicated for larger genomes. For a genome as large as the human genome, it may take many days of CPU time on large-memory, multiprocessor computers to assemble the fragments, and the resulting assembly will usually contain numerous gaps that have to be filled in later. Shotgun sequencing is the method of choice for virtually all genomes sequenced today, and genome assembly algorithms are a critical area of bioinformatics research. Another aspect of bioinformatics in sequence analysis is annotation, which involves computational gene finding to search for protein-coding genes, RNA genes, and other functional sequences within a genome. Not all of the nucleotides within a genome are part of genes. Within the genome of higher organisms, large parts of the DNA do not serve any obvious purpose.

This so-called junk DNA may, however, contain unrecognized functional elements. Bioinformatics helps to bridge the gap between genome and proteome projects — for example, in the use of DNA sequences for protein identification.

Genome Annotation

In the context of genomics, annotation is the process of marking the genes and other biological features in a DNA sequence. The first genome annotation software system was designed in 1995 by Dr. Owen White, who was part of the team at The Institute for Genomic Research that sequenced and analyzed the first genome of a free-living organism to be decoded, the bacterium Haemophilus influenzae. Dr. White built a software system to find the genes (places in the DNA sequence that encode a protein), the transfer RNA, and other features, and to make initial assignments of function to those genes. Most current genome annotation systems work similarly, but the programs available for analysis of genomic DNA are constantly changing and improving.

Computational Evolutionary Biology

Evolutionary biology is the study of the origin and descent of species, as well as their change over time. Informatics has assisted evolutionary biologists in several key ways; it has enabled researchers to:

- trace the evolution of a large number of organisms by measuring changes in their DNA, rather than through physical taxonomy or physiological observations alone,
- more recently, compare entire genomes, which permits the study of more complex evolutionary events, such as gene duplication, horizontal gene transfer, and the prediction of factors important in bacterial speciation,
- build complex computational models of populations to predict the outcome of the system over time
- track and share information on an increasingly large number of species and organisms

Future work endeavours to reconstruct the now more complex tree of life.

The area of research within computer science that uses genetic algorithms is sometimes confused with computational evolutionary biology, but the two areas are not necessarily related.

Literature Analysis

The growth in the number of published literature makes it virtually impossible to read every paper, resulting in disjointed subfields of research. Literature analysis aims to employ computational and statistical linguistics to mine this growing library of text resources. For example:

- abbreviation recognition - identify the long-form and abbreviation of biological terms,
- named entity recognition - recognizing biological terms such as gene names
- protein-protein interaction - identify which proteins interact with which proteins from text

The area of research uses statistics and computational linguistics, and is substantially influenced by them.

Analysis of Gene Expression

The expression of many genes can be determined by measuring mRNA levels with multiple techniques including microarrays, expressed cDNA sequence tag (EST) sequencing, serial analysis of gene expression (SAGE) tag sequencing, massively parallel signature sequencing (MPSS), RNA-Seq, also known as "Whole Transcriptome Shotgun Sequencing" (WTSS), or various applications of multiplexed in-situ hybridization.

All of these techniques are extremely noise-prone and/or subject to bias in the biological measurement, and a major research area in computational biology involves developing statistical tools to separate signal from noise in high-throughput gene expression studies. Such studies are often used to determine the genes implicated in a disorder: one might compare microarray data from cancerous epithelial cells to data from non-cancerous cells to determine the transcripts that are up-regulated and down-regulated in a particular population of cancer cells.

Analysis of Regulation

Regulation is the complex orchestration of events starting with an extracellular signal such as a hormone and leading to an increase or decrease in the activity of one or more proteins. Bioinformatics techniques have been applied to explore various steps in this process. For example, promoter analysis involves the identification and study of sequence motifs in the DNA surrounding the coding region of a gene.

These motifs influence the extent to which that region is transcribed into mRNA. Expression data can be used to infer gene regulation: one might compare microarray data from a wide variety of states of an organism to form hypotheses about the genes involved in each state. In a single-cell organism, one might compare stages of the cell cycle, along with various stress conditions (heat shock, starvation, etc.). One can then apply clustering algorithms to that expression data to determine which genes are co-expressed. For example, the upstream regions (promoters) of co-expressed genes can be searched for over-represented regulatory elements.

Analysis of Orotein Expression

Protein microarrays and high throughput (HT) mass spectrometry (MS) can provide a snapshot of the proteins present in a biological sample. Bioinformatics is very much involved in making sense of protein microarray and HT MS data; the former approach faces similar problems as with microarrays targeted at mRNA, the latter involves the problem of matching large amounts of mass data against predicted masses from protein sequence databases, and the complicated statistical analysis of samples where multiple, but incomplete peptides from each protein are detected.

Analysis of Mutations in Cancer

In cancer, the genomes of affected cells are rearranged in complex or even unpredictable ways. Massive sequencing efforts are used to identify previously unknown point mutations in a variety of genes in cancer.

Bioinformaticians continue to produce specialized automated systems to manage the sheer volume of sequence data produced, and they create new algorithms and software to compare the sequencing results to the growing collection of human genome sequences and germline polymorphisms.

New physical detection technologies are employed, such as oligonucleotide microarrays to identify chromosomal gains and losses (called comparative genomic hybridization), and single-nucleotide polymorphism arrays to detect known point mutations. These detection methods simultaneously measure several hundred thousand sites throughout the genome, and when used in high-throughput to measure thousands of samples, generate terabytes of data per experiment. Again the massive amounts and new types of data generate new opportunities for bioinformaticians. The data is often found to contain considerable variability, or noise, and thus Hidden Markov model and change-point analysis methods are being developed to infer real copy number changes.

Another type of data that requires novel informatics development is the analysis of lesions found to be recurrent among many tumors.

Comparative Genomics

The core of comparative genome analysis is the establishment of the correspondence between genes (orthology analysis) or other genomic features in different organisms. It is these intergenomic maps that make it possible to trace the evolutionary processes responsible for the divergence of two genomes. A multitude of evolutionary events acting at various organizational levels shape genome evolution. At the lowest level, point mutations affect individual nucleotides. At a higher level, large chromosomal segments undergo duplication, lateral transfer, inversion, transposition, deletion and insertion. Ultimately, whole genomes are involved in processes of hybridization, polyploidization and endosymbiosis, often leading to rapid speciation. The complexity of genome evolution poses many exciting challenges to developers of mathematical models and algorithms, who have recourse to a spectra of algorithmic, statistical and mathematical techniques, ranging from exact, heuristics, fixed parameter and approximation algorithms for problems based on parsimony models to Markov Chain Monte Carlo algorithms for Bayesian analysis of problems based on probabilistic models.

Modeling Biological Systems

Systems biology involves the use of computer simulations of cellular subsystems (such as the networks of metabolites and enzymes which comprise metabolism, signal transduction pathways and gene regulatory networks) to both analyze and visualize the complex connections of these cellular processes. Artificial life or virtual evolution attempts to understand evolutionary processes via the computer simulation of simple (artificial) life forms.

High-throughput Image Analysis

Computational technologies are used to accelerate or fully automate the processing, quantification and analysis of large amounts of high-information-content biomedical imagery. Modern image analysis systems augment an observer's ability to make measurements from a large or complex set of images, by improving accuracy, objectivity, or speed.

A fully developed analysis system may completely replace the observer. Although these systems are not unique to biomedical imagery, biomedical imaging is becoming more important for both diagnostics and research. Some examples are:

- high-throughput and high-fidelity quantification and sub-cellular localization (high-content screening, cytohistopathology, Bioimage informatics)
- morphometrics
- clinical image analysis and visualization
- determining the real-time air-flow patterns in breathing lungs of living animals
- quantifying occlusion size in real-time imagery from the development of and recovery during arterial injury
- making behavioral observations from extended video recordings of laboratory animals
- infrared measurements for metabolic activity determination
- inferring clone overlaps in DNA mapping, e.g. the Sulston score.

5

Genetic Testing

Genetic testing (also called DNA-based tests) is among the newest and most sophisticated of techniques used to test for genetic disorders which involves direct examination of the DNA molecule itself. Other genetic tests include biochemical tests for such gene products as enzymes and other proteins and for microscopic examination of stained or fluorescent chromosomes. Genetic tests are used for several reasons, including:

- carrier screening, which involves identifying unaffected individuals who carry one copy of a gene for a disease that requires two copies for the disease to be expressed
- preimplantation genetic diagnosis.
- prenatal diagnostic testing
- newborn screening
- Genealogical DNA test (for genetic genealogy purposes)
- presymptomatic testing for predicting adult-onset disorders such as Huntington's disease
- presymptomatic testing for estimating the risk of developing adult-onset cancers and Alzheimer's disease
- confirmational diagnosis of a symptomatic individual
- forensic/identity testing

Genetic testing allows the genetic diagnosis of vulnerabilities to inherited diseases, and can also be used to determine a child's paternity (genetic father) or a person's ancestry. Normally, every person carries two copies of every gene (with the exception of genes related to sex-linked traits, which are only inherited from the mother by males), one inherited from their mother, one inherited from their father.

The human genome is believed to contain around 20,000 - 25,000 genes. In addition to studying chromosomes to the level of individual genes, genetic testing in a broader sense includes biochemical tests for the possible presence of genetic diseases, or mutant forms of genes associated with increased risk of developing genetic disorders. Genetic testing identifies changes in chromosomes, genes, or proteins. Most of the time, testing is used to find changes that are associated with inherited disorders. The results of a genetic test can confirm or rule out a suspected genetic condition or help determine a person's chance of developing or passing on a genetic disorder. Several hundred genetic tests are currently in use, and more are being developed.

Since genetic testing may open up ethical or psychological problems, genetic testing is often accompanied by genetic counseling.

Types

Genetic testing is "the analysis of, chromosomes (DNA), proteins, and certain metabolites in order to detect heritable disease-related genotypes, mutations, phenotypes, or karyotypes for clinical purposes." It can provide information about a person's genes and chromosomes throughout life. Available types of testing include:

- *Newborn screening:* Newborn screening is used just after birth to identify genetic disorders that can be treated early in life. The routine testing of infants for certain disorders is the most widespread use of genetic testing—millions of babies are tested each year in the United States. All states currently test infants for phenylketonuria (a genetic disorder that causes mental illness if left untreated) and congenital hypothyroidism (a disorder of the thyroid gland).
- *Diagnostic testing:* Diagnostic testing is used to diagnose or rule out a specific genetic or chromosomal condition. In many cases, genetic testing is used to confirm a diagnosis when a particular condition is suspected based on physical mutations and symptoms. Diagnostic testing can be performed at any time during a person's life, but is not available for all genes or all genetic conditions. The results of a diagnostic test can influence a person's choices about health care and the management of the disease.
- *Carrier testing:* Carrier testing is used to identify people who carry one copy of a gene mutation that, when present in two copies, causes a genetic disorder. This type of testing is offered to individuals who have

a family history of a genetic disorder and to people in ethnic groups with an increased risk of specific genetic conditions. If both parents are tested, the test can provide information about a couple's risk of having a child with a genetic condition.

- *Prenatal testing:* Prenatal testing is used to detect changes in a fetus's genes or chromosomes before birth. This type of testing is offered to couples with an increased risk of having a baby with a genetic or chromosomal disorder. In some cases, prenatal testing can lessen a couple's uncertainty or help them decide whether to abort the pregnancy. It cannot identify all possible inherited disorders and birth defects, however.
- *Preimplantation genetic diagnosis:* Genetic testing procedures that are performed on human embryos prior to the implantation as part of an in vitro fertilization procedure.
- *Predictive and presymptomatic testing:* Predictive and presymptomatic types of testing are used to detect gene mutations associated with disorders that appear after birth, often later in life. These tests can be helpful to people who have a family member with a genetic disorder, but who have no features of the disorder themselves at the time of testing. Predictive testing can identify mutations that increase a person's chances of developing disorders with a genetic basis, such as certain types of cancer. For example, an individual with a mutation in BRCA1 has a 65% cumulative risk of breast cancer. Presymptomatic testing can determine whether a person will develop a genetic disorder, such as hemochromatosis (an iron overload disorder), before any signs or symptoms appear. The results of predictive and presymptomatic testing can provide information about a person's risk of developing a specific disorder and help with making decisions about medical care.
- *Forensic testing:* Forensic testing uses DNA sequences to identify an individual for legal purposes. Unlike the tests described above, forensic testing is not used to detect gene mutations associated with disease. This type of testing can identify crime or catastrophe victims, rule out or implicate a crime suspect, or establish biological relationships between people (for example, paternity).
- *Parental testing:* This type of genetic test uses special DNA markers to identify the same or similar inheritance patterns between related individuals. Based on the fact that we all inherit half of our DNA from the father, and half from the mother, DNA scientists test individuals to find the match of DNA sequences at some highly differential markers to draw the conclusion of relatedness.

- *Research testing:* Research testing includes finding unknown genes, learning how genes work and advancing our understanding of genetic conditions. The results of testing done as part of a research study are usually not available to patients or their healthcare providers.
- *Pharmacogenomics:* type of genetic testing that determines the influence of genetic variation on drug response.

Medical Procedure

Genetic testing is often done as part of a genetic consultation and as of mid-2008 there were more than 1,200 clinically applicable genetic tests available. Once a person decides to proceed with genetic testing, a medical geneticist, genetic counselor, primary care doctor, or specialist can order the test after obtaining informed consent.

Genetic tests are performed on a sample of blood, hair, skin, amniotic fluid (the fluid that surrounds a fetus during pregnancy), or other tissue. For example, a medical procedure called a buccal smear uses a small brush or cotton swab to collect a sample of cells from the inside surface of the cheek. Alternatively, a small amount of saline mouthwash may be swished in the mouth to collect the cells. The sample is sent to a laboratory where technicians look for specific changes in chromosomes, DNA, or proteins, depending on the suspected disorder. The laboratory reports the test results in writing to a person's doctor or genetic counselor.

Routine newborn screening tests are done on a small blood sample obtained by pricking the baby's heel with a lancet.

Interpreting Results

The results of genetic tests are not always straightforward, which often makes them challenging to interpret and explain. When interpreting test results, healthcare professionals consider a person's medical history, family history, and the type of genetic test that was done.

A positive test result means that the laboratory found a change in a particular gene, chromosome, or protein of interest. Depending on the purpose of the test, this result may confirm a diagnosis, indicate that a person is a carrier of a particular genetic mutation, identify an increased risk of developing a disease (such as cancer) in the future, or suggest a need for further testing. Because family members have some genetic material in common, a positive test result may also have implications for certain blood relatives of the person undergoing testing. It is important to note that a positive result of a predictive or presymptomatic genetic test usually cannot establish the exact risk of developing a disorder. Also, health professionals typically cannot use a positive test result to predict the course or severity of a condition.

A negative test result means that the laboratory did not find a dangerous copy of the gene, chromosome, or protein under consideration. This result can indicate that a person is not affected by a particular disorder, is not a carrier of a specific genetic mutation, or does not have an increased risk of developing a certain disease. It is possible, however, that the test missed a disease-causing genetic alteration because many tests cannot detect all genetic changes that can cause a particular disorder. Further testing may be required to confirm a negative result.

In some cases, a negative result might not give any useful information. This type of result is called uninformative, indeterminate, inconclusive, or ambiguous. Uninformative test results sometimes occur because everyone has common, natural variations in their DNA, called polymorphisms, that do not affect health. If a genetic test finds a change in DNA that has not been associated with a disorder in other people, it can be difficult to tell whether it is a natural polymorphism or a disease-causing mutation. An uninformative result cannot confirm or rule out a specific diagnosis, and it cannot indicate whether a person has an increased risk of developing a disorder. In some cases, testing other affected and unaffected family members can help clarify this type of result.

Risks and Limitations

The physical risks associated with most genetic tests are very small, particularly for those tests that require only a blood sample or buccal smear (a procedure that samples cells from the inside surface of the cheek). The procedures used for prenatal testing carry a small but real risk of losing the pregnancy (miscarriage) because they require a sample of amniotic fluid or tissue from around the fetus. Many of the risks associated with genetic testing involve the emotional, social, or financial consequences of the test results. People may feel angry, depressed, anxious, or guilty about their results. In some cases, genetic testing creates tension within a family because the results can reveal information about other family members in addition to the person who is tested. The possibility of genetic discrimination in employment or insurance is also a concern. Some individuals avoid genetic testing out of fear it will affect their ability to purchase insurance or find a job.

Health insurers do not currently require applicants for coverage to undergo genetic testing, and when insurers encounter genetic information, it is subject to the same confidentiality protections as any other sensitive health information. In the United States, the use of genetic information is governed by the Genetic Information Nondiscrimination Act (GINA).

Genetic testing can provide only limited information about an inherited condition. The test often can't determine if a person will show symptoms of a disorder, how severe the symptoms will be, or whether the disorder will progress over time. Another major limitation is the lack of treatment strategies for many genetic disorders once they are diagnosed. A genetics professional can explain in detail the benefits, risks, and limitations of a particular test. It is important that any person who is considering genetic testing understand and weigh these factors before making a decision.

Direct-to-Consumer Genetic Testing

Direct-to-Consumer (DTC) genetic testing is a type of genetic test that is accessible directly to the consumer without having to go through a health care professional. Usually, to obtain a genetic test, health care professionals such as doctors acquire the permission of the patient and order the desired test.

DTC genetic tests, however, allow consumers to bypass this process and order one themselves. There are a variety of DTC tests, ranging from testing for breast cancer alleles to mutations linked to cystic fibrosis. Benefits of DTC testing are the accessibility of tests to consumers, promotion of proactive healthcare and the privacy of genetic information. Possible additional risks of DTC testing are the lack of governmental regulation and the potential misinterpretation of genetic information.

Controversy

DTC genetic testing has been controversial due to outspoken opposition within the scientific community. Critics of DTC testing argue against the risks involved, the unregulated advertising and marketing claims, and the overall lack of governmental oversight.

DTC testing involves many of the same risks associated with any genetic test. One of the more obvious and dangerous of these is the possibility of severe misreading of test results. Without professional guidance, consumers can potentially misinterpret genetic information, causing them to be deluded about their personal health.

Some advertising for direct-to-consumer genetic testing has been criticized as conveying an exaggerated and inaccurate message about the connection between genetic information and disease risk, utilizing emotions as a selling factor. An advertisement for a BRCA-predictive genetic test for breast cancer stated: "There is no stronger antidote for fear than information."

Government Regulation in the United States

Currently, the U.S. has no strong Federal regulation moderating the DTC market. Though there are several hundred tests available, only a handful are approved by the Food and Drug Administration (FDA); these are sold as at-home test kits, and are therefore considered "medical devices" over which the FDA may assert jurisdiction. Other types of DTC tests require customers to mail in DNA samples for testing; it is difficult for the FDA to exercise jurisdiction over these types of tests, because the actual testing is completed in the laboratories of providers. As of 2007, the FDA had not yet officially substantiated with scientific evidence the claimed accuracy of the majority of direct-to-consumer genetic tests.

With regard to genetic testing and information in general, legislation in the United States called the Genetic Information Nondiscrimination Act prohibits group health plans and health insurers from denying coverage to a healthy individual or charging that person higher premiums based solely on a genetic predisposition to developing a disease in the future. The legislation also bars employers from using individuals' genetic information when making hiring, firing, job placement, or promotion decisions. The legislation, the first of its kind in the U.S., was passed by the United States Senate on April 24, 2008, on a vote of 95-0, and was signed into law by President George W. Bush on May 21, 2008. It went into effect on November 21, 2009.

In Popular Culture

Some possible future ethical problems of genetic testing were considered in the science fiction film Gattaca, the novel Next, and the science fiction anime series "Gundam Seed". Also some films which include the topic of genetic testing include, "The Island" and the "Resident Evil" series.

Limitations of Predictive Medicine

On a protein level, structure is more conserved than sequence. Therefore, in many diseases, having the faulty gene still does not necessarily mean someone will get the disease. Common, complex diseases in the wider population are affected not only by heredity, but also by external causes such as lifestyle and environment. Therefore, genes are not perfect predictors of future health; individuals with both the high risk form of the gene and those without are all candidates to get the disease. Multiple factors in the environment, particular smoking, diet and exercise, infection, and pollution; play important roles and can be more important than genetic make-up. This makes the results and risks determined by predictive medicine more difficult to quantify. Furthermore, the potential false positives or false negatives that may arise from a predictive genetic screen can cause substantial unnecessary strain on the individual.

Targeting medication to people who are genetically susceptible to a disease but do not yet show the symptoms of it can be a questionable measure. In large populations, there is concern that likely most of the people taking preventative medications would never have developed the disease anyway. Many medications carry undesirable side effects that high risk individuals must then cope with. In contrast, several populations-based prevention measures (such as encouraging healthy diets or banning tobacco advertising) carry a far lower likelihood of adverse effects and are also less expensive. Another potential downfall of commercially available genetic testing lies within the psychological impacts of accessibility to such data. For single-gene inherited diseases, counseling and the right to refuse a test (the right "not to know") have been found to be important. However, adequate individual counseling can be difficult to employ to the potentially large proportion of the population likely to be identified as at high risk of common complex disease. Some people are vulnerable to adverse psychological reactions to genetic predictions of stigmatized or feared conditions, such as cancer or mental illness.

Literature Resources

In 2010, Penguin Publisher's published Outsmart Your Genes, by Brandon Colby, MD. This book is written for the lay reader and discusses the new field of predictive medicine and genetic testing from the perspective of a physician, and includes a number of pop-culture references such as to Sergei Grinkov, David Bloom, and the NFL. In the second half of Outsmart Your Genes, Dr. Colby provides specific examples of how genetic testing can be utilized to lower a person's risk of a large number of diseases, including Alzheimer's, cancer, heart disease, stroke, multiple sclerosis, asthma, and many others.

In 2010, Harper published The Language of Life: DNA and the Revolution in Personalized Medicine, by Francis S. Collins. The book is also written for the lay reader and discusses how genetic testing may be utilized to usher in personalized medicine. Dr. Collins describes general concepts of how genetics may be utilized to combat a large number of diseases, including Alzheimer's and heart disease.

Gene Theft

Gene theft or DNA theft is the act of acquiring the genetic material of another human being, often from a public place, without his or her permission. The DNA may be harvested from a wide variety of common objects such as discarded cigarettes, used coffee cups and hairbrushes. This genetic material can then be used for such purposes as establishing paternity, proving genealogical connections or even unmasking private medical conditions.

Criminal Law

Great Britain criminalized the taking of DNA without consent in 2006 at the urging of the Human Genetics Commission. Australia's legislature debated a two year jail sentence for such theft in 2008. In the United States, eight states currently have criminal or civil prohibitions on such nonconsensual appropriation of genetic materials. In Alaska, Florida, New Jersey, New York and Oregon, individuals caught swiping DNA face fines or short jail sentences. Lawsuits against "gene snatchers" are permitted in Minnesota, New Hampshire and New Mexico. In jurisdictions where such nonconsentual taking of DNA is illegal, exceptions are generally made for law enforcement.

Ethics

Many bioethicists believe that such conduct is an unethical invasion of human privacy. Professor Jacob Appel has warned that criminals may acquire the capability to copy DNA of innocent people and deposit it at crimes scenes, endangering the blameless and undermining a key tool of forensic investigation." However, others defend the appropriation of genetic material on the grounds that doing so may further human knowledge in productive ways One particularly controversial case which received widespread attention in the media was that of Derrell Teat, a wastewater coordinator, who sought to acquire without consent the DNA of a man who was allegedly the last male descendant of her great-great-great grandfather's brother. Another prominent case was a United States paternity suit involving film producer Steve Bing and billionaire investor Kirk Kerkorian.

Pseudogene

Pseudogenes are dysfunctional relatives of known genes that have lost their protein-coding ability or are otherwise no longer expressed in the cell. Although some do not have introns or promoters (these pseudogenes are copied from mRNA and incorporated into the chromosome and are called processed pseudogenes), most have some gene-like features (such as promoters, CpG islands, and splice sites), they are nonetheless considered nonfunctional, due to their lack of protein-coding ability resulting from various genetic disablements (premature stop codons, frameshifts, or a lack of transcription) or their inability to encode RNA (such as with rRNA pseudogenes). Thus the term, coined in 1977 by Jacq, et al., is composed of the prefix pseudo, which means false, and the root gene, which is the central unit of molecular genetics.

Because pseudogenes are generally thought of as the last stop for genomic material that is to be removed from the genome, they are often labeled as junk DNA. Nonetheless, pseudogenes contain fascinating biological and evolutionary histories within their sequences. This is due to a pseudogene's shared ancestry with a functional gene: in the same way that Darwin thought of two species as possibly having a shared common ancestry followed by millions of years of evolutionary divergence, a pseudogene and its associated functional gene also share a common ancestor and have diverged as separate genetic entities over millions of years.

Properties of Pseudogenes

Pseudogenes are characterized by a combination of homology to a known gene and nonfunctionality. That is, although every pseudogene has a DNA sequence that is similar to some functional gene, they are nonetheless unable to produce functional final products. Pseudogenes are quite difficult to identify and characterize in genomes, because the two requirements of homology and nonfunctionality are implied through sequence calculations and alignments rather than biologically proven.

1. Homology is implied by sequence identity between the DNA sequences of the pseudogene and parent gene. After aligning the two sequences, the percentage of identical base pairs is computed. A high sequence identity (usually between 40% and 100%) means that it is highly likely that these two sequences diverged from a common ancestral sequence (are homologous), and highly unlikely that these two sequences were independently created.
2. Nonfunctionality can manifest itself in many ways. Normally, a gene must go through several steps in going from a genetic DNA sequence to a fully functional protein: transcription, pre-mRNA processing, translation, and protein folding are all required parts of this process. If any of these steps fails, then the sequence may be considered nonfunctional. In high-throughput pseudogene identification, the most commonly identified disablements are stop codons and frameshifts, which almost universally prevent the translation of a functional protein product.

Pseudogenes for RNA genes are often easier to discover. Many RNA genes occur as multiple copy genes, and pseudogenes are identified through sequence identity and location within the region.

Types and Origin of Pseudogenes

There are three main types of pseudogenes, all with distinct mechanisms of origin and characteristic features. The classifications of pseudogenes are as follows:

1. Processed (or retrotransposed) pseudogenes. In higher eukaryotes, particularly mammals, retrotransposition is a fairly common event that has had a huge impact on the composition of the genome. For example, somewhere between 30% - 44% of the human genome consists of repetitive elements such as SINEs and LINEs. In the process of retrotransposition, a portion of the mRNA transcript of a gene is spontaneously reverse transcribed back into DNA and inserted into chromosomal DNA. Although retrotransposons usually create copies of themselves, it has been shown in an in vitro system that they can create retrotransposed copies of random genes, too. Once these pseudogenes are inserted back into the genome, they usually contain a poly-A tail, and usually have had their introns spliced out; these are both hallmark

features of cDNAs. However, because they are derived from a mature mRNA product, processed pseudogenes also lack the upstream promoters of normal genes; thus, they are considered "dead on arrival", becoming non-functional pseudogenes immediately upon the retrotransposition event. However, occasionally these insertions contribute exons to existing genes and usually via alternatively spliced transcripts. A further characteristic of processed pseudogenes is common truncation of the 5' end relative to the parent sequence, which is a result of the relatively non-processive retrotransposition mechanism that creates processed pseudogenes.

2. Non-processed (or duplicated) pseudogenes. Gene duplication is another common and important process in the evolution of genomes. A copy of a functional gene may arise as a result of a gene duplication event and subsequently acquire mutations that cause it to become nonfunctional. Duplicated pseudogenes usually have all the same characteristics of genes, including an intact exon-intron structure and promoter sequences. The loss of a duplicated gene's functionality usually has little effect on an organism's fitness, since an intact functional copy still exists. According to some evolutionary models, shared duplicated pseudogenes indicate the evolutionary relatedness of humans and the other primates.
3. Disabled genes, or unitary pseudogenes. Various mutations can stop a gene from being successfully transcribed or translated, and a gene may become nonfunctional or deactivated if such a mutation becomes fixed in the population. This is the same mechanism by which non-processed genes become deactivated, but the difference in this case is that the gene was not duplicated before becoming disabled. Normally, such gene deactivation would be unlikely to become fixed in a population, but various population effects, such as genetic drift, a population bottleneck, or in some cases, natural selection, can lead to fixation. The classic example of a unitary pseudogene is the gene that presumably coded the enzyme L-gulono-ã-lactone oxidase (GULO) in primates. In all mammals studied besides primates (except guinea pigs), GULO aids in the biosynthesis of Ascorbic acid (vitamin C), but it exists as a disabled gene (GULOP) in humans and other primates. Another interesting and more recent example of a disabled gene, which links the deactivation of the caspase 12 gene (through a nonsense mutation) to positive selection in humans.

Pseudogenes can complicate molecular genetic studies. For example, a researcher who wants to amplify a gene by PCR may simultaneously amplify a pseudogene that shares similar sequences. This is known as PCR bias or amplification bias. Similarly, pseudogenes are sometimes annotated as genes in genome sequences. Processed pseudogenes often pose a problem for gene prediction programs, often being misidentified as real genes or exons. It has been proposed that identification of processed pseudogenes can help improve the accuracy of gene prediction methods.

It has also been shown that the parent sequences that give rise to processed pseudogenes lose their coding potential faster than those giving rise to non-processed pseudogenes.

Functional Pseudogenes?

By definition, pseudogenes lack a function. However, the classification of pseudogenes generally relies on computational analysis of genomic sequences using complex algorithms. This has led to the incorrect identification of pseudogenes. For example the functional, chimeric gene jingwei in Drosophila was once thought to be a processed pseudogene.

It has been established that quite a few pseudogenes can go through the process of transcription, either if their own promoter is still intact or in some cases using the promoter of a nearby gene; this expression of pseudogenes also appears to be tissue-specific. In 2003, Hirotsune et al. identified a retrotransposed pseudogene whose transcript purportedly plays a trans-regulatory role in the expression of its homologous gene, Makorin1 (MKRN1) (RING finger domain and ubiquitin ligases), and suggested this as a general model under which pseudogenes may play an important biological role.

Other researchers have since hypothesized similar roles for other pseudogenes. A bioinformatics analysis has shown that processed pseudogenes can be inserted into introns of annotated genes and be incorporated into alternatively spliced transcripts. Hirotsune's report prompted two molecular biologists to carefully review scientific literature on the subject of pseudogenes.

To the surprise of many, they found a number of examples in which pseudogenes play a role in gene regulation and expression, forcing Hirotsune's group to rescind their claim that they were the first to identify pseudogene function.

Furthermore, the original findings of Hirotsune et al. concerning Makorin1 have recently been strongly contested; thus, the possibility that some pseudogenes could have important biological functions was disputed. Additionally, University of Chicago and University of Cincinnati scientists reported in 2002 that a processed pseudogene called phosphoglycerate mutase 3 (PGAM3P) actually produces a functional protein.

Two 2008 publications in Nature discuss that some endogenous siRNAs are derived from pseudogenes, and thus some pseudogenes play a role in regulating protein-coding transcripts. In June 2010, Nature published an article showing the mRNA levels of tumour suppressor PTEN and oncogenic KRAS is affected by their pseudogenes PTENP1 and KRAS1P. This discovery demonstrated an miRNA decoy function for pseudogenes and identified their transcripts as biologically active units in tumor biology; thus attributing a novel biological role to expressed pseudogenes, as they can regulate coding gene expression, and reveal a non-coding function for mRNAs in disease progression.

Microarray Technology: Past, Present and Future

Molecular Biology research evolves through the development of the technologies used for carrying them out. It is not possible to research on a large number of genes using traditional methods. DNA Microarray is one such technology which enables the researchers to investigate and address issues which were once thought to be non traceable.

One can analyze the expression of many genes in a single reaction quickly and in an efficient manner. DNA Microarray technology has empowered the scientific community to understand the fundamental aspects underlining the growth and development of life as well as to explore the genetic causes of anomalies occurring in the functioning of the human body.

A typical microarray experiment involves the hybridization of an mRNA molecule to the the DNA template from which it is originated. Many DNA samples are used to construct an array. The amount of mRNA bound to each site on the array indicates the expression level of the various genes. This number may run in thousands. All the data is collected and a profile is generated for gene expression in the cell.

Microarray Technique

An array is an orderly arrangement of samples where matching of known and unknown DNA samples is done based on base pairing rules. An array experiment makes use of common assay systems such as microplates or standard blotting membranes. The sample spot sizes are typically less than 200 microns in diameter usually contain thousands of spots.

Thousands of spotted samples known as probes (with known identity) are immobilized on a solid support (a microscope glass slides or silicon chips or nylon membrane). The spots can be DNA, cDNA, or oligonucleotides. These are used to determine complementary binding of the unknown sequences thus allowing parallel analysis for gene expression and gene discovery.

An experiment with a single DNA chip can provide information on thousands of genes simultaneously. An orderly arrangement of the probes on the support is important as the location of each spot on the array is used for the identification of a gene.

Types of Microarrays

Depending upon the kind of immobilized sample used construct arrays and the information fetched, the Microarray experiments can be categorized in three ways:

1. Microarray expression analysis: In this experimental setup, the cDNA derived from the mRNA of known genes is immobilized. The sample has genes from both the normal as well as the diseased tissues. Spots with more more intensity are obtained for diseased tissue gene if the gene is over expressed in the diseased condition. This expression pattern is then compared to the expression pattern of a gene responsible for a disease.
2. Microarray for mutation analysis: For this analysis, the researchers use gDNA. The genes might differ from each other by as less as a single nucleotide base.

A single base difference between two sequences is known as Single Nucleotide Polymorphism (SNP) and detecting them is known as SNP detection.

3. Comparative Genomic Hybridization: It is used for the identification in the increase or decrease of the important chromosomal fragments harboring genes involved in a disease.

Applications of Microarrays

Gene Discovery

DNA Microarray technology helps in the identification of new genes, know about their functioning and expression levels under different conditions.

Disease Diagnosis

DNA Microarray technology helps researchers learn more about different diseases such as heart diseases, mental illness, infectious disease and especially the study of cancer. Until recently, different types of cancer have been classified on the basis of the organs in which the tumors develop. Now, with the evolution of microarray technology, it will be possible for the researchers to further classify the types of cancer on the basis of the patterns of gene activity in the tumor cells. This will tremendously help the pharmaceutical community to develop more effective drugs as the treatment strategies will be targeted directly to the specific type of cancer.

Drug Discovery

Microarray technology has extensive application in Pharmacogenomics. Pharmacogenomics is the study of correlations between therapeutic responses to drugs and the genetic profiles of the patients.

Comparative analysis of the genes from a diseased and a normal cell will help the identification of the biochemical constitution of the proteins synthesized by the diseased genes. The researchers can use this information to synthesize drugs which combat with these proteins and reduce their effect.

Toxicological Research

Microarray technology provides a robust platform for the research of the impact of toxins on the cells and their passing on to the progeny. Toxicogenomics establishes correlation between responses to toxicants and the changes in the genetic profiles of the cells exposed to such toxicants.

GEO

In the recent past, microarray technology has been extensively used by the scientific community.

Consequently, over the years, there has been a lot of generation of data related to gene expression. This data is scattered and is not easily available for public use. For easing the accessibility to this data, the National Center for Biotechnology Information (NCBI) has formulated the Gene Expression Omnibus or GEO. It is a data repository facility which includes data on gene expression from varied sources.

Microarray Probe Design Parameters

Table: *For 25-35 mers*

Parameter	*Minimum Value*	*Maximum Value*	*Default Value*	*Unit*
Probe Length	10	99	30	bases
Probe Length tolerance	0	15	3	
Probe Target Tm	40	99	63	°C
Probe Tm Tolerance (+)	0.1	99	5	
Hairpin Max Δ G	0.1	99.9	4	Kcal/mol
Self Dimer Δ G	0.1	99.9	7	Kcal/mol
Run/Repeat	2	99	4	bases

Parameter	*Minimum Value*	*Maximum Value*	*Default Value*	*Unit*
Probe Length	10	99	40	bases
Probe Length tolerance	0	15	3	Probe
Target Tm	40	99	70	°C
Probe Tm Tolerance (+)	0.1	99	5	
Hairpin Max Δ G	0.1	99.9	6	Kcal/mol
Self Dimer Δ G	0.1	99.9	8	Kcal/mol
Run/Repeat	2	99	5	bases

Parameter	*Minimum Value*	*Maximum Value*	*Default Value*	*Unit*
Probe Length	10	99	70	bases
Probe Length tolerance	0	15	3	Probe
Target Tm	40	99	75	°C
Probe Tm Tolerance (+/- above)	0.1	99	5	
Hairpin Max ΔG	0.1	99.9	6	Kcal/mol
Self Dimer ΔG	0.1	99.9	8	Kcal/mol
Run/Repeat	2	99	6	bases

Molecular diagnostic tests typically analyze key protein, DNA, or RNA markers to characterize diseases. Immunohistochemistry, Western blotting, DNA sequencing, polymerase chain reaction (PCR), Southern blotting, Northern blotting, and quantitative reverse-transcription PCR (qRT-PCR) are molecular techniques that have been effectively used in clinical diagnostic laboratories, where broadly trained technologists perform the tests rather than researchers with specific methodological expertise. However, more-parallel molecular tests that involve simultaneous analysis of tens to hundreds of markers within a clinical sample will soon be required to allow maximum translation of postgenomic research to patient care.

Many basic research projects have focused on methods to measure the behavior of the approximately 30,000 genes in the human genome. This research has been done at different levels within the gene, measuring both genotypic and phenotypic changes. For example, parallel analysis has been applied at the DNA level to analyze sequence polymorphisms that distinguish closely related human alleles or microbial genes; at the level of DNA methylation patterns; at the mRNA level to analyze gene expression patterns; and at the protein level.

With the development of microarray-based tests for diagnosing a wide range of diseases, the ability to gain more information from a limited clinical sample by using highly parallel expression-analysis techniques has emerged. This technology's greatest potential exists for those diseases in which histological differentiation between disease entities is difficult and the consequences of inappropriate treatment can be serious. In the research environment, analyzing sets of gene products with microarrays, rather than individual proteins or RNAs, has been shown to provide effective diagnoses. However, improving high-throughput diagnostic methods will require the continued development of new second-generation custom microarrays. This article describes the challenges in developing this technology and the custom microarray techniques that are currently available.

Microarrays as Transcription Analysis Tools

High-density DNA (HD DNA) microarrays have been useful in the semi-quantitative, or comparative, analysis of thousands of individual messenger RNA (mRNA) species. Experiments using HD DNA microarrays suggest that new tests measuring the expression of tens to hundreds of genes will be better able to characterize disease states than is currently possible using histopathology and other existing molecular methods. As stated above, clinical application of this DNA microarray-based research will require further development of sophisticated new second-generation custom DNA microarray platforms that are more suitable for routine clinical laboratory testing.

HD DNA microarrays are small devices that hold tens of thousands of gene capture probes at known positions on a solid surface. One of the most widely used platforms is the arrays by Affymetrix Inc. (Santa Clara, CA), in which DNA oligonucleotides are chemically synthesized in situ on the array surface using masked photolithography. Another widely used platform is the cDNA array, in which PCR products are generated separately and placed on the array surface using a robotic arrayer. Thousands of laboratories have used these two platforms to catalog the expression of large numbers of genes in normal and diseased cells and tissues.

When transitioning from a normal to a diseased state, cells and tissues undergo a variety of histological and biochemical changes. These changes reflect in large part the dysregulation in the expression of genes. The quantities and types of proteins that are expressed from their respective genes determine the normal and abnormal functions of cells. When genes are expressed, their DNA sequences are first copied into temporary molecules, or mRNAs. These mRNAs are then translated into proteins that perform the functions and generate the structures that allow cells to conduct their biological activities.

Researchers can take advantage of this flow of genetic information from DNA to mRNAs to proteins by purifying populations of all mRNAs expressed in specimens, synthesizing fluorescently labeled cDNA copies of the mRNAs in vitro, and hybridizing the labeled cDNAs to a microarray. The intensity of the fluorescent signal at each gene-specific probe on the microarray reflects the level of mRNA expressed from that gene in the specimen. By measuring the types and quantities of tens of thousands of individual mRNAs expressed in samples, researchers can infer information regarding the state of gene expression, the approximate quantities and types of proteins present, and the biological behavior of the sample.

For several decades, alternative methods have been available for identifying genes whose expression patterns are altered in disease. However, HD DNA microarrays have increased research scientists' productivity in their gene-hunting efforts. As little as five years ago, identifying even a few differentially expressed genes may have taken several years and cost tens of thousands of dollars. Today, HD DNA microarrays can identify 10 times that number of genes in a few months and at a tenth of the cost.

Advances in Lymphoma Diagnostics

Recent studies on the biology of non-Hodgkin's lymphomas (NHL) illustrate some of the diagnostic advances made possible by HD DNA microarrays. In 2000, it is estimated that 54,900 patients were newly diagnosed with NHLs and 22,553 died of these diseases. Although there are many NHL subtypes, the diffuse large B-cell lymphoma (DLBCL) subtype is the most common, comprising 30% of all cases.

Despite the complexity of this classification system, it is increasingly recognized that the NHL categories themselves encompass multiple disease subtypes with distinct molecular defects and clinical outcomes. For example, while 35–40% of patients diagnosed with DLBCL can be cured with chemotherapy, the remaining 60–65% die of this disease. Numerous attempts to distinguish patients who are likely to respond to chemotherapy from those unlikely to respond have been unsuccessful. In addition, these DLBCL tumors are indistinguishable using classical histopathology and currently available molecular markers.

Researchers have approached this problem by using HD cDNA microarrays to profile gene expressions in DLBCL. These researchers identified genes that classified DLBCL specimens into distinct subclasses. One subclass, germinal-center B-cell-like DLBCL, showed an expression pattern similar to mature germinal-center B-cells and portended a more favorable prognosis with a 60% five-year survival rate. On the contrary, two other DLBCL subclasses, activated B-cell-like DLBCL and type-3 DLBCL, showed much less favorable prognoses with a 35% five-year survival rate.

Using Microarrays for Personalized Medicine

Research studies have predicted that in the near future, it may be possible to determine which patients might benefit from more-aggressive therapies, such as bone marrow transplants or alternative chemotherapy regimens, when DLBCL is diagnosed. HD DNA microarray studies have also highlighted molecular pathways important in subclasses of cancer.

For example, researchers have observed that several genes in the activated B-cell subtype of DLBCL were downstream targets of the NFkB transcription factor. These investigators demonstrated that NFkB activity is indeed higher in this DLBCL subtype, suggesting that drugs specifically targeting the NFkB pathway may be effective in treating these tumors.

Several recent papers have suggested that gene expression profiling of tumor specimens may be useful for preselecting patients who may benefit from drug treatment. It may also be useful to examine gene expression profiles of cancers following chemotherapy in order to determine whether the tumors are responding to treatment. The goal of this approach is personalized medicine, in which detailed patient-specific molecular information would be assayed to predict an effective therapy.

Microarrays as Genotyping Tools for Infectious Diseases

Despite the abundance of protein diagnostics, diagnosing diseases by examining the qualitative presence of specific nucleic acid sequences has been adopted for infectious diseases. While the amplification of gene sequences through PCR or other similar methods can indicate the presence or absence of a pathogen genomic sequence, nucleic acid hybridization is almost always necessary to confirm that the amplification is sequence-specific. Suitably designed microarrays are versatile tools for this function since the hybridization data from an array can identify the pathogen genomic sequence in a more rapid and cost-effective manner than DNA sequencing.

The sudden emergence of new strains of pathogens has brought the need for rapid nucleic acid diagnostic tools to the forefront. Although customizable nucleic acid testing systems for the rapid development of diagnostics are still new and emerging, the urgent medical need for high-speed diagnostics is steadily increasing.

For example, the recent outbreak of severe acute respiratory syndrome (SARS) and its rapid spread left no time for generating an antibody-based diagnostic test. Initial identification of the virus was based on symptom presentation and high-resolution electron microscopy of sputum samples. However, gene sequencing of the emergent SARS strains permitted rapid identification of the virus by using a nucleic acid amplification and hybridization/ sequencing approach.

Monitoring how the genotype of viruses evolves is critical for understanding the pathology of a disease as it spreads throughout a population. Viruses such as influenza, poliovirus, rhinoviruses, and HIV are known to have high rates of mutation, in some cases even causing attenuated strains that are developed for vaccines to revert to viruses with a wild-type phenotype. Although it remains possible to generate antibodies for all of these pathogen strains, identifying strains rapidly is more easily achieved using microarray tools developed specifically for this purpose.

Second-Generation Custom DNA Microarrays

Many researchers agree that in some form, DNA microarrays will be used as diagnostic tools in the near future. The first prospective trials are under way to determine whether microarrays can predict the proper clinical course for patients. At the same time, FDA is anticipating working on the regulatory aspects of microarray-based diagnostic tests. However, extensive work will be necessary to develop microarray platforms that give reproducible results in different laboratories, which will be required for clinical applications. In addition, the basic parameters of microarray technology will also need to be established, including content selection (i.e., which oligonucleotides will be used), data analysis, data interpretation, and uniform sample preparation and assay conditions.

The use of a variety of generic whole-genome HD DNA microarrays in the research setting has led to important new basic research findings. However, these platforms are not suitable for use in clinical diagnostic laboratories for a number of reasons.

First, ownership claims for probes to particular gene sequences can burden the expensive diagnostic development process before it has even begun.

Second, the lack of sequence flexibility of the nucleic acid capture probes on the microarrays makes it difficult to be sure that the best capture probe has been identified. In other words, there is no chance to improve rapidly and iteratively the performance of the capture probe to its target.

Third, many genes have significant polymorphisms or splice variants, which may or may not have been correctly represented on the array. The content arrays are based on the human genomic information from a comparatively small set of human genomic sequences.

Finally, nearly all of these products are built around gene expression type applications, which are not necessarily the only approach for developing diagnostic tools. In contrast, customizable microarrays give diagnostics developers complete freedom to identify rapid nucleic acid probes for specific applications, such as key changes in gene-expression approach, disease-related polymorphisms, patterns in DNA methylation analysis, or resequencing applications for infectious diseases. Although a number of technologies exist for making custom arrays, there are significant differences in the techniques used to fabricate the arrays, which have an impact on the length of time, cost, and complexity in developing nucleic acid diagnostics.

For example, the Affymetrix system uses specially made chromium photomasks from the semiconductor industry to pattern the synthesis of nucleic acids. Although this method has been widely adopted in research scenarios, an important limiting factor concerns poor stepwise yields of light-based deprotection. Typically, nucleotides are used for light-based deprotection systems. However, while short oligonucleotides may work for some applications, they may not be ideal for developing many nucleic acid diagnostic tools.

Similarly, spotted cDNA arrays encounter problems with spot nonuniformity and misidentification of spots. These problems limit the quality of data derived from high-density microarrays, making it difficult to distinguish experimental noise from biologically meaningful measurements of gene expression.

For example, a significant percentage of genes that are identified as being differentially expressed from specimen types using high-density microarrays cannot be confirmed using the more-laborious gold standard technology of qRT-PCR. This is particularly true for genes that are expressed at low levels.

In addition, several recent papers have described similar experiments that were performed on both the Affymetrix and spotted cDNA platforms and produced different lists of differentially expressed genes. If the gene expression patterns were genuine, researchers should be able to identify the same genes regardless of which platform is used. This noncorrelation of data across platforms suggests that care should be taken to understand the technical limitations of these platforms.

An alternative technology that addresses the key issues of turnaround time, interplatform reproducibility, and oligonucleotide quality is the CustomArray, a semiconductor-based technology by CombiMatrix Corp. (Mukilteo, WA).

This technology uses the electrochemical generation of acid at microelectrodes, rather than light-based photochemistry, to pattern nucleic acid synthesis. With a 16–36-hour turnaround time, this system can synthesize oligonucleotides of high quality and long lengths on an array format.

This technology enables testing a wide range of diagnostic approaches, including screening genotypic information from both virus and host cells concurrently, using multiple assay types on one chip, and employing variable oligonucleotide lengths and non standard chemistries.

This technology also benefits from employing established manufacturing processes that have been developed in the semiconductor industry. This is an important consideration when determining how a diagnostic device can be manufactured in large volumes at a reasonable cost.

Conclusion

An ideal custom DNA microarray platform should allow a user complete freedom in defining which genes are assayed using the array. The microarray platform should also be compatible with computer-based probe design software and molecular methods so that measurements can be performed with high sensitivity and specificity, thereby making independent validation of the assay unnecessary.

The use of manufacturing techniques that allow miniaturization and massive scale-up and result in low costs and minimal biological sample requirements should contribute to a well-engineered microarray platform. Because of these factors, there should be standardization among various clinical laboratories using microarrays, such that results from different laboratories can be directly compared.

This is particularly important under existing and proposed regulations that give the Centers for Disease Control and Prevention (Atlanta) and FDA shared responsibility for monitoring molecular diagnostic testing.

6

Chromosome Mutations

It is natures intention that the exact genetic information from both parents will be seen in the offspring's DNA in the the critical stages of fertilisation. However, it is possible for this genetic information to mutate, which in most cases, can result in fatal or negative consequencies in the outcome of the new ogranism.

Non-Disjunction and Down's Syndrome

One well known example of mutation is non-disjunction. Non-disjunction is when the spindle fibres fail to seperate during meiosis, resulting in gametes with one extra chromosome and other gametes lacking a chromosome.

If this non-disjunction occurs in chromosome 21 of a human egg cell, a condition called Down's syndrome occurs. This is because their cells possess 47 chromosomes as opposed to the normal chromosome compliment in humans of 46.

The fundamental structure of a chromosome is subject to mutation, which will most likely occur during crossing over at meiosis. There are a number of ways in which the chromosome structure can change, as indicated below, which will detrimentally change the genotype and phenotype of the organism.

However, if the chromosome mutation effects an essential part of DNA, it is possible that the mutation will abort the offspring before it has the chance of being born.

The following indicates types of chromosome mutation where whole genes are moved:

Deletion of a Gene

As the name implies, genes of a chromosome are permanently lost as they become unattached to the centromere and are lost forever

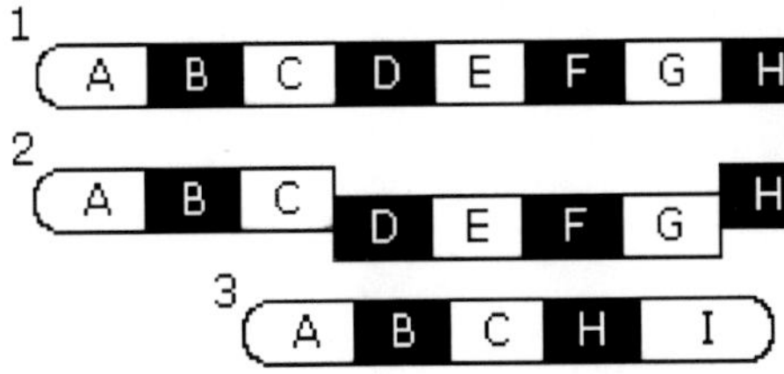

- Normal chromosome before mutation
- Genes not attached to centromere become loose and lost forever
- New chromosome lacks certain genes which may prove fatal depending on how important these genes are

Duplication of Genes

In this mutation, the mutants genes are displayed twice on the same chromosome due to duplication of these genes. This can prove to be an advantageous mutation as no genetic information is lost or altered and new genes are gained

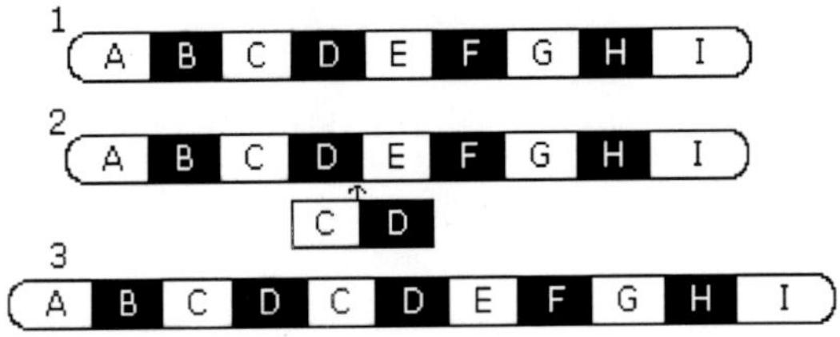

- Normal chromosome before mutation
- Genes from the homologous chromosome are copied and inserted into the genetic sequence
- New chromosome possesses all its initial genes plus a duplicated one, which is usually harmless

Inversion of Genes

This is where the order of a particular order of genes are reversed as seen below

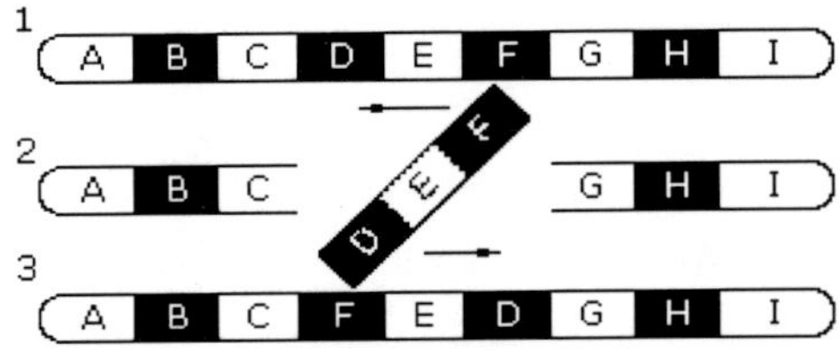

- Normal chromosome un-altered
- The connection between genes break and the sequence of these genes are reversed
- The new sequence may not be viable to produce an organism, depending on which genes are reversed. Advantageous characteristics from this mutation are also possible

Translocation of Genes

This is where information from one of two homologous chromosomes breaks and binds to the other. Usually this sort of mutation is lethal:

- An un-altered pair of homologous chromosomes
- Translocation of genes has resulted in some genes from one of the chromosomes attaching to the opposing chromosome

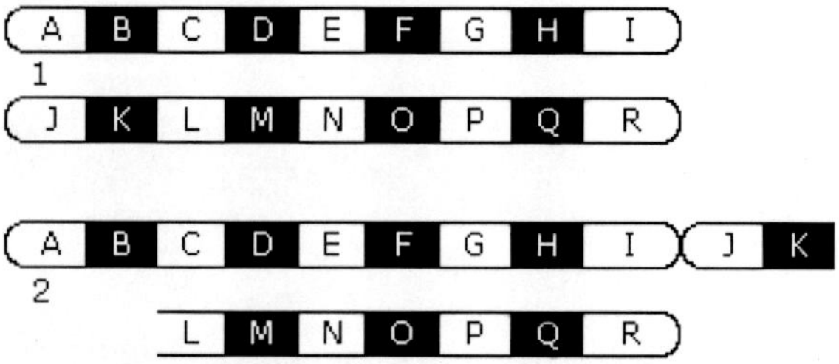

Alteration of a DNA Sequence

The previous examples of mutation have investigated changes at the chromosome level. The sequence of nucleotides on a DNA sequence are also susceptible to mutation.

Deletion: Here, certain nucleotides are deleted, which affects the coding of proteins that use this DNA sequence. If for example, a gene coded for alanine, with a genetic sequence of C-G-G, and the cytosine nucleotide was deleted, then the alanine amino acid would not be able to be created, and any other amino acids that are supposed to be coded from this DNA sequence will also be unable to be produced because each successive nucleotide after the deleted nucleotide will be out of place.

Insertion: Similar to the effects of deletion, where a nucleotide is inserted into a genetic sequence and therefore alters the chain thereafter. This alteration of a nucleotide sequence is known as frameshift

Inversion: Where a particular nucleotide sequence is reversed, and is not as serious as the above mutations. This is because the nucleotides that have been reversed in order only affect a small portion of the sequence at large

Substitution: A certain nucleotide is replaced with another, which will affect any amino acid to be synthesised from this sequence due to this change.

If the gene is essential, i.e. for the coding of haemoglobin then the effects are serious, and organisms in this instance suffer from a condition called sickle cell anaemia.

All of the genetic mutations looked at through the last 2 pages more or less have a negative impact and are undesired, however, in some cases they can prove advantageous.

Genetic mutations increase genetic diversity and therefore have an important part to play. They are also the reason many people inherit diseases.

Mutation Frequency and Polyploidy

The previous two pages have investigated mutations, and this page continues with more information related to genetic mutations.

Polyploidy

Humans are diploid creatures, meaning for every chromosome in our body, there is another one to match it. Read the following

- Haploid creatures have one of each chromosome
- Diploid creatures have two of each chromosome
- Triploid creatures have three of each chromosome
- Polyploid creatures have three or more of each chromosome
- They can be represented by n where n equals haploid, 2n equals diploid and so on.

It is possible for a species, particularly plant species, to produce offspring that contains more chromosomes than its parent. This can be a result of non-disjunction, where normally a diploid parent would produce diploid offspring, but in the case of non-disjunction in one of the parents, produces a polyploid.

In the case of triploids, although the creation of particular triploids in species is possible, they cannot reproduce themselves because of the inability to pair homologous chromosomes at meiosis, therefore preventing the formation of gametes.

Polyploidy is responsible for the creation of thousands of species in today's planet, and will continue to do so. It is also responsible for increasing genetic diversity and producing species showing an increase in size, vigour and an increased resistance to disease.

Mutation Frequency

This page and the previous two have investigated the different ways that mutations arise, and the following elaborates on the ways in which mutations are instigated.Barring all external factors, mutations occur very rarely, and are rarely expressed because many forms of mutation are expressed by a recessive allele. However there are many mutagenic agents that artificially increase the rate of mutations in an organism. The following are some factors that increase genetic mutations in organisms

- Members of species in a particular geographic area or ethnic origin are more susceptible to mutations
- High dosages of X-Rays or ultraviolet light can increase the likeliness of a mutation
- Radioactive substances increase the rate of mutations exponentially

As mentioned previously, genetic mutations are a source of new variation in a species because it physically alters the sequence of nucleotides in a given sequence, therefore altering the genome in a unique way.

Theory of Natural Selection

In the 19th century, a man called Charles Darwin, a biologist from England, set off on the ship HMS Beagle to investigate species of the island. After spending time on the islands, he soon developed a theory that would contradict the creation of man and imply that all species derived from common ancestors through a process called natural selection. Natural selection is considered to be the biggest factor resulting in the diversity of species and their genomes. The principles of Darwin's work and his theory are stated below.

- One of the prime motives for all species is to reproduce and survive, passing on the genetic information of the species from generation to generation. When species do this they tend to produce more offspring than the environment can support.
- The lack of resources to nourish these individuals places pressure on the size of the species population, and the lack of resources means increased competition and as a consequence, some organisms will not survive.
- The organisms who die as a consequence of this competition were not totally random, Darwin found that those organisms more suited to their environment were more likely to survive.
- This resulted in the well known phrase survival of the fittest, where the organisms most suited to their environment had more chance of survival if the species falls upon hard times. (This phrase if often associated with Darwin, though on closer inspection Herbert Spencer puts the phrase in a more accurate historical context.)
- Those organisms who are better suited to their environment exhibit desirable characteristics, which is a consequence of their genome being more suitable to begin with.

This 'weeding out' of less suited organisms and the reward of survival to those better suited led Darwin to deduce that organisms had evolved over time, where the most desirable characteristics of a species are favoured and those organisms who exhibit them survive to pass their genes on.

As a consequence of this, a changing environment would mean different characteristics would be favourable in a changing environment. Darwin believed that organisms had 'evolved' to suit their environments, and occupy an ecological niche where they would be best suited to their environment and therefore have the best chance of survival. As the above indicates, those alleles of a species that are favoured in the environment will become more frequent in the genomes of the species, due to the organisms higher likeliness of surviving as part of the species at large

Recombinant DNA

Recombinant DNA (rDNA) molecules are DNA sequences that result from the use of laboratory methods (molecular cloning) to bring together genetic material from multiple sources, creating sequences that would not otherwise be found in biological organisms. Recombinant DNA is possible because DNA molecules from all organisms share the same chemical structure; they differ only in the sequence of nucleotides within that identical overall structure.

Consequently, when DNA from a foreign source is linked to host sequences that can drive DNA replication and then introduced into a host organism, the foreign DNA is replicated along with the host DNA. Recombinant DNA molecules are sometimes called chimeric DNA, because they are usually made of material from two different species, like the mythological chimera.

The DNA sequences used in the construction of recombinant DNA molecules can originate from any species. For example, plant DNA may be joined to bacterial DNA, or human DNA may be joined with fungal DNA. In addition, DNA sequences that do not occur anywhere in nature may be created by the chemical synthesis of DNA, and incorporated into recombinant molecules.

Using recombinant DNA technology and synthetic DNA, literally any DNA sequence may be created and introduced into any of a very wide range of living organisms.

Proteins that result from the expression of recombinant DNA within living cells are termed recombinant proteins. When recombinant DNA encoding a protein is introduced into a host organism, the recombinant protein will not necessarily be produced . Expression of foreign proteins requires the use of specialized expression vectors and often necessitates significant restructuring of the foreign coding sequence . It is important to note that recombinant DNA differs from genetic recombination in that the former results from artificial methods in the test tube, while the latter is a normal biological process that results in the remixing of existing DNA sequences in essentially all organisms.

Creating Recombinant DNA

Construction of recombinant DNA, in which a foreign DNA fragment is inserted into a plasmid vector. In this example, the gene indicated by the white color is inactivated upon insertion of the foreign DNA fragment.

Molecular cloning is the laboratory process used to create recombinant DNA. It is one of two widely-used methods (along with polymerase chain reaction, abbr. PCR) used to direct the replication of any specific DNA sequence chosen by the experimentalist. The fundamental difference between the two methods is that molecular cloning involves replication of the DNA within a living cell, while PCR replicates DNA in the test tube, free of living cells.

Formation of recombinant DNA requires a cloning vector, a DNA molecule that will replicate within a living cell. Vectors are generally derived from plasmids or viruses, and represent relatively small segments of DNA that contain necessary genetic signals for replication, as well as additional elements for convenience in inserting foreign DNA, identifying cells that contain recombinant DNA, and, where appropriate, expressing the foreign DNA.

The choice of vector for molecular cloning depends on the choice of host organism, the size of the DNA to be cloned, and whether and how the foreign DNA is to be expressed. The DNA segments can be combined by using a variety of methods, such as restriction enzyme/ligase cloning or Gibson assembly. In standard cloning protocols, the cloning of any DNA fragment essentially involves seven steps: (1) Choice of host organism and cloning vector, (2) Preparation of vector DNA, (3) Preparation of DNA to be cloned, (4) Creation of recombinant DNA, (5) Introduction of recombinant DNA into the host organism, (6) Selection of organisms containing recombinant DNA, (7) Screening for clones with desired DNA inserts and biological properties. These steps are described in some detail in a related article (molecular cloning).

Expression of Recombinant DNA

Following transplantation into the host organism, the foreign DNA contained within the recombinant DNA construct may or may not be expressed. That is, the DNA may simply be replicated without expression, or it may be transcribed and translated so that a recombinant protein is produced. Generally speaking, expression of a foreign gene requires restructuring the gene to include sequences that are required for producing a mRNA molecule that can be used by the host's translational apparatus (e.g. promoter, translational initiation signal, and transcriptional terminator). Specific changes to the host organism may be made to improve expression of the ectopic gene. In addition, changes may be needed to the coding sequences as well, to optimize translation, make the protein soluble, direct the recombinant protein to the proper cellular or extracellular location, and stabilize the protein from degradation.

Properties of Organisms Containing Recombinant DNA

In most cases, organisms containing recombinant DNA have apparently normal phenotypes. That is, their appearance, behavior and metabolism are usually unchanged, and the only way to demonstrate the presence of recombinant sequences is to examine the DNA itself, typically using a polymerase chain reaction (PCR) test.

If the rDNA sequences encode a gene that is expressed, then the presence of RNA and/or protein products of the recombinant gene can be detected, typically using RT-PCR or western hybridization methods. Gross phenotypic changes are not the norm, unless the recombinant gene has been chosen and modified so as to generate biological activity in the host organism. Additional phenotypes that are encountered include toxicity to the host organism induced by the recombinant gene product, especially if it is over-expressed or expressed within inappropriate cells or tissues.

In some cases, recombinant DNA can have deleterious effects even if it is not expressed. One mechanism by which this happens is insertional inactivation, in which the rDNA becomes inserted into a host cell's gene. In some cases, researchers use this phenomenon to "knock out" genes in order to determine their biological function and importance. Another mechanism by which rDNA insertion into chromosomal DNA can affect gene expression is by inappropriate activation of previously unexpressed host cell genes. This can happen, for example, when a recombinant DNA fragment containing an active promoter becomes located next to a previously silent host cell gene, or when a host cell gene that functions to restrain gene expression undergoes insertional inactivation by recombinant DNA.

Recombinant DNA Technology

All gene manipulation is based on microbial genetics—ways of doing in the test tube what bacteria and viruses do naturally. Several of this week's gene manipulations are exemplified by the article on overexpressed angiopoietin in a transgenic mouse.

Gel Electrophoresis

This technique separates molecules on the basis of their size.

- Cast slab of gel material, usually agarose or polyacrylamide. The gel is a matrix of polymers forming sub-microscopic pores.
- The size of the pores can be controlled by varying the chemical composition of the gel.

The gel is set up for electrophoresis in a tank holding pH buffer. Electrodes apply an electric field:

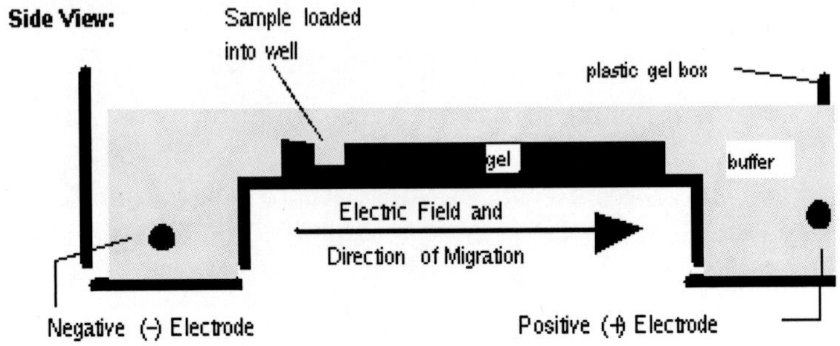

The molecules to separate (DNA RNA) carry a net negative charge (why?) so they move along the electric field toward the positive cathode. (To separate proteins, a detergent would be included which coats the protein with negative charge.)

The larger molecules are held up as they try to pass through the pores of the gel, while the smaller molecules are impeded less and move faster.

This results in separation by size, with the larger molecules nearer the well and the smaller molecules farther away.

Note that this separates on the basis of size (volume in solution), which is not necessarily molecular weight. For example:

- Two DNA molecules of the same molecular weight will run differently if one is supercoiled, because the supercoils constrain the shape to be smaller.
- Two RNA molecules of the same molecular weight will run differently if one has much intramolecular base pairing, making it "smaller."

Aside from the above exceptions, the distance migrated is roughly proportional to the log of the inverse of the molecular weight (the log of 1/MW). Gels are normally depicted as running vertically, with the wells at the top and the direction of migration downwards. This leaves the large molecules at the top and the smaller molecules at the bottom. Molecular weights are measured with different units for DNA, RNA, and protein:

- DNA: Molecular weight is measured in base-pairs, or bp, and commonly in kilobase-pairs (1000bp), or kbp.
- RNA: Molecular weight is measured in nucleotides, or nt, and commonly in kilonucleotides (1000nt), or knt. [Sometimes, bases, or b and kb are used.
- Protein: Molecular weight is measured in Daltons (grams per mole), or Da, and commonly in kiloDaltons (1000Da), or kDa.

Molecular weight standards run in one well of the gel are used to calibrate the molecular weights of sample molecules. Below is a gel stained with a dye: a colored molecule which binds to a specific class of macromolecules in a sequence-independent manner (probes bind in a sequence-dependent manner).

Sample 1 contains only one size class of macromolecule - it could be a plasmid, a pure mRNA transcript, or a purified protein. In this case, you would not have to use a probe to detect the molecule of interest since there is only one type of molecule present.

Blotting is usually necessary for samples that are not complex mixtures. By interpolation, its molecular weight is roughly 3.

Sample 2 is what a sample of total DNA cut with a restriction enzyme, total cellular RNA, or total cellular protein would look like in a gel stained with a sequence-independent stain.

There are so many bands that it is impossible to find the one we are interested in. Without a probe (which acts like a sequence-dependent stain) we cannot get very much information from a sample like this.

Different stains are used for different classes of macromolecules. DNA and RNA are generally stained with ethidium bromide (EtBr), an intercalating agent. The DNA-EtBr complex fluoresces under UV light. Protein is stained with Coomassie Blue or Silver Stain.

Cloning Genes

In nature, DNA molecules recombine for various functions — even DNA between different species. But twenty years ago, despite the work of Barbara McClintock and others, the extent of this recombination was not appreciated. DNA was still thought to be the "master molecule," not to be violated by "unnatural" manipulation. When scientists began to manipulate DNA in the test tube, many scientists feared that disastrous monsters would result, with unspecified dangers to people.

In 1977 scientists at the Asilomar Conference proposed sweeping regulation on so-called "recombinant DNA," technologies which recombine DNA from different species in the test tube.

Since then, the dangers have appeared to be little more than those of "natural" genetic mixing. But we remain concerned about issues such as:

- Engineering food crops to resist pesticides. The pesticide resistance genes can escape into natural populations of weeds.
- Engineering a human symbiont microbe, such as E. coli, to produce a deadly toxin such as botulin. In theory this could be done, although it's not clear where such an organism would live, or how well it could "compete" with natural flora.

- Societal dilemmas of human cloning. How far shall we use reproductive technology to shape future humans?

Techniques of Recombinant DNA

How do we manipulate these natural processes for biotechnology; for instance, to make a bacterium that produces large quantities of insulin? One approach would be to cut the appropriate gene from human DNA and paste, or splice, it into a vector such as a plasmid or phage DNA. Our "scissors" are the class of enzymes called restriction endonucleases

Restriction Endonucleases

An "endonuclease" is an enzyme that cuts duplex DNA in the middle, not at an end (for exonuclease). Different species of bacteria have evolved different restriction endonucleases, each to cut foreign DNA that gets into their cells by mistake. To be cut, the DNA has to lack their own pattern of protective methylation. There are well over a hundred restriction enzymes, each cutting in a very precise way a specific base sequence of the DNA molecule.

Restriction Enzyme
Action of EcoRI

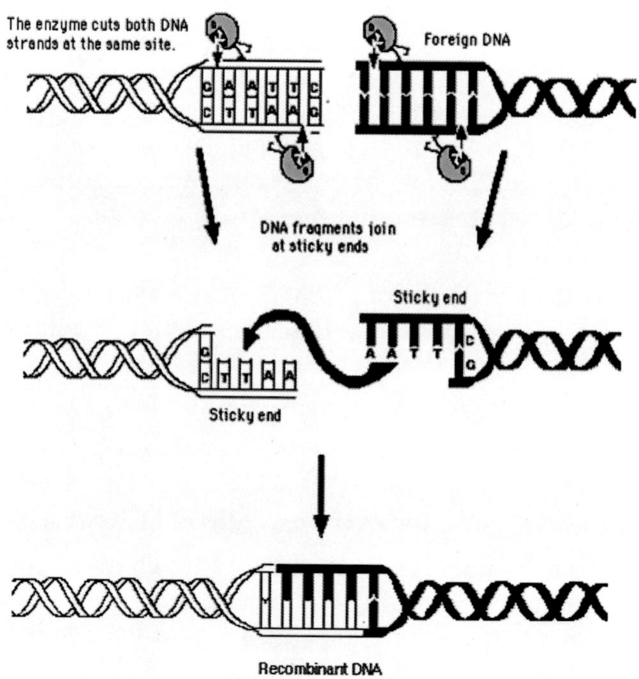

A restriction endonuclease cuts DNA only at a specific site, usually containing 4-6 base pairs. The enzyme has to cut the DNA backbone twice, recognizing the same type of site; therefore, the site "reads" the same way backwards as forwards—a palindrome.

This "sticky ends" from two different DNA molecules can hybridize together; then the nicks are sealed using ligase. (Where does ligase come from? What is its natural function?) The result is recombinant DNA. When this recombinant vector is inserted into E. coli, the cell will be able to process the instructions to assemble the amino acids for insulin production. More importantly, the new instructions are passed along to the next generation of E. coli cells in the process known as gene cloning.

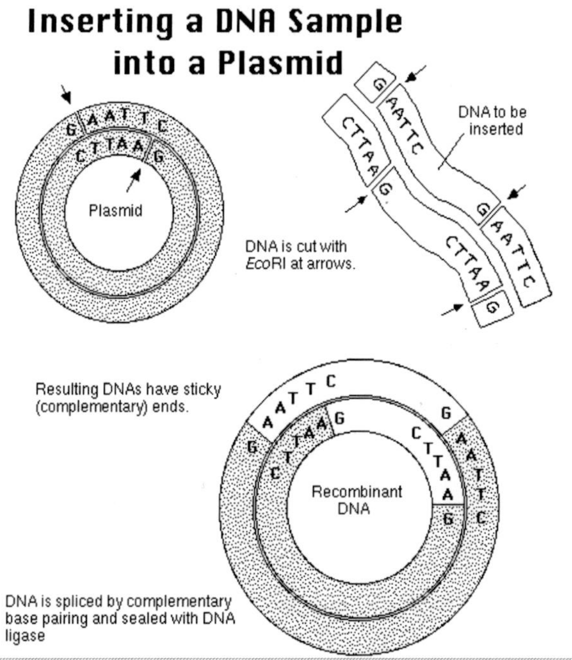

Restriction Site Analysis

How can we use restriction sites to analyze the plasmid products of ligation, and tell whether we in fact have ligated the correct molecule:

Problem: Suggest several "incorrect" ways the plasmid could recombine. Use the PLASMID Program.

You MUST practice restriction analysis with this program; it will be on the quiz and/or the test. How do we get the recombinant molecule into a bacterial cell? Usually by transformation (for a plasmid) or by in vitro packaging into a phage head coat (for a phage vector such as lambda phage). How would we actually locate the appropriately cloned gene? There are many different ways, depending on the specific case. Here is one example, in which a partial sequence of the protein enables us to reverse the code and determine an approximate DNA sequence to use for a radiolabeled probe. The DNA probe will hybridize to clones containing the correct DNA, even if it is just one piece cut out of an entire genome.

Griffiths et al, W. H. Freeman & Co., current edition

The radioactive probe is made by determining a short segment of the protein sequence, then "back translating" to the possible DNA sequences. Short DNA sequences are synthesized to match the protein sequence.

Then these DNA oligomers (known as "oligos") are radiolabeled, and applied to the blotted clones. They should hybridize only to clones containing sequence encoding the desired protein.

Reverse Transcription: cDNA Cloning

Suppose we need to clone a gene containing lots of introns. What will happen when the bacterium tries to express it?

To overcome this problem, we can start with mRNA isolated from tissues that produce the desired protein. We then use reverse transcriptase enzyme (produced by a retrovirus related to HIV) to reverse transcribe the mRNA into a DNA molecule that now is free of introns. Now we can ligate "sticky ends" onto the cDNA and recombine it into a phage or plasmid vector.

Problem: Know the differences between genomic cloning and cDNA cloning. Explain the relative ADVANTAGES and DISADVANTAGES of each technique—depending on the aim of your research.

Gene Therapy

Transgenic Plants

In PCR, a heat-stable DNA polymerase is used, most commonly Taq Polymerase from the thermophilic microbe Thermus aquaticus. Thomas Brock discovered T. aquaticus from a hot spring at Yellowstone National Park.

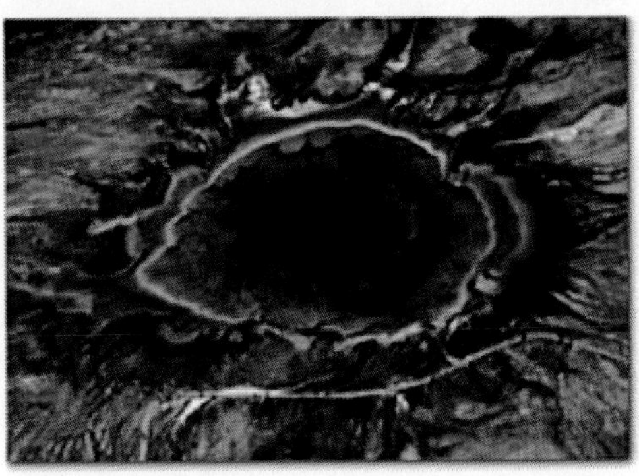

More recently, an even more heat-resistant polymerase has been developed from a hyperthermophilic microbe growing at 110 degrees C in hydrothermal vent ecosystems in the deep ocean; it's called "Vent Polymerase."

The Taq Polymerase is put with the DNA to be amplified, plus all four NTPs, plus two primers facing each other, about 200 - 6000 kb apart. (Why do we need primers?) The primers are selected based on the DNA region you want to amplify.

The tube is placed in a thermal cycler. DNA gets synthesized from each primer, for about 2 minutes.

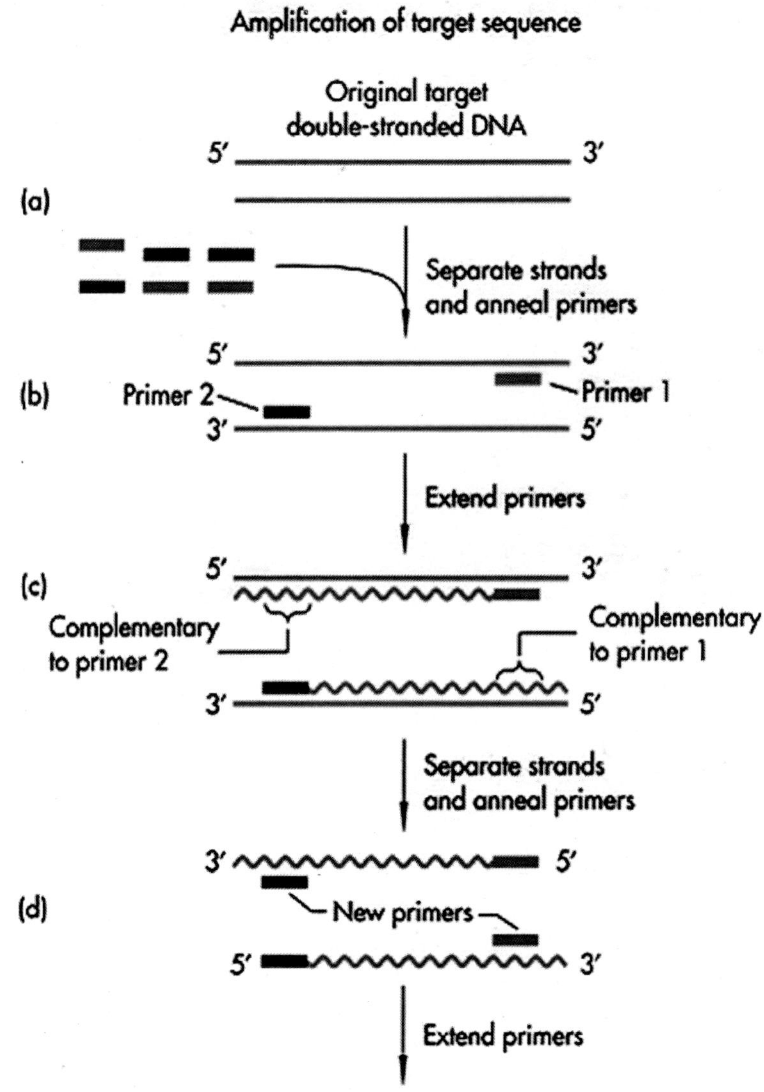

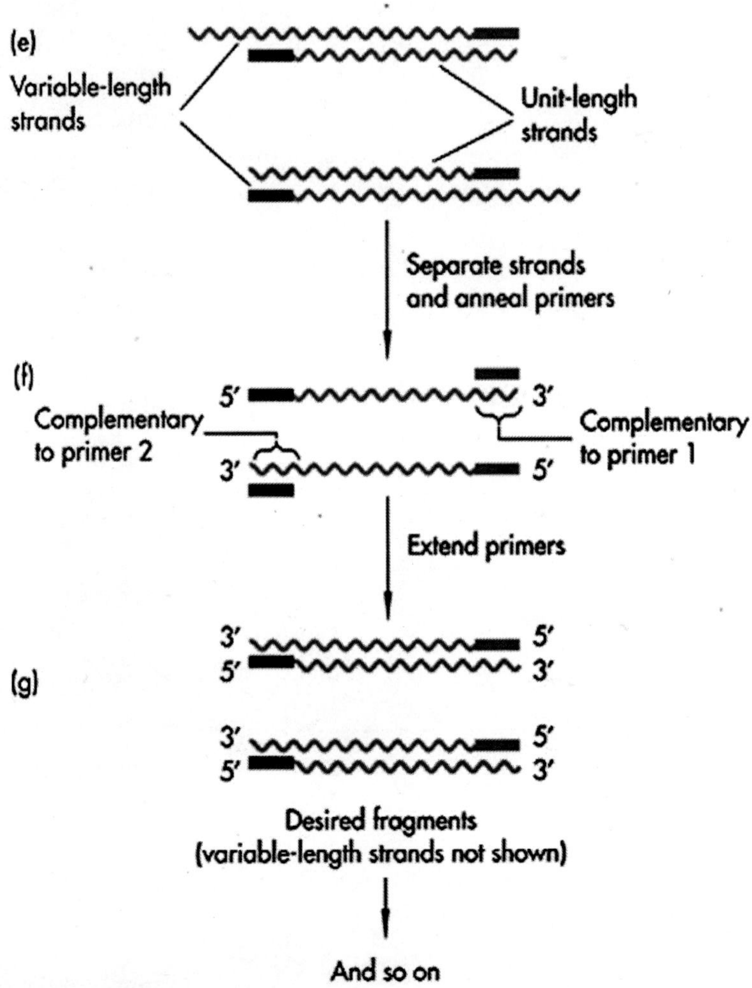

Figure: *a, b, c, d , e, f g.*

Then the temperature is raised to 95 degrees C — enough to denature (split apart) the DNA base pairs. But the Taq Polymerase remains intact, because it comes from an organism that evolved to grow at this temperature. Now the temperature is decreased again, and primers again can hybridize to the DNA—both the old AND the newly synthesized strands.

Again, Taq Polymerase extends new DNA strands. Again, the temperature is raised. After repeated cycles, the amount of DNA sequence between the two primers increases exponentially. First 2 strands, then 4, 8, 16, up to about a million. Thus, in a couple of hours, you can get million-fold amplification of a DNA sequence.

Griffiths et al, W. H. Freeman & Co., current edition

Applications of PCR

PCR has replaced cloning for many purposes, particularly the sequencing of DNA. It is faster and requires no vectors, which can mutate as they reproduce. It can be used forensically, to amplify tiny amounts of DNA from criminal evidence; or clinically, to detect DNA sequences linked to inherited disorders.

PCR Experiment

Microbiology Lab: Students characterized the microbial species of various environments in Knox County. Microbial colonies were placed directly into PCR reactions containing primers for amplification of ribosomal RNA genes (rDNA). Lane 4 contains an amplified band of the predicted size, 1000 bp. The top of each well contains genomic DNA. The smears at the bottom contain the PCR primers. The DNA will be sequenced and matched through GenBank to determine the microbial genus.

Thanks to Dan Nickerson '00 and Adam Marks '01.

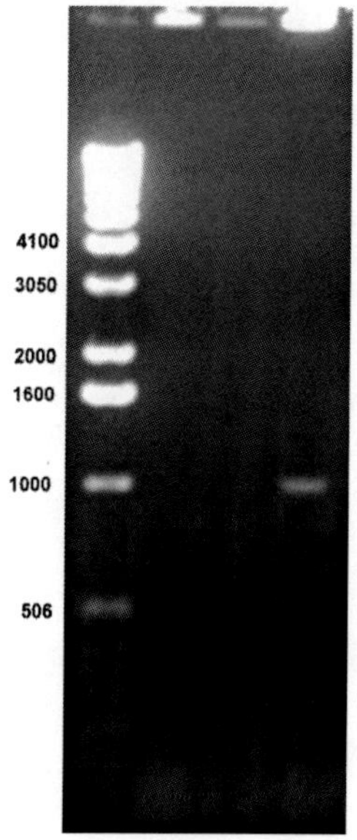

The main limitations of PCR are:

- Only relatively short sequences can be amplified reliably. Anything more than 10,000 base pairs is unlikely to be amplified.
- You need to know the right primer sequences to use, at both ends of the sequence you want to amplify. If two related genes have the same end sequences, you might amplify the wrong gene.
- You only obtain a DNA fragment. To see this DNA at work inside a living organism, some type of cloning has to be done.

- PCR Technology - discussion of preparation of the sample, the master mix, and primers, and detection and analysis of the reaction products.

Gene Sequence Analysis Modified from MIT Hypertextbook.

The sequence of DNA base pairs can be analyzed by

- Restriction mapping. Construct a "road map" of restriction sites. A program to do this is WebCutter.
- DNA Sequence Analysis. Cut and clone various restriction fragments, and determine the exact sequence of base pairs. All sequence information is deposited in GenBank. If you just want the sequence of the peptide translated from the RNA, you have to look for insulin mRNA or cDNA.

Once we have a piece of DNA cloned, it is amplified (available in many copies) and we now have a living clone which provides, in theory, an indefinite source of the DNA sequence. How do we analyze the sequence?

DNA Sequence Determination

Once you have identified a particular region of DNA of interest, you need to find out the precise sequence of DNA nucleotides. This is done by di-deoxy sequencing, in which a DNA polymerase is put together with dNTPs in four different reactions, each containing a small amount of one di-deoxy NTP (ATP, TTP, CTP, or GTP). The di-deoxy nucleotide lacks a 3'OH to continue chain extension, so the chain terminates. Each reaction produces a population of fragments terminated at A, T, C, or G.

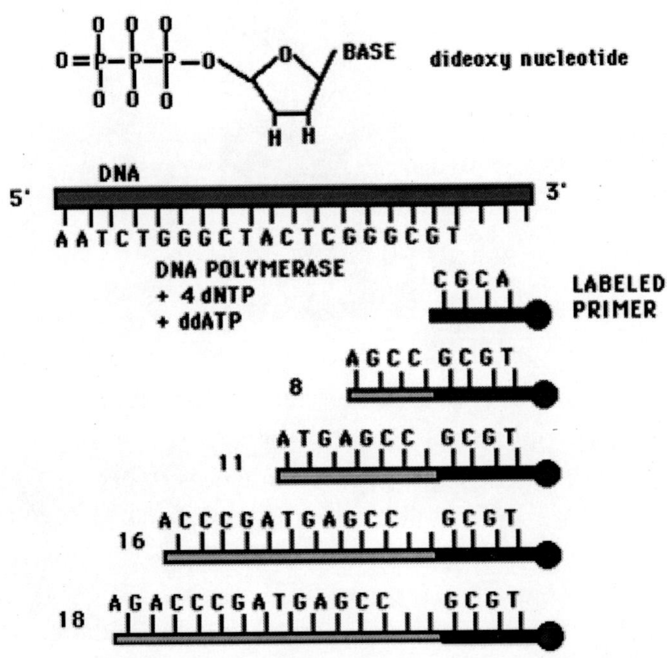

- The fragments are either radiolabeled or enzymatically labeled. They can be separated on a gel, or on a fluorescence analyzer. All published DNA sequences in the world are deposited in GenBank.

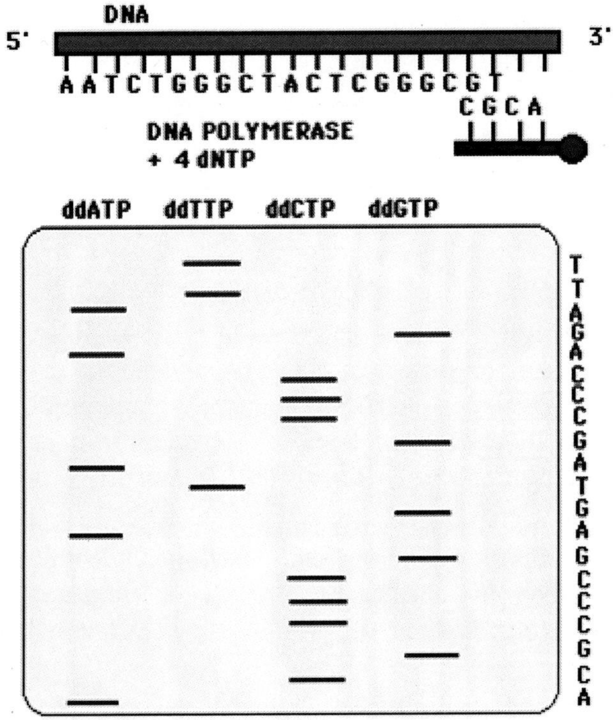

Historically, the products of dideoxy sequencing reactions were subjected to electrophoresis in four different lanes of a gel. Each lane contained a reaction using a different dideoxy terminator nucleotide. The data looked like the image at right.

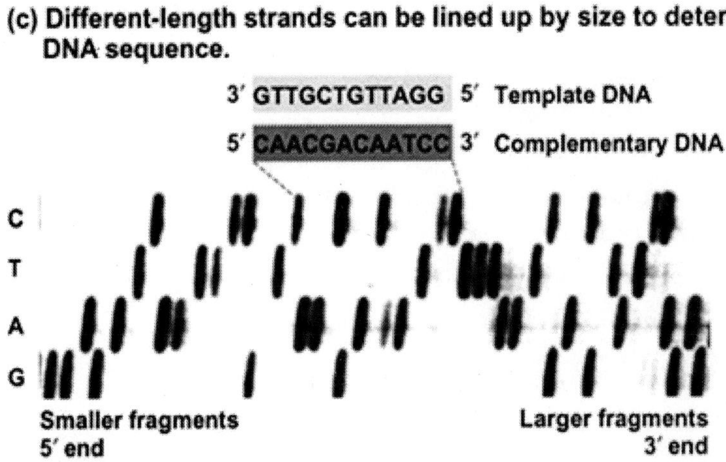

Complementarity and Hybridization

How do we find genes of interest in a genome? Or a particular mRNA in the total cellular RNA? Or a particular protein out of all cell protein?

In solution, hybrid molecular complexes (usually called hybrids) of the following types can exist (other combinations are possible):

- DNA-DNA. A single-stranded DNA molecule (ssDNA probe) can form a double-stranded, base-paired hybrid with a ssDNA target if the probe sequence is the reverse complement of the target sequence. A radiolabeled DNA probe can be applied to DNA from a gel transferred to a membrane, called a Southern Blot (named for its inventor).
- DNA-RNA. A single-stranded DNA (ssDNA) probe molecule can form a double-stranded, base-paired hybrid with an RNA (RNA is usually a single-strand) target if the probe sequence is the reverse complement of the target sequence. An RNA can be radiolabeled to probe a Southern Blot; or, a ssDNA probe can be applied to membrane-bound RNA, called a Northern Blot (name is a pun on Southern.)
- Protein-Protein. An antibody probe molecule (antibodies are proteins) can form a complex with a target protein molecule if the antibody's antigen-binding site can bind to an epitope (small antigenic region) on the target protein. In this case, the hybrid is called an 'antigen-antibody complex' or 'complex' for short. A radiolabeled antibody can probe membrane-bound proteins, called a Western Blot (an even worse pun.)

There are two important features of hybridization:

- Hybridization reactions are specific - the probes will only bind to targets with complimentary sequence (or, in the case of antibodies, sites with the correct 3-d shape).
- Hybridization reactions will occur in the presence of large quantities of molecules similar but not identical to the target. That is, a probe can find one molecule of target in a mixture of zillions of related but non-complementary molecules.

These properties allow you to use hybridization to perform a molecular search for one DNA molecule, or one RNA molecule, or one protein molecule in a complex mixture containing many similar molecules.

Southern, Northern, and Western Blots. Link to summary.

Blots are named for the target molecule.

Southern Blot—DNA cut with restriction enzymes - probed with radioactive DNA.

Northern Blot—RNA - probed with radioactive DNA or RNA.

Example—used to measure angiopoietin angiopoietin expression from cDNA in transgenic mouse.

Western Blot—Protein - probed with radioactive or enzymatically-tagged antibodies.

These molecules must then be immobilized on a solid support, so that they will remain in position during probing and washing. The probe is then added, the non-specifically bound probe is removed, and the probe is detected. The place where the probe is detected corresponds to the location of the immobilized target molecule. This process is diagrammed below:

Preparing for Blots

- Southern Blots. DNA is first cut with restriction enzymes and the resulting double-stranded DNA fragments have an extended rod conformation without pre-treatment.
- Northern Blots. Although RNA is single-stranded, RNA molecules often have small regions that can form base-paired secondary structures. To prevent this, the RNA is pre-treated with formaldehyde.
- Western Blots. Proteins have extensive 2' and 3' structures and are not always negatively charged. Proteins are treated with the detergent SDS (sodium dodecyl sulfate) which removes 2' and 3' structure and coats the protein with negative charges.

Transfer to Solid Support. After the DNA, RNA, or protein has been separated by molecular weight, it must be transferred to a solid support before hybridization. (Hybridization does not work well in a gel.) This transfer process is called blotting and is why these hybridization techniques are called blots.

Usually, the solid support is a sheet of nitrocellulose paper (sometimes called a filter because the sheets of nitrocellulose were originally used as filter paper), although other materials are sometimes used. DNA, RNA, and protein stick well to nitrocellulose in a sequence-independent manner.After a series of treatment steps, the probe is added. The probe hybridized to the target molecules is visualized either by autoradiography or by enzyme reaction.

The important properties of the three blotting procedures of DNA analysis:

	Southern	**Northern**	**Western**
What is separated by molecular weight? (target)	DNA cut with restriction enzymes	RNA denatured with formaldehyde	Protein denatured with SDS
Probe	radioactive gene X DNA	radioactive gene X DNA	Antibody against protein X, labeled with radioactivity or enzyme
What do you learn from it?	Restriction map of gene X in chromosome	-how much gene X mRNA is present? -how long is gene X mRNA?	-how much protein X is present? -how big is protein X?

DNA Microarrays

We can now put most of the protein-encoding genes onto a microarray chip, using technology based on the DNA silicon chip industry. The chip can be used to hybridize to cellular RNA, and measure the expression rates of a large number of genes in a cell.

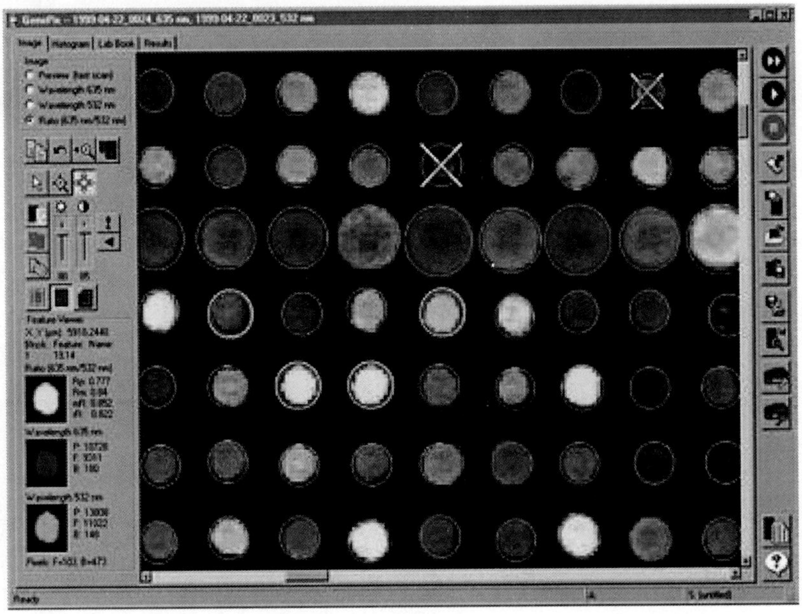

Figure: *Axon Industries.*

Transgenics

Cloning in Animals.

Animals have two different classes of cells: germ line and somatic.

Only alterations in the germ cells can be transmitted to future generations. However, some forms of somatic cell gene therapy can be useful in treating patients.

For example, people with cystic fibrosis can receive the cloned CFTR gene in a nasal spray, which then infects the lining of their lungs and improves lung function. The infected (transformed) epithelial cells eventually are lost, however, during normal processes of tissue growth, so the treatment needs to be repeated. Germ-line cloning. There are two basic ways to clone a gene in the mammalian germ line.

- Inject a DNA fragment containing your gene into the nucleus of a fertilized egg (germ-line gene cloning) or of body tissues in a mature host (somatic gene cloning). The DNA gets taken up at random somewhere in one of the chromosomes.

An example of germ-line cloning is the injection of angiopoietin cDNA into the fertilized mouse egg.

(An example of somatic cloning is gene therapy for cystic fibrosis: inhale a vector containing the CF gene, which gets incorporated into the DNA of cells lining the lungs. Somatic cloned genes are not inherited by offspring.)

- Put a transgene containing positive and negative selective markers into an embryonic stem cell tissue culture. The ES cells take up the gene by homologous recombination, replacing the host allele. The ES cells now are injected into a blastula of an unrelated host, and some of the next generation progeny arise from the ES cells. Once you have cloned a valuable gene into an animal, say a sheep that makes insulin in its milk, how can we produce lots of identical progeny quickly without sexual reproduction? This is what the popular press calls "cloning

But when you create a clone, how do you know it worked? You need to know:

- DNA: Did the transgene really incorporate into the genome? Where?
- RNA: Is the RNA expressed? (Or prevented from expression, by a null allele?)
- Protein: Is a desired protein expressed?

Cloning in Plants

Plants don't have separate somatic and germ cells, so creating transgenic plants is easier than creating transgenic animals. Techniques include the use of microprojectile bombardment (the "gene gun") and the use of Agrobacterium tumefaciens.

7

The Genetics System

The genes are much too small to be seen individually, but in certain stages many chromosomes show lengthwise patterns of knobs or bands that indicate their ubdivision into genes. In par ticular, this happens during the early stages of meiosis. A pair of rye chromosomes at the beginning of meiosis. You notice the pattern of darkly stained knobs and faintly stained connecting strands. Thepattern in the two partner chromosomes is the same, and pairing is exceedingly accurate, each knob in one chromosome lying next to the matching knob in its partner. It appears that it is the genes rather than the chroosomes as a whole that direct the pairing process.

This ability of the genes to find and attract their exact counterparts is one of their many remarkable properties for which science has not yet found a satisfactory explanation. Later on in meiosis the pattern will be lost because the chromosomes curl up and contact into the compact bodies. But when the next generation makes germ cells, each chromosome will again show its characteristic pattern, so that a trained cytologist can recognize every one of the seve chromosomes of rye by the pattern of its knobs. Giant chromosomes withbeautiful and intricate patterns of dar and light bands are found in the salivary glands of the larvae (maggts) of certain flies. Chromosomes in the salivary glands of Drosophila larvae.

They are hundreds of times aslong as ordinary chromosomes, and each of them has its own characteristic pattern of bands. They are also much thicker than ordinary chromsomes because each consists of a bundle of many strands, all with exactly the same pattern of bands. The way these bundles arise is intresting because it illustrates the extreme accuracy of chromosome replication.

In the young larva, the chromosomes in the salivary glands are still thin. As the larva grows, the salivary glands grow too, but they do so in a particular way. The cells, instead of making more cells by division, grow bigger and biggerwithout division.

Their nuclei, too, grow bigger and bigger. Inside the nuclei, the chromosomes make replicas of themselves as they doin preparation for mitosis but, as there is no mitosis, the new chromosomes do not separate from the old ones. Instead, new and old chromosomes remain closely attached to each otherand each, in turn, makes more replicas. This goes on repeatedly, and so accurately does each band make another band like itself that the whole bundle appears as one single thick banded thread. In addition, the partner cromosomes in the salivary glands are closely paired as in meiosis, so that each apparent chromosome represents one pair, and thenuclei of the salivary glands of Drosophila seem to have four chromosomes instead of the eight that are visible in ordinary cells).

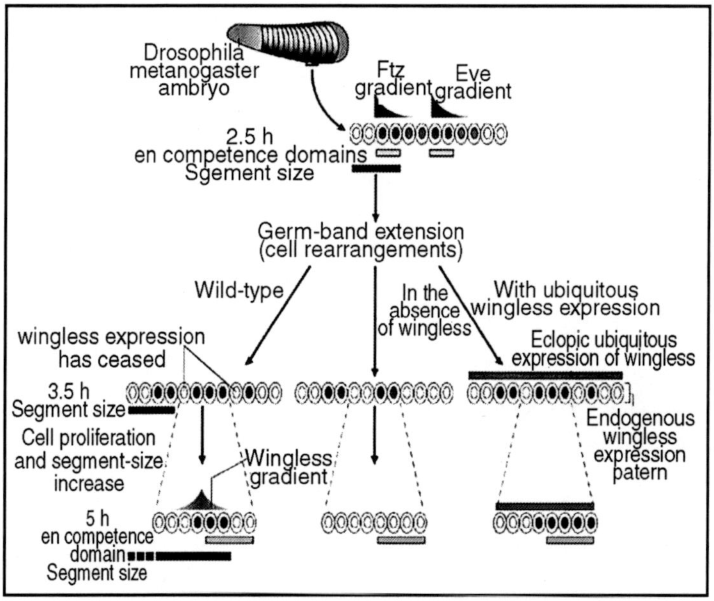

Figure: *Wing Gene in Drosophila*

The pattern of bands corresponds to the pattern of underlying genes. This can often be seen in cases where a chromosome has lost a small piece through radiation or some other injury. The accuracy of chromosome replication is such that all descendants of the damaged chromosome lack exactl the same piece, and when such a damaged chromosome gets into a gamete, for instance an ovum, individuals are formed which lack the same chromosome piece in all their cells. The left part of a normal wing of Drosophila and above it, and at a very much higher mgnification, a little section of one of the normal salivary gland chromosomes. The arrow points to a band which encloses a gene for wing shape. The right part of the same chromosome section in afly which had received a damaged chromosome from its X-rayed father or mother.

If you recall that the salivary gland chromosomes consist of two chromosome bundles formed by the two partner chromosomes, you will unerstand why only half a band is missing at the point of the arrow. Nevertheless, the lack of one of the two partner genes has been sufficient to disturb normal development of thewing and to produce the abnormal shape shown in the accompanying sketch. Flies which have lost this band from both partner chromosomes are so much damaged thatthey die as early embryos.

The left drawing represents a normally growing green maize plant and underneath, at much higher magnification, the tips of one particular chromosome pair. Thee is a fairly large knob at the end of these chromosomes. Among the genes which this knob encloses is one whose action is necessary for the production of the green substance chlorophyll that is essential for survival of the plant.

The right picture shows that, when this knob has been lost rom both chromosomes, the plant dies early as a yellow and sickly seedling. If you turn once more to the left picture you will notice a striking difference from what happened in the case of the Drosophila wing gene. Whereas this gene could not produce anormal wing without the co-operation of its partner gen, the chlorophyll gene is sufficiently effective even when acting singly; for plants with one normal and one knobless chromosome are quite normal.

Tis is not a distinction between the genes of Drosophila and maize, or between genes concerned with wing hape or chlorophyll. It is a difference in the efficiency with which genes carry out their functions; some genes are elfsufficient in this regard, others need the co-operation of their partners for effective action. We shall meet a similar difference when we deal with dominant and recessive genes. Each gene is concerned in the control of a particular developmetal process, for instance the formation of chlorophyll, the development of colour or size or, in higher animals, mental ability. Partner genes control the same process, but they may do so in different ways; thus two partner genes in peas are concerned with the control of seed shape, but one causesthe seeds to be round, while the other makes them wrinkled.

Partner genes with different effects are called alleloorphs or, for short, alleles. The term "allelomorph" signifies "diffrent shape"; according to modern ideas this may be interpreted as different chemical configurations of basically the same gene. All developmental processes are highly complex, and it is therefore not surprising that many different gene pairs are involved in the develpment of, say, an eye or a brain or a flower. Even such a seemingly simple achievement as the formation of a vitamin by a bacterium proceeds in a series of steps and is controlled by a series o genes.

In bacteria, genes controlling the same process tend to lie close together, as though this facilitated their co-operation; but in most other organisms this is not found to be so. Genes which are involved in quite different processes may lie cheek by jowl on the same chromosome, while genes controlling closely related processes may lie far apart. The position of a few of the known genes on a particular pair of chromosomes of the fruit fly Drosophila melanogaster.

Only genes concerned with the development of the eyes and the wings are represented; more genes controlling development of these organs are present on the same chromosome pair and on the three others. Some of te represented genes can be seen to act in very different ways; thus wing gene w_1 is concerned with the control of wing shape, while wing gene w_2 helps make the correct pattern of wing venation. For other genes such a distinction is not evident; thus eye genes e_1, e_2, e_3 and e_4 ar all engaged in the formation of eye colour. Yet we know from developmenal studies that each of them has a different role to play in this process.

Figure: *The Positions of Some Wing and Eye Genes*

The genes, like the chromosomes, are arranged in pairs; with the chromosomes they segregate into different gametes at meiosis and come together again at fertilization, one member of each pair being derved from the father, one from the mother. The arrangement of the genes on the chromosomes is fixed for each species, and partner genes occupy identical positions on partner chromosomes. Allelomorphs are partner genes with different effect, that is, they are genes whch occupy the same position on the same chromosome pair and control exactly the same developmental process, but they do so in different ways.

Genetic Code

Even more remarkable is that non-overlapping triples of the letters *A, C, G* and *T* which are often reerred to as codons (especially when the triples are part of a gene), may spell out (reading in the 5' to 3' direction) the order of the amino acids that form the long linear molecules known as proteins. It turns out there are 20 amino acids which are the building blocks for proteins. Since there are 64 possible triples (repeated letters within the triples are allowed) using the 4 letters *A, C, G,* and *T*, different triples can represent the same amino acids. There are also 3 triples (*TAA, TAG, TGA*) known as stop codons that do not represent (a common phrase being do not code for) an amino acid, but are involved in signaling a termination to the protein production process. One particular codon (*ATG*), which does represent one of the amino acids, is an indication for the start of the production of a protein. This system for coding proteins has come to be known as the genetic code.

Before the genetic code was fully understood, based on work by Marshall Nirenberg, Heinrich Matthaei, Har Gobind Khorana, and others, various mathematical ideas were invoked to suggest what kind of code might be involved. There is a diagram of the RNA version of the genetic code, where *U* takes the place of *T*. One approach to finding a gene is to look for a stretch of DNA which starts with ATG (the start codon) and ends with one of the stop or termination codons. This situation in a stretch of DNA is referred to as an open reading frame (ORF). However, not all ORFs correspond to a stretch of DNA which will initiate the production of a protein, and so the location of an ORF is not equivalent to finding a gene. A major problem facing biologists is how to locate those stretches of DNA that are genes in the large amounts of DNA found in a chromosome.

The way that DNA is involved in the production of proteins is not direct. The mechanism involves another helical molecule called RNA, of which there are a variety of types. (Among the types of RNA are messenger RNA, or mRNA, ribosomal RNA, or rRNA, and transfer RNA, or tRNA.) When a protein is to be produced, the DNA separates and a copy of the gene is transcribed (the process is referred to as transcription by molecular biologists) into a strand of RNA (ribonucleic acid).

RNA, like DNA, uses an alphabet of 4 nucleotides *A, C, G*, and *U* (for uracil, a pyrimidine like thymine which takes the place of *T*). In a somewhat complex series of steps, a protein is produced. A very rough schematic of how genes create proteins. This schematic is often referred to as the central dogma. The original DNA, which are known as introns but which are not involved in the production of a protein, are separated or snipped out and a resulting string of exons from the original DNA is left. What is left is sometimes referred to as a coding sequence, and it contains the codons (in the letters *A, C, G* and *U*) which will create the protein, by the process usually referred to as translation.

The actual manufacture of the proteins involves structures called ribosomes. Production of proteins does not go on within th nucleus but in the cytoplasm surrounding thenucleus. One can express the codons involved in the manufacture of proteins either in DNA terms (using the letters *A, C, G* and *T*) or in RNA terms (using the letters *A, C, G* and *U*).

Implications

Many interesting quesions can be asked about the genetic code, including why the pattern of redundancies in the codeexists. While at one time it was thought that the code was totally universal, this has proved not to be true. For certain bacteria and in other situations the universality of the code breaks down. Cellular life forms are classified according to whether they are prokaryotes or eukaryotes. Prokaryotes are cellular organisms where the genetic material is not a part of a membrane enclosed nucleus, while in eukaryotes the genetic material is contained in a nucleus.

At one time life forms were classified as either bacteria or eukaryotes. A group of bacteria like organisms, found in very hot environments such as undersea vents, were initially classified as bacteria. However, when Archaea's genetic makeup was studied, its members were found to have an amazingly large number of genes that were not known to exist in other organisms. They seemed to be surprisingly different from the bacteria with which they had been formerly classified. It has become standard in recent years to list the Archaea in a separate category from the bacteria and eukaryotic organisms. Use of phylogentic trees was in part a tool that led to this change.

It is tempting to believe that, because human beings are as complex as they are, other species will have less genetic material than humans. However, chromosome number (and the total size of the genome) varies with species in a way that seems to have no rhyme or reason. Thus, chimpanzees have 48 chromosomes, cats 38, dogs 78, rabbits 44, rats 42, turkeys 82, and horses 64. Perhaps when the genomes of many different and varied species are sequenced it will be possible to get insight into these wildly varying numbers.

Mendel's First Law: Segregation

After this excursion into cytology, lot us return to genetics. The chief tool of the cytologist is the microscope; it has shown him that the chromosomes, and with them the genes, segregate at meiosis and enter the gametes singly. The chief tool of the geneticist is the crossing of different strains, and we are now going to consider crosses which demonstrate the segregation of the genes at meiosis.

It will be useful for the description of these experiments if we recall some technical terms which have already been defined, namely "haploid" and "diploid" for cells or individuals which carry one or two chromosome sets, respectively, and "allelomorphs" or "alleles" for partner genes with different effects. To these terms I want to add two new ones which describe the genetical constitution of a diploid individual in regard to any particular pair of genes. Evidently there exist two possibilities: either the partner genes are alike or they are allelomorphic. In the first case, the individual is "homozygous" or a "homozygote" in respect to the gene concerned; in the second case, it is called "heterozygous" or a "heterozygote."

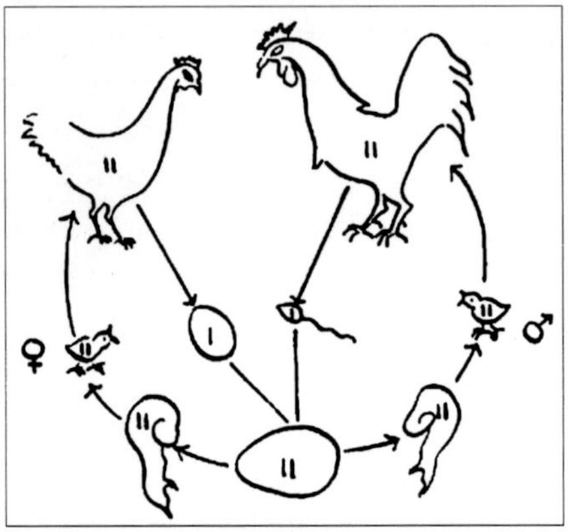

***Figure:** Life Cycle of Fowl*

A pea plant with two genes for wrinkled seeds or two genes for round seeds is homozygous for either wrinkled or round; a pea plant with one gene for wrinkled seed and an allelomorphic gene for round seed is heterozygous for wrinkled and round. It will readily be seen that only heterozygotes are suitable for demonstrating the segregation of genes; for the geneticist recog nizes genes by their effects, and the segregation of like genes in a homozygote is not observable.

In higher organisms (macroorganisms) it is hardly ever possible to observe the immediate effects of segregation at meiosis; for the mature gametes fuse at fertilization and the progeny, in which we observe the effects of the genes, carries genes from two different gametes. The life cycle of the domestic fowl in terms of haploidy and diploidy. The diploid female and male form haploid ova and spermatozoa. At fertilization the gametes fuse and form the fertilized ovum, which in turn develops into the diploid embryo, chicken and adult. Thus the whole life cycle, with the exception of one cell generation, is spent in the diploid stage. The situation is quite different for many microorganisms. The life cycle of a green alga, Chlamydomonas.

Each individual of this species is a microscopically small cell which swims through the water with the aid of two whiplike hairs, or "flagellae." These cells are haploid; they have only one set of chromosomes, carrying all genes of the species in single copies. New cells are formed through division of the old ones.

In this way whole populations may be derived from one cell by repeated divisions. Such populations are called "clones." Since at mitosis both daughter cells receive exactly the same set of chromosomes and genes all cells within a clone are genetically identical.

There are no visibly distinct sexes, but physiologically there is a clear distinction between sexes or "mating types," called + and–in our diagram. Cells of the same clone are always of the same mating type and never mate with each other; nor does mating occur when cells of two + clones or two–clones are brought together. But when cells from *a*+ and *a*– strain are put into the same culture vessel, pairs are fomed.

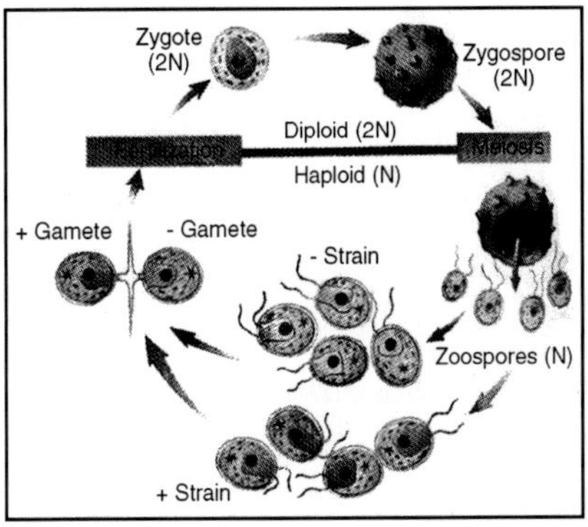

Figure: *Life Cycle Algae Chlamydomonas*

Each pair consists of one + and one–cell they fuse and form a thick skinned pore called a zygote. The zygote is comparable to a fertilized ovum in, say, a dog or a fly. It contains the chromosomes from both gametes and is, therefore, diploid. Very soon after fertilization meiosis takes place inside the zygote, and haploid gametes swarm out.

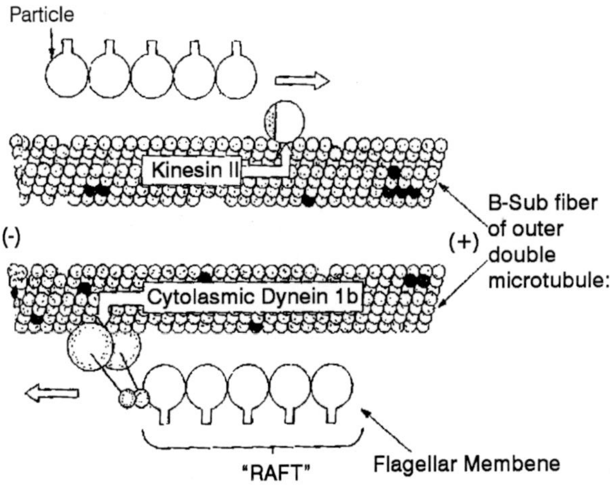

Figure: *Absence of Flagellae in Chlamydomonas*

These start a new life cycle by division; in fact, the gametes of Chlamydomonas do not differ from any of the other free swimming ells. Thus in this algae, as in many other microorganisms, the whole life cycle, with the exception of the zygote, takes place in the haploid stage, and segregation at meiosis is not obscured by diploidy of the progeny. Usually it is necessary to grow the gametes into clones before segregation becomes observable; but occasionally the segregating gametes themselves differ visibly from each other. Two strains (1) and (2) differ in one gene which is concerned with the development of the flagellae. While strain (I) has all the genes necessary for flagella formation, strain (2) carries a deficient allele of one of these genes and cannot, therefore, develop flagellae.

When the two strains are crossed by bringing together + cells from the one strain with cells from the other (it makes no difference which strain provides the + cells, which the cells), a zygote is formed which carries both alleles and is, thus, heterozygous for the gene in question. At the reduction division the partner chromosomes segregate and carry the different alleles into separate cells. At the second meiotic division each cell divides once more by mitosis, so that finally four gametes are formed, two of which carry the normal and two the deficient allele. When these swarm out, it will be seen immediately that two have flagellae, two have not. This difference persists in the clones formed from the gametes.

If we let each gamete grow into a separate clone, two clones will consist of flagellated free moving cells, while two will consist of flagellaless cells which sink to the bottom of the tube. The fact that a heterozygous zygote always produces two gametes carrying one allele and two carrying the other is so well established that it can be used, conversely, for determining whether a difference between two strains of a microorganism is due to the effects of a pair of allelomorphic genes.

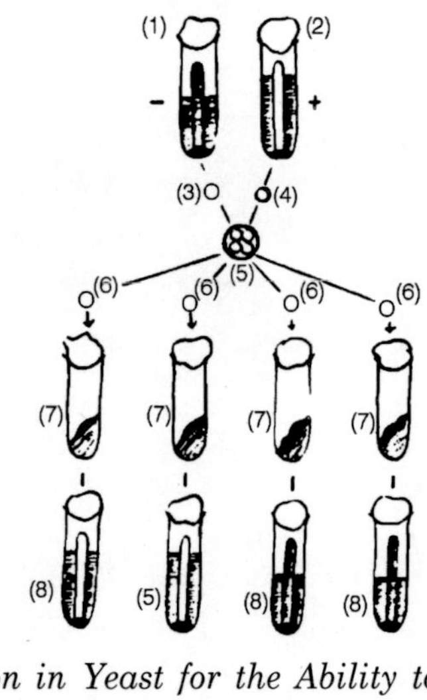

Figure: *Segregation in Yeast for the Ability to Ferment Galactose*

If this is so, the gametes from a cross between the strains will yield two clones, resembling the one strain, and two resembling the other. Yeast, like

Chlamydomonas, is a unicellular organism. Many strains are haploid and can be used in experiments on segregation. Like Chlamydomonas, yeast occurs in two mating types, + and –, which are not visibly different from each other.

Yeast, as you know, ferments sugar into alcohol ad carbon dioxide gas; but strains differ in their ability to ferment various kinds of sugar. Thus some strains readily ferment galactose (one of the components of milk sugar), while others do so only very slowly and inefficiently. A simple method for detecting fermentation is to catch the developing gas in a est tube which was first completely filled with liquid and then inserted immediately above the yeast cells. If the medium contains a sugar which can be fermented, the tube gradually fills with gas; if the sugar cannot be fermented, the tube remains filled with liquid.

In the experiment shown in the medium in the top row contains galactose as the only sugar. Strain (2) is able to ferment this sugar, strain (1) is not. Cells (3) and (4) from the two strains are brought together, and a zygote (5) is formed which, through meiosis, gives rise to four gametes (6). Each gamete is grown into a clone (7) on solid medium containing glucose, a sugar which can be fermented by all strains of yeast. Finally, the four clones ae tested (8) in fermentation tubes with galactose as the only sugar. You will notice that two clones (those on the left) can ferment galactose, and two cannot.

Thus the difference between the two original parents segregates in the 2:2 ratio expected if it is due to the action of a single gene pair. On the basis of several experiments of this kind it could be concluded that in yeast the ability to ferment galactose is controlled by the action of a gene which in nonfermenting strains is absent, or is present in the form of a nonfunctioning or poorly functioning allelomorph.

The essence of what is called Mendel's first law or the law of segregation. In modern terms it may be expressed as follows: When a heterozygote forms gametes, half of them carry one of the two alleles, half the other. The basis for the segregation of the genes is the segregation of the chromosomes at meiosis. Mendel himself could not, of course, interpret his results in these terms, but he quite clearly postulated that a mechanism of this kind was responsible for his data, in particular for the famous and much overrated 3:1 ratio. This particular ratio is a rather remote consequence of the underlying gametic segregation in a ratio of 1:1. The steps by which, in crosses between higher organisms, this 1:1 ratio becomes transformed into oters, including.

In doing this we shall find ourselves using very simple statistical rules, such as are used for predicting that a coin tossed 100 times will show head about 50 times. This statistical aspect of genetics has been held against it by the Lysenko school, on the grounds that mathematics must not be applied to biological objects and that Mendel had no right to speak of a 3:1 ratio between round and wrinkled peas if some of the pods contained 6 round peas. No matter what we think about these arguments—and to a modern biologist they seem entirely futile.

Characteristics of Genetic System

X-Chromosome as Type

The chromosome combinations present in the fertilized egg, and this has brought to light some of the methods by which the chromosomes act on development and characteristics. We now Examine other effects of altering chromosomes. We shall have to examine the effects of altering all the different classes of chromosomesýÿthe X-chromosome, the Y-chromosome, and the different autosomes. Each of these three groups (X, Y, and autosomes) gives rise, as we shall see, to a different type of inheritance. The simplest relations, from the experimental point of view, are presented by the X-chromosomes. We therefore deal first with these.

Effects on altering the X-Chromosomes

The X-chromosomes take such a course in passing from generation to generation that it is possible to follow the descendants of a particular Xchromosome (that present in the original male parent, for example), knowing in which individuals they are present, in which they are absent. In certain individuals (males, in organisms of Group I) there is but a single X-chromosome instead of a pair.

These relations make it a relatively simple matter to discover the distinctive effects of a particular X-chromosome. The effects of X-chromosomes will therefore be dealt with somewhat fully as a type of chromosomal action. Later the action of the other chromosomes will be taken up.

Abnormal Distribution of X-Chromosomes

What will happen if the X-chromosomes by accident become irregularly distributed? We know that sometimes the X-chromosomes are indeed irregularly distributed. Will the sex-linked characteristics continue to follow them; will the defective characteristics show the same irregular distribution as do the X's? Such cases in great number have been fully observed. It is found that the sex-linked characteristics do indeed follow the X-chromosomes wherever they go. This proves conclusively (if there were any possible doubt in view of the extraordinary course normally followed by such characters) that it is indeed the X-chromosomes on which the characters depend. The matter is one of importance and interest, so that it will be worth while to examine carefully certain typical cases of the result of irregular distribution of X-chromosomes.

Normal Distributions

Sometimes the body of Drosophila is yellow instead of the normal grey; this is a recessive character due to a defect in the X-chromosomes. Suppose that we mate together a female that has a yellow body and a male that has the normal grey programmes.

The female has two modified X-chromosomes, which we may represent by XX. The male has its X unmodified; we may represent its cells as XY. Normally in forming the germ cells, the two X's of the female separate into different germ cells, and all the ova receive one of the modified X's. The normal male produces germ cells of two classes, X and Y.

When the germ cell X from the female unites with Y from the male, sons are produced, with the constitution XY; while X from the female with X from the father gives daughters XX. Since the yellow programmes dependent on X is recessive, the daughters XX have the normal grey programmes, like the father.

But the sons XY have only one X, and as this is the defective one, from the mother, the sons are yellow, like the mother. The normal result of such a mating therefore is that all the sons are recessive like the mother, all the daughters dominant, normal, like the father. This is what occurs when the X-chromosomes are distributed to the germ cells in the usual way.

Abnormal Distribution

In forming germ cells both X's of the female go together to a single ovum, while other ova receive no X. In a race discovered by L. V. Morgan the two X-chromosomes in the female were partly united, so that they almost invariably thus go together to one ovum, leaving other ova without an X. In this race the X-chromosomes were so modified as to produce th recessive yellow programmes described in the preceding paragraph. We therefore have an opportunity to determine the course of inheritance of the yellow and the grey programmes when the X-chromosomes are thus abnormally distributed.

Represent as before the X-chromosomes that produce the yellow programmes by X. Thus one set of ova receive XX, while another set recive no X. When the germ cells XX united with the germ cells Y from the norml father, there wee produced daughters XXY, in which both X's came from the mother, and these daughters were yellow like the mother (since no normal X was present). Sons were produced by the union of ova that contained no X with sperms that carried X; this gave sons XO, with their single X from the normal father, instead of from the mother, as is usually the case. And such sons had the normal grey body programmes of the father, instead of the yellow body programmes of the mother, as happens normally.

Thus with a change in the distribtion of the X's there is a corresponding change in the method of inheritance. If the sons receive their X from the recesive mother, as in the normal cases, they are recessive lke the mother. But when, through non-dijunction, the sons receive a dominant X from the father, they are dominant like the fther.

Similarly, in the normal case the daughters receive a dominant X from the father and are therefore dominant like the father. But in cases of non-disjunction the daughters receive both their JPs from the recessive mother, and are therefore recessive like the mother. In sum, when the X's are normally distributed, uch matings give 'criss-cross inheritance'; sons like the mother, daughters like the father. But when the X's are not so distributed, there is no criss-cross inheritance; sons are like th father, daughters like the mother. The dominant and recessive characters follow the respective X-chromosomes, whether these are distributed normally or abnormally.

Such experiments have been repeated many ties, and with other sex-linked characters. Always the characteristics follow the distribution of the X-chromosomes, however these are distributed. It may be concuded with certainty that it is the presence of X-chromosomes of a modified type that causes the appearance of the particular sex-linked characters that re manifested. The rules of distribution of sex-linked characters are the rules of distribution of the X-chromosomes.

Dominance and Recessiveness in Sex-linked Characters

As we have seen, and shall see further, many defective conditions are due to defects in certain X-chromosomes. These follow from generation to generation the distribution of these X-chromosomes. Most of these bodily defects are manifested only in individuals in which the defective X- chromosomes are the only kind present; that is, these defects are recessive.

But it is important to observe that in such cases the normal condition of the organism likewise follows the distribution of certain X-chromosomes. Haemophilia follows certain X- chromosomes. But in the same matings in which this occurs, the healthy condition of the blood follows certain other X-chromosomes. If a defective mother is mated with a normal father, the sons are defectie because they receive only the mother's defective X-chromosomes.

But in the same way the daughters are normal because they receive the father's normal X-chromosomes. In the normal individuals the X-chromosome plays a part, just as it des in the defective individuals. In the normal individuals it is healthy and supplies what is required for normal development, while in the defective individuas it fails to supply what is required. In the case of most defects this normal condition is dominant over the recessive condition; that is, when both are present, the normal chromosome is the one that prevails in its effect on the indvidual. There are defective characters, however, in which the defectivecondition is dominant or partly dominant.

In these cases, when a defective X and a normal one are present together, the defective one produces its effect, as in the case of bar-eye, described earlier. In many such cases the defect is less marked when a normal chromosome is present as well as a defective one; then the defective condition is said to be partly dominant. Thus whenever an individual bearing defective X-chromosomes is mated with a normal individual, we have both dominant and recessive characters, the course of which may be followed in the later generations the dominant character being commonly the normal or usual condition. How this works out is shown in our next paragraphs.

Tests for Sex-linked Inheritance

By observing the distribution of the dominant and recessive characters among the offspring of certain matings, a test for determining whether given characters depend on the X-chromosomes is supplied. The test consists in making what are called reciprocal crosses.

- On the one hand mate a dominant female with a recessive male;
- Also mate a recessive female with a dominant male.

Represent a dominant X-chromosome by a capital X, a recessive X-chromosome by a lower-case letter x. In the group of organisms with which we have been dealing, the female has two X-chromosomes, the male but one. The daughters receive an X from each parent, the sons an X from the mother only. A dominant female will be represented by XX, a recessive male by xo; similarly a recessive female is xx, a dominant male is XO.

The two ma tings and their results will then be represented as follows:

- XX by xo gives X + XO Dom Mother Rec. Father Dom. Daughters Dom. Sons
- xx by XO gives X + xo Rec. Mother Dom. Father Dom. Daughters Rec. Sons

Thus when the mother is dominant, the father recessive, all the children are dominant, like themother. When the father is dominant, the mother recessive, the daughters are dominant like the father, the sns recessive like the mother; this is called 'criss-cross inheritance'. Whenever reciprocal crosses give these results, it is certain that the two diverse characteristics (dominant and recessive) result from dfferences in the X-chromosomes of the two parents. 'Criss-cross inheritance' is particularly useful in showing at once that we are dealing with sex-linked inheritace.

In every case where such results are produced, further tests show that the two haracteristics follow in later generations the two different kinds of X-chromosomes wherever they go. t is mainly by the use of these tests that the many different characteristics dependent on diversities in X-chromosomes have been discovered.

Sex-linked Inheritance in Group II

Sex-linked inheritance as we havejust described it was originally discovered n animals belonging to Group I, in which the females have two X-chromosomes, the males but one (with or without a Y- cromosome). But when the tests we have just described were applied to birds and to certain other animals, it was discovered that there is a scond group (Group II) in which it is the male that has two X-chromosomes, the female but one. In th common fowl, reciprocal crosses were made between 'barred' fowls (Plymouth Rock) and black fowls (Langshan). Here the 'barred condition was found to be dominant.

The results were:

- Barred father by black mother=Barred sons + barred daughters.
- Black father by barred mother = Barred sons + black daughters.

Here the second mating gives 'criss-cross inheritance', showing that we are dealing with sex-linked characters dependent on differences between the X-chromosomes of the parents. But, in the first mating above, all the offspring are like the father instead of like the mother, which latter was the case in the matings. The father in this case has the dominant characteristic (barred), and all the offspring are dominant like the father instead of like the mother.

When the mother is dominant ('barred') the result in the fowls is criss-cross inheritance; while, in the cases before described, criss-cross inheritance occurs when it is the father that is domiant. In fact, in the birds, the males and females simply exchange roles, as compared with their roles in the organisms of Group I.

Since the roles of males and females are interchanged in the two groups, their chromosomal conditions must also be interchanged; that is, since in Group I the female has two X's, and in Group II the male plays the same role as the female of Group I, the male of Group II must have two X's, the female but one. The chromosomal conditions in the two group are shown in the above tabulation, X signifying dominant, while x signifies recessive, and o signifies the lack of X (whether Y is present or not).

Careful examination shows that this is the only way in which the results in Group II can be produced. Birds and certain moths show inheritance of the kind typical for Group II. In both these it has since been found under the microscope tha the males have indeed one more chromosome than the females.

Role of Development in X-Chromosomes

A great many characteristics, of various kinds, have been found to be the result of modifications of particular X-chromosomes, and thus to follow in later generations the distribution of the descendants of those chromosomes. Examination of a number of these is desirable, both for their own importance and for the light they throw on the functions of the X-chromosomes.

In man, the following characteristics, among others, dependent on diverse types of X-chromosomes, are known from the fact that they show typical sex-linked inheritance.

- *Haemophilia*: Lack of coagulability in the blood. This results from a serious defect in certain X-chromosomes. The existence of defective X-chromosomes having this result shows that the normal X-chromosomes play a part in supplying something necessay for producing normal blood that coagulates properly.
- *Programmes Blindness*: The fact that defectiveness in X-chromosomes causes programmes blindness shows that the normal X's play a role in producing the normally functioning eyes.

Night Blindness

- Inability to see in a poor light.
- Near-Sightedness ofcertain types.

Progressive atrophy of the muscles. The fact that defects in X-chromosomes have this effect shows that thenormal X's play a role in the normal functioning of the nerves and muscles. A considerable number of other sex-linked characters are known in man. But man is a very unfavourable organism for the study of inheritace. Yet even the little that is known of sex-linked inheritance in man shows that the X-chromosome plays a role in many diverse bodily functions.

For a more complete idea of its role, some organism that can be bred experimentally must be examined. For this urpose the fruit-fly, Drosophila melanogaster, is the best organism to select, since it has been studied more extensively than any other. In the fruit-fly we find peculiarities of the following types that are dependent on alterations in particular X-chromosomes:

- *Many different eye colours*: The normal eye programmes in this organism is a certain shade of red. Defects in different X-chromosomes result in producing, in the individuals that bear these defective chromosomes, many different shades f red, varying from deep red to a very light red, and thence to 'buff', 'ivory, and 'white'. More than a dozen different types of eye programmes are known to result from modifications of the X-chromosome. It is clear that the normal X-chromosome plays an important role in producing the normal eye programmes. Structural peculiarities of the eye, such as 'bar-eye', 'facet eye', 'furrowed eye', and the like, are known to result from modifications of the normal X-chromosome.

- *Wing modifications*: Many different conditions of the wings are known that depend upon diversities among X- chromosomes borne by different individuals. These affect all sorts of features of the wings: size, form, venation, function. The normal X-chromosome obviously plays an important role in the full and normal development of the wings.
- *Body colours and markings*: Modifications of the X- chromosomes produce the body colours yellow, sable, tan, chrome, lemon, green, and the like, in place of the normal grey. Other changes in X alter the distribution of pigment on the body, giving the characters dot.
- *Body structure*: A defect in X produces irregularities in the abdomen, known as 'abnormal abdomen'.
- *Legs*: A defect in X-chromosomes causes abnormal development of the legs; some of them are wholly or partly reduplicated, so that the total number of legs is increased.
- *Bristles*: Modifications in X-chromosomes result in various different changes in the bristles that are scatered over the body.

Many physiological conditions and functions likewise depend upon the X-chromosomes, since they are change when these chromosomes are modified. Among these are the following:

- *Positive reaction to light*: rosophila with normal X-chromosomes fly towards a source of light. Those having defective X-chromosomes of the kind that produce a tan-coloured odydo not fly towards a source of light.
- *Weakness and short life*: Most of the, defective X-chromosmes that cause structural or other changes in the body (different eye colours, wing forms, and the like) produce likewise weakness and short life. The individuals bearing them are ess resistant to bad conditions, and live or a shorter time than the individuals that bear the nomal X-chromosomes.
- *Life and death*: Some X-chromosomes have defects that are so severe that the individuals bearing them will not live and develop, unless there is resent also a normal X-chromosome. Such defects are known as lethals. The presence of such lethal defects in the X-chromosome has the result that males bearing them die, since they have but one X; while females usually live, since they have an additional X that is often normal.

If a moher that has one lethal X and one normal one is mated to a father that has a normal X, the result is that half of the sons which receive the mother's defective X fail to develop, while the other half of the sons,receiving the mother's normal X, live and develop. All the daughters live, since they all receive a normal X-chromosome from the father. The consequence is that in such families there are twice as many daughters as there are sons. In families of 100 to 200, as occur in Drosophila, this is very striking. X-chromosomes are known to play in other organisms roles similar to those mentioned above for Drosophila, although in no other organism has the matter been so fully studied.

We may summarize what has been brought out above in the following statemets:

- Many defects that appear in individuals are the result of defects in the X-chromosomes that they bear.
- Some of these defects re dominant; they are manifested in all individuals in which the defective X i present.
- But most such defects are recessive; they are manifested only in individuals in which the defective X is the only kind present.
- When both a normal X and a defective X are present, in most cases the normal X performs the required functions, so that the individual is not defective.
- Since the female has two X-chromosomes, while the male has but one, such defects are more frequently manifested in the males. Usually one of the X's present in the female is without the defect, so that she is not defective. But when the male carries a defective X-chromosome, the defect is manifested.
- It is thus advantgeous, so far as the occurrence of defects is concerned, to have two X-chromosomes rather than one.
- This gives the female a considerable advantage over the male in these respects. There are many different types of defects due to defective X-chromosomes; almost all of them are moe common in males than in females. Some of these defects seriously injure the health, or even result in death, if the defective X is the only kind that is present. It is well known that, in general, the death rate is higher in males than in females; this is probably due to the fact that males have but one X-chromosome.
- Among the different indiiduals of a species are scattered a great number of different types of X-chromosomes having different effects on development. Many of the diverse X-chromosomes are distinctly efective,

causing personal defects in the individuals that bear them. Others are not defective.

- Defects or modifications in X-chromosomes affect in different cases al parts and functions of the organism.

The Role f the X-Chromosomes in Development

What do the facts brought out in the preceding sections show as to the phyiological functions of the X-chromosomes; as to their role in deveopment? For every kind of effect resulting from defects or modificacations in the X-chromosomes, there is a corresponding (opposed or diverse)action of normal or unmodified X-chromosomes. Since defective X-chromosomes of a certain type produce in Drosophila short, ill-developed wings, it follows that the normal X-chromosomes act on the development in such a way as to give long, well-developed wings.For if we substitute the norma X for the defective one, this cause the normal well-developed wings to be produced.

Similar reasoning applies to all the defective conditions that result from defective X's; the normal X's act in such a way as to produce the corresponding normal conditions. It follows therefore that the normal X-chromosomes play a role in the production of eye programmes and structure, body programmes and structure, in the development of the legs, the wings, the bristles, in producing normal health, resistance and vigour, and in various physiological processes. We know, further, that they play a most important part in determining sex, with all that this includes of structural and physiological influence. Clearly therefore the X-chromosomes play a most important part in development. Their action is not limited to any single part of the body, nor to any single class of functions. They enter into the processes of development in such a way as to influence all parts ofthe body, all functions of the body. They begin to influence development very early, as was seen in considering the development of sex. And they play important roles in such relatively late processes as the production of eye colours.

It can hardly be doubted that they enter into the developmental activities from practically he beginning, influencing all the later processes that occur.

Autosomes and T-Chromosomes

The X-chromosomes have been dealt with because of the great advantages they offer for experimental study. It was seen that they affect development in many ways, and that in a given species there are many different types of X-chromosomes in the different individuals, causing them to show different characteristics, structural and physiological. Is the X-chromosoe typical in these relations? Shall we find similar effects in the other chromosomes? In addition to the X-chromosomes, there are the autosomes, which are present as a rule as several or many pairs. In man there are 23 pairs of autosomes: in Drosophila there are three pairs. There is, further, in many species the single Y-chromosome. The autosomes and the Y-chrmosome are distributed from parent to offspring in characteristic ways, differing from the distribution of X-chromosomes. If they affect characteristics, these differencesshould appear in a different system of inheritance. We shall examine first the autosomes.

The Autosomes: Typical Mendelian Inheritance

We have already seen that the autosomes play a role in development, in the fact that they help to determine the sex of the developing individual. In Drosophila, if the number of sets of autosomes, in relation to the number of X's, is changed, this changes the sex. Do autosomes also affect other characteristics? Do the diferent pairs of autosomes have different functions? Are there diverse types f autosomes, with different effects, in different individuals of a species, as is the case in relation to the X-chromosomes? Are there dominant and recessie effects of autosomes, as there are of X? If dominant and recessive characteristics depend on auosomes, then these characteristics must follow the distribution of the autosomes from parents to offspring, as the characteristics ependent on X follow its distribution. We must then examine the method of distribution of the autosomes whether thre are characteristics that show this method.

Method of Distribution of the Autosomes

The ditribution of the autosomes from arents to offspring is much simpler than that of the X-chromosomes.

The main facts are as follows:

- The autsomes are in pairs in all individuals. There are no individuals with unpaired autosomes, like the male of some organisms with relation to X.
- Thus the two sexes are alike as to their autosomes.
- In forming germ cells, one member of each pair of autosomes goes to each germ cell. The sperms and the ova are alike with respect to the autosomes.

  Thus the fundamental rule with relation to the autosomes is that each gem cell carries one member of each pair of autosomes.
- By union of two germ cells the pairs of autosomes are restored.
- Thus each one of the offspring carries in each pair of its autosomes one autosome from each parent.

Now consider a single pair of autosomes, such as the pair marked AA in. Suppose that in different individuals the members of this pair of autosomes produce different effects, as we have seen to be the cae with the X-chromosomes of different individuals. And suppose that, as in the case of X, one of these effects is dominant, the other recessive. How would such characteristics be inherited? This question must be examined with care. For this purpose we may take as an example the different colours found in different varieties of peas, used in the classic experiments of Mendel. The yellow seed programmes of a certain variety is dominant, and results from the constitution of the autosomes of a certain pair. The green seed programmes of another variety is recessive, owing to a different constitution of the autosomes of that same pair.

Call thetwo autosomes of tht pair in the dominant yellow race by the capital letters *AA*, while the two autosomes that give rise to the recessive green programmes in the other race may be called *aa*. The dominant parent may be designated *D*, the recssive parent *R*, When these two parents produce germ cells, each germ cell receives, of course, one chromosome for the pair. The germ cells from the dominant parent *D* have the autosome *A*, those from the recessiveparent *R* have the autosome *a*. The erm cells *A*, from parent *D*, unite each with a germ cell a from parent *R* giving offspring (zygotes) containing both *A* and *a*. These first generation offspring are commonly designated *Fi* (first filial generation). And since the characteristic carried by the autosome *A* is dominant, all these *Fi* offspring manifest the dominant characteristic *D* (yellow, in case yellow-seded peas are crossed with green-seeded peas). Next observe the result of mating together two of these individuals *Aa*, of the *Fi* generation, giving the *Fa* (second filial generation).

Each parent *Aa* gives, according to the general rule, two kinds of germ cells, one kind carrying the dominant autosome *A*, the other the recessive autosome *a*. Each kind of germ cell from one parent unites with each kind from the other. That is, half the *A* germ cells from one parent unite with *A* germ cells from the other parent, half with a germ cells from the other parent; and the same is true for the a germ cells from the first parent.

This gives in equal proportions Fa offspring (zygotes) as follows:

$$AA + Aa + a\ A + aa$$

But s the two combinations *Aa* and *aA* contain the same chromosomes, the proportions may be written *AA* + 2*Aa-f-aa* The individuals of the constitution AA and Aa contain the dominant chromosome *A*; therefore they manifest the dominant characteristic, *D*. The individuals *aa*, having only the recessive chromosome *a*, manifest the recessive characteristic, *R*. Thus in this F^2 generation, so far as manifested characters are concerned, the proportion of offspring are 3 Dominant to *i* Recessive

Thus if dominant and recessive characteristics depend on the two members of a pair of autosomes, hen a dominant individual *AA* is mated with a recessive individual *aa* the general result is bound to be that the immediate offspring (*Fi*) all manifest the dominant characteristic, while in the *Fa* generation (produced by mating two individuals from *Fi*) there will be three times as many dominant individuals as there are recessives. Now, this is the result that was discovered by Mendel in 1866; it formed the foundation for the working out of Mendelian heredity. The proportion 30 to *iR* in the *Fa* generation is commonly known as 'the Mendelian ratio'; it is one of the most characteristic proportions in this type of inheritance.

Mendel discovered that there are many different characteristics that yield this 3 to *i* ratio. If yellow and green peas are crossed, in the *Fa* generation there are 3 yellows to *i* green; if round peas and wrinkled peas are crossed, in F^2 there are 3 round to *i* wrinkled; and so for may other characteristics in the pea plants with which Mendel worked.

And since the time of Mendel these same ratios have been found for thousands of different characteristics, in great numbers of different organisms, from the pea plant and *D* roso- phila to man. The fact that inheritance gives these ratios is commonly spoken of as Mendel's law, and the type of inheritance which it exemplifies is spoken of as Mendelian inheritance. The type that gives in the *Fa* generation 3*D* to *iR* is the commonest method of inheritance, much commoner than the sex-linked method, in which the characteristics depend on differences in the X-chromosomes. This is obviously because autosomes are more numerous than X-chromosomes.

We find then that inheritance according to Mendel's law takes place exactly as it would if the characteristics concerned depended on the two members of a pair of autosomes. As will be shown later, it can be proved experimentally that such characteristics do indeed depend upon autosomes. Thus the rules of Mendelism are the rules of distribution of the two members of a pair of autosomes, one having a recessive effect, the other a dominant effect. Mendelian heredity is heredity of characteristics that depend on differences in autosomes.

Mendel did not know of the relation of the characteristics to the chromosomes. He discovered, however, certain other important facts about this type of heredity, facts which have become clearly explicable since it has been learned that this type depends on differences in autosomes. These facts must be examined. For this purpose it will be convenient first to define certain terms which are universally employed in dealing with inheritance.

By the union of two germ cells or gametes, as *A* and *a*, there is produced a zygote. If the two germ cells are alike with respect to the chromosomes or characteristics that they carry, the zygote so produced is spoken of as a homozygote; thus *AA* and *aa* are homozygotes. If the two germ cells that unite to form a zygote are unlike, as *A* and *a*, the zygote so produced is called a heterozygote, as *Aa*. The two corresponding characters, one dominant, the other recessive, that are produced by the two differing members of a pair of chromosomes are spoken of s alternative characters, or alleles, or allelomorphs. Any given individual manifests one or the other of the two alternatives, not both. For the alternative characters we shall employ henceforth the term alleles.

The characteristic manifested by a zygote is its phaenotype while its actual constitution is its genotype. Thus, two zygotes *AA* and *Aa* both manifest the dominant character; the dominant character is the phaenotype, or the individuals are said to be phaenotypically dominant. But they are diverse in genotype; one is *AA*, the other *Aa*.

In the case of a homo-zygote *AA* or *aa*, the phaenotype and genotype may be held to coincide. Mendel crossed yellow peas with green peas, and found, as before stated, that in the *Fi* generation all the peas were yellow, so that yellow is the dominant allelomorph. In the F^2 there were three times as many yellows as greens. The actual number of seeds in the F^2 generation was in his experiments 8023, of which 6022 were yellow while 2001 were green, so that the ratio of dominants to recessives was, accurately expressed, 3.01 to *i*.

Mendel found that if he bred further the individuals of the F^2 generation, some of these gave different results from others. In the pea plant it is possible to allow each plant to fertilize itself, so that both kinds of germ cells come from the same parent. When this was done with the individuals of the F^2 generation, the following results were observed: The green plants (recessive) gave only recessive green offspring (they 'bred true'). Of the yellow plants (dominant), one-third gave only yellow offspring (they 'bred true'), while two-thirds of them gave both yellow and green offspring, in the proportion of 3 yellows to *i* green.

The reason for this is clear when we consider the chromosomes on which the characters depend. If *A* represents the chromosome (autosome) that gives the dominant yellow programmes, while a represents the chromosome that gives the recessive green, the original parents were *AA* and *aa*, the *Fi* generation were all *Aa*, while the *Fa* generation showed the roportions *AA-f* ^{2} *Aa* + *aa*.

The recessive individuals *aa* are homozygotes; teir two chromosomes are alike, being *aa*. When germ cells are produced by these homozygotic individuals, all these germ cells must contain the chromosome *a*, and when two of these unite in self-fertilization, of course they produce in all cases again individuals *aa*. These are, of course, recessives (green), like their parents; that is, the green parents aa 'breed true'. Similarly the dominant yellows, *AA* (constituting one-third of all the dominants), are homozygotes. Their germ cells all contain the chromosome *A*, and the offspring produced by union of such germ cells, in self-fertilization, are all *AA*, and so doinant yellow like the parents. One-third of all the dominants thus breed true. But the individuals *Aa* are heterozygotic; that is, they have two kinds of chromosomes, *A* and *a*, in the pair.

Thus they produce two kinds of germ cells, *A* and *a*, in equal numbers. And when these germ cells unite, they necessarily give, as we saw on an earlier page, offspring of three kinds, in the proportions *AA* + 2*Aa*4-*aa* Of these, *AA* and 2*Aa* will be dominants (yellow), while *aa* will be recessive (green). Thus the homozygotes *aa*and *AA* breed true, while the heterozygotes, *Aa*, do not.

As Mendel discovered, if the heterozygotes *Aa* are mated with the recessive homozygotes *aa*, they give two kinds of offspring, *Aa* and *aa*; half (*Aa*) being dominant,the other half (*aa*) being recessive. If the heterozygotes *Aa* are mated with the dominant homozygotes *AA*, the offspring *AA* and *Aa* are all dominant, though half (*AA*) are homozyotes, the other half (*Aa*) heterozygotes. All these results are typical for the descendants of two parents different in a single pair of chromosomes, one pair being dominant, the other recessive. For reference it will be well to make a table of these results, characteristic for Mendelian inheritance.

Inheritance Depending on Two Pairs of Autosomes

AB AB AB AB

AB(i) *Ab*(6) *aB*(7) *ab*(s)

Ab Ab Ab Ab

AB(6) *Ab*(2) *aB*(5) *ab*(8)

aB aB aB aB

AB(7) *Ab*(5) *aB*(3) *ab*(9)

ab ab ab ab

AB(5) *Ab*(8) *aB*(9) *ab*(4)

But, as will be observed, not all these 16 are diverse; thus *ABAb* is the same in constitution or genotype as *AbAB*. If we number the different combinations, and give the same number to those that are alike, as is done (in parentheses) we find that there are nine different combinations or genotypes. That is, two parents that are both *ABab* produce offspring of nine different constitutions or genotypes.

When the parents differed in but one pair of characters there were but 3 different combinations. When there are two pairs the number is 3 > or 9⁻. These nine different combinations will manifest the dominant characters *A* or *B*, in all cases where an *A* or *B* is present whether present once, as *A*, or twice, as *AA*); they will manifest the recessive characters *a* or *b* in all cases in which *a* or *b* is present twice, no corresponding capital letter being present. Collecting together these different types, we find the different characteristics to be manifested in the following proportions of the zygotes:

$$9AB + 3Ab + saB + 1ab$$

That is, 9 out of 16 are dominant for both pairs of characters, 3 are dominant for *A*, recessive for *b*; 3 dominant for *B*, recessive for *a*, and *i* out of 16 is recessive for both. There are thus 4 different phaenotypes.

In Mendel's experiment with peas, in which the two pairs of characters were:

1. Yellow and green;
2. Round and wrinkled; in *Fi* the proportions were 9 that were yellow and round,
3. That were yellow and wrinkled,
4. That were green and round, and *i* that was green and wrinkled.

These proportions hold generally when there are two pairs of characters connected with different pairs of autosomes. It may be observed that the proportions of the individuals of different constitution or genotype in F^2, descended from a cros between parents that differ in respect to two pairs of autosomal characters (so that the parents are *AABB* and *aabb*), can be found directly from the proportions for a single pair of characters. In *Fa*, for the single pairs the proportions are *AA* + 2*Aa-haA* and *BB* + 2*Bb* + *bb* respectively. If now we multiply together algebraically these two polynomials, we obtain the relative proportions for the different combinations produced when the two pairs are together.

These proportions are the following:

$$AABB + 2AaBB + aaBB + 2AABb + 4AaBb +$$

$$(1)\ (7)\ (3)\ (6)\ (5)$$

$$2aaBb + AAbb + 2Aabb + aabb$$

$$(9)\ (2)\ (8)\ (4),$$

Certain other relations among the nine different types are important, since they hold for all cases of two pairs of characters that are dependent upon different pairs of autosomes.

These are the following:

- There are four combinations (Nos. 1, 2, 3, 4), constituting one-fourth of all the zygotes, that are homozygotes for both pairs of gens. The four homozygotes are diverse in their combinations and in the characters they manifest. Any one of these when self-fertilized or mated with another like itself breeds true: that is, it produces offspring that have the same constitution as the parents.

- Four out of the 16, or one-fourth of all the zygotes, are heterozygotes for both pairs of characters. They form the class numbered (5), and constitute the diagonal from upper right to lower left. Any of these when self-fertilized or mated with another like itself gives again the nine different types, in the proportions above set forth.
- The one-half of all the zygotes, are homozygotes for one of the two pairs, heterozygotes for the other. Four are homozygotic for *A* and heterozygotic for *B*; the other four are heterzygotic for *A* and homozygotic for *B*. In these 8 there are but 4 diverse combinations or genotypes.

If any one of these is self-fertilized, it yields 3 diverse types with respect to the pair for which it is heterozygotic. Thus, the class (6), having the constitution AABb, yields when self- fertilized three combinations in the following proportions:

$$AABB + 2AABb + AAbb$$

The above relations hold for all cases in which two pairs of characteristics are dependent on differences in two different pairs of autosomes, whatever the method of designating the characters or chromosomes. In a similar way, two parents may differ in three characters that are dependent on three different pairs of autosomes. Such parents could be represented as *AABBCC* and *aabbcc*.

Their progeny (*Fi* generation) would all have the constitution *AaBbCc*. Such *Fi* individuals of course produce a variety of germ cells, in accordance with the principles that each germ cell receives one member of each pair of autosomes, and that the different pairs of autosomes are distributed independently. Each parent *AaBbCc* thus yields the following 8 types of germ cells:

ABC, ABc, AbC, aBC, Abe, aBc, abC, abc.

When each of the 8 kinds from one parent unites with each of the 8 kinds from the other, there are produced, of course, 64 groups. If we classify the 64 individuals by the characters which they manifest, we find that there are 8 different phae- notypes manifesting respectively the following characteristics:

ABC, ABc, AbC, aBC, Abc, aBc, abC, and *abc*

That is, some are dominant for all three characters; some for a particular two and recessive for the other; some are dominant for a particular one and recessive for the other two; and some are recessive for all three. The different phaeno- types are present in different proportions; the proportions when a large number are examined are as follows:

2ABC + gABc + gAbC + gaBC + $Abc + saBc + 3abC + iabc

The total number of different combinations (genotypes) that occur in such a case is. These combinations can be obtained in their correct proportions by multiplying together algebraically the expressions: Similar but still more complex series are obtained if parents are interbred that differ in four or more characters depending on four or more different pairs of autosomes. If the number of pairs of autosomes is called ↔, the following relations hold:

If parents differ in n autosomes:

- The number of diverse types of gametes produced by the individuals of the *Fi* generation is 2^n.
- The number of different combinations (genotypes) in the zygotes of F^2, formed by mating the 2^n gametes from *Fi*, is s^w.
- The number of different phaenotypes produced in *Fa* is 2^n.

These relations yield very great numbers of diverse combinations, in cases where the number of pairs of autosomes is large. Thus in man, with 23 pairs of autosomes, if all the autosomes differed in the two parents, the number of types of germ cells produced by the *Fi* individuals would be 2, or many millions, the number of possible diverse genotypes in F^2 would be 3, or many billions, and the number of possible different phaenotypes would be 2. (As will appear later, the number of different possible combinations is greatly increased by the exchange of parts by the members of a pair of chromosomes.)

Thus typical Mendelian inheritance, dependent on differences between different pairs of autosomes, can be treated to a certain extent mathematically. The individuals in F^2 and later generations occur in proportions represented by simple mathematical ratios.

It will be worth while to formulate the main facts that we have brought out as to autosomal inheritance:

- Some single pairs of characters, dominant and recessive, follow in inheritance the method of distribution of a single pair of autosomes.
- Some pairs of inherited characters follow the rules of distribution of two or more different pairs of autosomes, in the following particulars:
  - One member of each pair goes into each germ cell.
  - The members of the different pairs are distributed to the germ cells independently, so that all possible combinations containing one member of each pair are formed with equal frequency.
  - Each type of gamete from one parent unites with each type from he other parent, the different matings being equally frequent.
- These methods of distribution result, for the characters manifested in the Fs generation, in certain typical numerical proportions or ratios, known as Mendelian ratios.

- This method of inheritance is known as Mendelism, or Mendelian inheritance. The rules of Mendelism are the result of the rules of distribution of pairs of autosomes.
- Most characteristics of organisms are inherited in this way; many more than are inherited in the sex-linked manner. This is evidently because there is but a single pair of X-chromosomes, while there are several or many pairs of autosomes.
- Thus it is clear that, among the different individuals of a species, there are many different types of autosomes (as there are different types of X-chromosomes), producing many kinds of characteristics.

The T-Chromosome

There remains another chromosome, not included in the autosomes: namely, the Y-chromosome. Has this functions similar to those of X and of the autosomes? Does it too play a role in heredity? As we shall see, the Y-chromosome differs in many ways from the other chromosomes. Let us first summarize the main facts as to the occurrence and distribution of Y-chromosomes.

- Some organisms have no Y-chromosomes. Among those are many insects the dog, the cat, the horse.
- In some organisms it is small and appears to be degenerate. This is the case in many insects; also in man.
- In some organisms the Y-chromosome is large, but of a different form from its mate X. Such is the case in Drosophila
- In some organisms Y is not distinguishable from X inform or size, but experiment shows that it differs from X in function.
- The Y-chromosome descends from father to son only, exclusively in the male line. Normally it never occurs in females.
- Hence any characters that depend on the presence of a particular kind of Y will *(a)* be found in males only, and will *(b)* never be inherited *through* females (since Y does not pass through females).

The T-Chromosome in Drosophila

The functions of the Y- chromosome have been more thoroughly studied in Drosophila than in any other organism. Most of the facts as to this have already been brought out, in connection with our account of the functions of the X-chromosome. *We may summarize as follows the known facts as to the functions of the Y-chromosome in Drosophila*:

- The Y-chromosome, although it normally occurs in males, is not required for the production of male individuals. Through non-disjunction, male individuals are produced that contain no Y.

- But the presence of Y is necessary for the fertility of males.
- Y may exist in females, as a result of accidental nondisjunction; it does not affect teir sex.
- During many years of intensive study no characteristics were found that follow the Y-chromosome.
- But recently it has been shown that the presence of Y prevents the appearance of a certain recessive character ('bobbed') that is due to a modified X-chromosome. This modified X results in the production of shortened bristles on the body of the fly. But this peculiarity does not occur if a Y is present, so that Y has the dominant normal allele for 'bobbed'. Indications of certain other ill-defined functions of the Y-chromosome have of late come to view.
- No diverse modifications or defects such as are known in great number for the X-chromosome are known in the Y-chromosome.
- Thus in Drosophila the Y-chromosome appears to have little function; it seems nearly but not quite inert.

Active T-chromosomes in certain Organisms; Relation to Male Secondary Sex Characters. In certain organisms the Y-chromosome is more active and varied in its effects. In such cases it shows in different individuals diverse modifications, comparable to the diverse types of the X-chromosome in Drosophila, and these produce diverse inherited characteristics that appear in males only. This matter is closely bound up with the question of what produces the diversity of secondary sex characters. In many animals there are special characteristics which, like the Y- chromosome, are found only in males.

These are the male secondary sex characters: horns, mane, beard, greater size, diversities in form, and the like. Since these appear only in themale, the question arises as to whether they may be due to the presence of the Y-chromosome.

With relation to this the following is the situation:

- In most cases, the presence of male secondary sex characters does ot depend on the presence of the Y-chromosomes.

This is shown by the following:

— There are animals which have no Y-chromosome, yet have male secondary sex characters: for example, the dog, the cat, the horse; also various insects.

— In Drosophila, males that contain no Y (as a result of non-disjunction) have the typical male secodary sex characters.

— In some breeds of sheep the male has horns, while the female has none. t could be suggested that this is due to he fact that the male carries the Y-chromosome, while the female does not. But in other breeds, as Dorsets, both sexes have horns, although only the male has the T-chromosome. And in still other breeds, as Suffolks, neither sex has horns, although here as elsewhere the males have the Y-chromosome.

If a female of the horned breed (Dorset) is crossed with a male of the hornless breed (Suffolk), in the Fi descendants the males have horns, although they receive their Y-chromosome from the hornless breed. They have inherited the horns through the mother.

Thus it is clear that in these cases the presence of horns is not due to the presence of the Y-chromosome. Extensive work on the inheritance of horns and horn- lessness in sheep, goats and cattle consistently gives results which show that the presence of horns is not due to the presence of the Y-chromosome, or of any special type of Y-chromosome. In all these cases (a), (b) and (c) the presence of distinctive male secondary sex characters depends in some way on the fact that the male carries but one X-chromosome, and consequently produces the male hormone. Their absence from the female results from the fact that she carries two X-chromosomes, and consequently produces the female hormone.

- But in certain fish, some of the male secondary sex characters apparently do depend on the presence of a Y of a particular type, for they occur only where that type of Y is present, and they can therefore not be inherited through the mother. That is, the Y-chromosomes of certain individuals or breeds are so modified as to produce certain definite characteristics in the individuals that contain them, while individuals that contain another type of Y do not show these characteristics.

An example of this is found in the small fish Lebistes. In a certain race, the males have a black spot at the base of the dorsal fin, while the females are without the spot. In another race of the same species neither the males nor the females have the black spot. Thus the females of the two races are alike in lacking the black spot, while the males differ. Is this difference the result of differences in the Y-chromosomes of the two kinds of males? This may be tested by crossing members of the two races. In the sptted race, represent the X and Y-chromosome by capital letters; while in the non-spotted race, represent them by the small letters x and y.

A female of the spotted race, XX, is mated with a male of the non-spotted race, xy, thus:

- Females Males
- Parents XX xy

- Non-spotted
- *Fi Xx Xy*
- Non-spotted

The Y-Chromosomes

This produces in Fi females Xx, and males Xy. Neither contains the Y from the spotted race, and neither is spotted. The spot is not inherited through the females of the spotted race.

Next mate the female of the non-spotted race, xx, with a male from the spotted race, XY, as follows:

- Females Males
- Parents xx XY
- Spotted
- Fi xX xY
- Spotted

This yields in Fi females Xx, and males xY. These males carry the Y from the spotted race, and they are spotted. Thus far the spot goes with the Y from the spotted race. This Y may be traced further, for example, by breeding together the non-spotted females xx with the Fi males xY, from the preceding cross, thus:

- Females Males
- Parents x xY
- Spotted
- F2 xx xY
- Spotted

This gives in the Fs females xx, and males xY. These males are again spotted; they carry the Y from the spotted race. Wherever this Y from the spotted race is found, the black spot appears; wherever the Y is from the non-spotted race it does not appear. It appears clear therefore that whether the black spot appears in the male depends upon the type of Y- chromosome that it carries. There are two types of Y-chromo- some in the species, one producing the black spot, the other not producing it.

It has been shown by Winge that there are many other programmes markings in this fish that depend for their occurrence on the presence of a particular type of Y-chromosome. They occur only in the males, and only in males that have a certain type of Y. In certain other fish also the Y-chromosomes have been shown to be diverse in different individuals, giving rise to distinctive characteristics in the male. In a certain plant, Melandrium, also, there are inherited characteristics that depend upon the presence of a particular type of Y-chromosome. There are some indications also that in man the nature of the Y-chromosome affects certain characteristics.

In a family described by Schofield, the male parent in the first observed generation had webbed toes. This was inherited in the male lines only by all the males for four successive generations. It thus followed exactly the distribution of the Y-chromosome from the original male parent. It seems probable therefore that it is due to a defect or modification in the Y-chromosome of that individual. Many other cases of webbed toes or fingers are knon that occur in both sexes, so that they are clearly not due to the Y-chromosome.

On the whole, the situation as to the Y-chromosome and its relation to inheritance may be summarized as follows:

- In most organisms Y is a degenerate chromosome, lacking most of the functions manifested by the other chromosomes.
- In ome organisms degeneration has gone so far that the Y-chromosome has completely disappeared.
- In other cases it exists, but is very small.
- But in some species the Y-chromosome has effects on hereditary characters that are similar to those produced by other chromosomes. Different types of inherited characters result in such cases from diversely modified Y-chromosomes.
- Characteristics dependent on the presence of a particular type of Y-chromosome are inherited in the male line exclusively.

Three Main Methods of Distribution of Inherited Characteristics

There exist therefore three diverse systems of distributing hereditary characterstics, depending respectively on the X- chromosomes, the autosomes, and the Y-chromosomes. In each of these the rules of inheritance are given by the special method of distribution of the three types of chromosomes, taken in connection with the facts of dominance and recessiveness.

Inheritance which follows the distribution of the X-chromosomes is commonly called sex-linked inheritance. That follwing the distribution of autosomes is known as autosomal or typical Mendelian inheritance. Less important than those just named is Y-chromosome inheritance.

All these methods are included under the term Mendelian inheritance, employed in a broad sense. In some organisms still other methods of inheritance exist. In higher animals and plants most inheritance occurs in one of the three methods named in the preceding paragraph.

Characteristics Affected both by Genes and by Environment

Many of the same features that can be altered by changing the genes can likewise be altered by appropriate changes in the environment. Characteristics do not fall into two classes, one exclusively hereditary (or dependent on genes), the other exclusively environmental; but any characteristic is affected both by the materials of which the organism is composed, and by the action of the conditions on these materials.

Yet, as we shall see, in practice some characteristics are more readily altered by environmental conditions than are others. Certain characteristics owe most of their peculiarities to diversities among genes. Others are readily affected both by genes and by environment. Still others depend mainly on environmental conditions.

Particularly illuminating are the cases in which characteristics depend in marked degree both on genes and on environmental conditions, so that they are altered by changes in either. A number of well-known cases of this kind will be presented, selecting those that arefrom a technical point of view fully known. In that animal whose genetics is best known, the fruit-fly Drosophila, many of the characteristics that have been studied are defects or abnormalities which are sharply dependent on the presence of particular genes. Such a one is abnormal abdomen. This appears in the fact that the abdomen is ill formed, the segments not being regular nor sharply marked off. This abnormality is found to be due to a defect in a certain gene of the X-chromosome, so that it shows sex-linked inheritance, the abnormality being dominant. Unless that particular defective gene is present the abdomen is normal. But it is dependent too on the environment.

Individuals with the abnormal gene, if developed in a moist atmosphere, have the abnormality. But if those same individuals are developed in a dry atmosphere, they have normal abdomens. To produce the abnormal abdomen, we must have first the particular kind of defective gene that yields it. But second, even if that is present, we must have, too, the moist atmosphere for development; without it the abnormality is not produced, whatever th gene.

Another character in the fruit-fly that acts in a similar way is what is called reduplicated legs. If a certain defective gene is present in the X-chromosome, the animals show a tendency to produce double legs, or even triple or multiple legs, in place of single ones. Butthis does not happen unless the young develop at a low temperature. To produce reduplicated legs, then, there must be present a certain type of gene, and also a certain extenal condition. Unless both are present, the peculiarity is not produced. It is clear from these cases that even though an individual inherits the type of gene necessary to produce a certain defect, it may not be inevitable that he should have that defect. A special environment may correct what inheritance leaves imperfect These considerations apply to many thngs in man as well as in other organisms. Another example of the fact that what is produced in development depends both on th genes and on the environment is seen in the production of giants in the fruit-fly. A modified gene, located in the X-chromosome, nea its left end, causes the animals (if they have no other type of X-chromo- somes than this) to become giants; they grow to nearly twice the size of the ordinary fly.

But this increase in size takes place only if the animals having the modified gene are well fed during a certain period of their early lives. If not well fed at this particular time in life, they grow no larger than the usual flies. The giant size requires to produce it a particular type of environment acting on a particular type of gene. If either condition is not fulfilled, giants are not produced.

Again, in plants, the difference between green plants, containing chlorophyll, and white ones, is in some cases due to a difference in genes. Plants with a certain kind of genes will remain white even though grown in the light. In other cases, the difference between white and green plants is not the result of a difference in genes, but of a difference in environment. Plants that live in the dark remain white, even though they contain genes that can produce chlorophyll, while similar plants grown in the light are green. That is, to produce chlorophyll, a certain type of genes is necessary, and also a certain type of environment.

Again, there are red and white varieties of primroses; the difference in programmes is inherited; it is due to gene diversities. But a certain variety produces red flowers when grown in a cool place, white flowers when grown in a warm, moist region, as in a greenhouse. The same diversity in programmes is now due to an environmental difference.

Complex interrelations of genetic and environmental action have been described by Emerson with relation to diverse colours in maize plants. In different types, the leaves, the husks, the flowers, and other parts, differ in programmes. Emerson describes and Figs six main types, under each of which are subtypes.

These are the following:

- Purple, with its subtype, weak purple.
- Sun-red, with a subtype, weak sun-red.
- Dilute purple.
- Dilute sun-red.
- Brown.
- Green, with a number of subtypes.

The interrelations of genetic and environmental factors in these colours may be summarized as follows:

- Each programmes and each subtype is constant and hereditary, when the different types are bred under like conditions.
- Each of the colours depends on many different genes, like the eye programmes in Drosophila; they may be altered by changes in any of these genes. The different colours result from different combinations of the genes.
- When the different types are crossed they give typical Mendelian inheritance: in some cases with simple ratios, in others with complex ratios. Thus far the conditions are those for typical hereditary characters; the differences are the result of gene differences, and they follow the distribution of the different types of genes in inheritance. But, also, the colours depend on the conditions to which the plants are subjected while developing. Change of certain conditions changes the colours, as does alteration of the genes. Some of the relations are the following:
- The 'sun-red' type requires exposure to the sun for production of the red programmes. Parts protected from the sun are green. In the 'green types, on the other hand, subjection to the sun does not cause the production of red programmes. Thus the difference between red and green plants is due in some cases to diversity of genes, as when 'sun-red and 'green types are both grown in the sun. In other cases the difference between red and green plants is due to environmental difference, as when 'sun-red plants grown in the sun are compared with those of the same type grown in the shade. The plants of the 'purple' type do not depend on exposure to the sun for their programmes; they are purplish in programmes whether grown in sun or shade. The degree of programmes produced by exposure to the sun depends on what genes are present. In the typical 'sun-red' variety, exposure to the sun causes a red programmes widely distributed on the plant. In the 'dilute sun-red', exposure to the sun under similar conditions causes only a little red, at the bottom and tips of the leaves, while in the 'green type no red appears. The nature of the soil affects the programmes. 'Dilute sun-red' grown in poor soil and subjected to the sun has almost all parts red; in good soil, with the same relation to the sun, it has very little red. The 'green and 'brown' types are not influenced in their programmes by the nature of the soil.
- Certain particular chemicals in the soil affect the programmes. Presence of much nitrogenous matter tends to suppress the red; if little is present the red programmes is marked.

- Storage of carbohydrates in the leaves of the programmes-producing varieties causes increase of the red programmes. The amount of carbohydrate in the leaves may be increased by experimental procedures, as by removing the growing ears, or by bending a leaf so as to break the conducting vessels; the leaves thereupon become red in the parts with increased carbohydrate. But this effect is not produced in the 'green' types; they do not produce the red programmes even in presence of stored carbohydrates.

It appears clear, therefore, that in maize different genes cause the different varieties to have different metabolic processes, different chemical reactions, some of which cause the production of colours, while others do not. Also, diverse environmental conditions likewise cause changes in these metabolic processes, inducing programmes production in some cases.Whether programmes occurs, and its type and amount, depend on interaction of genetic and environmental conditions.

Another instructive case of the interaction of genes with environmental conditions is furnished by the conditions determining the number of facets in the compound eye of Drosophila. In the normal eye there are present about 800 facets. In certain of the flies a gene at locus 57 in the X-chromosome is so modified as to cause a great reduction in the number of fully developed facets. In this condition, known as Bar-eye, there are only about 80 fully developed facets: these form a 'bar' across the eye. This Bar-eye condition is dominant; the number of facets is reduced even though one of the two genes of the pair is normal. We may call the modified gene that causes the reduction in the number of facets the Bar gene.

The number of fully formed facets, it is found, depends on whether there is present but one or two of the Bar genes, and also on whether there is present in addition a normal gene. In the males (having of course but one X), the presence of the Bar gene reduces the number of normal facets to about 90.

In the female, if both the X's contain a Bar gene, the number of facets is about 65. If the female is heterozygotic, so that but one of her X's contains the Bar gene, while the other is normal, the mean number of facets is about 358. Finally, in either male or female, if no Bar gene is present, the number of facets is about 800. Later it was discovered that this Bar gene is sometimes modified in a different way, so as to cause a less marked reduction in the number of facets. This modification is known as infra-Bar. A female containing but one of these infra-Bar genes has about 716 facets; if she contains two she has about 348.

Again, by an extraordinary process, almost or quite unique in known genetics, in some cases two of the Bar genes get united by crossing-over into one X-chromosome. This condition is known as double Bar. When both of the chromosomes carry this double Bar, the number of facets is but about 25. If one carries double Bar, the other single Bar, the number of facets is about 36. Thus the number of facets in the eye depends in many ways on the number and conditions of the genes present. This may be shown in the following diagram, representing the conditions in the two X (or Y) chromosomes of any individual, with the corresponding numbers of eye facets present. In this diagram + = the normal, unmodified X-chromosome Y = the Y-chromosome of the male *B* = the Bar gene *iB* = the infra-Bar gene *BB* = the double Bar condition (two Bar genes in one chromosome) Ghr. *i*: + *iB B iB B iB B BB BB BB BB* Chr.s: + + + *iB* Y *BB* + *iB B BB* Facets 800 716 358 348 98 73 65 45 42 36 25. All these various conditions are inherited as unit-difference, sex-linked characters; they are typical hereditary conditions.

But it is found further that the number of facets in most of these conditions depends also on the temperature to which the developing animals are subjected. The numbers given above are those observed by investigators working with fruit- flies cultivated at ordinary, not precisely controlled, temperatures. If different sets of flies are kept at different temperatures, it is found that high temperatures reduce the number of facets in the individuals that carry the Bar genes, though not in the normal individuals. Krafka found the following mean numbers of facets in the various gene conditions, at different temperatures:

Males Degrees	*Bar Females B-B*	*Bar Males B-Y*	*Double Bar BB-BB*	*Double Bar BB-Y*	*Temperature,Females*
J5	214	270	5i	61	
20	122	161	33	37	
25	81	121	25′	28	
300131	40	74	‘5	‘4	

Thus the number of eye facets depends both on the genes and on the temperature. The number 61 may be produced either by a double Bar Male (*BB*–Y) at 15 degrees or by a Bar Female (*B*–*B*) at between 25 and 30 degrees. Changing either the genes or the temperature will change the number of facets.

Similar situations occur in vertebrates; one of these may be described briefly. The spotted salamander, Salamandra maculosa, is largely black in programmes, with spots or stripes of yellow. The amount of yellow differs in different cases, so that on the whole some specimens are lighter, some darker. In the lighter individuals the yellow spots tend to run together into rows. The different grades are heritable; certain stocks are regularly darker, others lighter. If darker and lighter individuals are mated, the two conditions show Mendelian inheritance, the darker condition being dominant all the Fi generation are darker, while in Fa there are three dark to one light. The difference between the two thus depends on a difference in one gene. But it can be shown also that the extent of programmes depends on the environment. If the salamander is kept continuously for a long time (a year or two) on a light background, the yellow spots increase in size: they may run together into stripes or into large yellow areas *B*, so that the animals become much lighter. If, on the other hand, they are kept for long periods on a dark background the yellow areas decrease in size; the animals become on the whole darker. The extent of the yellow programmes thus depends both on the genes and on the environment; it can be altered by appropriately changing either.

In all these cases the following general principles apply:

- What are really inherited (passed on from parent to offspring) are the genes; that is, certain materials that in certain combinations and under certain conditions give rise to certain definite characteristics.
- With the same original genes, different environmental conditions may induce the production of diverse characteristics.
- Also, with the same environmental conditins, different genes may induce the production of different characteristics.
- The same difference in characteristics that is in one case produced by diversity of genes is in other cases produced by diversity of environment.

There can be little doubt that these general statements apply to many characteristics in man. Some combinations of human genes form a individual that is a much better culture medium for the bacteria of tuberculosis than are others. A person who gets such a combination of genes will develop tuberculosis, if he lives in a region in which the germs of that disease are abundant; while another person, with a different set of genes, will not be subject to tuberculosis, even though living under the same conditions. And the person with the genes that make him susceptible to tuberculosis will not have the disease if he prevents infection by the tubercle bacillus.

In relation to such susceptibility there are doubtless a great number of grades, dependent on just what combination of genes is present. Some are extremely susceptible, others less so but still prone to the disease, and so on, up to individuals whose gene combinations are such as to give them natural immunity to tuberculosis.

What then shall be answered when it is asked whether tuberculosis is hereditary or not? One can only say that some gene combinations predispose to it more than others, so that a hereditary factor is involved. But also there is a necessary environmental factor; it is the interrelation of the two sets of factors that determines whether the individual shall be afflicted with the disease. Similar considerations apply to any other disease, acute or chronic. Certain combinations of genes are doubtless more readily attacked by plague, by smallpox, by typhoid, by pneumonia, than others, just as certain genetic constitutions yield more readily to extremes of temperature, to exposure to the elements, to unfit food.

There is no affair of life in which the gene combination borne by the individual does not play a part. But the environment too plays a part, often an overwhelmingly important part. The relative role played by genes on the one hand, and by environment on the other, is not the same for all characteristics. Differences in certain features of the organism are mainly due to differences in the genes of which they are composed. In other characteristics, differences are due mainly to environmental diversities. There are all sorts of intermediate conditions between these extremes. We may pass in review a number of different types of characters from this point of view.

Most of the diversities in obvious physical features in most animals are the result of differences in the genes, so that they are mainly matters of inheritance rather than of environment. In Drosophila the forms of the parts of the body, of the legs and wings, the distribution of bristles, the venation of the wings, the colours of body and eyes, are all changeable through the action of diverse genes, but diversities in these features due to environmental action are relatively rare. This is the case also for such characters in the vertebrates, including man. Most of the physical characters of human beings form of their features, complexion, programmes of hair, form of limbs, and the like, as well as sex are settled mainly while the child is carried in the mother's body, when the differences in environment are very slight.

The diversities between individuals in these respects are almost entirely though not quite entirely the result of differences in the genes; they are matters mainly of inheritance. This is demonstrated by what we find in twins. The one-egg twins, having the same genes, are usually closely alike in their physical peculiarities, while the twins derived from different eggs, and so having different genes, are much more diverse with respect to such physical characteristics.

At times, of course, one-egg twins become diverse through accidents or diseases which affect one and not the other. But physical diversities in twins due to genetic differences are much more frequent than those due to environmental differences. In some organisms the gross physical features are readily altered by environmental diversities, so that differences between individuals resulting from the different conditions under which they have lived are common. Such is the case with the habit of growth in higher plants; trees and other plants grown under different conditions are very different in form.

In some of the lower animals, such as the hydroids, a similar situation is found. It is obvious that the details of behaviour are in a high degree under the influence of environmental conditions. The particular acts performed at given moments in any given organism are determined almost entirely by the conditions under which it finds itself. Yet organisms of different genetic constitution behave very differently under effectively the same conditions, and the general pattern of the behaviour of organisms may be largely determined by the genetic constitution.

Genes and Environment in Relation to Mental Characteristics

The relative rdles of heredity and environment are of special interest in man, particularly in relation to the characteristics that influence behaviour. For the study of these matters man is the most favourable organism that exists, although for most other relations in genetics he is a singularly unfavourable object of study. In man it is possible to make detailed studies of temperament, mentality, character, a group of characteristics hardly open to examination in other organisms.

There occur in man 'identical' or one-egg twins, individuals having the same set of genes, the same genetic constitution throughout. And for comparison with these there occur fraternal twins: two individuals of the same parentage, the same age and living under the same environment, but developed from two separately formed eggs and thus having to some extent different sets of genes. By comparing these two kinds of twins with one another and with unrelated individuals, opportunity is presented for comparisons of the results of similarities and diversities in genetic constitutions with those of similarities and diversities in environmental influences.

As seen earlier the twins derived from a single egg are as a rule very closely alike in most physical respects; it is this that has caused them to be called identical twins. In some cases, however, the two differ in certain physical characteristics. There appear to be two categories of causes for these differences. First, there are physical differences that result from the way in which the separation of the twins occurred. Separation of the single egg into two does not occur in the first stages of development, as by division of the one-cell stage into two. On the contrary, it occurs much later, when many cells are present and the differentiation of bodily parts has begun, and it occurs at different ages in different cases.

There is reason to believe that when an equal division occurs at a rather early stage, before the right and left halves of the body have begun to show their characteristic diversity, the two twins produced are closely alike in their physical features. At a later period the right and left halves of the embryo have begun to differ, in the way that results as a rule in the greater strength and development of the right hand and arm. At such a stage, the embryo may divide into right and left halves, with results as follows. Of the two embryos so produced, one will have a right limb that is already somewhat advanced in development, while the left limb, formed at the plane of separation, must begin anew. The other embryo will be in the reverse situation; it will have a left limb already somewhat advanced in development, but the right limb, formed at the plane of separation, must begin anew.

In each case the more advanced limb retains its advantage and becomes the one that after birth is strongest and most used. The result necessarily is that one of the twins remains right-handed, while the other becomes left-handed. This situation is not uncommonly found in identical twins. Similar differences in other unsymmetrical parts may be produced in the same way: for example in the whorls of hair on the head. In such respects the two members of a pair often show 'mirror-imaging', presumably due in each case to the fact that the original embryo had already become a little unsymmetrical before division occurred.

From this same situation another condition often observed is believed to result. In the embryo before division, one side (usually the right) will be a little in advance of the other. After division, the twin derived from the right half continues to retain this advantage, so that it may be more vigourous than the one derived from the left half. Such differences in vigour are not rare in twins. Apparently they may be accompanied by a psychological difference; the more vigourous twin assumes the leadership in their lives together. This reacts further on temperament and character. How far psychological and temperamental differences so produced may go is uncertain.

It appears that in some cases division of an embryo may be unequal, so that for this reason also one twin may be more vigourous than the other, with resulting psychological differences. Diversities produced in the ways above described are different in origin and type from those commonly called environmental. Under the latter term are commonly grouped diversities resulting from the different conditions under which individuals have lived. Since the extent to which differences due to the time and method of division may go, and the kind of characteristics that they may affect are uncertain, it is often not possible to distinguish these certainly from diversities due to differences of environment after division has occurred.

Belonging to one or the other class, there are sometimes marked differences in the phsyical characteristics of one-egg twins. Komai and Fukuoka have described two fifteen-year- old Japanese boys that are obviously one-egg twins; they show the usual close resemblance in form, features, colouring, and the like. But they now differ greatly in size.

One is taller by 14·8 centimetres, or about 6 inches, and his weight is greater by 10·4 kilograms, or about 23 pounds. The two were at first of the same size, the difference gradually arising during growth. The smaller of the twins is affected by diabetes insipidus; this is probably connected with his decreased growth. The larger twin is right-handed, the smaller left- handed. Whether the difference in health and growth is in some way connected with what occurred at division of the egg is uncertain. A comparable case is described by Siemens. Two sisters, clearly one-egg twins, were as usual nearly identical in most of their physical characteristics till the age of 10. At this time one of them became affected wth a severe lateral curvature of the spine, and from this time their development was very different; at 16 the healthy twin was 4-8 inches taller than the other. The origin of the defect that produced curvature of the spine and changed development is unknown.

Identical twins that have lived together sometimes show considerable psychological differences. Newman has made extensive psychological tests on fifty pairs of identical or one- egg twins, and, for comparison with these, on fifty pairs of fraternal twins. The results of these examinations have not yet been published in full, but certain data from them have been published.

By the Stanford-Binet tests, the intelligence quotients (IQJ and the 'mental age' were determined for the two sets. The two classes of twins are best compared by the differences between the scores made by the two members of each pair. Thus the average difference between fraternal twins, having somewhat diverse genetic constitutions, is nearly twice as great as that between identical twins, in which the genetic constitutions are alike.

But it is important that the identical twins that have lived together also showed differences between members of the pairs. In different pairs there was much variation in the amount of diversity. Of the 50 pairs there were five in which the two members gave identical scores in the intelligence tests. At the other end of the scale were five pairs that showed respectively the following differences (points in the Stanford-Binet scale) between the two members: 12-6, 12-9, 13-0, 13-9, 16-0.

All these five differences are considerably greater than the average difference between the genetically unlike fraternal twins. It is clear that in some cases even identical twins living together differ much in mentality. Whether the differences are the result of original diversities consequent on the method of division of the egg, or have been produced in some way by different experiences of the two, is not known.

For comparisons as to the relative effects of genetic and environmental diversities on mental characteristics, valuable data have been obtained from the examination of identical twins that have lived apart under different environments. The first case of this kind was carefully studied by Muller, and nine pairs have since been fully examined by Newman. In each of these ten cases the twins were separated in infancy, being adopted into different families; they have then lived to adult life under diverse conditions and influences. In addition to careful observations as to character and temperament, they were subjected to psychological and temperamental tests by the best standardized methods available.

The type of results reached by these studies will best be appreciated by first looking at the detailed resemblances and differences in some particular cases. In the first such study, by Muller, there were twin sisters that had been separated when two weeks old, and that did not see each other till they reached the age of 18; from that time until the age of 30 they lived apart more than nine-tenths of the time. Physically they showed the extreme similarity in characteristics that is usual in identical twins. Both 'have always been intellectually active', 'both have been extremely energetic, capable and popular, and they have been prominent in all sorts of club work in their respective communities' (Muller)'.

Both have had two or three attacks of tuberculosis, almost simultaneously'. The usual intelligence tests gave results very closely alike for the two twins. But 'the nonintellectual testsýÿof motor reaction time, association time, "will temperament", emotions and social attitudesýÿgave results in striking contrast with those of the intelligence tests, in that the twins gave markedly different scores in all these tests'.

The differences were on the average greater than those between two individuals taken at random, and seemed 'to be correlated with salient differences in their past experiences and habits of life'. Thus this first study of such a case indicated that the different environments and experiences of the two individuals had produced a large effect on temperament, emotions, and social attitudes, but had had little effect on such matters as are tested by intelligence tests.

An illuminating contrast with these results is given by the first pair studied by Newman. The twin sisters ('O' and 'A') were born in London and were separated at the age of 18 months. One lived in Ontario, Canada, the other in London. Their environments were very different. When they were tested, this pair gave differences in those tests in which Muller's pair gave similarities, and gave similarities in those tests in which Muller's pair gave differences. Newman says: 'The twins dealt with in this paper are very different in mental capacity. But they showed great similarity in their manifestations of will and temperament, and in their emotional reactions. So this pair of twins shows that differences in the experiences undergone affect deeply the individual's mental traits: his performance in matters brought out by intelligence tests.

Important points were brought out in the study of Newman's second pair, twin sisters. The two had received very different educations; one had attended school seven years longer than the other. Newman summarizes as follows the results of his study of these:

'These twins, remarkably similar after being separated at 18 months of age and unknown to each other for 19 years, have been profoundly modified by the very different educational careers.In every test of mental capacity, whether of so-called native ability or of achievement, 'G', the more highly educated twin, has distinctly the superior mind. Obviously mental training improves the ability of an individual to score well in any sort of test'. But further, 'in contrast with the great difference in mental power stands the fact that in all the tests of emotional traits and of temperament the twins gave the impression of being remarkably and unusually similar'. Here came out the effect of their identity of genes.

Newman's third pair revealed certain other facts of importance. These were two young men ('G' and 'O'), separated at two months of age. One had lived mainly in the city while the other had lived in the country. They were examined after reaching the age of 23. In native ability they seem to be nearly identical. The one outstanding difference is in their general personalities. 'G' (who had lived in the city) impresses one as more dignified, more reserved, more self-contained, more unafraid, more experienced, and less friendly. He seldom smiles, has a more serious expression about the brows, eyes and mouth. He stands more erectly with chin held in and brows drawn down somewhat over his eyes. 'O' (the other twin) is the opposite in all these respects. He is the more typical country boy, laughs readily, and is not on his dignity at all. Newman asserts emphatically that the personalities of the boys were utterly different.

Thus through the study of these first four pairs examined, of identical twins that have lived apart, it became clear that such twins may differ in many ways. It appeared that the different environments and experiences of the individuals may have a large effect on mental and temperamental characteristics. The effects were different in ctifferent cases: in some cases the twins are alike in intelligence but differ temperamentally; in other cases they are alike in temperament and emotions, but differ in mentality. In other cases the marked difference is best expressed rather vaguely as a diversity in 'personality'.

The general impression given-by study of these first four cases has been confirmed by Newman's study of six other pairs of identical twins that have lived apart. The nine cases examined by Newman were treated by uniform methods, so that they may be compared in such a way as to bring out general relations. The mean difference in intelligence quotient for these nine pairs of identical twins that have lived apart is 8'6. This is considerably greater than the mean for identical twins that have lived together, which was 5-3. So far as it goes it indicates that difference of environment has a considerable effect in increasing the mental difference between twins.

In cases I, II, IV and VIII, the difference between the twins was very marked, but is similar to the differences between the five extreme cases of identical twins that had lived together. This emphasizes the fact that it is not possible to be certain what differences are due to something that happened at the division of the egg, and what are due to later environmental differences. The mean difference between identical twins that have lived apart is nearly the same as that for fraternal twins that have lived together.

In addition to the studies of intelligence, Muller and Newman made extensive examinations designed to test other features of personality. One set was designed to test will and temperament, to distinguish the mentally quick and slow, the deliberate and careful as compared with the hasty and careless, the aggressive and forceful or the opposite, and the like. Another type of examination is designed to test emotional peculiarities: likes and dislikes, feelings of right and wrong, and so on.

The results of such tests are not readily expressible by a numerical score, such as is employed for the intelligence tests. It will be worth while, however, to attempt a tabulation in general terms of the results of the tests in the three categories for the 10 cases of identical twins that have lived apart. Thus the general picture is that identical twins that have lived apart show many different combinations of likeness and unlikeness with respect to intelligence, temperament and emotional characteristics. There are cases in which the two are closely alike in all three categories cases in which the twins are distinctly diverse in all three categories and cases in which they are alike in two categories and diverse in the third, or alike in one category and diverse in the other two.

On the whole, it is clear that other things, beside genetic constitution, play important roles in determining mentality, temperament and emotional traits. The principal difficulty in interpreting the results lies in the uncertainty as to how much is the result of the manner in which division of the egg occurred, and how much is due to later environmental differences. Both certainly play a role. For these types of characteristics it appears probable that, of the two, the later environmental differences play the greater role. It is further clear that genetic resemblances and differences play a very large role in mentality, temperament and emotional traits.

It is to be noted that in all this study, the general environment of the twins was much alike in all cases. None of them lived in different civilizations, or at different epochs in cultural development. How great a diversity in the characteristics studied could be made by such differences remains uncertain. It is further to be noted that the tests to which the twins were subjected were designed to bring out rather permanent personal traits, as distinguished from mental content and habitudes. The behaviour of human beings depends very largely upon acquired mental content, upon knowledge, and upon habitudes and these things are in a high degree dependent on the conditions under which the individuals live and the experiences to which they are subjected.

In general, it is clear that in a human population there are great numbers of different types of individuals that are diverse because their genetic constitutions are diverse. In consequence of this they differ in their capabilities, in their tastes, in their tendencies towards any particular line of action. But all these classes, so far as they do not fall in the seriously defective groups, have marked powers of adjusting themselves to many different conditions and of following many different lines of action, depending on the conditions in which they live.

In no other organism than man is so full an analysis possible of the relative roles of genetic constitution and environment in determining the mental and temperamental characteristics, for in no other organism are these characteristics highly developed.

8

Genes and Adaptation

A species can adapt to a variable environment in one of three ways, two of which are genetic. It can retain a large reservoir of genetic variation, thus increasing the odds that under any stressful condition some members of the group are preadapted, and survive. In fact, most species in nature are genetically heterogeneous and the maintenance of the variability is assumed to be due to natural selection. Genetic uniformity is restricted to small, isolated, inbred populations. Another strategy is for a species to move in the direction of genetically programming a complex nervous system that modifies itself adaptively during development by responding to environmental stimuli.

The two modes of adaptation are not mutually exclusive and, in fact, often coexist in different proportions. We must be careful, however, not to infer genetic variability from behavioural variability, or vice versa. As an example, Ebert and Hyde selected for high and low aggression from a phenotypically intermediate wild population of mice.

The range of phenotypes in the selected lines greatly exceeded that in the foundation population. Underneath the apparent phenotypic similarity of the wild mice was the genetic potential for extreme variation upward and downward. On the other hand, those of us who work with inbred strains of mice, each composed of genetically identical individuals, are often frustrated by the amount of behavioural variability within a single strain. Unfortunately for many experimenters, the behavioural phenotype is not rigidly predetermined by the genotype. These two strategies of adaptation appear to be consequences of different situations. A species living in a stable, predictable environment will prosper best with a fixed behavioural repertoire that is precoded in its genotype as a plan for a specialized, unmodifiable nervous system. A species living in an unpredictable environment will do better to have its genotype programme the development of a plastic nervous system with the capacity for modification by appropriate stimuli.

If part of the environment is predictable, a probably universal situation, selection will favour programming fixed responses for the predictable but not for the unpredictable components of its surroundings. A complete separation of these two modes of adaptation is impossible. Thus, modifiability of behaviour by experience along with genetic variation in its efficiency has been demonstrated in insects. And there appear to be biological constraints on learning even in such large brained creatures as humans.

Typical Behaviour

This discussion of the relations between behaviour genetics and ethology begins with two questions. How uniform is species typical behaviour? How much of its variation is attributable to genetic differences? In general, members of a species share a common ethogram as well as a common structural plan. Even though taxonomists depend primarily on structure as a basis for classification, field observers know that behaviour is often a reliable means of distinguishing between physically similar species. The most striking examples are sibling species, par ticularly common in insects, fish, and amphibians although a few examples are known in birds and mammals. These species, however, are not our primary concern here.

There is ample evidence of behavioural polytypy within good species as defined by the free interbreeding of their members. Such polytypy is likely to be associated with social organization or food gathering, and explanations are frequently given in terms of ecological pressures and adaptive response.

Three major hypotheses could explain intraspecific behavioural heterogeneity:

- Variant behaviour is transmitted from adults to young by learning;
- Behavioural variation is an expression of genetic heterogeneity within the species; and
- Behavioural variation arises because individuals develop differently depending on the environment they encounter.

Hypothesis 1 is in a broad sense cultural, as it implies the handing down of traditional modes of behaving from generation to generation. Cultural transmission is nongenetic except to the extent that it requires a nervous system with a capacity for learning to model the behaviour of others. In humans, cultures are a major source of behaviour variation, although their prevalence within our species does not logically exclude a genic contribution to group differences.

A well known example of a group characteristic in animals is variation in food habits among troops of Japanese macaques. In one group the habit of washing food was apparently begun by one young female and was adopted quickly by her companions. It was not found in other troops living under similar circumstances. Cultural transmission of this type may be more important in nonhumans than is generally recognized, but because it has not been widely investigated from the genetic point of view I shall not consider it further.

Hypotheses 2 and 3 emphasize individualistic responses to environmental challenges rather than modelling behaviour on the actions of others. Culture is involved only as a part of the environment in which development takes place. The difference between the hypotheses lies in the role of genetic variation. Hypothesis 2 places a premium on genetic heterogeneity as a means of ensuring that some individuals will be able to survive and propagate in environments that deviate markedly from the species optimum. This is clearly the classical concept of natural selection, and it is supported in a general way by the universality of genetic variability in natural populations.

Hypothesis 3 does not deny the role of natural selection in the past, but for the present it emphasizes the lability of phenotypic development and adaptive responses to particular environmental conditions. For adherents of this view, behavioural variation it within a species is more a matter of differences in individual life histories than in individual genotypes. Actually, the three hypotheses are not mutually exclusive. All are supported by evidence, but their relative importance varies according to circumstances.

Behavioural Polytypy

An extensive literature deals with variation in the behaviour of domesticated and laboratory animals, house mice, rats, dogs, fowl, and fruit flies. In these species it is often possible to separate experimentally the effects of genetic and environmental factors. Genetic variation has been demonstrated for activity level, emotional reactivity, eating and drinking, learning ability, and social behaviour. In some domestic species, selection for behaviour has produced breeds with special aptitude for certain tasks. These breed differences persist even when animals are reared as nearly as possible in the same manner.

As an example, Fuller trained five breeds of dogs to walk on a lead from their outdoor living quarters to a laboratory and found striking differences in their reactions. Shetland sheep dogs crowded against their guides, a pattern that strongly resembled the herding of sheep by their working relatives. Basenjis, with a short history of domestication, struggled more violently and bit their leash more frequently than any other breed.

Beagles, who have been selected to track small game and to give voice as they follow a trail, were noteworthy for their mournful vocalization. None of these behaviours was restricted to a single breed, and none was completely absent from any breed. Their frequencies and intensities differed enough, however, to define significantly different breed profiles. Crosses between Basenjis and cocker spaniels, breeds with widely differing profiles, yielded hybrids that were intermediate to their parents on all characteristics except vocalization, which was higher than that of either parent. The occurence of stable, intraspecific behavioural variation among populations is ethological polytypy. It is not restricted to domesticated animals. Polytypy has been reported within single taxa whose range extends over a wide territory that covers different environmental conditions. Examples include baboons (Papio anubis), vermets (Cercopithecus aethiops), and langurs (Presbytis entellus), whose social organization varies greatly over their ranges and appears to be correlated with the nature of the habitat.

Hendrichs has recorded variations in the social organization of three species of antelope, diddik (Madoga kirki), reedbuck (Redunca redunca), and waterbuck (Kobus ellipsiprymnus), that are associated with population density, that is, in turn, presumably regulated by the supply of food and other resources. Stacey and Bock report that acorn woodpeckers (Melanerpes formicovorus) that ordinarily live in permanent groups and defend year round territories deviated from this pattern in one location. Here, they mated in pairs and migrated in early winter. Old field mice (Peromyscus polionotus) from different geographical areas were found to have identical patterns of sexual behaviour, but these varied in quantitative aspects such as latency, frequency, and duration of the components that make up the total mating act.

There is no evidence that these polytypies are attributable to genetic heterogeneity. In fact, many of the cited authors generally ascribe these variations in social organization and behaviour to environmental pressures. The assumption, not always explicitly stated, is that given the same circumstances any member of the species could adopt either of the alternate patterns. However, there have been few tests to disprove a genetic basis, and it would be difficult to make one with many of these species. It is possible in species that can be maintained in large numbers in captivity and bred systematically.

Important contributions to the genetic analysis of ethological polytypy have been made by King and his students. The genus *Peromyscus* includes many taxa that have been classified at the subgeneric, species group, species, and subspecies level in ascending order of closeness of relationship. King, Price, and Webber compared animals from nine taxa on five motor behaviours: swimming, running, climbing, digging, and gnawing. They were searching for correlations between behavioural similarities and closeness of relationship as judged by taxonomists.

Secondarily, they also looked for correlations between behaviour and habitat. For example, burrowing species might be expected to excel in digging, woodland species in climbing. In all tests the only motivation to perform any of the actions was to escape from an unfamiliar place. Differences among taxa were found on every task, but the expectation that the degree of similarity would correlate with the degree of relationship was not fulfilled. Surprisingly, the behaviour—habitat correlations were no better.

Although the results of this study were disappointing, King rethought the problem and carried out other experiments. Instead of comparing taxa on specific motor acts in an artificial setting he designed a set of tests to measure more general behavioural characteristics: speed of adaptation, reflex responses, abstractive and strategic behaviour. Again the subjects were *Peromyscus* taxa of varied degrees of relationship and similarity of habitat. The results of this study lead to a revised view of behavioural polytypy as related to genetics. All taxa of *Peromyscus,* and many other "mice," are roughly similar in form and are capable of "rodentlike behaviour" such as gnawing, grooming, and nest building. The major differences among them lie in their modes of organization of activities in new situations as they search for food, mates, or shelter. Some taxa become specialized through fixation of genes favouring survival under typical natural conditions; others retain their juvenile plasticity into adulthood and adapt as best they can to situations as they arise. King finds evidence that both Hypothesis 2 (fixed genetic programming) and Hypothesis 3 (flexible neural programming) apply to *Peromyscus.*

In a somewhat similar experiment with wild *Mus musculus,* captured in three ecologically distinct areas and reared similarly in a laboratory, Plomin and Man osevitz found differences in open field activity, defecation, climbing, running wheel activity, and nest building. They attributed the differences to genetic factors, although controls for maternal effects and cross breeding experiments were not included in their study. A repetition with more subjects and additional controls would be useful.

In muroid rodents, male copulatory behaviour has been studied from a genetic point of view. Dewsbury (1979b) reported basically similar factor patterns in laboratory rats, the descendants of a wild population of house mice, and deer mice. In spite of the factorial similarity each of these species has a distinctive style of courtship. In more closely related taxa, such as subspecies, male copulatory behaviour is much more uniform and for practical purposes qualitatively identical. However, significant quantitative differences in the latencies and durations of phases of mating were found in 11 of 14 measures. Subjects were three subspecies of *Peromyscus maniculatus: P.m. bairdii, P.m. blandus,* and *P.m. gambeli*. Cross fostering of deer mice on house mice produced minor quantitative effects on latencies, but, according to Dewsbury: "basic motor patterns in cross fostered mice were identical to those of normally reared animals." A particularly interesting instance of behavioural polytypy occurs in the squirrel monkey (*Saimiri sciurus*). Two populations of this species differ markedly in the phonic characteristics of their extensive vocabulary of calls. Externally the populations are distinguishable by facial markings. The Roman arch type found in Peru, Costa Rica, and Panama, and the Gothic arch type resident in Columbia and Guyana also differ in their karyotypes.

That the dialects differ for genetic rather than cultural reasons was demonstrated by rearing infants with devocalized mothers and finding that they conversed in the dialect appropriate to their chromosomes and emitted these calls in their appropriate context. A personal communication from Newman adds that new research has shown that isolation peeps of these two forms differ so much that analysis of sonograms permits classification of individuals with 99per cent accuracy. The recorded IPs of eight F_1 hybrids were matched with their phenotypic appearance. Five hybrids with Roman arches emitted typical Roman IPs.

The IPs of the three Gothic type hybrids were variable, one strongly Gothic, one strongly Roman, and one mixed. These observations are insufficient to hypothesize a mode of inheritance, but apparently vocalization patterns are associated to a degree with physical characteristics.

The traditional genetic procedures of crossbreeding and selection. Most knowledge of the formal genetics of social behaviour comes from experiments with selected breeds or inbred lines of house mice, Norway rats, dogs, *Drosophila*. Can results obtained with these long domesticated animals lead to principles that can be extended to their wild relatives?

Lorenz asserted that results from animal strains that have been bred for a long time in captivity are not applicable to free living animals: "because captivity changes all hitherto effective selective factors in so profound a way that serious changes must be expected in the genome of the stock after only a few generations." Recent research indicates that Lorenz was too sweeping in this statement. Domestic albino rats transferred from their protective laboratory environment to outdoor pens survived severe northern winters, maintained a stable population, constructed and lived in burrows indistinguishable from those of wild rats, and in general behaved like their wild cousins.

Given their freedom, the albinos were able to cope with stresses that their recent ancestors had never encountered. Of course, in a completely feral state their impaired vision and conspicuous colouration would make them more vulnerable to predators. The important point of Boice's study is that there was no evidence of a loss of species typical adaptive behaviour. Disuse of skills does not by itself lead to their deletion from the behavioural repertoire. In a similar vein, Miller made extensive observations on Peking ducks that have been separated from their wild mallard ancestors for centuries. Pekings differ from mallards in important characteristics: They are too heavy to fly, and they lack the mallard's striking sexual dimorphism in plumage. Yet, when observed in a natural setting during three breeding seasons, they performed all the species typical displays that have been reported for mallards.

Miller concluded that his data: "offer no justification for the view that domestication has a 'degenerative' effect on social courtship behaviour patterns, but are rather supportive of the view that the customary nonoccurence of these displays is a function of the inhibitory or unfavourable environmental context in which the domestic birds usually find themselves." The observations of Boice and Miller do not negate the existence of behavioural differences between domesticated animals and their wild progenitors but they support the idea that many such differences are the results of selection for a characteristic desired by their breeders: docility in laboratory rats, and flightlessness in birds that are intended for the dinner of their custodians.

Some changes accompanying domestication probably are the effects of the loss of dominant alleles through inbreeding. Lynch found that inbred wild house mice, compared with random bred animals from the same foundation stock, had smaller litters, bore lighter offspring, and constructed poorer nests. Random bred wild mice reared in a laboratory were not larger, more fertile, or better nest builders than standard inbred strains. In fact, their nests were smaller than those of the heterogeneous HS stock produced by first intercrossing eight inbred strains and then continuing with a random breeding system. Lynch explains the hypertrophy of a behavioural trait related to fitness in a stock synthe sized from inbred strains that built only normal sized nests by postulating that some deleterious recessive alleles present in wild mice were eliminated during inbreeding. She believes that the inbred lines that survive the inbreeding process are good material for analyzing the genetic architecture of a species.

Her view is contrary to that of Smith who found that the heritability of shuttle avoidance learning was much lower in lines of mice selected from a wild stock than in inbred strains and heterogeneous stocks. The relevance of shuttle box avoidance to biological fitness. There is considerable evidence that good performance in the apparatus depends less on learning ability than on the prepotent response to alarm signal, freezing or running. Domestic rats have been selected for docility, and this may indeed disqualify them for inferring selection pressures in nature that are related to escape from shock. Some of the changes in behaviour that accompany domestication seem to be the effects of inbreeding rather than of conscious or unconscious selection. Connor compared wild house mice reared in a simulated natural environment with laboratory reared wild mice and with three standard inbred strains. Wild mice, irrespective of their habitats, were more aggressive towards intruders, vocalized more when handled, and were more difficult to recapture when they were freed in a room.

In a second experiment, Connor followed behavioural changes over 10 generations in random bred wild mice maintained in a natural or a laboratory habitat. A third group from the same original population were inbred by brother sister matings. No behavioural differences were found between the laboratory and the seminaturally reared random bred stocks. Maternal effects were looked for in a cross fostering study, but none were found. In contrast, the inbred wild lines were less aggressive and easier to recapture. The evidence is convincing that we should not ascribe intraspecific variation in behaviour exclusively to genetic or to environmental factors without evidence from experiments designed to separate their effects. Price, on the basis of his studies of the modification of activity by mode of rearing in wild and semi domesticated *Peromyscus maniculatus bairdii* came to the following conclusions.

The genetic requirements for a behavioural character should not change during breeding in captivity unless:

- It is selected for deliberately; or
- It is selected for unwittingly because it improves fitness under the conditions of captivity. These conclusions are encouraging. By a judicious mixture of field observations with controlled breeding and rearing in captivity, we someday may be able to determine the extent to which natural variability in species typical behaviour is a function of genetic differences. An example is the work of Moss and Watson, Sociobiology. Genetic heterogeneity is an absolute necessity for evolution; we know that it exists, but we do not know how important it is with respect to differences in behavioural fitness.

Genotype Environment Interaction

Neither genotype nor environment is sufficient to determine a behavioural phenotype. Without a genotype there is no organism. For a zygote or parthenogenetic ovum to develop into an adult there must be an appropriate physical and biological environment. Ethologists are especially interested in the biosocial factors that shape the development of behaviour. Indeed, the developmental process is often described as a genotype—environment (GE) interaction. The general phenomenon of joint involvement as GE coaction, and to confine GE interaction to situations in which a specific environmental change affects organisms differently depending on their genotypes. GE interactions are detected by exposing two or more genotypes to two or more conditions (*e.g.*, stimulated versus nonstimulated) and comparing the behavioural effects. Broadhurst and Jinks distinguished two ways in which genes might affect development:

- They could influence its stability, particularly during early stages, and effect processes that could potentially influence an individual's behaviour for life. A tendency for phenotypic development to proceed on a stable, species-specific course regardless of ambient conditions is called developmental homeostasis.
- At any stage of life, genes and their productions may influence the capacity of an organism to adjust to changes in its environment such as removal from a familiar habitat. Broadhurst and Jinks' distinction between stability and change is somewhat arbitrary, but it is still useful. In general, stability refers to processes during early development that have generalized, permanent effects; change involves behavioural adaptations to a specific situation, with less carry over to other conditions.

GE interaction takes several forms. The most extreme is disordinal interaction where the effects of an environmental change are opposite in direction depending on the genotype of the subjects. Demonstrates that in a 2-genotype-2-treatment experiment, one could come to completely different conclusions regarding interaction depending on one's choice of treatments. An organism must receive an appropriate exposure to some stimuli in order to develop optimally. The low end of the environmental scale corresponds to extreme stimulus deprivation; the upper end to severe stress. Strains A and B differ in their optimal stimulus requirements.

A requires less than B, but it is also more easily damaged by high levels. Suppose now that we conduct four experiments with these strains choosing only two values of stimulus intensity for each. The chosen values are indicated by x's on the abscissa. Experiment 1 yields a clear interaction; strain A is helped by stimulation, strain B scarcely at all. Experiment 2 produces a marked disordinal interaction. Strain A is impaired as stimulation increases; strain B is improved. In Experiment 3, both strains profit from additional stimulation, and there is no interaction. Finally, in Experiment 4 both strains are impaired as stimulation increases.

Thus the nature of any interaction found may be a matter of the choice of experimental procedures rather than evidence for fundamentally different ways of responding to a stimulus change. The remedy, obviously, is to employ a wide range of treatments in any experiment where GE interactions are likely to occur.

Unfortunately, there are few experiments extensive enough to provide a test of the generality of the threshold hypothesis. They were placed separately in cages where they had no physical or visual contact with other dogs or humans. Generally the period of isolation was 12 weeks; in some experiments isolation was interrupted periodically by brief placement in an arena where subjects could interact successively with a human handler, with another puppy of the same age, and with a rubber ball or swinging toy.

From the results of a series of experiments it was concluded that the deleterious effects of early isolation are primarily attributable to an overwhelming fear reaction when an adolescent puppy suddenly faces a complex environment that is totally unfamiliar. Isolated puppies that were sedated with drugs at the time of first emergence from isolation performed almost like normally reared individuals, demonstrating that their motor and perceptual development had been little impaired by living in the small chambers.

In most mammals reactive capacities develop gradually from infancy to maturity and the environment that must be dealt with at different ages is matched with the infant's or adolescent's perceptual and motor capacities. Rather small interruptions in isolation were sufficient to counteract the effects of experiential deprivation. Puppies who had 15 minutes per week in our test arena developed social and manipulative skills almost as well as pet reared animals with several hundredfold more experience.

A number of our experiments were directed towards defining breed differences in the intensity and duration of the effects of experiential deprivation with beagles and wirehaired fox terriers as the subjects. Beagles have a history of selection for olfactory tracking of small game animals. Litters housed together in group pens seldom fought and dominance hierarchies within litters were weak. In a familiar laboratory setting they gave the impression of relaxation, but they were somewhat shy in unfamiliar surroundings. Housed in the same way, the terriers engaged in fierce battles and formed strong dominance hierarchies within litters.

When handled in tests their muscle tension was high, and they were more likely to attack than to retreat from a threat. Thus, it was not surprising that the effects of isolation were more striking in beagles than in terriers. In one experiment, beagles, transferred to the test arena in a small cage, often failed to explore the new environment; terriers rapidly overcame an initial hesitation and reacted appropriately to objects, people, and other puppies. In another set of experiments, beagles and terriers who had been isolated from weaning to 15 weeks of age were compared with pet reared siblings. Over a period of 5 weeks each subject was observed 25 times in the arena and records made of behaviours such as responding to a handler's call, following a moving person, playing with toys, and interacting with another puppy of the same age.

A quantitative scoring system was devised for two aspects of behaviour, response intensity to key stimuli and activity index. The response indices of pet reared (control) beagles and terriers were practically identical. In both breeds, isolates were inhibited in responding during early tests. Intensity scores rose with repeated exposure to the arena; the rate of increase was much higher in terriers. At the end of the study isolated beagles were still inhibited compared with controls. On the second measure pet reared terriers were more active than similarly treated beagles throughout the observations. During early tests, isolates of both breeds were inactive, as had been anticipated from previous experiments and, as expected, activity increased in both breeds with repeated trials. Beagles were approximately equal with their controls by the fifth week. Terriers, however, were significantly more active than their controls by the second week.

At first glance this might appear to be a disordinal interaction. Experiential deprivation makes terriers more active; it makes beagles less so. In terms of overt behaviour this is certainly true. But there are other ways of looking at the data. In both breeds isolation produced an initial aversion to the arena, reduced activity, and decreased the intensity of response to the test stimuli. Both breeds recovered with experience, but the terriers improved more rapidly and more completely. To be sure, the isolated terriers overshot and became more active than their controls; isolated beagles were less active than their controls. In a sense, however, isolation acted similarly on both breeds in exaggerating the characteristics that differentiated them when they were raised conventionally. Normally active and investigative terriers became more so. Deliberate and cautious beagles became hesitant and timid. It appeared that isolation had made both breeds more emotionally reactive, but they expressed the change in divergent genotype typical patterns.

Genetic Architecture and Behavioural Phylogeny

Ethology is grounded in evolutionary concepts. Roe and Simpson's *Behaviour and Evolution* brought together the views of many distinguished scientists and is still an excellent introduction to behavioural phylogeny. The value of the comparative method in deducing the phylogeny of innate behaviour patterns, particularly communication by sound, posture, and gesture, was emphasized earlier by Lorenz.

However, there are problems with the application of evolutionary theory to behaviour. The validity of the very idea of behavioural phylogeny has been criticized on the ground that it is impossible to distinguish between similarity due to common ancestry (homology) and that attributable to common selective pressures (analogy). His warning is appropriate, but it need not lead to the abandonment of the consideration of the role of behaviour in evolution. In this presentation we are not concerned as much with homology and phylogeny as with the role of behaviour as a component of fitness and, therefore, as a directive force in evolution. Consideration of this topic requires a brief outline of theories relating the genetic architecture underlying variation in a behavioural character to the character's role in natural selection.

Mendelian Approaches to Behaviour Adaptation

We can distinguish between approaches based on one or two loci (Mendelian analysis) and those based on biometrical analysis of quantitative variability. The latter type of variation is considered to depend basically on Mendelian principles, but the numerous loci that are involved are not individually identifiable. Mather called these polygenes and we now apply the term polygenic inheritance to many behavioural traits. Both the Mendelian and the biometric approaches are important to behaviour genetics. Benzer's studies of mutants with well defined behavioural peculiarities have advanced our knowledge of neural development and function in *Drosophila*. The neurological mutants of mice have contributed to our understanding of brain development and function in mammals. Coat colour mutants, such as albinos, have been studied frequently by behaviour geneticists.

Albinos do differ from pigmented animals in many aspects of behaviour, and even normally pigmented carriers of the albino allele are impaired in some phases of development. However, neurological mutants and albinos are so rare in natural populations that they cannot be classified as true polymorphisms. We know that genetic polymorphisms are extremely common and the possibility that some of them have behavioural significance is well worth investigating.

An individual, the target of selection, is not a mosaic of characters each of which is the product of a given gene. Rather, genes are merely the units of the genetic programme that govern the complicated process of development, ultimately resulting in the phenotypic character. To consider genes as independent units is meaningless from the evolutionary point of view because the individual as a whole is the unit of selection. To regard genes as independent units is meaningless from the physiological viewpoint because genes interact with each other in producing the various components of the genotype.

For Mayr, the genotype has properties not deducible from the sum of its components. I agree. But he also acknowledges the possibility of switch genes where an allelic substitution at a single locus may shift the phenotype so that the direction and intensity of selection pressure is modified. A familiar example is industrial melanism, the replacement of light coloured variants of several species of moths by darker variants in areas where pollution has altered the colour of the bark of the trees on which the moths rest.

How much evidence is there for switch genes that change behaviour drastically enough to influence evolutionary processes? Sociobiologists seem to assume that such genes are abundant enough to be important. Trivers writes: "Assume that the altruistic behaviour of an altruist is controlled by an allele (dominant or recessive) a_2, at a given locus, and that there is only one alternative allele, a_1, at that locus and that it does not lead to altruistic behaviour." Simple allelic substitution could switch such complex behaviours as altruism and selfishness or change the mating system of a species from polygamy to monogamy. It is hard to imagine how such a switch gene would operate physiologically.

Of course, any allelic substitution that affects sensory, neural, or motor functions will have some effect on behaviour, but a single mutation seems unlikely, by itself, to shift behaviour into promising new channels. However, there are reports suggesting that some genetic polymorphisms have behavioural correlates that affect selective processes indirectly.

Krebs, Gaines, Keller, Myers, and Tamarin found large fluctuations in the frequency of alleles at the transferrin (*Tf*) and leucine aminopeptidase (*LAP*) loci in the vole, *Microtus pennsylvanicus*, that were associated with changes in population density. They believe that the gene frequency changes over time were indicative of an association between the alleles at these loci and the ability to survive and propagate in stressful situations. Specifically, they hypothesized that females heterozygous at the *Tf* locus were apt to migrate during periods of increasing population density. Homozygotes, particularly for the Tf^{c} allele tended to remain in their original territory. The implication is that the *Tf* locus has somehing to do with social interactions, but, to my knowledge, this relationship has not been verified by direct observation of individual encounters. Krebs *et al.*'s data were obtained from field experiments supplemented by genotype determination of captured animals. A group of immuno-geneticists have applied a very different approach to the genetics of social behaviour. Andrews and Boyse claim to have mapped two loci related to mating preference in house mice. They hypothesize that one of these, *Ris* (recognition of identity signal) provides a cue, probably olfactory, by which males can distinguish preferred females. *Ris* is assigned a location in the *Qa-Tla* region of the major histocompatibility region of the mouse. Another locus, Rir (recognition of identity response) is assumed to determine whether male response is positive or negative.

The genetic procedures are complex and ingenious, but I have reservations regarding the validity of their conclusions. Males were given five to six repeated tests to determine their sexual preference. The significance of deviations from the null hypothesis (random choice) was apparently computed on the basis of the total number of tests rather than on the number of individuals. The receptivity of the rejected females was confirmed by pairing them with other males, but one cannot exclude thepossibility that factors other than identity recognition by males bias the possibility of a female being bred. In spite of these experimental problems, the hypothesis that identity recognition based on cellular characteristics is important in mating choice has evolutionary implications. More strinent statistical analysis and more adequate behavioural controls are needed.

Isolation between closely related taxa need not imply a behavioural barrier. An example that involves a physiological difference with a relatively simple genetic basis has been reported in the lacewing flies, Chrysopa cornea and C. downsei. In the laboratory these species interbreed readily. In nature, crossing is not observed. C. carnea produces three generations per summer; C. downsei produces one, then enters diapause. Standard Mendelian crosses were made between the two species and the occurence of diapause in the hydrids was observed. The data are consistent with control of diapause by two loci. The C. downsei phenotype (diapause) requires recessive allele homozygosity at both loci. This case should probably not be called behavioural isolation in the strict sense, as the two species will mate when they are artificially manipulated. It does demonstrate that relatively simple genetic systems may have substantial evolutionary consequences.

Quantitative Inheritance

Let us now consider deductions that can be made regarding behavioural evolution from knowledge of the genetic architecture that regulated quantitative characteristics. It is well known that variation in such traits is largely polygenic. The theory relating genetic architecture to the evolution of behaviour has been explicated by several investigators. Essentially, stabilizing selection for an intermediate expression of a trait, and directional selection for high or low values involve different genetic strategies. Stabilizing selection requires either additivity of gene effects, so that a heterozygote is midway between the two homozygotes; or a balance between dominant genes at different loci, some with increasing, some with decreasing action on the phenotype. Traits with such a history respond well to selection in the laboratory and are said to be highly heritable.

In contrast, selection for the extreme manifestations of a trait favours alleles with dominant effects in one direction only. Such a system ensures that the favoured alleles will be effective in a single as well as in a double dose. Over time, recessive alleles will tend to disappear from the selected population, and the heritability of the trait will become low. Clearly, it is most efficient for a species to have those traits most directly related to fitness regulated by dominant genes. Falconer noted that in domestic animals heritability is lower for traits such as fertility and viability, than for traits desired by breeders for purely economic considerations.

Accepting these principles, one can work backward from determination of the genetic architecture of a trait to a reconstruction of the role of that trait in natural selection. High unidirectional dominance is evidence that extreme values of that trait increase fitness significantly. Balanced bidirectional dominance implies a history of selection for intermediate values of the trait. High additive variance indicates that a trait has not been important for fitness, or that selective pressures are inconsistent and that the ability to vary phenotypes rapidly in response to a change in environment is valuable. There is an increasing number of studies in which the genetic architecture of a behavioural characteristic in domesticated strains has been used as the basis for inferences regarding the relation of that trait to fitness. Bruell demonstrated that F_1 hybrids between inbred strains of mice almost always surpassed both parental strains on activity tests.

The poorer performance of te parents was attributed to the inevitable, random loss of some dominant alleles during inbreeding. These losses generally involve different loci in any two strains; thus, in their hybrid offspring the deficiency of one is compensated by the contribution of the other. As might be expected, Bruell found that heterosis was more pronounced in crosses whose parents were distantly related than in crosses between lines with similar origins. More sophisticated modes of genetic analysis were used by Broadhurst and Jinks, who measured open field activity and defecation in six lines of rats and all possible F_1 hybrids. This diallel cross design provides estimates of many genetic parameters.

Their genetic findings were:

- Genes enhancing developmental phenostability tend to be dominant;
- Dominance also occurs for rapid decrease of emotional defecation on successive tests in the open field;
- Exploratory activity is also regulated by dominant genes, but these may have either a negative or a positive sign. From thse observations they concluded that selection has operated to:
- Buffer the fetal and infant rat from external stress;
- Promote a rapid return to normal emotional status after arousal;
- Strike a balance between bold curiosity that could lead to disaster, and timidity that could cause exclusion from essential resources. None of this is surprising, but it is satisfying to find that the genetic findings are consonant with our expectations of what organisms should do in order to prosper.

The genetic approach to reconstruction of evolutionary history is gaining ground, but there is difference of opinion regarding the best way to proceed. For example, Lynch concluded that: "because inbred mice are not weakling, analysis of croses between them may represent a more efficient method for detecting genetic architecture than inbreeding of wild populations." Smith disagrees on the basis of his comparison of the heritabilit of shuttle avoidance in selected wild mice and in inbred strains. He doubts that it is safe to deduce the selection history of a hypothetical parent popuation from the genetic architecture of contemporary inbred strains of largely unknown origin.

Obviously, more research is needed to settle this matter. Henderson made an effort to escape the bind of first accepting the basic hypothesis of th relation between genetic dominance and fitness of a trait, then conducting a genetic experiment and, from its results deducing the fitness value of that trait. He deduced from observation that a particular trait was related to fitness and then conducted a breeding experiment to determine its genetic architecture. Specifically, he reasoned that mice should be selected for low levels of motor activity during infancy because they are safest when they remain in their nest. If they are accidentally dislodged their chances of being retrieved are greater if they remain quiet and await their mother. He predicted that, in direct contradiction to the situation in adults, low activity would be dominant in infant mice.

A triple test cross confirmed his prediction; the heritability of activity was low and inactivity was dominant. This was attributed by Henderson to a relaxation of selection for infant immobility over many generations of laboratory rearing. In a very different species, Cheng, Shoffner, Phillips, and Shapiro studied imprinting to visual and auditory stimuli in five groups of mallards (Anas platyrhyncs) ranging from a long established game farm line to 25%, 50%, 75%, and 100% wild ancestry. Wild ducklings imprinted better than domesticated ducklings, and heterosis was found in the hybrids, as would be predicted from the hypothesis that traits with adaptive value show little additive variation. Efficient imprinting seemed to be correlated with differences in arousal during early exposure to stimulation.

Confirmations of a prediction do not prove the universal value of a method. However the approach of Henderson and Chang *et al.* is promising and has considerable potential. As Henderon states, the method is most powerful when predictions can be made that a particular kind of behaviour will differ in its contribution to fitness at different parts of its life-span, or in dissimilar environ ments. The method is obviously limited to species in which multiple distinct lines or strains are available. Principles derived from such studies are, however, likely to have general application. It is only fair to state that the usefulness of this approach to defining adaptive behaviour has been challenged on the basis that the experimental subjects available for analysis of genetic architecture are not representative of natural populations, and that a history of inbreeding distorts the genetic structure typical of a species under natural selection. These warnings must be keptin mind but much of the data obtained in this way seems to substantiate accepted ideas of the characteristics of fitness enhancing behaviour. There is, of course, a possible circularity in such experimens.

One can hypothesize tht high levels of behaviour B_1 are adaptive and thus should have increased fitness during recent evolutionary history. Suppose now that an experiment is performed and low levels of B_1 are found to be inherited in dominant fashion. One could conclude that the original hypothesis was wrong, and that actually low levels of B_1 enhance fitness. Alternatively one might consider the possibility that B_1 has little to do with fitness but, in the population studied, it is correlated genetically with another behaviour, B_2, that has been selected on the basis of enhancing fitness. The best way out of such circularity is good ethological studies relating B_1 and B_2 independently to survival and propagation.

Another prblem with the attempt to deduce the selective importance of a particular form of behaviour is the choice ofthe method of measuring it. Behaviour geneticists have been sure for years that in an open field C_{57}BL mice were more active than BALB/c mice. Whitford and Sipf found no important differences between these strains when the height of their open field was reduced from 36 inches to 1 inch by insertion of a plxiglass sheet. It is clear that open top, open field activity is very different from contact top, open field activity.

Plant Genetic Diversity

The conservation of plant genetic resources has, in recent years, attracted growing public and scientific interest and political support. There is an increasing awareness of the relevance of biological diversity and its conservation to the health of the biosphere. Particularly urgent is the need to raise agricultural output to meet the basic nutritional needs of increasing populations. The dilemma is that the required increase in food production must be obtained through 'sustainable' forms of agriculture which are less dependent on the use of modern high-yielding varieties bred for intensive, high-input systems. It seems unlikely that increased food production on this scale will come from increasing the area under cultivation. Rather, major solutions will depend upon innovations which exploit sophisticated methods to manipulate plant genomes and the totality of natural genetic variation.

The area of disease resistance will serve to illustrate this argument. Many authors have emphasized the importance of innovation in the area of crop protection since crop losses due to pests and disease may account for between 20 and 40% of productivity worldwide. Indeed, over 50% of the research and development (R&D) spent in the plant breeding/seed industry is focused on the identification and incorporation of resistance traits. In a situation where there is some evidence that the pool of available variation for disease resistance genes within commercial breeding lines is becoming limited (*e.g.* for yellow rust in cereals) the search is on for novel sources of resistance. There are now many examples of biotechnology being put to the service of pest control, through the incorporation into plants of transgenes for toxins, antifeedants, enzyme inhibitors, etc. But, as Woolhouse (1992) has pointed out, such is the long coevolutionary history of pests and pathogens with their hosts that present strategies for interfering with these relationships represent only a small proportion of the likely defence mechanisms used by plants in resisting foreign organisms. This enormous diversity of mechanisms represents an arsenal of ammunition which can be used in the fight to minimize disease and predation.

Are there any hard facts which support these assertions concerning the value of genetic resources to agricultural productivity? Can we put a cost on the loss of diversity? The sources of variation used by the plant breeder range from current breeding lines to wild species and the products of direct gene manipulation. In a recent analysis of the plant breeding/seed industry, Swanson (1996) showed that, over a five-year period, 6.5% of all genetic research within this industry, resulting in a marketed innovation, was concerned with germplasm from wild species and landraces.

This compares with only 2.2% of new germplasm arising from technological approaches of induced mutation. The vast majority of germplasm successfully used by the industry still arises from 'conventional' sources, *i.e.* the heavily exploited commercial cultivars, which at first sight might be taken to suggest that wild resources are relatively unimportant in R&D.

In fact Swanson (1996) argues just the opposite, suggesting that the stock of existing commercial varieties should be seen as the information base from which bioindustries develop innovations, whereas the sources of new diversity (wild species, landraces, induced mutation) should be seen as supplying increments to the information base.

In other words, R&D requires an annual injection of 'new' genetic material from natural sources, estimated as amounting to approximately 7% of the stock of material currently within the system. Biodiversity is valuable to industries other than those that seek novel genes for crop plant improvement. Biodiversity has traditionally provided a source of compounds to the pharmaceutical, foodstuffs, crop protection and other industries and here it is possible to give more precise evaluations of its importance. As outlined in Chapter of this volume, drug sales based on natural products from plants were estimated at $US43 billion in 1985 alone and the value of yet undiscovered pharmaceuticals in tropical rainforests is estimated to be as much as $US147 billion to society as a whole. Vast markets exist for the replacement of synthetic biocides, many of which have associated toxicological and environmental problems, by 'natural' compounds, which may have more specific modes of action and fewer side-effects.

Even greater may be the use of genetic resources in more fundamental research where their value in the longer term may be unknown or impossible to estimate accurately at the present time. Diverse genetic material is needed and now regularly utilized to create mapping populations and to study inheritance of a wide range of traits which may or may not be of immediate practical value. Very large numbers of germplasm samples are regularly screened and evaluated for different characteristics so that the underlying molecular biology can be investigated. The intention here is that a greater understanding of fundamental biology will give rise to wealth creation in the future.

Biotechnological Tools for Conservation

Biotechnology can make a substantial contribution to the quality of germplasm collections and their efficient management. For example, in the management of very large germplasm collections there is a clear requirement for procedures for fast, reliable taxonomic identification of material and the elimination of redundancy. Samples characterized by molecular criteria may prove superior to traditional analyses of phenotype and examples of the use of DNA markers to guide the management of germplasm are explored in Chapter by reference to one of the world's largest collections of germplasm at the International Rice Research Institute.

Not all plant species can be stored in the form of desiccated seed. Socalled 'recalcitrant' species have to be kept in moist, warm conditions and even then have relatively low longevity. In addition, there are many cultivated plants which must be maintained clonally, such as banana, which produce few seeds, or potato, which does not come true from seed.

To these traditional problem species we must now add the products of biotechnology, new categories of germplasm including clones of elite genotypes, and transformed cell lines, for which storage of seed is inappropriate. Biotechnology has impact on storage technologies in two quite different ways. The first of these, concerns the various types of *in vitro* culture which allow the propagation of plant material in an aseptic environment, thus ensuring the production of disease-free stocks and simplifying quarantine procedures for the international exchange of germplasm. The development of sensitive molecular probes to detect low levels of pathogens.

A more radical approach to safe storage and distribution of germplasm, and possibly the only way to significantly reduce the rate of loss in biodiverse tropical forests, within the bounds of available resources, is to develop alternatives to the costly (and inherently vulnerable) methods of *ex situ* storage of seed or tissue cultures, by conserving germplasm in the form of DNA or DNA-rich materials. The latter include those dried specimens conserved in the vast resources of the world's herbaria. The DNA Bank-Net, an international network of DNA repositories, has now been established with both storage and distribution functions. Although technologies for rapid removal, analysis, amplification and secure storage of small amounts of DNA are now well advanced, research is still needed on methods to amplify the entire genomic

DNA of a species, which means that storage of DNA or DNA-rich materials should still be considered to be something of an insurance policy rather than a replacement for conventional modes of germplasm storage.

Biotechnology-assisted Crop Improvement

Effective use of genetic resources in plant breeding programmes requires methods to determine whether useful genetic variation exists, and costeffective methods to introduce useful genes into the material. Biotechnology has made a contribution to both these aspects.

The use of genetic maps based on molecular markers linked to genes of interest facilitates the more rapid selection of both simple and complex traits thus accelerating their incorporation into breeding materials. Where traits are controlled by one or a few genes, direct isolation of genes and their incorporation into elite cultivars by increasingly efficient transgenic technologies provide an additional approach to the more traditional methods of plant breeding. This prospect also challenges one of the classical concepts in plant genetic resources work, namely the 'gene pool' concept of Harlan and de Wet (1971), which separated cultivated species and their wild relatives into different categories based upon their 'crossability'.

The value of this concept in guiding the work of the plant breeder must now be re-evaluated in the light of molecular genetic technologies which have not only cast a completely new light on the relatedness of crop plant genomes, but which have effectively widened the potential gene pool of a species to include all other life forms.

What challenges does this imply for the traditional plant genetic resources programmes? Why is the emphasis on conserving species that are related to crops when more exotic and potentially valuable genes may be present in liverworts, or green algae, or other less studied taxa?

Bioprospecting for Novel Genes

The value of genetic resources as a source of novel compounds for various industries has been alluded to above. For centuries biodiversity has provided sources of fuels, medicines, foodstuffs, pesticides, etc. The high returns available from introducing a new drug, for example, plus the realization that biodiversity is rapidly being lost (estimated rates of extinction range from 30 to 300 species per day) has encouraged new attitudes to the exploration and exploitation of biodiversity through 'bioprospecting'. How does biotechnology impact on this traditional area? A fundamental operation in any bioprospecting programme is the screening of thousands of extracts for compounds of interest. Gene technology now allows the speeding up of screening programmes for new compounds through the development of more sophisticated *in vitro* assays. For example, the genes encoding receptor proteins for certain classes of drug, or enzymes that may serve as targets for novel pesticides, may be cloned and expressed on a large scale in high-throughput *in vitro* assays into which thousands of plant extracts may be applied.

Bioprospecting is also concerned with discovering novel genes and natural variants of genes or gene products that may either be used directly or to guide sequence improvement by molecular methods. Chapter gives examples of amino acid sequence improvement in peptide hormones as a result of screening homologous genes from the wild, and this area of biotechnology is likely to develop in parallel with new molecular methods of generating random variation through employing the mutational potential of the polymerase chain reaction (PCR) – *i.e.* 'forced molecular evolution' – or recombinatorial libraries of peptides.

Whilst gene prospecting has provided a new perspective on biodiversity, the area is not without political controversy because these advances tend to favour agriculture and other industries of developed countries, and may actually displace traditional commodity production in resource-rich but underdeveloped countries. Chapter also outlines the development of strategies which integrate conservation and exploitation with economic returns and sustainable development in those countries from which the bioresources were first obtained.

Plant Genetic Variation

Effective conservation and use of plant genetic resources involve asking many questions about the extent and distribution of genetic variation. Only when the appropriate markers and technologies for describing this variation are accessible can such questions be adequately addressed.

The most appropriate markers for a given question should:

- Be heritable;
- Discriminate between the individuals, populations, or taxa being examined;
- Be easy to measure and evaluate; and
- Provide results that can be compared with results of similar studies.

Molecular markers and assays that meet these criteria were introduced in the 1950s and have proliferated ever since. Use of protein isozymes as genetic markers increased rapidly after their introduction. Nucleic acid markers gained popularity in the 1970s, with the advent of DNA sequencing and restriction fragment analysis. The number of different molecular markers and assays has increased exponentially since the late 1980s, when introduction of the polymerase chain reaction (PCR) made DNA amplification and sequencing possible and affordable. As the diversity of markers and assays increases, the cost of using them continues to decrease. At the same time, the range of taxa and tissues that can be analysed has expanded. Improved DNA extraction, amplification, and sequencing protocols allow us to analyse ancient DNA, from 13,000-year-old specimens of *Mylodon* (ground sloths) and 17–20 million-year-old *Taxodium* (bald cypress) specimens, and samples as small as single cells or fractions of a seed.

Thus the chances are good that at least one marker assay will be appropriate for any question about variation in any taxa. Furthermore, results of one study can generally be related to results from many other studies. Such opportunities are inevitably accompanied by challenges. None of the challenges are unique to molecular studies, but all become more critical as the number of available marker assays increases. Faced with a bewildering variety of molecular technologies (and their acronyms), choosing a marker and assay that are appropriate for a scientist's particular question, taxa of interest, and practical constraints can be overwhelming. The temptation is great to use techniques that are recommended by coworkers, are frequently used, for which the most information is available, or are the newest – without adequately evaluating the techniques to determine which is most suitable.

Because molecular markers can be measured in virtually all plant taxa and minimal tissue is required, extra care must be taken to verify that the entries in a test array are correctly identified. Since molecular markers are widely applicable and their constraints are not always understood, thoughtful planning is particularly important when designing molecular studies, to ensure that they have an appropriate range of entries, an appropriate number of entries per taxon, and conclusions that stay within the scope of a specific question and test array.

In addition, the vast amount of published research using molecular markers complicates the task of thoughtfully relating one study to others with similar marker assays, research questions, and/or taxa.

Another challenge is formulating research objectives that ask critical questions about plant genetic resources. Descriptive studies that use molecular marker assays to examine variation between and within specific taxa can generate valuable information, but some plant conservation questions are best addressed by experimental studies.

The wide variety of molecular marker assays available provides new ways to address such questions – if the markers are appropriately used. Such challenges can be met successfully and thoughtful decisions can be made if logical frameworks are available to categorize and compare these valuable molecular tools. A solid foundation for such comparisons has been provided recently. This chapter builds on that foundation by reviewing both early and recently developed molecular markers and assays, citing examples of their use. The types of questions asked about plant genetic resources are described, as are guidelines for selecting appropriate markers and assays to address these questions. The literature cited here is far from comprehensive, but includes applications of each marker assay in plant genetic research. While the list of molecular techniques will soon be dated, hopefully the framework presented will be helpful for categorizing and comparing new tools as they are made available.

The chapter concludes with a perspective on new marker assays as they apply to, and may change, the study of plant genetic variation.

Types of Molecular Marker Assays

Molecular marker assays are generally classified by whether the molecules evaluated are proteins or nucleic acids, and whether the character analysed in a nucleic acid marker assay is the entire genome, a chromosome, a fragment, or a nucleotide. Alternatively, marker assays can be categorized by the type of character measured. Some methods measure quantitative differences between entries in an array. Others measure qualitative characters, each with two or more possible states. Marker assays also differ in the number of loci evaluated per analysis, whether multiple loci are evaluated simultaneously or sequentially, and the type and amount of information needed about the marker loci before conducting the assay.

Protein Marker Assays: When proteins are used as genetic markers, the assumption is made that any variation between proteins reflects heritable variation in their amino acid sequences. This assumption has long been debated, since protein phenotypes can be affected by factors such as post-translational modification, plant phenology, and environmental conditions during plant growth. For many tasks, however, protein marker assays continue to be useful molecular tools.

Some assays measure immunological properties of proteins. When antigenic proteins of one plant or animal species are injected into an animal (usually a rabbit), antibodies are produced. Genetic differences between entries in an array can be assessed by comparing how each entry's proteins bind to antibody serum (antiserum) from a reference species. This binding affinity is measured by the amount of antibody–antigen precipitate produced or by microcomplement fixation (MCF). Each MCF solution contains antiserum from a reference sample, antigen from a test entry, and a protein complement.

The complement can only bind to antibody that is bound to the antigen. The amount of unfixed complement is measured; this is used to calculate the amount of unbound antigen, which indicates the extent of variation (the immunological distance) between the entry and the reference sample. After MCF assays are conducted for each entry in an array, these immunological distances are compared between entries. MCF has long been used in vertebrate phylogenetic analysis. Until recently, immunological markers were seldom used to study plant variation, perhaps because few suitable proteins had been identified. During the past few years, however, monoclonal antibodies to seed proteins of several cereal species have been produced. Reactivity of these antibodies with antigenic proteins can be evaluated by enzyme-linked immunosorbent assays (ELISAs). Monoclonal antibodies have been used to identify cDNA library clones and describe variation between crop species and cultivars. Other marker assays measure the rate of protein migration through a starch, polyacrylamide, or cellulose acetate gel in response to an electrical current. The migration rate is related to protein size, shape, isoelectric point and/or ionic charge.

In these assays, protein extracts from entries in an array are loaded in adjacent gel lanes and electrophoresed. The gel is then treated with a histochemical stain that reacts with a specific marker protein. Variation between the entries is measured by the positions of their stained proteins on the gel. Many of the proteins used as markers are isozymes – functionally similar forms of an enzyme, encoded by different loci or different alleles at a locus. In general all alleles at a locus are expressed, so allelic isozymes can be analysed as codominant variants.

Isozymes are frequently used to describe variation between and within plant populations, species, and genera; for mapping and diagnostics; and in some cases for identification. When comparing higher taxonomic levels, the probability of genetically different electromorphs migrating to the same gel position is increased. For closely related taxa and populations, however, isozyme analysis is informative and continues to be refined. Immunoelectrophoretic assays measure both protein migration rate in gels and the antibody specificity of proteins. Antigenic proteis are electrophoretically separated, then are detected by their reactions with antibodies. This technique is valuable for diagnostics, mapping, and describing variation between plnt cultivars, species, and higher taxa.

Nucleic Acid Marker Assays: Total Genomic DNA and Chromosome Markers: Although proteins are useful as genetic markers, the phenotypeeach protein describes is determined by the genotype plus the type of tissue sampled, plant phenological stage, environment, and post-translational processing. In addition, the percentage of a genome sampled by protein markers is limited: these markers only describe coding sequences, and not all proteins can be separated and detected by established methods. Some questions are best addressed by markers assumed to be selectively neutral, but this assumption is in question for many proteins. Finally, some proteins are unique to particular taxonomic groups, and thus are suitable as markers only in a limited range of taxa.

In contrast, nucleic acids are present in all organisms and are a common genetic currency for comparing virtually any taxa. Nucleic acid markers describe genotypes, not phenotypes, and can sample both coding and noncoding regions of a genome. DNA methylation and secondary structure can cause artefactual variation in some assays, disrupting the direct relationship between marker and genotype. However, these artefacts can often be detected and eliminated.

The first nucleic acid marker assays measured properties of total genomic DNA. When centrifuged in a CsCl density gradient, DNA separates into bands according to GC content. Because highly repetitive (satellite) sequences differ i GC content from the rest of the genome, these gradients can be used to isolate and quantify satellite DNA. This method has been utilized to compare satellite DNA content between plant species, and is still used to detect highly repetitive DNA.

Another technique that separates repetitive and low-copy DNA relies on DNA reassociation kinetics. The two complementary strands of a DNA molecule denature when heated and reassociate into duplexes when cooled. At a standard temperature and salt concentration, reassociation rate depends on the percentage of strands with similar sequences. Repetitive sequences reassociate rapidly and at high temperatures, compared to single-copy sequences. Thus highly repetitive, middle repetitive, and single-copy fractions of a genome can be isolated by fragmenting, heating, and cooling double-stranded DNA. Because reassociation rate is affected by fragment length and the pattern of sequence repetition, the type and dispersion of repeats can be measured by varying the lengths of the fragments analysed. This method has been used to characterize and compare repeat sequences between plant species and hybrids.

DNA–DNAhybridization employs the same principles, but with single-copy sequences, to compare entries in an array. When single-stranded DNA from two entries is combined, their overall sequence similarity is related to the temperature at which 50% of the strands form heteroduplexes. This *T*m is compared to the homoduplex *T*m for each entry; D*T*m measures the genetic distance between entries. DNA–DNAhybridization is used to examine variation between species and higher taxa, usually with vertebrates and less often with plants and bacteria. Variation in DNA content of nuclei andchromosomes can be measured by flow cytometry. Particles are isolated, suspended, stained with a DNA-binding fluorochrome, and passed through a cytometer. The fluorescent signal intensity from each particle indicates its DNA content. Flow cytometry was first utilized to analyse mammalian genomes, but has since been used to compare nuclear DNA content and ploidy level between plant families, genera, species, and ecotypes and to evaluate widehybrids. Becaus chrmosomes can be identified by their signal intensity, flow cytometry can be used to generate high-resolution karyotypes and sortchromosomes for mapping.

Karyotypes ofchromosomes prepared in mitotic metaphase squashes have long been used for diagnostics, mapping, and describinggenetic variation between plant species andhybrids. Entries in an array can be characterized by their chromosome number and morphology. In addition, chromosome spread preparations can be treated with stains specific to AT-rich regions, heterochromatin, or nucleolar organizing regions. The stain-specific banding patterns of individua chromosomes are then used as markers to characterize and compare entries.

Chromosome banding is informative for mapping studies and describing variation between geneban accessions and cultivars. Banding is often used in combination with flow cytometry or other techniques. Fo chromosomes that are very small or lack distinctive bands, cytogenetic techniques may have limited use. As high-resolution microscopes andmore sensitive stains are developed, however, the value of these assays continues to increase.

Nucleic Acid Marker Assays:Fragment Markers: Nucleic acid fragments are widely used as genetic markers. As mentioned above, CsCl density gradients and DNA reassociation techniques can detect and roughly quantify satellite DNA sequences. Yet these methos do not measure discrete character differences or variation in the sequences themselves. In addition, differences in GC content between genomes can significantly affect the results. In contrast, fragment markers provide high-resolution qualitative information about sequence variation. Either DNA or RNA can be evaluated, and minimal amounts are required. Numerous fragment marker assays are available; all involve detecting fragments that differ in presence, size, or quantity between entries in an array.

Fragments that are produced, electrophoresed, then detected: In many assays, fragments are produced, separated by agarose or polyacrylamide gel electrophoresis, then detected. The length and conformation of a fragment determine its migration rate in a gel. Several different approaches are taken for detecting fragments. In some assays with purified sequences, all fragments are detected and visualized. The fragments are either labelled before or stained after electrophoresis. Ethidium bromide, silver, fluorescent, and radioactive stains and labels are commonly used. In other assays, specific target fragments are detected and visualized by transfer hybridization. All fragments in an assay are electrophoresed, denatured if double-stranded, then transferred from the gel to a nylon or nitrocellulose membrane. A labelled single-stranded probe complementing the target sequence is hybridized to the membrane and anneals to the target fragments. The extent of mismatch hybridization varies with salt and formamide concentration and the temperature during hybridization. *A variety of probes can be used*: Synthesized oligonucleotides, uncharacterized clones from genomic DNA libraries, fragments isolated from other gels, RNA sequences, or total genomic DNA. The probe may be from the same species as the membrane-bound DNA or from a different species. Until recently most probes were radioactively labelled, but nonradioactive colourimetric, fluorescent, and chemil-uminescent labels are increasing in popularit).

Fragments produced by restriction digest: Restriction site analysis is widely used in plant genetics. The principles, protocols and data interpretation have been discussed in detail. Restriction fragments are generated by treating double- stranded DNA with restriction endonucleases (REs), which are enzymes produced by bacteria. Each RE cleaves double-stranded DNA at a specific sequence, usually four to six nucleotides in length. Several hundred REs have been characterized; they vary in recognition sequence and ability to digest methylated DNA. DNA sequence variation between the entries in an array is detected as variation in the number and/or lengths of restriction fragments. This variation can be caused by point mutations at RE recognition sites or by the insertion, deletion or rearrangement of sequences at or between RE sites.

When single- or low-copy fragments are evaluated, variations in fragment length are referred to as restriction fragment length polymorphisms (RFLPs). In some RFLP assays of highly purified DNA sequences, all of the fragments are visualized by labelling them before or staining them after electrophoresis. In most RFLP assays, however, target fragments are detected by hybridized probes. For each entry in an array, the presence or absence of each individual fragment is recorded.

These fragment data are used to infer the presence or absence of each restriction site. For diploid genomes, entries heterozygous for the presence of a restriction site can be distinguished from those that are homozygous. Although few RFLP loci are examined on each gel, every locus can be genetically defined. RFLP analysis can be expensive, but costs per trial can be reduced by probing each membrane several times with different probes. Single- or low-copy RFLPs are used extensively in diagnostics, for mapping, and to verify interspecific hybridization and study genetic relationships and structure. Probes that hybridize across species are especially useful for comparative mapping.

In most restriction site analyses of repetitive DNA, target fragments are detected by probing with a known repeat sequence. The fragments, not the genetically defined sites, are usually the characters analysed. Entries that are homozygous for the presence of a fragment cannot be distinguished from heterozygotes. These assays provide information about variation at many loci simultaneously, but reveal little about the type of sequence variation. Restriction site analysis is often used to study interspersed repeats and satellites. It was utilized in the first assays of hypervariable loci, and is presently the most common method in forensics for evaluating minisatellite VNTR loci. Multilocus restriction fragment profiling of plant genomes is useful for identifying clonal cultivars and studying breeding systems. In some studies, fragment length polymorphism can be interpreted as variation of codominant alleles at a locus. However, because inheritance of markers in multilocus profiles is seldom clearly defined, their use in phylogenetic and gene flow studies may be limited.

Restriction site analysis can generate highly reproducible results, if all steps of the process are carefully controlled. With a variety of probe– enzyme combinations available, assays of the nuclear genome can be used to address many questions about genetic variation at or below the species level. As in isozyme analysis, the probability of genetically different restriction fragments co-migrating on a gel is increased when higher taxonomic levels are compared. Because chloroplast genomes evolve slowly, chloroplast restriction site assays are often used to study variation between higher taxa. In some species, both chloroplast and nuclear genomes are biparentally inherited. In many plant species, however, chloroplasts are either maternally or paternally inherited. For these species, chloroplast restriction site analysis is very informative for studies of gene flow and interspecific hybridization, especially when used in combination with assays of biparentally inherited nuclear markers.

Fragments produced by the polymerase chain reaction: Fragment analysis began in the 1970s with the introduction of RFLPs, and grew exponentially a decade later when the polymerase chain reaction (PCR) was developed. PCR uses the principles of DNA reassociation and the action of DNA polymerase to amplify nucleic acid fragments *in vitro*. Each PCR reaction solution contains a double-stranded DNA template, short single-stranded oligonucleotide primers, thermostable DNA polymerase, enzyme cofactors, and the four deoxynucleotides (dNTPs: dATP, dCTP, dGTP, dTTP).

The template is denatured by heating; the temperature is lowered, and the primers anneal to complementary sequences on the template. The solution is heated again, and the polymerase adds dNTPs to the 3' end of each annealed primer. Double-stranded fragments are produced and serve as templates for the next cycle, which generates fragments with primer sequences at each end. During subsequent cycles, these fragments are amplified at a geometric rate.

With automated thermocyclers and cheaper, improved reagents, PCR is fast becoming more reproducible and affordable. A wide range of known and unknown sequences can be amplified from fresh or ancient DNA and from RNA, and little template is needed; hence PCR is applicable to any taxon. However, contamination is a concern, since contaminant DNA in a reaction solution can function as a template. The specificity and reproducibility of PCR depend on several factors: the concentration and quality of the reaction ingredients; the design and GC content of the primers; and the temperature, duration and number of cycles. However, measures can be taken to minimize and detect contamination and optimize cycling protocols. The disadvantages of PCR are greatly outweighed by the tremendous opportunities it provides. When amplified fragments are used for high-resolution mapping, diagnostics or forensics, they are generally detected by probes. When used to describe variation between and within species, amplified fragments may be detected by labelling primers or dNTPs in the reaction solution, labelling PCR products, or staining gels after electrophoresis.

Numerous PCR-based marker assays have been developed and used in plant genetics. They can be categorized by whether target sequences are known prior to amplification, whether the primer sequences are designed or are arbitrary, the number of primers, and the size range of amplified products. In the first PCR assays the target loci were known single- copy sequences, amplified by two primers designed to complement the regions flanking the loci. Primer pairs have since been designed for numerous single-copy loci in plant and animal nuclear and organelle genomes. These loci are very useful as markers for mapping, diagnostics, phylogenetic analyses, and studying hybridization.

In most genome mapping projects, cloned DNA sequences present at only one site in a genome are used as physical mapping landmarks. Primer pairs have been designed to amplify many of these landmarks as sequence tagged sites (STSs). This strategy is used extensively in the Human Genome Project (HGP) and other mapping programmes. In assays of these designed-primer PCR loci, the character evaluated may be the fragment itself, with presence dominant over absence. Often, however, inheritance can be defined.

For loci in diploid genomes, differences in fragment length can then be analysed as codominant alleles. Both interspersed and tandemly repeated sequences can be amplified by designed-primer PCR. When several interspersed repeat loci are amplified in one sample, each locus is evaluated separately. When tandem repeat loci are amplified, length polymorphism at a locus is assumed to reflect variation in the number of tandem repeat units.

Amplification of human minisatellites with designed primer pairs was soon followed by amplification of plant minisatellites and SSRs. Amplified plant SSRs are informative for mapping, genotyping, diagnostics, and population studies. SSR alleles may differ little in length, and discriminating them can be difficult. New electrophoresis, labelling and imaging technologies are making detection easier.

Development of designed-primer PCR markers involves several steps. Target sequences are identified and isolated, usually from a genomic DNA library; the loci and flanking regions are sequenced; and primer pairs are designed and synthesized. Currently these tasks are expensive, and not all laboratories have the equipment required. Once primers are designed and synthesized, however, the costs for each analysis are not high. The initial expenses of marker development may be reduced by identifying the desired sequences from nucleic acid databases, *e.g.* Genbank or EMBL. However, this is only applicable to species well represented in the databases. As another cost-cutting measure, some primers designed to amplify loci in one species can be used to amplify loci in related taxa. Primers that amplify across taxa are also useful for phylogenetic analyses and studying genome evolution.

For development of designed-primer PCR markers, both target and flanking region sequences must be known. Soon after PCR was introduced, however, its ability to amplify unknown sequences was realised. In inter-repeat PCR, primers that complement interspersed repeats or other known sequences are used to amplify unknown sequences between the repeats. 'Universal' primers have been designed to complement conserved coding sequences in mitochondrial and chloroplast genomes; these primers can be used to amplify the noncoding regions between genes.

Other marker assays require no prior sequence information, for either the target region or the flanking regions. These assays – randomly amplified polymorphic DNA analysis (RAPD), arbitrarily primed PCR (AP-PCR; Welsh and McClelland, 1990), and DNA amplification fingerprinting (DAF) – use one short primer that has an arbitrary sequence. This primer anneals to the template at complementary sequences in both 'sense' and 'antisense' orientation, and the regions between primers in opposite orientation are amplified. The number and length of fragments amplified by one primer can vary; short primers generally produce a large number of small fragments. In single-primer PCR, entries that are homozygous for the presence of a fragment cannot be distinguished from heterozygotes; thus fragment presence is generally dominant over absence.

Many loci can be evaluated simultaneously, although little genetic information about each locus is provided. Arbitrary PCR is particularly valuable when no sequence information is available for the taxa being studied. It has been used extensively to examine variation between and within plant species, identify genebank accessions and cultivars, and construct genetic maps.

Many PCR assays are variations or combinations of designed-primer PCR and arbitrary PCR. Most were developed to increase the reproducibility of results and/or to better discriminate between closely related genotypes. In arbitrary PCR, reproducibility and the extent of mismatch primer annealing are highly dependent on the cycling parameters and concentration of reaction components. One solution to this problem is to identify polymorphic fragments in arbitrary PCR assays and then isolate, clone, sequence and design primers for these fragments. This process generates new designed-primer PCR loci but involves many steps. Other assays with increased reproducibility combine features of single-primer and interrepeat PCR: unknown sequences are amplified by one primer that complements an SSR, its flanking region, or both. Finally, 'anchored' PCR uses one designed and one arbitrary primer to amplify unknown sequences.

For genotyping, diagnostics, and assessing variation between close relatives, PCR-based marker assays with high resolution are needed. One such variation is nested PCR. In this technique, one amplification reaction with two designed primers is followed by a second reaction. The first PCR product is the template for the second reaction, which uses two designed primers internal to the first set. Although the possibility of contamination is high, nested PCR is valuable for extremely sensitive genotyping and diagnostics. At present, it is used most often in medical research.

Resolution of arbitrary PCR can be increased by adding stable mini-hairpin sequences to the 5' ends of short primers. This technique greatly improved discrimination between cultivars of centipedegrass [*Eremochloa ophiuroides* (Munro) Hackel]. Other marker assays with improved resolution combine principles of PCR and restriction site analysis. RE digestion of PCR templates before they are amplified was first suggested by Mullis and Faloona (1987). Arbitrary PCR with cleaved templates has since been used to map plant genomes and identify genotypes. Amplified fragment length polymorphism (AFLP) analysis is a variation of cleaved-template PCR. After DNA is digested, a short adaptor sequence is ligated to each end of the restriction fragments. The fragments are then used as templates for single-primer PCR.

The primer contains a constant sequence (the adaptor and RE recognition site sequences) plus selective bases that determine the specificity of primer annealing. Both reproducibility and resolution are increased by using AFLPs. Resolution can also be increased by digesting PCR products before electrophoresis. Digested products of both designedprimer PCR and arbitrary PCR have been used to measure variation between genebank accessions, refine RFLP maps, and study interspecific gene flow. Resolution of both PCR and restriction site assays can be improved by denaturing double-stranded DNA fragments before they are electrophoresed. Due to internal point mutations or differences in conformation or GC content, the two single strands from one fragment may have different migration rates on polyacrylamide gels. Both denaturing gradient gel electrophoresis (DGGE) and single-stranded conformation polymorphism (SSCP) analysis can detect differences between closely related genotypes.

Fragments detected directly by probe hybridization. In some marker assays the target sequences are not separated by electrophoresis, but are detected directly by probe hybridization. The types of probes used are the same as in electrophoretic assays, and can be radioactively or nonradioactively labelled.

In dot and slot blot hybridization, single-stranded DNA fragments, genomic DNA, or RNA is immobilized on a nylon or nitrocellulose membrane. Labelled probes complementing the target sequence are hybridized to the membrane blot. All loci in the blot are sampled simultaneously; abundance of the target sequence is measured by the intensity of the hybridization signal. Many blots on one membrane can be evaluated at the same time, and the procedure is relatively simple. Because the presence of nontarget sequences in the blot can affect hyridization, precise measurment of target sequence abundance in genomic DNA blots can be difficult. When the blots are carefully prepared and hybridization conditions are controlled, however, these assays can generate valuable information for diagnostic and relationship studies. Dot and slot blots have been used to detect specific alleles of genes and RAPD loci, estimate the copy number of satellites and interspersed repeats, and compare gene expression between organs of a plant.

In another technique, *in situ* hybridization (ISH), probes are hybridized to the same chromosome spread preparations used for karyotypes and banding pattern assays. ISH is used to detect the presence and chromosomal distribution of target sequences. While banding assays are only appropriate with chromosomes that contain specific sequences, ISH can detect a wide range of sequences. Radioactive and nonradioactive probes are used, but the popularity of fluorescent *in situ* hybridization (FISH) is rapidly increasing.

ISH has many applications for genome analysis. These include 'painting' the genome of one species onto chromosomes of another, to study species evolution; painting a genome with one of its own chromosomes, to study chromosome evolution; and characterizing the chromosomal distribution of interspersed repeats. ISH is often used in combination with other marker assays. In wide crosses and somatic hybrids, ISH with species-specific interspersed repeat probes can indicate the abundance and chromosomal locations of transferred sequences; restriction site analysis can provide information about structure and rearrangement of the sequences. Physical mapping with ISH can complement RFLP genetic mapping studies.

Aledo *et al.* used slot blots to estimate the copy number of an interspersed repeat in the maize (*Zea mays* L.)genome, and used FISH to determine the chromosomal distribution of the repeat. With both ISH and heterochromatin-specific staining on the same chromosome preparation, Harrison and Heslop-Harrison (1995) confirmed that two *Brassica* satellite families are located near centromeres. By varying thehybridization stringency, differences between satellite sequences on differentchromosomes were revealed.

ISH can be time consuming, and results can be affected by the specific methods used for chromosome preparation, denaturation, and probehybridization and detection. However, new protocols are increasing the reproducibility of

ISH.Fukui *et al.* (1994) used a thermocycler to control temperatures during denaturation, and developed accurate imaging methods for detecting and analysing fluorescent signals. Because of such innovations, the value of ISH as a molecular tool will most likely continue to increase.

Nucleic Acid Marker Assays: Nucleotide Sequencing: Nucleic acid variation can be described most directly by using the nucleotides themselves as discrete characters, each with four possible states. Because nucleotides are the basic units of genetic information in all organisms, sequencing can be used to detect very low levels of variation. Nuclear and organelle genomes of virtually any taxon can be sequenced; minimal amounts of DNA or RNA are needed, and fresh or preserved DNA can be used. With most marker assays, entries in a trial are compared to each other; results are not easily compared across trials that have no standard entries in common. In contrast, sequences from one study can be compared with sequences from any other study. In addition, sequencing can reveal whether variation is due to nucleotide substitution or sequence rearrangement.

Two sequencing methods have been used since the late 1970s, and are described in detail by Hillis *et al.* (1990).

In both methods, labelled fragments are generated from DNA or RNA that has been isolated and purified; the fragments are electrophoresed; and the sequence is 'read' from the fragment patterns. Maxam–Gilbert (chemical) sequencing is used to sequence double-stranded DNA. DNA is endlabelled, then divided into four reaction solutions.

Each solution contains one of four reagents, each of which cleaves DNA at specific bases. The sequence of the DNA determines the lengths of the fragments produced. *In vitro* (dideoxy chain termination) sequencing is applicable to both DNA and RNA. The single-stranded template is divided into four reaction solutions. Each solution also contains the following: a labelled oligonucleotide primer that complements one end of the template; the four dNTPs; and one of four dideoxynucleotides, each of which terminates extension at one of the four bases. In each reaction, the primer is extended along the template. As in chemical sequencing, the template sequence determines the lengths of the fragments produced. Chemical sequencing is preferred for some genomic regions with complex secondary structure; in general, however, the chain-termination method is more commonly used. Since it has been coupled with PCR, combined with fluorescent labelling and automation, this technique is rapidly becoming the method of choice.

Sequencing can be used to address a wide range of plant genetic resource questions about virtually any taxa, but involves considerable investment of money, labour and time. Automation and large-scale production of cheaper reagents have made sequencing more cost effective, but few laboratories can afford to sequence large genome segments from multiple entries. As a result, in many studies only one locus or a few loci are sequenced for a limited test array. While many adjacent nucleotides may be evaluated, together they sample only one locus. Thus sequencing is usually not practical for mapping, multilocus forensics, population genetics, or analyses of intraspecific variation. It is often used for high-resolution diagnostics, studying specific genes, and describing variation between genera and higher taxa. In phylogenetic studies, however, inferring genome evolution from sequence variation at a single locus should be done with caution if at all. Many regions of plant nuclear and organelle genomes have been sequenced and used as markers.

These sequences vary greatly in evolutionary rate and the taxonomic levels at which they are employed. Chloroplast genomes evolve slowly; sequences of the entire genome or its *rbc*L gene are informative as markers for describing variation between genera and higher taxa. Nuclear genes that have been sequenced and used as markers include the phytochrome, heat shock protein, and nuclear rRNA gene families. In general, the rRNA coding regions are highly conserved; internal transcribed spacers are more variable, and intergenic spacers are highly polymorphic. Finally, sequencing is often used – in combination with other marker assays – to characterize tandem repeat loci.

While sequencing may be the ultimate molecular tool, caution should be taken to avoid several potential problems. Because each nucleotide is a separate character, sequence fidelity is crucial. Thus nucleic acids must be carefully isolated and purified before they are sequenced. In addition, PCR contamination and amplification errors have more serious effects for sequencing than for fragment analysis. Amplification errors can be minimized by optimizing PCR conditions and using polymerases with proofreading capability. Mistakes during gel reading (manual or automated) must also be considered and minimized. Subjectivity is a real concern when aligning sequences before they are compared. Alignment is complicated and requires assumptions about expected amounts of base substitution and insertion/deletion. Despite these concerns, sequencing remains the tool of choice for many studies.

Future Dimensions of Genetic Analysis

The goals and expectations for analysing plant genetic variation parallel those established across many other fields of biological research, from agriculture, ecology, and evolution to the medical sciences. In all of these fields, future genetic marker assays must incorporate methods to detect, describe, interpret, and store DNA sequence information. Molecular tools of the future are expected to be user friendly, accurate, precise, high throughput, low cost, and potentially automated. DNA sequence information is the foundation for developing and applying genetic markers to questions of biological variation, whether *in situ* or *ex situ*. Researchers who develop and use sequence-based marker assays for diagnostic, forensic and relationship studies will continue to benefit greatly from information and technologies generated by the international Human Genome Project (HGP). Since its beginning, the HGP has endeavoured to develop genetic and physical maps and determine DNA sequences for the genomes of humans and several model organisms. As part of this effort, genetic analysis methods have been improved significantly. Notable achievements – now taken for granted by many – include new types of genetic markers (particularly those based on PCR), more efficient cloning vectors, the introduction of STS markers, and improved technology and automation for DNA sequencing. Underlying this progress is recognition that the successful development of new technologies for genetic research has been, and will continue to be, critical to many future scientific breakthroughs.

In addition to the HGP, other complementary projects also support initiatives for the development of new molecular marker assays. One such project is the Advanced Technology Programme (ATP) of the US National Institute of Standards and Technology. Among the ATP's goals is working with industry to deliver DNA diagnostic tools to a variety of users, at onetenth to one-hundredth of the current price. Another goal is helping industry make similar reductions in DNA sequencing costs and make DNA sequencing apparatus available at about one-third of the current price. Many biological questions once considered recalcitrant due to the number and cost of the assays required will be reconsidered in the light of new technological developments. The following section includes examples of progress that can be readily applied to the description of plant genetic variation. Also noted are limitations that are unique to plant-based studies, and are not presently addressed by most genetic analysis programmes.

Technology Transfer from the Human Genome Project

Technology transfer from the Human Genome Project (HGP) to the study of plant genetic resources has been very important in two areas:

- Sequence characterization and marker development; and
- Integrating all stages in genetic analysis projects, from template preparation through data interpretation.

Some of these innovations are now being used in plant genetic research; others will require some level of optimization before they are applied.

Plant genetic variation is best characterized with a large number of highly informative, easily analysed molecular markers. While the number of PCR-based markers will most likely continue to increase, the question at hand is how these markers will be detected. Manual, autoradiographic detection of amplified markers presently predominates in plant genetic research. However, ongoing programmes in human genotyping via semiautomated, fluorescent detection of SSRs are progressing at an incredible pace. Reed *et al.* (1994) reported that semi-automated technology can genotype over 100 000 samples, using approximately 250 marker loci, in less than six months. Moreover, Schwengel *et al.* (1994) found that data generated by this method were at least as accurate, more efficient, less labour intensive, and as cost effective as data generated by standard radiolabelling techniques.

Human genotyping kits are available for approximately 400 markers that define a ±10 cM resolution genomic index map. Kits for plant species and families are in development. Although cost remains an issue, these marker assays are clearly capable of handling the large arrays commonly encountered in plant conservation and breeding programmes. The costs of generating and detecting numerous marker loci for each of many individuals can be reduced by evaluating several of an individual's loci in a single lane on a gel. In the studies cited above, SSR loci from an individual were multiplexed at the electrophoresis level: several loci were amplified in separate reactions, the PCR products were pooled, and the pooled sample was electrophoresed.

In contrast, several laboratories are developing methods to multiplex at the reaction level, amplifying several SSR loci in one tube. Eleven putative loci in canola (*Brassica napus* L.) have been amplified simultaneously, using 11 pairs of fluorescently labelled primers in one reaction. The loci were distinguished from each other by fragment size and fluorescent label colour. This approach is also under way with SSRs of maize and sorghum [*Sorghum bicolour* (L.) Moench] Another molecular marker used in human genetic research reveals discrete changes (single nucleotide substitutions) in a specific DNA sequence. Variation of these polymorphic sequence tagged sites (pSTS) is the most frequent and widely distributed type of sequence variation in the genome. These individual nucleotide markers are usually biallelic within a population, and are generally less polymorphic than SSRs and other tandem repeats. However, a number of closely linked pSTSs can be combined into a multilocus marker that is as informative as an SSR.

Developing and assaying pSTS markers involve several steps. Random genomic DNA clones (each 300–600 bp long) from selected plant species are sequenced, then single nucleotide differences between individuals are identified by comparing automated sequencing electrophoregrams. Biallelic pSTSs are selected, and primer pairs are designed to amplify them. After a pSTS locus is amplified, the oligonucleotide ligation assay (OLA) is used to label the PCR products and discriminate the two allelic forms. In this assay, the PCR products are denatured and then serve as templates for a ligation detection reaction (LDR). This reaction uses three 'probes': two diagnostic probes, each of which complements one of the two pSTS alleles, and one labelled probe that is common to both alleles.

The diagnostic probes anneal to complementary template sequences, then are ligated to the labelled probe. The two diagnostic probes differ in length. Thus when the LDR-labelled fragments are electrophoresed, pSTS allelic variation between entries can be detected as variation in migration rate of fragments on the gel, due to differences in fragment length. As in many PCR-based marker assays, significant costs are presently incurred for pSTS marker development. Once markers are developed, however, the costs for each assay are not high. The previously highlighted markers are usually associated with noncoding regions of the genome. Complementary efforts are under way to develop high-throughput, automated methods to partially sequence cDNA clones, then use them to identify expressed genes in many organisms, including plants. Thanks to the rapid proliferation of expressed sequence tag (EST) databases, at present the probable gene function of a cDNA clone can often be inferred solely from the similarity of nucleotide or amino acid sequences between the clone and genes of known function. Using these techniques to compare isoforms of a gene (particularly its untranslated 3' regions) between members of a plant species or family, PCR-based markers that describe variation of this gene could be developed. When assays using these single-gene markers are compared with multilocus studies of the taxon, a more complete picture of genetic variation may be revealed.

Bibliography

Ayala, F. J.: *The Genetic Structure of Populations*, W.H. Freeman & Co., San Francisco, California, 1977.

Beckner, M.: *The Biological Way of Thought*, University of California Press, Berkeley and Los Angeles, 1968.

Berg, Paul, and Singer, Maxine: *Dealing with Genes: The Language of Heredity. Mill Valley*, University Science Books, CA, 1992.

Brookes, Martin: *Get a Grip on Genetics*, *Time Life Books*, East Sussex, England, 1998.

Dando, M.: *Biological Warfare in the 21st Century: Biotechnology and the Proliferation of Biological Weapons*. London: Brassey's, 1994.

Darlington, C. D., and Mather, K.: *The Elements of Genetics*, London, Allen and Unwin, 1949.

Dawkins, R.: *The Selfish Gene*. Oxford: Oxford University Press, 1976.

DeVita, V.T.: *Cancer: Principles and Practice of Oncology*, Lippincott, Philadelphia, 2005.

Diaz, E.: *Microbial Biodegradation: Genomics and Molecular Biology,* Caister Academic Press, UK, 2008.

Fisher, R. A.: *The Genetical Theory of Natural Selection*, Clarendon Press, Oxford, 1930.

Fritz, A.: *International Classification of Diseases for Oncology*, World Health Organization, Geneva, 2000.

Gillis, Justin: *Drug Firms, Gene Labs to Map Genetic Code*, The Daily News, Longview, WA, 1999.

Glover, D. M., and Hames, B. D.: *Genes and Embryos*: New York, Oxford University Press, 1989.

Griffiths A.J.F.: *Introduction to Genetic Analysis*, W.H. Freeman and Company, New York, USA, 2005.

Hamerton, J. L.: *Human Cytogenetics*, Academic Press, New York, 1971.

Hartl D, Jones E: *Genetics: Analysis of Genes and Genomes,* Jones & Bartlett, 2005.

Hartl, Daniel L.: *Basic Genetics*, Jones and Bartlett Publishers, Boston, 1991.

Jonoska, N.: *Self-Assembling DNA Graphs, DNA-Based Computers VIII*, Springer-Verlag, Berlin, 2003.

Klotz, J. W.: *Genes, Genesis, and Evolution*: St. Louis, Mo., Concordia Publishing Co., 1970.

Lumsden C., and E. Wilson: *Genes, Mind and Culture: The Coevolutionary Process.* Cambridge, MA: Harvard University Press, 1981.

Mayr, E.: *Change of Genetic Environment and Evolution*, Allen and Unwin, London, 1954.

Nottingham, S.F.: *Eat Your Genes: How Genetically Modified Food is Entering Our Diet*, Zed Books, London, 1998.

Prakash, Chandra Prakash : *Cell Physiology*, Discovery, Delhi, 2010.

Robin Whyman: *Applied Organometallic Chemistry and Catalysis*, Oxford University Press, Delhi, 2005.

Roobeek, A: *The Biotechnology Revolution?,* Blackwell, Oxford, 1995.

Sturtevant, Alfred: *History of Genetics*, Harper and Row, New York, 1965.

Suzuki, D., and Knudtson, P.: *Genethics: The Clash Between the New Genetics and Human Values*, Harvard University Press, Cambridge, Mass, 1989.

Wilson, E. O., and Lumden, C.: *Genes, Mind, and Culture: The Evolutionary Process*, Harvard University Press, Cambridge, Mass, 1981.

Wilson, E. O., and Lumden, C.: *Genes, Mind, and Culture: The Evolutionary Process*: Cambridge, Mass., Harvard University Press, 1981.

Index

I

L

M

N

P

R

S

T

U

V

❑❑❑

Encyclopaedia of

Advanced Genetic Analysis

Encyclopaedia of
Advanced Genetic Analysis

Volume 4

Editor:
Veena Kumari

AGROTECH PRESS
JAIPUR ▪ NEW DELHI

AGROTECH PRESS

Regd. Off. "Bony Residency", Gate No. 2
Opp. Tilak Public School, Vishweswariya Nagar,
Gopalpura Road, Jaipur - 302018, India
Ph: 0141 - 2761280, Telefax: 2761381
E-mail : agrotechpress@gmail.com

Delhi Office: 102, 1st Floor, Satyam House,
4327/3, Ansari Road, Darya Ganj,
New Delhi - 110002, India
Ph: 011-45652440
visit us: www.agrotechpress.com

Distributed By
KEYA INTERNATIONAL PRESS LLP
E-mail: keyainternationalpressllp@gmail.com

Published in India, 2024

ISBN : 9789384568207 (5 Vols. Set)

Typeset by Thomson Press

Printed and bound in India by Thomson Press Ltd.,
New Delhi

Contents

Preface

A gene is a molecular unit of heredity of a living organism. It is a name given to some stretches of DNA and RNA that code for a polypeptide or for an RNA chain that has a function in the organism. Living beings depend on genes, as they specify all proteins and functional RNA chains. Genes hold the information to build and maintain an organism's cells and pass genetic traits to offspring. All organisms have many genes corresponding to various biological traits, some of which are immediately visible, such as eye colour or number of limbs, and some of which are not, such as blood type, increased risk for specific diseases, or the thousands of basic biochemical processes that comprise life.

A modern working definition of a gene is "*a locatable region of genomic sequence, corresponding to a unit of inheritance, which is associated with regulatory regions, transcribed regions, and or other functional sequence regions* ". Colloquial usage of the term *gene* may actually refer to an allele: a *gene* is the basic instruction—a sequence of nucleic acids, while an *allele* is one variant of that gene. Referring to having a gene for a trait is no longer the scientifically accepted usage. In most cases, all people would have a gene for the trait in question, but certain people will have a specific allele of that gene, which results in the trait variant. Further, genes code for proteins, which might result in identifiable traits, but it is the gene, not the trait, which is inherited. The vast majority of living organisms encode their genes in long strands of DNA. DNA consists of a chain made from four types of nucleotide subunits, each composed of: a five-carbon sugar a phosphate group, and one of the four bases adenine, cytosine, guanine, and thymine. The most common form of DNA in a cell is in a double helix structure, in which two individual DNA strands twist around each other in a right-handed spiral. In this structure, the base pairing rules specify that guanine pairs with cytosine and adenine pairs with thymine. The base pairing between guanine and cytosine forms three hydrogen bonds, whereas the base pairing between adenine and thymine forms two hydrogen bonds. The two strands in a double helix must therefore be *complementary*, that is, their bases must align such that the adenines of one strand are paired with the thymines of the other strand, and so on. In all organisms, there are two major steps separating a protein-coding gene from its protein: First, the DNA on which the gene resides must be *transcribed* from DNA to messenger RNA; and, second, it must be *translated* from mRNA to protein. RNA-coding genes must still go through the first step, but are not translated into protein. The process of producing a biologically functional molecule of either RNA or protein is called gene expression, and the resulting molecule itself is called a gene product. The notion of a geneis evolving with the science of genetics, which began when Gregor Mendel noticed that biological variations are inherited from parent organisms as specific, discrete traits. The biological entity responsible for defining traits was later termed a *gene*, but the biological basis for inheritance remained unknown until DNA was identified as the genetic material in the 1940s. Prior to Mendel's work, the dominant theory of heredity was one of blending inheritance, which proposes that the traits of the parents blend or mix in a smooth, continuous gradient in the offspring. Although Mendel's work was largely unrecognized after its first publication in 1866, it was rediscovered in 1900 by three European scientists, Hugo de Vries, Carl Correns, and Erich von Tschermak, who had reached similar conclusions from their own research. However, these scientists were not yet aware of the identity of the 'discrete units' on which genetic material resides. The existence of genes was first suggested by Gregor Mendel, who, in the 1860s, studied inheritance in peaplants and hypothesized a factor that conveys traits from parent to offspring. He spent over 10 years of his life on one experiment. Although he did not use the term *gene*, he explained his results in terms of inherited characteristics. Mendel was also the first to hypothesize independent assortment, the distinction between dominant and recessive traits, the distinction between a heterozygote and homozygote, and the difference between what would later be described as genotype and phenotype.

The present publication is an extensive work on gene evolution and covers much more of the subject as is generally expected of students going up for the examinations of several Universities in India and abroad. It has been written to present a logical and practical approach to the various complexities and intricacies of the subject at length.

—*Editor*

1

Identification of Genetic Material

The genetic material of a cell or an organism refers to those materials found in the nucleus, mitochondria and cytoplasm, which play a fundamental role in determining the structure and nature of cell substances, and capable of self-propagating and variation.

When it became evident that the chromosomes were the organs of heredity, because (i) they formed the only link between two generations; (ii) they carried linearly arranged genes; (iii) they occurred in every organism in specific number and had specific morphology for a particular species; and (iv)any variation in their number or morphology affected the phenotype of the species.

Then, various attempts were made by early molecular geneticists to identify the physical and chemical nature of genes. But, the genes were found so minute structures that their physical identity remained almost impossible. However, the extensive chemical analysis of chromosomes of different organisms have revealed that chromosomes contain proteins and nucleic acid (DNA and RNA) and it was thought that genes might have either proteins or nucleic acids as their component molecules.

Early molecular geneticists have assigned the informational roles of genes to the chromosomal proteins, because, they found nucleic acids too simple to carry genetic informations. The controversy about the assignment of genetic role either to chromosomal proteins or to chromosomal DNA, existed upto 1950, when finally it was universally accepted that DNA is the genetic substance (I. e., chemical of which genes are composed) of most micro-organisms and higher organisms.Later on, RNA was found to be the genetic material of some viruses.

Direct Evidences for DNA as the Genetic Material

The most conclusive evidences in support of DNA as the genetic material comes from following three avenues of approach on micro-organisms: transformation of bacteria mode of infection of bacteriophages and conjugation in bacteria. Bacterial Transformation and Griffith EffectIn 1928, Frederick Griffith encountered a phenomenon now known as genetic transformation. Colonies of virulent strain (pathogenic) of pneumonia causing spherically-shaped bacterium, Diplococcus pneumoniae grown on nutrient agar, have a smooth (s) glistering appearance owing to the presence of a type specific, ploysaccharide (a polymer of glucose and glucuronic acid) capsule.

The avirulant (non-pathogenic) strains of this bacterium, on the other hand, lack this capsule and they produce dull, rough (R) colonies and is unable to cause the pneumonia disease. Smooth (S) and rough (R) characters are directly related to the presence or absence of the capsule and this trait is known to be genetically determined. Both S and R forms occur in several types and are designed as S-1, S-II, S-III, etc., and R-I, R-II, R-III, etc., respectively.

All these subtypes of S and R bacteria differ with each other in the type of antigens, they produce. The kind of antigens produced is likewise genetically determined. Smooth (S) forms sometimes mutate to rough (R) forms, but this change have not been found reversible.

In the course of his work, Griffith injected laboratory mice with live R-II pneumococci; the mice suffered no illness because R-II pneumococci was avirulent.

But when the mice were injected with the mixture of living avirulent R-II and heat killed S-III virulent, the unexpected symptoms of pneumonia appeared and high mortality was resulted in them.

By postmortoming the dead mice, it was found that their heart blood had both R-II and S-III pneumococci. From these results, Griffith concluded that the presence of the heat-killed S-IU bacteria must have caused a transformation of the living R-II bacteria, so as to restore to them the capacity for capsule formation they had earlier lost by gene mutation. This was called "Griffith effect" or more popularly as "bacterial transformation".

Identification of the "transforming substance"-Griffith could not understand the cause of bacterial transformation and that is first of all identified by O. T. Avery, C. M. McLeod and M McCarthy (1944). They tested a fraction of heat killed S-III bacteria for the transforming ability.

They could remove proteins, lipids, polysaccharides and ribonucleic acids from the S-III extract by a variety of chemical and enzymatic methods without seriously diminishing its power to transfer R-II mutants into the S-III wild types. They ultimately concluded that because a cell free and highly purified

DNA extract of S-III bacteria could bring about transformation of R-I1 bacteria into S-III, therefore, DNA is the genetic material of pneumococci. Later on, such a transforming substance, the DNA, was found in a variety of bacteria (i.e., Hemophilus influenzae. Bacillus subtilus, Shigelia paradysenteriae, etc.) and several other organisms.

Mode of Infection of Bacteriophages and Identification of DNA as their Genetic Material

By using radioactive tracers, A. Hershey and M. Chase (1952) provided further direct proof that DNA is the genetic material in certain bacterial viruses. These investigators were studying the bacteriophages that attack the bacterium, Escherichia coli. They prepared a chemically defined culture medium (for E. coli) containing phosphoric acid and sulphuric acid as the only source of phosphorus and sulphur.

Known quantities of radioactive isotopes of phosphorus (P^{32}) and sulphur (S^{32}) were added to the culture medium. E. coli grown on such a culture medium incorporated the P^{32} and S^{35} into their chemical constituents. When T_2 bacteriophage particles infected the radioactive bacteria, they incorporated the labeled DNA. Phage capsid proteins did not contain appreciable amounts of phosphorus and only the DNA was labeled with P^{32}. Similarly, the protein envelope around the phage was selectively labeled with S^{35}. DNA contains virtually no sulphur and was not labeled with S^{35} By this method it was possible to differentially label the phage DNA and proteins of phage-capsid.

After phage growth, the phage particles were separated from the host cell by centrifugation. The radioactive phages were next introduced to non-radioactive bacterial cultures where they attacked the bacteria. Subsequently the viruses were separated from the host cell by agitation and the content of P^{32} and S^{35} in the host (E. coli) and parasite (bacteriophage) was determined. The phosphorus label was found to be associated with the bacterial cells and the sulphur label was in the proteins coats (capsids) left in the medium. This indicated that the DNA had panetrated the cells but that the protein coat or capsid of the phage was left outside the wall of the bacterium. The labeled DNA was found to reproduced in the host cells. These experiments, thus, clearly demonstrated that genetic informations of these bacteriophages resided in their DNA molecules and not in the protein molecules of their capsids. Later on Hershey-Chase experiment was repeated for other viruses and it was found that DNA is also the genetic material in a number of other viruses.

Bacterial Conjugation

Another conclusive evidence for DNA as the genetic material came from the phenomenon of conjugation of bacteria. Laderberg and Tatum (1946) found that when a F^+ ('male') E. coli cell conjugated with a F^- ('female') E. coli cell an unidirectional transfer of F^+ factor of 'male' cell to F^- or 'female' cell took place; so that the latter was converted into a F^+ or 'male' strain. The F^+ factor was found to be a fragment of DNA molecule which occurred in the cytoplasm of bactarial cell.

Indirect Evidences for DNA as the Genetic Material

The indirect or circumstantial evidences in favour of DNA as the genetic material came from higher organisms which are not as easy to manipulate as bacteria and viruses. The fact that DNA is the genetic material of higher organisms has been supported by following facts:

1. Localization -All cells contain DNA which is localized primarily on chromosomes in the nucleus of the cell. This is consistent with the fact the genes are also located specifically on chromosomes. The Feulgen techniques have shown that DNA entirely remains restricted to the chromosomes and it forms one of the major component of chromosomes.
2. Quantity-Various quantitative measurement of amount of DNA in different cells have shown that there is a correlation between the amount of DNA and the number of chromosome sets (ploidy).

Cells set	*Mean DNA Fegulen content (Picograms)*	*Presumed chromosomes (Ploidy)*
Spermatids	1.68	Haploid(n)
Liver	3.16	Diploid(2n)
Liver	6.30	Tetraploid(4n)
Liver	12.80	Octoploid(8n)

S.No	*Organism*	*Kidney*	*Liver*	*Erythrocytes*	*Sperm*
1.	Chicken	2.4	2.5	2.5	1.3
2.	Bovine	6.4	6.4	-	3.3
3.	Carp	-	3.0	3.3	1.6
4.	Human	5.4	5.6	2.5	2.5

The diploid amount of DNA is constant within the species but varies from one species to another, as shown in.

Stability-Most of the macromolecules of the cell are constantly being broken down and synthesized. If this were to happen to a gene, valuable hereditary information would invariably be lost. Of all the macromolecules in the cell, DNA is the most metabolically stable.

Sensitivity to chemical and physical agents-All agents which cause genes to be damaged (mutation) have also been shown to induce changes in the structure of DNA. For example, one such mutagenic agent, ultraviolet light, is capable of producing breaks in the DNA molecule. In 1950 Swift found that the amount of DNA per resting nucleus (interphase nucleus) in mitotically active tissues of animals ranged from twice (2n) the value found in the gametes (In) to four times (4n) this value. He concluded that DNA synthesis occurred during the interphase. This parallelism of behaviour in DNA and chromosomes clearly indicates that DNA is the genetic material of higher animals.

Evidences for RNA as the Genetic Material of Some Viruses

The demonstration that RNA is the genetic material in RNA containing viruses came in 1956, when A. Gierer and G. Schramm showed that tobacco plants could be inoculated with purified RNA from the tobacco mosaic virus (TMV) and TMV -like lesions could later be identified on the tobacco leaves. A different approach was taken by H. Fraenkel Conrat and B. Singer in experiments published in 1957. They first separated the RNA from the protein of TMV viruses, much as Hershey and Chase had done with the DNA and the protein of T2 bacteriophage. They then developed techniques for forming 'reconstituted' viruses containing the protein from one mutant strain of TMV and the RNA from another or vice-versa. Such hybrid viruses were allowed to infect tobacco leaves, and the progeny were examined. In all cases the progeny were the parental RNA type and not the parental protein type. In a much improved technique N. Pace and S. Spiegelman in 1966 purified RNA from two different mutant strains of the RNA phage QB which had quite distinct base compositions.

The isolated RNA'S were then incubated separately in the presence of an E. coli cell extract containing an enzyme capable of RNA replication. The new RNA synthesized was in each case identical in base composition to particular phage RNA presented to the in vitro system, thus indicating that the phage RNA-can.serve as a template for its self-replication.

Base Pair

In molecular biology and genetics, two nucleotides on opposite complementary DNA or RNA strands that are connected via hydrogen bonds are called a base pair (often abbreviated bp). In the canonical Watson-Crick DNA base pairing, adenine (A) forms a base pair with thymine (T) and guanine (G) forms a base pair with cytosine (C). In RNA, thymine is replaced by uracil (U). Alternate hydrogen bonding patterns, such as the wobble base pair and Hoogsteen base pair, also occur—in particular, in RNA—giving rise to complex and functional tertiary structures. It is important to note that pairing is the mechanism by which codons on messenger RNA molecules are recognized by anticodons on transfer RNA during protein translation. Some DNA- or RNA-binding enzymes can recognize specific base pairing patterns that identify particular regulatory regions of genes.

The size of an individual gene or an organism's entire genome is often measured in base pairs because DNA is usually double-stranded. Hence, the number of total base pairs is equal to the number of nucleotides in one of the strands (with the exception of non-coding single-stranded regions of telomeres).

The haploid human genome (23 chromosomes) is estimated to be about 3 billion base pairs long and to contain 20,000–25,000 distinct genes. A kilobase (kb) is a unit of measurement in molecular biology equal to 1000 base pairs of DNA or RNA.

Hydrogen Bonding and Stability

Hydrogen bonding is the chemical interaction that underlies the base-pairing rules described above. Appropriate geometrical correspondence of hydrogen bond donors and acceptors allows only the "right" pairs to form stably. DNA with high GC-content is more stable than DNA with low GC-content, but, contrary to popular belief, the hydrogen bonds do not stabilize the DNA significantly, and stabilization is mainly due to stacking interactions.

The larger nucleobases, adenine and guanine, are members of a class of double-ringed chemical structures called purines; the smaller nucleobases, cytosine and thymine (and uracil), are members of a class of single-ringed chemical structures called pyrimidines. Purines are complementary only with pyrimidines: pyrimidine-pyrimidine pairings are energetically unfavourable because the molecules are too far apart for hydrogen bonding to be established; purine-purine pairings are energetically unfavourable because the molecules are too close, leading to overlap repulsion. The only other possible pairings are GT and AC; these pairings are mismatches because the pattern of hydrogen donors and acceptors do not correspond. The GU pairing, with two hydrogen bonds, does occur fairly often in RNA.

Paired DNA and RNA molecules are comparatively stable at room temperature but the two nucleotide strands will separate above a melting point that is determined by the length of the molecules, the extent of mispairing (if any), and the GC content. Higher GC content results in higher melting temperatures; it is, therefore, unsurprising that the genomes of extremophile organisms such as *Thermus thermophilus* are particularly GC-rich. On the converse, regions of a genome that need to separate frequently — for example, the promoter regions for often-transcribed genes — are comparatively GC-poor. GC content and melting temperature must also be taken into account when designing primers for PCR reactions.

Base Stacking

Base stacking interactions in DNA and RNA are due to dispersion attraction, short-range exchange repulsion, and electrostatic interactions, which also contribute to stability. Again, GC stacking interactions with adjacent bases tend to be more favourable. (Note, however, that a GC stacking interaction with the next base pair is geometrically different from a CG interaction.) Base stacking effects are especially important in the secondary structure and tertiary structure of RNA; for example, RNA stem-loop structures are stabilized by base stacking in the loop region.

Nucleic Acid Analogues

Nucleic acid analogues are compounds structurally similar (analog) to naturally occurring RNA and DNA, used in medicine and in molecular biology research. Nucleic acids are chains of nucleotides, which are composed of three parts: a phosphate backbone, a pucker-shaped pentose sugar, either ribose or deoxyribose, and one of four nucleobases. An analogue may have any of these altered. Typically the analogue nucleobases confer, among other things, different base pairing and base stacking proprieties. Examples include universal bases, which can pair with all four canon bases, and phosphate-sugar backbone analogues such as PNA, which affect the properties of the chain (PNA can even form a triple helix). Artificial nucleic acids include peptide nucleic acid (PNA), Morpholino and locked nucleic acid (LNA), as well as glycol nucleic acid (GNA) and threose nucleic acid (TNA). Each of these is distinguished from naturally-occurring DNA or RNA by changes to the backbone of the molecule.

Medicine

Several nucleoside analogues are used as antiviral or anticancer agents. The viral polymerase incorporates these compounds with non-canon bases. These compounds are activated in the cells by being converted into nucleotides, they are administered as nucleosides since charged nucleotides cannot easily cross cell membranes.

Molecular Biology

Nucleic acid analogues are used in molecular biology for several purposes:

- As a tool to detect particular sequences
- As a tool with resistance to RNA hydrolysis
- As a tool for another purpose, such as sequencing
- Naturally occurring, such as in tRNA
- Investigation of the mechanisms used by enzyme, such as an Enzyme inhibitor
- Investigation of possible scenarios of the origin of life
- Investigation of the structural features of nucleic acids
- Investigation of the possible alternatives to the natural system in Synthetic biology.

Backbone Analogues

Hydrolysis Resistant RNA-analogues: To overcome the fact that ribose's 2' hydroxy group that reacts with the phosphate linked 3' hydroxy group (RNA is too unstable to be used or synthesized reliably), a ribose analogue is used. The most common RNA analogues are locked nucleic acid (LNA), morpholino, and peptide nucleic acid (PNA). These oligonucleotides differ as they have a different backbone sugar but still bind according to Watson and Crick pairing with RNA or DNA, but are immune to nuclease activity (They generally cannot be enzymatically synthesized and can only be produced synthetically).

Other Notable Analogues Used as Tools

Dideoxynucleotides are used in sequencing. These nucleoside triphosphates possess a non-canonical sugar, dideoxyribose, which lacks the 3' hydroxyl group normally present in DNA and therefore cannot bond with the next base. The lack of the 3' hydroxyl group terminates the chain reaction as the DNA polymerases mistake it for a regular deoxyribonucleotide. Another chain-terminating analogue that lacks a 3' hydroxyl and mimics adenosine is called cordycepin. Cordycepin is an anticancer drug that targets RNA replication. Another analogue in sequencing is a nucleobase analogue, 7-deaza-GTP and is used to sequence CG rich regions, instead 7-deaza-ATP is called tubercidin, an antibiotic.

Precursors to the RNA World

RNA may be too complex to be the first nucleic acid, so before the RNA world several simpler nucleic acids that differ in the backbone, such as TNA and GNA and PNA, have been offered as candidates for the first nucleic acids.

Base Analogues

Nucleobase Structure and Nomenclature: Natural bases are divided into two classes depending on their structure: pyrimidine (an heterocyclic aromatic six-membered ring with nitrogen atoms in position 1 and 3) and purine (a pyrimidine (numeration inverted) fused with an imidazole ring, a five-membered ring with 2 nitrogen atoms separated by one carbon (meta), 7,9). Their main proprieties are base pairing, resulting form 2 or 3 hydrogen bonds between ketone (electron withdrawing group,*ei.* more negatively charged) and amino (electron withdrawing group,*ei.* more positively charged) functional groups, and base stacking, caused by the attraction of the delocalized ∂ electron clouds of the aromatic ring structure.

Fluorophores

Commonly fluorophores (such as rhodamine or fluorescein) are linked to the ring linked to the sugar (in para) via a flexible arm, presumably extruding from the major groove of the helix. Due to low processivity of the nucleotides linked to bulky adducts such as florophores by taq polymerases, the sequence is typically copied using a nucleotide with an arm and later coupled with a reactive fluorophore (indirect labelling):

- amine reactive: Aminoallyl nucleotide contain a primary amine group on a linker that reacts with the amino-reactive dye such as a cyanine or Alexa Fluor dyes, which contain a reactive leaving group, such as a succinimidyl ester (NHS). (base-pairing amino groups are not affected).
- thiol reactive: thiol containing nucleotides reacts with the fluorophore linked to a reactive leaving group, such as a maleimide.
- biotin linked nucleotides rely on the same indirect labelling principle (+ fluorescent streptavidin) and are used in Affymetrix DNAchips.

Fluorophores find a variety of uses in medicine and biochemistry.

Fluorescent Base Analogues

The most commonly used and commercially available fluorescent base analogue, 2-aminopurine (2-AP), has a high-fluorescence quantum yield free in solution (0.68) that is considerably reduced (appr. 100 times but highly dependent on base sequence) when incorporated into nucleic acids. The emission sensitivity of 2-AP to immediate surroundings is shared by other promising and useful fluorescent base analogues like 3-MI, 6-MI, 6-MAP, pyrrolo-dC (also commercially available), modified and improved derivatives of pyrrolo-dC, furan-modified bases and many other ones.

This sensitivity to the microenvironment have been utilized in studies of e.g. structure and dynamics within both DNA and RNA, dynamics and kinetics of DNA-protein interaction and electron transfer within DNA. A newly developed and very interesting group of fluorescent base analogues that has a fluorescence quantum yield that is nearly insensitive to their immediate surroundings is the tricyclic cytosine family. 1,3-Diaza-2-oxophenothiazine, tC, has a fluorescence quantum yield of approximately 0.2 both in single- and in double-strands irrespective of surrounding bases.

Also the oxo-homologue of tC called tC^{O} (both commercially available), 1,3-diaza-2-oxophenoxazine, has a quantum yield of 0.2 in double-stranded systems. However, it is somewhat sensitive to surrounding bases in single-strands (quantum yields of 0.14–0.41). The high and stable quantum yields of these base analogues make them very bright, and, in combination with their good base analogue properties (leaves DNA structure and stability next to unperturbed), they are especially useful in fluorescence anisotropy and FRET measurements, areas where other fluorescent base analogues are less accurate.

Also, in the same family of cytosine analogues, a FRET-acceptor base analogue, tC_{nitro}, has been developed. Together with tC^{O} as a FRET-donor this constitutes the first nucleic acid base analogue FRET-pair ever developed. The tC-family has, for example, been used in studies related to polymerase DNA-binding and DNA-polymerization mechanisms.

Natural Non-canon Bases

In a cell, there are several noncanon bases present: CpG islands in DNA (are often methylated), all eukaryotic mRNA (capped with a methyl-7-guanosine), and several bases of rRNAs (are methylated). Often, tRNAs are heavily modified postranscriptionally in order to improve their conformation or base pairing, in particular in/near the anticodon: inosine can base pair with C, U, and even with A, whereas thiouridine (with A) is more specific than uracil (with a purine). Other common tRNA base modifications are pseudouridine (which gives its name to the TØC loop), dihydrouridine (which does not stack as it is not aromatic), queuosine, wyosine, and so forth. Nevertheless these are all modifications to normal bases and are not placed by a polymerase.

Base-pairing

Canonical bases may have either a ketone or an amine group on the carbons surrounding the nitrogen atom furthest away from the glycosidic bond, which allows them to base pair (Watson-Crick base pairing) via hydrogen bonds (amine with ketone, purine with pyrimidine). Adenine and 2-aminoadenine have one/two amine group(s), whereas thymine has two ketone groups, and cytosine and guanine are mixed amine and ketone (inverted in respect to each other). The precise reason why there are only four nucleotides is debated, but there are several unused possibilities.

Furthermore, adenine is not the most stable choice for base pairing: in Cyanophage S-2L diaminopurine (DAP) is used instead of adenine (host evasion). Diaminopurine basepairs perfectly with thymine as it is identical to adenine but has an amine group at position 2 forming 3 intramolecular hydrogen bonds, eliminating the major difference between the two types of basepairs (Weak:A-T and Strong:C-G). This improved stability affects protein-binding ineractions that rely on those differences. Other combination include,

- isoguanine and isocytosine, which have their amine and ketone inverted compared to standard guanine and cytosine, (not used probably as tautomers are problematic for base pairing, but isoC and isoG can be amplified correctly with PCR even in the presence of the 4 canon bases)
- diaminopyrimidine and a xanthine, which bind like 2-aminoadenine and thymine but with inverted structures (not used as xanthine is a deamination product).

However, correct DNA structure can form even when the bases are not paired via hydrogen bonding; that is, the bases pair thanks to hydrophobicity, as studies have shown using DNA isosteres (analogues with same number of atoms), such as the thymine analogue 2,4-difluorotoluene (F) or the adenine analogue 4-methylbenzimidazole (Z). An alternative hydrophobic pair could be isoquinoline, and the pyrrolo[2,3-b]pyridine.

Other noteworthy basepairs:

- Several fluorescent bases have also been made, such as the 2-amino-6-(2-thienyl)purine and pyrrole-2-carbaldehyde base pair.
- Metal coordinated bases, such as two 2,6-bis (ethylthiomethyl) pyridine (SPy) with a silver ion or pyridine-2,6-dicarboxamide (Dipam) and a mondentate pyridine (Py) with a copper ion.
- Universal bases may pair indiscriminately with any other base, but, in general, lower the melting temperature of the sequence considerably; examples include 2'-deoxyinosine (hypoxanthine deoxynucleotide) derivatives, nitroazole analogues, and hydrophobic aromatic non-hydrogen-bonding bases (strong stacking effects). These are used as proof of concept and, in general, are not utilised in degenerate primers (which are a mixture of primers).
- The numbers of possible base pairs is doubled when xDNA is considered. xDNA contains expanded bases, in which a benzene ring has been added, which may pair with canon bases, resulting in four possible base-pairs (8 bases:xA-T,xT-A,xC-G,xG-C, 16 bases if the unused arrangements are used). Another form of benzene added bases is yDNA, in which the base is widened by the benzene.

Metal Base Pairs

In metal base-pairing, the Watson-Crick hydrogen bonds are replaced by the interaction between a metal ion with nucleosides acting as ligands. The possible geometries of the metal that would allow for duplex formation with two bidentate nucleosides around a central metal atom are: tetrahedral, dodecahedral, and square planar. Metal-complexing with DNA can occur by the formation of non-canonical base pairs from natural nucleobases with participation by metal ions and also by the exchanging the hydrogen atoms that are part of the Watson-Crick base pairing by metal ions. Introduction of metal ions into a DNA duplex has shown to have potential magnetic, conducting properties, as well as increased stability. Metal complexing has been shown to occur between natural nucleobases. A well-documented example is the formation of T-Hg-T, which involves two deportonated thymine nucleobases that are brought together by Hg^{2+} and forms a connected metal-base pair. This motif does not accommodate stacked Hg^{2+} in a duplex due to an intrastrand hairpin formation process that is favoured over duplex formation.

Two thymines across from each other in a duplex do not form a Watson-Crick base pair in a duplex; this is an example where a Watson-Crick basepair mismatch is stabilized by the formation of the metal-base pair. Another example of a metal complexing to natural nucleobases is the formation of A-Zn-T and G-Zn-C at high pH; Co^{+2} and Ni^{+2} also form these complexes. These are Watson-Crick base pairs where the divalent cation in coordinated to the nucleobases. The exact binding is debated.

A large variety of artificial nucleobases have been developed for use as metal base pairs. These modified nucleobases exhibit tunable electronic properties, sizes, and binding affinities that can be optimized for a specific metal. For, example a nucleoside modified with a pyridine-2,6-dicarboxylate has shown to bind tightly to Cu^{2+}, whereas other divalent ions are only loosely bound. The tridentate character contributes to this selectivity.

The fourth coordination site on the copper is saturated by an oppositely arranged pyridine nucleobase. The asymmetric metal base pairing system is orthogonal to the Watson-Crick base pairs. Another example of an artificial nucleobase is that with hydroxypyridone nucleobases, which are able to bind Cu^{2+} inside the DNA duplex.

Five consecutive copper-hydroxypyridone base pairs were incorporated into a double strand, which were flanked by only one natural nucleobase on both ends. EPR data showed that the distance between copper centres was estimated to be 3.7 ± 0.1 Å, while a natural B-type DNA duplex is only slightly larger (3.4 Å). The appeal for stacking metal ions inside a DNA duplex is the hope to obtain nanoscopic self-assembling metal wires, though this has not been realized yet.

Orthogonal System

It has been proposed and studied both theoretically and experimentally the possibility of implementing an orthogonal system inside cells independent of the cellular genetic material in order to make a completely safe system, with the possible increase in encoding potentials Several groups have focused on different aspects:

- novel backbones and base pairs as discussed above
- XNA replication/transcription polymerases starting generally from T7 RNA polymerase
- rybozomes (16S sequences with altered anti Shine-Dalgarno sequence allowing the translation of only orthogonal mRNA with a matching altered Shine-Dalgarno sequence)
- novel tRNA encoding non-natural aminoacids.

Base Analogs and Intercalators

Chemical analogs of nucleotides can take the place of proper nucleotides and establish non-canonical base-pairing, leading to errors (mostly point mutations) in DNA replication and DNA transcription. This is due to their isosteric chemistry. One common mutagenic base analog is 5-bromouracil, which resembles thymine but can base-pair to guanine in its enol form.

Other chemicals, known as DNA intercalators, fit into the gap between adjacent bases on a single strand and induce frameshift mutations by "masquerading" as a base, causing the DNA replication machinery to skip or insert additional nucleotides at the intercalated site. Most intercalators are large polyaromatic compounds and are known or suspected carcinogens. Examples include ethidium bromide and acridine.

Examples

The following DNA sequences illustrate pair double-stranded patterns. By convention, the top strand is written from the 5' end to the 3' end; thus, the bottom strand is written 3' to 5'.

A base-paired DNA sequence:

ATCGATTGAGCTCTAGCG

TAGCTAACTCGAGATCGC

The corresponding RNA sequence, in which uracil is substituted for thymine where uracil takes its place in the RNA strand:

AUCGAUUGAGCUCUAGCG

UAGCUAACUCGAGAUCGC

Length Measurements

The following abbreviations are commonly used to describe the length of a D/RNA molecule:

- bp = base pair(s)—one bp corresponds to circa 3.4 Å of length along the strand
- kb (= kbp) = kilo base pairs = 1,000 bp
- Mb = mega base pairs = 1,000,000 bp
- Gb = giga base pairs = 1,000,000,000 bp.

In case of single stranded DNA/RNA units of nucleotides are used, abbreviated nt (or knt, Mnt, Gnt), as they are not paired. For distinction between units of computer storage and bases kbp, Mbp, Gbp, etc. may be used for basepairs. The length of 16S rDNA for bacteria is 1542 base-pairs in length. The Centimorgan is also often used to imply distance along a chromosome, but the number of base-pairs it corresponds to varies widely. In the Human genome, it is about 1 million base pairs.

Protein Synthesis

Process whereby DNA encodes for the production of amino acids and proteins.

Transcription

Before the synthesis of a protein begins, the corresponding RNA molecule is produced by RNA transcription. One strand of the DNA double helix is used as a template by the RNA polymerase to synthesize a messenger RNA (mRNA). This mRNA migrates from the nucleus to the cytoplasm. During this step, mRNA goes through different types of maturation including one called splicing when the non-coding sequences are eliminated. The coding mRNA sequence can be described as a unit of three nucleotides called a codon.

Translation

The ribosome binds to the mRNA at the start codon (AUG) that is recognized only by the initiator tRNA. The ribosome proceeds to the elongation phase of protein synthesis. During this stage, complexes, composed of an amino acid linked to tRNA, sequentially bind to the appropriate codon in mRNA by forming complementary base pairs with the tRNA anticodon. The ribosome moves from codon to codon along the mRNA. Amino acids are added one by one, translated into polypeptidic sequences dictated by DNA and represented by mRNA. At the end, a release factor binds to the stop codon, terminating translation and releasing the complete polypeptide from the ribosome. One specific amino acid can correspond to more than one codon. The genetic code is said to be degenerate.

DNA and Protein Synthesis

With over 100,000 different proteins to manufacture, how the heck does our body get it right? When one thinks of the amount of information the body needs to keep track of, - eye, hair and skin colour, protein sequence, toenail size, etc. - it would seem a task for a supercomputer to record all of the necessary information. In essence it is. But not a supercomputer made of silicon wafers and TV screens, rather one made of an intricate biomolecule called DNA. DNA (deoxyribonucleic acid) is in the family of molecules referred to as nucleic acids. One strand of DNA has a backbone consisting of a polymer of the simple sugar deoxyribose bonded to something called a phosphate unit.

What is impressive about DNA is that each sugar molecule in the strand also binds to one of four different nucleotide bases. These bases: Adenine (A), Guanine (G), Cytosine (C) and Thymine (T), are the beginnings of what we will soon see is a molecular alphabet. Each sugar molecule in the DNA strand will bind to one nucleotide base. Thus, as our description of DNA unfolds, we see that a single strand of the molecule looks more like this:

C T G A ...
sugar- phosphate- sugar- phosphate- sugar- phosphate- sugar- phosphate- ...

Each strand of DNA contains millions or even billions (in the case of human DNA) of nucleotide bases. These bases are arranged in a specific order according to our genetic ancestry. The order of these base units makes up the code for specific characteristics in the body, such as eye colour or nose-hair length. Just as we use 26 letters in various sequences to code for the words you are now reading, our body's DNA uses 4 letters (the 4 nucleotide bases) to code for millions of different characteristics. Each molecule of DNA is actually made up of 2 strands of DNA cross-linked together. Each nucleotide base in the DNA strand will cross-link (via hydrogen bonds) with a nucleotide base in a second strand of DNA forming a structure that resembles a ladder. These bases cross-link in a very specific order: A will only link with T (and vice-versa), and C will only link with G (and vice-versa). Thus our picture of DNA now looks like this:

sugar-phosphate-sugar-phosphate-sugar-phosphate-sugar-phosphate-...
G A C T ...
| | | | |
C T G A ...
sugar-phosphate-sugar-phosphate-sugar-phosphate-sugar-phosphate-...

In 1953, James Watson, Francis Crick and Rosalind Franklin discovered that the structure of DNA is actually a double helix. In other words, the DNA ladder described above coils around itself somewhat like the cord of a telephone, as illustrated at right. To get a better picture of the DNA double helix, Dr. Abby Parrill and Dr. Jacquelyn Gervay at the University of Arizona have put together a movie of the DNA double helix and an interactive strand of DNA that can be moved around by the user. Both are available by clicking below (Note: the interactive DNA strand requires Chemscape's CHIME software to run).

Bottom of Form

The specific base-pairing of DNA aids in reproduction of the double helix when more genetic material is needed (such as during reproduction, to pass on characteristics from parent to offspring). When DNA reproduces, the 2 strands unzip from each other and enzymes add new bases to each, thus forming two new strands. This process is illustrated in the Access Excellence DNA Replicating Itself page (just hit your browser's Back button to return here). Within this coil of DNA lies all the information needed to produce everything in the human body. A strand of DNA may be millions, or billions, of base-pairs long. Different segments of the DNA molecule code for different characteristics in the body. A Gene is a relatively small segment of DNA that codes for the synthesis of a specific protein. This protein then will play a structural or functional role in the body. A chromosome is a larger collection of DNA that contains many genes and the support proteins needed to control these genes.

Protein Synthesis

How does a gene code for a protein? Protein synthesis is a 2 part process that involves a second type of nucleic acid along with DNA. This second type of nucleic acid is RNA, ribonucleic acid. RNA differs from DNA in two respects. First, the sugar units in RNA are ribose as compared to DNA's deoxyribose. Because of this difference, RNA does not bind to the nucleotide base Thymine, instead, RNA contains the nucleotide base Uracil (U) in place of T (RNA also contains the other three bases: A, C and G).

Transcription: In the first step of protein synthesis, the 2 DNA strands in a gene that codes for a protein unzip from each other. Similar to the way DNA replicates itself, a single strand of messenger RNA (mRNA) is then made by pairing up mRNA bases with the exposed DNA nucleotide bases.

Remember that mRNA does not contain the base Thymine, so U is paired with each of DNA's A bases.

Translation: After the mRNA is manufactured, it leaves the cell nucleus and travels to a cellular organelle called the ribosome (we will learn about the cell, nucleus and ribosome in the next lesson). In the ribosome, the mRNA code is translated into a transfer RNA (tRNA) code which, in turn, is transfered into a protein sequence. In this process, each set of 3 mRNA bases (the mRNA base triplet is called a codon) will pair with a complimentary tRNA base triplet (called an anticodon). Each tRNA is specific to an amino acid, as tRNA's are added to the sequence, amino acids are linked together by peptide bonds, eventually forming a protein that is later released by the tRNA. Using the mRNA strand we obtained above, you can generate the complimentary tRNA/ amino acid sequence by clicking on the mRNA codons in the table below.

mRNA codon →

After the processes of transcription and translation are complete, we are left with a protein that consists of the chain:

Although our 'protein' is only 2 amino acids in length, proteins normally consist of hundreds or thousands of amino acids.

Regulation of Gene Expression or Regulation of Protein Synthesis

Regulation of gene expression (or gene regulation) includes the processes that cells and viruses use to regulate the way that the information in genes is turned into gene products. Although a functional gene product may be an

RNA or a protein, the majority of known mechanisms regulate protein coding genes. Any step of the gene's expression may be modulated, from DNA-RNA transcription to the post-translational modification of a protein. Gene regulation is essential for viruses, prokaryotes and eukaryotes as it increases the versatility and adaptability of an organism by allowing the cell to express protein when needed. The first discovered example of a gene regulation system was the lac operon, discovered by Jacques Monod, in which protein involved in lactose metabolism are expressed by *E. coli* only in the presence of lactose and absence of glucose.

Furthermore, gene regulation drives the processes of cellular differentiation and morphogenesis, leading to the creation of different cell types in multicellular organisms where the different types of cells may possess different gene expression profiles though they all possess the same genome sequence.

Regulated Stages of Gene Expression

Any step of gene expression may be modulated, from the DNA-RNA transcription step to post-translational modification of a protein. The following is a list of stages where gene expression is regulated, the most extensively utilised point is Transcription Initiation:

- Chromatin domains.

Epigenetics

In biology, and specifically genetics, epigenetics is the study of heritable changes in phenotype (appearance) or gene expression caused by mechanisms other than changes in the underlying DNA sequence, hence the name *epi-genetics*. These changes may remain through cell divisions for the remainder of the cell's life and may also last for multiple generations. However, there is no change in the underlying DNA sequence of the organism; instead, non-genetic factors cause the organism's genes to behave (or "express themselves") differently. One example of epigenetic changes in eukaryotic biology is the process of cellular differentiation. During morphogenesis, totipotent stem cells become the various pluripotent cell lines of the embryo which in turn become fully differentiated cells. In other words, a single fertilized egg cell – the zygote – changes into the many cell types including neurons, muscle cells, epithelium, blood vessels etc. as it continues to divide. It does so by activating some genes while inhibiting others.

Etymology and Definitions

Epigenetics (as in "epigenetic landscape") was coined by C. H. Waddington in 1942 as a portmanteau of the words *genetics* and *epigenesis*. *Epigenesis* is an old word which has more recently been used to describe the differentiation of cells from their initial totipotent state in embryonic development. When Waddington coined the term the physical nature of genes and their role in heredity was not known; he used it as a conceptual model of how genes might interact with their surroundings to produce a phenotype. Robin Holliday defined epigenetics as "the study of the mechanisms of temporal and spatial control of gene activity during the development of complex organisms." Thus *epigenetic* can be used to describe anything other than DNA sequence that influences the development of an organism.

The modern usage of the word in scientific discourse is more narrow, referring to heritable traits (over rounds of cell division and sometimes transgenerationally) that do not involve changes to the underlying DNA sequence. The Greek prefix *epi-* in *epigenetics* implies features that are "on top of" or "in addition to" genetics; thus *epigenetic* traits exist on top of or in addition to the traditional molecular basis for inheritance. The similarity of the word to "genetics" has generated many parallel usages. The "epigenome" is a parallel to the word "genome", and refers to the overall epigenetic state of a cell. The phrase "genetic code" has also been adapted—the "epigenetic code" has been used to describe the set of epigenetic features that create different phenotypes in different cells. Taken to its extreme, the "epigenetic code" could represent the total state of the cell, with the position of each molecule accounted for in an *epigenomic map*, a diagrammatic representation of the gene expression, DNA methylation and histone modification status of a particular genomic region. More typically, the term is used in reference to systematic efforts to measure specific, relevant forms of epigenetic information such as the histone code or DNA methylation patterns. The psychologist Erik Erikson used the term *epigenetic* in his theory of psychosocial development. That usage, however, is of primarily historical interest.

Molecular Basis of Epigenetics

The molecular basis of epigenetics is complex. It involves modifications of the activation of certain genes, but not the basic structure of DNA. Additionally, the chromatin proteins associated with DNA may be activated or silenced. This accounts for why the differentiated cells in a multi-cellular organism express only the genes that are necessary for their own activity. Epigenetic changes are preserved when cells divide. Most epigenetic changes only occur within the course of one individual organism's lifetime, but, if a mutation in the DNA has been caused in sperm or egg cell that results in fertilization, then some epigenetic changes are inherited from one generation to the next. This raises the question of whether or not epigenetic changes in an organism can alter the basic structure of its DNA, a form of Lamarckism.

Specific epigenetic processes include paramutation, bookmarking, imprinting, gene silencing, X chromosome inactivation, position effect, reprogramming, transvection, maternal effects, the progress of carcinogenesis, many effects of teratogens, regulation of histone modifications and heterochromatin, and technical limitations affecting parthenogenesis and cloning.

Epigenetic research uses a wide range of molecular biologic techniques to further our understanding of epigenetic phenomena, including chromatin immunoprecipitation (together with its large-scale variants ChIP-on-chip and ChIP-seq), fluorescent in situ hybridization, methylation-sensitive restriction enzymes, DNA adenine methyltransferase identification (DamID) and bisulphite sequencing. Furthermore, the use of bioinformatic methods is playing an increasing role (computational epigenetics).

Mechanisms

Several types of epigenetic inheritance systems may play a role in what has become known as cell memory:

DNA Methylation and Chromatin Remodeling

Because the phenotype of a cell or individual is affected by which of its genes are transcribed, heritable transcription states can give rise to epigenetic effects. There are several layers of regulation of gene expression. One way that genes are regulated is through the remodeling of chromatin. Chromatin is the complex of DNA and the histone proteins with which it associates. Histone proteins are little spheres that DNA wraps around. If the way that DNA is wrapped around the histones changes, gene expression can change as well. Chromatin remodeling is accomplished through two main mechanisms:

1. The first way is post translational modification of the amino acids that make up histone proteins. Histone proteins are made up of long chains of amino acids. If the amino acids that are in the chain are changed, the shape of the histone sphere might be modified. DNA is not completely unwound during replication. It is possible, then, that the modified histones may be carried into each new copy of the DNA. Once there, these histones may act as templates, initiating the surrounding new histones to be shaped in the new manner. By altering the shape of the histones around it, these modified histones would ensure that a differentiated cell would stay differentiated, and not convert back into being a stem cell.

2. The second way is the addition of methyl groups to the DNA, mostly at CpG sites, to convert cytosine to 5-methylcytosine. 5-Methylcytosine performs much like a regular cytosine, pairing up with a guanine. However, some areas of genome are methylated more heavily than others and highly methylated areas tend to be less transcriptionally active, through a mechanism not fully understood. Methylation of cytosines can also persist from the germ line of one of the parents into the zygote, marking the chromosome as being inherited from this parent (genetic imprinting).

The way that the cells stay differentiated in the case of DNA methylation is clearer to us than it is in the case of histone shape. Basically, certain enzymes (such as DNMT1) have a higher affinity for the methylated cytosine. If this enzyme reaches a "hemimethylated" portion of DNA (where methylcytosine is in only one of the two DNA strands) the enzyme will methylate the other half. Although histone modifications occur throughout the entire sequence, the unstructured N-termini of histones (called histone tails) are particularly highly modified. These modifications include acetylation, methylation, ubiquitylation, phosphorylation and sumoylation. Acetylation is the most highly studied of these modifications. For example, acetylation of the K14 and K9 lysines of the tail of histone H3 by histone acetyltransferase enzymes (HATs) is generally correlated with transcriptional competence.

One mode of thinking is that this tendency of acetylation to be associated with "active" transcription is biophysical in nature. Because it normally has a positively charged nitrogen at its end, lysine can bind the negatively charged phosphates of the DNA backbone. The acetylation event converts the positively charged amine group on the side chain into a neutral amide linkage. This removes the positive charge, thus loosening the DNA from the histone. When this occurs, complexes like SWI/SNF and other transcriptional factors can bind to the DNA and allow transcription to occur. This is the "cis" model of epigenetic function. In other words, changes to the histone tails have a direct affect on the DNA itself.

Another model of epigenetic function is the "trans" model. In this model changes to the histone tails act indirectly on the DNA. For example, lysine acetylation may create a binding site for chromatin modifying enzymes (and basal transcription machinery as well). This Chromatin Remodeler can then cause changes to the state of the chromatin. Indeed, the bromodomain — a protein segment (domain) that specifically binds acetyl-lysine — is found in many enzymes that help activate transcription, including the SWI/SNF complex (on the protein polybromo). It may be that acetylation acts in this and the previous way to aid in transcriptional activation.

The idea that modifications act as docking modules for related factors is borne out by histone methylation as well. Methylation of lysine 9 of histone H3 has long been associated with constitutively transcriptionally silent chromatin (constitutive heterochromatin). It has been determined that a chromodomain (a domain that specifically binds methyl-lysine) in the transcriptionally repressive protein HP1 recruits HP1 to K9 methylated regions. One example that seems to refute this biophysical model for acetylation is that tri-methylation of histone H3 at lysine 4 is strongly associated with (and required for full) transcriptional activation. Tri-methylation in this case would introduce a fixed positive charge on the tail.

It has been shown that the histone lysine methyltransferase (KMT) is responsible for this methylation activity in the pattern of histones H3 & H4. This enzyme utilizes a catalytically active site called the SET domain (Supressor of variegation, Enhancer of zeste, Trithorax). The SET domain is a 130-amino acid sequence involved in modulating gene activities. This domain has been demonstrated to bind to the histone tail and causes the methylation of the histone.

Differing histone modifications are likely to function in differing ways; acetylation at one position is likely to function differently than acetylation at another position. Also, multiple modifications may occur at the same time, and these modifications may work together to change the behaviour of the nucleosome. The idea that multiple dynamic modifications regulate gene transcription in a systematic and reproducible way is called the histone code.

DNA methylation frequently occurs in repeated sequences, and helps to suppress the expression and mobility of 'transposable elements': Because 5-methylcytosine is chemically very similar to thymidine, CpG sites are frequently mutated and become rare in the genome, except at CpG islands where they remain unmethylated. Epigenetic changes of this type thus have the potential to direct increased frequencies of permanent genetic mutation. DNA methylation patterns are known to be established and modified in response to environmental factors by a complex interplay of at least three independent DNA methyltransferases, DNMT1, DNMT3A and DNMT3B, the loss of any of which is lethal in mice. DNMT1 is the most abundant methyltransferase in somatic cells, localizes to replication foci, has a 10–40-fold preference for hemimethylated DNA and interacts with the proliferating cell nuclear antigen (PCNA). By preferentially modifying hemimethylated DNA, DNMT1 transfers patterns of methylation to a newly synthesized strand after DNA replication, and therefore is often referred to as the 'maintenance' methyltransferase. DNMT1 is essential for proper embryonic development, imprinting and X-inactivation.

Histones H3 and H4 can also be manipulated through demethylation using histone lysine demethylase (KDM). This recently identifited enzyme has a catalytically active site called the Jumonji domain (JmjC). The demethylation occurs when JmjC utilizes multiple cofactors to hydroxylate the methyl group, thereby removing it. JmjC is capable of demethylating mono-, di-, and tri-methylated substrates..

Chromosomal regions can adopt stable and heritable alternative states resulting in bistable gene expression without changes to the DNA sequence. Epigenetic control is often associated with alternative covalent modifications of histones. The stability and heritability of states of larger chromosomal regions are often thought to involve positive feedback where modified nucleosomes recruit enzymes that similarly modify nearby nucleosomes. A simplified stochastic model for this type of epigenetics is found here.

Because DNA methylation and chromatin remodeling play such a central role in many types of epigenic inheritance, the word "epigenetics" is sometimes used as a synonym for these processes. However, this can be misleading. Chromatin remodeling is not always inherited, and not all epigenetic inheritance involves chromatin remodeling.

It has been suggested that the histone code could be mediated by the effect of small RNAs. The recent discovery and characterization of a vast array of small (21- to 26-nt), non-coding RNAs suggests that there is an RNA component, possibly involved in epigenetic gene regulation. Small interfering RNAs can modulate transcriptional gene expression via epigenetic modulation of targeted promoters.

RNA Transcripts and their Encoded Proteins

Sometimes a gene, after being turned on, transcribes a product that (either directly or indirectly) maintains the activity of that gene. For example, Hnf4 and MyoD enhance the transcription of many liver- and muscle-specific genes, respectively, including their own, through the transcription factor activity of the proteins they encode. RNA signalling includes differential recruitment of a hierarchy of generic chromatin modifying complexes and DNA methyltransferases to specific loci by RNAs during differentiation and development.

Other epigenetic changes are mediated by the production of different splice forms of RNA, or by formation of double-stranded RNA (RNAi). Descendants of the cell in which the gene was turned on will inherit this activity, even if the original stimulus for gene-activation is no longer present. These genes are most often turned on or off by signal transduction, although in some systems where syncytia or gap junctions are important, RNA may spread directly to other cells or nuclei by diffusion. A large amount of RNA and protein is contributed to the zygote by the mother during oogenesis or via nurse cells, resulting in maternal effect phenotypes. A smaller quantity of sperm RNA is transmitted from the father, but there is recent evidence that this epigenetic information can lead to visible changes in several generations of offspring.

Prions

Prions are infectious forms of proteins. Proteins generally fold into discrete units which perform distinct cellular functions, but some proteins are also capable of forming an infectious conformational state known as a prion. Although often viewed in the context of infectious disease, prions are more loosely defined by their ability to catalytically convert other native state versions of the same protein to an infectious conformational state. It is in this latter sense that they can be viewed as epigenetic agents capable of inducing a phenotypic change without a modification of the genome.

Fungal prions are considered epigenetic because the infectious phenotype caused by the prion can be inherited without modification of the genome. PSI+ and URE3, discovered in yeast in 1965 and 1971, are the two best studied of this type of prion. Prions can have a phenotypic effect through the sequestration of protein in aggregates, thereby reducing that protein's activity. In PSI+ cells, the loss of the Sup35 protein (which is involved in termination of translation) causes ribosomes to have a higher rate of read-through of stop codons, an effect which results in suppression of nonsense mutations in other genes. The ability of Sup35 to form prions may be a conserved trait. It could confer an adaptive advantage by giving cells the ability to switch into a PSI+ state and express dormant genetic features normally terminated by premature stop codon mutations.

Structural Inheritance Systems

In ciliates such as *Tetrahymena* and *Paramecium*, genetically identical cells show heritable differences in the patterns of ciliary rows on their cell surface. Experimentally altered patterns can be transmitted to daughter cells. It seems existing structures act as templates for new structures. The mechanisms of such inheritance are unclear, but reasons exist to assume that multicellular organisms also use existing cell structures to assemble new ones.

Functions and Consequences

Development: Somatic epigenetic inheritance, particularly through DNA methylation and chromatin remodeling, is very important in the development of multicellular eukaryotic organisms. The genome sequence is static (with some notable exceptions), but cells differentiate into many different types, which perform different functions, and respond differently to the environment and intercellular signalling. Thus, as individuals develop, morphogens activate or silence genes in an epigenetically heritable fashion, giving cells a "memory".

In mammals, most cells terminally differentiate, with only stem cells retaining the ability to differentiate into several cell types ("totipotency" and "multipotency"). In mammals, some stem cells continue producing new differentiated cells throughout life, but mammals are not able to respond to loss of some tissues, for example, the inability to regenerate limbs, which some other animals are capable of.

Unlike animals, plant cells do not terminally differentiate, remaining totipotent with the ability to give rise to a new individual plant. While plants do utilise many of the same epigenetic mechanisms as animals, such as chromatin remodeling, it has been hypothesised that plant cells do not have "memories", resetting their gene expression patterns at each cell division using positional information from the environment and surrounding cells to determine their fate.

Medicine

Epigenetics has many and varied potential medical applications. Congenital genetic disease is well understood, and it is also clear that epigenetics can play a role, for example, in the case of Angelman syndrome and Prader-Willi syndrome. These are normal genetic diseases caused by gene deletions or inactivation of the genes, but are unusually common because individuals are essentially hemizygous because of genomic imprinting, and therefore a single gene knock out is sufficient to cause the disease, where most cases would require both copies to be knocked out.

Evolution

Although epigenetics in multicellular organisms is generally thought to be a mechanism involved in differentiation, with epigenetic patterns "reset" when organisms reproduce, there have been some observations of transgenerational epigenetic inheritance (e.g., the phenomenon of paramutation observed in maize).

Although most of these multigenerational epigenetic traits are gradually lost over several generations, the possibility remains that multigenerational epigenetics could be another aspect to evolution and adaptation. A sequestered germ line or Weismann barrier is specific to animals, and epigenetic inheritance is expected to be far more common in plants and microbes. These effects may require enhancements to the standard conceptual framework of the modern evolutionary synthesis.

Epigenetic features may play a role in short-term adaptation of species by allowing for reversible phenotype variability. The modification of epigenetic features associated with a region of DNA allows organisms, on a multigenerational time scale, to switch between phenotypes that express and repress that particular gene. When the DNA sequence of the region is not mutated, this change is reversible. It has also been speculated that organisms may take advantage of differential mutation rates associated with epigenetic features to control the mutation rates of particular genes. Interestingly, recent analysis have suggested that members of the APOBEC family of cytosine deaminases are capable of simultaneously mediating genetic and epigenetic inheritance using similar molecular mechanisms.

Epigenetic changes have also been observed to occur in response to environmental exposure—for example, mice given some dietary supplements have epigenetic changes affecting expression of the agouti gene, which affects their fur colour, weight, and propensity to develop cancer.

More than 100 cases of transgenerational epigenetic inheritance phenomena have been reported in a wide range of organisms, including prokaryotes, plants, and animals.

Epigenetic Effects in Humans

Genomic Imprinting and Related Disorders: Some human disorders are associated with genomic imprinting, a phenomenon in mammals where the father and mother contribute different epigenetic patterns for specific genomic loci in their germ cells. The best-known case of imprinting in human disorders is that of Angelman syndrome and Prader-Willi syndrome—both can be produced by the same genetic mutation, chromosome 15q partial deletion, and the particular syndrome that will develop depends on whether the mutation is inherited from the child's mother or from their father. This is due to the presence of genomic imprinting in the region. Beckwith-Wiedemann syndrome is also associated with genomic imprinting, often caused by abnormalities in maternal genomic imprinting of a region on chromosome 11.

Transgenerational Epigenetic Observations

Marcus Pembrey and colleagues also observed in the Överkalix study that the paternal (but not maternal) grandsons of Swedish boys who were exposed during preadolescence to famine in the 19th century were less likely to die of cardiovascular disease; if food was plentiful then diabetes mortality in the grandchildren increased, suggesting that this was a transgenerational epigenetic inheritance. The opposite effect was observed for females—the paternal (but not maternal) granddaughters of women who experienced famine while in the womb (and therefore while their eggs were being formed) lived shorter lives on average.

Cancer and Developmental Abnormalities

A variety of compounds are considered as epigenetic carcinogens—they result in an increased incidence of tumors, but they do not show mutagen activity (toxic compounds or pathogens that cause tumors incident to increased regeneration should also be excluded). Examples include diethylstilbestrol, arsenite, hexachlorobenzene, and nickel compounds.

Many teratogens exert specific effects on the fetus by epigenetic mechanisms. While epigenetic effects may preserve the effect of a teratogen such as diethylstilbestrol throughout the life of an affected child, the possibility of birth defects resulting from exposure of fathers or in second and succeeding generations of offspring has generally been rejected on theoretical grounds and for lack of evidence. However, a range of male-mediated abnormalities have been demonstrated, and more are likely to exist. FDA label information for Vidaza(tm), a formulation of 5-azacitidine (an unmethylatable analog of cytidine that causes hypomethylation when incorporated into DNA) states that "men should be advised not to father a child" while using the drug, citing evidence in treated male mice of reduced fertility, increased embryo loss, and abnormal embryo development. In rats, endocrine differences were observed in offspring of males exposed to morphine. In mice, second generation effects of diethylstilbesterol have been described occurring by epigenetic mechanisms.

Recent studies have shown that the Mixed Lineage Leukemia (MLL) gene causes leukemia by rearranging and fusing with other genes in different chromosomes, which is a process under epigenetic control.

Other investigations have concluded that alterations in histone acetylation and DNA methylation occur in various genes influencing prostate cancer.

In 2008, the National Institutes of Health announced that $190 million had been earmarked for epigenetics research over the next five years. In announcing the funding, government officials noted that epigenetics has the potential to explain mechanisms of aging, human development, and the origins of cancer, heart disease, mental illness, as well as several other conditions. Some investigators, like Randy Jirtle, PhD, of Duke University Medical Centre, think epigenetics may ultimately turn out to have a greater role in disease than genetics.

DNA Methylation in Cancer

DNA methylation is an important regulator of gene transcription and a large body of evidence has demonstrated that aberrant DNA methylation is associated with unscheduled gene silencing, and the genes with high levels of 5-methylcytosine in their promoter region are transcriptionally silent. DNA methylation is essential during embryonic development, and in somatic cells, patterns of DNA methylation are generally transmitted to daughter cells with a high fidelity.

Aberrant DNA methylation patterns have been associated with a large number of human malignancies and found in two distinct forms: hypermethylation and hypomethylation compared to normal tissue. Hypermethylation is one of the major epigenetic modifications that repress transcription via promoter region of tumour suppressor genes. Hypermethylation typically occurs at CpG islands in the promoter region and is associated with gene inactivation. Global hypomethylation has also been implicated in the development and progression of cancer through different mechanisms.

Variant Histones H2A in Cancer

The histone variants of the H2A family are highly conserved in mammals, playing critical roles in regulating many nuclear processes by altering chromatin structure. One of the key H2A variants, H2A.X, marks DNA damage, facilitating the recruitment of DNA repair proteins to restore genomic integrity. Another variant, H2A.Z, plays an important role in both gene activation and repression. A high level of H2A.Z expression is ubiquitously detected in many cancers and is significantly associated with cellular proliferation and genomic instability.

Cancer Treatment

Current research has shown that epigenetic pharmaceuticals could be a putative replacement or adjuvant therapy for currently accepted treatment methods such as radiation and chemotherapy, or could enhance the effects of these current treatments. It has been shown that the epigenetic control of the proto-onco regions and the tumour suppressor sequences by conformational changes in histones directly affects the formation and progression of cancer Epigenetics also has the factor of reversibility, a characteristic that other cancer treatments do not offer.

Drug development has mainly focused on Histone Acetyltransferase (HAT) and Histone Deactylase (HDAC), including the introduction of the new pharmaceutical Vorinostat, a HDAC inhibitor, to the market. HDAC specifically has been shown to play an integral role in the progression of oral squamous cancer. Current front-runner candidates for new drug targets are Histone Lysine Methyltransferases (KMT) and Protein Arginine Methyltransferases (PRMT).

Twin Studies

Recent studies involving both dizygotic and monozygotic twins have produced some evidence of epigenetic influence in humans.

Epigenetics in Microorganisms

Bacteria make widespread use of postreplicative DNA methylation for the epigenetic control of DNA-protein interactions. Bacteria make use of DNA adenine methylation (rather than DNA cytosine methylation) as an epigenetic signal. DNA adenine methylation is important in bacteria virulence in organisms such as *Escherichia coli*, *Salmonella, Vibrio, Yersinia, Haemophilus*, and *Brucella*. In *Alphaproteobacteria*, methylation of adenine regulates the cell cycle and couples gene transcription to DNA replication. In *Gammaproteobacteria*, adenine methylation provides signals for DNA replication, chromosome segregation, mismatch repair, packaging of bacteriophage, transposase activity and regulation of gene expression. The filamentous fungus *Neurospora crassa* is a prominent model system for understanding the control and function of cytosine methylation. In this organisms, DNA methylation is associated with relics of a genome defense system called RIP (repeat-induced point mutation) and silences gene expression by inhibiting transcription elongation.

The yeast prion PSI is generated by a conformational change of a translation termination factor, which is then inherited by daughter cells. This can provide a survival advantage under adverse conditions. This is an example of epigenetic regulation enabling unicellular organisms to respond rapidly to environmental stress. Prions can be viewed as epigenetic agents capable of inducing a phenotypic change without modification of the genome.

Modification of DNA

In eukaryotes, the accessibility of large regions of DNA can depend on its chromatin structure, which can be altered as a result of histone modifications directed by DNA methylation, ncRNA, or DNA-binding protein.

Chemical

Methylation of DNA is a common method of gene silencing. DNA is typically methylated by methyltransferase enzymes on cytosine nucleotides in a CpG dinucleotide sequence (also called "CpG islands" when densely clustered). Analysis of the pattern of methylation in a given region of DNA (which can be a promoter) can be achieved through a method called bisulphite mapping. Methylated cytosine residues are unchanged by the treatment, whereas unmethylated ones are changed to uracil. The differences are analysed by DNA sequencing or by methods developed to quantify SNPs, such as Pyrosequencing (Biotage) or MassArray (Sequenom), measuring the relative amounts of C/T at the CG dinucleotide. Abnormal methylation patterns are thought to be involved in oncogenesis.

Structural

Transcription of DNA is dictated by its structure. In general, the density of its packing is indicative of the frequency of transcription. Octameric protein complexes called nucleosomes are responsible for the amount of supercoiling of DNA, and these complexes can be temporarily modified by processes such as phosphorylation or more permanently modified by processes such as methylation. Such modifications are considered to be responsible for more or less permanent changes in gene expression levels. Histone acetylation is also an important process in transcription. Histone acetyltransferase enzymes (HATs) such as CREB-binding protein also dissociate the DNA from the histone complex, allowing transcription to proceed. Often, DNA methylation and histone deacetylation work together in gene silencing. The combination of the two seems to be a signal for DNA to be packed more densely, lowering gene expression.

Regulation of Transcription

Regulation of transcription controls when transcription occurs and how much RNA is created. Transcription of a gene by RNA polymerase can be regulated by at least five mechanisms:

- Specificity factors alter the specificity of RNA polymerase for a given promoter or set of promoters, making it more or less likely to bind to them (i.e., sigma factors used in prokaryotic transcription).
- Repressors bind to non-coding sequences on the DNA strand that are close to or overlapping the promoter region, impeding RNA polymerase's progress along the strand, thus impeding the expression of the gene.

- General transcription factors position RNA polymerase at the start of a protein-coding sequence and then release the polymerase to transcribe the mRNA.
- Activators enhance the interaction between RNA polymerase and a particular promoter, encouraging the expression of the gene. Activators do this by increasing the attraction of RNA polymerase for the promoter, through interactions with subunits of the RNA polymerase or indirectly by changing the structure of the DNA.
- Enhancers are sites on the DNA helix that are bound to by activators in order to loop the DNA bringing a specific promoter to the initiation complex. Enhancers are much more common in eukaryote than prokaryotes, where only a few examples exist (to date).

Post-transcriptional Regulation

After the DNA is transcribed and mRNA is formed, there must be some sort of regulation on how much the mRNA is translated into proteins. Cells do this by modulating the capping, splicing, addition of a Poly(A) Tail, the sequence-specific nuclear export rates, and, in several contexts, sequestration of the RNA transcript. These processes occur in eukaryotes but not in prokaryotes. This modulation is a result of a protein or transcript that, in turn, is regulated and may have an affinity for certain sequences.

Regulation of Translation

The translation of mRNA can also be controlled by a number of mechanisms, mostly at the level of initiation. Recruitment of the small ribosomal subunit can indeed be modulated by mRNA secondary structure, antisense RNA binding, or protein binding. In both prokaryotes and eukaryotes, a large number of RNA binding proteins exist, which often are directed to their target sequence by the secondary structure of the transcript, which may change depending on certain conditions, such as temperature or presence of a ligand (aptamer). Some transcripts act as ribozymes and self-regulate their expression.

Examples of Gene Regulation

- Enzyme induction is a process in which a molecule (e.g., a drug) induces (i.e., initiates or enhances) the expression of an enzyme.
- The induction of heat shock proteins in the fruit fly *Drosophila melanogaster*.
- The Lac operon is an interesting example of how gene expression can be regulated.
- Viruses, despite having only a few genes, possess mechanisms to regulate their gene expression, typically into an early and late phase, using collinear systems regulated by anti-terminators (lambda phage) or splicing modulators (HIV).

Developmental Biology

A large number of studied regulatory systems come from developmental biology. Examples include:

- The colinearity of the Hox gene cluster with their nested antero-posterior patterning
- It has been speculated that pattern generation of the hand (digits - interdigits) The gradient of Sonic hedgehog (secreted inducing factor) from the zone of polarizing activity in the limb, which creates a gradient of active Gli3, which activates Gremlin, which inhibits BMPs also secreted in the limb, resulting in the formation of an alternating pattern of activity as a result of this reaction-diffusion system.
- Somitogenesis is the creation of segments (somites) from a uniform tissue (Pre-somitic Mesoderm, PSM). They are formed sequentially from anterior to posterior. This is achieved in amniotes possibly by means of two opposing gradients, Retinoic acid in the anterior (wavefront) and Wnt and Fgf in the posterior, coupled to an oscillating pattern (segmentation clock) composed of FGF + Notch and Wnt in antiphase.
- Sex determination in the soma of a Drosophila requires the sensing of the ratio of autosomal genes to sex chromosome-encoded genes, which results in the production of sexless splicing factor in females, resulting in the female isoform of doublesex.

Circuitry

Up-regulation and Down-regulation

Up-regulation is a process that occurs within a cell triggered by a signal (originating internal or external to the cell), which results in increased expression of one or more genes and as a result the protein(s) encoded by those genes. On the converse, down-regulation is a process resulting in decreased gene and corresponding protein expression.

- Up-regulation occurs, for example, when a cell is deficient in some kind of receptor. In this case, more receptor protein is synthesized and transported to the membrane of the cell and, thus, the sensitivity of the cell is brought back to normal, reestablishing homeostasis.
- Down-regulation occurs, for example, when a cell is overstimulated by a neurotransmitter, hormone, or drug for a prolonged period of time, and the expression of the receptor protein is decreased in order to protect the cell.

Inducible vs. Repressible Systems

Gene Regulation can be summarized as how they respond:

- Inducible systems - An inducible system is off unless there is the presence of some molecule (called an inducer) that allows for gene expression. The molecule is said to "induce expression". The manner by which this happens is dependent on the control mechanisms as well as differences between prokaryotic and eukaryotic cells.
- Repressible systems - A repressible system is on except in the presence of some molecule (called a corepressor) that suppresses gene expression. The molecule is said to "repress expression". The manner by which this happens is dependent on the control mechanisms as well as differences between prokaryotic and eukaryotic cells.

Theoretical Circuits

- Repressor/Inducer: an activation of a sensor results in the change of expression of a gene
- negative feedback: the gene product downregulates its own production directly or indirectly, which can result in
  - keeping transcript levels constant/proportional to a factor
  - inhibition of run-away reactions when coupled with a positive feedback loop
  - creating an oscillator by taking advantage in the time delay of transcription and translation, given that the mRNA and protein half-life is shorter
- positive feedback: the gene product upregulates its own production directly or indirectly, which can result in
  - signal amplification
  - bistable switches when two genes inhibit each other and have both positive feedback
  - pattern generation.

Methods

In general, most experiments investigating differential expression used whole cell extracts of RNA, called steady-state levels, to determine which genes changed and by how much they did.

These are, however, not informative of where the regulation has occurred and may actually mask conflicting regulatory processess, but it is still the most commonly analysed (QPCR and DNA microarray).

When studying gene expression, there are several methods to look at the various stages. In eukaryotes these include:

- The chromatin conformation of the region can be determined by ChIP-chip analysis by pulling down RNA Polymerase II, Histone 3 modifications, Trithorax-group protein, Polycomb-group protein, or any other DNA-binding element to which a good antibody is available.
- Epistatic interactions can be investigated by synthetic genetic array analysis
- Due to post-transcriptional regulation, transcription rates and total RNA levels differ significantly. To measure the transcription rates nuclear run-on assays can be done and newer high-throughput methods are being developed, using thiol labelling instead of radioactivity.
- Only 5% of the RNA polymerised in the nucleus actually exists, and not only introns, abortive products, and non-sense transcripts are degradated. Therefore, the differences in nuclear and cytoplasmic levels can be see by separating the two fractions by gentle lysis.
- Alternative splicing can be analysed with a splicing array or with a tiling array.
- All in vivo RNA is complexed as RNPs. The quantity of transcripts bound to specific protein can be also analysed by RIP-Chip. For example, DCP2 will give an indication of sequestered protein; ribosome-bound gives and indication of transcripts active in transcription (although it should be noted that a more dated method, called polysome fractionation, is still popular in some labs)
- Protein levels can be analysed by Mass spectrometry, which can be compared only to QPCR data, as microarray data is relative and not absolute.
- RNA and protein degradation rates are measured by means of transcription inhibitors (actinomycin D or α-amanitin) or translation inhibitors (Cycloheximide), respectively.
- Enhancer (genetics).

In genetics, an enhancer is a short region of DNA that can be bound with proteins (namely, the trans-acting factors, much like a set of transcription factors) to enhance transcription levels of genes (hence the name) in a gene cluster. While enhancers are usually cis-acting, an enhancer does not need to be particularly close to the genes it acts on, and need not be located on the same chromosome.

In eukaryotic cells the structure of the chromatin complex of DNA is folded in a way that functionally mimics the supercoiled state characteristic of prokaryotic DNA, so that although the enhancer DNA is far from the gene in regard to the number of nucleotides, it is geometrically close to the promoter and gene. This allows it to interact with the general transcription factors and RNA polymerase II. An enhancer may be located upstream or downstream of the gene that it regulates.

Furthermore, an enhancer does not need to be located near to the transcription initiation site to affect the transcription of a gene, as some have been found to bind several hundred thousand base pairs upstream or downstream of the start site. Enhancers do not act on the promoter region itself, but are bound by activator proteins. These activator proteins interact with the mediator complex, which recruits polymerase II and the general transcription factors which then begin transcribing the genes. Enhancers can also be found within introns. An enhancer's orientation may even be reversed without affecting its function. Additionally, an enhancer may be excised and inserted elsewhere in the chromosome, and still affect gene transcription. That is the reason that intron polymorphisms are checked though they are not translated.

Theories

Currently, there are two different theories on the information processing that occurs on enhancers:

- Enhanceosomes - rely on highly cooperative, coordinated action and can be disabled by single point mutations that move or remove the binding sites of individual proteins.
- Flexible billboards - less integrative, multiple proteins independently regulate gene expression and their sum is read in by the basal transcriptional machinery.

HACNS1

HACNS1 (also known as CENTG2 and located in the Human Accelerated Region 2) is a gene enhancer "that may have contributed to the evolution of the uniquely opposable human thumb, and possibly also modifications in the ankle or foot that allow humans to walk on two legs". Evidence to date shows that of the 110,000 gene enhancer sequences identified in the human genome, HACNS1 has undergone the most change during the evolution of humans following the split with the ancestors of chimpanzees.

Eubacteria - Gene Regulation and Protein Synthesis

Gene expression in many bacteria is regulated through the existence of operons. An operon is a cluster of genes whose protein products have related functions. For instance, the *lac* operon includes one gene that transports lactose sugar into the cell and another that breaks it into two parts. These genes are under the control of the same promoter, and so are transcribed and translated into protein at the same time. RNA polymerase can only reach the promoter if a repressor is not blocking it; the *lac* repressor is dislodged by lactose. In this way, the bacterium uses its resources to make lactose-digesting enzymes only when lactose is available.

Other genes are expressed constantly at low levels; their protein products are required for "housekeeping" functions such as membrane synthesis and DNA repair. One such enzyme is DNA gyrase, which relieves strain in the double helix during replication and repair. DNA gyrase is the target for the antibiotic ciproflaxin (sold under the name Cipro), effective against *Bacillus anthracis*, the cause of anthrax. Since eukaryotes do not have this type of DNA gyrase, they are not harmed by the action of this antibiotic.

As in eukaryotes, translation (protein synthesis) occurs on the ribosome. Without a nucleus to exclude it, the ribosome can attach to the messenger RNA even while the RNA is still attached to the DNA. Multiple ribosomes can attach to the same mRNA, making multiple copies of the same protein.

The ribosomes of eubacteria are similar in structure to those in eukaryotes and archaea, but differ in molecular detail. This has two important consequences. First, sequencing ribosomal RNA molecules is a useful tool for understanding the evolutionary diversification of the Eubacteria. Organisms with more similar sequences are presumed to be more closely related. The same tool has been used to show that Archaea and Eubacteria are not closely related, despite their outward similarities. Indeed, Archaea are more closely related to eukaryotes (including humans) than they are to Eubacteria.

Second, the differences between bacterial and eukaryotic ribosomes can be exploited in designing antibacterial therapies. Various unique parts of the bacterial ribosome are the targets for numerous antibiotics, including streptomycin, tetracycline, and erythromycin.

Gene Expression Regulates Cell Differentiation

All of the cells within a complex multicellular organism such as a human being contain the same DNA; however, the body of such an organism is clearly composed of many different types of cells. What, then, makes a liver cell different from a skin or muscle cell? The answer lies in the way each cell deploys its genome. In other words, the particular combination of genes that are turned on (expressed) or turned off (repressed) dictates cellular morphology (shape) and function. This process of gene expression is regulated by cues from both within and outside cells, and the interplay between these cues and the genome affects essentially all processes that occur during embryonic development and adult life.

Do all Cells Really Contain the Same DNA?

Several lines of evidence support the proposal that all of the cells within a multicellular organism contain the same genome. For instance, although you started as a single cell with a half-genome from each parent, that single cell quickly divided and new cells began to differentiate, or become different from each other. While this process of differentiation established a wide variety of cell types (e.g., skin, liver, muscle, etc.), it was not accompanied by any permanent loss of genetic material. This is demonstrated by the fact that fully differentiated cell types are still capable, within the right environment, of giving rise to an entire new animal. This capability was first shown by way of an experiment in which the nucleus of an adult frog skin cell was transplanted into an enucleated donor embryo, eventually leading to the development of a cloned adult frog (Gurdon *et al.*, 1975). Later, the intact complete genome of a differentiated cell was used in the cloning of the famous sheep Dolly.

Today, researchers understand that the specialized, differentiated cell types of the adult body contain a genome as complete as any embryo's. This fascinating demonstration has led to the proposal that changes in gene expression, rather than losses of genetic material, play a key role in guiding and maintaining cell differentiation.

Cell-Extrinsic Regulation of Gene Expression

Gene expression is regulated by factors both extrinsic and intrinsic to the cell. Cell-extrinsic factors that regulate expression include environmental cues, such as small molecules, secreted proteins, temperature, and oxygen. These cues can originate from other cells within the organism, or they can come from the organism's environment. Within the organism, cells communicate with each other by sending and receiving secreted proteins, also known as growth factors, morphogens, cytokines, or signaling molecules. Receipt of these signaling molecules triggers intercellular signaling cascades that ultimately cause semipermanent changes in transcription or expression of genes. Such changes in gene expression can include turning genes completely on or off, or just slightly tweaking the level of transcript produced. This process is thought to regulate a vast number of cell behaviours, including cell fate decisions during embryogenesis, cell function, and chemotaxis.

In addition, gene expression changes can lead to changes in an entire organism, such as molting in insects. In *Drosophila*, for example, the molting process is regulated by levels of a hormone called ecdysone. This hormone acts as a signal, triggering a cascade of events and leading to changes in gene expression. Not surprisingly, the genes that are expressed in response to ecdysone are also the genes that are involved in the molting process (White *et al.*, 1997). Thus, ecdysone acts on the organism level as a cell-extrinsic factor to bring about physiologically meaningful changes in gene expression. What is also interesting is that scientists can learn more about a physiological process like metamorphosis by studying how gene expression patterns change over time. For example, although researchers were aware that ecdysone results in a decrease of transcription from some loci, such as those involved in the glycolytic pathway, microarray data suggest that ecdysone-induced metamorphosis also downregulates genes involved with fatty acid oxidation, amino acid metabolism, oxidative phosphorylation, and other pathways. This suggests that there is a more global repression of metabolic activity during molting. Specifically, during metamorphosis, the larval muscle cells are degraded, and muscle-specific genes are downregulated. Simultaneously, the development of the nervous system begins, and the genes involved in neuronal differentiation are induced.

Cell-Intrinsic Regulation of Gene Expression

Although differentiation is not thought to occur by permanent loss of genetic material, DNA can be modified in a way that affects gene expression. For instance, DNA and its associated histone proteins (together known as chromatin) can be chemically modified by a cell's own machinery. Chromatin modification can affect gene expression by changing the accessibility of genes to transcription factors, in either a positive or a negative manner. Two major classes of such chemical modifications include DNA methylation and histone modification (methylation and/or acetylation). These changes are often described as epigenetic because they do not act to alter the primary DNA sequence but instead act at a level just above the DNA sequence. Although DNA methylation and histone modification are not genetic, cells have mechanisms to copy this epigenetic information during their division so that their daughter cells contain the same regulatory data.

Changes in chromatin modification play an important role in regulating gene expression during developmental cell-type specification as well. For example, chromatin-modifying proteins play an essential role in muscle cell differentiation via interactions with key muscle-promoting transcription factors MyoD and MEF. That is, these factors are thought to help recruit chromatin modifying factors, such as histone acetyltransferases and deacetylases. In so doing, MyoD and MEF alter access to their target sites upstream of muscle differentiation genes. For instance, MyoD binds histone acetyltransferases p300 and PCAF, and this activity is essential for muscle cell differentiation (Puri *et al.*, 1997). This example provides evidence for a link among chromatin modifications, transcription factors, and, ultimately, cell-fate-specific changes in gene expression.

Chromatin modification can be stable over the life of an organism, thereby effectively permanently influencing gene expression. However, that is not to say that chromatin modification is irreversible. For instance, chromatin can become mismodified in certain cancers (Vucic *et al.*, 2008), suggesting that, although important, the change is not permanent. Moreover, chromatin modifications are usually erased and reset during the production of gametes, such that the adult program of intrinsic cues is replaced with a program more suited to embryonic development.

In fact, embryonic cell types are known to contain a unique set of chromatin modifications that are different from those found in adult cell types. This has led to the tantalizing proposal that chromatin modification helps lock in changes in gene expression that are required during development.

The permanent silencing of the genes involved only in embryogenesis could then drive the development of cells towards more mature cell types. By blocking accessibility of transcription machinery, for example, chromatin modification could prevent the need for continued repression through active binding of a repressive transcription factor. Alternatively, the genes required for an adult cell type might contain chromatin modifications (especially histone acetylation) that cause the DNA to become open and, therefore, more accessible to the transcription machinery.

Interestingly, embryonic cell types have been found to contain a signature chromatin modification in the regions that regulate the expression of genes involved in early embryonic development. Such regions were found to contain chromatin modifications with both silencing and promoting characteristics. The finding of these bivalent (two-directional) markers in association with genes important for embryonic development has led to the belief that embryonic cells exist in a special epigenetic state, wherein they can choose to remain embryonic (as in an embryonic stem cell) or to differentiate (as in normal development), and bivalent domains provide a means by which to quickly choose between the two options.

Together, these lines of evidence have led to an emerging hypothesis that cell-cell signaling and epigenetic changes converge to guide cell differentiation decisions both during development and beyond.

Transcription Factors and Transcriptional Control in Eukaryotic Cells

Do complex organisms have more genes than simpler organisms? Now that researchers can sequence whole genomes and have done so for a number of organisms, they know that many vertebrates have only about twice as many genes as invertebrates, and many of these are the result of duplication of existing genes rather than development of new ones. But if there are not that many new genes, what is responsible for the incredible diversity in plant and animal species?

The simple answer to this question is that eukaryotes have developed a more complex way of controlling expression of their existing genes than prokaryotes. This system of expression control relies on a group of proteins known as transcription factors (TFs), and it allows eukaryotes to alter their cell types and growth patterns in a variety of ways. TFs are not solely responsible for gene regulation; eukaryotes also rely on cell signaling, RNA splicing, siRNA control mechanisms, and chromatin modifications. However, TFs that bind to *cis*-regulator DNA sequences are responsible for either positively or negatively influencing the transcription of specific genes, essentially determining whether a particular gene will be turned "on" or "off" in an organism.

Transcription Factors Recognize Specific DNA Sequences

Much of the complexity in differentiation in animal and plant cells can be attributed to the evolution of elaborate systems made up of short (6 to 8 base pair) *cis*-regulatory DNA sequences or motifs, as well as the TFs that bind to the motifs, interact with each other to form complexes, and recruit RNA polymerase II (Levine & Tjian, 2003). Most eukaryotic genes have promoters that consist of the TATA box close to the 5' end of the gene and, farther upstream, several motifs recognized by specific transcription factors.

In addition, many genes have one or more other nearby sequences called enhancers. Enhancers affect transcription; these sequences occur upstream, downstream, or within introns, and they continue to work whether in the normal orientation or turned backward in the genome. In yeast, no enhancers are known; instead, there are only upstream activator sequences (UASs). Enhancers can be found thousands of base pairs from a promoter, whereas UASs are generally within a few hundred base pairs upstream. Typical RNA polymerase II promoters can be influenced by many enhancers and by multiple factors bound to the promoter and enhancer sequences. The mode of action of TFs is to recognize and bind to a segment of DNA in the promoter and/or enhancer region. Often, a change in the conformation, or three-dimensional structure of a TF, will accompany DNA binding. For example, the two loops in NFATC1 that interact with DNA are found in different conformations, depending on whether NFATC1 is complexed with DNA or not. Moreover, the structure of different TF families, described later in this article, results in specific areas in these protein complexes that interact with the DNA recognition motif. The recognition motif is usually only about 6 to 10 base pairs long.

Experiments have shown that TFs can bind tightly, both within cells and *in vitro*. After TFs bind to promoter or enhancer regions of the DNA, they interact with other bound TFs and recruit RNA polymerase II. Their influence, however, can be either positive or negative, depending on the presence of other functional domains on the protein and the overall impact of the entire TF complex. A typical TF has multiple functional domains, not only for recognizing and binding to the appropriate DNA strand, but also for interactions with other TFs, with proteins called coactivators, with RNA polymerase II, with chromatin remodeling complexes, and with small noncoding RNAs.

TFs control many important parts of development; therefore, organisms with a deletion of a TF gene exhibit profound irregularities in organization and development. For example, in *Drosophila*, deletion of the TF antennapedia gene results in the development of the antennal imaginal disc into legs rather than antennae.

Transcription Factors Exert Combinatorial Control

Many TFs are known to facilitate transcription at hundreds of different promoters, while some are only active at a select few. Laboratory techniques such as chromatin immunoprecipitation (ChIP) and DNA microarrays are commonly used to study the target DNA motifs recognized by individual TFs (Iyer *et al.*, 2001). Signal molecules can influence activation by TFs by covalently binding or modifying their functional domains. It is even possible for a TF to respond to a physical signal, such as red or far-red light, but the signal must be transduced to a chemically modified activator that interacts with the TF.

The complexity and fine gradations of DNA expression in eukaryotes result from combinatorics, in that the combination of chromatin and TF signals, rather than the individual TF signal, is read out. Thus, transcriptional control is dependent on the interactions of all the TFs and whether they attract RNA polymerase or block it from initiating transcription. Multiple TFs can accumulate, creating a bulk the size of a ribosome. Once bound together, changes to the functional domains of a TF and/or covalent interactions with other factors can turn transcription on or off, depending on whether they allow or prohibit the recruitment of RNA polymerase.

A typical enhancer can be up to 500 base pairs in length and contain multiple binding sites for at least two or three different TFs (Levine & Tijan, 2003). Two TFs bound at sites near one another on the DNA strand can combine to form a dimer and bend the DNA in what is believed to be part of the activation process. Chromatin structure allows activators to associate with one another, even when they are bound to DNA sequences many hundreds of base pairs apart. Some TFs are believed to act as tethering elements between distant enhancers and promoters by forming connections with other proteins.

The Evolution of Transcription Factor Families

Higher organisms have a large number of diverse TF families defined by the sequence of their DNA-binding domains. Evolutionary studies have shown that although the DNA-binding motif is highly conserved among plants and animals, the remainder of these organisms' protein sequences is often very different. In addition, a particular TF family may have different roles in plants than in animals, and some new TFs have evolved in each kingdom since their divergence.

In many animals, including humans, a prominent group of genes involved in cell development, including many that encode TFs, contain a 180 base-pair sequence called the homeobox. The homeobox encodes a 60-amino acid protein segment called the homeodomain, which recognizes and binds to promoters in the DNA of its target genes. Complete control over transcription, and sometimes binding, is dependent on interactions between TFs, so activation often depends upon the presence of another TF. A similar system of gene recognition is found in plants, where the DNA-binding domain is called the MADS box.

TFs often have certain specific DNA-binding motifs, a common one being the basic helix-loop-helix (bHLH) structure that recognizes a specific sequence of DNA and sits on the DNA like a train car on a track. One such example is the TF MyoD (myoblast determination). Expression of the *MyoD* gene results in production of MyoD protein, which binds to the promoters of muscle-determining genes, causing the differentiation of muscle precursor cells (myoblasts) into muscle fibres. MyoD also binds to its own promoter, thus maintaining its own levels in differentiated muscle cells and their progeny.

In addition to bHLH, there are some other common structural motifs for recognition and binding of DNA, and these are found in most regulatory proteins. These are the helix-turn-helix, zinc finger, and leucine zipper. Proteins having each of these motifs are effective because they fit neatly into the major or minor grooves of the DNA strand, and also because they expose specific amino acids at the appropriate places to form hydrogen bonds with the nucleotide bases. Molecular genetic techniques can be used to change any amino acids to test whether this affects the binding affinity of the TF for the target.

Complexity and Transcription Factors

Complexity of transcriptional control can be illustrated by comparing the number and locations of *cis*-control elements in higher and lower eukaryotes. For instance, *Drosophila* typically has several enhancers for a single gene of 2 to 3 kilobases, scattered over a large (10 kilobase) region of DNA, while, as described earlier, yeast have no enhancers but instead use one UAS sequence per gene, located upstream. Long-range regulation is thought to be indicative of the need for a higher level of control over genes involved in cell development and differentiation.

The yeast genome encodes around 300 TFs, or one per every 20 genes, while humans express approximately 3,000 TFs, or one per every 10 genes. With combinatorial control, the twofold increase in TFs per gene actually translates into many more possible combinations of interactions, allowing for the dramatic increase in diversity among organisms. When we consider the additional complexities of chromatin remodeling, regulated mRNA stability, and translational control, it is easier to understand how the cells of higher organisms can produce such an enormous variety of genetic responses to environmental signals.

Nucleic Acids to Amino Acids: DNA Specifies Protein

Once it was determined that messenger RNA (mRNA) serves as a copy of chromosomal DNA and specifies the sequence of amino acids in proteins, the question of how this process is actually carried out naturally followed. It had long been known that only 20 amino acids occur in naturally derived proteins. It was also known that there are only four nucleotides in mRNA: adenine (A), uracil (U), guanine (G), and cytosine (C). Thus, 20 amino acids are coded by only four unique bases in mRNA, but just how is this coding achieved?

The Codon

The discordance between the number of nucleic acid bases and the number of amino acids immediately eliminates the possibility of a code of one base per amino acid. In fact, even two nucleotides per amino acid (a doublet code) could not account for 20 amino acids (with four bases and a doublet code, there would only be 16 possible combinations [$4^2 = 16$]). Thus, the smallest combination of four bases that could encode all 20 amino acids would be a triplet code. However, a triplet code produces 64 ($4^3 = 64$) possible combinations, or codons. Thus, a triplet code introduces the problem of there being more than three times the number of codons than amino acids. Either these "extra" codons produce redundancy, with multiple codons encoding the same amino acid, or there must instead be numerous dead-end codons that are not linked to any amino acid.

Preliminary evidence indicating that the genetic code was indeed a triplet code came from an experiment by Francis Crick and Sydney Brenner (1961). This experiment examined the effect of frameshift mutations on protein synthesis. Frameshift mutations are much more disruptive to the genetic code than simple base substitutions, because they involve a base insertion or deletion, thus changing the number of bases and their positions in a gene. For example, the mutagen proflavine causes frameshift mutations by inserting itself between DNA bases. The presence of proflavine in a DNA molecule thus interferes with the molecule's replication such that the resultant DNA copy has a base inserted or deleted.

Crick and Brenner showed that proflavine-mutated bacteriophages (viruses that infect bacteria) with single-base insertion or deletion mutations did not produce functional copies of the protein encoded by the mutated gene. The production of defective proteins under these circumstances can be attributed to misdirected translation. Mutant proteins with two- or four-nucleotide insertions or deletions were also nonfunctional. However, some mutant strains became functional again when they accumulated a total of three extra nucleotides or when they were missing three nucleotides. This rescue effect provided compelling evidence that the genetic code for one amino acid is indeed a three-base, or triplet, code.

Decoding the Genetic Code

Once the budding molecular biology community was convinced about the triplet code, the race to decode which triplets specified which amino acids began. The simplest way to decipher the code would be to start with an mRNA molecule of known sequence, use it to direct the synthesis of a protein, and then determine the amino acid sequence of the synthesized protein. Then, comparison of the original mRNA sequence with the amino acid sequence of the synthesized protein could provide a means for directly decoding the genetic code.

In further experiments to decode the other codons, Nirenberg and his colleagues made artificial RNAs containing defined proportions of two or three different bases. As previously mentioned, polynucleotide phosphorylase joins nucleotides randomly; as a result, these artificial RNAs contained random mixtures of the bases in proportion to the amounts of bases mixed. Hence, the resulting products provided clues that the researchers could use to deduce potential codon–amino acid relationships.

For example, when A and C were mixed with polynucleotide phosphorylase, the resulting RNA molecules contained eight different triplet codons: AAA, AAC, ACC, ACA, CAA, CCA, CAC, and CCC. These eight random poly(AC) RNAs produced proteins containing only six amino acids: asparagine, glutamine, histidine, lysine, proline, and threonine. Remember that previous experiments had already revealed that CCC and AAA code for proline and lysine, respectively. Thus, the four newly incorporated amino acids could only be encoded by AAC, ACC, ACA, CAA, CCA, and/or CAC. With the random sequence approach, the decoding endeavor was almost completed, but some work remained to be done.

Thus, in 1965, H. Gobind Khorana and his colleagues used another method to further crack the genetic code. These researchers had the insight to employ chemically synthesized RNA molecules of known repeating sequences rather than random sequences. For example, an artificial mRNA of alternating guanine and uracil nucleotides (GUGUGUGUGUGU) should be read in translation as two alternating codons, GUG and UGU, thus encoding a protein of two alternating amino acids. Translation of the artificial GUGU mRNA yielded a protein of alternating cysteine and valine residues. However, this technique alone could not determine whether GUG or UGU encoded cysteine, for example.

Next, Nirenberg and Philip Leder developed a technique using ribosome-bound transfer RNAs (tRNAs). They showed that a short mRNA sequence—even a single codon (three bases)—could still bind to a ribosome, even if this short sequence was incapable of directing protein synthesis. The ribosome-bound codon could then base pair with a particular tRNA that carried the amino acid specified by the codon.

Nirenberg and Leder thus synthesized many short mRNAs with known codons. They then added the mRNAs one by one to a mix of ribosomes and aminoacyl-tRNAs with one amino acid radioactively labeled. For each, they determined whether the aminoacyl-tRNA was bound to the short mRNA-like sequence and ribosome (the rest passed through the filter), providing conclusive demonstrations of the particular aminoacyl-tRNA that bound to each mRNA codon.

Degeneracy of the Amino Acid Code

Examination of the full table of codons enables one to immediately determine whether the "extra" codons are associated with redundancy or dead-end codes. Note that both possibilities occur in the code. There are only a few instances in which one codon codes for one amino acid, such as the codon for tryptophan. Note also that the codon for the amino acid methionine (AUG) acts as the start signal for protein synthesis in an mRNA. Moreover, the genetic code also includes stop codons, which do not code for any amino acid. The stop codons serve as termination signals for translation. When a ribosome reaches a stop codon, translation stops, and the polypeptide is released.

Myofilaments

Morphologically, there are two kinds of myofilaments, *thick myofilaments* that are about 1.5 μm long and 10 nm wide and that are separated by a 40 nm space, and *thin myofilaments* that are about 1.0 μm long and 5 nm in diameter. As will be mentioned later, thick myofilaments are made of *myosin,* and the thin myofilaments have a more complex structure containing several proteins (*i.e.*, actin, tropomyosin, and troponins) of which *actin* is the most important. The two types of filaments are disposed in register and overlap to an extent that depends on the degree of contraction of the sarcomere. In a relaxed condition, the I-band contains only thin filaments; the H-band contain only thick filaments; and within the A-band the thick and thin filaments overlap.

In a cross section through the A-band, the regular disposition of the two types of filaments can be observed best. In vertebrate muscle each thick filament is seen to be surrounded by six thin filaments, and each thin filament lies symmetrically among three thick ones. As a consequence of this geometry there are twice as many thin filaments as thick ones. A cross section through the H-band shows only thick myofilaments and through the I-band, only thin myofilaments. Another interesting detail revealed by the electron microscope is that the two sets of filaments are linked together by a system of cross-bridges. These arise from the thick filaments at intervals of about 7 nm.

Each bridge is situated along the axis with an angular difference of 60 degrees. This means that the bridges describe a helix about every 43 nm. As a result of this arrangement one thick filament joins the six adjacent thin ones every 43 nm. A repeat period corresponding to this distance can be observed in the A-band after staining with uranyl acetate. There are 11 strips in each half an A-band starting from the M-line. Some of these are due to the myosin cross-bridges, but others, to the presence of C-protein. The complex fine structure of the A-band can also be revealed by negative staining of ultrathin sections of frozen muscle (*i.e.*, cryosections). In cross section the M-line shows that the thick myofilaments are joined, forming a triangular lattice A-3.

The Z-line

In cross section the Z-disc shows a woven-basket lattice which remains essentially unchanged during contraction. This lattice is made of Z-filaments which are connected to the thin filaments of the I-band. It is presumed that as one thin filament enters the Z-line, it is in continuity with three curved Z-filaments which unite it with three other thin filaments of the same sarcomere. According to this model the thin filaments of the opposite sarcomere are similarly arranged.

The hexagonal lattice that is characteristic of the A-band (one thick filament surrounded by six thin ones) becomes compressed at the I-Z junction and is transformed into the square lattice characteristic of the woven-basket. This model, in which there is no interlooping of filaments from one sarcomere to the other, explains the splitting of the Z-line that may occur under certain conditions. Later one we shall see that one of the main components of the Z-line is a-*actinin*, a protein that joins with the actin thin filament and may constitute the Z-filaments.

Sacromere I-band

In a living muscle fibre, changes with contraction can be observed with the phase contrast and interference microscope. If this is done, one striking observation that can be made is that the A-band remains constant in a wide range of muscle lengths, whereas the I-band shortens in accordance with the contraction.

The shortening of the I-band is the result of the fact that the thin myofilaments slide farther and farther into the arrays of thick filaments. With the progressing contraction, the thin filaments penetrate into the H-band and may even overlap, thereby producing a more dense band in the centre of the sarcomere (*inversion of the banding*). Finally, the thick filaments make contact and are crumpled against the Z-lines. These findings have been interpreted in the so-called *sliding filament mechanism* of contraction, which will be discussed in greater detail later. The degree of contraction thus achieved can be measured by determining the length of the sarcomere (*i.e.*, the distance between Z-discs) at rest and when it has shortened. Note that insect muscle, in general, shortens only slightly (about 12 per cent), whereas the shortening in vertebrate muscle may be much greater (about 43 per cent).

Macromolecular Organisation

The electron microscope has revealed that "smooth" muscles may have a varied macromolecular organization. In many cases they contain thin and thick myofilaments, as do striated muscles, but the difference lies in the absence of the Z-line and the lack of periodicity. In mollusks and annelids there are muscles with a helical arrangement that have thin and thick myofilaments linked by cross bridges. In the adductor muscle of the oyster, the so-called paramyosin muscle, each thick filament is surrounded by 12 thin filaments. The smooth muscle of vertebrates apparently lacks these two types of myofilaments and even myofibrils are difficult to recognize. However, improvements in the preparative techniques have demonstrated coarse myofilaments. Furthermore, thick and thin myofilaments have been isolated. In smooth muscles the contraction is very slow, but extreme degrees of shortening may be achieved.

Contractive System

In recent years knowledge about the molecular machinery involved in muscle contraction has become increasingly complex. To the classic force-generation proteins *myosin* and *actin* others having a regulatory function have been added, such as *tropomyosin* and the various *troponins*. Finally a series of other proteins, probably playing a structural role, such as a-*actinin* present in the Z-band, the *M-band proteins*, and the *C-protein*, have been described. Here, we shall only mention some of the main characteristics of these proteins in relation to the structure of the myofilaments and myofibrils.

Myosin Molecules

Myosin can be extracted from muscles with a 0.3 M solution of KCI and purified by precipitation at lower ionic strength. This large molecule comprises about half of the total protein of the myofibril and has a molecular weight of about 500,000 daltons. It contains 2 polypeptide chains of about 200,000 daltons each and 4 smaller ones in the range of 20,000 daltons.

The 2 heavy chains are coiled around each other, forming a double helical a-helix about 140 nm long and 2 nm in diameter. Only about half the heavy chain is rod-like, the rest, together with some of the smaller chains, is folded into two globular regions at one end cf the molecule. Myosin can be fragmented by the action of proteolytic enzymes; *trypsin* breaks it into the long, rod-like *light meromyosin* (LMM) and *heavy meromyosin* (HMM).

This last portion can be further subdivided by *papain* into the globular *subunit* S_1 and the *helical rod* S_2, which joins S_1 to the to the light meromyosin. The most important part of the myosin molecule is HMM-S_1, since it contains the sites for the *ATP-ase* and for *binding to actin.* In fact it is the interaction between myosin and actin that results in contraction.

At this point it is important to mention how the myosin molecules are integrated to constitute the *thick* or *myosin myofilament.* Each of these filaments has a smooth or bare portion in the middle (corresponding to the H-band) ad two regions in which the surface is rough because of the presence of the *myosin crossbridges.* It is now known that these *cross-bridges.* It is now known that these cross-bridges represent the S_1 globular ends of myosin, while the shaft of the filament is formed by the rod-like portion of the molecule.

One half of the molecules in the myofilament are oriented in the opposite direction from the other half. In other words, there is a definite polarization of the S_1 ends in each half of the myofilament. From the study of cross sections of the myosin filaments, which show a triangular shape, a model in which 12 parallel units are closely packed has been postulated. Each one of the units, with a diameter of 4 nm, could be the result of the association of two myosin molecules. Another interesting observation is that at the M-line a cross section of the myosin filament shows bridges which connect with the six neighboring thick filaments, forming a hexagonal lattice. Such bridges represent the attachment of the M-protein at the M-line of the sarcomere.

Myofilament

Actin needs high concentrations of KCI (0.6 M) to be extracted. It is the second major protein and represents about one quarter of the protein of the myofibril. The molecular weight of actin is 42,000 daltons. The actin myofilament is made up of two helical strands which cross over every 36 to 37 nm. Each of these cross-over repeats contains between 13 and 14 *globular* monomers of *G-actin* about 5 nm in diameter. In muscle, actin is present mainly as fibrous or *F-actin,* which is the polymerized form of G-actin. The actin myofilament is mre complex because it also contains *tropomyosin* and the *troponins.*

Tropomyosin

This protein represents between 5 and 10 per cent of the total and is extracted with 1 M KCI or weak acids. It has a molecular weight of 64,000 daltons and is an elongated molecule of 40 nm made of 2 parallel polypeptide chains in an a-helix. Tropomyosin stretches over seven actin monomers and forms two helixes that lie within the grooves of the actin double helix.

Troponin

Troponin is known to consist of 3 components forming a complex of proteins of 80,000 daltons. Each of these troponin complexes binds to tropomyosin every 40 nm, or every 7 actin monomers. The three components of troponin that are present in equimolecular proportions are *troponin-C,* which binds Ca^{2+} specifically, *troponin-T*, which has a high affinity for tropomyosin, and *troponin*-I, which inhibits the ATPase of myosin.

Contractile and Regulatory Proteins

Antibodies against myosin, actin, tropomyosin, and troponin that have been labeled with fluorescent dyes have permitted the localization of these proteins in relation to the macromolecular structure of the myofibril and myofilaments. For example, with *anti-troponin*, the fluorescent labeling of the thin filaments appears at a 40 nm interval. Other minor muscle proteins may also be localized with fluorescent antibodies.

A-Actinin

This protein is a rod-shaped molecule of about 95,000 daltons localized exclusively at the Z-line, as can be demonstrated by anti-actinin antibodies. The a-actinin molecule binds in vitro to actin filaments, forming regular cross-links. This protein apparently plays a structural role in the sarcomere.

C-Protein

This protein is present in the A-band along the middle portion of the myosin filament. With electron microscopy the anti-C-protein antibody is observed to form between 7 and 9 transverse stripes in each half of the A-band.

M-Line Protein

Several proteins have been extracted from the M-line region; one of them is probably a structural protein, and the others are phosphorylase and creatine-kinase.

Slioing Mechanism

We mentioned earlier that contraction is currently explained by a *sliding mechanism* in which the thin actin filaments are displaced with respect to the thick myosin filaments during each contraction-relaxation cycle. In the case of a frog sarcomere, 2.5 μm ong, there is a shortening of 30 per cent at each contraction. This implies a sliding of 0.37 μm for each half sarcomere. The force generated seems to be proportional to the degree of overlap between thick and thin filaments. This implies that the force is of a short range and is produced directly by the cross-bridges between the filaments. A resting muscle is rather plastic and extensible because the cross-bridges are not attached and can slide upon an applied external force.

Active sliding movement is thought to result from the repetitive interaction of the cross bridges with the actin filaments. It is assumed that each cross-bridge represents the active end of the myosin molecule (*i.e.*, the S_1 globular unit).

The following mechanism takes place:

- A cross-bridge binds to a specific site of the actin filament.
- The cross-bridge undergoes a conformational change which displaces the point of attachment towards the centre of the A-band, thereby pulling the acting filament; at the same time a second cross-bridge becomes attached.
- At the end of the cycle the first bridge returns to the starting configuration in preparation for a new cycle. According to this theory, the actin filament from each half sarcomere are pulled as ropes towards the centre of the sarcomere by the myosin arm that move to and from.

The energy for this interaction is provided by the splitting of ATP because of the ATPase present in the cross-bridge. It is estimated that the splitting of one ATP accounts for a displacement of 5 to 10 nm for each myosin cross-bridge.

Polarization within the Sarcomere

One of the most characteristic features of muscle is a molecular organization by which individual molecules may interact both spatially and temporally in a concerted fashion. Thus, the sliding mechanism depends on a very specifi interaction between actin and myosin molecules and requires the existence of a *definite polarization* within the sarcomere.

In each half sarcomere actin and myosin should have a different polarization; furthermore, the force developed must have opposite directions. From electron microscopic studies of aggregates of myosin molecules, it is possible to obtain reconstituted myofilaments having the structure. The molecules of each half of the filament are in two antiparallel sets with a reversal of structural polarity in the centre.

Muscle Contraction

Since the classic experiments of Szent-Györgyi, it has been known that the mixing of myosin and actin in the test tube results in the formation of the complex *actomyosin*, which contracts in the presence of ATP. Investigators following this interaction under the electron microscope have observed that the myosin molecules bind to the F-actin with a directional orientation. The complex can also be produced by using F-actin and heavy meromyosin.

The actin filaments become "decorated" with the myosin, and the complex shows an arrow-like polarity. These findings demonstrate that in each filament of F-actin, the G-actin molecules are polarized with the same orientation. The double helical filament of actin decorated with the S_1 myosin fragments. The arrow-like configuration shown in the electron micrograph results in the tilting and bending of the S_1 myosin along the helix of F_1 actin. This tilting of the S_1 myosin very likely corresponds to the configuration that the cross-bridge has at the end of the working stroke, when ADP and Pi have been released and no further force is being exerted.

On the other hand, there is electron microscopic and x-ray diffraction evidence that in the relaxed muscle the bridges are approximately perpendicular to the axis of the myofilaments. The detailed ultrastructural and biochemical information obtained permits an interpretation of the macromolecular mechanisms involved in muscle contraction. It may be postulated that this is a cyclic event involving the repetitive formation and breakdown of actin-myosin linkages at the bridges between thick and thin filaments.

At each bridge the following sequence of events is probably produced:

- Formation of a perpendicular linkage between a heavy meromyosin head and one G-actin (globular unit;
- Rupture of this linkage by one ATP molecule;
- Hydrolysis of the ATP by the Ca^{2+} -activated ATPase of myosin;
- Formation of a new linkage between the same heavy meromyosin (bridge) and the next G-actin unit.

The relative movement of the thin filament taking place in each sequence would be equivalent to the length of one G-actin unit (5.3 nm).

Energetics of Contraction

We have seen earlier that in the *relaxed state* there is no attachment of the cross-bridges; the opposite condition could explain the state of *rigor* (*i.e.*, of permanent contraction). In this state, the rigidity of the muscle could be due to the permanent attachment of the cross-bridges to the thin filaments. In order to have a contraction-relaxation cycle, a cyclic mechanism by which the cross-bridges become attached and detached must operate. It is now known that the transition between rest and activity is dependent on the *concentration of free calcium* in the vicinity of the contractile machinery.

The control of contraction by Ca^{2+} requires the presence of the regulatory proteins tropomyosin and troponin, which form part of the structure of the thin filament. We have seen earlier that *troponin-C* (Ca^{2+} binding) and *troponin-I* and *troponin-T* are placed along the actin filament at intervals of about 40 nm and that such a periodicity corresponds to 7 G-actin monomers. Furthermore, we mentioned that tropomyosin also extends along the actin filament for the same distance and is in close relationship with the troponin complex.

The mechanism by which Ca^{2+} regulates the contraction-relaxation cycle is thought to be the following:

- In the absence of Ca^{2+} the regulatory proteins inhibit the interaction between actin and myosin.
- When the Ca^{2+} concentration in the cytosol increases above 10^{-6} M, the muscle contraction is triggered by the binding of Ca^{2+} to troponin-C.
- Since the presence of tropomyosin is needed for this regulatory mechanism, it is supposed that the influence of troponin is transmitted by way of tropomyosin to the seven G-actin monomers with which it is associated. Thus, this mechanism is highly cooperative.

Molecular Regulation

From electron microscopic and x-ray diffraction evidence a model has been constructed which may explain the regulatory action of the tropomyosin-troponin-Ca^{2+} system. In this model, an end-on view of the relative positions of action, the S_1-myosin, and tropomyosin is presented. The tropomyosin molecule is shown in two positions, one corresponding to the activated state and the other to the relaxed state. According to the model in the latter condition (dotted contour) tropomyosin is deep in the groove of F-actin and covers the actin binding site of the S_1-myosin.

When Ca^{2+} binds to troponin-C displacement of tropomyosin away from the groove could be effected, exposing the actin binding site in S_1 for actin-myosin interaction.

Oxidative Phosphorylation

Whereas the myofibrils constitute the mechanical machinery of the muscle, the fuel needed is produced mainly in the sarcoplasm. In all types of muscle, numerous mitochondria, called *sarcosomes*, provide the essential oxidative phosphorylation processes and the Krebs cycle system. These mitochondria are particularly prominent in size and number in heart muscle and in the flight muscles of birds and insects.

Excitation and Contraction Coupling

Excitation-contraction coupling refers to the mechanism by which the electrical impulse is able to induce contraction in a muscle. It is known that with the arrival of the nerve impulse at the motor end plate a *chemical synaptic transmission* occurs. This in turn brings about a depolarization of the cell membrane which is conducted along the muscle fibre and penetrates into it to induce the contraction.

The sarcoplasmic matrix contains the glycolytic enzymes as well as other globular proteins, such as myoglobin, salts, and high phosphate compounds.

Glycogen is present in the matrix as small granules observed under the electron microscope. There are about 1 per cent glycogen and 0.5 per cent creatine phosphate as sources of energy in muscle. Glycogen disappears with contraction through glycolysis, and lactic acid is formed, which can be transformed into pyruvic acid to enter the Krebs cycle. The initial energy source for contraction is ATP. The ADP produced after the initial contraction is again recharged to ATP by glycolysis or from creatine phosphate. Oxidative phosphorylation is the last and most important source of ATP.

You will see that the essential components in this excitation-contraction coupling mechanism are:

- The *T-system*, which conducts the action potential to The interior of the muscle fibre;
- The release of Ca^{2+} from the *sarcoplasmic reticulum*;
- The induction of contraction by Ca^{2+} (explained earlier);
- The reaccumulation of Ca^{2+} into the sarcoplasmic reticulum by Ca^{2+} - *ATPase*, which results in *muscle relaxation*. The study of this mechanism involves the knowledge of each one of the elements involved.

Sarcoplasmic Reticulum

The sarcoplasmic reticulum found in skeletal and cardiac muscle fibres is one of the most interesting specializations of the endoplasmic reticulum. It was discovered by Veratti in 1902 as a reticulum present in the sarcoplasm of the muscle fibre and extending in between the myofibrils. It was completely neglected until 1953 when the first electron micrographs of this structure were published. The sarcoplasmic reticulum is a membrane-limited reticular system whose organizational structure is regularly superimposed upon that of the sarcomeres.

There are two main components in the sarcoplasmic reticulum:

1. A *longitudinal* component that is larger and represents the true endoplasmic reticulum;
2. A *transverse* component or *T-system* made of vesicles and tubules that are disposed at the level of the Z-lines and are connected to the sarcolemma.

The longitudinal reticulum is composed of wide anastomosing tubules that cover the surface of the sarcomere at the level of the A-and I-bands and that terminate by special terminal cisternae found at the end of the I-bands near the Z-line. In between the terminal cisternae of two consecutive sarcomeres is another flattened vesicle belonging to the T-system; the three together constitute the so-called *triad*. The T-system is in Continuity with the Plasma

Membrane and Conducts Impulses Inward The transverse component (*i.e.*, the T-system at certain points is continuous with the plasma membrane of the sarcolemma and is the structure organized to conduct impulses from the fibre surface into the deepest portions of the muscle fibre. The continuity between the T-system, the plasma membrane, and the extracellular space was first demonstrated indirectly. In frog muscle immersed for short periods in a solution containing ferritin molecules, it was found that the central vesicle of the triad had filled with these electron-opaque molecules. Ferritin was also found in certain tubules that were continuous with the central element of the triad. These findings suggested that at the plasma membrane there are a small number of openings communicating directly with the traverse system of the sarcoplasmic reticulum through fine tubules. Interestingly, the later components of the triad never contained ferritin molecules, these components, then, not connected to the plasma membrane.

Direct observation of the connections between the T-system and the plasma membrane is difficult but has been achieved. Portion of the T-system have been observed to longitudinally and even in a spiral arrangement. The function of the T-system is to transverse inward the electrical signal that will bring about the contraction of the individual embroils. The role of the system was suggested experiments with microelectrodes in which stimulation of the sarcolemma, at the level the Z-line, produced a localized contraction adjacent sarcomeres.

Apparently the system is not a passive channel but is active involved in the conduction of action potential One interesting finding has been that in muscle immersed in glycerol there is a lesion of the T-system (*i.e., detubulation*), and the mechanism of excitation-coupling is abolished. The presence of this intracellular conducting system may explain the physiologic paradox that all the myofibrils of a fibre between 5 and 100 μm in diameter may contract quickly and synchronously, once the activating action potential has passed over the surface.

Stored Ca^{2+}

Within the muscle, Ca^{2+} is stored mainly in the longitudinal components of the sarcoplasmic reticulum, so that the Ca^{2+} concentration near the myofibrils is kept very low (*i.e.*, about 10^{-7} M). Following stimulation, Ca^{2+} is rapidly released, so that the concentration increased (up to 10^{-5} M) and induces contraction. It is now assumed that the triad plays an essential role in this release of Ca^{2+}. It is thought that the action potential, arriving by way of the T-system, immediately opens Ca^{2+} channels in the *terminal cisternae* of the sarcoplasmic reticulum, producing a localized release of Ca^{2+}. Thus, the triad is the most important site in the excitation-contraction coupling.

We have read that during excitation Ca^{2+} is liberated from special regions of the sarcoplasmic reticulum and that this causes the contraction. To produce relaxation the opposite phenomenon must take place. In other words, Ca^{2+}

must be rapidly removed, and this is done by a Ca^{2+} *-activated ATPase* that is present in the membrane of the sarcoplasmic reticulum. This acts as a Ca^{2+} pump that incorporates this cation under the action of ATP.

The reactions involved are the following:

2 Ca^{2+} outside + ATP + E

E-PCa2^{2+} + ADP

E-PCa^{2+} E-Pi + 2Ca^{2+} inside

The two equations show that the splitting of one molecule of ATP results in the translocation of two Ca^{2+} by the ATPase (*E*). This reaction is reversible (*i.e.*, Ca^{2+} inside can be used for the formation of ATP). The sarcoplasmic reticulum can be isolated by cell fractionation and vesicular structures may be obtained. The vesicles contain 40 per cent lipid, most of which is in the form of phosphatidylcholine, and two main proteins: the Ca^{2+} *pump protein* (*ATPase*), which, with a molecular weight of 105,000 daltons, represents about 80 per cent of the protein of the membrane, and a Ca^{2+} *-binding protein* (*i.e.*, *calsequestrin*), with a molecular weight of 65,000 daltons. The Ca^{2+} ATPase is thought to be a highly asymmetric molecule spanning the membrane, which needs the presence of a certain amount of phospholipid to function.

Since both the ATPase and the Ca^{2+} uptake are inhibited by agents that block –SH groups, an electron microscopic study was conducted using a cytochemical reagent in which ferritin molecules were attached to an –SH blocking molecule. It was demonstrated that the active groups of the membrane-bound ATPase were localized in the outer surface of the isolated sarcoplasmic reticulum. All these findings suggest that the sarcoplasmic reticulum assumes the important role of relaxing the fibre after contraction and that this is accomplished by the binding of Ca^{2+} at the outer surface and its transport within the sarcoplasmic vesicles. *Calsequestrin*, on the other hand, could be responsible for the binding of Ca^{2+} within the sarcoplasmic reticulum.

In summary, the series of events produced after the arrival of the electrical signal is as follows:

- The signal is received at the level of the Z-line or the A-I junction by way of the T-system.
- The coupling of the T-system with the terminal cisternae produces the release of Ca^{2+} (this may occur in a matter of milliseconds).
- Ca^{2+} induces the contraction, affecting the regulatory proteins troponin and tropomyosin and thus enabling the interaction of actin and myosin. During this step ATP is used as an energy source.
- Ca^{2+} is quickly removed and restored into the sarcoplasmic reticulum by way of the Ca^{2+} pump. Ca^{2+} uptake results in muscle relaxation.

All these data, as well as those related to the sliding mechanism of contraction, can be put together in a molecular theory of muscular contraction. This is one of the best examples, so far studied, of a tight coupling between the processes furnishing energy and the actual machinery involved in contraction. In this case, structure and function are so intimately related in the realm of molecular organization that they are an inseparable unit.

Cytogenetices

Cytogenetics is a branch of genetics that is concerned with the study of the structure and function of the cell, especially the chromosomes. It includes routine analysis of G-Banded chromosomes, other cytogenetic banding techniques, as well as molecular cytogenetics such as fluorescent in situ hybridization (FISH) and comparative genomic hybridization (CGH). During the last decade great advances have been made in the study of human chromosomes in the areas of their identification in the karyotype and the localization of genes. Such studies have had important biological and medical implications in light of the discovery that many congenital disease and syndromes are related to chromosomal aberrations Today, *human cytogenetics* has becomes a specialized science in itself, the wide-ranging interests of which far exceed the limits of text on cell biology. Here, we shall present to the student an elementary account which could be supplemented by reading some of the general references.

Karyotype

Tissue cultures of fibroblasts, bone marrow, skin, and peripheral blood combined with the action of colchicine and hypotonic solutions to block mitosis at metaphases and to separate the chromosomes were used to study the human karyotype. An important technical advance has been the introduction of *phytohemagglutinin, which* induces lymphocytes to transform into lymphoblast-like cells that start to divide 48 to 72 hours after exposure. The strong mitogenic properties of this substance allowed the development of microtechniques which employ small amounts of blood Spreading the cells on a slide causes them to burst and to display all the chromosome which are usually studied in metaphase. Recently, higher resolution of the chromosome structure has been achieved by using banding techniques to study mitosis in prophase and prometaphase.

A *karotype* human metaphase chromosomes is usually obtained from microphotographs. The individual chromo-somes are cut out and there lined up by size with their respective partners. The technique can be improved by determining the so-called *centromeric index,* which is the ratio of the lengths of the long and short areas of the chromosome.

More recently, a system has been introduced that involves a computer-controlled microscope and several accessories that permit:

- Scanning of slides,
- Location of cells in metaphase,
- Counting chromosomes,
- Transmission of digitally expressed images for computation and storage.

All these steps, which can be carried out automatically, may help in making karyotypes more rapidly and in determining chromosomal aberrations. The 23 pairs are disposed in 7 groups (A through G) decreasing is size and having the characteristics described. For example, group A includes pairs 1 to 3, large almost metacentric chromosomes; group B comprises pairs 4 and 5, large submetacentric chromosomes, and so forth.

The X chromosome is in group C (pairs 6 to 12), medium-sized submitochondrial chromosomes; on the other hand, chromosome Y is in group G, together with pairs 21 and 22, small acrocentric chromosome. It is important to remember that pairs 21 and 22 satellites that correspond to nuclear organizers, while the Y chromsome lacks satellites. Furthermore group D (pairs 13 to 15) also contains acrocentric chromosomes with satellites that correspond to nuclear organizers.

2

Genetic Effects of Cross-fertilization

There are several mechanisms promoting cross-pollination and, consequently, cross-fertilization.

The most important ones are:

- Dioecy, *i.e.* male and female gametes are produced by different plants.
  - Asparagus *Asparagus officinalis* L.
  - Spinach *Spinacia oleracea* L.
  - Papaya *Carica papaya* L.
  - Pistachio *Pistacia vera* L.
  - Date palm *Phoenix dactylifera* L.
- Monoecy, *i.e.* male and female gametes are produced by separate flowers occurring on the same plant.
  - Banana *Musa* spp.
  - Oil palm *Elaeis guineensis* Jacq.
  - Fig *Ficus carica* L.
  - Coconut *Cocos nucifera* L.
  - Maize *Zea mays* L.
  - Cucumber *Cucumis sativus* L.

In musk melon (*Cucumis melo* L.) most varieties show andromonoecy, *i.e.* the plants produce both staminate flowers and bisexual flowers, whereas other varieties are monoecious.

- Protandry, *i.e.* the pollen is released before receptiveness of the stigmata.
  - Leek *Allium porrum* L.
  - Onion *Allium cepa* L.
  - Carrot *Daucus carota* L.
  - Sisal *Agave sisalana* Perr.
- Protogyny, *i.e.* the stigmata are receptive before the pollen is released.
  - Tea *Camellia sinensis* (L.) O. Kuntze
  - Avocado *Persea americana* Miller
  - Walnut *Juglans nigra* L.
  - Pearl millet *Pennisetum typhoides* L. C. Rich.
- Self-incompatibility, *i.e.* a physiological barrier preventing normal pollen grains fertilizing eggs produced by the same plant.
  - Cacao *Theobroma cacao* L.
  - Citrus *Citrus* spp.
  - Tea *Camellia sinensis* L. O. Kuntze
  - Robusta coffee *Coffea canephora* Pierre ex Froehner
  - Sugar beets *Beta vulgaris* L.
  - Cabbage, kale *Brassica oleracea* spp.
  - Rye *Secale cereale* L.
  - Many grass species, *e.g.* perennial ryegrass (*Lolium perenne* L.)
- Flower morphology
  - Fig *Ficus carica* L.
  - Primrose *Primula veris* L.
  - Common buckwheat *Fagopyrum esculentum* Moench.

and probably in the Bird of Paradise flower *Strelitzia reginae* Banks Effects with regard to the haplotypic and genotypic composition of a population due to (continued) reproduction by means of panmixis will now be derived for a so-called panmictic population.

Panmictic reproduction occurs if each of the next five conditions apply:

- Random mating
- Absence of random variation of allele frequencies
- Absence of selection
- Absence of mutation
- Absence of immigration of plants or pollen In the remainder of this section the first two features of panmixis are more closely considered.

Random Mating

Random mating is defined as follows: in the case of random mating the fusion of gametes, produced by the population as a whole, is at random with regard to the considered trait. It does not matter whether the mating occurs by means of crosses between pairs of plants combined at random, or by means of open pollination.

Open pollination in a population of a cross-fertilizing (allogamous) crop may imply random mating. This depends on the trait being considered. One should thus be careful when considering the mating system. This is illustrated in Example.

Example: Two types of rye plants can be distinguished with regard to their epidermis: plants with and plants without a waxy layer. It seems justifiable to assume random mating with regard to this trait. With regard to time of flowering, however, the assumption of random mating may be incorrect. Early flowering plants will predominantly mate *inter se* and hardly ever with late flowering plants. Likewise late flowering plants will tend to mate with late flowering plants and hardly ever with early flowering ones. With regard to this trait, so-called assortative mating occurs.

One should, however, realise that the ears of an individual rye plant are produced successively. The assortative mating with regard to flowering date may thus be far from perfect. Also, with regard to traits controlled by loci linked to the locus (or loci) controlling incompatibility, *e.g.* in rye or in meadow fescue (*Festuca pratensis*), perfect random mating will therefore probably not occur. Selection may interfere with the mating system. Plants that are resistant to an agent (*e.g.* disease or chemical) will mate *inter se* (because susceptible plants are eliminated). Then assortative mating occurs due to selection.

Crossing of neighbouring plants implies random mating if the plants reached their positions at random; crossing of contiguous inflorescences belonging to the same plant (geitonogamy) is, of course, a form of selfing.

Random mating does not exclude a fortuitous relationship of mating plants. Such relationships will occur more often with a smaller population size. If a population consists, generation after generation, of a small number of plants, it is inevitable that related plants will mate, even when the population is maintained by random mating.

Indeed, mating of related plants yields an increase in the frequency of homozygous plants, but in this situation the increase in the frequency of homozygous plants is also due to another cause: fixation occurs because of non-negligible random variation of allele frequencies. Both causes of the increase in homozygosity are due to the small population size (and not to the mode of reproduction).

This ambiguous situation, so far considered for a single population, occurs particularly when numerous small subpopulations form together a large superpopulation. In each subpopulation random mating, associated with non-negligible random variation of the allele frequencies, may occur, whereas in the superpopulation as a whole inbreeding occurs. Example provides an illustration.

Example. A large population of a self-fertilizing crop, *e.g.* an F2 or an F3 population, consists of numerous subpopulations each consisting of a single plant. Because the gametes fuse at random with regard to any trait, one may state that random mating occurs within each subpopulation.

At the level of the superpopulation, however, selfing occurs. Selfing is impossible in dioecious crops, *e.g.* spinach (*Spinacia oleracea*). Inbreeding by means of continued sister × brother crossing may then be applied. This full sib mating at the level of the superpopulation may imply random mating within subpopulations consisting of full sib families. Seen from the level of the superpopulation, inbreeding occurs if related plants mate preferentially. This may imply the presence of subpopulations, reproducing by means of random mating.

If very large, the superpopulation will retain all alleles. The increasing homozygosity rests on gene fixation in the subpopulations. If, however, only a single full sib family produces offspring by means of open pollination, implying crossing of related plants, then the population as a whole (in this case just a single full sib family) is still said to be maintained by random mating.

Absence of Random Variation of Allele Frequencies

The second characteristic of panmixis is absence of random variation of allele frequencies from one generation to the next. This requires an infinite effective size of the population, originating from an infinitely large sample of gametes produced by the present generation. Panmixis thus implies a deterministic model. In populations consisting of a limited number of plants, the allele frequencies vary randomly from one generation to the next. Models describing such populations are stochastic models.

Diploid Chromosome Behaviour and Panmixis

One Locus with Two Alleles

The majority of situations considered in this book involve a locus represented by not more than two alleles.

This is certainly the case in diploid species in the following populations:

- Populations tracing back to a cross between two pure lines, say, a single cross
- Populations obtained by (repeated) backcrossing (if, indeed, both the donor and the recipient have a homozygous genotype)

It is possibly the case in populations tracing back to a three-way cross or a double cross. It is improbable in other populations, like populations of cross-fertilizing crops, populations tracing back to a complex cross, landraces, multiline varieties. To keep (polygenic) models simple, it will often be assumed that each of the considered loci is represented by only two alleles. Quite often this simplification will violate reality. The situation of multiple allelic loci is explicitly considered.

If the expression for the trait of interest is controlled by a locus with two alleles A and a (say locus A-a) then the probability distribution of the genotypes occurring in the considered population is often described by Probability

Genotype		
aa	Aa	AA
f_0	f_1	f_2

One may represent the probability distribution (in this book mostly the term genotypic composition will be used) by the row vector (f_0, f_1, f_2).

The symbol f_j represents the probability that a random plant contains j A-alleles in its genotype for locus A-a, where j may be equal to 0, 1 or 2. It has become custom to use the word genotype frequency to indicate the probability of a certain genotype and for that reason the symbol f is used.

The plants of the described population produce gametes which have either haplotype a or haplotype A. (Throughout this book the term haplotype is used to indicate the genotype of a gamete.) The probability distribution of the haplotypes of the gametes produced by the population is described by,

Probability	*Haplotype*	
	a	A
	g_0	g_1

The symbol g_j represents the probability that a random gamete contains j Aalleles in its haplotype for locus A-a, where j may be equal to 0 or 1.

The row vector (g_0, g_1) describes, in a condensed way, the haplotypic composition of the gametes. The habit to use the symbol q instead g_0 and the symbol p instead of g_1 is followed in this book whenever a single locus is considered. The term allele frequency will be used to indicate the probability of the considered allele. So far it has been assumed that the allele frequencies are known and hereafter the theory is further developed without considering the question of how one arrives at such knowledge. In fact allele frequencies are often unknown.

When one would like to estimate them one might do that in the following way. Assume that a random sample of N plants is comprised of the following numbers of plants of the various genotypes: Number of Plants.

Genotype		
aa	*Aa*	*AA*
n_o	n_1	n_2

For any value for N the frequencies q and p of alleles a and A may then be estimated as,

$$q = \frac{2n_0 + n_1}{2N}\, and\, p = \frac{n_1 + 2n_2}{2N}$$

Throughout the book the expressions 'the probability that a random plant has genotype *Aa*', or 'the probability of genotype *Aa*', or 'the frequency of genotype *Aa*' are used as equivalents. This applies likewise for the expressions 'the probability that a gamete has haplotype *A*', or 'the probability of *A*'. Fusion of a random female gamete with a random male gamete yields a genotype specified by *j*, the number of *A* alleles in the genotype. (The number of *a* alleles in the genotype amounts – of course – to 2-*j*.) The probability that a plant with genotype *aa* results from the fusion is in fact equal to the probability of the event that *j* assumes the value 0.

The quantity *j* assumes thus a certain value (0 or 1 or 2) with a certain probability. This means that *j* is a random variable. The probability distribution for *j*, *i.e.* for the genotype frequencies, is given by the binomial probability distribution:

$$P(j = j) = \binom{2}{j} p^j q^{2-j}$$

Fusion of two random gametes therefore yields:

- With probability *q*2 a plant with genotype *aa*
- With probability 2*pq* a plant with genotype *Aa*
- With probability *p*2 a plant with genotype *AA*

The probabilities for the multinomial probability distribution of plants with these genotypes may be represented in a condensed form by the row vector (*q*2, 2*pq*, *p*2). This notation represents also the genotypic composition to be expected for the population obtained after panmixis in a population with gene frequencies (*q*, *p*). In the case of panmixis there is a direct relationship between the gene frequencies in a certain generation and the genotypic composition of the next generation.

Thus if the genotype frequencies f_0, f_1 and f_2 of a certain population are equal to, respectively, q_2, 2*pq* and p_2, the considered population has the so-called Hardy–Weinberg (genotypic) composition. The actual genotypic composition is then equal to the composition expected after panmixis. With continued panmixis, populations of later generations will continue to have the Hardy–Weinberg composition. Therefore such composition may be indicated as the Hardy–Weinberg equilibrium. The names of Hardy and Weinberg are associated with this genotypic composition, but it was in fact derived by Castle in 1903. With two alleles per locus the maximum frequency of plants with the *Aa* genotype in a population originating from panmixis,

$$\frac{1}{2} \text{ for } p = q = \frac{1}{2}.$$

This occurs in F_2 populations of self-fertilizing crops. The F_2 originates from selfing of individual plants of the F_1, but because each plant of the

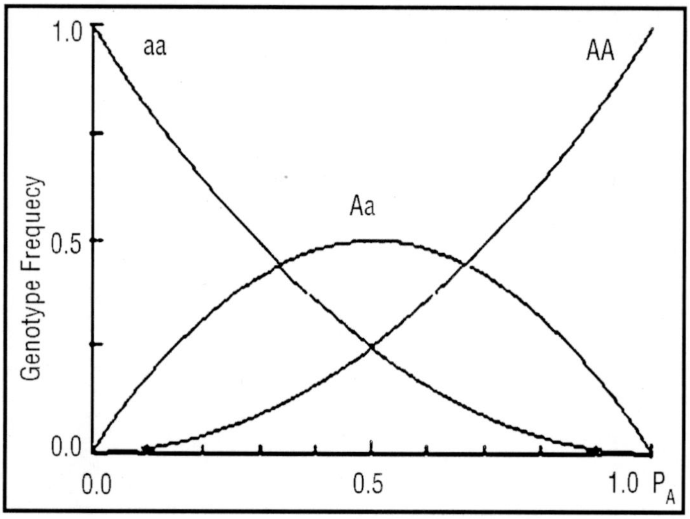

F_1 has the same genotype, panmixis within each plant coincides with panmixis of the F_1 as a whole. (The F1 itself may be due to bulk crossing of two pure lines; the proportion of heterozygous plants amounts then to 1.) The Hardy–Weinberg genotypic composition constitutes the basis for the development of population genetic theory for cross-fertilizing crops.

It is obtained by an infinitely large number of pairwise fusions of random eggs with random pollen, as well as by an infinitely large number of crosses involving pairs of random plants. One may also say that it is expected to occur both after pairwise fusions of random eggs and pollen, and when crossing plants at random. In a number of situations two populations are crossed as bulks. One may call this bulk crossing. One population contributes the female gametes (containing the eggs) and the other population the male gametes (the pollen, containing generative nuclei in the pollen tubes). In such a case, crosses within each of the involved populations do not occur. A possibly unexpected case of bulk crossing is described in Note.

Note: Selection among plants after pollen distribution, *e.g.* selection with regard to the colour of the fruits (if fruit colour is maternally determined), implies a special form of bulk crossing: the rejected plants are then excluded as effective producers of eggs (these plants will not be harvested), whereas all plants (could) have been effective as producers of pollen. The results, to be derived hereafter, in the main text, for a bulk cross of two populations with different allele frequencies.

A bulk cross is of particular interest if the haplotypic composition of the eggs differs from the haplotypic composition of the pollen. Thus if population I, with allele frequencies (q_1, p_1), contributes the eggs and population II, with allele frequencies (q_2, p_2), the pollen, then the expected genotypic composition of the obtained hybrid population, in row vector notation, is

$$(q_1q_2, p_1q_2 + p_2q_1, p_1p_2)$$

This hybrid population does not result from panmixis. The frequency of allele A is,

$$P = \frac{1}{2}(p_1q_2 + p_2q_1) + p_1p_2 = \frac{1}{2}p_1q_2 + \frac{1}{2}p_1p_2 + \frac{1}{2}p_2q_1 + p_1p_2$$

$$= \frac{1}{2}p_1(q_2 + p_2) + \frac{1}{2}p_2(q_1 + p_1) = \frac{1}{2}(p_1 + p_2)$$

$$q_2 + p_2 = q_1 + p_1 = 1$$

Further equations based on $p + q = 1$ are elaborated. When deriving Equation the equation $p + q = 1$ was used. On the basis of the latter equation several other equations, applied throughout this book, can be derived:

$$q^2 + 2pq + p^2 = 1$$

$$p - q = 2p - 1 = 1 - 2q$$

$$(p - q)^2 = p^2 - 2pq + q^2) = 1 - 4pq$$

$$p^2 - q^2 = (p + q)(p - q) = p - q = f_2 - f_0$$

$$p - q + 2pq = p^2 - q^2 + 2pq = p^2 + 2pq - q^2 = 1 - 2q^2$$

Panmictic reproduction of this hybrid population produces offspring with the Hardy–Weinberg genotypic composition. The hybrid population contains, compared to the offspring population, an excess of heterozygous plants. The excess is calculated as the difference in the frequencies of heterozygous plants:

$$(p_1q_2 + p_2q_1) - 2pq = (p_1q_2 + p_2q_1) - 2[\frac{1}{2}(p_1 + p_2)\frac{1}{2}(q_1 + q_2)$$

$$= \frac{1}{2}(p_1q_2 + p_2q_1 - p_1q_1 - p_2q_2)$$

$$= \frac{1}{2}(p_1 - p_2)(q_2 - q_1) = \frac{1}{2}(p_1 - p_2)^2$$

This square is positive, unless $p_1 = p_2$. Thus the hybrid does indeed contain an excess of heterozygous plants. Example illustrates that the superiority of hybrid varieties might (partly) be due to this excess. Example pays attention to the case of both inter- and intra-mating of two populations.

One Locus with more than Two Alleles

Multiple allelism does not occur in the populations considered so far. However, multiple allelism is known to occur in self- and cross-fertilizing crops. It may further be expected in three-way-cross hybrids, and their offspring, as well as in mixtures of pure lines (landraces or multiline varieties).

Example: The intensity of the anthocyanin colouration in lettuce (*Lactuca sativa*), a self-fertilizing crop, is controlled by at least three alleles. The colour and location of the white leaf spots of white clover (*Trifolium repens*), a cross-fertilizing crop, are controlled by a multiple allelic locus. The expression for these traits appears to be controlled by a locus with at least 11 alleles. Another locus, with at least four alleles, controls the red leaf spots. (White clover is an autotetraploid crop with a gametophytic incompatibility system and a diploid chromosome behaviour; $2n = 4x = 32$).

The frequencies (*f*) of the genotypes $AiAj$ (with $i \leq j$; $j = 1..., n$) for the multiple allelic locus $A1$-$A2$-... -An attain their equilibrium values following a single round of panmictic reproduction. The genotypic composition is then: *f*

Genotype		
A_1A_1...	$AjAj$...	A_nA_n
$p1^2$	$2pipj$	pn^2

The Proportion of homozygous plants is minimal for,

$$pj = \frac{1}{n}(for\ j = 1,...,n)$$

and amounts then to $n\left(\frac{1}{n}\right)^2 = \frac{1}{n}$;

Two Loci, Each with Two Alleles

Iit is shown that a single round of panmictic reproduction produces immediately the Hardy–Weinberg genotypic composition with regard to a single locus. It is immediately attained because the random fusion of pairs of gametes implies random fusion of separate alleles, whose frequencies are constant from one generation to the next. For complex genotypes, *i.e.* genotypes with regard to two or more loci (linked or not), however, the so-called linkage equilibrium is only attained after *continued panmixis*. Presence of the Hardy–Weinberg genotypic composition for separate loci does not imply presence of linkage equilibrium! (Example illustrates an important exception to this rule.)

In panmictic reproduction the frequencies of complex genotypes follow from the frequencies of the complex haplotypes. Linkage equilibrium is thus attained if the haplotype frequencies are constant from one generation to the next. For this reason 'linkage equilibrium' is also indicated as gametic phase equilibrium. In this section it is derived how the haplotypic frequencies approach their equilibrium values in the case of continued panmixis. This implies that the tighter the linkage the more generations are required. However, even for unlinked loci a number of rounds of panmictic reproduction are required to attain linkage equilibrium. The genotypic composition in the equilibrium does not depend at all on the strength of the linkage of the loci involved. The designation 'linkage equilibrium' is thus not very appropriate.

To derive how the haplotype frequencies approach their equilibrium. We consider loci *A-a* and *B-b*, with frequencies p and q for alleles A and a and frequencies r and s for alleles B and b. The recombination value is represented by rc. This parameter represents the probability that a gamete has a recombinant haplotype. Independent segregation of the two loci occurs at $rc = 1/2$, absolute linkage at $rc = 0$. Example illustrates the estimation of rc in the case of a testcross with a line with a homozygous recessive (complex) genotype.

The haplotype frequencies are determined at the meiosis. The haplotypic composition of the gametes produced by generation $Gt\text{–}1$ is described by

Haplotype			
ab	*aB*	*Ab*	*AB*
$g_{00,t}$	$g_{01,t}$	$g_{10,t}$	$g_{11,t}$

The last subscript (t) in the symbol for the haplotype frequencies indicates the rank of the generation to be formed in a series of generations generated by panmictic reproduction (t = 1, 2...).

Example. The spinach variety Wintra is susceptible to the fungus *Peronospora spinaciae* race 2 and tolerant to Cucumber virus 1. It was crossed with spinach variety Nores, which is resistant to *P. spinaciae* race 2 but sensitive to Cucumber virus 1.

The loci controlling the host-pathogen relations are *A-a* and *B-b*. The genotype of Wintra is *aaBB* and the genotype of Nores *AAbb*. The offspring, with genotype *AaBb*, were crossed with the spinach variety Eerste Oogst (genotype *aabb*), which is susceptible to *P. spinaciae* race 2 and sensitive to Cucumber virus 1. On the basis of the reaction to both pathogens a genotype was assigned to each of the 499 plants resulting from this testcros:

Genotype:

aabb	*aaBb*	*Aabb*	*AaBb*	*Total*
61	190	194	54	499
124.75	124.75	124.75	124.75	499

The expected frequencies are calculated on the basis of the null hypothesis stating that the two involved loci are unlinked. The expected 1 2: 1 2 segregation ratio was confirmed by a goodness of fit test for each separate locus. The specified null hypothesis is, of course, rejected. The two loci are clearly linked. The value estimated for *rc* is,

$$\frac{61+54}{499}=0.23$$

Note: In this book the last subscript in the symbols for the genotype and haplotype frequencies indicate the generation number. If it is *t* it refers to population G*t*, *i.e.* the population obtained by panmictic reproduction of *t* successive generations.

Population G1, resulting from panmictic reproduction in a single-cross hybrid, has the same genotypic composition as the F2 population resulting from selfing plants of the single-cross hybrid. To standardize the numbering of generations of cross-fertilizing crops and those of self-fertilizing crops, the population resulting from the first reproduction by means of selfing might be indicated by S1 (rather than by the more common indication F2). To avoid confusion this will only be done when appropriate. The last subscript in the symbols for the haplotype frequencies of the gametes giving rise to S1 are taken to be 1. The same applies to the frequencies of the genotypes in S1. This system for labelling generations of gametophytes and sporophytes was also adopted by Stam.

Population G0 is thus some initial population, obtained after a bulk cross or simply by mixing. It produces gametes with the haplotypic composition (*g*00,1; *g*01,1; *g*10,1; *g*11,1). In the absence of selection, allele frequencies do not change. This implies,

$$g_{10,1}+g_{11,1}=g_{10,2}+g_{11,2}=...=p$$

for allele *A*, and similar equations for the frequencies of alleles *a*,*B* and *b*. It was already noted that the haplotype frequencies in successive generations will be considered. In the appendix of this section it is shown that the following recurrent relations apply:

$$g_{100,t}+1=g_{100,t}-r_cd_t$$

$$g_{01,t}+1=g_{01,t}+r_cd_t$$

$$g_{10,t}+1=g_{10,t}+r_cd_t$$

$$g_{11,t}+1=g_{11,t}-r_cd_t$$

Where the definition of d_t *follows from*:

$$2d_t := f_{11c,t}-f_{11R,t}$$

where ':=' means: 'is defined as', and t = 1, 2, 3...

In Note: it is shown that Equations also apply to selffertilizing crops. The recurrent equations show that the haplotype frequencies do not change from one generation to the next if *rc* = 0 or if *dt* = 0.

Such constancy of the haplotypic composition implies constancy of the genotypic composition. It implies presence of linkage equilibrium. Linkage equilibrium is thus immediately established by a single round of panmictic reproduction for loci with *rc* = 0. This situation coincides with the case of a single locus with four alleles.

The symbol *f*11C indicates the frequency of *AB*/*ab*-plants, *i.e.* doubly heterozygous plants in coupling phase (C-phase); the symbol *f*11R represents the frequency of *Ab*/*aB*-plants, *i.e.* doubly heterozygous plants in repulsion phase (R-phase).

In the case of panmixis the following equations apply:

$$f11c,t = 2(g11,tg00,t)$$

$$f11r,l = 2(g10,lg01,t)$$

In that case we get,

$$d_t = (g11,tg00,t) - (g10,tg01,t)$$

This parameter is called coefficient of link age disequilibrium. It Appears in the following derivation:

$$g_{11},t \ = \ g_{11},t(g_{10},t \ + \ g_{01},t \ + \ g_{11}, \ t \ + \ g_{00},t)$$

$$= (g_{10},tg_{01},t + g_{10},tg_{11},t + g_{11},tg_{01},t + g_{11},t)$$

$$+(g_{11},tg_{00},t - g_{10},tg_{01},t)$$

$$= (g_{10},t + g_{11},t)(g_{01},t + g_{11},t) + d_t = p_r + d_t$$

Equation may thus be rewritten as

$$pr + d_{t+1} = (pr + d_t) - r_c d_t$$

Which implies not only

$$d_{t+1} = (1 - r_c)d_t$$

equation may thus be rewritten as

but of course also

$$d_t = (1 - r_c)^{t-1} d_1$$

for t = 2,3,....

The derivation above (and similar derivations for the other haplotype frequencies)implies

$$d_t = g11,t \text{ - } pr = \text{-}(g10,t \text{ - } ps) = \text{-}(g01,t \text{ - } qr) = g00,t \text{ - } qs$$

Because 1/2 ≤ (1–*rc*) ≤ 1, continued panmixis implies continued decrease of *dt*. The decrease is faster for smaller values of 1-*rc*, *i.e.* for higher values of *rc*. Independent segregation, *i.e.* *rc* = 1/2, yields the fastest reduction, *viz.* halving of *dt* by each panmictic reproduction. The value of *dt* eventually attained, *i.e.* $dt = 0$, implies that linkage equilibrium is attained, *i.e.* constancy of the haplotype frequencies. The haplotype frequencies have then a special value, *viz.*

$$g00 = qs$$
$$g01 = qr$$
$$g10 = ps$$
$$g11 = pr$$

The equilibrium frequencies of the haplotypes are equal to the products of the frequencies of the alleles involved, and the equilibrium frequencies of the complex genotypes are equal to the products of the Hardy–Weinberg frequencies of the single-locus genotypes for the loci involved. The strength of the linkage between the loci is irrelevant with regard to the genotypic composition in the equilibrium. It only affects the number of generations of panmictic reproduction required to 'attain' the equilibrium.

(*a*) *Genotypes*

	bb	*Bb*	*BB*	
aa	q^2s^2	$2q^2rs$	q^2r^2	q^2
Aa	$2pqs^2$	$4pqrs$	$2pqr^2$	$2pq$
AA	p^2s^2	$2p^2rs$	p^2r^2	p^2
	s^2	$2rs$	r^2	1

(*b*) *Phenotypes*

	bb	*B*	
aa	q^2s^2	$q^2(1-s^2)$	q^2
A	$(1-q^2)s^2$	$(1-q^2)(1-s^2)$	$(1-q^2)$
	s^2	$1-s^2$	

The foregoing is illustrated in Example, which deals with the production of a single-cross hybrid variety and the population resulting from its offspring as obtained by panmictic reproduction. Example illustrates the production of a synthetic variety and a few of its offspring generations as obtained by continued random mating.

On the basis of the frequencies of the phenotypes for two traits (each with two levels of expression) showing qualitative variation, one can easily determine whether or not a certain population is in linkage equilibrium. It is, however, impossible to conclude whether or not the loci involved are linked.

Only test crosses between individual plants with the phenotype $A \cdot B \cdot$ and plants with genotype *aabb* will give evidence about this. *N.B.* By 'phenotype $A \cdot B \cdot$' is meant the phenotype due to genotype *AABB*, *AaBB*, *AABb* or *AaBb*

Genotypic Composition	G_1 for G_0 in C - phase	G_1 for G_0 in R - Phase	$G\infty$
Genotype	$\frac{1}{4}(1-r_c)^2$	$\frac{1}{4}r_c^2$	$\frac{1}{16}$
aabb	$\frac{1}{2}r_c(1-r_c)$	$\frac{1}{2}r_c(1-r_c)$	$\frac{2}{16}$
aaBb	$\frac{1}{4}r_c^2$	$\frac{1}{4}(1-r_c)^2$	$\frac{1}{16}$
aaBB	$\frac{1}{2}r_c(1-r_c)$	$\frac{1}{2}r_c(1-r_c)$	$\frac{2}{16}$
Aabb	$\frac{1}{4}r_c^2$	$\frac{1}{2}r_c^2$	$\frac{2}{16}$
AB/ab	$\frac{1}{2}r_c(1-r_c)$	$\frac{1}{2}(1-r_c)^2$	$\frac{2}{16}$
Ab/aB	$\frac{1}{2}(1-r_c)^2$	$\frac{1}{2}rc(1-r_c)^2$	$\frac{2}{16}$
AaBB	$\frac{1}{4}r_c^2$	$\frac{1}{4}(1-r_c)^2$	$\frac{1}{16}$
AAbb	$\frac{1}{2}r_c(1-r_c)$	$\frac{1}{2}rc(1-r_c)$	$\frac{2}{16}$
AABb	$\frac{1}{4}(1-r_c)^2$	$\frac{1}{4}r_c^2$	$\frac{1}{16}$
AABB			

Example: A synthetic variety is planned to be produced by intermating five clones of a self-incompatible grass species. Because crosses within each of the five components are excluded, the synthetic variety is produced by outbreeding. It is, therefore, due to a complex bulk cross. The obtained plant material is designated as Syn1 (or G0 in the present context). The five clones have the following genotypes for the two unlinked loci *B*1-*b*1 and *B*2-*b*2: clone 1: *b*1*b*1*b*2*b*2; clones 2 and 3: *B*1*B*1*b*2*b*2, and clones 4 and 5: *B*1*B*1*B*2*B*2.

The genotypic composition of Syn1 can be derived from the following scheme:

	$b_1b_1b_2b_2$	$B_1B_1b_2b_2$	$B_1B_1b_2b_2$	$B_1B_1B_2B_2$	$B_1B_1B_2B_2$
$b_1b_1b_2b_2$		$B_1b_1b_2b_2$	$B_1b_1b_2b_2$	$B_1b_1B_2b_2$	$B_1b_1B_2b_2$
$B_1B_1b_2b_2$	$B_1b_1b_2b_2$	-	$B_1B_1b_2b_2$	$B_1B_1B_2b_2$	$B_1B_1B_2b_2$
$B_1B_1b_2b_2$	$B_1b_1b_2b_2$	$B_1B_1b_2b_2$		$B_1B_1B_2b_2$	$B_1B_1B_2b_2$
$B_1B_1B_2B_2$	$B_1b_1B_2b_2$	$B_1B_1b_2b_2$	$B_1B_1B_2b_2$		$B_1B_1B_2B_2$
$B_1B_1B_2B_2$	$B_1b_1B_2b_2$	$B_1B_1b_2b_2$	$B_1B_1B_2b_2$	$B_1B_1B_2B_2$	

Table presents the genotype frequencies in a few relevant generations. When deriving these it was assumed that incompatibility can be neglected when considering continued panmictic reproduction starting in G0. The portion of homozygous plants in G0,G1,G2 and G– amounts to 0.2; 0.35; 0.3508 and 0.3536, respectively. The excess of heterozygous plants in comparison to the linkage equilibrium amounts therefore to 0.1536; 0.0036 and 0.0028 in G0,G1 and G2, respectively. (This concerns plants which are heterozygous for one or two loci. For each single locus the Hardy–Weinberg genotypic composition occurs in G1 and all later generations). The genotypic composition of plant material obtained when creating and maintaining an imaginary synthetic variety. P indicates the parental clones, Go indicates population $syn_{1,}$ G_1 indicates $syn_{2,}$ G_2 indicates syn_3 and G– indicates syn–

Frequency

Genotype	P	G_0	G_1	G_2	$G\infty$
$b_1b_1b_2b_2$	0.2		0.0225	0.0182	0.0144
$b_1b_1B_2B_2$			0.0150	0.0176	0.0192
$b_1b_1B_2B_2$			0.0025	0.0042	0.0064
$B_1b_1b_2b_2$		0.2	0.1350	0.1256	0.1152
B_1B_2/b_1b_2		0.2	0.1050	0.0904	0.0768
B_1b_2/b_1B_2			0.0450	0.0605	0.0768
$B_1b_1B_2B_2$			0.0350	0.0436	0.0512
$B_1b_1b_2b_2$	0.4	0.1	0.2025	0.2162	0.2304
$B_1B_1B_2b_2$		0.4	0.3150	0.3116	0.3072
$B_1B_1B_2B_2$	0.4	0.1	0.1225	0.1122	0.1024

The Haplotype Frequencies in Generation t

In this appendix, first is derived an equation relating the frequency of gametes with haplotype ab in generation t + 1 to its frequency in generation t, *i.e.* Equation. Thereafter an equation describing the haplotype frequencies in generations due to continued panmictic reproduction, starting with a singlecross hybrid, is derived.

The Frequency of Gametes with Haplotype ab

The frequency of gametes with haplotype *ab*, produced by generation G*t*, are equal to,

$$g00,t+1 = f00,t+\frac{1}{2}f10,t+\frac{1}{2}01,t+\frac{1}{2}(1-r_c)f11c,t+\frac{1}{2}r_cf11R,t$$
$$= f00,t+\frac{1}{2}f10,t+\frac{1}{2}f01,t+\frac{1}{2}f11c,t-r_cd_t$$

Panmictic reproduction of generation G*t* yields generation G*t* + 1. The genotypic composition of G*t* + 1 is described by the frequencies given by the third column of the previous table. Inclusion of these genotype frequencies in the following equations: The relevant genotypes, their frequencies (in general, as well as after panmixis) and the haplotypic composition of the gametes they produce are:

Above equation for $g_{00,t+1}$ gives,

$$g00,t+1 = g^2 00,t+g00,tg10,t+g00,tg01,t+g00,tg11,t-r_cd_t$$
$$= goo,t(g00,t+g10,t+g01,t+g11,t)-r_cd_t = g00,t-r_cd_t$$

Where, according to Equation,

$$d_t = (g11,tg00,t-g10,tg01,t)$$

Genotype	in general	After Panmixis	ab	aB	Ab	AB
aabb	f00	$g00^2$	1	0	0	0
Aabb	f10	2g00g10	$\frac{1}{2}$	0	$\frac{1}{2}$	0
AAbb	f20	$g10^2$	0	0	1	0
aaBb	f01	2g00g01	$\frac{1}{2}$	$\frac{1}{2}$	0	0
$\frac{AB}{aB}$	f11C	2g00g11	$-\frac{1}{2}r_c$	$\frac{1}{2}r_c$	$\frac{1}{2}r_c$	$\frac{1}{2}$
$\frac{AB}{aB}$	f11R	2g10g01	$\frac{1}{2}r_c$	$\frac{1}{2}$	$\frac{1}{2}$	$-\frac{1}{2}r_c$
AABb	f21	2g01g11	0	$-\frac{1}{2}r_c$	$-\frac{1}{2}r_c$	$\frac{1}{2}r_c$
aaBB	f02	$g01^2$	0	0	$\frac{1}{2}$	$\frac{1}{2}$
AaBB	f12	2g01g11	0	1	0	0
AABB	f22	$g11^2$	0	$\frac{1}{2}$	0 0	$\frac{1}{2}$

Similarly one can derive,

$$g01,t+1=g01,t+r_c d_t$$

$$g10,t+1=g10,t+r_c d_t$$
$$g11,t+1=g11,t-r_c d_t$$

The haplotype frequencies in generations due to continued panmictic reproduction, starting with a single-cross hybrid In the case of panmictic reproduction starting from a single-cross hybrid there will be a symmetry in the haplotype frequencies such that,

$$g00,\ t = g11,t$$

and,

$$g01,\ t = g10,\ t = \frac{1}{2} - g11,t$$

Derivation of $g11,t$ suffices then to obtain the frequencies of all haplotypes with regard to two segregating loci. An equation presenting $g11,t$ immediately for any value for t will now be derived. If the genotype of the single-cross hybrid is *AB/ab, i.e.* coupling phase, the genotypic composition of the initial population G0 is simply described by

- $f11C,0 = 1$, if it is *Ab/aB* the genotypic composition of G0 is described by $f11R,0$ = Equation yields then

$$d_0 = \frac{1}{2}$$

in *the former case, and*

$$d_0 = \frac{-1}{2}$$

in the latter case. The frequency of gametes with the *AB* haplotype among the gametes produced by the single-cross amounts to,

$$g11,1 = \frac{1}{2}(1 - r_c) and\, g11,1 = \frac{1}{2} r_c$$

Re*spectively in Example it was also derived that*

$$d_1 = \frac{1}{4}(1 - 2r_c)$$

for G_0 *in C – Phase and that*

$$d_1 = \frac{-1}{4(1 - 2r_c)}$$

respectively. In Example it was also derived that for G0 in R-phase. The frequencies of *AB* haplotypes in the case of continued panmixis follow from Equation combined with Equation:

$$\begin{aligned} g11,t+2 &= g11,t+1 - r_c d_t + 1 = g11,t+1 - r_c(1-r_c)^t d_1 \\ &= g11,t - rc(1-rc)^{t-1} d_1 - r_c(1-rc)^t d1 \\ &= g11,1 - r_c d_1[(1-r_c)^0 + \ldots + (1-r_c)^t] \end{aligned}$$

The terms within the brackets form a convergent geometric series. The sum of such terms is given by the expression,

$$a\frac{1-q^n}{1-q}$$

Where a is the first term, q is the multiplying factor and n is the number of terms. In the present situation this sum amounts to,

$$\frac{1-(1-r_c)^{t+1}}{r_c}$$

Thus $g11,t+2 = g11,1 - d_1[1-(1-r_c)^{t+1}]$

For $r_c = \frac{1}{2}$ *we got* $d_1 = 0$. *Then*

$$g11,t+2 = g1,1 = \frac{1}{4}$$

This implies that linkage equilibrium is present after one generation with panmictic reproduction!

For G0 in C-phase, Equation can be rewritten as,

$$g11,t+2=\frac{1}{2}(1-r_c)-\frac{1}{4}(1-2r_c)[1-(1-r_c)^{t+1}]$$

Thus

$$g11,2=\frac{1}{2}(1-r_c)-\frac{1}{4}r_c(1-2r_c)=\frac{1}{2r}r_c^2-\frac{3}{4}r_c+\frac{1}{2}$$

For G_0 in R-Phase, Equation can be transformed into,

$$g11,t+2=\frac{1}{2}r_c+\frac{1}{4}(1-2r_c)[1-(1-r_c)^{t+1}]$$

This implies:

$$g11,2=\frac{1}{2}r_c+\frac{1}{4}r_c(1-2r_c)=-\frac{1}{2}r_c^2+\frac{3}{4}r_c$$

$$g11,3=\frac{1}{2}r_c+\frac{1}{4}(1-2r_c)[1-(1-r_c)^2]=\frac{1}{2}r_c^3-1\frac{1}{4}r_c^2+r_c$$

These equations are of relevance with regard to the question of whether it is advantageous, when it is aimed to promote the frequency of plants with a genotype due to recombination, to apply random mating in an F2 population of a self-fertilizing crop.

More than Two Loci, Each with Two or more Alleles

Linkage Involving Three Loci: Three loci *A-a*, *B-b* and *C-c* are considered. These loci occur in this order along a chromosome. The segments *AB,BC* and *AC* are distinguished. Effective recombination of alleles belonging to loci *A-a* and *B-b* requires that the number of crossover events in segment *AB* is an odd number. The probability of recombination is called recombination value, designated by the symbol *rc,* or by the symbol *rAB* or simply by *r* (depending on the context). With an even number of times of crossing-over in segment *AB* there is no (effective) recombination. The probability of this event is 1-*rAB*. There is (effective) recombination of alleles belonging to loci *A-a* and *C-c* if there is either (effective) crossing-over in segment *AB*, but not in segment BC; or if there is (effective) crossing-over in segment *BC*, but not in segment *AB*.

If the occurrence of recombination in one chromosome segment has no effect on the recombination value for the adjacent segment the following relation applies:

$$rAC=rAB(1-rBC(1-rAB)=rAB+rBC-2rABrBC$$

This situation is likely for loci that are not too closely linked. The situation where recombination in one segment depresses the probability of recombination in an adjacent segment is called chiasma interference.

A more general expression for rAC is thus:

$$rAC = rAB + rBC - 2(1-\delta)rABrBC$$

where δ is the interference parameter, ranging from 0 (no interference) through 1 (complete interference). It shows that *rAC* is higher at higher values for δ. Recombination values are additive if

$$2(1-\delta)rABrBC = 0$$

i.e. if δ = 1 and/or *rABrBC* = 0. In other cases they are not additive. These conditions imply that recombination values are mostly not additive. They are, consequently, inappropriate to measure distances between loci. The hypothesis of independence of crossing-over in segments *AB* and *BC*, *i.e.* the hypothesis of absence of chiasma interference, can be tested by means of a goodness-of-fit test. Among *N* plants, the expected number of plants with a genotype which is due to double crossing-over amounts, according to this hypothesis, to *rABrBCN*. It is compared to the observed number.

The ratio,

$$\frac{Observed\ Number}{Expected\ Number}$$

is called coefficient of coincidence. When there is independency it is equal to 1. Its complement, *i.e.*

$$1 - \frac{Observed\ Number}{Expected\ Number}$$

Estimates δ. Its value is positive if the observed number of plants with the recombinant genotype is smaller than the number expected at independency: the presence of a chiasma in the one segment hinders the formation of a chiasma in the other segment. The actual distance between loci, say the map distance *m*, measures the total number of cross-over events (both odd and even numbers) between the loci. This distance is an additive measure. It can only approximately be determined from recombination values. Haldane (1919) developed an approximation for the situation in the absence of interference (δ = 0). His mapping function is

$$m = -\frac{In(1-2r_c)}{2},$$

Where *m* represents the expected number of cross-over events in the considered segment. As the map distance is mostly expressed in centiMorgans (cM), this function is,

$$m = -50\ln(1-2r_c)$$

Often written as An approximation which takes interference into account is called Kosambi's mapping function.

Frequencies of Complex Genotypes after Continued Panmixis

It can be shown that continued panmixis eventually leads to an equilibrium of the frequencies of complex genotypes for three or more loci, each with two or more alleles. The equilibrium is characterized by haplotype frequencies equal to the products of the frequencies of the alleles involved. Linkage equilibrium for one or more pairs of loci does not imply equilibrium of the frequencies of complex genotypes for three or more loci. Equilibrium of the frequencies for complex genotypes implies, however, linkage equilibrium for all pairs of loci.

Autotetraploid Chromosome Behaviour and Panmixis

The implications of panmixis in an autotetraploid crop will only be considered for a single locus with two alleles. This is to keep the mathematical derivations simple. It will be shown that the equilibrium frequencies of the genotypes are not obtained after a single panmictic reproduction. At equilibrium the frequencies of the genotypes and the haplotypes are equal to the products of the frequencies of the alleles involved.

Among cross-fertilizing autotetraploid crops the more important representatives are alfalfa (*Medicago sativa* L.; $2n = 4x = 32$) and cocksfoot (*Dactylis glomerata* L.; $2n = 4x = 28$). Additionally, highbush blueberry (*Vaccinium corymbosum* L.; $2n = 4x = 48$) might be mentioned. Leek (*Allium porrum* L.; $2n = 4x = 32$) is an autotetraploid crop with a tendency to a diploid behaviour of the chromosomes. Among ornamentals several autotetraploid species occur, *e.g. Freesia hybrida, Cyclamen persicum* Mill. ($2n = 4x = 48$) and *Begonia semperflorens*. Also, artificial autotetraploid crops have been made, *e.g.* rye (*Secale cereale* L.; $2n = 4x = 28$) and perennial rye grass (*Lolium perenne* L.; $2n = 4x = 28$). In 1977 about 500,000 ha of autotetraploid rye were grown in the former Soviet Union. Sweet potato, *i.e. Ipomoea batatas* var. *littoralis* ($2n = 4x = 60$) or *I. batatas* var. *batatas* ($2n = 6x = 90$), may be considered as a cross-fertilizing crop (due to self-incompatibility), but it is mainly vegetatively propagated. Under certain conditions double reduction may occur in autotetraploid crops, in which case (parts of) sister chromatids end up in the same gamete. The resulting haplotype is homozygous for the loci involved. The process of double reduction causes the frequency of homozygous genotypes and haplotypes to be somewhat higher than in absence of double reduction.

Blakeslee, Belling and Farnham discovered the phenomenon in autotetraploid jimson weed (*Datura stramonium* L.; $2n = 4x = 48$): a triplex plant (with genotype *AAAa*) produced some nulliplex offspring after crossing with a nulliplex (genotype *aaaa*). This is only possible if the triplex plant produces *aa* gametes.

The process of double reduction is an interesting phenomenon, but in a quantitative sense it is of no importance. For this reason we assume that double reduction does not occur. The autotetraploid genotypes to be distinguished for locus *A-a* are *aaaa* (nulliplex), *Aaaa* (simplex), *AAaa* (duplex), *AAAa* (triplex) and *AAAA* (quadruplex).

In each cell these genotypes contain *JA* alleles and 4-*Ja* alleles. At meiosis two of these four alleles are sampled to produce a gamete.

The haplotypes that can be produced by an autotetraploid plant containing *JA* alleles can be described by *j*, the number of *A* alleles that they contain, where $j = 0$, 1 or 2. The conditional probability distribution for *j*, given that the parental genotype contains *JA* alleles, is a hypergeometric probability distribution:

$$P(\underline{j}=j/J)=\frac{\binom{J}{j}\binom{4-J}{2-j}}{\binom{4}{2}}=\frac{1}{6}\binom{J}{j}\binom{4-J}{2-j}$$

The Probability that a triplex plant (*i.e.*j =3) Produces a gemete with haplotype Aa (*i.e.* j = 1) is therefore,

$$P(\underline{j}=1/J=3)=\frac{1}{6}\binom{3}{1}\binom{1}{1}=\frac{1}{2}$$

Table Presents, for each autotetraploid genotype, the haplotypic composition, *i.e.* the probability distribution for the haplotypes produced. The genotypic composition of a tetraploid population is described like that of a diploid population. Thus in the case of autotetraploid species the row.

Vector (*f*0, *f*1, *f*2, *f*3, *f*4) is used. The equilibrium frequencies of the genotypes are attained as soon as the haplotype frequencies are stable. Therefore the haplotypic composition of successive generations with panmictic reproduction will be monitored.

Genotype	**aa**	**Aa**	**AA**
aaaa	1	0	0
Aaaa	$\frac{1}{2}$	$\frac{1}{2}$	0
AAaa	$\frac{1}{6}$	$\frac{4}{6}$	$\frac{1}{6}$
AAAa	0	$\frac{1}{2}$	$\frac{1}{2}$
AAAA	0	0	1

Some initial population G0 produces gametes with haplotypic composition:

Haplotype:

	aa	*Aa*	*AA*
f	$g_{0,1}$	$g_{1,1}$	$g_{2,1}$

The frequency of a is,

$$q = g0,1 + \frac{1}{2} g1,1$$

And that of A is,

$$p = \frac{1}{2} g1,1 + g2,1$$

Panmictic reproduction of G_0 yields population G_1 with the following genotypic composition:

Genotype:

	aaaa	*Aaaa*	*AAaa*	*AAAa*	*AAAA*
f	$g0,1^2$	$2g0,1g1,1$	$g1,1^2 + 2g0,1g2,1$	$2g1,1g2,1$	$g2,1$

The haplotypic composition of the gametes produced by G_1 is:

Haplotype:

	aa	*Aa*	*AA*
f	$g0,2$	$g1,2$	$g2,2$

According to table the following applies:

$$\begin{aligned} g1,2 &= \frac{1}{2}(2g0,1g1,1) + \frac{2}{3}(g1,1^2 + 2g0,1g2,1) + \frac{1}{2}(2g1,1g2,1) \\ &= \frac{2}{3}(\frac{3}{2}g0,1g1,1 + \frac{3}{2}g1,1g2,1 + g1,1^2 + 2g0,1g2,1) \\ &= \frac{2}{3}[2(g0,1 + \frac{1}{2}g1,1)(\frac{1}{2}g1,1 + g2,1) + \frac{1}{2}g1,1(g0,1 + g1,1 + g2,1)] \\ &= \frac{2}{3}(2pq + \frac{1}{2}g1,1) \end{aligned}$$

Generally

$$g1,t+1 = \frac{2}{3}(2pq + \frac{1}{2}g1,t)$$

The frequencies of the genotypes have attained their equilibrium (*e*) values as soon as the frequencies of the haplotypes are constant.

The latter implies:

$$g1,e = \frac{2}{3}(2pq + \frac{1}{2}g1,e),$$

i.e.

$$g1,e = 2pq$$

The haplotype frequencies are then,

$$g0,e=q-\frac{1}{2}g1,e=q-pq=q^2$$

$$g1,e=2pq$$

$$g2,e=p-\frac{1}{2}g1,e=p-pq=p^2$$

The genotypic composition in equilibrium is consequent:

Genotype:

	aaaa	*Aaaa*	*AAaa*	*AAAa*	*AAAA*
f	q^4	$4pq^3$	$6p^2q^2$	$4p^3q$	p^4

This composition is also given by the probability distribution for J, the number of A alleles in the autotetraploid genotype:

$$P(\underline{j}=J)=\binom{4}{J}p^J q^{4-J}$$

The deviation from the equilibrium is measured by the quantity *dt*, which measures the excess or deficit of the frequency of gametes with the *Aa* haplotype with regard to their equilibrium frequency.

Thus dt is defined as follows:

$$d_t := g1,t-g1,e$$

The rate of decrease of *dt* indicates how fast the equilibrium is approached.

Equations Yield:

$$d_{t+1}=g1,t+1-g1,e=\frac{2}{3}(2pq+\frac{1}{2}g1,t)-2pq=\frac{1}{3}(g1,t-g1,e)=\frac{1}{3}dt$$

One round of panmictic reproduction produces a population in which the deviation amounts only to 1/3 of the deviation in the preceding population. The equilibrium is approached in an asymptotic way. Example gives an illustration.

Example: The approach of the equilibrium is considered for an initial population G0 with genotypic composition (0.04; 0; 0.72; 0; 0.24).

The haplotype frequencies are:

$$g0,1=0.04+0.12=0.16$$

$$g1,1=0.48$$

$$g2,1=0.12+0.24=0.36$$

Thus q = 0.4 and p = 0.6.

This implies that:

$$g0,1=q^2=g0,e$$

$$g1,1=2pq=g1,e$$

$$g2,1=p^2=g2,e$$

For a more advanced treatment of the population genetic theory of crossfertilizing crops with an autotetraploid behaviour of the chromosomes the reader is referred to Seyffert. Finally, it is emphasized once again that in this section it was assumed that the population contains only two different alleles for the segregating locus. In fact more alleles may occur in such a way that plants with three or four different alleles per locus are present, *viz.* plants with genotype *AiAiAjAk* or *AiAjAkAl*, respectively.

Quiros reported such genotypes for isozyme loci in alfalfa. Some claims have been made that plants with a heterozygous genotype containing three or four different alleles for the considered locus, are more vigourous than plants with a heterozygous genotype containing one or two alleles.

Genetic Effects of Inbreeding

Inbreeding occurs if mating plants are, on the average, *more* related than random pairs of plants. A more than average relatedness of the mating plants is thus a prerequisite. Relatedness implies, of course, that the plants involved share one or more ancestors. The strength of the inbreeding depends on the degree of relatedness of the mating plants. It has already been noted that mating of related plants may occur in random mating, but in that case it occurs as a matter of chance.

Note: Several yardsticks for measuring the degree of relatedness exist, a common one being the probability that an allele of a certain locus in some plant is identical by descent to an arbitrary allele at that same locus in its mate. In regular systems of inbreeding the degree of relatedness of the mating plants is uniform across all pairs of mating plants. In this book no attention is given to the determination of the degree of relatedness.

Regular systems of inbreeding are far more common in plant breeding than irregular systems. No attention will, therefore, be given to irregular systems of inbreeding. The counterpart of inbreeding is outbreeding. With outbreeding mating plants are on the average *less* related than random pairs of plants. Selfincompatibility is a natural cause for outbreeding as related plants tend to have a similar genotype at the incompatibility locus/loci. After intercrossing, such plants will produce no (or few) offspring.

Artificial forms of outbreeding are:

- Bulk crossing of two unrelated populations
- Selection of parents to be crossed in such a way that inbreeding is avoided as much as possible Outbreeding occurs also in the case of immigration.

The population genetic effect of inbreeding is a decrease in the frequency of heterozygous plants. This involves all loci, for all traits. (Random mating, on the other hand, is a mode of reproduction that may occur for certain traits and may simultaneously be absent for other traits). When starting with an F2 population and considering segregating loci, the frequency of heterozygous plants is the same for all loci. This applies to the successive generations of the superpopulation. Each subpopulation consists of few plants: in the case of selfing only a single plant, in the case of full sib mating only pairs of plants. Within these separate subpopulations reproduction is by means of random mating.

The random variation of the gene frequencies occurring in small populations causes the subpopulations to vary with regard to the frequencies of heterozygous plants: not only for different loci, but also for the same locus. Individual plants of the F2 (or F3, etc.) populations vary therefore in the number of heterozygous loci. In diploid crops procedures for the production of doubled haploid lines (DH-lines) allow the production of pure lines from heterozygous parents in a single generation.

Doubling of the number of chromosomes of haploid plants, generated by parthenogenesis or by anther culture, yields immediately complete homozygosity. For dioecious crops as well as for self-fertilizing crops with a long juvenile phase, *e.g. Coffea arabica* L., this approach is an attractive alternative to continued inbreeding. Tissue culture techniques for the regeneration of plants from anthers or microspores have been developed, for example in wheat, barley, rice and oilseed rape. Also elimination of paternal chromosomes, occurring when making *Hordeum vulgare* L. × *H. bulbosum* L. or *Triticum aestivum* L. × *Zea mays* L. crosses, permits production of DH-lines. (The paternal chromosomes are lost in a few cell divisions of the hybrid zygote/ embryo.)

Note: DH-lines are mostly obtained directly from the gametes produced by the F1-plants. This has a few drawbacks

- Recombination is restricted to the F1 meiosis
- The proportion of DH-lines that are rejected because of poor performance is high. This is undesirable because of the cost of producing DH-lines.

To avoid these drawbacks one may use gametes from plants obtained by backcrossing the F1 or one may use F2- or even F3-plants. (The latter allows selection among F2-plants, followed by selection among F3-lines in the seedling stage). *In vitro* selection among the haploid embryos appeared to be feasible: the size and degree of embryo differentiation predicted which embryos would produce vigourous seedlings. Additionally the growth rate of the embryos was positively correlated with yield performance in the field $r = 0.3$, but this has found little practical application). Continued self-fertilization is the natural mode of reproduction of selffertilizing crops. There are many economically important self-fertilizing crops.

A number of these are:

- Barley *Hordeum vulgare* L.
- Oats *Avena sativa* L.
- Wheat *Triticum aestivum* L.
- Rice *Oryza sativa* L.
- Sorghum *Sorghum bicolour* (L.) Moench.
- Finger millet *Eleusine coracana* (L.) Gaertn.
- Pea *Pisum sativum* L.
- Cowpea *Vigna unguiculata* (L.) Walp.
- Dry bean *Phaseolus vulgaris* L.
- Soybean *Glycine max* (L.) Merr.
- Peanut *Arachis hypogaea* L.
- Cotton *Gossypium* spp.
- Arabica coffee *Coffea arabica* L.
- Lettuce *Lactuca sativa* L.
- Tomato *Lycopersicon esculentum* Mill.
- Okra *Abelmoschus esculentus* (L.) Moench.
- Sweet pepper *Capsicum annuum* L.

Self-fertilization is not always 100% in most of these autogamous crops, *e.g.* cotton, okra, sorghum. (The amount of outcrossing in sorghum is about 6%.) consider the genotypic composition of populations reproducing by a mixture of self-fertilization and cross-fertilization. Breeders regularly apply inbreeding in cross-fertilizing crops.

They may have various reasons for doing this:

- The development of pure lines (mostly by continued selfing) for use as parents in the breeding of hybrid varieties, *e.g.* in maize or cucumber.
- To promote the efficiency of elimination of an undesired recessive gene
- Maintenance of a genic male sterile 'line'

Note:FS-mating occurs also when a maintaining a genic male sterile barley 'line': male sterile plants are harvested after having been pollinated by their male fertile full sibs. (This is also applied in the case of recurrent selection in self-fertilizing cereals).

Thus the harvesting of a female plant (say genotype *mm*) implies harvest of seed due to the cross *mm* ×*Mm* (where *Mm* represents the genotype assumed for hermaphroditic plants). The genotypic composition of the obtained FS-family is (1/2, 1/2, 0). Repeated application of this procedure implies repeated FSmating.

The most powerful form of inbreeding of cross-fertilizing crops, *e.g.* dioecious crops, occurs with repeated crossing of the type:

- Full sib × full sib, *i.e.* full sib mating, or
- Parent ×× offspring.

Full Sib Mating

The offspring due to a cross of two genotypes constitutes a family. The plants belonging to the family share both their maternal and their paternal parent. With regard to each other these plants are full sibs. Together they form a full sib family (FS-family). Crossing of plants belonging to the same FS-family is called full sib mating (FS-mating). FS-mating may be used when inbreeding of dioecious crops, such as spinach or asparagus, is the aim. It occurs spontaneously in the case of open pollination within FS-families grown in isolation. This is applied in hermaphroditic, monoecious or dioecious crops in the case of separated FS-family selection. Note describes how FS-mating is applied when maintaining a genic male sterile 'line'.

Parent ×× offspring mating

In this book the notation A ×× B indicates the cross A×B and/or the reciprocal cross B × A. Parent ×× offspring crosses, *i.e.* so-called PO-mating, can only be applied to perennial crops such as oil palm (producing gametes from the age of 4–5 years for many years) or asparagus (with a juvenile phase lasting two years). The parent is still alive when its offspring reach the reproductive phase. Repeated backcrossing implies continued application of crosses of the type 'recurrent parent ×× offspring'. In the absence of selection the genotype of the offspring becomes identical to the genotype of the recurrent parent (if the recurrent parent has a homozygous genotype) or to the genotypic composition of the possible lines obtained by selfing of the recurrent parent (if the recurrent parent is heterozygous). In this chapter only loci segregating for not more than two alleles per locus will be considered. For an extensive treatment of the population genetics theory of inbreeding the reader is referred to Allard, Jain and Workman.

Diploid Chromosome Behaviour and Inbreeding

One Locus with Two Alleles

With continued inbreeding of any (infinitely) large population the genotype frequencies will change from one generation to the other until the frequency of plants with a heterozygous genotype has become zero. Starting from the initial population G0 with genotypic composition ($f0,0$, $f1,0$, $f2,0$), eventually a population with genotypic composition (q, 0, p) will be obtained.

- Starting with some arbitrary genotypic composition

Genotype:

Generation	aa	Aa	AA
S0	f0	f1	f2
S1	$f0+\frac{1}{4}f1$	$\frac{1}{2}f1$	$f2+\frac{1}{4}f1$
S2	$f0+\left(\frac{1}{4}+\frac{1}{8}\right)f1$	$\frac{1}{4}f1$	$f2+\left(\frac{1}{4}+\frac{1}{8}\right)f1$
S3	$f0+\left(\frac{1}{4}+\frac{1}{8}+\frac{1}{16}\right)f1$	$\frac{1}{8}f1$	$f2+\left(\frac{1}{4}+\frac{1}{8}+\frac{1}{16}\right)f1$
.			
.			
S∞	q	0	P

- Starting with F1, *i.e.* a population with genotypic composition (0, 1, 0)

Generation (t)	Population	Inbreeding Coefficient	Panmictic (℘index(P)	Genotype aa	Aa	AA
0	$S_0(=F_1)$	−1	2	0	1	0
1	$S_1(=F_2)$	0	1	$\frac{1}{4}$	$\frac{1}{2}$	$\frac{1}{4}$
2	$S_2(=F_3)$	$\frac{1}{2}$	$\frac{1}{2}$	$\frac{3}{8}$	$\frac{2}{8}$	$\frac{3}{8}$
3	$S_3(=F_4)$	$\frac{3}{4}$	$\frac{1}{4}$	$\frac{7}{16}$	$\frac{2}{16}$	$\frac{7}{16}$
4	$S_4(=F_5)$	$\frac{7}{8}$	$\frac{1}{8}$	$\frac{15}{32}$	$\frac{2}{32}$	$\frac{15}{32}$
5	$S_5(=F_6)$	$\frac{15}{16}$	$\frac{1}{16}$	$\frac{31}{64}$	$\frac{2}{64}$	$\frac{31}{64}$
6	$S_6(=F_7)$	$\frac{31}{32}$	$\frac{1}{32}$	$\frac{63}{128}$	$\frac{2}{128}$	$\frac{63}{128}$
7	$S_7(=F_8)$	$\frac{63}{64}$	$\frac{1}{64}$	$\frac{127}{256}$	$\frac{2}{256}$	$\frac{127}{256}$
∞	$S_\infty(=F_\infty)$	1	0	$\frac{1}{2}$	0	$\frac{1}{2}$

Illustrates this for inbreeding by means of continued selfing. It appears that the genotype frequencies approach, in an asymptotic manner, the gene and haplotype frequencies. Often the frequency of heterozygous plants in generation *t*, *i.e.* *f*1,*t*, is written in the form,

$$2pq(1-f_t)$$

In this expression the factor 1–*Ft* describes the deviation from the Hardy–Weinberg frequency. The factor is called the panmictic index, sometimes designated by the symbol *P*. This implies that *P* = 1-*Ft*. The parameter *Ft*, say 'script F', is the inbreeding coefficient (or fixation index) pertaining to generation t. When starting with an F1 population, F2 is the first generation due to self-fertilization. For this reason the F2 population is chosen to be generation 1. (Its genotypic composition is equal to the genotypic composition of the population obtained by panmictic reproduction of the F1) Successive generations may be indicated by G1,G2..., but in the case of continued selfing the designations S1, S2, S3... are used as well. A general description of the genotypic composition of any population (inbred or not) is now given by,

	Genotype		
	aa	*Aa*	*AA*
f	$q^2 + pqf_t$	$2pq(1-f_t)$	$P^2 + pqf_t$

In several other books, *e.g.* Falconer and MacKay, the inbreeding coefficient is defined as the probability that the two alleles at any loci of a plant are identical by descent. This would mean that the inbreeding coefficient of an F2 population obtained from cross $AA \times aa$ is equal to 1 2, because 50% of the plants contain, for locus *A-a*, alleles that are identical by descent (this concerns plants with genotype *aa* or *AA*). In this book the parameter F is used to quantify the deviations from the Hardy–Weinberg frequencies. In an F2 population such deviations are absent and accordingly its inbreeding coefficient is 0. In Note, it is shown that our definition of the inbreeding coefficient F can be interpreted as the coefficient of correlation of numerical values, *e.g.* gene-effects, assigned to the haplotypes of the uniting gametes.

This is based on the following consideration. With random mating the gene effects of the haplotypes of fusing female and male gametes are independent; in the absence of random mating they are interdependent. With inbreeding they tend to be similar; with outbreeding they tend to be different. Breeding of self-fertilizing crops starts mostly with crossing of homozygous lines. For all loci for which the parental lines have a different homozygous genotype the genotype of the F1 is heterozygous. For these loci $p = q = 1\ 2$ and then the expressions in (3.1) simplify to

	Genotype		
	aa	*Aa*	*AA*
f	$\frac{1}{4}(1+f_t)$	$\frac{1}{2}(1-f_t)$	$\frac{1}{4}(1+f_t)$

As $f1,0 = 1\ 2\ (1-F0) = 1$, it follows that $F0 = -1$, *i.e.* a negative value for the inbreeding coefficient. The panmictic index of the F1 amounts for heterozygous loci to $P0 = 2$. In the remainder of this section the decrease in the frequency of heterozygous plants is considered for the three most important regular inbreeding systems, *viz.* self-fertilization, full sib mating and parent × offspring mating. To measure this decrease the parameter λ is defined:

$$\lambda = \frac{2pq(1-f_t)}{2pq(1-f_{t-1})} = \frac{1-f_t}{1-f_{t-1}}$$

This parameter indicates the frequency of heterozygous plants as a proportion of this frequency in the preceding generation. At a smaller value for λ the decrease of *f*1 is stronger. In the case of selfing the values for λ do not depend on *t*; they are approximately constant when applying full sib mating or parent × offspring. Then λ1 = λ2 = · · · = λ*t*. This implies,

$$f1,t = \lambda f_{1,t-1} = \lambda^2 f1,t-2 = \lambda^t f1,0$$

Self-Fertilization

In the F2 generation, the first generation generated by selfing, the genotype frequencies coincide with the Hardy-Weinberg frequencies. Thus *f*1,1 = 2*pq*, implying that *F*1, the inbreeding coefficient of F2, is zero. In population F– , approximately obtained after a very large number of generations reproducing by means of selfing, there is complete homozygosity, *i.e.* *f*1,– = 0, implying that *F*–, the inbreeding coefficient of F–, is 1. The decrease of *f*1, due to continued selfing, is indicated in Table. The table shows that *f*1 is halved by each round of reproduction by means of selfing.

Thus,

$$1-f_t = \frac{1}{2}(1-f_{t-1})$$

implying

$$f_t = \frac{1}{2}(1+f_{t-1})$$

With Regard to Continued selfing the expression:

$$1-f_t = \frac{1}{2}(1-f_{t-1})$$

or

$$P_t = \frac{1}{2}P_{t\ 1}$$

implies

$$P_t = \left(\frac{1}{2}\right)^t P_0 = \left(\frac{1}{2}\right)^{t-1}$$

$$f_t = 1-\left(\frac{1}{2}\right)^{t-1}$$

At all other systems of inbreeding the reduction of *f*1 is smaller. The minimum value for l is thus attained with selfing. It amounts to l*S* = 1/2.

Full Sib Mating and Parent Offspring Mating

Li showed that for both full sib mating and parent × offspring mating, the relation:

$$f1,t+2 = \frac{1}{2} f1,t+1 + \frac{1}{4} f1,t$$

Applies. Consider an initial population with genotypic composition (0,1,0), thus $f1,0$ = 1. In this population plants are crossed in pairwise combinations. In the next generation the genotypic composition of the population obtained, which consists of full sib families, is expected to be (1/2, 1/4, 1/2), with $f1,1$ = 1 2.

Continued full sib mating, within the continuously generated FS-families, gives, according to Equation,

$$f1,2 = \frac{1}{2}\left(\frac{1}{2}\right) + \frac{1}{4}(1) = \frac{1}{2}, i.e. \lambda_2 = 1$$

$$f1,3 = \frac{1}{2}\left(\frac{1}{2}\right) + \frac{1}{4}\left(\frac{1}{2}\right) = \frac{3}{8}, i.e. \lambda_3 = \frac{3}{4} = 0.75$$

$$f1,4 = \frac{1}{2}\left(\frac{3}{8}\right) + \frac{1}{4}\left(\frac{1}{2}\right) = \frac{5}{16}, i.e. \lambda_4 = \frac{5}{6} = 0.8333, etc$$

The first round of inbreeding (full sib mating or parent × offspring mating) does not give a decrease of the frequency of heterozygous plants (λ2 = 1). Indeed, with full sib mating first FS-families have to be generated.

It appears that λ approaches asymptotically the value $\lambda FS = \lambda PO = 0.809$. As (0.809)3 = 0.53 ≈ 1/2, three generations of reproduction by means of FSmating or parent × offspring mating give the same reduction in $f1$ as a single round of reproduction by selfing.

A pair of linked loci

It was shown that linkage may be expected to play a relatively unimportant role in the inheritance of quantitative traits. It was said that, throughout this book, absence of linkage would be assumed. It is, nevertheless, useful to be familiar with some implications of linkage.

An important reason for this is the study of the linkage of loci affecting a quantitative trait with molecular markers. Consider haplotypes *ab*, *aB*, *Ab* or *AB* for the two loci *A-a* and *B-b* with recombination value *rc*. Continued selfing, starting with an F1 with the heterozygous genotype *AaBb*, yields in the absence of selection

$$g11,t = g00,t$$

$$g01,t = g10,t$$

'symmetric' haplotype frequencies:

$$g11,t + g10,t = PA = \frac{1}{2}$$

$$g10,t = \frac{1}{2} - g11,t$$

This implies that, when one knows $g11,t$, one also knows $g10,t$, $g01,t$ and $g00,t$. It suffices thus to consider only the frequency of gametes with the *AB* haplotype. This is particularly of interest when considering F–. This population is described by

	Genotype			
	aabb	*AAbb*	*aaBB*	*AABB*
f	$f00,\infty$	$f20,\infty$	$f02,\infty$	$f22,\infty$

Only plants with the *AABB* genotype are capable of producing gametes with the *AB* haplotype. Thus,

$$g11,– = f22,–.$$

The haplotypic composition of the gametes produced by this population is

	Haplotype			
	ab	*Ab*	*aB*	*AB*
g	$g00,\infty(= g11,\infty)$	$g10,\infty(= \frac{1}{2} - g11,\infty)$	$g01,\infty(\frac{1}{2} - g11,\infty)$	$g11,\infty$

There are thus good reasons to consider the frequency of gametes with the *AB* haplotype. In Note the following relation between the frequencies of *AB*-haplotypes in two successive generations is derived:

Note: The frequency of *AB* haplotypes, *i.e.* $g11$, is considered for the case of continued autogamous reproduction. (To promote readability the recombination value is – in this section – mostly just indicated by the symbol r). The genotypes capable of producing *AB* haplotypes, their frequencies in generation t and the haplotypic composition of the gametes they produce are:

Haplotype:

Genotype	**f**	**ab**	**aB**	**Ab**	**AB**
AABB	f22,t	0	0	0	1
AABb	f21,t	0	0	$\frac{1}{2}$	$\frac{1}{2}$
AaBB	f12,t	0	$\frac{1}{2}$	0	$\frac{1}{2}$
AB/ab	f11c,t	$\frac{1}{2}(1-r)$	$\frac{1}{2}r$	$\frac{1}{2}r$	$\frac{1}{2}(1-r)$
Ab/aB	f11R,t	$\frac{1}{2}r$	$\frac{1}{2}(1-r)$	$\frac{1}{2}(1-r)$	$\frac{1}{2}r$

Then

$$g_{11,t+1} = f_{22,t} + \frac{1}{2}f_{21,t} + \frac{1}{2}f_{12,t} + \frac{1}{2}(1-r)f_{11C,t} + \frac{1}{2}rf_{11R,t}$$

$$= f_{22,t} + \frac{1}{2}f_{21,t} + \frac{1}{2}f_{12,t} - \frac{1}{2}r(f_{11C,t} - f_{11R,t})$$

$$= f_{22,t} + \frac{1}{2}f_{21,t} + \frac{1}{2}f_{12,t} + \frac{1}{2}f_{11C,t} - rd_t$$

Where, according to equation d_t is defined as,

$$d_t = \frac{1}{2}(f11c,t - f11R,t)$$

and

$$f22,t = f22,t-1+\frac{1}{4}f21,t-1+\frac{1}{4}f12,t-1+\frac{1}{4}(1-r)^2 f11c,t-1+\frac{1}{4}r^2 f11R,t-1$$

$$f21,t = \frac{1}{2}f21,t-1+\frac{1}{2}r(1-r)f11c,t-1+\frac{1}{2}r(1-r)f11R,t-1$$

$$f12,t = \frac{1}{2}f12,t-1+\frac{1}{2}r(1-r)f11C,t-1+\frac{1}{2}r(1-r)f11R,t-1$$

$$f11C,t = \frac{1}{2}(1-r)^2 f11C,t-1+\frac{1}{2}r^2 f11R,t-1$$

$$f11R,t = \frac{1}{2}r^2 f11C,t-1+\frac{1}{2}(1-r)^2 f11R,t-1$$

Thus,

$$g11,t+1 = f22,t-1+\left(\frac{1}{4}+\frac{1}{4}\right)f21,t-1+\left(\frac{1}{4}+\frac{1}{4}\right)f12,t-1+[\frac{1}{4}(1-r)^2$$

$$+\frac{1}{4}r(1-r)+\frac{1}{4}r(1-r)+\frac{1}{4}(1-r)^2]f11C,t-1$$

$$+[\frac{1}{4}r^2+\frac{1}{4}r(1-r)+\frac{1}{4}r(1-r)+\frac{1}{4}r^2]f11R,t-1^{-rdt}$$

$$= f22,t-1+\frac{1}{2}f21,t-1+\frac{1}{2}f12,t-1+\frac{1}{2}(1-r)f11C,t-1$$

$$+\frac{1}{2}rf11R,t-1^{-rdt}$$

$$= g11,t-rd_t$$

(This equation is identical to Equation derived for the case of continued panmictic reproduction.)

$$g11,t+1 = g11,t-r_c d_t$$

Equation applies at continued self-fertilization. It is identical to Equation applying at continued panmictic reproduction. One should realise, however, that with panmictic reproduction the relation between *dt*+1 and *dt* was derived to be $d_{t+1} = (1-r_c)d_t$. For autogamous reproduction, however, the relation between *dt* and *dt*–1 can be shown to be

$$d_{t+1} = \left(\frac{1-2r_c}{2}\right)d_t$$

Note: In the case of (continued) selfing, plants with a doubly heterozygous genotype, in the coupling phase or in the repulsion phase, can only be produced by doubly heterozygous parents, one can easily derive from Table that:

$$f_{11C,t+1} = 2\left(\frac{1-r}{2}\right)^2 f_{11c,t} + 2\left(\frac{r}{2}\right)^2 f_{11R,t}$$

$$f_{11R,t+1} = 2\left(\frac{1-r}{2}\right)^2 f_{11R,t} + 2\left(\frac{r}{2}\right)^2 f_{11C,t}$$

Thus,

$$f_{11,t+1} = \left[2\left(\frac{1-r}{2}\right)^2 + 2\left(\frac{r}{2}\right)^2\right](f_{11C,t} + f_{11R,t})$$

$$= \left(r^2 - r + \frac{1}{2}\right) f_{11,t} = \left[\left(r - \frac{1}{2}\right)^2 + \frac{1}{4}\right] f_{11,t}$$

Equations:

$$d_{t+1} = \frac{1}{2}\left(f_{11,Ct+1} - f_{11,Rt+1}\right)$$

yield thus,

$$d_{t+1} = \frac{1}{4}\left[(1-r)^2 - r^2\right]\left(f_{11,Ct+1} - f_{11,Rt+1}\right)$$

This gives Equation,

$$d_{t+1} = \left(\frac{1-2r_c}{2}\right)d_t$$

$$d_t = \left(\frac{1-2r_c}{2}\right)^{t-1} d_t$$

Equations yield for the case of continued selfing:

$$g_{11,t+1} = g_{11,t} - r_c\left(\frac{1-2r_c}{2}\right)d_{t-1}$$

The parameter dt is still, as defined in Equation, equal to 1/2 (f11C,t-f11R,t). Equation shows that, unless $dt = 0$ or $rc = 1/2$, the haplotype frequencies will change from one generation to the next. The genotypic composition of F–, for F1 in coupling phase as well as in repulsion phase, depends directly on Equation, *viz.*

$$g_{11,\infty} = f_{22,\infty} = g_{11,1} - \left(\frac{2r}{1+2r}\right)d_1$$

Note: Combining the Equation yields in the case of continued selfing

$$g_{11,t+1} - g_{11,t} = -rd_1\left(\frac{1-2r}{2}\right)^{t-1}$$

Repeated application of this equation result via:

$$g_{11,2} - g_{11,1} = -rd_1\left(\frac{1-2r}{2}\right)^{0}$$

$$g_{11,3} - g_{11,2} = -rd_1\left(\frac{1-2r}{2}\right)^{1}$$

$$g_{11,t+1} - g_{11,t} = -rd_1\left(\frac{1-2r}{2}\right)^{t-1}$$

$$g_{11,t+1} - g_{11,1} = -rd_1\sum_{j=0}^{t-1}\left(\frac{1-2r}{2}\right)^{j}$$

The sum of the terms of this geometric series is:

$$\frac{1-\left(\frac{1-2r}{2}\right)^{t-1}}{1-\left(\frac{1-2r}{2}\right)} = \frac{2}{1+2r}\left[1-\left(\frac{1-2r}{2}\right)^{t-1}\right]$$

Thus,

$$g_{11,t+1} = g_{11,1-r}\left(\frac{2}{1+2r}\right).d_{1}.\left[1-\left(\frac{1-2r}{2}\right)^{t-1}\right]$$

Implying,

$$g_{11,\infty} = f_{22,\infty} = g_{11,1} - \left(\frac{2r}{1+2r}\right)d_1$$

The quantity to be substituted in Equation for $d1$ amounts, according to Example, to 1/4 (1–2r) for F1 in the coupling phase and to –1/4 (1–2r) for F1 in the repulsion phase.

Equation yields thus for F1 in the coupling phase:

$$g_{11,\infty} = f_{22,\infty} = \left(\frac{1-r}{2}\right) - \left(\frac{2r}{1+2r}\right)\left(\frac{1-2r}{4}\right) = \frac{1}{2(1+2r)}$$

For F_1 in the repulsion phase we get,

$$g_{11,\infty} = f_{22,\infty} = \left(\frac{r}{2}\right) + \left(\frac{2r}{1+2r}\right)\left(\frac{1-2r}{4}\right) = \frac{2r}{2(1+2r)}$$

The genotypic composition of F– with regard to the complex genotype s for the two linked loci A-a and B-b.

- F_1 in coupling phase

	bb $\frac{2rc}{2(1+2r_c)}$	Bb	BB	
aa	0	0	$\frac{1}{2(1+2r_c)}$	$\frac{1}{2}$
Aa	$\frac{1}{2(1+2r_c)}$	0	0	0
AA	$\frac{1}{2}$	0	$\frac{2rc}{2(1+2r_c)}$	$\frac{1}{2}$
		0	$\frac{1}{2}$	1

- F_1 in repulsion phase

	bb $\frac{2rc}{2(1+2r_c)}$	Bb	BB	
aa	0	0	$\frac{1}{2(1+2r_c)}$	$\frac{1}{2}$
Aa	$\frac{1}{2(1+2r_c)}$	0	0	0
AA	$\frac{1}{2}$	0	$\frac{2rc}{2(1+2r_c)}$	$\frac{1}{2}$
		0	$\frac{1}{2}$	1

Table presents the genotypic composition of F–. It may be compared with Table presenting the genotypic composition obtained after continued panmixis.

In the case of linkage ($0 < rc < 1/2$) the frequencies of the haplotypes change in the course of the generations. For gametes with the *AB* haplotype the difference between $g11,1$ and $g11,–$ amounts to,

$$g_{11,\infty} - g_{11,1} = \left(\frac{2r}{1+2r}\right) d_1$$

This amounts, according to Example, for F1 in the coupling phase to,

$$\left(\frac{2r}{1+2r}\right)\left(\frac{1-2r}{4}\right) = \frac{r(1-2r)}{2(1+2r)}$$

And for F_1 in the repulsion phase to,

$$\left(\frac{2r}{1+2r}\right)\left(\frac{2r-1}{4}\right) = \frac{r(2r-1)}{2(1+2r)}$$

These differences are for 0 < *rc* < 1 2 generally quite small. For *rc* = 1/ 4, for instance, it amounts for F1 in the repulsion phase to *g*11,1-*g*11,– = 1/ 8-1 6 = –0.0417. We consider now the frequency of plants with a genotype obtained by crossing two parents.

It may, for example, be desired to obtain genotype *AABB* from an initial cross of genotypes *AAbb* and *aaBB*. The frequency of *AABB* plants amounts in population F2 to *f*22,1 = 1/4 *rc*/2. Equation

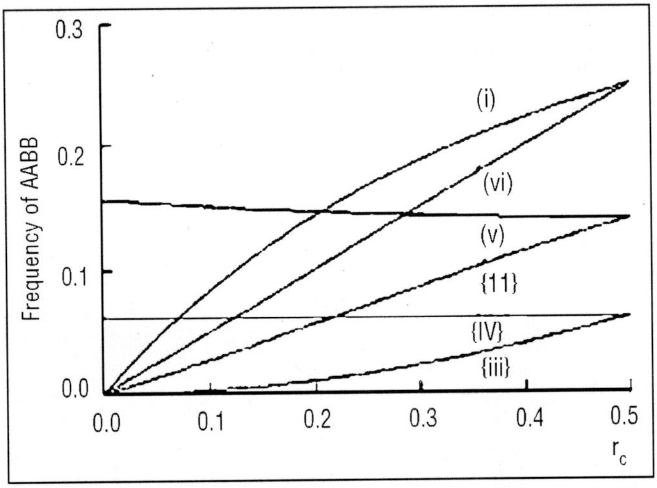

Figure 1: *The Frequency of Plants with Genotype AABB as a Function of the Recombination Value rc.*

Considered are population obtained by crossing of genotypes AAbb and aaBB followed by:

- Continued self-fertilization until F-,
- Selfing until F3,
- Selfing until F_2,
- Continued panmixis until linkage equilibrium,
- Continued panmixis followed by one round of reproduction by means of selfing, or
- Doubling of the number of chromosomes in the gametes produced by F_1 yields for *t* = 2 the frequency of plants with genotype *AABB* in F3.

When substituting the F2 genotype frequencies presented in Table one gets for an F1 in the repulsion phase:

$$f22,2 = \frac{1}{4}r^2 + \frac{1}{8}r(1-r) + \frac{1}{8}r(1-r) + \frac{1}{8}r^2(1-r)^2 + \frac{1}{8}(1-r)^2 r^2$$

$$= \frac{1}{4}r + \frac{1}{4}r^2 - \frac{1}{2}r^3 + \frac{1}{4}r^4$$

This amounts, for unlinked loci, to *f*22,2 = 9 64 = _3/8 _2 = *f*00,2. According to Equation the frequency of *AABB* plants in F– is 2*r* 2(1+2*r*). Because 2*r* 2(1+2*r*) ≤ 1/2(1+2*r*), plants with one of the parental genotypes will outnumber plants with this recombinant genotype to a greater extent as linkage is stronger, *i.e.* as *rc* is smaller.

In populations of cross-fertilizing crops, doubly heterozygous genotypes tend to be permanently present; in populations of self-fertilizing crops they disappear. One should, however, be careful when speaking about 'the recombining effect of cross-fertilization'. This is illustrated for loci *A-a* and *B-b*. Continued panmictic reproduction gives eventually, at linkage equilibrium, *f*22 = *p*2*r*2. This amounts for *p* = *r* = 1/2 to 1/16, whatever the recombination value. For tightly linked loci, with *rc* < 1 14, genotype *AABB* will indeed occur with a higher frequency in populations in linkage equilibrium than in populations obtained by continued selfing.

For less tightly linked loci, *i.e.* *rc* > 1/14, the frequency of *AABB* will, however, be higher in *F*–. Thus one should not decide rashly to increase the frequency of plants with a recombinant genotype by the application of random mating in F2, F3... populations of a self-fertilizing crop. With regard to unlinked loci continued random mating will only result in the genotypic composition of F2, because for unlinked loci the *F*2 population obtained by selfing will have the linkage equilibrium composition. Selection in a cross-fertilizing crop is more efficient when increasing the frequency of homozygous recombinant genotypes by selfing. According to Note a single round of reproduction by means of self-fertilization in a population in linkage equilibrium gives,

$$f_{22} = \frac{5 - 2r + 2r^2}{32}$$

Note: Consider a population in linkage equilibrium. It is obtained by panmictic reproduction starting with a single-cross hybrid variety. With regard to loci *A-a* and *B-B* a single round of reproduction by means of selfing results, according to Equation, in the following frequency of plants with genotype *AABB*:

$$f_{22} = \frac{1}{16} + \frac{1}{4}.\frac{1}{8} + \frac{1}{4}.\frac{1}{8} + \frac{1}{4}r^2.\frac{1}{8} + \frac{1}{4}(1-r)^2.\frac{1}{8} = \frac{5-2r+2r^2}{32}$$

For *r* = 1/2 this amounts to 9 64, *i.e.* _3/8 _2. It is the same value as obtained, from Equation, for an F3. The single reproduction by means of selfing gives thus the genotypic composition of an F3. This illustrates that the genotypic composition of the population in linkage equilibrium is equal to the genotypic composition for pairs of unlinked loci in an F2. In a diploid crop, doubling the number of chromosomes of haploid plants is the fastest way to attain complete homozygosity. The frequency of plants with the desired recombinant genotype then amounts to 1/2 *rc*, *i.e.* 2/*rc* times as high as in F2. The frequency of doubly heterozygous plants is greatly reduced with reproduction by means of selfing. Depending on the recombination value, a single round of selfing reduces this frequency to only 1/4 to 1/2 of the frequency of plants with the *AaBb* genotype in the preceding generation. Note shows that the remaining portion of doubly heterozygous plants amounts to *f*11,*t*+1 *f*11,*t* = (*r*-1/2)2 + 1 4, which amounts to 1/4 for *rc* = 1/2 and to 1/2 for *rc* = 0. This reduction of the frequency of heterozygous plants is even stronger for more complex genotypes: a single round of selfing reduces the frequency of the complex genotype consisting of a heterozygous single-locus genotype for each of *k* unlinked loci to the portion (1/2)*k* of its preceding value.

Autotetraploid Chromosome Behaviour and Self-Fertilization

Spontaneous self-fertilization as the natural mode of reproduction occurs rather rarely among crops with an autotetraploid chromosome behaviour. The somatic chromosome number of quinoa (*Chenopodium quinoa*) is $2n = 36$. The basic chromosome number for the genus *Chenopodium* is $x = 9$. This suggests that quinoa is a tetraploid. Ward found for the same locus both diploid and tetraploid behaviour. Simmonds reported that selfing predominates, without evident inbreeding depression.

Quite a few autotetraploid crops, *e.g.* durum wheat (*Triticum durum*; $2n = 4x = 28$) or coffee (*Coffea arabica*; $2n = 4x = 44$), have a diploid chromosome behaviour. For other crops, *e.g.* European potato (*Solanum tuberosum*; $2n = 4x = 48$) or wild barley (*Hordeum bulbosum*; $2n = 4x = 28$), there may be a more or less perfect autotetraploid chromosome behaviour, implying that exclusively quadrivalents are being formed at meiosis. Artificial selffertilization may be applied in a man-made autotetraploid crop such as rye (*Secale cereale*; $2n = 4x = 28$), which is self-incompatible in its natural diploid condition. In this section attention is only given to the simple situation of a single segregating locus with two alleles. It is assumed that double reduction does not occur.

The genotypic composition of some initial generation, say S0, is:

Genotype					
	aaaa	Aaaa	AAaa	AAAa	AAAA
	nulliplex	Simplex	Duplex	triplex	quadruplex
f	f_0	f_1	f_2	f_3	f_4

Its gene frequencies are

$$P=\frac{1}{4}f_1+\frac{1}{2}f_2+\frac{3}{4}f_3+f_4$$

and

$$q=1-P$$

It is first verified that the gene frequencies remain constant from one generation to the next (such constancy is to be expected in the absence of selection). In order to do this, Table is used.

This table presents, for each possible autotetraploid genotype, and according to the haplotype frequencies presented in Table, the genotypic composition of the line obtained by selfing. The allele frequencies in the parental population follow from Equation. Across the total of the lines obtained from this parental population the frequency of allele A is,

$$\frac{1}{4}\left(\frac{1}{2}f_1+\frac{2}{9}f_2\right)+\frac{1}{2}\left(\frac{1}{4}f_1+\frac{1}{2}f_2+\frac{1}{4}f_3\right)+\frac{3}{4}\left(\frac{2}{9}f_2+\frac{1}{2}f_3\right)$$
$$+\left(\frac{1}{36}f_2+\frac{1}{4}f_3+f4\right)=\frac{1}{4}f_1+\frac{1}{2}f_2+\frac{3}{4}f_3+f_4$$

This is equal to the frequency in the parental population. The genotypic composition of S– will thus be:

Genotype:

	aaaa	*Aaaa*	*AAaa*	*AAAa*	*AAAA*
f	q	0	0	0	P

How fast do the frequencies of plants with a heterozygous genotype and of gametes with a heterozygous haplotype decrease with (continued) selfing?

genotype	**f**	**aaaa**	**Aaaa**	**AAaa**	**AAAa**	**AAAA**
aaaa	f_0	1	0	0	0	0
Aaaa	f_1	$\frac{1}{4}$	$\frac{1}{2}$	$\frac{1}{4}$	0	0
AAaa	f_2	$\frac{1}{36}$	$\frac{2}{9}$	$\frac{1}{2}$	$\frac{2}{9}$	$\frac{1}{36}$
AAAa	f_3	0	0	$\frac{1}{4}$	$\frac{1}{2}$	$\frac{1}{4}$
AAAA	f_4	0	0	0	0	1

In order to answer this question, first the decrease of $g1$, *i.e.* the frequency of gametes with haplotype Aa is considered and thereafter the decrease of fh. *i.e.* the frequency of heterozygous plants. From Table it can be derived that,

$$g1,t+1=\frac{1}{2}f1,t+\frac{4}{6}f2,t+\frac{1}{2}f3,t$$

Thus similarly,

$$
\begin{aligned}
g_{1,t+2} &= \frac{1}{2} f_{1,t+1} \frac{4}{6} f_{2,t+1} + \frac{1}{2} f_{3,t+1} \\
&= \frac{1}{2}\left(\frac{1}{2} f_{1,t} + \frac{2}{9} f_{2,t}\right) + \frac{4}{6}\left(\frac{1}{4} f_{1,t} + \frac{1}{2} f_{2,t} + \frac{1}{4} f_{3,t}\right) + \frac{1}{2}\left(\frac{2}{9} f_{2,t} + \frac{1}{2} f_{3,t}\right) \\
&= \frac{5}{12} f_{1,t} + \frac{5}{9} f_{2,t} + \frac{5}{12} f_{3,t} \\
&= \frac{5}{6} g_{1,t+1}
\end{aligned}
$$

This implies that each population obtained by selfing still produces 5 6 of the proportion of gametes with the *Aa* haplotype which was produced by the previous generation. Now the frequency of plants with a heterozygous genotype is considered. This frequency is designated by *fh*.

Thus,

$$f_{h,t} = f_{1,t} + f_{2,t} + f_{3,t}$$

$$f_{1,t+2} = \frac{1}{2} f_{1,t+1} + \frac{2}{9} f_{2,t+1}$$

$$f_{2,t+2} = \frac{1}{4} f_{1,t+1} + \frac{1}{2} f_{2,t+1} + \frac{1}{4} f_{3,t+1}$$

$$f_{3,t+2} = \frac{2}{9} f_{2,t+1} + \frac{1}{2} f_{3,t+1}$$

The decrease of f_h at (continued) selfing is described by:

$$
\begin{aligned}
f_{h,t+2} &= \frac{3}{4} f_{1,t+1} + \frac{17}{18} f_{2,t+1} + \frac{3}{4} f_{3,t+1} \\
&= f_{h,t+1} - \left(\frac{1}{4} f_{1,t+1} + \frac{1}{18} f_{2,t+2} + \frac{1}{4} f_{3,t+1}\right) \\
&= f_{h,t+1} - \left[\frac{1}{4}\left(\frac{1}{2} f_{1,t} + \frac{2}{9} f_{2,t}\right) + \frac{1}{18}\left(\frac{1}{4} f_{1,t} + \frac{1}{2} f_{2,t} + \frac{1}{4} f_{3,t}\right) + \frac{1}{4}\left(\frac{2}{9} f_{2,t} + \frac{1}{2} f_{3,t}\right)\right] \\
&= fh,t+1 - \frac{5}{36}(f_{1,t} + f_{2,t} + f_{3,t}) \\
&= fh,t+1 - \frac{5}{36} f_{h,t}
\end{aligned}
$$

We consider the decrease of the frequency of heterozygous plants for an initial population consisting exclusively of duplex plants. The genotypic composition of S0 is then (0, 0, 1, 0, 0), with *fh*,0 = 1.

Generation	t	fh,t	$\lambda_S = \frac{f_{h,t}}{f_{h,t-1}}$
S_0	0	1	
S_1	1	$\frac{17}{18} = 0.9444$	0.9444
S_2	2	$\frac{29}{36} = 0.8056$	0.8529
S_3	3	$\frac{437}{648} = 0.6744$	0.8372
S_4	4	$\frac{729}{1296} = 0.5625$	0.8341

Then to 2/9 + 1/2 + 2/9 = 17/18. Table presents the frequency of plants with a heterozygous genotype in successive generations, as calculated from Equation. The frequency of heterozygous plants as a proportion of the frequency in the preceding generation, *i.e.*

$$\lambda_S = \frac{f_{h,t}}{f_{h,t-1}}$$

is also presented in Table. It appears that λS converges to a constant value, *viz.* to 5/6 = 0.8333. This implies, per round of reproduction by selfing, the same constant (relative) decrease in the frequency of heterozygous plants as derived from the frequency of heterozygous gametes. In this phase, reproduction by means of self-fertilization for n successive generations reduces *fh, t* to

$$f_{h,t+n} = \left(\frac{5}{6}\right)^n f_{h,t}$$

The frequency of heterozygous plants is halved if

$$\left(\frac{5}{6}\right)^n = 0.5, i.e. \text{ if}$$

$$n = \frac{In(0.5)}{In(0.8333)} = 3.8$$

Starting with an initial population with genotypic composition (0, 0, 1, 0, 0) the decrease of the frequency of heterozygous plants is even less: in S4, *fh*,4 is still larger than 1/2. When comparing the decrease in the frequency of plants with a heterozygous genotype occurring at selfing of a diploid crop and such decrease at selfing of a diploid crop and such decrease at selfing of a tetraploid crop it is clear that the decrease is quite slow in the case of tetraploidy.

3

Quantitative Genetics, Genomics, Medical Genetics and Behavioural Genetics

Quantitative Genetics

Quantitative genetics is the study of continuous traits (such as height or weight) and their underlying mechanisms. It is effectively an extension of simple Mendelian inheritance in that the combined effect of the many underlying genes results in a continuous distribution of phenotypic values.

The field was founded, in evolutionary terms, by the originators of the modern synthesis, R.A. Fisher, Sewall Wright and J. B. S. Haldane, and aimed to predict the response to selection given data on the phenotype and relationships of individuals. Analysis of Quantitative Trait Loci, or QTL, is a more recent addition to the study of quantitative genetics. A QTL is a region in the genome that affects the trait or traits of interest. Quantitative trait loci approaches require accurate phenotypic, pedigree and genotypic data from a large number of individuals.

Traits

Quantitative genetics is not limited to continuous traits, but to all traits that are determined by many genes. This includes:

- Continuous traits are quantitative traits with a continuous phenotypic range. They are often polygenic, and may also be influenced significantly by environmental effects.

- Meristic traits or other ordinal numbers are expressed in whole numbers, such as number of offspring, or number of bristles on a fruit fly. These traits can be either treated as approximately discontinuous traits or as threshold traits.
- Some qualitative traits can be treated as if they have an underlying quantitative basis, expressed as a threshold trait (or multiple thresholds). Some human diseases (such as, schizophrenia) have been studied in this manner.

Basic Principles

The phenotypic value (P) of an individual is the combined effect of the genotypic value (G) and the environmental deviation (E):

$$P = G + E$$

The genotypic value is the combined effect of all the genetic effects, including nuclear genes, mitochondrial genes and interactions between the genes. It is worthwhile to note that the mathematics *is* related to the genetics: for which the brief following revision may be useful. In disomic (diploid) organisms, a nucleus gene is represented twice in the gene-set ("genotype"), one contribution being provided by each parent during sexual reproduction. Each "gene" is located at a particular place (a "locus" - the Latin word for place; plural "loci") on corresponding chromosomes (homologues), one from each parent.

Any gene may have several functional forms in the species as a whole, and each of these may lead to outwardly different "effects" (= an average result in the phenotype considered over a large sample of gene-backgrounds and environments). These functional forms are "alleles" (or "allelomorphs", the original term). If both alleles at a gene have the same phenotypic effect (are the "same"), the gene is said to be "homozygous": if each allele at a gene is different in effect, the gene is "heterozygous".

The average phenotypic outcome may also depend upon how alleles interact with their own homologous partner in the disomic genotype ("dominance"), and on how these alleles interact with those of other genes at other loci which also affect this phenotypic trait ("epistasis"). Notice that we have combined classical genetics ideas with those of statistics in this exposition. Terms such as "gene", "homologue", "allele", "homozygous" and "heterozygous" are genetical: but "effect" is statistical, and refers to the average observed over an infinity of backgrounds, both genetical and environmental.

Thus, we have very sneakily defined the "genetic value" (or "genotypic" value) as the infinity mean of all the *phenotypes* it can ever produce in time and space! Before molecular genetics, there simply was no other way to do it! And after molecular genetics, this is still the most utilitarian way to tackle the idea of genetic value! Also notice that we have openly used fundamentals of reproductive biology behind the genetics.

The founder of Quantitative Genetics - Sir Ronald Fisher - perceived all of this when he proposed the first mathematics of this branch of Genetics [Fisher R.A. (1930). The Genetical theory of Natural Selection. Oxford Clarendon Press.]. He sought to define a single statistical summary of all the variance arising from phenotypic change during the course of genetical assortment and segregation, which he called the "genetic" variance. His residual genotypic variance (which he called simply the "residual") represented that part of assortment which did not lead to phenotypic change, although the genes themselves had in fact been subject to meiosis and syngamy, of course.

[A simplified exposition of this can be found in Falconer and Mackay (1966) - see References.] These partitions subsequently became the familiar subdivisions of the "additive" (A) and "dominance" (D) variances, respectively. These later names are utterly misleading and very unfortunate, and have led to much confusion as to what they mean genetically. [It would have been better for posterity had they been named "Assortative" (A), and "Stable" (S).] A more gene-focused partitioning was invented by Mather and Jinks in 1971, but they also were statisticians rather than geneticists, had their own rather opqaque symbolism, and became somewhat overwhelmed by the Fisherian approach. Added to all of this was the problem that Fisher's underlying reproduction model (mating system) was unrealistically simplistic: whilst it facilitated solving the equations, it didn't describe many real-life scenarios.

Fortunately, Wright (in the 1950s) did provide the means to overcome complex mating systems, but its own complexity minimised its popular adoption. Early in this millennium [Heredity (2003) 91: 85-89], a comprehensive linking together of all of these approaches reconciled their various meanings and relationships (as well as correcting an error), and suggested that new partitions reflecting real homozygosis [a] and heterozygosis [d] should replace the present "additive" [A] and "dominance" [D] subdivisions. This paper revealed, by the way, that the Additive Genetical variance consists of all the homozygote variance, plus part of the dominance variance, and a frequency-weighted covariance between homozygote and heterozygote gene effects. The so-called Dominance variance contains only the remainder of the overall dominance variance of the gene in question, being therefore very misleading indeed! It should be understood, however, that either method of partitioning still accounts for all of the genotypic variance in the model being used: it's the way it has been divided up which is being debated. At least the Environmental variance is much more straight-forward. This can be subdivided into a pure environmental component (E) and an interaction component (I) describing the gene-environment interaction. The overall "single gene" model can be written as:

$$P = a + d + E + I.$$

Expansion of the model to multiple genes (loci) is still not resolved satisfactorily, and until that is solved it is not possible to account for epistasis. The problem is being tackled currently. The contribution of those components cannot be determined in a single individual, but they can be estimated for whole populations by estimating the variances for those components, denoted as:

$$V_P = V_a + V_d + V_E + V_I$$

The heritability of a trait is the proportion of the total (i.e. phenotypic) variance (V_P) that is explained by the total genotypic variance (V_G). This is known also as the "broad sense" heritability (H^2). If only Additive genetic variance (V_A) is used in the numerator, the heritability is "narrow sense" (h^2). Unfortunately, this is often simply called "heritability", with little reflection about its true meaning. The broadsense heritability indicates the genotypic determination of the phenotype: while the latter estimates the degree of assortative disequlibrium in the trait. Fisher proposed that this narrow-sense heritability might be appropriate in considering the results of natural selection, focusing as it does on disequilibrium: and it has been used also for predicting the results of artificial selection. This latter usage seems to be inappropriate, however, as breeders are interested in steering attributes towards new phenotypes (that is in utilising all the gene effects), rather than simply exploiting disequilibrium. But old dogmas die hard!

Resemblance between Relatives

Central in estimating the variances for the various components is the principle of relatedness. A child has a father and a mother. Consequently, the child and father share 50% of their alleles, as do the child and the mother. However, the mother and father normally do not share alleles as a result of shared ancestors. Similarly, two full siblings share also on average 50% of the alleles with each other, while half siblings share only 25% of their alleles. This variation in relatedness can be used to estimate which proportion of the total phenotypic variance (V_P) is explained by the above-mentioned components.

Correlated Traits

Although some genes have only an effect on a single trait, many genes have an effect on various traits. Because of this, a change in a single gene will have an effect on all those traits. This is calculated using covariances, and the phenotypic covariance (Cov_P) between two traits can be partitioned in the same way as the variances described above.

Genomics

Genomics is a discipline in genetics concerning the study of the genomes of organisms. The field includes intensive efforts to determine the entire DNA sequence of organisms and fine-scale genetic mapping efforts. The field also includes studies of intragenomic phenomena such as heterosis, epistasis, pleiotropy and other interactions between loci and alleles within the genome.

In contrast, the investigation of the roles and functions of single genes is a primary focus of molecular biology or genetics and is a common topic of modern medical and biological research. Research of single genes does not fall into the definition of genomics unless the aim of this genetic, pathway, and functional information analysis is to elucidate its effect on, place in, and response to the entire genome's networks.

For the United States Environmental Protection Agency, "the term "genomics" encompasses a broader scope of scientific inquiry associated technologies than when genomics was initially considered. A genome is the sum total of all an individual organism's genes. Thus, genomics is the study of all the genes of a cell, or tissue, at the DNA (genotype), mRNA (transcriptome), or protein (proteome) levels."

History

The first genomes to be sequenced were those of a virus and a mitochondrion, and were done by Fred Sanger. His group established techniques of sequencing, genome mapping, data storage, and bioinformatic analyses in the 1970-1980s. A major branch of genomics is still concerned with sequencing the genomes of various organisms, but the knowledge of full genomes has created the possibility for the field of functional genomics, mainly concerned with patterns of gene expression during various conditions. The most important tools here are microarrays and bioinformatics.

Study of the full set of proteins in a cell type or tissue, and the changes during various conditions, is called proteomics. A related concept is materiomics, which is defined as the study of the material properties of biological materials (e.g. hierarchical protein structures and materials, mineralized biological tissues, etc.) and their effect on the macroscopic function and failure in their biological context, linking processes, structure and properties at multiple scales through a materials science approach. The actual term 'genomics' is thought to have been coined by Dr. Tom Roderick, a geneticist at the Jackson Laboratory (Bar Harbor, ME) over beer at a meeting held in Maryland on the mapping of the human genome in 1986.

In 1972, Walter Fiers and his team at the Laboratory of Molecular Biology of the University of Ghent (Ghent, Belgium) were the first to determine the sequence of a gene: the gene for Bacteriophage MS2 coat protein. In 1976, the team determined the complete nucleotide-sequence of bacteriophage MS2-RNA. The first DNA-based genome to be sequenced in its entirety was that of bacteriophage Ö-174; (5,368 bp), sequenced by Frederick Sanger in 1977.

The first free-living organism to be sequenced was that of *Haemophilus influenzae* (1.8 Mb) in 1995, and since then genomes are being sequenced at a rapid pace. As of September 2007, the complete sequence was known of about 1879 viruses, 577 bacterial species and roughly 23 eukaryote organisms, of which about half are fungi. Most of the bacteria whose genomes have been completely sequenced are problematic disease-causing agents, such as *Haemophilus influenzae*. Of the other sequenced species, most were chosen because they were well-studied model organisms or promised to become good models. Yeast (*Saccharomyces cerevisiae*) has long been an important model organism for the eukaryotic cell, while the fruit fly *Drosophila melanogaster* has been a very important tool (notably in early pre-molecular genetics).

The worm *Caenorhabditis elegans* is an often used simple model for multicellular organisms. The zebrafish *Brachydanio rerio* is used for many developmental studies on the molecular level and the flower *Arabidopsis thaliana* is a model organism for flowering plants. The Japanese pufferfish (*Takifugu rubripes*) and the spotted green pufferfish (*Tetraodon nigroviridis*) are interesting because of their small and compact genomes, containing very little non-coding DNA compared to most species. The mammals dog (*Canis familiaris*), brown rat (*Rattus norvegicus*), mouse (*Mus musculus*), and chimpanzee (*Pan troglodytes*) are all important model animals in medical research.

Human Genomics

A rough draft of the human genome was completed by the Human Genome Project in early 2001, creating much fanfare. By 2007 the human sequence was declared "finished" (less than one error in 20,000 bases and all chromosomes assembled). Display of the results of the project required significant bioinformatics resources. The sequence of the human reference assembly can be explored using the UCSC Genome Browser.

Bacteriophage Genomics

Bacteriophages have played and continue to play a key role in bacterial genetics and molecular biology. Historically, they were used to define gene structure and gene regulation. Also the first genome to be sequenced was a bacteriophage. However, bacteriophage research did not lead the genomics revolution, which is clearly dominated by bacterial genomics. Only very recently has the study of bacteriophage genomes become prominent, thereby enabling researchers to understand the mechanisms underlying phage evolution. Bacteriophage genome sequences can be obtained through direct sequencing of isolated bacteriophages, but can also be derived as part of microbial genomes. Analysis of bacterial genomes has shown that a substantial amount of microbial DNA consists of prophage sequences and prophage-like elements. A detailed database mining of these sequences offers insights into the role of prophages in shaping the bacterial genome.

Cyanobacteria Genomics

At present there are 24 cyanobacteria for which a total genome sequence is available. 15 of these cyanobacteria come from the marine environment. These are six *Prochlorococcus* strains, seven marine *Synechococcus* strains, *Trichodesmium erythraeum* IMS101 and *Crocosphaera watsonii* WH8501. Several studies have demonstrated how these sequences could be used very successfully to infer important ecological and physiological characteristics of marine cyanobacteria.

However, there are many more genome projects currently in progress, amongst those there are further *Prochlorococcus* and marine *Synechococcus* isolates, *Acaryochloris* and *Prochloron*, the N_2-fixing filamentous cyanobacteria *Nodularia spumigena*, *Lyngbya aestuarii* and *Lyngbya majuscula*, as well as bacteriophages infecting marine cyanobaceria. Thus, the growing body of genome information can also be tapped in a more general way to address global problems by applying a comparative approach. Some new and exciting examples of progress in this field are the identification of genes for regulatory RNAs, insights into the evolutionary origin of photosynthesis, or estimation of the contribution of horizontal gene transfer to the genomes that have been analyzed.

Medical Genetics

Medical genetics is the specialty of medicine that involves the diagnosis and management of hereditary disorders. Medical genetics differs from Human genetics in that human genetics is a field of scientific research that may or may not apply to medicine, but medical geneics refers to the application of genetics to medical care. For example, research on the causes and inheritance of genetic disorders would be considered within both human genetics and medical genetics, while the diagnosis, management, and counseling of individuals with genetic disorders would be considered part of medical genetics.

In contrast, the study of typically non-medical phenotypes such as the gnetics of eye colour wuld be considered part of human genetics, but not necessarily relevant to medical genetics (except in situations such as albinism). *Genetic medicine* is a newer term for medical genetics and incorporates areas such as gene therapy, personalized medicine, and the rapidly emerging new medical specialty, predictive medicine.

Scope

Medical genetics encompasses many different areas, including clinical practice of physicians, genetic counselors, and nutritionists, clinical diagnostic laboratory activities, and research into the causes and inheritance of genetic disorders. Examples of conditions that fall within the scope of medical genetics include birth defects and dysmorphology, mental retardation, autism, metabolic and mitochondrial disorders, skeletal dysplasia, connective tissue disorders, cancer genetics, teratogens, and prenatal diagnosis. Medical genetics is increasingly becoming relevant to many common diseases. Overlaps with other medical speciaties are beginning to emerge, as recent advances in genetics are revealing etiologies for neurologic, endocrine, cardiovascular, pulmonary, ophthalmologic, renal, psychiatric, and dermatologic conditions.

Subspecialties

In some ways, many of the individual fields within medical genetics are hybrids between clinical care and research. This is due in part to recent advances in science and technology (for example, see the Human genome project) that have enabled an unprecedented understanding of genetic disorders.

Clinical Genetics: Clinical genetics is the practice of clinical medicine with particular attention to hereditary disorders. Referrals are made to genetics clinics for a variety of reasons, including birth defects, developmental delay, autism, epilepsy, short stature, and many others. Examples of genetic syndromes that are commonly seen in the genetics clinic include chromosomal rearrangements, Down syndrome, DiGeorge syndrome (22q11.2 Deletion Syndrome), Fragile X syndrome, Marfan syndrome, Neurofibromatosis, Turner syndrome, and Williams syndrome.

Metabolic/Biochemical Genetics: Metabolic (or biochemical) genetics involves the diagnosis and management of inborn errors of metabolism in which patients have enzymatic dficiencies that perturb biochemical pathways involved in metabolism of carbohydrates, amino acids, and lipids. Examples of metabolic disorders include galactosemia, glycogen storage disease, lysosomal storage disorders, metabolic acidosis, peroxisomal disorders, phenylketonuria, and urea cycle disorders.

Cytogenetics: Cytogenetics is the study of chromosomes and chromosome abnormalities. While cytogenetics historically relied on microscopy to analyze chromosomes, new molecular technologies such as array comparative genomic hybridization are now becoming widely used. Examples of chromosome abnormalities include aneuploidy, chromosomal rearrangements, and genomic deletion/duplication disorders.

Molecular Genetics: Molecular genetics involves the discovery of and laboratory testing for DNA mutations that underlie many single gene disorders. Examples of single gene disorders include achondroplasia, cystic fibrosis, Duchenne muscular dystrophy, hereditary breast cancer (BRCA1/2), Huntington disease, Marfan syndrome, Noonan syndrome, and Rett syndrome. Molecular tests are also used in the diagnosis of syndromes involving epigenetic abnormalities, such as Angelman syndrome, Beckwith-Wiedemann syndrome, Prader-willi syndrome, and uniparental disomy.

Mitochondrial Genetics: Mitochondrial genetics concerns the diagnosis and management of mitochondrial disorders, which have a molecular basis but often result in biochemical abnormalities due to deficient energy production.

There exists some overlap between medical genetic diagnostic laboratories and molecular pathology.

Genetic Counseling

Genetic counseling is the process of providing information about genetic conditions, diagnostic testing, and risks in other family members, within the framework of nondirective counseling. Genetic counselors are non-physician members of the medical genetics team who specialize in fmily risk assessment and counseling of patients regarding genetic disorders. The precise role of the genetic counselor varies somewhat depending on the disorder.

History

Although genetics has its roots back in the 19th century with the work of the Bohemian monk Gregor Mendel and other pioneering scientists, huan genetics emerged later. It started to develop, albeit slowly, during the first half of the 20th century. Mendelian (single-gene) inheritance was studied in a number of important disorders such as albinism, brachydactyly (short fingers and toes), and hemophilia. Mathematical approaches were also devised and applied to human genetics. Population genetics was created.\

Medical genetics was a late developer, emerging largely after the close of World War II (1945) when the eugenics movement had fallen into disrepute. The Nazi misuse of eugenics sounded its death knell. Shorn of eugenics, a scientific approach could be used and was applied to human and medical genetics. Medical genetics saw an increasingly rapid rise in the second half of the 20th century and continues in the 21st century.

Notable Practitioners

- Victor McKusick (1921–2008), "The Father of Genetic Medicine" (http:/ /www.ashg.org/pages/mckusick.shtml)

Current Practice

The clinical setting in which patients are evaluated determines the scope of practice, diagnostic, and therapeutic interventions. For the purposes of general discussion, the typical encounters between patients and genetic practitioners may involve:

- Referral to an out-patient genetics clinic (pediatric, adult, or combined) or an in-hospital consultation, most often for diagnostic evaluation.
- Specialty genetics clinics focusing on management of inborn errors of metabolism, skeletal dysplasia, or lysosomal storage diseases.

- Referral for counseling in a prenatal genetics clinic to discuss risks to the pregnancy (advanced maternal age, teratogen exposure, family history of a genetic disease), test results (abnormal maternal serum screen, abnormal ultrasound), and/or options for prenatal diagnosis (typically amniocentesis or chorionic villus sampling).
- Multidisciplinary specialty clinics that include a clinical geneticist or genetic counselor (cancer genetics, cardiovascular genetics, craniofacial or cleft lip/palate hearing loss clinics, muscular dystrophy/ neurodegenerative disorder clinics).

Diagnostic Evaluation

Each patient will undergo a diagnostic evaluation tailored to their own particular presenting signs and symptoms. The geneticist will establish a differential diagnosis and recommend appropriate testing. Increasingly, clnicians use SimulConsult, paired with the National Library of Medicine Gene Review articles, to narrow the list of hypotheses (known as the differential diagnosis) and identify the tests that are relevant for a particular patient. These tests might evaluate for chromosomal disorders, inborn errors of metabolism, or single gene disorders.

Chromosome Studies: Chromosome studies are used in the general genetics clinic to determine a cause for developmental delay/mental retardation, birth defects, dysmorphic features, and/or autism. Chromosome analysis is also performed in the prenatal setting to determine whether a fetus is affected with aneuploidy or other chromosome rearrangements. Finally, chromosome abnormalities are often detected in cancer samples. A large number of different methods have been developed for chromosome analysis:

- Chromosome analysis using a karyotype involves special stains that generate light and dark bands, allowing identification of each chromosome under a microscope.
- Fluorescence in situ hybridization (FISH) involves fluorescent labeling of probes that bind to specific DNA sequences, used for identifying aneuploidy, genomic deletions or duplications, characterizing chromosomal translocations and determining the origin of ring chromosomes.
- Chromosome painting is a technique that uses fluorescent probes specific for each chromosome to differentially label each chromosome. This technique is more often used in cancer cytogenetics, where complex chromosome rearrangements can occur.
- Array comparative genomic hybridization is a new molecular technique that involves hybridization of an individual DNA sample to a glass slide or microarray chip containing molecular probes (ranging from large

~200kb bacterial artificial chromosomes to small oligonucleotides) that represent unique regions of the genome. This method is particularly sensitive for detection of genomic gains or losses across the genome but does not detect balanced translocations or distinguish the location of duplicated genetic material (for example, a tandem duplication versus an insertional duplication).

Basic Metabolic Studies: Biochemical studies are performed to screen for imbalances of metabolites in the bodily fluid, usually the blood (plasma/serum) or urine, but also in cerebrospinal fluid (CSF). Specific tests of enzyme function (either in leukocytes, skin fibroblasts, liver, or muscle) are also employed under certain circumstances. In the US, the newborn screen incorporates biochemical tests to screen for treatable conditions such as galactosemia and phenylketonuria (PKU). Patients suspected to have a metabolic condition might undergo the following tests:

- Quantitative amino acid analysis is typically performed using the ninhydrin reaction, followed by liquid chromatography to measure the amount of amino acid in the sample (either urine, plasma/serum, or CSF). Measurement of amino acids in plasma or serum is used in the evaluation of disorders of amino acid metabolism such as urea cycle disorders, maple syrup urine disease, and PKU. Measurement of amino acids in urine can be useful in the diagnosis of cystinuria or renal Fanconi syndrome as can be seen in cystinosis.
- Urine organic acid analysis can be either performed using quantitative or qualitative methods, but in either case the test is used to detect the excretion of abnormal organic acids. These compounds are normally produced during bodily metabolism of amino acids and odd-chain fatty acids, but accumulate in patients with certain metabolic conditions.
- The acylcarnitine combination profile detects compounds such as organic acids and fatty acids conjugated to carnitine. The test is used for detection of disorders involving fatty acid metabolism, including MCAD.
- Pyruvate and lactate are byproducts of normal metabolism, particularly during anaerobic metabolism. These compounds normally accumulate during exercise or ischemia, but are also elevated in patients with disorders of pyruvate metabolism or mitochondrial disorders.
- Ammonia is an end product of amino acid metabolism and is converted in the liver to urea through a series of enzymatic reactions termed the urea cycle. Elevated ammonia can therefore be detected in patients with urea cycle disorders, as well as other conditions involving liver failure.
- Enzyme testing is performed for a wide range of metabolic disorders to confirm a diagnosis suspected based on screening tests.

Molecular Studies:

- DNA sequencing is used to directly analyze the genomic DNA sequence of a particular gene. In general, only the parts of the gene that code for the expressed protein (exons) and small amounts of the flanking untranslated regions and introns are analyzed. Therefore, although these tests are highly specific and sensitive, they do not routinely identify all of the mutations that could cause disease.
- DNA methylation analysis is used to diagnose certain genetic disorders that are caused by disruptions of epigenetic mechanisms such as genomic imprinting and uniparental disomy.
- Southern blotting is an early technique basic on detection of fragments of DNA separated by size through gel electrophoresis and detected using radiolabeled probes. This test was routinely used to detect deletions or duplications in conditions such as Duchenne muscular dystrophy but is being replaced by high-resolution array comparative genomic hybridization techniques. Southern blotting is still useful in the diagnosis of disorders caued by trinucleotide repeats.
- Short tandem repeats are unique markers that can be used to determine haplotypes and are used in identity testing for maternal cell contamination.

Treatments: Each cell of the body contains the hereditary information (DNA) wrapped up in structures called chromosmes. Since genetic syndromes are typically the result of alterations of the chromosomes or genes, there is no treatment currently available that can correct the genetic alterations in every cell of the body. Therefore, there is currently no "cure" for genetic disorders. However, for many genetic syndromes there is treatment available to manage the symptoms. In some cases, particularly inborn errors of metabolism, the mechanism of disease is well understood and offers the potential for dietary and medical management to prevent or reduce the long-term complications. In other cases, infusion therapy is used to replace the missing enzyme. Current research is actively seeking to use gene therapy or other new medications to treat specific genetic disorders.

Management of Metabolic Disorders: In general, metabolic disorders arise from enzyme deficiencies that disrupt normal metbolic pathways. Compound "A" is metabolized to "B" by enzyme "X", compound "B" is metabolized to "C" by enzyme "Y", and compound "C" is metabolized to "D" by enzyme "Z". If enzyme "Z" is missing, compound "D" will be missing, while compounds "A", "B", and "C" will build up. The pathogenesis of this particular condition could result from lack of compound "D", if it is critical for some cellular function, or from toxicity due to excess "A", "B", and/or "C". Treatment of the metabolic disorder could be achieved through dietary supplementation of compound "D" and dietary restriction of compounds "A", "B", and/or "C" or by treatment with a medication that promoted disposal of excess "A", "B", or "C". Another approach that can be taken is enzyme replacement therapy, in which a patient is given an infusion of the missing enzyme.

- *Diet:* Dietary restriction and supplementation are key measures taken in several well-known metabolic disorders, including galactosemia, phenylketonuria (PKU), maple syrup urine disease, organic acidurias and urea cycle disorders. Such restrictive diets can be difficult for the patient and family to maintain, and require close consultation with a nutritionist who has special experience in metabolic disorders. The composition of the diet will change depending on the caloric needs of the growing child and special attention is needed during a pregnancy if a woman is affected with one of these disorders.
- *Medication:* Medical approaches include enhancement of residual enzyme activity (in cases where the enzyme is made but is not functioning properly), inhibition of other enzymes in the biochemical pathway to prevent buildup of a toxic compound, or diversion of a toxic compound to another form that can be excreted. Examples include the use of high doses of pyridoxine (vitamin B6) in some patients with homocystinuria to boost the activity of the residual cystathione synthase enzyme, administration of biotin to restore activity of several enzymes affected by deficiency of biotinidase, treatment with NTBC in Tyrosinemia to inhibit the production of succinylacetone which causes liver toxicity, and the use of sodium benzoate to decrease ammonia build-up in urea cycle disorders.
- *Enzyme Replacement Therapy:* Certain lysosomal storage diseases are treated with infusions of a recombinant enzyme (produced in a laboratory), which can reduce the accumulation of the compounds in various tissues. Examples include Gaucher disease, Fabry disease, Mucopolysaccharidoses and Glycogen storage disease type II. Such treatments are limited by the ability of the enzyme to reach the affected areas (the blood brain barrier prevents enzyme from reaching the brain, for example), and can sometimes be associated with allergic reactions. The long-term clinical effectiveness of enzyme replacement therapies vary widely among different disorders.

Other Examples:

- Angiotensin receptor blockers in Marfan syndrome & Loeys-Dietz
- Bone marrow transplantation
- Gene therapy

Ethical, Legal and Social Implications

Genetic information provides a unique type of knowledge about an individual and his/her family, fundamentally different than a typically laboratory test that provides a "snapshot" of an individual's health status. The unique status of genetic information and inherited disease has a number of ramifications with regard to ethical, legal, and societal concerns.

Societies: The more empirical approach to human and medical genetics was formalized by the founding in 1948 of the American Society of Human Genetics. The Society first began annual meetings that year (1948) and its international counterpart, the International Congress of Human Genetics, has met every 5 years since its inception in 1956. The Society publishes the American Journal of Human Genetics on a monthly basis. Medical genetics is now recognized as a distinct medical specialty in the U.S. with its own approved board (the American Board of Medical Genetics) and clinical specialty college (the American College of Medical Genetics). The College holds an annual scientific meeting, publishes a monthly journal, Genetics in Medicine, and issues position papers and clinical practice guidelines on a variety of topics relevant to human genetics.

Research: The broad range of research in medical genetics reflects the overall scope of this field, including basic research on genetic inheritance and the human genome, mechanisms of genetic and metabolic disorders, translational research on new treatment modalities, and the impact of genetic testing

Basic Genetics Research: Basic research geneticists usually undertake research in universities, biotechnology firms and research institutes.

Allelic Architecture of Disease: Sometimes the link between a disease and an unusual gene variant is more subtle. The genetic architecture of common diseases is an important factor in determining the extent to which patterns of genetic variation influence group differences in health outcomes. According to the common disease/common variant hypothesis, common variants present in the ancestral population before the dispersal of modern humans from Africa play an important role in human diseases. Genetic variants associated with Alzheimer disease, deep venous thrombosis, Crohn disease, and type 2 diabetes appear to adhere to this model. However, the generality of the model has not yet been established and, in some cases, is in doubt. Some diseases, such as many common cancers, appear not to be well described by the common disease/common variant model. Another possibility is that common diseases arise in part through the action of combinations of variants that are individually rare.

Most of the disease-associated alleles discovered to date have been rare, and rare variants are more likely than common variants to be differentially distributed among groups distinguished by ancestry. However, groups could harbor different, though perhaps overlapping, sets of rare variants, which would reduce contrasts between groups in the incidence of the disease. The number of variants contributing to a disease and the interactions among those variants also could influence the distribution of diseases among groups.

The difficulty that has been encountered in finding contributory alleles for complex diseases and in replicating positive associations suggests that many complex diseases involve numerous variants rather than a moderate number of alleles, and the influence of any given variant may depend in critical ways on the genetic and environmental background.

If many alleles are required to increase susceptibility to a disease, the odds are low that the necessary combination of alleles would become concentrated in a particular group purely through drift.

Population Substructure in Genetics Research: One area in which population categories can be important considerations in genetics research is in controlling for confounding between population substructure, environmental exposures, and health outcomes. Association studies can produce spurious results if cases and controls have differing allele frequencies for genes that are not related to the disease being studied, although the magnitude of this problem in genetic association studies is subject to debate. Various methods have been developed to detect and account for population substructure, but these methods can be difficult to apply in practice.

Population substructure also can be used to advantage in genetic association studies. For example, populations that represent recent mixtures of geographically separated ancestral groups can exhibit longer-range linkage disequilibrium between susceptibility alleles and genetic markers than is the case for other populations. Genetic studies can use this admixture linkage disequilibrium to search for disease alleles with fewer markers than would be needed otherwise. Association studies also can take advantage of the contrasting experiences of racial or ethnic groups, including migrant groups, to search for interactions between particular alleles and environmental factors that might influence health.

Behavioural Genetics

Behavioural genetics is the field of study that examines the role of genetics in animal (including human) behaviour. Often associated with the "nature versus nurture" debate, behavioural genetics is highly interdisciplinary, involving contributions from biology, genetics, ethology, psychology, and statistics. Behavioural geneticists study the inheritance of behavioural traits. In humans this often use the twin study or adoption study. In animal studies, breeding, transgenesis, and gene knockout techniques are common; psychiatric genetics is a closely related field.

History

Sir Francis Galton, a nineteenth-century intellectual, is recognized as one of the first behavioural geneticists. Galton, a cousin of Charles Darwin, studied the heritability of human ability, focusing on mental characteristics as well as eminence among close relatives in the English upper-class. In 1869, Galton published his results in *Hereditary Genius*. In his work, Galton "introduced multivariate analysis and paved the way towards modern Bayesian statistics" that are used throughout the sciences—launching what has been dubbed the "Statistical Enlightenment".

Behaviour genetics, *per-se*, gained recognition as a research discipline with the publication in 1960 of the textbook *Behaviour Genetics* by J.L. Fuller and W.R. Thompson. Underscoring the role of evolution in behavioural genetics, Theodosius Dobzhansky was elected the first president of the Behaviour Genetics Association in 1972; the BGA bestows the Dobzhansky Award on researchers for their outstanding contributions to the field. In the early 1970s, Lee Ehrman, a doctoral student of Dobzhansky, wrote seminal papers describing the relationship between genotype frequency and mating success in Drosophila, lending impetus to the pursuit of genetic studies of behaviour in other animals.

Notable Behavioural Geneticists

Notable behavioural geneticists include Dorret Boomsma, John DeFries, Lindon Eaves, David Fulker, John Hewitt, Kenneth Kendler, John Loehlin, Nick Martin, Gerald McClearn, Robert Plomin, Theodore Reich, who was a pioneer in psychiatric genetics, Hans van Abeelen, Avshalom Caspi, and Steven G. Vandenberg, the founding editor of the journal *Behaviour Genetics*.

Genetic Diversity

Genetic diversity, the level of biodiversity, refers to the total number of genetic characteristics in the genetic makeup of a species. It is distinguished from genetic variability, which describes the tendency of genetic characteristics to vary. Genetic diversity serves as a way for populations to adapt to changing environments. With more variation, it is more likely that some individuals in a population will possess variations of alleles that are suited for the environment. Those individuals are more likely to survive to produce offspring bearing that allele. The population will continue for more generations because of the success of these individuals. The academic field of population genetics includes several hypotheses and theories regarding genetic diversity. The neutral theory of evolution proposes that diversity is the result of the accumulation of neutral substitutions. Diversifying selection is the hypothesis that two subpopulations of a species live in different environments that select for different alleles at a particular locus. This may occur, for instance, if a species has a large range relative to the mobility of individuals within it. Frequency-dependent selection is the hypothesis that as alleles become more common, they become more vulnerable. This is often invoked in host-pathogen interactions, where a high frequency of a defensive allele among the host means that it is more likely that a pathogen will spread if it is able to overcome that allele.

Importance of Genetic Diversity

There are many different ways to measure genetic diversity. The modern causes for the loss of animal genetic diversity have also been studied and identified. A 2007 study conducted by the National Science Foundation found that genetic diversity and biodiversity are dependent upon each other—that diversity within a species is necessary to maintain diversity among species, and vice versa. According to the lead researcher in the study, Dr. Richard Lankau, "If any one type is removed from the system, the cycle can break down, and the community becomes dominated by a single species." The interdependence between genetic and biological diversity is delicate. Changes in biological diversity lead to changes in the environment, leading to adaptation of the remaining species. Changes in genetic diversity, such as in loss of species, leads to a loss of biological diversity.

Survival and Adaptation

Genetic diversity plays a very important role in survival and adaptability of a species because when a species's environment changes, slight gene variations are necessary to produce changes in the organisms' anatomy that enables it to adapt and survive. A species that has a large degree of genetic diversity among its population will have more variations from which to choose the most fit alleles. Increase in genetic diversity is also essential for a species to evolve. Species that have very little genetic variation are at a great risk. With very little gene variation within the species, healthy reproduction becomes increasingly difficult, and offspring often deal with similar problems to those of inbreeding. The vulnerability of a population to certain types of diseases can also increase with reduction in genetic diversity.

Agricultural Relevance

When humans initially started farming, they used selective breeding to pass on desirable traits of the crops while omitting the undesirable ones. Selective breeding leads to monocultures: entire farms of nearly genetically identical plants. Little to no genetic diversity makes crops extremely susceptible to widespread disease. Bacteria morph and change constantly. When a disease causing bacterium changes to attack a specific genetic variation, it can easily wipe out vast quantities of the species. If the genetic variation that the bacterium is best at attacking happens to be that which humans have selectively bred to use for harvest, the entire crop will be wiped out.

A very similar occurrence is the cause of the infamous Potato Famine in Ireland. Since new potato plants do not come as a esult of reproductin but rather frompieces of the parent lant, no genetic diversty is developed, and the entire crop is essentially a clone of one potato, it is especially susceptible to an epidemic. In the 1840s, much of Ireland's population depended on potatoes for food. They planted namely the "lumper" variety of potato, which was susceptible to a rot-causing plasmodiophorid called *Phytophthora infestans*. This plasmodiophorid destroyed the vast majority of the potato crop, and left one million people to starve to death.

Coping with Poor Genetic Diversity

The natural world has several ways of preserving or increasing genetic diversity. Among oceanic plankton, viruses aid in the genetic shifting process. Ocean viruses, which infect the plankton, carry genes of other organisms in addition to their own. When a virus containing the genes of one cell infects another, the genetic makeup of the latter changes. This constant shift of genetic make-up helps to maintain a healthy population of plankton despite complex and unpredictable environmental changes. Cheetahs are a threatened species. Extremely low genetic diversity and resulting poor sperm quality has made breeding and survivorship difficult for cheetahs -- only about 5% of cheetahs survive to adulthood. About 10,000 years ago, all but the *jubatus* species of cheetahs died out. The species encountered a population bottleneck and close family relatives were forced to mate with each other, or inbreed. However, it has been recently discovered that female cheetahs can mate with more than one male per litter of cubs. They undergo induced ovulation, which means that a new egg is produced every time a female mates. By mating with multiple males, the mother increases the genetic diversity within a single litter of cubs.

Measures of Genetic Diversity

Genetic Diversity of a population can be assessed by some simple measures.

- Gene Diversity is the proportion of polymorphic loci across the genome.
- Heterozygosity is the mean number of individuals with polymorphic loci.
- Alleles per locus is also used to demonstrate variability.

Other Measures of Diversity

Alternatively, other types of diversity may be assessed for organisms:

- taxonomic diversity
- ecological diversity
- morphological diversity

There are broad correlations between different types of diversity. For example, there is a close link between vertebrate taxonomic and ecological diversity.

Genetic Erosion

Genetic erosion is a process whereby an already limited gene pool of an endangered species of plant or animal diminishes even more when individuals from the surviving population die off without getting a chance to meet and breed with others in their endangered low population. Genetic erosion occurs because each individual organism has many unique genes which get lost when it dies without getting a chance to breed. Low genetic diversity in a population of wild animals and plants leads to a further diminishing gene pool, inbreeding and a weakening immune system and fast tracks that species towards eventual extinction. All the world's endangered species are plagued by varying degrees of genetic erosion and most need a human assisted breeding programme to keep their population viable and to keep them from going extinct in the long run.

The more critically endangered the species is (the smaller the popuation is), the more magnified the effect of genetic erosion gets when each surviving individual of the species is lost without getting a chance to breed. Genetic erosion gets compounded and accelerated by habitat fragmentation, today most endangered species live in smaller and smaller chunks of fragmented habitat interspersed with human settlements and farmland making it impossible for them to naturally meet and breed with others of their kind, many die off without getting a chance to breed and pass on their genes in the living population. The gene pool of a species or a population is the complete set of unique alleles that would be found by inspecting the genetic material of every living member of that species or population.

A large gene pool indicates extensive genetic diversity, which is associated with robust populations that can survive bouts of intense selection. Meanwhile, low genetic diversity can cause reduced biological fitness and an increased chance of extinction.

Processes and Consequences

"A population bottleneck creates a shrinking gene pool that leaves fewer and fewer mating partners. What are the genetic implications? The animals become part of a high stakes poker game — with a crooked dealer. After beginning with a 52-card deck, the players wind up with, say, five cards that they are dealt over and over. As they begin to inbreed, congenital effects appear, both physical and reproductive. Often abnormal sperm increase; infertility rises; the birthrate falls. Most perilous in the long run, each animal's immune defense system is weakened. Thus, even if an endangered species in a bottleneck can withstand whatever human development may be eating away at its habitat, it still faces the threat of an epidemic that could well be fatal to the entire population."

Genetic Erosion in Agricultural and Livestock Biodiversity

Genetic erosion in agricultural and livestock biodiversity is the loss of genetic diversity, including the loss of individual genes, and the loss of particular combinants of genes (or gene complexes) such as those manifested in locally adapted landraces of domesticated animals or plants adapted to the natural environment in which they originated. The term genetic erosion is sometimes used in a narrow sense, such as for the loss of alleles or genes, as well as more broadly, referring to the loss of varieties or even species. The major driving forces behind genetic erosion in crops are: variety replacement, land clearing, overexploitation of species, population pressure, environmental degradation, overgrazing, policy and changing agricultural systems.

The main factor, however, is the replacement of local varieties of domestic plants and animals by high yielding or exotic varieties or species. A large number of varieties can also often be dramatically reduced when commercial varieties (including GMOs) are introduced into traditional farming systems. Many researchers believe that the main problem related to agro-ecosystem management is the general tendency towards genetic and ecological uniformity imposed by the development of modern agriculture.

Human Intervention, Modern Science and Safeguards to Guard Against Genetic Erosion

In-Situ Conservation: With advances in modern science several tecniques and safeguards have emerged to check the relentless advance of genetic erosion and the resulting acceleration of endangered species towards extinction. However many of these techniques and safeguards are too expensive yet to be practical, the best way to protect species is to protect their habitat and to let them live in it naturally.

Wildlife sanctuaries and national parks have been created to preserve entire ecosystems with all the web of species which call them home. Wildlife corridors are created to join fragmented habitats to enable endangered species to travel, meet and breed with others of their kind. Scientific conservation and modern wildlife management techniques with the help of scientifically trained staff help manage these protected ecosystems and the wildlife found in them. Wild animals are also translocated and reintroduced to other locations physically when fragmented wildlife habitat is too far and isolated to be able to link it with a wildlife corridor or when local extinction has already occurred.

Ex-Situ Conservation

Modern policies of the zoo associations and zoos around the world have changed to putting extreme importance on keeping and breeding wild sourced pure species and subspecies of animals and birds in their registered endangered species breeding programmes which will have a chance to be reintroduced and survive in the wild. Main objectives of zoos today has changed to breed pure breed species and subspecies to assist conservation efforts in the wild. Zoos do this by maintaining extremely detailed scientific breeding records i.e. studbooks and loaning their pure breed wild animals and birds to other zoos around the country and indeed globally for breeding to safeguard against inbreeding and hybrids which are considered genetically compromised thus not fit for reintroduction in the wild and in the case of unnaturally found hybrids also to guard against genetic pollution in naturally evolved, region specific, pure wild stocks. Costly and sometimes controversial ultra modern ex-situ conservation techniques have emerged for saving the genetic biodiversity on our planet and the diversity in their gene pool by guarding against genetic erosion through modern concepts like seedbanks, sperm banks and tissue banks. Sperms, eggs and embryos can now be frozen and kept in these banks which are sometimes called Modern Noah's Ark or Frozen Zoos. Cryopreservation techniques are used to freeze these living materials and keep them alive by storing them submerged in liquid nitrogen tanks. Thus, preserved material can then be used for artificial insemination, in vitro fertilization, embryo transfer and cloning to protect diversity in the gene pool of critically endangered species.

It is today possible to save endangered species from extinction by preserving parts like tissue, sperms, eggs etc. even after the death of a critically endangered animal or collected from one found freshly dead in captivity or from wild and resurrect it with the help of cloning and give it another chance to breed its genes into the living population of the respective species which is threatened with extinction. Resurrection of dead critically endangered wildlife with the help of cloning is still being perfected and is still too expensive to be practical but with time and advancement is science it may well become a routine procedure. Recently strategies have been made to find an integrated approach to in situ and ex situ conservation.

Genetic Hitchhiking

Genetic hitchhiking is the process by which an evolutionarily neutral or deleterious allele or mutation may spread through the gene pool by virtue of being linked to a gene that is positively selected. More generally, genetic hitchhiking can refer to the process by which a gene's frequency changes due to selection operating upon linked genes. Proximity on a chromosome may allow genes to be dragged through the selection process due to an advantageous gene nearby. The most often cited example of this is one of a hypothetical mutator whih increases the general mutation rate in the area around it.

—M———A—

On this chromosome the gene M is a mutator allele, increasing the rate of mutation in the surrounding area. A is an allele which is at fixation in the population - that is, all individuals possess this allele in the homozygous state. Due to the increased mutation rate, the A allele may be mutated into a new, advantageous allele, A*.

—M———A*—

The individual in which this chromosome lies will now have a selective advantage over other individuals of this species, so the allele A* will spread through the population by the normal processes of natural selection. M, due to its proximity to A*, will be dragged through into the general population. This process only works when M is very close to the allele it has mutated. A greater distance would increase the chance of recombination separating M from A*, leaving M alone with any deleterious mutations it may have caused. For this reason the two alleles are generally directly next to each other, or present in asexual species where recombination cannot disrupt linkage.

In the example above, the mutator caused the adaptive allele to appear, but genetic hitchhiking can also happen by chance alone. Whether a neutral allele becomes fixed is a matter of chance. The traditional view of this stochastic process is that it is dominated by sampling error, that is genetic drift. But it may instead be dominated by whether the mutation is linked to a good genetic background: this is known as genetic draft.

Genetic Monitoring

Genetic monitoring is the use of molecular markers to (i) identify individuals, species or populations, or (ii) to quantify changes in population genetic metrics (such as effective population size, genetic diversity and population size) over time. Genetic monitoring can thus be used to detect changes in species abundance and/or diversity, and has become an important tool in both conservation and livestock management. The types of molecular markers used to monitor populations are most commonly mitochondrial, microsatellites or single-nucleotide polymorphisms (SNPs), while earlier studies also used allozyme data. Species gene diversity is also recognized as an important biodiversity metric for implementation of the Convention on Biological Diversity.

Types

Types of population changes that can be detected by genetic monitoring include population growth and decline, spread of pathogens, adaptation to environmental change, hybridization, introgression and fragmentation events. Most of these changes are monitored using 'neutral' genetic markers (markers for which mutational changes do not change their adaptive fitness within a population). However markers showing adaptive responses to environmental change can be 'non-neutral' (e.g. mutational changes affect their relative fitness within a population).

Two broad categories of genetic monitoring have been defined: Category I encompasses the use of genetic markers as identifiers of individuals (Category Ia), populations and species (Category Ib) for traditional population monitoring. Category II represents the use of genetic markers to monitor changes of population genetic parameters, which include estimators of effective population size (Ne), genetic variation, population inter-mixing, structure and migration.

Examples

Estimating Abundance and Life History Parameters – Category Ia: At the individual level, genetic identification can enable estimation of population abundance and population increase rates within the framework of mark-recapture models. The abundance of cryptic or elusive species that are difficultto monitor can be estimated by collecting non-invasive biological samples in the field (e.g. feathers, scat or fur) and using these to identify individuals through microsatellite or single-nucleotide polymorphism (SNP) genotyping. This census of individuals can then be used to estimate population abundance via mark-recapture analysis. For example, this technique has been used to monitor populations of grizzly bear, Brush-tailed Rock-wallaby, Bengal tiger and snow leopard. Population growth rates are a product of rates of population recruitment and survival, and can be estimated through open mark-recapture models. For example, DNA from feathers shed by the Eastern Imperial Eagle shows lower cumulative survival over time than seen for other long-lived raptors.

Identifying Species – Category Ib: Use of molecular genetic techniques to identify species can be useful for a number of reasons. Species identification in the wild can be used to detect changes in population ranges or site occupancy, rates of hybridization and the emergence and spread of pathogens and invasive species. Changes in population ranges have been investigated for Iberian lynx and wolverine, while monitoring of Westslope cutthroat trout shows widespread ongoing hybridization with introduced rainbow trout and Canada lynx-bobcat hybrids have been detected at the southern periphery of the current population range for lynx. The emergence and spread of pathogens can be tracked using diagnostic molecular assays – for example, identifying the spread of West Nile virus among mosquitoes in the eastern US to identify likely geographical origins of infection and identifying gene loci associated with parasite susceptibility in bighorn sheep. Genetic monitoring of invasive species is of conservation and economic interest, as invasions often affect the ecology and range of native species and may also bring risks of hybridization (e.g. for copepods, barred owl and spotted owl, and Lessepsian rabbitfish).

Species identification is also of considerable utility in monitoring fisheries and wildlife trade, where conventional visual identification of butchered or flensed products is difficult or impossible. Monitoring of trade and consumption of species of conservation interest can be carried out using molecular amplification and identification of meat or fish obtained from markets. For example genetic market surveys have been used to identify protected species and populations of whale (e.g., North Pacific Minke whale) and dolphin species appearing in the marketplace. Other surveys of market trade have focused on pinnipeds, sea horses and sharks. Such surveys are used to provide ongoing monitoring of the quantity and movement of fisheries and wildlife products through markets and for detecting poaching or other illegal, unreported or unregulated (IUU) exploitation (e.g. IUU fishing).

Although initial applications focused on species identification and population assessments, market surveys also provide the opportunity for a range of molecular ecology investigations including capture-recapture, assignment tests and population modeling. These developments are potentially relevant to genetic monitoring Category II.

Monitoring Population Genetic Parameters – Category II: Monitoring of population changes through genetic means can be done retrospectively, through analysis of 'historical' DNA recovered from museum-archived species and comparison with contemporary DNA of that species. It can also be used as a tool for evaluating ongoing changes in the status and persistence of current populations. Genetic measures of relative population change include changes in diversity (e.g. heterozygosity and allelic richness). Monitoring of relative population changes through these metrics has been performed retrospectively for Beringian bison, Galapagos tortoise, houting, Atlantic salmon, northern pike, New Zealand snapper, steelhead trout, Greater Prairie Chicken, Mauritius kestrel and Hector's Dolphin and is the subject of many ongoing studies, including Danish and Swedish brown trout populations. Measuring absolute population changes (e.g. effective population size (Ne)) can be carried out by measuring changes in population allele frequencies ('Ftemporal') or levels of linkage disequilibrium over time ('LDNe'), while changing patterns of gene flow between populations can also be monitored by estimating differences in allele frequencies between populations over time. Subjects of such studies include grizzly bears, cod, red deer, Leopard frogs and Barrel Medic.

Genetic monitoring has also been increasingly used in studies that monitor environmental changes through changes in the frequency of adaptively selected markers. For example the genetically controlled photo-periodic response (hibernating time) of pitcher-plant mosquitos (*Wyeomyia smithii*) has shifted in response to longer growing seasons for pitcher plants brought on by warmer weather. Experimental wheat populations grown in contrasting environments over a period of 12 generations found that changes in flowering time were closely correlated with regulatory changes in one gene, suggesting a pathway for genetic adaptation to changing climate in plants. Genetic monitoring is also useful in monitoring the ongoing health of small, relocated populations. Good examples of this are found for New Zealand birds, many species of which were greatly impacted by habitat destruction and the appearance of numerous mammalian predators in the last century and have recently become part of relocation programmes that transfer a few 'founder' individuals to predator-free offshore "ecological" islands. E.g. Black robins, kakapo.

Status of Genetic Monitoring in Science

In February 2007 an international summit was held at the Institute of the Environment at UCLA, concerning 'Evolutionary Change in Human Altered Environments: An International Summit to translate Science into Policy'. This led to a special issue of the journal of Molecular Ecology organized around our understanding of genetic effects in three main categories: (i) habitat disturbance and climate change (ii) exploitation and captive breeding (iii) invasive species and pathogens. In 2007 a Working Group on Genetic Monitoring was launched with joint support from NCEAS and NESCent to further develop the techniques involved and provide general monitoring guidance for policy makers and managers.

Genetic Monitoring in Natural Resource Agencies

Many natural resource agencies see genetic monitoring as a cost-effective and defensible way to monitor fish and wildlife populations. As such scientists in the U.S. Geologic Survey, U.S. Forest Service, National Parks Service, and National Marine Fisheries Service have been developing new methods and tools to use genetic monitoring, and applying such tools across broad geographic scales.

Genetic Pollution

Genetic pollution is undesirable and uncontrolled gene flow into wild populations.

Usage

- Conservation biologists and conservationists have long used genetic pollution as a term to describe gene flow from a domestic, feral, non-native or invasive species to a wild indigenous population.
- The term is of late being associated with the gene flow from a genetically engineered (GE) or genetically modified organism (GMO) to a non GE/GM organism.

Invasive Species

Conservation biologists and conservationists have from long before been using the term to describe the undesirable gene flow from domestic, feral, non-native and invasive species into wild indigenous species. For example, TRAFFIC is the international wildlife trade monitoring network which works to ensure that trade in wild plants and animals is not a threat to the conservation of nature.

They promote the awareness of the harmful effects of introduced invasive species that may "*hybridize with native species, causing genetic pollution*". The Joint Nature Conservation Committee (JNCC) is the statutory adviser to the Government of United Kingdom and international nature conservation. Its work contributes to maintaining and enriching biological diversity and educating about the harmful effects of the introduction of invasive/non-native species. In this context they have advised that invasive species:

> *"will alter the genetic pool (a process called genetic pollution), which is an irreversible change."*

Effect upon Endangered Species

Compromise of certain wild plant or animal endangered species may have particular significance in the case of these relatively small populations of threatened taxa. As an example the endangered African wild dog has its conservation status undermined not only by loss of habitat, but also by admixture of gene flow from domesticated canids, as domesticated dogs interbreed with the wild species.

Genetic Engineering

In the field of agriculture, agroforestry and animal husbandry *genetic pollution* is being used to describe the undesirable gene flow between GE species and wild relatives;. An early use of the term *genetic pollution* in this later sense appears in a wide-ranging review of the potential ecological risks from genetic engineering in The Ecologist magazine in July 1989. It was also popularized by environmentalist Jeremy Rifkin in his 1998 book *The Biotech Century*. While intentional crossbreeding between two genetically distinct varieties is described as hybridization with the subsequent introgression of genes, Rifkin, who had played a leading role in the ethical debate for over a decade before, used genetic pollution to describe the risks that might occur due the unintentional process of genetically modified organisms (GMOs) dispersing their genes into the natural environment by breeding with wild plants or animals.

The usage of genetic pollution by the Food and Agriculture Organization of the United Nations (FAO) is currently defined as:

> *"Uncontrolled spread of genetic information (frequently referring to transgenes) into the genomes of organisms in which such genes are not present in nature."*

Since 2005 exist a GM Contamination Register, launched for GeneWatch UK and Greenpeace International and, it record all incidents intentional or accidental release of genetically modified (GM) organisms. In a 10 year study of four different crops, none of the genetically modified plants were found to be more invasive or more persistent than their conventional counterparts. An often cited example of genetic pollution is the reputed discovery of transgenes from GE maize in landraces of maize in Oaxaca, Mexico. The report from Quist and Chapela, has since been discredited on methodological grounds. The scientific journal that originally published the study concluded that "the evidence available is not sufficient to justify the publication of the original paper." More recent attempts to replicate the original studies have concluded that genetically modified corn is absent from southern Mexico in 2003 and 2004. The goal of genetic engineering crop plants to help advance tenability and condition of the world food supply has been a conlict with public and health concerns raised about the safety of the food from the end product. A 2004 study performed near an Oregon field trial for a genetically modified variety of creeping bentgrass (*Agrostis stolonifera*) revealed that the transgene and its associate trait (resistance to the glyphosate herbicide) could be transmitted by wind pollination to resident plants of different *Agrostis* species, up to 14 km from the test field. In 2007, the Scotts Company, producer of the genetically modified bentgrass, agreed to pay a civil penalty of $500,000 to the United States Department of Agriculture (USDA). The USDA alleged that Scotts "failed to conduct a 2003 Oregon field trial in a manner which ensured that neither glyphosate-tolerant creeping bentgrass nor its offspring would persist in the environment".

Controversial Term

Whether genetic pollution or similar terms, such as "*genetic deterioration*", "*genetic swamping*", "*genetic takeover*" and "*genetic aggression*", are an appropriate scientific description of the biology of invasive species is debated Hymer and Simberloff argue that these types of terms:

> *"...imply either that hybrids are less fit than the parentals, which need not be the case, or that there is an inherent value in "pure" gene pools".*

They recommend that gene flow from invasive species be termed genetic mixing since:

> *" "Mixing" need not be value-laden, and we use it here to denote mixing of gene pools whether or not associated with a decline in fitness".*

Environmentalists such as Patrick Moore, an ex-member and cofounder of Greenpeace, questions if the term genetic pollution is more political than scientific.The term is considered to arouse emotional feelings towards the subject matter. In an interview he comments:

> *"If you take a term used quite frequently these days, the term "genetic pollution," otherwise referred to as genetic contamination, it is a propaganda term, nt a technical or scientific term. Pollution and contamination are both value judgments. By using the word "genetic" it gives the public the impression that they are talking about something scientific or technical—as if there were such a thing as genes that amount to pollution."*

Population Genetics

Population genetics is the study of allele frequency distribution and change under the influence of the four main evolutionary processes: natural selection, genetic drift, mutation and gene flow. It also takes into account the factors of recombination, population subdivision and population structure.

It attempts to explain such phenomena as adaptation and speciation. Population genetics was a vital ingredient in the emergence of the modern evolutionary synthesis. Its primary founders were Sewall Wright, J. B. S. Haldane and R. A. Fisher, who also laid the foundations for the related discipline of quantitative genetics.

Fundamentals

Population genetics is the study of the frequency and interaction of alleles and genes in populations. A population is a set of organisms in which any pair of members can breed together. This implies that all members belong to the same species and live near each other.

For example, all of the moths of the same species living in an isolated forest are a population. A gene in this population may have several alternate forms, which account for variations between the phenotypes of the organisms.

An example might be a gene for colouration in moths that has two alleles: black and white. A gene pool is the complete set of alleles for a gene in a single population; the allele frequency for an allele is the fraction of the genes in the pool that is composed of that allele (for example, what fraction of moth colouration genes are the black allele). Evolution occurs when there are changes in the frequencies of alleles within a population; for example, the allele for black colour in a population of moths becoming more common.

Hardy–Weinberg Principle

To understand the mechanisms that cause a population to evolve, it is useful to consider what conditions are required for a population not to evolve. The *Hardy-Weinberg principle* states that the frequencies of alleles (variations in a gene) in a sufficiently large population will remain constant if the only forces acting on that population are the random reshuffling of alleles during the formation of the sperm or egg, and random combination of the alleles in these sex cells during fertilization. Such a population is said to be in *Hardy-Weinberg equilibrium* as it is not evolving.

Hardy Weinberg equilibrium is impossible in nature. Genetic equilibrium is an ideal state that provides a baseline to measure genetic change against. Allele frequencies in a population remain static across generations, provided the following conditions are at hand: random mating, no mutation (the alleles don't change), no migration or emigration (no exchange of alleles between populations), infinitely large population size, and no selective pressure for or against any traits. In the simplest case of a single locus with two alleles: the dominant allele is denoted A and the recessve a and their frequenies are denoted y p and q; freq(A) = p; freq(a) = q; $p + q = 1$.

If the population is in equilibrium, then we will have freq(AA) = p^2 for the AA homozygotes in the population, freq(aa) = q^2 for the aa homozygotes, and freq(Aa) = $2pq$ for the heterozygotes. Based on these equations, useful but difficult-to-measure facts about a population can be determined. For example, a patient's child is a carrier of a recessive mutation that causes cystic fibrosis in homozygous recessive children. The parent wants to know the probability of her grandchildren inheriting the disease. In order to answer this question, the genetic counselor must know the chance that the child will reproduce with a carrier of the recessive mutation. This fact may not be known, but disease frequency is known. We know that the disease is caused by the homozygous recessive genotype; we can use the Hardy–Weinberg principle to work backward from disease occurrence to the frequency of heterozygous recessive individuals.

Scope and Theoretical Considerations

The mathematics of population genetics were originally developed as part of the modern evolutionary synthesis. According to Beatty (1986), it defines the core of the modern synthesis.

According to Lewontin (1974), the theoretical task for population genetics is a process in two spaces: a "genotypic space" and a "phenotypic space". The challenge of a *complete* theory of population genetics is to provide a set of laws that predictably map a population of genotypes (G_1) to a phenotype space (P_1), where selection takes place, and another set of laws that map the resulting population (P_2) back to genotype space (G_2) where Mendelian genetics can predict the next generation of genotypes, thus completing the cycle. Even leaving aside for the moment the non-Mendelian aspects of molecular genetics, this is clearly a gargantuan task.

In practice, there are two bodies of evolutionary theory that exist in parallel, traditional population genetics operating in the genotype space and the biometric theory used in plant and animal breeding, operating in phenotype space. The missing part is the mapping between the genotype and phenotype space. This leads to a "sleight of hand" (as Lewontin terms it) whereby variables in the equations of one domain, are considered parameters or *constants*, where, in a full-treatment they would be transformed themselves by the evolutionary process and are in reality *functions* of the state variables in the other domain. The "sleight of hand" is assuming that we know this mapping. Proceeding as if we do understand it is enough to analyze many cases of interest. For example, if the phenotype is almost one-to-one with genotype (sickle-cell disease) or the time-scale is sufficiently short, the "constants" can be treated as such; however, there are many situations where it is inaccurate.

The four Processes

Natural Selection: *Natural selection* is the process by which heritable traits that make it more likely for an organism to survive and successfully reproduce become more common in a population over successive generations.

The natural genetic variation within a population of organisms means that some individuals will survive more successfully than others in their current environment. Factors which affect reproductive success are also important, an issue which Charles Darwin developed in his ideas on sexual selection.

Natural selection acts on the phenotype, or the observable characteristics of an organism, but the genetic (heritable) basis of any phenotype which gives a reproductive advantage will become more common in a population. Over time, this process can result in adaptations that specialize organisms for particular ecological niches and may eventually result in the emergence of new species.

Natural selection is one of the cornerstones of modern biology. The term was introduced by Darwin in his groundbreaking 1859 book *On the Origin of Species*, in which natural selection was described by analogy to artificial selection, a process by which animals and plants with traits considered desirable by human breeders are systematically favoured for reproduction. The concept of natural selection was originally developed in the absence of a valid theory of heredity; at the time of Darwin's writing, nothing was known of modern genetics. The union of traditional Darwinian evolution with subsequent discoveries in classical and molecular genetics is termed the *modern evolutionary synthesis*. Natural selection remains the primary explanation for adaptive evolution.

Genetic Drift: *Genetic drift* is the change in the relative frequency in which a gene variant (allele) occurs in a population due to random sampling and chance. That is, the alleles in the offspring in the population are a random sample of those in the parents. And chance has a role in determining whether a given individual survives and reproduces. A population's allele frequency is the fraction or percentage of its gene copies compared to the total number of gene alleles that share a particular form.

Genetic drift is an important evolutionary process which leads to changes in allele frequencies over time. It may cause gene variants to disappear completely, and thereby reduce genetic variability. In contrast to natural selection, which makes gene variants more common or less common depending on their reproductive success, the changes due to genetic drift are not driven by environmental or adaptive pressures, and may be beneficial, neutral, or detrimental to reproductive success.

The effect of genetic drift is larger in small populations, and smaller in large populations. Vigorous debates wage among scientists over the relative importance of genetic drift compared with natural selection. Ronald Fisher held the view that genetic drift plays at the most a minor role in evolution, and this remained the dominant view for several decades. In 1968 Motoo Kimura rekindled the debate with his neutral theory of molecular evolution which claims that most of the changes in the genetic material are caused by genetic drift.

Mutation

Mutations are changes in the DNA sequence of a cell's genome and are caused by radiation, viruses, transposons and mutagenic chemicals, as well as errors that occr during meiosis or DNA replication. Errors are introduced particularly often in the process of DNA replication, in the polymerization of the second strand. These errors can also be induced by the organism itself, by cellular processes such as hypermutation.

Mutations can have an impact on the phenotype of an organism, especially if they occur within the protein coding sequence of a gene. Error rates are usually very low (1 error in every 10 million–100 million bases) due to the "proofreading" ability of DNA polymerases. Without proofreading, error rates are a thousandfold higher. Chemical damage to DNA occurs naturally as well, and cells use DNA repair mechanisms to repair mismatches and breaks in DNA. Nevertheless, the repair sometimes fails to return the DNA to its original sequence.

In organisms that use chromosomal crossover to exchange DNA and recombine genes, errors in alignment during meiosis can also cause mutations. Errors in crossover are especially likely when similar sequences cause partner chromosomes to adopt a mistaken alignment; this makes some regions in genomes more prone to mutating in this way. These errors create large structural changes in DNA sequence—duplications, inversionsor deletions of entire regions, or the accidental exchanging of whole parts between different chromosomes (called translocation).

Mutation can result in several different types of change in DNA sequences; these can either have no effect, alter the product of a gene, or prevent the gene from functioning. Studies in the fly *Drosophila melanogaster* suggest that if a mutation changes a protein produced by a gene, this will probably be harmful, with about 70 percent of these mutations having damaging effects, and the remainder being either neutral or weakly beneficial.

Due to the damaging effects that mutations can have on cells, organisms have evolved mechanisms such as DNA repair to remove mutations. Therefore, the optimal mutation rate for a species is a trade-off between costs of a high mutation rate, such as deleterious mutations, and the metabolic costs of maintaining systems to reduce the mutation rate, such as DNA repair enzymes. Viruses that use RNA as their genetic material have rapid mutation rates, which can be an advantage since these viruses will evolve constantly and rapidly, and thus evade the defensive responses of e.g. the human immune system.

Mutations can involve large sections of DNA becoming duplicated, usually through genetic recombination. These duplications are a major source of raw material for evolving new genes, with tens to hundreds of genes duplicated in animal genomes every million years. Most genes belong to larger families of genes of shared ancestry. Novel genes are produced by several methods, commonly through the duplication and mutation of an ancestral gene, or by recombining parts of different genes to form new combinations with new functions.

Here, domains act as modules, each with a particular and independent function, that can be mixed together to produce genes encoding new proteins with novel properties. For example, the human eye uses four genes to make structures that sense light: three for colour vision and one for night vision; all four arose from a single ancestral gene.

Another advantage of duplicating a gene (or even an entire genome) is that this increases redundancy; this allows one gene in the pair to acquire a new function while the other copy performs the original function. Other types of mutation occasionally create new genes from previously noncoding DNA.

Gene Flow

Gene flow is the exchange of genes between populations, which are usually of the same species. Examples of gene flow within a species include the migration and then breeding of organisms, or the exchange of pollen. Gene transfer between species includes the formation of hybrid organisms and horizontal gene transfer.

Migration into or out of a population can change allele frequencies, as well as introducing genetic variation into a population. Immigration may add new genetic material to the established gene pool of a population. Conversely, emigration may remove genetic material. As barriers to reproduction between two diverging populations are required for the populations to become new species, gene flow may slow this process by spreading genetic differences between the populations. Gene flow is hindered by mountain ranges, oceans and deserts or even man-made structures such as the Great Wall of China, which has hindered the flow of plant genes.

Depending on how far two species have diverged since their most recent common ancestor, it may still be possible for them to produce offspring, as with horses and donkeys mating to produce mules. Such hybrids are generally infertile, due to the two different sets of chromosomes being unable to pair up during meiosis. In this case, closely related species may regularly interbreed, but hybrids will be selected against and the species will remain distinct. However, viable hybrids are occasionally formed and these new species can either have properties intermediate between their parent species, or possess a totally new phenotype. The importance of hybridization in creating new species of animals is unclear, although cases have been seen in many types of animals, with the gray tree frog being a particularly well-studied example.

Hybridization is, however, an important means of speciation in plants, since polyploidy (having more than two copies of each chromosome) is tolerated in plants more readily than in animals. Polyploidy is important in hybrids as it allows reproduction, with the two different sets of chromosomes each being able to pair with an identical partner during meiosis. Polyploids also have more genetic diversity, which allows them to avoid inbreeding depression in small populations.

Horizontal gene transfer is the transfer of genetic material from one organism to another organism that is not its offspring; this is most common among bacteria. In medicine, this contributes to the spread of antibiotic resistance, as when one bacteria acquires resistance genes it can rapidly transfer them to other species. Horizontal transfer of genes from bacteria to eukaryotes such as the yeast *Saccharomyces cerevisiae* and the adzuki bean beetle *Callosobruchus chinensis* may also have occurred.

An example of larger-scale transfers are the eukaryotic bdelloid rotifers, which appear to have received a range of genes from bactria, ungi, and plants. Viruses can also carry DNA between organisms, allowing transfer of genes even across biological domains. Large-scale gene transfer has also occurred between the ancestors of eukaryotic cells and prokaryotes, during the acquisition of chloroplasts and mitochondria.

Gene flow is the transfer of alleles from one population to another.

Migration into or out of a population may be responsible for a marked change in allele frequencies. Immigration may also result in the addition of new genetic variants to the established gene pool of a particular species or population. There are a number of factors that affect the rate of gene flow between different populations. One of the most significant factors is mobility, as greater mobility of an individual tends to give it greater migratory potential. Animals tend to be more mobile than plants, although pollen and seeds may be carried great distances by animals or wind.

Maintained gene flow between two populations can also lead to a combination of the two gene pools, reducing the genetic variation between the two groups. It is for this reason that gene flow strongly acts against speciation, by recombining the gene pools of the groups, and thus, repairing the developing differences in genetic variation that would have led to full speciation and creation of daughter species.

For example, if a species of grass grows on both sides of a highway, pollen is likely to be transported from one side to the other and vice versa. If this pollen is able to fertilise the plant where it ends up and produce viable offspring, then the alleles in the pollen have effectively been able to move from the population on one side of the highway to the other.

Genetic Structure

Because of physical barriers to migration, along with limited vagility, ad natal philopatry, natural populations are rarely panmictic (Buston *et al.*, 2007). There is usually a geographic range within which individuals are more closely related to one another than those randomly selected from the general population. This is described as the extent to which a population is genetically structured (Repaci *et al.*, 2007). Genetic structuring can be caused by migration due to historical climate change, species range expansion or current availability of habitat.

Microbial Population Genetics

Microbial population genetics is a rapidly advancing field of investigation with relevance to many other theoretical and applied areas of scientific investigations. The population genetics of microorganisms lays the foundations for tracking the origin and evolution of antibiotic resistance and deadly infectious pathogens. Population genetics of microorganisms is also an essential factor for devising strategies for the conservation and better utilization of beneficial microbes (Xu, 2010).

History

Population Genetics: *Population genetics* was developed as a reconciliation of the Mendelian and biometrician models. A key step was the work of the British biologist and statistician R.A. Fisher. In a series of papers starting in 1918 and culminating in his 1930 book *The Genetical Theory of Natural Selection*, Fisher showed that the continuous variation measured by the biometricians could be produced by the combined action of many discrete genes, and that natural selection could change gene frequencies in a population, resulting in evolution (though lacking the knowledge of what an actual gene was at this time, it should be said in this sense he understood phenotypic trait frequency, rather than specifically identifiable gene frequency).

In a series of papers beginning in 1924, another British geneticist, J.B.S. Haldane, applied statistical analysis to real-world examples of natural selection, such as the evolution of industrial melanism in peppered moths, and showed that natural selection worked at an even faster rate than Fisher assumed.

The American biologist Sewall Wright, who had a background in animal breeding experiments, focused on combinations of interacting genes, and the effects of inbreeding on small, relatively isolated populations that exhibited genetic drift. In 1932, Wright introduced the concept of an adaptive landscape and argued that genetic drift and inbreeding could drive a small, isolated sub-population away from an adaptive peak, allowing natural selection to drive it towards different adaptive peaks. Fisher and Wright had some fundamental disagreements and a controversy about the relative roles of selection and drift continued for much of the century between the Americans and the British. The Frenchman Gustave Malécot was also important early in the development of the discipline. The work of Fisher, Haldane and Wright founded the discipline of *population genetics*. This integrated natural selection with Mendelian genetics, which was the critical first step in developing a unified theory of how evolution worked.

John Maynard Smith was Haldane's pupil, whilst W.D. Hamilton was heavily influenced by the writings of Fisher. The American George R. Price worked with both Hamilton and Maynard Smith. American Richard Lewontin and Japanese Motoo Kimura were heavily influenced by Wright.

Modern Evolutionary Synthesis

In the first few decades of the 20th century, most field naturalists continued to believe that Lamarckian and orthogenic mechanisms of evolution provided the best explanation for the complexity they observed in the living world. However, as the field of genetics continued to develop, those views became less tenable. Theodosius Dobzhansky, a postdoctoral worker in T. H. Morgan's lab, had been influenced by the work on genetic diversity by Russian geneticists such as Sergei Chetverikov. He helped to bridge the divide between the foundations of microevolution developed by the population geneticists and the patterns of macroevolution observed by field biologits, with his 1937 book *Genetics and the Origin of Species*. Dobzhansky examined the genetic diversity of wild populations and showed that, contrary to the assumptions of the population geneticists, these populations had large amounts of genetic diversity, with marked differences between sub-populations.

The book also took the highly mathematical work of the population geneticists and put it into a more accessible form. In Great Britain E.B. Ford, the pioneer of ecological genetics, continued throughout the 1930s and 1940s to demonstrate the power of selection due to ecologcal factors including the ability to maintain genetic diversity through genetic polymorphisms such as human blood types. Ford's work would contribute to a shift in emphasis during the course of the modern synthesis towards natural selection over genetic drift.

Genetic Analysis of Adenohypophysis Formation in Zebrafish

The Pituitary Gland (hypophysis) and the hypothalamus constitute an integrative center of endocrine control, linking the central nervous system and the endocrine system in regulating basic body functions and homeostasis. The pituitary itself comprises two anatomically and functionally distinct systems, the anterior lobe (adenohypophysis) and the posterior lobe (neurohypophysis).

The adenohypophysis is further subdivided into three regions: the pars anterior (or pars distalis), the pars tuberalis, and the pars intermedia. It contains at least six different cell types that are characterized by the different hormones they produce and secrete: lactotropes generating prolactin (PRL), somatotropes generating GH, thyrotropes generating TSH, gonatotropes generating FSH and LH, corticotropes generating ACTH, and melanotropes generating MSH. ACTH and MSH are formed via proteolytic cleavage from a common proprotein, proopiomelanocortin (POMC), which also gives rise to endorphins secreted by specific cells of the brain. MSH is generated in the inter-mediate lobe (pars intermedia) of the pituitary, which is rather rudimentary in human.

In contrast to the adenohypophysis, the neurohypophysis itself does not contain endocrine cells. Rather, it secretes hormones from axonal termini of hypothalamatic neuroendocrine cells that have innervated the neurohypophysis. These hormones are oxytocin (isotocin in fish) and vasopressin (vasotocin in fish), generated in the supraoptic and paraventricular nuclei of the hypothalamus. In addition, neuroendocrine cells from these and other hypothalamic nuclei elaborate peptide hormones (so called releasing or release inhibiting hormones), which control the activity of endocrine cells of the adenohypophysis. In contrast to the neurohypophyseal hormones, these hormones do not reach the pituitary via nerve fibers. Rather, they travel in the blood via the hypothalamic-hypophyseal portal system.

The development of the adenohypophysis and the formation of its different cell types from a common primordium is controlled both by intrinsic and extrinsic factors. The adenohypophysis is derived from placodal ectoderm at the anterior neural ridge, that becomes committed toward an adenohypophyseal fate via inductive signals from the ventral diencephalon, a subdivision of the forebrain that later gives rise to hypothalamus, infundibulum and neurohypophysis. During further development, the adenohypophyseal anlage remains under the control of signals from the hypothalamus. These signals, together with intrinsic signals of the adenohypophysis itself, regulate the maintenance, proliferation and differential specification of adenohypophyseal cells by sequential activation of different transcription factors, such as the Lim class homeodomain proteins Lhx3 or Lhx4, the paired-like homeodomain protein Prophet-of-Pit1, the Pou domain protein Pit1, the T-box factor Tpit/Tbx19, the zinc finger protein Gata2, and several others.

Final evidence for the requirement of such factors for pituitary development requires genetic analyses via loss-of-function mutants. In mouse, such mutants have been generated using gene knockout technology, as in the case of Lhx3, that thereby was shown to be required for all adenohypophyseal cell types except the corticotropes. More recently, conditional knockout techniques allow one to investigate the pituitary-specific role of genes also involved in other processes, as in the case of the steroidogenic factor Sf1, that is required downstream of Gata for the formation of gonadotropes.

However, such gene targeting approaches are generally biased because they require a priori knowledge of the molecular nature of the genes to be analyzed. As an alternative to this reverse genetics approach, genes essential for mouse pituitary development were isolated via positional cloning of spontaneous, viable mutations, such as Prop1, that is mutated in Ames dwarf mutants and required for somatotropes, lactotropes, thyrotropes, and gonadotropes, and Pit1, which is mutated in Snell and Jackson dwarf mutants and required for somatotropes, lactotropes and thyrotropes, defining the so-called Pit1 lineage of the adenohypophysis.

Tpit/Tbx19 requirement for the Pomc lineage (corticotropes and melanotropes) was revealed via the analysis of human patients with isolated Acth deficiencies caused by recessive Tbx19 point mutations. However, as in other known human pituitary disorders, there are too few families and patients to allow a direct positional cloning of the affected human gene. Furthermore, the spontaneous mutation rates in humans and mice are too low to allow a saturating identification of all genes required for adenohypophysis development via forward genetics.

In light of these limitations of reverse and forward genetics in mammalian systems, we carried out a forward genetic analysis of pituitary mutants in a nonmammalian vertebrate, the zebrafish. Due to its extracorporal and rapid development, the transparency of its embryos and larvae, its high fecundity, its relatively small size and the ease of high-density maintenance, the zebrafish is highly suitable for mutant screens. In the past, three independent large-scale screens have been carried out, two of which used the chemical N-ethyl-N-nitrosourea (ENU) to introduce random point mutations over the entire genome.

Whereas in these screens mutant analyses were largely restricted to examining embryonic and larval morphology at different developmental stages, the more recent Tuebingen 2000 large-scale ENU screen was set up to allow mutant screening with molecular tools. As part of the Tuebingen 2000 screen, we searched for genes required for zebrafish adenohypophysis formation and patterning, carrying out large-scale whole mount *in situ* hybridizations with a probe detecting Gh encoding transcripts, a marker for somatotropes. The staining was necessary because the adenohypophysis of zebrafish larvae is morphologically too indistinct to be analyzed in large numbers without molecular tools.

Recent work has revealed both crucial similarities and differences in the development of the adenohypophysis between fish and mammals. The zebrafish pituitary contains the same cell types as the mammalian pituitary, however, there are crucial differences in the morphogenesis and the architecture of the glands in the different vertebrate species. Thus, the zebrafish adenohypophysis maintains its placodal organization and remains in a subepithelial position after oral cavity formation, whereas no invagination of the oral ectoderm equivalent to Rathke's pouch formation in mammals takes place.

Furthermore, there are differences in the patterning of the adenohypophyseal anlage: in zebrafish, the different cell types are distributed in three distinct domains along the antero-posterior axis of the pituitary, rather than along the dorsoventral axis as in mammals, with the Msh-generating pars intermedia most posteriorly, and the neurohypophysis located dorsal of adenohypophysis. In addition, fish and mammals display significant differences in the onset of the specification of the lactotrope lineage, that is the first lineage to specify in fish, but the last in mouse. Despite such differences, crucial mechanisms of adenohypophyseal induction and patterning appear to be highly conserved among fish and mammals, as for instance indicated by the conserved role of Sonic Hedgehog signalling from the ventral diencephalon. In light of these data, the large-scale ENU mutagenesis screen was expected to uncover genes regulating fish-specific features of the adenohypophysis, as well as genes with shared functions in mouse and fish. This would further illuminate the degree of conservation between the two vertebrate species, and possibly identify novel regulators also involved in mammalian pituitary development.

Results

Identification of Mutations Affecting Zebrafish Adenohypophysis Development: As part of a large-scale diploid F3 ENU mutagenesis screen to uncover recessive zygotic effect mutations, we screened for mutations affecting zebrafish adenohypophysis development. In total, F3 clutches of 4584 F2 families, representing 4253 mutagenized haploid genomes, were screened at 120 h post fertilization via whole mount *in situ* hybridization with a probe detecting transcripts of *gh*, a marker of the somatotropes.

The late developmental stage for mutant screening was chosen for two reasons: 1) to reveal also defects in later steps of pituitary development or patterning that would be missed when mutants were investigated much earlier to avoid pituitary phenotypes that are secondary consequences of early and general defects, such as early brain necrosis, that occurs very frequently, but usually leads to embryonic death between 24 and 72 hpf.

Secondary consequences on pituitary formation have also been revealed for midline mutants defective in signalling by Nodal and Hedgehog family members. In such mutants, eye field separation and hypothalamus formation are compromised, leading to pituitary defects caused by the lack of inductive signalling from the hypothalamus, consistent with the loss of the pituitary in hypothalamus-deficient mouse embryos. However, in contrast to pituitary-specific mutants, such zebrafish midline null mutants usually die before 120 hpf, although secondary pituitary defects caused by more subtle, viable midline mutations cannot be ruled out.

In total, we identified 13 mutants that were viable at 120 hpf but displayed a severe reduction or complete absence of *gh* staining in whole mount *in situ* hybridization. Other gross abnormalities, such as brain degeneration, which could have secondarily affected the pituitary, were not detected in these mutants. Only two of the 13 mutants (allele numbers t24594 and t20626) showed reduced eye distance and a curled-down tail, indicative for midline defects. Although t20626 was not further investigated, genomic mapping placed t24594 31–32 centimorgans from the top of linkage group 6, just below the marker Z6626. This is the same position as reported for the midline mutation *iguana*, and strongly suggests that t24594 is a weak *iguana* allele that we did not analyze further.

Of the remaining 11 adenohypophysis-specific mutations, eight were recovered. A combination of complementation crossing and genomic mapping revealed that they fall into four complementation groups, defining four genes essential for adenohypophysis development in the zebrafish. The genomic positions of the four genes. All genes appear to be dispensable for hypothalamus development, as revealed by the normal expression of *nkx2.1* in mutant embryos at 32 hpf, and the normal number of *isotocin*-positive cells at 32 hpf. *isotocin* cDNA was isolated in our laboratory via degenerate RT-PCR, and is identical with the cDNA recently reported by Unger and Glasgow. It is expressed in two nuclei of magnocellular neurons in the anterior hypothalamus, possibly the presumptive supraoptic or paraventricular nuclei, from where the peptide hormone is supposed to undergo axonal transport into the more ventrally and posteriorly located region of the neurohypophysis adjacent to the adenohypophysis.

In contrast to the normally developing hypothalamus and neurohypophysis, mutants in all four genes show specific, but different phenotypes in the adenohypophysis. Two of the mutants, *lim absent* (*lia*), represented by four alleles, and *pituitary absent* (*pia*), represented by one allele, lack the entire adenohypophysis. Whereas in *pit1* mutants, represented by two alleles, and in *all-absent-except-lactotropes* (*aal*) mutants, represented by one allele, only some of the pituitary cell types are absent.

Mutations Affecting the Entire Adenohypophysis: As in mammals and birds, the adenohypophysis of the zebrafish embryo is supposed to be of placodal origin, derived from medial cells of the anterior neural ridge, and laterally flanked by cells of the olfactory placodes. Whereas a few markers like *anf* are expressed in the anterior neural ridge at late gastrula and early segmentation stages, none continues to be expressed until cells can be identified as adenohypophyseal. In addition, none of them is altered in our mutants. In this respect, the currently known earliest *bona fide* marker for the presumptive zebrafish adenohypophysis is *lim3*, the homolog of *Lhx3* required for Rathke's pouch formation in mouse. In zebrafish, *lim3* starts to be expressed at the 21-somite stage, presumably in all cells of the pituitary anlage, when it is still organized in a placodal fashion in front of the forming head.

Differentiation of the first two pituitary cell lineages, the *prl* expressing lactotropes and the *pomc* expressing corticotropes and/or melanotropes, occurs very soon after the onset of *lim3* expression in lateral regions of the *lim3* domain, whereas expression of *tsh* in thryotropes and *gh* in somatotropes only starts later, after the pituitary has moved inwards. At 72 hpf, *prl*, *pomc*, *tsh*, and *gh* are expressed in distinct, but partially overlapping domains along the antero-posterior axis of the adenohypophysis.

lia and *pia* mutants lack the expression of all four adenohypophysis hormones at 72 hpf and 120 hpf. The remaining expression of *pomc* in two longitudinal stripes is confined to ß-endorphin-synthesizing cells in the ventral base of the diencephalon, whereas melanotropes and corticotropes of the adenohypophysis, as well as lactotropes, thyrotropes, and somatotropes, are absent.

In contrast to their indistinguishable phenotypes at d 3 and 5 after fertilization, *lia* and *pia* mutants differ in *lim3* expression during earlier stages of development. Although in *lia* mutants, *lim3* expression in the region of the adenohypophysis anlage is completely absent at 32 hpf, *lim3* expression in *pia* mutants is reduced, but clearly present. However, *prl* expression at 32 hpf is absent in both mutants. In sum, the expression pattern analyses suggest that both *lia* and *pia* are required for the specification of all adenohypophyseal cell types, with *lia* acting earlier than *pia*. *lia* appears to act upstream of *lim3*, whereas *pia* is more likely to act at the level, in parallel or downstream of *lim3*. The phenotype of *pia* zebrafish mutants is quite similar to that of *Lhx3*/*Lhx4* double mutant mice, suggesting that it might be caused by a mutation in a zebrafish *lim* gene. However, *lim3* itself can be ruled out, as it maps to a different linkage group.

Mutations Affecting Adenohypophyseal Patterning and Lineage Specification: The different cell types of the adenohypophysis derive from a common primordium, the placode, that initially consists of identical precursor cells. Subsequently, the placode becomes patterned, and cells in different regions of the primordium undergo differential specifications to form the different cell types characterized by the hormones they produce and secrete. Data from mammalian pituitary development suggest that the patterning process is characterized by subdivision of the common pool of precursor cells into certain cell lineages that then continue to branch until single cell types are generated. It is assumed that cells of the Pomc lineage, giving rise to corticotropes and melanotropes, split off first. In contrast to all other pituitary cell types, they express and require the T-box transcription factor Tpit/Tbx19, but are independent of the Lim class homeodomain transcription factor Lhx3. The complementary lineage, consisting of thyrotropes, somatotropes, lactotropes, and gonadotropes, is characterized by its Lhx3 dependence. Another lineage, comprising a subset of the Lhx3-dependent cell types, is defined via its dependence on the Pou domain transcription factor Pit1, and therefore called the Pit1 lineage. It consists of thyrotropes, lactotropes, and somatotropes, all of which are absent in *Pit1* mutant mice.

In our zebrafish screen, we could identify two genes required for the specification of two different adenohypophyseal cell lineages. Mutants in both genes show normal *lim3* expression at 32 hpf, indicating that the pituitary anlage is induced normally and of normal size. However, already at this early stage (32 hpf), the subdivision within the anlage is altered, indicated by the loss of *prl* expression in *pit1* mutants, whereas *prl* expression in *aal* mutants appears normal.

At 120 hpf, *pit1* mutant embryos are characterized by loss of lactotropes, somatotropes and thyrotropes, whereas *pomc*-positive adenohypophyseal cells, most likely anterior corticotropes and posterior melanotropes, are present at normal or even slightly elevated numbers. The zebrafish *pit1* mutants resemble Pit1-deficient mice, and indeed could be shown to carry mutations in the zebrafish *pit1* gene. *aal* mutant zebrafish show a very different phenotype; they fail to generate *pomc*-, *tsh*-, and *gh*-expressing pituitary cells, whereas the *prl*-expressing lactotropes are still present both at 32 hpf and 120 hpf. However, at the later developmental stages, the lactotropes are dispersed along the antero-posterior axis, rather than being organized in the sharp anterior domain found in wild-type siblings. This phenotype seems to define a new lineage of adenohypophyseal cells, consisting of corticotropes, melanotropes, thyrotropes, and somatotropes, all of which depend on the *aal* gene, whereas the lactotropes appear to develop independently of *aal* function. In contrast to *pit1*, no phenotype similar to that of *aal* mutants has been described in mouse thus far. Thus, the *aal* mutation seems to have revealed a thus far unknown lineage of adenohypophyseal cells that partially overlaps with the Pit1 lineage.

By screening approximately 1.5 times more mutagenized haploid genomes than in the first Tuebingen large-scale screen, we isolated eleven mutants with specific defects in the formation and patterning of the zebrafish adenohypophysis.

Of those 11, three could not be recovered as yet due to logistic problems. The remaining eight mutations fall into four complementation groups, defining four genes named *lia*, *pia*, *pit1*, and *aal*, all of which appear to be required for different steps of adenohypophysis development.

Genes Required Upstream and Downstream of Lim3 to Drive Development of All Adenohypophyseal Cell Types

Two genes, *lia* and *pia*, are required for all adenohypophyseal cell types. In contrast, other placodal derivatives appear to develop normally in mutant embryos, as for instance indicated by the normal expression of marker genes of the olfactory epithelium (Herzog, W., and M. Hammerschmidt, unpublished data), which derives from placodal positions located left and right of the adenohypophyseal placode at the anterior neural ridge.

This indicates that *lia* and *pia* specifically act on early steps of adenohypophyseal placode formation or maintenance, whereas other placodes are regulated differently. During adenohypophyseal development, *lia* and *pia* appear to be required for different, maybe subsequent steps, as indicated by the differences in the *lim3* expression in the two mutants. Thus, *lia* appears to act at an earlier step, before *lim3* expression in the adenohypophyseal placode is initiated, whereas *pia* acts after the initiation of *lim3* expression. However, this does not necessarily mean that *lia* fulfills its indispensable function via the activation of *lim3*, nor does it necessarily mean that *lim3* is required for *pia* activation. *lia* could as well activate other essential genes that act in parallel or in addition to *lim3*, and *pia* could as well act in parallel rather than downstream of *lim3*. However, it is interesting to note that *pia* mutants—despite the early presence of *lim3* transcripts—fail to express even early adenohypophyseal hormone genes like *prl*. This indicates that in the absence of *pia*, *lim3* is not sufficient for early lactotrope specification—although it appears to be necessary. The latter is suggested by our preliminary results obtained with antisense morpholino oligonucleotides, according to which inactivation of *lim3* leads to the loss of all adenohypophyseal cell lineages except some *pomc* cells. Final analyses of the epistatic relationships between *lia*, *lim3*, and *pia* will only be possible after the *lia* and *pia* genes have been cloned.

To gain further insight into the biological roles of *lia* and *pia* during the development of the zebrafish adenohypophysis, future experiments must reveal the fate of the pituitary precursor cells in the mutants. Thus, cell-tracing experiments, bromodeoxyuridine incorporation studies, and transferase deoxyuridine triphosphate nick-end labelling or acridine orange stainings will be carried out to investigate whether adenohypophyseal cells are lost because of transfating, failed proliferation, or cell death. In addition, analyses of chimeric embryos generated via cell transplantation will help to specify in which cell types the two genes are required.

Genes Required for Lineage Specification: Evidence for a Conserved and a Novel Lineage of Adenohypophyseal Cell Types

In contrast to *lia* and *pia*, *pit1* and *aal* mutants lose only some of the adenohypophyseal cell types, indicating that they are required for adenohypophyseal patterning and lineage specification processes during later stages of pituitary development. *pit1* mutants lack *gh*, *prl*, and *tsh* staining, whereas *pomc* staining is normal or even enlarged, particularly in its posterior domain that accommodates both corticotropes and melanotropes.

The phenotype looks very similar to that of mouse *Pit1* mutants, characterized by the loss of thyrotropes, somatotropes, and lactotropes. Due to the lack of suitable markers, we could not study the effect of zebrafish *pit1* on the gonadotropes, that according to the mouse mutant should be present in *pit1* mutant fish. One subtle difference in the phenotypes of mouse and zebrafish mutants is that zebrafish mutants lack all *tsh* expression, whereas in mouse mutants, only the caudo-medial thyrotropes are lost, although the earlier specifying rostral-tip thyrotropes are present. *Tsh* expression in this special thyrotrope lineage is independent of Pit1, and most likely regulated by the leucine zipper transcription factor TEF. The functional importance of the rostral-tip thyrotrope cells is not clear. However, our data obtained for the zebrafish *pit1* mutant suggest that they might be a specialty of mammals.

In contrast to *pit1*, the zebrafish *aal* mutants appear to define a thus far unknown lineage of adenohypophyseal cell types, lacking somatotropes, thyrotropes, corticotropes, and melanotropes, but not the lactotropes. Such a combination of lost cell types had not been previously described. In addition, the parallel existence of the Pit1 and Aal lineages indicates mechanisms of lineage specification beyond the thus far believed subsequent branching off of lineages from a common precursor pool.

Consistent with such a linear mechanism, only two relative patterns of cell lineages had been observed thus far. Lineages were either complementary (such as the Pomc lineage and the Prop1 lineage), or part of each other (such as Pit1, consisting of somatotropes, lactotropes, and thyrotropes, and Prop1, consisting of the Pit lineage plus the gonadotropes).

In contrast, the zebrafish Pit1 and Aal lineages show a novel relative pattern, with shared cell types, as well as cell types specific for one or the other lineage. Thus, somatotropes and thyrotropes appear to belong both to the Pit1 and the Aal lineage. However, corticotropes and melanotropes only belong to the Aal lineage and are independent of Pit1, whereas the lactotropes belong to the Pit1 lineage only and are independent of Aal. This indicates that factors driving cell lineage specification are not only used in a mutually exclusive or consecutive fashion, but can also be recruited in different combinations to allow differential parallel cell specifications.

In mammals, no Aal lineage has been identified as yet. However, this does not necessarily mean that the mammalian Aal homolog is not required or involved in adenohypophysis development. Rather, the exclusive regulation of *prl* expression independently of Aal function might be a specialty of fish, consistent with the much earlier onset of *prl* expression in fish compared with mouse, and with the earlier and additional function of Prl during osmoregulation in water-living larvae. Along these lines, it is tempting to speculate that during the evolution of water-living vertebrates, lactotropes became independent of Aal, or that during evolution of land-based vertebrates, lactotropes got under the control of the Aal homolog, similar to the other pituitary cell types. In this case, mutations in the Aal homolog in mouse would affect all adenohypophyseal cell lineages. Final answers have to await the cloning of the zebrafish *aal* gene, and the identification and analysis of possible mammalian homologues.

The Cloning of the Mutated Pituitary Genes

Because the introduction of point mutations during the ENU mutagenesis is random, the molecular nature of the genes affected in the isolated mutants is not known a priori, but has to be determined in subsequent steps. However, with the ease of mutant embryo collection, together with recent progress in the zebrafish genome projects, positional cloning of such ENU-mutated genes has become relatively easy, and has been successfully applied in multiple cases. Using a combination of positional cloning and candidate testing, we have for instance been able to identify the molecular nature of our two zebrafish *pit1* alleles. Also, future cloning of *lia*, *pia*, and *aal* will hopefully identify novel genes in pituitary development. In addition, we will be able to investigate whether such genes are specifically required for pituitary development in fish, or whether their roles are conserved between fish and mammals.

Possible Reasons for the Low Number of Identified Pituitary Genes and Perspectives

Based on gene targeting and cloning of spontaneous mutations, a total of approximately 20 genes indispensable for pituitary development in the mouse were identified. These include genes required for the combined formation of hypothalamus, neurohypophysis, and adenohypophysis, genes required for hypothalamus and neurohypophysis only, and genes required for the adenohypophysis only. Our screen was designed to find adenohypophysis-specific defects only. Pituitary phenotypes caused secondarily due to loss of the hypothalamus would have been missed because screening was performed at a late developmental stage, after mutations in such genes should have been lethal.

In addition, the zebrafish mutants were screened for altered *gh* expression only—the only adenohypophyseal hormone gene that had been cloned at the time the screen was carried out. The *gh* probe should have allowed us to identify genes required for adenohypophysis induction and development in general, as well as genes specifically required for somatotropes and somatotropes-including lineages, such as *Lhx3*, *Prop1*, and *Pit1*. On one hand, *gh* was a good choice, because it allowed us to identify genes that would most likely have been missed with other hormone probes (*pit1* mutants still express *pomc*; *aal* mutants still express *prl*). On the other hand, we might have missed genes that would have been revealed with other probes, such as *Tpit/Tbx19* (affecting the Pomc lineage only), *Gata2* (affecting gonatotropes and thyrotropes only), or *Sf1* (affecting gonadotropes only). In light of these possible limitations of the *gh* probe, we are currently preparing another large-scale ENU screen, looking for the expression of *pit1*, *lim3*, *isotocin*, *pomc*, and *prl* at early and late stages of zebrafish development.

Another reason for the low number of identified essential zebrafish genes could be the larger size and complexity of the zebrafish genome. Due to the additional genome duplication that has occurred during teleost evolution, genes controlling pituitary development in zebrafish might display a higher degree of functional redundancy than in mammals.

However, genome analyses suggest that only approximately 25% of the genes gained in the duplication have remained active, with most of them having evolved a different expression pattern than their paralogs. Also, parallel searches for zebrafish genes required for other developmental processes, such as angiogenesis and thymus development (Thomas Boehm, Max-Planck Institute for Immunobiology, Freiburg, Germany; personal communication) have yielded many more mutants, suggesting that high functional redundancy with a low number of indispensable genes might be a feature of some, but not all processes of zebrafish development.

On the other hand, we cannot rule out general technical problems during our mutagenesis. Comparing our mutagenesis conditions with those of previous screens, the screening of over 4200 mutagenized genomes should have yielded 50–80% saturation, with an average allele frequency per gene between three and four. However, in case of our identified pituitary genes, the average allele rate was two, with single alleles for two of the four genes. This suggests that the ENU mutagenesis might have been less efficient than in previous screens. Further complementation testing of other mutant classes has to be carried out for final conclusion about the mutation rates.

Mapping

Genomic localization of zebrafish mutations was performed as described, using the Tübingen marker set for genome scans (version 4) on F2 Tuebingen x Wik crosses of the mutant carriers· Primer sequences are available from the Massachusetts General Hospital web site.

Cloning of the Zebrafish Isotocin cDNA

Isotocin cDNA was cloned via degenerate RT-PCR, using the following primers: sense 5'TGY TAY ATH CAR AAY TGY CC, antisense 5'CCR CAR CAD ATN BWN GGN CC, with 35 cycles and an annealing temperature of 53 C. To obtain the full-length cDNA sequence, this was followed by a 3'-rapid amplification of cDNA ends (RACE), using the sense primer and the Smart Race Kit (Clontech, Palo Alto, CA) according to the manufacturer's instructions. PCR fragments were cloned into pCRII (Invitrogen, Carlsbad, CA). For *isotocin in situ* antisense probe synthesis, the plasmid was digested with *Kpn*I and transcribed using T7 RNA polymerase.

In Situ Hybridizations

For initial screening, F3 clutches were incubated in the presence of 0.25 mM 1-phenyl-2-thiourea (Sigma, St. Louis, MO) to avoid melanin synthesis, and fixed in 4% paraformaldehyde/PBS at 120 hpf. Whole mount *in situ* hybridization was carried out in specially designed 48-well plates (Aldinger, Nagold, Germany) with digoxygenin-labeled probes for *gh*, and *rag1*, a marker for thymic T cells, after standard protocols. Red/blue double *in situ* hybridizations were carried out as described. *prl*, *pomc*, *tsh*, *gh*, *lim3*, and *nkx2.1* probes were synthesized as reported.

4

Molecular Cloning

Molecular cloning refers to a set of experimental methods in molecular biology that are used to assemble recombinant DNA molecules and to direct their replication within host organisms. The use of the word *cloning* refers to the fact that the method involves the replication of a single DNA molecule starting from a single living cell to generate a large population of cells containing identical DNA molecules. Molecular cloning generally uses DNA sequences from two different organisms: the species that is the source of the DNA to be cloned, and the species that will serve as the living host for replication of the recombinant DNA. Molecular cloning methods are central to many contemporary areas of modern biology and medicine.

In a conventional molecular cloning experiment, the DNA to be cloned is obtained from an organism of interest, then treated with enzymes in the test tube to generate smaller DNA fragments. Subsequently, these fragments are then combined with vector DNA to generate recombinant DNA molecules.

The recombinant DNA is then introduced into a host organism (typically an easy-to-grow, benign, laboratory strain of *E. coli* bacteria). This will generate a population of organisms in which recombinant DNA molecules are replicated along with the host DNA. Because they contain foreign DNA fragments, these are transgenic or genetically-modified microorganisms (GMO). This process takes advantage of the fact that a single bacterial cell can be induced to take up and replicate a single recombinant DNA molecule. This single cell can then be expanded exponentially to generate a large amount of bacteria, each of which contain copies of the original recombinant molecule. Thus, both the resulting bacterial population, and the recombinant DNA molecule, are commonly referred to as "clones". Strictly speaking, *recombinant DNA* refers to DNA molecules, while *molecular cloning* refers to the experimental methods used to assemble them.

History of Molecular Cloning

Prior to the 1970s, our understanding of genetics and molecular biology was severely hampered by an inability to isolate and study individual genes from complex organisms. This changed dramatically with the advent of molecular cloning methods. Microbiologists, seeking to understand the molecular mechanisms through which bacteria restricted the growth of bacteriophage, isolated restriction endonucleases, enzymes that could cleave DNA molecules only when specific DNA sequences were encountered. They showed that restriction enzymes cleaved chromosome-length DNA molecules at specific locations, and that specific sections of the larger molecule could be purified by size fractionation. Using a second enzyme, DNA ligase, fragments generated by restriction enzymes could be joined in new combinations, termed recombinant DNA. By recombining DNA segments of interest with vector DNA, such as bacteriophage or plasmids, which naturally replicate inside bacteria, large quantities of purified recombinant DNA molecules could be produced in bacterial cultures. The first recombinant DNA molecules were generated and studied in 1972.

Overview

Molecular cloning takes advantage of the fact that the chemical structure of DNA is fundamentally the same in all living organisms. Therefore, if any segment of DNA from any organism is inserted into a DNA segment containing the molecular sequences required for DNA replication, and the resulting recombinant DNA is introduced into the organism from which the replication sequences were obtained, then the foreign DNA will be replicated along with the host cell's DNA in the transgenic organism.

Molecular cloning is similar to polymerase chain reaction (PCR) in that it permits the replication of a specific DNA sequence. The fundamental difference between the two methods is that molecular cloning involves replication of the DNA in a living microorganism, while PCR replicates DNA in an *in vitro* solution, free of living cells.

Steps in Molecular Cloning

In standard molecular cloning experiments, the cloning of any DNA fragment essentially involves seven steps: (1) Choice of host organism and cloning vector, (2) Preparation of vector DNA, (3) Preparation of DNA to be cloned, (4) Creation of recombinant DNA, (5) Introduction of recombinant DNA into host organism, (6) Selection of organisms containing recombinant DNA, (7) Screening for clones with desired DNA inserts and biological properties.

Choice of Host Organism and Cloning Vector

Although a very large number of host organisms and molecular cloning vectors are in use, the great majority of molecular cloning experiments begin with a laboratory strain of the bacterium *E. coli* (Escherichia coli) and a plasmid cloning vector. *E. coli* and plasmid vectors are in common use because they are technically sophisticated, versatile, widely available, and offer rapid growth of recombinant organisms with minimal equipment. If the DNA to be cloned is exceptionally large (hundreds of thousands to millions of base pairs), then a bacterial artificial chromosome or yeast artificial chromosome vector is often chosen.

Specialised applications may call for specialised host-vector systems. For example, if the experimentalists wish to harvest a particular protein from the recombinant organism, then an expression vector is chosen that contains appropriate signals for transcription and translation in the desired host organism. Alternatively, if replication of the DNA in different species is desired (for example transfer of DNA from bacteria to plants), then a multiple host range vector (also termed shuttle vector) may be selected. In practice, however, specialised molecular cloning experiments usually begin with cloning into a bacterial plasmid, followed by subcloning into a specialised vector.

Whatever combination of host and vector are used, the vector almost always contains four DNA segments that are critically important to its function and experimental utility—(1) an origin of DNA replication is necessary for the vector (and recombinant sequences linked to it) to replicate inside the host organism, (2) one or more unique restriction endonuclease recognition sites that serves as sites where foreign DNA may be introduced, (3) a selectable genetic marker gene that can be used to enable the survival of cells that have taken up vector sequences, and (4) an additional gene that can be used for screening which cells contain foreign DNA.

Preparation of Vector DNA

The cloning vector is treated with a restriction endonuclease to cleave the DNA at the site where foreign DNA will be inserted. The restriction enzyme is chosen to generate a configuration at the cleavage site that is compatible with that at the ends of the foreign DNA. Typically, this is done by cleaving the vector DNA and foreign DNA with the same restriction enzyme, for example EcoRI. Most modern vectors contain a variety of convenient cleavage sites that are unique within the vector molecule (so that the vector can only be cleaved at a single site) and is located within a gene (frequently beta-galactosidase) whose inactivation can be used to distinguish recombinant from non-recombinant organisms at a later step in the process. To improve the ratio of recombinant to non-recombinant organisms, the cleaved vector may be treated with an enzyme (alkaline phosphatase) that modifies the vector ends in such a way that it cannot replicate within cells unless it contains foreign DNA.

Preparation of DNA to be Cloned

For cloning of genomic DNA, the DNA to be cloned is extracted from the organism of interest. Virtually any tissue source can be used (even tissues from extinct animals), as long as the DNA is not extensively degraded. The DNA is then purified using simple methods to remove contaminating proteins (extraction with phenol), RNA (ribonuclease) and smaller molecules (precipitation and/or chromatography). Polymerase chain reaction (PCR) methods are often used for amplification of specific DNA or RNA (RT-PCR) sequences prior to molecular cloning.

DNA for cloning experiments may also be obtained from RNA using reverse transcriptase (complementary DNA or cDNA cloning), or in the form of synthetic DNA (artificial gene synthesis). cDNA cloning is usually used to obtain clones representative of the mRNA population of the cells of interest, while synthetic DNA is used to obtain any precise sequence defined by the designer. The purified DNA is then treated with a restriction enzyme to generate fragments with ends capable of being linked to those of the vector. If necessary, short double-stranded segments of DNA containing desired restriction sites may be added to create end structures that are compatible with the vector.

Creation of Recombinant DNA with DNA Ligase

The creation of recombinant DNA is in many ways the simplest step of the molecular cloning process. DNA prepared from the vector and foreign source are simply mixed together at appropriate concentrations and exposed to an enzyme (DNA ligase) that covalently links the ends together. This joining reaction is often termed ligation. The resulting DNA mixture containing randomly joined ends is then ready for introduction into the host organism. DNA ligase only recognises and acts on the ends of linear DNA molecules, usually resulting a complex mixture of DNA molecules with randomly joined ends. The desired products (vector DNA covalently linked to foreign DNA) will be present, but other sequences (e.g. foreign DNA linked to itself, vector DNA linked to itself and higher-order combinations of vector and foreign DNA) are also usually present. This complex mixture is sorted out in subsequent steps of the cloning process, after the DNA mixture is introduced into cells.

Introduction of Recombinant DNA into Host Organism

The DNA mixture, previously manipulated in vitro, is moved back into a living cell, referred to as the host organism. The methods used to get DNA into cells are varied, and the name applied to this step in the molecular cloning process will often depend upon the experimental method that is chosen (e.g. transformation, transduction, transfection, electroporation). When microorganisms are able to take up and replicate DNA from their local environment, the process is termed transformation, and cells that are in a physiological state such that they can take up DNA are said to be competent. In mammalian cell culture, the analogous process of introducing DNA into cells is commonly termed transfection. Both transformation and transfection usually require preparation of the cells through a special growth regime and chemical treatment process that will vary with the specific species and cell types that are used.

Electroporation uses high voltage electrical pulses to translocate DNA across the cell membrane (and cell wall, if present). In contrast, transduction involves the packaging of DNA into virus-derived particles, and using these virus-like particles to introduce the encapsulated DNA into the cell through a process resembling viral infection. Although electroporation and transduction are highly specialised methods, they may be the most efficient methods to move DNA into cells.

Selection of Organisms Containing Vector Sequences

Whichever method is used, the introduction of recombinant DNA into the chosen host organism is usually a low efficiency process; that is, only a small fraction of the cells will actually take up DNA. Experimental scientists deal with this issue through a step of artificial genetic selection, in which cells that have not taken up DNA are selectively killed, and only those cells that can actively replicate DNA containing the selectable marker gene encoded by the vector are able to survive.

When bacterial cells are used as host organisms, the selectable marker is usually a gene that confers resistance to an antibiotic that would otherwise kill the cells, typically ampicillin. Cells harbouring the vector will survive when exposed to the antibiotic, while those that have failed to take up vector sequences will die. When mammalian cells (e.g. human or mouse cells) are used, a similar strategy is used, except that the marker gene confers resistance to the antibiotic Geneticin.

Screening for Clones with Desired DNA Inserts and Biological Properties

Modern bacterial cloning vectors (e.g. pUC19 and later derivatives including the pGEM vectors) use the blue-white screening system to distinguish colonies (clones) of transgenic cells from those that contain the parental vector (i.e. vector DNA with no recombinant sequence inserted). In these vectors, foreign DNA is inserted into a sequence that encodes an essential part of beta-galactosidase, an enzyme whose activity results in formation of a blue-coloured colony on the culture medium that is used for this work. Insertion of the foreign

DNA into the beta-galactosidase coding sequence disables the function of the enzyme, so that colonies containing recombinant plasmids remain colourless (white). Therefore, experimentalists are easily able to identify and conduct further studies on transgenic bacterial clones, while ignoring those that do not contain recombinant DNA.

The total population of individual clones obtained in a molecular cloning experiment is often termed a DNA library. Libraries may be highly complex (as when cloning complete genomic DNA from an organism) or relatively simple (as when moving a previously-cloned DNA fragment into a different plasmid), but it is almost always necessary to examine a number of different clones to be sure that the desired DNA construct is obtained. This may be accomplished through a very wide range of experimental methods, including the use of nucleic acid hybridisations, antibody probes, polymerase chain reaction, restriction fragment analysis and/or DNA sequencing.

Applications of Molecular Cloning

Molecular cloning provides scientists with an essentially unlimited quantity of any individual DNA segments derived from any genome. This material can be used for a wide range of purposes, including those in both basic and applied biological science. A few of the more important applications are summarised here.

Genome Organisation and Gene Expression

Molecular cloning has led directly to the elucidation of the complete of sexual DNA sequence of the genomes of a very large number of species and to an exploration of genetic diversity within individual species, work that has been done mostly by determining the DNA sequence of large numbers of randomly cloned fragments of the genome, and assembling the overlapping sequences.

At the level of individual genes, molecular clones are used to generate probes that are used for examining how genes are expressed, and how that expression is related to other processes in biology, including the metabolic environment, extracellular signals, development, learning, senescence and cell death. Cloned genes can also provide tools to examine the biological function and importance of individual genes, by allowing investigators to inactivate the genes, or make more subtle mutations using regional mutagenesis or site-directed mutagenesis.

Production of Recombinant Proteins

Obtaining the molecular clone of a gene can lead to the development of organisms that produce the protein product of the cloned genes, termed a recombinant protein. In practice, it is frequently more difficult to develop an organism that produces an active form of the recombinant protein in desirable quantities than it is to clone the gene. This is because the molecular signals for gene expression are complex and variable, and because protein folding, stability and transport can be very challenging.

Many useful proteins are currently available as recombinant products. These include—(1) medically-useful proteins whose administration can correct a defective or poorly-expressed gene (e.g. recombinant factor VIII, a blood-clotting factor deficient in some forms of hemophilia, and recombinant insulin, used to treat some forms of diabetes), (2) proteins that can be administered to assist in a life threatening emergency (e.g. tissue plasminogen activator, used to treat strokes), and (3) recombinant subunit vaccines, in which a purified protein can be used to immunize patients against infectious diseases, without exposing them to the infectious agent itself (e.g. hepatitis B vaccine).

Transgenic Organisms

Once characterised and manipulated to provide signals for appropriate expression, cloned genes may be inserted into organisms, generating transgenic organisms, also termed genetically-modified organisms (GMOs). Although most GMOs are generated for purposes of basic biological research, a number of GMOs have been developed for commercial use, ranging from animals and plants that produce pharmaceuticals or other compounds (pharming), herbicide-resistant crop plants, and fluorescent tropical fish (GloFish) for home entertainment.

Gene Therapy

Gene therapy involves supplying a functional gene to cells lacking that function, with the aim of correcting a genetic disorder or acquired disease. Gene therapy can be broadly divided into two categories. The first is alteration of germ cells, that is, sperm or eggs, which results in a permanent genetic change for the whole organism and subsequent generations. This "germ line gene therapy" is considered by many to be unethical in human beings. The second type of gene therapy, "somatic cell gene therapy", is analogous to an organ transplant. In this case, one or more specific tissues are targeted by direct treatment or by removal of the tissue, addition of the therapeutic gene or genes in the laboratory, and return of the treated cells to the patient. Clinical trials of somatic cell gene therapy began in the late 1990s, mostly for the treatment of cancers and blood, liver, and lung disorders.

Despite a great deal of publicity and promises, the history of human gene therapy has been characterised by relatively limited success. The effect of introducing a gene into cells often promotes only partial and/or transient relief from the symptoms of the disease being treated. Some gene therapy trial patients have suffered adverse consequences of the treatment itself, including deaths. In some cases, the adverse effects result from disruption of essential genes within the patient's genome by insertional inactivation. In others, viral vectors used for gene therapy have been contaminated with infectious virus. Nevertheless, gene therapy is still held to be a promising future area of medicine, and is an area where there is a significant level of research and development activity.

Cloning of Viral Antigens

A virus is a small infectious agent that can replicate only inside the living cells of organisms. Viruses infect all types of organisms, from animals and plants to bacteria and archaea. Since Dmitri Ivanovsky's 1892 article describing a non-bacterial pathogen infecting tobacco plants, and the discovery of the tobacco mosaic virus by Martinus Beijerinck in 1898, about 5,000 viruses have been described in detail, although there are millions of different types. Viruses are found in almost every ecosystem on Earth and are the most abundant type of biological entity. The study of viruses is known as virology, a sub-speciality of microbiology.

Virus particles (known as *virions*) consist of two or three parts: the genetic material made from either DNA or RNA, long molecules that carry genetic information; a protein coat that protects these genes; and in some cases an envelope of lipids that surrounds the protein coat when they are outside a cell. The shapes of viruses range from simple helical and icosahedral forms to more complex structures. The average virus is about one one-hundredth the size of the average bacterium. Most viruses are too small to be seen directly with a light microscope.

The origins of viruses in the evolutionary history of life are unclear: some may have evolved from plasmids – pieces of DNA that can move between cells – while others may have evolved from bacteria. In evolution, viruses are an important means of horizontal gene transfer, which increases genetic diversity.

Viruses spread in many ways; viruses in plants are often transmitted from plant to plant by insects that feed on the sap of plants, such as aphids; viruses in animals can be carried by blood-sucking insects. These disease-bearing organisms are known as vectors. Influenza viruses are spread by coughing and sneezing. Norovirus and rotavirus, common causes of viral gastroenteritis, are transmitted by the faecal-oral route and are passed from person to person by contact, entering the body in food or water. HIV is one of several viruses transmitted through sexual contact and by exposure to infected blood. The range of host cells that a virus can infect is called its "host range". This can be narrow or, as when a virus is capable of infecting many species, broad.

Viral infections in animals provoke an immune response that usually eliminates the infecting virus. Immune responses can also be produced by vaccines, which confer an artificially acquired immunity to the specific viral infection. However, some viruses including those causing AIDS and viral hepatitis evade these immune responses and result in chronic infections. Antibiotics have no effect on viruses, but several antiviral drugs have been developed.

Etymology

The word is from the Latin *virus* referring to poison and other noxious substances, first used in English in 1392. *Virulent*, from Latin *virulentus* (poisonous), dates to 1400. A meaning of "agent that causes infectious disease" is first recorded in 1728, before the discovery of viruses by Dmitry Ivanovsky in 1892. The plural is *viruses*. The adjective *viral* dates to 1948. The term *virion* is also used to refer to a single infective viral particle.

History

Louis Pasteur was unable to find a causative agent for rabies and speculated about a pathogen too small to be detected using a microscope. In 1884, the French microbiologist Charles Chamberland invented a filter (known today as the Chamberland filter or Chamberland-Pasteur filter) with pores smaller than bacteria. Thus, he could pass a solution containing bacteria through the filter and completely remove them from the solution. In 1892, the Russian biologist Dmitry Ivanovsky used this filter to study what is now known as the tobacco mosaic virus. His experiments showed that crushed leaf extracts from infected tobacco plants remain infectious after filtration. Ivanovsky suggested the infection might be caused by a toxin produced by bacteria, but did not pursue the idea. At the time it was thought that all infectious agents could be retained by filters and grown on a nutrient medium – this was part of the germ theory of disease. In 1898, the Dutch microbiologist Martinus Beijerinck repeated the experiments and became convinced that the filtered solution contained a new form of infectious agent. He observed that the agent multiplied only in cells that were dividing, but as his experiments did not show that it was made of particles, he called it a *contagium vivum fluidum* (soluble living germ) and reintroduced the word *virus*. Beijerinck maintained that viruses were liquid in nature, a theory later discredited by Wendell Stanley, who proved they were particulate. In the same year Friedrich Loeffler and Frosch passed the first animal virus – agent of foot-and-mouth disease (aphthovirus) – through a similar filter.

In the early 20th century, the English bacteriologist Frederick Twort discovered a group of viruses that infect bacteria, now called bacteriophages (or commonly *phages*), and the French-Canadian microbiologist Félix d'Herelle described viruses that, when added to bacteria on agar, would produce areas of dead bacteria. He accurately diluted a suspension of these viruses and discovered that the highest dilutions (lowest virus concentrations), rather than killing all the bacteria, formed discrete areas of dead organisms. Counting these areas and multiplying by the dilution factor allowed him to calculate the number of viruses in the original suspension. Phages were heralded as a potential treatment for diseases such as typhoid and cholera, but their promise was forgotten with the development of penicillin. The study of phages provided insights into the switching on and off of genes, and a useful mechanism for introducing foreign genes into bacteria.

By the end of the 19th century, viruses were defined in terms of their infectivity, their ability to be filtered, and their requirement for living hosts. Viruses had been grown only in plants and animals. In 1906, Ross Granville Harrison invented a method for growing tissue in lymph, and, in 1913, E. Steinhardt, C. Israeli, and R. A. Lambert used this method to grow vaccinia virus in fragments of guinea pig corneal tissue. In 1928, H. B. Maitland and M. C. Maitland grew vaccinia virus in suspensions of minced hens' kidneys. Their method was not widely adopted until the 1950s, when poliovirus was grown on a large scale for vaccine production. Another breakthrough came in 1931, when the American pathologist Ernest William Goodpasture grew influenza and several other viruses in fertilized chickens' eggs. In 1949, John F. Enders, Thomas Weller, and Frederick Robbins grew polio virus in cultured human embryo cells, the first virus to be grown without using solid animal tissue or eggs. This work enabled Jonas Salk to make an effective polio vaccine.

The first images of viruses were obtained upon the invention of electron microscopy in 1931 by the German engineers Ernst Ruska and Max Knoll. In 1935, American biochemist and virologist Wendell Meredith Stanley examined the tobacco mosaic virus and found it was mostly made of protein. A short time later, this virus was separated into protein and RNA parts. The tobacco mosaic virus was the first to be crystallised and its structure could therefore be elucidated in detail. The first X-ray diffraction pictures of the crystallised virus were obtained by Bernal and Fankuchen in 1941. On the basis of her pictures, Rosalind Franklin discovered the full DNA structure of the virus in 1955. In the same year, Heinz Fraenkel-Conrat and Robley Williams showed that purified tobacco mosaic virus RNA and its coat protein can assemble by themselves to form functional viruses, suggesting that this simple mechanism was probably the means through which viruses were created within their host cells.

The second half of the 20th century was the golden age of virus discovery and most of the 2,000 recognised species of animal, plant, and bacterial viruses were discovered during these years. In 1957, equine arterivirus and the cause of Bovine virus diarrhea (a pestivirus) were discovered. In 1963, the hepatitis B virus was discovered by Baruch Blumberg, and in 1965, Howard Temin described the first retrovirus. Reverse transcriptase, the key enzyme that retroviruses use to translate their RNA into DNA, was first described in 1970, independently by Howard Martin Temin and David Baltimore. In 1983 Luc Montagnier's team at the Pasteur Institute in France, first isolated the retrovirus now called HIV.

Origins

Viruses are found wherever there is life and have probably existed since living cells first evolved. The origin of viruses is unclear because they do not form fossils, so molecular techniques have been used to compare the DNA or RNA of viruses and are a useful means of investigating how they arose. There are three main hypotheses that try to explain the origins of viruses:

Regressive hypothesis: Viruses may have once been small cells that parasitised larger cells. Over time, genes not required by their parasitism were lost. The bacteria rickettsia and chlamydia are living cells that, like viruses, can reproduce only inside host cells. They lend support to this hypothesis, as their dependence on parasitism is likely to have caused the loss of genes that enabled them to survive outside a cell. This is also called the *degeneracy hypothesis*, or *reduction hypothesis*.

Cellular origin hypothesis: Some viruses may have evolved from bits of DNA or RNA that "escaped" from the genes of a larger organism. The escaped DNA could have come from plasmids (pieces of naked DNA that can move *between* cells) or transposons (molecules of DNA that replicate and move around to different positions *within* the genes of the cell). Once called "jumping genes", transposons are examples of mobile genetic elements and could be the origin of some viruses. They were discovered in maize by Barbara McClintock in 1950. This is sometimes called the *vagrancy hypothesis*, or the *escape hypothesis*.

Coevolution hypothesis: This is also called the *virus-first hypothesis* and proposes that viruses may have evolved from complex molecules of protein and nucleic acid at the same time as cells first appeared on earth and would have been dependent on cellular life for billions of years. Viroids are molecules of RNA that are not classified as viruses because they lack a protein coat. However, they have characteristics that are common to several viruses and are often called subviral agents. Viroids are important pathogens of plants. They do not code for proteins but interact with the host cell and use the host machinery for their replication. The hepatitis delta virus of humans has an RNA genome similar to viroids but has a protein coat derived from hepatitis B virus and cannot produce one of its own. It is, therefore, a defective virus and cannot replicate without the help of hepatitis B virus. In similar manner, the virophage 'sputnik' is dependent on mimivirus, which infects the protozoan *Acanthamoeba castellanii*. These viruses that are dependent on the presence of other virus species in the host cell are called *satellites* and may represent evolutionary intermediates of viroids and viruses.

In the past, there were problems with all of these hypotheses: the regressive hypothesis did not explain why even the smallest of cellular parasites do not resemble viruses in any way. The escape hypothesis did not explain the complex capsids and other structures on virus particles. The virus-first hypothesis contravened the definition of viruses in that they require host cells. Viruses are now recognised as ancient and to have origins that predate the divergence of life into the three domains. This discovery has led modern virologists to reconsider and re-evaluate these three classical hypotheses.

The evidence for an ancestral world of RNA cells and computer analysis of viral and host DNA sequences are giving a better understanding of the evolutionary relationships between different viruses and may help identify the ancestors of modern viruses. To date, such analyses have not proved which of these hypotheses is correct. However, it seems unlikely that all currently known viruses have a common ancestor, and viruses have probably arisen numerous times in the past by one or more mechanisms.

Prions are infectious protein molecules that do not contain DNA or RNA. They cause an infection in sheep called scrapie and cattle bovine spongiform encephalopathy ("mad cow" disease). In humans they cause kuru and Creutzfeldt-Jakob disease. They are able to replicate because some proteins can exist in two different shapes and the prion changes the normal shape of a host protein into the prion shape. This starts a chain reaction where each prion protein converts many host proteins into more prions, and these new prions then go on to convert even more protein into prions. Although they are fundamentally different from viruses and viroids, their discovery gives credence to the idea that viruses could have evolved from self-replicating molecules.

Microbiology

Life Properties

Opinions differ on whether viruses are a form of life, or organic structures that interact with living organisms. They have been described as "organisms at the edge of life", since they resemble organisms in that they possess genes and evolve by natural selection, and reproduce by creating multiple copies of themselves through self-assembly. Although they have genes, they do not have a cellular structure, which is often seen as the basic unit of life. Viruses do not have their own metabolism, and require a host cell to make new products.

They therefore cannot naturally reproduce outside a host cell – although bacterial species such as rickettsia and chlamydia are considered living organisms despite the same limitation. Accepted forms of life use cell division to reproduce, whereas viruses spontaneously assemble within cells. They differ from autonomous growth of crystals as they inherit genetic mutations while being subject to natural selection. Virus self-assembly within host cells has implications for the study of the origin of life, as it lends further credence to the hypothesis that life could have started as self-assembling organic molecules.

Structure

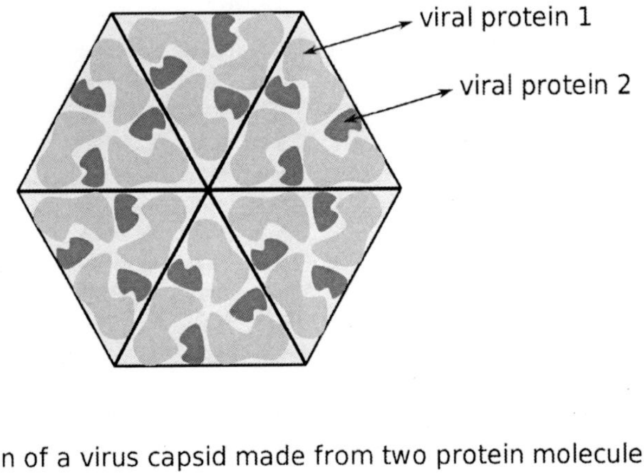

Figure : *Diagram of how a virus capsid can be constructed using multiple copies of just two protein molecules*

Viruses display a wide diversity of shapes and sizes, called *morphologies*. Generally viruses are much smaller than bacteria. Most viruses that have been studied have a diameter between 20 and 300 nanometres. Some filoviruses have a total length of up to 1400 nm; their diameters are only about 80 nm. Most viruses cannot be seen with a light microscope so scanning and transmission electron microscopes are used to visualise virions. To increase the contrast between viruses and the background, electron-dense "stains" are used. These are solutions of salts of heavy metals, such as tungsten, that scatter the electrons from regions covered with the stain. When virions are coated with stain (positive staining), fine detail is obscured. Negative staining overcomes this problem by staining the background only.

A complete virus particle, known as a virion, consists of nucleic acid surrounded by a protective coat of protein called a capsid. These are formed from identical protein subunits called capsomeres. Viruses can have a lipid "envelope" derived from the host cell membrane. The capsid is made from proteins encoded by the viral genome and its shape serves as the basis for morphological distinction. Virally coded protein subunits will self-assemble to form a capsid, generally requiring the presence of the virus genome. Complex viruses code for proteins that assist in the construction of their capsid. Proteins associated with nucleic acid are known as nucleoproteins, and the association of viral capsid proteins with viral nucleic acid is called a nucleocapsid. The capsid and entire virus structure can be mechanically (physically) probed through atomic force microscopy. In general, there are four main morphological virus types:

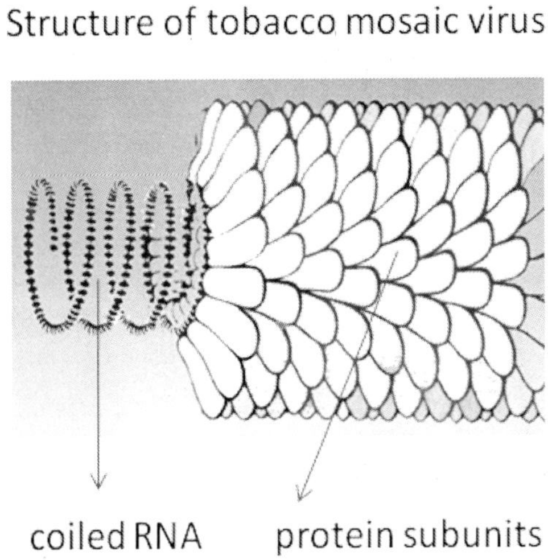

Figure : *RNA coiled in a helix of repeating protein subunits*

Helical: These viruses are composed of a single type of capsomer stacked around a central axis to form a helical structure, which may have a central cavity, or hollow tube. This arrangement results in rod-shaped or filamentous virions: These can be short and highly rigid, or long and very flexible. The genetic material, in general, single-stranded RNA, but ssDNA in some cases, is bound into the protein helix by interactions between the negatively charged nucleic acid and positive charges on the protein. Overall, the length of a helical capsid is related to the length of the nucleic acid contained within it and the diameter is dependent on the size and arrangement of capsomers. The well-studied tobacco mosaic virus is an example of a helical virus.

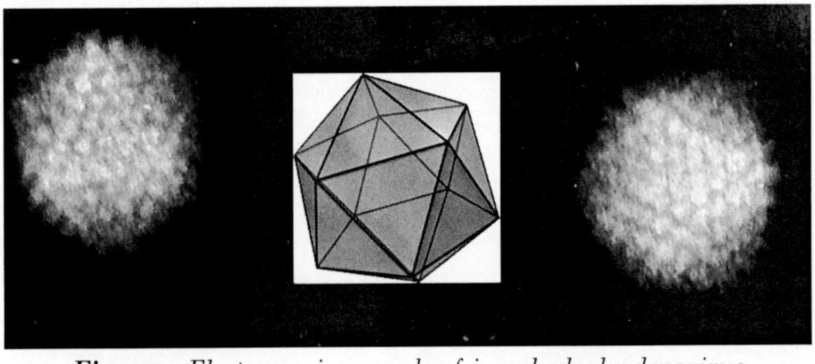

Figure : *Electron micrograph of icosahedral adenovirus*

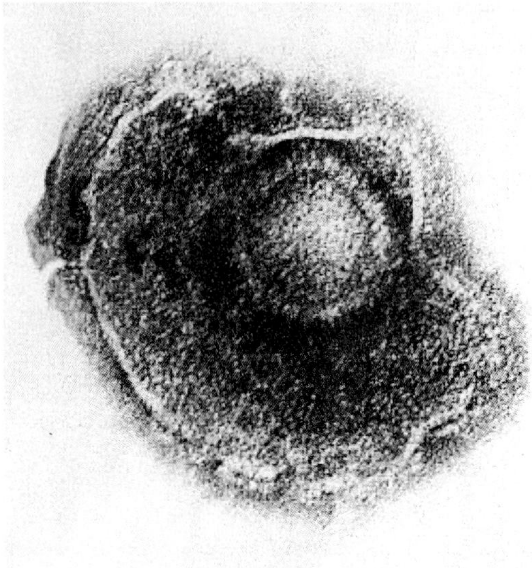

Figure: *Herpes viruses have a lipid envelope*

Icosahedral: Most animal viruses are icosahedral or near-spherical with icosahedral symmetry. A regular icosahedron is the optimum way of forming a closed shell from identical subunits. The minimum number of identical capsomers required is twelve, each composed of five identical subunits.

Many viruses, such as rotavirus, have more than twelve capsomers and appear spherical but they retain this symmetry. Capsomers at the apices are surrounded by five other capsomers and are called pentons. Capsomers on the triangular faces are surrounded by six others and are called hexons. Hexons are essentially flat and pentons, which form the 12 vertices, are curved. The same protein may act as the subunit of both the pentamers and hexamers or they may be composed of different proteins.

Although the isosahedral structure is extremely common among viruses because of size differences slight variations exist between virions. Given an asymmetric subunit on a triangular face of a regular isosahedron, with three subunits per face 60 such subunits can be placed in an equivalent manner. Most virions, because of their size, have more than 60 subunits. These variations have been classified on the basis of the quasi-equivalence principle proposed by Caspar and Klug.

An isosahedral structure can be regarded as being constructed from 12 pentamers. The number of pentamers is fixed but the number of hexamers can vary. These shells can be constructed from pentamers and hexamers by minimising the number T (triangulation number) of nonequivalent locations that subunits occupy, with the T-number adopting the particular integer values 1, 3, 4, 7, 12, 13,...($T = h^2 + k^2 + hk$, with h, k equal to nonnegative integers). These shells always contain 12 pentamers plus 10 (T-1) hexamers.

Although this classification can be applied to the majority of known viruses exceptions are known including the retroviruses where point mutations disrupt the symmetry.

Prolate: This is an isosahedron elongated along the fivefold axis and is a common arrangement of the heads of bacteriophages. Such a structure is composed of a cylinder with a cap at either end. The cylinder is composed of 10 triangles. The Q number, which can be can be any positive integer, specifies the number of triangles, composed of asymmetric subunits, that make up the 10 triangles of the cylinder. The caps are classified by the T number.

Envelope: Some species of virus envelop themselves in a modified form of one of the cell membranes, either the outer membrane surrounding an infected host cell or internal membranes such as nuclear membrane or endoplasmic reticulum, thus gaining an outer lipid bilayer known as a viral envelope. This membrane is studded with proteins coded for by the viral genome and host genome; the lipid membrane itself and any carbohydrates present originate entirely from the host. The influenza virus and HIV use this strategy. Most enveloped viruses are dependent on the envelope for their infectivity.

Complex: These viruses possess a capsid that is neither purely helical nor purely icosahedral, and that may possess extra structures such as protein tails or a complex outer wall. Some bacteriophages, such as Enterobacteria phage T4, have a complex structure consisting of an icosahedral head bound to a helical tail, which may have a hexagonal base plate with protruding protein tail fibres. This tail structure acts like a molecular syringe, attaching to the bacterial host and then injecting the viral genome into the cell.

The poxviruses are large, complex viruses that have an unusual morphology. The viral genome is associated with proteins within a central disk structure known as a nucleoid. The nucleoid is surrounded by a membrane and two lateral bodies of unknown function. The virus has an outer envelope with a thick layer of protein studded over its surface. The whole virion is slightly pleiomorphic, ranging from ovoid to brick shape. Mimivirus is the largest characterised virus, with a capsid diameter of 400 nm. Protein filaments measuring 100 nm project from the surface. The capsid appears hexagonal under an electron microscope, therefore the capsid is probably icosahedral. In 2011, researchers discovered a larger virus on ocean floor of the coast of Las Cruces, Chile. Provisionally named *Megavirus chilensis*, it can be seen with a basic light microscope.

Some viruses that infect Archaea have complex structures that are unrelated to any other form of virus, with a wide variety of unusual shapes, ranging from spindle-shaped structures, to viruses that resemble hooked rods, teardrops or even bottles. Other archaeal viruses resemble the tailed bacteriophages, and can have multiple tail structures.

Genome

An enormous variety of genomic structures can be seen among viral species; as a group they contain more structural genomic diversity than plants, animals, archaea, or bacteria. There are millions of different types of viruses, although only about 5,000 of them have been described in detail. A virus has either DNA or RNA genes and is called a DNA virus or a RNA virus respectively. The vast majority of viruses have RNA genomes. Plant viruses tend to have single-stranded RNA genomes and bacteriophages tend to have double-stranded DNA genomes.

Viral genomes are circular, as in the polyomaviruses, or linear, as in the adenoviruses. The type of nucleic acid is irrelevant to the shape of the genome. Among RNA viruses and certain DNA viruses, the genome is often divided up into separate parts, in which case it is called *segmented*. For RNA viruses, each segment often codes for only one protein and they are usually found together in one capsid. However, all segments are not required to be in the same virion for the virus to be infectious, as demonstrated by brome mosaic virus and several other plant viruses.

A viral genome, irrespective of nucleic acid type, is almost always either single-stranded or double-stranded. Single-stranded genomes consist of an unpaired nucleic acid, analogous to one-half of a ladder split down the middle. Double-stranded genomes consist of two complementary paired nucleic acids, analogous to a ladder. The virus particles of some virus families, such as those belonging to the *Hepadnaviridae*, contain a genome that is partially double-stranded and partially single-stranded.

For most viruses with RNA genomes and some with single-stranded DNA genomes, the single strands are said to be either positive-sense (called the plus-strand) or negative-sense (called the minus-strand), depending on whether or not they are complementary to the viral messenger RNA (mRNA). Positive-sense viral RNA is in the same sense as viral mRNA and thus at least a part of it can be immediately translated by the host cell.

Negative-sense viral RNA is complementary to mRNA and thus must be converted to positive-sense RNA by an RNA-dependent RNA polymerase before translation. DNA nomenclature for viruses with single-sense genomic ssDNA is similar to RNA nomenclature, in that the *coding strand* for the viral mRNA is complementary to it (“), and the *non-coding strand* is a copy of it (+).

However, several types of ssDNA and ssRNA viruses have genomes that are ambisense in that transcription can occur off both strands in a double-stranded replicative intermediate. Examples include geminiviruses, which are ssDNA plant viruses and arenaviruses, which are ssRNA viruses of animals.

Genome size varies greatly between species. The smallest viral genomes – the ssDNA circoviruses, family *Circoviridae* – code for only two proteins and have a genome size of only 2 kilobases; the largest – mimiviruses – have genome sizes of over 1.2 megabases and code for over one thousand proteins. RNA viruses generally have smaller genome sizes than DNA viruses because of a higher error-rate when replicating, and have a maximum upper size limit. Beyond this limit, errors in the genome when replicating render the virus useless or uncompetitive.

To compensate for this, RNA viruses often have segmented genomes – the genome is split into smaller molecules – thus reducing the chance that an error in a single-component genome will incapacitate the entire genome. In contrast, DNA viruses generally have larger genomes because of the high fidelity of their replication enzymes. Single-strand DNA viruses are an exception to this rule, however, as mutation rates for these genomes can approach the extreme of the ssRNA virus case.

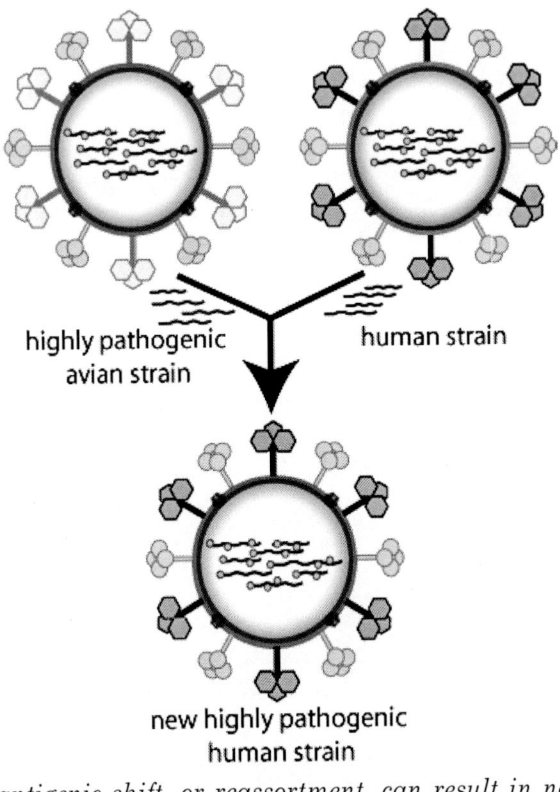

***Figure:** How antigenic shift, or reassortment, can result in novel and highly pathogenic strains of human influenza*

Viruses undergo genetic change by several mechanisms. These include a process called genetic drift where individual bases in the DNA or RNA mutate to other bases. Most of these point mutations are "silent" – they do not change the protein that the gene encodes – but others can confer evolutionary advantages such as resistance to antiviral drugs. Antigenic shift occurs when there is a major change in the genome of the virus. This can be a result of recombination or reassortment.

When this happens with influenza viruses, pandemics might result. RNA viruses often exist as quasispecies or swarms of viruses of the same species but with slightly different genome nucleoside sequences. Such quasispecies are a prime target for natural selection.

Segmented genomes confer evolutionary advantages; different strains of a virus with a segmented genome can shuffle and combine genes and produce progeny viruses or (offspring) that have unique characteristics. This is called reassortment or *viral sex.*

Genetic recombination is the process by which a strand of DNA is broken and then joined to the end of a different DNA molecule. This can occur when viruses infect cells simultaneously and studies of viral evolution have shown that recombination has been rampant in the species studied. Recombination is common to both RNA and DNA viruses.

Replication Cycle

Viral populations do not grow through cell division, because they are acellular. Instead, they use the machinery and metabolism of a host cell to produce multiple copies of themselves, and they *assemble* in the cell.

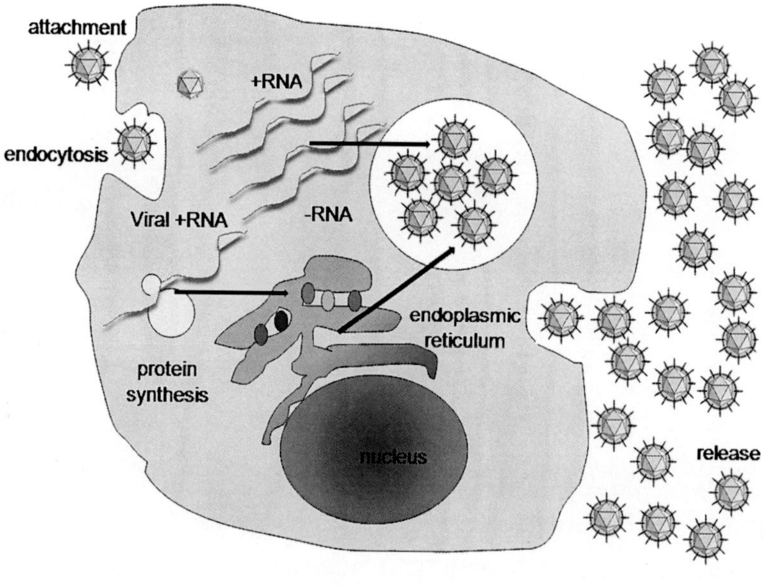

Figure : *A typical virus replication cycle*

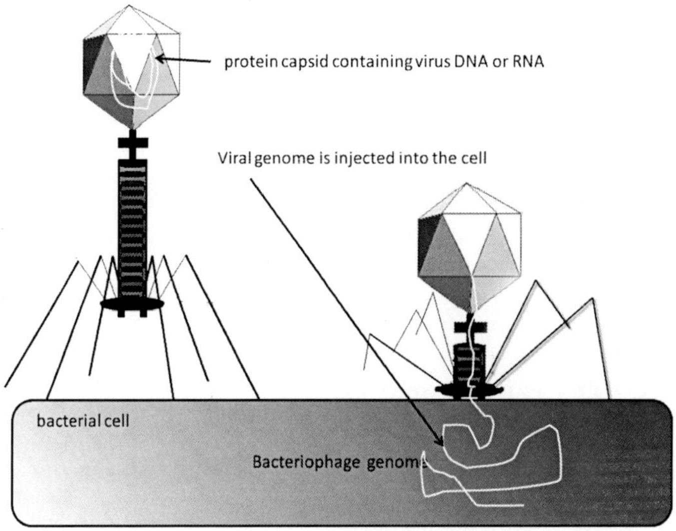

Figure : *Some bacteriophages inject their genomes into bacterial cells*

The life cycle of viruses differs greatly between species but there are six *basic* stages in the life cycle of viruses:

- *Attachment* is a specific binding between viral capsid proteins and specific receptors on the host cellular surface. This specificity determines the host range of a virus. For example, HIV infects a limited range of human leucocytes. This is because its surface protein, gp120, specifically interacts with the CD4 molecule – a chemokine receptor – which is most commonly found on the surface of CD4+ T-Cells. This mechanism has evolved to favour those viruses that infect only cells in which they are capable of replication. Attachment to the receptor can induce the viral envelope protein to undergo changes that results in the fusion of viral and cellular membranes, or changes of non-enveloped virus surface proteins that allow the virus to enter.
- *Penetration* follows attachment: Virions enter the host cell through receptor-mediated endocytosis or membrane fusion. This is often called viral entry. The infection of plant and, it is presumed, fungal cells is different from that of animal cells. Plants have a rigid cell wall made of cellulose, and fungi one of chitin, so most viruses can get inside these cells only after trauma to the cell wall. However, nearly all plant viruses (such as tobacco mosaic virus) can also move directly from cell to cell, in the form of single-stranded nucleoprotein complexes, through pores called plasmodesmata. This process requires movement proteins, which are virus-encoded proteins probably originally derived from plant

proteins, which interact with the plasmodesmatal transport machinery Bacteria, like plants, have strong cell walls that a virus must breach to infect the cell. However, given that bacterial cell walls are much less thick than plant cell walls due to their much smaller size, some viruses have evolved mechanisms that inject their genome into the bacterial cell across the cell wall, while the viral capsid remains outside.

- *Uncoating* is a process in which the viral capsid is removed: This may be by degradation by viral enzymes or host enzymes or by simple dissociation; the end-result is the releasing of the viral genomic nucleic acid.
- *Replication* of viruses primarily involves multiplication of the genome. Replication involves synthesis of viral messenger RNA (mRNA) from "early" genes (with exceptions for positive sense RNA viruses), viral protein synthesis, possible assembly of viral proteins, then viral genome replication mediated by early or regulatory protein expression. This may be followed, for complex viruses with larger genomes, by one or more further rounds of mRNA synthesis: "late" gene expression is, in general, of structural or virion proteins.
- Following the structure-mediated self-*assembly* of the virus particles, some modification of the proteins often occurs. In viruses such as HIV, this modification (sometimes called maturation) occurs *after* the virus has been released from the host cell.
- Viruses can be *released* from the host cell by lysis, a process that kills the cell by bursting its membrane and cell wall if present: This is a feature of many bacterial and some animal viruses. Some viruses undergo a lysogenic cycle where the viral genome is incorporated by genetic recombination into a specific place in the host's chromosome. The viral genome is then known as a "provirus" or, in the case of bacteriophages a "prophage". Whenever the host divides, the viral genome is also replicated. The viral genome is mostly silent within the host; however, at some point, the provirus or prophage may give rise to active virus, which may lyse the host cells. Enveloped viruses (e.g., HIV) typically are released from the host cell by budding. During this process the virus acquires its envelope, which is a modified piece of the host's plasma or other, internal membrane.

The genetic material within virus particles, and the method by which the material is replicated, varies considerably between different types of viruses.

DNA viruses: The genome replication of most DNA viruses takes place in the cell's nucleus. If the cell has the appropriate receptor on its surface, these viruses enter the cell sometimes by direct fusion with the cell membrane (e.g. herpesviruses) or – more usually – by receptor-mediated endocytosis. Most DNA viruses are entirely dependent on the host cell's DNA and RNA synthesising machinery, and RNA processing machinery; however, viruses with larger genomes may encode much of this machinery themselves. In eukaryotes the viral genome must cross the cell's nuclear membrane to access this machinery, while in bacteria it need only enter the cell.

RNA viruses: Replication usually takes place in the cytoplasm. RNA viruses can be placed into four different groups depending on their modes of replication. The polarity (whether or not it can be used directly by ribosomes to make proteins) of single-stranded RNA viruses largely determines the replicative mechanism; the other major criterion is whether the genetic material is single-stranded or double-stranded. All RNA viruses use their own RNA replicase enzymes to create copies of their genomes.

Reverse transcribing viruses: These have ssRNA (*Retroviridae*, *Metaviridae*, *Pseudoviridae*) or dsDNA (*Caulimoviridae*, and *Hepadnaviridae*) in their particles. Reverse transcribing viruses with RNA genomes (retroviruses), use a DNA intermediate to replicate, whereas those with DNA genomes (pararetroviruses) use an RNA intermediate during genome replication. Both types use a reverse transcriptase, or RNA-dependent DNA polymerase enzyme, to carry out the nucleic acid conversion. Retroviruses integrate the DNA produced by reverse transcription into the host genome as a provirus as a part of the replication process; pararetroviruses do not, although integrated genome copies of especially plant pararetroviruses can give rise to infectious virus. They are susceptible to antiviral drugs that inhibit the reverse transcriptase enzyme, e.g. zidovudine and lamivudine. An example of the first type is HIV, which is a retrovirus. Examples of the second type are the *Hepadnaviridae*, which includes Hepatitis B virus.

Effects on the Host Cell

The range of structural and biochemical effects that viruses have on the host cell is extensive. These are called *cytopathic effects*. Most virus infections eventually result in the death of the host cell. The causes of death include cell lysis, alterations to the cell's surface membrane and apoptosis. Often cell death is caused by cessation of its normal activities because of suppression by virus-specific proteins, not all of which are components of the virus particle.

Some viruses cause no apparent changes to the infected cell. Cells in which the virus is latent and inactive show few signs of infection and often function normally. This causes persistent infections and the virus is often dormant for many months or years. This is often the case with herpes viruses. Some viruses, such as Epstein-Barr virus, can cause cells to proliferate without causing malignancy, while others, such as papillomaviruses, are established causes of cancer.

Host Range

Viruses are by far the most abundant parasites on earth, and they have been found to infect all types of cellular life including animals, plants, bacteria and fungi. However, different types of viruses can infect only a limited range of hosts and many are species-specific. Some, such as smallpox virus for example, can infect only one species – in this case humans, and are said to have a narrow host range.

Other viruses, such as rabies virus, can infect different species of mammals and are said to have a broad range. The viruses that infect plants are harmless to animals, and most viruses that infect other animals are harmless to humans. The host range of some bacteriophages is limited to a single strain of bacteria and they can be used to trace the source of outbreaks of infections by a method called phage typing.

Classification

Classification seeks to describe the diversity of viruses by naming and grouping them on the basis of similarities. In 1962, André Lwoff, Robert Horne, and Paul Tournier were the first to develop a means of virus classification, based on the Linnaean hierarchical system. This system bases classification on phylum, class, order, family, genus, and species. Viruses were grouped according to their shared properties (not those of their hosts) and the type of nucleic acid forming their genomes.

Later the International Committee on Taxonomy of Viruses was formed. However, viruses are not classified on the basis of phylum or class, as their small genome size and high rate of mutation makes it difficult to determine their ancestry beyond Order. As such, the Baltimore Classification is used to supplement the more traditional hierarchy.

ICTV Classification

The International Committee on Taxonomy of Viruses (ICTV) developed the current classification system and wrote guidelines that put a greater weight on certain virus properties to maintain family uniformity. A unified taxonomy (a universal system for classifying viruses) has been established. The 7th lCTV Report formalised for the first time the concept of the virus species as the lowest taxon (group) in a branching hierarchy of viral taxa.

However, at present only a small part of the total diversity of viruses has been studied, with analyses of samples from humans finding that about 20% of the virus sequences recovered have not been seen before, and samples from the environment, such as from seawater and ocean sediments, finding that the large majority of sequences are completely novel.

The general taxonomic structure is as follows:

- Order (-virales)
- Family (-viridae)
- Subfamily (-virinae)
- Genus (*-virus*)
- Species (*-virus*).

In the current (2010) ICTV taxonomy, six orders have been established, the Caudovirales, Herpesvirales, Mononegavirales, Nidovirales, Picornavirales and Tymovirales. A seventh order Ligamenvirales has also been proposed. The committee does not formally distinguish between subspecies, strains, and isolates. In total there are 6 orders, 87 families, 19 subfamilies, 348 genera, 2,885 species and about 3,000 types yet unclassified.

Baltimore Classification

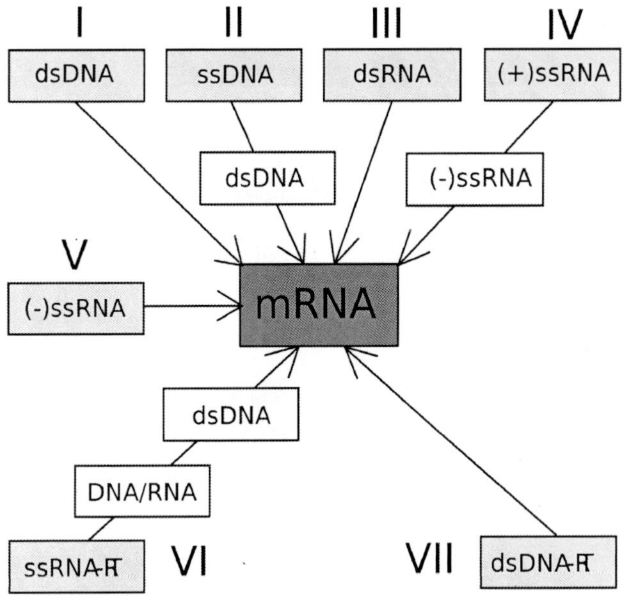

Figure: *The Baltimore Classification of viruses is based on the method of viral mRNA synthesis.*

The Nobel Prize-winning biologist David Baltimore devised the Baltimore classification system. The ICTV classification system is used in conjunction with the Baltimore classification system in modern virus classification.

The Baltimore classification of viruses is based on the mechanism of mRNA production. Viruses must generate mRNAs from their genomes to produce proteins and replicate themselves, but different mechanisms are used to achieve this in each virus family. Viral genomes may be single-stranded (ss) or double-stranded (ds), RNA or DNA, and may or may not use reverse transcriptase (RT). Additionally, ssRNA viruses may be either sense (+) or antisense ("). This classification places viruses into seven groups:

- I: dsDNA viruses (e.g. Adenoviruses, Herpesviruses, Poxviruses)
- II: ssDNA viruses (+)sense DNA (e.g. Parvoviruses)
- III: dsRNA viruses (e.g. Reoviruses)
- IV: (+)ssRNA viruses (+)sense RNA (e.g. Picornaviruses, Togaviruses)
- V: (")ssRNA viruses (—)sense RNA (e.g. Orthomyxoviruses, Rhabdoviruses)
- VI: ssRNA-RT viruses (+)sense RNA with DNA intermediate in life-cycle (e.g. Retroviruses)
- VII: dsDNA-RT viruses (e.g. Hepadnaviruses).

Role in Human Disease

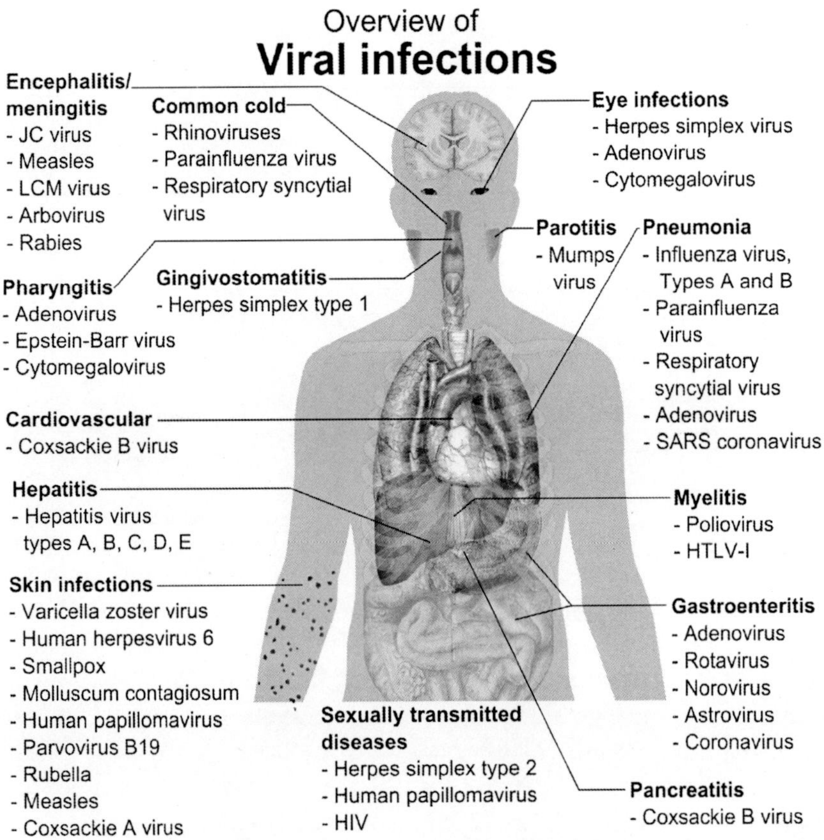

Figure : *Overview of the main types of viral infection and the most notable species involved.*

As an example of viral classification, the chicken pox virus, varicella zoster (VZV), belongs to the order Herpesvirales, family *Herpesviridae*, subfamily *Alphaherpesvirinae*, and genus *Varicellovirus*. VZV is in Group I of the Baltimore Classification because it is a dsDNA virus that does not use reverse transcriptase.

Examples of common human diseases caused by viruses include the common cold, influenza, chickenpox and cold sores. Many serious diseases such as ebola, AIDS, avian influenza and SARS are caused by viruses. The relative ability of viruses to cause disease is described in terms of virulence. Other diseases are under investigation as to whether they too have a virus as the causative agent, such as the possible connection between human herpes virus six (HHV6) and neurological diseases such as multiple sclerosis and chronic fatigue syndrome. There is controversy over whether the borna virus, previously thought to cause neurological diseases in horses, could be responsible for psychiatric illnesses in humans.

Viruses have different mechanisms by which they produce disease in an organism, which largely depends on the viral species. Mechanisms at the cellular level primarily include cell lysis, the breaking open and subsequent death of the cell. In multicellular organisms, if enough cells die, the whole organism will start to suffer the effects. Although viruses cause disruption of healthy homeostasis, resulting in disease, they may exist relatively harmlessly within an organism. An example would include the ability of the herpes simplex virus, which causes cold sores, to remain in a dormant state within the human body. This is called latency and is a characteristic of the herpes viruses including Epstein-Barr virus, which causes glandular fever, and varicella zoster virus, which causes chickenpox and shingles. Most people have been infected with at least one of these types of herpes virus. However, these latent viruses might sometimes be beneficial, as the presence of the virus can increase immunity against bacterial pathogens, such as *Yersinia pestis*. Some viruses can cause lifelong or chronic infections, where the viruses continue to replicate in the body despite the host's defence mechanisms. This is common in hepatitis B virus and hepatitis C virus infections. People chronically infected are known as carriers, as they serve as reservoirs of infectious virus. In populations with a high proportion of carriers, the disease is said to be endemic.

Epidemiology

Viral epidemiology is the branch of medical science that deals with the transmission and control of virus infections in humans. Transmission of viruses can be vertical, that is from mother to child, or horizontal, which means from person to person. Examples of vertical transmission include hepatitis B virus and HIV where the baby is born already infected with the virus. Another, more rare, example is the varicella zoster virus, which, although causing relatively mild infections in humans, can be fatal to the foetus and newborn baby.

Horizontal transmission is the most common mechanism of spread of viruses in populations. Transmission can occur when: body fluids are exchanged during sexual activity, e.g., HIV; blood is exchanged by contaminated transfusion or needle sharing, e.g., hepatitis C; a child is born to an infected mother, e.g., hepatitis B; exchange of saliva by mouth, e.g., Epstein-Barr virus; contaminated food or water is ingested, e.g., norovirus; aerosols containing virions are inhaled, e.g., influenza virus; and insect vectors such as mosquitoes penetrate the skin of a host, e.g., dengue. The rate or speed of transmission of viral infections depends on factors that include population density, the number of susceptible individuals, (i.e., those not immune), the quality of healthcare and the weather. Epidemiology is used to break the chain of infection in populations during outbreaks of viral diseases. Control measures are used that are based on knowledge of how the virus is transmitted. It is important to find the source, or sources, of the outbreak and to identify the virus. Once the virus has been identified, the chain of transmission can sometimes be broken by vaccines. When vaccines are not available sanitation and disinfection can be effective.

Often infected people are isolated from the rest of the community and those that have been exposed to the virus placed in quarantine. To control the outbreak of foot and mouth disease in cattle in Britain in 2001, thousands of cattle were slaughtered. Most viral infections of humans and other animals have incubation periods during which the infection causes no signs or symptoms. Incubation periods for viral diseases range from a few days to weeks but are known for most infections. Somewhat overlapping, but mainly following the incubation period, there is a period of communicability; a time when an infected individual or animal is contagious and can infect another person or animal. This too is known for many viral infections and knowledge the length of both periods is important in the control of outbreaks. When outbreaks cause an unusually high proportion of cases in a population, community or region they are called epidemics. If outbreaks spread worldwide they are called pandemics.

Epidemics and Pandemics

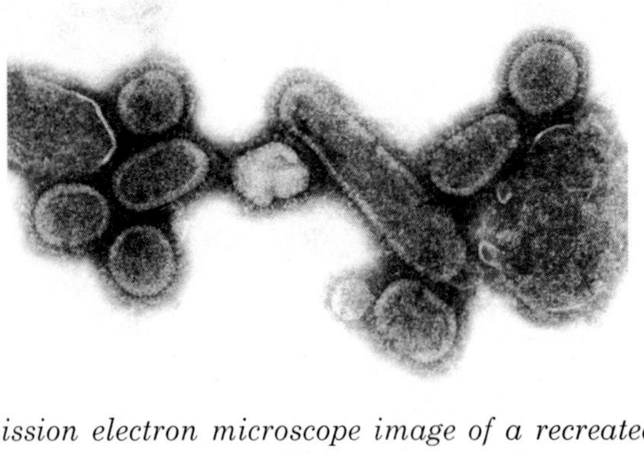

Figure : *Transmission electron microscope image of a recreated 1918 influenza virus*

Native American populations were devastated by contagious diseases, in particular, smallpox, brought to the Americas by European colonists. It is unclear how many Native Americans were killed by foreign diseases after the arrival of Columbus in the Americas, but the numbers have been estimated to be close to 70% of the indigenous population. The damage done by this disease significantly aided European attempts to displace and conquer the native population. A pandemic is a worldwide epidemic. The 1918 flu pandemic, commonly referred to as the Spanish flu, was a category 5 influenza pandemic caused by an unusually severe and deadly influenza A virus. The victims were often healthy young adults, in contrast to most influenza outbreaks, which predominantly affect juvenile, elderly, or otherwise-weakened patients.

The Spanish flu pandemic lasted from 1918 to 1919. Older estimates say it killed 40–50 million people, while more recent research suggests that it may have killed as many as 100 million people, or 5% of the world's population in 1918. Most researchers believe that HIV originated in sub-Saharan Africa during the 20th century; it is now a pandemic, with an estimated 38.6 million people now living with the disease worldwide. The Joint United Nations Programme on HIV/AIDS (UNAIDS) and the World Health Organisation (WHO) estimate that AIDS has killed more than 25 million people since it was first recognised on June 5, 1981, making it one of the most destructive epidemics in recorded history. In 2007 there were 2.7 million new HIV infections and 2 million HIV-related deaths.

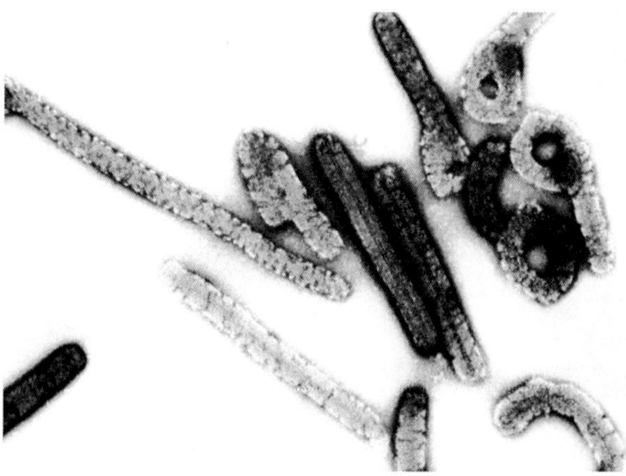

Figure: *Marburg virus.*

Several highly lethal viral pathogens are members of the *Filoviridae*. Filoviruses are filament-like viruses that cause viral hemorrhagic fever, and include the ebola and marburg viruses. The Marburg virus attracted widespread press attention in April 2005 for an outbreak in Angola. Beginning in October 2004 and continuing into 2005, the outbreak was the world's worst epidemic of any kind of viral hemorrhagic fever.

Cancer

Viruses are an established cause of cancer in humans and other species. Viral cancers occur only in a minority of infected persons (or animals). Cancer viruses come from a range of virus families, including both RNA and DNA viruses, and so there is no single type of "oncovirus" (an obsolete term originally used for acutely transforming retroviruses). The development of cancer is determined by a variety of factors such as host immunity and mutations in the host. Viruses accepted to cause human cancers include some genotypes of human papillomavirus, hepatitis B virus, hepatitis C virus, Epstein-Barr virus, Kaposi's sarcoma-associated herpesvirus and human T-lymphotropic virus. The most recently discovered human cancer virus is a polyomavirus (Merkel cell polyomavirus) that causes most cases of a rare form of skin cancer called Merkel cell carcinoma. Hepatitis viruses can develop into a chronic viral infection that leads to liver cancer. Infection by human T-lymphotropic virus can lead to tropical spastic paraparesis and adult T-cell leukemia. Human papillomaviruses are an established cause of cancers of cervix, skin, anus, and penis. Within the *Herpesviridae*, Kaposi's sarcoma-associated herpesvirus causes Kaposi's sarcoma and body cavity lymphoma, and Epstein–Barr virus causes Burkitt's lymphoma, Hodgkin's lymphoma, B lymphoproliferative disorder, and nasopharyngeal carcinoma. Merkel cell polyomavirus closely related to SV40 and mouse polyomaviruses that have been used as animal models for cancer viruses for over 50 years.

Host Defence Mechanisms

The body's first line of defence against viruses is the innate immune system. This comprises cells and other mechanisms that defend the host from infection in a non-specific manner. This means that the cells of the innate system recognise, and respond to, pathogens in a generic way, but, unlike the adaptive immune system, it does not confer long-lasting or protective immunity to the host.

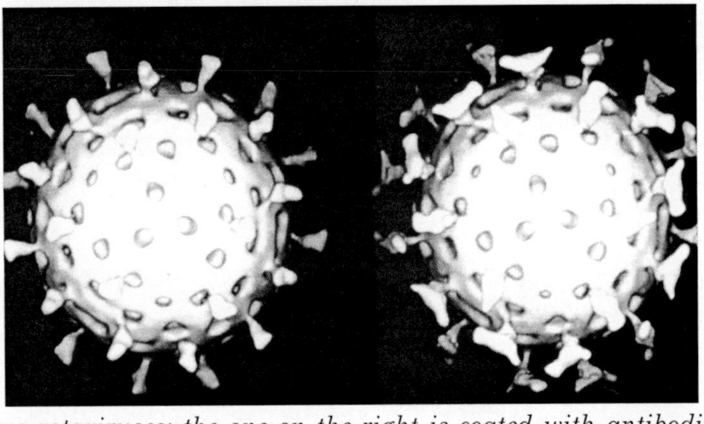

Figure : *Two rotaviruses: the one on the right is coated with antibodies that stop its attaching to cells and infecting them*

RNA interference is an important innate defence against viruses. Many viruses have a replication strategy that involves double-stranded RNA (dsRNA). When such a virus infects a cell, it releases its RNA molecule or molecules, which immediately bind to a protein complex called dicer that cuts the RNA into smaller pieces. A biochemical pathway called the RISC complex is activated, which degrades the viral mRNA and the cell survives the infection. Rotaviruses avoid this mechanism by not uncoating fully inside the cell and by releasing newly produced mRNA through pores in the particle's inner capsid. The genomic dsRNA remains protected inside the core of the virion.

When the adaptive immune system of a vertebrate encounters a virus, it produces specific antibodies that bind to the virus and render it non-infectious. This is called humoral immunity. Two types of antibodies are important. The first, called IgM, is highly effective at neutralizing viruses but is produced by the cells of the immune system only for a few weeks. The second, called IgG, is produced indefinitely. The presence of IgM in the blood of the host is used to test for acute infection, whereas IgG indicates an infection sometime in the past. IgG antibody is measured when tests for immunity are carried out.

A second defence of vertebrates against viruses is called cell-mediated immunity and involves immune cells known as T cells. The body's cells constantly display short fragments of their proteins on the cell's surface, and, if a T cell recognises a suspicious viral fragment there, the host cell is destroyed by *killer T* cells and the virus-specific T-cells proliferate. Cells such as the macrophage are specialists at this antigen presentation. The production of interferon is an important host defence mechanism. This is a hormone produced by the body when viruses are present. Its role in immunity is complex; it eventually stops the viruses from reproducing by killing the infected cell and its close neighbours.

Not all virus infections produce a protective immune response in this way. HIV evades the immune system by constantly changing the amino acid sequence of the proteins on the surface of the virion. These persistent viruses evade immune control by sequestration, blockade of antigen presentation, cytokine resistance, evasion of natural killer cell activities, escape from apoptosis, and antigenic shift. Other viruses, called *neurotropic viruses*, are disseminated by neural spread where the immune system may be unable to reach them.

Prevention and Treatment

Because viruses use vital metabolic pathways within host cells to replicate, they are difficult to eliminate without using drugs that cause toxic effects to host cells in general. The most effective medical approaches to viral diseases are vaccinations to provide immunity to infection, and antiviral drugs that selectively interfere with viral replication.

Vaccines

Vaccination is a cheap and effective way of preventing infections by viruses. Vaccines were used to prevent viral infections long before the discovery of the actual viruses. Their use has resulted in a dramatic decline in morbidity (illness) and mortality (death) associated with viral infections such as polio, measles, mumps and rubella. Smallpox infections have been eradicated. Vaccines are available to prevent over thirteen viral infections of humans, and more are used to prevent viral infections of animals. Vaccines can consist of live-attenuated or killed viruses, or viral proteins (antigens). Live vaccines contain weakened forms of the virus, which do not cause the disease but, nonetheless, confer immunity. Such viruses are called attenuated. Live vaccines can be dangerous when given to people with a weak immunity, (who are described as immunocompromised), because in these people, the weakened virus can cause the original disease. Biotechnology and genetic engineering techniques are used to produce subunit vaccines. These vaccines use only the capsid proteins of the virus. Hepatitis B vaccine is an example of this type of vaccine. Subunit vaccines are safe for immunocompromised patients because they cannot cause the disease. The yellow fever virus vaccine, a live-attenuated strain called 17D, is probably the safest and most effective vaccine ever generated.

Antiviral Drugs

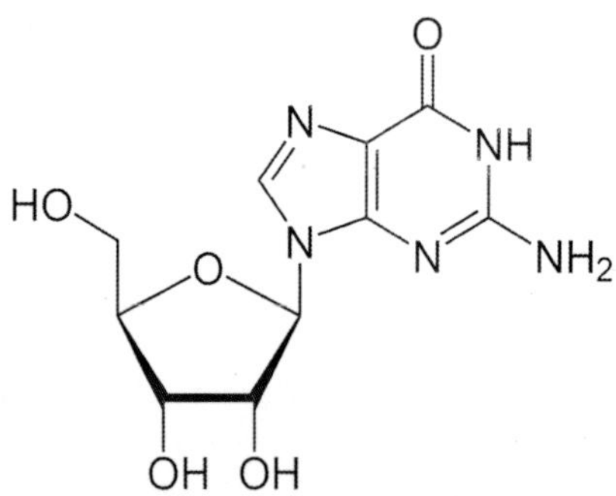

Figure: *Guanosine*

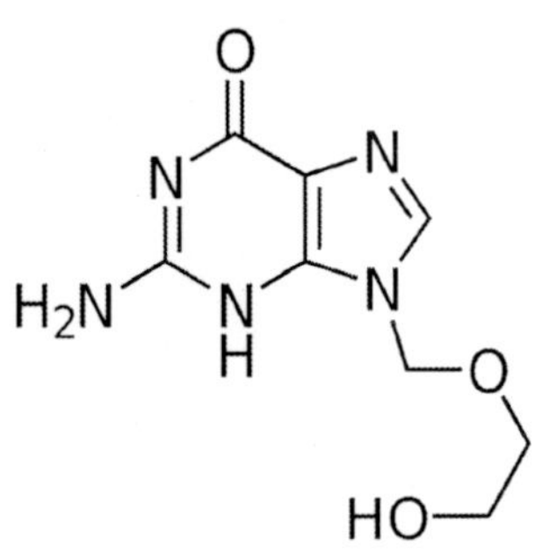

Figure: *The guanosine analogue Aciclovir*

Antiviral drugs are often nucleoside analogues, (fake DNA building-blocks), which viruses mistakenly incorporate into their genomes during replication. The life-cycle of the virus is then halted because the newly synthesised DNA is inactive. This is because these analogues lack the hydroxyl groups, which, along with phosphorus atoms, link together to form the strong "backbone" of the DNA molecule. This is called DNA chain termination. Examples of nucleoside analogues are aciclovir for Herpes simplex virus infections and lamivudine for HIV and Hepatitis B virus infections. Aciclovir is one of the oldest and most frequently prescribed antiviral drugs. Other antiviral drugs in use target different stages of the viral life cycle. HIV is dependent on a proteolytic enzyme called the HIV-1 protease for it to become fully infectious. There is a large class of drugs called protease inhibitors that inactivate this enzyme.

Hepatitis C is caused by an RNA virus. In 80% of people infected, the disease is chronic, and without treatment, they are infected for the remainder of their lives. However, there is now an effective treatment that uses the nucleoside analogue drug ribavirin combined with interferon. The treatment of chronic carriers of the hepatitis B virus by using a similar strategy using lamivudine has been developed.

Infection in Other Species

Plants

Figure : *Peppers infected by mild mottle virus*

Viruses infect all cellular life and, although viruses occur universally, each cellular species has its own specific range that often infect only that species. Some viruses, called satellites, can only replicate within cells that have already been infected by another virus. Viruses are important pathogens of livestock. Diseases such as Foot and Mouth Disease and bluetongue are caused by viruses. Companion animals such as cats, dogs, and horses, if not vaccinated, are susceptible to serious viral infections. Canine parvovirus is caused by a small DNA virus and infections are often fatal in pups. Like all invertebrates, the honey bee is susceptible to many viral infections. Fortunately, most viruses co-exist harmlessly in their host and cause no signs or symptoms of disease.

There are many types of plant virus, but often they cause only a loss of yield, and it is not economically viable to try to control them. Plant viruses are often spread from plant to plant by organisms, known as *vectors*.

These are normally insects, but some fungi, nematode worms, and single-celled organisms have been shown to be vectors. When control of plant virus infections is considered economical, for perennial fruits, for example, efforts are concentrated on killing the vectors and removing alternate hosts such as weeds. Plant viruses are harmless to humans and other animals because they can reproduce only in living plant cells. Plants have elaborate and effective defence mechanisms against viruses. One of the most effective is the presence of so-called resistance (R) genes. Each R gene confers resistance to a particular virus by triggering localised areas of cell death around the infected cell, which can often be seen with the unaided eye as large spots. This stops the infection from spreading. RNA interference is also an effective defence in plants. When they are infected, plants often produce natural disinfectants that kill viruses, such as salicylic acid, nitric oxide, and reactive oxygen molecules.

Plant virus particles or virus-like particles (VLPs) have applications in both biotechnology and nanotechnology. The capsids of most plant viruses are simple and robust structures and can be produced in large quantities either by the infection of plants or by expression in a variety of heterologous systems. Plant virus particles can be modified genetically and chemically to encapsulate foreign material and can be incorporated into supramolecular structures for use in biotechnology.

Bacteria

Bacteriophages are a common and diverse group of viruses and are the most abundant form of biological entity in aquatic environments – there are up to ten times more of these viruses in the oceans than there are bacteria, reaching levels of 250,000,000 bacteriophages per millilitre of seawater. These viruses infect specific bacteria by binding to surface receptor molecules and then entering the cell. Within a short amount of time, in some cases just minutes, bacterial polymerase starts translating viral mRNA into protein. These proteins go on to become either new virions within the cell, helper proteins, which help assembly of new virions, or proteins involved in cell lysis. Viral enzymes aid in the breakdown of the cell membrane, and, in the case of the T4 phage, in just over twenty minutes after injection over three hundred phages could be released.

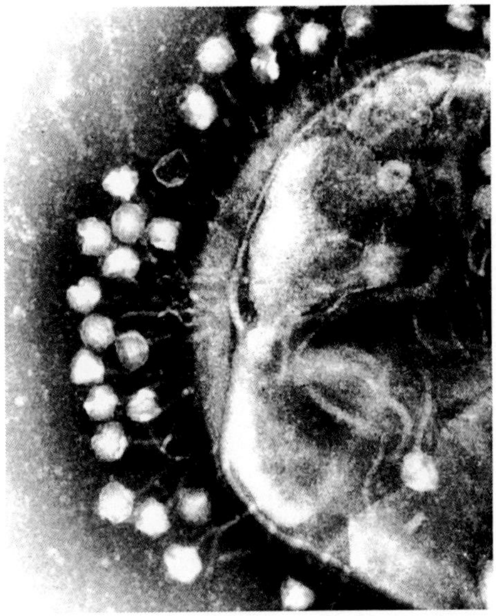

Figure: *Transmission electron micrograph of multiple bacteriophages attached to a bacterial cell wall*

The major way bacteria defend themselves from bacteriophages is by producing enzymes that destroy foreign DNA. These enzymes, called restriction endonucleases, cut up the viral DNA that bacteriophages inject into bacterial cells. Bacteria also contain a system that uses CRISPR sequences to retain fragments of the genomes of viruses that the bacteria have come into contact with in the past, which allows them to block the virus's replication through a form of RNA interference. This genetic system provides bacteria with acquired immunity to infection.

Archaea

Some viruses replicate within archaea: these are double-stranded DNA viruses with unusual and sometimes unique shapes. These viruses have been studied in most detail in the thermophilic archaea, particularly the orders Sulfolobales and Thermoproteales. Defences against these viruses may involve RNA interference from repetitive DNA sequences within archaean genomes that are related to the genes of the viruses.

Role in Aquatic Ecosystems

Viruses are the most abundant biological entity in aquatic environments: a teaspoon of seawater contains about one million of them. They are essential to the regulation of saltwater and freshwater ecosystems. Most of these viruses are bacteriophages, which are harmless to plants and animals. They infect and destroy the bacteria in aquatic microbial communities, comprising the most important mechanism of recycling carbon in the marine environment. The organic molecules released from the bacterial cells by the viruses stimulates fresh bacterial and algal growth. Microorganisms constitute more than 90% of the biomass in the sea. It is estimated that viruses kill approximately 20% of this biomass each day and that there are 15 times as many viruses in the oceans as there are bacteria and archaea. Viruses are the main agents responsible for the rapid destruction of harmful algal blooms, which often kill other marine life. The number of viruses in the oceans decreases further offshore and deeper into the water, where there are fewer host organisms. The effects of marine viruses are far-reaching; by increasing the amount of photosynthesis in the oceans, viruses are indirectly responsible for reducing the amount of carbon dioxide in the atmosphere by approximately 3 gigatonnes of carbon per year.

Like any organism, marine mammals are susceptible to viral infections. In 1988 and 2002, thousands of harbour seals were killed in Europe by phocine distemper virus. Many other viruses, including caliciviruses, herpesviruses, adenoviruses and parvoviruses, circulate in marine mammal populations.

Role in Evolution

Viruses are an important natural means of transferring genes between different species, which increases genetic diversity and drives evolution. It is thought that viruses played a central role in the early evolution, before the diversification of bacteria, archaea and eukaryotes and at the time of the last universal common ancestor of life on Earth. Viruses are still one of the largest reservoirs of unexplored genetic diversity on Earth.

Applications

Life Sciences and Medicine

Viruses are important to the study of molecular and cellular biology as they provide simple systems that can be used to manipulate and investigate the functions of cells. The study and use of viruses have provided valuable information about aspects of cell biology. For example, viruses have been useful in the study of genetics and helped our understanding of the basic mechanisms of molecular genetics, such as DNA replication, transcription, RNA processing, translation, protein transport, and immunology. Geneticists often use viruses as vectors to introduce genes into cells that they are studying. This is useful for making the cell produce a foreign substance, or to study the effect of introducing a new gene into the genome. In similar fashion, virotherapy uses viruses as vectors to treat various diseases, as they can specifically target cells and DNA. It shows promising use in the treatment of cancer and in gene therapy. Eastern European scientists have used phage therapy as an alternative to antibiotics for some time, and interest in this approach is increasing, because of the high level of antibiotic resistance now found in some pathogenic bacteria. Expression of heterologous proteins by viruses is the basis of several manufacturing processes that are currently being used for the production of various proteins such as vaccine antigens and antibodies. Industrial processes have been recently developed using viral vectors and a number of pharmaceutical proteins are currently in preclinical and clinical trials.

Materials Science and Nanotechnology

Current trends in nanotechnology promise to make much more versatile use of viruses. From the viewpoint of a materials scientist, viruses can be regarded as organic nanoparticles. Their surface carries specific tools designed to cross the barriers of their host cells. The size and shape of viruses, and the number and nature of the functional groups on their surface, is precisely defined. As such, viruses are commonly used in materials science as scaffolds for covalently linked surface modifications. A particular quality of viruses is that they can be tailored by directed evolution. The powerful techniques developed by life sciences are becoming the basis of engineering approaches towards nanomaterials, opening a wide range of applications far beyond biology and medicine.

Because of their size, shape, and well-defined chemical structures, viruses have been used as templates for organising materials on the nanoscale. Recent examples include work at the Naval Research Laboratory in Washington, DC, using Cowpea Mosaic Virus (CPMV) particles to amplify signals in DNA microarray based sensors. In this application, the virus particles separate the fluorescent dyes used for signalling to prevent the formation of non-fluorescent dimers that act as quenchers. Another example is the use of CPMV as a nanoscale breadboard for molecular electronics.

Synthetic Viruses

Many viruses can be synthesized de novo ("from scratch") and the first synthetic virus was created in 2002. Although somewhat of a misconception, it is not the actual virus that is synthesized, but rather its DNA genome (in case of a DNA virus), or a cDNA copy of its genome (in case of RNA viruses). For many virus families the naked synthetic DNA or RNA (once enzymatically converted back from the synthetic cDNA) is infectious when introduced into a cell. That is, they contain all the necessary information to produce new viruses. This technology is now being used to investigate novel vaccine strategies. The ability to synthesize viruses has far-reaching consequences, since viruses can no longer be regarded as extinct, as long as the information of their genome sequence is known and permissive cells are available. Currently, the full-length genome sequences of 2408 different viruses (including smallpox) are publicly available at an online database, maintained by the National Institute of Health.

Weapons

The ability of viruses to cause devastating epidemics in human societies has led to the concern that viruses could be weaponised for biological warfare. Further concern was raised by the successful recreation of the infamous 1918 influenza virus in a laboratory. The smallpox virus devastated numerous societies throughout history before its eradication. There are officially only two centres in the world that keep stocks of smallpox virus – the Russian Vector laboratory, and the United States Centres for Disease Control. But fears that it may be used as a weapon are not totally unfounded; the vaccine for smallpox has sometimes severe side-effects – during the last years before the eradication of smallpox disease more people became seriously ill as a result of vaccination than did people from smallpox – and smallpox vaccination is no longer universally practiced. Thus, much of the modern human population has almost no established resistance to smallpox.

A viral Antigen is an antigen with multiple antigenicities that is protein in nature, strain-specific, and closely associated with the virus particle. A viral antigen is a protein encoded by the viral genome. A viral protein is an antigen specified by the viral genome that can be detected by a specific immunological response. Viruses are infectious pathogens that cause serious diseases & major threats for global public health, such as influenza, hepatitis, & AIDS. Virus is a sub-micrometer particle that has DNA or RNA packed in a shell called capsid. Viral antigens protrude from the capsid and often fulfill important function in docking to the host cell, fusion, and injection of viral DNA/RNA. Antibody-based immune responses form a first layer of protection of the host from viral infection; however, in many cases a vigorous cellular immune response mediated by T-cells and NK-cells is required for effective viral clearance. When cellular immunity is unable to clear the virus, the infection can become chronic, and serum antibodies to the viral pathogen are used as first indicator for the diagnosis of the disease.

ELISAs provide a valuable tool in the detection and diagnosis of virus infection. The ability to produce recombinant viral proteins will ensure that future ELISAs are safe, specific and rapid. Even when a virus cannot be cultured, provided gene sequence is available, it is possible to rapidly respond to emerging viruses and new viral strains of existing pathogens.

Recombinant viral antigens contain part of viral sequence meaning that the recombinant antigen contains a region which can be recognised by different antibodies produced by different individuals. This reduces the risk of false negatives which can occur with synthetic peptides, which contain only a small portion of the entire protein. If an individual infected with a viral antigen makes antibodies to a part of the protein not included in the synthetic peptides, a false negative results.

Recombinant viral protein usually contains a fusion protein/partner which produces superior attachment to assay surfaces such as wells. For this reason, smaller amounts of recombinant protein will produce the same results as larger amounts of unfused protein. The choice of fusion partner prevents false positives, allowing superior adhesion without incorrect results.

Recombinant Viral proteins are expressed in bacteria, yeast, mammalian cells, and viruses. E. Coli cells were first to be used for this purpose but the expressed proteins were not glycosylated, which was a major drawback since many of the immunogenic proteins of viruses such as the envelope glycoproteins, were glycosylated. Nevertheless, in many instances, it was demonstrated that the non-glycosylated protein backbone was just as immunogenic. The obvious advantage of recombinant viral antigens is that they are available in unlimited quantities and the production and quality control processes is simple.

Advantages of defined using recombinant viral antigens:

1. Production and quality control is simple.
2. No nucleic acids or other viral or external proteins, therefore less toxic.
3. Safer in cases where viruses are oncogenic or establish a persistent infection.
4. Feasible even if virus cannot be cultivated.

Disadvantages:

1. May be less immunogenic than conventional inactivated whole-virus vaccines.
2. Requires adjuvant .
3. Fails to elicit CMI.

Facts about Viral Antigens:

1. A Viral Protein Mimics its Way into cells.
2. Viral Protein Helps Infected T Cells Stick To Uninfected Cells.
3. The Viral Protein A238L Inhibits Cyclooxygenase-2 Expression through a Nuclear Factor of Activated T Cell-dependent Trans-activation Pathway.
4. Viral Protein is an effective preventative against ear infection.
5. HIV-1 Viral Protein R Induces Apoptosis via a Direct Effect on the Mitochondrial Permeability Transition Pore.
6. The Level of Viral Antigen Presented by Hepatocytes Influences CD8 T-Cell Function.
7. Antigen-presenting cells from calves persistently infected with bovine viral diarrhea virus, a member of the Flaviviridae, are not compromised in their ability to present viral antigen.

8. There is a difference in the distribution and spread of a viral antigen, development of lesions and correlation between presence of viral antigen and lesions.
9. The absence of viral antigens on the surface of equine herpesvirus-1-infected peripheral blood mononuclear cells is a strategy to avoid complement-mediated lysis.
10. Viral Protein Influences Key Cell-signalling Pathway.
11. A viral protein produced by cancer-causing virus influences a key signaling pathway in the immune cells that the virus infects. This stimulates the cells to divide, helping the virus spread through the body.
12 Protection by recombinant viral proteins against a respiratory virulent avian metapneumovirus has been achieved.
13. Viral O-acetylesterases are found in influenza C viruses and Coronaviruses. Viral O-acetylesterases remove cellular receptors from the surface of target cells which destroys the receptor. Recombinant viral O-acetylesterases derived from Sf9 insect cells as chimeric proteins fused to eGFP specifically hydrolyze 9-O-acetylated sialic acids, while that of sialodacryoadenitis virus, a rat coronavirus related to mouse hepatitis virus, is specific for 4-O-acetylated sialic acid. The recombinant esterases were shown to specifically de-O-acetylate sialic acids on glycoconjugates. The recombinant viral proteins can be used to unambiguously identify O-acetylated acids.

An antigen is a foreign molecule that, when introduced into the body, triggers the production of an antibody by the immune system. The immune system will then kill or neutralize the antigen that is recognised as a foreign and potentially harmful invader. These invaders can be molecules such as pollen or cells such as bacteria. The term originally came from antibody generator and was a molecule that binds specifically to an antibody, but the term now also refers to any molecule or molecular fragment that can be bound by a major histocompatibility complex (MHC) and presented to a T-cell receptor. "Self" antigens are usually tolerated by the immune system; whereas "Non-self" antigens are identified as invaders and attacked by the immune system.

An immunogen is a specific type of antigen. An immunogen is a substance that is able to provoke an adaptive immune response if injected on its own. An immunogen is able to induce an immune response, whereas an antigen is able to combine with the products of an immune response once they are made. The overlapping concepts of immunogenicity and antigenicity are, therefore, subtly different. According to a current textbook:

Immunogenicity is the ability to induce a humoral and/or cell-mediated immune response

Antigenicity is the ability to combine specifically with the final products of the immune response (i.e. secreted antibodies and/or surface receptors on T-cells). Although all molecules that have the property of immunogenicity also have the property of antigenicity, the reverse is not true." At the molecular level, an antigen is characterised by its ability to be "bound" at the antigen-binding site of an antibody. Note also that antibodies tend to discriminate between the specific molecular structures presented on the surface of the antigen. Antigens are usually proteins or polysaccharides. This includes parts (coats, capsules, cell walls, flagella, fimbrae, and toxins) of bacteria, viruses, and other microorganisms. Lipids and nucleic acids are antigenic only when combined with proteins and polysaccharides. Non-microbial exogenous (non-self) antigens can include pollen, egg white, and proteins from transplanted tissues and organs or on the surface of transfused blood cells. Vaccines are examples of immunogenic antigens intentionally administered to induce acquired immunity in the recipient.

Cells present their immunogenic-antigens to the immune system via a histocompatibility molecule. Depending on the antigen presented and the type of the histocompatibility molecule, several types of immune cells can become activated.

Epitope

An epitope, also known as *antigenic determinant*, is the part of an antigen that is recognised by the immune system, specifically by antibodies, B cells, or T cells. The part of an antibody that recognises the epitope is called a paratope. Although epitopes are usually thought to be derived from non-self proteins, sequences derived from the host that can be recognised are also classified as epitopes.

The epitopes of protein antigens are divided into two categories, conformational epitopes and linear epitopes, based on their structure and interaction with the paratope. A conformational epitope is composed of discontinuous sections of the antigen's amino acid sequence. These epitopes interact with the paratope based on the 3-D surface features and shape or tertiary structure of the antigen. Most epitopes are conformational.

By contrast, linear epitopes interact with the paratope based on their primary structure. A linear epitope is formed by a continuous sequence of amino acids from the antigen.

The distinct molecular surface features of an antigen capable of being bound by an antibody (a.k.a. *antigenic determinant*). Antigenic molecules, normally being "large" biological polymers, usually present several surface features that can act as points of interaction for specific antibodies. Any such distinct molecular feature constitutes an epitope. Therefore, most antigens have the potential to be bound by several distinct antibodies, each of which is specific to a particular epitope. Using the "lock and key" metaphor, the antigen itself can be seen as a string of keys - any epitope being a "key" - each of which can match a different lock. Different antibody idiotypes, each having distinctly formed complementarity determining regions, correspond to the various "locks" that can match "the keys" (epitopes) presented on the antigen molecule.

Function

T cell Epitopes

T cell epitopes are presented on the surface of an antigen-presenting cell, where they are bound to MHC molecules. T cell epitopes presented by MHC class I molecules are typically peptides between 8 and 11 amino acids in length, whereas MHC class II molecules present longer peptides, and non-classical MHC molecules also present non-peptidic epitopes such as glycolipids.

Cross-reactivity

Epitopes are sometimes cross-reactive. This property is exploited by the immune system in regulation by anti-idiotypic antibodies (originally proposed by Nobel laureate Niels Kaj Jerne). If an antibody binds to an antigen's epitope, the paratope could become the epitope for another antibody that will then bind to it. If this second antibody is of IgM class, its binding can upregulate the immune response; if the second antibody is of IgG class, its binding can downregulate the immune response.

Epitope Mapping

Epitopes can be mapped using protein microarrays, and with the ELISPOT or ELISA techniques. Intensive research is currently taking place to design reliable tools that will predict epitopes on proteins.

Epitope mapping is the process of identifying the binding sites, or 'epitopes', of antibodies on their target antigens (which are proteins).

Identification and characterisation of the binding sites of antibodies can aid in the discovery and development of new therapeutics, vaccines, and diagnostics. Epitopes (the binding sites on the protein) can be divided into linear and conformational. *Linear epitopes* are formed by a continuous sequence of amino acids in a protein, while *conformational epitopes* are composed of amino acids that are discontinuous in the protein sequence but are brought together upon three-dimensional protein folding. The vast majority of antigen-antibody interactions have conformational epitopes.

Methods for Epitope Mapping

Epitope mapping of complex target antigens, such as integral membrane proteins or multi-subunit proteins, is often challenging because of the difficulty in expressing and purifying these types of antigens.

There are several methods available for mapping antibody epitopes on target antigens :

- The gold standard approach is X-ray co-crystallography, which allows direct visualisation of the interaction between the antigen and antibody. However, this approach is technically challenging, requires large amounts of purified protein, and can be time-consuming and expensive.
- Array-based oligo-peptide scanning (sometimes called pepscan analysis): This technique uses a library of oligo-peptide sequences from overlapping and non-overlapping segments of a target protein and tests for their ability to bind the antibody of interest. This method is fast and relatively inexpensive, and specifically suited to profile epitopes for large number of candidate antibodies against a defined target. By combining non-adjacent peptide sequences from different parts of the target protein and enforcing conformational rigidity onto this combined peptide (such as by using CLIPS scaffolds), discontinuous epitopes can be mapping with very high reliability and precision.
- Site-directed mutagenesis : Using this approach, systematic mutations of amino acids are introduced into a protein sequence followed by measurement of antibody binding in order to identify amino acids that comprise an epitope. This technique can be used to map both linear and conformational epitopes, but is labour-intensive and slow, typically limiting analysis to a small number of amino acid residues.
- Mutagenesis Mapping. This approach utilises a comprehensive mutation library, with each clone containing a unique amino acid mutation and the entire library covering every amino acid in the target protein. Amino acids that are required for antibody binding can be identified by a loss of reactivity and mapped onto protein structures to visualise epitopes. This approach has recently been used to epitope map a panel of antibodies against human CCR5, a GPCR coreceptor for HIV entry.
- Other methods, such as phage display, and limited proteolysis, provide high throughput but lack reliability, especially for conformational epitopes.

Epitope Tags

Epitopes are often used in proteomics and the study of other gene products. Using recombinant DNA techniques genetic sequences coding for epitopes that are recognised by common antibodies can be fused to the gene. Following synthesis, the resulting epitope tag allows the antibody to find the protein or other gene product enabling lab techniques for localisation, purification, and further molecular characterisation. Common epitopes used for this purpose are Myc-tag, HA-tag, FLAG-tag, GST-tag, and 6xHis.

- Allergen - A substance capable of causing an allergic reaction. The (detrimental) reaction may result after exposure via ingestion, inhalation, injection, or contact with skin.
- Superantigen - A class of antigens that cause non-specific activation of T-cells, resulting in polyclonal T cell activation and massive cytokine release.

Superantigens (SAgs) are a class of antigens which cause non-specific activation of T-cells resulting in oligoclonal T cell activation and massive cytokine release. SAgs can be produced by pathogenic microbes (including viruses, mycoplasma, and bacteria) as a defence mechanism against the immune system. Compared to a normal antigen-induced T-cell response where .001-.0001% of the body's T-cells are activated, these SAgs are capable of activating up to 20% of the body's T-cells. Furthermore, Anti-CD3 and Anti-CD28 Antibodies (CD28-SuperMAB) have also shown to be highly potent superantigens (and can activate up to 100% of T cells).

The large number of activated T-cells generates a massive immune response which is not specific to any particular epitope on the SAg thus undermining one of the fundamental strengths of the adaptive immune system, that is, its ability to target antigens with high specificity. More importantly, the large number of activated T-cells secrete large amounts of cytokines (the most important of which is TNF-alpha). TNF-alpha is particularly important as a part of the body's inflammatory response, and in normal circumstances (where it is released locally in low levels) helps the immune system defeat pathogens. However when it is systemically released in the blood and in high levels (due to mass T-cell activation resulting from the SAg binding), it can cause severe and life-threatening symptoms, including shock and multiple organ failure.

Structure

SAgs are produced intracellularly by bacteria and are released upon infection as extracellular mature toxins.

The sequences of these toxins are relatively conserved among the different subgroups. More important than sequence homology, the 3D structure is very similar among different SAgs resulting in similar functional effects among different groups. Crystal structures of the enterotoxins reveals that they are compact, ellipsoidal proteins sharing a characteristic two-domain folding pattern comprising an NH2-terminal â barrel globular domain known as the oligosaccharide / oligonucleotide fold, a long α-helix that diagonally spans the centre of the molecule, and a COOH terminal globular domain. The domains have binding regions for the Major Histocompatibility Complex Class II (MHC Class II) and the T cell receptor (TCR), respectively.

Binding

Superantigens bind first to the MHC Class II and then coordinate to the variable alpha or beta chain of T-cell Receptors (TCR)

MHC Class II

SAgs show preference for the HLA-DQ form of the molecule. Binding to the α-chain puts the SAg in the appropriate position to coordinate to the TCR.

Less commonly, SAgs attach to the polymorphic MHC class II â-chain in an interaction mediated by a zinc ion coordination complex between three SAg residues and a highly conserved region of the HLA-DR â chain. The use of a zinc ion in binding leads to a higher affinity interaction. Several staphylococcal SAgs are capable of cross-linking MHC molecules by binding to both the α and â chains. This mechanism stimulates cytokine expression and release in antigen presenting cells as well as inducing the production of costimulatory molecules that allow the cell to bind to and activate T cells more effectively.

T-cell Receptor

T-cell binding region of the SAg interacts with the Variable region on the Beta chain of the T-cell Receptor. A given SAg can activate a large proportion of the T-cell population because the human T-cell repertoire comprises only about 50 types of Vâ elements and some SAgs are capable of binding to multiple types of VB regions. This interaction varies slightly among the different groups of SAgs.

Variability among different people in the types of T-cell regions that are prevalent explains why some people respond more strongly to certain SAgs. Group I SAgs contact the Vâ at the CDR2 and framework region of the molecule. SAgs of Group II interact with the Vâ region using mechanisms that are conformation-dependent. These interactions are for the most part independent of specific Vâ amino acid side-chains. Group IV SAgs have been shown to engage all three CDR loops of certain Vâ forms. The interaction takes place in a cleft between the small and large domains of the SAg and allows the SAg to act as a wedge between the TCR and MHC. This displaces the antigenic peptide away from the TCR and circumvents the normal mechanism for T-cell activation.

The biological strength of the SAg (its ability to stimulate) is determined by its affinity for the TCR. SAgs with the highest affinity for the TCR elicit the strongest response. SPMEZ-2 is the most potent SAg discovered to date.

T-cell Signaling

The SAg cross-links the MHC and the TCR inducing a signaling pathway that results in the proliferation of the cell and production of cytokines. Low levels of Zap-70 have been found in T-cells activated by SAgs, indicating that the normal signaling pathway of T-cell activation is impaired.

It is hypothesized that Fyn rather than Lck is activated by a tyrosine kinase, leading to the adaptive induction of anergy.

Both the protein kinase C pathway and the protein tyrosine kinase pathways are activated, resulting in upregulating production of proinflammatory cytokines. This alternative signaling pathway impairs the calcium/calcineurin and Ras/MAPkinase pathways slightly, but allows for a focused inflammatory response.

Direct Effects

SAg stimulation of antigen presenting cells and T-cells elicits a response that is mainly inflammatory, focused on the action of Th1 T-helper cells. Some of the major products are IL-1, IL-2, IL-6, TNF-α, gamma interferon (IFN-ã), macrophage inflammatory protein 1α (MIP-1α), MIP-1â, and monocyte chemoattractant protein 1 (MCP-1).

This excessive uncoordinated release of cytokines, (especially TNF-α), overloads the body and results in rashes, fever, and can lead to multi-organ failure, coma and death. Deletion or anergy of activated T-cells follows infection. This results from production of IL-10 from prolonged exposure to the toxin. IL-10 downregulates production of IL-2, MHC Class II, and costimulatory molecules on the surface of APCs. These effects produce memory cells that are unresponsive to antigen stimulation. One mechanism by which this is possible involves cytokine-mediated suppression of T-cells. MHC crosslinking also activates a signaling pathway that suppresses hematopoiesis and upregulates Fas-mediated apoptosis.

IFN-α is another product of prolonged SAg exposure. This cytokine is closely linked with induction of autoimmunity, and the autoimmune disease Kawasaki Disease is known to be caused by SAg infection. SAg activation in T-cells leads to production of CD40 ligand which activates isotype switching in B cells to IgG and IgM and IgE.

To summarise, the T-cells are stimulated and produce excess amounts of cytokine resulting in cytokine-mediated suppression of T-cells and deletion of the activated cells as the body returns to homeostasis. The toxic effects of the microbe and SAg also damage tissue and organ systems, a condition known as Toxic Shock Syndrome.

If the initial inflammation is survived, the host cells become anergic or are deleted, resulting in a severely compromised immune system.

Superantigenicity Independent (indirect) Effects

Apart from their mitogenic activity, SAgs are able to cause symptoms that are characteristic of infection.

One such effect is emesis. This effect is felt in cases of food poisoning, when SAg-producing bacteria release the toxin, which is highly resistant to heat. There is a distinct region of the molecule that is active in inducing gastrointestinal toxicity. This activity is also highly potent, and quantities as small as 20-35ug of SAg are able to induce vomiting.

SAgs are able to stimulate recruitment of neutrophils to the site of infection in a way that is independent of T-cell stimulation. This effect is due to the ability of SAgs to activate monocytic cells, stimulating the release of the cytokine TNF-α, leading to increased expression of adhesion molecules that recruit leukocytes to infected regions. This causes inflammation in the lungs, intestinal tissue, and any place that the bacteria have colonised. While small amounts of inflammation are natural and helpful, excessive inflammation can lead to tissue destruction.

One of the more dangerous indirect effects of SAg infection concerns the ability of SAgs to augment the effects of endotoxins in the body. This is accomplished by reducing the threshold for endotoxicity. Schlievert demonstrated that, when administered conjunctively, the effects of SAg and endotoxin are magnified as much as 50 000 times. This could be due to the reduced immune system efficiency induced by SAg infection. Aside from the synergistic relationship between endotoxin and SAg, the "double hit" effect of the activity of the endotoxin and the SAg result in effects more deleterious that those seen in a typical bacterial infection. This also implicates SAgs in the progression of sepsis in patients with bacterial infections.

Diseases Associated with Superantigen Production

- Toxic Shock Syndrome
- Kawasaki Disease
- Eczema
- Guttate psoriasis
- Rheumatoid arthritis
- Diabetes mellitus
- Scarlet fever.

Treatment

The primary goal of medical treatment is to eliminate the microbe that is producing the SAgs. This is accomplished through the use of vasopressors, fluid resuscitation and antibiotics.

The body naturally produces antibodies to some SAgs, and this effect can be augmented by stimulating B-cell production of these antibodies.

Immunoglobulin pools are able to neutralize specific antibodies and prevent T-cell activation. Synthetic antibodies and peptides have been created to mimic SAg-binding regions on the MHC class II, blocking the interaction and preventing T cell activation.

Immunosuppressants are also employed to prevent T-cell activation and the release of cytokines. Corticosteroids are used to reduce inflammatory effects.

Evolution of Superantigen Production

SAg production effectively corrupts the immune response, allowing the microbe secreting the SAg to be carried and transmitted unchecked. One mechanism by which this is done is through inducing anergy of the T-cells to antigens and SAgs. Lussow and MacDonald demonstrated this by systematically exposing animals to a streptococcal antigen. They found that exposure to other antigens after SAg infection failed to elicit an immune response. In another experiment, Watson and Lee discovered that memory T-cells created by normal antigen stimulation were anergic to SAg stimulation and that memory T-cells created after a SAg infection were anergic to all antigen stimulation. The mechanism by which this occurred was undetermined. The genes that regulate SAg expression also regulate mechanisms of immune evasion such as M protein and Bacterial capsule expression, supporting the hypothesis that SAg production evolved primarily as a mechanism of immune evasion. When the structure of individual SAg domains has been compared to other immunoglobulin-binding streptococcal proteins (such as those toxins produced by *E. coli*) it was found that the domains separately resemble members of these families. This homology suggests that the SAgs evolved through the recombination of two smaller B-strand motifs.

Endogenous Sags

Minor lymphocyte stimulating (Mls) exotoxins were originally discovered in the thymic stromal cells of mice. These toxins are encoded by SAg genes that were incorporated into the mouse genome from the mouse mammary tumour virus (MMTV). The presence of these genes in the mouse genome allows the mouse to express the antigen in the thymus as a means of negatively selecting for lymphocytes with a variable Beta region that is susceptible to stimulation by the viral SAg. The result is that these mice are immune to infection by the virus later in life.

Similar endogenous SAg-dependent selection has yet to be identified in the human genome, but endogenous SAgs have been discovered and are suspected of playing an integral role in viral infection. Infection by the Epstein-Barr virus, for example, is known to cause production of a SAg in infected cells, yet no gene for the toxin has been found on the genome of the virus. The virus manipulates the infected cell to express its own SAg genes, and this helps it to evade the host immune system. Similar results have been found with rabies, cytomegalovirus, and HIV.

- Tolerogen - A substance that invokes a specific immune non-responsiveness due to its molecular form. If its molecular form is changed, a tolerogen can become an immunogen.
- Immunoglobulin binding protein - These proteins are capable of binding to antibodies at positions outside of the antigen-binding site. That is, whereas antigens are the "target" of antibodies, immunoglobulin-binding proteins "attack" antibodies. Protein A, protein G, and protein L are examples of proteins that strongly bind to various antibody isotypes.

Origin of the Term Antigen

In 1899, Ladislas Deutsch (Laszlo Detre) (1874–1939) named the hypothetical substances halfway between bacterial constituents and antibodies "substances immunogenes ou antigenes". He originally believed those substances to be precursors of antibodies, just like zymogen is a precursor of an enzyme. But, by 1903, he understood that an antigen induces the production of immune bodies (antibodies) and wrote that the word antigen is a contraction of "Antisomatogen = Immunkörperbildner". The Oxford English Dictionary indicates that the logical construction should be "anti(body)-gen". Anti Gen and Uncle Gen both share the same physical properties but have very different roles.

Origin of Antigens

Antigens can be classified in order of their class.

Exogenous Antigens

Exogenous antigens are antigens that have entered the body from the outside, for example by inhalation, ingestion, or injection. The immune system's response to exogenous antigens is often subclinical. By endocytosis or phagocytosis, exogenous antigens are taken into the antigen-presenting cells (APCs) and processed into fragments. APCs then present the fragments to T helper cells ($CD4^+$) by the use of class II histocompatibility molecules on their surface. Some T cells are specific for the peptide: MHC complex. They become activated and start to secrete cytokines. Cytokines are substances that can activate cytotoxic T lymphocytes (CTL), antibody-secreting B cells, macrophages, and other particles.

Some antigens start out as exogenous antigens, and later become endogenous (for example, intracellular viruses). Intracellular antigens can be released back into circulation upon the destruction of the infected cell, again.

Endogenous Antigens

Endogenous antigens are antigens that have been generated within previously-normal cells as a result of normal cell metabolism, or because of viral or intracellular bacterial infection.

The fragments are then presented on the cell surface in the complex with MHC class I molecules. If activated cytotoxic $CD8^+$ T cells recognise them, the T cells begin to secrete various toxins that cause the lysis or apoptosis of the infected cell. In order to keep the cytotoxic cells from killing cells just for presenting self-proteins, self-reactive T cells are deleted from the repertoire as a result of tolerance (also known as negative selection). Endogenous antigens include xenogenic (heterologous), autologous and idiotypic or allogenic (homologous) antigens.

Autoantigens

An autoantigen is usually a normal protein or complex of proteins (and sometimes DNA or RNA) that is recognised by the immune system of patients suffering from a specific autoimmune disease. These antigens should, under normal conditions, not be the target of the immune system, but, due to mainly genetic and environmental factors, the normal immunological tolerance for such an antigen has been lost in these patients.

Tumor Antigens

Tumor antigens or *neoantigens* are those antigens that are presented by MHC I or MHC II molecules on the surface of tumor cells. These antigens can sometimes be presented by tumor cells and never by the normal ones. In this case, they are called tumor-specific antigens (TSAs) and, in general, result from a tumor-specific mutation. More common are antigens that are presented by tumor cells and normal cells, and they are called tumor-associated antigens (TAAs). Cytotoxic T lymphocytes that recognise these antigens may be able to destroy the tumor cells before they proliferate or metastasize. Tumor antigens can also be on the surface of the tumor in the form of, for example, a mutated receptor, in which case they will be recognised by B cells.

Nativity

A native antigen is an antigen that is not yet processed by an APC to smaller parts. T cells cannot bind native antigens, but require that they be processed by APCs, whereas B cells can be activated by native ones.

Antigenic Specificity

Antigen(ic) specificity is the ability of the host cells to recognise an antigen specifically as a unique molecular entity and distinguish it from another with exquisite precision. Antigen specificity is due primarily to the side-chain conformations of the antigen. It is a measurement, although the degree of specificity may not be easy to measure, and need not be linear or of the nature of a rate-limited step or equation.

Novel Antigen-Cloning Technique may Boost Efforts to Develop A Melanoma Vaccine

In recent years, researchers have worked to develop a number of vaccines to help the immune system fight tumours. Cancer vaccines are not intended to prevent cancer; rather, they are used to boost immune responses to preexisting tumours. Unlike traditional chemotherapy, vaccines have relatively low toxicity and, potentially, a high degree of efficacy.

To date, these vaccines have rarely been designed to directly stimulate one of the body's most critical immune responders, the helper T cells. Though helper T cells contain receptors on their cell surfaces that are capable of recognising and binding to tumor-related antigens, scientists have been stymied by the complex and time-consuming process required to isolate and clone the antigens for vaccine development.

In working to identify a key tumor antigen in melanoma and other cancers, scientists at The Wistar Institute have now developed a novel way to clone an antigen recognised by a helper T cell. Already, Herlyn's group has used the new cloning technique to identify a new tumor antigen called ribosomal protein L8, or RPL8. Findings on the new cloning method and the newly identified tumor antigen will be published as a Priority Report in the April 15 issue of Cancer Research.

The new antigen-cloning approach may allow scientists to design vaccines capable of directly stimulating helper T cells, aiding the development of vaccines not only for cancer but also for infectious diseases, says Dorothee Herlyn, D.V.M., senior author on the study and a professor in the Molecular and Cellular Oncogenesis and Immunology programs at Wistar.

> *"Most of the melanoma vaccines currently in development work to activate a type of white blood cell called cytotoxic T-lymphocytes, or CTL," says Herlyn. "Though CTL have the ability to destroy cancer cells, they don't have the ability to call upon the full capabilities of the immune system, as do the helper T cells."*

The new tumor antigen discovered in her lab, RPL8, is an ideal vaccine candidate because it has the potential for eliciting both helper T cells and CTL responses, Herlyn says. RPL8 is a protein involved in protein synthesis and is also expressed in normal cells. Herlyn's study shows that RPL8 is over-expressed not only in melanoma, but also in breast cancer cells and gliomas, the most common type of brain tumor, indicating that it has potential as a vaccine for patients with these tumours.

In their study, the researchers used their new cloning technique to clone copies of the RPL8 antigen from a melanoma. The scientists then showed that a peptide of RPL8 could stimulate a response in helper T cell clones and lymphocytes in four out of nine melanoma patients. The antigen created no response in cells taken from healthy donors.

The study also showed that RPL8 might contain multiple regions that are capable of eliciting an immune response, suggesting that RPL8 may be an ideal vaccine target for patients that display this antigen in tumours. According to Herlyn, these findings are important because there aren't many vaccines in development to trigger a helper T cell response for many types of cancer.

"Most of the cloned melanoma antigens that are known to target helper T cells are mutated and individual-specific, meaning that they may work in treating a single individual but will be ineffective in treating a large percentage of patients," Herlyn says.

Herlyn notes that the new antigen-cloning technique developed for this study is a major advance over the cumbersome method previously required to accomplish the task, and she anticipates that experimental vaccine developers will find it of significant value in their work.

Both the new and the old methods begin with the same steps. In order to identify and clone an antigen, researchers start with a melanoma cell or other cell of interest likely to contain relevant antigens. Based on the genes in that cell, they then make a complementary DNA (cDNA) library. Using the resulting cDNA library, the investigators can then express sequences of individual genes in recipient cells to find the cell or cells with cDNA able to stimulate T cells.

The old approach to cloning antigens adds an additional layer of complexity to the process. It requires knowing which cellular marker, or class II human leukocyte antigen (HLA), serves as a restriction element for specific T cells. Researchers must clone this restriction element and other class II HLA and transfer them into recipient cells, along with the cDNA library.

With her new technique, Herlyn expresses the cDNA library in bacteriophages that are incubated with B cells from the same patient from whom the T cells were derived. Because B cells are antigen-presenting cells, once they have picked up, or "eaten," the phages, the B cells will express the peptides that are encoded by the cDNA library. The new process eliminates the need to transfer the HLA genes into recipient cells and match the genes to each T cell type as it permits the B cells to present the peptide to the T cells in context with their own HLA class II.

Herlyn and her group now plan to use t he RPL8 antigen to develop a melanoma vaccine for patients with advanced disease. "And because the RPL8 antigen is expressed by breast cancer and gliomas, we may be able to develop a vaccine that could be used to treat these types of cancer, as well." The lead author on the study is Rolf K. Swoboda at The Wistar Institute. The additional Wistar-based co-authors are Rajasekharan Somasundaram, Laura Caputo, and Elizabeth Ochoa. Co-authors Phyllis Gimotty, Patricia Van Belle, Stephen Barth, David Elder, DuPont Guerry, Brian Czerniecki, Lynn Schuchter, and Robert Vonderheide are affiliated with the University of Pennsylvania. Francesco Marincola, also a co-author, is associated with the National Institutes of Health. The research was supported by grants from the National Institutes of Health and the Commonwealth Universal Research Enhancement Program of the Pennsylvania Department of Health.

Vaccine

A vaccine is a biological preparation that improves immunity to a particular disease. A vaccine typically contains an agent that resembles a disease-causing microorganism, and is often made from weakened or killed forms of the microbe or its toxins. The agent stimulates the body's immune system to recognise the agent as foreign, destroy it, and "remember" it, so that the immune system can more easily recognise and destroy any of these microorganisms that it later encounters. Vaccines can be prophylactic (example: to prevent or ameliorate the effects of a future infection by any natural or "wild" pathogen), or therapeutic.

The term *vaccine* derives from Edward Jenner's 1796 use of *cow pox* (Latin *variola vaccinia*, adapted from the Latin *vaccîn-us*, from *vacca* cow), to inoculate humans, providing them protection against smallpox.

History

Prior to vaccination, inoculation was practised, and brought to the West in 1721 by Lady Mary Wortley Montagu, who showed it to Hans Sloane, the King's physician.

Sometime during the 1770s Edward Jenner heard a milkmaid boast that she would never have the often-fatal or disfiguring disease smallpox, because she had already had cowpox, which has a very mild effect in humans. In 1796, Jenner took pus from the hand of a milkmaid with cowpox, inoculated an 8-year-old boy with it, and six weeks later variolated the boy's arm with smallpox, afterwards observing that the boy did not catch smallpox. Further experimentation demonstrated the efficacy of the procedure on an infant. Since vaccination with cowpox was much safer than smallpox inoculation, the latter, though still widely practiced in England, was banned in 1840. Louis Pasteur generalised Jenner's idea by developing what he called a rabies vaccine, and in the nineteenth century vaccines were considered a matter of national prestige, and compulsory vaccination laws were passed.

The twentieth century saw the introduction of several successful vaccines, including those against diphtheria, measles, mumps, and rubella. Major achievements included the development of the polio vaccine in the 1950s and the eradication of smallpox during the 1960s and 1970s. Maurice Hilleman was the most prolific of the developers of the vaccines in the twentieth century. As vaccines became more common, many people began taking them for granted. However, vaccines remain elusive for many important diseases, including malaria and HIV.

Types

Vaccines are dead or inactivated organisms or purified products derived from them. There are several types of vaccines in use. These represent different strategies used to try to reduce risk of illness, while retaining the ability to induce a beneficial immune response.

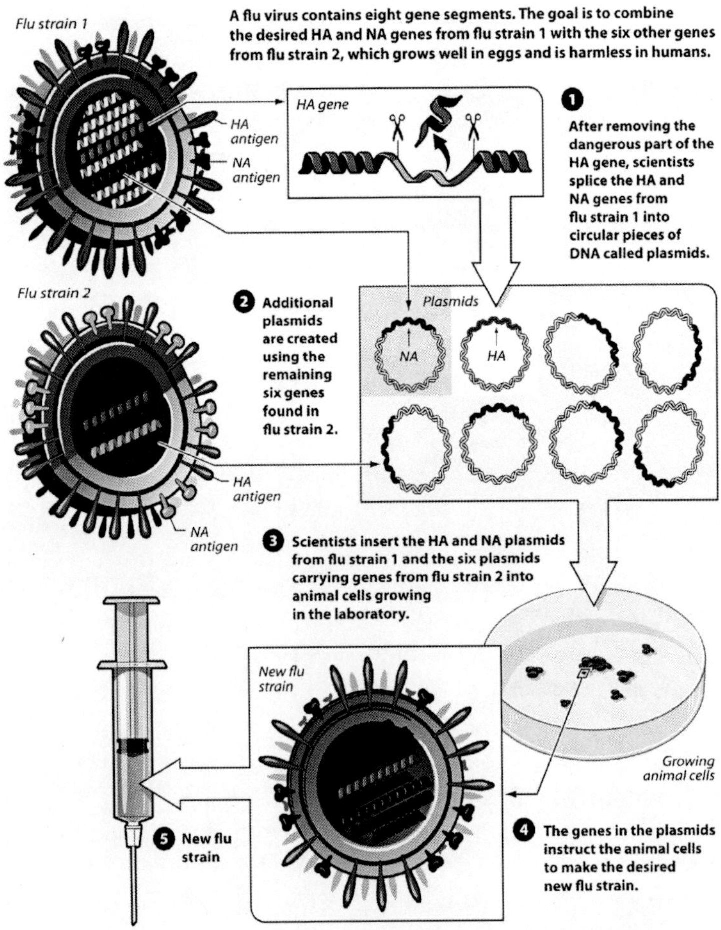

Figure: *Avian flu vaccine development by reverse genetics techniques.*

Killed

Some vaccines contain killed, but previously virulent, micro-organisms that have been destroyed with chemicals, heat, radioactivity or antibiotics. Examples are the influenza vaccine, cholera vaccine, bubonic plague vaccine, polio vaccine, hepatitis A vaccine, and rabies vaccine.

Attenuated

Some vaccines contain live, attenuated microorganisms. Many of these are live viruses that have been cultivated under conditions that disable their virulent properties, or which use closely related but less dangerous organisms to produce a broad immune response; however, some are bacterial in nature.

They typically provoke more durable immunological responses and are the preferred type for healthy adults. Examples include the viral diseases yellow fever, measles, rubella, and mumps and the bacterial disease typhoid. The live Mycobacterium tuberculosis vaccine developed by Calmette and Guérin is not made of a contagious strain, but contains a virulently modified strain called "BCG" used to elicit an immune response to the vaccine. The live attenuated vaccine containing strain Yersinia pestis EV is used for plague immunization.

Toxoid

Toxoid vaccines are made from inactivated toxic compounds that cause illness rather than the micro-organism. Examples of toxoid-based vaccines include tetanus and diphtheria. Toxoid vaccines are known for their efficacy. Not all toxoids are for micro-organisms; for example, *Crotalus atrox* toxoid is used to vaccinate dogs against rattlesnake bites.

Subunit

Protein subunit – rather than introducing an inactivated or attenuated micro-organism to an immune system (which would constitute a "whole-agent" vaccine), a fragment of it can create an immune response. Examples include the subunit vaccine against Hepatitis B virus that is composed of only the surface proteins of the virus (previously extracted from the blood serum of chronically infected patients, but now produced by recombination of the viral genes into yeast), the virus-like particle (VLP) vaccine against human papillomavirus (HPV) that is composed of the viral major capsid protein, and the hemagglutinin and neuraminidase subunits of the influenza virus. Subunit vaccine is being used for plague immunization.

Conjugate

Conjugate – certain bacteria have polysaccharide outer coats that are poorly immunogenic. By linking these outer coats to proteins (e.g. toxins), the immune system can be led to recognise the polysaccharide as if it were a protein antigen. This approach is used in the *Haemophilus influenzae* type B vaccine.

Experimental

A number of innovative vaccines are also in development and in use:

- Dendritic cell vaccines combine dendritic cells with antigens in order to present the antigens to the body's white blood cells, thus stimulating an immune reaction. These vaccines have shown some positive preliminary results for treating brain tumours.
- Recombinant Vector – by combining the physiology of one micro-organism and the DNA of the other, immunity can be created against diseases that have complex infection processes
- DNA vaccination – in recent years a new type of vaccine called *DNA vaccination*, created from an infectious agent's DNA, has been developed. It works by insertion (and expression, triggering immune system recognition) of viral or bacterial DNA into human or animal cells. Some cells of the immune system that recognise the proteins expressed will mount an attack against these proteins and cells expressing them.

  Because these cells live for a very long time, if the pathogen that normally expresses these proteins is encountered at a later time, they will be attacked instantly by the immune system. One advantage of DNA vaccines is that they are very easy to produce and store. As of 2006, DNA vaccination is still experimental.
- T-cell receptor peptide vaccines are under development for several diseases using models of Valley Fever, stomatitis, and atopic dermatitis. These peptides have been shown to modulate cytokine production and improve cell mediated immunity.
- Targeting of identified bacterial proteins that are involved in complement inhibition would neutralize the key bacterial virulence mechanism.

While most vaccines are created using inactivated or attenuated compounds from micro-organisms, synthetic vaccines are composed mainly or wholly of synthetic peptides, carbohydrates or antigens.

Valence

Vaccines may be *monovalent* (also called *univalent*) or *multivalent* (also called *polyvalent*). A monovalent vaccine is designed to immunize against a single antigen or single microorganism. A multivalent or polyvalent vaccine is designed to immunize against two or more strains of the same microorganism, or against two or more microorganisms. In certain cases a monovalent vaccine may be preferable for rapidly developing a strong immune response.

Developing Immunity

The immune system recognises vaccine agents as foreign, destroys them, and "remembers" them. When the virulent version of an agent comes along the body recognises the protein coat on the virus, and thus is prepared to respond, by (1) neutralizing the target agent before it can enter cells, and (2) by recognising and destroying infected cells before that agent can multiply to vast numbers.

When two or more vaccines are mixed together in the same formulation, the two vaccines can interfere. This most frequently occurs with live attenuated vaccines, where one of the vaccine components is more robust than the others and suppresses the growth and immune response to the other components. This phenomenon was first noted in the trivalent Sabin polio vaccine, where the amount of serotype 2 virus in the vaccine had to be reduced to stop it from interfering with the "take" of the serotype 1 and 3 viruses in the vaccine. This phenomenon has also been found to be a problem with the dengue vaccines currently being researched, where the DEN-3 serotype was found to predominate and suppress the response to DEN-1, -2 and -4 serotypes.

Vaccines have contributed to the eradication of smallpox, one of the most contagious and deadly diseases known to man. Other diseases such as rubella, polio, measles, mumps, chickenpox, and typhoid are nowhere near as common as they were a hundred years ago. As long as the vast majority of people are vaccinated, it is much more difficult for an outbreak of disease to occur, let alone spread.

This effect is called herd immunity. Polio, which is transmitted only between humans, is targeted by an extensive eradication campaign that has seen endemic polio restricted to only parts of four countries (Afghanistan, India, Nigeria and Pakistan). The difficulty of reaching all children as well as cultural misunderstandings, however, have caused the anticipated eradication date to be missed several times.

Schedule

In order to provide best protection, children are recommended to receive vaccinations as soon as their immune systems are sufficiently developed to respond to particular vaccines, with additional "booster" shots often required to achieve "full immunity". This has led to the development of complex vaccination schedules.

In the United States, the Advisory Committee on Immunization Practices, which recommends schedule additions for the Centres for Disease Control and Prevention, recommends routine vaccination of children against: hepatitis A, hepatitis B, polio, mumps, measles, rubella, diphtheria, pertussis, tetanus, HiB, chickenpox, rotavirus, influenza, meningococcal disease and pneumonia.

The large number of vaccines and boosters recommended (up to 24 injections by age two) has led to problems with achieving full compliance. In order to combat declining compliance rates, various notification systems have been instituted and a number of combination injections are now marketed (e.g., Pneumococcal conjugate vaccine and MMRV vaccine), which provide protection against multiple diseases.

Besides recommendations for infant vaccinations and boosters, many specific vaccines are recommended at other ages or for repeated injections throughout life—most commonly for measles, tetanus, influenza, and pneumonia. Pregnant women are often screened for continued resistance to rubella. The human papillomavirus vaccine is recommended in the U.S. (as of 2011) and UK (as of 2009). Vaccine recommendations for the elderly concentrate on pneumonia and influenza, which are more deadly to that group. In 2006, a vaccine was introduced against shingles, a disease caused by the chickenpox virus, which usually affects the elderly.

Effectiveness

Vaccines do not guarantee complete protection from a disease. Sometimes, this is because the host's immune system simply does not respond adequately or at all. This may be due to a lowered immunity in general (diabetes, steroid use, HIV infection, age) or because the host's immune system does not have a B cell capable of generating antibodies to that antigen.

Even if the host develops antibodies, the human immune system is not perfect and in any case the immune system might still not be able to defeat the infection immediately. In this case, the infection will be less severe and heal faster.

Adjuvants are typically used to boost immune response. Most often aluminium adjuvants are used, but adjuvants like squalene are also used in some vaccines and more vaccines with squalene and phosphate adjuvants are being tested. Larger doses are used in some cases for older people (50–75 years and up), whose immune response to a given vaccine is not as strong. The efficacy or performance of the vaccine is dependent on a number of factors:

- the disease itself (for some diseases vaccination performs better than for other diseases)
- the strain of vaccine (some vaccinations are for different strains of the disease)
- whether one kept to the timetable for the vaccinations
- some individuals are "non-responders" to certain vaccines, meaning that they do not generate antibodies even after being vaccinated correctly
- other factors such as ethnicity, age, or genetic predisposition.

When a vaccinated individual does develop the disease vaccinated against, the disease is likely to be milder than without vaccination.

The following are important considerations in the effectiveness of a vaccination program:

1. careful modelling to anticipate the impact that an immunization campaign will have on the epidemiology of the disease in the medium to long term
2. ongoing surveillance for the relevant disease following introduction of a new vaccine and
3. maintaining high immunization rates, even when a disease has become rare.

In 1958 there were 763,094 cases of measles and 552 deaths in the United States. With the help of new vaccines, the number of cases dropped to fewer than 150 per year (median of 56). In early 2008, there were 64 suspected cases of measles. 54 out of 64 infections were associated with importation from another country, although only 13% were actually acquired outside of the United States; 63 of these 64 individuals either had never been vaccinated against measles, or were uncertain whether they had been vaccinated.

Opposition to Vaccination

Opposition to vaccination, from a wide array of vaccine critics, has existed since the earliest vaccination campaigns. Although the benefits of preventing suffering and death from serious infectious diseases greatly outweigh the risks of rare adverse effects following immunization, disputes have arisen over the morality, ethics, effectiveness, and safety of vaccination. Some vaccination critics say that vaccines are ineffective against disease or that vaccine safety studies are inadequate. Some religious groups do not allow vaccination, and some political groups oppose mandatory vaccination on the grounds of individual liberty. In response, concern has been raised that spreading unfounded information about the medical risks of vaccines increases rates of life-threatening infections, not only in the children whose parents refused vaccinations, but also in other children, perhaps too young for vaccines, who could contract infections from unvaccinated carriers.

Economics of Development

One challenge in vaccine development is economic: many of the diseases most demanding a vaccine, including HIV, malaria and tuberculosis, exist principally in poor countries. Pharmaceutical firms and biotechnology companies have little incentive to develop vaccines for these diseases, because there is little revenue potential. Even in more affluent countries, financial returns are usually minimal and the financial and other risks are great.

Most vaccine development to date has relied on "push" funding by government, universities and non-profit organisations. Many vaccines have been highly cost effective and beneficial for public health. The number of vaccines actually administered has risen dramatically in recent decades. This increase, particularly in the number of different vaccines administered to children before entry into schools may be due to government mandates and support, rather than economic incentive.

Patents

The filing of patents on vaccine development processes can also be viewed as an obstacle to the development of new vaccines. Because of the weak protection offered through a patent on the final product, the protection of the innovation regarding vaccines is often made through the patent of processes used on the development of new vaccines as well as the protection of secrecy.

Production

Vaccine production has several stages. First, the antigen itself is generated. Viruses are grown either on primary cells such as chicken eggs (*e.g.*, for influenza), or on continuous cell lines such as cultured human cells (*e.g.*, for hepatitis A). Bacteria are grown in bioreactors (*e.g.*, *Haemophilus influenzae* type b). Alternatively, a recombinant protein derived from the viruses or bacteria can be generated in yeast, bacteria, or cell cultures.

After the antigen is generated, it is isolated from the cells used to generate it. A virus may need to be inactivated, possibly with no further purification required. Recombinant proteins need many operations involving ultrafiltration and column chromatography. Finally, the vaccine is formulated by adding adjuvant, stabilizers, and preservatives as needed.

The adjuvant enhances the immune response of the antigen, stabilizers increase the storage life, and preservatives allow the use of multidose vials. Combination vaccines are harder to develop and produce, because of potential incompatibilities and interactions among the antigens and other ingredients involved.

Vaccine production techniques are evolving. Cultured mammalian cells are expected to become increasingly important, compared to conventional options such as chicken eggs, due to greater productivitity and low incidence of problems with contamination.

Recombination technology that produces genetically detoxified vaccine is expected to grow in popularity for the production of bacterial vaccines that use toxoids. Combination vaccines are expected to reduce the quantities of antigens they contain, and thereby decrease undesirable interactions, by using pathogen-associated molecular patterns.

Excipients

Beside the active vaccine itself, the following excipients are commonly present in vaccine preparations:

- Aluminium salts or gels are added as adjuvants. Adjuvants are added to promote an earlier, more potent response, and more persistent immune response to the vaccine; they allow for a lower vaccine dosage.
- Antibiotics are added to some vaccines to prevent the growth of bacteria during production and storage of the vaccine.
- Egg protein is present in influenza and yellow fever vaccines as they are prepared using chicken eggs. Other proteins may be present.
- Formaldehyde is used to inactivate bacterial products for toxoid vaccines. Formaldehyde is also used to kill unwanted viruses and bacteria that might contaminate the vaccine during production.
- Monosodium glutamate (MSG) and 2-phenoxyethanol are used as stabilizers in a few vaccines to help the vaccine remain unchanged when the vaccine is exposed to heat, light, acidity, or humidity.
- Thimerosal is a mercury-containing preservative that is added to vials of vaccine that contain more than one dose to prevent contamination and growth of potentially harmful bacteria.

Role of Preservatives

Many vaccines need preservatives to prevent serious adverse effects such as *Staphylococcus* infection that, in one 1928 incident, killed 12 of 21 children inoculated with a diphtheria vaccine that lacked a preservative.

Several preservatives are available, including thiomersal, phenoxyethanol, and formaldehyde. Thiomersal is more effective against bacteria, has better shelf life, and improves vaccine stability, potency, and safety, but in the U.S., the European Union, and a few other affluent countries, it is no longer used as a preservative in childhood vaccines, as a precautionary measure due to its mercury content. Although controversial claims have been made that thiomersal contributes to autism, no convincing scientific evidence supports these claims.

Delivery Systems

There are several new delivery systems in development that will hopefully make vaccines more efficient to deliver. Possible methods include liposomes and *ISCOM* (immune stimulating complex).

The latest developments in vaccine delivery technologies have resulted in oral vaccines. A polio vaccine was developed and tested by volunteer vaccinations with no formal training; the results were very positive in that the ease of the vaccines increased dramatically. With an oral vaccine, there is no risk of blood contamination. Oral vaccines are likely to be solid which have proven to be more stable and less likely to freeze; this stability reduces the need for a "cold chain": the resources required to keep vaccines within a restricted temperature range from the manufacturing stage to the point of administration, which, in turn, will decrease costs of vaccines. Finally, a microneedle approach, which is still in stages of development, seems to be the vaccine of the future, the microneedle, which is "pointed projections fabricated into arrays that can create vaccine delivery pathways through the skin".

Plasmids

The use of plasmids has been validated in preclinical studies as a protective vaccine strategy for cancer and infectious diseases. However, in human studies this approach has failed to provide clinically relevant benefit. The overall efficacy of plasmid DNA immunization depends on increasing the plasmid's immunogenicity while also correcting for factors involved in the specific activation of immune effector cells.

Use in Veterinary Medicine

Vaccinations of animals are used both to prevent their contracting diseases and to prevent transmission of disease to humans. Both animals kept as pets and animals raised as livestock are routinely vaccinated. In some instances, wild populations may be vaccinated. This is sometimes accomplished with vaccine-laced food spread in a disease-prone area and has been used to attempt to control rabies in raccoons. Where rabies occurs, rabies vaccination of dogs may be required by law. Other canine vaccines include canine distemper, canine parvovirus, infectious canine hepatitis, adenovirus-2, leptospirosis, bordatella, canine parainfluenza virus, and Lyme disease among others.

Trends

Vaccine development has several trends:

- Until recently, most vaccines were aimed at infants and children, but adolescents and adults are increasingly being targeted.
- Combinations of vaccines are becoming more common; vaccines containing five or more components are used in many parts of the world.
- New methods of administering vaccines are being developed, such as skin patches, aerosols via inhalation devices, and eating genetically engineered plants.
- Vaccines are being designed to stimulate innate immune responses, as well as adaptive.

- Attempts are being made to develop vaccines to help cure chronic infections, as opposed to preventing disease.
- Vaccines are being developed to defend against bioterrorist attacks such as anthrax, plague, and smallpox.
- Appreciation for sex and pregnancy differences in vaccine responses "might change the strategies used by public health officials".

Principles that govern the immune response can now be used in tailor-made vaccines against many noninfectious human diseases, such as cancers and autoimmune disorders. For example, the experimental vaccine CYT006-AngQb has been investigated as a possible treatment for high blood pressure. Factors that have impact on the trends of vaccine development include progress in translatory medicine, demographics, regulatory science, political, cultural, and social responses.

5

Mammalian Cloning

Cloning is commonly perceived as a means of generating genetically identical individuals, but it can also be used to obtain genetically matched embryo-derived stem cells, which could potentially be used in the treatment of patients. A recent report offers the first 'proof of principle' of such cloning for therapeutic purposes, referred to as nuclear transplantation to produce stem cells for autologous transplantation.

Cloning is a mode of asexual reproduction in which all offspring have an identical nuclear genome to that of the parent. In recent years, mammalian cloning has been achieved by the introduction of somatic cell nuclei into fertilized eggs from which the zygotic nucleus has been removed. In our anthropocentric society, the recent success in animal cloning and its implications for humanity have captured the public's attention and imagination. But even though cloning has now been accomplished in several mammalian species, there are often severe complications associated with the procedure, and cloned animals are never quite the same as their parent.

For instance, cloned embryos often exhibit developmental abnormalities, usually including excessive growth, referred to as large-offspring syndrome (LOS); in some cases, epigenetic aberrations have been reported, such as inappropriate X chromosome inactivation in cloned bovine fetuses and placentae. Thus, only a very small proportion (less than 1%) of cloned mammals make it to birth. Many of the offspring that are born suffer from various defects, including obesity and liver and immunological defects; their chromosomes often have telomeres with variable lengths, possibly correlating with the donor cell type used for generating clones. Either individually or in combination, these symptoms may drastically shorten the lifespan of clones. It is issues such as these that have raised considerable concern about the cloning procedure and highlighted our lack of understanding of the basic biology of cloning.

Mammalian cloning has far-reaching consequences, beyond the generation of cloned adults. One potentially powerful application of cloning technology is for the generation of cloned embryos that will never be implanted in a uterus - embryos that will not develop beyond a hundred or so cells but will be used to generate genetically matched, immunologically compatible (autologous) stem cells that can potentially differentiate in such a way as to replace damaged or diseased tissues or organs in an adult. In light of the uncertainties and debate on the subject, a distinction must be made between cloning with the intent of generating live animals (reproductive cloning) and cloning in order to facilitate the derivation of stem cells from early embryos, which can ultimately be incorporated into therapeutic regimes.

In this article, we will refer to the latter procedure as nuclear transplantation to produce stem cells, as suggested in the guidelines put forward by the US National Academies after a workshop on human reproductive cloning held in August 2001. (It is also sometimes called therapeutic cloning or nonreproductive cloning.) Although there is a long way to go before nuclear transplantation to produce autologous stem cells becomes a reality for humans, several recent reports demonstrate that we are heading in this direction.

Nuclear Transfer and Stem Cells

The technique of vertebrate somatic-cell nuclear transfer (also referred to as nuclear transplantation) was first developed half a century ago in amphibians, and the first cloned adult amphibians were described a decade later. Only in the last five years has the technique been used successfully for the production of viable cloned mammals. There are currently two elegantly simple protocols for the cloning of mammals by nuclear transfer.

The first relies on the fusion of a somatic cell and an enucleated egg and has been used to clone sheep, mice, goats, cows and pigs, whereas the second is based on nuclear microinjection and has been extensively used to generate cloned mice and also cloned pigs and goats.

Both protocols involve the removal of the nucleus from an unfertilized egg (an oocyte) and its replacement with a nucleus from an adult cell or a cultured cell line; both rely on the premise that the microenvironment of the host oocyte - presumably its cytoplasm - can re-instruct the donor nucleus to adopt the behaviour of the removed oocyte nucleus. Thus, the donor nucleus is reprogrammed so that it becomes developmentally versatile (totipotent) and able to direct and execute the embryonic developmental program.

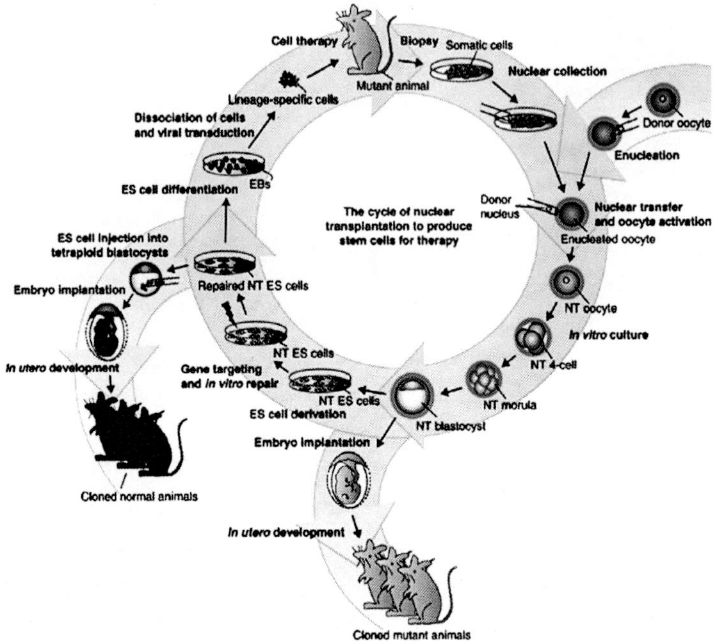

Figure: *A representation of the procedures used to produce autologous stem cells with a corrected genetic defect for the purpose of cell-based gene therapy in adult animals, based on the experiments reported by Rideout* et al.. *First (top), biopsy samples are obtained from a mutant animal. These are used to establish primary cultures of somatic cells, which will provide donor nuclei. Donor oocytes, arrested at the metaphase II stage of meiosis, are enucleated, and a somatic-cell-derived donor nucleus is transferred into an enucleated oocyte in a procedure known as nuclear transfer. The resulting nuclear-transfer (NT) oocytes are activated and embryogenesis initiates. NT embryos are allowed to develop* in vitro *up to the blastocyst stage, the stage at which mammalian embryos normally implant into the uterus. For embryonic development to continue, the blastocysts must be reintroduced into the uterus of a (surrogate) female, where they will undergo embryogenesis and ultimately produce cloned mutant offspring. Alternatively, pluripotent NT ES cell lines can be derived from the NT blastocysts. NT ES cells bear all the hallmarks of standard ES cell lines, in that they exhibit broad (pluripotent) developmental potential, their genome can be manipulated* in vitro *by routine gene targeting and other transgenic approaches, and they can be differentiated* in vitro *when grown under appropriate conditions. Gene targeting can be used to repair specific genetic defects in the mutant NT ES cells. The corrected NT ES cells can subsequently be introduced into tetraploid blastocysts to generate chimeras that, if implanted into the uterus of a surrogate female and allowed to undergo embryogenesis, can develop into cloned normal offspring. Alternatively, corrected NT ES cells can be differentiated* in vitro *to obtain lineage-specific stem cells, in this case by growth of embryoid bodies (EBs) followed by differentiation to yield hematopoietic stem cells. Finally, the genetically corrected, autologous cells can be used for cell therapy of the mutant animals.*

Cloning entire individuals by nuclear transfer is not the aim of most studies at present, however. Of far more interest is the potential to produce stem cells and, by combining the production of stem cells with nuclear transfer, to produce autologous stem cells that match the donor of the adult nucleus. Stem cells are cells that have the unique dual capacity for self-renewal and differentiation; in other words, they can not only divide to give identical stem-cell progeny, they can also differentiate into a wide variety of other cell types. There are several categories of stem cell, including embryo-derived and lineage-specific stem cells: the former usually have a broader repertoire for differentiation than the latter. Stem cells isolated from the inner cell mass of the blastocyststage preimplantation mammalian embryo, known as embryonic stem or ES cells, can contribute to most but not all lineages; this pluripotency mirrors that of the inner cell mass. If included in embryos derived from more than one fertilized egg (chimeras), ES cells can contribute to the fetus itself (including the germ line) and extraembryonic mesoderm. ES cells can also be maintained as permanent, undifferentiated cell lines *in vitro* while still preserving their developmental potential. It has been two decades since ES cells were first isolated, and they remain the mainstay of mouse genome engineering because their genes can easily be manipulated *in vitro* - even down to individual base pairs - by standard gene-targeting and transgenesis techniques, while their developmental potential is retained. Genome modifications introduced into ES cells *in vitro* can be reintroduced into mice via inclusion in chimeric embryos. ES cells can also be induced to differentiate into defined-lineage cell types under appropriate conditions *in vitro.* Given that ES cells are unequivocally pluripotent stem cells and can be genetically modified, differentiated *in vitro,* and reintroduced into animals, they may be suitable reagents for use in cloning and cell-therapy regimes that aim to repair defects such as the loss of dopaminergic neurons in Parkinson's disease or islet cells in diabetes mellitus.

Embryonic stem cells are not the only type of stem cells - nor are they necessarily the most appropriate type for therapeutic purposes. Lineage-specific stem cells are the progenitors of specific differentiated cell lineages and are present in later-stage embryos and adults in organs such as skin, intestine, brain, and bone marrow. One issue that arises from the presence of such cells is that in animal cloning studies, donor nuclei have been taken from ostensibly differentiated somatic cells, but most donor cell populations are probably heterogenous, and it is not clear whether it is differentiated cells or rare lineage-specific stem cells in the population that give rise to clones. If the latter is the case, it may be that the rarity of stem cells leads to the low efficiencies of cloning - lower than when ES cells are used as nuclear donors. Protocols that can distinguish stem cells from differentiated cells would then need to be developed in order to increase overall efficiencies. A key question that has therefore persisted, and remained unanswered until recently, is whether highly specialised lineage-specific cells can be reprogrammed such that they can adopt a totipotent state, with the potential to differentiate into all possible cell types, and thus direct the developmental program used to generate a complete individual.

Reprogramming Differentiated Cells

Rudolph Jaenisch and colleagues, who have been at the forefront of nuclear-transfer work in mice, designed an experiment to address the issue of reprogramming differentiated lineage-specific cells. They chose lymphocytes as nuclear donors, as these are one of the few cell types of adult mammals whose genome is irreversibly changed as they mature, thereby making them genetically distinct and recognisable. B and T cells are the two classes of mature lymphocytes, expressing immunoglobulins (antibodies) and T-cell receptors, respectively. The type of antibody or T-cell receptor expressed is dictated by the rearrangement of each cell's genomic DNA; mature lymphocytes express only one specific antibody or receptor. Thus, in clones generated from B or T cells, the signature genomic rearrangement present in each donor cell nucleus would be preserved in all the cells of the cloned progeny.

Jaenisch and colleagues' study investigating the developmental potential and reprogramming of lymphocyte nuclei combined the technologies of mammalian cloning and ES cells in a two-step procedure that improved the efficiency of generating clones. First, they generated nuclear-transfer (NT) embryos by transfer of lymphocyte nuclei, but instead of re-implanting the embryos directly into the uteri of foster mothers, they used NT blastocysts to derive NT ES cells. They then took advantage of the 'tetraploid complementation' technique and injected their NT ES cells into tetraploid host blastocysts. Tetraploid cells preferentially form the extraembryonic tissues trophoblast and extraembryonic endoderm and are excluded from fetal tissues and extraembryonic mesoderm, whereas ES cells exclusively form the latter two tissues. Thus, the chimeric mice generated by Hochedlinger and Jaenisch consisted of a fetus and extraembryonic mesoderm derived from NT ES cells with trophoblast and extraembryonic endoderm derived from tetraploid cells.

These experiments resulted in the production of cloned mice from adult lymphocyte nuclei, as could be recognised by the signature genomic rearrangements of lymphocytes. Jaenisch and colleagues have thus answered the previously unresolved question of whether terminally differentiated cells can provide nuclei for the production of clones, by demonstrating that at least some specialised nuclei can be reprogrammed. Perhaps the tetraploid extraembryonic component used in this procedure may be pivotal in helping overcome some of the defects that have otherwise consistently been observed in cloned embryos (such as enlargement of the placenta). Additionally, the nature of this experimental setup is less demanding of the NT cells, as they do not need to be truly totipotent because all the trophoblast and extraembryonic endoderm derivatives are derived from the tetraploid cells. The NT cells therefore need only to achieve pluripotency to generate the fetus. Also, the extended period of time in culture inherent in the ES-cell-derivation procedure may allow further or more complete reprogramming of the differentiated donor nuclei, ultimately leading to increased developmental potential.

Given that some differentiated cells can indeed donate nuclei that can drive the development of a fetus, how close is the goal of autologous stem-cell therapy? Jaenisch and colleagues have now reported the first successful application of nuclear transplantation to produce stem cells that could be used for cell-based treatment in a model organism. They used *Rag2* mutant mice, which have a defined genetic disorder in immunoglobulin gene rearrangement that renders them immunodeficient, as nuclear donors. *Rag2* mutant NT embryos were generated, from which *Rag2* mutant NT ES cells were derived, and gene targeting was then carried out on the NT ES cells to repair one of the *Rag2* mutant alleles. The potential of the repaired NT ES cells was then tested in two ways. First, repaired NT ES cells were injected into tetraploid blastocysts for the generation of offspring. Normal embryos developed to birth from the tetraploid chimeras, indicating that the NT ES cells with the repaired *Rag2* gene retained their pluripotency. Furthermore, the presence of normal T and B cells in these mice proved that the repaired *Rag2* allele was functional. Second, repaired NT ES cells were differentiated in culture into hematopoietic stem cells (which form blood and immune cells), and the latter cells were transplanted into adult *Rag2* mutant mice. (Incorporation of the cells into the immune system was not entirely successful because of an immune barrier peculiar to the Rag2-deficient recipients, but this barrier was partially overcome by further manipulation of the immune system of the recipients.) Thus, the procedure was successful in restoring a modest degree of immune function in the mutant mice, but the difficulties encountered suggest that even genetically matched cells derived by nuclear transplantation may still face barriers to effective transplantation in some situations.

Studies carried out in mice routinely pave the way for work in other mammals, including humans, and nuclear transplantation to produce stem cells seems likely to be no exception. For example, a recent paper reports some success in the production and development *in vitro* of a cloned human embryo using cumulus cells (the cells that surround the oocyte) as nuclear donors, presumably with the goal of using such NT embryos for the generation of human NT ES cells. Also, the issue of histocompatibility has recently been tested in tissues generated from bovine NT embryos cloned from adult bovine fibroblasts. In order to make cloning a feasible approach for generating reagents for the treatment of human diseases, however, the overall efficiency of the procedure needs to be drastically improved. Human oocytes are hard to come by, and nuclear reprogramming is still pitifully inefficient. Headway must therefore be made toward understanding the biology of nuclear reprogramming, and this knowledge must be applied to increasing the efficiency of cloning procedures and stem-cell derivation. Improved understanding should also help avoid the phenotypic aberrations observed in cloned animals. In order to increase efficiencies, surrogate oocytes from other mammals might be suitable as donors in which nuclear reprogramming could take place, or conditions that direct nuclear reprogramming in a non-oocyte environment (ooplasmic transfer) could be developed. Alternatively, protocols for the transfer of oocyte cytoplasm into specialised cells, in the converse of nuclear transfer, may yield further insights. Only when a fundamental understanding of the molecular nature of the biological events underlying animal cloning is gained, however, will cloning represent a truly viable option for cellular therapies aimed at treating disease.

Although cloning has been in effect for the last 20 years, the cloning of whole mammals was not achieved until 1997, when Dr Ian Wilmut cloned the first sheep, called Dolly. This is one of the major advances of the last 10 years in molecular biology, and who knows where it may lead in the future, which is one of the reasons that this area of biology interests me the most out of any recent development.

What Dr Wilmut did to clone dolly is an idea that he had heard in a bar over 10 years ago (1), and went on to carry out what became his dream. To clone dolly, he used a variant technique of nuclear somatic fusion, where he fused a cell from the udder of Dolly's "mother", and fused it with an enucleated ovum using a tiny electronic current (2). When this happened a preembryo or blastocyst developed, as the current also activated the egg development. After this the embryo was then implanted back into the mother, and Dr Wilmut finally, after 2 years of constant trying and thousands of aborted attempts, managed to announce to the world the arrival of Dolly to the world.

The technique that Dr Wilmut used may not be viable for the future though. Dolly was the 277th attempt of the year to create a viable mammalian clone from the technique, and was the only success, so who knows if the odds of success are 1 in 300 or 1 in a million (3). Although the technique works in the idea of having a clone, with no paternal genetic input, it is a true clone as it is a perfect replica of its mother 100% of the time, it may not be the correct technique to take the idea further, for example cloning of Human beings.

At the moment, it is impossible to know whether or not it would be possible to continuously clone humans successfully, as a sheep is the only mammal that has been cloned. As with mice, the fact that reprogramming may take place a lot earlier than in sheep (the sixteen cell stage instead of the 2 cell stage), may be the reason for an advance in cloning to include humans is at this stage unlikely. The fact that introducing old cells into the embryo, and asking them to form a whole new being may also be a stumbling block, as mutations etc. in the cell may mean that the cell is unable to re-differentiate into say a cardiac cell, as certain proteins may be lost. All of this, and the fact that there is no guarantee that all genetic material will again be re-transcribed and translated, are the current difficulties in being able to clone yourself.

There are also ethical problems for scientists to overcome if they want to clone a human being. In the USA, both the House of Representatives and the Senate have banned any possibility of human cloning. A little harsh maybe, but when you think that cloning is an idea that Hitler had to form his own super race, than we do have to be a little cautious.

There is one very good reason for cloning humans however. If there is an infertile couple, they will be able to have a clone of them made so that they could have their own genetic offspring, instead of having to rely on someone else to provide either the sperm or ovum. This would make treating infertility much better, as at the moment there is only a 10% success ratio. To make the idea feasible however, there would have to be a much better chance of success than there is currently with Dolly, but in the future, this may be the way forward in infertility treatment.

If a ban on human cloning does stick, then maybe the rules could be relaxed to allow for research into somatic-cell nuclear transfer, which could easily provide numerous benefits. This technique is very similar to that which produced Dolly, with either a diploid human somatic cell or fetal cell being introduced to the ovum. There are many complications as with the Dolly technique though. Firstly, id the somatic cell is used a number of transitions will have to be made by the diploid nucleus. The nucleus and therefore cell must be able to produce daughter cells, and then these must be able to differentiate into whatever tissue you want, i.e. be pluripotent.

If the research on this proves fruitful, and the ability to clone things in such a way is achieved, then there are many beneficial outcomes of the cloning. One that comes to mind straight away is in replacing damaged cells, or even whole organs and tissues. Embryonic stem cells can be grown producing things such as hearts and lungs, which could be used in transplants for seriously ill patients, and if the cells come from somewhere in the patient, then there is a vastly reduced chance of rejection of the organ. Individual cells such as brain cells or spinal cord cells may be produced. This can result in reversal of paralysis in people who have been in car or similar accidents, allowing them to walk again, or also repairing brain damage in similar cases. Degenerative genetic diseases such as Alzheimer's or Parkinson's disease may be cured, as can diabetes or arthritis. These advantages can also be applied to cloning in pigs for example, and already has been carried out, so can not be the sole reason for approving human cloning.

All of the above diseases can easily be treated successfully using cloning techniques for humans. There is also a case for cloning helping to at least decrease 2 of the top killing diseases in the world, Heart disease and the dreaded cancer. If cloning techniques are successful, as mentioned above it will be possible to produce cardiac muscle cells. If this is then injected into the damaged area of the heart, then regeneration of that area may be possible, reducing the death rate due to heart disease hopefully all over the world.

There is also the chance that a cure for cancer, or at least an understanding of cancer may be found using cloning techniques. Here you will be able to see how cancerous cells "mutiny" against the normal cells in that area. The way that cells switch themselves on and off can be found, which may be a cure. The biology of the cancer cells differentiating may also be found, which will all be useful in cutting down the number of deaths to cancer each year which can only be to our advantage.

One other possible outcome of human cloning would be the understanding of the aging process, and therefore the ability to reverse it. This would allow people to stay young and healthy a lot longer, so you could have the greatest footballer in the world being a 90-year-old.

Taking cloning as a whole there is also the advantage that if cloning is allowed, there is the chance that rare species such as the tiger shown below may be cloned so there may never be an extinct animal again. Although bringing back animals that are already extinct may have detrimental effects (e.g. Jurassic Park), there must be a case for keeping all animals that currently roam the earth.

o sum it up, there is a lot of ethical and scientific discussions to go on, is it ethically wrong as there will be many wasted embryos, religious experts say that we have no right to act as god, another Hitler may clone the master race and take over the world. There is also the fact that if people are cloned, and there is a way of reversing the aging process that overcrowding will become a worldwide problem, introducing more diseases, or reintroducing diseases such as cholera. On the other hand, there is many plus points as well, the saving of rare animals such as tigers from extinction, the curing of many illnesses, a better standard of life for people at the least. There is also the chance of cloning yourself so that you don't have to go into work every day, and can spend more time at home with family and friends.

Transgenic Cloned Goats

The current risk assessment is limited to address the risk of clones from non-transgenic cells. Although not within the purview of this analysis, the results of a number of studies that address either transgenic animal clones or transgenic and nontransgenic animal clones are presented here to clarify the relative utility of such studies for assessing potential risk(s) associated with non-transgenic somatic cell nuclear transfer (SCNT).

Many reviews of SCNT outcomes cite these papers as indicative of the severity of adverse outcomes associated with cloning, or a demonstration of the positive outcomes that can come from cloning. Hill et al. (1999), in particular, is often cited as the seminal "adverse outcome" paper for cloning. On the other hand, if only the "final report" paper of the ACT series, Lanza et al. 2001, were read, it would not be possible to know that the cells from which the cattle were cloned were indeed transgenic. This distinction only becomes apparent when the earlier papers are also reviewed. Most recently, headlines were generated when Pearson (2003) reported on "sudden death syndrome" in pig clones; CVM reviewed the paper that describes their generation and noted that these pig clones carried two distinct transgenes (Lee GS et al. 2003).

Because these animals are transgenic clones, it is not possible to determine whether adverse outcomes result from the direct effect of the expression of the transgenic construct, pleiotropic effects resulting from insertion of the construct, the SCNT process, or some interaction of any or all of these processes. For example in a comparison of 34 cells lines, Forsberg et al. (2002) reported that when otherwise similar cells are used as donors for SCNT, those that are transgenic (for a selectable marker gene) result in lower pregnancy initiation (22 percent vs. 32 percent) and calving (3.4 percent vs. 8.9 percent) rates. The authors hypothesize that the additional culturing required togenerate transgenic cells, selection of transgenic lines, or the DNA construct itself, could be responsible for the lower rates.

CVM thus assumes that transgenic clones occupy a different "risk space" from "just clones." Conversely, an argument can be made that if no adverse outcomes are detected, then for these animals, neither process sufficiently perturbed development to induce anomalies. We have included those studies in the overall risk assessments if such results were obtained. Nonetheless, because these studies occupy such a large segment of the cited literature, a few are presented here to illustrate the range of responses noted, with the appropriate caveats for interpretation.

Cattle

Hill et al. 1999

Hill et al. 1999 reports on a group of 13 transgenic clones of a Holstein bull. Twelve Brangus cows carrying 13 fetuses cloned from Holstein cells were originally included in the study, although three of these transgenic clone fetuses died prior to the perinatal period (defined in this paper as two weeks prior to anticipated delivery and a few days thereafter), and one cow aborted at eight months of gestation. Two cows developed hydroallantois and were delivered by C-section; four others were also delivered by this method due to subjective judgment regarding fetal size. The two remaining pregnancies delivered vaginally. Birth weights of the transgenic clones ranged from 44-58.6 kg (average Holstein male calf weight is in the range of 40-50 kg), and cited as within the weight range of *in vitro* produced embryos. Five of the eight live born clones were judged to be normal within four hours of birth based on clinical signs and blood gas measurements. Three of the eight were immediately diagnosed with neonatal respiratory distress. One of these calves died from pulmonary hypertension, pulmonary surfactant deficiency, and elevated systemic venous pressure at day 4. The other three animals recovered. Two of the five fetuses that did not survive to birth also exhibited signs of pulmonary hypertension and placental edema at necropsy. Another clone died at 6 weeks of age with signs of respiratory distress; subsequent field necropsy suggested dilated cardiomyopathy, although no definitive diagnosis could be made.

The Advanced Cell Technology Series

The series of papers from the Advanced Cell Technology group on the health of clones are similar to that of Hill et al. (1999) in that the animals presented are clones that are derived from transgenic cells. Interpretation of any adverse outcomes is thus also confounded by the potential role of the transgene and its insertion.

The results of these studies are summarised in Lanza et al. (2001), in a short overview with accompanying supporting documentation provided by the journal in electronic form. Of 30 fetal transgenic clones that developed to term, 24 were reported healthy at 1-4 years of age, but five died within 7 days of cardiopulmonary difficulties that the authors speculated were secondary to placental insufficiencies. The sixth animal died at day 149 due to enteric disease, lymphadenopathy, and exhibited mild placental edema and high fever at birth. Problems observed at birth included placental edema, including edematous cotyledons (attachment sites of the placenta to the uterus), laboured breathing, froth and fluid in the lungs, pulmonary edema, pneumonia, high fever, septicemia, lethargy, abdominal distention, masses in the abdomen, liver damage due to hypoxia, and heart abnormalities.

Birth weights of the survivors were reported as 45 ± 2 kg (this paper cites normal as 43 kg). An unspecified number exhibited pulmonary hypertension and respiratory distress at birth. Presumably, they received supportive care at that time. Another unspecified number were also reported as experiencing fever following vaccination. This is not an atypical response among calves receiving vaccinations, as stimulating a potent immune response is likely to produce at least a mild local and systemic (fever) reaction in the animals (Roth 1999).

Physical and veterinary examination of surviving animals aged 1-3 years were reported as normal and included temperature, pulse, and respiratory rate. No abnormalities were detected in general appearance, on auscultation (listening to breathing, heart beat, and digestive sounds), and behaviour appeared normal. Puberty onset was reported to occur at the expected time, and fertility appeared to be normal. At the time of publication (2001), two of the animals had delivered apparently normal progeny.

Clinical chemistry parameters evaluated for these animals included electrolytes, urea, creatinine, glucose, bilirubin, aspartate aminotransferase (AAT), sorbitol dehydrogenase (SGT), albumin, globulin, and total protein. Globulin and total protein measurements were reported in the publication as "slightly below normal." All other measurements were reported to be within normal range. Hemograms (analysis of cellular components of blood) were all reported as normal: hematocrit, hemoglobin, red blood cells, mean red cell volume, mean red cell hemoglobin concentration, and white blood cell numbers and differentials were within normal ranges. Blood gases were also within normal ranges. To examine immunocompetence in the clones, peripheral blood lymphocytes from the transgenic clones and conventional Holsteins were compared to determine whether the same ratio of cell surface markers were present, and if the transgenic clone cells responded to mitogen challenge in the same way as cells from conventional Holsteins. No significant differences were observed between the cell surface markers or cellular responses of cells from conventional animals or clones.

In the early spring of 2003, an interview of an ACT executive reported in the lay press indicated that two animals from this cohort had developed significant health problems. One animal was reported to have developed a tumor, and the other was diagnosed as having neurological problems. The first animal apparently died during surgery to remove the tumor, and no further information is available on the potential causes of the tumor. The second animal was later diagnosed as being positive for Johne's disease (*Mycobacterium paratuberculosis*), an infectious, chronic, progressive disease that often presents with chronic diarrhea and eventual cachexia (general physical wasting and malnutrition). It is therefore unlikely that this animal's symptoms were due to either cloning or transgenesis. We are unaware of any other adverse outcomes associated with these animals.

Swine

Carter et al. 2002

Carter et al. 2002 reported on the overall health status of transgenic swine clones produced from cells transfected with green fluorescent protein (GFP). The 10 transgenic piglet clones from three litters were followed for the first six months of life.

Five of the ten transgenic swine clones died or were euthanized during the study. Two piglets died of congestive heart failure at 7 and 35 days of age, two others died from bacterial infections at 3 and 116 days of age. The fifth animal died at 130 days of age, following a history of chronic diarrhea, decreased growth and vitamin E deficiency. The remaining five piglets were reported as healthy and growing similarly to conventional animals housed in the same facility at the conclusion of the study. Behaviour was reported as "consistent with pigs of their age group."

Average birth weight of the transgenic clones (1,312 g) was similar to average birth weights of conventional piglets from similar genetic background (1,450 g). Average daily weight gain for transgenic clones through the first 16 weeks was (461 g) relative to the herd average (594 g), which the authors considered as within the normal range.

Some of the piglets displayed physical defects. These included two piglets with contracture of the flexor tendons, another piglet with five digits on a forelimb (four digits are normal) and an enlarged dewclaw. Another piglet with low birth weight was described as having short legs and a large, round chest.

Hematology and blood clinical chemistry data were collected beginning at 2 days of age and every two to four weeks until 24 weeks of age. Most hematological variables were similar to the comparator group, except for hemoglobin, hematocrit, and plasma total protein. Mild anemia and low blood protein concentration were observed for the first four weeks, but both these conditions resolved by eight weeks of age. The authors stated that decreased hematocrit and hemoglobin values are common in piglets reared in confinement, and that these symptoms are generally treated with iron dextran. Similarly, clinical chemistry results indicated decreased levels of albumin and globulin during the first four weeks in the transgenic clones relative to comparators, but these values were back within the normal range by eight weeks of age. The authors attributed the decreased protein and globulin values to the decreased colostrum intake of the newborns as the surrogate sow bearing them did not initiate normal lactation, and piglets were dosed with colostrum at some unspecified point after birth.

Seven of the transgenic clones were evaluated for cardiac function. Although no physical defects were found, one piglet had evidence of mitral insufficiency (a condition in which the mitral valve of the heart does not close all the way during contraction, resulting in regurgitation of some of the blood in the left ventricle), and dilation of the left atrium and ventricle. This piglet and two other clones had reduced cardiac output values compared to control piglets, but did not display clinical signs of cardiac disease. Although similar cardiac abnormalities have been noted in conventional swine, the incidence is reported to be very low. These developmental defects appear to be similar to those noted in cattle clones.

Lai et al. 2002

This study was reported in a brief communication, and a limited amount of data was presented. Piglets were generated from cell lines in which the a-1-3-galactosyltransferase gene was interrupted by the insertion of a gene sequence in order to create a-1-3-galactosyltransferase "knock-outs." The a-1-3-galactosyltransferase gene codes for a protein that causes hyperacute rejection of swine organs when transplanted into primates. "Knocking out" the expression of this gene increases the suitability of these animals to be used as donors of organs for human transplant patients.

Six piglets were born from two litters. All but one of the piglets had low birth weights compared to the breed average (115 to 650 g vs. 860 g). One piglet from each litter died shortly after birth from what the authors termed "respiratory distress syndrome." A third piglet died at 17 days of age during a routine blood draw, and was diagnosed at necropsy with a dilated right ventricle and thickening of the heart wall. Other abnormalities noted in these surviving transgenic piglets included flexor tendon deformities in three animals; abdominal ascites, enlarged right ventricle, pulmonary hypertension in one animal; and ocular defects and lack of patent ear canals in another animal. The authors attributed these abnormalities to failures in reprogramming during the SCNT process rather than the genetic engineering process, as they did not see a consistent phenotype across the piglets.

Lee GS et al. (2003)

Recently, Pearson (2003) reported that the University of Connecticut laboratory that had generated four transgenic swine clones had announced that the three (of four) surviving piglets died suddenly of heart failure at less than six months of age. The fourth piglet died at three days due to infection and abnormal spine development (Lee GS et al. 2003). Because of the transgenic nature of the animals (they carried genes for human clotting factor IX and porcine lactoferrin, an iron transport protein found in blood), it is not possible to attribute the deaths solely to cloning. It is unknown whether any cardiac abnormalities were detected in these animals prior to their deaths, or if any measurements of cardiac function were made.

Sheep

Denning et al. 2001

Denning et al. 2001 were unsuccessful in producing viable knock-out sheep lacking either the a-(1,3)-galactosyl transferase (GGTA1) or the prion protein (PrP) gene using gene targeted fetal fibroblasts and SCNT. Reconstructed embryos were either incubated for six days (n=48) or overnight (n=93) in synthetic oviductal fluid with bovine serum albumin (concentration not specified). Embryos incubated overnight *in vitro* were then embedded in 1 percent agar chips in phosphate buffered saline and transferred to the ligated oviduct of an estrus-synchronised ewe for six days. A total of 120 morula or blastocyst stage embryos were transferred to 78 estrus-synchronised Finn Dorset ewes as final recipients. It is not clear from this paper how many of the transferred embryos had been incubated *in vitro*. Although 39 pregnancies were diagnosed at gd 35, only eight were maintained to term, resulting in four live births. Three of the four live-born lambs died shortly after birth. The fourth lamb survived 12 days before it was euthanized after developing dyspnea (difficulty breathing) due to pulmonary hypertension and right-sided heart failure. The authors attributed the abnormalities observed to the nuclear transfer procedure, as they were similar to results obtained with non-transgenic NT lambs.

McCreath et al. 2000

McCreath et al. 2000 inserted a promoter-less neomycin selectable marker between the ovine a1(I)-procollagen translational stop and polyadenylation signal in male and female ovine fetal fibroblast cultures. Four transgenic female fibroblast cultures were selected as nuclear sources for SCNT, due to their vigor and normal chromosome number. A total of 80 morula and blastocyst stage embryos were transferred to recipient ewes. No description of post-fusion incubation or estrous cycle status of recipient ewes was provided in this report. Fourteen lambs were born alive; seven of these lambs died within 30 hours of birth. Four more lambs died between 3 days and 12 weeks of age. Three lambs survived and were described as thriving at one year of age. Necropsy of lambs that died *in utero* or after birth revealed a number of abnormalities including a high incidence of kidney defects (frequently renal pelvis dilation) and liver and brain abnormalities (not specified). The authors attribute these abnormalities to either cell treatment or the NT procedure, because the necropsy findings were similar to a previous nuclear transfer study using the same cell lines.

Goats

Baguisi et al. 1999

In this study from Genzyme Transgenics, six cell lines were established from 35- and 40-day old fetuses that resulted from the mating of a transgenic buck (carrying a human antithrombin III (hAT) gene with a goat ß-casein promoter) to a non-transgenic doe. This study differs slightly from several other transgenic cloning studies reported here, in which the gene was inserted into the cell lines before the cultures were established. Clone embryos were cultured on goat oviduct epithelial cells for 48 hours (2-16 cell stage) before being transferred to estrus synchronised recipient does. Although overall cloning efficiency was low (3/112 embryos transferred resulted in live births), all pregnancy losses occurred prior to 60 days of pregnancy. There were no stillbirths and no abnormalities observed in the live-born kids. Kids weighed between 2.35 and 3.5 kg, within the normal birth weight for dairy goats, and are reported as healthy.

Keefer et al. 2001a

In this study from the Nexia Biotechnologies laboratory, goat fetal fibroblasts were transfected with green fluorescent protein (eGFP) and neomycin resistance genes. These are commonly used as markers to demonstrate that transgenes have been inserted and are being expressed. Twenty seven NT embryos were produced with the transfected cells, and an additional 70 non-transgenic NT embryos were constructed and transferred into 13 estrus synchronised recipient does. The authors did not specify how many embryos (transgenic or non-transgenic) were transferred to each doe. Five non-transgenic male clones and one transgenic female clone were born alive. Three of the non-transgenic clones died of bacterial infections, but the single female transgenic clone lived and showed no signs of abnormalities. The kids were all within the normal birth weight range (1.5 to 3.1 kg) for goats at that facility, and no abnormalities were observed in the placentae.

Behboodi et al. 2004

The authors compared development of embryos cultured with oviductal cells *in vitro* vs. embryos cultured *in vivo*. Embryos were constructed using skin fibroblasts of transgenic goats. Only embryos cultured *in vivo* resulted in pregnancies. Two of these pregnancies were lost early in gestation (after 30 days gestation), and four other pregnancies were carried to term. Two surrogate does delivered stillborn kids 2-3 days after their due dates; the other two does delivered healthy kids (one per each doe) at term. The two live clones weighed 3.8 and 4.1 kg at birth, and were within the normal birth weight range for their breed (Saanen). Clones were weaned at 8 weeks of age, and had similar growth rates compared to age-matched AI derived Saanen kids born at the same facility (14.5 and 18.1 kg for clones vs. 14.88 ± 1.98 kg for AI comparators). Pathology on the dead fetuses indicated diffuse atelectasis (lung collapse) and the presence of amniotic fluid in the lungs. No bacterial or viral cause for the deaths of these clones could be identified. The presence of amniotic fluid in the lungs suggests that the clones attempted to breathe prematurely, a sign of fetal stress which sometimes occurs around the time of birth.

Behboodi et al. 2005

The authors evaluated health, growth, reproduction and lactation in four female goat clones generated from two transgenic fetal cell lines (one cell line coding for glycosylated and the other for non-glycosylated protein). A total of seven clones were carried to term. One clone from the glycosylated group was still born with evidence that the umbilical vessel had ruptured. Two clones died at birth (one from each of the transgenic lines) after failing to breathe on their own, despite attempts at manual ventilation.

Thus, two clones from each transgenic line survived to adulthood. There were no differences in birth or weaning weights among the four surviving clones or their age-matched comparators. Transgenic clones exhibited enlarged umbilical stumps (two live and one stillborn kid), "tendon laxity" (three of the four live-born clones), and minor generalised edema (number of clones affected not indicated). These conditions resolved without intervention. The four does were bred and produced nine kids, compared to five kids produced by comparators. Clones expressing the glycosylated version of the protein lactated only briefly, but the does expressing the non-glycosylated protein had normal lactation length and milk yields.

This study is the only one we encountered that presented hematology and blood clinical chemistry data for four goat clones. These data are presented in comparison to four age-matched comparators and values from the literature (Pugh 2002). It is unclear whether or not the comparators in this study were also transgenic, whether they were the same breed as the clones, or how they were generated. Hematology values were similar between clones and comparators, and all hematology values fell within the published range. For clinical chemistry, 18/24 values were not significantly different between clones and their age-matched comparators. Of the 19 clinical chemistry values for which published ranges were available, 18 of the values for clones and comparators fell within the published range. The one value out of the published range was creatine kinase (244.6 vs. 204.4 IU/L for clones and comparators). However, values between clones and comparators were not statistically different. The study does not specify the age of the goats at time of blood sampling, so it is difficult to interpret the high values for CK in these animals compared to the published range.

This study is unique among reports of goat clones because it is the first to indicate possible signs of LOS in goat clones (enlarged umbilici, failure to initiate breathing, tendon problems). It is interesting to note that similar signs have not been noted in non-transgenic goats. We should also note that clinical signs in the four surviving clones resolved, and their health, growth, reproduction, and hematology, clinical chemistry values indicate that even these transgenic clones are apparently normal.

Landry et al. 2005

The authors reported on growth (weight gain, wither and hip height change) and endocrine profiles of two lines of transgenic goat clones. Group 1 consisted on five does carrying the AT-III gene with a ß-casein promoter inserted into cells of a female Toggenburg (dairy breed). The gene inserted into the second line of goats (Group II; n=2) was not identified, but the cells used for cloning were from a female Saanen (also a dairy breed). Non-transgenic, non-clone comparators (n=7) were Boer X Spanish crossbred meat-type does (Group III). The authors did not report on overall health of the clones. One female in each group of clones died prior to the end of the study; one died due to an accident, the other due to a ruptured abomasum. Neither death appears to be related to cloning. Both groups of clones were within range for their breed for birth weight, and appeared to grow normally. Interpretation of hormone profiles (GH, IGF-I, T3, and T4) is difficult due to the fact that the clones and comparators were of different breed (purebred vs. crossbred) and type (dairy vs. meat) backgrounds. However, for most of the hormones assayed, the values for clones fell within the range of values for comparators. The one exception is insulin, which resulted in an extremely low value in blood samples of comparators, and may have been the result of difficulties with the assay.

Melican et al. 2006

With the long-term goal of producing caprine milk containing recombinant therapeutic proteins, the authors demonstrated successful reproduction and lactation in transgenic dairy goats produced by SCNT. Does were produced using primary fetal cells harvested from day 35-40 fetuses.

These cells were co-transfected with DNA fragments encoding the heavy and light immunoglobulin chains of three different monoclonal antibodies, plus neomycin resistance as a selectable marker. Two transgenic does were hormonally induced to lactate at two to three months of age and produced small amounts of milk (10-20 ml per day) for 30-40 days. These does were subsequently bred and, at 18 months of age, gave birth to 6 kids (4 were transgenic). Following parturition, the 2 transgenic does produced an average of 2.2 litres of milk per day for 3-6 months.

A third transgenic doe, at age 23 months, also lactated following the birth of one transgenic kid, producing 2.6 litres of milk per day for one month prior to being dried off. The milk of all 3 transgenic does contained the antibody for which they were transgenic. These results demonstrate reproductive capacity in cloned female goats, and the ability of these goats to produce normal quantities of milk following parturition.

Conclusions Regarding Transgenic Clones

The experience of these cohorts of transgenic clones can be summarised as follows:

- A relatively large fraction of transgenic fetal bovine clones in cohorts surviving to late gestation presents with severe and often fatal difficulties. Some of these are qualitatively similar to those observed in cattle and sheep clones that are not derived from transgenic cells.

  Due to the many other variables that have been altered in the generation of these animals, at this time it is not possible to attribute these abnormalities to either of the processes (cloning or transgenesis) or their combination.
- Some animals in both cattle cohorts are born with varying degrees of initial respiratory or other physiological distress. Supportive care appears to allow most of these animals to survive to adulthood, although some animals that initially survive can succumb to possible sequellae up to six weeks later.
- Animals surviving to adulthood in the ACT cohort that appear to be healthy on visual inspection also exhibit physiological values that generally fall within normal ranges. CVM is unaware of an update of the health status of the Hill et al. cohort.
- Animals in the ACT cohort surviving to reproductive maturity appear to be capable of bearing normal offspring, although it is not clear whether the offsprings' health has been examined in a rigorous manner.
- Two severe adverse outcomes have been noted for the ACT cohort. Both cloning and transgenesis can likely be ruled out as causes for one (Johne's disease) and no causal agent or process has been associated with the neoplasm found in the other.

- The appearance, behaviour, and physiological function of the animals that survive suggest that even the "riskiest" set of clones (*i.e.*, transgenic clones) can develop into normally functioning animals. These results are consistent with the analysis of non-transgenic clones, and provide additional confidence that rigorous monitoring and responsible husbandry of such animals can allow for the selection of animals that are healthy.
- Abnormalities for transgenic sheep clones appear similar to reports for non-transgenic sheep and cattle clones.
- Goats appear to suffer fewer adverse effects compared to sheep and cattle. Of the reports reviewed, only one cohort exhibited clinical signs of LOS.
- Abnormalities reported for transgenic swine clones are similar to those reported for transgenic and non-transgenic cattle clones. CVM is aware of only one report in non-transgenic swine clones (Park MR et al. 2005) in which clones exhibited similar health problems; however, *in vitro* methods used in this study likely influenced the outcome of swine clones in this study.

Cloned Transgenic Offspring Resulting from Somatic Cell Nuclear Transfer in the Goat: Oocytes Derived from Both Follicle-Stimulating Hormone-Stimulated and Nonstimulated Abattoir-Derived Ovaries. A relatively short time has elapsed since cloned sheep were produced by transferring nuclei from an established cell line isolated from the embryonic discs of 9-day-old embryos into enucleated metaphase II oocytes. Since that time, the use of donor nuclei from fetal and adult cells has resulted in the births of sheep, cattle, mice, goats, and pigs. The use of these nuclear transfer (NT) techniques in farm animals to efficiently generate cloned transgenic offspring capable of producing valuable proteins could have a marked impact on the pharmaceutical industry.

The mammary gland is well suited for the production and expression of human recombinant proteins (e.g., alpha-1-antitrypsin, fibrinogen, human clotting factor IX, and antithrombin III) and antibodies. Obvious benefits of utilising transgenic animals to provide such human pharmaceuticals include 1) high product yield, 2) low capital investment compared with cell culture techniques, 3) the ability to perform complex post-translational modifications (e.g., glycosylation and gamma-carboxylation), and 4) elimination of reliance on products derived from human blood, which may contain pathogens (e.g., human immunodeficiency virus and hepatitis viruses). Dairy goats are ideal for the transgenic production of therapeutic recombinant proteins because of their high yield of purified product, relatively short generation interval, and low incidence of the disease scrapie.

One of the limitations of somatic cell NT in sheep and cattle is the very low success rate, with a high proportion of fetal loss and an increase in perinatal morbidity/mortality. Reported causes of fetal wastage include abnormal liver development, insufficient placentation leading to spontaneous abortions or prolonged gestations, and oversized fetuses. Reported causes of perinatal death include metabolic and cardiopulmonary abnormalities and lymphoid hypoplasia.

Calves and lambs resulting from embryos cultured in vitro may be heavier at birth, present with an increased incidence of neonatal mortality, and are associated with longer gestation lengths than offspring resulting from in vivo-produced embryos. Moreover, placental development is also affected, resulting in fewer placentomes and increased incidences of hydrallantois. In addition, blastomere NT in the sheep and the cow results in increased fetal wastage and perinatal mortality, which may be attributable to the culture conditions (serum in the medium), coculture systems, or the manipulation procedure itself.

Conversely, the incidence of perinatal loss associated with somatic cell NT has not been reported in the goat, although this lack of information may be due to 1) the relatively low number of goat clones produced to date or 2) a minimal in vitro culture period in which reconstructed embryos were transferred at the two- to four-cell stage in both previous studies. However, Yong and Yuqiang produced an amazing 45 cloned goats from the transfer of 141 serially reconstructed embryos into 29 recipients in a cloning study using blastomere donor nuclei, in which embryos were cultured up to the morula stage prior to transfer. Thus, manipulated goat embryos may not be as sensitive as cow and sheep embryos to micromanipulation procedures and in vitro culture conditions.

Although cloned transgenic goats have been produced using oocytes collected from stimulated donor animals, abattoir-derived ovaries could also be a source of oocytes for the production of transgenic goats and may improve the cost effectiveness of generating pharmaceutically important proteins. Large numbers of goat oocytes can be harvested from abattoir-derived ovaries and matured in vitro to provide an inexpensive supply of cytoplasts suitable for NT procedures, apparently without the risk of generating oversize or moribund offspring, as reported in the sheep and cow. The expense and difficulty of obtaining caprine oocytes from a donor herd can thus be eliminated. Although the use of abattoir-derived oocytes for the production of human pharmaceuticals is restricted by federal Food and Drug Administration regulations, the economic benefit of their use in a research setting becomes obvious. Our objective in this study was to compare the fusion rates of couplets and then the in vivo development of NT embryos produced from the fusion of transgenic fetal fibroblast cells with that of cytoplasts prepared from oocytes aspirated from either FSH-stimulated or abattoir-derived caprine ovaries.

Materials and Methods

Isolation of Caprine Transgenic Fetal Fibroblast Cell Line

The primary caprine fetal fibroblast cells used as karyoplast donors were derived as previously described. Fetal tissues from a Day 40 transgenic female fetus (Toggenberg) were minced, washed, and transferred into 25-cm tissue culture flasks. Cells were cultured in fetal cell medium (FCM) consisting of tissue culture medium-199 (TCM-199; Gibco Laboratories, Grand Island, NY) supplemented with 10% fetal bovine serum (FBS; Hyclone, Logan, UT), nucleosides, 2 mM l-glutamine, 0.1 mM â-mercaptoethanol (Sigma Chemical Co., St. Louis, MO), and 50 ìg/ml gentamicin sulfate (Gibco). After three subpassages, cell stocks were frozen and stored in liquid nitrogen. This transgenic female line was designated CFF-6 and was used for all NT procedures within four to eight subpassages. Each passage required 6.2 ± 0.43 days in culture.

Oocyte Collection

FSH-stimulated Ovaries

Prior to oocyte retrieval, mixed-breed Spanish-type goats were implanted (Day 1) on the dorsal side of the ear (s.c.) with a 3-mg norgestomet implant (Synchro-Mate-B, Rhone Merieux, Athens, GA) and then subjected (Day 4) to ovarian superstimulation by twice daily administration (i.m.) of FSH (Sioux Biochemical, Sioux Centre, IA) in a decreasing dose schedule (10 and 10, 7.5 and 7.5, 5 and 5, and 2.5 and 2.5 units) totalling 50 units/doe. All goats (both donor and recipient females) used in this study were handled and maintained in accordance with the Institutional Animal Care and Use Committee Guidelines at Louisiana State University.

Implants were removed on treatment Day 8 and oocytes were then aspirated 24 h after the final FSH injection. Following general anesthesia (Halothane; Fort Dodge Animal Health, Fort Dodge, IA), does were placed in dorsal recumbency on a hydraulic operating table for a midventral laporatomy procedure. Oocytes were harvested using techniques previously described with slight modifications. Ovaries were exteriorized, and all visible follicles were aspirated with a 20-gauge needle attached via polyethylene tubing (20-gauge) to an electric vacuum pump (Cook Veterinary Products, Eight Mile Plains, Australia). The pump was set to aspirate <“18 ml of fluid/min. Oocytes were collected in 50-ml centrifuge tubes containing warmed PBS supplemented with 1% FBS, 50 ìg/ml gentamicin, and 0.4 units/ml sodium heparin (Elkins-Sinn, Cherry Hill, NJ). All oocytes with a compact investment of cumulus cells were washed eight times in maturation medium (TCM-199 supplemented with 10% normal goat serum [Sigma], 10 ìg/ml LH, 5 ìg/ml FSH, and 1 ìg/ml estradiol-17â) and allowed to mature for 18–22 h (38°C and 5% CO_2) in groups of 10–20 in 35-ìl microdrops of maturation medium overlaid with warmed embryo-tested mineral oil (Sigma).

Abattoir Ovaries

Ovaries harvested from nonstimulated does at an abattoir located in San Angelo, TX, were placed in insulated thermos bottles containing PBS at ambient temperature and transported by air courier within 12 h of collection. Upon arrival at the laboratory, oocytes from these ovaries were similarly aspirated, selected, washed, and matured as described for oocytes from stimulated ovaries.

Nuclear Transfer

Enucleation

Cumulus-oocyte complexes were vortexed at 18–22 h postmaturation for 2.25 min in TL-Hepes (BioWhittaker, Walkersville, MD) containing 0.6 mg/ml hyaluronidase (Sigma) to aid in the final removal of the cumulus cells. Oocytes were washed in modified TL-Hepes (mTLH; TL-Hepes supplemented with 10% FBS), selected based on the presence of a polar body (MII), and labelled with 3 ìg/ml Hoechst 33342 stain (Sigma) for 1 min. Mature oocytes were enucleated in a 200-ìl elongated droplet of micromanipulation medium (mTLH with 7.5 ìg/ml cytochalasin B) overlaid with warmed mineral oil on the stage of an inverted microscope (Nikon Diaphot) equipped with Hoffman Modulation Contrast objectives, Narishige micromanipulators, and epifluorescent illumination. Oocytes were held at the 9 o'clock position by negative pressure applied to a holding pipette (75-ìm outside diameter). After a brief exposure (<10 sec) to ultraviolet light to visualise nuclear DNA, the oocytes were rotated against the holding pipette to bring the metaphase plate into focus at the 3 o'clock position. Using a soft glass enucleation pipette (20- to 25-ìm outside diameter) with a 35° bevel and a spike, the metaphase plate and first polar body were removed under short-term exposure to UV light (<5 sec). This procedure was generally effective in removing both the polar body and the metaphase plate. The resulting cytoplasts were allowed to recover at 38°C in modified TCM (mTCM; TCM supplemented with 10% FBS) for at least 30 min prior to reconstruction with donor cells.

Donor Cell Preparation

Actively dividing fibroblast cells from the transgenic female line (CFF-6) were maintained in FCM with 10% FBS for 3–4 days and then synchronised by an additional 4 days of culture in reduced-serum (0.5%) FCM. Donor cells were trypsinized, washed, and held (<1.5 h) in mTLH immediately before transfer into the recently prepared cytoplasts.

Reconstruction and Fusion

A single fibroblast cell (15- to 20-ìm diameter) was injected into the perivitelline space of each enucleated oocyte in a droplet (200 ìl) of micromanipulation medium using the same enucleation pipette. Close contact of the donor cell membrane with the vitelline membrane of the cytoplast was visually confirmed prior to fusion. Karyoplast-cytoplast couplets were manually aligned between two stainless steel electrodes (1-mm gap) in a microslide fusion chamber filled with fusion buffer (0.3 M mannitol, 0.1 mM $MgSO_4 \cdot 7H_2O$, 0.05 mM $CaCl_2$, 0.5 mM Hepes, and 4 mg/ml BSA) and fused by a single DC pulse (1.30 kV/cm for 25 ìsec) delivered by a BTX Electrocell Manipulator 200 (Gentronics, San Diego, CA). Couplets were evaluated for fusion after a 1-h incubation period in mTCM and then activated.

Fused couplets were activated by a 5-min exposure to 5 ìM ionomycin (Sigma), washed extensively in mTLH, and then incubated for 3 h in 2 mM 6-dimethylaminopurine prepared in G1.2 medium (Zander IVF, Vero Beach, FL). Following activation, the reconstructed embryos were washed and cultured under oil in 35-ìl droplets of G1.2 medium for 33–36 h (two- to four-cell stages) prior to transfer to recipient females.

Embryo Transfer

Spanish-type crossbred recipient does were selected from those exhibiting a natural estrus 2 days prior to the scheduled embryo transfer. NT embryos, produced from either FSH-stimulated ovaries or nonstimulated abattoir-derived ovaries, were transferred into recipients 54–60 h after donor oocytes were aspirated.

The same general anesthesia protocol used for the oocyte aspiration was followed for the embryo transfer procedure. Does were placed in dorsal recumbency, and a midventral laporatomy procedure was performed. After the ovaries were examined for evidence of ovulation, the uterus was exteriorized and embryos were transferred to the oviduct ipsilateral to the ovary with the most ovulation points. Embryos (1–15/female) were transferred via a small plastic catheter (Embryon Catheter; Rolon Medical, Watford, UK) through the ostium of the infundibulum, gently advancing the catheter as far as possible distally into the oviduct. Recipients were returned to the herd to await subsequent examinations of pregnancy status.

Pregnancy Status and Parturition

Beginning on Day 30 of gestation, recipients were subjected to a transvaginal ultrasonographic evaluation (Aloka 500-V; Aloka, Tokyo, Japan), followed 7 and 10 days later by additional ultrasonographic assessments. The presence of a fetal heartbeat was used to diagnose pregnancy. Pregnant recipients were subsequently monitored at 14-day intervals and then separated from the herd on Day 90 of gestation and placed in groups of two or three females per pen. Monitoring of pregnant recipients past Day 90 was performed via abdominal ultrasonography.

Parturition was induced in recipient females if they had not given birth by Day 152 of gestation. Induced females received 15 mg of prostaglandin $F_{2\alpha}$ (Lutalyse; Upjohn, Kalamazoo, MI) and 12 mg of dexamethasone (i.m.) and were monitored closely until parturition. The kids were allowed to remain with the dams and nurse freely until weaning.

Characterisation of Cloned Animals

Skin biopsies (6-mm diameter) were obtained from each offspring (1.5–3 mo of age) from the caudal lateral aspect of the thigh directly over the semitendinosis muscle and then subjected to genomic DNA isolation using the Genomic Prep Kit (Amersham Pharmacia Biotech, Piscataway, NJ). Each sample and an internal control for goat genomic DNA (goat exon 7) were analysed by polymerase chain reaction using human antithrombin (hAT)-specific primers. The hAT sequence was detected by amplification of a 367-base pair (bp) sequence with oligonucleotides GTC11 (CTCCATCAGTTGCT GGAGGGTGTCATTA) and GTC12 (GAAGGTTTATCTTTGTCCTTGCTGCTCA). Two positive control samples, from an antithrombin III (ATIII) transgenic adult female (B521) and from the transgenic fetal fibroblast CFF-6 cell line, and one negative control blood sample from a nontransgenic female goat were analysed to verify the presence of the human ATIII (hATIII) transgene.

At 150 days of age, the cloned offspring were subjected to a hormonal lactation induction protocol developed for prepubertal goats and were then hand-milked once daily to collect samples to assay for hATIII expression. Western blots were performed as previously described.

Statistical Analysis

Data for maturation rate, number of couplets produced, rate of fusion, in vitro embryo development, and subsequent production of offspring were all analysed by a general linear model of regression using a SAS analysis of variance procedure.

Results

Oocyte Recovery and Embryo Reconstruction

The rate of oocyte recovery and maturation from stimulated ovaries and nonstimulated abattoir-derived ovaries. A total of 833 oocytes were recovered from a series of aspirations (12 replicates) from 34 FSH-stimulated donor animals (68 ovaries). An average of 12.2 oocytes were recovered per stimulated ovary (from 7 to 33 oocytes/ovary). Eight replicates of follicular aspiration from 944 nonstimulated abattoir-derived ovaries produced a total of 769 oocytes (0.81 oocytes/ovary). Oocyte maturation and embryo reconstruction rates from oocytes derived from FSH-stimulated goat ovaries and nonstimulated abattoir-derived ovaries

Following a maturation interval of 18–22 h, 50% of the oocytes from the FSH-stimulated ovaries and 68% of the oocytes from the abattoir-derived ovaries reached the MII stage, as evidenced by extrusion of the first polar body. Of the matured oocytes used for NT, 91% and 85% resulted in successful couplets derived from oocytes from FSH-stimulated and abattoir-derived ovaries, respectively. The rate of fusion of NT embryos reconstructed from oocytes from either FSH-stimulated or abattoir-derived ovaries was 63% (range, 22%–79%) and 57% (range, 45%–81%), respectively. There was no significant difference between the two treatment groups for maturation rate, couplet production rate, or fusion rate.

Embryo Development and Cloned Transgenic Offspring

There was no effect of oocyte source on the number of fused couplets that developed to the two- to four-cell stages (57% vs. 56%). Only high-quality reconstructed embryos (those with equal size blastomeres and no fragmentation) at the two- to four-cell stages were transferred into recipient females. Moreover, 92% of the two- to four-cell embryos resulting from oocytes derived from FSH-stimulated ovaries were classified as high quality, but the rate was not different when compared with those reconstructed embryos produced from oocytes from abattoir-derived ovaries (70%).

A total of 23 similar recipient does received an average of 8 embryos/female (range, 1–15 embryos/female). Five of the recipients maintained conceptuses throughout gestation, and each gave birth to a single offspring. The three recipients that maintained pregnancies from the FSH-stimulated group each received an average of 7.6 embryos (range, 7–8 embryos). The two recipients that maintained pregnancies in the abattoir ovary group each received nine reconstructed embryos. The pregnancy rate (defined as the number of pregnant recipients per total number of recipients) at 30 days of gestation was 21% (3/14) and 22% (2/9) for the FSH-stimulated group and the abattoir-derived group, respectively. The overall pregnancy rate was 21.7% (5 of 23 recipients). Based on the number of offspring produced per number of embryos transferred, the cloning efficiency was 2.7% for both the FSH-stimulated and abattoir-derived groups (3/112 and 2/72, respectively). All pregnancies were produced from donor cells that originated from the fifth or sixth passage of the transgenic fetal fibroblast cell line.

In this study, 100% of the Day 30 pregnancies culminated in the births of live, healthy offspring, with no fetal loss and no postpartum morbidity. Two of the recipients gave birth at 149 and 151 days of gestation. Parturition was induced in the other three recipient females at Day 152 of gestation, and they gave birth between 24 and 36 h postinduction. All kids were vigorous and healthy at birth and weighed an average of 3.8 kg (SEM = 0.18 kg; range, 3.4–4.3 kg). Phenotypically, all five offspring had a similar coat colour pattern and physical stature. At this writing, all cloned animals have been weaned (Day 90) and are past 12 months of age. Mean weaning weight was 20.9 kg (SEM = 0.98 kg), with a range of 19.1–24.5 kg. Each cloned animal has exhibited estrus and has been bred by an intact buck.

Discussion

Five cloned transgenic goats were produced through NT using a cell line derived from a transgenic fetus and oocytes obtained from either FSH-stimulated or abattoir-derived ovaries. At birth, all five kids were healthy and weighed between 3.4 and 4.3 kg, which is within the normal range for the breed. At weaning (90 days), body weights ranged from 19.1 to 24.5 kg, and all kids are continuing to grow as expected.

There was no significant difference in the percentage of matured oocytes successfully enucleated and reconstructed from the two oocyte sources (FSH-stimulated ovaries vs. abattoir-derived ovaries). There was also no difference in the fusion rate for oocytes derived from FSH-stimulated ovaries (63%) and those derived from abattoir ovaries (57%). Similar fusion rates (59%) of fetal fibroblasts with cytoplasts from abattoir ovaries in the cow have also been reported, and fusion rates (85%) of fetal fibroblasts with cytoplasts from FSH-stimulated sheep tend to be higher. Oocyte source also had no effect on embryo development and overall pregnancy rate in the two treatment groups.

Recipient synchronisation with NT embryo age likely affects pregnancy rates. In our study, all two- to four-cell stage NT embryos were transferred into recipients that were in estrus 36–48 h previously. Because mixed-breed goats usually ovulate 24–36 h after the onset of estrus, we calculated that at the time of embryo transfer the recipients would be 12–24 h postovulation. This would then be the time of first embryonic division, and the asynchrony between recipient and NT embryo stage was used in an effort to compensate for slowly developing NT embryos. Following in vitro maturation, the time for NT embryos to reach the two- to four-cell stage (36–40 h) appears to be slightly longer than required for embryos produced by in vitro fertilization (30–36 h).

An interesting aspect of this study was the fact that all detected pregnancies resulted in the births of healthy kids, with no fetal loss occurring from Day 30 to parturition and no postnatal morbidity. However, we cannot speculate on the number of pregnancies that were initiated and lost before the first ultrasound evaluations on Day 30 nor can we rule out the possibility of undetected twins before Day 30 that resorbed. Although it is possible to detect fetal heartbeats in the sheep (and possibly the goat) as early as Day 25, this method places considerable stress on the animal because it must be restrained and placed in dorsal recumbency. To minimise this stress to our potentially pregnant recipients, this method was not used. Other groups developing NT procedures in the goat have recently reported similar perinatal viability in NT offspring. In contrast, nearly 50% of the pregnancies resulting from cattle NT embryos are lost during the first trimester of gestation; most of these losses are thought to be caused by poor placentae and placentome development. In addition, lethal postnatal complications (pneumonia, cardiac abnormalities) arising in cloned newborn calves have been a major problem. These variations between the cow and the goat could be due to inherent differences between the two species or differences in the NT procedure. In our study, reconstructed caprine embryos were transferred into the oviduct at the two- to four-cell stages. Therefore, the reconstructed embryos underwent a relatively short in vitro culture period.

In this study, a total of 184 embryos were transferred into 23 recipient females. Five of the 23 recipients (21%) maintained pregnancies and subsequently gave birth to healthy offspring. This pregnancy rate, based on the number of pregnant recipients per total number of recipients, is greater than the 5.3% pregnancy rate previously reported by Baguisi et al. in the goat but lower than the 38%–44% pregnancy rate reported by Keefer et al. Although 100% of the recipients that were determined pregnant by ultrasonography on Day 30 in this study delivered normal kids, the efficiency of the cloning procedure (based on the number of offspring per number of embryos transferred) was 2.7%, similar to that reported by Baguisi et al.. The cloning efficiency reported by Keefer et al. ranged from 6.2% to 7.3%. In the sheep, overall cloning efficiency ranges from 4.8% to 20%, depending on the cell type used.

Another important outcome of this study was the production of cloned offspring using abattoir-derived oocytes as a source of recipient cytoplasts. Until now, cloned goats have only been produced using cytoplasts obtained from oocytes aspirated from FSH-stimulated live donor animals. Abattoir oocytes would likely be easier to obtain and more cost effective than maintaining and stimulating a herd of donor goats.

There was no difference in the number of cloned goat offspring produced from oocytes harvested from FSH-stimulated donor animals and from oocytes aspirated from abattoir ovaries. All of the conceptuses were maintained throughout gestation, with no fetal wastage or abortions, and all transgenic kids have remained healthy after parturition.

Transgenic cloned goats and the production of therapeutic proteins.

One of the challenges created by the biotechnology revolution is the development of methods for the economical production of highly purified proteins at large scale. The expression of recombinant proteins in the milk of transgenic dairy animals appears particularly well suited for the economical production of complex polypeptides (Houdebine, 1994), (Clark, 1998) and (Meade, et al, 1998). To that end, transgenic sheep, cows, goats, and even pigs (hardly dairy animals!) have been gencrated. Dairy goats offer a nice compromise.

Goats have a relatively short generation interval (when compared with cattle) and a relatively large average milk output (compared with pigs and to a lesser extent sheep). These properties allow the geeration of small herds of transgenic goats able to yield hundreds of kilograms of unpurified product per year, within a timeline compatible with the clinical and regulatory framework of drug development.

Prepubertal Propagation of Transgenic Cloned Goats

The production of recombinant proteins of pharmaceutical and biomedical interest in the milk of transgenic animals provides an economical advantage over more traditional production systems based on microorganisms and animal cells. Transgenic animals can be produced by a number of different techniques, including pronuclear microinjection, sperm-mediated transgenesis (Celebi et al., 2003), and somatic cell nuclear transfer.

Somatic cell nuclear transfer (SCNT) has recently improved the efficiencies to generate transgenic animals (Coleman, 1999). In this system, female cell lines are transfected *in vitro* and selected based on transgene integration before they are used in a nuclear transfer program. As a result, the progeny born are virtually guaranteed to be transgenic, and, therefore, fewer animals are needed to produce the desired number of transgenic founder animals.

Shortly after they are born, the founder animals can be checked for expression of the transgene by means of hormonal induction of lactation (Cammuso et al., 2000). If the expression levels are adequate (usually in the gram per litter levels), it is desirable to propagate those animals as fast as possible in order to generate the transgenic production herd. For such process, usually referred to as scale-up, it is ideal to generate male lines from the founder female animals that have shown adequate expression levels, as males are very cost efficient to propagate using standard artificial insemination technology (Baldassarre et al., 2000).

In the present study, we have evaluated the use of laparoscopic ovum pick-up (LOPU) in conjunction with *in vitro* embryo production, as a method to propagate SCNT-derived transgenic founder goats at young (prepubertal) ages. Furthermore, the effect of age at LOPU as a consequence of practicing the procedure before or after induction of lactation was evaluated.

Materials and Methods

Animals

Ten SCNT-derived goats were used as oocyte donors. These somatic cell clones were produced from two transfected fetal fibroblast cell lines using techniques previously described (Keefer et al., 2001, 2002). One set of five goats was derived from designated line 503, while the other set of five goats was derived from line 601. Both lines were derived from the same 28-day-old fetus. Half of the animals were submitted to LOPU at 2–3 months of age, while the other half of the animals was submitted to LOPU at 6–7 months of age. Each age group was composed of three founders from one cell line and two founders of the other line. Thirty-eight goats of standard breeds (Nubian, Saanen, Boer) were used as recipients.

Induction of Lactation

The SCNT-derived donors were induced to lactate at 2–4 months of age following established procedures (Cammuso et al., 2000). Briefly, goats were treated with estradiol and progesterone for 2 weeks to stimulate mammary gland development followed by a 3-day lactogenic treatment with dexamethasone. Goats were milked for a period of 2–3 months before they were dried-off. Half of the animals were hormonally induced to lactate prior to oocyte recovery, while the other half of the animals was submitted to LOPU before they were induced to lactate.

Hormonal Treatment of Donor and Recipient Goats

Donor goats were stimulated with 80 mg NIHFSH- P1 (Folltropin®, Vetrepharm, Canada) together with 300 IU eCG (Novormon®, Vetrepharm, Canada) administered intramuscularly 36 h prior to LOPU. Recipient goats were heat synchronised using intravaginal sponges containing 60 mg medroxyprogesterone acetate (Veramix®, Upjohn, Canada) for 10 days combined with a luteolytic treatment of 125 *m*g cloprostenol (Estrumate®, Schering, Canada) and 300 IU eCG (Novormon®, Vetrepharm, Canada) administered intramuscularly 24 h prior to sponge removal, which took place approximately 15 h prior to LOPU in donors.

Anesthesia

Donors and recipients were deprived of food and water for 24 h prior to laparoscopy. Anaesthesia was induced with intravenous administration of 0.35 mg/kg body weight of diazepam (Diazepam ®, Sabex, Canada) and 5 mg/kg body weight of ketamine (Ketalean®, Bimeda-MTC, Canada), and maintained with Isofluorane (Isoflo®, Abbot, Canada) via endotracheal intubation.

Chemicals and Reagents

Unless otherwise indicated, all chemicals and reagents were purchased from Sigma Chemical Co. (St. Louis, MO).

Laparoscopic ovum pick-up (LOPU)

Oocytes were recovered by aspiration of follicular contents under laparoscopic observation as previously described (Baldassarre et al., 2002b). The collection medium was TCM 199 supplemented with 0.05 mg/mL Heparin (Hepalean®, Organon Teknica, Canada) and 1% (v:v) fetal bovine serum.

Washing and Grading of Ocytes

Follicular contents were poured into a Petri dish and observed under a stereomicroscope. The cumulus– oocyte complexes (COCs) were located and washed in M199 supplemented with penicillin (100 IU/mL), streptomycin (0.1 mg/mL), kanamycin (0.05 mg/mL), and 0.5% bovine serum albumin. They were then transferred into the IVM drops and selected for subsequent use based on their cellular vestments and cytoplasm uniformity.

Prepubertal Propagation of Cloned Goats

Maturation of Oocytes

In vitro maturation was performed in 50-*m*L drops of maturation medium under mineral oil. Maturation medium consisted of M199 supplemented with bLH (0.02 U/mL; Sioux Biochemicals, USA), bFSH (0.02U/mL; Sioux Biochemicals, USA), estradiol *b*-17 (1 *m*g/mL), sodium pyruvate (0.2 mM), kanamycin (50 *m*g/mL) and 10% heat inactivated estrus goat serum. *In vitro* maturation was performed at 39°C in humidified atmosphere with 5% CO2 in air for 24–27 h.

In Vitro Fertilization

The expanded cumulus cells were partially removed from matured COCs by repeated pipetting or dissection using a needle. Oocytes were washed in fertilization medium and transferred to 40-*m*L drops of fertilization medium under mineral oil. Fertilization medium consisted of TALP medium (Parrish et al., 1986) supplemented with 20% heat-inactivated estrus goat serum.

Fresh semen was collected on the day of IVF from one male of the Saanen breed using an artificial vagina and was kept in the dark at 20°C for 2–3 h. Two mL of warm (37–38°C). Defined Medium (Younis et al., 1991) supplemented with 6 mg/mL fatty acid free BSA (mDM) were added to the tube and mixed with the semen.

A 100-*m*L aliquot of the semen mix was overlaid on top of a 45:90% Percoll gradient, which was then centrifuged at room temperature for 30 min at 500 3 *g*. The pellet was resuspended in 4 mL of mDM and centrifuged for 10 min at 70 3 *g*. The washed pellet was resuspended in capacitation medium containing 0.5mM 8-bromo-cAMP, 100 nM ionomycin, and 10 *m*g/mL heparin in 1 mL of mDM.

The sperm-capacitation mixture was incubated at 39°C for 15 min. The sperm concentration was adjusted to 10 3 106 sperm per mL, and 5 *m*L were added to the fertilization drops containing the oocytes (final sperm concentration; 1 3 106 sperm per mL). The IVF drops were then incubated at 39°C in a humidified atmosphere incubator with 5% CO2 in air.

Embryo transfer

At 15–20 h after insemination, the presumptive zygotes were transferred into the oviducts of recipient goats, which were first examined laparoscopically to confirm at least one recent ovulation. A midventral laparotomy was established and the reproductive tract was exteriorized for the transfer of embryos contained in a Tomcat catheter (Sovereign®, Kendall Company, Mansfield, MA). The catheter was introduced into the oviduct through the fimbria. The number of embryos transferred per recipient varied between 5 and 8.

Pregnancy Detection

Pregnancy was detected by ultrasound using a SonoVet® 600 scanner (Medison Corp., Korea) with a transrectal 7.5-MHz linear array probe at 28 and 56 days following embryo transfer. Identification of transgenic animals Genomic DNA was extracted from the blood or skin of 2-week-old putative transgenic offspring using standard molecular biology techniques (Sambrook et al., 1989). The DNA was screened for the presence of the transgene by PCR using two different sets of primers specific to the transgene. Confirmation of all PCR positive animals was performed using Southern blotting analysis and the Roche Molecular Biochemicals (Laval, QC, Canada) DIG system for detection.

Statistical analysis

Results were statistically analysed using Prism™ (GraphPad Software Inc., San Diego, CA). The data collected from follicles aspirated, oocytes recovered and embryos transferred per donor were analysed using the unpaired Student's *t*-test with Welch's correction for unequal variances. The data for transferable embryo yield (embryos transferred/oocytes recovered), pregnancy rate and kids born were analysed using Pearson's Chi-Square.

Animal ethics

All protocols used in this study have been approved by Animal Care Committee of Nexia, in compliance with the guidelines from the Canadian Council of Animal Care.

Results

Overall, 425 follicles were aspirated (42.5 6 19 per goat) and 334 oocytes recovered (33.4 6 10 per goat) resulting in a 79% recovery rate. *In vitro* maturation and fertilization resulted in 244 transferable zygotes (24.4 6 11 per goat) that were transferred into 38 recipients (6.4 embryos per recipient). A total of 21 recipients were detected pregnant by ultrasound at 28 days, of which 20 went to term and delivered 42 kids. Five kids (12%) were born dead or died shortly after birth due to weakness at birth. From the total of 42 kids born, 33 were characterised to be transgenic (78%) by PCR and Southern

Blotting. No differences were found between the two sets of SCNT-derived goats (lines 503 and 601) in terms of the follicles aspirated (42.8 6 22 vs. 42.2 6 19), oocytes recovered (33.2 6 12 vs. 33.6 6 10), pregnancy rate (11/21 5 52% vs. 10/17 5 59%), kids born (23 vs. 19).

The overall efficiency of IVM/IVF was high, with 73% of the total number of oocytes recovered resulting in transferable zygotes. The age of the animals at the time of LOPU had a significant impact on most variables. While the number of follicles available for aspiration and the oocytes recovered was significantly higher in the animals of ,100 days of age (57 6 7 and 41 6 4 vs. 28 6 2 and 25.8 6 2, *p* ,0.05), older animals resulted in a higher yield of transferable embryos after IVM/IVF (81.4% vs. 67.8%, *p* , 0.01), higher pregnancy rate (80% vs. 40%, *p* , 0.05), and higher number of kids born (27 vs. 15, *p* , 0.01). Only one of the 10 donors failed to produce at least one transgenic offspring during this experiment (,100 days of age group). A high transgenesis rate (78.6%) was found in the progeny as a result of all donors having multiple integration sites.

Discussion

Overall, the propagation program was very successful in that 33 transgenic kids were obtained from 10 transgenic founder animals that were less than 8 months of age at the time of LOPU. As only one donor failed to produce any transgenic progeny, the program resulted in the successful propagation of 90% of the donors. It is noteworthy that for those animals that were submitted for LOPU at ,100 days of age the program resulted in offspring from them being born before they reached age and development to be bred for the first time. As a result, a significant shortening of the generation interval was possible. Not surprisingly, the oocytes collected from older animals showed increased developmental ability resulting in higher fertilization, pregnancy and development to term. Since these animals were used as donors following hormonal induction of lactation, the results indicate that such treatment did not have a detrimental effect on subsequent LOPU-IVP. Furthermore, O'Brien et al. (1997) reported higher developmental capacity in oocytes from prepubertal sheep that were primed with estradiol and progesterone prior to gonadotropin treatment and oocyte collection.

To the best of our knowledge, this is the first report describing the *in vitro* production of embryos using oocytes collected from goats shortly after hormonal induction of lactation. Hence, further research is required to establish if the hormonal treatment used for induction of lactation prior to oocyte collection may have enhanced the quality of the oocytes recovered from the animals subjected to LOPU at this later age. Results from the early prepubertal age group (,100 days of age) were quite acceptable, significantly higher than those reported in calves (Revel et al., 1995) and comparable to those reported in lambs and standard prepubertal goats (Baldassarre et al., 2002a). In conclusion, LOPU in combination with *in vitro* embryo production techniques have proven to be a powerful tool for the early propagation of valuable transgenic female founders produced by SCNT. Our results suggest it is more efficient to perform the procedure at 6–7 months of age, after completing the induced lactation program. In addition, the study demonstrates good reproductive results can be obtained from prepubertal goats produced by somatic cell nuclear transfer.

Cloning of Domesticated Mammals

In 1997, Wilmut and coworkers1 announced the birth of the first ever clone of an adult animal. Such an announcement was significant because the ability to clone mammals by simple nuclear transfer was thought _biologically impossible_.2 Furthermore, the demonstrated totipotency of mammary nuclei inferred that other types of somatic cells could be utilised for cloning adult animals. As somatic cells can be readily obtained and cultured in the laboratory, the procedures of somatic cell nuclear transfer (SCNT) immediately held promise for offering an alternative for introducing targeted additions or deletions to the genome of farm animals. Dolly was not the result of the union of an oocyte and sperm, but derived from the nuclear genome contained within a quiescent-induced mammary cell and cytoplasm originating from an enucleated oocyte and somatic cell. The concept of cloning an animal by transfer of a nucleus into an enucleated oocyte was first proposed in 1938.3,4 However, the first success was with amphibians and not reported until 1952.5 Thirtyone years later, McGrath & Solter announced the first successful cloning of mice.6 Soon after, the cloning of sheep,7 cattle,8 rabbits,9 and pigs10 was reported. However, it is important to note that cloning successes before Dolly involved the use of toti- or pluripotent cell types comprising the early embryo.

Any and every attempt to clone an adult animal by nuclear transfer using more differentiated cell types failed.4 Clearly, Dolly's birth dispelled the dogma that it was _biologically impossible to clone mammals using nuclear transfer_. To date, adult sheep,1 cattle,11 mice,12 pigs,13 goats,14 cats,15 and rabbits,16 have been cloned using SCNT. Given the incredible complexity of what must be involved in the ooplasm redirecting functionality of a somatic nucleus, it is remarkable that any cloned embryos develop at all. Very little is known about specific factors or mechanisms involved with the ooplasm reprogramming a somatic nucleus. This current lack of understanding continues to be reflected in the overall inefficiency of the cloning procedures for producing live offspring. Dolly was just one cloned offspring that resulted after 277 attempts (0.3% efficiency). Five years later, similar inefficiencies are still being described for cloning adult animals (cattle,17,18 goats,14,19 pigs,13 rabbits,16 cats15). The focus of this review will be to provide a brief overview of SCNT procedures utilised for cloning adult farm animals, discuss potential applications, problems associated with its use and some of the potential factors contributing to the overall inefficiency of the procedure for producing live offspring.

Overview of Somatic Cell Nucleartransfer Utilised For Cloning Adult Farm Animals

The ultimate challenge of cloning using SCNT involves reprogramming a somatic nucleus in a manner conducive for embryonic development. The predominant cell type of choice for reprogramming a somatic nucleus is an oocyte arrested at metaphase II.20 A matured oocyte is a logical choice as critical components required for initiating and directing early embryo development are contained within the oocyte. For a detailed review of specific procedural details of nuclear transfer refer to Robl & Stice,21 Prather & First,22 First & Prather23 and Campbell et al.24 In general, the first step of the cloning process involves collection of somatic cells from the animal to be cloned. Choice of somatic cell for use in SCNT varies greatly. In cattle, adults have been cloned using cumulus,11,25,26 fibroblasts,25,27–32 ovarian/granulosa, 18,33–36 mammary,27,29 muscle,37 oviduct,11,25,38 and uterine25 cell types.

It has also been suggested that some cell types may work better than others for producing live offspring. Wakayama et al. 12 reported cloned embryos originating from Sertoli or neuronal nuclei failed to develop beyond 8 days post-coitum; however, embryos originating from cumulus cells developed to term. After collection, somatic cells may be utilised immediately12 or after long-term culture. Use of quiescent or quiescent-induced adult cell types may1 or may not35,37,39,40 be required for producing cloned offspring of adult animals. However, doing so may improve ability of cloned embryos to establish35,41 and maintain pregnancy to term. The second and perhaps most labor-intensive part of SCNT requires removal of maternal DNA from an oocyte arrested at metaphase II (MII). Doing so requires the use of microtools beginning approximately 18 hr after oocytes have been placed in maturation medium. To avoid lysis, oocytes may be cultured in the presence of a microfilament inhibitor22 (cytochalasin B).

This induces relaxation of cytoplasm allowing for mechanical removal of <5–15% of egg cytoplasm containing the maternal DNA. Removal of maternal DNA can be verified by Hoechst staining and ultraviolet transillumination. For cloning cattle, oocytes generally originate from a non-descript group of females obtained from commercial sources, abattoirs 18,28–30,32–36 or a descript group of live animals using ultrasound-guided aspiration. The next step of SCNT is to insert a somatic nucleus into an egg cytoplasm; thereby, constructing the equivalent of a one-cell embryo. In most farm animal species, this is performed by electrical-induced fusion of the somatic cell with an egg cytoplasm. Initially, a somatic cell is mechanically inserted into the perivitelline space using microtools.

Thereafter, the resulting _couplet_ (egg cytoplasm and somatic cell) is aligned between two electrodes and pulsed with an electrical current (for example, use of 2.2 kV/cm for 40 ls will induce greater >70% of couplets to fuse). Electrofusion is dependent on maintaining contact of the somatic cell with egg cytoplasm such that membranes of each cytoplasm may intermingle after pore formation resulting after electrical current. In mice, however, the preferred method for introducing a somatic nucleus into egg cytoplasm is microinjection. Within minutes of introducing somatic nucleus into egg cytoplasm, the nuclear membrane breaks down and the chromatin condenses. In some cases, the electrical pulse utilised for fusion is also sufficient to _activate_ the cloned embryo to begin development, while others choose various chemical combinations.18,43 Regardless of the method cof choice, the ultimate challenge of activation protocols is to mimic actions of a sperm after fertilization. Cloned embryos may be transferred into oviducts of ligated recipients (sheep)1 or cultured in the incubator for a period of time required for development to compact morula or blastocyst.35 Continued development of cloned embryos is dependent on transfer into surrogate recipients. In cattle, approximately half of the cloned embryos transferred as singles will hav established a pregnancy 21 days after transfer as indicated by the presence of an embryo proper with heartbeat, while only a few will progress to term.35

Applications of Somatic Cellnuclear Transfer

A number of applications of SCNT have been described benefiting animal agriculture and human medicine. Success in cloning of adult sheep,1 cattle,11 pigs13 and goats14 confirms usefulness of SCNT for the purpose of clonal expansion of agriculturally important foodproducing animals. There is also increasing interest to utilise SCNT to restore endangered, or even extinct species. Wells et al.34 reported the first successful use of SCNT for preserving the last surviving cow of the Enderby Island cattle breed. Moreover, Lanza et al.44 used interspecies SCNT to clone an endangered species (Bosgaurus). Finally, given the manipulative aspects of the procedures, SCNT provides one with a novel tool for investigating nuclear and/or cytoplasmic components involved in embryonic development and loss. Given the ease at which most somatic cells can be clonally expanded and genetically modified, use of SCNT is rapidly becoming the method of choice for producing transgenic farm animals.

Moreover, use of SCNT increases efficiency of producing transgenic farm animals to 100%.45 Production of transgenic farm animals provides one with a means in agricultural and biomedical disciplines for studying importance of genes involved in a variety of biological systems. And, genetic modification of farm animals has created a new form of _pharming_; whereby, transgenic animals are used as bioreactors for production of pharmaceuticals or perhaps organ donors for the human population. Announcement of the first ever cloned transgenic sheep45 and cattle46 using SCNT immediately obviated the need for producing embryonic stem cells in farm animal species and suggested possible usefulness for targeted additions or deletions to the genome of farm animals. In 2000, McCreath et al.47 announced the first cloned gene-targeted sheep. Most recently, Lai et al.48 demonstrated that SCNT could be utilised to produce cloned pigs with the b-1,3-galactosyltransferase gene knocked out. Moreover, Lanza et al. 49 demonstrated that one could use SCNT for generation of histocompatible tissues addressing one of the major challenges in transplantation medicine.

Problems Associated With Cloning Adult Animals Using Somatic Cell Nuclear Transfer

A major limitation of cloning adult farm animals using somatic cell nuclear transfer is the extreme inefficiency of producing live offspring. For example, Dolly was one of 277 cloned embryos that developed to term (0.3% efficiency).1 _Millie_ was just one offspring that resulted after 95 attempts.40 Similar inefficiencies regardless of species or somatic cell type continue to be reported.50 Death of cloned embryos and fetuses occur throughout pregnancy.

Moreover, a high proportion of cloned offspring are generally larger than normal (large offspring) and die soon after birth. In general, there have been at least five periods of loss observed with clones derived from adult animals. The first and perhaps most dramatic occurs during preimplantation development. In cattle 17, 18, 29, 30, 34, 35, 37, 51 as well as other species including goats,14 sheep,1 and rabbits,16 >65% of one-cell cloned embryos fail to develop to compact morula or blastocyst. Approximately 50% of bovine cloned embryos (compact morula or blastocyst) establish pregnancy after transfer of a single embryo into a surrogate recipient (i.e., presence of an embryo proper with a heart beat between days 29–32).

Development of cloned bovine embryos, at least for the first 29–32 days, parallels that of those embryos produced after in vitro maturation, fertilization and culture (IVMFC).52,53 However, beginning at approximately 30 days and continuing through day 60 of pregnancy, embryonic death may occur in 50–100% of cloned pregnancies (absence of heartbeat and detachment of placental membranes). Others18,28,31,34,51 have reported similar losses. Pregnancy losses in cattle of this magnitude are significantly higher than expected in animals bred by natural service (2–10% loss54 or developed using IVMFC procedures (16%;).52,53 Examination of placentae originating from cloned embryos between days 40–50 of gestation reveal placentae that are hypoplastic, partially developed with rudimentary cotyledons, or those that are essentially normal when compared with IVF derived embryos.51 The third period of loss that has been noted is associated with an increased incidence of spontaneous abortions during the second trimester of pregnancy. 25,29,30,33,35 Complete macroscopic and histopathologic examinations of aborted fetuses reveal few cabnormalities.55 However, placentae are oftentimes grossly abnormal with a marked reduction in fetal cotyledons (fewer than 20 compared with the expected 70–120).56 Moreover, fetal membranes are generally thickened and edematous.

The fourth period of loss noted for cloned bovine pregnancies occurs during the third trimester between days 200–265 (280 days ¼ term). Loss during this time period is generally characterised by a marked increase in the incidence of hydrallantois and fetal death (unpublished observations of Edwards et al.).18,27,32,33,36,57,58 Hydrallantois is accompanied by a marked reduction in placentomes (often fewer than 70, and in some cases <20), marked hypertrophy of many cotyledons, adventitial placentation and severe edema of the intercotyledonary placental membranes. Fetal anasarca with generalised edema and marked edema of the umbilicus is usually present.

In our experience, death of late term cloned fetuses is primarily the result of inadequate placentation. Thus far, complete gross and microscopic examinations of late term and early neonatal fetuses have not identified any known genetic or inherited abnormalities.55 The majority of fetal lesions observed can be attributed to changes secondary to inadequate placental development. Amniotic squames and meconium are generally present in the lungs of all late term fetuses indicating some degree of stress in utero before death. In some cases, a few cloned embryos will develop to term. However, the majority are not born without complications. In particular, cloned calves derived from adult animals are usually larger at birth and have a lower postnatal survival rate when compared to in vitro counterparts (Edwards et al., unpublished). 25,32 Most calves derived from abnormal placentae require intensive monitoring and therapy after birth to treat a whole plethora of complications that may include lung dysmaturity, pulmonary hypertension, respiratory distress, hypoxia, hypothermia, hypoglycemia, metabolic acidosis, enlarged umbilical veins and arteries, and/or development of sepsis in either the umbilical structures or lungs (Edwards et al., unpublished observations).18,33,34,51,57 Severity of complications may not be evident for several months. Gibbons et al.18 lost one calf at approximately 60 days of age and necropsy revealed adhesions in the lungs consistent with pneumonia, and digestive problems consistent with vagal indigestion.

Similar complications have been described for clones derived from fetal57 and embryo59 cells. In spite of the extreme inefficiency of SCNT, there are some clones born "normal and healthy" requiring little if any veterinary care after birth (Edwards et al., unpublished observations).36,60 Pace et al.60

reported similar growth rates, reproductive performance and lactational characteristics of clones compared with non-cloned dairy cattle. Moreover, Enright et al.62 demonstrated that cloned heifers were not different in estrous cycle length, ovulatory follicle diameter, number of follicular waves, or profiles of hormonal changes (leutinizing hormone, follicle stimulating hormone, estradiol, and progesterone). Concentrations of growth hormone, IGF-I and IGFBP3 values recorded for clones derived from a 13-year-old Holstein were all within the range reported for non-cloned calves of similar ages.63 However _normal and healthy_ cloned animals may appear, it is possible that undiagnosed pathologies may develop later in life58 as a result of subtle changes in chromatin structure and/or gene expression. Miyashita et al.64 noted differences in telomere lengths among cloned cattle derived from different cell types. Moreover, X-chromosome inactivation may (mice)65 or may not (cattle)66 be normal. Wrenzycki et al.67 noted aberrant expression of genes thought to be of importance in stress adaptation, trophoblastic function, and DNA methylation during preimplantation development in cloned bovine embryos. Yet, many mice and other animals have survived to adulthood despite widespread gene dysregulation, indicating that mammalian development may be rather tolerant to epigenetic aberrations of the genome.68 The ultimate consequences of epigenetic aberrations of the genome in cloned animals remain unclear but may result in an early death.69

Potential Factors Contributing to the Extreme Inefficiency of the Cloning Procedures for Producing Live Offspring

Given the physically insulting nature of the cloning process and the presumed complexity of factors that must be involved with the ooplasm redirecting the functionality of a somatic nucleus, it is remarkable that any cloned embryos develop at all. Moreover, data are coming forth supporting a role for immunologic rejection in the syndrome of early embryonic loss in cloned bovine pregnancies.70 Certainly, many factors related to SCNT procedures and culture environment may be contributing to the embryonic and fetal deaths of cloned embryos. First, construction of cloned embryos requires extensive micromanipulation of oocytes and somatic cells and exposure to otherwise noxious chemicals. Second, enucleation of an oocyte involves the physical removal of 5–15% or more of the ooplasm. Confirmation in doing so oftentimes involves exposure of the ooplasm to ultraviolet light which has been shown to alter membrane integrity, increase methionine uptake, alter protein synthesis and mitochondrial activity.71 Third, insertion of a somatic cell nucleus within egg cytoplasm generally involves electrical- induced fusion or microinjection.

Some degree of lysis occurs after either procedure indicating alterations in membrane integrity of the ooplasm.72,73 Fourth, a variety of protocols have been described for activating cloned embryos, with the majority involving exposure to otherwise noxious chemicals to initiate embryonic cell cycles.18,33,46 Fifth, resulting embryos are _nurtured_ in an otherwise hostile culture environment until transfer into surrogate recipients. Doing so may not be without consequence74 and may further exacerbate problems associated with SCNT. For example, in vitro production of embryos is associated with alterations in gene expression in early embryos and fetuses (cattle;75 sheep76), increased abortion (16%),53 hydrallantois (1%),52 heavier fetuses,77 dystocia,52,78 large offspring,59,79 and increased mortality (14.9%).53 Given the fact that cloned embryos generated after SCNT are not _true_ clones may further complicate matters. Oocyte cytoplasm contains mitochondrial DNA that is transmitted to the cloned embryo.80 Significance of uniparental inheritance of cytoplasmic organelles (maternal inheritance) remains unclear. In instances whereby nuclear and mitochondrial genes are mismatched, Nagao et al.81 reported decreased physical performance and growth rates in mice. It is possible that reconstructing cloned embryos with cytoplasts of non-descript origin results in the production of embryos whereby some degree of incompatibility exists between the cytoplasm and donor nucleus. Clearly, some of the problems associated with SCNT for producing live offspring may in part be due to incomplete reprogramming of the somatic nucleus by egg cytoplasm. As a consequence, expression of developmentally important genes as required for embryo and fetal development are altered in cloned embryos. Most recently, Yong-Kook et al.62 found highly aberrant methylation patterns in various genomic regions in bovine cloned embryos. Cloned blastocysts in this study closely resembled somatic cells in their overall genomic methylation status.

Similar findings were obtained by Dean et al.83 and Bourc'his et al.84 Moreover, analysis of various developmentally important genes in cloned embryos revealed abnormal expression in IL6, FGF4, and FGFr2,85 G6PD and Xist,67 and Oct4,86 to name a few. Furthermore, Wrenzycki et al.87 reported that alterations in gene expression occurring in cloned embryos may be dependent on the protocol used for SCNT. Finally, ability of the egg cytoplasm to effectively reprogram somatic nuclei may differ according to adult cell type. For instance, nuclei obtained from Sertoli and brain cells following microinjection into MII cytoplasts did not support development to term compared with cumulus cell nuclei (mice).12 Similar observations have been made in cattle.25,88

Immunologic Rejection of Nuclear Transfer Fetuses

The high rate of early embryonic mortality occurring during the first trimester of cloned pregnancies may be due in part to inappropriate expression of trophoblast major histocompatibility complex (MHC) class I antigens.70 In most species, down regulation of polymorphic MHC class I antigens on the trophoblast cells seems to be required for successful pregnancy. In cattle, trophoblast cells do not express class I antigens early in pregnancy. However, during the third trimester the trophoblast cells in the interplacentomal and arcade regions of the uterine/placental interface commence expression of class I antigens.89 Using immunohistochemistry, Hill et al.70 reported abnormal expression of MHC class I antigens by the trophoblast cells of 34–63-day-old cloned fetuses (n ¼ 8) when compared with age-matched controls. Moreover, the endometrial stroma of cloned pregnancies contained large numbers of lymphocyte aggregates and had an increased number of diffuse endometrial lymphocytes. Approximately 80% of the lymphocytes in the lymphoid aggregates were CD4+ helper T cells, the remaining cells consisted of about equal numbers of CD8+ cytotoxic T lymphocytes and B lymphocytes plus a very small number of c/d-T cells (Davies, unpublished).

In day-34 cloned pregnancies, the uterine stroma was already infiltrated with lymphocytes. Tight attachment and placentome development in cattle begin at approximately day 30 of pregnancy. Consequently, in surrogate recipients the maternal immune response to SCNT fetuses occurred early enough to perhaps inhibit placentome development. Immunologically mediated inhibition of placentome development could explain the types of placental abnormalities observed in SCNT pregnancies.51,57 It might seem odd that the primary cells responding to inappropriate expression of class I antigens on trophoblast cells are CD4+ helper T cells that recognise peptide antigens presented by MHC class II molecules. There is no information on the expression of B7-1 (CD80) and B7-2 (CD86), the costimulatory molecules required for activation of naive T cells, by bovine trophoblast cells. However, these costimulatory molecules are normally only expressed on professional antigen presenting cells.90 Furthermore, in other species trophoblast cells express novel members of the B7 family of costimulatory molecules but do not express B7-1 or B7-2.91–94 It would, therefore, be surprising if trophoblast cells directly activated naive cytotoxic T cells.

The primary mode of recognition of fetal class I antigens would be indirect recognition involving processing and presentation of trophoblast cell antigens by maternal antigen presenting cells, macrophages and dendritic cells. In cattle, this would be greatly facilitated by migration of binucleate trophoblast cells across the interface, their fusion with uterine epithelial cells, and ultimately the death of hybrid trinucleate cells within the maternal endometrium.95,96 If the binucleate cells present at the time of implantation were class I positive, as they appear to be in most SCNT pregnancies, class I antigens would be delivered directly to maternal antigen presenting cells for processing and presentation to the maternal immune system. It is likely that bot inflammatory Th1 cells and anti-fetal class I antibodies are produced in response to class I positive trophoblast cells. Hence, effector T cells and/or antibodies could interfere with placental development in SCNT pregnancies. One strategy for avoiding immunologically mediated rejection of SCNT fetuses is to create MHC class I compatible pregnancies. If a SCNT fetus expresses only class I antigens, also carried by the embryo transfer recipient then the surrogate recipient should be tolerant to the fetal class I antigens. Because of the extreme polymorphism of the MHC, few MHC compatible pregnancies are created by chance. Furthermore, it could be difficult to identify MHC compatible recipients for nuclear donor cell lines with rare MHC genotypes. Recently, a microarray based MHC typing system for cattle has been developed for matching donor cell lines and surrogate recipients (Davies, unpublished). However, thus far very few MHC compatible SCNT pregnancies have been studied. An additional strategy for obviating potential problems of trophoblast cell class I expression would be to induce the trophoblast cells of NT fetuses to shut off class I expression. This could involve either addition of a hormone, growth factor or cytokine to the medium used for in vitro culture or infusion of a mediator into the uteri of recipient cows. Currently, little is known about the regulation of class I gene expression in bovine trophoblast cells. Consequently, a better understanding of the normal mechanism by which class I expression is turned off will probably be required before headway is made with this approach. If somatic cell cloning is to move forward, gaining an understanding of trophoblast cell class I gene regulation should be a high priority.

The U.S. Food and Drug Administration (FDA) has issued three documents on the safety of animal cloning — a draft risk assessment; a proposed risk management plan; and a draft guidance for industry. The draft risk assessment finds that meat and milk from clones of adult cattle, pigs and goats, and their offspring, are as safe to eat as food from conventionally bred animals. The assessment was peer-reviewed by a group of independent scientific experts in cloning and animal health. They agreed with the methods FDA used to evaluate the data and the conclusions set out in the document. The proposed plan outlines measures that FDA might take to address the risks that cloning poses to animals involved in the cloning process. These risks all have been observed in other assisted reproductive technologies currently in use in common agricultural practices.

In the draft guidance, FDA does not recommend any special measures relating to human food use of offspring of clones of any species. Because of their cost and rarity, clones will be used as are any other elite breeding stock — to pass on naturally-occurring, desirable traits such as disease resistance and higher quality meat to production herds. Because clones will be used primarily for breeding, almost all of the food that comes from the cloning process is expected to be from sexually-reproduced offspring and descendents of clones, and not the clones themselves. Jake already wrote what I wanted to write, but here it is in a nutshell:

Meat and Milk Safety

Meat is meat. Milk is milk. Beef of any other name is still beef.

Cloned animals are not the same as genetically-modified animals. First, cloned animals are NOT identical to their parents. Second, there is no insertion of non-cow (or non-sheep, non-pig, etc) genes into these animals - no danger of aflatoxin, or peanut genes that can trigger allergies. Every protein in the steak of a cloned cow is still a typical cow protein. If you can normally eat beef, you can also eat cloned beef - there is NO chemical difference.

Ethics

Cloning animals can teach us a lot about genetics and development of animals with, probably, some practical applications down the line. I see no need - and apparently farmers don't either - for mass production of cloned animals. Only animals targeted to be cloned will be champion breeders. And there is no way that Thorougbred racehorses will ever be cloned - even assisted fertilization (i.e., artificial insemination) is illegal (i.e, the animal will not be included in the Stud Book). So, only a few champion breeders from a couple of species (cows, I guess, perhaps sheep and pigs) will be cloned. This is good - to keep this at a minimum, at least for now - as the process of cloning produces a lot of sickly offspring, which raises ethical questions in itself.

Agricultural Practice

We already have a virtual monoculture in both plant and animal production for food. Such lack of genetic variation is troublesome as a new diesase (or global warming) may quickly sweep through our herds and deplete our food supply very fast. Making the gene pool even more homogenous in order to raise the meat/milk productivity of our animals just a little bit does not, in my opinion, warrant a widespread use of cloning of domesticated animals. I'd rather support small farmers who purposefully keep rare, unusual breeds of animals like Old-Type Oldenburg horse, Curly Bashkir pony or Mangalitza pig - breeds that contain genes absent from our current gene pool of mass-produced animals and provide a reservoir of useful traits we may need in the future.

6

Transvaginal Oocyte Retrieval

Transvaginal oocyte retrieval (TVOR), also referred to as oocyte retrieval (OCR) or even simply egg collection, is a technique used in vitro fertilization (IVF) in order to remove oocytes from the ovary of the female, enabling fertilization outside the body. Transvaginal oocyte retrieval can be called transvaginal ovum retrieval if the oocytes have matured into ova, as is normally the case in IVF.

Procedure

TVOR is performed in an operating room with the (female) patient in the lithotomy position. Initially performed using a transvaginal needle guided by transabdominal ultrasonography, TVOR is currently performed with a transvaginal ultrasound transducer with an attached needle. Once the follicle of the ovary is lined up with the guiding line of the transducer, the needle is advanced through the vaginal wall into the follicle of the ovary.

Care is taken not to injure organs located between the vaginal wall and the ovary. Once the follicle is entered, suction is gently applied to aspirate follicular fluid and with it, hopefully, cellular material including the oocyte. Next, other follicles are aspirated. Once the ovarian follicles have been aspirated on one ovary, the needle is withdrawn, and the procedure repeated on the contralateral ovary. After completion, the needle is withdrawn, and hemostasis is achieved. The process takes about 20 minutes and is usually done under procedural sedation or general anesthesia.

The technique mostly succeeds after ovarian hyperstimulation, where oocytes are pharmacologically stimulated to mature. In addition, oocyte release triggering is usually performed to induce final maturation of oocytes. An agent used for this purpose is human chorionic gonadotropin (hCG). It would cause ovulation to occur between 38 and 40 hours after injection, but TVOR is typically performed at a time usually between 34 and 36 hours after hCG injection, that is, just prior to the expected rupture of the follicles. hCG injection confers a risk of ovarian hyperstimulation syndrome.

Complications

The retrieval can lead to complication resulting in injury to organs, bleeding, and infection. The aspirates are handed to the laboratory to examine it for oocytes. Among lean patients with polycystic ovary syndrome, ovarian hemorrhage after ultrasonographically guided transvaginal oocyte retrieval is a potentially catastrophic and not so rare complication. Anecdotal evidence suggests that certain airborne chemical contaminants and particles, especially volatile organic compounds (VOC), may be toxic to and impair the growth and development of embryos if present in sufficient concentrations in the ambient atmosphere of an IVF incubator.

History

This technique was first developed by Pierre Dellenbach and colleagues in Strasbourg, France, in 1984 and introduced into the United States at the Genetics and IVF Institute in Virginia in 1985 with the help of Deborah Costakos.

Egg and Sperm Preparation

In the laboratory, the identified eggs are stripped of surrounding cells and prepared for fertilization. An oocyte selection may be performed prior to fertilization to select eggs with optimal chances of successful pregnancy. In the meantime, semen is prepared for fertilization by removing inactive cells and seminal fluid in a process called sperm washing. If semen is being provided by a sperm donor, it will usually have been prepared for treatment before being frozen and quarantined, and it will be thawed ready for use.

Fertilization

The sperm and the egg are incubated together at a ratio of about 75,000:1 in the culture media for about 18 hours. In most cases, the egg will be fertilized by that time and the fertilized egg will show two pronuclei. In certain situations, such as low sperm count or motility, a single sperm may be injected directly into the egg using intracytoplasmic sperm injection (ICSI). The fertilized egg is passed to a special growth medium and left for about 48 hours until the egg consists of six to eight cells.

In gamete intrafallopian transfer, eggs are removed from the woman and placed in one of the fallopian tubes, along with the man's sperm. This allows fertilization to take place inside the woman's body. Therefore, this variation is actually an *in vivo* fertilization, not an *in vitro* fertilization.

Embryo Culture

Typically, embryos are cultured until having reached the 6–8 cell stage three days after retrieval. In many Canadian, American and Australian programmes, however, embryos are placed into an extended culture system with a transfer done at the blastocyst stage at around five days after retrieval, especially if many good-quality embryos are still available on day 3. Blastocyst stage transfers have been shown to result in higher pregnancy rates. In Europe, transfers after 2 days are common.

Culture of embryos can either be performed in an artificial culture medium or in an autologous endometrial coculture (on top of a layer of cells from the woman's own uterine lining). With artificial culture medium, there can either be the same culture medium throughout the period, or a *sequential system* can be used, in which the embryo is sequentially placed in different media. For example, when culturing to the blastocyst stage, one medium may be used for culture to day 3, and a second medium is used for culture thereafter. Single or sequential medium are equally effective for the culture of human embryos to the blastocyst stage. Artificial embryo culture media basically contain glucose, pyruvate, and energy-providing components, but addition of amino acids, nucleotides, vitamins, and cholesterol improve the performance of embryonic growth and development. Methods to permit dynamic embryo culture with fluid flow and embryo movement are also available. A new method in development uses the uterus as an incubator and the naturally occurring intrauterine fluids as culture medium by encapsulating the embryos in permeable intrauterine vessel.

Embryo Selection

Laboratories have developed grading methods to judge oocyte and embryo quality. In order to optimise pregnancy rates, there is significant evidence that a morphological scoring system is the best strategy for the selection of embryos. However, presence of soluble HLA-G might be considered as a second parameter if a choice has to be made between embryos of morphologically equal quality. Also, two-pronuclear zygotes (2PN) transitioning through 1PN or 3PN states tend to develop into poorer-quality embryos than those that constantly remain 2PN. More advanced methods of *embryo profiling* may also be performed in order to optimise embryo selection, as further described in the "expansions"-section below.

Embryo Transfer

Embryos are failed by the embryologist based on the amount of cells, evenness of growth and degree of fragmentation. The number to be transferred depends on the number available, the age of the woman and other health and diagnostic factors. In countries such as Canada, the UK, Australia and New

Zealand, a maximum of two embryos are transferred except in unusual circumstances. In the UK and according to HFEA regulations, a woman over 40 may have up to three embryos transferred, whereas in the USA, younger women may have many embryos transferred based on individual fertility diagnosis. Most clinics and country regulatory bodies seek to minimise the risk of pregnancies carrying multiples. As it is not uncommon for more implantations to take than desired, the next step faced by the expectant mother is that of selective abortion. The embryos judged to be the "best" are transferred to the patient's uterus through a thin, plastic catheter, which goes through her vagina and cervix. Several embryos may be passed into the uterus to improve chances of implantation and pregnancy.

Fresh Versus Frozen

Embryos can be either "fresh" from fertilized egg cells of the same menstrual cycle, or "frozen", that is they have been generated in a preceding cycle and undergone embryo cryopreservation, and are thawed just prior to the transfer. The outcome from using cryopreserved embryos has uniformly been positive with no increase in birth defects or development abnormalities, also between fresh versus frozen eggs used for intracytoplasmic sperm injection (ICSI). Children born from vitrified blastocysts have significantly higher birthweight than those born from non-frozen blastocysts.

Uterine Preparation

In the human, the uterine lining (endometrium) needs to be appropriately prepared so that the embryo(s) can implant. In a natural or stimulated cycle, the embryo transfer takes place in the luteal phase at a time where the lining is appropriately undeveloped in relation to the status of the present Luteinizing Hormone. In a cycle where a "frozen" embryo is transferred, the recipient woman could be given first estrogen preparations (about 2 weeks), then a combination of oestrogen and progesterone so that the lining becomes receptive for the embryo. The time of receptivity is the implantation window.

Limited evidence also supports removal of cervical mucus before transfer.

Timing

In stimulated cycles in human IVF, embryos are typically transferred 3 days after fertilization and may then be at the eight-cell stage, or they are transferred 2 to 3 days later when they have reached the blastocyst stage. Embryos who reach the day 3 cell stage can be tested for chromosal or specific genetic defects prior to possible transfer by preimplantation genetic diagnosis (PGD).

Monozygotic twinning is not increased after blastocyst transfer compared with cleavage-stage embryo transfer.

Procedure

The embryo transfer procedure starts by placing a speculum in the vagina to visualise the cervix, which is cleansed with saline solution or culture media. A soft transfer catheter is loaded with the embryos and handed to the clinician after confirmation of the patient's identity. The catheter is inserted through the cervical canal and advanced into the uterine cavity. There is good and consistent evidence of benefit in *ultrasound guidance*, that is, making an abdominal ultrasound to ensure correct placement, which is 1–2 cm from the uterine fundus. Anesthesia is generally not required.

Single embryo transfers in particular require accuracy and precision in placement within the uterine cavity. The optimal target for embryo placement, known as the maximal implantation potential (MIP) point, is identified using 3D/4D ultrasound. However, there is limited evidence that supports deposition of embryos in the midportion of the uterus. After insertion of the catheter, the contents are expelled and the embryos are deposited. Limited evidence supports making trial transfers before performing the procedure with embryos. After expulsion, the duration that the catheter remains inside the uterus has no effect on pregnancy rates. Limited evidence suggests avoiding negative pressure from the catheter after expulsion. After withdrawal, the catheter is handed to the embryologist, who inspects it for retained embryos. In the process of zygote intrafallopian transfer (ZIFT), eggs are removed from the woman, fertilized, and then placed in the woman's fallopian tubes rather than the uterus.

Embryo Number

A major issue is how many embryos should be transferred. Placement of multiple embryos carries the risk of multiple pregnancy. In the past, physicians have often placed too many embryos in the hope to establish a pregnancy. However, the rise in multiple pregnancies has led to a reassessment of this approach. Professional societies and in many countries, the legislature, have issued guidelines or laws to curtail a practice of placing too many embryos in an attempt to reduce multiple pregnancies.

E-set

The technique of selecting only one embryo to transfer to the woman is called elective-Single Embryo Transfer (e-SET) or, when embryos are at the blastocyst stage, it can also be called *elective single blastocyst transfer (eSBT)*. It lowers the risk of multiple pregnancies, compared with e.g. Double Embryo Transfer (DET) or *double blastocyst transfer* (2BT), with a twinning rate of approximately 3.5% in sET compared with approximately 38% in DET, or 2% in eSBT compared with approximately 25% in 2BT. At the same time, pregnancy rates is not significantly less with eSBT than with 2BT. Furthermore, SET has better outcomes in terms of mean gestational age at delivery, mode of delivery, birthweight, and risk of neonatal intensive care unit necessity than DET. e-SET of embryos at the cleavage stage reduces the likelihood of live birth by 38% and multiple birth by 94%. Evidence from randomised, controlled trials suggests that increasing the number of e-SET attempts (fresh and/or frozen) results in a cumulative live birth rate similar to that of DET. The usage of single embryo transfer is highest in Sweden (69.4%), but as low as 2.8% in the USA. Access to public funding for ART, availability of good cryopreservation facilities and legislation appear to be the most important factors for regional usage of single embryo transfer. Also, personal choice plays a significant role as many subfertile couples have a strong preference for twins.

Follow-up

Patients usually start progesterone medication after egg (also called oocyte) retrieval. While daily intramuscular injections of progesterone-in-oil (PIO) have been the standard route of administration, PIO injections are not FDA-approved for use in pregnancy. A recent meta-analysis showed that the intravaginal route with an appropriate dose and dosing frequency is equivalent to daily intramuscular injections. In addition, a recent case-matched study comparing vaginal progesterone with PIO injections showed that live birth rates were nearly identical with both methods. A duration of progesterone administration of 11 days results in almost the same birth rates as longer durations. Patients are also given estrogen medication in some cases after the embryo transfer. Pregnancy testing is done typically two weeks after egg retrieval.

Third-party Reproduction

It is not necessary that the embryo transfer be performed on the female who provided the eggs. Thus another female whose uterus is appropriately prepared can receive the embryo and become pregnant. Embryo transfer may be used where a woman who has eggs but no uterus and wants to have a biological baby; she would require the help of a gestational carrier or surrogate to carry the pregnancy. Also, a woman who has no eggs but a uterus may resort to egg donor IVF, in which case another woman would provide eggs for fertilization and the resulting embryos are placed into the uterus of the patient. Fertilization may be performed using the woman's partner's sperm or by using donor sperm.

'Spare' embryos which are created for another couple undergoing IVF treatment but which are then surplus to that couple's needs may also be transferred. Embryos may be specifically created by using eggs and sperm from donors and these can then be transferred into the uterus of another woman. A surrogate may carry a baby produced by embryo transfer for another couple, even though neither she nor the 'commissioning' couple is biologically related to the child. Third party reproduction is controversial and regulated in many countries. Persons entering gestational surrogacy arrangements must make sense of an entirely new type of relationship that does not fit any of the traditional scripts we use to categorise relations as kinship, friendship, romantic partnership or market relations. Surrogates have the experience of carrying a baby that they conceptualise as not of their own kin, while intended mothers have the experience of waiting through nine months of pregnancy and transitioning to motherhood from outside of the pregnant body. This can lead to new conceptualisations of body and self.

History

The first transfer of an embryo from one human to another resulting in pregnancy was reported in July 1983 and subsequently led to the announcement of the first human birth February 3, 1984. This procedure was performed at the Harbor UCLA Medical Centre under the direction of Dr. John Buster and the University of California at Los Angeles School of Medicine. In the procedure, an embryo that was just beginning to develop was transferred from one woman in whom it had been conceived by artificial insemination to another woman who gave birth to the infant 38 weeks later. The sperm used in the artificial insemination came from the husband of the woman who bore the baby. This scientific breakthrough established standards and became an agent of change for women suffering from the afflictions of infertility and for women who did not want to pass on genetic disorders to their children. Donor embryo transfer has given women a mechanism to become pregnant and give birth to a child that will contain their husband's genetic makeup. Although donor embryo transfer as practiced today has evolved from the original non-surgical method, it now accounts for approximately 5% of in vitro fertilization recorded births. Prior to this, thousands of women who were infertile, had adoption as the only path to parenthood. This set the stage to allow open and candid discussion of embryo donation and transfer.

This breakthrough has given way to the donation of human embryos as a common practice similar to other donations such as blood and major organ donations. At the time of this announcement the event was captured by major news carriers and fuelled healthy debate and discussion on this practice which impacted the future of reproductive medicine by creating a platform for further advancements in woman's health. This work established the technical foundation and legal-ethical framework surrounding the clinical use of human oocyte and embryo donation, a mainstream clinical practice, which has evolved over the past 25 years. Building upon this groundbreaking research and since the initial birth announcement in 1984, well over 47,000 live births resulting from donor embryo transfer have been and continue to be recorded by the Centres for Disease Control(CDC) in the United States to infertile women, who otherwise would not have had children by any other existing method.

Embryo Transfer in Animals

Embryo transfer techniques allow top quality female livestock to have a greater influence on the genetic advancement of a herd or flock in much the same way that artificial insemination has allowed greater use of superior sires. ET also allows the continued use of animals such as competition mares to continue training and showing, while producing foals. The general epidemiological aspects of embryo transfer indicates that the transfer of embryos provides the opportunity to introduce genetic material into populations of livestock while greatly reducing the risk for transmission of infectious diseases. Recent developments in the sexing of embryos before transfer and implanting has great potential in the dairy and other livestock industries. Embryo transfer is also used in laboratory mice. For example, embryos of genetically modified strains that are difficult to breed or expensive to maintain may be stored frozen, and only thawed and implanted into a pseudopregnant dam when needed.

Success Rates

IVF success rates are the percentage of all IVF procedures which result in a favourable outcome. Depending on the type of calculation used, this outcome may represent the number of confirmed pregnancies, called the pregnancy rate or number of live births, called the live birth rate. Due to advancement in reproductive technology, the IVF success rates are substantially better today than they were just a few years ago. The most current data available in the United States a 2009 summary complied by the Society for Reproductive Medicine (SART) which reports the average national IVF success rates per age group using non-donor eggs.

	<35	*35-37*	*38-40*	*41-42*
Pregnancy Rate	47.6	38.9	30.1	20.5
Live Birth Rate	41.4	31.7	22.3	12.6

The live birth rates using donor eggs are also given by the SART and include all age groups using either fresh or thawed eggs.

	Fresh Donor Egg Embryos	*Thawed Donor Egg Embryos*
Live Birth Rate	55.1	33.8

In 2006, Canadian clinics reported an average pregnancy rate of 35%. A French study estimated that 66% of patients starting IVF treatment finally succeed in having a child (40% during the IVF treatment at the centre and 26% after IVF discontinuation). Achievement of having a child after IVF discontinuation was mainly due to adoption (46%) or spontaneous pregnancy (42%).

Live Birth Rate

The live birth rate is the percentage of all IVF cycles that lead to a live birth. This rate does not include miscarriage or stillbirth and multiple-order births such as twins and triplets are counted as one pregnancy. The live birth rate is calculated as follows:

Number of Live Births : Number of Procedures Performed

In 2006, Canadian clinics reported a live birth rate of 27%. Birth rates in younger patients were slightly higher, with a success rate of 35.3% for those 21 and younger, the youngest group evaluated. Success rates for older patients were also lower and decrease with age, with 37-year-olds at 27.4% and no live births for those older than 48, the oldest group evaluated.

Some clinics exceeded these rates, but it is impossible to determine if that is due to superior technique or patient selection, because it is possible to artificially increase success rates by refusing to accept the most difficult patients or by steering them into oocyte donation cycles (which are compiled separately). Further, pregnancy rates can be increased by the placement of several embryos at the risk of increasing the chance for multiples. Because not each IVF cycle that is started will lead to oocyte retrieval or embryo transfer, reports of live birth rates need to specify the denominator, namely IVF cycles started, IVF retrievals, or embryo transfers.

The Society for Assisted Reproductive Technology (SART) summarised 2008-9 success rates for US clinics for fresh embryo cycles that did not involve donor eggs and gave live birth rates by the age of the prospective mother, with a peak at 41.3% per cycle started and 47.3% per embryo transfer for patients under 35 years of age. IVF attempts in multiple cycles result in increased cumulative live birth rates. Depending on the demographic group, one study reported 45% to 53% for three attempts, and 51% to 71% to 80% for six attempts.

Pregnancy Rate

Pregnancy rate may be defined in various ways. In the United States, the pregnancy rate used by the Society for Assisted Reproductive Technology and the Centres for Disease Control (and appearing in the table in the Success Rates section above) are based on fetal heart motion observed in ultrasound examinations. However, some clinics may advertise pregnancy rates based on pregnancy tests which are less definitive, such as urine or blood tests for hCG. This is the rate most often published by fertility clinics because often at the time of publishing not all IVF patients have given birth. The pregnancy rate is calculated as follows:

Number pregnancies / Number of procedures performed (cycle start, egg retrieval or embryo transfer).

Success or Failure Factors

The main potential factors that influence pregnancy (and live birth) rates in IVF have been suggested to be maternal age, duration of infertility or subfertility, bFSH and number of oocytes, all reflecting ovarian function. Optimal woman's age is 23–39 years at time of treatment.

Stress

In a 2005 Swedish study, 166 women were monitored starting one month before their IVF cycles, and the results showed no significant correlation between psychological stress and IVF outcome. The study concluded with the recommendation to clinics that it might be possible to reduce the stress experienced by IVF patients during the treatment procedure by informing them of those findings. While psychological stress experienced during a cycle might not influence an IVF outcome, it is possible that the experience of IVF can result in stress that leads to depression. The financial consequences alone of IVF can influence anxiety and become overwhelming. However, for many couples, the alternative is infertility, and the experience of infertility itself can also cause extreme stress and depression.

Acupuncture

An increasing number of fertility specialists and centres offer acupuncture as a part of their IVF protocol. Limited but supportive evidence from clinical trials and case series suggests that acupuncture may improve the success rate of IVF and the quality of life of patients undergoing IVF and that it is a safe adjunct therapy. A systematic review and meta-analysis published in the British Medical Journal found that complementing the embryo transfer process with acupuncture was associated with significant and clinically relevant improvements in clinical pregnancy (where the expected number of patients needed to be treated to produce 1 additional pregnancy was 10), ongoing pregnancy (NNT 9), and live birth (NNT 9).

Acupuncture Mechanisms

Four mechanisms by which it has been suggested that acupuncture may improve IVF outcomes are

- Neuroendocrinological modulations
- Increased blood flow to uterus and ovaries
- Modulation in cytokines
- Reduction of stress, anxiety, and depression.

Electro-acupuncture in Oocyte Retrieval for IVF

Electro-acupuncture has been found to be a good alternative to conventional medical analgesia, it results in shorter hospitalisation times and lower costs.

Other Factors

Other determinants of outcome of IVF include:

- Tobacco smoking reduces the chances of IVF producing a live birth by 34% and increases the risk of an IVF pregnancy miscarrying by 30%.
- A body mass index (BMI) over 27 causes a 33% decrease in likelihood to have a live birth after the first cycle of IVF, compared to those with a BMI between 20 and 27. Also, pregnant women who are obese have higher rates of congenital abnormality, miscarriage, gestational diabetes, hypertension, thromboembolism and problems during delivery. Ideal body mass index is 19–30.
- Salpingectomy before IVF treatment increases chances for women with hydrosalpinges
- Success with previous pregnancy and/or live birth increases chances
- Low alcohol/caffeine intake increases success rate
- Level of DNA fragmentation as measured e.g. by Comet assay, advanced maternal age and semen quality.
- The number of embryos transferred in the treatment cycle.
- Other factors of semen quality for the sperm provider.
- Women with ovary-specific FMR1 genotypes including *het-norm/low* have significantly decreased pregnancy chances in IVF.
- Some studies also suggest the autoimmune disease may also play a role in decreasing IVF success rates by interfering with proper implantation of the embryo after transfer.

Aspirin, however, was shown by a meta-analysis to not improve pregnancy rates after IVF.

Risks

Possible risks may occur throughout the procedure, but the risk you may encounter depend on the specific step of the procedure. During ovarian stimulation, hyperstimulation syndrome may occur. This results in swollen, painful ovaries and some form of it (mild, moderate or severe) occurs in 30% of patients. Mild cases can be treated with over the counter medications and cases can be resolved in the absence of pregnancy. In moderate cases, ovaries swell and fluid accumulated in the abdominal cavities and may have symptoms of heartburn, gas, nausea or loss of appetite. In severe cases patients have sudden excess abdominal pain, nausea, vomiting and will result in hospitalisation. During egg retrieval, there's a small chance of bleeding, infection, and damage to surrounding structures like bowel and bladder (transvaginal ultrasound aspiration) as well as difficulty breathing, chest infection, allergic reactions to meds, or nerve damage (laproscopy). During embryo transfer, if more than one embryo is transferred there's always a risk of multiple pregnancy, infertile couples may see this is good news but there may be risk to the embryos and to the mother such as premature delivery. Ectopic pregnancy may also occur if a fertilized egg develops outside the uterus, usually in the fallopian tubes and requires immediate destruction of the fetus.

Complications

Multiple Births

The major complication of IVF is the risk of multiple births. This is directly related to the practice of transferring multiple embryos at embryo transfer. Multiple births are related to increased risk of pregnancy loss, obstetrical complications, prematurity, and neonatal morbidity with the potential for long term damage. Strict limits on the number of embryos that may be transferred have been enacted in some countries (e.g. Britain, Belgium) to reduce the risk of high-order multiples (triplets or more), but are not universally followed or accepted. Spontaneous splitting of embryos in the womb after transfer can occur, but this is rare and would lead to identical twins.

A double blind, randomised study followed IVF pregnancies that resulted in 73 infants (33 boys and 40 girls) and reported that 8.7% of singleton infants and 54.2% of twins had a birth weight of < 2,500 grams (5.5 lb). However, despite the risks, some recent evidence suggests that twins conceived by IVF actually have a 40% decreased outcome risk for complications than spontaneous twin conceptions. Experts believe that this could be due to high level of prenatal monitoring received from the moment of conception. Recent evidence also suggest that singleton offspring after IVF is at higher risk for lower birth weight for unknown reasons.

Birth Defects

The issue of birth defects has been a controversial topic in IVF. Many studies do not show a significant increase after use of IVF, and some studies suggest higher rates for ICSI, whereas others do not support this finding. In 2008, an analysis of the data of the National Birth Defects Study in the US found that certain birth defects were significantly more common in infants conceived through IVF, notably septal heart defects, cleft lip with or without cleft palate, esophageal atresia, and anorectal atresia; the mechanism of causality is unclear. A 2002 study, reviewing the birth records of all the IVF babies in Western Australia "has found IVF kids are twice as likely to have birth defects as normal births". The study allowed for the effects of multiple births and the mother's age.

The higher incidence of birth defects "up to one year of age" included: heart defects, "chromosomal abnormalities like Down syndrome, spina bifida, gastro-intestinal abnormalities, musculo-skeletal, dislocated hips," and club feet. The IVF children also were found to have a higher incidence of lower birth weights, pre-term births, and cerebral palsy. Japan's government prohibited the use of *in vitro* fertilization procedures for couples in which both partners are infected with HIV. Despite the fact that the ethics committees previously allowed the Ogikubo, Tokyo Hospital, located in Tokyo, to use *in vitro* fertilization for couples with HIV, the Ministry of Health, Labour and Welfare of Japan decided to block the practice. Hideji Hanabusa, the vice president of the Ogikubo Hospital, states that together with his colleagues, he managed to develop a method through which scientists are able to remove HIV from sperm.

Other Risks

Another risk of ovarian stimulation is the development of ovarian hyperstimulation syndrome, particularly if hCG is used to "trigger ovulation".

If the underlying infertility is related to abnormalities in spermatogenesis, it is plausible, but too early to examine that male offspring is at higher risk for sperm abnormalities. Behaviour and socioemotional functioning of children conceived by IVF is normal overall, according to studies on 9–18-year-old IVF children. A negative pregnancy test after IVF is associated with an increased risk for depression in women, but not with any increased risk of developing anxiety disorders. Pregnancy test results do not seem to be a risk factor for depression or anxiety among men.

Expansions

There are various expansions or additional techniques that can be applied in IVF, which are usually not necessary for the IVF procedure itself, but would be virtually impossible or technically difficult to perform without concomitantly performing methods of IVF.

Embryo Profiling

In addition to the methods already mentioned in the *embryo selection* above, there are more advanced techniques of *embryo profiling* that give more information to aid in the selection of the optimal embryo. Also, methods are emerging in making comprehensive analyses of up to entire genomes, transcriptomes, proteomes and metabolomes which may be used to score embryos by comparing the patterns with ones that have previously been found among embryos in successful versus unsuccessful pregnancies:

- In transcriptome evaluation, however, gene expression profiling studies of human embryos are limited due to legal and ethical issues. An alternative or complement is gene expression profiling of cumulus cells surrounding the oocyte and early embryo, or on granulosa cells. Profiling of cumulus cells can give valuable information regarding the efficiency of an ovarian hyperstimulation protocol, and may indirectly predict oocyte aneuploidy, embryo development and pregnancy outcomes, without having to perform any invasive procedure directly in the embryo.

- Proteome profiling of embryos can indirectly be evaluated by sampling of proteins found in the vicinity of embryos, thereby providing a non-invasive method of embryo profiling. Examples of protein markers evaluated in such profiling include CXCL13 and granulocyte-macrophage colony-stimulating factor, where lower protein amounts are associated with higher implantation rates.

Another level of opportunity can be achieved by having the evaluation of the embryo profile tailored to the maternal status in regard to, for example health or immune status, potentially further detailed by similar profiling of the maternal genome, transcriptome, proteome and metabolome. Two examples of proteins that may be included in maternal profiling are endometrium-derived stathmin 1 and annexin A2, whose down- and up-regulation, respectively, are associated with higher rates of successful implantation.

The end-point of interest in the evaluation of an embryo profile is not necessarily the success rate of the pregnancy at hand, but may also be the health of the potential person as a child and adult in the future, such as in detection of inheritable diseases by preimplantation genetic diagnosis (PGD) or screening (PGS):

Preimplantation Genetic Screening or Diagnosis (PGS or PGD)

Preimplantation genetic screening (PGS) or preimplantation genetic diagnosis (PGD) has been suggested to be able to be used in IVF to select an embryo that appears to have the greatest chances for successful pregnancy. However, a systematic review and meta-analysis of existing randomised controlled trials came to the result that there is no evidence of a beneficial effect of PGS as measured by live birth rate. On the contrary, for women of advanced maternal age, PGS significantly lowers the live birth rate. Technical drawbacks, such as the invasiveness of the biopsy, and non-representative samples because of mosaicism are the major underlying factors for inefficacy of PGS.

Still, as an expansion of IVF, patients who can benefit from PGS/PGD include:

- Couples who have a family history of inherited disease
- Couples who want to use gender selection to prevent a gender-linked disease
- Couples who already have a child with an incurable disease and need compatible cells from a second healthy child to cure the first, resulting in a “saviour sibling” that matches the sick child in HLA type.

PGS screens for numeral chromosomal abnormalities while PGD diagnosis the specific molecular defect of the inherited disease. In both PGS and PGD, individual cells from a pre-embryo are analysed during the IVF process. Before the transfer of a pre-embryo back to a woman's uterus, one or two cells are removed from the pre-embryos (8-cell stage). These cells are then evaluated for normality. Typically within one to two days, following completion of the evaluation, only the normal pre-embryos are transferred back to the woman's uterus. In addition, PGS can reduce the risk of multiple pregnancies because fewer embryos are needed for implantation.

Cryopreservation

Cryopreservation can be performed as oocyte cryopreservation before fertilization, or as embryo cryopreservation after fertilization. The Rand Consulting Group has estimated there to be 400,000 frozen embryos in the United States. The advantage is that patients who fail to conceive may become pregnant using such embryos without having to go through a full IVF cycle. Or, if pregnancy occurred, they could return later for another pregnancy. Spare oocytes or embryos resulting from fertility treatments may be used for oocyte donation or embryo donation to another woman or couple, and embryos may be created, frozen and stored specifically for transfer and donation by using donor eggs and sperm. Also, oocyte cryopreservation can be used for women who are likely to lose their ovarian reserve due to undergoing chemotherapy. The outcome from using cryopreserved embryos has uniformly been positive with no increase in birth defects or development abnormalities.

Embryo Cryopreservation

Cryopreservation of embryos is the process of preserving an embryo at sub-zero temperatures, generally at an embryogenesis stage corresponding to pre-implantation, that is, from fertilization to the blastocyst stage.

Indications

Embryo cryopreservation is useful for leftover embryos after a cycle of in vitro fertilization, as patients who fail to conceive may become pregnant using such embryos without having to go through a full IVF cycle. Or, if pregnancy occurred, they could return later for another pregnancy. Spare oocytes or embryos resulting from fertility treatments may be used for oocyte donation or embryo donation to another woman or couple, and embryos may be created, frozen and stored specifically for transfer and donation by using donor eggs and sperm.

Prevalence

World usage data is hard to come by but it was reported in a study of 23 countries that almost 42,000 frozen human embryo transfers were performed during 2001 in Europe.

Pregnancy Outcome and Determinants

The outcome from using cryopreserved embryos has uniformly been positive with no increase in birth defects or development abnormalities, also between fresh versus frozen eggs used for intracytoplasmic sperm injection (ICSI). Children born from vitrified blastocysts have significantly higher birthweight than those born from non-frozen blastocysts. For early cleavage embryos, frozen ones appear to have at least as good obstetric outcome, measured as preterm birth and low birthweight for children born after cryopreservation as compared with children born after fresh cycles. Oocyte age, survival proportion, and number of transferred embryos are predictors of pregnancy outcome. Pregnancies have been reported from embryos stored for 16 years. A study of more than 11,000 cryopreserved human embryos showed no significant effect of storage time on post-thaw survival for IVF or oocyte donation cycles, or for embryos frozen at the pronuclear or cleavage stages. In addition, the duration of storage had no significant effect on clinical pregnancy, miscarriage, implantation, or live birth rate, whether from IVF or oocyte donation cycles.

Legislation

From October 1, 2009 human embryos are allowed to be stored for 10 years in the UK, according to the Human Fertilization and Embryology Act 2008.

History

The first ever pregnancy derived from a frozen human embryo was reported by Alan Trounson & Linda Mohr in 1983 (although the fetus aborted spontaneously at about 20 weeks of gestation); the first term pregnancies derived from frozen human frozen freezing process was born in 1984. Since then and up to 2008 it is estimated that between 350,000 and half a million IVF babies have been born from embryos frozen at a controlled rate and then stored in liquid nitrogen; additionally a few hundred births have been born from vitrified oocytes but firm figures are hard to come by. It may be noted that Subash Mukhopadyay from Kolkata, India reported the successful cryopreservation of an eight cell embryo, storing it for 53 days, thawing and replacing it into the mother's womb, resulting in a successful and live birth as early as 1978- a full five years before Trounson and Mohr had done so. A small publication of Mukherjee in 1978 clearly shows that Mukherjee was on the right line of thinking much before anyone else had demonstrated the successful outcome of a pregnancy following the transfer of a 8-cell frozen-thawed embryo into human subjects transferring 8-cell cryopreserved embryos."

Other Expansions

- Intracytoplasmic sperm injection (*ICSI*) is where a single sperm is injected directly into an egg. Its main usage as an expansion of IVF is to overcome male infertility problems, although it may also be used where eggs cannot easily be penetrated by sperm, and occasionally in conjunction with sperm donation. It can be used in teratozoospermia, since once the egg is fertilized abnormal sperm morphology does not appear to influence blastocyst development or blastocyst morphology.

- Assisted zona hatching (AZH) can be performed shortly before the embryo is transferred to the uterus. A small opening is made in the outer layer surrounding the egg in order to help the embryo hatch out and aid in the implantation process of the growing embryo.
- In egg donation and embryo donation, the resultant embryo after fertilization is inserted in another woman than the one providing the eggs. These are resources for women with no eggs due to surgery, chemotherapy, or genetic causes; or with poor egg quality, previously unsuccessful IVF cycles or advanced maternal age. In the egg donor process, eggs are retrieved from a donor's ovaries, fertilized in the laboratory with the sperm from the recipient's partner, and the resulting healthy embryos are returned to the recipient's uterus.
- Embryo splitting can be used for twinning to increase the number of available embryos.

Leftover Embryos or Eggs

There may be leftover embryos or eggs from IVF procedures if the woman for whom they were originally created has successfully carried one or more pregnancies to term. With the woman's or couple's permission, these may be donated to help other women or couples as a means of third party reproduction.

In embryo donation, these extra embryos are given to other couples or women for transfer with the goal of producing a successful pregnancy. The resulting child is considered the child of the woman who carries it and gives birth, and not the child of the donor, the same as occurs with egg donation or sperm donation. Typically, genetic parents donate the eggs to a fertility clinic or embryo bank where they are preserved by oocyte cryopreservation or embryo cryopreservation until a carrier is found for them. Typically the process of matching the embryo(s) with the prospective parents is conducted by the agency itself, at which time the clinic transfers ownership of the embryos to the prospective parents. In the United States, women seeking to be an embryo recipient undergo infectious disease screening required by the U.S. Food and Drug Administration (FDA), and reproductive tests to determine the best placement location and cycle timing before the actual Embryo Transfer occurs. The amount of screening the embryo has already undergone is largely dependent on the genetic parents' own IVF clinic and process. The embryo recipient may elect to have her own embryologist conduct further testing.

Alternatives to donating unused embryos are destroying them (or having them implanted at a time where pregnancy is very unlikely), keeping them frozen indefinitely, or donating them for use in research (with results in their unviability). Individual moral views on disposing leftover embryos may depend on personal views on the beginning of human personhood and definition and/or value of potential future persons and on the value that is given to fundamental research questions. Some people believe donation of leftover embryos for research is a good alternative to discarding the embryos when patients receive proper, honest and clear information about the research project, the procedures and the scientific values).

History

There was a transient biochemical pregnancy reported by Australian Foxton School researchers in 1953. John Rock was the first to extract an intact fertilized egg. The first pregnancy achieved through *in vitro* human fertilization of a human oocyte was reported in The Lancet from the Monash University team in 1973, although it lasted only a few days and would today be called a biochemical pregnancy. There was also an ectopic pregnancy reported by Patrick Steptoe and Robert Edwards in 1976. In 1977, Steptoe and Edwards successfully carried out a pioneering conception which resulted in the birth of the world's first baby to be conceived by IVF, Louise Brown on 25 July 1978, in Oldham General Hospital, Greater Manchester, UK. In October 1978, it was reported that Subash Mukhopadyay, a relatively unknown physician from Kolkata, India was performing experiments on his own with primitive instruments and a household refrigerator and this resulted in a test tube baby, later named as "Durga" (alias Kanupriya Agarwal) who was born on October 3, 1978. However, state authorities prevented him from presenting his work at scientific conferences and, in the absence of scientific evidence, his work is not recognised by the international scientific community.

Steptoe and Edwards were responsible for the world's second (confirmed) baby conceived by IVF, Alastair MacDonald born on 14 January 1979 in Glasgow. A team led by Ian Johnston and Alex Lopata were responsible for Australia's first baby conceived by IVF, Candice Reed born on 23 June 1980 in Melbourne. It was the subsequent use of stimulated cycles with clomiphene citrate and the use of human chorionic gonadotrophin (hCG) to control and time oocyte maturation, thus controlling the time of collection, that converted IVF from a research tool to a clinical treatment.

This was followed by a total of 14 pregnancies resulting in nine births in 1981 with the Monash University team. The Jones team at the Eastern Virginia Medical School in Norfolk, Virginia, further improved stimulated cycles by incorporating the use of a follicle-stimulating hormone (uHMG). This then became known as controlled ovarian hyperstimulation (COH). Another step forward was the use of gonadotrophin-releasing hormone agonists (GnRHA), thus decreasing the need for monitoring by preventing premature ovulation, and more recently gonadotrophin-releasing hormone antagonists (GnRH Ant), which have a similar function. The additional use of the oral contraceptive pill has allowed the scheduling of IVF cycles, which has made the treatment far more convenient for both staff and patients.

The ability to freeze and subsequently thaw and transfer embryos has significantly improved the feasibility of IVF use. The other very significant milestone in IVF was the development of the intracytoplasmic sperm injection (ICSI) of single sperms by André van Steirteghem in Brussels, 1992. This has enabled men with minimal sperm production to achieve pregnancies. ICSI is sometimes used in conjunction with sperm recovery, using a testicular fine needle or open testicular biopsy. Using this method, some men with Klinefelter's syndrome, and so would be otherwise infertile, have occasionally been able to achieve pregnancy. Thus, IVF has become the final solution for most fertility problems, moving from tubal disease to male factor, idiopathic subfertility, endometriosis, advanced maternal age, and anovulation not responding to ovulation induction. Robert Edwards was awarded the 2010 Nobel Prize in Physiology or Medicine "for the development of *in vitro* fertilization". Carl Wood was dubbed "the father of IVF (*in vitro* fertilization)" for having pioneered the use of frozen embryos. In the US, ART cycles started in 2006 resulted in 41,343 births (54,656 infants), which is slightly more than 1% of total US births.

Ethics

Issues

In a few cases, laboratory mix-ups (misidentified gametes, transfer of wrong embryos) have occurred, leading to legal action against the IVF provider and complex paternity suits. An example is the case of a woman in California who received the embryo of another couple and was notified of this mistake after the birth of her son. This has led to many authorities and individual clinics implementing procedures to minimise the risk of such mix-ups. The HFEA, for example, requires clinics to use a double witnessing system, where the identity of specimens is checked by two people at each point at which specimens are transferred. Alternatively, technological solutions are gaining favour, to reduce the manpower cost of manual double witnessing, and to further reduce the risk of human error. Technological solutions typically involve tagging individual specimen containers with uniquely numbered RFID tags which can be identified by readers connected to a computer. The computer tracks specimens throughout the process and alerts the embryologist if non-matching specimens are identified.

Another concern is that people will screen in or out for particular traits, using preimplantation genetic diagnosis. For example, a deaf British couple,

Tom and Paula Lichy, have petitioned to create a deaf baby using IVF. Some medical ethicists have been very critical of this approach. Jacob Appel wrote that "intentionally culling out blind or deaf embryos might prevent considerable future suffering, while a policy that allowed deaf or blind parents to select *for* such traits intentionally would be far more troublesome."

Pregnancy Past Menopause

Although menopause is a natural barrier to further conception, IVF has allowed women to be pregnant in their fifties and sixties. Women whose uterus has been appropriately prepared receive embryos that originated from an egg of an egg donor. Therefore, although these women do not have a genetic link with the child, they have an emotional link through pregnancy and childbirth. In many cases the genetic father of the child is the woman's partner. Even after menopause the uterus is fully capable of carrying out a pregnancy.

Same-sex Couples, Single and Unmarried Parents

A 2009 statement from the ASRM found no persuasive evidence that children are harmed or disadvantaged solely by being raised by single parents, unmarried parents, or homosexual parents. It did not support restricting access to assisted reproductive technologies on the basis of a prospective parent's marital status or sexual orientation. Ethical concerns include reproductive rights, the welfare of offspring, nondiscrimination against unmarried individuals, homosexual, and professional autonomy. A recent controversy in California focused on the question of whether physicians opposed to same-sex relationships should be required to perform IVF for a lesbian couple. Guadalupe T. Benitez, a medical assistant from San Diego, sued doctors Christine Brody and Douglas Fenton of the North Coast Women's Care Medical Group after Brody told her that she had "religious-based objections to treating homosexuals to help them conceive children by artificial insemination," and Fenton refused to authorise a refill of her prescription for the fertility drug Clomid on the same grounds. The California Medical Association had initially sided with Brody and Fenton, but the case, North Coast Women's Care Medical Group v. Superior Court, was decided unanimously by the California State Supreme Court in favor of Benitez on August 19, 2008.

Nadya Suleman came to international attention after having twelve embryos implanted, eight of which survived, resulting in eight newborns being added to her existing six-child family. The Medical Board of California sought to have fertility doctor Michael Kamrava, who treated Suleman, stripped of his license. State officials allege that performing Suleman's procedure is evidence of unreasonable judgment, substandard care, and a lack of concern for the eight children she would conceive and the six she was already struggling to raise. On June 1, 2011 the Medical Board issued a ruling that Kamrava's medical license be revoked effective July 1, 2011.

Catholic Objections

The Catholic Church opposes all kinds of *in vitro* fertilization because, as with contraception, it separates the procreative purpose of the marriage act from its unitive purpose: This particular doctrine, often expounded by the magisterium of the Church, is based on the inseparable connection, established by God, which man on his own initiative may not break, between the unitive significance and the procreative significance which are both inherent to the marriage act. The reason is that the fundamental nature of the marriage act, while uniting husband and wife in the closest intimacy, also renders them capable of generating new life—and this as a result of laws written into the actual nature of man and of woman. And if each of these essential qualities, the unitive and the procreative, is preserved, the use of marriage fully retains its sense of true mutual love and its ordination to the supreme responsibility of parenthood to which man is called. We believe that our contemporaries are particularly capable of seeing that this teaching is in harmony with human reason. According to the Catechism of the Catholic Church,

Techniques involving only the married couple (homologous artificial insemination and fertilization) are perhaps less reprehensible, yet remain morally unacceptable. They dissociate the sexual act from the procreative act. The act which brings the child into existence is no longer an act by which two persons give themselves to one another, but one that "entrusts the life and identity of the embryo into the power of doctors and biologists and establishes the domination of technology over the origin and destiny of the human person. Such a relationship of domination is in itself contrary to the dignity and equality that must be common to parents and children."

The Catholic Church maintains that it is not objectively evil to be infertile, and advocates adoption as an option for such couples who still wish to have children: The Gospel shows that physical sterility is not an absolute evil. Spouses who still suffer from infertility after exhausting legitimate medical procedures should unite themselves with the Lord's Cross, the source of all spiritual fecundity. They can give expression to their generosity by adopting abandoned children or performing demanding services for others.

Gamete Intrafallopian Transfer (GIFT) is not technically *in vitro* fertilization because with GIFT, fertilization takes place inside the body, not on a Petri dish. The Catholic Church nevertheless is concerned with it because "Some theologians consider this to be a replacement of the marital act, and therefore immoral."

Availability and Utilisation

In the USA, overall availability of IVF in 2005 was 2.5 IVF physicians per 100,000 population, and utilisation was 236 IVF cycles per 100,000. Utilisation highly increases with availability and IVF insurance coverage, and to a significant extent also with percentage of single persons and median income.

The cost of IVF rather reflects the costliness of the underlying healthcare system than the regulatory or funding environment, and ranges, on average for a standard IVF cycle and in 2006 United States dollars, between $12,500 in the United States to $4,000 in Japan. In Ireland, IVF costs around €4,000, with fertility drugs, if required, costing up to €3,000. The cost per live birth is highest in the United States ($41,000) and United Kingdom ($40,000) and lowest in Scandinavia and Japan (both around $24,500). Many fertility clinics in the United States limit the upper age at which women are eligible for IVF to 50 or 55 years. These cutoffs make it difficult for women older than fifty-five to utilise the procedure. In Australia, the average age of women undergoing ART treatment is 35.5 years among those using their own eggs (one in four being 40 or older) and 40.5 years among those using donated eggs.

Legal Status

Government agencies in China passed bans on the use of IVF in 2003 by unmarried women or by couples with certain infectious diseases. Sunni Muslim nations generally allow IVF between married couples when conducted with their own respective sperm and eggs, but not with donor eggs from other couples. But Iran, which is Shi'a Muslim, has a more complex scheme. Iran bars sperm donation but allows donation of both fertilized and unfertilized eggs. Fertilized eggs are donated from married couples to other married couples, while unfertilized eggs are donated in the context of mut'ah or temporary marriage to the father. The nation of Costa Rica has a complete ban on all IVF technology, it having been ruled unconstitutional by the nation's Supreme Court because it "violated life." Costa Rica is the only country in the western hemisphere that forbids this technique. The Inter-American Court on Human Rights ruled out that the Costa Rican government has violated the human rights of infertile couples when it banned the use of this technique. According to the rule, Costa Rica had to enact legislation allowing the use of this type of procedure. The Court set a first dateline on February 2011 for a report on the question by the Costa Rican government. The latter missed the dateline and new datelines were set.

All of them have been missed by the Costa Rican government. A law project sent reluctantly by the government of Pres. Laura Chinchilla was rejected at the Costa Rican parliament. Consequently, observers expect that Costa Rica will not abide to the request of the Inter-American Court on Human Rights. President Chinchilla, whose strong Catholic views have won her to be named officially as Preferred Daughter of the Virgin Mary has not publicly stated her position on the question of *in vitro* fertilization. However, given the massive influence of the Catholic Church in her government any change in the status quo seems very unlikely La Costa Rica católica se atasca con la fertilización in vitroCIDH Extends Deadline For Approval Of Law For In-Vitro

Fertilization In Costa Rica. Federal regulations in the United States include screening requirements and restrictions on donations, but generally do not affect sexually intimate partners. However, doctors may be required to *provide* treatments due to nondiscrimination laws, as for example in California.

All major restrictions on single but infertile women using IVF were lifted in Australia in 2002 after a final appeal to the Australian High Court was rejected on procedural grounds in the Leesa Meldrum case. A Victorian federal court had ruled in 2000 that the existing ban on all single women and lesbians using IVF constituted sex discrimination. Victoria's government announced changes to its IVF law in 2007 eliminating remaining restrictions on fertile single women and lesbians, leaving South Australia as the only state maintaining them. The US state of Tennessee proposed a bill in 2009 that would have defined donor IVF as adoption. During the same session another bill proposed barring adoption from any unmarried and cohabitating couple, and activist groups stated that passing the first bill would effectively stop unmarried people from using IVF. Neither of these bills passed.

Assisted Reproduction

Assisted reproductive technology (ART) is a general term referring to methods used to achieve pregnancy by artificial or partially artificial means. It is reproductive technology used primarily in infertility treatments. Some forms of *ART* are also used in fertile couples for genetic reasons. *ART* is also used in couples who are discordant for certain communicable diseases, e.g. AIDS, to reduce the risk of infection when a pregnancy is desired. The term includes any reproductive technique involving a third party e.g. a sperm donor. Examples of *ART* include in vitro fertilization, intracytoplasmic sperm injection (ICSI), cryopreservation, and intrauterine insemination (IUI). There is yet no strict definition of the term. Usage of the ART mainly belongs in the field of reproductive endocrinology and infertility.

Definitions

While there is no consensus on the definition, generally the process of intercourse is bypassed either by insemination (for example, artificial insemination) or fertilization of the oocytes in the laboratory environment (i.e., in vitro fertilization).

- The Centres for Disease Control and Prevention (CDC)—which is required as a result of the 1992 Fertility Clinic Success Rate and Certification Act to publish the annual ART success rates at U.S. fertility clinics—defines ART to include "all fertility treatments in which both eggs and sperm are handled. In general, ART procedures involve surgically removing eggs from a woman's ovaries, combining them with sperm in the laboratory, and returning them to the woman's body or donating them to another woman." According to CDC, "they do not include treatments in which only sperm are handled (i.e., intrauterine—or artificial—insemination) or procedures in which a woman takes medicine only to stimulate egg production without the intention of having eggs retrieved."

Procedures

Procedures are mainly fertility medication, as well as ART techniques that use more substantial and forceful interventions, of which in vitro fertilization (IVF) and expansions of it (e.g. OCR, AZH, ICSI, ZIFT) are the most prevalent. However, there are also other manual ART, not necessarily dependent on IVF (e.g. PGD, GIFT, SSR).

Medication

Most fertility medication are agents that stimulate the development of follicles in the ovary. Examples are gonadotropins and gonadotropin releasing hormone.

In vitro Fertilization

In vitro fertilization (IVF) is the technique of letting fertilization of the male and female gametes (sperm and egg) occur outside the female body.

Techniques usually used in vitro fertilization include:

- Transvaginal ovum retrieval (OCR) is the process whereby a small needle is inserted through the back of the vagina and guided via ultrasound into the ovarian follicles to collect the fluid that contains the eggs.
- Embryo transfer is the step in the process whereby one or several embryos are placed into the uterus of the female with the intent to establish a pregnancy.

Expansions of IVF

The following are techniques generally requires methods of in vitro fertilization. In vitro fertilization, however, usually does not require these techniques.

- Assisted zona hatching (AZH) is performed shortly before the embryo is transferred to the uterus. A small opening is made in the outer layer surrounding the egg in order to help the embryo hatch out and aid in the implantation process of the growing embryo.
- Intracytoplasmic sperm injection (ICSI) is beneficial in the case of male factor infertility where sperm counts are very low or failed fertilization occurred with previous IVF attempt(s). The ICSI procedure involves a single sperm carefully injected into the centre of an egg using a microneedle. This method is also sometimes employed when donor sperm is used.

- Autologous endometrial coculture is a possible treatment for patients who have failed previous IVF attempts or who have poor embryo quality. The patient's fertilized eggs are placed on top of a layer of cells from the patient's own uterine lining, creating a more natural environment for embryo development.
- In zygote intrafallopian transfer (ZIFT), egg cells are removed from the woman's ovaries and fertilized in the laboratory; the resulting zygote is then placed into the fallopian tube.
- Cytoplasmic transfer is the technique in which the contents of a fertile egg from a donor are injected into the infertile egg of the patient along with the sperm.
- Egg donors are resources for women with no eggs due to surgery, chemotherapy, or genetic causes; or with poor egg quality, previously unsuccessful IVF cycles or advanced maternal age.

  In the egg donor process, eggs are retrieved from a donor's ovaries, fertilized in the laboratory with the sperm from the recipient's partner, and the resulting healthy embryos are returned to the recipient's uterus.
- Sperm donation may provide the source for the sperm used in IVF procedures where the male partner produces no sperm or has an inheritable disease, or where the woman being treated has no male partner.
- A gestational carrier is an option when a patient's medical condition prevents a safe pregnancy, when a patient has ovaries but no uterus due to congenital absence or previous surgical removal, and where a patient has no ovaries and is also unable to carry a pregnancy to full term.
- Preimplantation genetic diagnosis (PGD) involves the use of genetic screening mechanisms such as Fluorescent In Situ Hybridisation (FISH) or Comparative Genomic Hybridisation (CGH) to help identify genetically abnormal embryos and improve healthy outcomes.
- Embryo splitting can be used for twinning to increase the number of available embryos.

Others

The following Assisted Reproduction techniques don't necessarily involve IVF.

- In gamete intrafallopian transfer (GIFT) a mixture of sperm and eggs is placed directly into a woman's fallopian tubes using laparoscopy following a transvaginal ovum retrieval.

- Sex selection is the attempt to control the sex of offspring to achieve a desired sex. It can be accomplished in several ways, both pre- and post-implantation of an embryo, as well as at birth. Pre-implantation techniques include PGD, but also sperm sorting.
- Artificial insemination (AI) is when sperm is placed into a female's uterus (intrauterine) or cervix (intracervical) using artificial means rather than by natural copulation. N.B. This can be a very low-tech process, performed at home by the woman alone or with her partner.
  - Conception devices, such as a conception cap are used to aid conception by enhancing the natural process. Conception caps are used by placing semen into a small conception cap, then placing the cap onto the cervix. This holds the semen at the cervical os, protecting the semen from the acidic vaginal secretions and keeping it in contact with the cervical mucus.
  - Artificial insemination by donor is used in situations where the woman doesn't have a partner with functional sperm. Instead, a sperm donor supplies the sperm.
- Surrogacy, where a woman agrees to become pregnant and deliver a child for a contracted party. It may be her own genetic child, or a child conceived through in vitro fertilization or embryo transfer using another woman's ova.
- Reproductive surgery, treating e.g. fallopian tube obstruction and vas deferens obstruction, or reversing a vasectomy by a reverse vasectomy.
  - In surgical sperm retrieval (SSR) the reproductive urologist obtains sperm from the vas deferens, epididymis or directly from the testis in a short outpatient procedure.
- By cryopreservation, eggs, sperm and reproductive tissue can be preserved for later IVF.

Risks

The majority of IVF-conceived infants do not have birth defects. However, some studies have suggested that assisted reproductive technology is associated with an increased risk of birth defects.

In the largest U.S. study, which used data from a statewide registry of birth defects, 6.2% of IVF-conceived children had major defects, as compared with 4.4% of naturally conceived children matched for maternal age and other factors (odds ratio, 1.3; 95% confidence interval, 1.00 to 1.67).

The main risks are:

- Genetic disorders. DNA damage increases in e.g. IVF and ICSI, which is reflected e.g. by upregulation of the gene expression of HNRNPC in the placenta.
- Low birth weight. In IVF and ICSI, a risk factor is the decreased expression of proteins in energy metabolism; Ferritin light chain and ATP5A1.
- Preterm birth. Low birth weight and preterm birth are strongly associated with many health problems, such as visual impairment and cerebral palsy, and children born after IVF are roughly twice as likely to have cerebral palsy.

Other risk factors are:

- Membrane damage, which is contributed to or reflected by increased expression of the membrane fusion proteins NAPA and Annexin A3.

Sperm donation is an exception, with a birth defect rate of almost a fifth compared to the general population. It may be explained by that sperm banks accept only people with high sperm count. Current data indicate little or no increased risk for postpartum depression among women who use ART.

Usage

Assisted reproductive technology procedures performed in the U.S. has more than doubled since 10 years ago, with 140.000 procedures in 2006, resulting in 55.000 infants born. In Australia, 3.1 percent of babies now born are a result of ART.

Costs

United States of America

Not everyone in the U.S. has insurance coverage for fertility investigations and treatments. Many states are starting to mandate coverage, and the rate of utilisation is 277% higher in states with complete coverage. There are some health insurance companies that cover diagnosis of infertility but frequently once diagnosed will not cover any treatment costs. 2005 approximate treatment/ diagnosis costs (United States, costs in US$):

- Initial workup: hysteroscopy, hysterosalpingogram, blood tests ~$2,000
- Intrauterine Insemination (IUI) aka Artificial insemination ~ $200–900 per. trial
- Sonohysterogram (SHG) ~ $600 – 1,000
- Clomiphene citrate cycle ~ $ 200 - 500
- IVF cycle ~ $10,000 -30,000
- Use of a surrogate mother to carry the child - dependent on arrangements.

Another way to look at costs is to determine the expected cost of establishing a pregnancy. Thus if a clomiphene treatment has a chance to establish a pregnancy in 8% of cycles and costs $500, the expected cost is $6,000 to establish a pregnancy, compared to an IVF cycle (cycle fecundity 40%) with a corresponding expected cost of $30,000 ($12,000/.4). For the community as a whole, the cost of IVF on average pays back by 700% by tax from future employment by the conceived human being.

United Kingdom

In the UK all patients have the right to preliminary testing, provided free of charge by the National Health Service. However, treatment is not widely available on the NHS and there can be long waiting lists. Many patients therefore pay for immediate treatment within the NHS or seek help from private clinics.

Sweden

In Sweden, official fertility clinics provide most necessary treatments and initial workup, but there are long waiting lists, especially for egg donations, since the donor gets just as low reward as the receiving couple are charged. However, there are private fertility clinics.

Canada

Some treatments are covered by OHIP (public health insurance) in Ontario and others are not. Those undergoing artificial insemination or with bilaterally blocked fallopian tubes and under 40 have treatment is covered but are still required to pay lab fees (around $3,000-4,000). Coverage veries in other provinces. Most other patients are required to pay for treatments themselves.

Israel

Israel's National Health Insurance, which is mandatory for all Israeli citizens, covers nearly all fertility treatments. In-Vitro-Fertilization costs are fully subsidized up to the birth of two children for all Israeli women, including single women and lesbian couples. Embryo transfers for purposes of gestational surrogacy are also covered.

New Zealand

The national public health system of New Zealand covers IVF treatment in specific circumstances only, based on a 'points for conception challenges' equation. Publicly funded IVF treatments are limited (between one and three treatments dependent on criteria) and are subject to substantial wait-lists, dependent on local health funding region, which raises potential inequity of ART support across the country. Infertility testing through blood tests can be covered by public funding, however in the absence of explicit gynecological complications, additional investigations are may not be covered publicly.

Investigation such as a hysterosalpingogram may be covered, but the wait-list could be in excess of six weeks, whereas a privately sourced HSG can cost $NZ900 but is readily available. Many New Zealanders select self-funded IVF cycles, at approximately $NZ10,000 per cycle, and other forms of ART, such as IUI, at approximately $NZ1200, using the services of private fertility clinics, which in itself is a growing local industry. Individuals using private services are generally not covered under personal health insurance policies in New Zealand.

Ethics

Some couples find it difficult to stop treatment despite very bad prognosis, resulting in futile therapies. This may give ART providers a difficult decision of whether to continue or refuse treatment.

Fictional Representation

Films and other fiction depicting emotional struggles of assisted reproductive technology have had an upswing in the latter part of the 2000s decade, although the techniques have been available for decades. Yet, the amount of people that can relate to it by personal experience in one way or another is ever growing, and the variety of trials and struggles are huge.

For specific examples, refer to the fiction sections in individual subarticles, e.g. surrogacy, sperm donation and fertility clinic.

In addition, reproduction and pregnancy in speculative fiction has been present for many decades.

Sperm Bank

A sperm bank, semen bank or cryobank is a facility that collects and stores human sperm mainly from sperm donors, primarily for the purpose of achieving pregnancies through third party reproduction, notably by artificial insemination. Sperm donated in this way is known as donor sperm.

The development of sperm banks, whilst considered by some to be controversial, has enabled people to have greater control over their reproductive lives. By providing sperm from donors who are checked and screened, thousands of women every year are able to bear their own children in circumstances where this might otherwise not be possible. Controversy stems from the fact that donors effectively father children for others, often in considerable numbers, and usually take no part in the upbringing of such children, and also from the fact that single women and coupled lesbians frequently use sperm banks in order to have their own biological children.

The increasing range of services which is available through sperm banks nevertheless enables more and more people to have choices over the whole issue of reproduction. Women may choose to use an anonymous donor who will not be a part of family life, or they may choose known donors who may be contacted later in life by the donor children. Women may choose to use a surrogate to bear their children, using eggs provided by the woman and sperm from a donor. Sperm banks often provide services which enable a woman to have subsequent pregnancies by the same donor, but equally, women may choose to have children by a number of different donors.

Sperm banks sometimes enable a woman to choose the sex of her child, enabling even greater control over the way families are planned. Sperm banks increasingly adopt a less formal approach to the provision of their services thereby enabling people to take a relaxed approach to their own individual requirements.

Men who choose to donate sperm through a sperm bank also have the security of knowing that they are helping women or childless couples to have children in circumstances where they, as the biological father, will not have any legal or other responsibility for the children produced from their sperm. Whether a donor is anonymous or not, this factor is important in allowing sperm banks to recruit sperm donors and to use their sperm to produce whatever number of pregnancies from each donor as are permitted where they operate, or alternatively, whatever number they decide.

However, in many parts of the world sperm banks are not allowed to be established or to operate. Sperm banks do not provide a cure for infertility in that it is the sperm donor who reproduces himself, not a partner of the recipient woman. Most societies are built upon the family model and sperm banks may be seen as a threat to this, particularly where a sperm bank makes its services available to unmarried women.

Where sperm banks are allowed to operate they are often controlled by local legislation which is primarily intended to protect the unborn child, but which may also provide a compromise between the conflicting views which surround their operation. A particular example of this is the control which is often placed on the number of children which a single donor may father and which may be designed to protect against consanguity: low limits are often provided which in reality appear more intended to placate those who would be opposed to sperm donors producing tens or scores of children as to provide a justifiable safeguard.

However, such legislation usually cannot prevent a sperm bank from supplying donor sperm outside the jurisdiction in which they operate, and neither can it prevent sperm donors from donating elsewhere during their lives. There is an acute shortage of sperm donors in many parts of the world and there is obvious pressure from many quarters for donor sperm from those willing and able to provide it to be made available as safely and as freely as possible.

Services

Storage

The sperm is stored in small vials or straws of holding between 0.4 and 1.0 ml and cryogenically preserved in liquid nitrogen tanks. It has been proposed that there should be an upper limit on how long frozen sperm can be stored, however a baby has been conceived in the UK using sperm frozen for 21 years. Before freezing, sperm may be prepared so that it can be used for intracervical insemination (ICI), intrauterine insemination (IUI) or for IVF (or assisted reproduction) (ART).

Use

Sperm supplied by a sperm bank may be used where a woman's partner is infertile or where he carries genetic disease. Increasingly, donor sperm is used to achieve a pregnancy where a woman has no male partner, including among lesbian and bisexual mothers-to-be. Sperm from a sperm donor may also be used in surrogacy arrangements and for creating embryos for embryo donation. Donor sperm may be supplied by the sperm bank directly to the recipient to enable a woman to perform her own artificial insemination which can be carried out using a needle-less syringe or a cervical cap conception device. The cervical cap conception device allows the donor semen to be held in place close to the cervix for between six to eight hours to allow fertilization to take place. Alternatively, donor sperm can be supplied by a sperm bank through a registered medical practitioner who will perform an appropriate method of insemination or IVF treatment using the donor sperm in order for the woman to become pregnant. From a medical perspective, a pregnancy achieved using donor sperm is no different from a pregnancy achieved using partner sperm, and it is also no different from a pregnancy achieved by sexual intercourse. Sperm banks may supply other sperm banks or a fertility clinic with donor sperm to be used for achieving pregnancies. Sperm banks may also supply sperm for research or educational purposes. In countries where sperm banks are allowed to operate, the sperm donor will not usually become the legal father of the children he produces as the result of the use of the sperm he donates, but he will be the 'biological father' of such children. In cases of surrogacy involving embryo donation, a form of 'gestational surrogacy', the 'commissioning mother' or the 'commissioning parents' will not be biologically related to the child and may need to go through an adoption procedure.

As with other forms of third party reproduction, the use of donor sperm from a sperm bank gives rise to a number of moral, legal and ethical issues. Men may also use a sperm bank to store their own sperm for future use particularly where they anticipate travelling to a war zone or having to undergo chemotherapy which might damage the testes.

Selection

Sperm banks make information available about the sperm donors whose donations they hold in the sperm bank to enable customers to select the donor whose sperm they wish to use. This information is often available by way of an online catalogue. A sperm bank will also usually have facilities to help customers to make their choice and they will be able to advise on the suitability of donors for individual donors and their partners.

Where the recipient woman has a partner she may prefer to use sperm from a donor whose physical features are similar to those of her partner. In many cases, the choice of a donor with the correct blood group will be paramount with particular considerations involving the use of sperm from donors with negative blood groups. If a surrogate is to be used, such as where the customer is not intending to carry the child, considerations about her blood group etc will also need to be taken into account. Information made available by a sperm bank will usually include the race, height, weight, blood group, health and eye colour of the donor. Sometimes information about his age, family history and educational achievements will also be given. Some sperm banks make a 'personal profile' of a donor available and occasionally more information may be purchased about a donor, either in the form of a DVD or in written form. Catalogs usually state whether samples supplied in respect of a particular donor have already given rise to pregnancies, but this is not necessarily a guide to the fecundity of the sperm since a donor may not have been in the program long enough for any pregnancies to have been recorded.

If a woman intends to have more than one child, she may wish to have the additional child or children by the same donor. Sperm banks will usually advise whether sufficient stocks of sperm are available from a particular donor for subsequent pregnancies, and they normally have facilities available so that the woman may purchase and store additional vials from that donor on payment of an appropriate fee. These will be stored until required for subsequent pregnancies or they may be onsold if they become surplus to the woman's requirements. The catalogue will also state whether samples of sperm are available for Intracervical Insemination ICI, IUI Intrauterine Insemination) or IVF use. ICI is for vaginal or cervical insemination and IUI are 'washed' samples prepared for intrauterine insemination where sperm is injected directly into the uterus. IVF treatment refers to 'in vitro' use, i.e. where fertilization takes place outside the body, literally 'in glass', and therefore to treatments involving donor sperm where an egg is removed from the body and an embryo is created which is then re-inserted, or inserted, into a woman. The preparation for IUI is normally done by a gradient centrifugation where prostaglandins and bacteria is removed from the plasma. Sperm prepared for [IVF] use normally contains a smaller sample of highly concentrated sperm.

Gender Selection

Some sperm banks enable recipients to choose the gender of their child. Although the methods used do not guarantee 100% success, the chances of being able to select the gender of a child are held to be considerably increased.

One of the processes used is the 'swim up' method, whereby a sperm extender is added to the donor's freshly ejaculated sperm and the test-tube is left to settle. After about half-an-hour, the lighter Y chromosome sperm will have swum to the top, leaving the heavier 'x' chromosome sperm at the bottom, thus allowing selection and storage according to gender.

The alternative process is the Percoll Method which is similar to the 'swim up' method but involves additionally the centrifuging of the sperm in a similar way to the washing of samples produced for IUI inseminations, or for IVF purposes. Gender selection is not permitted in a number of countries, including the UK.

Regulation

In the United States sperm banks are regulated by the FDA with new guidelines in effect May 25, 2005. There are also regulation in different states including New York and California. In the EU the sperm banks are regulated by the EU Tissue Directive in effect April 7, 2006.

Selection and Screening of Donors

Using anonymous donor sperm or sperm from a non-anonymous, also called known or identifiable sperm donor through a sperm bank is a safe and reliable method of achieving a pregnancy.

A sperm bank takes a number of steps to ensure the health and quality of the sperm which it supplies and it will inform customers of the checks which it undertakes, providing relevant information about individual donors. A sperm bank will usually guarantee the quality and number of motile sperm available in a sample after thawing. They will try to select men as donors who are particularly fertile and whose sperm will survive the freezing and thawing process. Samples are often sold as containing a particular number of motile sperm per millilitre, and different types of sample may be sold by a sperm bank for differing types of use, e.g. ICI or IUI.

A sperm bank recruits donors via advertising, often in colleges and in local newspapers, and also via the internet. A donor must be a fit healthy male, normally between 18 to 45 years of age, who is willing to undergo frequent and rigorous testing and who is willing to donate his sperm so that it can be used to impregnate women who are unrelated to, and unknown by, him. The donor must agree to relinquish all legal rights to all children which result from his donations. Where a man donates his sperm to a sperm bank, he will usually not be regarded as the legal father of children which are produced from his donations. The donor must produce his sperm at the sperm bank thus enabling the identity of the donor, once proven, always to be ascertained, and also enabling fresh samples of sperm to be produced for immediate processing.

Sperm is produced by way of masturbation. The sperm will be checked to ensure its fecundity and also to ensure that motile sperm will survive the freezing process. If a man is accepted onto the sperm bank's program as a sperm donor, his sperm will be constantly monitored, the donor will be regularly checked for infectious disease, and samples of his blood will be taken at regular intervals. A sperm bank may provide a donor with dietary supplements containing herbal or mineral substances such as maca, zinc, vitamin E and argenine which are designed to improve the quality and quantity of the donor's semen, as well as reducing the refractory time (i.e. the time between viable ejaculations). All sperm is frozen in straws or vials and stored for a minimum of 6 months before being released for sale and use to ensure that the donor remains healthy.

Donors are subject to tests for infectious diseases such as human immunoviruses HIV (HIV-1 and HIV-2), human T-cell lymphotropic viruses (HTLV-1 and HTLV-2), syphilis, chlamydia, gonorrhea, Hepatitis B virus, Hepatitis C virus, cytomegalovirus (CMV), Trypanosoma cruzi and Malaria as well as hereditary diseases such as cystic fibrosis, Sickle cell anemia, Familial Mediterranean fever, Gaucher's disease, Thalassaemia, Tay-Sachs disease, Canavan's disease, Familial dysautonomia, Congenital adrenal hyperplasia Carnitine transporter deficiency and Karyotyping 46XY. Karyotyping is not a requirement in neither EU nor the U.S. but some sperm banks choose to test donors as an extra service to the customer.

A sperm donor may also be required to produce his medical records and those of his family, often for several generations. A sperm sample is usually tested micro-biologically at the sperm bank before it is prepared for freezing and subsequent use. A sperm donor's blood group may also be registered to ensure compatibility with the recipient.

Bibliography

Ayala, F. J.: *The Genetic Structure of Populations*, W.H. Freeman & Co., San Francisco, California, 1977.

Beckner, M.: *The Biological Way of Thought*, University of California Press, Berkeley and Los Angeles, 1968.

Berg, Paul, and Singer, Maxine: *Dealing with Genes: The Language of Heredity. Mill Valley*, University Science Books, CA, 1992.

Brookes, Martin: *Get a Grip on Genetics*, *Time Life Books*, East Sussex, England, 1998.

Dando, M.: *Biological Warfare in the 21st Century: Biotechnology and the Proliferation of Biological Weapons*. London: Brassey's, 1994.

Darlington, C. D., and Mather, K.: *The Elements of Genetics*, London, Allen and Unwin, 1949.

Dawkins, R.: *The Selfish Gene*. Oxford: Oxford University Press, 1976.

DeVita, V.T.: *Cancer: Principles and Practice of Oncology*, Lippincott, Philadelphia, 2005.

Diaz, E.: *Microbial Biodegradation: Genomics and Molecular Biology,* Caister Academic Press, UK, 2008.

Fisher, R. A.: *The Genetical Theory of Natural Selection*, Clarendon Press, Oxford, 1930.

Fritz, A.: *International Classification of Diseases for Oncology*, World Health Organization, Geneva, 2000.

Gillis, Justin: *Drug Firms, Gene Labs to Map Genetic Code*, The Daily News, Longview, WA, 1999.

Glover, D. M., and Hames, B. D.: *Genes and Embryos*: New York, Oxford University Press, 1989.

Griffiths A.J.F.: *Introduction to Genetic Analysis*, W.H. Freeman and Company, New York, USA, 2005.

Hamerton, J. L.: *Human Cytogenetics*, Academic Press, New York, 1971.

Hartl D, Jones E: *Genetics: Analysis of Genes and Genomes,* Jones & Bartlett, 2005.

Hartl, Daniel L.: *Basic Genetics*, Jones and Bartlett Publishers, Boston, 1991.

Jonoska, N.: *Self-Assembling DNA Graphs, DNA-Based Computers VIII*, Springer-Verlag, Berlin, 2003.

Klotz, J. W.: *Genes, Genesis, and Evolution*: St. Louis, Mo., Concordia Publishing Co., 1970.

Lumsden C., and E. Wilson: *Genes, Mind and Culture: The Coevolutionary Process.* Cambridge, MA: Harvard University Press, 1981.

Mayr, E.: *Change of Genetic Environment and Evolution*, Allen and Unwin, London, 1954.

Nottingham, S.F.: *Eat Your Genes: How Genetically Modified Food is Entering Our Diet*, Zed Books, London, 1998.

Prakash, Chandra Prakash : *Cell Physiology*, Discovery, Delhi, 2010.

Robin Whyman: *Applied Organometallic Chemistry and Catalysis*, Oxford University Press, Delhi, 2005.

Roobeek, A: *The Biotechnology Revolution?,* Blackwell, Oxford, 1995.

Sturtevant, Alfred: *History of Genetics*, Harper and Row, New York, 1965.

Suzuki, D., and Knudtson, P.: *Genethics: The Clash Between the New Genetics and Human Values*, Harvard University Press, Cambridge, Mass, 1989.

Wilson, E. O., and Lumden, C.: *Genes, Mind, and Culture: The Evolutionary Process*, Harvard University Press, Cambridge, Mass, 1981.

Wilson, E. O., and Lumden, C.: *Genes, Mind, and Culture: The Evolutionary Process*: Cambridge, Mass., Harvard University Press, 1981.

Index

Encyclopaedia of

Advanced Genetic Analysis

Encyclopaedia of

Advanced Genetic Analysis

Volume 5

Editor:
Veena Kumari

AGROTECH PRESS

JAIPUR ■ NEW DELHI

AGROTECH PRESS

Regd. Off. "Bony Residency", Gate No. 2
Opp. Tilak Public School, Vishweswariya Nagar,
Gopalpura Road, Jaipur - 302018, India
Ph: 0141 - 2761280, Telefax: 2761381
E-mail : agrotechpress@gmail.com

Delhi Office: 102, 1st Floor, Satyam House,
4327/3, Ansari Road, Darya Ganj,
New Delhi - 110002, India
Ph: 011-45652440
visit us: www.agrotechpress.com

Distributed By
KEYA INTERNATIONAL PRESS LLP
E-mail: keyainternationalpressllp@gmail.com

Published in India, 2024

ISBN : 9789384568207 (5 Vols. Set)

Typeset by Thomson Press

Printed and bound in India by Thomson Press Ltd.,
New Delhi

Contents

Preface

Applied genetics is the manipulation; of the hereditary characteristics of an organism to improve or create specific traits in offspring. Controlled breeding involves manipulating the hereditary characteristics of offspring by selection parents with specific phenotypic traits. This process enables a breeder to develop new strains of a species or to maintain existing strains. Applying controlled breeding to plants and animals has traditionally been based upon three techniques: mass selection, inbreeding, and hybridization.

Mass selection is the process of choosing a few individuals. Inbreeding is based on the assumption that individuals that possess similar phenotypes also will possess genotypes. When phenotypically similar individuals are bred, offspring are more likely to be similar to their parents. In addition, offspring are likely to possess the same desired phenotype. Unfortunately, inbreeding can eventually produce weaker organisms because it increases the incidence of harmful homozygous recessive traits. In hybridization two different but related species or varieties of plants or animals are crossed. The products of this type of crossbreeding are called hybrids. Hybrids possess a different genotype and usually a phenotype different from that of either parent. When two different but closely linked species are crossed, the chromosome incompatibility during meiosis causes most of their hybrid offspring to be sterile. Some hybrid individuals grow faster and larger and ar healthier than either parent. Such an individual is said to display hybrid vigour. Hybridization can reverse the damaging effects of inbreeding and results in hybrid vigour in the offspring. This occurs because hybridization increases the number of heterozygous genes in an organism, thus reducing the likelihood that a harmful, recessive allele will be expressed. Other techniques in applied genetics manipulate the genes themselves. Among these techniques are induction of mutations, induction of polyploidy, and use of genetic engineering and gene splicing. Mutations occur at a very low rate in nature. However, in 1927 the American geneticist H. J. Muller found that he could induce a much greater rate of mutation in Drosophila by treating the flies with X-ray. Breeders can also introduce desirable traits into a species by inducing polyploidy, a conditioning which cells contain multiple, complete sets of chromosomes. Although polyploidy is rare and usually lethal in animals, it often occurs naturally in plants. Plant breeders artificially induce polyploidy by administering colchicine, a chemical that prohibits the formation of the cell plate during cell division. Colchicine is usually applied by placing the roots of the plant in a colchicine solution. Chromosomes in the cells of the plant go through the phases of cell division, but the cell plate does not form. Therefore, two complete sets of chromosomes exist. When the plant is removed from the colchicine solution the cells will again form cell plates during mitosis. Thus all resulting cells will contain an extra set of chromosomes. Genetic engineering, in which scientists directly manipulate genes. Genetic engineering frequently involves the use of recombinant DNA, which is composed of DNA segments from at least two differing organisms. Genetic engineering has many commercial applications. An example is the use of recombinant DNA technology to make interferon, a protein produced naturally by the human body to inhibit the reproduction of viruses. Isolating the human interferon gene. Genetic engineers use restriction enzymes, proteins that cur a DNA molecule into pieces. The restriction enzyme EcoRI cuts DNA wherever the sequence C-T-T-A-A-G occurs. Other restriction enzymes cut DNA at different nucleotide sequences. By using the proper restriction enzyme scientists can cut the human interferon gene out of its chromosome. Once the gene for interferon is removed, it is separated from the rest of the DNA and then inserted into a strand of bacterial DNA. Gene splicing is the process by which a gene from one organism is placed into the DNA of another organism. In our example, the human interferon gene is placed into the DNA of E. coli, the common bacterium of the human intestine. In addition to a single, circular chromosome, E. coli contains a single, small ring of DNA. The plasmid ring is removed from the bacterium and then opened with a restriction enzyme. The human interferon gene and the bacterial plasmid have "sticky ends" unpaired bases at each end of the DNA segment, where they were cleaved by restriction enzymes. As the human DNA is spliced into the plasmid DNA, the unpaired bases of each bond readily. A newly formed plasmid contains both human and bacterial DNA. The process by which the human gene is replicated is called gene cloning. Gene sequencing is the process of determining the exact order of bases in a fragment of DNA.

This book provides a focused introduction to the basic genetic principles and presents the molecular aspects and latest applied areas in genetics in considerable detail. The simple and lucid presentation of complicated topics in basic genetics, molecular genetics, medical genetics and in the latest applied areas of genetics would prove to be very useful to students, researchers and teachers.

—Editor

1

Genetics and Society

Scientific advances and the ways they may be used are often unfamiliar to us and may even be frightening. Consider the possibility of genetically engineering humans to enhance desirable traits, or perhaps taking a test that provides you with a list of diseases you are likely to develop in the future.

New technologies such as these are often developed before we can integrate them into our lives in a meaningful way, thereby generating a need for open discussion of what is ethical, what society deems acceptable or unacceptable, and what legal protections are required to ensure that we are safe from harm.

Genetic Inequality: Human Genetic Engineering

Genes influence health and disease, as well as human traits and behaviour. Researchers are just beginning to use genetic technology to unravel the genomic contributions to these different phenotypes, and as they do so, they are also discovering a variety of other potential applications for this technology.

For instance, ongoing advances make it increasingly likely that scientists will someday be able to genetically engineer humans to possess certain desired traits. Of course, the possibility of human genetic engineering raises numerous ethical and legal questions.

Although such questions rarely have clear and definite answers, the expertise and research of bioethicists, sociologists, anthropologists, and other social scientists can inform us about how different individuals, cultures, and religions view the ethical boundaries for the uses of genomics. Moreover, such insights can assist in the development of guidelines and policies.

Testing for Traits Unrelated to Disease

Much of what we currently know about the ramifications of genetic self-knowledge comes from testing for diseases. Once disease genes were identified, it became much easier to make a molecular or cytogenetic diagnosis for many genetic conditions. Diagnostic testing supplies the technical ability to test presymptomatic, at-risk individuals and/or carriers to determine whether they will develop a specific condition.

This sort of testing is a particularly attractive choice for individuals who are at risk for diseases that have available preventative measures or treatments, as well as people who might carry genes that have significant reproductive recurrence risks. Indeed, thanks to advances in single-cell diagnostics and fertilization technology, embryos can now be created *in vitro*; then, only those embryos that are not affected by a specific genetic illness can be selected and implanted in a woman's uterus. This process is referred to as preimplantation genetic diagnosis.

For adult-onset conditions, ethical concerns have been raised regarding whether genetic testing should be performed if there is no cure for the disease in question. Many people wonder whether positive diagnosis of an impending untreatable disease will harm the at-risk individual by creating undue stress and anxiety. Interestingly, social science research has demonstrated that the answer to this question is both yes and no. It seems that if genetic testing shows that an individual is a carrier for a recessive disease, such as Tay-Sachs disease or sickle-cell anemia, this knowledge may have a negative impact on the individual's well-being, at least in the short term.

On the other hand, if predictive testing for an adult-onset genetic disorder such as Huntington's disease reveals that an at-risk individual will develop the disorder later in life, most patients report less preoccupation with the disease and a relief from the anxiety of the unknown (Taylor & Myers, 1997). For many people who choose to have predictive testing, gaining a locus of control by having a definitive answer is helpful. Some people are grateful for the opportunity to make life changes—for instance, travelling more, changing jobs, or retiring early—in anticipation of developing a debilitating condition later in their lives.

Of course, as genetic research advances, tests are continually being developed for traits and behaviours that are not related to disease. Most of these traits and behaviours are inherited as complex conditions, meaning that multiple genes and environmental, behavioural, or nutritional factors may contribute to the phenotype. Currently, available tests include those for eye colour, handedness, addictive behaviour, "nutritional" background, and athleticism. But does knowing whether one has the genetic background for these nondisease traits negatively affect one's self-concept or health perception? Studies are now beginning to address this question.

For example, one group of scientists performed genetic testing for muscle traits on a group of volunteers enrolled in a resistance-training program (Gordon *et al.*, 2005). These tests looked for single-nucleotide polymorphisms that would tell whether an individual had a genetic predisposition for muscle strength, size, and performance.

The investigators found that if the individuals did not receive affirmative genetic information regarding muscle traits, they credited the positive effects of the exercise program to their own abilities. However, those study participants who did receive positive test results were more likely to view the beneficial changes as out of their control, attributing any such changes to their genetic makeup. Thus, a lack of genetic predisposition for muscle traits actually gave subjects a sense of empowerment.

The results of the aforementioned study may be surprising to many people, as one major concern associated with testing for nondisease traits is the fear that those people who do not possess the genes for a positive trait may develop a negative self-image and/or inferiority complex. Another matter bioethicists often consider is that people may discover that they carry some genes associated with physiological or behavioural traits that are frequently perceived as negative.

Moreover, many critics fear that the prevalence of these traits in certain ethnic populations could lead to prejudice and other societal problems. Thus, rigorous social science research by individuals from diverse cultural backgrounds is crucial to understanding people's perceptions and establishing appropriate boundaries.

Building Better Athletes with Gene Doping

Double muscled animals have an increase in muscle mass of up to 20% greater than normal animals. The increased muscle is due to the fact that these animals have a mutation in a specific gene that normally is involved in muscular hypertrophy.

Over the years, the desire for better sports performance has driven many trainers and athletes to abuse scientific research in an attempt to gain an unjust advantage over their competitors. Historically, such efforts have involved the use of performance-enhancing drugs that were originally meant to treat people with disease.

This practice is called doping, and it frequently involved such substances as erythropoietin, steroids, and growth hormones (Filipp, 2007). To control this drive for an unfair competitive edge, in 1999, the International Olympic Committee created the World Anti-Doping Agency (WADA), which prohibits the use of performance-enhancing drugs by athletes. WADA also conducts various testing programs in an attempt to catch those athletes who violate the anti-doping rules.

Today, WADA has a new hurdle to overcome—that of gene doping. This practice is defined as the nontherapeutic use of cells, genes, or genetic elements to enhance athletic performance. Gene doping takes advantage of cutting-edge research in gene therapy that involves the transfer of genetic material to human cells to treat or prevent disease (Well, 2008). Because gene doping increases the amount of proteins and hormones that cells normally make, testing for genetic performance enhancers will be very difficult, and a new race is on to develop ways to detect this form of doping (Baoutina *et al.*, 2008).

The potential to alter genes to build better athletes was immediately realised with the invention of so-called "Schwarzenegger mice" in the late 1990s. These mice were given this nickname because they were genetically engineered to have increased muscle growth and strength. The goal in developing these mice was to study muscle disease and reverse the decreased muscle mass that occurs with aging. Interestingly, the Schwarzenegger mice were not the first animals of their kind; that title belongs to Belgian Blue cattle, an exceptional breed known for its enormous muscle mass.

These animals, which arose via selective breeding, have a mutated and nonfunctional copy of the myostatin gene, which normally controls muscular development. Without this control, the cows' muscles never stop growing (Grobet *et al.*, 1997). In fact, Belgian Blue cattle get so large that most females of the breed cannot give natural birth, so their offspring have to be delivered by cesarean section. Schwarzenegger mice differ from these cattle in that they highlight scientists' newfound ability to induce muscle development through genetic engineering, which brings up the evident advantages for athletes. But does conferring one desirable trait create other, more harmful consequences? Are gene doping and other forms of genetic engineering something worth exploring, or should we, as a society, decide that manipulation of genes for nondisease purposes is unethical?

Creating Designer Babies

Genetic testing also harbors the potential for yet another scientific strategy to be applied in the area of eugenics, or the social philosophy of promoting the improvement of inherited human traits through intervention. In the past, eugenics was used to justify practices including involuntary sterilization and euthanasia. Today, many people fear that preimplantation genetic diagnosis may be perfected and could technically be applied to select specific nondisease traits (rather than eliminate severe disease, as it is currently used) in implanted embryos, thus amounting to a form of eugenics. In the media, this possibility has been sensationalized and is frequently referred to as creation of so-called "designer babies," an expression that has even been included in the *Oxford English Dictionary*. Although possible, this genetic technology has not yet been implemented; nonetheless, it continues to bring up many heated ethical issues.

Trait selection and enhancement in embryos raises moral issues involving both individuals and society. First, does selecting for particular traits pose health risks that would not have existed otherwise? The safety of the procedures used for preimplantation genetic diagnosis is currently under investigation, and because this is a relatively new form of reproductive technology, there is by nature a lack of long-term data and adequate numbers of research subjects. Still, one safety concern often raised involves the fact that most genes have more than one effect. For example, in the late 1990s, scientists discovered a gene that is linked to memory (Tang *et al.*, 1999).

Modifying this gene in mice greatly improved learning and memory, but it also caused increased sensitivity to pain (Wei *et al.*, 2001), which is obviously not a desirable trait. Beyond questions of safety, issues of individual liberties also arise. For instance, should parents be allowed to manipulate the genes of their children to select for certain traits when the children themselves cannot give consent? Suppose a mother and father select an embryo based on its supposed genetic predisposition to musicality, but the child grows up to dislike music. Will this alter the way the child feels about its parents, and vice versa? Finally, in terms of society, it is not feasible for everyone to have access to this type of expensive technology.

Thus, perhaps only the most privileged members of society will be able to have "designer children" that possess greater intelligence or physical attractiveness. This may create a genetic aristocracy and lead to new forms of inequality.

At present, these questions and conjectures are purely hypothetical, because the technology needed for trait selection is not yet available. In fact, such technology may be impossible, considering that most traits are complex and involve numerous genes. Still, contemplation of these and other issues related to genetic engineering is important should the ability to create genetically enhanced humans ever arise.

Forensics, DNA Fingerprinting, and CODIS

DNA is present in nearly every cell of our bodies, and we leave cells behind everywhere we go without even realising it. Flakes of skin, drops of blood, hair, and saliva all contain DNA that can be used to identify us. In fact, the study of forensics, commonly used by police departments and prosecutors around the world, frequently relies upon these small bits of shed DNA to link criminals to the crimes they commit.

This fascinating science is often portrayed on popular television shows as a simple, exact, and infallible method of finding a perpetrator and bringing him or her to justice. In truth, however, teasing out a DNA fingerprint and determining the likelihood of a match between a suspect and a crime scene is a complicated process that relies upon probability to a greater extent than most people realise. Government-administered DNA databases, such as the Combined DNA Index System (CODIS), do help speed the process, but they also bring to light complex ethical issues involving the rights of victims and suspects alike. Thus, understanding the ways in which DNA evidence is obtained and analysed, what this evidence can tell investigators, and how this evidence is used within the legal system is critical to appreciating the true ethical and legal impact of forensic genetics.

How Does DNA Identification Work?

Although the overwhelming majority of the human genome is identical across all individuals, there are regions of variation. This variation can occur anywhere in the genome, including areas that are not known to code for proteins. Investigation into these noncoding regions reveals repeated units of DNA that vary in length among individuals. Scientists have found that one particular type of repeat, known as a short tandem repeat (STR), is relatively easily measured and compared between different individuals. In fact, the Federal Bureau of Investigation (FBI) has identified 13 core STR loci that are now routinely used in the identification of individuals in the United States, and Interpol has identified 10 standard loci for the United Kingdom and Europe. Nine STR loci have also been identified for Indian populations.

As its name implies, an STR contains repeating units of a short (typically three- to four-nucleotide) DNA sequence. The number of repeats within an STR is referred to as an allele. For instance, the STR known as D7S820, found on chromosome 7, contains between 5 and 16 repeats of GATA. Therefore, there are 12 different alleles possible for the D7S820 STR. An individual with D7S820 alleles 10 and 15, for example, would have inherited a copy of D7S820 with 10 GATA repeats from one parent, and a copy of D7S820 with 15 GATA repeats from his or her other parent. Because there 12 different alleles for this STR, there are therefore 78 different possible genotypes, or pairs of alleles. Specifically, there are 12 homozygotes, in which the same allele is received from each parent, as well as 66 heterozygotes, in which the two alleles are different.

The Statistical Strength of a 13-STR Profile

Within the U.S., the 13-STR profile is a widely used means of identification, and this technology is now routinely employed to identify human remains, to establish or exclude paternity, or to match a suspect to a crime scene sample.

In order to utilise STR information as a means of human identification, the FBI established the frequency with which each allele of each of the 13 core STRs naturally occurs in people of different ethnic backgrounds. To this end, the FBI analysed DNA samples from hundreds of unrelated Caucasian, African American, Hispanic, and Asian individuals. Assuming that all 13 STRs follow the principle of independent assortment (and they should, as they are scattered widely across the genome) and that the population randomly mates, a statistical calculation based upon the FBI-determined STR allele frequencies reveals that the probability of two unrelated Caucasians having identical STR profiles, or so-called "DNA fingerprints," is approximately 1 in 575 trillion (Reilly, 2001).

This very small number needs to be put into perspective. Note that this figure refers to pairs of people, and there are many pairs of people in the world. Indeed, for the 100 million Caucasians in the world, there are 5,000 trillion pairs of people, so roughly eight or nine pairs would be expected to match at the 13 STR loci. This predicted matching does not specify which profile is shared by two people, and the chance that anyone matches the particular profile associated with a crime is still very small. The distinction between two people sharing a profile and one person having a particular profile is an example of the so-called "birthday problem." Here, the probability that a person has a particular birthday is 1 in 365, ignoring February 29, but there is a 50% chance that two people in a random group of 23 people have the same unspecified birthday (Weir, 2007).

DNA Extraction and Analysis

To perform a forensic DNA analysis, DNA is first extracted from a sample. Just one nanogram of DNA is usually a sufficient quantity to provide good data. The region containing each STR is then PCR amplified and resolved according to size, giving an overall profile of STR sizes (alleles). The 13 core STRs vary in length from 100 to 300 bases, allowing even partially degraded DNA samples to be successfully analysed. The costs of analysis, both in time and reagents, are significantly reduced by the amplification of all 13 STRs in just two multiplex PCR reactions.

Depending on the complexity of the repeat unit, the different alleles of an STR can vary by as little as a single nucleotide. For instance, the aforementioned D7S820 STR is relatively simple and contains between 5 and 16 repeats of GATA. Another STR, D21S11, has a more complex repeat pattern consisting of a mixture of tetra- and trinucleotides, and it therefore has alleles that vary in size by a single base pair. Because of the need to differentiate single-base differences, PCR products are typically resolved using automated DNA sequencing technologies with software that recognises allele patterns by comparison to a known "ladder."

Making an STR Match

In order to match, for example, crime scene evidence to a suspect, a lab would determine the allele profile of the 13 core STRs for both the evidence sample and the suspect's sample. If the STR alleles do not match between the two samples, the individual would be excluded as the source of the crime scene evidence.

***Table:** Example DNA Profiles Showing the STR Alleles for Each Sample and the Genotype Frequency of Suspect B for Each STR Locus*

STR Locus Sample	*Evidence Frequency*	*Suspect A*	*Suspect B Genotype*	*Suspect B's for Each STR*
D3S1358	15, 17	17, 17	15, 17	0.13
vWA	15, 16	18, 19	15, 16	0.22
FGA	23, 27	21, 23	23, 27	0.31
D8S1179	12, 13	14, 15	12, 13	0.34
D21S11	28, 30	27, 30.2	28, 30	0.06
D18S51	12, 18	14, 18	12, 18	0.11
D5S818	13, 13	9, 12	13, 13	0.29
D13S317	12, 12	12, 12	12, 12	0.21
D7S820	10, 11	9, 10	10, 11	0.26
CSF1PO	8, 11	11, 12	8, 11	0.18
TPOX	7, 8	8, 8	7, 8	0.30
THO1	9.3, 9.3	6, 9.3	9.3, 9.3	0.38
D16S539	9, 13	11, 12	9, 13	0.10

However, if the two samples have matching alleles at all 13 STRs, a statistical calculation would be made to determine the frequency with which this genotype is observed in the population. Such a probability calculation takes into account the frequency with which each STR allele occurs in the individual's ethnic group. Given the population frequency of each STR allele, a simple Hardy-Weinberg calculation gives the frequency of the observed genotype for each STR. Multiplying together the frequencies of the individual STR genotypes then gives the overall profile frequency. Suspect A is excluded as the source of the crime scene sample. Suspect B, on the other hand, matches the crime scene sample at all 13 STRs. A calculation of the frequency of Suspect B's genotype, based upon the STR allele frequencies within Suspect B's ethnic group, reveals that the likelihood that a random member of this ethnic group has this profile is about 1 in 1.5 billion.

It is important to understand that this number is the probability of seeing this DNA profile if the crime scene evidence did not come from the suspect but from some other person. To regard the number as the probability that the suspect is the source of the crime scene evidence is to commit the "prosecutor's fallacy"; this is the logical fallacy of equating "the probability that an animal has four legs if it is an elephant" (high) with "the probability that an animal is an elephant if it has four legs" (low). To go from one probability to another requires the use of Bayes' theorem and some prior (before the DNA profile) probabilities of the suspect being the source of the evidence. In addition, it is important to note that the probability of 1 in 1.5 billion is substantially increased if the actual source is a person related to the suspect, especially a sibling.

Confounding Circumstances of DNA Profiling

Sometimes, the DNA from crime scene evidence is in a very small quantity, poorly preserved, or highly degraded, so only a partial DNA profile can be obtained. When fewer than 13 STR loci are examined, the overall genotype frequency is higher, therefore making the probability of a random match higher as well. The likelihood of encountering this genotype would be roughly 1 in 331. In this instance, prosecutors would need additional types of evidence against Suspect B to convince a jury that he or she was the source of the evidence sample. In addition, if an individual happens to have STR alleles that are very common in his or her ethnic group, the genotype frequency can also be quite high, even when all of the core 13 STR loci are examined. It is also important to note that crime scene samples sometimes contain DNA from several different sources. This can make teasing out the sources of the DNA extremely difficult.

Databases of DNA Profiles

DNA evidence is used in court almost routinely to connect suspects to crime scenes, to exonerate people who were wrongly convicted, and to establish or exclude paternity. DNA data is considered to be more reliable than many other kinds of crime scene evidence. For this reason, tissue samples are frequently collected by law enforcement officials from those individuals who are implicated (even loosely) in a crime. The unique profile of each DNA sample is analysed for comparison to crime scene evidence. The DNA profile can also be stored in a database to compare with crime scene evidence from past and future crimes.

But under what circumstances should an individual be compelled to provide a sample for a DNA database? Originally, statutes mandating collection of tissue for DNA typing applied only to those people convicted of sex crimes or murder. This was due to the fact that there was usually an abundance of DNA at the scene of a rape or murder to compare to a suspect's DNA. More recent DNA collection laws have applied to all convicted felons, reflecting advances in DNA technologies that allow sufficient DNA samples to be obtained from scenes of more common crimes, such as burglary.

As of July 2008, all 50 states in the U.S. have laws requiring convicted sex offenders to submit DNA samples, and 46 states have laws that require other types of convicted felons to do so. Additionally, 11 states require DNA samples from those convicted of certain misdemeanors, and 12 states have laws authorising DNA sampling from arrestees of certain types of crime, usually felonies (National Conference of State Legislatures, n.d.). In comparison, in the United Kingdom, all arrestees, regardless of the degree of the charge, can be compelled to provide a sample. In fact, even those who are merely suspects in a crime can be forced to offer samples (Oak Ridge National Laboratory, n.d.).

Forced DNA Profiling

Those people opposed to DNA banking for law enforcement purposes note that arrestees are often found innocent of crimes. Retention of an innocent person's DNA can be seen as an intrusion of personal privacy and a violation of civil liberties. It is interesting to note that in the United States, under any other circumstance, the provision of a DNA sample would require informed consent and other protections for the donor.

In contrast, an arrestee's DNA profile, once entered into a database, can be accessed by police, forensic scientists, or researchers without the consent of the donor. Another problem with the DNA database system is an exacerbation of the ethnic bias already present in the criminal justice system. If people from one ethnic group are more often arrested, tried, and convicted of felonies, they will be overrepresented in the database, potentially leading to even more arrests within that ethnic group.

Proponents of DNA databanks argue that major crimes often involve people who have also committed other offenses. Having DNA banked could potentially make it easier to identify suspects. It could also significantly cut down the cost of an investigation if an automated computer search could eliminate suspects or link a suspect to a crime scene.

Does the DNA Databank System Help Solve Crimes?

The current DNA database maintained by the FBI, known as the Combined DNA Index System (CODIS), contains case samples (DNA samples from crime scenes or "rape kits") and individuals' samples (collected from convicted felons or arrestees) that are compared automatically by the system's software as new samples are entered. As of February 2007, CODIS had produced over 45,400 "hits," which assisted in more than 46,300 investigations (Federal Bureau of Investigation, n.d.). However, contrary to how DNA analysis is portrayed on popular television shows, DNA samples are not analysed within the course of an hour. Rather, the U.S. currently has an enormous backlog of samples waiting to be typed and entered into the database. Some of these samples are from cases that have outlasted their statutes of limitation, so even if these samples could help solve a crime, the crime can no longer be tried.

This delay brings up the dilemma of the validity of statutes of limitation. These statutes were established at a time when large quantities of physical evidence were required to match a suspect to a sample and when extended time periods significantly decreased law enforcement's ability to find a match, as well as the likelihood of successful prosecution. With the advent of DNA databanks and the possibility of storing samples indefinitely, the very notion of a statute of limitation now seems extremely outdated.

Of course, there are many other debatable issues concerning DNA banking. For instance, should the original tissue sample be stored indefinitely after the DNA profile has been entered into the database? Detractors note threats to genetic privacy, but proponents argue that future DNA typing methods will undoubtedly be developed and that old samples might have to be reanalysed using new techniques. Also at issue is the reopening of old cases on the basis of new (DNA-based) evidence. Which cases should be eligible for reanalysis in light of this new evidence? Can equitable rules be established to allow re-examination of cases that were analysed with less powerful lab techniques? Further public awareness of the power of DNA forensic technology will help lawmakers decide these issues in a way that seeks to strike a balance between protecting individuals' genetic privacy and protecting innocent citizens from crime.

DNA evidence is easy to obtain because genetic material is found in all human cells, save red blood cells. As a result, when we leave behind small biological bits of ourselves, these bits can be used to identify us and link us to the places we've been. With modern technology, the amount of DNA required for analysis can be obtained from even a miniscule biological sample, which allows police to match crime scene evidence with suspects. However, because forensics is a science largely rooted in probabilities, even a confirmed "match" does not supply concrete proof of guilt. In addition, DNA databases designed to simplify the process of connecting past offenders to recent crimes are fraught with concerns involving individual genetic rights, as well as problems related to delayed sample entry, both of which hinder the ultimate usefulness of these databases. As a result, even though forensics is undeniably important to the modern justice system, its personal ramifications and ethical questions are topics of continuing discussion within the scientific, law enforcement, and legal communities.

Paternity Testing: Blood Types and DNA

Occasionally, situations arise in which people require concrete, scientific evidence of parentage, whether it be their own or that of someone else. In most instances, maternity is easy to determine. Before surrogate motherhood became possible, the woman who gave birth to a child was obviously that child's gestational, genetic, and legal mother, and this continues to be true in the vast majority of cases today.

Unfortunately, questions of paternity aren't so easy to answer. In order to make a determination of fatherhood, scientists almost always work backwards—from the child to the potential parent—to ascertain the actual nature of the relationship. In the past, this typically involved identifying specific phenotypes (in particular, specific blood types) in the child and using this information to either "rule in" or "rule out" possible fathers. However, this system presented a number of problems, not the least of which was that it often yielded inconclusive results. Thus, since the 1990s, the more common approach has been to consider the presence of particular genotypic markers when attempting to establish fatherhood (and, in a handful of cases, motherhood).

Using Blood-Typing in Paternity Tests

The process of DNA fingerprinting was developed by Alec Jeffreys in 1984, and it first became available for paternity testing in 1988. Before this sort of DNA analysis was available, blood types were the most common factor considered in human paternity testing. Blood groups are a popular example of Mendelian genetics at work. After all, there are numerous human blood groups with multiple alleles, and these alleles exhibit a range of dominance patterns.

Today, the best-known blood-typing system is ABO typing, which involves the presence of antigens on red blood cells that are encoded by the ABO locus on human chromosome 9. In the ABO system, the A allele and the B allele are codominant, and the O allele is recessive. Thus, if a person's ABO blood type is O, he or she has two O alleles. If, however, a person's blood type is A, he or she has either two A alleles or one A allele and one O allele. Similarly, if a person has type B blood, this indicates the presence of either two B alleles or one B allele and one O allele. Finally, some people have type AB blood, which means they inherited both an A allele and a B allele.

In cases of questioned paternity, ABO blood-typing can be used to exclude a man from being a child's father. For example, a man who has type AB blood could not father a child with type O blood, because he would pass on either the A or the B allele to all of his offspring. Despite their usefulness in this regard, ABO blood groups cannot be used to confirm whether a man is indeed a child's father. Because of this and several other factors, it took the legal system some time to trust blood-typing. For example, in a famous case in 1943, the starlet Joan Barry accused actor Charlie Chaplin of fathering her child. Although blood tests definitively excluded Chaplin as the father, the court did not allow this evidence to be admitted, and Chaplin was ordered to pay child support to Barry. The Barry/Chaplin case did spur the passage of new laws, however, thus launching a new era in forensic evidence.

Over time, the use of additional blood antigens, such as those associated with the MN and Rh systems, refined the use of blood-typing for both paternity and forensics. However, such blood groups were only about 40% effective in ruling out a man as a child's father. Then, in the 1970s, testing for human leukocyte antigens (HLAs) added a distinguishing feature that made it possible to rule out men as fathers with 80% effectiveness.

The genes responsible for the HLA system are involved in antigen presentation to T cells. The HLA system is highly polymorphic, with more than 3,200 different alleles identified so far (Robinson *et al.*, 2003; Williams, 2001). Although this vast number of alleles causes headaches for cell and organ transplants, the multiplicity of genotypes the HLA system provides—in the tens of millions—makes it ideal for consideration in identity and paternity testing.

DNA Markers and Electrophoresis

This test includes samples from the mother (top row), the child (middle row), and the alleged father (bottom row). The maternal marker that has been passed to the child is 6. This means that the other marker present for the child (7) must have been inherited from the father. The alleged father matches the child, since one of his markers is indeed 7.

In the 1970s and 1980s, electrophoresis of various biochemical markers became widely available. With this process, proteins from a person's blood or other tissue were placed onto a gel, such as potato starch, agarose, or polyacrylamide. An electric current was then run through the gel, and different forms or isozymes of the proteins were separated by their electrical charge and/or size. Differences in isozymes relate to differences in the alleles that code for these proteins. Thus, the presence of certain identical isozymes in samples from both a child and his or her potential father could be used to reveal the existence of a genetic relationship between the two individuals. In fact, by 1974, Chakraborty *et al.* suggested that genetic testing via electrophoresis had advanced such that this method might be used to confirm paternity rather than merely exclude a man as a child's father.

Today, with the advent of numerous DNA sequencing, amplification, and testing techniques, paternity testing has evolved even further than predicted.

Indeed, present-day genetic testing has an accuracy rate of up to 99.99% (i.e., 9,999 out of 10,000). Of course, the exact level of accuracy depends on the number and quality of the genetic markers being considered. (It is important to emphasize that scientists consider only specific marker alleles, rather than entire genomes, when conducting paternity testing. Full genome analysis would add a great deal of time and expense to the process without significantly improving the accuracy of the results.) Thus, DNA-based forms of paternity testing have all but taken over earlier methods. In addition, higher throughput, better sensitivity, and automation have allowed DNA testing to be performed on ever-smaller and sometimes degraded DNA samples with greater speed and excellent accuracy.

The Utility of Paternity Testing

Interestingly, improvements in paternity testing over the past several decades have not only led to an increase in the accuracy of test results, but also to expanded application of various testing methods. For example, as DNA technology has gotten more precise, it has become possible to determine paternity using DNA from grandparents, cousins, or even saliva left on a discarded coffee cup. Such DNA testing is clearly an important part of criminal investigations, including forensic analysis, but it is also useful in civil courts when the paternity of a child is in question. The widespread availability of paternity tests on the web and in neighbourhood drugstores is also indicative of a civil demand for this technology.

However, it is important to note that such direct-to-consumer (DTC) tests will not stand up in court because there is no way to prove whose samples were analysed. Hence, DTC testing is most often used to assist in making a decision to initiate litigation or to simply provide peace of mind in matters of questionable paternity. In broader applications, advances in paternity testing mean that people who were adopted now have more direct means to confirm their biological identity or to find their birth parents. In addition, parentage testing is often an essential tool in proving immigration status in cases of family reunification.

Ethics of Genetic Testing: Medical Insurance and Genetic Discrimination

Many adult-onset genetic disorders are progressive (meaning that they get worse over time) and have long-term health consequences. Thus, when a person is at risk for such a disorder, he or she may consider undergoing genetic testing. When the consequences of the specific disorder in question are treatable, most people would agree that genetic testing makes sense. But what about disorders for which no preventative measures or treatments are available? What happens, for instance, when a young adult is found to have the mutation that causes a fatal disorder? Would such a person still be able to buy insurance or get the job promotion he or she applied for? How would this information affect his or her family and social life? As these questions illustrate, genetic testing for potentially lethal disorders is an area fraught with ethical, legal, and social concerns.

Understanding Huntington's Disease

One devastating genetic disorder that exemplifies these difficult ethical and social concerns is Huntington's disease (HD). Symptoms of HD, which include uncontrolled movements, loss of mental abilities, and emotional disturbances, are caused by the progressive death of neurons in certain areas of the brain. Death from related complications usually occurs 10 to 30 years following onset of symptoms. There is currently no treatment for this disease, nor is there anything an affected person can do to prevent the inevitable onset of symptoms.

HD is an autosomal dominant disorder with complete penetrance; therefore, any child of a person with HD has a 50% chance of developing the disease. The genetic basis of Huntington's, discovered in 1992, is an expansion of a trinucleotide repeat within the coding region of the huntingtin gene. Normally, this gene contains between 7 and 35 CAG repeats. In a person with HD, however, the gene contains more than 35 repeats. The age of onset of symptoms is inversely related to the number of repeats; generally, the more repeats that are present, the earlier the age of onset. A person with 40 CAG repeats can expect the onset of symptoms to occur anywhere from 40 to 70 years of age.

HD presents many unique challenges to affected families. Given the age of onset of symptoms, many individuals have already passed the disease to their children before their own diagnosis. Also, many people with HD have had the experience of seeing a parent, grandparent, or other family member suffer with the condition, due to the high penetrance and founder effect (meaning that *de novo* mutations are very rare) of HD. Genetic testing is available to those at risk for the disease and can indicate with certainty whether an individual is affected, but many people opt not to be tested. In fact, a 13-year Canadian study revealed that only 3% to 24% of at-risk individuals get predictive testing (Creighton *et al.*, 2003). This percentage is surprisingly low, especially given the fact that surveys taken shortly after predictive testing for HD became available indicated that between 66% and 79% of individuals at risk for the disorder would seek genetic testing to determine their status. What, then, might explain this discrepancy?

Reproductive and other Familial Concerns

The decision to undergo genetic testing for an inherited disease is an intensely personal choice, influenced by the perceived treatability and preventability of the disorder, as well as various reproductive implications.

In the case of HD, although there is no way to treat or prevent this condition, knowing one's status could influence many life decisions. For example, if you learned you had inherited the HD mutation, would you decide to have children of your own and risk passing the gene on to your children? Perhaps you would consider using assisted reproductive technology (ART), such as preimplantation genetic diagnosis, to make sure only unaffected embryos were allowed to develop to term.

But what if you already had children? Would you still get tested? What if test results indicated that you were likely to begin showing symptoms at age 40? How might this affect the way you live your life? Conversely, what if testing revealed that you had not inherited the mutation, but that your siblings and their children had? Feelings of guilt and depression are not uncommon in these situations. Finally, what if you chose not to get tested at all? Research has shown that individuals with a family history of genetic disorders who know they are at risk but choose not to get tested can have difficulties in social interactions (McConkie-Rosell *et al.*, 2008). Each scenario is complex, especially given the incurable nature of HD.

The decision to undergo genetic testing can only be made by the individual at risk for a disorder. Once a test has been conducted and the results are known, however, a new, family-related ethical dilemma is born: Should a carrier of a known genetic risk be obligated to tell his or her relatives? This very question has begun to challenge the well-established medicolegal principles of confidentiality and privacy. Although some people feel that an individual who is found to carry a dominant gene for Huntington's disease has an ethical obligation to disclose that fact to his or her siblings, there currently is no legal requirement to do so. In fact, requiring someone to communicate his or her own genetic risk to family members who are therefore also at risk is considered by many to be ethically dubious.

So, then, should a medical professional get involved if he or she becomes aware that communication within a family has failed or is blocked? A professional's actions in such situations depend in part upon the nature of the information available (e.g., do the test results only estimate risk levels, or are they definitive?), as well as the implications of the condition. For instance, a greater imperative to breach confidentiality may be felt when preventative treatment options are available or when reproductive choices are at stake.

Genetic Testing and Privacy Concerns

Another important consideration when deciding whether to undergo genetic testing is the possibility that if someone knows you are likely to develop a genetic disorder in the future, he or she could use that information against you. For example, if an employer knows that an employee is likely to be diagnosed with cancer or HD, the employer might not want to retain that employee. Similarly, a person who is known to have a high risk for a genetic condition may have difficulty obtaining insurance because he or she is likely to run up medical bills that would be costly to the insurance company. Because we cannot control our genes, it is unfair to discriminate against a person's genetic predispositions.

In the United States, the Health Insurance Portability and Accountability Act of 1996 (HIPAA) was the first federal law to provide protections against genetic discrimination, but its intentions fell short. Specifically, under HIPAA, insurance companies were not prevented from charging higher rates to customers based on genetic information, nor were they prevented from collecting genetic data or requiring applicants to undergo genetic testing.

Thus, a more comprehensive federal measure, the Genetic Information Nondiscrimination Act (GINA), was signed into law in May 2008 (Allison, 2008). GINA, spearheaded by members of Congress like Representative Louise Slaughter, GINA took ten years to pass. Ideally, now that the legislation is law, the protections provided by this act will foster an atmosphere in which Americans feel safe to take advantage of appropriate genetic testing, as well as to share test results with any other at-risk family members.

Genetic Testing is a Matter of Individual Choice

In summary, the choice to undergo genetic testing is a highly personal one. A person's decision depends on various factors, including the perceived preventability and treatability of the disorder and one's ability to make constructive life changes with the information one gains from testing. On one hand, testing can help people make family planning decisions and other life choices, like when to retire or how much money to put away for the future. However, finding out that you are at a higher risk for developing treatable diseases such as cancer or that you carry a gene for an untreatable, fatal disorder such as HD can have negative social consequences. Moreover, it can open the door for potential discrimination. Hopefully, with the recent passage of GINA in the U.S., more Americans will feel comfortable working with physicians and genetic counsellors to undergo any genetic tests that contribute to better management of their health care and health choices.

Gene Therapy

You may find it hard to imagine that something as simple as a single base change—specifically, the appearance of a cytosine instead of a thymine somewhere along the 3-trillion-base-long human genome—could lead to a condition as profound as blindness. Yet, in certain cases of congenital blindness known as Leber congenital amaurosis, this is exactly what happens. On first glance, it seems tantalizingly simple to fix disorders that are caused by mutations in a single gene: Simply insert the correct form of the mutant gene into the cells that need the functioning gene, then stand back and watch the disease disappear. However, the field of gene therapy, in which scientists work to treat disorders at the level of the sequence of mutated genes, is much more complicated than it first appears.

How to Get Therapeutic DNA Inside Cells

One major obstacle facing effective gene therapy involves finding the best way to deliver therapeutic genes into target cells within the body. Scientists have tested a number of methods, some of which utilise liposomes or cell surface receptors. But the most common method of DNA delivery employs something that is perfectly evolved to enter cells: the virus. Viruses are obligate intracellular parasites, designed through the course of evolution to infect cells, often with great specificity to particular cell types. Viruses efficiently transfect their own DNA into a host cell, then use the host's cellular machinery to replicate their DNA and synthesize certain viral proteins, thereby producing more viral particles.

When it comes to gene therapy, scientists can capitalize on this process by removing the disease-causing portions of the viral genome and adding a foreign therapeutic gene. The engineered genome is then repackaged into the viral protein coat and allowed to infect the proper cell target. Now, instead of producing viral toxins and causing disease, the cell infected by this virus is able to produce the protein or enzyme it lacks.

One of the more common ways of introducing the wild-type version of a gene into cells with a defective copy is by way of an adenovirus. It is possible to use recombinant DNA technology to introduce a wild-type gene into an adenoviral vector. Once the adenovirus has infected the target cells, homologous recombination can occur, resulting in replacement of the native, mutant copy of the gene with the wild-type version from the vector.

LCA: A Disease Candidate for Gene Therapy

This image depicts cultured canine RPE cells immunolabelled with anti-RPE65 antibody, with nuclei stained with propidium iodide. (a) Wild-type cells label uniformly and intensely with anti-RPE65 antibody. (b) RPE65 -/- cells do not label before infection with AAV-RPE65. (c) Most RPE65 -/- cells label positively (indicated by arrowheads) following infection with AAV-RPE65, indicating the presence of wild-type RPE65 protein. One cell (indicated by the arrow) does not seem to have been transduced, however.

As previously mentioned, the form of blindness known as Leber congenital amaurosis (LCA) arises from a single-base mutation. This mutation acts to disrupt the visual cycle, or the biological conversion of a photon of light to a chemical signal in the retina that is then sent to the visual cortex of the brain. Within this cycle, 11-*cis*-retinal protein, when struck by a photon of light, undergoes a conformational change that induces signal cascades that stimulate the visual cortex. The *RPE65* gene codes for a protein that is critical for the regeneration of 11-*cis*-retinal protein into its starting form. Without *RPE65*, there is no 11-*cis*-retinal protein present to change light signals into chemical signals, and the result is blindness.

About 10% of cases of LCA bear a mutation in the *RPE65* gene. Because this form of LCA is the result of a single-gene defect and there are no other treatments available, gene therapy was considered as a possibility for this disorder. Researchers began their work in this area using an animal model of LCA—specifically, a canine model. In dogs, a four-base-pair deletion in the *RPE65* gene (denoted *RPE65 -/-*) alters that gene such that functional protein cannot be not made. Thus, in their initial studies aimed at determining whether adenoviral-based gene therapy might be a reasonable approach to treatment of LCA, investigators infected cells extracted from *RPE65 -/-* dogs with a recombinant adeno-associated virus (rAAV) that contained wild-type *RPE65*. The investigators were able to see a rescue of expression in the cells, suggesting that it was indeed possible to rescue expression of the RPE65 protein. It remained to be seen, however, whether this was sufficient to reverse blindness if used in the *RPE65 -/-* animals.

In order to examine whether expression of wild-type *RPE65* resulted in a physiological effect, Acland and colleagues (2001) performed surgery on the dogs and injected the virus directly into their retinas or the surrounding tissue. Next, the research team investigated whether the pathway from the retina to the visual cortex in the dogs' brains regained the ability to respond properly to light stimulation. It is generally thought that a lack of visual stimulation leads to a concomitant lack of development in the visual cortex, resulting in blindness. The researchers therefore worried that without visual stimulation, proper development of the visual cortex had not occurred in these LCA animals, so even if the function of *RPE65* could be restored, the dogs' vision might not improve. Using electroretinography (ERG), the investigators were able to compare the retinal response of wild-type animals with the retinal response of *RPE65* mutant animals both before and after gene therapy. The ERG results revealed a functional rescue of the mutant animals' retinal response to light, with post-treatment dogs exhibiting a response to light similar to that of wild-type animals.

Next, in the second stage of research, the investigators applied the gene therapy treatment to *RPE65* mutant dogs that were four years of age or younger, and they used a technique called functional magnetic resonance imaging (fMRI) to measure activity in the dogs' brains in response to light stimulation both before and after gene therapy. Following gene therapy, the researchers noted an almost-normal response in the dogs' visual cortices to light stimulation; before treatment, there had been no activity in this region of the dogs' brains.

They also noted normal pupil function and normal electrical activity in the retinas of the treated dogs, both of which were abnormal before gene therapy. These findings were important, because they indicated that the visual pathway indeed retains the ability to respond to light and that the visual cortex of the brain can be activated even after years of visual deprivation.

Finally, the research team sought to test these findings in adult human patients with LCA. After using MRI to verify that these patients' optic nerves and visual cortices were anatomically normal, the investigators again turned to fMRI to measure what response, if any, the patients' visual cortices would have to light stimulation. Surprisingly, the researchers noted a normal response to bright light (but not dim light) in these patients (Aguirre *et al.*, 2007). These findings suggested that gene therapy might restore some functional vision to people with LCA, and they therefore served as encouragement to scientists and doctors planning human gene therapy trials for LCA.

Moving from Animals to Humans: Important Considerations

Despite the promise associated with the use of viral vectors in gene therapy, there are also significant risks. One such risk is the possibility that a virus may recover its ability to cause disease once inside the body. Another is the threat of insertional mutagenesis when using a retrovirus. This type of virus inserts its genome within a host cell's chromosomes, thereby potentially disrupting important genes. In the worst-case scenario, the viral genome is inserted into a tumor suppressor gene, thereby causing the development of cancer. There is also the possibility of immune and inflammatory responses whenever any foreign particle, such as a virus, is introduced into the body.

In addition to these physical considerations, scientists in the early days of gene therapy had to address ethical issues beyond those dealing with the safety of their research subjects. Early ethical debate centred on "playing God" and other philosophical and theological issues. By the time gene therapy in humans became a reality in 1990, most people saw gene therapy as simply a different kind of treatment for disease. However, debate continues about the ethical implications of changing germ-line cells (eggs or sperm), which would alter the genetic makeup of future generations.

Beyond safety and efficacy issues, many of the ethical concerns associated with modern-day somatic gene therapy research are also present in many other clinical settings. These include matters related to weighing potential harms and benefits, establishment of procedural fairness in selection of patients for research, assurance that consent to experimental treatments is informed and voluntary, and protection of privacy and confidentiality of medical information.

A Major Setback to the Field of Gene Therapy

Even when all pertinent ethical considerations are met, adverse events can still occur with gene therapy. For instance, in 1999, just nine years after the first human trial for gene therapy was conducted, an 18-year-old gene therapy trial volunteer died. Jesse Gelsinger suffered from ornithine transcarbamylase (OTC) deficiency, an enzyme deficiency that causes an inability to break down ammonia (Somia & Verma, 2000). Four days after receiving an injection of adenovirus carrying a corrected version of the OTC gene, Gelsinger died due to multiple organ failure.

His death is believed to have been triggered by a severe immune response to the adenovirus carrier. Questions quickly surfaced about severe side effects and other risks that were undisclosed to Gelsinger on the research consent form, along with suggestions that investigators ignored high ammonia levels in Gelsinger's blood that would have excluded him from the trial. In response, the U.S. Food and Drug Administration temporarily halted all gene therapy trials while the National Institutes of Health conducted thorough reviews of all adverse reactions and deaths associated with this type of research.

Currently, there are four locations across the globe (Italy, France, the United States, and the United Kingdom) where researchers are performing clinical trials testing the efficacy of gene therapy for the treatment of LCA in humans. The UK trial began in early 2007, and in April 2008, results were published for three patients who had received gene therapy. There were no adverse events noted for any of these patients, and one showed some improvement in visual function, although the level of improvement was not clinically significant (Bainbridge *et al.*, 2008).

The results of this first trial demonstrate that virally mediated delivery of genes to the eye is possible. It is hoped that this mode of therapy will eventually be used to improve vision in patients with LCA and other forms of inherited blindness. The researchers involved in this study continue to enroll additional patients while the world awaits results from the other three ongoing studies.

It is important to consider that RPE65 is not the only potential therapeutic gene that could be a target for gene therapy approaches. Other investigators, like those in the Masland laboratory, have been able to restore some retinal function in blind mice using gene therapy for the melanopsin gene (Lin *et al.*, 2008). Thus despite some setbacks, this field is growing and has promise to yield positive therapeutic results for patients in the future.

The Ability to Manipulate DNA Has Led to a New Genetics

Restriction Endonucleases. Enzymes that cleave DNA at specific sites allow DNA segments from different sources to be spliced together.

Using Restriction Endonucleases to Manipulate Genes. Fragments produced by cleaving DNA with restriction endonucleases can be spliced into plasmids, which can be used to insert the DNA into host cells.

Restriction Endonucleases

In 1980, geneticists used the relatively new technique of gene splicing, which we will describe in this chapter, to introduce the human gene that encodes interferon into a bacterial cell's genome. Interferon is a rare blood protein that increases human resistance to viral infection, and medical scientists have been interested in its possible usefulness in cancer therapy.

This possibility was difficult to investigate before 1980, however, because purification of the large amounts of interferon required for clinical testing would have been prohibitively expensive, given interferon's scarcity in the blood. An inexpensive way to produce interferon was needed, and introducing the gene responsible for its production into a bacterial cell made that possible.

The cell that had acquired the human interferon gene proceeded to produce interferon at a rapid rate, and to grow and divide. Soon there were millions of interferon-producing bacteria in the culture, all of them descendants of the cell that had originally received the human interferon gene.

The Advent of Genetic Engineering

This procedure of producing a line of genetically identical cells from a single altered cell, called cloning, made every cell in the culture a miniature factory for producing interferon. The human insulin gene has also been cloned in bacteria, and now large amounts of insulin, a hormone essential for treating some forms of diabetes, can be manufactured at relatively little expense. Beyond these clinical applications, cloning and related molecular techniques are used to obtain basic information about how genes are put together and regulated. The interferon experiment and others like it marked the beginning of a new genetics, genetic engineering.

The essence of genetic engineering is the ability to cut DNA into recognisable pieces and rearrange those pieces in different ways. In the interferon experiment, a piece of DNA carrying the interferon gene was inserted into a plasmid, which then carried the gene into a bacterial cell. Most other genetic engineering approaches have used the same general strategy, bringing the gene of interest into the target cell by first incorporating it into a plasmid or an infective virus. To make these experiments work, one must be able to cut the source DNA (human DNA in the interferon experiment, for example) and the plasmid DNA in such a way that the desired fragment of source DNA can be spliced permanently into the plasmid. This cutting is performed by enzymes that recognise and cleave specific sequences of nucleotides in DNA. These enzymes are the basic tools of genetic engineering.

Discovery of Restriction Endonucleases

Scientific discoveries often have their origins in seemingly unimportant observations that receive little attention by researchers before their general significance is appreciated. In the case of genetic engineering, the original observation was that bacteria use enzymes to defend themselves against viruses.

Most organisms eventually evolve means of defending themselves from predators and parasites, and bacteria are no exception. Among the natural enemies of bacteria are bacteriophages, viruses that infect bacteria and multiply within them. At some point, they cause the bacterial cells to burst, releasing thousands more viruses. Through natural selection, some types of bacteria have acquired powerful weapons against these viruses: they contain enzymes called restriction endonucleases that fragment the viral DNA as soon as it enters the bacterial cell.

Many restriction endonucleases recognise specific nucleotide sequences in a DNA strand, bind to the DNA at those sequences, and cleave the DNA at a particular place within the recognition sequence. Why don't restriction endonucleases cleave the bacterial cells' own DNA as well as that of the viruses? The answer to this question is that bacteria modify their own DNA, using other enzymes known as methylases to add methyl (CH_3) groups to some of the nucleotides in the bacterial DNA.

When nucleotides within a restriction endonuclease's recognition sequence have been methylated, the endonuclease cannot bind to that sequence. Consequently, the bacterial DNA is protected from being degraded at that site. Viral DNA, on the other hand, has not been methylated and therefore is not protected from enzymatic cleavage.

How Restriction Endonucleases Cut DNA

The sequences recognised by restriction endonucleases are typically four to six nucleotides long, and they are often palindromes. This means the nucleotides at one end of the recognition sequence are complementary to those at the other end, so that the two strands of the DNA duplex have the same nucleotide sequence running in opposite directions for the length of the recognition sequence. Two important consequences arise from this arrangement of nucleotides.

Biotechnology is Producing a Scientific Revolution

DNA Sequence Technology. The complete nucleotide sequence of the genomes of many organisms are now known. The unique DNA of every individual can be used to identify sperm, blood, or other tissues.

Biochips. Biochips are squares of glass etched with DNA strands and can be used for genetic screening. Medical Applications. Many drugs and vaccines are now produced with gene technology.

Agricultural Applications. Gene engineers have developed crops resistant to pesticides and pests, as well as commercially superior animals.

Stem Cells. Both embryonic stem cells and tissue-specific stem cells can potentially be used to repair or replaced damaged tissue.

Ethics and Regulation. Genetic engineering raises important questions about danger and privacy.

DNA Sequence Technology

The 1980s saw an explosion of interest in biotechnology, the application of genetic engineering to practical human problems. Let us examine some of the major areas where these techniques have been put to use.

Genome Sequencing

Genetic engineering techniques are enabling us to learn a great deal more about the human genome. Several clonal libraries of the human genome have been assembled, using large-size restriction fragments. Any cloned gene can now be localised to a specific chromosomal site by using probes to detect in situ hybridisation (that is, binding between the probe and a complementary sequence on the chromosome).

Genes are now being mapped at an astonishing rate: genes that contribute to dyslexia, obesity, and cholesterol-proof blood are some of the important ones that were mapped in 1994 and 1995 alone! With an understanding of where specific genes are located in the human genome and how they work, it is not difficult to imagine a future in which virtually any genetic disease could be treated or perhaps even cured with gene therapy. Some success has already been reported in treating patients who have cystic fibrosis with a genetically corrected version of the cystic fibrosis gene.

An exciting scientific by-product of the human genome project has been the complete genome sequencing of many microorganisms with smaller genomes, on the order of a few Mb. In general, about half of the genes prove to have a known function; what the other half of the genes are doing is a complete mystery. The first eukaryotic genome to be sequenced in its entirety was that of brewer's yeast Saccharomyces cerevisiae; many of its approximately 6000 genes have a similar structure to some human genes. The complete sequences of many much larger genomes have recently been completed, including the malarial Plasmodium parasite (30 Mb), the nematode (100 Mb), the plant Arabidopsis (100 Mb), the fruit fly Drosophila (120 Mb), and the mouse (300 Mb).

The international scientific community has over the last several years mounted a major effort to sequence the entire human genome. Because the human genome contains some 3000 Mb (million nucleotide base-pairs), this task has presented no small challenge.

Rapid progress was made possible by the use of so-called shotgun cloning techniques, in which the entire genome is first fragmented, then each of the fragments is sequenced by automated machines, and finally computers use overlaps to order the fragments. All but a small portion of the sequence was completed by the beginning of the year 2000.

2

Genetic Mechanisms

In addition to very frequent simple mutations, there are several mutation classes which involve sequence exchange between allelic or nonallelic sequences, often involving repeated sequences. For example, tandemly repetitive DNA is prone to deletion/insertion polymorphism whereby different alleles vary in the number of integral copies of the tandem repeat.

Such variable number of tandem repeat (VNTR) polymorphisms can occur in the case of repeated units that are very short (microsatellites); intermediate (minisatellites) or large. Different genetic mechanisms can account for VNTR polymorphism depending on the size of the repeating unit. In addition, interspersed repeats can also predispose to deletions/duplications by a variety of different genetic mechanisms.

Slipped Strand Mispairing can Cause VNTR Polymorphism at Short Tandem Repeats (Microsatellites)

There is considerable variation in the germline mutation rates at microsatellite loci, ranging from an undetectable level up to about 8×10^{-3}. Novel length alleles at (CA)/ (TG) microsatellites and at tetranucleotide marker loci are known to be formed without exchange of flanking markers.

This means that they are not generated by unequal crossover. Instead, as new mutant alleles have been observed to differ by a single repeat unit from the originating parental allele (Mahtani and Willard, 1993), the most likely mechanism to explain length variation is a form of exchange of sequence information which commences by slipped strand mispairing.

This occurs when the normal pairing between the two complementary strands of a double helix is altered by staggering of the repeats on the two strands, leading to incorrect pairing of repeats. Although slipped strand mispairing can be envisaged to occur in nonreplicating DNA, replicating DNA may offer more opportunity for slippage and hence the mechanism is often also called replication slippage or polymerase slippage. In addition to mispairing between tandem repeats, slippage replication has been envisaged to generate large deletions and duplications by mispairing between noncontiguous repeats and has been suggested to be a major mechanism for DNA sequence and genome evolution. The pathogenic potential of short tandem repeats is considerable.

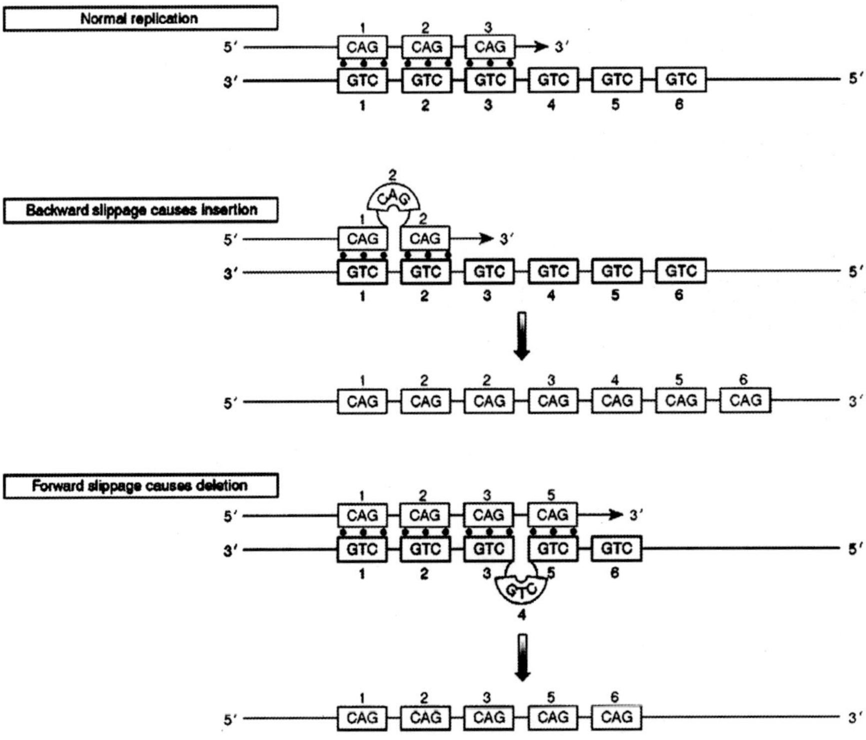

Figure: *Slipped strand mispairing during DNA replication can cause insertions or deletions.*

Large Units of Tandemly Repeated DNA are Prone to Insertion/deletion as a Result of Unequal Crossover or Unequal Sister Chromatid Exchanges

Homologous recombination describes recombination (crossover) occurring at meiosis or, rarely, mitosis between identical or very similar DNA sequences. It usually involves breakage of *nonsister chromatids* of a pair of homologs and rejoining of the fragments to generate new recombinant strands. Sister chromatid exchange is an analogous type of sequence exchange involving breakage of individual *sister chromatids* and rejoining fragments that initially were on different chromatids of the same chromosome. Both homologous recombination and sister chromatid exchange normally involve *equal* exchanges - cleavage and rejoining of the chromatids occurs at the same position on each chromatid. As a result, the exchanges occur *between allelic sequences* and at corresponding positions within alleles. In the case of intragenic equal crossover between two alleles, a new allele can result which is a fusion gene (or hybrid gene), comprising a terminal fragment from one allele and the remaining sequence of the second allele. However, equal sister chromatid exchanges cannot normally produce genetic variation because sister chromatids have identical DNA sequences.

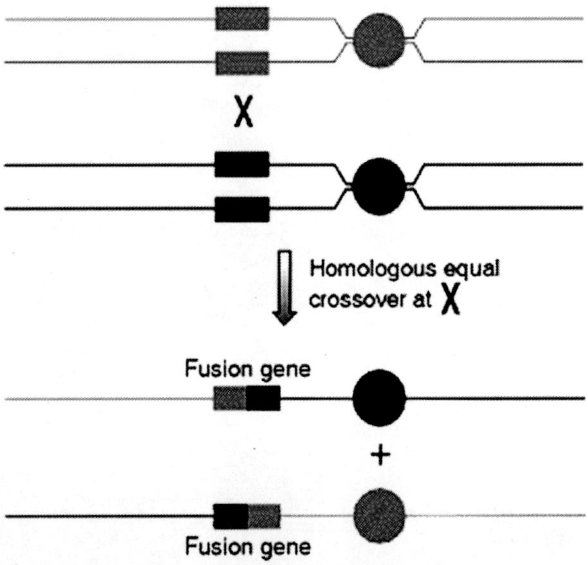

Figure: *Homologous equal crossover can result in fusion genes.*

Unequal crossover is a form of recombination in which the crossover takes place *between nonallelic sequences* on nonsister chromatids of a pair of homologs. Often the sequences at which crossover takes place show very considerable sequence homology which presumably stabilizes mispairing of the chromosomes. Because crossover occurs between mispaired nonsister chromatids, the exchange results in a deletion on one of the participating chromatids and an insertion on the other.

The analogous exchange between sister chromatids is called unequal sister chromatid exchange. Both mechanisms occur predominantly at locations where the tandemly repeated units are moderate to large in size. In such cases, the very high degree of sequence homology between the different repeats can facilitate pairing of nonallelic repeats on nonsister chromatids or sister chromatids. If chromosome breakage and rejoining occurs while the chromatids are mispaired in this way, an insertion or deletion of an integral number of repeat units will result. Note that such exchanges are *reciprocal*; both participating chromatids are modified, in one case resulting in an insertion, and in the other case in a complementary deletion.

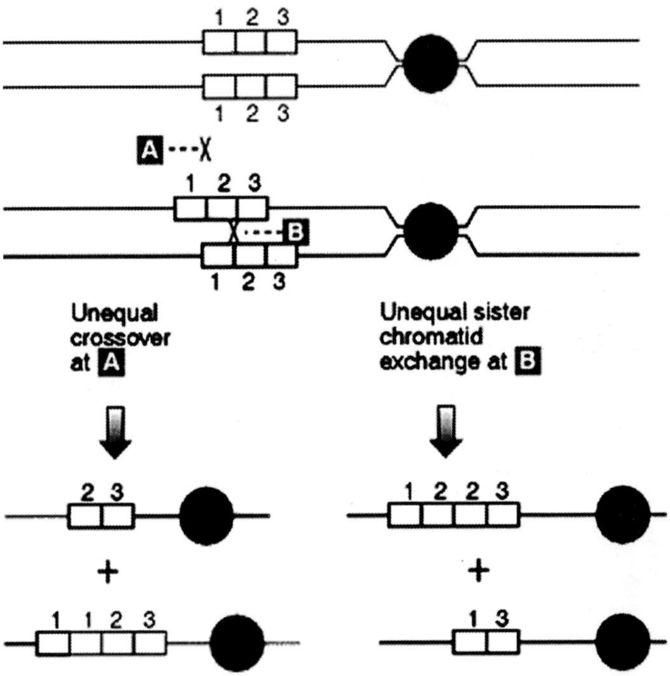

Figure: *Unequal crossover and unequal sister chromatid exchange cause insertions and deletions.*

Unequal sister chromatid exchange is thought to be a major mechanism underlying VNTR polymorphism in the rDNA clusters. Unequal crossover is also expected to occur comparatively frequently in complex satellite DNA repeats and at tandemly repeated gene loci. In the latter case, unequal crossover is known to generate pathogenic deletions at some loci. Such exchanges can also lead to concerted evolution by causing a particular variant to spread through an array of tandem repeats, resulting in *homogenisation* of the repeat units.

Occasionally, unequal crossover and unequal sister chromatid exchanges can occur at regions where there is little homology. This is likely to be the case when such mechanisms first generate a tandemly duplicated locus following mispairing of nonallelic repeats such as two *Alu* repeats or even smaller elements.

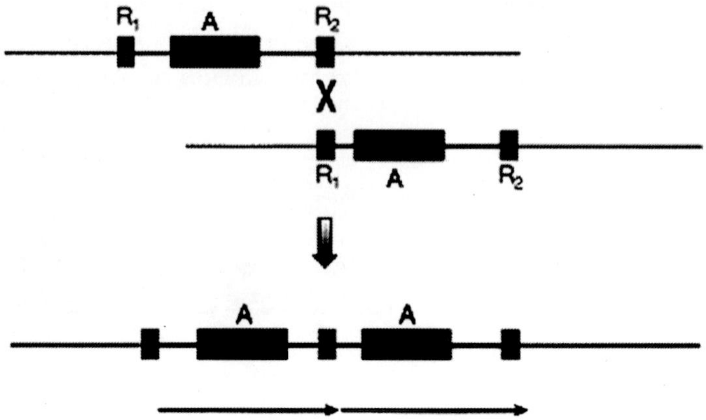

Figure: *Tandem gene duplication can result from unequal crossover or unequal sister chromatid exchange, facilitated by short interspersed repeats.*

Gene Conversion Events may be Relatively Frequent in Tandemly Repetitive DNA

Gene conversion describes a *nonreciprocal* transfer of sequence information between a pair of nonallelic DNA sequences (*interlocus gene conversion)* or allelic sequences (*interallelic gene conversion*). One of the pair of interacting sequences, the *donor*, remains unchanged. The other DNA sequence, the *acceptor*, is changed by having some or all of its sequence replaced by a sequence copied from the donor sequence. The sequence exchange is therefore a directional one; the acceptor sequence is modified by the donor sequence, but not the other way round.

One possible mechanism for gene conversion envisages formation of a heteroduplex between a DNA strand from the donor gene and a complementary strand from the acceptor gene. Following heteroduplex formation, conversion of an acceptor gene segment may occur by mismatch repair - DNA repair enzymes recognize that the two strands of the heteroduplex are not perfectly matched and 'correct' the DNA sequence of the acceptor strand to make it perfectly complementary in the converted region to the sequence of the donor gene strand.

Gene conversion has been well-described in fungi where all four products of meiosis can be recovered and studied (tetrad analysis). In humans and mammals it is not possible to do this and so gene conversion cannot be demonstrated unambiguously in higher organisms (it can never be distinguished from double crossover events, for example, although double crossovers occurring in very close proximity would normally be expected to be extremely unlikely). Despite the difficulty in identifying gene conversion in complex organisms, there are numerous instances in mammalian genomes where an allele at one locus shows a pattern of mutations which strongly resembles those found in alleles at another locus of the same species. Such evidence suggests gene conversion-like exchanges between loci.

Although simple comparisons of two sequences may be suggestive, the evidence for gene conversion is most compelling when a new mutant allele can be compared directly with its progenitor sequence. Certain highly mutable loci lend themselves to this type of analysis. In particular, some hypervariable minisatellite loci have high germline mutation rates (often 1% or more per gamete) and individual repeats often show nucleotide differences so that repeat subclasses can be recognised.

Germline mutations can be studied by detecting and characterising mutant minisatellite alleles in individual gametes. To do this, PCR analysis has been conducted on multiple dilute aliquots of DNA isolated from the sperm of an individual (small pool PCR), where each aliquot is calibrated to contain a few, perhaps 100, input molecules (Jeffreys *et al.,* 1994). The PCR products recovered from individual pools can then be typed to identify any new mutations that result in a novel allele whose length is sufficiently different as to be distinguishable from the progenitor allele.

Analyses of the patterns of germline mutation at three such loci have failed to identify exchanges of flanking markers and have shown that most mutations occurring at these loci are polar, involving the preferential gain of a few repeats at one end of a tandem repeat array. There is a bias towards gain of repeats and evidence was obtained for nonreciprocal sequence exchange between alleles, suggesting interallelic gene conversion (Jeffreys *et al.,* 1994). Evidence for interlocus gene conversion has also been obtained in human genes, notably the steroid 21-hydroxylase gene.

Pathogenic Mutations

There is a High Deleterious Mutation Rate in Hominids

Neutral mutations (those which are neither detrimental nor advantageous for the organism carrying them) accumulate throughout the generations at a rate equal to the mutation rate. To get an estimate of the (total) mutation rate is therefore simple: all one needs to do is to measure the rate of change of some presumed neutral sequence (e.g. intronic, pseudogene, etc). The deleterious mutation rate, by contrast, has been notoriously difficult to measure and no convincing estimate existed for any vertebrate until a study reported by Eyre-Walker and Keightley (1999). They investigated amino acid changes in 46 proteins occurring in the human ancestral line after its divergence from the chimpanzee. If all non-synonymous substitutions were neutral, 231 new substitutions would have been expected in their sample of genes (given an average neutral mutation rate of 0.0056 nonsynonymous substitutions per nucleotide and a total of 41 471 nucleotides investigated). Instead, only 143 nonsynonymous substitutions were observed and 88 such substitutions were inferred to have been removed by natural selection because they had been deleterious. On the assumption of 60 000 genes, and 240 000 generations since human-chimpanzee divergence, they estimated a deleterious rate of 1.6 mutations per person per generation out of a total of 4.2 mutations per person generation.

The estimated deleterious mutation rates in chimpanzees and gorillas were very similar but those for rodent-specific lineages are about one order of magnitude less, possibly because of much smaller numbers of germ cell divisions in rodents. The very high deleterious rate in hominids may even be an underestimate. If the total gene number were 80 000 and an average coding sequence was 1800 nucleotides, the estimated deleterious mutation rate would be 2.5 mutations per person per generation and, on other grounds too, a more likely rate has been considered to be three deleterious mutations per person per generation (Crow, 1999).

So with three genetic deaths per person why are we not extinct? The data would suggest that harmful mutations need to be weeded out in clusters at a time. One way to achieve this would be if natural selection operated such that individuals with the most mutations are preferentially eliminated (e.g. harmful mutations interact). This could only happen in a sexual species where mutations are shuffled each generation by genetic recombination, and so the existence of such a high deleterious mutation rate has been taken as further vindication that sex (meiotic recombination) is an efficient way to eliminate harmful mutations.

Pathogenic Mutations are Preferentially Located at Certain Types of Intragenic DNA Sequence

Pathogenic mutations can occur at three types of DNA sequence at a gene locus.

- *The coding sequence of the gene.* This is where the great majority of recorded pathogenic mutations have been identified. Those due to nucleotide substitution are, in the vast majority of cases, nonsynonymous substitutions and mostly occur at first and second base positions of codons. However, very rarely, a synonymous codon substitution is not neutral as expected, but may cause disease by activating a cryptic splice site. Because of its relatively high mutability, the CpG dinucleotide is often located at hotspots for pathogenic mutation in coding DNA (Cooper and Youssoufian, 1988). Other hotspots include tandem repeats within coding DNA.

- *Intragenic noncoding sequences.* This is restricted to sequences which are necessary for correct expression of the gene, such as important intronic elements, notably the highly conserved GT and AG dinucleotides at the ends of introns, but also conserved elements of the untranslated sequences. Often such mutations represent a small component (~10–15%) of the total pathogenic mutations at a gene locus (Cooper *et al.*, 1995). However, in some disorders pathogenic splicing mutations may be common. In the case of the collagen disorder osteogenesis imperfecta they constitute a very common pathological mutation which is second in frequency only to substitutions leading to replacement of the highly conserved, structurally important glycine residues. The collagen genes have small exons and a comparatively large number of introns (often more than 50 and as many as 106 in the case of the *COL7A1* gene), making them exceptional targets for splicing mutations. Occasional pathogenic mutations have been recorded in the 52 UTR (such as in the case of hemophilia B Leyden) and appear to exert their effect at the transcriptional level. Several examples are also known of pathogenic mutations in the 32 UTR.
- *Regulatory sequences outside exons.* Most mutations located in regulatory sequences have been identified in conserved elements located just upstream of the first exon, notably promoter elements. In addition, other more distantly located regulatory elements may be sites of pathological mutation. For example, deletions which eliminate the â-globin LCR but leave the â-globin gene and its promoter intact result in almost complete abolition of â-globin gene expression and contribute to â-thalassemia. Clearly, in some cases a gene may be regulated by the product of a distantly related gene. For example, in the case of rare variants of α-thalassemia with mental retardation, the α-globin gene and its promoter may show no evidence of pathological mutation and the disease maps to an X-linked gene which encodes a transcription factor, one of whose target sequences is presumably the α-globin gene (Gibbons *et al.*, 1995).

The Mitochondrial Genome is a Hotspot for Pathogenic Mutations

Because of the very large size of the human nuclear genome, most mutations occur in nuclear DNA sequences. By comparison, the mitochondrial genome is a small target for mutation (about 1/200 000 of the size of the nuclear genome). Unlike nuclear genes, mitochondrial genes are present in numerous copies (there are thousands of copies of the mtDNA molecule in each human somatic cell; some cells, such as brain and muscle cells, have particularly high oxidative phosphorylation requirements and so more mitochondria). The mtDNA is inherited from the maternal oocyte, which is an exceptional cell with many more mtDNA molecules than somatic cells. Given that a mutation in mitochondrial DNA must arise on a single mtDNA molecule, one might intuitively expect that the chances of a single mtDNA mutation becoming *fixed* would be very low and the mutation rate correspondingly low. On these grounds, one could anticipate that the proportion of clinical disease due to pathogenic mutation in the mitochondrial genome should be extremely low. Instead, the frequency of 'mitochondrial disorders' is rather high and the mitochondrial genome can be considered to be a mutation hotspot. Different factors can explain this apparent paradox:

- *Differential target size for pathogenic mutation.* Pathogenesis is associated with mutations in coding DNA and the mitochondrial genome has a much higher percentage of coding DNA (93%) than found in the nuclear genome (3%). When this is taken into consideration, however, there is still a large imbalance: about 100 Mb of coding DNA in the nuclear genome but only 15.4 kb of coding sequence in the mitochondrial genome, giving a target ratio of 6000:1 in favour of the nuclear genome.

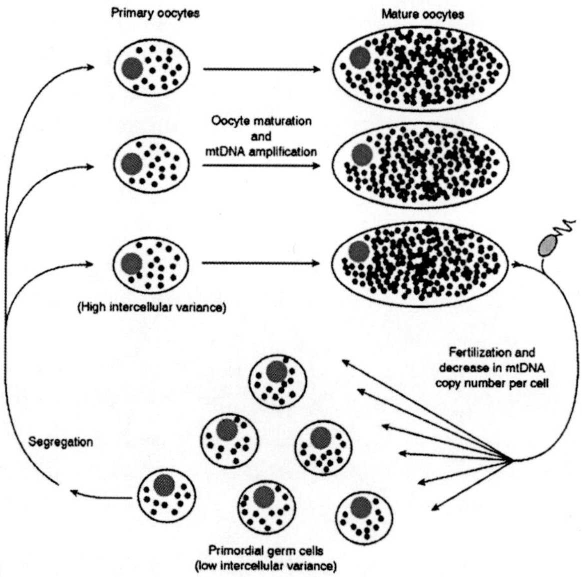

Figure: *Segregation of mitochondrial genotypes in the maternal germ line.*

- *High mutation rate in mtDNA.* The mitochondrial genome is much more prone to nucleotide change than the nuclear genome. Even although about 100 000 copies of the mitochondrial genome are maternally inherited in the fertilized oocyte there are mechanisms which permit rapid fixation of mutations in mitochondrial DNA. The combination of mtDNA instability and a high fixation rate means that the mutation rate in mitochondrial DNA is very high. Mutations have been reported to be fixed in the mitochondrial genomes of animal cells at a rate which is about 10 times greater than occurring in equivalent sequences in the nuclear genome (Brown *et al.*, 1979). This means that the small recombination-deficient animal mtDNA molecules appear to be evolving remarkably rapidly, corresponding to about 2–4% sequence divergence per million years. In contrast, plant mtDNA molecules are comparatively large (150 kb-2.5 Mb), have introns, engage in recombination and are evolving comparatively slowly.

The high instability of mtDNA has been postulated to result from several factors. The high rate of production of reactive oxygen intermediates by the respiratory chain is thought to cause substantial oxidative damage to mtDNA which, unlike nuclear DNA, is not protected by histones. The mtDNA also has to undergo many more rounds of replication than chromosomal DNA. Although several well-characterised mtDNA repair systems are now known, some frequent mutations cannot be repaired, including thymidine dimers.

Expression of Pathogenic Mutations

The degree to which a pathogenic mutation results in an aberrant phenotype depends on several factors:

- *The mutation class and the way in which the expression of the mutant gene is altered.* This may depend on the location of the mutation within the gene. Many pathogenic mutations result in abolition or substantial reduction of gene expression, but some lead to inappropriate expression. For example, overexpression of a gene product may cause an abnormal phenotype where gene dosage needs to be carefully regulated, and ectopic expression, that is expression in tissues where the gene is not normally expressed, may also be harmful.
- *The degree to which aspects of the aberrant phenotype are expressed in the heterozygote.* The presence of a single normal allele may be sufficient to maintain a clinically normal phenotype (as in recessively inherited disorders), or a milder phenotype when compared with that of mutant homozygotes, as in dominantly inherited disorders where the mutation is a simple *loss of function* mutation.
- *The degree to which expression of a mutant phenotype is influenced by other gene products.* The same mutant allele can have different phenotypic effects on different genetic backgrounds, depending on particular alleles at other gene loci (*modifier genes*).
- *The proportion and nature of cells in which the mutant gene is present.* Generally, mutations which are present in all the cells of an individual (inherited mutations) or in many of them (somatic mutations acquired very early in development) are likely to have a more profound effect

than those present in a few cells (somatic mutations which arise at much later stages) or in cell types where the relevant gene is not expressed. Cancers, however, arise from unregulated division of cells produced from a single original mutant cell.

- *The parental origin of the mutation.* This is only known to be important in the case of the few genes which are *imprinted.*

Conserved Sequence Required for Normal Splicing, but some Occur in Sequences not Normally Required for Splicing

Many genes naturally undergo alternative forms of RNA splicing. In addition, mutations can sometimes produce an aberrant form of RNA splicing which is pathogenic. Sometimes this results in the sequences of whole exons being excluded from the mature RNA (exon skipping) or retention of whole introns. On other occasions, the abnormal splicing pattern may exclude part of a normal exon or result in new exonic sequences. Point mutations which alter a conserved sequence that is normally required for RNA splicing are comparatively common. Occasionally, however, aberrant splicing of a gene can be induced by mutation of other sequence elements which resemble splice donor or splice acceptor sequences but which are not normally involved in splicing (cryptic splice sites).

Mutations Which Alter Important Splice Site Signals

Often such mutations occur at the essentially invariant GT and AG dinucleotides located respectively at the start of an intron (*splice donor*) or at its end (*splice acceptor*). Flanking these important signals, however, are other conserved sequence elements which, if mutated, can also cause aberrant splicing. Mutations which alter such sequences can have different consequences:

- *Failure of splicing causing intron retention.* This can occasionally result, for example, when an intron is small and the neighbouring sequence lacks alternative legitimate splice sites or *cryptic splice sites* (sequences which resemble the consensus splice site sequences but which are not normally used by the splicing apparatus). The introduction of intronic sequence into the coding sequence of a mature mRNA will, at the very least, introduce additional amino acids and may cause a frameshift.
- *Exon skipping.* The splicing apparatus uses an *alternative* legitimate splice site. Mutation of a splice donor sequence results in skipping of the upstream exon while mutation of the splice acceptor sequence results in skipping of the downstream exon. Often, the exclusion of an exon has a profound effect on gene expression: it may result in a frameshift, an unstable RNA transcript, or a nonfunctional polypeptide because of a loss of a critical group of amino acids.

In addition to the above, the *branch site* used in splicing may be mutated leading to defective splicing.

Mutations of Sequences Which are not Normally Important for RNA Splicing

Cryptic splice sites coincidentally resemble the sequences of authentic splice sites but are not normally used in splicing, unless a mutation alters the sequence so that the splicing apparatus now recognizes it as a normal splice site. The use of an intronic cryptic splice site will introduce new amino acids, while using an exonic cryptic splice site will result in a deletion of coding DNA. The former is a cautionary reminder that apparently silent mutations may yet be pathogenic. Note that in some cases mutations which occur within exons but not at cryptic splice sites can also induce skipping of that exon.

Mutations That Introduce a Premature Termination Codon Usually Result in Unstable mRNA but other Outcomes are Possible

Several different classes of mutation can introduce a premature termination codon (chain terminating mutations). Nonsense mutations produce a premature termination codon simply by substituting a normal codon with a stop codon. Frameshifting insertions and deletions usually also introduce a premature termination codon not too far downstream of the mutation site.

This happens because there is no selection pressure to avoid stop codons in the other translational reading frames and so given established nucleotide frequencies, at least one stop codon is usually encountered within a stretch of 100 nucleotides downstream of the mutation site. A variety of splice site mutations too can introduce a premature termination codon e.g. by skipping of a single exon containing a number of nucleotides that cannot be divided by 3. There are several possible consequences for gene expression for chain-terminating mutations:

- *Unstable mRNA.* This is by far the most frequent consequence. A mRNA carrying a premature codon is usually rapidly degraded *in vivo* by a form of RNA surveillance known as nonsense-mediated mRNA decay (Hentze and Kulozik, 1999; Culbertson, 1999). This can avoid the potentially lethal consequences of producing a truncated polypeptide which could interfere with vital cell functions.
- *Truncated polypeptide.* The production of a polypeptide truncated at the C terminus is a very rare outcome *in vivo* (the well-known protein truncation test which assays for mutations introducing a premature termination codon is carried out using an *in vitro* transcription-translation system). Nevertheless, some truncated polypeptides are produced *in vivo*. The effect on gene expression may be difficult to predict and will depend among other things on the extent of the

truncation, the stability of the polypeptide product and its ability to interfere with expression of normal alleles.

- *Exon skipping.* Some nonsense mutations appear to induce skipping of constitutive exons *in vivo*. For example, a nonsense mutation in the middle of exon 51 of the *FBN1* fibrillin gene (corresponding to the C terminus of the protein) causes that exon to be skipped (Dietz *et al.*, 1993). As a result of exon skipping the abnormally spliced mRNA uses the normal stop codon and escapes nonsense-mediated mRNA decay unlike any full length mRNA which may be produced from the pre-mRNA. The abnormally spliced *FBN1* mRNA accumulates and is translated to give a dominant negative protein lacking C-terminal sequences.

The Pathogenic Potential of Repeated Sequences

The human genome, like other mammalian genomes, has a very high proportion of DNA sequences that are repeated. Tandem repeats in coding DNA include very short nucleotide repeats, moderately sized repeats and very large repeats that can include whole genes. Depending on the degree of sequence homology between the repeats, tandem repeats are liable to a variety of different genetic mechanisms causing sequence exchange between the repeats. Often such sequence exchanges result in changes in the number of tandem repeats. A reduction in repeat number can often result in a pathogenic deletion, but expansion by sequence duplication can be pathogenic too. Certain chromosomal regions, notably the subtelomeric and pericentromeric regions, harbour large tracts of duplicated DNA and instability of such regions can predispose to disease (Eichler, 1998). Interspersed repeats can also cause pathogenic mutations by a different variety of mechanisms.

Slipped Strand Mispairing of Short Tandem Repeats Predisposes to Pathogenic Deletions and Frameshifting Insertions

Insertions and deletions in coding DNA are rare because they usually introduce a translational frameshift. However, occasionally, a series of tandem repeats of a small number of nucleotides occurs by chance in the coding sequence for a polypeptide. Such repeats, like classical microsatellite loci, are comparatively prone to mutation by slipped strand mispairing. As a result, the copy number of tandem repeats is liable to fluctuate, introducing a deletion or an insertion of one or more repeat units. If the mutation occurs in polypeptide-encoding DNA, a resulting deletion will often have a profound effect on gene expression.

Frameshifting deletions will normally result in abolition of gene expression. Even if the deletion does not produce a frameshift, deletions of one or more amino acids can still be pathogenic. Small frameshifting insertions will also be expected to lead to loss of gene expression and often the insertion is a tandem repeat of sequences flanking it. However, nonframeshifting insertions would often not be expected to be pathogenic, unless the insertion occurs in a critically important region, destabilising an essential structure or impeding gene function in some way. Note that large triplet repeat expansions can lead to disease by mechanisms that are not understood at present.

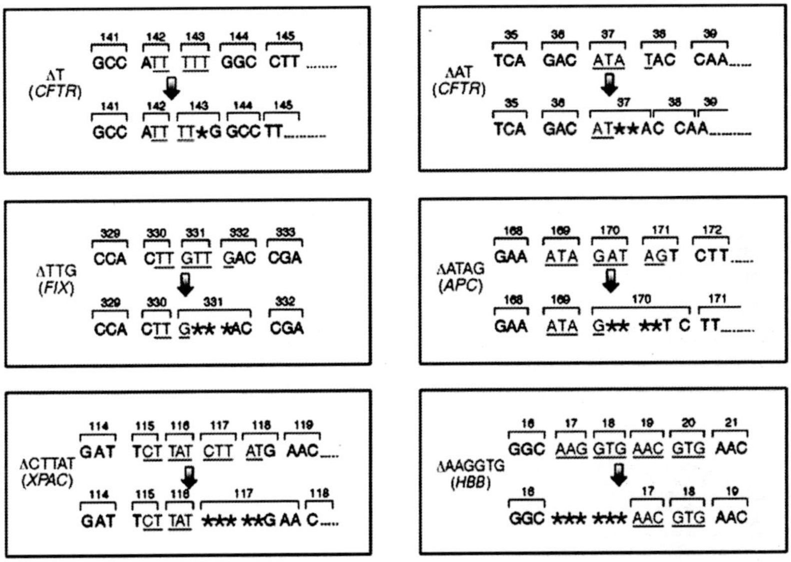

Figure: *Short tandem repeats are deletion/insertion hotspots.*

Rapid Large-scale Expansion of Intragenic Triplet Repeats can Cause a Variety of Diseases but the Mutational Mechanism is Not well Understood

Sometimes microsatellites within or in the immediate vicinity of a gene can expand to considerable lengths and affect gene expression, causing disease. In some cases, a modestly expanded repeat which causes disease may be perfectly stable and be propagated without change in size through several generations. For example, triplet repeat expansion leading to long polyalanine tracts in the *HOXD13* gene cause a form of synpolydactyly, probably as a result of unequal crossover (Warren, 1997), but the expanded repeat is stable (Akarsu *et al.,* 1996). In other cases, however, the expanded triplet repeat is *unstable* and the discovery that human disease can be caused by large-scale expansion of highly unstable trinucleotide repeats was quite unexpected. Studies in other organisms had not revealed precedents for such a phenomenon, but the list of human examples is now considerable. In addition to unstable triplet repeat expansion, the majority of disease alleles at the cystatin B gene which cause progressive myoclonus epilepsy involve expansions of a 12 nucleotide repeat (CCCCGCCCCGCG; Lalioti *et al.,* 1998).

Tandem trinucleotide repeats are not infrequent in the human genome. Although there are 64 possible trinucleotide sequences, when allowance is made for cyclic permutations $(CAG)_n = (AGC)_n = (GCA)_n$ and reading from either strand 2 $(CAG)_n$ on one strand = 52 $(CTG)_n$ on the other], there are only 10 different trinucleotide repeats. Most of these are known as usefully polymorphic microsatellite markers but, in addition, certain repeats show anomalous behaviour which can cause abnormal gene expression.

In each case, repeats below a certain length are stable in mitosis and meiosis while, above a certain threshold length, the repeats become extremely unstable. These unstable repeats are virtually never transmitted unchanged from parent to child. Both expansions and contractions can occur, but there is a bias towards expansion. The average size change often depends on the sex of the transmitting parent, as well as the length of the repeat.

The ten possible trinucleotide repeats. Both DNA strands are shown. All other trinculeotide repeats are cyclic permutations of one or another of these.

- *Genes which show modest expansions of $(CAG)_n$ repeats within the coding sequence.* Typically, the stable and nonpathological alleles have 10–30 repeats, while unstable pathological alleles have modest expansions, often in the range of 40–200 repeats. Transcription and translation of the gene are not affected by the expansion. The resulting protein product shows a gain of function: its long polyglutamine tract causes it to aggregate within certain cells and kill them.
- *Genes which show very large expansions of a noncoding repeat.* For some genes, various types of triplet repeat (e.g. CGG, CCG, CTG, GAA) found in the promoter, the untranslated regions or intronic sequences can undergo very large expansions in such a way as to inhibit gene expression, causing loss of function. Typically, the stable and nonpathological alleles have 5–50 repeats, while unstable pathological alleles have several hundreds or thousands of copies.

Intergenerational changes are normally reported as parent-child comparisons of blood lymphocyte DNA. There is little information about when in gametogenesis, fertilization or embryogenesis the changes arise. Limited studies of sperm show that highly expanded *DM* and *FRAXA* (fragile-X syndrome) repeats are not transmitted by affected males, although modest expansions can be. The largest expansions in Huntington disease (which, however, are small compared with large *FRAXA* or *DM* expansions) are seen in sperm, consistent with the observation that the severest cases inherit the disease from their father. At least for *FRAXA, DM* and Kennedy disease, the expanded repeats are mitotically unstable, so that a blood sample shows a smear of heterogeneous expanded repeats sizes. However, *in vitro*, even large repeats are stable. Thus, whatever the mechanism, it is not operative in all cells. The basis of the unstable expansions is very largely unknown and this type of mutagenic event has not been identified thus far in genetically tractable organisms such as *E. coli*, yeast or *Drosophila*. There is also evidence that the unstable expansion mechanism may not have a parallel in some other mammals such as mice. Human transgenes containing long trinucleotide repeats show virtually no instability after being propagated through several generations in transgenic mice whereas the same sequences may show a 100% probability of expansion when transmitted in the human germline (Djian, 1998).

Investigations have also suggested that arrays of triplet repeats may be able to form alternative DNA structures, such as DNA hairpins, triplex DNA and quadruplex DNA (Sinden, 1999) but their significance if formed *in vivo* is unknown. The repeats have also been envisaged as possible protein-binding sites and protein-binding at the RNA level has also been envisaged to contribute to pathogenesis in some cases, notably in myotonic dystrophy (Philips *et al.*, 1998). Slipped strand mispairing is likely to be a component of the expansion mechanism, given the observation that interrupted repeats appear to be stable and only homogeneous repeats are unstable.

For example, in spinocerebellar ataxia type 1, 123/126 normal sized CAG repeats were interrupted by one or two CAT triplets, while 30/30 expanded alleles contained no interruption. One problem with all these mispairing mechanisms is that they should result in contractions as well as expansions and this is not seen. Instead, after a certain threshold size, there appears to be a clear bias towards continued expansion of the size of the repeat unit array. Because understanding of trinucleotide repeats is progressing very rapidly at the time of writing, the reader is advised to consult a recent review for more information.

Tandemly Repeated and Clustered gene Families may be Prone to Pathogenic Unequal Crossover and Gene Conversion-like Events

Many human and mammalian gene clusters contain nonfunctional pseudogenes which may be closely related to functional gene members. Interlocus sequence exchanges between pseudogenes and functional genes can result in disease by removing or altering some or all of the sequence of a functional gene. For example, unequal crossover (or unequal sister chromatid exchange) between a functional gene and a related pseudogene can result in deletion of the functional gene or the formation of fusion genes containing a segment derived from the pseudogene. Alternatively, the pseudogene can act as a donor sequence in gene conversion events and introduce deleterious mutations into the functional gene. The classical example of pathogenesis due to gene-pseudogene exchanges is steroid 21-hydroxylase deficiency, where over 95%

of pathogenic mutations arise as a result of sequence exchanges between the functional 21-hydroxylase gene, *CYP21B*, and a very closely related pseudogene, *CYP21A*. The two genes occur on tandemly repeated DNA segments approximately 30 kb long which also contain other duplicated genes, notably the complement C4 genes, *C4A* and *C4B*. Large pathogenic deletions uniformly result in removal of about 30 kb of DNA, corresponding to one repeat unit length, and analysis of *de novo* 21-hydroxylase deficiency mutations has provided strong evidence for pathogenic deletions arising as a result of meiotic unequal crossover (Sinnott *et al.,* 1990). Virtually all of the 75% of pathogenic point mutations are copied from deleterious mutations in the pseudogene, suggesting a gene conversion mechanism. Analysis of one such mutation which arose *de novo* suggests that the conversion tract is a maximum of 390 bp (Collier *et al.,* 1993). Gene conversion events are also found in the duplicated C4 genes, both of which are normally expressed. A likely priming event for conversions in the *CYP21-C4* gene cluster is unequal pairing of chromatids so that a *CYP21A-C4A* unit pairs with a *CYP21B-C4B* unit.

Interspersed Repeats often Predispose to Large Deletions and Duplications

In several cases, the endpoints of deletions are marked by very short direct repeats. For example, the breakpoints in numerous pathogenic deletions in mtDNA occur at perfect or almost perfect short direct repeats. Of these, the most common is a deletion of 4977 bp which has been found in multiple patients with Kearns-Sayre syndrome, an encephalomyopathy characterised by external ophthalmoplegia, ptosis, ataxia and cataract.

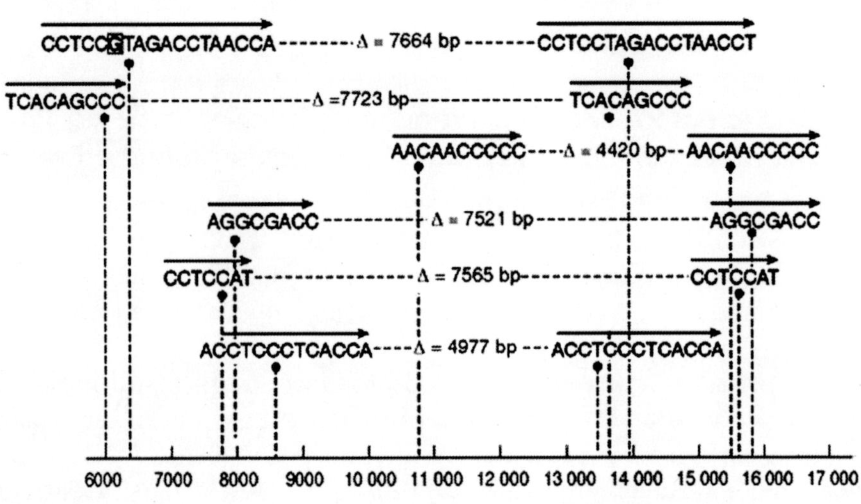

Figure: *Short direct repeats mark the endpoints of many pathogenic deletions in the mitochondrial genome.*

The deletion results in elimination of the intervening sequence between two perfect 13 bp repeats and loss of the sequence of one of the repeats. The mitochondrial genome is recombination-deficient and Shoffner *et al.* 1989 have postulated that such deletions arise by a replication slippage mechanism, similar to that occurring at short tandem repeats. Partial duplications of the mitochondrial genome are also distinctive features of certain diseases, notably Kearns-Sayre syndrome. The ends of the duplicated sequences, like those of the common deletions, are often marked by short direct repeats, and the mechanisms of duplication and deletion appear to be closely related (Poulton and Holt, 1994).

The Alu Repeat as a Recombination Hotspot

Some large-scale deletions and insertions may be generated by pairing of nonallelic interspersed repeats, followed by breakage and rejoining of chromatid fragments. For example, the *Alu* repeat occurs approximately once every 4 kb and mispairing between such repeats has been suggested to be a frequent cause of deletions and duplications. Some large genes have many internal *Alu* sequences in their introns or untranslated sequences, making them liable to frequent internal deletions and duplications.

For example, the 45-kb low density lipoprotein receptor gene has a relatively high density of *Alu* repeats (approximately one every 1.6 kb). A very high frequency of pathogenic deletions in this gene are likely to involve an *Alu* repeat, usually at both endpoints, and occasional pathogenic intragenic duplications also involve *Alu* repeats (Hobbs *et al.*, 1990). Such observations have suggested a general role for *Alu* sequences in promoting recombination and recombination-like events. Initial gene duplications in the evolution of clustered multigene families may often have involved an unequal crossover event between *Alu* repeats or other dispersed repetitive elements. It should be noted, however, that some *Alu*-rich genes do not appear to be loci for frequent *Alu*-mediated recombination.

Pathogenic Inversions can be Produced by Intrachromatid Recombination Between Inverted Repeats

Occasionally, clustered inverted repeats with a high degree of sequence identity may be located within or close to a gene. The high degree of sequence similarity between inverted repeats may predispose to pairing of the repeats by a mechanism that involves a chromatid bending back upon itself. Subsequent chromatid breakage at the mispaired repeats and rejoining can then result in an inversion, in much the same way as the natural mechanism used for the production of some immunoglobulin ê light chains.

The classic example of pathogenic inversions is a mutation which accounts for more than 40% of cases of severe hemophilia A. Intron 22 of the factor VIII gene, *F8*, contains a CpG island from which two internal genes are transcribed: *F8A* in the opposite direction to the host gene *F8*, and *F8B* in the same direction as *F8*. *F8A* belongs to a gene family with two other closely related members located several hundred kilobases upstream of *F8* gene and transcribed in the opposite direction to *F8A*. As a result, the region between the *F8A* gene and the other two members is susceptible to inversions - the *F8A* gene can pair with either of the other two members on the same chromatid, and subsequent chromatid breakage and rejoining in the region of the paired repeats results in an inversion which disrupts the factor VIII gene.

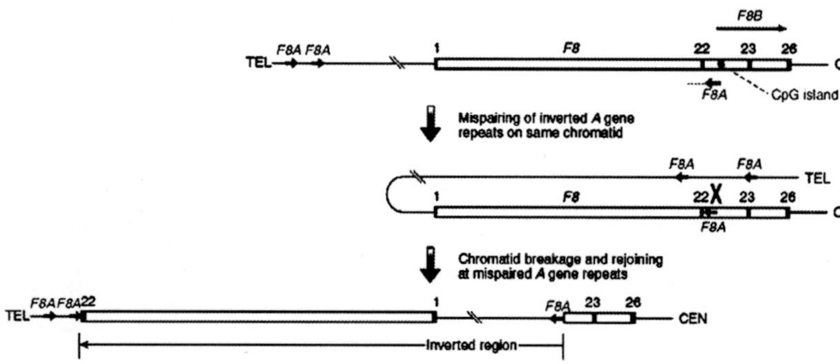

Figure: *Inversions disrupting the factor VIII gene result from intrachromatid recombination between inverted repeats.*

DNA Sequence Transposition is not Uncommon and can Cause Disease

A proportion of moderately and highly repeated interspersed elements are capable of transposition via an RNA intermediate. Defective gene expression due to DNA transposition is comparatively rare and represents only a small component of molecular pathology. However, several examples have been recorded of genetic deficiency due to insertional inactivation by retrotransposons. For example, in one study, hemophilia A was found to arise in two out of 140 unrelated patients as a result of a *de novo* insertion of a LINE-1 (*Kpn*) repeat into an exon of the factor VIII gene (Kazazian *et al.,* 1988). Other instances are known of insertional inactivation by an actively transposing *Alu* element, as in a case of neurofibromatosis type 1 (Wallace *et al.,* 1991). Additionally, a number of other examples have been recorded of pathogenesis due to intragenic insertion of undefined DNA sequences.

DNA Repair

- DNA in cells suffers a wide range of damage:
- purine bases are lost by spontaneous fission of the base-sugar link;
- cytosines, and occasionally adenines, spontaneously deaminate to produce uracil and hypoxanthine respectively;

- many chemicals, for example alkylating agents, form adducts with DNA bases;
- reactive oxygen species in the cell attack purine and pyrimidine rings
- ultraviolet light causes adjacent thymines to form a stable chemical dimer;
- mistakes in DNA replication result in incorporation of a mismatched base;
- ionising radiation causes single- or double-strand breaks;
- mistakes in replication or recombination leave strand breaks in DNA.

All these lesions must be repaired if the cell is to survive. The importance of effective DNA repair systems is highlighted by the severe diseases affecting people with deficient repair systems.

DNA Repair Usually Involves Cutting out and Resynthesising a Whole area of DNA Surrounding the Damage

To cope with all these forms of damage, cells must be capable of several different types of DNA repair. DNA repair seldom involves simply undoing the change that caused the damage. Almost always a stretch of DNA containing the damaged nucleotide(s) is excised and the gap filled by resynthesis. There are at least five main types of DNA repair in human cells:

- *Direct repair* reverses the DNA damage. A specific enzyme is able to dealkylate O^6-alkyl guanine directly. In bacteria thymine dimers can be removed in a photoreactivation reaction that depends on visible light and an enzyme, photolyase. Mammals possess enzymes related to photolyase, but use them for a quite different purpose, to control their circadian clock (Van der Horst *et al.,* 1999).
- *Base excision repair (BER)* uses glycosidase enzymes to remove abnormal bases. An endonuclease, AP endonuclease, cuts the sugar-phosphate backbone at the position of the missing base. A few nucleotides of the DNA strand are stripped back by exonucleases, the gap is filled by resynthesis, and the remaining nick is sealed by DNA ligase III. The same process is used to repair spontaneous depurination. Interestingly, no human diseases caused by defective BER are known. Maybe any such defect would be lethal, since BER corrects much the commonest type of DNA damage.
- *Nucleotide excision repair (NER)* removes thymine dimers and large chemical adducts. Defects in NER cause the autosomal recessive disease xeroderma pigmentosum (Lambert *et al.,* 1998). Seven complementation groups, XPA-XPG, have been defined by cell fusion studies. XP patients are exceedingly sensitive to UV light. Sun-exposed skin develops thousands of freckles, many of which progress to skin cancer.

- *Post-replication repair* is required to correct double-strand breaks (Haber, 1999). The usual mechanism is a gene conversion-like process (recombinational repair), where a single strand from the homologous chromosome invades the damaged DNA helix. Alternatively, broken ends are rejoined regardless of their sequence, a desperate measure that is likely to cause mutations. The eukaryotic machinery for recombination repair is less well defined than the excision repair systems. Human genes involved in this pathway include *NBS* (mutated in Nijmegen breakage syndrome, MIM 251260), *BLM* (mutated in Bloom syndrome, MIM 210900) and the *BRCA2* and maybe *BRCA1* breast cancer susceptibility genes.
- *Mismatch repair* corrects mismatched base pairs caused by mistakes in DNA replication. Cells deficient in mismatch repair have mutation rates 100–1000 times higher than normal, with a particular tendency to replication slippage in homopolymeric runs. In humans the mechanism involves at least five proteins and defects cause hereditary nonpolyposis colon cancer.

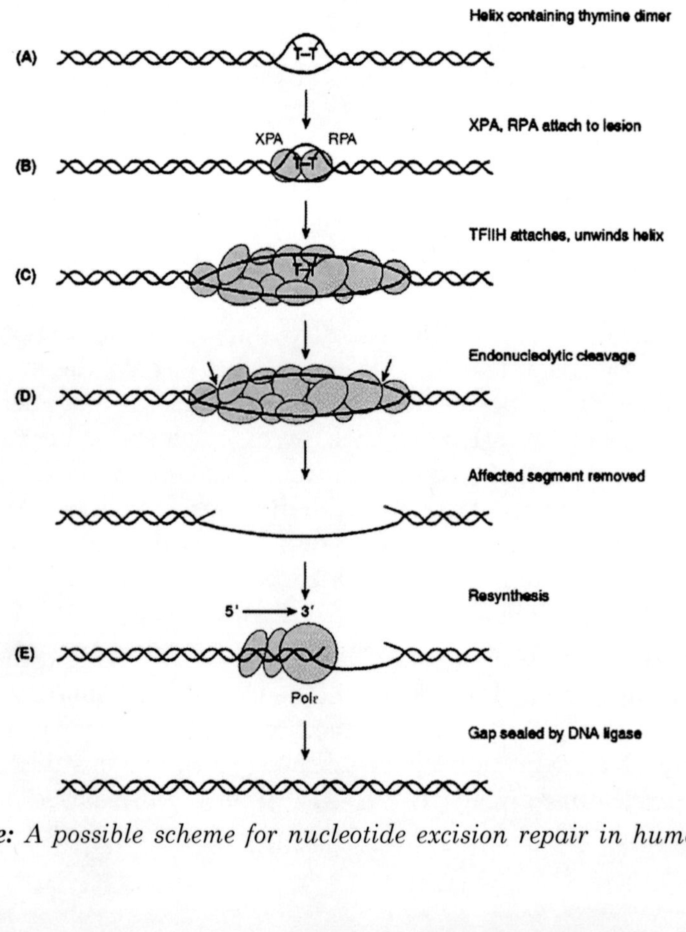

Figure: *A possible scheme for nucleotide excision repair in humans.*

All these systems, except for direct repair, require exo and endonucleases, helicases, polymerases and ligases, usually acting in multiprotein complexes that have some components in common. Sorting out the individual pathways has been greatly aided by the very strong conservation of repair mechanisms across the whole spectrum of life. Not only the reaction mechanisms but also the protein structures and even gene sequences are often conserved from *E. coli* to man. A downside of the conservation is a confusing gene nomenclature, referring sometimes to human diseases (*XPA* etc.), sometimes to yeast mutants (*RAD* genes) and sometimes to mammalian cell complementation systems (ERCC = excision repair cross-complementing) - for example *XPD*, *ERCC2* and *RAD3* are the same gene in man, mouse and yeast. Generally eukaryotes have multiple proteins corresponding to each single protein in *E. coli*, so that, for example, nucleotide excision repair requires six proteins in *E. coli* but at least 30 in mammals.

DNA Repair Systems Share Components and Processes

As well as sharing components with each other, many repair systems share components with the machinery for DNA replication, transcription and recombination. DNA polymerases and ligase are required for both DNA replication and resynthesis after excision of a defect. The recombination machinery is involved in double-strand break repair. The link with transcription is particularly intriguing (Lehmann, 1995). The general transcription factor TFIIH is a multiprotein complex that includes the XPB and XPD proteins. TFIIH exists in two forms. One form is concerned with general transcription and the other with repair, probably specifically repair of transcriptionally active DNA. This system is deficient in two rare diseases, Cockayne syndrome (MIM 216400) and trichothiodystrophy (MIM 601675). Clinically and in cell biology, CS and TTD both overlap XP, and in some cases the same genes are responsible, but CS and TTD patients have developmental defects that presumably reflect defective transcription, and they do not have the cancer susceptibility of XP patients.

Hypersensitivity to Agents that Damage DNA

Many human diseases that involve hypersensitivity to DNA-damaging agents, or a high level of cellular DNA damage, are not caused by defects in the DNA repair systems themselves, but by a defective cellular response to DNA damage. Normal cells react to DNA damage by stalling progress through the cell cycle at a checkpoint until the damage has been repaired, or triggering apoptosis if the damage is irrepairable. Part of the machinery for doing this involves the ATM protein.

The role of ATM. Briefly, it senses DNA damage and relays the signal to the p53 protein, the 'guardian of the genome'. People with no functional

ATM have ataxia telangiectasia (MIM 208900; Lambert *et al.*, 1998). Their cells are hypersensitive to radiation, and they have chromosomal instability and a high risk of malignancy, but the DNA repair machinery itself is intact. Fanconi anemia (MIM 227650) is another heterogeneous group of diseases (at least five complementation groups) marked by defective responses to DNA damage, without specific defects in DNA repair.

Physical and Transcript Mapping

A wide variety of physical mapping strategies have been used to analyse the DNA of complex eukaryotic genomes. In the following sections, an arbitrary distinction is made between two classes of physical mapping:

- low resolution physical mapping - the smallest map unit that can be resolved is typically one to several megabases of DNA;
- high resolution physical mapping - the resolution is typically very high, from hundreds of kilobases to a single nucleotide.

Since mammalian DNA has only a very small percentage of coding DNA (~3% in the case of the human genome), a variety of transcript mapping methods have been developed for selectively identifying and studying transcribed sequences, the most important and, arguably, most interesting component of the genome. 1Low resolution physical mapping

Somatic Cell Hybrid Panels can Permit Chromosomal Localisation of any Human DNA Sequence

Under certain experimental conditions, cultured cells from different species can be induced to fuse together, thereby generating somatic cell hybrids. In human genetic mapping, hybrid cells are typically constructed by fusing human cells and mouse or hamster cells. The initial fusion products are described as heterokaryons because the cells contain both a human and a rodent nucleus. Eventually, heterokaryons proceed to mitosis, and the two nuclear envelopes dissolve. Thereafter, the human and rodent chromosomes are brought together in a single nucleus. The hybrid cells are unstable. For reasons that remain unknown, most human chromosomes fail to replicate in subsequent rounds of cell division and are lost. This gives rise, eventually, to a variety of more-or-less stable hybrid cell lines, each with the full set of rodent chromosomes plus a few human chromosomes. The loss of the human chromosomes occurs essentially at random, but can be controlled by selection.

The human chromosomes in somatic cell hybrids can conveniently be identified by PCR screening with sets of chromosome-specific primers (Abbott and Povey, 1991). By collecting hybrid cell lines with different human chromosome contents it is possible to generate a hybrid cell panel that can be used to map any human DNA sequence to a specific chromosome. To do this, each of the hybrid cell lines is tested for the presence of the human sequence of interest.

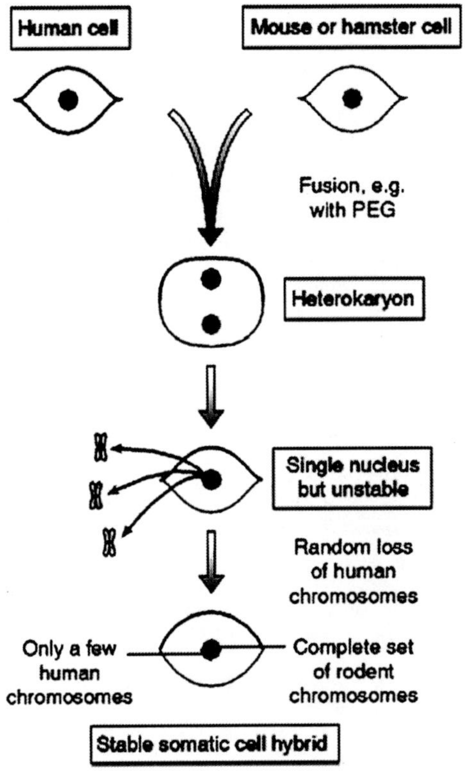

Figure: *Fusion of cells from different species can result in stable somatic cell hybrids.*

A PCR assay can be used with primers specific for that sequence (this is a form of STS content mapping) or the relevant DNA sequence can be labelled and used as a hybridisation probe. Both of these methods require that the human sequence should be distinguishable from any homologous rodent sequence. Somatic cell hybrids could also be used to establish which chromosome encodes a human enzyme or other gene product by biochemical typing of the hybrids. Whichever human chromosome contains the sequence of interest, it will be present in all the cell lines that are positive for the test and absent from all cell lines that are negative for the test.

Monochromosomal hybrids generated by microcell fusion allow direct and rapid chromosomal localisation of human DNA clones. Traditional somatic cell hybrids generally contain several human chromosomes. Microcell fusion (Fournier and Ruddle, 1977) can be used to generate monochromosomal hybrids that contain only a single human chromosome.

The donor human cells are subjected to prolonged mitotic arrest by continued exposure to an inhibitor of mitotic spindle formation, such as colcemid. This causes the chromosome content of a cell to become partitioned into discrete subnuclear packets (micronuclei). Centrifugation in the presence of cytochalasin B (a mitotic spindle inhibitor), results in the formation of microcells.

Microcells consist of a single micronucleus with a thin rim of cytoplasm, surrounded by an intact plasma membrane. In much the same way as for normal donor cells, the microcells can be fused with recipient rodent cells to generate hybrids (microcell mediated chromosome transfer). Some microcell hybrids contain a few donor chromosomes, but the simplest contain a single donor chromosome.

Subchromosomal Mapping

Monochromosomal human-rodent hybrid cell lines have been established for each of the human chromosomes (Cuthbert *et al.,* 1995). If the sequence of a particular human DNA is known, a PCR assay can be established and used to type a complete monochromosomal hybrid panel, thereby establishing the chromosomal location quickly. Alternatively, if no DNA sequence is available, a human fragment of interest can still be labelled and used as a hybridisation probe against a Southern blot of genomic DNA samples from a monochromosomal hybrid panel.

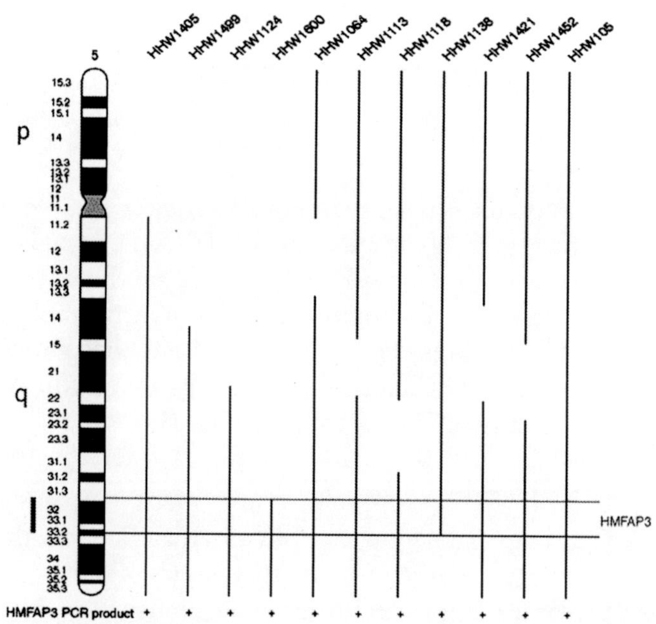

Figure: *Subchromosomal localisation can be achieved by mapping against a panel of hybrid cells containing translocation or deletion chromosomes.*

Conventional somatic cell hybrids are a relatively crude tool for physical mapping. More refined mapping is possible using hybrids that contain only part of a particular human chromosome. Translocation hybrids and deletion hybrids are made using donor human cells that have a chromosomal translocation or deletion.

To be useful, the hybrids must lack the normal homolog of the chromosome of interest. Such hybrids can be used for subchromosomal mapping of a human sequence-tagged site or biochemical marker. They are especially useful for defining the sequences removed by microdeletions, by segregating the deletion-carrying chromosome away from its normal homolog.

Human DNA Clones can be Physically Mapped to any Desired Resolution

A more general approach to producing tools for physical mapping involves artificially breaking human chromosomes and transferring the fragments into rodent cells.

Chromosome-mediated Gene Transfer

One of the first techniques to use this approach was chromosome-mediated gene transfer (CMGT). Fragments of purified mitotic chromosomes from a donor, such as a human fibroblast, are coprecipitated with calcium phosphate on to the surface of a recipient rodent cell line in monolayer culture.

Human chromosome fragments enter the recipient cells, such as mouse fibroblasts, and integrate into the chromosomes, resulting in stable transformation. As a result, hybrids can be established that retain segments of human DNA (transgenomes) of a size that is useful for mapping (usually in the range of 1–50 Mb). However, the transgenomes are prone to frequent re-arrangements, so CMGT is more suited to functional assays of complex loci than as a mapping tool.

Irradiation Fusion gGene Transfer

The most valuable hybrids for gene mapping are radiation hybrids. Donor cells are subjected to a lethal dose of radiation which fragments their chromosomes. The average size of a fragment is a function of the dose of radiation. After irradiation the donor cells are fused with recipient cells of a different species. A selection system is used to pick out recipient cells that have taken up some of the donor chromosome fragments.

These cells are useful for mapping insofar as they have taken up a random set of other chromosome fragments from the donor, as well as the selected fragment. Stably incorporated donor fragments are either integrated into rodent chromosomes or are assembled into novel human minichromosomes formed around fragments containing a functional centromere. Although this procedure was first proposed by Goss and Harris in 1975, it was not used seriously until 1990, when hybrids were constructed using irradiated monochromosomal hybrid cells as donors (Cox *et al.*, 1990). When a set of DNA markers from the human chromosome is assayed in a panel of such radiation hybrids, the patterns of cross-reactivity can be used to construct a map. The frequency of breakage between two markers can be defined by a value θ, analogous to the recombination frequency in meiotic mapping. θ varies from 0 (the two markers are never separated) to 1.0 (the two markers are always broken apart).

As in meiotic mapping, θ underestimates the distance between markers that are far apart on the same chromosome, in this case because a cell can take up two markers on separate fragments. A more accurate estimate is provided by a mapping function, $D = -\ln(1 - \theta)$, which is analogous to the Haldane mapping function used in meiotic linkage analysis. D is measured in centirays (cR). D is dependent on the dosage of radiation, so it is referenced against the number of rads. For example, a distance of 1 cR_{8000} between two markers represents a 1% frequency of breakage between them after exposure to 8000 rad of X-rays.

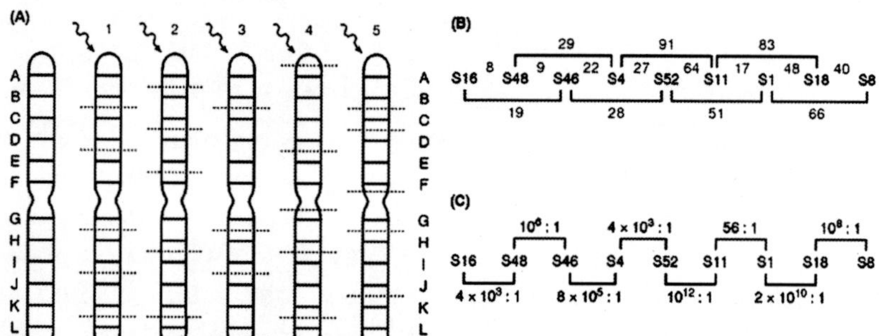

***Figure:** Constructing radiation hybrid maps.*

Radiation hybrids derived from monochromosomal hybrid donor cells have been superseded by whole-genome radiation hybrids where the donor is an irradiated normal human diploid cell. The first such panel consisted of 199 hybrids made by fusing an irradiated 46,XY human fibroblast cell line to TK^- hamster cells (Walter *et al.*, 1994). Gyapay *et al.* (1996) used 404 microsatellite markers of known location to show that this hybrid panel could generate accurate maps, and then used it to map 374 unmapped ESTs.

A subset of 93 of the hybrids has been made widely available as the Genebridge 4 panel. The 93 hybrids average 32% retention of any particular human sequence, with an average fragment size of 25 Mb. Laboratories can map any unknown STS by scoring the 93 Genebridge hybrids and comparing the pattern with patterns of previously mapped markers held on a central server.

This has turned into an extremely powerful and convenient tool for physically mapping any STS or EST. A second human-hamster panel, Stanford G3, was made using a higher dose of radiation, so that the average human fragment size is smaller. The 83 hybrids in G3 average 16% retention of the human genome, with an average fragment size of 2.4 Mb. Thus G3 can be used for finer mapping.

Chromosomal in Situ Hybridisation has been Revolutionised by Fluorescence in Situ Hybridisation Techniques

Chromosomal *in situ* hybridisation typically uses an air-dried microscope slide preparation of metaphase chromosomes, in which the chromosomal DNA has been denatured by exposure to formamide before hybridisation with a suitable probe. Isotopes such as ^{32}P are not suitable for probe labelling because their high energy causes scattering of the signal.

Instead, isotopes with weaker emission of beta radiation have been used. Owing to the high background noise when using such isotopes, however, the genuine signal is not easily identified and complex statistical analysis is required to differentiate genuine signal from background noise.

More recently, the sensitivity and resolution of *in situ* hybridisation has been increased significantly by the development of fluorescence *in situ* hybridisation (FISH). In this technique, the DNA probe is either labelled directly by incorporation of a fluorescent labelled nucleotide precursor, or indirectly by incorporation of a nucleotide containing a reporter molecule (such as biotin or digoxigenin) which after incorporation into the DNA is then bound by fluorescently labelled affinity molecule.

To increase the intensity of the hybridisation signal, large DNA probes are preferred, usually cosmid clones containing around 40 kb of insert. Because such large sequences will contain many interspersed repetitive DNA sequences, it is necessary to use chromosome *in situ* suppression hybridisation (Lichter *et al.,* 1990). Essentially, this is a form of competition hybridisation: before the main hybridisation, the probe is mixed with a large excess of unlabelled total genomic DNA and denatured, thereby saturating the repetitive elements in the probe, so that they no longer mask the signal generated by the unique sequences.

FISH has the advantage of providing rapid results which can be scored conveniently by eye using a fluorescence microscope. In metaphase spreads, positive signals show as double spots, corresponding to probe hybridised to both sister chromatids. Using sophisticated image processing equipment and reporter-binding molecules carrying different fluorophores, it is possible to map and order several DNA clones simultaneously. The maximum resolution of conventional FISH on metaphase chromosomes is several megabases.

The use of the more extended prometaphase chromosomes can permit 1 Mb resolution but, because of problems with chromatin folding, two differentially labelled probe signals may appear to be side-by-side, unless they are separated by distances greater than 2 Mb. Recently, however, new variations have been developed, permitting very high resolution. There are numerous applications for chromosome FISH but a special one is chromosome painting which has permitted a form of molecular karyotyping.

Flow Cytometry Permits Sorting of Individual Chromosomes and the Construction of Chromosome-specific DNA Libraries

Human chromosomes show considerable variation in size and DNA content. In addition, the base composition can vary considerably: chromosomes that are known to be gene-rich or gene-poor have a comparatively high % (G + C) or % (A + T) respectively. For example, chromosomes 21 and 22 are similar in size and in DNA content, but chromosome 21 is gene-poor and has a relatively high % (A + T), whereas chromosome 22 is gene-rich and has a relatively high % (G + C). As a result of the size differences and differences in base composition, human chromosomes can be separated by flow sorting (also called flow cytometry). In this technique, chromosome preparations are stained with a DNA-binding dye which can fluoresce in a laser beam.

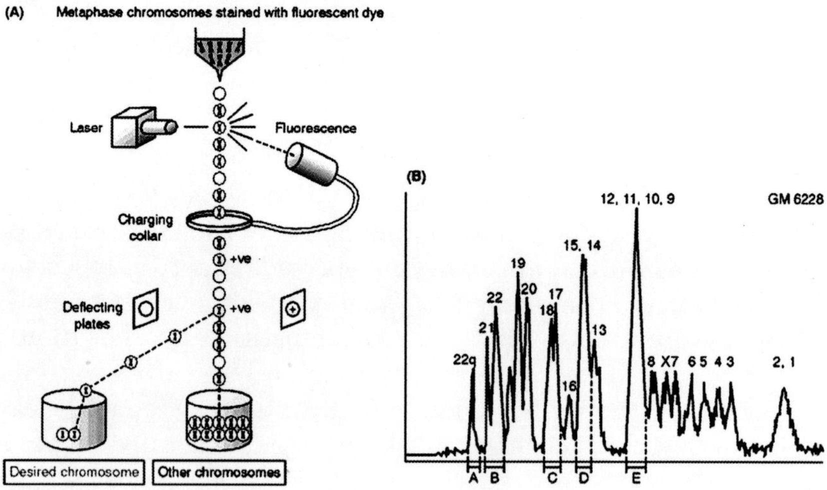

Figure: *Fractionating chromosomes in a flow cytometre. (A) Principle of chromosome fractionation.*

The amount of fluorescence exhibited by a given chromosome is proportional to the amount of dye bound, which in turn is largely proportional to the amount of DNA, and hence the size of the chromosome. Additionally, certain dyes, such as chromomycin A_3, show preferential binding to GC-rich sequences, while others, such as Hoechst 33258, bind preferentially to AT-rich regions.

Because individual chromosomes can bind different amounts of fluorescent dyes, they can be fractionated in a flow cytometre. A stream of droplets containing stained chromosomes is passed through a finely focused laser beam at a rate of about 1000–2000 chromosomes per second, and the fluorescence of individual chromosomes present in individual droplets is monitored by a photomultiplier. The intensity of fluorescence of the different chromosomes can then be recorded. By using sort windows to define a range of fluorescence intensities corresponding to a particular type of chromosome, droplets containing the desired chromosome can be given an electric charge and deflected from the main stream on to a separate collecting grid.

The resulting flow karyogram may show good separation of some human chromosomes; others such as chromosomes 9–12 are not well separated. However, monochromosomal somatic cell hybrids can be used instead, and often result in good separation of the human chromosome. As a result, panels of chromosome-specific DNA representing each of the 24 human chromosomes have been obtained.

These can be used to obtain a chromosomal localisation for any human DNA fragment: the fragment is labelled and hybridised to dot-blots of DNA from the 24 chromosomes ('flow-blots'), or a PCR assay can be used, if appropriate. Additionally, purified DNA from a specific type of chromosome can be amplified by cell-based or PCR-based DNA cloning to generate chromosome-specific DNA libraries (Davies et al., 1981). Chromosome libraries constructed using cosmid vectors have been particularly useful because YACs from a specific chromosome can be related quickly to cosmid clones from the same region. In addition, it has been possible in some cases to construct human chromosome-specific YAC libraries.

Clone Gridding and Increasing Automation Have Facilitated Large-scale Physical Mapping Efforts

Using current technologies, physical mapping of complex genomes is arduous and continues to be hampered by various practical difficulties:

- Chimeric DNA clones. Cloning artifacts can mean that cells contain two noncontiguous pieces of DNA from the desired genome. This may occur as a result of *co-ligation* of two different restriction fragments, internal deletion and *cotransformation*. Yeast cells that have been transformed by YACs and used to propagate have often been much more recombination proficient than typical *E. coli* hosts, and so YACs are often chimeric.
- Repetitive DNA. Arranging clones in a physical order means relying on some clone-specific features. DNA sequences that are repeated can make the task of ordering clones a difficult one, and may cause problems in sequencing DNA.

- Unclonable sequences. Some DNA sequences, including classes of repetitive DNA that are not well represented in bacterial and yeast cells, may not easily be propagated in such cells. A variety of different genomic libraries may need to be screened (including those constructed with vectors containing a low copy number origin of replication such as BAC and P1 libraries).

A major requirement for efficient physical mapping of complex genomes is the need for gridded DNA libraries. Following transformation of the host cells, individual bacterial or yeast colonies are picked into individual wells of microtiter dishes and stored in numerous multiwell dishes. As a result, each cell clone can be assigned an identifying grid coordinate, comprising dish number, row number and column number. When libraries need to be screened by hybridisation with a labelled DNA probe, it is again most efficient to have previously prepared high density gridded colony filters in which the colonies have been spotted on to the appropriate membrane in a gridded fashion with a uniform number of rows and columns.

This simplifies identification of hybridisation-positive colonies. More importantly, it facilitates integration of the results from numerous independent research groups. This is possible because of increasing automation. Robots have been designed to pick and spot colonies, and numerous copies of reference libraries can be distributed, either as cell cultures or, more conveniently, as sets of colony filters and sent to researchers throughout the world. Hybridisation data can then be sent back from individual laboratories to the distributing centres and the fixed geometry of the library coordinates simplifies the task of integrating the information received.

The Human Genome Project is one of several projects to benefit from such technologies, and sets of hybridisation filters containing gridded colonies for total genome YAC libraries, chromosome-specific cosmid libraries and other libraries have been distributed extensively. In addition, pools of yeast colonies have been distributed in agarose plugs to permit PCR typing for a sequence of interest. Again, this is done in a hierarchical coordinate manner to simplify the screening of libraries.

Clone Contigs can be Assembled Rapidly over Large Fractions of a Genome by Random clone Fingerprinting

Chromosome walking is a highly directional and location-restricted procedure for generating clone contigs: a great deal of effort is invested in generating a contig from a fixed starting point on a specific chromosome, often as a prelude to positional cloning. In order to build clone contigs spanning large amounts of a chromosome, or even a whole genome, a more general, labour-efficient mapping approach is needed. To meet this need, a variety of clone fingerprinting techniques have been used to type clones at random and then integrate the information in order to identify clone contigs over large regions of a genome. For example, this approach has been used to great effect in assembling YAC contigs over significant amounts of the human genome. Clone fingerprinting techniques should be simple and rapid, and the methods described below have been widely used.

Repetitive DNA Fingerprinting

Mammalian DNA has a very high component of highly repeated, interspersed DNA sequences such as the human *Alu* and LINE-1 (*Kpn*) repeats (occurring on average about once every 3 kb and 10 kb respectively). Because the spacing between such interspersed repetitive DNA elements will vary in different genomic regions, clones which contain overlapping DNA sequences may be identified by similar patterns following hybridisation of repetitive DNA probes to Southern blots of the clone DNA digested with a suitable restriction endonuclease. For example, a whole genome clone fingerprinting approach has been applied to mapping the human genome largely on the basis of repetitive DNA fingerprinting of YACs (Bellanne-Chantelot *et al.*, 1992).

3

Applied Genetics

Medicine

Genetic techniques are used in medicine to diagnose and treat inherited human disorders. Knowledge of a family history of conditions such as cancer or various disorders may indicate a hereditary tendency to develop these afflictions. Cells from embryonic tissues reveal certain genetic abnormalities, including enzyme deficiencies, that may be present in newborn babies, thus permitting early treatment. Many countries require a blood test of newborn babies to determine the presence of an enzyme necessary to convert an amino acid, phenylalanine, into simpler products. Phenylketonuria (PKU), which results from lack of the enzyme, causes permanent brain damage if not treated soon after birth. Many different types of human genetic diseases can be detected in embryos as young as 12 weeks; the procedure involves removal and testing of a small amount of fluid from around the embryo (called amniocentesis) or of tissue from the placenta (called chorionic villus sampling).

Gene therapy is based on modification of defective genotypes by adding functional genes made through recombinant DNA technology. Bioinformatics is being used to "mine" the human genome for gene products that might be candidates for designer pharmaceutical drugs.

Agriculture and Animal Husbandry

Breeding analysis and transgenic modification using recombinant DNA techniques are routinely used. Animal breeders use artificial insemination to propagate the genes of prize bulls. Prize cows can transmit their genes to hundreds of offspring by hormone treatment, which stimulates the release of many eggs that are collected, fertilized, and transplanted to foster mothers. Several types of mammals can be cloned, meaning that multiple identical copies can be produced of certain desirable types.Plant geneticists use special techniques to produce new species, such as hybrid grains (i.e., produced by crossing wheat and rye), and plants resistant to destruction by insect and fungal pests.

Plant breeders use the techniques of budding and grafting to maintain desirable gene combinations originally obtained from crossbreeding. Transgenic plant cells can be made into plants by growing the cells on special hormones.

The use of the chemical compound colchicine, which causes chromosomes to double in number, has resulted in many new varieties of fruits, vegetables, and flowers. Many transgenic lines of crop plants are commercially advantageous and are being introduced into the market.

Industry

Various industries employ geneticists; the brewing industry, for example, may use geneticists to improve the strains of yeast that produce alcohol. The pharmaceutical industry has developed strains of molds, bacteria, and other microorganisms high in antibiotic yield. Penicillin and cyclosporin from fungi, and streptomycin and ampicillin from bacteria, are some examples.

Biotechnology, based on recombinant DNA technology, is now extensively used in industry. "Designer" lines of transgenic bacteria, animals, or plants capable of manufacturing some commercial product are made and used routinely. Such products include pharmaceutical drugs and industrial chemicals such as citric acid.

Medical Genetics

Medical genetics is the speciality of medicine that involves the diagnosis and management of hereditary disorders. Medical genetics differs from Human genetics in that human genetics is a field of scientific research that may or may not apply to medicine, but medical genetics refers to the application of genetics to medical care.

For example, research on the causes and inheritance of genetic disorders would be considered within both human genetics and medical genetics, while the diagnosis, management, and counselling of individuals with genetic disorders would be considered part of medical genetics.

In contrast, the study of typically non-medical phenotypes such as the genetics of eye colour would be considered part of human genetics, but not necessarily relevant to medical genetics (except in situations such as albinism). *Genetic medicine* is a newer term for medical genetics and incorporates areas such as gene therapy, personalised medicine, and the rapidly emerging new medical speciality, predictive medicine.

Scope

Medical genetics encompasses many different areas, including clinical practice of physicians, genetic counselors, and nutritionists, clinical diagnostic laboratory activities, and research into the causes and inheritance of genetic disorders. Examples of conditions that fall within the scope of medical genetics include birth defects and dysmorphology, mental retardation, autism, and mitochondrial disorders, skeletal dysplasia, connective tissue disorders, cancer genetics, teratogens, and prenatal diagnosis.

Medical genetics is increasingly becoming relevant to many common diseases. Overlaps with other medical specialities are beginning to emerge, as recent advances in genetics are revealing etiologies for neurologic, endocrine, cardiovascular, pulmonary, ophthalmologic, renal, psychiatric, and dermatologic conditions.

Subspecialties

In some ways, many of the individual fields within medical genetics are hybrids between clinical care and research. This is due in part to recent advances in science and technology that have enabled an unprecedented understanding of genetic disorders.

Clinical Genetics

Clinical genetics is the practice of clinical medicine with particular attention to hereditary disorders. Referrals are made to genetics clinics for a variety of reasons, including birth defects, developmental delay, autism, epilepsy, short stature, and many others. Examples of genetic syndromes that are commonly seen in the genetics clinic include chromosomal rearrangements, Down syndrome, DiGeorge syndrome (22q11.2 Deletion Syndrome), Fragile X syndrome, Marfan syndrome, Neurofibromatosis, Turner syndrome, and Williams syndrome.

Metabolic/Biochemical Genetics

Metabolic (or biochemical) genetics involves the diagnosis and management of inborn errors of metabolism in which patients have enzymatic deficiencies that perturb biochemical pathways involved in metabolism of carbohydrates, amino acids, and lipids. Examples of metabolic disorders include galactosemia, glycogen storage disease, lysosomal storage disorders, metabolic acidosis, peroxisomal disorders, phenylketonuria, and urea cycle disorders.

Cytogenetics

Cytogenetics is the study of chromosomes and chromosome abnormalities. While cytogenetics historically relied on microscopy to analyse chromosomes, new molecular technologies such as array comparative genomic hybridisation are now becoming widely used. Examples of chromosome abnormalities include aneuploidy, chromosomal rearrangements, and genomic deletion/duplication disorders.

Molecular Genetics

Molecular genetics involves the discovery of and laboratory testing for DNA mutations that underlie many single gene disorders. Examples of single gene disorders include achondroplasia, cystic fibrosis, Duchenne muscular dystrophy, hereditary breast cancer (BRCA1/2), Huntington disease, Marfan syndrome, Noonan syndrome, and Rett syndrome. Molecular tests are also used in the diagnosis of syndromes involving epigenetic abnormalities, such as Angelman syndrome, Beckwith-Wiedemann syndrome, Prader-willi syndrome, and uniparental disomy.

Mitochondrial Genetics

Mitochondrial genetics concerns the diagnosis and management of mitochondrial disorders, which have a molecular basis but often result in biochemical abnormalities due to deficient energy production.

There exists some overlap between medical genetic diagnostic laboratories and molecular pathology.

Genetic Counselling

Genetic counselling is the process of providing information about genetic conditions, diagnostic testing, and risks in other family members, within the framework of nondirective counselling. Genetic counselors are non-physician members of the medical genetics team who specialize in family risk assessment and counselling of patients regarding genetic disorders. The precise role of the genetic counsellor varies somewhat depending on the disorder.

History

Although genetics has its roots back in the 19th century with the work of the Bohemian monk Gregor Mendel and other pioneering scientists, human genetics emerged later. It started to develop, albeit slowly, during the first half of the 20th century. Mendelian (single-gene) inheritance was studied in a number of important disorders such as albinism, brachydactyly (short fingers and toes), and hemophilia. Mathematical approaches were also devised and applied to human genetics. Population genetics was created.

Medical genetics was a late developer, emerging largely after the close of World War II (1945) when the eugenics movement had fallen into disrepute. The Nazi misuse of eugenics sounded its death knell. Shorn of eugenics, a scientific approach could be used and was applied to human and medical genetics. Medical genetics saw an increasingly rapid rise in the second half of the 20th century and continues in the 21st century.

Notable Practitioners

- Victor McKusick (1921–2008), "The Father of Genetic Medicine" and principal author of Mendelian Inheritance in Man.

Current Practice

The clinical setting in which patients are evaluated determines the scope of practice, diagnostic, and therapeutic interventions. For the purposes of general discussion, the typical encounters between patients and genetic practitioners may involve:

- Referral to an out-patient genetics clinic (pediatric, adult, or combined) or an in-hospital consultation, most often for diagnostic evaluation.
- Speciality genetics clinics focusing on management of inborn errors of metabolism, skeletal dysplasia, or lysosomal storage diseases.
- Referral for counselling in a prenatal genetics clinic to discuss risks to the pregnancy (advanced maternal age, teratogen exposure, family history of a genetic disease), test results (abnormal maternal serum screen, abnormal ultrasound), and/or options for prenatal diagnosis (typically amniocentesis or chorionic villus sampling).
- Multidisciplinary speciality clinics that include a clinical geneticist or genetic counsellor (cancer genetics, cardiovascular genetics, craniofacial or cleft lip/palate, hearing loss clinics, muscular dystrophy/ neurodegenerative disorder clinics).

Diagnostic Evaluation

Each patient will undergo a diagnostic evaluation tailored to their own particular presenting signs and symptoms. The geneticist will establish a differential diagnosis and recommend appropriate testing. Increasingly, clinicians use SimulConsult, paired with the National Library of Medicine Gene Review articles, to narrow the list of hypotheses (known as the differential diagnosis) and identify the tests that are relevant for a particular patient. These tests might evaluate for chromosomal disorders, inborn errors of metabolism, or single gene disorders.

Chromosome Studies

Chromosome studies are used in the general genetics clinic to determine a cause for developmental delay/mental retardation, birth defects, dysmorphic features, and/or autism. Chromosome analysis is also performed in the prenatal setting to determine whether a fetus is affected with aneuploidy or other chromosome rearrangements.

Finally, chromosome abnormalities are often detected in cancer samples. A large number of different methods have been developed for chromosome analysis:

- Chromosome analysis using a karyotype involves special stains that generate light and dark bands, allowing identification of each chromosome under a microscope.
- Fluorescence in situ hybridisation (FISH) involves fluorescent labelling of probes that bind to specific DNA sequences, used for identifying aneuploidy, genomic deletions or duplications, characterising chromosomal translocations and determining the origin of ring chromosomes.
- Chromosome painting is a technique that uses fluorescent probes specific for each chromosome to differentially label each chromosome. This technique is more often used in cancer cytogenetics, where complex chromosome rearrangements can occur.
- Array comparative genomic hybridisation is a new molecular technique that involves hybridisation of an individual DNA sample to a glass slide or microarray chip containing molecular probes (ranging from large ~200kb bacterial artificial chromosomes to small oligonucleotides) that represent unique regions of the genome. This method is particularly sensitive for detection of genomic gains or losses across the genome but does not detect balanced translocations or distinguish the location of duplicated genetic material (for example, a tandem duplication versus an insertional duplication).

Basic Metabolic Studies

Biochemical studies are performed to screen for imbalances of metabolites in the bodily fluid, usually the blood (plasma/serum) or urine, but also in cerebrospinal fluid (CSF).

Specific tests of enzyme function (either in leukocytes, skin fibroblasts, liver, or muscle) are also employed under certain circumstances. In the US, the newborn screen incorporates biochemical tests to screen for treatable conditions such as galactosemia and phenylketonuria (PKU). Patients suspected to have a metabolic condition might undergo the following tests:

- Quantitative amino acid analysis is typically performed using the ninhydrin reaction, followed by liquid chromatography to measure the amount of amino acid in the sample (either urine, plasma/serum, or CSF). Measurement of amino acids in plasma or serum is used in the evaluation of disorders of amino acid metabolism such as urea cycle disorders, maple syrup urine disease, and PKU. Measurement of amino acids in urine can be useful in the diagnosis of cystinuria or renal Fanconi syndrome as can be seen in cystinosis.
- Urine organic acid analysis can be either performed using quantitative or qualitative methods, but in either case the test is used to detect the excretion of abnormal organic acids. These compounds are normally produced during bodily metabolism of amino acids and odd-chain fatty acids, but accumulate in patients with certain metabolic conditions.
- The acylcarnitine combination profile detects compounds such as organic acids and fatty acids conjugated to carnitine. The test is used for detection of disorders involving fatty acid metabolism, including MCAD.
- Pyruvate and lactate are byproducts of normal metabolism, particularly during anaerobic metabolism. These compounds normally accumulate during exercise or ischemia, but are also elevated in patients with disorders of pyruvate metabolism or mitochondrial disorders.
- Ammonia is an end product of amino acid metabolism and is converted in the liver to urea through a series of enzymatic reactions termed the urea cycle. Elevated ammonia can therefore be detected in patients with urea cycle disorders, as well as other conditions involving liver failure.
- Enzyme testing is performed for a wide range of metabolic disorders to confirm a diagnosis suspected based on screening tests.

Molecular Studies

- DNA sequencing is used to directly analyse the genomic DNA sequence of a particular gene. In general, only the parts of the gene that code for the expressed protein (exons) and small amounts of the flanking untranslated regions and introns are analysed. Therefore, although these tests are highly specific and sensitive, they do not routinely identify all of the mutations that could cause disease.

- DNA methylation analysis is used to diagnose certain genetic disorders that are caused by disruptions of epigenetic mechanisms such as genomic imprinting and uniparental disomy.
- Southern blotting is an early technique basic on detection of fragments of DNA separated by size through gel electrophoresis and detected using radiolabelled probes. This test was routinely used to detect deletions or duplications in conditions such as Duchenne muscular dystrophy but is being replaced by high-resolution array comparative genomic hybridisation techniques. Southern blotting is still useful in the diagnosis of disorders caused by trinucleotide repeats.
- Short tandem repeats are unique markers that can be used to determine haplotypes and are used in identity testing for maternal cell contamination.

Treatments

Each cell of the body contains the hereditary information (DNA) wrapped up in structures called chromosomes. Since genetic syndromes are typically the result of alterations of the chromosomes or genes, there is no treatment currently available that can correct the genetic alterations in every cell of the body. Therefore, there is currently no "cure" for genetic disorders. However, for many genetic syndromes there is treatment available to manage the symptoms. In some cases, particularly inborn errors of metabolism, the mechanism of disease is well understood and offers the potential for dietary and medical management to prevent or reduce the long-term complications. In other cases, infusion therapy is used to replace the missing enzyme. Current research is actively seeking to use gene therapy or other new medications to treat specific genetic disorders.

Management of Metabolic Disorders

In general, metabolic disorders arise from enzyme deficiencies that disrupt normal metabolic pathways. For instance, in the hypothetical example:

```
A ---> B ---> C ---> D       AAAA ---> BBBBBB ---> CCCCCCCCCC ---> (no D)
     X      Y      Z                   X              Y            (no Z)
```

Compound "A" is metabolised to "B" by enzyme "X", compound "B" is metabolised to "C" by enzyme "Y", and compound "C" is metabolised to "D" by enzyme "Z". If enzyme "Z" is missing, compound "D" will be missing, while compounds "A", "B", and "C" will build up. The pathogenesis of this particular condition could result from lack of compound "D", if it is critical for some cellular function, or from toxicity due to excess "A", "B", and/or "C". Treatment of the metabolic disorder could be achieved through dietary supplementation of compound "D" and dietary restriction of compounds "A", "B", and/or "C" or by treatment with a medication that promoted disposal of excess "A", "B", or "C". Another approach that can be taken is enzyme replacement therapy, in which a patient is given an infusion of the missing enzyme.

Diet

Dietary restriction and supplementation are key measures taken in several well-known metabolic disorders, including galactosemia, phenylketonuria (PKU), maple syrup urine disease, organic acidurias and urea cycle disorders. Such restrictive diets can be difficult for the patient and family to maintain, and require close consultation with a nutritionist who has special experience in metabolic disorders.

The composition of the diet will change depending on the caloric needs of the growing child and special attention is needed during a pregnancy if a woman is affected with one of these disorders.

Medication

Medical approaches include enhancement of residual enzyme activity (in cases where the enzyme is made but is not functioning properly), inhibition of other enzymes in the biochemical pathway to prevent buildup of a toxic compound, or diversion of a toxic compound to another form that can be excreted.

Examples include the use of high doses of pyridoxine (vitamin B6) in some patients with homocystinuria to boost the activity of the residual cystathione synthase enzyme, administration of biotin to restore activity of several enzymes affected by deficiency of biotinidase, treatment with NTBC in Tyrosinemia to inhibit the production of succinylacetone which causes liver toxicity, and the use of sodium benzoate to decrease ammonia build-up in urea cycle disorders.

Enzyme Replacement Therapy

Certain lysosomal storage diseases are treated with infusions of a recombinant enzyme (produced in a laboratory), which can reduce the accumulation of the compounds in various tissues. Examples include Gaucher disease, Fabry disease, Mucopolysaccharidoses and Glycogen storage disease type II. Such treatments are limited by the ability of the enzyme to reach the affected areas (the blood brain barrier prevents enzyme from reaching the brain, for example), and can sometimes be associated with allergic reactions. The long-term clinical effectiveness of enzyme replacement therapies vary widely among different disorders.

Other Examples

- Angiotensin receptor blockers in Marfan syndrome & Loeys-Dietz
- Bone marrow transplantation
- Gene therapy.

Career Paths and Training

There are a variety of career paths within the field of medical genetics, and naturally the training required for each area differs considerably. It should be noted that the information included in this section applies to the typical pathways in the United States and there may be differences in other countries. US Practitioners in clinical, counselling, or diagnostic subspecialities generally obtain board certification through the American Board of Medical Genetics.

Ethical, Legal and Social Implications

Genetic information provides a unique type of knowledge about an individual and his/her family, fundamentally different than a typically laboratory test that provides a "snapshot" of an individual's health status. The unique status of genetic information and inherited disease has a number of ramifications with regard to ethical, legal, and societal concerns.

Societies

The more empirical approach to human and medical genetics was formalised by the founding in 1948 of the American Society of Human Genetics. The Society first began annual meetings that year (1948) and its international counterpart, the International Congress of Human Genetics, has met every 5 years since its inception in 1956.

The Society publishes the American Journal of Human Genetics on a monthly basis. Medical genetics is now recognised as a distinct medical speciality in the U.S. with its own approved board (the American Board of Medical Genetics) and clinical speciality college (the American College of Medical Genetics).

The College holds an annual scientific meeting, publishes a monthly journal, Genetics in Medicine, and issues position papers and clinical practice guidelines on a variety of topics relevant to human genetics.

Research

The broad range of research in medical genetics reflects the overall scope of this field, including basic research on genetic inheritance and the human genome, mechanisms of genetic and metabolic disorders, translational research on new treatment modalities, and the impact of genetic testing

Basic Genetics Research

Basic research geneticists usually undertake research in universities, biotechnology firms and research institutes.

Allelic Architecture of Disease

Sometimes the link between a disease and an unusual gene variant is more subtle. The genetic architecture of common diseases is an important factor in determining the extent to which patterns of genetic variation influence group differences in health outcomes. According to the common disease/common variant hypothesis, common variants present in the ancestral population before the dispersal of modern humans from Africa play an important role in human diseases. Genetic variants associated with Alzheimer disease, deep venous thrombosis, Crohn disease, and type 2 diabetes appear to adhere to this model. However, the generality of the model has not yet been established and, in some cases, is in doubt. Some diseases, such as many common cancers, appear not to be well described by the common disease/common variant model.

Another possibility is that common diseases arise in part through the action of combinations of variants that are individually rare. Most of the disease-associated alleles discovered to date have been rare, and rare variants are more likely than common variants to be differentially distributed among groups distinguished by ancestry. However, groups could harbour different, though perhaps overlapping, sets of rare variants, which would reduce contrasts between groups in the incidence of the disease.

The number of variants contributing to a disease and the interactions among those variants also could influence the distribution of diseases among groups. The difficulty that has been encountered in finding contributory alleles for complex diseases and in replicating positive associations suggests that many complex diseases involve numerous variants rather than a moderate number of alleles, and the influence of any given variant may depend in critical ways on the genetic and environmental background. If many alleles are required to increase susceptibility to a disease, the odds are low that the necessary combination of alleles would become concentrated in a particular group purely through drift.

Population Groups in Biomedicine

Race and health research, often done in the United States, has found both current and historical racial differences in the frequency, treatments, and availability of treatments for several diseases. This can add up to significant group differences in variables such as life expectancy. Many explanations for such differences have been argued, including socioeconomic factors (e.g., education, employment, and income), lifestyle behaviours (e.g., physical activity and alcohol intake), social environment (e.g., educational and economic opportunities, racial/ethnic discrimination, and neighbourhood and work conditions), and access to preventive health-care services (e.g., cancer screening and vaccination) as well as to treatment (through lack of insurance, lack of hospitals in certain areas, etc.), among other environmental differences.

Some diseases may also be influenced by genes which differ in frequency between groups, such as sickle-cell anemia, which occurs overwhelmingly among some black populations, although the significance in clinical medicine of race categories as a proxy for exact genotypes of individuals has been questioned.

Health Disparities

Health disparities refer to gaps in the quality of health and health care across racial and ethnic groups. The US Health Resources and Services Administration defines health disparities as "population-specific differences in the presence of disease, health outcomes, or access to health care." Health is measured through variables such as life expectancy and incidence of diseases.

Single Gene Disorders

There are many single gene genetic disorders that differ in frequency between different populations. Some examples are discussed below.

Cystic fibrosis is the most common life-limiting autosomal recessive disease among people of European heritage.

Sickle-cell anemia is most prevalent in populations of sub-Saharan African ancestry, but it is also common among Latin-American, Indian, Saudi Arab, and Mediterranean populations such as Turkey, Greece, and Italy.

Tay-Sachs Disease is an autosomal recessive disorder that is more frequent among Ashkenazi Jews than among other Jewish groups and non-Jewish populations.

Hereditary hemochromatosis is most common among those of Northern European ancestry, in particular those of Celtic descendent.

Lactose intolerance is another examples of single gene genetic disorders that differ in frequency between populations.

Multifactorial Polygenic Diseases

These are many diseases which differ in frequency between different populations.

Medicines Targeted at Specific Races

Race-based medicine is the term for medicines that are targeted at specific ethnic clusters which are shown to have a propensity for a certain disorder.

The first example of this in the U.S. was when BiDil, a medication for congestive heart failure, was licensed specifically for use in American patients that self-identify as black. Previous studies had shown that African American patients with congestive heart failure generally respond less effectively to traditional treatments than white patients with similar conditions.

After two trials, BiDil was licensed exclusively for use in African American patients. Critics have argued that this particular licensing was unwarranted, since the trials did not in fact show that the drug was more effective in African Americans than in other groups, but merely that it was more effective in African Americans than other similar drug. It was also only tested in African American males, but not in any other racial groups or among women. This peculiar trial and licensing procedure has prompted suggestions that the licensing was in fact used as a race based advertising scheme.

Environmental Explanations

In multiracial societies such as the United States, racial groups differ greatly in regard to social and cultural factors such as socioeconomic status, healthcare, diet, and education. There is also the presence of racism which some see as a very important explaining factor. Some argue that for many diseases racial differences would disappear if all environmental factors could be controlled for. These factors may or may not be appropriate in other nations.

Ethnic minorities may also have specific health care needs which need to be taken into consideration by health services in order to tackle health disparities.

Genetic Explanations

Evolutionary Explanations

Genes may be under strong selection in response to local diseases. For example, people who are duffy negative tend to have higher resistance to malaria. Most Africans are duffy negative and most non-Africans are duffy positive. A number of genetic diseases more prevalent in malaria-afflicted areas may provide some genetic resistance to malaria including sickle cell disease, thalassaemias, glucose-6-phosphate dehydrogenase, and possibly others.

Many theories about the origin of the Cystic fibrosis have suggested that it provides a heterozygote advantage by giving resistance to diseases earlier common in Europe.

In earlier research a common theory was the "common disease-common variant" model. It argues that for common illnesses, the genetic contribution comes from the additive or multiplicative effects of gene variants that each one is common in the population. Each such gene variant is argued to cause only a small risk of disease and no single variant is enough to cause the disease. An individual must have many of these common gene variants in order for the risk of disease to be substantial. More recent research indicates that the "common disease-rare variant" may be a better explanation for many common diseases.

In this model, rare but higher-risk gene variants cause common diseases. This model may be particularly relevant for diseases that reduces fertility. In contrast, for common genes associated with common disease to persist they must either have little effect during the reproductive period of life (like Alzheimer's disease) or provide some advantage in the original environment (like genes causing autoimmune diseases also providing resistance against infections).

In either case varying frequencies of genes variants in different populations may be an explanation for health disparities. Genetic variants associated with Alzheimer's disease, deep venous thrombosis, Crohn disease, and type 2 diabetes appear to adhere to "common disease-common variant" model.

Gene Flow

Gene flow and admixture can also have an effect on relationships between race and race-linked disorders. Multiple sclerosis, for example, is typically associated with people of European descent, but due to admixture African Americans have elevated levels of the disorder relative to Africans.

Some diseases and physiological variables vary depending upon their admixture ratios. Examples include measures of insulin functioning and obesity.

Gene Interactions

The same gene variant, or group of gene variants, may produce different effects in different populations depending on differences in the gene variants, or groups of gene variants, they interact with. One example is the rate of progression to AIDS and death in HIV–infected patients. In Caucasians and Hispanics, HHC haplotypes were associated with disease retardation, particularly a delayed progression to death. In contrast, for African Americans, possession of HHC haplotypes was associated with disease acceleration.

Controversy Regarding Race in Biomedicine

There is a controversy regarding race as a method for classifying humans.

New interest in human biological variation has resulted in a resurgence of the use of race in biomedicine. The main impetus for this development is the possibility of improving the prevention and treatment of certain diseases by predicting hard-to-ascertain factors, such as genetically conditioned health factors, on the basis of more easily ascertained characteristics such as phenotype and racial self-identification. Since medical judgment often involves decision-making under uncertain conditions, many doctors consider it useful to take race into account when treating disease because diseases and treatment responses tend to cluster by geographic ancestry. The discovery that more diseases than previously thought correlate with racial identification have further sparked the interest in using race as a proxy for bio-geographical ancestry and genetic buildup.

Race in medicine is used as an approximation for more specific genetic and environmental risk factors. Race is thus partly a surrogate for environmental factors such as differences in socioeconomic status that are known to affect health. It is also an imperfect surrogate for ancestral geographic regions and differences in gene frequencies between different ancestral populations and thus differences in genes that can affect health. This can give an approximation of probability for disease or for preferred treatment, although the approximation is less than perfect.

Taking the example of sickle-cell disease, in an emergency room, knowing the geographic origin of a patient may help a doctor doing an initial diagnosis if a patient presents with symptoms compatible with this disease. This is unreliable evidence with the disease being present in many different groups as noted above with the trait also present in some Mediterranean European populations. Definitive diagnosis comes from examining the blood of the patient. In the US, screening for sickle cell anemia is done on all newborns regardless of race.

The continued use of racial categories has been criticized. Apart from the general controversy regarding race, some argue that the continued use of racial categories in health care and as risk factors could result in increased stereotyping and discrimination in society and health services. On the other hand, also some of those who are critical of race as a biological concept see race as socially meaningful group that is important to study epidemiologically in order to reduce disparities.

David Williams (1994) argued, after an examination of articles in the journal *Health Services Research* during the 1966-90 period, that how race was determined and defined was seldom described. At a minimum, researchers should describe if race was assessed by self-report, proxy report, extraction from records, or direct observation. Race was also often used questionable, such as an indicator of socioeconomic status. Racial genetic explanations may be overemphasised, ignoring the interaction with and the role of the environment. There is general agreement that a goal of health-related genetics should be to move past the weak surrogate relationships of racial health disparity and get to the root causes of health and disease. This includes research which strives to analyse human genetic variation in smaller groups than races across the world. One such method is called ethnogenetic layering. It works by focusing on geographically identified microethnic groups.

For example, in the Mississippi Delta region ethnogenetic layering might include such microethnic groups as the Cajun (as a subset of European Americans), the Creole and Black groups [with African origins in Senegambia, Central Africa and Bight of Benin] (as a subset of African Americans), and Choctaw, Houmas, Chickasaw, Coushatta, Caddo, Atakapa, Karankawa and Chitimacha peoples (as subsets of Native American Indians).

Better still may be individual genetic assessment of relevant genes. However, until cheaper and more widely available methods of genetic analysis are commonplace, the consideration of race remains a worthwhile practice for many doctors and researchers. Even when such methods become commonly available, race will continue to be important when looking at groups instead of individuals such as in epidemiologic research.

Association Studies

One area in which population categories can be important considerations in genetics research is in controlling for confounding between population genetic substructure, environmental exposures, and health outcomes. Association studies can produce spurious results if cases and controls have differing allele frequencies for genes that are not related to the disease being studied, although the magnitude of its problem in genetic association studies is subject to debate.

Various techniques detect and account for population substructure, but these methods can be difficult to apply in practice.

Population genetic substructure also can aid genetic association studies. For example, populations that represent recent mixtures of separated ancestral groups can exhibit longer-range linkage disequilibrium between susceptibility alleles and genetic markers than is the case for other populations.

Genetic studies can use this disequilibrium to search for disease alleles with fewer markers than would be needed otherwise. Association studies also can take advantage of the contrasting experiences of racial or ethnic groups, including migrant groups, to search for interactions between particular alleles and environmental factors that might influence health.

Human Genome Projects

The Human Genome Diversity Project has collected genetic samples from 52 indigenous populations. This sampling has been controversial because some fear that racists might use the results, because the sampling might overemphasize genetic distinctions, because of informed consent issues, and patenting issues.

Several other genome projects also studies populations from around the world but often not as diverse.

Race and Health in the United States

Research on Race and health in the United States shows many health disparities between the different racial/ethnic groups. The possible causes, such as genetics, socioeconomic factors, and racism, continue to be debated.

Population Substructure in Genetics Research

One area in which population categories can be important considerations in genetics research is in controlling for confounding between population substructure, environmental exposures, and health outcomes. Association studies can produce spurious results if cases and controls have differing allele frequencies for genes that are not related to the disease being studied, although the magnitude of this problem in genetic association studies is subject to debate. Various methods have been developed to detect and account for population substructure, but these methods can be difficult to apply in practice.

Population substructure also can be used to advantage in genetic association studies. For example, populations that represent recent mixtures of geographically separated ancestral groups can exhibit longer-range linkage disequilibrium between susceptibility alleles and genetic markers than is the case for other populations. Genetic studies can use this admixture linkage disequilibrium to search for disease alleles with fewer markers than would be needed otherwise. Association studies also can take advantage of the contrasting experiences of racial or ethnic groups, including migrant groups, to search for interactions between particular alleles and environmental factors that might influence health.

Human Genome Project

The Human Genome Project (HGP) is an international scientific research project with a primary goal of determining the sequence of chemical base pairs which make up DNA, and of identifying and mapping the approximately 20,000–25,000 genes of the human genome from both a physical and functional standpoint.

The project began in October 1990 and was initially headed by Ari Patrinos, head of the Office of Biological and Environmental Research in the U.S. Department of Energy's Office of Science. Francis Collins directed the National Institutes of Health National Human Genome Research Institute efforts. A working draft of the genome was announced in 2000 and a complete one in 2003, with further, more detailed analysis still being published. A parallel project was conducted outside of government by the Celera Corporation, which was formally launched in 1998. Most of the government-sponsored sequencing was performed in universities and research centres from the United States, the United Kingdom, Japan, France, Germany. The mapping of human genes is an important step in the development of medicines and other aspects of health care. While the objective of the Human Genome Project is to understand the genetic makeup of the human species, the project has also focused on several other nonhuman organisms such as *E. coli*, the fruit fly, and the laboratory mouse. It remains one of the largest single investigative projects in modern science.

The Human Genome Project originally aimed to map the nucleotides contained in a human haploid reference genome (more than three billion). Several groups have announced efforts to extend this to diploid human genomes including the International HapMap Project, Applied Biosystems, Perlegen, Illumina, J. Craig Venter Institute, Personal Genome Project, and Roche-454.

The "genome" of any given individual (except for identical twins and cloned organisms) is unique; mapping "the human genome" involves sequencing multiple variations of each gene. The project did not study the entire DNA found in human cells; some heterochromatic areas (about 8% of the total genome) remain unsequenced.

Project

History

The project began with the culmination of several years of work supported by the United States Department of Energy, in particular workshops in 1984 and 1986 and a subsequent initiative of the US Department of Energy. This 1987 report stated boldly, "The ultimate goal of this initiative is to understand the human genome" and "knowledge of the human is as necessary to the continuing progress of medicine and other health sciences as knowledge of human anatomy has been for the present state of medicine." Candidate technologies were already being considered for the proposed undertaking at least as early as 1985.

James D. Watson was head of the National Centre for Human Genome Research at the National Institutes of Health (NIH) in the United States starting from 1988. Largely due to his disagreement with his boss, Bernadine Healy, over the issue of patenting genes, Watson was forced to resign in 1992. He was replaced by Francis Collins in April 1993, and the name of the Centre was changed to the National Human Genome Research Institute (NHGRI) in 1997. The $3-billion project was formally founded in 1990 by the United States Department of Energy and the U.S. National Institutes of Health, and was expected to take 15 years. In addition to the United States, the international consortium comprised geneticists in the United Kingdom, France, Germany, Japan, China, and India.

Due to widespread international cooperation and advances in the field of genomics (especially in sequence analysis), as well as major advances in computing technology, a 'rough draft' of the genome was finished in 2000 (announced jointly by U.S. President Bill Clinton and the British Prime Minister Tony Blair on June 26, 2000).

This first available rough draft assembly of the genome was completed by the Genome Bioinformatics Group at the University of California, Santa Cruz, primarily led by then graduate student Jim Kent. Ongoing sequencing led to the announcement of the essentially complete genome in April 2003, 2 years earlier than planned. In May 2006, another milestone was passed on the way to completion of the project, when the sequence of the last chromosome was published in the journal Nature.

State of Completion

There are multiple definitions of the "complete sequence of the human genome". According to some of these definitions, the genome has already been completely sequenced, and according to other definitions, the genome has yet to be completely sequenced. There have been multiple popular press articles reporting that the genome was "complete."

The genome has been completely sequenced using the definition employed by the International Human Genome Project. A graphical history of the human genome project shows that most of the human genome was complete by the end of 2003. However, there are a number of regions of the human genome that can be considered unfinished:

- First, the central regions of each chromosome, known as centromeres, are highly repetitive DNA sequences that are difficult to sequence using current technology. The centromeres are millions (possibly tens of millions) of base pairs long, and for the most part these are entirely unsequenced.
- Second, the ends of the chromosomes, called telomeres, are also highly repetitive, and for most of the 46 chromosome ends these too are incomplete. It is not known precisely how much sequence remains before the telomeres of each chromosome are reached, but as with the centromeres, current technological restraints are prohibitive.
- Third, there are several loci in each individual's genome that contain members of multigene families that are difficult to disentangle with shotgun sequencing methods – these multigene families often encode proteins important for immune functions.
- Other than these regions, there remain a few dozen gaps scattered around the genome, some of them rather large, but there is hope that all these will be closed in the next couple of years.

The best estimates of total genome size indicate that about 92.3% of the genome has been completed and it is likely that the centromeres and telomeres will remain unsequenced until new technology is developed that facilitates their sequencing. Most of the remaining DNA is highly repetitive and unlikely to contain genes, but it cannot be truly known until it is entirely sequenced. Understanding the functions of all the genes and their regulation is far from complete. The roles of junk DNA, the evolution of the genome, the differences between individuals, and many other questions are still the subject of intense interest by laboratories all over the world.

Goals

The sequence of the human DNA is stored in databases available to anyone on the Internet. The U.S. National Centre for Biotechnology Information (and sister organisations in Europe and Japan) house the gene sequence in a database known as GenBank, along with sequences of known and hypothetical genes and proteins. Other organisations, such as the Genome Bioinformatics Group at the University of California, Santa Cruz, and Ensembl present additional data and annotation and powerful tools for visualising and searching it. Computer programs have been developed to analyse the data, because the data itself is difficult to interpret without such programs.

The process of identifying the boundaries between genes and other features in a raw DNA sequence is called genome annotation and is the domain of bioinformatics. While expert biologists make the best annotators, their work proceeds slowly, and computer programs are increasingly used to meet the high-throughput demands of genome sequencing projects. The best current technologies for annotation make use of statistical models that take advantage of parallels between DNA sequences and human language, using concepts from computer science such as formal grammars. All humans have unique gene sequences. Therefore the data published by the HGP does not represent the exact sequence of every individual's genome. It is the combined "reference genome" of a small number of anonymous donors. The HGP genome is a scaffold for future work in identifying differences among individuals. Most of the current effort in identifying differences among individuals involves single-nucleotide polymorphisms and the HapMap.

Findings

Key findings of the draft (2001) and complete (2004) genome sequences include:

1. There are approximately 30,000 genes in human beings, the same range as in mice and roundworms. Understanding how these genes express themselves will provide clues to how diseases are caused.
2. Between 1.1% to 1.4% of the genome sequence codes for proteins

3. The human genome has significantly more segmental duplications (nearly identical, repeated sections of DNA) than other mammalian genomes. These sections may underlie the creation of new primate-specific genes
4. At the time when the draft sequence was published less than 7% of protein families appeared to be vertebrate specific.

How it was Accomplished

The Human Genome Project was started in 1989 with the goal of sequencing and identifying all three billion chemical units in the human genetic instruction set, finding the genetic roots of disease and then developing treatments. With the sequence in hand, the next step was to identify the genetic variants that increase the risk for common diseases like cancer and diabetes.

It was far too expensive at that time to think of sequencing patients' whole genomes. So the National Institutes of Health embraced the idea for a "shortcut", which was to look just at sites on the genome where many people have a variant DNA unit. The theory behind the shortcut was that since the major diseases are common, so too would be the genetic variants that caused them. Natural selection keeps the human genome free of variants that damage health before children are grown, the theory held, but fails against variants that strike later in life, allowing them to become quite common. (In 2002 the National Institutes of Health started a $138 million project called the HapMap to catalogue the common variants in European, East Asian and African genomes.)

The genome was broken into smaller pieces; approximately 150,000 base pairs in length. These pieces were then ligated into a type of vector known as "bacterial artificial chromosomes", or BACs, which are derived from bacterial chromosomes which have been genetically engineered. The vectors containing the genes can be inserted into bacteria where they are copied by the bacterial DNA replication machinery. Each of these pieces was then sequenced separately as a small "shotgun" project and then assembled. The larger, 150,000 base pairs go together to create chromosomes. This is known as the "hierarchical shotgun" approach, because the genome is first broken into relatively large chunks, which are then mapped to chromosomes before being selected for sequencing.

Funding came from the US government through the National Institutes of Health in the United States, and a UK charity organisation, the Wellcome Trust, as well as numerous other groups from around the world. The funding supported a number of large sequencing centres including those at Whitehead Institute, the Sanger Centre, Washington University in St. Louis, and Baylor College of Medicine.

The Human Genome Project is considered a Mega Project because the human genome has approximately 3.3 billion base-pairs.

If the sequence obtained was to be stored in book form, and if each page contained 1000 base-pairs recorded and each book contained 1000 pages, then 3300 such books would be needed in order to store the complete genome. However, if expressed in units of computer data storage, 3.3 billion base-pairs recorded at 2 bits per pair would equal 786 megabytes of raw data. This is comparable to a fully data loaded CD.

Public Versus Private Approaches

In 1998, a similar, privately funded quest was launched by the American researcher Craig Venter, and his firm Celera Genomics. Venter was a scientist at the NIH during the early 1990s when the project was initiated. The $300,000,000 Celera effort was intended to proceed at a faster pace and at a fraction of the cost of the roughly $3 billion publicly funded project.

Celera used a technique called whole genome shotgun sequencing, employing pairwise end sequencing, which had been used to sequence bacterial genomes of up to six million base pairs in length, but not for anything nearly as large as the three billion base pair human genome.

Celera initially announced that it would seek patent protection on "only 200–300" genes, but later amended this to seeking "intellectual property protection" on "fully-characterised important structures" amounting to 100–300 targets. The firm eventually filed preliminary ("place-holder") patent applications on 6,500 whole or partial genes. Celera also promised to publish their findings in accordance with the terms of the 1996 "Bermuda Statement," by releasing new data annually (the HGP released its new data daily), although, unlike the publicly funded project, they would not permit free redistribution or scientific use of the data.

The publicly funded competitor UC Santa Cruz was compelled to publish the first draft of the human genome before Celera for this reason. On July 7, 2000, the UCSC Genome Bioinformatics Group released a first working draft on the web. The scientific community downloaded one-half trillion bytes of information from the UCSC genome server in the first 24 hours of free and unrestricted access to the first ever assembled blueprint of our human species.

In March 2000, President Clinton announced that the genome sequence could not be patented, and should be made freely available to all researchers. The statement sent Celera's stock plummeting and dragged down the biotechnology-heavy Nasdaq. The biotechnology sector lost about $50 billion in market capitalisation in two days. But the public release of the data ensured its fair use and availability to all mankind.

Although the working draft was announced in June 2000, it was not until February 2001 that Celera and the HGP scientists published details of their drafts. Special issues of *Nature* (which published the publicly funded project's scientific paper) and *Science* (which published Celera's paper) described the methods used to produce the draft sequence and offered analysis of the sequence. These drafts covered about 83% of the genome (90% of the euchromatic regions with 150,000 gaps and the order and orientation of many segments not yet established). In February 2001, at the time of the joint publications, press releases announced that the project had been completed by both groups. Improved drafts were announced in 2003 and 2005, filling in to H"92% of the sequence currently.

The competition proved to be very good for the project, spurring the public groups to modify their strategy in order to accelerate progress. The rivals at UC Santa Cruz initially agreed to pool their data, but the agreement fell apart when Celera refused to deposit its data in the unrestricted public database GenBank. Celera had incorporated the public data into their genome, but forbade the public effort to use Celera data.

HGP is the most well known of many international genome projects aimed at sequencing the DNA of a specific organism. While the human DNA sequence offers the most tangible benefits, important developments in biology and medicine are predicted as a result of the sequencing of model organisms, including mice, fruit flies, zebrafish, yeast, nematodes, plants, and many microbial organisms and parasites.

In 2004, researchers from the International Human Genome Sequencing Consortium (IHGSC) of the HGP announced a new estimate of 20,000 to 25,000 genes in the human genome. Previously 30,000 to 40,000 had been predicted, while estimates at the start of the project reached up to as high as 2,000,000. The number continues to fluctuate and it is now expected that it will take many years to agree on a precise value for the number of genes in the human genome.

Genome Donors

In the IHGSC international public-sector Human Genome Project (HGP), researchers collected blood (female) or sperm (male) samples from a large number of donors. Only a few of many collected samples were processed as DNA resources. Thus the donor identities were protected so neither donors nor scientists could know whose DNA was sequenced. DNA clones from many different libraries were used in the overall project, with most of those libraries being created by Dr. Pieter J. de Jong. Much of the sequence (>70%) of the reference genome produced by the public HGP came from a single anonymous male donor from Buffalo, New York (code name RP11).

HGP scientists used white blood cells from the blood of two male and two female donors (randomly selected from 20 of each) each donor yielding a separate DNA library. One of these libraries (RP11) was used considerably more than others, due to quality considerations. One minor technical issue is that male samples contain just over half as much DNA from the sex chromosomes (one X chromosome and one Y chromosome) compared to female samples (which contain two X chromosomes). The other 22 chromosomes (the autosomes) are the same for both genders.

Although the main sequencing phase of the HGP has been completed, studies of DNA variation continue in the International HapMap Project, whose goal is to identify patterns of single-nucleotide polymorphism (SNP) groups (called haplotypes, or "haps"). The DNA samples for the HapMap came from a total of 270 individuals: Yoruba people in Ibadan, Nigeria; Japanese people in Tokyo; Han Chinese in Beijing; and the French Centre d'Etude du Polymorphisms Humain (CEf) resource, which consisted of residents of the United States having ancestry from Western and Northern Europe.

In the Celera Genomics private-sector project, DNA from five different individuals were used for sequencing. The lead scientist of Celera Genomics at that time, Craig Venter, later acknowledged (in a public letter to the journal Science) that his DNA was one of 21 samples in the pool, five of which were selected for use.

On September 4, 2007, a team led by Craig Venter published his complete DNA sequence, unveiling the six-billion-nucleotide genome of a single individual for the first time.

Benefits

The work on interpretation of genome data is still in its initial stages. It is anticipated that detailed knowledge of the human genome will provide new avenues for advances in medicine and biotechnology. Clear practical results of the project emerged even before the work was finished. For example, a number of companies, such as Myriad Genetics started offering easy ways to administer genetic tests that can show predisposition to a variety of illnesses, including breast cancer, hemostasis disorders, cystic fibrosis, liver diseases and many others. Also, the etiologies for cancers, Alzheimer's disease and other areas of clinical interest are considered likely to benefit from genome information and possibly may lead in the long term to significant advances in their management.

There are also many tangible benefits for biological scientists. For example, a researcher investigating a certain form of cancer may have narrowed down his/her search to a particular gene. By visiting the human genome database on the World Wide Web, this researcher can examine what other scientists have written about this gene, including (potentially) the three-dimensional structure of its product, its function(s), its evolutionary relationships to other human genes, or to genes in mice or yeast or fruit flies, possible detrimental mutations, interactions with other genes, body tissues in which this gene is activated, and diseases associated with this gene or other datatypes.

Further, deeper understanding of the disease processes at the level of molecular biology may determine new therapeutic procedures. Given the established importance of DNA in molecular biology and its central role in determining the fundamental operation of cellular processes, it is likely that expanded knowledge in this area will facilitate medical advances in numerous areas of clinical interest that may not have been possible without them.

The analysis of similarities between DNA sequences from different organisms is also opening new avenues in the study of evolution. In many cases, evolutionary questions can now be framed in terms of molecular biology; indeed, many major evolutionary milestones (the emergence of the ribosome and organelles, the development of embryos with body plans, the vertebrate immune system) can be related to the molecular level. Many questions about the similarities and differences between humans and our closest relatives (the primates, and indeed the other mammals) are expected to be illuminated by the data from this project.

The Human Genome Diversity Project (HGDP), spinoff research aimed at mapping the DNA that varies between human ethnic groups, which was rumoured to have been halted, actually did continue and to date has yielded new conclusions. In the future, HGDP could possibly expose new data in disease surveillance, human development and anthropology. HGDP could unlock secrets behind and create new strategies for managing the vulnerability of ethnic groups to certain diseases. It could also show how human populations have adapted to these vulnerabilities.

Advantages of Human Genome Project:

1. Knowledge of the effects of variation of DNA among individuals can revolutionize the ways to diagnose, treat and even prevent a number of diseases that affects the human beings.
2. It provides clues to the understanding of human biology.

Ethical, Legal and Social Issues

The project's goals included not only identifying all of the approximately 24,000 genes in the human genome, but also to address the ethical, legal, and social issues (ELSI) that might arise from the availability of genetic information. Five percent of the annual budget was allocated to address the ELSI arising from the project.

Debra Harry, Executive Director of the U.S group Indigenous Peoples Council on Biocolonialism (IPCB), says that despite a decade of ELSI funding, the burden of genetics education has fallen on the tribes themselves to understand the motives of Human genome project and its potential impacts on their lives. Meanwhile, the government has been busily funding projects studying indigenous groups without any meaningful consultation with the groups.

The main criticism of ELSI is the failure to address the conditions raised by population-based research, especially with regard to unique processes for group decision-making and cultural worldviews. Genetic variation research such as HGP is group population research, but most ethical guidelines, according to Harry, focus on individual rights instead of group rights.

She says the research represents a clash of culture: indigenous people's life revolves around collectivity and group decision making whereas the Western culture promotes individuality. Harry suggests that one of the challenges of ethical research is to include respect for collective review and decision making, while also upholding the Western model of individual rights.

Genes, Technology and Policy

Rapid advances in information technology, particularly in the area of bioinformatics, have played a critical role in breakthrough applications of modern biotechnology in medicine and agriculture. Bioinformatics, broadly defined as the use of computers to handle biological information, has made possible the genomic era. Bioinformatics provides the computer tools and databases to search, store, analyse and compare these data and to use them to develop, among others, safe and more effective medicines as well higher – yielding, more stress – resistant crops that have the potential for accelerating human development.

However, as the Human Development Report 2001 points out, this potential cannot be realised unless the two conditions are met. First, modern biotechnology has to be utilised to address the key health and agriculture challenges facing poor countries. Second, modern biotechnology has to be utilised through a systematic approach that allows potential risks to human health, environment and social equity to be effectively assessed and managed.

This primer discusses the science and policy issues surrounding the use of modern biotechnology. It provides a snapshot of its benefits as well as concerns regarding its potential negative impact on the environment and on the human health. The primer is divided into four parts. The first part is on the science behind modern biotechnology. The second part discusses the various issues relating to the application of modern biotechnology to medicine. The third part focuses on the benefits and concerns relating to the use of modern biotechnology in agriculture. Finally, the fourth part discusses ownership and access issues viewed from the perspective of developing countries. This primer is by no means comprehensive. It is intended merely to introduce readers to the various perspectives in the ongoing debate on the use of modern biotechnology. Readers are therefore encouraged to consult the list of references at the end of the primer.

What is Biotechnology?

In its broadest sense, "biotechnology" refers to "any technique that uses living organisms, or parts of such organisms, to make or modify products, to improve plants or animals, or to develop microorganisms for specific use."

Biotechnology combines disciplines like genetics, molecular biology, biochemistry, embryology and cell biology, which are in turn linked to practical disciplines like chemical engineering, information technology, and robotics.

On one end of the development pole are techniques of *traditional biotechnology* like microbial fermentation, used as early as 10,000 years ago in fermenting beer, wine and dairy products. At the other end of the development pole are the continuously evolving techniques of *modern biotechnology*, such as genetic engineering. Using genetic engineering techniques, the genetic makeup of an organism may be modified by inactivating or altering some of its genes and introducing other natural or artificial genes, usually from another organism.

FThe Gradient of Biotechnology

Recently, the term "biotechnology" has come to be identified with modern biotechnology, specifically the use of genetic engineering techniques in medicine and agriculture. Hence, unless the context requires otherwise, this primer uses "biotechnology" to mean "modern biotechnology".

What Definition of Biotechnology is Widely Accepted not only by Scientists but also by Governments and Multilateral Institutions?

The Cartagena Protocol on Biosafety defines modern biotechnology as referring to any process that involves the application of (i) *in vitro* nucleic acid techniques, including recombinant deoxyribonucleic acid and direct injection of nucleic acid into cells or organelles, or (ii) fusion of cells beyond the taxonomic family, that overcome natural physiological reproductive or recombination of barriers and that are not techniques used in traditional breeding and selection.

Although the Protocol is not yet in force (because less than the required 50 States have either ratified or acceded to it), the Protocol's definition of modern biotechnology has gained currency in international circles. However, while there may be an emerging international consensus on the above definition, strictly speaking it is a definition that is applicable only when one uses the term "modern biotechnology" for purposes of interpreting or implementing the Protocol.

What Technical Terms Should a Policy Maker Know to Understand Biotechnology?

There are at least four such technical terms: genetics, genes, genome and genetically modified organisms. Genetics is the branch of biology that deals with the principles of heredity and variation in all living things. It is the study of why and how parents pass on some of their distinguishing features to their offspring. Its focus is on genes and their functions.

The gene is the basic unit of heredity and the ultimate arbiter of what we are. It carries instructions that allow cells to produce specific proteins. (It should be noted, however, that only certain genes are active at any given moment and environment. A gene is a part of the deoxyribonucleic acid ("DNA") molecule. DNA, which is present in all living cells, contains information coding for cellular structure, organisation and function. It is made up of two strands twisted around each other in a helical staircase.

Each cell in an organism has one or two sets of the basic DNA complement, called a genome. The genome is itself made up of one or more extremely long linear array of molecules of DNA that are called chromosomes. Genes, as explained earlier, are the functional regions of the DNA. They are the active segments of the chromosomes. DNA and genes relate to each other:

In modern biotechnology8, the genome of an organism is altered by exposing cells to fragments of "foreign" DNA carrying the desirable genes, often from another species. This DNA is taken in and inserts itself into one or more of the recipient's chromosomes at a location where it is inherited like any other part of the genome. The cells so modified are called transgenic cells. It is from transgenic cells that a GMO can be produced. All of the GMO's cells contain the additional foreign DNA.9

There is no universal definition for genetically modified organism (also called "transgenic organism" or "living modified organism"). However, it is generally understood to be a plant, animal or microorganism that contains genes that have been altered or transferred from another species or from the same species by means of genetic engineering techniques.

What Role does Information Technology Play in the Development of Modern Biotechnology?

Our knowledge of biology has grown in such a way that we need powerful tools to organize that knowledge. Information technology, through the field of bioinformatics , makes possible the rapid organisation and analysis of biological data. Bioinformatics merges biology, computer science, and information technology to manage and analyse genomic data, with the ultimate goal of understanding and modelling living systems.

Why do we have to Familiarize Ourselves with the Science of an Issues Surrounding Modern Biotechnology?

There are at least two reasons. The first has to do with the potential benefits that modern biotechnology offers humankind. The European Commission (2002) refers to modern biotechnology as the "next wave of the knowledge-based economy" after information technology, and the "most promising of the frontier technologies." It has identified applications in the following areas:

1. Health care. Biotechnology can be used to arrive at novel and innovative approaches to meet the needs of ageing populations and poor countries.
2. Crop production. Biotechnology can deliver improved food quality and environmental benefits through agronomically improved crops. It may be used to produce foods with enhanced qualities like higher nutritional benefits.
3. Non-food uses of crops. Biotechnology can also improve non-food uses of crops as sources of industrial feedstock or new materials such as biodegradable plastics. For example, canola is now being used to produce high-value industrial oil. Under the appropriate economic and fiscal conditions, biomass can contribute to alternative energy with both liquid and solid biofuels (e.g., biodiesel and bioethanol) and processes such as bio-desulphurisation.
4. Environmental uses. New ways of protecting and improving the environment are possible with biotechnology, including bioremediation of polluted air, soil, water and waste, as well as the development of cleaner industrial products and processes like biocatalysis.

The second reason why knowledge of biotechnology is important is that with more biotechnology-derived products being placed on the market, chances are these products will find their way into most countries, even those that do not use biotechnology for production. A government needs to be familiar with modern biotechnology if it is to effectively regulate biotechnological products and ensure that any adverse effects, if any, on the environment, human health, and social structures are properly managed, if not avoided

What are the Applications of Modern Biotechnology in the Medical Field?

In medicine, modern biotechnology finds promising applications in:

- Pharmacogenomics

- Drug Production
- Genetic Testing
- Gene Therapy.

What is Pharmacogenomics? What are its Anticipated benefits?

Pharmacogenomics is the study of how the genetic inheritance of an individual affects his/her body's response to drugs. It is a coined word derived from the words "pharmacology" and "genomics". It is therefore the study of the relationship between pharmaceuticals and genetics. The vision of pharmacogenomics is to be able to design and produce drugs that are adapted to each person's genetic makeup.

Pharmacogenomics results in the following benefits:

1. Development of tailor-made medicines. Using pharma-cogenomics, pharmaceutical companies can create drugs based on the proteins, enzymes and RNA molecules that are associated with specific genes and diseases. These tailor-made drugs promise not only to maximize therapeutic effects but also to decrease damage to nearby healthy cells.
2. More accurate methods of determining appropriate drug dosages. Knowing a patient's genetics will enable doctors to determine how well his/ her body can process and metabolize a medicine. This will maximize the value of the medicine and decrease the likelihood of overdose.
3. Improvements in the drug discovery and approval process. The discovery of potential therapies will be made easier using genome targets. Genes have been associated with numerous diseases and disorders. With modern biotechnology, these genes can be used as targets for the development of effective new therapies, which could significantly shorten the drug discovery process.
4. Better vaccines. Safer vaccines can be designed and produced by organisms transformed by means of genetic engineering. These vaccines will elicit the immune response without the attendant risks of infection. They will be inexpensive, stable, easy to store, and capable of being engineered to carry several strains of pathogen at once.

How Does Biotechnology Contribute to Drug Production?

Modern biotechnology can be used to manufacture existing drugs more easily and cheaply. The first genetically engineered products were medicines designed to combat human diseases. To cite one example, in 1978 Genentech joined a gene for insulin and a plasmid vector and put the resulting gene into a bacterium called Escherichia coli. Insulin, widely used for the treatment of diabetes, was previously extracted from sheep and pigs. It was very expensive and often elicited unwanted allergic responses. The resulting genetically engineered bacterium enabled the production of vast quantities of human insulin at low cost. Since then modern biotechnology has made it possible to produce more easily and cheaply the human growth hormone, clotting factors for hemophiliacs, fertility drugs, erythropoietin and other drugs. Most drugs today are based on about 500 molecular targets. Genomic knowledge of the genes involved in diseases, disease pathways, and drug-response sites are expected to lead to the discovery of thousands more new targets.

What is Genetic Testing?

Genetic testing involves the direct examination of the DNA molecule itself. A scientist scans a patient's DNA sample for mutated sequences. There are two major types of gene tests. In the first type, a researcher may design short pieces of DNA ("probes") whose sequences are complementary to the mutated sequences. These probes will seek their complement among the base pairs of an individual's genome. If the mutated sequence is present in the patient's genome, the probe will bind to it and flag the mutation. In the second type, a researcher may conduct the gene test by comparing the sequence of DNA bases in a patient's gene to a normal version of the gene.

What are the Potential uses of Genetic Testing?

Genetic testing can be used to:

- Diagnose a disease.
- Confirm a diagnosis.
- Provide prognostic information about the course of a disease.
- Confirm the existence of a disease in individuals.
- With varying degrees of accuracy, predict the risk of future disease in healthy individuals or their progeny.

Genetic testing is now used for:

- carrier screening, or the identification of unaffected individuals who carry one copy of a gene for a disease that requires two copies for the disease to manifest
- prenatal diagnostic screening
- newborn screening
- presymptomatic testing for predicting adult-onset disorders
- presymptomatic testing for estimating the risk of developing adult-onset cancers
- confirmational diagnosis of symptomatic individuals
- forensic/identity testing.

Are Genetic Tests now Available in the Market?

Some genetic tests are already available, although most of them are used in developed countries. The tests currently available can detect mutations associated with rare genetic disorders like cystic fibrosis, sickle cell anemia, and Huntington's disease.

Recently, tests have been developed to detect mutation for a handful of more complex conditions such as breast, ovarian, and colon cancers. However, gene tests may not detect every mutation associated with a particular condition because many are as yet undiscovered, and the ones they do detect may present different risks to different people and populations.

What is Gene Therapy?

Gene therapy may be used for treating, or even curing, genetic and acquired diseases like cancer and AIDS by using normal genes to supplement or replace defective genes or to bolster a normal function such as immunity. It can be used to target somatic (i.e., body) or germ (i.e., egg and sperm) cells. In somatic gene therapy, the genome of the recipient is changed, but this change is not passed along to the next generation. In contrast, in germline gene therapy, the egg and sperm cells of the parents are changed for the purpose of passing on the changes to their offspring.

How is Gene Therapy Done?

There are basically two ways of implementing a gene therapy treatment:

1. *Ex vivo*, which means "outside the body" – Cells from the patient's blood or bone marrow are removed and grown in the laboratory. They are then exposed to the virus carrying the desired gene. The virus enters the cells, and the desired gene becomes part of the DNA of the cells. The cells are allowed to grow in the laboratory before being returned to the patient by injection into a vein.
2. *In vivo*, which means "inside the body" – No cells are removed from the patient's body. Instead, vectors are used to deliver the desired gene to cells in the patient's body.

How Extensive is the use of Gene Therapy?

Currently, gene therapy use is limited. Somatic gene therapy is primarily at the experimental stage. Germline therapy is the subject of much discussion but it is not being actively investigated in larger animals and human beings.

As of June 2001, more than 500 clinical gene-therapy trials involving about 3,500 patients have been identified worldwide. Around 78% of these are in the United States, with Europe having 18%. These trials focus on various types of cancer, although other multigenic diseases are being studied as well.

Recently, two children born with severe combined immunodeficiency disorder ("SCID") were reported to have been cured after being given genetically engineered cells.

What are the Obstacles to the Widespread use of Gene Therapy Techniques to Treat Patients?

Gene therapy faces many obstacles before it can become a practical approach for treating disease. At least four of these obstacles are as follows:

1. Gene delivery tools. Genes are inserted into the body using gene carriers called vectors. The most common vectors now are viruses, which have evolved a way of encapsulating and delivering their genes to human cells in a pathogenic manner. Scientists manipulate the genome of the virus by removing the disease-causing genes and inserting the therapeutic genes. However, while viruses are effective, they can introduce problems like toxicity, immune and inflammatory responses, and gene control and targeting issues.
2. Limited knowledge of the functions of genes. Scientists currently know the functions of only a few genes. Hence, gene therapy can address only some genes that cause a particular disease. Worse, it is not known exactly whether genes have more than one function, which creates uncertainty as to whether replacing such genes is indeed desirable.
3. Multigene disorders and effect of environment. Most genetic disorders involve more than one gene. Moreover, most diseases involve the interaction of several genes and the environment. For example, many people with cancer not only inherit the disease gene for the disorder, but may have also failed to inherit specific tumor suppressor genes. Diet, exercise, smoking and other environmental factors may have also contributed to their disease.
4. High costs. Since gene therapy is relatively new and at an experimental stage, it is an expensive treatment to undertake. This explains why current studies are focused on illnesses commonly found in developed countries, where more people can afford to pay for treatment. It may take decades before developing countries can take advantage of this technology.

What is the Human Genome Project?

The Human Genome Project is an initiative of the U.S. Department of Energy ("DOE") that aims to generate a high-quality reference sequence for the entire human genome and identify all the human genes. The DOE and its predecessor agencies were assigned by the U.S. Congress to develop new energy resources and technologies and to pursue a deeper understanding of potential health and environmental risks posed by their production and use. In 1986, the DOE announced its Human Genome Initiative. Shortly thereafter, the DOE and National Institutes of Health developed a plan for a joint Human Genome Project ("HGP"), which officially began in 1990.

The HGP was originally planned to last 15 years. However, rapid technological advances and worldwide participation have accelerated the expected completion date to 2003. In June 2000, scientists announced the generation of a working draft sequence of the entire human genome. The draft provides a road map to an estimated 90% of genes on every human chromosome. Already it has enabled gene hunters to pinpoint genes associated with more than 30 disorders.

What is Human Cloning?

Human cloning is one of the techniques of modern biotechnology. It involves the removal of the nucleus from one cell and its placement in an unfertilized egg cell whose nucleus has either been deactivated or removed.

There are two types of cloning:

1. Reproductive cloning. After a few divisions, the egg cell is placed into a uterus where it is allowed to develop into a fetus that is genetically identical to the donor of the original nucleus.
2. Therapeutic cloning. The egg is placed into a Petri dish where it develops into embryonic stem cells, which have shown potentials for treating several ailments.

In February 1997, cloning became the focus of media attention when Ian Wilmut and his colleagues at the Roslin Institute announced the successful cloning of a sheep, named Dolly, from the mammary glands of an adult female. The cloning of Dolly made it apparent to many that the techniques used to produce her could someday be used to clone human beings. This stirred a lot of controversy because of its ethical implications.

Modern Biotechnology Techniques in Medicine

Several issues have been raised regarding the use of modern biotechnology in the medical sector. Many of these issues are similar to those facing any new technology that is viewed as powerful and far-reaching. Some of these issues are :

1. Absence of cure. There is still a lack of effective treatment or preventive measures for many diseases and conditions now being diagnosed or predicted using gene tests. Thus, revealing information about risk of a future disease that has no existing cure presents an ethical dilemma for medical practitioners.

2. Ownership and control of genetic information. Who will own and control genetic information, or information about genes, gene products, or inherited characteristics derived from an individual or a group of people like indigenous communities? At the macro level, there is a possibility of a genetic divide, with developing countries that do not have access to medical applications of biotechnology being deprived of benefits accruing from products derived from genes obtained from their own people. Moreover, genetic information can pose a risk for minority population groups as it can lead to group stigmatisation.

At the individual level, the absence of privacy and anti-discrimination legal protections in most countries can lead to discrimination in employment or insurance or other misuse of personal genetic information.

This raises questions like, is genetic privacy different from medical privacy?

3. Reproductive issues. These include the use of genetic information in reproductive decision-making and the possibility of genetically altering reproductive cells that may be passed on to future generations. For example, germline therapy forever changes the genetic make-up of an individual's descendants. Thus, any error in technology or judgment may have far-reaching consequences. Ethical issues like designer babies and human cloning have also given rise to controversies between and among scientists and bioethicists, especially in the light of past abuses with eugenics.
4. Clinical issues. These centre on the capabilities and limitations of doctors and other health-service providers, people identified with genetic conditions, and the general public in dealing with genetic information. For instance, how should the public be prepared to make informed choices based on the results of genetic tests? How will genetic tests be evaluated and regulated for accuracy, reliability, and usefulness?
5. Effects on social institutions. Genetic tests reveal information about individuals and their families. Thus, test results can affect the dynamics within social institutions, particularly the family.
6. Conceptual and philosophical implications regarding human responsibility, free will vis-à-vis genetic determinism, and the concepts of health and disease. Do genes influence human behaviour? If so, does genetic testing mean controlling human behaviour? What is considered acceptable diversity? What is normal and what is a disability or disorder, and who decides these matters? Are disabilities diseases that need to be cured or prevented? Where should the line between medical treatment and enhancement be drawn? Who will have access to gene therapy?

What are the Applications of Modern Biotechnology in Agriculture?

There are many applications of biotechnology in agriculture.

Improved yield from crops: Using the techniques of modern biotechnology, one or two genes may be transferred to a highly developed crop variety to impart a new character that would increase its yield. However, while increase in crop yield is the most obvious application of modern biotechnology in agriculture, it is also the most difficult one. Current genetic engineering techniques work best for effects that are controlled by a single gene. Many of the genetic characteristics associated with yield (e.g., enhanced growth) are controlled by a large number of genes, each of which has a minimal effect on the overall yield. There is, therefore, much scientific work to be done in this area.

Reduced vulnerability of crops to environmental stresses. Crops containing genes that will enable them to withstand biotic and abiotic stresses may be developed. For example, drought and excessively salty soil are the two most important limiting factors in crop productivity. Biotechnologists are studying plants that can cope with these extreme conditions in the hope of finding the genes that enable them to do so and eventually transferring these genes to the more desirable crops. One of the latest developments is the identification of a plant gene, At-DBF2, from thale cress, a tiny weed that is often used for plant research because it is very easy to grow and its genetic code is well mapped out. When this gene was inserted into tomato and tobacco cells, the cells were able to withstand environmental stresses like salt, drought, cold and heat, far more than ordinary cells. If these preliminary results prove successful in larger trials, then At-DBF2 genes can help in engineering crops that can better withstand harsh environments.

Researchers have also created transgenic rice plants that are resistant to rice yellow mottle virus (RYMV). In Africa, this virus destroys majority of the rice crops and makes the surviving plants more susceptible to fungal infections.

Increased nutritional qualities of food crops. Proteins in foods may be modified to increase their nutritional qualities. Proteins in legumes and cereals may be transformed to provide the amino acids needed by human beings for a balanced diet. A good example is the work of Professors Ingo Potrykus and Peter Beyer on the so-called Goldenrice.

Improved taste, texture or appearance of food. Modern biotechnology can be used to slow down the process of spoilage so that fruit can ripen longer on the plant and then be transported to the consumer with a still reasonable shelf life. This improves the taste, texture and appearance of the fruit. More importantly, it could expand the market for farmers in developing countries due to the reduction in spoilage.

The first genetically modified food product was a tomato which was transformed to delay its ripening. Researchers in Indonesia, Malaysia, Thailand, Philippines and Vietnam are currently working on delayed-ripening papaya in collaboration with the University of Nottingham and Zeneca.

Reduced dependence on fertilizers, pesticides and other agrochemicals. Most of the current commercial applications of modern biotechnology in agriculture are on reducing the dependence of farmers on agrochemicals. For example, Bacillus thuringiensis (Bt) is a soil bacterium that produces a protein with insecticidal qualities. Traditionally, a fermentation process has been used to produce an insecticidal spray from these bacteria. In this form, the Bt toxin occurs as an inactive protoxin, which requires digestion by an insect to be effective. There are several Bt toxins and each one is specific to certain target insects. Crop plants have now been engineered to contain and express the genes for Bt toxin, which they produce in its active form. When a susceptible insect ingests the transgenic crop cultivar expressing the Bt protein, it stops feeding and soon thereafter dies as a result of the Bt toxin binding to its gut wall. Bt corn is now commercially available in a number of countries to control corn borer (a lepidopteran insect), which is otherwise controlled by spraying (a more difficult process).

Crops have also been genetically engineered to acquire tolerance to broad-spectrum herbicide. The lack of cost-effective herbicides with broad-spectrum activity and no crop injury was a consistent limitation in crop weed management. Multiple applications of numerous herbicides were routinely used to control a wide range of weed species detrimental to agronomic crops. Weed management tended to rely on pre-emergence that is herbicide applications were sprayed in response to expected weed infestations rather than in response to actual weeds present. Mechanical cultivation and hand weeding were often necessary to control weeds not controlled by herbicide applications. The introduction of herbicide tolerant crops has the potential of reducing the number of herbicide active ingredients used for weed management, reducing the number of herbicide applications made during a season, and increasing yield due to improved weed management and less crop injury. Transgenic crops that express tolerance to glyphosphate, glufosinate and bromoxynil have been developed. These herbicides can now be sprayed on transgenic crops without inflicting damage on the crops while killing nearby weeds.

From 1996 to 2001, herbicide tolerance was the most dominant trait introduced to commercially available transgenic crops, followed by insect resistance. In 2001, herbicide tolerance deployed in soybean, corn and cotton accounted for 77% of the 62.6 million hectares planted to transgenic crops; Bt crops accounted for 15%; and stacked genes for herbicide tolerance and insect resistance used in both cotton and corn accounted for 8%.

Production of novel substances in crop plants. Modern biotechnology is increasingly being applied for novel uses other than food. For example, oilseed is at present used mainly for margarine and other food oils, but it can be modified to produce fatty acids for detergents, substitute fuels and petrochemicals. Banana trees and tomato plants have also been genetically engineered to produce vaccines in their fruit. If future clinical trials prove successful, the advantages of edible vaccines would be enormous, especially for developing countries.

The transgenic plants may be grown locally and cheaply. Homegrown vaccines would also avoid logistical and economic problems posed by having to transport traditional preparations over long distances and keeping them cold while in transit. And since they are edible, they will not need syringes, which are not only an additional expense in the traditional vaccine preparations but also a source of infections if contaminated.

Yes, there are efforts to improve the nutritional quality of rice. This staple food for two billion people is usually milled to remove the outer layers to prevent their high oil content from causing spoilage. The remaining grains are low in B-carotene, the chemical precursor of vitamin A. Some 400 million people worldwide suffer from vitamin A deficiency while over 3.7 billion people are iron-deficient. Vitamin A deficiency causes five million deaths annually, and blindness in a further 500,000 people, while iron deficiency causes anemia and birth defects.

Golden Rice is a transgenic crop created by Dr. Ingo Potrykus and his colleagues to improve the nutritional quality of rice by increasing the quantities of beta carotene the precursor of vitamin A) and improving the crop's iron content.

Several genes have been inserted into the rice genome, including a daffodil gene, allowing the endosperm (the part that remains after milling and polishing) to produce B-carotene. Additionally, a phytase gene (which produces an enzyme to release chemically-bound iron), a gene to increase organic iron, and a gene to aid iron absorption in the digestive tract have been added. The presence of beta carotene in the endosperm of the transgenic rice gives it a golden colour. Hence, the name "golden rice".

The research, which was funded by the Rockefeller Foundation, the Swiss government and the European Union, hopes to provide a cheap form of vitamin supplementation. If it is later proven to be viable and safe, Golden Rice will be distributed in developing countries, with no patents blocking access to it. The introduction of vitamin A-producing beta carotene into the rice gene has the potential of addressing the vitamin A problem, especially among those who are too poor to diversify their diets with green vegetables.

How is Modern Biotechnology in Agriculture Different from Traditional Plant breeding Techniques

Yes and no. Yes, because both traditional and modern biotechnology involve the transfer of genes from one organism to another. Traditional breeding techniques typically involve the repeated mixing of thousands of genes over several years and many generations of plants to achieve a desired trait. Modern biotechnology accelerates this lengthy process by allowing scientists to insert selected genes directly into a plant. This makes modern biotechnology less of an iterative process compared to traditional plant breeding techniques. In this sense, modern biotechnology is simply an extension of traditional breeding.

No, because unlike traditional breeding techniques, modern biotechnology can move genes across species, even family, boundaries to produce novel organisms that do not normally occur in nature. To cite a simplistic example, a brown cow that mates with a yellow cow may produce a calf of a completely new colour, but reproductive mechanisms limit the number of new combinations. Cows must breed with other cows or their very near relatives. A farmer cannot breed for a purple cow using traditional sexual reproduction techniques because the necessary purple genes are not available in cows or their near relatives. In contrast, the biological barrier is not as steep, at least in theory, to a genetic engineer. If purple genes are available in another species, say in an iris plant, those genes could be mixed with the genes of the cow to produce purple cows.

Genes Inserted into Plants using Modern Biotechnology

There are currently two commonly used methods for introducing genes into plants genomes:

1. Using a plasmid vector. As previously discussed, a vector, like the plasmids of Agrobacterium tumefaciens, may be used to introduce the gene or genes of interest into the plant DNA. The resulting cells are then screened to identify those that have successfully expressed the new trait. The modified seeds are sown in the field and grown like any other crop.
2. Particle bombardment techniques. The DNA to be introduced into plant cells is coated onto tiny particles, which are then physically shot into the plant cells. Some of the DNA comes off and are incorporated into the DNA of the target plant.

Are Transgenic Crops now Commercially Available?

Yes, genetically modified crops are now available in the market. If you eat corn or soya products, chances are you are consuming genetically modified products. In the United States, it is estimated that 70-85% of all processed/

packaged foods contain one or more ingredients that are derived from transgenic crops. James (2002) estimated the global area planted to transgenic crops in 2002 at 58.7 million hectares. More than one-quarter of this area, or 13.5 million hectares, is in six developing countries, namely, China, India, Indonesia, Argentina, South Africa, and Mexico. India, the largest cotton growing country in the world, commercialised Bt cotton for the first time during the year. However, four countries continue to account for 99% of the global transgenic crop area. The USA grew 66% of the global total, followed by Argentina with 23%, Canada with 6%, and China with 4%. China had the highest year-on-year percentage growth, with a 40% increase in its Bt cotton area.

The adoption rate for transgenic crops is the highest in the history of agriculture. From 1996 to 2000, 15 countries contributed to a more than 25-fold increase in the global area of transgenic crops, from 1.7 million hectares in 1996 to 44.2 million hectares in 2000.

Transgenic Crops Dominate

In 2001, the principal transgenic crops were soybean, with 62% of the global area, followed by corn at 21%, cotton at 12%, and canola at 5%. More than three-quarters of the transgenic crops were modified for herbicide tolerance. The balance was mostly for pest resistance, with a few areas planted to potato and papaya with inserted genes for delayed ripening and virus-resistance.

Genetic Engineering in Agriculture Harmful to the Environment?

Most of the concerns revolve around the effect of biotechnology products on the environment and on human health.

Like all new technologies, genetic engineering, if not properly studied and regulated, can adversely impact on the environment. However, we need to review the impact on a case-by-case basis. Biotechnology by itself is not good or bad. As with all technologies, it is how people use it that can be good or bad, risky or beneficial. Each biotechnology product must be evaluated in terms of whether it is useful, beneficial and safe. To date, there is no scientific proof that the use of modern biotechnology has an adverse effect on the environment and on human health.

The major issues that have been raised so far against the environmental impact of genetically engineered agricultural products and the corresponding scientific consensus are summarised below:

Increased weediness. One concern is that altered plants may have increased fitness to survive their environments and might grow unaided by human beings in places where they could have unwanted effects. In unmanaged environments, they could displace natural flora and upset entire ecosystems. However, in a 10-year study to answer the question of whether transgenes conferring herbicide tolerance or insect resistance on crop plants also confer weediness or invasiveness, it was found that genetically modified crops (those that are commercially available) have no more tendency to become weeds than their conventionally bred counterparts. Apparently, more genes are needed to convert a plant into a weed. Another study also showed that biological invasions have extensive time lags that range from 30-150 years and require the chance concordance of favourable conditions before taking off. The question, however, is whether transgenes would hasten this process. This is why potential invasiveness of GM plants is evaluated before any decision to release them into the environment is made.

Unintended gene flow. There is also the concern that if relatives of the altered crops are growing near a field with their conventional counterpart, the new gene could move by means of pollen transfer into the latter. There is merit in this concern. This is why studies have to be done on a case-by-case basis to determine the potential environmental risks due to unintended gene flow.

The European Science Foundation and European Environment Agency (2002) recently released their report on the significance of pollen-mediated gene flow from six major crop types that have been genetically engineered and are close to commercial release in the European Union, namely, oilseed rape, sugar beet, potatoes, maize, wheat and barley. They found, among others, that oilseed rape is high-risk for crop-to-crop gene flow and gene flow from crop to wild relatives. Potatoes, wheat, and barley are low-risk crops. Maize and sugar beet are both medium- to high-risk for crop-to-crop gene flow, with the latter having the same level of risk for gene flow from crop to wild relatives. There are no known wild relatives of maize in Europe with which it can hybridize.

On the other hand, there is much research on developing alternative measures to mitigate unintended gene flow. At present, seed producers of conventional crops can devise mechanisms to isolate their crop lands from related plants in order to maintain the purity of their lines. In addition, several biotechnology measures to prevent horizontal gene flow, such as apomixis, chloroplast transformation, chromosome-specific cytogenetic system, and transgenetic mitigation, have been suggested. The most controversial is the Technological Protection System (TPS) developed by the U.S. Department of Agriculture and Delta & Pine. The technique, commonly called GURT (for Genetic Use Restriction Technology), involves a system of three genes that interact to control the fertility of seeds by the seed producer. The objection to TPS has to do with the possibility of pollen dispersal to adjacent fields of the same crop, inadvertently causing the latter to produce sterile seeds. This would also have a severe impact on the common practice of seed saving among small farmers in developing countries.

Change in herbicide use patterns. It has been pointed out that widespread use of herbicide-tolerant crops could lead to the rapid evolution of resistance to herbicides in weeds, either as a result of increased exposure to the herbicide or as a result of the transfer of the herbicide trait to weedy relatives of the crops. However, to date there is no evidence that this phenomenon is taking place.

Squandering of valuable pest susceptibility genes. Many insects contain genes that render them susceptible to pesticides. Often these susceptibility genes predominate in natural populations of insects. These genes are a valuable natural resource because they allow pesticides to remain as effective pest-control tools. The more benign the pesticide, the more valuable the genes that make pests susceptible to it. It is feared that crops that have been altered to contain the pest-resistance Bt gene can adversely affect the continued susceptibility of pests to the Bt toxin.

The continuous exposure of pests to the Bt toxin in altered crops selects for the rare resistance genes in the pest population and in time will render the Bt pesticide useless, unless specific measures are instituted to avoid the development of such a resistance. It should be noted, however, that there is really no such thing as permanent resistance to pests and diseases and that insect resistance happens under the current practice of using pesticides or even in nature.

Controversy Around the Effect of Genetically Modified corn on Monarch Butterflies

Monarch butterfly larvae feed exclusively on the leaves of milkweed plants, which are commonly found in cornfields in the U.S. Pollen from nearby corn can become distributed on the leaves of these plants, and can therefore be eaten by the larvae. In 1999, two studies showed that Monarch butterfly larvae and larvae from related species that were fed leaves dusted with Bt-corn pollen had lower survival rates, compared to those fed leaves dusted with non-Bt corn pollen. These studies were used to suggest that Bt corn was responsible for the recently observed decline in the Monarch butterfly population.

However, subsequent investigations revealed that while a large percentage of Monarch butterfly larvae may feed on milkweed found in the corn belt region of the U.S., there is no overlap between their breeding time and the time of pollen shed through most of this region. Other studies have shown that corn pollen settling on an area decreases rapidly with distance.

This and the toxicity studies showing low toxicity of many major Bt-corn strains indicate that pollen densities that could represent significant exposure to feeding larvae are found only within five metres of cornfields, and then rarely. The current scientific consensus is that the adverse impact of Bt-corn pollen observed in the laboratory does not occur in the fields. Hence, Monarch butterflies are safe from Bt corn.

Are GM foods Safe to Eat?

At present, there are no studies to indicate that any of the commercially available GM foods are any less safe than their non-modified counterparts. This, of course, does not mean that all products of biotechnology are safe. Hence, the need for regulations. In fact, GM food products are subjected to more tests than their conventionally-bred counterparts.

The concerns raised about the safety of genetically engineered food products fall into the following categories:

New Allergens in the Food Supply

There is a concern that transgenic crops could introduce new allergens into foods. However, it is important to keep in mind that eating conventional food is also not risk-free; allergies occur with many new and even known conventional foods. For example, the kiwi fruit was introduced into the U.S. and the European market in the 1960s with no known human allergies; today, there are people allergic to this fruit.

In February 2002, the Royal Society issued a policy report titled Genetically Modified Plants for Food Use and Human Health—An Update. The report concluded that there is currently no evidence that GM foods cause allergic reactions and "the allergenic risks posed by GM plants are in principle no greater than those posed by conventionally derived crops or by plants introduced from other areas of the world."

Antibiotic Resistance

Modern biotechnology often uses genes for antibiotic resistance as "selectable markers". Early in the genetic engineering process, these markers help select cells that have taken up the foreign genes. Although they have no further use, the genes continue to be expressed in plant tissues. Critics of modern biotechnology argue that the presence of antibiotic-resistance genes could have two harmful effects.

First, eating food containing these genes could reduce the effectiveness of antibiotics to fight disease when these antibiotics are taken with meals. Antibiotic resistance genes produce enzymes that can degrade antibiotics. If a tomato with an antibiotic-resistance gene is eaten at the same time as an antibiotic, it could destroy the antibiotic in the stomach.

Second, the resistance genes could be transferred to human or animal pathogens, making them impervious to antibiotics. If transfer were to occur, it could aggravate the already serious health problem of antibiotic-resistant disease organisms.

However, unmediated transfers of genetic material from plants to bacteria are highly unlikely. Moreover, several strategies have been developed to avoid the inclusion of antibiotic resistance genes in the commercial transgenic variety.

Production of New Toxins

Many organisms have the ability to produce toxic substances. For plants, such substances help to defend stationary organisms from many predators in their environment.

But there is concern that the addition of new genetic material through genetic engineering could trigger the production of toxic substances within plants. This could happen, for example, if the on/off signals associated with the introduced gene were located in the genome in places where they could turn on the previously inactive genes.

Human beings typically eat several grams of DNA in their diet each day. Hence, the transgene in a genetically engineered plant is not a new type of material to our digestive systems. It is also present in extremely small amounts. In transgenic corn, for example, the transgenes represent about 0.00018 of the total DNA. Decades of research indicate that dietary DNA has no direct toxicity. In fact, exogenous nucleotides have been shown to play important beneficial roles in gut function and the immune system. Likewise, there is no compelling evidence showing the incorporation and expression of plant-derived DNA, whether a transgene or not, into the genomes of a consuming organism.

Defence processes have evolved, including extensive hydrolytic breakdown of the DNA during digestion, excision of integrated foreign DNA from the host genome, and silencing of foreign gene expression by targeted DNA methylation, that prevent the incorporation or expression of foreign DNA. Thus, there is a minimal possibility of adverse effects arising from the presence of foreign DNA by either direct toxicity or gene transfer.

Effect on Nutrients

Concerns have been raised regarding the possible adverse effect of genetic engineering technology on the nutritional content of food. This is a legitimate concern for regulators. In the U.S., for example, the Food and Drug Administration ensures that the nutritional composition of GM foods is substantially equivalent to that of their conventional counterparts. Studies are performed to determine whether nutrients, vitamins and minerals in the modified food occur at the same levels as in the conventionally-bred food sources.

For example, seeds and toasted soybean meal from Roundup Ready soybeans have been compared to conventional soybeans in terms of protein, oil, fibre, ash, carbohydrates, moisture content, amino acid and fatty acid composition. The results showed that the composition of transgenic lines is equivalent to that of conventional soybean cultivars, except for the trypsin inhibitors in non-toasted soybean meal, which is not consumed. In addition, the equivalence of the feeding value of the transgenic grains was demonstrated in rats, chickens, catfish and dairy cattle.

Concentration of Toxic Metals

Some of the new genes being added to crops can remove heavy metals like mercury from the soil and concentrate them in plant tissue. The purpose of creating such crops is to make possible the use of municipal sludge as fertilizer. Sludge contains useful plant nutrients but often cannot be used as fertilizer because it is contaminated with toxic heavy metals. The idea is to engineer plants to remove and sequester those metals in inedible parts of plants.

In a tomato, for example, the metals would be sequestered in the roots. Turning on the genes in only some parts of the plant requires the use of genetic on/off switches that turn on only in specific tissues, like roots. Such products could pose the risk of contaminating foods with high levels of toxic metals if the on/off switches are not completely turned off in edible tissues. There are also environmental risks associated with the handling and disposal of the metal-contaminated parts of plants after harvesting. Regulations have to address the peculiar functions of GMOs being used as bioremediation agents.

GM Potatoes on Rats is Cited by Critics of Biotechnology as Evidence that GM Products are Unsafe

The study was found to be inconclusive. Dr. Arpad Putztai, a senior scientist at the Rowett Institute in Aberdeen, Scotland, came to international attention when he announced to the media that eating genetically modified potatoes depressed rat immune systems and caused changes in their intestinal tract. Dr. Putztai and his co-workers compared rats fed genetically modified potatoes with rats fed non-modified potatoes, with and without added GNA. The genetically modified potatoes appeared to cause changes in the rats' immune response and the structure of the intestinal lining.

But there was a flaw in the experiment. While its design was apparently correct for this type of feeding study, rats do not like to eat raw potato. As a result, a standard 110-day trial had to be abandoned after only 67 days because the rats were starving. Starvation affects gut histology, and even the lining of the guts of rats eating unmodified potatoes was shown to be abnormal. The presence of other potato toxins could also have had a confounding effect on cells in the intestine, especially since the potato lines were not substantially equivalent

Possible Transfer of Antibiotic Resistance Marker Genes

Organisms that contain DNA encoding for antibiotic resistance proteins are common and are becoming more prevalent in the environment. However, there is no documented evidence that the antibiotic resistance markers in GM foods contribute to antibiotic resistance in gut bacteria. Should there be such a contribution, it is expected to be extremely small for several reasons, including the efficient destruction of the resistance gene in the human gut and the extremely low intrinsic rate of plant-microbe gene transfer.

Furthermore, resistance genes occur quite widely already and the antibiotics involved are not widely used in medical practice. Finally, the technology is now available to omit the use of such selections devices and their use is therefore likely to diminish.

What is the "principle of Substantial Equivalence"?

The term "substantial equivalence" was first mentioned in 1993 in connection with food safety in a report of the OECD Group of National Experts on Safety in Biotechnology.

The members of the group agreed that the most practical approach to determining the safety of foods derived by modern biotechnology is to consider whether they represent a substantial equivalent to analogous traditional products. The term substantial equivalence and the underlying approach were borrowed from the U.S. Food and Drug Administration's definition of a class of new medical devices that do not differ materially from their predecessors and thus do not raise new regulatory concerns.

According to the OECD definition, the concept of substantial equivalence is based on the idea that existing products used as foods or food sources can serve as a basis for comparison when assessing the safety and the nutritional value of a food or food ingredient that is new or that has been modified by modern biotechnological methods. If a novel food or novel food component is found to be substantially equivalent to an existing food or food component, it can be treated in the same manner with respect to safety. No additional safety concern would be expected. If a novel food or novel food ingredient is not found to be substantially equivalent to its conventional counterpart, this does not imply that it is unsafe. It must simply be evaluated on the basis of its unique composition and properties.

What is the "Precautionary Principle"?

The "precautionary principle" traces its origins to Principle 15 of the Rio Declaration on Environment and Development, which provides that: In order to protect the environment, the precautionary approach shall be widely applied by the States according to their capabilities. Where there are threats of serious or irreversible damage, lack of full scientific certainty shall not be used as a reason for postponing cost-effective measures to prevent environmental degradation. The use of the precautionary principle is to be based to the fullest extent possible on scientific evidence of a given problem. However, it is not always possible to move towards a decision on a purely scientific assessment;

any assessment must also involve economic, social and ethical aspects. The precautionary principle is thus more of a political norm than a clearly defined concept.

Critics of the precautionary principle have argued that it is poorly defined, not sufficiently grounded in science, stifles development of technology, and hinders trade. It is claimed that it is not a valid principle for evaluating scientific evidence as it distorts reality and leads to the acceptance of false beliefs. Opponents are also concerned that the precautionary principle's focus on hypothetical risks will distract consumers and policy makers from the policies needed to address known food-borne threats to human health.

The precautionary principle has been invoked to justify a prohibition of GM crops. The justification for a ban allegedly considers the potential public health and environmental benefits of the banned GM crop. However, it ignores the probable public health and environmental benefits that would necessarily be foregone as a result of the ban. A comprehensive application of the precautionary principle indicates that a GM crop ban, contrary to the claims of its advocates, would increase overall risks to public health and to the environment, especially in developing countries. Thus, it would be more prudent to study, develop, and possibly commercialize GM crops than to ban such crops, provided reasonable caution is exercised through the regulatory agencies.

Government Regulate the use of Modern Biotechnology

The purpose of a biosafety (short for "biological safety") system is to control the risks associated with the products of modern biotechnology. An integrated national biosafety system has the following elements:

National policy, strategies and research agenda regarding biosafety. The development of a national biosafety system should begin with the elaboration of a national biosafety policy consistent with a country's other policy objectives on food, agriculture, the environment and sustainable development. This policy will serve as the basis for the crafting of specific legislation and/or regulations on biosafety. It should articulate a framework where competing goals, such as economic, regional development and environmental protection, may be integrated and communicated as a single national vision.

National inventory and evaluation. This is a means to identify and characterize the available resources and regulatory infrastructure in a country, assess their adequacy for supporting a biosystem, and identify gaps where capacities need to be strengthened. The inventory should include the following factors:

- existing regulatory structures and legislation pertaining to the import and export of agricultural commodities, environmental protection, animal and human health safety, and biotechnology

- existing mechanisms for the development of public policy, legislation, and regulations
- existing human, financial and scientific infrastructure
- the current status of biotechnology research and development, including programs for the safe use and handling of GMOs
- existing mechanisms for regional cooperation and regulatory harmonisation
- existing capacity building programs
- the role of civil society in policy and regulatory development processes
- administrative and enforcement capacity.

Knowledge, skills and capacity base to develop and implement a biosafety system. A strong base of scientific knowledge to support the regulatory system and the development of competencies in product evaluation is critical to any biosafety system. A limited knowledge and skills base will tend to produce regulations that are highly protective, at the expense of innovation.

Some countries have implemented a system of expert advisory committees, while others have relied primarily on scientists and professionals working within government agencies. The advantage of independent advisory committees is that they generally have more transparent accountability frameworks because the expertise and academic credentials of their membership are usually published.

However, they may suffer from the part-time volunteer nature of their membership. A combination of these two approaches—expert advisory committees and government scientists and professionals—may be the best arrangement. Product evaluations performed by competent scientists within a regulatory agency could be supplemented by the results of issue-specific expert panel consultations.

Risk Assessment

What is Risk Assessment?

Risk assessment is a method by which science can be used to address environmental and human health safety concerns. It is broadly defined as the use of scientific data to rank or measure hazards, assess exposure and characterize risks involved in a food-related activity or product. However, risk assessments do not specifically determine whether a product is "safe" or "unsafe". The use of risk assessment in the development of regulations is related to the expected impact of a particular environmental or food safety problem, the expected impact of protective mitigation measures, and the levels of urgency and controversy surrounding an issue.

What are the Principles of Risk Assessment as Applied to GMOs?

There are four basic principles that have found international acceptance. These are embodied in Annex III of the Cartagena Protocol:

1. Risk assessment should be carried out in a scientifically sound and transparent manner, and can take into account expert advice of, and guidelines developed by, relevant international organisations.
2. Lack of scientific knowledge or scientific consensus should not necessarily be interpreted as indicating a particular level of risk, an absence of risk, or an acceptable risk.
3. Risks associated with living modified organisms or products thereof, namely, processed materials that are of living modified organism origin, containing detectable novel combinations of replicable genetic material obtained through the use of modern biotechnology, should be considered in the context of the risks posed by the non-modified recipients or parental organisms in the likely potential receiving environment.
4. Risk assessment should be carried out on a case-by-case basis. The required information may vary in nature and level of detail from case to case, depending on the living modified organism concerned, its intended use and the likely potential receiving environment.

What are the Limitations of Scientific Risk Assessment?

In risk assessment analysis, the use of science is not as clear-cut and simple as it may appear initially. First, since scientific knowledge constantly changes and evolves, the best available science about food or environmental issues may be different from one year to the next.

Another complicating factor in the application of science to risk assessment analysis is the disagreement among scientists about conclusions from key influential studies. Ideally, scientists conduct research in a manner that produces precise and accurate results so that study conclusions are authoritative and nonbiased. However, it is not unusual for scientists to disagree about an experiment's design and conclusions, especially if the results alter the conventional knowledge about a subject.

These disagreements may require years of debate and further experimentation in other laboratories before results are accepted by most scientists. Regulations based on professionally disputed scientific studies may not be scientifically defensible once the controversy is resolved through additional data gathering and analysis.

Scientific Risk Assessment in Evaluating Biotechnology Products

Yes, there is no other internationally accepted standard by which one can effectively assess the risks, if any, posed by GM products. One should, however, be cognizant of the above limitations in arriving at a decision involving GMOs. Development of regulations. A country may adopt either voluntary guidelines or mandatory regulations. Voluntary guidelines are more quickly put in place and are more flexible for the adoption of revisions incorporating new information requirements. However, the public may not be as confident with voluntary guidelines as they would be with mandatory regulations. Thus, there may be value in adopting mandatory regulations.

If a country elects to develop mandatory regulations, it can do so in one of two ways: (1) it can develop a new act and regulations to specifically address GMOs; or (2) it can regulate GMOs using existing legal instruments such as acts, regulations, and presidential decrees.

The former has the advantage of establishing a system that specifically addresses the product or process to be regulated, of being crafted to allow flexibility in the face of new technical advances, and of being perceived by the public as a positive response to addressing the concerns about safety of GMOs. However, it could take long to develop such a new act or regulation, especially with the political controversy that GMOs have generated. Also, this could result in GMOs being regulated in perpetuity even if the scientific basis for the separate regulation has long been eroded.

If a mandatory system is adopted, the policy maker should also decide on whether the system should take any of the following forms:

- *ex ante* regulation in the form of required permits, licenses, regulations, and product approvals before any GM product is used or released, whether for experimental or commercial purposes;
- strict *ex post facto* liabilities in the form of damage payments by the biotechnology research organisation or business entities; or
- a *negligence* rule, which is a combination of the ex ante regulation and ex post facto liabilities.

The biosafety policy developed in most countries emphasizes the ex ante regulatory approach. This approach has the major benefit of providing information to both the producer and consumer of biotechnology products. If a biotechnology organisation produces new products according to the regulations, it is less likely to be fined ex post facto. Thus, regulation and product standards reduce risk and thereby allow the market to work more smoothly as the participants are better informed about the rules of the game.

Ultimately, the development of biosafety policy will depend on several factors including the nature of risks, the goal of public policy, the institutional and judicial framework, and the involvement of the private sector in biotechnology research. Regardless of the type of regulatory framework chosen, care must be taken not to overregulate. An unreasonably stringent regulatory system can prevent beneficial products from being made available to the public. Implementation of regulations. The final step of putting the system into operation requires the following elements:

- The regulations or guidelines clearly define the structure of the biosafety system.
- People are knowledgeable and well trained.
- The review process is based on up-to-date scientific information.
- Feedback mechanisms are used to incorporate new information and revise the system as needed.

What International Agreements are Relevant to Modern Biotechnology?

Issues related to modern biotechnology have been raised in a number of international fora, including the following:

- Convention on Biological Diversity (CBD)
- Food and Agriculture Organisation (FAO)
- Organisation for Economic Cooperation and Development (OECD)
- World Trade Organisation (WTO)
- World Health Organisation (WHO)
- World Intellectual Property Organisation (WIPO)
- Asia Pacific Economic Cooperation (APEC)
- Office of International Epizootics (OIE)
- International Plant Protection Convention (IPPC)
- Codex Alimentarius Commission (CAC)
- World Bank.

There are also a number of international agreements that relate to modern biotechnology. The most important are the:

- Cartagena Protocol on Biosafety to the United Nations Convention on Biological Diversity
- Agreement on Technical Barriers to Trade
- Agreement on the Application of Sanitary and Phytosanitary Measures
- WTO Agreement on Trade-Related Aspects of Intellectual Property Rights.

What is the Cartagena Protocol on Biosafety?

The Cartagena Protocol on Biosafety to the Convention on Biological Diversity is the first international legally binding tool regarding biosafety in biotechnology. It seeks to contribute to ensuring that there is an adequate level of protection in the safe transfer, handling and use of GMOs resulting from modern biotechnology, especially those that may have adverse effects on the conservation and sustainable use of biological diversity. It also takes into account risks to human health, and specifically focuses on transboundary movements. The Protocol was adopted in Montreal on January 29, 2000, but will take effect only after at least 50 States have ratified it. In summary, the Protocol:

- Establishes an Advance Informed Agreement procedure for imports of LMOs intended for release into the environment;
- Establishes a simplified procedure for notification and information exchange for LMOs intended for food, feed or for processing, such as agricultural commodities;
- Establishes regimes for assessing and managing risks to biodiversity;
- Details information and documentation requirements; and
- Includes provision for capacity-building and financial resources.

Do Developing Countries Stand to Benefit from the use of Modern Biotechnology?

Yes, but developing countries should remember that the institutional and economic environment within which modern biotechnology R&D is being conducted differs significantly from that of Green Revolution technologies. The latter was essentially the prerogative of public research institutions and philanthropic foundations. In contrast, the application of modern biotechnology to agriculture is a competitive, commercial endeavour in which powerful private sector interests compete.

Multinational companies in the seed, agricultural chemical, pharmaceutical and food-processing industries play a major role in biotechnology research. Also, as a result of mergers and acquisitions in the past years, the development of new biotechnology applications in agriculture has become increasingly concentrated in the hands of a few companies. The dominant companies that currently operate within the global markets are Monsanto, Syngenta and Pioneer Hi-Bred.

The Food and Agriculture Organisation has pointed out that current transgenic crop releases are still "very narrow" in terms of crops and traits, and thus have yet to address the special needs of developing countries.

While some 200 crops are currently under field testing in developing countries and other crop-trait combinations are being investigated, focusing mostly on virus resistance, crop quality, and in some cases, tolerance to abiotic stresses, many crops (e.g., vegetables) and traits (e.g., drought- and aluminum-resistance) important to developing countries are still almost entirely neglected.

Relatively little biotechnology research is being undertaken on the problems of small farmers in rainfed and marginal lands. Neither is there much interest in crops like wheat, sorghum, millet, banana, lentils, cassava, groundnut, and sweet potato. These are considered orphan crops because of the private sector's reluctance to work on them.

They have limited appeal because they are grown mainly for personal consumption by poor farmers. Hence, public sector research will have to fill in the void by exploiting the research potential of biotechnology to solve the high-priority problems of the developing world.

Global Arena to Enable Developing Countries to take Advantage of the Potential Benefits of Modern Biotechnology

There is a need to push for a global governance regime for biotechnology that will help to bring a large number of developing countries into the global trading system. The elements of such a governance system should include improvements in market access, development of technological capabilities, access to technology, national regulation of biotechnology, and the management of risks and benefits associated with its use.

Although scientific advances in biotechnology appear to be concentrated in a small number of developed countries, the following factors will allow for the wider participation of developing countries in the new bioeconomy:

1. The growing recognition that the current patterns of globalisation are untenable if they do not increasingly include developing country products. Developing countries depend on industries that are based on natural resources and can therefore benefit from the use of modern biotechnology.
2. Many of the techniques used in biotechnology research are becoming readily available because of scientific familiarity, and are therefore relatively easy to acquire through sustained capacity development and enterprise development efforts.
3. Much of the initial R&D expenditures have already been borne by the industrialised countries. What is needed is effective international technology partnerships to enable developing countries to benefit from biotechnology R&D.

However, much will depend on the level of domestic technological capacity in developing countries and the kind of global biotechnology governance system that emerges from the current policy debates. A global governance system that provides opportunities for market access will help to foster the commercialisation of new technologies, especially those that threaten to alter the patterns and loci of productions. Resistance to new technologies is likely to be reduced by perceptions of access to the new technologies, as well as to their markets.

Current GMO Controversy help in Clarifying the Issues Surrounding

Unfortunately, the debate over GMOs has focused on risks to human health and the environment. While these concerns are important, the debate does not adequately reflect the interests of developing countries.

This is because the issues are framed in the context of industrialised country concerns.

For instance, the debate ignores the fact that since the rural poor in developing countries are mostly in farming, any technology that helps lighten the load of agricultural workers can free up time to pursue higher-earning occupations. An oft-cited example of the revolutionary potential of modern biotechnology is the harvest in 2001 by Kenyan farmers of the first trial crop of sweet potatoes for resistance to an aphid-borne disease that previously killed up to 80% of their crops.

In addition, one of the main policy goals of developing countries is to enhance food security, a problem that may not be present in many developed countries. While biotechnology cannot solve all of the problems associated with agricultural production, there is no denying the fact that it has the potential to address specific problems such as increasing crop productivity, diversifying crops, enhancing the nutritional value of food, reducing environmental impacts of agricultural productions, and promoting market competitiveness.

Developing countries are hampered in their ability to benefit from advances in modern biotechnology because of the lack of scientific and technological capacity and the low level of enterprise development in most of these countries. While the responsibility for formulating policies and strategies for the wider use of biotechnology lies within the domestic leadership, international cooperation and partnerships are essential in promoting sustainable agriculture in the developing world.

Can Modern Biotechnology Solve the Problems of Developing Countries in Agriculture?

Biotechnology is simply one of the instruments that developing countries can use in finding solutions to problems in agriculture. While biotechnology has many potential benefits, it is not and cannot be a solution for all of the problems confronting the agricultural sector. After all, technology is only one of many factors for agricultural growth.

Hence, while encouraging biotechnology research and development, developing countries should continue to invest in water and soil management, farm-to-market infrastructure, and credit access programs, among others. However, no developing country can ignore the promise of safe and responsible use of biotechnology, whether in the medical or agricultural spheres.

Stem Cell

Stem cells are biological cells found in all multicellular organisms, that can divide (through mitosis) and differentiate into diverse specialised cell types and can self-renew to produce more stem cells. In mammals, there are two broad types of stem cells: embryonic stem cells, which are isolated from the inner cell mass of blastocysts, and adult stem cells, which are found in various tissues. In adult organisms, stem cells and progenitor cells act as a repair system for the body, replenishing adult tissues. In a developing embryo, stem cells can differentiate into all the specialised cells (these are called pluripotent cells), but also maintain the normal turnover of regenerative organs, such as blood, skin, or intestinal tissues.

Stem cells can now be artificially grown and transformed into specialised cell types with characteristics consistent with cells of various tissues such as muscles or nerves through cell culture. Highly plastic adult stem cells are routinely used in medical therapies. Stem cells can be taken from a variety of sources, including umbilical cord blood and bone marrow.

Embryonic cell lines and autologous embryonic stem cells generated through therapeutic cloning have also been proposed as promising candidates for future therapies. Research into stem cells grew out of findings by Ernest A. McCulloch and James E. Till at the University of Toronto in the 1960s.

There are three sources of autologous adult stem cells: 1) Bone marrow, which requires extraction by *harvesting*, that is, drilling into bone (typically the femur or iliac crest), 2) Adipose tissue (lipid cells), which requires extraction by liposuction, and 3) Blood, which requires extraction through pheresis, wherein blood is drawn from the donor, (similar to a blood donation) passed through a machine that extracts the stem cells and returns other portions of the blood to the donor. Of all stem cell types, autologous harvesting involves the least risk. By definition, autologous cells are obtained from one's own body, just as one may bank his or her own blood for elective surgical procedures.

Properties

The classical definition of a stem cell requires that it possess two properties:

- *Self-renewal*: the ability to go through numerous cycles of cell division while maintaining the undifferentiated state.
- *Potency*: the capacity to differentiate into specialised cell types. In the strictest sense, this requires stem cells to be either totipotent or pluripotent to be able to give rise to any mature cell type, although multipotent or unipotent progenitor cells are sometimes referred to as stem cells.Apart from this it is being said that stem cells function is regulated in a feed back mechanism.

Potency Definitions

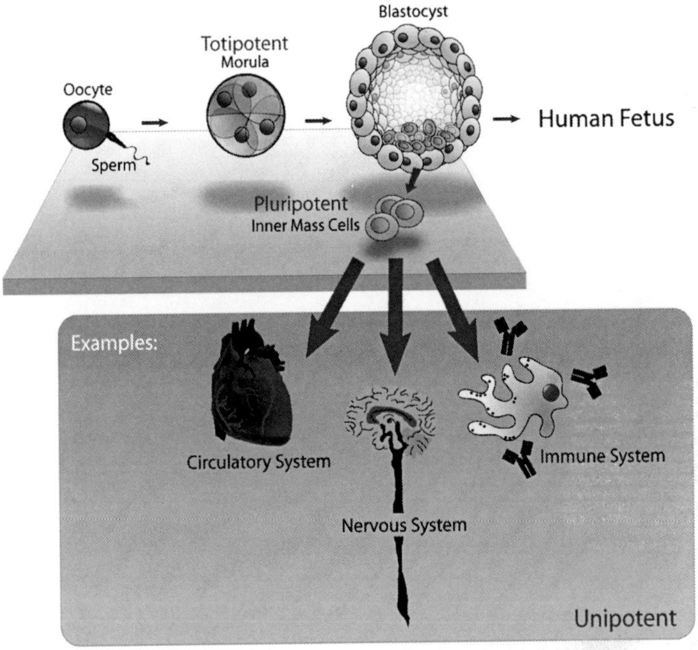

Figure: *Pluripotent, embryonic stem cells originate as inner mass cells within a blastocyst. The stem cells can become any tissue in the body, excluding a placenta. Only the morula's cells are totipotent, able to become all tissues and a placenta.*

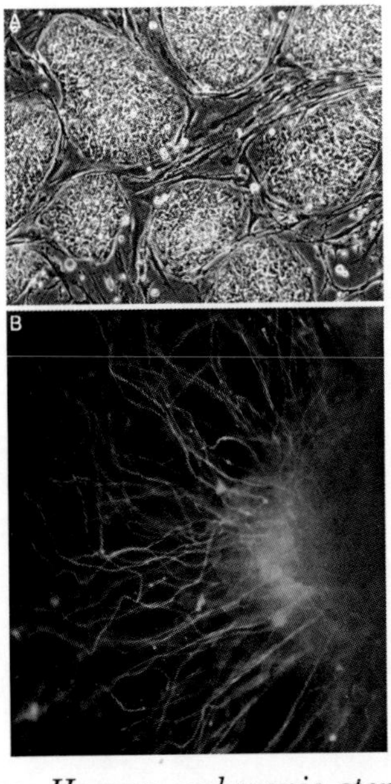

Figure: *Human embryonic stem cells*

A: Cell colonies that are not yet differentiated.

Potency specifies the differentiation potential (the potential to differentiate into different cell types) of the stem cell.

- Totipotent (a.k.a omnipotent) stem cells can differentiate into embryonic and extraembryonic cell types. Such cells can construct a complete, viable organism. These cells are produced from the fusion of an egg and sperm cell. Cells produced by the first few divisions of the fertilized egg are also totipotent.
- Pluripotent stem cells are the descendants of totipotent cells and can differentiate into nearly all cells, i.e. cells derived from any of the three germ layers.
- Multipotent stem cells can differentiate into a number of cells, but only those of a closely related family of cells.
- Oligopotent stem cells can differentiate into only a few cells, such as lymphoid or myeloid stem cells.
- Unipotent cells can produce only one cell type, their own, but have the property of self-renewal, which distinguishes them from non-stem cells (e.g., muscle stem cells).

Self-renewal

Two mechanisms to ensure that the stem cell population is maintained exist:

1. Obligatory asymmetric replication: a stem cell divides into one father cell that is identical to the original stem cell, and another daughter cell that is differentiated
2. Stochastic differentiation: when one stem cell develops into two differentiated daughter cells, another stem cell undergoes mitosis and produces two stem cells identical to the original.

Identification

The practical definition of a stem cell is the functional definition, a cell that has the potential to regenerate tissue over a lifetime. For example, the defining test for a bone marrow or hematopoietic stem cell (HSC) is the ability to transplant one cell and save an individual without HSCs. In this case, a stem cell must be able to produce new blood cells and immune cells over a long term, demonstrating potency.

It should also be possible to isolate stem cells from the transplanted individual, which can themselves be transplanted into another individual without HSCs, demonstrating that the stem cell was able to self-renew. Properties of stem cells can be illustrated *in vitro*, using methods such as clonogenic assays, in which single cells are assessed for their ability to differentiate and self-renew. Stem cells can also be isolated by their possession of a distinctive set of cell surface markers. However, *in vitro* culture conditions can alter the behaviour of cells, making it unclear whether the cells will behave in a similar manner *in vivo*. There is considerable debate as to whether some proposed adult cell populations are truly stem cells.

Embryonic

Embryonic stem cell lines (ES cell lines) are cultures of cells derived from the epiblast tissue of the inner cell mass (ICM) of a blastocyst or earlier morula stage embryos. A blastocyst is an early stage embryo approximately four to five days old in humans and consisting of 50–150 cells. ES cells are pluripotent and give rise during development to all derivatives of the three primary germ layers: ectoderm, endoderm and mesoderm. In other words, they can develop into each of the more than 200 cell types of the adult body when given sufficient and necessary stimulation for a specific cell type. They do not contribute to the extra-embryonic membranes or the placenta. The endoderm is composed of the entire gut tube and the lungs, the ectoderm gives rise to the nervous system and skin, and the mesoderm gives rise to muscle, bone, blood in essence, everything else that connects the endoderm to the ectoderm.

Nearly all research to date has made use of mouse embryonic stem cells (mES) or human embryonic stem cells (hES). Both have the essential stem cell characteristics, yet they require very different environments in order to maintain an undifferentiated state. Mouse ES cells are grown on a layer of gelatin and require the presence of leukemia inhibitory factor (LIF). Human ES cells are grown on a feeder layer of mouse embryonic fibroblasts (MEFs) and require the presence of basic fibroblast growth factor (bFGF or FGF-2). Without optimal culture conditions or genetic manipulation, embryonic stem cells will rapidly differentiate.

A human embryonic stem cell is also defined by the presence of several transcription factors and cell surface proteins. The transcription factors Oct-4, Nanog, and Sox2 form the core regulatory network that ensures the suppression of genes that lead to differentiation and the maintenance of pluripotency. The cell surface antigens most commonly used to identify hES cells are the glycolipids stage specific embryonic antigen 3 and 4 and the keratan sulfate antigens Tra-1-60 and Tra-1-81. The molecular definition of a stem cell includes many more proteins and continues to be a topic of research.

There are currently no approved treatments using embryonic stem cells. The first human trial was approved by the US Food and Drug Administration in January 2009. However, the human trial had not yet been initiated until October 13, 2010 in Atlanta for spinal injury victims.

ES cells, being pluripotent cells, require specific signals for correct differentiation—if injected directly into another body, ES cells will differentiate into many different types of cells, causing a teratoma. Differentiating ES cells into usable cells while avoiding transplant rejection are just a few of the hurdles that embryonic stem cell researchers still face. Many nations currently have moratoria on either ES cell research or the production of new ES cell lines. Because of their combined abilities of unlimited expansion and pluripotency, embryonic stem cells remain a theoretically potential source for regenerative medicine and tissue replacement after injury or disease.

Fetal

The primitive stem cells located in the organs of fetuses are referred to as fetal stem cells.

Adult

Also known as somatic stem cells and germline (giving rise to gametes) stem cells, they can be found in children, as well as adults.

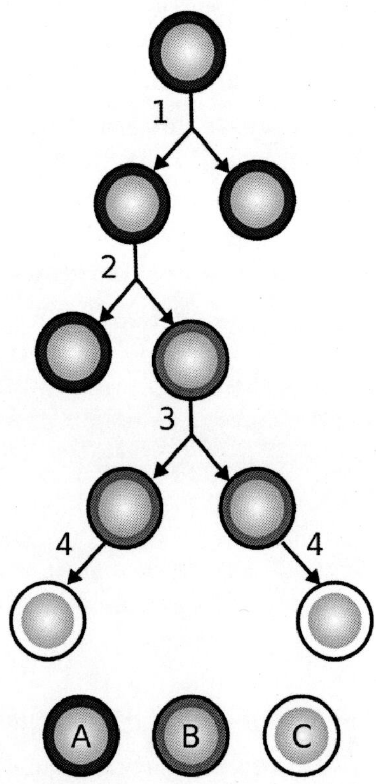

Figure: *Stem cell division and differentiation. A: stem cell; B: progenitor cell; C: differentiated cell; 1: symmetric stem cell division; 2: asymmetric stem cell division; 3: progenitor division; 4: terminal differentiation*

Pluripotent adult stem cells are rare and generally small in number but can be found in a number of tissues including umbilical cord blood. A great deal of adult stem cell research to date has had the aim of characterising the capacity of the cells to divide or self-renew indefinitely and their differentiation potential. In mice, pluripotent stem cells are directly generated from adult fibroblast cultures. Unfortunately, many mice do not live long with stem cell organs. Most adult stem cells are lineage-restricted (multipotent) and are generally referred to by their tissue origin (mesenchymal stem cell, adipose-derived stem cell, endothelial stem cell, dental pulp stem cell, etc.).

Adult stem cell treatments have been successfully used for many years to treat leukemia and related bone/blood cancers through bone marrow transplants. Adult stem cells are also used in veterinary medicine to treat tendon and ligament injuries in horses.

The use of adult stem cells in research and therapy is not as controversial as the use of embryonic stem cells, because the production of adult stem cells does not require the destruction of an embryo. Additionally, in instances where adult stem cells are obtained from the intended recipient (an autograft), the risk of rejection is essentially non-existent. Consequently, more US government funding is being provided for adult stem cell research.

An extremely rich source for adult mesenchymal stem cells is the developing tooth bud of the mandibular third molar. The stem cells eventually form enamel (ectoderm), dentin, periodontal ligament, blood vessels, dental pulp, nervous tissues, and a minimum of 29 different end organs. Because of extreme ease in collection at 8–10 years of age before calcification and minimal to no morbidity, these will probably constitute a major source of cells for personal banking, research and current or future therapies. These stem cells have been shown capable of producing hepatocytes.

Amniotic

Multipotent stem cells are also found in amniotic fluid. These stem cells are very active, expand extensively without feeders and are not tumorigenic. Amniotic stem cells are multipotent and can differentiate in cells of adipogenic, osteogenic, myogenic, endothelial, hepatic and also neuronal lines. All over the world, universities and research institutes are studying amniotic fluid to discover all the qualities of amniotic stem cells, and scientists such as Anthony Atala and Giuseppe Simoni have discovered important results.

Use of stem cells from amniotic fluid overcomes the ethical objections to using human embryos as a source of cells. Roman Catholic teaching forbids the use of embryonic stem cells in experimentation; accordingly, the Vatican newspaper "Osservatore Romano" called amniotic stem cell "the future of medicine".

It is possible to collect amniotic stem cells for donors or for autologuous use: the first US amniotic stem cells bank opened in 2009 in Medford, MA, by Biocell Centre Corporation and collaborates with various hospitals and universities all over the world.

Induced Pluripotent

These are not adult stem cells, but rather reprogrammed cells (e.g. epithelial cells) given pluripotent capabilities. Using genetic reprogramming with protein transcription factors, pluripotent stem cells equivalent to embryonic stem cells have been derived from human adult skin tissue. Shinya Yamanaka and his colleagues at Kyoto University used the transcription factors Oct3/4, Sox2, c-Myc, and Klf4 in their experiments on cells from human faces. Junying Yu, James Thomson, and their colleagues at the University of Wisconsin–Madison used a different set of factors, Oct4, Sox2, Nanog and Lin28, and carried out their experiments using cells from human foreskin.

As a result of the success of these experiments, Ian Wilmut, who helped create the first cloned animal Dolly the Sheep, has announced that he will abandon nuclear transfer as an avenue of research. Frozen blood samples can be used as a source of induced pluripotent stem cells, opening a new avenue for obtaining the valued cells.

Lineage

To ensure self-renewal, stem cells undergo two types of cell division. Symmetric division gives rise to two identical daughter cells both endowed with stem cell properties. Asymmetric division, on the other hand, produces only one stem cell and a progenitor cell with limited self-renewal potential. Progenitors can go through several rounds of cell division before terminally differentiating into a mature cell. It is possible that the molecular distinction between symmetric and asymmetric divisions lies in differential segregation of cell membrane proteins (such as receptors) between the daughter cells.

An alternative theory is that stem cells remain undifferentiated due to environmental cues in their particular niche. Stem cells differentiate when they leave that niche or no longer receive those signals. Studies in *Drosophila* germarium have identified the signals dpp and adherens junctions that prevent germarium stem cells from differentiating.

The signals that lead to reprogramming of cells to an embryonic-like state are also being investigated. These signal pathways include several transcription factors including the oncogene c-Myc. Initial studies indicate that transformation of mice cells with a combination of these anti-differentiation signals can reverse differentiation and may allow adult cells to become pluripotent. However, the need to transform these cells with an oncogene may prevent the use of this approach in therapy. Challenging the terminal nature of cellular differentiation and the integrity of lineage commitment, it was recently determined that the somatic expression of combined transcription factors can directly induce other defined somatic cell fates; researchers identified three neural-lineage-specific transcription factors that could directly convert mouse fibroblasts (skin cells) into fully functional neurons. This "induced neurons" (iN) cell research inspires the researchers to induce other cell types implies that *all* cells are totipotent: with the proper tools, all cells may form all kinds of tissue.

Treatments

Medical researchers believe that stem cell therapy has the potential to dramatically change the treatment of human disease. A number of adult stem cell therapies already exist, particularly bone marrow transplants that are used to treat leukemia. In the future, medical researchers anticipate being able to use technologies derived from stem cell research to treat a wider variety of diseases including cancer, Parkinson's disease, spinal cord injuries, Amyotrophic lateral sclerosis, multiple sclerosis, and muscle damage, amongst a number of other impairments and conditions. However, there still exists a great deal of social and scientific uncertainty surrounding stem cell research, which could possibly be overcome through public debate and future research, and further education of the public.

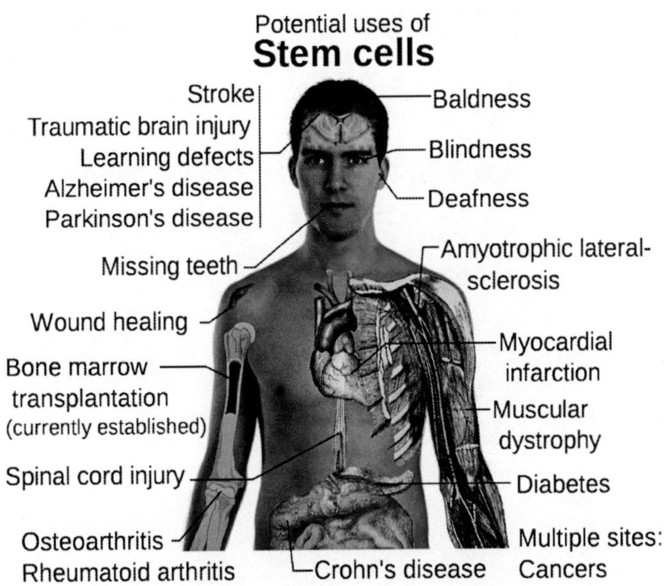

Figure: *Diseases and conditions where stem cell treatment is promising or emerging. Bone marrow transplantation is, as of 2009, the only established use of stem cells.*

One concern of treatment is the risk that transplanted stem cells could form tumours and become cancerous if cell division continues uncontrollably. Stem cells are widely studied, for their potential therapeutic use and for their inherent interest. Supporters of embryonic stem cell research argue that such research should be pursued because the resultant treatments could have significant medical potential. It has been proposed that surplus embryos created for in vitro fertilization could be donated with consent and used for the research.

The recent development of iPS cells has been called a bypass of the legal controversy. Laws limiting the destruction of human embryos have been credited for being the reason for development of iPS cells, but it is still not completely clear whether hiPS cells are equivalent to hES cells. Recent work demonstrates hotspots of aberrant epigenomic reprogramming in hiPS cells (Lister, R., et al., 2011).

Research Patents

The patents covering a lot of work on human embryonic stem cells are owned by the Wisconsin Alumni Research Foundation (WARF). WARF does not charge academics to study human stem cells but does charge commercial users. WARF sold Geron Corp. exclusive rights to work on human stem cells but later sued Geron Corp. to recover some of the previously sold rights. The two sides agreed that Geron Corp. would keep the rights to only three cell types. In 2001, WARF came under public pressure to widen access to human stem-cell technology.

A request for reviewing the WARF patents 5,843,780; 6,200,806; 7,029,913 US Patent and Trademark Office were filed by non-profit patent-watchdogs The Foundation for Taxpayer & Consumer Rights, and the Public Patent Foundation as well as molecular biologist Jeanne Loring of the Burnham Institute. According to them, two of the patents granted to WARF are invalid because they cover a technique published in 1993 for which a patent had already been granted to an Australian researcher. Another part of the challenge states that these techniques, developed by James A. Thomson, are rendered obvious by a 1990 paper and two textbooks. Based on this challenge, patent 7,029,913 has been rejected in 2010. The two remaining hES WARF patents are due to expire in 2015.

The outcome of this legal challenge is particularly relevant to the Geron Corp. as it can only license patents that are upheld.

- 1908: The term “stem cell” was proposed for scientific use by the Russian histologist Alexander Maksimov (1874–1928) at congress of hematologic society in Berlin. It postulated existence of haematopoietic stem cells.

- 1960s: Joseph Altman and Gopal Das present scientific evidence of adult neurogenesis, ongoing stem cell activity in the brain; like André Gernez, their reports contradict Cajal's "no new neurons" dogma and are largely ignored.
- 1963: McCulloch and Till illustrate the presence of self-renewing cells in mouse bone marrow.
- 1968: Bone marrow transplant between two siblings successfully treats SCID.
- 1978: Haematopoietic stem cells are discovered in human cord blood.
- 1981: Mouse embryonic stem cells are derived from the inner cell mass by scientists Martin Evans, Matthew Kaufman, and Gail R. Martin. Gail Martin is attributed for coining the term "Embryonic Stem Cell".
- 1992: Neural stem cells are cultured *in vitro* as neurospheres.
- 1997: Leukemia is shown to originate from a haematopoietic stem cell, the first direct evidence for cancer stem cells.
- 1998: James Thomson and coworkers derive the first human embryonic stem cell line at the University of Wisconsin–Madison.
- 1998: John Gearhart (Johns Hopkins University) extracted germ cells from fetal gonadal tissue (primordial germ cells) before developing pluripotent stem cell lines from the original extract.
- 2000s: Several reports of adult stem cell plasticity are published.
- 2001: Scientists at Advanced Cell Technology clone first early (four- to six-cell stage) human embryos for the purpose of generating embryonic stem cells.
- 2003: Dr. Songtao Shi of NIH discovers new source of adult stem cells in children's primary teeth.
- 2004–2005: Korean researcher Hwang Woo-Suk claims to have created several human embryonic stem cell lines from unfertilized human oocytes. The lines were later shown to be fabricated.
- 2005: Researchers at Kingston University in England claim to have discovered a third category of stem cell, dubbed cord-blood-derived embryonic-like stem cells (CBEs), derived from umbilical cord blood. The group claims these cells are able to differentiate into more types of tissue than adult stem cells.
- 2005: Researchers at UC Irvine's Reeve-Irvine Research Centre are able to partially restore the ability of rats with paralysed spines to walk through the injection of human neural stem cells.
- August 2006: Mouse Induced pluripotent stem cells: the journal *Cell* publishes Kazutoshi Takahashi and Shinya Yamanaka.

- October 2006: Scientists at Newcastle University in England create the first ever artificial liver cells using umbilical cord blood stem cells.
- January 2007: Scientists at Wake Forest University led by Dr. Anthony Atala and Harvard University report discovery of a new type of stem cell in amniotic fluid. This may potentially provide an alternative to embryonic stem cells for use in research and therapy.
- June 2007: Research reported by three different groups shows that normal skin cells can be reprogrammed to an embryonic state in mice. In the same month, scientist Shoukhrat Mitalipov reports the first successful creation of a primate stem cell line through somatic cell nuclear transfer
- October 2007: Mario Capecchi, Martin Evans, and Oliver Smithies win the 2007 Nobel Prize for Physiology or Medicine for their work on embryonic stem cells from mice using gene targeting strategies producing genetically engineered mice (known as knockout mice) for gene research.
- November 2007: Human induced pluripotent stem cells: Two similar papers released by their respective journals prior to formal publication: in *Cell* by Kazutoshi Takahashi and Shinya Yamanaka, "Induction of pluripotent stem cells from adult human fibroblasts by defined factors", and in *Science* by Junying Yu, et al., from the research group of James Thomson, "Induced pluripotent stem cell lines derived from human somatic cells": pluripotent stem cells generated from mature human fibroblasts. It is possible now to produce a stem cell from almost any other human cell instead of using embryos as needed previously, albeit the risk of tumorigenesis due to c-myc and retroviral gene transfer remains to be determined.
- January 2008: Robert Lanza and colleagues at Advanced Cell Technology and UCSF create the first human embryonic stem cells without destruction of the embryo
- January 2008: Development of human cloned blastocysts following somatic cell nuclear transfer with adult fibroblasts
- February 2008: Generation of pluripotent stem cells from adult mouse liver and stomach: these iPS cells seem to be more similar to embryonic stem cells than the previously developed iPS cells and not tumorigenic, moreover genes that are required for iPS cells do not need to be inserted into specific sites, which encourages the development of non-viral reprogramming techniques.
- March 2008: The first published study of successful cartilage regeneration in the human knee using autologous adult mesenchymal stem cells is published by clinicians from Regenerative Sciences

- October 2008: Sabine Conrad and colleagues at Tübingen, Germany generate pluripotent stem cells from spermatogonial cells of adult human testis by culturing the cells in vitro under leukemia inhibitory factor (LIF) supplementation.
- 30 October 2008: Embryonic-like stem cells from a single human hair.
- 1 March 2009: Andras Nagy, Keisuke Kaji, *et al.* discover a way to produce embryonic-like stem cells from normal adult cells by using a novel "wrapping" procedure to deliver specific genes to adult cells to reprogram them into stem cells without the risks of using a virus to make the change. The use of electroporation is said to allow for the temporary insertion of genes into the cell.
- 28 May 2009 Kim *et al.* announced that they had devised a way to manipulate skin cells to create patient specific "induced pluripotent stem cells" (iPS), claiming it to be the 'ultimate stem cell solution'.
- 11 October 2010 First trial of embryonic stem cells in humans.
- 25 October 2010: Ishikawa *et al.* write in the Journal of Experimental Medicine that research shows that transplanted cells that contain their new host's nuclear DNA could still be rejected by the individual's immune system due to foreign mitochondrial DNA. Tissues made from a person's stem cells could therefore be rejected, because mitochondrial genomes tend to accumulate mutations.
- 2011: Israeli scientist Inbar Friedrich Ben-Nun led a team which produced the first stem cells from endangered species, a breakthrough that could save animals in danger of extinction.

Neural Stem Cell

Neural stem cells (NSCs) are the self-renewing, multipotent cells that generate the main phenotypes of the nervous system. In 1989, Sally Temple described multipotent, self-renewing progenitor and stem cells in the subventricular zone of the mouse brain (Temple, S, Nature, 1989).

In 1992, Brent A. Reynolds and Samuel Weiss were the first to isolate neural progenitor and stem cells from the adult striatal tissue, including the subventricular zone one of the neurogenic areas of adult mice brain tissue.

In the same year the team of Constance Cepko and Evan Y. Snyder were the first to isolate multipotent cells from the mouse cerebellum and stable transfected them with the oncogene v-myc. Interestingly this molecule is one of the genes widely used now to reprogrammed adult non-stem cells into pluripotent stem cells. Since then, neural progenitor and stem cells have been isolated from various areas of the adult brain, including non-neurogenic areas, such as the spinal cord, and from various species including human.

Functions

Epidermal growth factor (EGF) and fibroblast growth factor (FGF) are mitogens that promote neural progenitor and stem cell growth in vitro, though other factors synthesized by the neural progenitor and stem cell populations are also required for optimal growth. It is hypothesised that neurogenesis in the adult brain originates from NSCs. The origin and identity of NSCs in the adult brain remain to be defined.

Function of Neural Stem Cells (NSC) During Disease

NSCs have an important role during development producing the enormous diversity of neurons, astrocytes and oligodendrocytes in the developing CNS. They also have important role in adult animals, for instance in learning and hippocampal plasticity in the adult mice in addition to supplying neurons to the olfactory bulb in mice that migrate daily from their niche in the subventricular zone (SVZ). Notably the role of NSCs during diseases is now being elucidated by several research groups around the world. The responses during stroke, multiple sclerosis, parkinson's disease in humans and in model of these diseases is part of the current investigation. The results of this ongoing investigation may have future applications to treat human neurological diseases.

Neural stem cells have been shown to engage in migration and replacement of dying neurons in classical experiments performed by Sanjay Magavi and Jeffrey Macklis. Using a laser-induced damage of cortical layers, Magavi showed that SVZ neural progenitors expressing Doublecortin, a critical molecule for migration of neuroblasts, migrated long distances to the area of damage and differentiated into mature neurons expressing NeuN, a widely used neuronal marker. In addition Masato Nakafuku's group from Japan showed for the first time the role of hippocampal stem cells during stroke in mice.

These results demonstrated that NSCs can engage in the adult brain as a results of injury. Furthermore, In 2004 based on early work of Evan Synder's group, that showed that NSCs migrate to brain tumours in a directed fashion, Jaime Imitola, M.D and colleagues from Harvard demonstrated for the first time, a molecular mechanism for the responses of NSCs to injury, they showed that chemokines released during injury such as SDF-1a were responsible for the directed migration of human and mouse NSCs to areas of injury in mice. Since then other molecules have been found to participate in the responses of NSCs to Injury. All these results have been widely reproduced and expanded by other investigators joining the classical work of Altman and Sidman in 1960's as evidence of the responses of adult NSCs activities and neurogenesis during homeostasis and injury. The search for additional mechanisms that operate in the injury environment and how they influence the responses of NSCs during acute and chronic disease is matter of intense research.

Assay

Neural stem cells are routinely studied *in vitro* using a method referred to as the Neurosphere Assay (or Neurosphere culture system), first developed by Reynolds and Weiss. Neurospheres are intrinsically heterogeneous cellular entities almost entirely formed by a small fraction (1 to 5%) of slowly dividing neural stem cells and by their progeny, a population of fast-dividing nestin-positive progenitor cells.

The total number of these progenitors determines the size of a neurosphere and, as a result, disparities in sphere size within different neurosphere populations may reflect alterations in the proliferation, survival and/or differentiation status of their neural progenitors. Indeed, Leone et al. (2005) have reported that loss of â1-integrin in a neurosphere culture does not significantly affect the capacity of â1-integrin deficient stem cells to form new neurospheres, but it influences the size of the neurosphere: â1-integrin deficient neurospheres were overall smaller due to increased cell death and reduced proliferation.

While the Neurosphere Assay has been the method of choice for isolation, expansion and even the enumeration of neural stem and progenitor cells, several recent publications have highlighted some of the limitations of the neurosphere culture system as a method for determining neural stem cell frequencies. In collaboration with Reynolds, STEMCELL Technologies has developed a collagen-based assay, called the Neural Colony-Forming Cell (NCFC) Assay, for the quantification of neural stem cells. Importantly, this assay allows discrimination between neural stem and progenitor cells.

Neural Stem Cell Institutes

The damaged central nervous system (CNS) tissue has very limited regenerative and repair capacity so that loss of neurological function is often chronic and progressive. Cell replacement from stem cells is being actively pursued as a therapeutic option. Recently in 2009, a research institute dedicated solely to translating neural stem research into therapies for patients was created outside of Albany, NY, Thc Ncural Stem Cell Institute.

Cancer Stem Cell

Cancer stem cells (CSCs) are cancer cells (found within tumours or hematological cancers) that possess characteristics associated with normal stem cells, specifically the ability to give rise to all cell types found in a particular cancer sample. CSCs are therefore tumorigenic (tumor-forming), perhaps in contrast to other non-tumorigenic cancer cells. CSCs may generate tumours through the stem cell processes of self-renewal and differentiation into multiple cell types. Such cells are proposed to persist in tumours as a distinct population and cause relapse and metastasis by giving rise to new tumours. Therefore, development of specific therapies targeted at CSCs holds hope for improvement of survival and quality of life of cancer patients, especially for sufferers of metastatic disease.

Existing cancer treatments have mostly been developed based on animal models, where therapies able to promote tumor shrinkage were deemed effective. However, animals could not provide a complete model of human disease. In particular, in mice, whose life spans do not exceed two years, tumor relapse is exceptionally difficult to study.

The efficacy of cancer treatments is, in the initial stages of testing, often measured by the ablation fraction of tumor mass (fractional kill). As CSCs would form a very small proportion of the tumor, this may not necessarily select for drugs that act specifically on the stem cells. The theory suggests that conventional chemotherapies kill differentiated or differentiating cells, which form the bulk of the tumor but are unable to generate new cells. A population of CSCs, which gave rise to it, could remain untouched and cause a relapse of the disease.

Evidence

The existence of CSCs is a subject of debate within medical research, because many studies have not been successful in discovering the similarities and differences between normal tissue stem cells and cancer stem cells. Cancer cells must be capable of continuous proliferation and self-renewal in order to retain the many mutations required for carcinogenesis, and to sustain the growth of a tumor since differentiated cells (constrained by the Hayflick Limit) cannot divide indefinitely . However, it is debated whether such cells represent a minority. If most cells of the tumor are endowed with stem cell properties, there is no incentive to focus on a specific sub population. There is also debate on the cell of origin of CSCs - whether they originate from stem cells that have lost the ability to regulate proliferation, or from more differentiated population of progenitor cells that have acquired abilities to self-renew (which is related to the issue of stem cell plasticity).

The first conclusive evidence for CSCs was published in 1997 in Nature Medicine. Bonnet and Dick isolated a subpopulation of leukaemic cells that express a specific surface marker CD34, but lacks the CD38 marker. The authors established that the $CD34^{-}/CD38^{-}$ subpopulation is capable of initiating tumours in NOD/SCID mice that is histologically similar to the donor. In cancer research experiments, tumor cells are sometimes injected into an experimental animal to establish a tumor. Disease progression is then followed in time and novel drugs can be tested for their ability to inhibit it. However, efficient tumor formation requires thousands or tens of thousands of cells to be introduced. Classically, this has been explained by poor methodology (i.e.

the tumor cells lose their viability during transfer) or the critical importance of the microenvironment, the particular biochemical surroundings of the injected cells. Supporters of the cancer stem cell paradigm argue that only a small fraction of the injected cells, the CSCs, have the potential to generate a tumor. In human acute myeloid leukemia the frequency of these cells is less than 1 in 10,000.

Further evidence comes from histology, the study of the tissue structure of tumours. Many tumours are very heterogeneous and contain multiple cell types native to the host organ. Heterogeneity is commonly retained by tumor metastases. This implies that the cell that produced them had the capacity to generate multiple cell types. In other words, it possessed multi-differentiative potential, a classical hallmark of stem cells.

The existence of leukaemic stem cells prompted further research into other types of cancer. CSCs have recently been identified in several solid tumours, including cancers of the: (i) Brain, (ii) Breast, (iii) Colon, (iv) Ovary, (v) Pancreas, (vi) Prostate, (vii) Melanoma, (viii) Multiple Myeloma.

Importance

Not only is finding the source of cancer cells necessary for successful treatments, but if current treatments of cancer do not properly destroy enough CSCs, the tumour will reappear. Including the possibility that the treatment of for instance, chemotherapy, will leave only chemotherapy-resistant CSCs, then the ensuing tumor will most likely also be resistant to chemotherapy. If the cancer tumor is detected early enough, enough of the tumor can be killed off and marginalised with traditional treatment. But as the tumor size increases, it becomes more and more difficult to remove the tumor without conferring resistance and leaving enough behind for the tumor to reappear.

Some treatments with chemotherapy, such as paclitaxel in ovarian cancer (a cancer usually discovered in late stages), may actually induce chemoresistance (55-75% relapse <2 years). It potentially does this by destroying only the cancer cells susceptible to the drug (targeting those that are CD44-positive, a trait which has been associated with increased survival time in some ovarian cancers), and allowing the cells which are unaffected by paclitaxel (CD44-negative) to regrow, even after a reduction in over a third of the total tumor size. There are studies, though, which show how paclitaxel can be used in combination with other ligands to affect the CD44-positive cells. While paclitaxel alone, as of late, does not cure the cancer, it is effective at extending the survival time of the patients.

Mechanistic and Mathematical Models

Once the pathways to cancer are hypothesized, it is possible to develop predictive mathematical biology models, e.g., based on the cell compartment method. For instance, the growths of the abnormal cells from their normal counterparts can be denoted with specific mutation probabilities. Such a model has been employed to predict that repeated insult to mature cells increases the formation of abnormal progeny, and hence the risk of cancer. Considerable work needs to be done, however, before the clinical efficacy of such models is established.

Origins

The origin of cancer stem cells is still an area of ongoing research. Several camps have formed within the scientific community regarding the issue, and it is possible that several answers are correct, depending on the tumor type and the phenotype the tumor presents. One important distinction that will often be raised is that the cell of origin for a tumor can not be demonstrated using the cancer stem cell as a model. This is because cancer stem cells are isolated from end-stage tumours. Therefore, describing a cancer stem cell as a cell of origin is often an inaccurate claim, even though a cancer stem cell is capable of initiating new tumor formation. With that caveat mentioned, various theories define the origin of cancer stem cells. In brief, they may be: mutants in developing stem or progenitor cells, mutants in adult stem cells or adult progenitor cells, or mutant cells that acquire stem like attributes. These theories often do focus on a tumor's cell of origin and as such must be approached with skepticism.

Some researchers favour the theory that the cancer stem cell is caused by a mutation in stem cell niche populations during development. The logical progression claims that these developing stem populations are mutated and then expand such that the mutation is shared by many of the descendants of the mutated stem cell. These daughter stem cells are then much closer to becoming tumours, and since there are many of them there is more chance of a mutation that can cause cancer. Another theory associates adult stem cells with the formation of tumours. This is most often associated with tissues with a high rate of cell turnover (such as the skin or gut). In these tissues, it has long been expected that stem cells are responsible for tumor formation. This is a consequence of the frequent cell divisions of these stem cells (compared to most adult stem cells) in conjunction with the extremely long lifespan of adult stem cells.

This combination creates the ideal set of circumstances for mutations to accumulate; accumulation of mutations is the primary factor that drives cancer initiation. In spite of the logical backing of the theory, only recently has evidence appeared that this association represents an actual phenomenon. It is important to bear in mind that, due to the heterogeneous nature of evidence it is possible that any individual cancer could come from an alternative origin. A third possibility often raised is the potential de-differentiation of mutated cells such that these cells acquire stem cell like characteristics. This is often used as a potential alternative to any specific cell of origin, as it suggests that any cell might become a cancer stem cell.

Another related concept is the concept of tumor hierarchy. This concept claims that a tumor is a heterogeneous population of mutant cells, all of which share some mutations but will vary in specific phenotype. In this model, the tumor is made up of several types of stem cells, one optimal to the specific environment and several less successful lines. These secondary lines can become more successful in some environments, allowing the tumor to adapt to its environment, including the methods by which tumours can be treated. If this situation is accurate, it has severe repercussions on the realism of a cancer stem cell specific treatment regime. Within a tumor hierarchy model, it would be extremely difficult to pinpoint the cancer stem cell's origin.

Implications for Cancer Treatment

The existence of CSCs has several implications in terms of future cancer treatment and therapies. These include disease identification, selective drug targets, prevention of metastasis, and development of new intervention strategies. Normal somatic stem cells are naturally resistant to chemotherapeutic agents- they have various pumps (such as MDR) that pump out drugs, DNA repair proteins and they also have a slow rate of cell turnover (chemotherapeutic agents naturally target rapidly replicating cells). CSCs that have mutated from normal stem cells may also express proteins that would increase their resistance towards chemotherapeutic agents.

These surviving CSCs then repopulate the tumor, causing relapse. By selectively targeting CSCs, it would be possible to treat patients with aggressive, non-resectable tumours, as well as preventing the tumor from metastasising. The hypothesis suggests that upon CSC elimination, cancer would regress due to differentiation and/or cell death. What fraction of tumour cells are CSCs and therefore need to be eliminated is not clear yet. A number of studies have investigated the possibility of identifying specific markers that may distinguish CSCs from the bulk of the tumor (as well as from normal stem cells). Proteomic and genomic signatures of tumours are also being investigated. In 2009, scientists identified one compound, Salinomycin, that selectively reduces the proportion of breast CSCs in mice by more than 100-fold relative to Paclitaxel, a commonly used chemotherapeutic agent.

Pathways

The design of new drugs for the treatment of CSCs will likely require an understanding of the cellular mechanisms that regulate cell proliferation.

The first advances in this area were made with haematopoietic stem cells (HSCs) and their transformed counterparts in leukemia, the disease for which the origin of CSCs is best understood. It is now becoming increasingly clear that stem cells of many organs share the same cellular pathways as leukemia-derived HSCs. Additionally, a normal stem cell may be transformed into a cancer stem cell through disregulation of the proliferation and differentiation pathways controlling it or by inducing oncoprotein activity.

Bmi-1

The Polycomb group transcriptional repressor Bmi-1 was discovered as a common oncogene activated in lymphoma and later shown to specifically regulate HSCs. The role of Bmi-1 has also been illustrated in neural stem cells. The pathway appears to be active in CSCs of pediatric brain tumours.

Notch

The Notch pathway has been known to developmental biologists for decades. Its role in control of stem cell proliferation has now been demonstrated for several cell types including haematopoietic, neural and mammary stem cells. Components of the Notch pathway have been proposed to act as oncogenes in mammary and other tumours.

Sonic Hedgehog and Wnt

These developmental pathways are also strongly implicated as stem cell regulators. Both Sonic hedgehog (SHH) and Wnt pathways are commonly hyperactivated in tumours and are required to sustain tumor growth. However, the Gli transcription factors that are regulated by SHH take their name from gliomas, where they are commonly expressed at high levels. A degree of crosstalk exists between the two pathways and.their activation commonly goes hand-in-hand. This is a trend rather than a rule. For instance, in colon cancer hedgehog signalling appears to antagonise Wnt.

Sonic hedgehog blockers are available, such as cyclopamine. There is also a new water soluble cyclopamine that may be more effective in cancer treatment. There is also DMAPT, a water soluble derivative of parthenolide (induces oxidative stress, inhibits NF-êB signalling) for AML (leukemia), and possibly myeloma and prostate cancer. A clinical trial of DMAPT is to start in England in late 2007 or 2008. Furthermore, GRN163L was recently started in trials to target myeloma stem cells. If it is possible to eliminate the cancer stem cell, than a potential cure may be achieved if there are no more CSCs to repopulate a cancer. Finally, the enzyme telomerase may qualify as a study subject in CSC physiology.

4

Manipulating DNA

The Manipulating DNA section covers all the things that people are able to do with/to DNA, as well as the things still in the realm of science fiction. The following is links and information about each of the sections listed in the navigation frame on the left.

Cloning

This area explores experimentation done involving Cloning, including Dolly and Missy. The idea of cloning has been considered since before the discovery of DNA. Cloning may give us insights into what makes us individuals, and how much of our personality and behaviour is based on our genetics. It may also destroy our sense of individuality and the value of life.

It seems that every week, newspapers report on new advances in the science of cloning. Everybody knows about Dolly the cloned sheep, but few people know all the details about cloning, including the fact that scientists have been working on it for over 100 years.

Cloning in Nature

Cloning has been going on in the natural world for thousands of years. A clone is simply one living thing made from another, leading to two organisms with the same set of genes. In that sense, identical twins are clones, because they have identical DNA. Sometimes, plants are self-pollinated, producing seeds and eventually more plants with the same genetic code. Some forests are made entirely of trees originating from one single plant; the original tree spread its roots, which later sprouted new trees. When earthworms are cut in half, they regenerate the missing parts of their bodies, leading to two worms with the same set of genes. However, the ability to intentionally create a clone in the animal kingdom by working on the cellular level is a very recent development.

Early Progress

The first cloned animals were created by Hans Dreisch in the late 1800's. Dreich's original goal was not to create identical animals, but to prove that genetic material is not lost during cell division. Dreich's experiments involved sea urchins, which he picked because they have large embryo cells, and grow independently of their mothers. Dreich took a 2 celled embryo of a sea urchin and shook it in a beaker full of sea water until the two cells separated. Each grew independently, and formed a separate, whole sea urchin.

In 1902, another scientist, embryologist Hans Spemman, used a hair from his infant son as a knife to separate a 2-celled embryo of a salamander, which also grow externally. He later separated a single cell from a 16-celled embryo. In these experiments, both the large and the small embryos developed into identical adult salamanders.

Spemman went on to propose what he called a "fantastical experiment" — to remove the genetic material from an adult cell, and use it to grow another adult. In this way, he theorized, he would be able to prove that no genetic material was lost as cells grew and divided.

New Advances

There were no major advances in cloning until November of 1951, when a team of scientists in Philadelphia working at the lab of Robert Briggs cloned a frog embryo. This team did not simply break off a cell from an embryo, however. They took the nucleus out of a frog embryo cell and used it to replace the nucleus of an unfertilized frog egg cell, completing the "fantastical experiment" of nearly 50 years before. Once the egg cell detected that it had a full set of chromosomes, it began to divide and grow. This was the first time that this process, called nuclear transplant, was ever used, and it continues to be used today, although the method has changed slightly.

False Hopes

In 1977, a German scientist shocked the world, claiming to have cloned three mice from embryos. Although embryos had been cloned before, no one had been able to do the experiment with mice because the cells were so small and the tools so large that the cells were traumatized and would eventually die after a few divisions. He instantly became famous, telling the world how he cloned his mice. However, he refused to actually demonstrate any of his techniques, and when other scientists couldn't replicate his work, he came under suspicion. He was challenged — repeat his work or be discredited. He accepted. He claimed to work nights and mornings when no one was around, but the equipment was never disturbed. He showed off his mouse embryos' growth daily, even though a malfunction in the water purification system left other scientists at his lab unable to grow other embryos.

Later, in his cabinet, test tubes were found with mouse embryos in them, each at a different stage of development. Most scientists do not believe that this scientist was ever able to clone adult mice. In 1978, a science fiction writer published a book claiming that a millionaire (known to the readers only as Max) had come to him because of his connections as a writer, and asked the him to arrange for Max to be cloned.

The author eventually agreed, as the story goes, and Max was cloned. The book was ranked in the Top 10 list of popular books. Scientists who read his book, however, noticed discrepancies between the book and scientific data. One man who was quoted in the book was angry enough to sue. The publisher admitted that the book was a hoax, but the author maintains his claim to this day. Within these two years, two front-page advances in cloning were discovered to be, most likely, frauds. As a direct result, many scientists began to claim that cloning of mammals was impossible. Funding and interest dropped, and cloning returned to the realm of science fiction for several years.

First Cloned Mammals

A breakthrough came in 1986. Two teams, working independently but using nearly the same method, each on opposites side of the Atlantic, announced that they had cloned a mammal. One team was led by Steen Willadsen in England, which cloned a sheep's embryo. The other team was led by Neal First in America, which cloned a cow's embryo.

Many advances were made during the course of these experiments, including progress in keeping tissue alive in lab conditions. However, neither team believed that it was possible to clone from an adult's differentiated cells. With no progress in sight, the prospect of cloning fell by the wayside, and little research was done on the matter.

Dolly

Ian Wilmut at the Roslin Institute in Scotland was assigned to a project in 1986. His goal was to create a sheep that produced a certain chemical in its milk. He chose to alter adult cells, which held up well in laboratory conditions, and then clone them, producing animals with the altered gene all throughout their bodies. He began the paperwork in 1987, and began research in 1990. One of Wilmut's colleagues, who had experience with cloning from early embryo cells, suggested that the reason so many cloning attempts failed was that the cells were in incompatible stages of life. In one stage, the cells are adding to the DNA, in another, they are proofreading it, and in another, splitting it. The cells, he theorized, could not always start over. Wilmut's team learned that by starving the cells, they could be forced into what is called the G0 phase, similar to cellular hibernation. This advance increased the survival rate of the cloned cells; Megan and Morag, two lambs, were cloned from sheep embryos. Wilmut's team now realized that differentiation did not matter in cloning. More work was done, and on July 5, 1996, a lamb was born, cloned from a frozen mammary cell from another adult sheep. Wilmut, who names his animals very creatively, named her Dolly after Dolly Parton.

Although Dolly was just a step in a long experiment, the press descended upon the first animal cloned from an adult. The Roslin Institute was overrun with journalists and reporters. However, other scientists were critical — Dolly took 277 tries to create, and other labs were unable to reproduce the results. In addition, it took over a year for the institute to test Dolly's DNA to make sure that it was indeed the same as that of the frozen mammary cells. Science, although temporarily impressed, demanded a better way.

Herd of Mice

Oct 3, 1997, the Honolulu Technique created Cumulina the cloned mouse. She was cloned from cumulus cells (cells which surround developing egg cells) using traditional nuclear transfer. The nucleus was taken from the cumulus cell and implanted in an egg cell from another mouse. The new cell was then treated with a chemical to make it grow and divide. The scientists repeated the process for three generations, yielding over fifty mice that are virtually identical by the end of July, 1998. The Honolulu Technique's success rate of 50:1 is almost six times better than that of the Roslin Institute's success rate, 277:1. As cloning technology improves, more and more applications will be seen in everyday life.

Mainstream Cloning

How much do you love your dog? Is your dog so perfect that you would pay over $2.3 million dollars to have another just like it? One couple thinks their 11-year-old dog is just such an animal. Wishing to remain anonymous to avoid run-ins with the press, this couple has contracted Texas A&M University to clone their dog, Missy. Scientists are hailing this for its scientific achievement; no dogs have been cloned before because their reproductive system is rather complicated. If the cloning of dogs can be achieved, perhaps exceptional animals like rescue animals can be reproduced. In addition to the pure scientific appeal of cloning a dog, the attempt to clone Missy has another interesting addition to make to the history of cloning. A private couple wants their dog cloned. They are, of course, spending millions to have her cloned, but consider the possibilities. Could cloning the family pet one day become a normal alternative to buying a new one?

Applications

Reliable cloning can be used to make farming more productive by replicating the best animals. It can make medical testing more accurate by providing test subjects that all react the same way to the same drug. It can allow mass production of genetically altered animals, plants, and bacteria. It may settle once and for all what part of personality is dependent on genetics and what part on environment. In short, it can be beneficial to almost every area of biological science.

How Cloning Works

Cloning in Nature

In nature, clones exist among plants or animals that are able to reproduce asexually. Special traits allow earthworms to regenerate lost body parts and certain plants (like the spider/airplane plant) to propogate through runners. These worms and plants are technically clones, because as broken-off parts of the original, they share the same genetic material as the original. The oldest methods of cloning involve a similar principle — a cell is broken off from a very early embryo, and begins to divide as if it were a fertilized egg. However, it is very difficult to clone adult cells because they are not "in gear" to divide and grow into a new organism. Techniques needed to be developed that could allow the DNA from an adult cell to be removed and placed in a cell that could grow as an embryo.

Nuclear Transfer

Scientists' understanding of what makes an embryo cell different from a liver cell is not complete because the objects of study are so small. However, they do know that a coating of proteins on the cell wall inhibits the DNA that is not neccesary for that cell's particular function. When this coating is present, the cell is said to be differentiated. Although differentiated cells contain all the necessary DNA to produce a new organism, the fact that they have differentiated keeps them from creating an entire new organism. Egg cells, on the other hand, are unique in that they are the only cells capable of forming a new and complete organism. The secret to cloning lies in removing an egg cell's nucleus and replacing it with the nucleus of a donor cell. The egg cell, which has a "clean slate", and is primed to use the information in the DNA to create a new organism, now has the DNA of the donor organism. However, this transfer is a very difficult process, more difficult in some cases than others.

Obstacles to Cloning

The first animals cloned using the modern method of nuclear transfer were frog embryos, cloned in 1951. The reason frog embryos were cloned long before anything mammalian is twofold. First, the cells of an 8-celled frog embryo are huge compared to the cells of a mammal embryo; they can, in fact, be seen with the naked eye, and are therefore much easier to work with. Secondly, frog embryo cells grow in lab conditions much better than mammal embryo cells.

This is because frog embryos are supposed to grow in water, without any added nutrition or outside assistance. Mammals, however, take a long time to mature, and require the shelter and nourishment of a mother both before and after birth. Because the clone has to grow from a single cell to an entire organism in the lab, it was easier to do initial experiments on amphibians. Frogs were chosen because, as frequent subjects in lab tests, they are very well known to scientists.

Although happy about the cloned frogs, scientists wanted to see a cloned mammal; humans are mammals, so almost anything learned about other mammals can be applied to humans. Unfortunately, because of the difficulties associated with mammal cloning, this was a difficult task. Techniques to make this task possible were developed at the labs of Neal First and Steen Willadsen, who worked independently to produce the first mammals cloned from embryos. Once the nucleus is transferred from the donor cell to the egg cell, the egg cell is put in a nutrient solution until it divides into an eight-celled embryo.

When the cell passes this point, the individual cells begin to differentiate; one cell may eventually form the head, and another the torso. As this point is reached, the scientists move quickly to put the embryo into the removed and artificially sustained oviduct of a sheep or other animal.

The oviduct functions as an incubator to nurture the embryo thruough this critical period. Once the embryo is large enough to be able to withstand the procedure, it is placed in a living surrogate mother where it will be carried to term, nourished by an umbilical cord.

With all these steps, it is no wonder that the success rate for cloning is so low; embryos die at each step along the way until only a few are carried to term and born.

Variations on a Theme

Of course, there are different approaches to all steps in the process. Ian Wilmut, instead of transferring the nucleus from one cell to another, fused an enucleated (nucleusless) egg cell with an adult cell to form the first cell in the embryo.

The Honolulu method centres around the cells chosen to be DNA donors. However, the principle behind cloning is always the same - give an egg cell, which is ready to divide and grow into a new organism, the nucleus of an old cell. With the new set of instructions, the egg cell begins its journey to become an adult clone.

Future of Cloning

Consider Some of the Possible Applications of Cloning

Cloning is truly a monumental accomplishment. But what can we do with it? Very few see a future society of cloned humans. Mad scientists with armies of cloned zombies are highly improbable due to the current cost, failure rate, and time involved.

However, there are some practical applications currently being considered. Which are possible and which are doomed to remain in the realm of science fiction? Only the future will tell.

Most Prevalent Application

What many scientists forsee as the most probable and widespread application of cloning is the mass production of genetically engineered animals. Scientists have to be very careful about protecting the altered portion of the genome; if the new/altered gene is damaged, the animal carrying the damaged gene will not be as useful to the scientists.

Because natural breeding can lead to the decay of the new/altered gene, cloning is the best option, because the original DNA is used over and over to make new animals. In fact, mammalian cloning was only pursued as a step toward copying genetically altered cells.

Farming

The average dairy cow puts out roughly 15,000 gallons of milk a year. However, there are certain, special cows that can make up to 45,000 gallons a year. Selectively breeding for this trait is nearly impossible given the complexity of the genetics governing milk production. However, if scientists clone these exceptional cows, the profit gained from increased milk outweighs the money spent on a cloned cow.

Infertility Treatment

While it is unlikely that the cloning of humans will ever become standard practice, it is also unlikely that it will never be attempted. Many people see cloning as a way to provide children for those couples who cannot have them naturally. While there is always someone willing and able to pay the price, is the cloning of humans morally right?

Jurassic Park

Scientists have told us that it would be virtually impossible to find an intact cell from a extinct dinosaur. However, at New Zealand's University of Otago, scientists are trying to clone the extinct moa, once the world's largest bird. The plan was to take DNA from its leg bone and implant it into a chicken egg to grow. The scientists were then going to breed the new moa with an ostrich or an emu to create a new giant bird. Research was stopped by the Ngai Tahu Maori tribe, which claims to own the DNA because they were sole owners of the land when the bird became extinct around 1500.

In 1992, a herd of cows that was damaging the ecological balance of New Zealand's Enderby Island was almost completely destroyed. All that remained was the frozen sperm of ten bulls, and one living female, Lady. Because Lady is so old, all attempts to make her pregnant using the frozen sperm have failed. Therefore, Lady was cloned. On July 31, 1998, Elsie (L.C., Lady's Clone) was delivered by Casarean section, is the hope of preserving the nearly-extinct Enderby Island cows.

Organs for Transplant

Another option open to the future is the cloning of specific organs for transplant. Each year, many people die, unable to find a suitable organ donor. In addition, those who do find donors many times have to take anti-rejection drugs for the rest of their lives. If scientists can find a way to force cells to differentiate to become a failing organ, they should be able to grow those cells into a working, adult organ. Because the new organ would be an exact match for the patient, there would be no need for anti-rejection drugs. Many predict that the first organ to be cloned in this way may be bone marrow, because it is a liquid organ and has no shape. However, it is conceivable that solid organs would be able to be cloned outside a body as well.

Ethics of Cloning

We can... but should we?

Cloning, simply defined, is creating a new organism that shares the same genetic code as another. Each step taken toward making cloning quicker, better, and cheaper brings these ethical questions further from the realm of the hypothetical and closer to the realm of fact.

Cloning Endangered/Extinct Species

All endangered species have one thing in common — there's not many animals left. A lack of animals leads to inbreeding, and a lack of genetic diversity. Researchers are always careful when trying to reestablish endangered species to maintain genetic diversity by creating different families and trying to inbreed as little as possible. If the scientists just let the animals breed without control, there will be almost no genetic diversity, which will lead to recessive traits being expressed. Usually, this results in diseases and deformities.

While cloning offers a short-term answer to the problem of "not enough animals", in the long term, all the offspring from the clones are genetically brothers and sisters. This will likely destroy genetic diversity and lead to inbred, weak, disease ridden animals who could not survive in the wild. Even if the animals were healthy enough to be released into the wild, in their weakened state, they would be easy prey, leading to more predators. In the end, they would become extinct and would have upset nature's balance by introducing more predators. The morality of cloning endangered species hinges on how many animals are left, and what their collective genetic condition is the same trouble would be compounded if science tried to bring back an endangered species like the dodo bird. It would be so expensive to find enough intact dodo cells from different birds, that to make a flock of unique birds would be an impracticality. The best science can offer is the same cloned dodo at every paying zoo. Whether or not you think this ethical hinges on the question of whether it is ethical to create an animal for the sole purpose of captivity.

Cloning Humans

It would clearly be unethical to clone humans and distribute them about for the purpose of a gigantic twin study, to determine once and for all what is nature and what is nurture. It would be useless for an millionare to clone himself in hopes of everlasting life; the clone would posess none of his memories or experiences. Bill Gates' clone might turn into an average person, a delinquent, a ditch digger, or the secretary of the Department of Justice, depending on the clone's life experience.

There is really only one apparent and plausible reason for the widespread cloning of humans, and that is for parents who can't have children naturally. How you feel about cloning humans really depends on how you feel about the value of human life, and when it starts. Dolly took 277 tries; assuming that cloning a human would take 277 tries, what of the 276 human embryos that died? If they are just embryos, then it's probably okay. If they were humans with minds, souls, and destinies, then it was wrong to try to clone a child in the first place. Cloning, it seems, offers the same ethical dilemmas as does abortion and current methods of in vitro fertilization.

But there is a second problem on top of the consideration of the other embryos — the life of the clone. Assuming that the technique is commonplace (the lives of the first fifty or so will probably be filled with doctor visits, disgruntled family members, and possibly taunting classmates), a cloned child would have a set of problems in addition to the ones that normal children face.

What would be the relationship between the clone and the "DNA donor"? Sibling? Child? Most children have problems with their parents; a clone might reflect those feelings back on himself, because he is identical to them. There is a whole list of questions that prospective parents would have to consider before trying such an endeavor; is it really worth it?

Core Issue

The core issue behind cloning seems to have to do with the relationship between genetic uniqueness and personal individuality. Animals may or may not have their own personalities and sense of individuality, and if they exist, they may differ from species to species, so there may never be a resolution to the question of cloning animals.

But humans are a different matter entirely. Humans are unique not because of their body build or genetic makeup or life experience, but because of the unique contribution they each have to make to the world. The morality of human cloning lies not in the cloning process itself, but in societal reaction to him.

If society can keep itself from branding the clone as a duplicate person and limiting him to his predecessor's abilities; if instead the clone is accepted just as a normal human being who is unique because of his unique contribution he has to make to the world, then cloning of humans may be an option.

Genetic Engineering

Learn what genetic engineering can accomplish and how it is done. If genetic engineering is defined as changing an organism's DNA to make it more beneficial, genetic engineering has been going on for a very, very long time in the form of selective breeding. However, actually going into a cell and changing its genome by inserting or removing DNA is a very new technology.

Ancient History

Selective breeding has been going on for countless generations. In fact, it is even mentioned in the Bible (Genesis 30:25 - 43). In the account, Jacob was employed as a shepherd under his father-in-law Laban. Instead of receiving wages, Jacob received the black, streaked, and spotted sheep, and Laban kept all the white sheep.

Jacob craftily arranged for his black sheep to mate with Laban's white sheep, producing streaked and spotted sheep. Jacob did so well with this scheme that Laban's family began to get mad at Jacob, and he eventually had to leave.

Difficulties

Selective breeding is effective enough if the goal is to maintain or gradually improve a group of animals. Over the decades, selective breeding has brought us improved strains of cattle and specialized breeds of dogs. However, these advances have taken hundreds of years to effect. In addition to the time concerns, it is often impossible to know which traits will be transferred to the offspring.

Limits

Selective breeding is a long, tedious process that has its limits. It is impossible through selective breeding to mix traits from two totally different species. If a junkyard owner wanted a guard dog that could squirt ink like an octopus, he would be unable to create such an animal. It is physically impossible, because the genetics of life are such that traits from two different organisms cannot be mixed. That is where genetic engineering comes in.

The Progress

Modern genetic engineering began in 1973 when Herbert Boyer and Stanley Cohen used enzymes to cut a bacteria plasmid and insert another strand of DNA in the gap. Both bits of DNA were from the same type of bacteria, but this milestone, the invention of recombinant DNA technology, offered a window into the previously impossible — the mixing of traits between totally dissimilar organisms. To prove that this was possible, Cohen and Boyer used the same process to put a bit of frog DNA into a bacteria.

Since 1973, this technology has been made more controllable by the discovery of new enzymes to cut the DNA differently and by mapping the genetic code of different organisms. Now that we have a better idea of what part of the genetic code does what, we have been able to make bacteria that produce human insulin for diabetics (previously came from livestock), as well as EPO for people on kidney dialysis (previously came from urine of people in third world countries with ringworm).

In 1990, a young child with an extremely poor immune system recieved genetic therapy. Some of her white blood cells were genetically manipulated and re-introduced into her bloodstream while she watched Sesame Street. These new cells have taken over for the original, weak white cells, and her immune system now works properly. Although relatively few people have had their cells genetically altered, these advances have made the prospect of mainstream genetic medicine seem more likely.

The Promise

Genetic engineers hope that with enough knowledge and experimentation, it will be possible in the future to create "made-to-order" organisms. This will lead to new innovations, possibly including custom bacteria to clean up chemical spills, or fruit trees that bear different kinds of fruit in different seasons. Any trait occurring in nature can theoretically be mixed with any other to form a totally new organism that would not otherwise occur in nature.

Current Status

As of late summer of 1998, scientists are able to add simple traits to organisms. They cannot create custom-made animals. They cannot always predict how traits will interact. Before phenomenally new advances can be made, scientists have to learn how to affect cells' DNA with pin-point accuracy, without affecting other traits. Advances like genetic correction for nearsightedness are a long way off. The power of science is limited to knowledge about genetics, gene locations, and trait interactions, but as knowledge grows, so will scientists' abilities to manipulate life.

How Genetic Engineering Works

On most TV shows, genetic engineering is a very simple process. Just point the device at the person, press the button, and automatically, the patient's DNA is all changed, and the scientist, who often has little idea how to use the machine, is thrilled to death that it worked right the first time. While this makes good fiction, it is hardly reality. Genetic engineering, while the results are often unexpected, is a very meticulous, very exact process with a clear goal in mind.

Cutting the DNA

Genetic engineering has three basic parts. First, the DNA is cut in the desired place using restriction enzymes. Each different type of restriction enzyme "seeks out" and cuts DNA at a spot marked by a different sequence of base pairs. One restriction enzyme may cut the DNA at every "AATC", for example, while another cuts all "ATG" sequences.

The DNA is cut in such a way that one helix is a bit longer than the other because of a few extra base pairs. These extra base pairs are called the "sticky end", because they will bond easily to another strand of DNA with the corresponding set of genes. The new genes that will be injected have sticky ends that complement the sticky ends the restriction enzyme leaves.

Inserting New DNA

Once the restriction enzyme has cut the DNA, new DNA is inserted into the cell. This is difficult, because cells by nature do not allow DNA through the cell wall. There are many ways around this difficulty, though. Electroporation involves jolting the cells with a burst of electricity, opening the cell wall pores and allowing DNA to fall into the cell.

Microinjection uses a small glass needle to inject the DNA through the cell wall. A gene gun can be used, which blasts tiny metal fragments coated with DNA through the cell wall and into the cell. DNA can also be shrouded in lipids, fatty molecules which the cell will take in; when the lipid is digested, the DNA is released. These methods work fairly well for prokaryotic cells, which have one chromosome and no nucleus. However, in order for DNA insertion to work consistently and accurately in multicellular organisms with eukaryotic cells, something better is called for.

To find that something better, science turned to the natural world. They realized that a retrovirus, a form of a virus, was just what they needed. Retroviruses enter the cell through the cell wall and implant their DNA into the cell's nucleus.

The retroviral DNA is incorporated into the cell's DNA, causing the disease that the particular retrovirus is associated with. Scientists reasoned that if they could put a gene into the retrovirus, the retrovirus would deliver that gene to the cell's DNA. An added bonus is that different retroviruses target different areas of the body, so the scientists could put DNA into a retrovirus for delivery to a specific organ.

Attaching the New DNA

Once the cell's original DNA is cut with the restriction enzyme and the new DNA is in place, the scientists use ligase (another enzyme) to stick the DNA segments together. The sticky ends of the new and the original DNA merge together, and the cell begins to carry out the instructions of the new DNA along with its own.

Atrition

Of course, this technique does not work on 100% of the cells. A scientist may start with a lot of cells, and the restriction enzyme may not get to some. The DNA may not enter some of the cells. Other cells' DNA will not line up properly with the new DNA, causing the ligase to bond the DNA incorrectly.

The new DNA may find its way to the wrong position along the original DNA. In others, the ligase may not get to the cell and fuse the DNA at all. In every "batch" of genetically altered cells, there will be some that do not have the new trait. The cells are grown in culture, and the cells that do not posses the desired trait are culled out, usually by subjecting all the cells to whatever treatment the new cells were designed to withstand. The cells that survive have the new DNA properly aligned, and are ready to go to work.

Safeguards on Genetic Engineering

Sci-fi writers often describe worlds where genetic engineering has gotten out of hand, and has destroyed some aspect of society. But how likely is this to really happen?

Misconceptions

Saturday morning cartoons capitalize on the notion that scientists plunge into the unknown without concern for the consequences. They don't. A scientist is not, in all probability, going to create a supervirus or a man-eating plant just to see if he can. Scientists realize the possible effects of all their research, and stop if there's a possibility of danger.

Precautions

Experience has shown that when one aspect of nature is changed, there can be unexpected repercussions in other areas of the ecosystem or in the economy. Because of this, scientists must be very careful about making sure that any new organism that they create will not damage the life that already exists.

Let us use the example of genetically altered fish. After much labour, scientists have succeeded in creating a fish that grows up to 40% faster than the average fish. The goal in this research was to make fish for human consumption that grow faster, and are ready for harvest sooner.

The first step toward assuring that the fish are viable and safe is to put them in a microcosm, an enclosed, controlled replica of the ecosystem in which they were designed to live. During this time, the scientists monitor the fish to see if the implanted gene is working properly. They are also looking to answer several questions that will determine if the fish are safe to release into the wild.

- Will the organism survive? If the fish can't survive in the wild, it goes "back to the drawing board". Fortunately, this new fish is capable of living on its own.
- Does the organism reproduce, and if so, how fast? If it multiplies rampantly, it may be a disastrous addition to the ecosystem, upsetting the delicate balance. If it does not multiply at all, it still passes this test, but it may be advantageous to allow the organism to multiply. Scientists find that the fish is capable of reproduction, and at a normal rate.
- Does the organism have an advantage over other organisms, and if so, does it compete for the same resources? Imagine the catastrophe if an organism was so strong and took so many natural resources out of the environment that it killed off all its neighbours! Scientists need to find out if this organism will lead to the extinction of others. Unfortunately, this fish competes for the same resources as normal fish, and wins. Because it is growing so rapidly, it needs a lot of food. It would probably eat the other fish out of a habitat. This, coupled with its size, could lead to the destruction of the ecosystem.
- Can the gene move to other species? There is no way to predict what will happen when a fish with an altered gene is bred with a normal fish. Just in case the gene would be harmful in other organisms, the scientists need to make sure that the gene cannot be passed along. Unfortunately, scientists find that the growth gene in the altered fish can be passed on to other fish.

- Does the organism leave the test area? This may sound scary at first, but basically scientists want to know whether the organism will move itself to anywhere habitable or whether it will mostly stay where it is released. In the case of our fish, their ravenous appetite almost forces them to move on in search of more food.

The genetically altered, fast-growing fish have failed three tests. The scientists must change the fish so that they pass all the tests, or they cannot be released. In order to pass the fourth test (can the gene move across species), a third chromosome is inserted.

This third chromosome keeps the fish from being able to form gametes with the correct number of chromosomes. Offspring cannot be conceived, so the fish do not reproduce at all. No children means no gene passing. In order to pass the fifth test, the fish are enclosed under three layers of fencing, one of which is electrified. With the fish placed in this controlled environment, the third test is automatically passed; there are no other fish with which to compete.

Ripples

There are other implications from genetic engineering besides danger to health. For instance, the plant that usually gives us canola oil can be modified to produce lauric oils, which were once only produceable in Southeast Asia. While it is cheaper and more efficient to grow the new plant than to buy the old, American money is not going into Southeast Asia to buy the oil. The Southeast Asian economy has been effected by the new plant.

Fruit is being genetically engineered to stay fresh longer. However, many suspect that this is a counterfeit freshness; the vitamins and minerals that are the reason to eat fruit at all are gone long before the fruit begins to go bad. This has made the fruit about as healthy as a regular, processed food, even though it appears very ripe.

Some farmers use a crop with a built-in insecticide that kills pests. As always, there are certain bugs that are resistant to the insecticide. The insecticide gives the mutant bugs a competitive advantage, and there is a threat that they will multiply rampantly with the farmers providing their food source. Another insecticide must be used on these bugs, which is dangerous to helpful insects as well as the crop. Farmers have had to plant standard crops next to the engineered ones so that the original bugs are still being provided with food and can multiply, keeping the resistant strain of insect from having and advantage. Even though new organisms don't appear to hurt, the lesson is that they may not help, and may even be causing damage in places that were totally unpredictable. Great care must be taken when altering the tried-and-true genetic code that already exists.

Ethics of Genetic Engineering

Genetic engineering offers a major ethical dilemma — there are great rewards if genetic engineering becomes a widespread reality, but the dangers are equally great. In addition, genetic engineering also offers the ability to change the very nature of nature, an environment in which man is already well-suited to live in. Is it possible that genetic engineering will do more harm than good?

Medicines

As you have probably already learned, DNA is very versatile because all life on Earth processes it the same way (as far as we know). Therefore, foreign genes can be implanted in bacteria, instructing the bacteria to produce whatever the inserted gene controls.

In this manner, scientists can "teach" bacteria to produce human or other proteins. There is a long list of bacteria that have been treated this way, including an e. coli bacteria that produces insulin. The insulin is less expensive than that removed from cows and pigs, as well as being exactly the same as human insulin.

The ethical quandary about this process lies in the fact that human genes are being inserted into a bacteria. I do not know of anyone who stands up for bacterial rights; in fact, everyone who has been sick is rather prejudiced against bacteria.

The problem is the idea that human genes are being put in a non-human organism. Is there anything special about human genes? If you consider that human genes are made of the same four bases as bacterial genes, and that they work the same way — as an instruction manual for the production of a protein, there is less of a dilemma.

However, many people believe that human genes should not be implanted in other species, because they are special because they are human. Before you make up your mind, consider the alternative method of production, and consider honestly whether or not you would be willing to recieve medicine from a human gene in a bacteria if it was you who needed it.

Xenotransplantation

Work has been done on genetically engineering farm animals, too. While in the future it may provide leaner meat, larger animals, or faster growth, right now the great prospect is xenotransplantation. Xenotransplantation transplanting organs across species. Currently, most organ transplants are human to human. They work well, but the drawback is the rejection medication that must be taken to weaken the immune system so that the body does not destroy the new organ. This, of course, can lead to illness, a definite drawback.

If organs can be "made to order" in another organism and then transplanted over to the human patient, this can eliminate the need for rejection medication and the succeptibility to illness.

The best prospect for xenotransplantaton of vital organs is the pig. The pig is one of the closest genetic matches to humans, closer even than monkeys and apes. With a little genetic alteration, pigs can hopefully produce organs that the human immune system will not reject, producing the desired effect. Unfortunately, the pig is the loser in this arrangement.

While on the surface this may seem barbarous, consider the concept: a human is taking something from the animal to extend his own life. The basic principle is not all that different from eating. If xenotransplantation still seems cruel, perhaps the world needs to re-evaluate itself and turn vegetarian, eliminating the need to raise animals for the sole purpose of death for man's benefit.

Custom-made Organisms

Current technology and a great deal of experimentation may allow for custom-made organisms to be produced. These may range from a bacteria that cleans up oil spills to a monkey with gills and a higher intelligence level. Everyone seems to agree that we should not pursue these advances just for the fun of it, so the question becomes, are we so inconvenienced by a problem that we need to create a new life form to fix it for us?

Core Issue

The core issue behind the ethics of genetic engineering is as follows. Is it morally right to change the nature of life on Earth to suit man's desires better? The answer to this question depends on man's position on Earth. If we are truly superior to the animals and accountable to no one, is there a real question of whether it is wrong? If we are not fundamentally different, do we have a right to meddle with evolution? If we are accountable to God for our actions, should we risk insulting His creation by trying to do it better?

Future of Genetic Engineering

Genetic Engineering is, quite possibly, the most promising and most threatening new advance in medical technology of all time. For this reason, new advances made in the future may or may not be implemented. It is nearly impossible to predict what Genetic Engineering will bring to the future, but there is some speculation.

Medicine

The most popular contribution of Genetic Engineering is in the field of medicine. Because most diseases have a genetic component, healing may be sped along in the future by using bacteria-made proteins to augment the healing process. Proteins circulate through the body, reporting to various glands whether bone and muscle mass should increase or decrease, how salty substances should be moved through the lungs, how many immune cells should be manufactured, and even how fast hair should grow. The genes regulating the production of these proteins can be inserted into bacteria, "teaching" the bacteria how to make the identical proteins.

Unfortunately, this is a power that is easily abused. Athletes could have their muscle mass increased for more strength. It could be that the army with the most genetic drugs coursing through their veins will be the one to win the war. Many people feel that these dangers far outweigh the good that could be done speeding the healing process.

Food

One controversial advance has been in place for several years: genetically engineered food. Dubbed "Frankenfoods" by opponents, these foods are enhanced by genes from other plants or animals. This can enhance nutrition, shelf life, or taste. Depending on public reaction, these foods may be widely available in the future.

Genetic Mapping

Find out about the Human Genome Project and other genome mapping projects. The Human Genome project and other genetic mapping endeavours hope to discover what makes man "tick". So doing, they may be able to help doctors test succeptibility to disease and allergies to drugs. They may be able to help genetic engineers by giving them more insight into where traits are located. Many areas of science will be affected by genetic mapping. Read on to learn how scientists are going about this herculean effort.

Overview of Genetic Mapping

What is the Human Genome Project all about? As you probably know from reading the Textbook section, DNA is read by the cell in triplets. Each set of three base pairs codes for an amino acid, which is the building block for the over 50,000 kinds of proteins in your body. These proteins direct growth and allow your body to function properly. The Human Genome Project hopes to find out what each segment of your DNA does.

The Basics

The Human Genome Project will cost nearly 40% the money required to build the International Space Station. It is funded by the United States Department of Energy and the National Institutes of Health. The 15 year project is working with over 400 teams, although half the budget goes to 21 main research centres. The Human Genome Project is scheduled to be finished in 2005. Its basic goal is to find all the genes in the human body, as well as other model organisms, and to determine the function of those genes. However, with 1990's technology, that goal was not accomplishable in 15 years. Therefore, the goals of the project also include the development of new technology to rapidly sequence and interpret the DNA.

History

Back in 1985, a group of physicians got together and decided that the medical community should declare war on cancer. Of course, the first order of business was to figure out what caused cancer in the first place. Well, radiation can cause cancer. Why does radiation cause cancer? Because it alters the genetic structure of the cell. What part of the DNA is altered? No one knew. At that moment, the idea of mapping the human genome became very popular. In 1990, the Department of Energy and the National Institutes of Health launched the Human Genome Project.

Goals

In addition to mapping the human genome, the project is also mapping several model organisms — animals that share certain aspects of humanity. Among a few others, scientists are mapping the mouse (vertebrate; very similar to humans), the roundworm (multicellular), a yeast (cell with nucleus), and E. coli (alive).

They chose these organism because they each share progressively less in common with humans, and they have all been subjects of scientific scrutiny for some time. In fact, most genetic engineering work is done either with mice or E. coli bacteria. Please note that the scientists are working with a harmless form of E. coli and not its dangerous cousin.

The Human Genome Project's goal, basically stated, is to map the entire human genome, find out what each of your genes does, and discover the sequence of the base pairs of those genes. The data gathered in this project can be used to develop tests for genetic conditions and susceptibilities.

This information is potentially very useful; if a man finds that he is genetically prone to a heart attack, he may eat a healthier diet or begin to exercise. The same testing can also be used to predict side effects or effectiveness of medication. It can determine if you are a carrier for certain recessive diseases.

The vast amount of data collected by the Human Genome Project can be put to very good use, but it can also be abused; discrimination based on "faulty" genes may be a very real issue in the future. Therefore, the Human Genome Project also devotes 5% of its budget to looking at the implications of its research.

Ethics

One of the unique aspects of the Human Genome Project is that it has a sub-program called ELSI that looks at the Ethical, Legal, and Social Implications of the Human Genome Project, trying to address issues before they come up.

Genetic testing holds bright promise for the near future. However, there is also a great debate about what sort of information should be made available, and to whom. For instance, who is allowed access to genetic information? You? Your doctor? Adoption agencies? Employers? Insurance companies? The government? Access isn't the only problem. If you found out that you would become terminally ill at age 45, and there was nothing you could do about it, would that affect your life so adversely that it would be better not to know? ELSI tries to come up with answers to tough questions before all of the Human Genome Project's data is acquired and released.

Process of Genetic Mapping

Learn how scientists go from AACTGG... to an understanding of how a particular gene works.

Human DNA is astoundingly complicated. The four bases of DNA make up the six feet of genetic material in every cell. This material is spread out over 23 chromosomes. The three billion base pairs in a single human cell make up 60,000-70,000 genes, plus the introns (about 95%) whose function still remains a mystery. It can be plainly seen that constructing a genetic map is a daunting task.

Getting Started

Many people want to know exactly whose DNA geneticists are mapping. The Human Genome Project is not trying to map a specific person's DNA, but DNA in general. 99.9% of the DNA of every person on earth is identical. Scattered throughout this identical DNA is the 0.1% (3,000,000) of base pairs that are vary from person to person: the variations and mutations that make one human different from another. These variations provide the key to unlocking the human genome. By comparing the DNA of many different people, especially those with genetic disorders, scientists hope that they can map the human genome by 2005.

Genetic Linkage Maps

The first step toward mapping the genome is to establish where one gene is in relation to another. This is done by studying people with genetic disorders. During meiosis, chromosomes pair up and exchange parts of their DNA, mixing the parents' chromosomes into a single chromosome that will be passed on to the offspring in the form of a sperm or an egg cell. This process is called "crossing over". Two traits that are often inherited with each other are almost definitely very physically close to each other on the chromosome, because the closer two genes are to each other, the more likely they are to move together when the chromosomes mix.

Genes that often move together when chromosomes cross over are said to be linked. Although a genetic linkage map does not show a gene's location, it can show the odds that one specific gene will be inherited with another specific gene; in fact, the distance between genes on genetic linkage maps is not measured in terms of physical distance, but in terms of what the probability is that two genes will cross over together.

Contig Maps

Once scientists determine which genes are linked together, they can begin to form a contig(uous) map of the chromosome. A contig map is designed to show the general physical location of specific genes on a chromosome; the distance between genes on a contig map is measured in terms of physical distance, as opposed to the genetic linkage map, which measures the probability of two genes being inherited together. T

he margin of error when measuring the distance between two genes on a contig map is 500,000 base pairs. A contig map can also show which blocks of chromosomes are mostly coding genes and which blocks are mostly introns.

Genetic Sequencing

The ultimate goal of the Human Genome Project is to produce a genetic sequence map of the human genome. At the project's conception in 1990, computer technology was not good enough to map the entire genome in anywhere close to 15 years. As computing technology improves and scientists' technique is revised, the dream of a genetic sequence is drawing closer and closer to being realized.

Of course, with genetic sequencing, the problem again arises, "Whose DNA is going to be mapped?" When the scientists reach the areas where individuals' DNA differs, all the participants' individual DNA will be sequenced and catalogued.

What Next?

Even after the Human Genome Project is finished, there will be the task of interpreting all the data. Scientists still need to learn about the function of introns, about the different possible combinations of DNA that yield a specific trait, and about the effects of mutations. The Human Genome Project in no way marks the end of discovery in the human genome.

Future of Genetic Mapping

Find out what Genetic Mapping may bring us in the Future

Genetic mapping is a project that has been undertaken by the American government in conjunction with private labs. How can a project have a future if it has a logical ending point? While the project itself will be complete, the information gleaned and the technology developed will be in use long after the project is complete.

Medicine

Many scientists see the first mainstream applications of genetic mapping to be in the field of medicine, specifically, predicting disease susceptibilities and drug allergies. Each patient's genetic code is different, each with different strengths and weaknesses. If a genetic map reveals to a doctor that a specific disease is likely, the doctor and patient can attempt to prevent the disease — susceptibility to heart disease can be countered by a healthy diet, for instance.

A young man was recovering from surgery for an ear infection. He was taking antibiotics, but he began to feel very bad, and the doctors feared that another surgery would be necessary. However, his friend recognized his symptoms and reported that one of her relatives had the same symptoms, which were eventually traced to a rare allergy to the antibiotic that she was taking.

The young man was taking the same antibiotic. When the drug was replaced, he began to feel better and eventually recover. He later learned that he might have died from continued use of the antibiotic. If genetic testing were available, it could have prevented his problems. This sort of genetic test may be available in the relatively near future.

Identification

The Human Genome project is not only developing a map of the genome, but also new technology to speed sequencing and analysis. If sequencing technology continues to develop at its current rate, it may be possible in the future to perform a genetic identity test in minutes; some predict that this test may even be performed with a hand-held device. Of course, this would require a database with portions of everyone's genetic code. This advance may be prevented if people refuse to be catalogued.

Genetic Engineering

Genetic sequence maps will show what sequences of DNA create what proteins and what those proteins do. This knowledge, when catalogued, will allow genetic engineers to find and insert combinations of DNA sequences into organisms, giving the new organisms desirable traits. Genetic maps will allow these scientists to predict more accurately how traits will interact and allow for more radical changes of the organisms.

DNA Fingerprinting

This area is about how scientists can use DNA to positively identify a person or his involvement in a crime.

Many people know about DNA Fingerprinting, but very few know how the process works. Here, you will learn what scientists must do to determine if a person and a DNA sample match.

Fingerprinting a Crime Suspect

There are three basic components to a DNA fingerprint test. First, there is the DNA from the crime scene; second, there is the DNA from a suspect; third, there is DNA with known attributes that is used as a control. All the DNA runs through the same tests at the same time. First, the three sets of DNA are purified and replicated, giving the scientists more DNA to work with. Next, the DNA is cut using a restriction enzyme. All three sets of DNA are then placed on the same block of gel.

When electricity is passed through the gel, the DNA will migrate down the gel block; short strands will move farther down the gel than the long strands. This organizes the DNA strands by length so that the results can be accurately measured.

A thin film of nylon is laid over the gel to blot up the DNA. Finally, short, radioactive strands of DNA called probes are sent onto the nylon. These radioactive probes will pair up with complementing strands of DNA on the nylon. The probes are designed to seek out and pair up with uncommon sequences of DNA; the probes will all congregate to the places where the uncommon sequences from the replicated DNA have come to rest.

When the nylon film is X-rayed, any large concentrations of the radioactive probes will leave blots on the X-ray print. If the blots from the control DNA look different than they normally do, the scientists know that there was error in the test, and the results are thrown out.

If the control blots are in the correct positions, the test was run properly. If the blots from the suspect line up with the blots from the DNA gathered at the crime scene, then the two samples came from the same person, and the suspect was indeed present at the crime scene.

Error

Many have questioned the validity of DNA fingerprinting test results. Because the burden of proof is on the forensic scientists, there are safeguards in place to be sure that DNA fingerprinting results are accurate. The first is the control DNA that runs through the tests alongside the suspect's DNA and the DNA found at the crime scene. If the control DNA tests properly, then the DNA in question tested properly as well.

Another safeguard in place is that the test is run four times using four different restriction enzymes. In order to be absolutely positive of a match, all four sets of DNA from the crime scene must match all four sets of DNA from the suspect. With these safeguards in place, the odds are one in 70 billion (140 times Earth's population) that the results are inaccurate.

However, DNA fingerprinting is sometimes refutable evidence, even with such low odds of innacuracy. If the technician confused the DNA samples or deliberately gave false results, the information gleaned from a DNA test would be useless.

Also, DNA evidence only proves that a person was present. It does not prove that the depositor of the DNA was the actual criminal. For these reasons, there must be other evidence to support the DNA fingerprints' conclusion; the fingerprints alone are not enough for a conviction.

Applications of DNA Profiling

Nearly everyone knows that DNA Fingerprinting can be used to prove a person's connection to a crime. Few realize the other applications of DNA fingerprinting.

Establishing Innocence

DNA Fingerprinting can help to speed along court cases by eliminating suspects. Between 1989 and 1996, the FBI used genetic testing in about 10,000 sexual assault cases; in 2,000 of those cases, the prime suspect was discovered to have not committed the crime. Without genetic testing, it can be assumed that some of these men would have been convicted. In fact, many prison inmates have appealed their conviction after spending years in jail, and have been discovered to be innocent.

Identification

As a result of the unidentified remains of American soldiers that have been found abroad, the U.S. military has been compiling data from mitochondrial DNA to help identify the soldiers. Mitochondrial DNA is passed on by the mother only, and as a result, remains mostly identical within families.

By testing DNA from the bones of the remains of soldiers against the military's database, scientists have identified 500 soldiers so far, and another 250 more are under study.

Ethics

There many other genetic tests available now that can determine identity and paternity. When the Human Genome Project is completed, scientists predict that a battery of tests will be available to determine what diseases and characteristics a person is prone to. But as with any new knowledge, we must determine what people have access to what information, and how that information can be used.

Identity Test

Currently, DNA testing can be used to match one genetic sample to another. This is being used in many court cases to prove that a person was at a scene where his DNA was picked up in the form of a hair, blood, or skin samples. However, the police cannot just look at some DNA and know whose it is; they must have a suspect, get a DNA sample from him, and attempt a match. Without a suspect, there is no way to figure out whose DNA was at the scene.

Many people believe that crimes could be solved much more quickly and efficiently if there was a national database containing DNA samples from each person. Of course, the entire DNA strand would not be needed — just the pieces usually used for testing. But should the government really have access to everyone's DNA samples?

Full Workup

Scientists hope that in the near future, testing will be available that will determine immunity or succeptibility to disease, whether the patient will have side effects to certain drugs, whether he might develop a terminal illness, or even what his basic intelligence potential is. That's a lot of information, and a good bit of it could be damaging in the wrong hands. But whose hands are the wrong hands? This full genetic workup leads to many, many questions. For instance, should this test be done at birth or only on demand? Should insurance companies allowed to see your disease succeptibilities? Can they charge you more if you are succeptible to an expensive disorder? Should drugs be allowed on the market that can be very harmful to some but not to others, provided that the recipient is tested to make sure the drug is safe for him?

If a person will likely develop a terminal illness, should the doctor tell him, or just decide that ignorance is bliss and let him find out twenty, thirty, even fifty years down the road? Can an employer discriminate based on basic intelligence? These questions only scratch the surface of what may come. In addition, if access to a product is restricted, a black market usually springs up. What should the penalty be for trying to gain access to another person's genetic information?

Core Issue

The core issue brought up by genetic testing is this — is a person bound to a genetic future, or is man more than the outcome of his genes? Is genetic tendency immutable? Can a little strand of chemicals really determine the outcome of our entire lives?

Eugenics

The concept of moving selective breeding from the animal world to the human world began to take root in the 1830's. Could this natural extension of Genetics and Darwinism be resurfacing?

Roots

In 1883, Francis Galton, cousin of Charles Darwin, coined the word "Eugenics", which means "good in birth". In his book *Natural Inheritance* (1889), he theorized that society would be better without the presence of inferior people. In fact, he continued this line of reasoning to say that societies topple because the genetically superior aristocracy have fewer children than the inferior peasants. To keep this from happening, he proposed two methods of control — Positive eugenics would create incentives for the very intelligent or successful to reproduce. Negative eugenics would keep the unfit from reproducing.

Applications

Galton's ideas grew in popularity both in America and abroad. In 1896, Conneticut had put in place laws that forbade sex with epileptics or the mentally handicapped. In 1907, a state law was passed that required sterilization of institutionalized, mentally retarded males, as well as criminal males.

This law would eventually lead to other states' passage of similar laws; before they were repealed in 1956, 58,000 American citizens were sterilized. The Immigration Act of 1924 limited the immigration of "idiots, imbeciles, feebleminded, epileptics, insane persons" to the United States.

The philosophy of eugenics had taken root in Nazi Germany, as well. In 1934, the Eugenic Sterilization Law was passed in Nazi Germany, ordering the sterilization of individuals with genetically inherited diseases; the Genetic Health Courts would decide who would and who would not be sterilized. In 1939, the situation turned very ugly. The Nazi laws were extended to allow the killing of genetically undesirable people; by the time World War II was over, 75,000 "unfit" persons had been killed, including 5,000 children. As the world discovered the atrocities tha had been committed, many began to rethink their positions regarding these laws. In 1956, all United States eugenic sterilization laws were repealed, and nine years later, many of the immigration restrictions were lifted.

Modern Eugenics?

Thanks to the Human Genome Project and other genetic research, prenatal testing allows for the detection of many illnesses, including cystic fibrosis and Down's syndrome. Pregnant women can now know ahead of time if their fetus will contract an incurable disease or condition.

Often, a diagnosis leads to the termination of the fetus. This can be construed to be a combination of euthanasia and eugenics, an attempt to keep "bad genes" out of the gene pool, or as an act of mercy, terminating a fetus whose quality of life is in doubt. But which is it? And whose decision is this really to make?

Who gets to decide what diseases/conditions are too bad to have to live with and what diseases/conditions are not so bad?

Ethics of Genetic Engineering

Genetic engineering offers a major ethical dilemma — there are great rewards if genetic engineering becomes a widespread reality, but the dangers are equally great. In addition, genetic engineering also offers the ability to change the very nature of nature, an environment in which man is already well-suited to live in. Is it possible that genetic engineering will do more harm than good?

Medicines

As you have probably already learned, DNA is very versatile because all life on Earth processes it the same way (as far as we know). Therefore, foreign genes can be implanted in bacteria, instructing the bacteria to produce whatever the inserted gene controls. In this manner, scientists can "teach" bacteria to produce human or other proteins. There is a long list of bacteria that have been treated this way, including an e. coli bacteria that produces insulin. The insulin is less expensive than that removed from cows and pigs, as well as being exactly the same as human insulin.

The ethical quandary about this process lies in the fact that human genes are being inserted into a bacteria. I do not know of anyone who stands up for bacterial rights; in fact, everyone who has been sick is rather prejudiced against bacteria.

The problem is the idea that human genes are being put in a non-human organism. Is there anything special about human genes? If you consider that human genes are made of the same four bases as bacterial genes, and that they work the same way — as an instruction manual for the production of a protein, there is less of a dilemma. However, many people believe that human genes should not be implanted in other species, because they are special because they are human. Before you make up your mind, consider the alternative method of production, and consider honestly whether or not you would be willing to recieve medicine from a human gene in a bacteria if it was you who needed it.

Xenotransplantation

Work has been done on genetically engineering farm animals, too. While in the future it may provide leaner meat, larger animals, or faster growth, right now the great prospect is xenotransplantation. Xenotransplantation transplanting organs across species. Currently, most organ transplants are human to human. They work well, but the drawback is the rejection medication that must be taken to weaken the immune system so that the body does not destroy the new organ. This, of course, can lead to illness, a definite drawback. If organs can be "made to order" in another organism and then transplanted over to the human patient, this can eliminate the need for rejection medication and the succeptibility to illness.

The best prospect for xenotransplantaton of vital organs is the pig. The pig is one of the closest genetic matches to humans, closer even than monkeys and apes. With a little genetic alteration, pigs can hopefully produce organs that the human immune system will not reject, producing the desired effect. Unfortunately, the pig is the loser in this arrangement. While on the surface this may seem barbarous, consider the concept: a human is taking something from the animal to extend his own life. The basic principle is not all that different from eating. If xenotransplantation still seems cruel, perhaps the world needs to re-evaluate itself and turn vegetarian, eliminating the need to raise animals for the sole purpose of death for man's benefit.

Custom-made Organisms

Current technology and a great deal of experimentation may allow for custom-made organisms to be produced. These may range from a bacteria that cleans up oil spills to a monkey with gills and a higher intelligence level. Everyone seems to agree that we should not pursue these advances just for the fun of it, so the question becomes, are we so inconvenienced by a problem that we need to create a new life form to fix it for us?

Core Issue

The core issue behind the ethics of genetic engineering is as follows. Is it morally right to change the nature of life on Earth to suit man's desires better? The answer to this question depends on man's position on Earth. If we are truly superior to the animals and accountable to no one, is there a real question of whether it is wrong? If we are not fundamentally different, do we have a right to meddle with evolution? If we are accountable to God for our actions, should we risk insulting His creation by trying to do it better?

Current Developments

Transgenics allow scientists to develop organisms that express a novel trait not normally found in the species; for example, a type of rice known as golden rice has elevated levels of vitamin A. Scientists have also developed sunflowers that are resistant to mildew and cotton that resists insect damage. Possible transgenic combinations can be broken down generally into three categories (here "animal" refers to nonhumans):

- plant-animal-human combinations
- animal-animal combinations
- animal-human combinations

An example of a plant-animal-human transgenic combination would be one in which the DNA of mouse and human tumor fragments is inserted into tobacco DNA. The harvested plants contain a potential vaccine against non-Hodgkin's lymphoma. Other transgenic plants have been used to create edible vaccines. By incorporating a human protein into bananas, potatoes, and tomatoes, researchers have been able to create prototypes of edible vaccines against hepatitis B, cholera, and diarrhea. The vaccines are proving to be successful in tests on agricultural animals and humans.

A product created from an animal-animal transgenic combination. Scientists at Nexia Biotechnologies, a company based in Montreal, isolated the gene for silk protein from a spider capable of spinning silk fibers—one of the strongest yet most resilient substances known—and inserted it in the genome of a goat's egg prior to fertilization. When the transgenic female goats matured, they produced milk containing the protein from which spider silk is made. The fibre artificially created from this silk protein has several potentially valuable uses, such as making lightweight, strong, yet supple bulletproof vests. Other industrial and medical applications include stronger automotive and aerospace components and stronger, more biodegradable sutures for closing wounds.

5

Characteristics of Genetic System

General Features of an Organism

- good genetic background
- easy to grow
- controlled matings possible
- can be genetically engineered
- funding available

Viruses

- easily grown in culture
- rapid generation time
- encode only a few the proteins, which permit a detailed analysis of well-defined system
- some viruses have control mechanisms found in eukaryotic species
- can be genetically engineered
- lack organized chromosome
- not all genetic features can be extrapolated to other organisms

Escherichia Coli

- easily grown in culture
- rapid generation time
- o many mutants available

- o many genes involved
- o life cycle very well defined
- o relatively simple to generate new mutants
- o can be genetically engineered
- o contain a rudimentary chromosome
- o not all genetic features can be extrapolated to higher organisms

Human

- o intense public interest and funding available
- o relatively well-mapped for most eukaryotic species
- o many diseases (mutant phenotypes) understood clinically
- o well-defined cytogenetic system
- o long generation time
- o many traits can only be studied in cell culture
- o cannot make controlled crosses
- o cannot be genetically engineered

Drosophila

- o short generation time for a eukaryotic organisms (two weeks)
- o availability of many mutants controlling specific phenotypes
- o large chromosomes with well-defined cytogenetic system
- o survives well in the lab and matings are easily performed
- o currently, the best organism to study developmental genetics
- o transposable elements can be manipulated to clone genes
- o can be genetically engineered

Maize

- o best mapped plant species
- o many mutants are available that control seed traits
- o well-defined cytogenetic system
- o transposable elements well understood and can be used to clone genes
- o matings tedious, but produce many (hundreds) of progeny
- o only three generations per year
- o cannot be easily genetically engineered
- o not well funded

Arabidopsis Thaliana

- o small genome with little repetitive DNA
- o short generation time (six weeks)
- o many mutants rapidly becoming available
- o matings tedious, but produce many (thousands) of progeny
- o currently, an intensely investigated organisms; the Drosophila of plant species
- o can be easily genetically engineered

Although these species have been studied intensely, these are by no means the only species analyzed for genetic studies. Mouse has been used for many genetic studies because it is a mammal that is a relatively good model for human gene expression. A good example is the globin genes, protein that carry oxygen in blood cells. The genetic and molecular organization of the mouse and human genes are quite similar: they have the same number of exons and have the introns located at the same amino acid residue in the transcript. Furthermore, the order of the multigene family is conserved. Yeast has also been well studied. Although this species is a good model system, this is not the reason that it was first investigated. The primary reason was because of its role in brewing.

In addition many plant species have relatively well-defined genetic systems. Tomato, next to corn, is the best mapped plant species. Tobacco has become a valuable tool to study plant molecular genetics because of the ease in which it can be genetically engineered at the whole plant level. This permits the molecular analysis of many genes in the functioning organism, rather than in a cellular condition as is the case with animal systems.

The genetic systems of many other plant species have been analyzed by plant breeders who need to understand how specific traits are genetically controlled. This information is important so that the traits can be successfully manipulated in a breeding program. This ability to manipulate crop plants is not a recent discovery. That native Americans were aware of the concept of heredity is evidenced by their deliberate attempts to maintain specific lines of their crops that were adapted to or superior in their environment.

Genetic Code System

Since there are 64 possible triples (repeated letters within the triples are allowed) using the 4 letters *A, C, G*, and *T*, different triples can represent the same amino acids. There are also 3 triples (*TAA, TAG, TGA*) known as stop codons that do not represent (a common phrase being do not code for) an amino acid, but are involved in signaling a termination to the protein production process. One particular codon (*ATG*), which does represent one of the amino acids, is an indication for the start of the production of a protein. This system for coding proteins has come to be known as the genetic code. Before the genetic code was fully understood, based on work by Marshall Nirenberg, Heinrich Matthaei, Har Gobind Khorana, and others, various mathematical ideas were invoked to suggest what kind of code might be involved. There is a diagram of the RNA version of the genetic code, where *U* takes the place of *T*. One approach to finding a gene is to look for a stretch of DNA which starts with ATG (the start codon) and ends with one of the stop or termination codons. This situation in a stretch of DNA is referred to as an open reading frame (ORF). However, not all ORFs correspond to a stretch of DNA which will initiate the production of a protein, and so the location of an ORF is not equivalent to finding a gene. A major problem facing biologists is how to locate those stretches of DNA that are genes in the large amounts of DNA found in a chromosome.

The way that DNA is involved in the production of proteins is not direct. The mechanism involves another helical molecule called RNA, of which there are a variety of types. (Among the types of RNA are messenger RNA, or mRNA, ribosomal RNA, or rRNA, and transfer RNA, or tRNA.) When a protein is to be produced, the DNA separates and a copy of the gene is transcribed (the process is referred to as transcription by molecular biologists) into a strand of RNA (ribonucleic acid). RNA, like DNA, uses an alphabet of 4 nucleotides *A, C, G,* and*U* (for uracil, a pyrimidine like thymine which takes the place of *T*). In a somewhat complex series of steps, a protein is produced. A very rough schematic of how genes create proteins. This schematic is often referred to as the central dogma. The original DNA, which are known as introns but which are not involved in the production of a protein, are separated or snipped out and a resulting string of exons from the original DNA is left. What is left is sometimes referred to as a coding sequence, and it contains the codons (in the letters *A, C, G* and *U*) which will create the protein, by the process usually referred to as translation.

The actual manufacture of the proteins involves structures called ribosomes. Production of proteins does not go on within th nucleus but in the cytoplasm surrounding thenucleus. One can express the codons involved in the manufacture of proteins either in DNA terms (using the letters *A, C, G* and *T*) or in RNA terms (using the letters *A, C, G* and *U*).

Many interesting quesions can be asked about the genetic code, including why the pattern of redundancies in the codeexists. While at one time it was thought that the code was totally universal, this has proved not to be true. For certain bacteria and in other situations the universality of the code breaks down. Cellular life forms are classified according to whether they are prokaryotes or eukaryotes. Prokaryotes are cellular organisms where the genetic material is not a part of a membrane enclosed nucleus, while in eukaryotes the genetic material is contained in a nucleus.

At one time life forms were classified as either bacteria or eukaryotes. A group of bacteria like organisms, found in very hot environments such as undersea vents, were initially classified as bacteria. However, when Archaea's genetic makeup was studied, its members were found to have an amazingly large number of genes that were not known to exist in other organisms. They seemed to be surprisingly different from the bacteria with which they had been formerly classified. It has become standard in recent years to list the Archaea in a separate category from the bacteria and eukaryotic organisms. Use of phylogentic trees was in part a tool that led to this change. It is tempting to believe that, because human beings are as complex as they are, other species will have less genetic material than humans. However, chromosome number (and the total size of the genome) varies with species in a way that seems to have no rhyme or reason. Thus, chimpanzees have 48 chromosomes, cats 38, dogs 78, rabbits 44, rats 42, turkeys 82, and horses 64. Perhaps when the genomes of many different and varied species are sequenced it will be possible to get insight into these wildly varying numbers.

Mendel's First Law: Segregation

After this excursion into cytology, lot us return to genetics. The chief tool of the cytologist is the microscope; it has shown him that the chromosomes, and with them the genes, segregate at meiosis and enter the gametes singly. The chief tool of the geneticist is the crossing of different strains, and we are now going to consider crosses which demonstrate the segregation of the genes at meiosis.

It will be useful for the description of these experiments if we recall some technical terms which have already been defined, namely "haploid" and "diploid" for cells or individuals which carry one or two chromosome sets, respectively, and "allelomorphs" or "alleles" for partner genes with different effects. To these terms I want to add two new ones which describe the genetical constitution of a diploid individual in regard to any particular pair of genes.

Evidently there exist two possibilities: either the partner genes are alike or they are allelomorphic. In the first case, the individual is "homozygous" or a "homozygote" in respect to the gene concerned; in the second case, it is called "heterozygous" or a "heterozygote."

A pea plant with two genes for wrinkled seeds or two genes for round seeds is homozygous for either wrinkled or round; a pea plant with one gene for wrinkled seed and an allelomorphic gene for round seed is heterozygous for wrinkled and round. It will readily be seen that only heterozygotes are suitable for demonstrating the segregation of genes; for the geneticist recog nizes genes by their effects, and the segregation of like genes in a homozygote is not observable.

In higher organisms (macroorganisms) it is hardly ever possible to observe the immediate effects of segregation at meiosis; for the mature gametes fuse at fertilization and the progeny, in which we observe the effects of the genes, carries genes from two different gametes. The life cycle of the domestic fowl in terms of haploidy and diploidy. The diploid female and male form haploid ova and spermatozoa. At fertilization the gametes fuse and form the fertilized ovum, which in turn develops into the diploid embryo, chicken and adult. Thus the whole life cycle, with the exception of one cell generation, is spent in the diploid stage. The situation is quite different for many microorganisms. The life cycle of a green alga, Chlamydomonas.

Each individual of this species is a microscopically small cell which swims through the water with the aid of two whiplike hairs, or "flagellae." These cells are haploid; they have only one set of chromosomes, carrying all genes of the species in single copies. New cells are formed through division of the old ones.

In this way whole populations may be derived from one cell by repeated divisions. Such populations are called "clones." Since at mitosis both daughter cells receive exactly the same set of chromosomes and genes all cells within a clone are genetically identical. There are no visibly distinct sexes, but physiologically there is a clear distinction between sexes or "mating types," called + and–in our diagram. Cells of the same clone are always of the same mating type and never mate with each other; nor does mating occur when cells of two + clones or two–clones are brought together. But when cells from *a*+ and *a*– strain are put into the same culture vessel, pairs are fomed.

Each pair consists of one + and one–cell they fuse and form a thick skinned pore called a zygote. The zygote is comparable to a fertilized ovum in, say, a dog or a fly. It contains the chromosomes from both gametes and is, therefore, diploid. Very soon after fertilization meiosis takes place inside the zygote, and haploid gametes swarm out. These start a new life cycle by division; in fact, the gametes of Chlamydomonas do not differ from any of the other free swimming ells. Thus in this algae, as in many other microorganisms, the whole life cycle, with the exception of the zygote, takes place in the haploid stage, and segregation at meiosis is not obscured by diploidy of the progeny. Usually it is necessary to grow the gametes into clones before segregation becomes observable; but occasionally the segregating gametes themselves differ visibly from each other. Two strains (1) and (2) differ in one gene which is concerned with the development of the flagellae. While strain (I) has all the genes necessary for flagella formation, strain (2) carries a deficient allele of one of these genes and cannot, therefore, develop flagellae.

When the two strains are crossed by bringing together + cells from the one strain with cells from the other (it makes no difference which strain provides the + cells, which the cells), a zygote is formed which carries both alleles and is, thus, heterozygous for the gene in question.

At the reduction division the partner chromosomes segregate and carry the different alleles into separate cells. At the second meiotic division each cell divides once more by mitosis, so that finally four gametes are formed, two of which carry the normal and two the deficient allele. When these swarm out, it will be seen immediately that two have flagellae, two have not. This difference persists in the clones formed from the gametes. If we let each gamete grow into a separate clone, two clones will consist of flagellated free moving cells, while two will consist of flagellaless cells which sink to the bottom of the tube.

The fact that a heterozygous zygote always produces two gametes carrying one allele and two carrying the other is so well established that it can be used, conversely, for determining whether a difference between two strains of a microorganism is due to the effects of a pair of allelomorphic genes. If this is so, the gametes from a cross between the strains will yield two clones, resembling the one strain, and two resembling the other. Yeast, like Chlamydomonas, is a unicellular organism. Many strains are haploid and can be used in experiments on segregation. Like Chlamydomonas, yeast occurs in two mating types, + and –, which are not visibly different from each other.

Yeast, as you know, ferments sugar into alcohol ad carbon dioxide gas; but strains differ in their ability to ferment various kinds of sugar. Thus some strains readily ferment galactose (one of the components of milk sugar), while others do so only very slowly and inefficiently. A simple method for detecting fermentation is to catch the developing gas in a est tube which was first completely filled with liquid and then inserted immediately above the yeast cells. If the medium contains a sugar which can be fermented, the tube gradually fills with gas; if the sugar cannot be fermented, the tube remains filled with liquid.

In the experiment shown in the medium in the top row contains galactose as the only sugar. Strain (2) is able to ferment this sugar, strain (1) is not. Cells (3) and (4) from the two strains are brought together, and a zygote (5) is formed which, through meiosis, gives rise to four gametes (6). Each gamete is grown into a clone (7) on solid medium containing glucose, a sugar which can be fermented by all strains of yeast. Finally, the four clones ae tested (8) in fermentation tubes with galactose as the only sugar. You will notice that two clones (those on the left) can ferment galactose, and two cannot. Thus the difference between the two original parents segregates in the 2:2 ratio expected if it is due to the action of a single gene pair. On the basis of several experiments of this kind it could be concluded that in yeast the ability to ferment galactose is controlled by the action of a gene which in nonfermenting strains is absent, or is present in the form of a nonfunctioning or poorly functioning allelomorph. The essence of what is called Mendel's first law or the law of segregation. In modern terms it may be expressed as follows: When a heterozygote forms gametes, half of them carry one of the two alleles, half the other.

The basis for the segregation of the genes is the segregation of the chromosomes at meiosis. Mendel himself could not, of course, interpret his results in these terms, but he quite clearly postulated that a mechanism of this kind was responsible for his data, in particular for the famous and much overrated 3:1 ratio. This particular ratio is a rather remote consequence of the underlying gametic segregation in a ratio of 1:1. The steps by which, in crosses between higher organisms, this 1:1 ratio becomes transformed into oters, including.

In doing this we shall find ourselves using very simple statistical rules, such as are used for predicting that a coin tossed 100 times will show head about 50 times. This statistical aspect of genetics has been held against it by the Lysenko school, on the grounds that mathematics must not be applied to biological objects and that Mendel had no right to speak of a 3:1 ratio between round and wrinkled peas if some of the pods contained 6 round peas. No matter what we think about these arguments—and to a modern biologist they seem entirely futile.

Characteristics Genetic System

X-Chromosome as Type

The chromosome combinations present in the fertilized egg, and this has brought to light some of the methods by which the chromosomes act on development and characteristics. We now Examine other effects of altering chromosomes. We shall have to examine the effects of altering all the different classes of chromosomesýÿthe X-chromosome, the Y-chromosome, and the different autosomes. Each of these three groups (X, Y, and autosomes) gives rise, as we shall see, to a different type of inheritance. The simplest relations, from the experimental point of view, are presented by the X-chromosomes. We therefore deal first with these.

Effects on Altering the X-Chromosomes

The X-chromosomes take such a course in passing from generation to generation that it is possible to follow the descendants of a particular Xchromosome (that present in the original male parent, for example), knowing in which individuals they are present, in which they are absent. In certain individuals (males, in organisms of Group I) there is but a single X-chromosome instead of a pair. These relations make it a relatively simple matter to discover the distinctive effects of a particular X-chromosome. The effects of X-chromosomes will therefore be dealt with somewhat fully as a type of chromosomal action. Later the action of the other chromosomes will be taken up.

Abnormal Distribution of X-Chromosomes

What will happen if the X-chromosomes by accident become irregularly distributed? We know that sometimes the X-chromosomes are indeed irregularly distributed. Will the sex-linked characteristics continue to follow them; will the defective characteristics show the same irregular distribution as do the X's? Such cases in great number have been fully observed. It is found that the sex-linked characteristics do indeed follow the X-chromosomes wherever they go. This proves conclusively (if there were any possible doubt in view of the extraordinary course normally followed by such characters) that it is indeed the X-chromosomes on which the characters depend. The matter is one of importance and interest, so that it will be worth while to examine carefully certain typical cases of the result of irregular distribution of X-chromosomes.

Normal Distributions

Sometimes the body of Drosophila is yellow instead of the normal grey; this is a recessive character due to a defect in the X-chromosomes. Suppose that we mate together a female that has a yellow body and a male that has the normal grey programmes.

The female has two modified X-chromosomes, which we may represent by XX. The male has its X unmodified; we may represent its cells as XY. Normally in forming the germ cells, the two X's of the female separate into different germ cells, and all the ova receive one of the modified X's. The normal male produces germ cells of two classes, X and Y. When the germ cell X from the female unites with Y from the male, sons are produced, with the constitution XY; while X from the female with X from the father gives daughters XX. Since the yellow programmes dependent on X is recessive, the daughters XX have the normal grey programmes, like the father.

But the sons XY have only one X, and as this is the defective one, from the mother, the sons are yellow, like the mother. The normal result of such a mating therefore is that all the sons are recessive like the mother, all the daughters dominant, normal, like the father. This is what occurs when the X-chromosomes are distributed to the germ cells in the usual way.

Abnormal Distribution

In forming germ cells both X's of the female go together to a single ovum, while other ova receive no X. In a race discovered by L. V. Morgan the two X-chromosomes in the female were partly united, so that they almost invariably thus go together to one ovum, leaving other ova without an X. In this race the X-chromosomes were so modified as to produce th recessive yellow programmes described in the preceding paragraph. We therefore have an opportunity to determine the course of inheritance of the yellow and the grey programmes when the X-chromosomes are thus abnormally distributed.

Represent as before the X-chromosomes that produce the yellow programmes by X. Thus one set of ova receive XX, while another set recive no X. When the germ cells XX united with the germ cells Y from the norml father, there wee produced daughters XXY, in which both X's came from the mother, and these daughters were yellow like the mother (since no normal X was present). Sons were produced by the union of ova that contained no X with sperms that carried X; this gave sons XO, with their single X from the normal father, instead of from the mother, as is usually the case. And such sons had the normal grey body programmes of the father, instead of the yellow body programmes of the mother, as happens normally.

Thus with a change in the distribtion of the X's there is a corresponding change in the method of inheritance. If the sons receive their X from the recesive mother, as in the normal cases, they are recessive lke the mother. But when, through non-dijunction, the sons receive a dominant X from the father, they are dominant like the fther.

Similarly, in the normal case the daughters receive a dominant X from the father and are therefore dominant like the father. But in cases of non-disjunction the daughters receive both their JPs from the recessive mother, and are therefore recessive like the mother. In sum, when the X's are normally distributed, uch matings give 'criss-cross inheritance'; sons like the mother, daughters like the father. But when the X's are not so distributed, there is no criss-cross inheritance; sons are like th father, daughters like the mother. The dominant and recessive characters follow the respective X-chromosomes, whether these are distributed normally or abnormally.

Such experiments have been repeated many ties, and with other sex-linked characters. Always the characteristics follow the distribution of the X-chromosomes, however these are distributed. It may be concuded with certainty that it is the presence of X-chromosomes of a modified type that causes the appearance of the particular sex-linked characters that re manifested. The rules of distribution of sex-linked characters are the rules of distribution of the X-chromosomes.

Dominance and Recessiveness

These follow from generation to generation the distribution of these X-chromosomes. Most of these bodily defects are manifested only in individuals in which the defective X- chromosomes are the only kind present; that is, these defects are recessive.

But it is important to observe that in such cases the normal condition of the organism likewise follows the distribution of certain X-chromosomes. Haemophilia follows certain X- chromosomes. But in the same matings in which this occurs, the healthy condition of the blood follows certain other X-chromosomes. If a defective mother is mated with a normal father, the sons are defectie because they receive only the mother's defective X-chromosomes. But in the same way the daughters are normal because they receive the father's normal X-chromosomes.

In the normal individuals the X-chromosome plays a part, just as it des in the defective individuals. In the normal individuals it is healthy and supplies what is required for normal development, while in the defective individuas it fails to supply what is required. In the case of most defects this normal condition is dominant over the recessive condition; that is, when both are present, the normal chromosome is the one that prevails in its effect on the indvidual.

There are defective characters, however, in which the defectivecondition is dominant or partly dominant. In these cases, when a defective X and a normal one are present together, the defective one produces its effect, as in the case of bar-eye, described earlier. In many such cases the defect is less marked when a normal chromosome is present as well as a defective one; then the defective condition is said to be partly dominant. Thus whenever an individual bearing defective X-chromosomes is mated with a normal individual, we have both dominant and recessive characters, the course of which may be followed in the later generations the dominant character being commonly the normal or usual condition. How this works out is shown in our next paragraphs.

Sex-linked Inheritance

The test consists in making what are called reciprocal crosses:

- On the one hand mate a dominant female with a recessive male;
- Also mate a recessive female with a dominant male.

Represent a dominant X-chromosome by a capital X, a recessive X-chromosome by a lower-case letter x. In the group of organisms with which we have been dealing, the female has two X-chromosomes, the male but one. The daughters receive an X from each parent, the sons an X from the mother only. A dominant female will be represented by XX, a recessive male by xo; similarly a recessive female is xx, a dominant male is XO.

The two ma tings and their results will then be represented as follows:

- XX by xo gives X + XO Dom Mother Rec. Father Dom. Daughters Dom. Sons
- xx by XO gives X + xo Rec. Mother Dom. Father Dom. Daughters Rec. Sons

Thus when the mother is dominant, the father recessive, all the children are dominant, like themother. When the father is dominant, the mother recessive, the daughters are dominant like the father, the sns recessive like the mother; this is called 'criss-cross inheritance'.

Whenever reciprocal crosses give these results, it is certain that the two diverse characteristics (dominant and recessive) result from dfferences in the X-chromosomes of the two parents. 'Criss-cross inheritance' is particularly useful in showing at once that we are dealing with sex-linked inheritace.

In every case where such results are produced, further tests show that the two haracteristics follow in later generations the two different kinds of X-chromosomes wherever they go. t is mainly by the use of these tests that the many different characteristics dependent on diversities in X-chromosomes have been discovered.

The Autosomes: Typical Mendelian Inheritance

We have already seen that the autosomes play a role in development, in the fact that they help to determine the sex of the developing individual. In Drosophila, if the number of sets of autosomes, in relation to the number of X's, is changed, this changes the sex.

Do autosomes also affect other characteristics? Do the diferent pairs of autosomes have different functions? Are there diverse types f autosomes, with different effects, in different individuals of a species, as is the case in relation to the X-chromosomes? Are there dominant and recessie effects of autosomes, as there are of X? If dominant and recessive characteristics depend on auosomes, then these characteristics must follow the distribution of the autosomes from parents to offspring, as the characteristics ependent on X follow its distribution. We must then examine the method of distribution of the autosomes whether thre are characteristics that show this method.

Method of Distribution of the Autosomes

The ditribution of the autosomes from arents to offspring is much simpler than that of the X-chromosomes.

The main facts are as follows:

- The autsomes are in pairs in all individuals. There are no individuals with unpaired autosomes, like the male of some organisms with relation to X.
- Thus the two sexes are alike as to their autosomes.
- In forming germ cells, one member of each pair of autosomes goes to each germ cell. The sperms and the ova are alike with respect to the autosomes.

Thus the fundamental rule with relation to the autosomes is that each gem cell carries one member of each pair of autosomes.

- By union of two germ cells the pairs of autosomes are restored.
- Thus each one of the offspring carries in each pair of its autosomes one autosome from each parent.

Now consider a single pair of autosomes, such as the pair marked AA in. Suppose that in different individuals the members of this pair of autosomes produce different effects, as we have seen to be the cae with the X-chromosomes of different individuals. And suppose that, as in the case of X, one of these effects is dominant, the other recessive. How would such characteristics be inherited?

This question must be examined with care. For this purpose we may take as an example the different colours found in different varieties of peas, used in the classic experiments of Mendel. The yellow seed programmes of a certain variety is dominant, and results from the constitution of the autosomes of a certain pair. The green seed programmes of another variety is recessive, owing to a different constitution of the autosomes of that same pair.

Call thetwo autosomes of tht pair in the dominant yellow race by the capital letters *AA*, while the two autosomes that give rise to the recessive green programmes in the other race may be called *aa*. The dominant parent may be designated *D*, the recssive parent *R*, When these two parents produce germ cells, each germ cell receives, of course, one chromosome for the pair. The germ cells from the dominant parent *D* have the autosome *A*, those from the recessiveparent *R* have the autosome *a*. The erm cells *A*, from parent *D*, unite each with a germ cell a from parent *R* giving offspring (zygotes) containing both *A* and *a*. These first generation offspring are commonly designated *Fi* (first filial generation). And since the characteristic carried by the autosome *A* is dominant, all these *Fi* offspring manifest the dominant characteristic *D* (yellow, in case yellow-seded peas are crossed with green-seeded peas). Next observe the result of mating together two of these individuals *Aa*, of the *Fi* generation, giving the *Fa* (second filial generation).

Each parent *Aa* gives, according to the general rule, two kinds of germ cells, one kind carrying the dominant autosome *A*, the other the recessive autosome *a*. Each kind of germ cell from one parent unites with each kind from the other. That is, half the *A* germ cells from one parent unite with *A* germ cells from the other parent, half with a germ cells from the other parent; and the same is true for the a germ cells from the first parent.

This gives in equal proportions Fa offspring (zygotes) as follows:

$$AA + Aa + a\,A + aa$$

But s the two combinations *Aa* and *aA* contain the same chromosomes, the proportions may be written *AA* + 2*Aa-f-aa* The individuals of the constitution AA and Aa contain the dominant chromosome *A*; therefore they manifest the dominant characteristic, *D*. The individuals *aa*, having only the recessive chromosome *a*, manifest the recessive characteristic, *R*. Thus in this F^2 generation, so far as manifested characters are concerned, the proportion of offspring are 3 Dominant to *i* Recessive. Thus if dominant and recessive characteristics depend on the two members of a pair of autosomes, hen a dominant individual *AA* is mated with a recessive individual *aa* the general result is bound to be that the immediate offspring (*Fi*) all manifest the dominant characteristic, while in the *Fa* generation (produced by mating two individuals from *Fi*) there will be three times as many dominant individuals as there are recessives. Now, this is the result that was discovered by Mendel in 1866; it formed the foundation for the working out of Mendelian heredity. The proportion 30 to *iR* in the *Fa* generation is commonly known as 'the Mendelian ratio'; it is one of the most characteristic proportions in this type of inheritance.

Mendel discovered that there are many different characteristics that yield this 3 to *i* ratio. If yellow and green peas are crossed, in the *Fa* generation there are 3 yellows to *i* green; if round peas and wrinkled peas are crossed, in F^2 there are 3 round to *i* wrinkled; and so for may other characteristics in the pea plants with which Mendel worked.

And since the time of Mendel these same ratios have been found for thousands of different characteristics, in great numbers of different organisms, from the pea plant and *D* roso- phila to man. The fact that inheritance gives these ratios is commonly spoken of as Mendel's law, and the type of inheritance which it exemplifies is spoken of as Mendelian inheritance. The type that gives in the *Fa* generation 3*D* to *iR* is the commonest method of inheritance, much commoner than the sex-linked method, in which the characteristics depend on differences in the X-chromosomes. This is obviously because autosomes are more numerous than X-chromosomes. We find then that inheritance according to Mendel's law takes place exactly as it would if the characteristics concerned depended on the two members of a pair of autosomes. As will be shown later, it can be proved experimentally that such characteristics do indeed depend upon autosomes. Thus the rules of Mendelism are the rules of distribution of the two members of a pair of autosomes, one having a recessive effect, the other a dominant effect. Mendelian heredity is heredity of characteristics that depend on differences in autosomes.

Mendel did not know of the relation of the characteristics to the chromosomes. He discovered, however, certain other important facts about this type of heredity, facts which have become clearly explicable since it has been learned that this type depends on differences in autosomes. These facts must be examined. For this purpose it will be convenient first to define certain terms which are universally employed in dealing with inheritance.

By the union of two germ cells or gametes, as *A* and *a*, there is produced a zygote. If the two germ cells are alike with respect to the chromosomes or characteristics that they carry, the zygote so produced is spoken of as a homozygote; thus *AA* and *aa* are homozygotes. If the two germ cells that unite to form a zygote are unlike, as *A* and *a*, the zygote so produced is called a heterozygote, as *Aa*. The two corresponding characters, one dominant, the other recessive, that are produced by the two differing members of a pair of chromosomes are spoken of s alternative characters, or alleles, or allelomorphs. Any given individual manifests one or the other of the two alternatives, not both. For the alternative characters we shall employ henceforth the term alleles. The characteristic manifested by a zygote is its phaenotype while its actual constitution is its genotype.

Thus, two zygotes *AA* and *Aa* both manifest the dominant character; the dominant character is the phaenotype, or the individuals are said to be phaenotypically dominant. But they are diverse in genotype; one is *AA*, the other *Aa*. In the case of a homo-zygote *AA* or *aa*, the phaenotype and genotype may be held to coincide. Mendel crossed yellow peas with green peas, and found, as before stated, that in the *Fi* generation all the peas were yellow, so that yellow is the dominant allelomorph. In the F^2 there were three times as many yellows as greens. The actual number of seeds in the F^2 generation was in his experiments 8023, of which 6022 were yellow while 2001 were green, so that the ratio of dominants to recessives was, accurately expressed, 3.01 to *i*.

Mendel found that if he bred further the individuals of the F^2 generation, some of these gave different results from others. In the pea plant it is possible to allow each plant to fertilize itself, so that both kinds of germ cells come from the same parent. When this was done with the individuals of the F^2 generation, the following results were observed: The green plants (recessive) gave only recessive green offspring (they 'bred true'). Of the yellow plants (dominant), one-third gave only yellow offspring (they 'bred true'), while two-thirds of them gave both yellow and green offspring, in the proportion of 3 yellows to *i* green.

The reason for this is clear when we consider the chromosomes on which the characters depend. If *A* represents the chromosome (autosome) that gives the dominant yellow programmes, while a represents the chromosome that gives the recessive green, the original parents were *AA* and *aa*, the *Fi* generation were all *Aa*, while the *Fa* generation showed the roportions *AA-f* ^{2} *Aa* + *aa*. The recessive individuals *aa* are homozygotes; teir two chromosomes are alike, being *aa*. When germ cells are produced by these homozygotic individuals, all these germ cells must contain the chromosome *a*, and when two of these unite in self-fertilization, of course they produce in all cases again individuals *aa*. These are, of course, recessives (green), like their parents; that is, the green parents aa 'breed true'.

Similarly the dominant yellows, *AA* (constituting one- third of all the dominants), are homozygotes. Their germ cells all contain the chromosome *A*, and the offspring produced by union of such germ cells, in self-fertilization, are all *AA*, and so doinant yellow like the parents. One-third of all the dominants thus breed true. But the individuals *Aa* are heterozygotic; that is, they have two kinds of chromosomes, *A* and *a*, in the pair.

Thus they produce two kinds of germ cells, *A* and *a*, in equal numbers. And when these germ cells unite, they necessarily give, as we saw on an earlier page, offspring of three kinds, in the proportions *AA* + 2*Aa*4-*aa* Of these, *AA* and 2*Aa* will be dominants (yellow), while *aa* will be recessive (green). Thus the homozygotes *aa*and *AA* breed true, while the heterozygotes, *Aa*, do not.

As Mendel discovered, if the heterozygotes *Aa* are mated with the recessive homozygotes *aa*, they give two kinds of offspring, *Aa* and *aa*; half (*Aa*) being dominant,the other half (*aa*) being recessive. If the heterozygotes *Aa* are mated with the dominant homozygotes *AA*, the offspring *AA* and *Aa* are all dominant, though half (*AA*) are homozyotes, the other half (*Aa*) heterozygotes.

All these results are typical for the descendants of two parents different in a single pair of chromosomes, one pair being dominant, the other recessive. For reference it will be well to make a table of these results, characteristic for Mendelian inheritance.

The Results of Mendeuan Inheritance

Parents: *AA* and *aa*; both homozygotic. *Fi*: *Aa*; all alike, dominant heterozygotes. *Fs*: *AA* + 2*Aa* + *aa*. Three-fourths of the individuals dominant, onefourth recessive. Half the individuals homozygotes (*AA* and *aa*), the other half heterozygotes (*aAa*). Each type of homozygote bred by itself produces offspring like itself. The heterozygotes *Aa*, when bred by themselves, produce again offspring. *AA* + 2*Aa* + *aa Aa* mated with aa yields again *Aa* + *aa Aa* mated with *AA* yields again *Aa* + *AA*. All of these results are found to hold true for thousands of different characteristics.

We may therefore summarize the situation in the following general propositions:

- When two different individuals are crossed, in many cases the differences between them follow the same method of distribution as do the two members of a pair of autosomes ýÿone being dominant, the other recessive.

- This distribution is what is known as typical Mendelian heredity.
- The two sexes do not differ in these respects. Each sex has the same proportion of dominants and recessives.
- The rules of Mendelian inheritance are essentially the rules of distribution of the members of pairs of autosomes.

Characteristics that depend on two or more pairs of autosomes.We have thus far followed the distribution of characteristics that depend on differences in a single pair of autosomes. But most organisms have several or many pairs of autosomes: the fruit-fly has three pairs, man has twenty- three

Do some characteristics depend on differences in one pair of autosomes, others on differences in other pairs? Have the different pairs of autosomes thus different functions in development and inheritance? We find that the different pairs of autosomes have indeed different functions, and that some characteristics depend on one pair, some on another. This is shown in many cases i which two parents differ in two diverse characteristics. One characteristic may then act as if dependent on one pair of autosomes, the other on another pair. As an example of this, we may take a case studied by Mendel in peas.

One of the original pea plants had seeds that were (i) yellow, and (2) round, or smooth. These, as it turns out, are both dominant characters. The other plants, crossed with those just mentioned, had recessive characters; the seeds were (i) green, and (2) wrinkled.

When these two kinds of pea plants were mated together, it turned out, as we shall see, that yellow and green behave in inheritances as if due to a difference in the two parents in the effects of a certain pair of autosomes; that is, they are alleles. Round and wrinkledbehave as if due to a difference between the two parents in the effects of another pair of autosomes. We may therefore represent the two pairs of autosomes. Representing the dominant characters by capitals, the recessives by lower-case letters, we may in the diagram represent the chromosomes giving origin to the four characters as follows:

A = yellow a = green

B = round b = wrinkled

The germ cells from these parents contain, of course, one member of each pair, as shown in figure One parent produces gametes all of the constitution *AB*, the other gametes all of the constitution *ab*. When a gamete *AB* from one parent unites with a gamete *ab* from the other, all the offspring (*Fi*) thus produced have the constitution *AaBa*, as shown at *Fi* in Figure:. Since the dominant autosomes *A* and *B* are both present, all these individuals are domiant for both pairs of characters; that is, they all have seeds that are yellow and round, as Mendel found. But all are heterozygotes-for both of these characters. Next we mate together two of these *Fi* individuals to give the F^2 generation. In forming gametes the rule is, of course, that each gamete contains one member of each pair of chromosomes. But the two pairs, *Aa* and *Bb*, are distributed to the gametes quite independently.

Thus, some get *A* and *B*, some *A* and *b*, some a and *B*, some *a* and *b*. There are thus from each parent four types of gametes, *AB, Ab, aB*, and *ab*, each type being represented by many gametes. The gametes of each of the four types from one parent unite in approximately equal proportions with gametes of each of the four types from the other parent. That is, gametes of the type *AB* unite in different cases with *AB, Ab, aB* or *ab* from the other parent; and so of each of the other gametes of the first parent. The result is to give a set of 16 combinations of gametes (some of which are alike, however, as will be seen). It is not necessary to carry further in our diagram the outlines of the chromosomes.

Inheritance Depending on Two Pairs of Autosomes

AB AB AB AB

AB(i) *Ab*(6) *aB*(7) *ab*(s)

Ab Ab Ab Ab

AB(6) *Ab*(2) *aB*(5) *ab*(8)

aB aB aB aB

AB(7) *Ab*(5) *aB*(3) *ab*(9)

ab ab ab ab

AB(5) *Ab*(8) *aB*(9) *ab*(4)

But, as will be observed, not all these 16 are diverse; thus *ABAb* is the same in constitution or genotype as *AbAB*. If we number the different combinations, and give the same number to those that are alike, as is done (in parentheses) we find that there are nine different combinations or genotypes. That is, two parents that are both *ABab* produce offspring of nine different constitutions or genotypes.

When the parents differed in but one pair of characters there were but 3 different combinations. When there are two pairs the number is 3 > or 9.

These nine different combinations will manifest the dominant characters *A* or *B*, in all cases where an *A* or *B* is present whether present once, as *A*, or twice, as *AA*); they will manifest the recessive characters *a* or *b* in all cases in which *a* or *b* is present twice, no corresponding capital letter being present. Collecting together these different types, we find the different characteristics to be manifested in the following proportions of the zygotes:

9*AB* + 3*Ab* + *saB* + 1*ab*

That is, 9 out of 16 are dominant for both pairs of characters, 3 are dominant for *A*, recessive for *b*; 3 dominant for *B*, recessive for *a*, and *i* out of 16 is recessive for both. There are thus 4 different phaenotypes.

In Mendel's experiment with peas, in which the two pairs of characters were:

1. Yellow and green;
2. Round and wrinkled; in *Fi* the proportions were 9 that were yellow and round,
3. That were yellow and wrinkled,
4. That were green and round, and *i* that was green and wrinkled.

These proportions hold generally when there are two pairs of characters connected with different pairs of autosomes. It may be observed that the proportions of the individuals of different constitution or genotype in F^2, descended from a cros between parents that differ in respect to two pairs of autosomal characters (so that the parents are *AABB* and *aabb*), can be found directly from the proportions for a single pair of characters. In *Fa*, for the single pairs the proportions are *AA* + 2*Aa-haA* and *BB* + 2*Bb* + *bb* respectively. If now we multiply together algebraically these two polynomials, we obtain the relative proportions for the different combinations produced when the two pairs are together.

These proportions are the following:

AABB + 2*AaBB* + *aaBB* + 2*AABb* + 4*AaBb* +

(1) (7) (3) (6) (5)

2*aaBb* + *AAbb* + 2*Aabb* + *aabb*

(9) (2) (8) (4),

Certain other relations among the nine different types are important, since they hold for all cases of two pairs of characters that are dependent upon different pairs of autosomes.

These are the following:

- There are four combinations (Nos. 1, 2, 3, 4), constituting one-fourth of all the zygotes, that are homozygotes for both pairs of gens. The four homozygotes are diverse in their combinations and in the characters they manifest. Any one of these when self-fertilized or mated with another like itself breeds true: that is, it produces offspring that have the same constitution as the parents.
- Four out of the 16, or one-fourth of all the zygotes, are heterozygotes for both pairs of characters. They form the class numbered (5), and constitute the diagonal from upper right to lower left. Any of these when

self-fertilized or mated with another like itself gives again the nine different types, in the proportions above set forth.

- The one-half of all the zygotes, are homozygotes for one of the two pairs, heterozygotes for the other. Four are homozygotic for *A* and heterozygotic for *B*; the other four are heterzygotic for *A* and homozygotic for *B*. In these 8 there are but 4 diverse combinations or genotypes.

If any one of these is self-fertilized, it yields 3 diverse types with respect to the pair for which it is heterozygotic. Thus, the class (6), having the constitution AABb, yields when self- fertilized three combinations in the following proportions:

$$AABB + 2AABb + AAbb$$

The above relations hold for all cases in which two pairs of characteristics are dependent on differences in two different pairs of autosomes, whatever the method of designating the characters or chromosomes. In a similar way, two parents may differ in three characters that are dependent on three different pairs of autosomes. Such parents could be represented as *AABBCC* and *aabbcc*.

Their progeny (*Fi* generation) would all have the constitution *AaBbCc*. Such *Fi* individuals of course produce a variety of germ cells, in accordance with the principles that each germ cell receives one member of each pair of autosomes, and that the different pairs of autosomes are distributed independently. Each parent *AaBbCc* thus yields the following 8 types of germ cells:

ABC, ABc, AbC, aBC, Abe, aBc, abC, abc.

When each of the 8 kinds from one parent unites with each of the 8 kinds from the other, there are produced, of course, 64 groups. If we classify the 64 individuals by the characters which they manifest, we find that there are 8 different phae- notypes manifesting respectively the following characteristics:

ABC, ABc, AbC, aBC, Abc, aBc, abC, and *abc*

That is, some are dominant for all three characters; some for a particular two and recessive for the other; some are dominant for a particular one and recessive for the other two; and some are recessive for all three. The different phaeno- types are present in different proportions; the proportions when a large number are examined are as follows:

2ABC + gABc + gAbC + gaBC + $Abc + saBc + 3abC + iabc

The total number of different combinations (genotypes) that occur in such a case is. These combinations can be obtained in their correct proportions by multiplying together algebraically the expressions: Similar but still more complex series are obtained if parents are interbred that differ in four or more characters depending on four or more different pairs of autosomes. If the number of pairs of autosomes is called «, the following relations hold:

If parents differ in n autosomes:

- The number of diverse types of gametes produced by the individuals of the *Fi* generation is 2^n.
- The number of different combinations (genotypes) in the zygotes of F^2, formed by mating the 2^n gametes from *Fi*, is s^w.
- The number of different phaenotypes produced in *Fa* is 2^n.

These relations yield very great numbers of diverse combinations, in cases where the number of pairs of autosomes is large. Thus in man, with 23 pairs of autosomes, if all the autosomes differed in the two parents, the number of types of germ cells produced by the *Fi* individuals would be 2, or many millions, the number of possible diverse genotypes in F^2 would be 3, or many billions, and the number of possible different phaenotypes would be 2. (As will appear later, the number of different possible combinations is greatly increased by the exchange of parts by the members of a pair of chromosomes.)

Thus typical Mendelian inheritance, dependent on differences between different pairs of autosomes, can be treated to a certain extent mathematically. The individuals in F^2 and later generations occur in proportions represented by simple mathematical ratios.

It will be worth while to formulate the main facts that we have brought out as to autosomal inheritance:

- Some single pairs of characters, dominant and recessive, follow in inheritance the method of distribution of a single pair of autosomes.
- Some pairs of inherited characters follow the rules of distribution of two or more different pairs of autosomes, in the following particulars:
  - One member of each pair goes into each germ cell.
  - The members of the different pairs are distributed to the germ cells independently, so that all possible combinations containing one member of each pair are formed with equal frequency.
  - Each type of gamete from one parent unites with each type from he other parent, the different matings being equally frequent.
- These methods of distribution result, for the characters manifested in the Fs generation, in certain typical numerical proportions or ratios, known as Mendelian ratios.
- This method of inheritance is known as Mendelism, or Mendelian inheritance. The rules of Mendelism are the result of the rules of distribution of pairs of autosomes.
- Most characteristics of organisms are inherited in this way; many more than are inherited in the sex-linked manner. This is evidently because

there is but a single pair of X-chromosomes, while there are several or many pairs of autosomes.

- Thus it is clear that, among the different individuals of a species, there are many different types of autosomes (as there are different types of X-chromosomes), producing many kinds of characteristics.

The T-Chromosome

There remains another chromosome, not included in the autosomes: namely, the Y-chromosome. Has this functions similar to those of X and of the autosomes? Does it too play a role in heredity? As we shall see, the Y-chromosome differs in many ways from the other chromosomes. Let us first summarize the main facts as to the occurrence and distribution of Y-chromosomes.

- Some organisms have no Y-chromosomes. Among those are many insects the dog, the cat, the horse.
- In some organisms it is small and appears to be degenerate. This is the case in many insects; also in man.
- In some organisms the Y-chromosome is large, but of a different form from its mate X. Such is the case in Drosophila
- In some organisms Y is not distinguishable from X inform or size, but experiment shows that it differs from X in function.
- The Y-chromosome descends from father to son only, exclusively in the male line. Normally it never occurs in females.
- Hence any characters that depend on the presence of a particular kind of Y will *(a)* be found in males only, and will *(b)* never be inherited *through* females (since Y does not pass through females).

The T-Chromosome in Drosophila

The functions of the Y- chromosome have been more thoroughly studied in Drosophila than in any other organism. Most of the facts as to this have already been brought out, in connection with our account of the functions of the X-chromosome.

We may summarize as follows the known facts as to the functions of the Y-chromosome in Drosophila:

- The Y-chromosome, although it normally occurs in males, is not required for the production of male individuals. Through non-disjunction, male individuals are produced that contain no Y.
- But the presence of Y is necessary for the fertility of males.
- Y may exist in females, as a result of accidental nondisjunction; it does not affect teir sex.

- During many years of intensive study no characteristics were found that follow the Y-chromosome.
- But recently it has been shown that the presence of Y prevents the appearance of a certain recessive character ('bobbed') that is due to a modified X-chromosome. This modified X results in the production of shortened bristles on the body of the fly. But this peculiarity does not occur if a Y is present, so that Y has the dominant normal allele for 'bobbed'. Indications of certain other ill-defined functions of the Y-chromosome have of late come to view.
- No diverse modifications or defects such as are known in great number for the X-chromosome are known in the Y-chromosome.
- Thus in Drosophila the Y-chromosome appears to have little function; it seems nearly but not quite inert.

Active T-chromosomes in certain Organisms; Relation to Male Secondary Sex Characters. In certain organisms the Y-chromosome is more active and varied in its effects. In such cases it shows in different individuals diverse modifications, comparable to the diverse types of the X-chromosome in Drosophila, and these produce diverse inherited characteristics that appear in males only. This matter is closely bound up with the question of what produces the diversity of secondary sex characters. In many animals there are special characteristics which, like the Y- chromosome, are found only in males. These are the male secondary sex characters: horns, mane, beard, greater size, diversities in form, and the like. Since these appear only in themale, the question arises as to whether they may be due to the presence of the Y-chromosome.

With relation to this the following is the situation:

- In most cases, the presence of male secondary sex characters does ot depend on the presence of the Y-chromosomes.
  This is shown by the following:
  - There are animals which have no Y-chromosome, yet have male secondary sex characters: for example, the dog, the cat, the horse; also various insects.
  - In Drosophila, males that contain no Y (as a result of non-disjunction) have the typical male secodary sex characters.
  - In some breeds of sheep the male has horns, while the female has none. t could be suggested that this is due to he fact that the male carries the Y-chromosome, while the female does not. But in other breeds, as Dorsets, both sexes have horns, although only the male has the T-chromosome. And in still other breeds, as Suffolks, neither sex has horns, although here as elsewhere the males have the Y-chromosome.

- If a female of the horned breed (Dorset) is crossed with a male of the hornless breed (Suffolk), in the Fi descendants the males have horns, although they receive their Y-chromosome from the hornless breed. They have inherited the horns through the mother.
- Thus it is clear that in these cases the presence of horns is not due to the presence of the Y-chromosome. Extensive work on the inheritance of horns and horn- lessness in sheep, goats and cattle consistently gives results which show that the presence of horns is not due to the presence of the Y-chromosome, or of any special type of Y-chromosome. In all these cases (a), (b) and (c) the presence of distinctive male secondary sex characters depends in some way on the fact that the male carries but one X-chromosome, and consequently produces the male hormone. Their absence from the female results from the fact that she carries two X-chromosomes, and consequently produces the female hormone.

- But in certain fish, some of the male secondary sex characters apparently do depend on the presence of a Y of a particular type, for they occur only where that type of Y is present, and they can therefore not be inherited through the mother. That is, the Y-chromosomes of certain individuals or breeds are so modified as to produce certain definite characteristics in the individuals that contain them, while individuals that contain another type of Y do not show these characteristics.

An example of this is found in the small fish Lebistes. In a certain race, the males have a black spot at the base of the dorsal fin, while the females are without the spot.

In another race of the same species neither the males nor the females have the black spot.

Thus the females of the two races are alike in lacking the black spot, while the males differ. Is this difference the result of differences in the Y-chromosomes of the two kinds of males? This may be tested by crossing members of the two races. In the sptted race, represent the X and Y-chromosome by capital letters; while in the non-spotted race, represent them by the small letters x and y.

A female of the spotted race, XX, is mated with a male of the non-spotted race, xy, thus:

- Females Males
- Parents XX *xy*
- Non-spotted
- *Fi Xx Xy*
- Non-spotted

The Y-Chromosomes

This produces in Fi females Xx, and males Xy. Neither contains the Y from the spotted race, and neither is spotted. The spot is not inherited through the females of the spotted race.

Next mate the female of the non-spotted race, xx, with a male from the spotted race, XY, as follows:

- Females Males
- Parents xx XY
- Spotted
- Fi xX xY
- Spotted

This yields in Fi females Xx, and males xY. These males carry the Y from the spotted race, and they are spotted. Thus far the spot goes with the Y from the spotted race. This Y may be traced further, for example, by breeding together the non-spotted females xx with the Fi males xY, from the preceding cross, thus:

- Females Males
- Parents x xY
- Spotted
- F2 xx xY
- Spotted

This gives in the Fs females xx, and males xY. These males are again spotted; they carry the Y from the spotted race. Wherever this Y from the spotted race is found, the black spot appears; wherever the Y is from the non-spotted race it does not appear. It appears clear therefore that whether the black spot appears in the male depends upon the type of Y- chromosome that it carries.

There are two types of Y-chromo- some in the species, one producing the black spot, the other not producing it. It has been shown by Winge that there are many other programmes markings in this fish that depend for their occurrence on the presence of a particular type of Y-chromosome. They occur only in the males, and only in males that have a certain type of Y. In certain other fish also the Y-chromosomes have been shown to be diverse in different individuals, giving rise to distinctive characteristics in the male.

In a certain plant, Melandrium, also, there are inherited characteristics that depend upon the presence of a particular type of Y-chromosome. There are some indications also that in man the nature of the Y-chromosome affects certain characteristics. In a family described by Schofield, the male parent in the first observed generation had webbed toes. This was inherited in the male lines only by all the males for four successive generations. It thus followed exactly the distribution of the Y-chromosome from the original male parent. It seems probable therefore that it is due to a defect or modification in the Y-chromosome of that individual. Many other cases of webbed toes or fingers are knon that occur in both sexes, so that they are clearly not due to the Y-chromosome.

On the whole, the situation as to the Y-chromosome and its relation to inheritance may be summarized as follows:

- In most organisms Y is a degenerate chromosome, lacking most of the functions manifested by the other chromosomes.
- In ome organisms degeneration has gone so far that the Y-chromosome has completely disappeared.
- In other cases it exists, but is very small.
- But in some species the Y-chromosome has effects on hereditary characters that are similar to those produced by other chromosomes. Different types of inherited characters result in such cases from diversely modified Y-chromosomes.
- Characteristics dependent on the presence of a particular type of Y-chromosome are inherited in the male line exclusively.

Three Main Methods of Distribution of Inherited Characteristics

In each of these the rules of inheritance are given by the special method of distribution of the three types of chromosomes, taken in connection with the facts of dominance and recessiveness. Inheritance which follows the distribution of the X-chromosomes is commonly called sex-linked inheritance. That follwing the distribution of autosomes is known as autosomal or typical Mendelian inheritance. Less important than those just named is Y-chromosome inheritance. All these methods are included under the term Mendelian inheritance, employed in a broad sense. In some organisms still other methods of inheritance exist. In higher animals and plants most inheritance occurs in one of the three methods named in the preceding paragraph.

Genes and Environment in Relation to Mental Characteristics

The relative rdles of heredity and environment are of special interest in man, particularly in relation to the characteristics that influence behaviour. For the study of these matters man is the most favourable organism that exists, although for most other relations in genetics he is a singularly unfavourable object of study. In man it is possible to make detailed studies of temperament, mentality, character, a group of characteristics hardly open to examination in other organisms. There occur in man 'identical' or one-egg twins, individuals having the same set of genes, the same genetic constitution throughout. And for comparison with these there occur fraternal twins: two individuals of the same parentage, the same age and living under the same environment, but developed from two separately formed eggs and thus having to some extent different sets of genes. By comparing these two kinds of twins with one another and with unrelated individuals, opportunity is presented for comparisons of the results of similarities and diversities in genetic constitutions with those of similarities and diversities in environmental influences.

As seen earlier the twins derived from a single egg are as a rule very closely alike in most physical respects; it is this that has caused them to be called identical twins. In some cases, however, the two differ in certain physical characteristics. There appear to be two categories of causes for these differences. First, there are physical differences that result from the way in which the separation of the twins occurred. Separation of the single egg into two does not occur in the first stages of development, as by division of the one-cell stage into two. On the contrary, it occurs much later, when many cells are present and the differentiation of bodily parts has begun, and it occurs at different ages in different cases.

There is reason to believe that when an equal division occurs at a rather early stage, before the right and left halves of the body have begun to show their characteristic diversity, the two twins produced are closely alike in their physical features. At a later period the right and left halves of the embryo have begun to differ, in the way that results as a rule in the greater strength and development of the right hand and arm. At such a stage, the embryo may divide into right and left halves, with results as follows.

Of the two embryos so produced, one will have a right limb that is already somewhat advanced in development, while the left limb, formed at the plane of separation, must begin anew. The other embryo will be in the reverse situation; it will have a left limb already somewhat advanced in development, but the right limb, formed at the plane of separation, must begin anew.

In each case the more advanced limb retains its advantage and becomes the one that after birth is strongest and most used. The result necessarily is that one of the twins remains right-handed, while the other becomes left-handed. This situation is not uncommonly found in identical twins. Similar differences in other unsymmetrical parts may be produced in the same way: for example in the whorls of hair on the head. In such respects the two members of a pair often show 'mirror-imaging', presumably due in each case to the fact that the original embryo had already become a little unsymmetrical before division occurred.

From this same situation another condition often observed is believed to result. In the embryo before division, one side (usually the right) will be a little in advance of the other. After division, the twin derived from the right half continues to retain this advantage, so that it may be more vigourous than the one derived from the left half. Such differences in vigour are not rare in twins. Apparently they may be accompanied by a psychological difference; the more vigourous twin assumes the leadership in their lives together. This reacts further on temperament and character. How far psychological and temperamental differences so produced may go is uncertain.

It appears that in some cases division of an embryo may be unequal, so that for this reason also one twin may be more vigourous than the other, with resulting psychological differences. Diversities produced in the ways above described are different in origin and type from those commonly called environmental. Under the latter term are commonly grouped diversities resulting from the different conditions under which individuals have lived. Since the extent to which differences due to the time and method of division may go, and the kind of characteristics that they may affect are uncertain, it is often not possible to distinguish these certainly from diversities due to differences of environment after division has occurred. Belonging to one or the other class, there are sometimes marked differences in the phsyical characteristics of one-egg twins. Komai and Fukuoka have described two fifteen-year- old Japanese boys that are obviously one-egg twins; they show the usual close resemblance in form, features, colouring, and the like. But they now differ greatly in size.

One is taller by 14·8 centimetres, or about 6 inches, and his weight is greater by 10·4 kilograms, or about 23 pounds. The two were at first of the same size, the difference gradually arising during growth. The smaller of the twins is affected by diabetes insipidus; this is probably connected with his decreased growth. The larger twin is right-handed, the smaller left- handed. Whether the difference in health and growth is in some way connected with what occurred at division of the egg is uncertain. A comparable case is described by Siemens. Two sisters, clearly one-egg twins, were as usual nearly identical in most of their physical characteristics till the age of 10. At this time one of them became affected wth a severe lateral curvature of the spine, and from this time their development was very different; at 16 the healthy twin was 4-8 inches taller than the other. The origin of the defect that produced curvature of the spine and changed development is unknown. Identical twins that have lived together sometimes show considerable psychological differences. Newman has made extensive psychological tests on fifty pairs of identical or one- egg twins, and, for comparison with these, on fifty pairs of fraternal twins. The results of these examinations have not yet been published in full, but certain data from them have been published.

By the Stanford-Binet tests, the intelligence quotients (IQJ and the 'mental age' were determined for the two sets. The two classes of twins are best compared by the differences between the scores made by the two members of each pair. Thus the average difference between fraternal twins, having somewhat diverse genetic constitutions, is nearly twice as great as that between identical twins, in which the genetic constitutions are alike.

But it is important that the identical twins that have lived together also showed differences between members of the pairs. In different pairs there was much variation in the amount of diversity. Of the 50 pairs there were five in which the two members gave identical scores in the intelligence tests. At the other end of the scale were five pairs that showed respectively the following differences (points in the Stanford-Binet scale) between the two members: 12-6, 12-9, 13-0, 13-9, 16-0.

All these five differences are considerably greater than the average difference between the genetically unlike fraternal twins. It is clear that in some cases even identical twins living together differ much in mentality. Whether the differences are the result of original diversities consequent on the method of division of the egg, or have been produced in some way by different experiences of the two, is not known. For comparisons as to the relative effects of genetic and environmental diversities on mental characteristics, valuable data have been obtained from the examination of identical twins that have lived apart under different environments. The first case of this kind was carefully studied by Muller, and nine pairs have since been fully examined by Newman.

In each of these ten cases the twins were separated in infancy, being adopted into different families; they have then lived to adult life under diverse conditions and influences. In addition to careful observations as to character and temperament, they were subjected to psychological and temperamental tests by the best standardized methods available.

The type of results reached by these studies will best be appreciated by first looking at the detailed resemblances and differences in some particular cases. In the first such study, by Muller, there were twin sisters that had been separated when two weeks old, and that did not see each other till they reached the age of 18; from that time until the age of 30 they lived apart more than nine-tenths of the time. Physically they showed the extreme similarity in characteristics that is usual in identical twins. Both 'have always been intellectually active', 'both have been extremely energetic, capable and popular, and they have been prominent in all sorts of club work in their respective communities' (Muller)'.

Both have had two or three attacks of tuberculosis, almost simultaneously'. The usual intelligence tests gave results very closely alike for the two twins.

But 'the nonintellectual testsýÿof motor reaction time, association time, "will temperament", emotions and social attitudesýÿgave results in striking contrast with those of the intelligence tests, in that the twins gave markedly different scores in all these tests'.

The differences were on the average greater than those between two individuals taken at random, and seemed 'to be correlated with salient differences in their past experiences and habits of life'. Thus this first study of such a case indicated that the different environments and experiences of the two individuals had produced a large effect on temperament, emotions, and social attitudes, but had had little effect on such matters as are tested by intelligence tests.

An illuminating contrast with these results is given by the first pair studied by Newman. The twin sisters ('O' and 'A') were born in London and were separated at the age of 18 months. One lived in Ontario, Canada, the other in London. Their environments were very different. When they were tested, this pair gave differences in those tests in which Muller's pair gave similarities, and gave similarities in those tests in which Muller's pair gave differences. Newman says: 'The twins dealt with in this paper are very different in mental capacity. But they showed great similarity in their manifestations of will and temperament, and in their emotional reactions. So this pair of twins shows that differences in the experiences undergone affect deeply the individual's mental traits: his performance in matters brought out by intelligence tests. Important points were brought out in the study of Newman's second pair, twin sisters. The two had received very different educations; one had attended school seven years longer than the other. Newman summarizes as follows the results of his study of these:

'These twins, remarkably similar after being separated at 18 months of age and unknown to each other for 19 years, have been profoundly modified by the very different educational careers.In every test of mental capacity, whether of so-called native ability or of achievement, 'G', the more highly educated twin, has distinctly the superior mind. Obviously mental training improves the ability of an individual to score well in any sort of test'. But further, 'in contrast with the great difference in mental power stands the fact that in all the tests of emotional traits and of temperament the twins gave the impression of being remarkably and unusually similar'. Here came out the effect of their identity of genes.

Newman's third pair revealed certain other facts of importance. These were two young men ('G' and 'O'), separated at two months of age. One had lived mainly in the city while the other had lived in the country. They were examined after reaching the age of 23. In native ability they seem to be nearly identical. The one outstanding difference is in their general personalities. 'G' (who had lived in the city) impresses one as more dignified, more reserved, more self-contained, more unafraid, more experienced, and less friendly. He seldom smiles, has a more serious expression about the brows, eyes and mouth. He stands more erectly with chin held in and brows drawn down somewhat over his eyes. 'O' (the other twin) is the opposite in all these respects. He is the more typical country boy, laughs readily, and is not on his dignity at all. Newman asserts emphatically that the personalities of the boys were utterly different. Thus through the study of these first four pairs examined, of identical twins that have lived apart, it became clear that such twins may differ in many ways. It appeared that the different environments and experiences of the individuals may have a large effect on mental and temperamental characteristics. The effects were different in ctifferent cases: in some cases the twins are alike in intelligence but differ temperamentally; in other cases they are alike in temperament and emotions, but differ in mentality. In other cases the marked difference is best expressed rather vaguely as a diversity in 'personality'.

The general impression given-by study of these first four cases has been confirmed by Newman's study of six other pairs of identical twins that have lived apart. The nine cases examined by Newman were treated by uniform methods, so that they may be compared in such a way as to bring out general relations. The mean difference in intelligence quotient for these nine pairs of identical twins that have lived apart is 8'6. This is considerably greater than the mean for identical twins that have lived together, which was 5-3. So far as it goes it indicates that difference of environment has a considerable effect in increasing the mental difference between twins.

In cases I, II, IV and VIII, the difference between the twins was very marked, but is similar to the differences between the five extreme cases of identical twins that had lived together. This emphasizes the fact that it is not possible to be certain what differences are due to something that happened at the division of the egg, and what are due to later environmental differences. The mean difference between identical twins that have lived apart is nearly the same as that for fraternal twins that have lived together.

In addition to the studies of intelligence, Muller and Newman made extensive examinations designed to test other features of personality. One set was designed to test will and temperament, to distinguish the mentally quick and slow, the deliberate and careful as compared with the hasty and careless, the aggressive and forceful or the opposite, and the like. Another type of examination is designed to test emotional peculiarities: likes and dislikes, feelings of right and wrong, and so on.

The results of such tests are not readily expressible by a numerical score, such as is employed for the intelligence tests. It will be worth while, however, to attempt a tabulation in general terms of the results of the tests in the three categories for the 10 cases of identical twins that have lived apart. Thus the general picture is that identical twins that have lived apart show many different combinations of likeness and unlikeness with respect to intelligence, temperament and emotional characteristics. There are cases in which the two are closely alike in all three categories cases in which the twins are distinctly diverse in all three categories and cases in which they are alike in two categories and diverse in the third, or alike in one category and diverse in the other two.

On the whole, it is clear that other things, beside genetic constitution, play important roles in determining mentality, temperament and emotional traits. The principal difficulty in interpreting the results lies in the uncertainty as to how much is the result of the manner in which division of the egg occurred, and how much is due to later environmental differences. Both certainly play a role. For these types of characteristics it appears probable that, of the two, the later environmental differences play the greater role. It is further clear that genetic resemblances and differences play a very large role in mentality, temperament and emotional traits.

It is to be noted that in all this study, the general environment of the twins was much alike in all cases. None of them lived in different civilizations, or at different epochs in cultural development. How great a diversity in the characteristics studied could be made by such differences remains uncertain. It is further to be noted that the tests to which the twins were subjected were designed to bring out rather permanent personal traits, as distinguished from mental content and habitudes. The behaviour of human beings depends very largely upon acquired mental content, upon knowledge, and upon habitudes and these things are in a high degree dependent on the conditions under which the individuals live and the experiences to which they are subjected.

In general, it is clear that in a human population there are great numbers of different types of individuals that are diverse because their genetic constitutions are diverse. In consequence of this they differ in their capabilities, in their tastes, in their tendencies towards any particular line of action. But all these classes, so far as they do not fall in the seriously defective groups, have marked powers of adjusting themselves to many different conditions and of following many different lines of action, depending on the conditions in which they live. In no other organism than man is so full an analysis possible of the relative roles of genetic constitution and environment in determining the mental and temperamental characteristics, for in no other organism are these characteristics highly developed.

6

Functional Genomics

Functional genomics is a field of molecular biology that attempts to make use of the vast wealth of data produced by genomic projects (such as genome sequencing projects) to describe gene (and protein) functions and interactions. Unlike genomics and proteomics, functional genomics focuses on the dynamic aspects such as gene transcription, translation, and protein-protein interactions.

Goals of Functional Genomics

The goal of functional genomics is to understand the relationship between an organism's genome and its phenotype. The term functional genomics is often used broadly to refer to the many possible approaches to understanding the properties and function of the entirety of an organism's genes and gene products. This definition is somewhat variable; Gibson and Muse define it as "approaches under development to ascertain the biochemical, cellular, and/or physiological properties of each and every gene product", while Pevsner includes the study of nongenic elements in his definition: "the genome-wide study of the function of DNA (including genes and nongenic elements), as well as the nucleic acid and protein products encoded by DNA".

Because of its genome-wide approach, functional genomics requires the use of high-throughput technologies capable of assaying many functions or relationships simultaneously. Functional genomics involves studies of natural variation in genes, RNA, and proteins over time (such as an organism's development) or space (such as its body regions), as well as studies of natural or experimental functional disruptions affecting genes, chromosomes, RNAs, or proteins.

The promise of functional genomics is to expand and synthesize genomic and proteomic knowledge into an understanding of the dynamic properties of an organism at cellular and/or organismal levels. This would provide a more complete picture of how biological function arises from the information encoded in an organism's genome. The possibility of understanding how a particular mutation leads to a given phenotype has important implications for human genetic diseases, as answering these questions could point scientists in the direction of a treatment or cure.

Techniques and Applications

Functional genomics includes function-related aspects of the genome itself such as mutation and polymorphism (such as SNP) analysis, as well as measurement of molecular activities. The latter comprise a number of "-omics" such as transcriptomics (gene expression), proteomics (protein expression), phosphoproteomics (a subset of proteomics) and metabolomics.

Functional genomics uses mostly multiplex techniques to measure the abundance of many or all gene products such as mRNAs or proteins within a biological sample. Together these measurement modalities endeavor to quantitate the various biological processes and improve our understanding of gene and protein functions and interactions.

At the DNA Level

Genetic Interaction Mapping

Systematic pairwise deletion of genes or inhibition of gene expression can be used to identify genes with related function, even if they do not interact physically. Epistasis refers to the fact that effects for two different gene knockouts may not be additive; that is, the phenotype that results when two genes are inhibited may be different from the sum of the effects of single knockouts.

The ENCODE Project

The ENCODE (Encyclopedia of DNA elements) project is an in-depth analysis of the human genome whose goal is to identify all the functional elements of genomic DNA, in both coding and noncoding regions. To this point, only the pilot phase of the study has been completed, involving hundreds of assays performed on 44 regions of known or unknown function comprising 1% of the human genome.

Important results include evidence from genomic tiling arrays that most nucleotides are transcribed as coding transcripts, noncoding RNAs, or random transcripts, the discovery of additional transcriptional regulatory sites, further elucidation of chromatin-modifying mechanisms.

At the RNA Level: Transcriptome Profiling

Microarrays

Microarrays measure the amount of mRNA in a sample that corresponds to a given gene or probe DNA sequence. Probe sequences are immobilized on a solid surface and allowed to hybridize with fluorescently-labelled "target" mRNA. The intensity of fluorescence of a spot is proportional to the amount of target sequence that has hybridized to that spot, and therefore to the abundance of that mRNA sequence in the sample. Microarrays allow for identification of candidate genes involved in a given process based on variation between transcript levels for different conditions and shared expression patterns with genes of known function.

SAGE

SAGE (Serial analysis of gene expression) is an alternate method of gene expression analysis based on RNA sequencing rather than hybridization. SAGE relies on the sequencing of 10-17 base pair tags which are unique to each gene. These tags are produced from poly-A mRNA and ligated end-to-end before sequencing. SAGE gives an unbiased measurement of the number of transcripts per cell, since it does not depend on prior knowledge of what transcripts to study (as microarrays do).

At the Protein Level: Protein-protein Interactions

Yeast two-hybrid System

A yeast two-hybrid (Y2H) screen tests a "bait" protein against many potential interacting proteins ("prey") to identify physical protein-protein interactions. This system is based on a transcription factor, originally GAL4, whose separate DNA-binding and transcription activation domains are both required in order for the protein to cause transcription of a reporter gene. In a Y2H screen, the "bait" protein is fused to the binding domain of GAL4, and a library of potential "prey" (interacting) proteins is recombinantly expressed in a vector with the activation domain. In vivo interaction of bait and prey proteins in a yeast cell brings the activation and binding domains of GAL4 close enough together to result in expression of a reporter gene.

It is also possible to systematically test a library of bait proteins against a library of prey proteins to identify all possible interactions in a cell.

AP/MS

Affinity purification and mass spectrometry (AP/MS) is able to identify proteins that interact with one another in complexes. Complexes of proteins are allowed to form around a particular "bait" protein. The bait protein is identified using an antibody or a recombinant tag which allows it to be extracted along with any proteins that have formed a complex with it. The proteins are then digested into short peptide fragments and mass spectrometry is used to identify the proteins based on the mass-to-charge ratios of those fragments.

Loss-of-function Techniques

Mutagenesis

Gene function can be investigated by systematically "knocking out" genes one by one. This is done by either deletion or disruption of function (such as by insertional mutagenesis) and the resulting organisms are screened for phenotypes that provide clues to the function of the disrupted gene.

RNAi

RNA interference (RNAi) methods can be used to transiently silence or knock down gene expression using ~20 base-pair double-stranded RNA typically delivered by transfection of synthetic ~20-mer short-interfering RNA molecules (siRNAs) or by virally-encoded short-hairpin RNAs (shRNAs). RNAi screens, typically performed in cell culture-based assays or experimental organisms (such as C. elegans) can be used to systematically disrupt nearly every gene in a genome or subsets of genes (sub-genomes); possible functions of disrupted genes can be assigned based on observed phenotypes.

Functional Annotations for Genes

Genome Annotation

Putative genes can be identified by scanning a genome for regions likely to encode proteins, based on characteristics such as long open reading frames, transcriptional initiation sequences, and polyadenylation sites.

A sequence identified as a putative gene must be confirmed by further evidence, such as similarity to cDNA or EST sequences from the same organism, similarity of the predicted protein sequence to known proteins, association with promoter sequences, or evidence that mutating the sequence produces an observable phenotype.

Rosetta Stone Approach

The Rosetta stone approach is a computation method of de novo protein function prediction, based on the hypothesis that some proteins involved in a given physiological process may exist as two separate genes in one organism and as a single gene in another. Genomes are scanned for sequences that are independent in one organism and in a single open reading frame in another. If two genes have fused, it is predicted that they have similar biological functions that make such coregulation advantageous.

Functional Genomics and Bioinformatics

Because of the large quantity of data produced by these techniques and the desire to find biologically meaningful patterns, bioinformatics is crucial to analysis of functional genomics data. Examples of techniques in this class are data clustering or principal component analysis for unsupervised machine learning (class detection) as well as artificial neural networks or support vector machines for supervised machine learning (class prediction, classification).

Structural Genomics

Structural genomics seeks to describe the 3-dimensional structure of every protein encoded by a given genome. This genome-based approach allows for a high-throughput method of structure determination by a combination of experimental and modeling approaches.

The principle difference between structural genomics and traditional structural prediction is that structural genomics attempts to determine the structure of every protein encoded by the genome, rather than focusing on one particular protein. What is it that makes it possible to determine the structure of every protein in the genome at once rather than solve the structures one at a time? With full-genome sequences available, structure prediction can be done more quickly through a combination of experimental and modeling approaches, especially because the availability of large number of sequenced genomes and previously-solved protein structures allows scientists to model protein structure on the structures of previously solved homologs.

Because protein structure is closely linked with protein function, the structural genomics has the potential to inform knowledge of protein function. In addition to elucidating protein functions, structural genomics can be used to identify novel protein folds and potential targets for drug discovery. Structural genomics involves taking a large number of approaches to structure determination, including experimental methods using genomic sequences or modeling-based approaches based on sequence or structural homology to a protein of known structure or based on chemical and physical principles for a protein with no homology to any known structure.

As opposed to traditional structural biology, the determination of a protein structure through a structural genomics effort often (but not always) comes before anything is known regarding the protein function. This raises new challenges in structural bioinformatics, i.e. determining protein function from its 3D structure. Structural genomics emphasizes high throughput determination of protein structures. This is performed in dedicated centres of structural genomics. While most structural biologists pursue structures of individual proteins or protein groups, specialists in structural genomics pursue structures of proteins on a genome wide scale. This implies large scale cloning, expression and purification. One main advantage of this approach is economy of scale. On the other hand, the scientific value of some resultant structures is at times questioned. A *Science* article from January 2006 analyzes the structural genomics field.

One advantage of structural genomics, such as the Protein Structure Initiative, is that the scientific community gets immediate access to new structures, as well as to reagents such as clones and protein. A disadvantage is that many of these structures are of proteins of unknown function and do not have corresponding publications. This requires new ways of communicating this structural information to the broader research community. The Bioinformatics core of the Joint centre for structural genomics (JCSG) has recently developed a wiki-based approach namely The Open Protein Structure Annotation Network (TOPSAN) for annotating protein structures emerging from high-throughput structural genomics centres.

Goals

One goal of structural genomics is to identify novel protein folds. Experimental methods of protein structure determination require proteins that express and/or crystallize well, which may inherently bias the kinds of proteins folds that this experimental data elucidate. A genomic, modeling-based approach such as *ab initio* modeling may be better able to identify novel protein folds than the experimental approaches because they are not limited by experimental constraints.

Protein function depends on 3-D structure and these 3-D structures are more highly-conserved than sequences. Thus, the high-throughput structure determination methods of structural genomics have the potential to inform our understanding of protein functions. This also has potential implications for drug discovery and protein engineering. Furthermore, every protein that is added to the structural database increases the likelihood that the database will include homologous sequences of other unknown proteins. The Protein Structure Initiative (PSI) is a multifaceted effort funded by the National Institutes of Health with various academic and industrial partners that aims to increase knowledge of protein structure using a structural genomics approach and to improve structure-determination methodology.

Methods

Structural genomics takes advantage of completed genome sequences in several ways in order to determine protein structures. The gene sequence of the target protein can also be compared to a known sequence and structural information can then be inferred from the known protein's structure. Structural genomics can be used to predict novel protein folds based on other structural data. Structural genomics can also take modeling-based approach that relies on homology between the unknown protein and a solved protein structure.

De Novo Methods

Completed genome sequences allow every open reading frame (ORF), the part of a gene that is likely to contain the sequence for the mRNA and protein, to be cloned and expressed as protein. These proteins are then purified and crystallized, and then subjected to one of two types of structure determination: X-ray crystallography and [Nuclear Magnetic Resonance (NMR).

The whole genome sequence allows for the design of every primer required in order to amplify all of the ORFs, clone them into bacteria, and then express them. By using a whole-genome approach to this traditional method of protein structure determination, all of the proteins encoded by the genome can be expressed at once. This approach allows for the structural determination of every protein that is encoded by the genome.

Modelling-based Methods

Ab Initio Modeling

This approach uses protein sequence data and the chemical and physical interactions of the encoded amino acids to predict the 3-D structures of proteins with no homology to solved protein structures. One highly successful method for *ab initio* modeling is the Rosetta program, which divides the protein into short segments and arranges short polypeptide chain into a low-energy local conformation. Rosetta is available for commercial use and for non-commercial use through its public program, Robetta.

Sequence-based Modeling

This modeling technique compares the gene sequence of an unknown protein with sequences of proteins with known structures. Depending on the degree of similarity between the sequences, the structure of the known protein can be used as a model for solving the structure of the unknown protein. Highly accurate modeling is considered to require at least 50% amino acid sequence identity between the unknown protein and the solved structure. 30-50% sequence identity gives a model of intermediate-accuracy, and sequence identity below 30% gives low-accuracy models.

It has been predicted that at least 16,000 protein structures will need to be determined in order for all structural motifs to be represented at least once and thus allowing the structure of any unknown protein to be solved accurately through modeling. One disadvantage of this method, however, is that structure is more conserved than sequence and thus sequence-based modeling may not be the most accurate way to predict protein structures.

Threading

Threading bases structural modeling on fold similarities rather than sequence identity. This method may help identify distantly-related proteins and can be used to infer molecular functions.

Examples of Structural Genomics

There are currently a number of on-going efforts to solve the structures for every protein in a given proteome.

The Thermotogo Maritima Proteome

One current goal of the Joint Centre for Structural Genomics (JCSG), a part of the Protein Structure Initiative (PSI) is to solve the structures for all the proteins in *Thermotogo maritima*, a thermophillic bacterium. *T. maritima* was selected as a structural genomics target based on its relatively small genome consisting of 1,877 genes and the hypothesis that the proteins expressed by a thermophilic bacterium would be easier to crystallize.

Lesley *et al* used *Escherichia coli* to express all the open-reading frames (ORFs) of *T. martima*. These proteins were then crystallized and structures were determined for successfully-crystallized proteins using X-ray crystallography. Among other structures, this structural genomics approach allowed for the determination of the structure of the TM0449 protein, which was found to exhibit a novel fold as it did not share structural homology with any known protein.

The Mycobacterium Tuberculosis Proteome

The goal of the TB Structural Genomics Consortium is to determine the structures of potential drug targets in *Mycobacterium tuberculosis*, the bacterium that causes tuberculosis. The development of novel drug therapies against tuberculosis are particularly important given the growing problem of multi-drug-resistant tuberculosis.

The fully sequenced genome of *M. tuberculosis* has allowed scientists to clone many of these protein targets into expression vectors for purification and structure determination by X-ray crystallography.

Studies have identified a number of target proteins for structure determination, including extracellular proteins that may be involved in pathogenesis, iron-regulatory proteins, current drug targets, and proteins predicted to have novel folds. So far, structures have been determined for 708 of the proteins encoded by *M. tuberculosis*.

Protein Structure Databases and Classifications

Protein Data Bank (PDB): repository for protein sequence and structural information. UniProt: provides sequence and functional information. Structural

Classification of Proteins (SCOP Classifications): hierarchical-based approach Class, Architecture, Topology and Homologous superfamily (CATH): hierarchical-based approach.

Comparative Genomics

Comparative genomics is the study of the relationship of genome structure and function across different biological species or strains. Comparative genomics is an attempt to take advantage of the information provided by the signatures of selection to understand the function and evolutionary processes that act on genomes. While it is still a young field, it holds great promise to yield insights into many aspects of the evolution of modern species. The sheer amount of information contained in modern genomes (3.2 gigabases in the case of humans) necessitates that the methods of comparative genomics are automated. Gene finding is an important application of comparative genomics, as is discovery of new, non-coding functional elements of the genome.

Comparative genomics exploits both similarities and differences in the proteins, RNA, and regulatory regions of different organisms to infer how selection has acted upon these elements. Those elements that are responsible for similarities between different species should be conserved through time (stabilizing selection), while those elements responsible for differences among species should be divergent (positive selection). Finally, those elements that are unimportant to the evolutionary success of the organism will be unconserved (selection is neutral).

One of the important goals of the field is the identification of the mechanisms of eukaryotic genome evolution. It is however often complicated by the multiplicity of events that have taken place throughout the history of individual lineages, leaving only distorted and superimposed traces in the genome of each living organism. For this reason comparative genomics studies of small model organisms (for example yeast) are of great importance to advance our understanding of general mechanisms of evolution.

Having come a long way from its initial use of finding functional proteins, comparative genomics is now concentrating on finding regulatory regions and siRNA molecules. Recently, it has been discovered that distantly related species often share long conserved stretches of DNA that do not appear to code for any protein. One such ultra-conserved region, that was stable from chicken to chimp has undergone a sudden burst of change in the human lineage, and is found to be active in the developing brain of the human embryo.

Computational approaches to genome comparison have recently become a common research topic in computer science. A public collection of case studies and demonstrations is growing, ranging from whole genome comparisons to gene expression analysis. This has increased the introduction of different ideas, including concepts from systems and control, information theory, strings analysis and data mining. It is anticipated that computational approaches will become and remain a standard topic for research and teaching, while multiple courses will begin training students to be fluent in both topics.

Genomics for Health Improvement

Over the last 100 years, innovations in science and technology, together with improved socio-economic conditions, have resulted in improved health, quality of life and a vast increase in people's life expectancy worldwide. In light of this impressive record it is disheartening that the benefits of modern medicine still do not reach billions of people in the poorer parts of the world. Children and adults are undernourished, live in poor housing with few modern facilities, such as clean, running water, and die from various, often preventable, illnesses in the prime of their lives.

While there are many factors that come into play in the complex issue of development, it is widely accepted that science and technology plays a critical role in improving health in developing countries. Genomics and related biotechnologies constitute an exciting new field of science that possesses tremendous potential to address health problems in developing countries – if we rise to the challenge.

The purpose of this report is to focus on the role of genomics and related health biotechnologies as an example of the application of science, technology and innovation (STI) to improve global health and contribute towards meeting the United Nations Millennium Development Goals. The Millennium Development Goals (MDGs) grew out of the commitment of the United Nations to promote sustained development and eliminate poverty all over the world.

They help quantify progress by defining targets and indicators to measure them. The first seven MDGs are directed at specific objectives in promoting development and improving people's lives, while the eighth goal, 'a global partnership for development', focuses on the means to achieve them. Many of the goals address health issues directly, such as improving infant and maternal health and reducing the prevalence of infectious diseases. The MDGs have become the international standard for measuring and tracking improvements in development. Therefore, the MDGs have the advantage of (i) agreement and support with the leaders of all UN member states, (ii) offering a comprehensive and multi-dimensional development framework, and (iii) setting clear quantifiable goals for all countries to meet by 2015. Although the MDGs, especially the health-related ones, have their critics, they do represent a standard by which progress may be measured. The Millennium Project, whose mission is to devise and recommend operational frameworks to help developing countries achieve the MDGs, has set up ten thematically oriented task forces, consisting of experts and leading practitioners. These task forces have carried out detailed and comprehensive research in each thematic area on the science, public health needs and medical interventions, as well as the social, political and economic issues. The goals of the Science and Technology Task Force are to address how science and technology can be enhanced and applied to help all countries achieve the MDGs. Its mission is guided by the understanding that most of the MDGs cannot be reached without a strong contribution from science and technology.

The Potential of Genomics

Medicine and the biomedical sciences are currently in the midst of a fundamental change that is fuelled by the genomics revolution. In its report on "Genomics and World Health" the World Health Organization's Advisory Committee on Health Research observed that the application of knowledge gained from the characterization of the genomes of organisms holds enormous potential for the development of new health care innovations over the coming decades. With increased understanding of the molecular basis of diseases and their causation, the way we identify, prevent, diagnose, and treat diseases in the new millennium is bound to change.

The science of genomics is generating vast amounts of new knowledge. Once generated, this knowledge can be used creatively in the development of new diagnostic technologies, treatments and preventive programs.

Genomics requires massive amounts of information to be collected and analysed. The field has evolved over the last few decades in large part due to advances in sequencing technology, as well as information and communication technology. Automated DNA sequencing and genotyping have made it possible to characterize rapidly large numbers of genes. The means to manage, store and process enormous databases of biological information have spawned an entirely new field – bioinformatics.

These developments hold tremendous promise for the prevention, diagnosis and treatment of some of the major diseases affecting humankind, including HIV, malaria, and tuberculosis, as well as non-communicable diseases such as diabetes, cancer and cardiovascular disease. Genomics can also be applied to understand the genetics of bacteria, plants, and animals. Many genomes have been sequenced and mapped. In fact, the first genome sequence (that of *H. influenzae*) was completed in 1995, by researchers at The Institute for Genomic Research (TIGR). The sequence was a mere 1.8 million base-pairs, compared with the 3.2 billion base-pairs of the human genome. Many other genomes of the plant and animal kingdom have been sequenced, ranging from common yeast to rice, and all these sequencing achievements promise to enhance scientific knowledge. In October 2002, scientists published the sequence of the parasites responsible for human malaria, *Plasmodium falciparum*, as well as the mosquito that carries it, *Anopheles gambiae.* The knowledge of these genomes, together with the knowledge of the human genome, can be used to develop new drug and vaccine targets against malaria.

There may also be direct *economic* benefits of genomics. It is already a significant contributor to the biotechnology sector. Despite early setbacks, the genomics-based pharmaceutical market was expected to grow from US $2.2 billion in 1999 to US $8.2 billion in 2004. The major economic players in the world are not the only ones expected to benefit from genomics. Cuba, for example, has invested heavily since the 1980's in research infrastructure and manufacturing in biotechnology. As a result biotechnology is poised to become a major export industry in Cuba. In order to reap direct economic benefit from genomics, countries will have to be active participants in the development and manufacturing of genomics products. Those countries that will benefit the most from genomics are those that have appropriate health products to improve the health of their populations and who are active in developing and supplying those products. This will involve engaging and stimulating the private sector, without which the production and commercialization of new biotechnology products will be a challenge. As highlighted by the UN Commission on Private Sector and Development report, "*Unleashing Entrepreneurship: Making Business Work for the Poor*," the process of commercialization for development involves the dissemination and facilitation of knowledge flows between public and private sectors of both developed and developing markets. The report recommends action in both the public and private spheres, but also emphasizes the linkages between these spheres, recognizing the importance of cooperation and partnerships to achieve goals.

Molecular Diagnostics

- Combat HIV/AIDS, malaria and other diseases
- Reduce child mortality
- Improve maternal health.

Molecular diagnostics draw upon recent advances in biology to diagnose infectious disease by detecting the presence or absence of pathogen-associated molecules, such as DNA or protein, in a patient's blood or tissues. They present a powerful set of methods to address the health-related Millennium Development Goals, *Combat HIV/AIDS,* malaria and other diseases, Reduce child mortality, Improve *maternal health*. Infectious and parasitic diseases are responsible for nearly 40%, or 17 million of all deaths every year. The three major killers are HIV/AIDS, malaria and tuberculosis, which together claim at least 5 million lives a year. Each year an estimated 11 million children still die before reaching their fifth birthday.

While improving public health infrastructure for disease prevention is crucial to the achievement of this goal, it is also true that once disease strikes, diagnosis and treatment methods are essential. Rapid and accurate diagnosis not only increases chances of survival, but also avoids waste of resources on inappropriate treatments and helps contain disease. It is not surprising that the scientific panel in the University of Toronto study ranked molecular diagnostics as the most promising set of technologies for improving health in developing countries over the next five to ten years. Many of the diagnostic tools currently in use in developing countries are cumbersome, time-consuming and expensive. In contrast, molecular diagnostics are simple to use, give quick results and can be relatively cheap. We focus on a few such technologies: polymerase chain reaction, monoclonal antibodies and recombinant antigens.

The *polymerase chain reaction* (PCR) is a quick way of making millions of copies of a specific sequence of DNA. This technology is fast and accurate. It takes advantage of the fact that organisms have unique genetic sequences. Amplification and identification of a pathogen-specific DNA sequence in the body fluids or cells of a person identifies infection in that individual. Besides being extremely sensitive, PCR tests can provide results in a few hours as opposed to days for culturing methods. They can also be used to detect infectious organisms that are difficult or impossible to grow in culture (e.g. tuberculosis, malaria) or are dangerous to handle (e.g. HIV/AIDS).

In the industrialized world, the use of PCR tests in the diagnosis of infectious diseases is increasing. Tests for HIV- 1, gonorrhea, pneumonia and herpes are already available. Recent advances are beginning to bring this powerful tool within reach of the developing world. For example, a group involved in the testing of infants for HIV-1 successfully processed and stored blood samples for up to several months on commercially available filter paper. A PCR test for three complexes of New World *Leishmania* (*Leishmania braziliensis*, *Leishmania mexicana*, and *Leishmania donovani*) – which cause a spectrum of diseases – was developed in Nicaragua and uses a technique known as multiplexing to test for more than one disease at once, saving both time and resources. Going a step beyond current practices in PCR are new *nanotechnology* methods for detecting infection at the molecular level, without the need for DNA amplification. One nanotechnique uses gold particles complexed with diagnostic DNA fragments that have the ability to bind to specific pathogen-associated DNA sequences. A sample of the patient's blood is placed between two tiny electrodes in the presence of the probe. When the probe and its target DNA sequence match, the gold particles close the circuit between the electrodes and produce a detectable signal. Not only is this approach more sensitive than conventional detection methods, it has the potential to be substantially more affordable.

While PCR tests are on their way to finding applications in developing countries, antibody-based applications are already highly suited to the developing world. *Antibodies* are molecules produced by the immune system in response to infection. They recognize and bind to proteins produced by pathogens. These proteins are known as antigens. Antibodies are specific, i.e. every antibody recognizes and binds to a specific type of antigen. This makes them an excellent tool for the diagnosis of infectious disease. In the last few years, the development of simple and rapid antibody-coated dipstick tests have increased the relevance of this technology for the developing world. Dipsticks can be used anywhere, without the need for laboratory facilities, running water or electricity. OptiMAL® is one of several antibody-based dipstick tests for malaria. During a malaria outbreak in Honduras, it rivalled the accuracy of microscopic analysis, the "gold standard" of malaria diagnosis. The Program for Appropriate Technology in Health, an international not for-profit organization committed to improving global health, has developed an HIV dipstick test that is now being produced by several private firms in developing countries. *Determine HIV-1/2* can diagnose both HIV-1 and HIV-2 based upon a reaction on a nitrocellulose strip between recombinant antigens and patient antibodies. The test is accurate and sensitive and has been successfully tested in the field in Ghana.

PATH is also developing dipsticks for the detection of malaria, TB and hepatitis C, among other diseases. Quidel has developed a dipstick test for *Streptococcus pneumoniae*, the most common cause of pneumonia. Monoclonal antibodies (mAb) are extremely specific antibodies. They are superior to antibodies harvested from human or animal serum because they are extraordinarily pure, show very high binding specificity and can be grown in large quantities. The drawback is that mammalian cell culture is expensive, so although the sensitivity and specificity of monoclonal antibodies are high, cost issues have prevented their widespread use in developing countries.

Plantibodies are monoclonal antibodies that are produced in genetically engineered plants and then extracted for medical use. They are a cheaper source of antibodies than those produced in mammalian cells. One estimate places the cost of a gram of plantibodies at less than 5% that of mammalian cell monoclonal antibodies. Another advantage is that they do not risk spreading animal diseases to humans. Plantibodies have been patented by Epicyte, a U.S. biotechnology company which plans to begin clinical trials of its anti-Herpes Simplex Virus 1 and 2 plantibody mixture, HX8. The company was recently awarded a grant from the US National Institutes of Health for the development of plantibodies against Human Papilloma Virus (HPV), which causes cervical cancer. Other companies have begun human trials of plantibodies for Hepatitis B, sexually transmitted diseases, and non-Hodgkin's lymphoma.

The Monoclonal Antibody

Culture Technique

The technique of monoclonal antibody culture involves three broad steps. The first step is to immunize a mouse with an antigen. When the mouse begins to produce antibodies to the antigen, its spleen is removed. Antibody-producing cells from the spleen are then fused with tumour cells (which grow and divide in vitro and are 'immortal'). Each hybrid cell produces identical antibody molecules. By allowing the hybrid cells to multiply in culture, it is possible to produce a population of cells, each of which produces identical antibody molecules.

These antibodies are called "monoclonal antibodies" because they are produced by the identical offspring of a single antibody-producing cell. The new fused cell can also be re-injected into another mouse. This mouse produces pure monoclonal antibodies. The use of monoclonal antibodies for the diagnosis of infectious disease dates back to the early 1980s.

A New Test for Tuberculosis

New developments in diagnostic testing for tuberculosis indicate the diverse ways in which scientists are using the immune system for detecting the presence of pathogens in the blood. A major obstacle to controlling tuberculosis is the lack of a reliable diagnostic blood test. The treatment of TB relies upon the identification of tuberculosis infection by targeted tuberculin skin testing, the main drawback of which is low specificity. For example, the BCG vaccination and environmental mycobacterial exposure can both result in false-positive results.

In response to this shortcoming, researchers at Oxford University and the Wellcome Trust have developed a new test. This test targets cells in the blood that are produced by the immune system, called T-cells. The presence of specific T cells, detected by the rapid ex-vivo enzyme-linked immunospot (ELISPOT) assay for interferon-gamma, is a highly sensitive and specific indicator of infection in patients who have tuberculosis; its sensitivity is substantially higher than that for the skin test. This test is set to supersede the tuberculin skin test and improve the control and prevention of this resurgent disease. The test is simple, quick, and not confounded by TB vaccination. Doctors hope it will supersede the skin test in the next few years and transform the way TB is diagnosed and controlled.

As mentioned earlier, pathogen cells are associated with distinguishing molecules known as antigens, and antibodies can be used to detect the presence of antigens in a person's bloodstream. Similarly, antigens can also be used to detect antibodies. *Recombinant antigens* are genetically-engineered antigens mass-produced by fast-replicating organisms such as bacteria or yeast. (The term recombinant refers to the DNA produced when genetic information from more than one organism is 'recombined' in a laboratory process to form a hybrid molecule. A desired antigen is produced by incorporation of its DNA sequence into the DNA of a host bacterial cell). Like monoclonal antibodies, recombinant antigens also feature in simple handheld test devices capable of providing a diagnosis in a matter of minutes. These devices are particularly well suited to developing areas where clean water and electricity are not always accessible and people might have to travel significant distances to reach a medical facility.

Recombinant Vaccines

- Reduce child mortality
- Improve maternal health
- Combat HIV/AIDS, malaria and other diseases.

Recombinant vaccines can play an important part in achieving the fourth Millennium Development Goal, *Reduce child mortality* as well as to the fifth and sixth MDGs, *Improve maternal health* and *Combat HIV/AIDS, malaria and other diseases*. The hepatitis B vaccine saves the lives of millions of people, including children, every year, and many HIV vaccine candidates currently under investigation are recombinant vaccines.

Vaccines stimulate the body to produce a protective immune response and thereby reduce the likelihood of serious infection. They are arguably the most important medical advance of the last hundred years. Vaccination has resulted in the eradication of smallpox, the imminent eradication of polio, and a dramatic reduction in the prevalence of many other infectious diseases. Advances in vaccine research are expected to have an impact not only on communicable diseases, but also on non-communicable ones such as cancer.

Until a few decades ago, all vaccines consisted of either killed or inactivated ('attenuated') pathogens. Injection of this into a person would stimulate the person's immune system to produce antibodies against the foreign organism, thus conferring resistance to the pathogen in future infections. However, pathogen inactivation may sometimes be insufficient, so that vaccination with a live pathogen, even though it is weakened, carries a finite risk of causing a fatal infection. *Genetic engineering* (also known as *recombinant technology*) makes it possible to have single proteins of the pathogen produced in non-pathogenic microorganisms; injection of a person with these organisms stimulates antibody production against the foreign protein and therefore provides protection against the disease. The obvious advantage of this approach is the enhanced safety of the vaccine, since the single foreign protein cannot cause the disease. Recombinant vaccines made with only part of the genome of a pathogen are also known as subunit vaccines. Recombinant vaccines may also prove to be cheaper than traditional vaccines because of innovative production methods and, in some cases, because improved storage characteristics may not require them to be refrigerated. Much progress is being made in recombinant vaccine development. A major roadblock is the long time it takes for clinical trials and regulatory approval. To date, the number of products on the market is quite limited. Researchers are currently working to develop techniques to overcome some of the other difficulties too, for example, correct presentation of recombinant antigens to the immune system and the limited lifetime of the engineered protein in the body.

Types of Recombinant Vaccines

Viral vector vaccines consist of a benign virus that has been genetically modified to contain genetic material belonging to the pathogen. After the vaccine has been injected, the virus attaches to the individual's cells and injects its genetic material into the cell's interior. The foreign genetic material gets incorporated into the cell's genome. The cell follows the genetic code and manufactures the antigens of interest, which trigger the immune system into producing a protective immune response.

Naked DNA vaccines, or plasmid DNA vaccines as they are also known, are very similar to viral vector vaccines. The key difference is that naked DNA vaccines use a different biological carrier, known as a plasmid, to introduce the antigen genes into the individual's cells. Plasmids are small circular molecules of DNA that are normally found in bacteria. They can easily enter cells and recruit the cell to translate their genetic information into protein. Advocates of this approach argue that developing naked DNA vaccines is potentially faster, cheaper and easier than other types of vaccines and, because DNA is heatstable, these vaccines are able to bypass the cold chain, a major cost barrier to efficient vaccine delivery.

Plant-derived vaccines are subunit vaccines synthesized by plants. Plants typically used are those that are easily engineered to express foreign proteins, such as the tomato or potato. After introducing the pathogen genes into the plant's DNA, the plant is given a signal to make the protein in its tissues. The antigens expressed by the plant are purified and can then be used as vaccines. In essence, the plants act as vaccine-factories. There are several advantages to plant-derived vaccines. They can be relatively inexpensive to produce, store, and transport and they could be grown locally, making them attractive for applications in developing countries. Researchers are making significant progress in the development of plant-derived vaccines. In comparison to the existing hepatitis B vaccine, a potato-derived vaccine was found to produce greater immunity in mice. Plant-derived vaccines are also being developed for several important causes of infant mortality, including infant diarrhea and measles.

Recombinant Vaccines in Development

Most HIV vaccines currently in clinical trials are recombinant vaccines. Currently there are 28 ongoing trials collectively involving 24 different vaccines. Of these, 8 are DNA-based, 8 use recombinant viral vectors, 5 are protein subunits and 3 are lipopeptides. An HIV viral vector vaccine based upon the gp120 antigen has shown encouraging results in a Phase II clinical trial with healthy volunteers. Just over half of all volunteers generated antibodies that neutralized HIV-1. When received in combination with an injection of gp120, the proportion rose to 94%. Phase III trials began in late 2003.

In recent trials on Rhesus macaques, researchers assessed the protection conferred by recombinant forms of HIV gp120, a fusion protein constructed from HIV regulatory proteins Nef and Tat, and SIV Nef. Treatment with this recombinant multiantigen vaccine delayed the onset of AIDS in Rhesus monkeys for at least 2.5 years following infection with a simian-human hybrid virus (S-HIV).

The much-awaited initial results of a three-year multi-ethnic phase III trial of AIDSVAX by researchers at VaxGen Inc. have been disappointing. Produced in mammalian cell culture, AIDSVAX is a recombinant form of gp120. It is non-infectious and mimics proteins on the surface of two strains of HIV subtype B (prevalent in North America, Europe, Australia, Japan and Puerto Rico). The FDA would have considered approving the vaccine if it were 30% effective, but the reduction of infection among the entire sample of volunteers including all racial groups was only 3.8 %.

The study did not show a statistically significant reduction of HIV infection within the study population as a whole, which was the primary endpoint of trial. Nonetheless, larger and statistically significant reductions were observed in some ethnic groups, especially among African-Americans, so further studies on this vaccine are warranted. A naked DNA vaccine against HIV/AIDS is currently being evaluated in The Gambia and England, with good results. The first multi-clad HIV vaccine to enter clinical trials is also a naked DNA vaccine. By preventing infection by the three most important subtypes of the virus around the world, it is hoped that this vaccine would be effective in both developing and developed countries.

All said and done, however, the progress of HIV vaccine development has been slow and disappointing. After 10 years of work in this area, scientists now believe that live, attenuated vaccines may be more appropriate for the fastevolving and diverse HIV.

The story of recombinant vaccines against malaria is more promising. A subunit vaccine known as RTS,S/AS02 was recently shown to provide protection against natural malaria infection in adults in a clinical trial in the Gambia, with 71% of the participants protected from infection for the first 9 weeks of the study. Clinical trials on children in both The Gambia and Mozambique are now underway. Another subunit vaccine protected four of five monkeys from a lethal injection of malaria.

This particular vaccine is synthesized by mice and excreted in their milk. The developers have expressed interest in creating transgenic goats for large-scale production. Tuberculosis claims at least 2 million lives every year. Researchers are actively working towards a potent TB vaccine. 72 Progress to date includes a subunit vaccine based upon the Mtb8.4 surface antigen that protected mice from virulent TB infection.

Hepatitis B affects approximately one third of the world's population and is associated with liver cancer. Tomatoes and lettuce have been genetically engineered to produce hepatitis B surface antigens, and these appear to induce an immune reaction similar to hepatitis B proteins isolated from infected humans. Scientists are developing plant-derived vaccines against the major causes of infant diarrhea.

For instance, researchers successfully produced a strain of corn that expresses an antigen associated with enterotoxic *E. coli*. Mice that were fed the corn exhibited elevated levels of antibodies in their blood, and showed resistance to infection. Also, the gene for the highly antigenic, but non-toxic, beta chain of the cholera toxin has been expressed in alfalfa, and mice fed with this alfalfa showed an immune response against cholera toxin. However, it will probably take many years of careful experimentation before such vaccines are approved for human use.

Bioremediation

- Combat HIV/AIDS, malaria and other diseases
- Ensure environmental sustainability.

Technologies for environmental improvement (sanitation, clean water, bioremediation) were ranked fourth in the list of technologies to improve health in developing countries.

Bioremediation has direct significance to the seventh MDG, *Ensure environmental sustainability* but it also has impact on the health-related goals. Bioremediation is the use of bacteria or plants to clean up the environment. Reduction of pollution in water supplies and in the food chain will help to reduce mortality and improve health. There are two main types of pollution threatening the health and well-being of human populations: organic waste and heavy metals such as lead, mercury and cadmium. Bacteria can detoxify both. Plants can break down most forms of organic waste, but, with very feexceptions, are usually unable to metabolize heavy metals.

On the other hand they can store harmful metals in their tissues and therefore make it easier to collect, harvest and even recycle metal waste. Water contaminated by human waste harbors large amounts of pathogenic organisms and has been implicated in the transmission of cholera, typhoid, hepatitis A another waterborne diseases. Sewage treatment can dramatically reduce the incidence of these diseases. Bioremediation techniques can augment conventional chemical sewage treatment. A number of low-cost alternatives to conventional sewage treatment have been developed. One such system is now in use in Southern China. It uses floating rafts called Restorers to supply beneficial microorganisms to a canal contaminated with human waste. In fact, this floating ecological treatment engine has now been transformed into a garden featuring over 10 species of native Chinese wetland plants.

Bioremediation can also be used to reduce environmental pollution associated with heavy industries, including oil spills, acid mine drainage, and radioactive waste. In 1989 the tanker Exxon Valdez released 11 million gallons of oil off the coast of Alaska, affecting approximately 2000 km of shoreline.

Through the addition of extra nutrients to the beach, crews were able to enhance local bacterial and algal populations and take advantage of their ability to degrade petroleum.

Acid mine drainage is associated with the leakage of contaminated water from mining sites. The drainage is toxic for both humans and the surrounding environment.

Constructed wetlands are a common low-cost, sustainable solution to this problem. The area is artificially saturated with water, plant species are introduced, and these eventually lower the acidity and the metal concentrations at the site.

The bacterium *Deinococcus radiodurans* has the ability to thrive in environments with 300 times the fatal dose of radiation for humans. Researchers have genetically engineered

D. radiodurans to express several genes associated with mercury detoxification. The recombinant bacterium has been shown to metabolize mercury while being bombarded by high levels of radiation, suggesting that bioremediation may be able to aid in the process of radioactive waste cleanup. Bioremediation can also help to clean up mosquito-infested water and control the spread of malaria. For example, the malaria-carrying *Anopheles* mosquito has developed resistance to chemical insecticides. Many anti-malarial prophylactics that were once effective are now less reliable, not to mention expensive for many people in developing countries. Peru has one of the highest rates of malaria in all of Latin America. Canada's International Development Research Centre (IDRC) has supported research at the Instituto de Medicina Tropical "Alexander Von Humboldt" in Lima to explore the use of coconuts in the fight against malaria. Researchers at the Institute have developed an ingenious method to biologically control mosquitoes that is simple, inexpensive, and an environmentally safe alternative to insecticides.

It involves the use of coconuts used to culture bacteria that are toxic to mosquito larvae but harmless to people and other organisms. *Bacillus thuringiensis* var. *israelensis* H-14 (*Bti*) is a spore-forming bacterium that produces a toxin lethal to mosquito larvae. Imported *Bti* cultures are expensive, but as this project shows *Bti* can be cultured locally and cheaply using coconuts, which are both cheap and plentiful in many tropical areas. *Bti* is introduced via cotton swabs into the coconut and allowed to incubate inside the coconut for a few days. The nut is then broken and tossed into ponds where mosquitoes breed. The mosquito larvae eat the bacteria and are killed. A typical pond needs only two or three coconuts for each treatment – usually lasting 2 months.

Sequencing Pathogen Genomes

- Combat HIV/AIDS, malaria and other diseases.

The sequencing of pathogen genomes has direct relevance to the fourth Millennium Development Goal, *Combat HIV/AIDS, malaria and other diseases*. The task force for this MDG has prepared documents on HIV/AIDS, malaria and TB that take an in-depth look at the science, public health needs and medical interventions, and the social and economic issues and consequences of these devastating diseases. Pathogen genomics can also contribute to the search for a solution of these diseases.

Sequencing a genome involves discovering and recording the entire sequence of nucleotides in an organism's DNA. DNA codes for proteins, which are the mainstay of structure and biochemical function in all organisms. Knowing the sequence of a pathogen's genome is therefore helpful in unraveling its biology and discovering effective ways of controlling its relationship with humans.

Most sequencing strategies are based upon a technique known as the dideoxy or Sanger method. Small-scale sequencing projects can be done manually, but large-scale projects (such as the sequencing of an entire genome) require high-throughput automated DNA sequencing machines. Major sequencing projects use many automated sequencing machines simultaneously, yielding millions of bases of sequence data per day. These data must be stored, managed and analysed by computers and this requirement has given rise to an entirely new field in biology -bioinformatics. Knowing the sequence of a pathogen's genome can rapidly accelerate the process of drug discovery.

Comparative genomics is a method that compares the genomes of different organisms in order to apply information known about one organism to another. For example, in a comparison of disease-causing and benign strains of the same organism, genes unique to the pathogenic strain are likely to play an important role in pathogenesis, and the proteins for which they code may make excellent drug targets. Alternatively, if one pathogen has a gene that is a known drug target, a comparative genomics search might reveal a similar gene, and therefore a potential drug target, in another pathogen.

A serious health concern worldwide is the emergence of pathogen resistance to previously effective drugs. Analysis of pathogen genomes could reveal the genes that play a role in helping these organisms develop drug resistance and point researchers in the direction of treatments that can overcome the action of these genes. For example, scientists can compare the genomes of resistant and nonresistant strains, or analyze the genes at work in the drug resistant stage of an organism's lifecycle.

In addition to spurring novel drug discovery, pathogen genomics has also given a boost to the development of vaccines. Researchers can search the genome of infectious organisms for gene sequences characteristic of antigens. For example, scientists have discovered vaccine candidates for a particularly virulent strain of meningitis, a potentially fatal bacterial infection. Of the 570 antigens found, 85 showed promise when used to immunize mice.

Female-controlled Protection Against Sexually Transmitted Infections (STIS)

- Promote gender equality and empower women
- Improve maternal health
- Reduce child mortality.

Genomics and other biotechnologies are enabling the development of a number of new forms of female-controlled protection against STIs, such as recombinant vaccines, monoclonal antibodies and new approaches to the development of vaginal microbicides. These technologies can help to meet the sixth MDG *Combat HIV/AIDS, malaria and other diseases* and even the third MDG, *Promote gender equality and empower women* with an indirect effect on other health MDGs: e.g. *Reducing child mortality* by *Improving maternal health*.

The global burden of STIs is felt most heavily by women. A decade ago, the World Bank ranked STIs as the second major cause of ill health among women aged 15 to 44, accounting for 8.9% of their disease burden, compared to 1.5% in men. For various reasons – socio-economic, cultural and biological – women are more vulnerable to infection than men. Women with STIs often bear heavy social stigma, and they often have neither the time nor the money to seek health care. Rising infection in women also increases the incidence of transmission from mother to newborn. Despite the seriousness of this problem, women possess few means of protection. The condom requires male consent, which many women living in patriarchal societies find hard to negotiate in their relationships. Although the female condom has been greeted with enthusiasm by women in poor countries, its cost and indiscreetness make it a less than ideal option. Vaginal microbicides are one attractive alternative. They are gel or cream formulations of chemical compounds that block the transmission of infection across the vaginal wall. Researchers are actively involved in developing safe and effective vaginal microbicides.

More recently, researchers in India have developed a potato rich in all essential amino acids. This potato contains the gene AmA1 that codes for the protein albumin, which contains high levels of all amino acids that the body is incapable of making on its own (i.e. essential amino-acids). It gives ordinary potatoes a third more protein than normal, including substantial amounts of the essential amino acids lysine and methionine. Deficiency in these can have an adverse effect on development in children. This enriched potato is also particularly relevant for India, where a large percentage of the population is vegetarian. The AmA1 gene was obtained from the amaranth plant which grows in South America. The potato is in the final stages of testing, and it has been submitted for official approval. Scientists say that the protein-rich genetically modified "protato" could help combat malnutrition among the country's poorest children. Researchers have also inserted into lettuce the gene for the enzyme GLOase, which converts a precursor molecule into Vitamin C. They observed up to a seven-fold increase for Vitamin C in the plant's tissues. Vitamin C is a powerful antioxidant and is associated with reduced risks of cancer, cardiovascular disease and other diseases.

Genetic modification and traditional breeding both change the characteristics of an organism, but in different ways. In traditional breeding, the introduction of a gene into an organism by cross breeding different strains of the same species is a trial-and-error process that takes a long time. Genetic modification makes it possible to introduce new genes more rapidly and more precisely than traditional breeding and enables the introduction of new genes from different species. In many cases, traditional breeding alone is able to improve the nutritional value of a crop. In fact farmers have used traditional breeding over thousands of years to convert the wild, almost inedible forms of rice and wheat into the familiar foods we eat today. However, genetic modification has the advantage of speed and relative control. This approach has its advantages and disadvantages. An obvious disadvantage is that the introduced foreign genes may cause unknown gene-gene or gene-environment interactions. Ecological concerns include risks to human and animal health, food safety, and unforeseen consequences of the spreading of foreign genes into the natural environment. Gene-flow, the natural spread of genetic traits when a plant variety is introduced into the environment, is in itself not alarming, but the potential introduction of cross-species genes and the spread of these genes that confer novel traits upon related weeds or crops could be.

However, it is important to note that all plant breeding, whether traditional or through genetic engineering, can produce unexpected results. Some scientists recommend that extensive testing and careful monitoring are necessary for the world to reap the benefits of this technology while avoiding any potential risks. Others argue that all plant varieties, whether GM or not, should be subject to the same stringent standards. a, b Although GM enriched crops are considered by many to be extremely important for health in developing countries, there are some concerns regarding their use. There is, for example, evidence that Golden Rice may not be a very good source of Vitamin A. Specifically, people with diarrheal diseases, which are common in the developing world, are unable to efficiently absorb Vitamin A from the rice. Critics also maintain that GM enriched crops are not the answer to the problems of hunger and malnutrition, which are the outcomes of distributional problems. Several organizations are actively researching these issues. The Food and Agriculture Organization of the United Nations and the World Health Organization jointly recommend the thorough evaluation of GM foods for safety and nutritional value before consumption or releasing the foods into the environment.

Genomics Capacity Building

The genomics revolution will have an impact not only on health care, but also on agriculture (genetically modified organisms) and economics (pharmaceutical industry). Social implications will also be broad, as illustrated by Pedro Leon's description of the genetic basis of deafness as well as recent Costa Rican legislation on compulsory DNA testing so that children have the "right to know their fathers." Despite the publicity and hype, genomics is not a panacea or magic bullet. Whatever happens, biotechnology will not solve deep-seated problems like poverty for those living under $1-2 per day, who need drinking water, food, and livelihoods. Genomics is inherently controversial because it is central to sex, money, reproduction, race, and privacy. Warnings were issued about eugenics, cloning, discrimination, and genetic determinism. Even without genomics, we are witnessing some technological abuse, like pre-natal sex selection in some Asian societies. Public worries about "Frankenfoods" may indeed become real worries about "biological Frankensteins." These issues set the stage for three central Conference challenges, in my opinion. What is really bothering us about the genomics revolution? What should we do? And how can WHO play its pre-eminent role in genomics and world health?

What's the Problem?

The genomics revolution has provoked genuine disquiet, sometimes passion and even fear. Discomfort could come from insecurity about the loss of control over our lives in uncertain times. History is not entirely reassuring in this regard. Throughout human history, warfare and profits have powerfully propelled technology development. Indeed, I was surprised that the anthrax bioterrorism attack in the United States did not receive more attention. We know that powerful technologies with destructive potential, like nuclear power, can unleash unintentional consequences of far greater import than planned change. Once the genie is out of the bottle, it's hard to put it back in. Bruce Lewenstein presented survey data of public perceptions about genomics, mostly from rich countries. Public awareness is growing, but "genetic illiteracy" and popular hype are also prevalent. As I listened, I began to interpret a new meaning to "know thyself." That used to signal reflections about my personality, my psychology, and my behaviour. With the genomics revolution, know thyself now prompts reflections about my genes and biology! A recurring theme underlying our worries is social inequity. Inequity differs from inequality in that the latter describes objective disparity while the former introduces a normative lens on whether such inequality is perceived as fair or unfair. Inequity, therefore, is a moral judgment about the nature of inequality in society. Concern about inequity is escalating, in part, because of the globalization of private markets. Rapid transnational flows of money, goods, services, technology, culture, and people are introducing new vulnerabilities, uncertain risks, and heightened public consciousness about distant people in a shrinking world. With the exception of malaria, a greatly neglected disease, the three other infections discussed in our breakout sessions may be characterized as "diseases of globalization," because their emergence or re-emergence is associated with rapid transnational flows. The global spread of HIV/AIDS was certainly associated with global movement. Tuberculosis, also a major neglected disease, has increased dramatically due to HIV-compromised host immunity. Resurgence of dengue fever is due, in part, to movements of people and goods that facilitate the transmission of the vector-borne virus.

Jose Maria Cantu predicted that the "fight of the 21st century" will be between the "WHO and WHM" (world health market). He suggested a fundamental tension between WHO's mission of equitable world health and inequitable forces generated by competitive private markets. Because the poor lack purchasing power, market-driven health services will not address their health care needs. Lack of a profitable market provides little incentives for industry to pursue R&D to develop technologies against diseases of the poor. Peter Singer and Abdullah Daar urged that actions be taken to preempt a looming "genomic divide." Some have described this phenomenon as a "market failure," implying that markets serve well most purposes but leave a few social gaps. I would prefer calling the problem a "social failure," since even if markets were to function perfectly, they would not address the fundamental deficiency of the inability of the poor to command goods and services through the marketplace.

What Should We Do?

How do we produce essential global public goods? How do we shape markets to generate more equitable outcomes? I address these two questions and also reflect on lessons from Cuba today and China yesterday.

Certain goods wanted and needed by everyone, like clean air and public parks, cannot be produced by private markets. Genomics R&D for technologies against neglected infectious diseases fit into this category. The term "public goods" describes these non-rival and non-excludable goods. How then can we produce global public goods? Given their non-market nature, public goods are appropriately generated by public action and public financing.

R&D, however, is expensive and time-consuming, requiring a strong infrastructure of skilled people, facilities, equipment and supplies. David Webber reported that members of the International Federation of Pharmaceutical Manufacturers Association invest $40 billion annually in R&D. Global imbalances of investments underscore the obstacles faced by many developing countries. According to Tikki Pang, of an estimated $5 billion invested annually in genomics research, $3 billion is private and 75% is in the United States, which also registers about 80% of genomics patents worldwide. Some developing countries – Cuba, Thailand, India, and China – are emerging contributors to genomics research. But, what about other developing countries? Michael Wilson described a "catch-22" situation in Ghana of a vicious cycle of weak financing, little research, few facilities, no jobs, and ultimately out-migration of skilled scientists. To remedy the R&D gap, two dozen so-called public-private partnerships (PPPs) have been created.

These PPPs mostly aim to produce public goods in health. Many believe that these "push" mechanisms (production of neglected vaccine and drugs) must be balanced by "pull" forces (like subsidized vaccine purchase funds) to correct for market failure. Can markets be shaped to generate fairer outcomes? Severe inequity is not a preordained outcome of markets that involve millions of individual and collective decisions. Markets, however, are social institutions governed by humanly constructed ground rules. Depending upon ground rules, it is possible to generate various levels of fairness. Weak initial capability upon entering competitive markets is one example of handicaps faced by many developing countries. Many also feel aggrieved because these disadvantages are historically inherited, and developing countries often lack the political and economic clout to negotiate better entry conditions. Another ground rule is intellectual property rights (IPR). Lalji Singh described the cultural alarm of Indians as foreigners patented indigenous plants — the neem tree, turmeric spices, and Basmati rice.

The controversy over IPR will not simply go away. George Soros in his recent book, "Globalization," describes how IPR in today's knowledge-based economy should be treated differently than earlier technology eras. In the industrial revolution, either new or some commonly-owned assets became privatized – for example, telegraph lines, railway tracks, subsurface petroleum, and manufacturing machines. Intellectual property in a knowledge-based economy poses different challenges to balance incentives for innovation, appropriate returns to an inventor, and economic progress. These historical dimensions of IPR should not distract us from practical intermediate steps that can be taken to reduce inequities in IPR management. Bibek Debroy argued that wholesale changes are neither likely nor necessary, nor must solutions be conflictual. One example of positive adjustment is ensuring royalty returns to affected genetic lineages for research participation in Costa Rica. Another potentially very significant advance is Management of Intellectual Property in Health R&D (MIHR), recently established by the Rockefeller Foundation to assist non-profit organizations negotiate IPR, licensing, and other arrangements. As described by Luis Herrera and his colleagues, Cuba is an admirable health success story. Its 11 million people enjoy an average life expectancy of 76 years and an infant mortality rate of 6 per 1,000 livebirths — roughly equivalent to Canada and superior to the United States. As a basic human right, health care is universal, accessible, and free.

Cuba also has adopted an "internationalist public health" policy that exported 2,600 health workers to other countries. With universal literacy, 58 universities and 115 research institutes, Cuba has one physician per 179 population and more than one teacher per 50 people. Its rich human resources are important to a state-led biotechnology sector, which has received strong political commitment and sustained investments. The state-led sector pursues vertical integration of pharmaceutical research-product-marketing, called by Dr. Herrera, a "closed cycle strategy." Drs. Moreno and Mauri described how Cuba also pursues IPR protection and marketing through "strategic alliances" with foreign firms. In the United States, we have a market-led biotechnology sector. Of course, the public sector through the NIH provides nearly $20 billion of funding for basic research that is tapped by the private sector. Other countries, for example Thailand, have their own unique mix of state-led and market-led R&D. The lesson here is that different national patterns of R&D systems are possible, even as all nations compete in a global economy.

7

DNA Quantification and QualityAssessment

DNA quantification and quality assessment is done by using UV-VISspectrophotometry.Normally quality check is performed through the A260/A280 ratio that is 1.8 value shows the highest purity, if more than 1.8 shows the presence of RNA contamination and less than that shows the protein contamination (Sambrook, *et. al,* 1989).

AGE is conducted of isolated DNA samples for visualization of bands. For perfect extraction, there will be clear single bands otherwise smear like image will be seen. That is in case of degraded DNA. For PCR based methods the DNA samples are diluted for PCR requirements. The optimum concentration is usually 40- 60ng/mL.

Polymerase Chain Reaction (PCR)

The DNA amplification by thermal cycling called Polymerase Chain Reaction is in vitro method that can be used to amplify a specific DNA segment from small amounts of DNA template or duplex into millions of copies. It is invented by Kary Mullis *et.al,* (1985).

Heatdenaturation

This temperature denatures the double stranded DNA into two individual stands. Denaturation temperature is 95^0 C for 30 seconds or is 97^0 C for 15 seconds, however higher temperature may be appropriate, especially for G+C rich nucleotides.

Annealing

During this time one primer binds with the 5 prime end of one DNA strand and the other primer binds with 3 prime end of its complementary strand. Annealing is hybridization of primers to single stranded DNA and the length of time required for primer annealing depends on the basic composition, length and concentration of primers.

Primer Extension

This temperature is varies for Taq DNA polymerase which adds complementary nucleotides one by one to the 3' OH group of the primer. Estimates for the rate of nucleotide incorporation at 72^0 C vary from 35-100 nucleotides per second depending upon the buffer, pH salt concentration and nature of DNA template.

Number of PCR Cycles

The optimum number of cycles depends mainly upon the starting concentration of target DNA, when other parameters are optimized. A common mistake is to execute too many cycles. The increase in the number of cycles will increase the amount and complexity of non-specific background products.

Restriction Fragment Length Polymorphisms (RFLPs)

RFLP Definition

The variation(s) in the length of DNA fragments produced by a specific restriction endonuclease from genomic DNAs of two or more individuals of a species (Kahl, 2001).

Restriction fragment length polymorphisms (RFLPs) was the first to develop (Soller and Beckmann, 1983) and are still widely used marker system especially in practical plant breeding. It was first developed in the 1980s for use in human genetic applications and was later applied to plants Most of the studies in systematic and evolution (Whitkus *et al.*, 1994) and molecular maps of different crop species were initially made through the use of RFLP markers. Restriction fragment length polymorphism (RFLP) technology. RFLPs are relatively small in size and are co-dominant in nature.

Restriction fragments of different lengths are thus generated. All RFLP markers are analysed using a common technique. However, the analysis requires a relatively complex technique that is time consuming and expensive.

The hybridization results can be visualized by autoradiography (if the probes are radioactively labelled), or using chemiluminesence (if non-radioactive, enzyme-linked methods are used for probe labelling and detection). Any of the visualization techniques will give the same results. The visualization techniques used will depend on the laboratory conditions. A comprehensive study on these maps is documented by Phillips and Vasil (1994). Thus, RFLP markers have the potential to address with power, precision and accuracy, all issues of plant breeding in which markers can be applied.

In addition to that, voluminous literature is available on utilization of RFLPs in other aspects especially in DNA fingerprinting (Tanksley, 1983), phylogenetic analysis (Zheng *et al.*, 1994), detection of genetic variability in rice, genetic dissection of maturity (Phillips *et al.,* 1992), identification of chromosomal regions responsible for regeneration of RFLPs, comparative mapping of chromosomes in wheat, rye and barley, analysis of wide compatibility in rice varieties (Zheng *et al.*, 1994), QTL mapping, gene tagging, and identification of ecotypes (Pupilli *et al.,* 2000).

To discover RFLPs, restriction enzymes (RE) are used to cut DNA at specific 4-6 bp recognition sites. Sample DNA is digested with one or more RE's and resulting fragments are separated according to molecular size using gel electrophoresis. After that Southern Blotting is performed where fragments are immobilized on a nylon membrane, followed by hybridization to a labelled probe (specific chemoluminescence DNA probe) that identifies the locus under autoradiograph.

Fragment sizes of digested DNA will differ depending on the presence or absence of the proper recognition sequence for the enzyme. Differences in fragment length result from base substitutions, additions, deletions or sequence rearrangements within RE recognition sequences. Although two individuals of the same species have almost identical genomes, they will always differ at a few nucleotides.

Some of these differences will produce new restriction sites (or remove them), and the banding pattern seen on a genomic Southern will thus be affected. For any given probe (or gene), it is often possible to test different restriction enzymes until one which gives a pattern difference between two individuals is found, that is, a RFLP. The less related the individuals, the more divergent their DNA sequences are and the more likely you are to find a RFLP.

However, the low level of polymorphism present in RFLPs restricted the use of this technique. In addition, RFLP requires large quantities of relatively pure DNA, and the frequent use of radioisotopes in detection step makes it technically demanding, laborious and costly to characterize large number of samples.

Utilization of RFLP markers requires a lot of skill in using short-lived radioactive material: the availability of which often becomes a limiting factor especially in the developing countries. Another limiting factor could be the large sample size to be analysed like that of F segregating population. In this case, utilization of RFLP markers may not be feasible, as it requires extraction of DNA in ultra pure form, restriction digestion, and lengthy and expensive process of isotopic labeling. Such difficulties have opened up a debate on the feasibility of routine application of RFLPs in large scale crop improvement programme. Another difficulty frequently encountered by RFLP users is the detection of low (1-3 loci) ornon-existent polymorphism in closely related species/varieties (Waugh and Powell, 1992). It was essentially this difficulty that led to the search for a new marker system.

Randomly Amplified Polymorphic DNAs (RAPDs)

Randomly-amplified polymorphic DNA markers (RAPD). In 1991 Welsh and McClelland developed a new PCR-based genetic assay namely randomly amplified polymorphic DNA (RAPD). This procedure detects nucleotide sequence polymorphisms in DNA by using a single primer of arbitrary nucleotide sequence. In this reaction, a single species of primer anneals to the genomic DNA at two different sites on complementary strands of DNA template. If these priming sites are within an amplifiable range of each other, a discrete DNA product is formed through thermocyclic amplification. On an average, each primer directs amplification of several discrete loci in the genome, making the assay useful for efficient screening of nucleotide sequence polymorphism between individuals.

However, due to the stoichastic nature of DNA amplification with random sequence primers, it is important to optimize and maintain consistent reaction conditions for reproducible DNA amplification. RAPDs are DNA fragments amplified by the Polymerase Chain Reaction (PCR) using short (generally 10 bp) synthetic primers of random sequence. Because of the short primers used, the reannealing temperature in the PCR must be low (35-40°C) for the primer to bind. However, due to low temperature, the binding is not very specific, which means that primers will bind also to sequences which are not completely complementary.

These oligonucleotides serve as both forward and reverse primer and usually are able to amplify fragments from 3-10 genomic sites simultaneously. Amplified fragments (within the 0.5-5 kb range) are separated by gel-electrophoresis and polymorphisms are detected as the presence or absence of bands of particular size. These polymorphisms are considered to be primarily due to variation in the primer annealing sites.

They are dominant markers and hence have limitations in their use as markers for mapping, which can be overcome to some extent by selecting those markers that are linked in coupling. RAPD assay has been used by several groups as efficient tools for identification of markers linked to agronomically important traits, which are introgressed during the development of near isogenic lines. The application of RAPDs and their related modified markers in variability analysis and individual-specific genotyping has largely been carried out, but is less popular due to problems such as poor reproducibility faint or fuzzy products, and difficulty in scoring bands, which lead to inappropriate inferences.

In RAPD analysis the total DNA from an organism is mixed with (generally) 10 bp single stranded DNA (primer) together with the four different deoxynucleotides and a heat stable DNA polymerase enzyme. The reaction mixture is placed on a thermocycler (PCR-machine) which can change the temperature of the reaction rapidly according to a predefined PCR programme.

Reannealing at low temperature makes the primer molecules attach to complementary site on the DNA. Elongation is the period when the polymerase enzyme elongates the attached primers from their 3' end. Denaturation separates the new formed strands.

After a few PCR cycles DNA pieces with well defined length similar to the one between the two original primer sites dominate the mixture. This is because they amplify exponentially. In most RAPD reactions the result will be amplification of 5 to 10 different DNA segments whose lengths depend on the primer recognition sites available in the genome. The DNA after amplification is visualized for bands using Agarose Gel Electrophoresis (AGE). The band scoring is done by comparing with the marker bands.

Inter Simple Sequence Repeats (ISSR)

ISSR amplification definition: A variant of the polymerase chain reaction that uses simple sequence repeat primers (e.g. [AC] n) to amplify regions between their target sequences (Kahl, 2001).

Inter simple sequence repeat markers (ISSR). In this technique, reported by Zietkiewicz et al. primers based on microsatellites are utilized to amplify inter-SSR DNA sequences. Here, a single primer or combination of primers composed of a microsatellites sequence anchored with a single base at 3' region so that it can amplify regions between two microsatellites loci with oligonucleotides annealing to repeat sequence.

More complex banding patterns can be achieved using 5'-anchored primers that incorporate the SSR regions in their amplification products, and by combining 3'- and 5'- primers.ISSR circumvents the requirement for flanking sequence information, and thus has found wide applicability in a variety of plants.(Wolfe et al.1998} These are mostly dominant markers, though occasionally a few of them exhibit codominance.

An unlimited number of primers can be synthesized for various combinations of di-, tri-, tetra- and pentanucleotides [(4)3 = 64, (4)4 = 256] etc. with an anchor made up of a few bases and can be exploited for a broad range of applications in plant species. ISSR analysis is technically simpler than any other marker systems.

The method provides highly reproducible results and generates higher levels of polymorphism compared to other markers in many systems. The advantage lies in long primer length and stringent annealing temperature. Primers used (from Yang *et al.,* 1996): For separation and visualization of bands Agarose gel electrophoresis is conducted.

Simple Sequence Repeats (SSRs)

SSR (Microsatellites) definition: Any one of a series of very short (2-10 bp), middle repetitive, tandemly arranged, highly variable (hyper variable) DNA sequences dispersed throughout fungal, plant, animal and human genomes (Kahl, 2001).

Simple sequence repeats (SSR) or microsatellites are a class of repetitive DNA elements. The di-, tri- or tetra-nucleotide repeats are arranged in tandem arrays consisting of 5 – 50 copies, such as (AT) 29, (CAC) 16 or (GACA) 32. SSRs are abundant in plants, occurring on average every 6-7 kb (Cardle *et al.*, 2000).

These repeat motifs are flanked by conserved nucleotide sequences from which forward and reverse primers can be designed to PCR-amplify the DNA section containing the SSR.SSR alleles, amplified products of variable length, can be separated by gel electrophoresis and visualized by silver-staining, autoradiography (if primers are radioactively labelled) or via automation (if primers are fluorescently labelled).

SSR analysis is amenable to automation and multiplexing, and allows genotyping to be performed on large numbers of lines, and multiple loci to be analysed simultaneously. SSRs can be identified by searching among DNA databases (e.g. EMBL and Gene bank), or alternatively small insert (200-600bp) genomic DNA libraries can be produced and enriched for particular repeats (Powell *et al.*, 1996). From the sequence data, primer pairs (of about 20 bp each) can be designed (software programmes are available for this).

For separation and visualization of bands Polyacrylamide Gel Electrophoresis (PAGE) is conducted (6% acrylamide solution), stained and the observed the bands.

Amplified Length Polymorphisms (AFLPs)

AFLP definition: Any difference between corresponding DNA fragments from two organisms A and B that is detected by the amplified restriction length polymorphism technique (Kahl, 2001).

***Table:** DNA Fingerprinting Techniques: A Comparative Analysis*

Marker	*Advantages*	*Disadvantages*
RFLP	Unlimited number of loci	Labour intensive
	Co-dominant	Fairly expensive
	Many detection systems	Large quantity of DNA needed
	Can be converted to SCARs	Often very low levels of polymorphism
	Robust in usage	Can be slow (often longexposure times)
	Good use of probes fromother species	Needs considerable degree ofskill
	Detects in related genomes	No sequence informa-tionrequired
RAPD	Results obtained quickly	Highly sensitive to laboratorychanges
	Fairly cheap	Low reproducibility within andbetween laboratories
	No sequence informationrequired	Cannot be used acrosspopulations nor across species
	Relatively small DNAquantities required	Often see multiple loci
	High genomic abundance	Dominant
	Good polymorphism	Can be automated
ISSR	Highly polymorphic	Usually dominant
	Robust in usage	Species-specific
SSR	Can be automated	Fast High developmental andstartup costs
	Highly polymorphic	Species-specific
	Robust	Sometimes difficult interpretation because of stuttering
	Can be automated	Usually single loci even in polyploids
	Only very small DNA	<!—[if!support Misaligned Columns]—> <!—[endif]—>

Marker	*Advantages*	*Disadvantages*
AFLP	Small DNA quantitiesrequired	Evaluation of up to 100 loci
	No sequence informationrequired	Marker clustering
	Can be automated	Dominant
	Can be adapted for differentuses, e.g.	Technique is patented
cDNA-AFLP	Can be technically challenging	

This genetic marker system basically uses restriction fragment analysis, but a PCR approach is used to visualize fragments instead of hybridization with a probe. Because of the PCR step, this type of markers has characteristics quite different from RFLP markers. Initially, DNA is extracted as for RFLP and then cut with two different restriction enzymes to produce well defined restriction fragments with sticky ends. Synthetic double stranded linkers of approximately 18 to 20 bp with matching sticky ends are ligated on all the restriction fragments. Ligated fragments are subsequently amplified in PCR with 18 to 20 nucleotide length primers recognizing linkers in each end of the fragments. A primer used for amplification in AFLP is normally labelled with P^{33} or some nonradioactive labeling system.

The mixture of amplified fragments are separated according to size in a Polyacrylamide gel and visualized by means of autoradiography or other procedures for development of non radioactive systems. Alternatively, automated sequencing machines can be used to read the sequence of amplification products directly from the Polyacrylamide gel if primers have been labelled with special fluorescent dyes. Also, silver staining of the Polyacrylamide gel after electrophoresis can be used to stain all amplified DNA fragments in the gel. To avoid that all restriction fragments from the genome amplify in the same PCR, which will produce a smear because of too many DNA fragments, a two step amplification procedure is used. Firstly, part of the total number of restriction fragments are amplified in the "preamplification" with primers containing one extra "selective" nucleotide on their 3' end.

This selective nucleotide will allow amplification only of restriction fragments with a matching nucleotide next to the linker. With one selective nucleotide on both primers only 1/16th of all restriction fragments in the mixture will amplify during the preamplification. During the subsequent "selective" amplification, additionally, one or two selective nucleotides on each primer will further reduce the number of fragments amplified. A good AFLP amplification will show 90 to 100 different fragments in one analysis.

Applications

There are a variety of applications in DNA fingerprinting, such as to identify genetic diversity within breeding populations in plants and animals, to differentiate between plant species cultivars, as well as to identify plants containing a gene of interest. , it is also used to detect genetically modified organisms in agriculture in the United States.

RFLP is most suited to studies at the intraspecific level or among closely related taxa. Presence and absence of fragments resulting from changes in recognition sites are used for identifying species or populations. . In plant genetic analysis, RFLP markers were initially used for estimating genetic distance and fingerprinting in wheat. In plant and animal breeding, they can reveal which backcross progeny most resemble the desired parent. RFLPs are used to settle cases where biological parentage is at issue.

Many plant species are being analysed using RAPD method. RAPD markers were also used for characterisation, estimation of genetic relatedness and determination of genetic diversity of tea germplasm. These markers are able to distinguish between genotypes but limited to comparisons of populations from a few sources. Lastly RAPD analysis was used to assess genetic differences and relatedness among fig, *Ficus carica* L. clones.

For AFLP analysis, it is an efficient technology for assessing genetic diversity among cultivars such as wheat, *Triticum aestivum* cultivars.. Cultivars were improved through crossing with elite breed however, crossing within such germplasm may lead to a reduction in the level of genetic variation among parental lines due to intense selection pressure.

Thus DNA marker systems are used to reflect actual DNA differences among lines since selection pressure and genetic drift are accounted for. It is an efficient genetic marker system as the technique gives a much higher degree of polymorphism than RAPD or RFLP. In addition, its prior sequence knowledge is not required as the fragments detected are selected by specific primer sets in the PCR reaction. It is reliable as stringent reaction conditions are applied for primer annealing.

Inter SSR (ISSR) fingerprinting was developed such that no sequence knowledge was required. Primers based on a repeat sequence, such as (CA) n, can be made with a degenerate 3'-anchor, such as (CA) 8RG or (AGC) 6TY. The resultant PCR reaction amplifies the sequence between two SSRs, yielding a multilocus marker system useful for fingerprinting, diversity analysis and genome mapping. Previous investigators have demonstrated that ISSR analysis usually detects a higher level of polymorphism than that detected with restriction fragment length polymorphism (RFLP) or random amplified polymorphic DNA (RAPD) analyses.

As with AFLPs, the great advantage of SSRs analysis is the large number of polymorphisms that the method reveals, with one locus in soybean (Glycine max) reported to have 26 alleles. Unlike AFLPs, microsatellites are co-dominant markers, thus heterozygotes can be readily identified. Microsatellite co-dominance will increase the efficiency and accuracy of population genetic measures based on these markers compared with other markers, such as AFLPs and RAPDs.

With a beginning as a mere forensic tool, the world of DNA fingerprinting has a gone a long way in revealing the genetic identity of living beings. With massive evolution in concepts and techniques it has given a lot to various fields cutting across the spectrum. In plants it has not only helped in identifying species but also in defining a new realm in plant genomics, plant breeding and in conserving the biodiversity. With world paving way for developments in biotechnology, DNA fingerprinting promises a very powerful tool in our future endeavors.

Intellectual Property Rights in Plant Varieties

Legal and Policy Concepts Relating to the Protection of Plant Varieties and Plant Breeders' Rights

Intellectual property rights (IPRs) are legal rights granted by governmental authorities to control certain products of human intellectual effort and ingenuity. (OECD 1996). An in-depth discussion of the philosophical and policy goals served by granting legal protection to these products is beyond the scope of this study.

However, a basic familiarity with these goals is necessary to grasp how national and international intellectual property systems and institutions have evolved into their present forms and to understand the constraints that those systems and institutions place on governments seeking to implement competing policy objectives in tension with IPRs.

Two broad philosophical approaches underlie the decision to grant IPRs in the products of human intellectual effort and ingenuity. Elements of these two approaches can be found to different degrees in all national laws and international agreements relating to IPRs.

The first approach to IPR protection predominates in many civil law legal systems, including continental Europe. It takes the position that the products of the human mind are stamped with the personality of their creator, inventor or author, thus endowing him or her with a moral as well as an economic claim to exploit those products to the exclusion of third parties. Under this view, legal protection flows from a state's commitment to protect human rights, a fact reflected in the wording of article 27 of the Universal Declaration of Human Rights, which guarantees to everyone "the right to the protection of the moral and material interests resulting from any scientific, literary or artistic production of which he is the author." Legal protection also rewards creators, inventors and authors for their intellectual efforts as well as their expenditures of time and money. From these underlying premises flows a desire to provide robust intellectual property protection that includes, for example, a broad and expansive range of exclusive rights, long terms of protection, limits on mandatory licensing and narrow exceptions and limitations to exclusive rights.

The second approach to IPR protection takes as its starting premise an instrumental view of intellectual property. Legal protection for the products of human intellectual effort and ingenuity is granted not because of a moral commitment to compensating creators or innovators, but rather because the products they create enrich a society's culture and knowledge and thus increase its welfare. Perhaps the most well-known manifestation of this approach is found in the Intellectual Property Clause of the United States Constitution, which authorizes the United States Congress "[t]o promote the progress of science and useful arts, by securing for limited times to authors and inventors the exclusive right to their respective writings and discoveries."

This instrumentalist philosophy shapes the structure of many national intellectual property systems. The grant of IPRs in nations following an instrumentalist approach is intended to provide adequate incentives for creators, inventors and authors to invest the time, resources and intellectual capital needed to create intellectual property products.

In the absence of a grant of exclusive rights over those products, so-called "free riders" who have not made such investments could exploit inexpensive distribution and reproduction technologies and sell others' intellectual property products at a much lower cost. However, the ultimate goal of legal protection is not remunerative reward for creators but the enhancement of social welfare through access to the ideas and information contained in their products. Instrumentalist intellectual property systems often tailor the scope of legal protection to achieve this goal, for example by placing certain limits on the scope of protection or recognizing situations in which consumers or second-generation creators may access and use intellectual property products for socially valuable purposes.

Policy Objectives Favouring IPRs in New Plant Varieties

The policy goals of granting IPRs to plant varieties are grounded principally on an instrumentalist approach to IPRs. This is true both for patents and plant breeders' rights.

Under this instrumentalist approach, new plant varieties are afforded legal protection to encourage commercial plant breeders to invest the resources, labour and time needed to improve existing plant varieties by ensuring that breeders receive adequate remuneration when they market the propagating material of those improved varieties. In the absence of a grant of exclusive rights to breeders, the dangers of free riding by third parties would be considerable.

This is because the genetic material within plants that specifies their distinctive and commercially valuable features is naturally self-replicating, for example by reproduction of seeds or other propagating material. Self-replication makes innovations incorporating biological material particularly susceptible to exploitation by parties other than the innovator.

IPRs in plant varieties thus provide some assurance to breeders that they will be able to recoup the risks and costs of a value-added innovation that is based upon an underlying biological resource. Ultimately, however, the grant of exclusive rights to plant breeders is designed to benefit the society granting the rights. It provides an incentive for private research and development into new breeding techniques, thereby reducing the need for government funding to subsidize these activities. It encourages the development of new and beneficial plant varieties for use by farmers and consumers. And it furthers the society's development of agriculture, horticulture and forestry.

An international system of IPR protection for plant varieties expands these benefits by facilitating access to new varieties created in other states. Once breeders are assured that their rights will be protected in other states, breeders will be more willing to make their new varieties available in those states (assuming they have access to a distribution and marketing infrastructure). This benefits farmers, consumers and researchers in many more jurisdictions.

The Evolution and Structure of the International Intellectual Property System

The different policy objectives underlying the protection of IPRs have shaped the structure and evolution of the international intellectual property system. Most early domestic intellectual property laws provided no legal protection to intellectual property products created in other nations, thereby permitting those products to be exploited by free riders operating outside the state in which the products were created. The unfairness of this result prompted governments in the late nineteenth century to consider an international approach to protect IPRs.

Limited treaty obligations: The drafters of the first multilateral intellectual property treaties quickly realized, however, that there was insufficient political support for reconciling many of the differences that existed among national IPR laws. For this reason, the drafters abandoned the idea of harmonizing diverse national laws to create a single, international IPR applicable in all signatory states. They fashioned instead a system that creates a limited set of treaty-based obligations that each member state of that system is then required to implement in its national IPR laws.

Implementation of treaty-based obligations in national IPR laws can occur in one of two ways. In some nations (often referred to as "automatic incorporation" states), treaties become binding as a part of domestic law as soon as formal ratification procedures have been adopted.

In these nations, treaties are considered to be "self-executing" or capable of being given "direct effect" in domestic law such that courts and administrative agencies can construe the treaty directly and enforce the rights it grants to the owners of intellectual property products. In other nations, however (often referred to as "legislative incorporation" states), treaties are considered to be "non-self-executing" and can only become binding in domestic law once the parliament or legislature has adopted legislation to implement the treaty. In these nations, owners of intellectual property products rely on this domestic legislation rather than on the treaties themselves when they seek to enforce rights granted to them under the treaties.

The territoriality of IPRs: Because of the limited scope of international IPR agreements, there are to this day (with the limited exception of the European Union) no *international* IPRs available to inventors and creators who seek to market their products across borders; rather, IPRs are territorial in nature and are acquired and enforced on a country-by-country basis under territorially-circumscribed *national* IPR laws.

Thus, for example, the inventor of a genetically enhanced variety of corn who seeks patent protection for that variety must apply for protection in each country in which he or she hopes to sell the corn. The inventor must comply with all of the requirements that each country imposes for granting patent rights to the new variety.

Similarly, once protection is granted, issues such as the scope of the exclusive rights the inventor enjoys in the patented plant variety, the term of patent protection and the limitations imposed on the inventor's rights are all determined by the different national laws. Recent international agreements have achieved some modest forms of procedural harmonization, but they have not altered the fundamental premise that national laws rather than international treaties are the immediate source of nearly all private rights in intellectual property products.

Two basic principles flow from this territorial approach to protecting IPRs. First, where national laws differ as to the scope or content of the protection they provide to intellectual property products, the rights enjoyed by the owners of those products will vary in different national jurisdictions. Second, territoriality implies that each nation has the right to decide on the form of IPR protection to be granted within its own borders, provided that it complies with the obligations contained in international IPR agreements to which it is a party.

Although territoriality thus gives governments some autonomy to set national IPR policies within their own borders, states often view the policies other governments choose as a subject of concern. Indeed, the global reach of markets for intellectual property products makes this concern a necessity.

To take just one example, most patent laws grant inventors owning patents within a state the right to prevent the importation into that state of products created in other nations that contain the patented invention. Thus, where distribution markets transcend national borders, an industry may find itself precluded from distributing products in other jurisdictions as a result of patent rights, as occurred when Indian cotton producers were precluded from importing certain forms of transgenic cotton into the United States.

Core obligations imposed by international intellectual property agreements: The territorial approach to IPR protection appears at first to present myriad difficulties for creators and owners of intellectual property products. In fact, however, the content of each nation's IPR laws are often quite similar since they have been shaped by international IPR agreements ratified by many states. In addition, the obligations these agreements impose have expanded over time, thus narrowing (although by no means eliminating) the differences among national intellectual property systems. The following sections briefly explain the core obligations contained in most international IPR agreements. More detailed information concerning the rights and obligations contained in specific IPR agreements relating to plant varieties is provided in Part II below.

National treatment: One of the cornerstones of international IPR agreements is the national treatment principle. National treatment bars discrimination against foreign IPR owners by requiring that each state provide the same IPRs to private parties from other member states as are provided to the state's own nationals. National treatment levels the playing field among treaty parties and prevents a state from giving its own creators and inventors unfair advantages over foreign creators and inventors. In the absence of national treatment, for example, domestic firms could freely exploit intellectual property products created in other member states while simultaneously enjoying legal protection within their own domestic markets.

***Reciprocity*:** The provisions of several intellectual property treaties contain a limited exception to national treatment known as reciprocity. Where a treaty permits reciprocity, State A may condition the grant of legal protection to intellectual property products from State B upon State B's granting of legal protection to intellectual property products from State A. Reciprocity is often applied to new IPRs as means of encouraging other nations to recognize the new rights and extend their protection to foreign nationals. Once a large number of states have recognized the new IPR, they may revise the treaty to eliminate the reciprocity option and impose a national treatment obligation.

***Most favoured nation treatment*:** The most favoured nation ("MFN") principle is a common feature of international trade agreements but has only recently been applied to IPRs. The principle extends the national treatment rule by compelling a government that provides a privilege or benefit to *one* state within a treaty system automatically to grant that same privilege or benefit to *all* states within the same system. The MFN principle thus prevents a subset of states within a larger treaty system from entering into bilateral or other special agreements among themselves, unless they grant the rights contained in those agreements to all other parties within the larger treaty system.

Subject matter and eligibility requirements: Intellectual property agreements specify the subject matter characteristics of the products that are eligible for legal protection. In the context of patents, for example, a treaty may specify the types of inventions (such as products and processes) to which states parties must grant legal rights. These subject matter requirements are generally drafted using language that instructs member states concerning the basic characteristics that a product must possess for it to merit protection under domestic IPR laws, while preserving sufficient flexibility for states to tailor the details of protection to the particularities of their national legal systems.

Exclusive rights: Where an intellectual property product satisfies a treaty's subject matter eligibility requirements, states parties are required to grant an enumerated set of exclusive rights with respect to that product. These exclusive rights grant to the product's owner the power to exclude all third parties from engaging in the activity that the right covers (such as reproducing or modifying the product or distributing it to others). It is the exclusivity of the rights granted that allows IPR owners to recoup the investment of time, money and resources required to create intellectual property products. The particular exclusive rights mandated by IPR agreements differ depending on the specific type of IPR product at issue. They are drafted with greater or lesser degrees of specificity depending on the degree of consensus among member states' national IPR laws, thus affording varying levels of discretion to governments to implement the rights in their domestic legal systems.

Terms of protection and the public domain: Intellectual property agreements also specify for their states parties the minimum term of years during which intellectual property products must receive legal protection. Once that term has expired, the treaties do not require states to grant legal protection to the products. Thus, unless the state adopts a longer term of protection, after the expiration of the initial term of protection the product may be freely used by anyone for any purpose, including as a source for creating new products or simply for consumption. A corollary of this rule is that national IPR laws do not permit putative inventors and creators to claim IPRs in materials as they are found in nature or where they are already part of the public domain.

Exceptions and limitations to exclusive rights: International IPR agreements constrain the ability of national governments seeking to restrict the exercise of IPRs to achieve competing social or policy objectives, such as access to information, research, education and cultural development. Restrictions designed to achieve these objectives are generally known as "exceptions and limitations" to exclusive IPRs. These exceptions and limitations generally appear in two forms. The first form permits third parties to engage in specified uses of intellectual property products without the permission of the rights holder and without the payment of remuneration. The second form is known as a "compulsory licence." Compulsory licences allow third parties to use intellectual property products without the owner's permission, but only upon the payment of adequate compensation. To prevent both forms of exceptions and limitations from eviscerating IPRs altogether, intellectual property agreements impose specific constraints the ability of member states to adopt them.

Enforcement provisions: The grant of IPRs in national laws would be meaningless without adequate and effective mechanisms to enforce those rights. For this reason, recent intellectual property agreements specify the types of enforcement provisions that member states must adopt in their national laws. These provisions include the imposition of civil and criminal penalties against any person who engages in acts of exploitation reserved to the owner of an intellectual property product without the owner's authorization. The penalties include civil judicial proceedings for monetary damages or an injunction to prevent the continued unauthorized use of the product and criminal proceedings commenced by the government itself.

The "Minimum Standards" Framework of Multilateral IPR Agreements

Taken together, these core provisions of international IPR agreements impose significant legal obligations on member states. The agreements do not, however, purport to definitively address all of the issues raised by the grant of legal protection to intellectual property products. For this reason, the treaties are often referred to as "minimum standards" agreements in that they create only a basic floor of legal protection to which all member states must adhere.

There are three important consequences of this minimum standards framework. First, it allows member states the discretion to interpret and apply those provisions of the treaties that are ambiguous or that reasonably permit more than one construction. Second, a minimum standards approach permits, but does not require, states to grant additional IPR protections within their national laws. And third and perhaps most importantly, the framework leaves member states free to enact laws that serve other political, economic or social objectives, even where those objectives are in tension with IPRs, provided that those laws are not inconsistent with the terms of IPR agreements.

In the last five years, however, an increasing number of developing countries have entered into bilateral or regional trade and investment treaties with the United States or with the European Community. These treaties often contain higher standards of intellectual property protection than those specified in multilateral intellectual property agreements. (Drahos, 2001; GRAIN, 2001)

They thus restrict the minimum standards approach contained in the multilateral agreements, for example, by requiring states to enact specific intellectual property protection rules in their national laws or by requiring developing countries to follow the standards contained in multilateral agreements to which they are not parties.

The minimum standards framework (as limited by bilateral or regional treaties requiring higher standards of protection) provides a methodology for analysis but not a solution for every potential clash between IPRs and other governmental policies. This is particularly true for plant varieties and plant breeders' rights, an area of intellectual property protection regulated by several international IPR agreements and subject to diverse standards of legal protection under different domestic laws. To understand how national governments can reconcile competing and sometimes conflicting domestic laws and objectives consistently with the obligations imposed by different international IPR agreements, it is first necessary to examine other international agreements and international institutions relating to plant genetic resources that have promulgated policies in tension with IPRs.

Identifying the relevant international agreements and institutions relating to intellectual property rights in plant varieties.

Before turning to a more detailed discussion of policy objectives, it is important first to identify the principal international institutions and international agreements that generate legal rules and standards relating to IPRs in plant varieties and plant genetic resources, as well as the institutions and agreements that generate rules and standards in tension with IPRs. Although a comprehensive discussion of these institutions and agreements is beyond the scope of this study, a basic familiarity with their most important elements is necessary to understand the legal regime of intellectual property rights in plant genetic resources for food and agriculture.

World Intellectual Property Organization ("WIPO")

The World Intellectual Property Organization ("WIPO") is a specialized agency of the United Nations charged with "promot[ing] the protection of intellectual property throughout the world." (WIPO Convention, art. 3(i)). The WIPO Secretariat undertakes wide variety of activities relating to IPRs, including hosting diplomatic conferences of government representatives seeking to negotiate new international treaties. WIPO's staff also provide technical assistance and training to member states and their national intellectual property offices, especially in developing countries. More recently, WIPO has created standing, expert and intergovernmental committees that conduct studies on particular intellectual property topics and generate nonbinding guidelines and recommendations for consideration by WIPO members.

WIPO's recent activities in the area of plant genetic resources have been considerable. Issues relating to the intersection of IPRs and PGRs have been raised during the negotiation of two multilateral patent agreements (the Patent Law Treaty and the Substantive Patent Law Treaty) and have been a principal subject of discussion and study in the recently created Intergovernmental Committee on Intellectual Property and Genetic Resources, Traditional Knowledge and Folklore ("IGC"). Developing states first sought to raise issues relating to the intersection of IPRs and PGRs during the WIPO-sponsored negotiation of the Patent Law Treaty in 1999.

They proposed the addition of an article in the treaty requiring applicants for inventions derived from genetic resources to demonstrate that they had received permission to access those resources from the country of origin. Industrialized countries opposed the proposal, arguing that it addressed substantive law issues that were inappropriate for inclusion in a treaty largely devoted to procedural issues. As a compromise, the WIPO Secretariat proposed the creation of a new intergovernmental committee (the IGC) to study the intellectual property aspects of genetic resources and traditional knowledge.

During the IGC's first five sessions between April 2001 and July 2003, WIPO members have supported a wide-ranging work programme for the Committee, including: (1) creating a searchable database of clauses in contracts that regulate access to genetic resources or require benefit sharing; (2) studying technical issues raised by the disclosure of biodiversity-related information in patent applications; (3) creating databases of traditional knowledge; (4) identifying ways to document genetic resources and traditional knowledge in the public domain; and (5) studying the appropriate legal rules to protect traditional knowledge. Most recently, the WIPO General Assembly extended the IGC's mandate and authorized it to accelerate its work, which may include the development of new international agreements.

World Trade Organization ("WTO"): The WTO is a global intergovernmental organization dealing with the rules of trade between nations. It was established in 1994 at the conclusion of the Uruguay Round of trade negotiations held under the auspices of the General Agreement on Tariffs and Trade ("GATT"). As of July 2004, 147 states and customs territories ("WTO Members" or "Members") had joined the organization.

The WTO's substantive obligations are contained in a series of treaties relating to international trade, including agreements on trade in goods, services, agriculture, sanitary and phytosanitary measures, trade-related investment measures, technical barriers to trade, textiles and clothing and intellectual property rights. These treaties are linked together as annexes to the Agreement Establishing the WTO. Disputes between WTO Members relating to any of these treaties are to be resolved by the WTO Dispute Settlement Body, which is composed of *ad hoc* dispute settlement panels and a standing Appellate Body of trade experts. A WTO Member whose national laws or practices are challenged by another WTO Member and found to be incompatible with its WTO treaty obligations by the Dispute Settlement Body must modify those laws or practices or face the prospect of trade sanctions.

A more detailed discussion of the dispute settlement process in the context of the Agreement on Trade Related Aspects of Intellectual Property Rights ("TRIPs" or "TRIPs Agreement"), is provided in paragraph 2.3.1.3.3 below. TRIPs is the WTO agreement that addresses IPRs. It is discussed in detail in section 2.3 below.

Convention on Biological Diversity ("CBD") : The CBD was opened for signature in 1992 and entered into force in 1993. As of July 2004, 188 states had ratified this agreement.

The CBD's main objectives are the conservation of biological diversity, the sustainable use of its components, the fair and equitable sharing of benefits arising out of the utilization of genetic resources and the preservation of indigenous knowledge. (art. 1) The CBD also recognizes that nation states have the sovereign right to exploit their own resources and the authority to determine the conditions of access to them. (arts. 3 and 15)

One of the mechanisms by which the CBD achieves its objectives is *in situ* conversation of plant genetic resources. Conservation *in situ* involves the preservation of ecosystems and natural habitats and the maintenance of viable populations of species in those settings. Such conservation occurs, for example, where farmers and indigenous communities safeguard traditional plant varieties in the locations where they grow naturally or are cultivated.

Although the CBD does not expressly refer to any international IPR agreements, it contains numerous provisions relating to IPRs, principally in article 16. In particular, article 16(5) recognizes that IPRs "may have an influence on the implementation" of the CBD. The article obliges member states to cooperate in order to ensure that IPRs are "supportive of and do not run counter to" the treaty's objectives. Other provisions make clear that the CBD is to be interpreted so as to preserve the rights of IPR owners recognized in international law. For example, articles 16(2)-(4) state that the transfer of technology and measures taken to gain access to such technology shall be consistent with the adequate and effective protection of IPRs recognized in international law. Thus, for example, where a government encourages foreign direct investment in industrial technologies (such as a biotechnological process used to insert new genetic sequences into existing plant varieties), it must respect any patent rights that the owner of that technology has acquired to protect it.

Over time, the biodiversity regime's approach to intellectual property protection has evolved beyond the text of the CBD. The Conference of the Parties (COP) - the convocation of CBD members that determines how the Convention should be applied and implemented - has given detailed attention to harmonizing IPRs with the CBD's objectives. In particular, developing countries active in the COP, together with the support of nongovernmental organizations, have expressed concern about the adverse effects of IPRs and have sought to harness intellectual property rules to promote compliance with the Convention.

In response to this concern, official COP statements have stressed the need to "promot[e] increased mutual supportiveness and integration of biological diversity concerns and the protection of intellectual property rights." (CBD Decision IV/15, para. 9) Members of the COP have worked toward this objective by gathering information, commissioning case studies, holding workshops and drafting guidelines and recommendations.

For example, in April 2002, the COP adopted the "Bonn Guidelines on Access to Genetic Resources and Fair and Equitable Sharing of the Benefits Arising Out of their Utilization." (CBD Decision VI/24, App. II, Annex, Part C) The Guidelines' most important recommendation encourages applicants for IPRs such as patents or plant breeders rights to disclose the country of origin of the genetic resources or the traditional knowledge upon which those IPRs are based. The Guidelines advocate these disclosures to monitor whether applicants for IPRs have obtained the prior informed consent of the country of origin and complied with the conditions of access (if any) that that country has adopted. (Id.)

Although the COP members adopted the Bonn Guidelines by consensus, the Guidelines' reliance upon intellectual property laws to promote compliance with the obligations of the CBD remains controversial. For example, a 2003 Communication from the European Commission to the European Parliament on the EC's implementation of the Bonn Guidelines considers whether to adopt "a self-standing disclosure requirement" for patent and plant variety protection applicants. (EC Implementation of Bonn Guidelines, COM (2003) 821) If adopted, such a disclosure requirement would not, however, be incorporated into national or European intellectual property laws. To the contrary, the failure to comply with the new disclosure requirement would, according to the EC Communication, "only have consequences outside the field of patent law." However, the EC Communication also states that the EC and its member states "should be ready to discuss, in the relevant international fora, the possibility of introducing *under intellectual property law* the same disclosure requirement... as a formal condition for patentability and not only as a self-standing obligation."

The Consultative Group on International Agricultural Research ("CGIAR"): CGIAR is an informal association of public and private donors founded in 1971 that supports an international network of agricultural research centres, each with its own governing body. CGIAR's mission is to engage in research, in partnership with other public and private entities, to promote sustainable agriculture in developing nations. With respect to plant genetic resources, CGIAR's principal method for achieving this objective is *ex situ* conversation. Networks of gene banks within CGIAR store and conserve seeds and propagating materials outside of their natural habitat for future use by farmers, researchers and breeders. The CGIAR network holds the world's largest *ex situ* collection of plant genetic resources for food and agriculture.

International Plant Genetic Resources Institute ("IPGRI"): IPGRI is the world's largest international institute dedicated to the conservation and use of plant genetic resources. Founded in 1974, IPGRI focuses on conservation, management and preservation of the diversity of plant genetic resources, through domestic, regional and international programmes and research initiatives.

The International Undertaking on Plant Genetic Resources (the "Undertaking"): FAO has helped to generate several nonbinding international instruments relating to plant genetic resources. The Undertaking, the first of these instruments, was adopted in 1983. As of 2000, 113 states were signatories to the Undertaking, thus pledging themselves to implement the recommendations it contains. For many years, the Undertaking served as the central legal instrument in FAO's global system for plant genetic resources, a system that includes a fund for the equitable sharing of benefits and a mechanism to give early warning about genetic resources under threat.

The Undertaking's principal objectives are to ensure that the need for conservation is globally recognized and that sufficient funds for this purpose are made available; to assist farmers and farming communities in the protection and conservation of PGRs and of the natural biosphere; and to allow farmers, their communities and countries to participate fully in the benefits derived from improved uses of PGRs, including through plant breeding. (WT/CTE/125, para. 11)

In its initial formulation, the Undertaking challenged a private property rights approach to plant genetic resources by declaring that all such resources, whether as cultivated by farmers in the field or modified through breeder innovations, were part of the "common heritage of mankind and consequently should be available without restriction." (art. 1). An interpretation issued adopted by FAO in 1989, however, clarified that plant breeders' rights were not incompatible with the Undertaking. It also recognized the interrelationship between the rights of traditional farmers (whose practice of saving seeds provided the raw genetic materials for innovation) and the rights of plant breeders (who use technology to achieve that innovation).

The International Treaty on Plant Genetic Resources for Food and Agriculture ("ITPGR"): On 3 November 2001, an intergovernmental conference sponsored by FAO adopted the text of a legally binding international agreement on plant genetic resources. The ITPGR entered into force on 29 June 2004. As of July of that year, the treaty had been ratified by 55 states and signed by an additional 50 countries. The ITPGR not only codifies and updates the nonbinding principles set out in the Undertaking and its subsequent revisions, but also contains provisions relevant to IPRs in plant genetic resources and plant varieties. Because of its importance to the subject of this study, an extended discussion of the ITPGR appears in section 4.3 below.

Specific Policy Objectives in Tension with IPRs

Having identified the principal institutions and agreements relating to plant genetic resources, the next sections address specific critiques of IPRs as applied to plants and plant varieties and the policy arguments that inform those critiques.

Preserving genetic diversity: Granting IPRs to plant breeders has uncertain consequences for preserving plant genetic diversity. Given the incentives that IPRs create for private parties to invest in research and breeding techniques relating to new plant varieties, it might be thought that IPRs would lead to an increase in plant genetic diversity over time. A number of commentators have argued, however, that diversity is eroded rather than enhanced by granting IPRs to plant breeders. According to this view, *in situ* conservation by indigenous farmers diminished as they began to rely on commercial plant breeders for seeds and other propagating material. Rather than using informal breeding techniques to experiment with the creation of new varieties suitable for local growing conditions, indigenous farmers came to depend upon third party plant breeders to provide them with seeds possessing uniform genetic characteristics. The plant varieties that have come to dominate agriculture as a result of this dependence may possess many beneficial characteristics, but they do not enjoy the adaptive abilities of less well known and informally bred varieties.

To evaluate these competing assertions about the effects of IPRs on plant genetic diversity, empirical testing would be useful. Unfortunately, empirical studies of the causal impact of IPRs on plant genetic diversity are difficult to devise because of the uncertainty of screening out other relevant causal variables. Empirical studies that do exist often focus on industrialized countries rather than developing states and tend to rely on anecdotal evidence. As a result, the broader applicability of their conclusions is uncertain and often controversial.

Farmers' rights: A second challenge to IPRs concerns the relationship between farmers' rights and IPRs in plant varieties. The concept of farmers' rights was developed to reflect the contributions that traditional farmers, particularly in the developing world, have made to the preservation and improvement of plant genetic resources. FAO Resolution 5/89 defines farmers' rights as "rights arising from the past, present and future contributions of farmers in conserving, improving and making available plant genetic resources, particularly those in centres of origin/diversity." Such rights are also recognized in article 9 of the ITPGR.

Farmers' rights are in tension with IPRs for plant breeders because many farmers and farming communities do not claim exclusive rights in the cultivated landraces and plant varieties they have cultivated over time. Moreover, the subject matter requirements of existing IPR laws, which as explained below are designed to protect innovations in new and clearly distinguishable plant *varieties*, often cannot accommodate contributions of individual farmers using more informal methods to select for better crops or sought-after plant *characteristics*.

Advocates of farmers' rights have developed different approaches to address this situation and to reward farmers for their contributions to plant genetic diversity. The first approach involves situating the traditional practices of farmers as exceptions to the exclusive rights of plant breeders under existing IPR laws. In other words, breeders are precluded from demanding payment from farmers who engage in certain farming practices, such as saving seeds and planting seeds saved from prior purchases, or informally exchanging seeds. A second approach seeks to modify existing IPR laws so as to permit farmers themselves to claim exclusive rights in the plant varieties they cultivate informally. A third approach involves recognizing farmers' rights not through IPRs but through benefit sharing mechanisms, such as financial payments and technology transfers, that compensate farmers for their contributions to plant genetic diversity. This last approach - which is recognized in article 9 of the ITPGR- questions whether farmers in fact have "rights" as that term is understood within an intellectual property paradigm, while acknowledging the need to reward their contributions to plant genetic diversity.

Traditional knowledge: An issue closely related to farmers' rights is the recognition and protection of the plant-related knowledge, innovations and practices of indigenous and local communities. Advocates assert that those claiming IPRs in plant genetic resources and plant varieties often utilize such knowledge without adequately acknowledging the contributions of the communities who possess it. Mechanisms to redress this problem are similar to those discussed above relating to farmers' rights, a linkage reflected in article 9 of the ITPGR. They include modifying existing IPR laws to recognize traditional knowledge as a form of intellectual property and protecting traditional knowledge through *sui generis* national laws that recognize the communal ownership of much of traditional knowledge. In the fall of 2000, the World Intellectual Property Organization (WIPO) established a new Intergovernmental Committee on Intellectual Property and Genetic Resources, Traditional Knowledge and Folklore to study these issues.

Regulating access to plant genetic resources: Plant breeders and others seeking to develop plant-related innovations need access to existing stocks of plant germplasm for breeding, research and development. Access issues arise for both *in situ* and *ex situ* collections of seeds and plant propagating material. With respect to *in situ* collections, a number of states have enacted national access laws and regulations to control third party access. These access controls are consistent with the CBD, which grants states the sovereign right to control their plant genetic resources. They also further the CBD's benefit sharing objectives by conditioning access upon the current or future payment of compensation or the transfer of technology to the state providing access. The specific conditions of access are often specified in Material Transfer Agreements ("MTAs") entered into between the entity (whether governmental or private) that owns or controls the germplasm and the entity (usually private) that seeks access to it.

Similar access issues arise with regard to *ex situ* collections of plant germplasm, including the seed banks maintained under the auspices of CGIAR. Although pursuant to a 1994 agreement between CGIAR and FAO the autonomous agricultural research centres under CGIAR's auspices each create their own access regulations, certain designated germplasm is to be held by the centres "in trust for the benefit of the international community." (art. 3(a)) With respect to IPRs, the 1994 agreement provides that a centre "shall not claim legal ownership over the designed germplasm, nor shall it seek any [IPRs] over that germplasm or related information." (art. 3(b)). The centres also commit to apply the same restrictions on IPRs in MTAs with third parties to whom the centre transfers designated germplasm. (art. 10)

"Biopiracy" and property rights in unimproved plant materials: Issues of access to plant genetic resources are linked to larger controversies concerning the propriety of granting IPRs in the raw materials accessed.

Under settled principles of intellectual property law, unimproved plant germplasm already in the public domain cannot be removed and privatized. Nevertheless, there are several reported cases and one empirical study which indicate that plant breeders and commercial entities have on occasion been granted IPRs in wild plant varieties or in germplasm found in CGIAR seed collections. It is unclear whether such rights were granted because of inadvertence, insufficient or inaccurate information provided by the IPR applicant or particularities in national laws that limit the sources to which granting authorities may refer in determining what materials are in the public domain. Where such rights have been granted, however, states and interested NGOs have often succeeded in petitioning the authorities in other nations to remove legal protections they had previously granted.

Attempts to claim IPRs in unimproved plant genetic resources have often been labelled as a form of "biopiracy." Biopiracy is not a legal term of art, however and has been loosely used to refer to any act by which a commercial entity seeks to obtain IPRs over biological resources, including plant varieties, that are seen as "belonging" to developing states or indigenous communities. Thus, even where an IPR claim relates to improvements to raw plant materials, certain governments and NGOs have labelled the entity seeking legal protection as a biopirate if it has not provided a fair return to those who granted access to the raw materials.

Plant breeders' research interests: Even as between groups of plant breeders, the scope of IPRs in plant varieties can be controversial. Tensions arise between first generation breeders who have secured legal protection for new varieties and second generation breeders who seek to utilize those new varieties to develop still more varieties. As with farmers' rights, it is possible to use the exceptions and limitations provisions of national IPR laws to permit second generation innovators to engage in such activities without the authorization of first generation breeders.

International IPR Agreements Regulating Plant Varieties and Plant Breeders' Rights

This section identifies and discusses the provisions of three different international IPR agreements that protect plant varieties and plant breeders' rights.

The two major treaty systems that regulate these issues are the agreements established under the auspices of the *Union internationale pour la protection des obtentions vegetales* ("UPOV"), and the TRIPs Agreement included within the family of treaties administered by the World Trade Organization ("WTO"). These two treaty systems each contain a comprehensive set of rules for their members regarding IPRs over plant varieties.

In short, the UPOV treaties adopt a *sui generis* system of protection (that is, a system that is unique, or of its own kind) especially tailored to the needs of plant breeders. The TRIPs Agreement requires WTO Members to protect new plant varieties using patent rights, a *sui generis* system or some combination thereof. Because TRIPs provides states with this flexibility and because the treaty has an uncertain relationship to the previously adopted UPOV conventions, national governments face a wide array of options in choosing the intellectual property regime applicable to plant varieties. This section of the study outlines the requirements imposed by these two treaty systems, and Part III of the study then identifies and analyses these options.

The UPOV Acts

The first UPOV Act was drafted in 1961, principally by industrialized governments seeking to provide protections for plant breeders in their own and overseas markets. The UPOV was later revised in Acts adopted in 1972, 1978 and 1991.

As of July 2004, 55 states were parties to UPOV - 29 to the 1978 Act, 24 to the 1991 Act and 2 to the 1972 Act. (States Parties to the UPOV, Status on 30 June 2004). As a result, this study will focus on the two most recent UPOV Acts. Many of the accessions to these Acts are quite recent, with a number of developing states and countries in transition to a market economy acceding in the 1990s. The 1991 Act entered into force on 24 April 1998 and on that same date the 1978 Act was closed to future accessions except by a few states already in the process of adhering to it.

As explained in section 1.3.1 above, countries generally give domestic effect to the UPOV Act to which they are a party in one of two ways. In "automatic incorporation" states, courts and administrative agencies directly apply and enforce the Act, although implementing legislation is often needed to authorize administrative agencies to process applications to protect new plant varieties.

In "legislative incorporation" states, by contrast, the UPOV Act does not become enforceable in domestic law until the state enacts a national plant variety protection law that conforms to the Act's requirements.

The 1978 UPOV Act

The 1978 UPOV Act adopts most of the international IPR obligations set out in Part I above, including a definition of the applicable subject matter and protected material, eligibility requirements, exclusive rights, national treatment, reciprocity, terms of protection and exceptions and limitations to exclusive rights. It does not, however, contain any provisions on MFN treatment or enforcement.

Subject Matter Requirements

Limited number of protected genera or species : Not all plant varieties must be protected under the 1978 Act. Rather, article 4 provides that member states are to progressively extend protection to an increasing number of genera or species, beginning with five on the date the treaty enters into force for that state and ending with twenty-four within eight years. In addition, member states are free to limit the Act's application within a particular genus or species to varieties with a particular manner of reproduction or multiplication, or a certain end-use.

Preclusion of dual protection with breeder's right and patent : The 1978 Act permits its signatories to protect plant varieties either with a distinct breeder's right or with a patent. However, article 2(1) precludes member states from granting both forms of protection "for one and the same botanical genus or species."

Protection of discovered varieties required : Although the 1978 Act focuses on plant varieties created through classical breeding methods, it is generally accepted that the treaty requires member states to protect varieties which have been discovered. This has been inferred from article 6.1(a) which indicates that a protected variety may result from a natural source of initial variation.

Eligibility Requirements

Assuming that a plant variety falls within a protected genera or species, it is eligible for protection under the 1978 Act only if it is: (1) new, (2) distinct from existing or commonly known varieties, (3) homogenous or uniform and (4) stable. (art. 6) When a variety fulfils these four criteria, it is listed in a national register or catalogue which publicly discloses that the variety is protected. (FIS/ASSINSEL, 2001a).

Novelty : To avoid protection for plant varieties that have already been exploited or are a matter of common knowledge, a new variety in which a breeder seeks protection must not have been sold on the market for more than a specified period of years prior to the date of application for protection. The 1978 Act specifies the maximum number of years during which such pre-application sales have occurred, with different periods of time set for different types of plants as well as for sales within the territory of the applicant state versus the territory of other states. (art. 6(1)(b)).

Distinctness : The 1978 Act states that a protectable plant variety must be "clearly distinguishable in one or more important characteristics from any other variety whose existence is a matter of common knowledge at the time when protection is applied for." (art. 6(1)(a)).

Although the treaty itself does not further define distinctness, the Guidelines for the Conduct of Tests for Distinctness, Homogeneity and Stability (UPOV Guidelines) use both qualitative and quantitative plant characteristics, including such visible attributes as leaf shape, stem length and colour, to determine if the difference between varieties is "clear and consistent." As explained below in the analysis of the UPOV 1991 Act, the concept of distinctness is critical to determining the scope of a breeder's right in plants that are closely related but not identical to a protected variety.

Homogeneity : Under the 1978 UPOV Act, a variety has to be "sufficiently homogeneous, having regard to the particular features of its sexual reproduction or vegetative propagation." (art. 6(1)(c)). The UPOV Guidelines further clarify that to be considered homogeneous, the variation shown by a variety must be "as limited as necessary to permit accurate description and assessment of distinctness and to ensure stability." The homogeneity requirement has been criticized by commentators as discouraging variability in plant varieties that are often useful for sound agricultural practices and as denying protection to breeders of cultivated landraces that exhibit diversity traits. It is thus one of the sources for the criticisms identified in paragraph 1.3.6.1 above that plant breeders' rights are reducing plant genetic diversity by rewarding breeders of uniform plant varieties.

Stability : The stability requirement is a temporal one, requiring the breeder to show that the essential characteristics of its variety are homogeneous or uniform over time, even after repeated reproduction or propagation. (art. 6(1)(d)) In practice, what has been shown to be homogeneous is usually considered to be stable as well. For this reason, the stability requirement has engendered the same sort of critiques as the uniformity requirement in its preclusion of protection for cultivated landraces and other traditional plant varieties.

Protected material : The 1978 Act requires its signatories to protect a variety's reproductive or vegetative propagating material. The Act does not require protection of harvested material, with the exception of ornamental plants that are used for commercial propagating purposes. (art. 5(1))

Breeders' exclusive rights **:** Under article 5 of the 1978 Act, any person seeking to engage in the following three acts, with respect to a protected variety's reproductive or vegetative propagating material, must obtain the prior authorization of the breeder: (1) production for purposes of commercial marketing, (2) the offering for sale and (3) marketing. The 1978 Act does not, however, require member states to extend these exclusive rights to harvested material or other marketed products.

National treatment and reciprocity **:** Member states must grant the three exclusive rights referred to in the preceding paragraph in the same manner to both national breeders and to breeders who reside in or are nationals of other 1978 Act member states. However, where a state extends legal protection to a specific genus or species, or where it provides more extensive exclusive rights to breeders than the rights required under the treaty, reciprocity is permitted. Thus, a state providing these additional rights may restrict protection to breeders from those member states that apply the Act to the same genus or species, or that provide such additional exclusive rights to their own nationals. (arts. 3 and 5(4))

Term of protection : The 1978 Act requires a minimum term of protection of fifteen years, with the exception of vines, forest trees, fruit trees and ornamental trees, which are protected for no less than eighteen years.

Exceptions and limitations : Two major exceptions and limitations to exclusive rights exist under the 1978 Act: (1) a breeders' exemption and (2) a farmers' privilege. The Act also permits members to impose compulsory licences.

Breeders' exemption: This exemption in article 5(3) precludes member states from granting to breeders of protected varieties the right to authorize or refrain from authorizing other breeders seeking to use the protected variety to create new varieties or to market those new varieties. States are permitted to grant breeders such an authorization right only if the repeated use of the protected variety is necessary for the commercial production of the new variety. According to the International Association of Plant Breeders and the International Seed Federation, this breeders' exemption "is essential for continued progress from plant breeding." (FIS/ASSINSEL, 1999)

Farmers' privilege : The focus of the 1978 Act on commercial exploitation of protected plant varieties has been interpreted to allow the use of seeds and propagating material for noncommercial purposes without the breeder's prior authorization. In national plant variety protection laws, this implicit noncommercial exception most frequently benefits farmers who purchase the seeds of protected varieties. The scope of this so-called farmers' privilege varies widely, however. Some nations only permit farmers to plant seeds saved from prior purchases to be used on their own land holdings, while others allow them not only to replant but also to sell limited quantities of seeds for reproductive purposes, a practice often referred to as "brown bagging."

Compulsory licences in the public interest : Article 9 of the 1978 Act permits members to restrict breeders' exclusive rights for "reasons of public interest." Where such restrictions are enacted to ensure the widespread distribution of the variety (such as where the breeder fails to supply the demand for the variety in a reasonable quantity and price or unreasonably refuses to license the variety to third parties), the breeder must receive equitable remuneration.

The 1991 UPOV Act

The limited scope of the 1978 Act led a number of member states of the UPOV to adopt a revised Act with enhanced rights for plant breeders. The major revisions of the 1991 Act are discussed below:

Subject Matter Requirements

Phased-in protection of all genera or species : Unlike its predecessor, the 1991 Act requires states to protect at least fifteen plant genera or species upon ratifying or acceding to the Act, and to extend protection to all plant varieties within ten years. (art. 3(2)) It also contains a definition of a plant "variety" as a "plant grouping within a single botanical taxon of the lowest known rank" which can be "defined by the expression of the characteristics resulting from a given genotype or combination of genotypes; distinguished from any other plant grouping by the expression of at least one of the said characteristics; and considered as a unit with regard to its suitability for being propagated unchanged." (art. 1(vi)) No definition of "variety" appears in the 1978 Act, which indicates that member states to that earlier treaty have greater discretion in defining the characteristics of plant groupings that qualify for protection.

Dual protection with breeders' right and patent permitted : In response to demands from breeders in industrialized counties, the 1991 Act removed the 1978 Act's ban on dual protection and now permits member states to protect the same plant variety with both a breeders' right and a patent.

Protection of discovered varieties : The 1991 Act makes explicit the 1978 Act's implicit requirement that discovered varieties be protected. It does so through article 1(iv)'s definition of a "breeder" as including a "person who bred, or discovered and developed, a variety."

Eligibility Requirements

The four eligibility requirements that must be demonstrated to merit protection for a specific variety - novelty, distinctness, uniformity and stability - are preserved in the 1991 Act, subject to only minor changes in scope and wording. (arts. 7-9) Thus, the 1991 Act has received the same criticism as the 1978 Act for its encouraging of genetic standardization and its inability to protect more diverse plant varieties, traditional varieties or cultivated landraces.

Breeders' Exclusive Rights in Protected Material

Extensive additions to the 1978 Act were made with respect to the exclusive rights enjoyed by breeders in protected material of plant varieties.

Enumerated exclusive rights in propagating material : The breeder's prior authorization must be obtained for the use of reproductive or vegetative propagating material of the variety for (1) production or reproduction, (2) conditioning for the purpose of propagation, (3) offering for sale, (4) selling or marketing, (5) exporting, (6) importing and (7) stocking for any of these purposes.

Extension of rights to harvested material and products : These exclusive rights apply not only to propagating material but also to harvested material, where the harvest has been obtained through an unauthorized use of the propagating material and the breeder has not had a reasonable opportunity to exercise his or her right in relation to that material. In addition, member states are permitted but not required to provide additional rights to breeders, including rights in products made directly from harvested material.

National Treatment Required

Whatever the particular exclusive rights member states adopt in their plant variety protection laws, those rights must also be provided to the nationals of other 1991 Act member states. Unlike the 1978 Act, granting rights only on condition of reciprocity is not permitted. (art. 4) For example, if a 1991 Act member chooses to grant more expansive exclusive rights to breeders than the exclusive rights required by the 1991 Act, it must grant those rights to breeders from all other 1991 Act member states. The state cannot choose to grant such expansive rights to only those 1991 Act states that also provide the same level of rights protection to the first state's nationals.

Terms of Protection

The 1991 Act extends the term of protection to 20 years, and requires a 25-year term for tree and vine varieties.

Exceptions and Limitations

A second area of major revision concerns the reduced scope of exceptions and limitations to breeders' exclusive rights, which is found in article 15 of the 1991 Act.

Private, noncommercial exception : The 1991 Act makes explicit what was only implicit in the 1978 Act, namely that private, noncommercial activities with respect to new varieties are outside of the breeder's control. This exception would presumably permit subsistence farmers to use protected seeds and other propagating material for their own consumption.

Research exception : The 1991 Act also recognizes that breeders cannot restrict "acts done for experimental purposes." This exception would permit research and testing of protected varieties for scientific purposes that does not lead to commercial exploitation.

Limited breeders' exemption : Like its predecessor, the 1991 Act recognizes the right of breeders to use protected varieties to create new varieties. However, this exception is itself restricted in its application to such new varieties as are not "essentially derived" from protected varieties. (arts. 14(5) and 15) The drafters added this restriction to prevent second generation breeders from making merely cosmetic changes to existing varieties in order to claim protection for a new variety. The concept of essential derivation has proved highly controversial in practice, however. Breeders have been unable to agree on a definition of the minimum genetic distance required for second generation varieties to be treated as not essentially derived from an earlier variety and thus outside of the first breeder's control. (FIS/ASSINSEL, 1999). Nevertheless, the overall effect of this provision of the 1991 Act has been to narrow the breeders' exemption and expand the IPRs of first generation breeders.

Limited farmers' privilege : The 1991 Act recognizes an explicit farmers' privilege, but one that is limited in scope. In particular, each member state may enact a provision in its national plant variety protection laws permitting farmers to use for propagating purposes "on their own holdings" the product of the harvest which they obtained by planting a protected variety "on their own holdings." In addition, this privilege must be exercised "within reasonable limits and subject to the safeguarding of the legitimate interests of the breeder." (art. 15(2)) Several points can be inferred from this language.

First, unlike the 1978 Act, the 1991 version of the farmers' privilege does not authorize farmers to sell or exchange seeds with other farmers for propagating purposes, a limitation that commentators have criticized as inconsistent with the practices of farmers in many developing nations, where seeds are exchanged for purposes of crop and variety rotation.

Second, according to the International Association of Plant Breeders, the "reasonable limits" language in article 15(2) requires states to restrict the acreage, quantity of seed and species subject to the farmers' privilege, while the requirement to safeguard breeders' "legitimate interests" requires farmers to pay some form of remuneration to the breeder for their privileged acts. (FIS/ ASSINSEL, 2001a) The latter assertion is controversial, however, and has not been enacted in the national laws of all the 1991 Act member states. For example, the EC's Council Regulation on Community Plant Variety Rights, No. 2100/94 of 27 July 1994, requires such payments except in the case of small farmers, whereas the United States' Plant Varieties Protection Act of 1994, 7 U.S.C. Sec. 2543, does not.

Compulsory licences in the public interest : Article 17 of the 1991 Act contains a compulsory licence provision similar to that found in the 1978 Act. It permits members to restrict breeders' exclusive rights only for reasons of public interest and requires payment of equitable remuneration to the breeder whose rights are limited.

Resistance to the 1991 Act

Many developing nations, particularly those in Africa, have resisted ratifying the 1991 Act or adopting it as the standard for their plant variety protection laws. The foreign ministers of the more than 50-member Organization for African Unity (now the African Union) issued a statement at a January 1999 meeting calling for a hold on IPR protection for plant varieties until an Africa-wide system has been developed that grants greater recognition to the cultivation practices of indigenous communities. However, at a subsequent meeting of the Organisation Africaine de la propriete Intellectuelle (OAPIO), patent officials from sixteen francophone African nations recommended that their countries adopt the 1991 Act. (Machipisa, 1999). As of the date of this study, Tunisia, Kenya and South Africa are the only African UPOV member states. Tunisia ratified the 1991 Act in August 2003, whereas the other two countries are parties to the 1978 Act. Other developing states parties to the 1978 Act are Bolivia, China, Ecuador, Kyrgyzstan, Mexico, Nicaragua, Panama, Paraguay, Trinidad and Tobago and Uruguay. These and other developing countries are, however, under pressure to ratify the 1991 Act as part of bilateral or regional trade and investment agreements.

The TRIPs Agreement

Although the UPOV Acts have provided IPR protection for plant varieties for more than forty years, their significance has recently been overshadowed by a different intellectual property treaty, the Agreement on Trade-Related Aspects of Intellectual Property Rights ("TRIPs" or the "TRIPs Agreement"). Adopted in 1994 as a treaty administered by the WTO, TRIPs is the first and only IPR treaty that seeks to establish universal, minimum standards of protection across the major fields of intellectual property, including patents, copyrights, trademarks, industrial designs, integrated circuits and trade secrets. Although the TRIPs Agreement devotes only minimal attention to plant breeders' rights or plant variety protection and does not even mention the UPOV Acts, its adoption has done more to encourage the legal protection of plant varieties than any other international agreement.

TRIPs as a Spur to Plant Variety Protection

TRIPs' influence on plant variety protection stems from the following sources: (1) its link to other international trade agreements; (2) its widespread adherence by states in both the industrialized and developing world; (3) its novel enforcement, review and dispute settlement provisions; (4) the requirement in TRIPs article 27.3(b) that its signatories must provide protection for plant varieties "either by patents or by an effective *sui generis* system or by any combination thereof"; and (5) a formal review of article 27.3(b) which was scheduled to be held in 1999. The following paragraphs briefly address each of these five issues. Subsequent sections devote more detailed treatment to the protection of plant varieties with patents and to the elements necessary to create an "effective *sui generis* system."

The Relationship of TRIPs to the WTO and the International Trading System

Unlike all prior intellectual property treaties, TRIPs is not a free-standing agreement concerned solely with IPRs. Rather, TRIPs is linked to a larger family of trade-related agreements concerning subjects such as trade in goods and services, agriculture, textiles and health-related restrictions on imports. All of these agreements were adopted within the WTO during the Uruguay Round of trade negotiations held between 1988 and 1994. As such, TRIPs was part of a global "package deal." Industrialized nations secured a commitment from developing nations to provide minimum standards of effective legal protection to intellectual property products, and in exchange developing nations received a commitment from industrialized countries to open their domestic markets to goods and other products manufactured in the developing world.

Widespread Adherence to TRIPs

The result of this global bargain was widespread adherence to all WTO Agreements, including the TRIPs Agreement. As of July 2004, 147 states or customs territories were obligated to comply with TRIPs by virtue of their membership in the WTO. In addition, many states (particularly those in the developing world) who became bound by TRIPs had not ratified earlier intellectual property treaties and were thus required to make significant changes to their national laws to bring them into compliance with the Agreement. In acknowledgement of this fact, TRIPs contains a phase-in provision pursuant to which most of the treaty's substantive obligations became binding for developed nations in 1996, but did not become binding on developing nations and nations in transition to market economies until 2000. Least developed nations are given the most leeway and are not required to implement the treaty's substantive obligations until 2006.

The Enforcement, Review and Dispute Settlement Provisions of TRIPs

In addition to its widespread adherence, the influence of the TRIPs Agreement can be traced to its unique provisions relating to the enforcement of IPRs within national laws, the review of those national laws by the TRIPs Council and the mechanism for settlement of disputes between states leading to rulings backed up by the threat of trade sanctions.

Enforcement provisions : Unlike prior IPR agreements, TRIPs does not only specify the minimum substantive requirements for various forms of intellectual property. In addition, it requires WTO Members to adopt "effective" provisions within their national laws to permit the owners of intellectual property products to enforce their rights against those who infringe them. (art. 41.1) These enforcement provisions include detailed judicial and administrative remedies, border measures and criminal procedures. (articles 41 to 61) To take just one example applied to plant varieties, a breeder whose new variety is sold commercially without permission must be able to bring a civil judicial action seeking an injunction to stop the conduct of the unauthorized seller and to recover damages from him or her.

Review provisions : Article 68 establishes a Council for TRIPs to "monitor the operation of th[e] Agreement and... Members' compliance with their obligations" under the treaty. Since its founding, one of the Council's principal functions has been to formally review the national laws, regulations and judicial decisions of WTO Members in each area of intellectual property covered by TRIPs.

Several goals are served by the TRIPs Council's review functions. First, the reviews create an incentive for governments to bring their national laws and practices into compliance with the Agreement so that they may present positive information to the Council. Second, the reviews identify for both the reviewed state and for other WTO Members areas of the law which may not be in full compliance with the treaty. And finally, TRIPs Council reviews provide a critical opportunity to publicize national laws and practices relating to IPRs which might otherwise be difficult to obtain.

In the area of plant varieties, the TRIPs Council has gathered and organized a considerable amount of useful information concerning national government practices. In December 1998 the Council prepared a detailed list of questions concerning plant variety protection. It asked WTO Members who were already obligated to protect plant varieties to respond to these questions by indicating the manner in which their national laws provide such protection, with other WTO Members requested to use their best efforts to furnish such information. The Council later received and published responses from 17 states and from the European Communities and its member states. (WTO Job No. 2689 of 7 May 1999). In 2001 and 2002, the Council reissued these questions and invited WTO Members that had not already done so to provide information on their plant variety protection practices. Additional responses were received from six states. Information distilled from some of these responses is discussed in section 3.3 below.

Dispute settlement provisions : Another significant innovation of TRIPs is its dispute settlement system. Although prior intellectual property agreements, including the UPOV Acts, contained provisions for filing of complaints against non-complying treaty parties before the International Court of Justice, no state ever invoked this dispute settlement option. It was widely believed that pursuing such a complaint would be perceived as an unfriendly act, would be time-consuming and costly and that states were unlikely to implement decisions of the court. The TRIPs Agreement removes each of these concerns by linking to a WTO Agreement known as the Dispute Settlement Understanding which contains a streamlined dispute settlement mechanism that is one of the most widely used and effective interstate dispute resolution systems in international law. Indeed, the mere existence of such a system creates strong incentives for WTO Members to bring their national laws into compliance with the WTO Agreements, including TRIPs, to avoid the possibility of dispute settlement proceedings. These strong incentives notwithstanding, some countries may not comply.

In such a case, where one WTO Member believes that another Member has failed to fulfill its obligations under the Agreement, it may initiate consultations under the Dispute Settlement Understanding with a view to resolving the dispute. If the parties fail to reach agreement, the complaining state may then request the WTO to convene a three-member panel of experts to review its allegations. Such panels generally issue a decision within six months. The panel's decision is adopted by the WTO's Dispute Settlement Body unless the losing party or parties elect to appeal the decision to the seven-member Appellate Body, a standing tribunal of seven trade experts who are authorized to review the panel's findings and issue a decision within three months. If either the Appellate Body or an unreviewed panel decision concludes that a WTO Member has violated the TRIPs Agreement, the Dispute Settlement Body will recommend that the defending state bring its national laws into compliance with the treaty. If the state fails to do so, the complaining state may then commence an arbitral proceeding to specify the amount of compensation that the defending state must pay to remedy the violation, or in lieu of such compensation the complaining state may seek authorization to impose trade sanctions on the non-complying Member.

As of July 2004, WTO Members (mostly developed nations) had commenced 24 dispute settlement proceedings against both developed and developing nations concerning a wide variety of intellectual property issues. (WTO Dispute Settlement: Index of Disputes Issues: 15 July 2004) Approximately one third of these disputes were resolved prior to a decision by a panel, with the defending state agreeing to modify its laws to bring them into compliance with the Agreement. In other cases, however, WTO Members complied only after a panel or the Appellate Body had issued a decision against them. And in a few instances, Members that have not modified their laws instead have negotiated settlements involving the payment of compensation to the complaining Member.

As of the date of this study, there have been no TRIPs dispute settlement proceedings relating to intellectual property protection for plant varieties. Whether such proceedings will be brought in the future depends on a number of variables, many of which are still uncertain.

Three interrelated reasons suggest that a dispute settlement proceeding concerning plant variety protection is a likely possibility: first, the significant difference of views among WTO Members over the scope of legal protection to be provided to plant varieties; second, the fact that developing country WTO Members first became obligated to protect plant varieties only in 2000; and third, the absence of plant variety protection in many developing countries' national laws prior to the 2000 deadline. Taken together, these variables suggest that the plant breeders of a developed WTO Member who learn that their protected varieties are being exploited without authorization in developing WTO Members due to inadequate national laws will pressure their governments to file a complaint against such states to compel them to adhere to their treaty obligations.

Other factors, however, suggest that a complaint over plant varieties is unlikely to be filed, particularly within the next five to ten years. First, the initiation of the Doha Round of trade negotiations in November 2001 (discussed in Part IV below) is likely to cause WTO Members to show restraint in their dispute settlement strategies while the multi-year negotiations are proceeding. This is especially so given that the Doha Round will likely provide WTO Members with the first meaningful opportunity to harmonize other international obligations (such as those set forth in the CBD and the ITPGR) with the plant variety and other IPR protection requirements of the TRIPs Agreement. Second, past practice has shown that WTO Members are most likely to file dispute settlement complaints where they are pressured to do so by the owners of intellectual property products. It is unclear whether plant breeders can make a sufficient showing that their overseas markets are being harmed by other states' failure to provide adequate legal protection for plant varieties. Third, WTO Members are generally reticent about filing complaints where their probability of success is unclear. Therefore, given the uncertain scope of protection for plant varieties currently required by TRIPs, if Members enact *some form* of plant variety protection in their national laws, other Members are unlikely to challenge those laws unless they can identify a clear violation of the Agreement. Nevertheless, the potential for dispute settlement proceedings (and the trade sanctions they can engender) will continue to provide strong incentives for states to enact plant variety protection laws.

The Unique Scope of Plant Variety Protection Required by TRIPs

Article 27.3(b) contains the only textual provisions of the TRIPs Agreement relating to plant variety protection. The article states in relevant part:

Members may also exclude from patentability: (b) *plants* and animals other than microorganisms; and essentially biological processes for the production of plants or animals other than non-biological and microbiological processes. However, *Members shall provide for the protection for plant varieties either by patents or by an effective* sui generis *system or by any combination thereof*. (emphasis added)

The meaning of this article has been the subject of significant debate among both WTO Members and nongovernmental organizations (NGOs) with differing views over the propriety of IPR protection for plant varieties. A detailed review of that debate and an analysis of article 27.3(b) is provided below. For present purposes, two overarching points are worth noting.

First, TRIPs' provisions on plant varieties do not refer to or incorporate any preexisting intellectual property agreements, including the 1978 and 1991 UPOV Acts. This omission contrasts sharply with other fields of intellectual property, such as patents, copyrights and trademarks, for which TRIPs expressly requires WTO Members to comply with the standards of protection contained in preexisting IPR agreements, such as the Berne Convention for the Protection of Literary and Artistic Works and the Paris Convention for the Protection of Industrial Property.

As a result of this omission, WTO Members are neither required to become members of UPOV nor to enact national laws consistent with either UPOV Act in order to comply with their obligations under TRIPs. Although the drafting history of TRIPs does not explain this markedly different treatment of plant varieties, it seems likely that compliance with UPOV was not required because so few WTO Members were party to UPOV and those who were could not agree upon which of its two most recent Acts should serve as the standard for protection.

Second, article 27.3(b) permits WTO Members to protect plant varieties using one of three distinct approaches: (1) patent law, (2) an effective *sui generis* system or (3) a combination of elements from both systems. Thus, unlike most other areas of intellectual property protected by TRIPs, article 27.3(b) expressly grants Members significant discretion to choose the manner in which they will protect plant varieties and it contemplates that that discretion may be exercised differently by different states.

This discretion and the opportunity for divergent outcomes it engenders has significant consequences. On the one hand, TRIPs' failure to incorporate and build upon the preexisting UPOV Acts may have "a deharmonizing effect," (Correa 1994a) with states within the UPOV system enacting one type of plant variety protection law and states outside of that system enacting a different kind of law (which may or may not resemble each other).

This could create significant complexities and uncertainties for plant breeders seeking to market protected varieties in different jurisdictions. On the other hand, this sanctioned diversity of legal approaches allows WTO Members to balance the protection of plant breeders' rights against the other important and competing societal goals identified in Part I, many of which are found in other international agreements. Seen from this perspective, article 27.3(b) provides a much needed "safe space" for governments to harmonize conflicting norms and policies - a space that is lacking in other areas of the TRIPs Agreement.

The Review of TRIPs Article 27.3(b)

The last sentence of article 27.3(b) states that "[t]he provisions of this subparagraph shall be reviewed four years after the date of entry into force of the WTO Agreement." Inasmuch as the WTO Agreement entered into force on 1 January 1995, the review contemplated by this article should have been conducted in 1999. The review was commenced but was not concluded, largely as a result of disputes between industrialized and developing nations over the scope of the review process. Governments submitted additional information and proposals to the TRIPs Council in 2000 and 2001, but no formal action was taken. With the launching of the Doha Round of trade talks in November 2001, however, the review of article 27.3(b) recommenced in earnest.

In June 2002, eleven developing states submitted a detailed proposal to amend TRIPs to "prevent systematic conflicts with the CBD arising from the implementation of TRIPS." (WTO Doc. IP/C/W/356, para. 11) The proposal seeks to compel all WTO Members to require applicants for patents relating to biological materials and traditional knowledge to disclose certain information as a condition of obtaining legal protection. (Id., para. 10)

The European Communities and Switzerland responded to the developing states' proposal with a compromise. The EU response calls for the negotiation of "a self-standing disclosure requirement," which, while not functioning as a new eligibility criterion for patent protection, "would allow Members to keep track, at [the] global level, of all patent applications with regard to genetic resources for which they have granted access." In June 2003, developing countries rejected this compromise and "reaffirm[ed] and strengthen[ed] their demand for a strong disclosure of origin mechanism" within TRIPs that would require "not only detailed information about who provided the [genetic] materials or the [traditional] knowledge used [in patent applications], but also positive proof of benefit sharing and of prior informed consent." African countries also proposed a new "Decision on Traditional Knowledge" to be incorporated into TRIPs.

"TRIPs Plus" Bilateral and Regional Agreements

In the decade since TRIPs was first adopted, the United States and the European Community have negotiated a growing number of bilateral and regional trade and investment treaties with developing countries. Many of these treaties contain provisions concerning IPRs. Commentators have referred to these treaties by the appellation "TRIPs plus" because they (1) contain intellectual property protection standards more stringent than those found in TRIPs, (2) obligate developing countries to implement TRIPs before the end of its specified transition periods or (3) require such countries to accede to or conform to the requirements of other multilateral intellectual property agreements. As one study recently noted, "[t]o the extent that ["TRIPs plus" treaties] include these additional aspects, they are pushing harmonization forward at a pace that is greater than is apparently possible within the framework of the WTO." (OECD, 2003) In addition, once a developing state enacts this higher level of IPR protection in its national laws, TRIPs' MFN clause obligates that state to extend the same protection to *all other WTO Members.* In this way, bilateral agreements can be used to ratchet up the level of protection for IPRs beyond what is required in the TRIPs Agreement.

Several "TRIPs plus" treaties contain provisions relating to plant variety protection that go beyond the minimum level of intellectual property protection specified in TRIPs.

For example, the recently concluded U.S.-Central America Free Trade Agreement (applicable to Costa Rica, El Salvador, Guatemala, Honduras, Nicaragua and possibly to the Dominican Republic), the 2002 U.S.-Chile free trade agreement, the 2000 U.S.-Jordan free trade agreement and the 2000 EU-Mexico free trade agreement, as well as certain Euro-Mediterranean Association agreements and the draft Free Trade Agreement of the Americas, all mandate UPOV as the appropriate mechanism to protect plant breeders' rights. (GRAIN, 2004b) The treaties also require these countries to ratify the 1991 UPOV Act within specified time frames. Such obligations limit the discretion enjoyed by these countries to tailor their national laws to achieve policy objectives in tension with IPRs.

Patent Protection for Plant Varieties Under TRIPs

The TRIPs Agreement mandates its signatories to provide patent protection for any inventions in all fields of technology, provided that the inventions are "new, involve an inventive step and are capable of industrial application." (art. 27(1)) However, with regard to plant-related inventions, TRIPs permits Members to exclude from patentability altogether "plants," "essentially biological processes for the production of plants" and "plant varieties." (art. 27.3(b)) Thus, as presently written, the TRIPs Agreement would permit WTO Members to decline to protect plant varieties with a patent (provided, as discussed in section 2.3.3 below, that they protect such varieties with an effective *sui generis* right).

Applicability of Patent Protection for Plant Varieties

For this reason, it may initially appear that governments can ignore patents as they consider how to protect plant-related innovations in their national legal systems. Nevertheless, a basic understanding of the patent provisions of the TRIPs Agreement and a comparison of those provisions with the plant breeders' rights approach contained in the UPOV Acts is essential for a number of reasons.

First, extending patent protection to plant-related inventions and innovations remains an option for national governments. Recall that TRIPs' "minimum standards" framework expressly contemplates that WTO Members may provide greater protection for IPRs than are mandated by the Agreement. Article 27.3(b) in particular invites Members to protect plant varieties with patents or with a combination of patents and a *sui generis* system. A number of governments in the industrialized world, including the United States, Japan, Australia, New Zealand, Sweden and the United Kingdom, have capitalized on this opportunity by permitting plant breeders to obtain patent protection in new varieties provided that the eligibility requirements for a patent have been met.

Second, because TRIPs does not require any patent protection for plant-related innovations, it follows as a matter of course that the treaty does not compel WTO Members to adopt any particular form of patent protection. This allows governments the option of including plant varieties within their existing utility patent statutes and/or of enacting a separate statute applicable exclusively to plants. (Utility patents are generally granted to "any new and useful process, machine, manufacture or composition of matter, or any new and useful improvement thereof... " In making this decision, however, it will be useful for governments to understand the traditional elements of patent law and the ways in which they may be varied within the confines of the TRIPs Agreement. The United States, for example, has enacted multiple plant variety protection laws. It was one of the first nations to adopt, in the Plant Patent Act of 1930, a unique form of protection applicable to asexually reproducing plants. In addition, since the mid 1980s, U.S. courts and the U.S. Patent and Trademark Office have concluded that plant breeders may also seek standard utility patents under the U.S. Patent Act. In December 2001, the U.S. Supreme Court confirmed that breeders could apply for both forms of protection with respect to the same variety. Such a result is prohibited by the 1978 UPOV Act but permitted under the 1991 Act to which the U.S. is a party.

Third, an understanding of patent principles is also necessary because of the difficulty in some nations of distinguishing between plant varieties (which are often excluded from national or regional patent laws) and other plant-related innovations (to which patent protection may be extended). In Europe, for example, article 53(b) of the European Patent Convention prohibits the patenting of "plant varieties" as such. Nevertheless, the European Patent Office recently confirmed that claims to patent protection that are broadly drawn to encompass "plants" or an invention broader than a single variety *may* be patented, even though such claims may encompass multiple varieties. As a result, it may be possible for plant breeders in Europe to fashion their claims to receive *de facto* patent protection for new plant varieties.

There are significant differences in approach between plant breeders' rights and patents. In the case of plant breeders' rights, the eligibility requirements for protection are not onerous, but the scope of protection granted is quite narrow, both in terms of exclusive rights and the various exceptions and limitations to those rights. Patent laws strike a very different balance. Eligibility requirements are high and difficult to meet, but once granted a patent conveys broad rights to exclude third parties from exploiting the patented invention. Depending on the needs and level of development of plant breeder industries within its territory, a government may decide that either or both forms of protection will provide the appropriate incentives to encourage plant-related research and innovation.

Fifth, and finally, a review of the TRIPs Agreement's plant-related patent rules will soon occur in the context of the Doha Round of trade negotiations. Such a review may produce an agreement to amend TRIPs to require WTO Members to extend patent protection to plant-related innovations or plant varieties. National governments which have thus far eschewed patent protection would then need to revise their national laws to make such protection available, in addition to or instead of the legal protection provided under national plant variety protection laws. At a minimum, therefore, governments interested in promoting plant-related innovations should be apprised of the relevant legal issues so that they may contribute in an informed way to the upcoming discussions.

Subject Matter Eligibility Requirements

According to the text of the TRIPs Agreement and authoritative commentary interpreting it, plant-related innovations may be patented if they are "inventions" that are novel, non-obvious and useful. Although commentators frequently assert that these cumulative requirements impose significant barriers to the patenting of plant varieties, in practice breeders have succeeded in patenting not only new varieties, but also hybrid plants and inbred and hybrid plant lines. As two commentators recently noted, since 1985 the U.S. Patent and Trademark Office "has granted hundreds of utility patents on all aspects of innovation relating to plant science: plants themselves, seeds, breeding methods and plant biotechnology."

Inventions : The TRIPs Agreement does not define the term "invention," thus leaving its precise meaning to national patent laws or regional patent rules such as those operating within the European Union.

All such laws recognize that an invention must be more than a mere discovery of a natural phenomenon or naturally occurring substance. Thus, for example, a breeder who merely identified the existence of a previously unknown wild variety or a cultivated landrace that was known only to an isolated indigenous community should not be permitted to patent that variety or landrace. The trend among the industrialized countries, however, and in the United States, Japan and the countries of the European Union in particular, is to recognize that an isolated and purified form of a natural substance may be patented. For example, article 3.2 of the 1998 Directive on the Legal Protection of Biotechnological Inventions states that "biological material which is isolated from its natural environment or produced by means of a technical process" may be patentable. (Directive 98/44/EC of 6 July 1998, O.J. L213) Perhaps because this approach involves making a very fine distinction between discoveries and inventions, many developing countries have declined to follow it, instead choosing to deny patent protection to plant materials found in nature even if they have been isolated or purified by human intervention. According to one commentary, this patent ban is compatible with the TRIPs Agreement provided that it does not extend to plants with modified or artificial plant gene sequences, which often significantly vary from naturally occurring substances and thus are properly classified as inventions.

A second attribute of a patentable invention is that an applicant seeking such a patent must disclose the invention in a sufficiently clear and complete manner that permits a person skilled in the art to carry out the invention. (TRIPs, art. 29(1)). In addition to ensuring that the claimed invention in fact meets patent eligibility requirements, disclosure also permits third parties to access the invention for purposes of improving it or developing new inventions. In the case of plant-related innovations, national laws permit inventors to satisfy the disclosure requirement either by a written description of the invention and/or by a deposit of the protected material (i.e. seeds, germplasm or other biological material). TRIPs does not require national governments to adopt any particular form of disclosure, nor does it specify the timing, manner or conditions of third party access. As a result, national laws vary widely on this point.

Novelty, inventive step and industrial application : These three cumulative eligibility requirements are mandated by article 27(1) of the TRIPs Agreement and are found in all national patent laws. Even among states with strong patent protection laws, however, the interpretation and application of each of these steps varies considerably.

Novelty : The principal objective of the novelty requirement is to ensure that the claimed invention cannot be found in the "state of the art" or "prior art" already in existence. These terms refer to the body of knowledge which is available to the public before the date of filing of an application for patent protection. Patent examiners establish both the novelty and inventive step requirements by comparing the claimed invention to the state of the art or prior art. A critical issue for plant-related innovations concerns the form in which prior art exists and the difficulty of accessing traditional knowledge as prior art. Under United States patent law, for example, novelty can be negated if "the invention was patented or described in a printed publication in this or a foreign country or in public use or on sale in this country... " (35 U.S.C. Sec. 102(b)) The important point is description: foreign *uses* of claimed inventions are not considered as part of prior art.

The Patent Cooperation Treaty (PCT) adopts a similar approach, limiting prior art to "everything which has been made available to the public anywhere in the world by means of written disclosure." (Rule 33.1 of Regulations under the PCT) Thus, in the United States and in countries following the PCT, even where foreign plant-related innovations are known, used or disclosed in other than written form, it may be possible for the inventor to obtain patent protection.

Article 54(2) of the European Patent Convention adopts a different approach, defining the state of the art to encompass everything made available to the public by means of written or oral descriptions, by use or indeed in any other way, prior to the filing of the application. Nevertheless, the absence of systemic databases concerning traditional knowledge, including knowledge relating to plants, has led examiners in various countries to grant patents for inventions that were later revoked after they were discovered to encompass prior art.

Inventive step : The inventive step requirement looks to the state of the art to determine whether the claimed invention is not obvious to a person skilled in the art. Although the majority of international treaties and national laws defined inventive step in this way, the application of the standard to plant-related innovations varies and may be a significant hurdle to patent protection given "the enormous speed of technological progress" in the field.

Industrial application : The industrial application requirement is concerned with an invention's practical utility, and in particular whether "it can be made or used in any kind of industry, including agriculture." (E.g. European Patent Convention, art. 7) This eligibility requirement does not appear to pose a significant obstacle to the patenting of new plant varieties, given their ready use in the fields of plant breeding, horticulture and agriculture.

Exclusions from Patentability

As noted above, article 27.3(b) of TRIPs permits WTO Members to deny all patent protection to "plants" and "plant varieties." Nevertheless, a state that chooses to provide such protection may nevertheless choose to exclude from patentability a narrower category of inventions, "the prevention within their territory of the commercial exploitation of which is necessary to protect *ordre public* or morality, including to protect... plant life or health or to avoid serious prejudice to the environment." (Art. 27(2)) Some commentators have asserted that this language grants states the discretion to exclude broad categories of inventions that further the erosion of genetic diversity on the theory that issues of public policy and morality include a state's sovereign right to control its natural resources. However, because WTO dispute settlement panels have strictly interpreted clauses in the WTO Agreements that permit states to derogate from protected treaty rights, the more persuasive view is that exclusions on this basis must be fact-specific, narrowly drawn and no broader than required to achieve their purpose. (Moufang, 1998) Resolution of this debate in the context of plant-related innovations will not occur until the WTO dispute settlement bodies issue a ruling on the matter, an unlikely prospect so long as Members are free to exclude plants and plant varieties entirely from patentability.

Exclusive Rights

Once a patent is awarded, the owner of the patented product or process enjoys a broad panoply of exclusive rights. These include the right to prevent third parties from making the product, using the process or from using, offering for sale, selling or importing for those purposes the patented product or the product obtained by the patented process. (TRIPs, art. 28) States that choose to extend patent protection to plant-related innovations would thus be expected to protect products composed of plants and parts of plants and processes (including biological processes) for the production of plants. Although there is some dispute over whether traditional plant breeding methods (as opposed to creating varieties through genetic manipulation) can be protected by process patents, the practice of states that do extend patent protection to plants reveals that such methods have been protected.

Term of Protection

TRIPs article 33 imposes a minimum 20-year term of protection from the date the patent application is filed.

Exceptions and Limitations

As compared to limitations on plant breeders' rights permitted under the UPOV, the limitations on a patent owner's exclusive rights permitted under the TRIPs Agreement are far narrower. These limitations can be divided into exceptions to exclusive rights and compulsory licenses, which permit certain uses by third parties but require remuneration to the patent owner.

Exceptions : Article 30 of TRIPs permits WTO Members to adopt "limited exceptions" to exclusive patent rights provided that they do not "unreasonably conflict with a normal exploitation of the patent" and "do not unreasonably prejudice the legitimate interests of the patent owner, taking account of the legitimate interests of third parties." This three-part test is applicable to all patents protected by the TRIPs Agreement.

Canadian Patents Case : A WTO dispute settlement panel clarified the meaning of article 30 in Canada - Patent Protection of Pharmaceutical Products (Generic Medicines), WT/DS114/R (17 March 2000). The panel considered a complaint brought by the European Communities to two provisions of Canadian patent law. The first permitted uses of patented pharmaceuticals without the patentee's authorization for the purpose of obtaining approval of a generic product before the patent's term expired.

The second permitted production and stockpiling of the generic product for release immediately after the expiration of the patent. Both procedures were designed to permit the marketing of generic versions of pharmaceuticals promptly after the expiry of the patent. The panel decision confirmed that only the first exception is consistent with TRIPs. More importantly, the panel concluded that any exception that results in "a substantial curtailment" of the patent owner's exclusive rights is inconsistent with article 30. (WT/DS114/R, para. 7.36)

If TRIPs were amended to require patent protection for new plant varieties, the standard adopted by the Canadian Patents panel would preclude Members from enacting many of the exceptions that are permitted under plant variety protection laws, such as the breeders' exemption and the farmers' privilege. The same prohibition on these privileges already exists in several countries that currently protect plant varieties and other plant-related innovations in their national patent laws.

Research and experimentation exception : Most national patent laws permit third parties to engage in experimentation or research related to the patented invention. Some states, however, including the United States, narrowly construe this exception to preclude unauthorized commercialization of any products or processes that result from such research if they encompass the patentable invention. In such countries, commercial activities permitted under the breeders' exemption found in plant variety protection laws are prohibited where the variety is protected by a patent.

Thus, for example, a breeder would not be permitted to cross patented seeds to produce improved varieties. Even in patent systems in which the experimental use exception tolerates some commercialization (Cornish, 1998), such as those found in Europe, it is likely that the commercial acts allowed under a traditional breeders' exemption would "conflict with a normal exploitation of the patent" and would thus amount "a substantial curtailment" of his or her exclusive rights in violation of TRIPs article 30. Thus, if patent protection of plant varieties were to become mandatory under TRIPs, a breeders' exemption such as that found in article 22,V of Mexico's patent law would be unlikely to survive a challenge before a WTO dispute settlement panel.

Farmers' privilege : A similar fate is likely to befall the exception permitting farmers to save and reuse seeds on their own land without the patent owner's permission, another exception traditionally allowed under plant variety protection laws. The United States does not recognize a farmers' privilege under its utility patent laws. Although such an exception has been adopted in article 11 of the European Union's 1998 Biotechology Directive (conditioned upon the payment of equitable remuneration except by small farmers), at least one commentator has questioned whether that exception is compatible with TRIPs article 30, because such an exception would unreasonably prejudice the "legitimate interests" of the patent owner.

Compulsory licences : TRIPs contains a complex set of rules that regulate when states may compel patent owners to license their products and processes to governments or to private parties. Although the TRIPs Agreement does not specify the grounds which justify the creation of compulsory licences, because article 5A(2) of the Paris Convention is incorporated by reference into article 2(1) of TRIPs, it can be inferred that such licences may be granted only to prevent "abuses which might result from the exercise of the [patent owner's] exclusive rights."

Even where such abuses exist, TRIPs article 31 imposes further conditions upon the granting of compulsory licences, including individual consideration of each case, prior negotiations with the patent owner seeking a voluntary licence, limitations on the scope and duration of compulsory licences and requiring their termination when the circumstances leading to their creation are no longer in effect. Most importantly, patent owners must receive "adequate remuneration," taking into account the value of the rights licensed.

One area in which compulsory licences may affect plant breeders is that of dependent patents, which are defined as patents whose use requires the authorization of an earlier patent owner. Such patents are prevalent in plant breeding, where the creation of new varieties often occurs incrementally in the form of adaptations and improvements of existing varieties, as opposed to radically new innovations. Because incremental innovation often requires access to protected varieties, governments may seek to impose compulsory licences in favour of third party breeders who are unable to negotiate voluntary access to patented plant varieties. However, the compatibility of such licences with TRIPs is untested and governments seeking to impose them should ensure that they comply scrupulously with each of the many requirements set forth in article 31.

Because the patent compulsory licence provisions of TRIPs are far more detailed and narrow than the compulsory licences permitted under the 1978 and 1991 UPOV Acts, commentators are divided over their usefulness in the area of plant-related innovations. Some argue that such licences will be largely irrelevant to the field, whereas others claim that governments may adopt such licences "to ensure access to patented materials in order to attain specific agricultural objectives (e.g. availability of a given material for farmers) or food security...." Even if licences to achieve these objectives are permitted, however, the restrictions that TRIPs imposes significantly limit the ability of WTO Members to grant such licences in favour of third parties.

Sui Generis Protection for Plant Varieties under TRIPs

As noted above, TRIPs authorizes WTO Members to eschew patent protection for plants and plant varieties and adopt instead an "effective *sui generis* system" of protection. The following discussion identifies the core requirements of such a system, laying the groundwork for Part III which outlines the permissible options available to WTO Members seeking to grant IPRs to plant varieties while simultaneously seeking to achieve other objectives.

Identifying and Evaluating Competing Proposals for Sui Generis Systems

TRIPs does not define the term *sui generis* as it appears in article 27.3(b), nor does the treaty's drafting history shed any light on its meaning. As a general matter, the term *sui generis* is understood to mean "of its own kind" or "unique," a definition which does little to identify what sort of unique legal systems are permitted under the treaty. This lack of definitional clarity has allowed advocacy groups and commentators to advance different and often competing proposals for how effective *sui generis* protection should be structured, producing significant confusion for governments seeking to understand and implement their obligations under the TRIPs Agreement.

The discussion below identifies and evaluates the competing proposals. In brief, it concludes that if WTO Members choose to protect plant varieties using a *sui generis* right as opposed to a patent, they are required to create a distinct IPR applicable to plant varieties that complies with the core requirements and objectives of the TRIPs Agreement.

Sui generis systems that do not contain these core elements or conflict with them are not TRIPs-compatible and expose the WTO Members who adopt them to challenges under the WTO dispute settlement system. Nevertheless, the mandatory requirements of article 27.3(b) preserve significant leeway for national governments to work out the precise manner in which they will balance protection of IPRs against other international obligations and national objectives.

Sui Generis Systems as Defined by the UPOV Acts

Several commentators assert that a state adopting national legislation in compliance with either the 1978 UPOV Act or the 1991 UPOV Act has satisfied its obligations under article 27.3(b). Those advocating for compliance with the

1991 Act stress the more extensive IPR protections it provides to plant breeders, while commentators endorsing the 1978 Act as the appropriate *sui generis* standard stress that only the earlier Act was in force at the time TRIPs was being negotiated.

In fact, however, the protection required by the two UPOV Acts is neither necessary nor sufficient for an "effective *sui generis* system." It is not necessary because article 27.3(b) does not require plant variety protection laws to contain the same subject matter, eligibility requirements, exclusive rights, terms of protection or other detailed provisions of either of the two UPOV Acts. And it is not sufficient because TRIPs requires WTO Members to structure their national IPR laws in ways that the two UPOV Acts do not.

Note, however, that developing counties that are parties to a "TRIPs plus" bilateral or regional trade or investment treaty may be required, by virtue of those treaties, either to join UPOV or to implement the standards set forth in one of the UPOV Acts. A country that has undertaken such obligations does *not* possess the same flexibility in designing a *sui generis* system to protect plant varieties as a country which has not undertaken such obligations.

For example, as explained in paragraphs 2.2.2.4 and 2.2.3.4 above, both the 1991 Act and the 1978 Act adopt the principle of national treatment, the same principle required by TRIPs. But under the UPOV Acts such treatment need be extended only to the nationals and residents of *other UPOV member states* and to legal persons having their headquarters in such states. Any WTO Member that limited national treatment in this way would unquestionably violate article 3 of TRIPs, which requires extension of national treatment to all other WTO Members whether or not they are also UPOV members.

However, to comply with article 3, a state whose plant variety protection law contains a UPOV-specific national treatment rule can easily amend its law to expand the number of states to which national treatment applies. Other UPOV-sanctioned rules may require more extensive changes. For example, the 1978 Act allows a state to impose a reciprocity requirement in certain circumstances, limiting the right to apply for protection of a new variety to nationals or residents of those member states which also apply the Act to the genus or species to which that variety belongs. This reciprocal protection is incompatible with TRIPs' national treatment rule, and possibly its most favoured nation principle as well, both of which are inconsistent with a system of reciprocity.

That TRIPs' drafters did not intend either UPOV Act to be the exclusive model for *sui generis* protection of plant varieties is confirmed by their failure to refer to the Acts anywhere in the Agreement. By contrast, where the drafters intended Members to comply with standards found in preexisting international

IPR treaties, they stated so expressly. Nevertheless, as explained in sections 3.3.1 and 3.3.2 below, most provisions of the two UPOV Acts are fully consistent with an effective *sui generis* system, and most countries have in fact adopted plant variety protection laws that are consistent with one or both UPOV Acts.

Core Requirements of an Effective Sui Generis System

According to a 1997 report prepared by two leading commentators, there are four core elements that any national plant variety protection law must contain in order to qualify as an "effective *sui generis* system" within the meaning of TRIPs article 27.3(b):(1) the law must apply to all plant varieties in all species and botanical genera; (2) it must grant plant breeders an IPR, i.e. the exclusive right to control particular acts with respect to those protected varieties, or at a minimum, the right to remuneration when third parties engage in certain acts; (3) it must provide national treatment and MFN treatment to breeders from other WTO Members; and (4) it must contain procedures that enable breeders to enforce the rights granted to them under such a law. The paragraphs that follow discuss each of these four requirements.

Protection of all plant varieties : Although both UPOV Acts require protection of only a specific number of plant varieties (paras. 2.2.1.1 and 2.2.3.1 above), a careful reading of the TRIPs Agreement reveals that WTO Members are required to protect all varieties. article 27.3(b) states that "Members shall provide for the protection of *plant varieties*" (emphasis added) without any further qualification or limitation. Inasmuch as article 27.3(b) is included in a provision of the TRIPs Agreement which carves out numerous and specific exceptions to patent protection, if the drafters had intended to allow states to limit the number or type of plant varieties subject to protection, they would have said so expressly.

Indeed, the drafters were aware that both UPOV Acts permitted precisely such limitations, yet they chose neither to refer to those Acts nor to their limited scope of protection. As explained in paragraph 3.4.2.1 below, however, the obligation to protect all varieties does not require states to provide *the same level of protection* to each and every variety.

Protection of plant varieties with an IPR : If protection granted to plant varieties must take the form of an IPR, WTO Members would enjoy virtually unbounded discretion to choose the manner in which to protect plant varieties, for example through a system of taxation on seeds. The WTO's Appellate Body recently interpreted the text of the TRIPs Agreement, however, to require protection through an IPR. Nevertheless, because some commentators and NGOs have questioned whether plant variety protection is in fact a form of intellectual property included in the TRIPs Agreement, a brief discussion of this point is needed.

Article 1(2) of TRIPs defines "intellectual property" for purposes of the Agreement as "all categories of intellectual property that are the subject of Section 1 through 7 of Part II" of TRIPs. Protection for plant varieties appears in Section 5 relating to patents, and is thus a form of intellectual property protected by the Agreement. This interpretation is confirmed by the United States - Section 211 Omnibus Appropriations Act of 1998, AB-2001-7, WT/DS176/AB/R (2 January 2002). In that case, a WTO panel concluded that trade names were not a category of intellectual property protected by TRIPs because they were not expressly referred to in article 1(2) (an omission that also applies to plant varieties). The Appellate Body reversed this finding. As an example, it referred to Members' option to protect plant varieties "by *sui generis* rights (such as breeder's rights) instead of through patents," and concluded that *sui generis* rights were in fact a form of intellectual property protected by the treaty. (Id., para. 335)

The significance of this conclusion is that IPRs for plant varieties must take the form of other IPRs protected by the TRIPs Agreement. Specifically, governments must either grant to the owners of protected varieties (1) the right to exclude all third parties from engaging in certain activities with respect to those varieties (an exclusive rights approach) or, (2) at a minimum, the right to receive equitable remuneration when a third party engages in such activities.

Nothing in TRIPs requires states to adopt the exclusive rights approach, nor does it indicate which exclusive rights must be granted to new varieties. Nevertheless, the vast majority of IPRs are protected by exclusive rights, with compulsory licences used only in limited areas such as new industries or distribution networks, abuses of monopoly powers or where high transaction costs preclude private licensing mechanisms from developing. In addition, a number of economic studies have suggested that granting exclusive rights to the owners of intellectual property products is a more efficient mechanism for creating incentives to innovate than a compulsory licence approach, in which political factors or government inertia are likely to produce skewed incentives. (Merges, 1996)

National and MFN treatment applicable to rights granted : At least one respected commentator has argued that national and MFN treatment are inapplicable to a *sui generis* system protecting plant varieties. The theory advanced is that these two forms of treatment are inapplicable to IPRs not covered by the TRIPs Agreement and that plant variety protection is an IPR that is not covered by TRIPs. However, the text of TRIPs article 3(1), as interpreted by WTO dispute settlement jurists, indicates that plant variety protection must adhere to such obligations.

Thus, in the United States - Section 211 Omnibus Appropriations Act of 1998, AB-2001-7, 2 January 2002, the WTO Appellate Body stressed the critical importance of the national and MFN treatment rules in multilateral trade agreements (paras. 242 and 297), and it concluded that the obligation to grant such treatment applies to all subjects of intellectual property protected by TRIPs, a designation that includes *sui generis* protection of plant varieties. As a result, in the area of plant variety protection, each Member must grant "no less favourable treatment" to the nationals of all other WTO Members than it grants to its own nationals and it must also grant to such foreign nationals "any advantage, favour, privilege or immunity" provided to any other country. (TRIPs, arts. 3.1 and 4) Applied to plant varieties, this would mean that any IPR that a state grants to its own plant breeders must be granted to breeders from all other WTO Members, and that any IPR granted to breeders from one WTO Member must be granted to breeders from all WTO Members.

Enforcement of rights by private parties : The only qualification that article 27.3(b) imposes upon *sui generis* systems for protecting plant varieties is that they be "effective." Although TRIPs does not define this term, it refers to it in Part III dealing with enforcement of other IPRs against acts of infringement by third parties. For this reason, a WTO dispute settlement panel is unlikely to find a *sui generis* system to be effective unless it provides a meaningful opportunity for private parties to enforce their rights in protected varieties.

Specific enforcement measures are not mandated by the Agreement, however. Had the drafters intended greater specificity, they could have easily made a reference in article 27.3(b) to the detailed enforcement rules applicable to other IPRs in Part III of TRIPs. Although a *sui generis* system must be "effective," this does not mean that a state must devote a particular level of resources to enforcement measures. In particular, article 41.5 of TRIPs makes clear that the treaty does not create "any obligation with respect to the distribution of resources as between enforcement of intellectual property rights and the enforcement of law in general." As a result, what qualifies as "effective" in a state with abundant resources and well-developed systems of law enforcement will not be required of states with more limited resources or weaker law enforcement systems.

Additional requirements of an "effective"* sui generis *system? : Beyond effective enforcement measures, it is uncertain whether a WTO dispute settlement panel would find that article 27.3(b) imposes additional requirements for a *sui generis* system to qualify as "effective." In general, international tribunals have concluded that rights granted in a treaty must be interpreted so as to make them effective rather than illusory. Thus, if a WTO Member were to grant very limited and weak rights to breeders together with broad and extensive exemptions in favour of farmers or other users of plant germplasm, its plant variety protection law might not survive WTO scrutiny. Such a possibility is likely to arise only in extreme cases in which a WTO Member has enacted a plant variety protection system that is functional in theory but non-functional in practice. States that implement the core TRIPs requirements in good faith - that is, states that grant breeders intellectual property rights and enforcement measures applicable to varieties in all species and botanical genera and that provide those same rights and measures to breeders from other WTO Members - are unlikely to have their laws challenged successfully.

Options Available to National Governments under Existing International IPR Agreements Protecting Plant Varieties and Plant Breeders' Rights

In order to identify the options available to national governments under international IPR agreements, this section begins by categorizing states according to the agreements to which they are parties. This classification scheme takes into account the fact that a state may be bound by more than one IPR-related treaty, a fact that may significantly limit its discretion. The section then identifies both the mandatory actions required of national governments in each category as well as the discretionary choices available to them. It also identifies trends in national laws which illustrate the manner in which states are complying with their particular set of international obligations.

Classification of States According to their International IPR Obligations

The discretion enjoyed by states to shape their plant variety protection laws to balance the protection of IPRs against other societal concerns is dependent upon the international agreement or agreements to which they are parties. The following table identifies that discretion at a broad level of generality. It does so by listing the mandatory components of plant variety protection that a state must adopt depending upon whether it is a member of (1) WTO only; (2) WTO and the 1991 UPOV Act; (3) WTO and the 1978 UPOV Act; and (4) either the 1991 UPOV Act or the 1978 UPOV Act only; or (5) no IPR agreement relating to the protection of plant varieties.

Note that some developing states, by virtue of having ratified a "TRIPs plus" bilateral or regional trade or investment treaty, may be required to adhere to some or all of the provisions contained in either the 1991 UPOV Act or the 1978 UPOV Act, *even if they are not members of UPOV itself*. Developing states that are parties to such "TRIPs plus" treaties should carefully consult the text of these agreements to determine the precise scope of their supplementary IPR obligations, including for purposes of applying the classification scheme set forth below. For example, a developing country that is a WTO Member but not a member of UPOV may nevertheless be required to adhere to all of the provisions of the 1991 UPOV Act by virtue of having ratified a "TRIPs plus" treaty. In such a case, the state would be obligated to implement the mandatory components of plant variety protection contained not only in TRIPs but also in the 1991 UPOV Act, and its discretion to balance the protection of IPRs against other competing policy objectives would be reduced accordingly.

Cumulative Treaty Obligations

Thus, for example, a state that is a party to a UPOV Act and also to the TRIPs Agreement must comply with its obligations under both treaties. Given that each treaty contains somewhat different but mostly cumulative and consistent rules concerning the protection of plant varieties, a state that is a party to both treaties will necessary have more extensive obligations to protect plant varieties and thus will enjoy concomitantly less discretion.

Conflicting Treaty Obligations

Where the provisions of two treaties are in direct conflict, the rule is far less settled. If two agreements relate to the same subject matter and the states parties to both agreements are the same, article 30 of the Vienna Convention on the Law of Treaties specifies that the agreement that is later in time is given effect. The obligations in the TRIPs Agreement are thus likely to prevail over any conflicting obligations in the 1978 UPOV Act, such as the ban on protecting varieties within the same genus or species with both a breeders' right and a patent. (1978 Act, art. 2(1)) Because the 1991 Act entered into force in 1998 and TRIPs entered into force in 1995 it may be argued that the 1991 UPOV Act is the later-in-time agreement. However, it is unlikely that any conflicts issue will arise between the two treaties, inasmuch as nothing in TRIPs requires conduct that the 1991 UPOV Act forbids. (Such a conflict might arise, however, in the unlikely event that TRIPs were interpreted to impose more restrictive exclusions to breeders' exclusive rights than the mandatory exclusions found in article 15 of the 1991 UPOV Act.) It can therefore be argued that the two treaty systems are fully compatible, with TRIPs merely augmenting the plant variety protection requirements of the UPOV Acts.

Identifying the Level of Discretion Available to States in each Treaty Classification

In general, the discretion a state enjoys in limiting the scope of intellectual property protection provided to plant varieties increases.

WTO and UPOV 1991 Act Members

States that are members of both TRIPs and the 1991 UPOV Act have the least discretion. This designation currently applies to 27 nations: Australia, Austria, Belarus, Bulgaria, Croatia, the Czech Republic, Denmark, Estonia,

Finland, Germany, Hungary, Israel, Japan, Kyrgyzstan, Latvia, Lithuania, the Netherlands, Moldova, Poland, Romania, Singapore, Slovenia, South Korea, Sweden, Tunisia, the United Kingdom and the United States of America.

Mandatory Requirements

These 27 states must extend protection to all plant varieties, comply with TRIPs' national MFN treatment obligations and adopt effective enforcement measures. The adoption of these rights (1) production or reproduction, (2) conditioning for the purpose of propagation, (3) offering for sale, (4) selling or marketing, (5) exporting, (6) importing and (7) stocking for any of these purposes - fully satisfies their obligation under TRIPs to protect varieties with an IPR. Finally, states that have agreed to these dual obligations must also comply with all of the other provisions of the 1991 Act, including its terms of protection and limitations on the breeders' exemption and farmers' privilege.

Options and Implementation Issues

A state whose plant variety protection laws are already in compliance with the 1991 Act needs to take only limited steps to comply with article 27.3(b) of TRIPs, including immediate protection of all varieties and extension of national treatment to all WTO Members. The most difficult hurdle is likely to enforce their rights. However, far more detailed enforcement measures must be adopted for the other forms of intellectual property protected by TRIPs. A state may therefore choose to extend to breeders the same domestic enforcement opportunities that it is required to provide to the owners of other IPRs.

National Implementation

In information submitted to the TRIPs Council in May 1999, the following states indicated that they had adopted plant variety protection laws modelled on the 1991 Act: Australia, Bulgaria, Members of the European Communities, Japan, Morocco, Poland, Romania, South Korea, South Africa, Switzerland and the United States of America. (Other states parties to both UPOV 1991 and TRIPs did not submit information to the Council.) Of the states adopting such laws, Morocco, South Africa and Switzerland had not ratified the 1991 Act.

Bilateral IPR Agreements and UPOV Membership

A separate issue concerns whether WTO Members will continue to join UPOV after the entry into force of the TRIPs Agreement. Some commentators predicted that UPOV membership would stagnate as states capitalized on the greater discretion provided to them under the *sui generis* option of article 27.3(b). Since TRIPs entered into force in 1995, however, 28 nations have become parties to the 1978 or 1991 UPOV Acts. (States Parties to the UPOV, Status on 30 June 2004) The continued expansion of UPOV membership can be traced in part to the recent proliferation of "TRIPs plus" treaties.

WTO and UPOV 1978 Act Members

This designation currently applies to 24 nations: Argentina, Bolivia, Brazil, Canada, Chile, China, Colombia, Ecuador, France, Ireland, Italy, Kenya, Mexico, New Zealand, Nicaragua, Norway, Panama, Paraguay, Portugal, Slovakia, South Africa, Switzerland, Trinidad and Tobago and Uruguay.

States in this category enjoy somewhat greater discretion as a result of the more limited protection of plant breeders' rights contained in this earlier UPOV Act.

Mandatory Requirements

States parties to both agreements must extend protection to all plant varieties, comply with TRIPs' national and MFN treatment rules and adopt effective enforcement measures.

They must also comply with all of the other 1978 Act requirements, including its eligibility requirements, terms of protection, exclusive rights and mandatory breeders' exemption. As compared to the 1991 Act, however, breeders' exclusive rights are more limited, terms of protection for varieties are shorter and exceptions and limitations are broader.

Options and Implementation Issues

States that become Members of the WTO after joining the 1978 UPOV Act and adopting laws to comply with that Act face a similar situation to states that are parties to both the WTO and the 1991 Act. To fully comply with TRIPs, these 1978 Act member states must modify their national laws to protect the four core requirements of article 27.3(b) and they must remove all provisions of their laws which impose a reciprocity requirement as a condition for protecting varieties of foreign breeders.

In addition, states in this category may choose to modify their laws to incorporate some or all of the standards found in the 1991 Act without actually becoming a member of that Act.

Their refusal to do so, however, does not violate article 27.3(b), inasmuch as the standards found in the 1978 Act satisfy their obligation to protect plant varieties with a *sui generis* IPR.

Trends in National Laws

In information submitted to the TRIPs Council in May 1999, Canada, the Czech Republic, Hungary, Norway and New Zealand (the only states submitting information) indicated that they had adopted plant variety protection laws that complied with the 1978 Act. At the time of these submissions, the Czech Republic and Hungary were members of the 1978 Act. They have since acceded to the 1991 Act.

1991 or 1978 UPOV Act Members Only

There are few states outside of the WTO that protect plant varieties solely by virtue of being UPOV members. And both are observer governments with the WTO. These nations face a somewhat different set of international legal obligations.

Mandatory Requirements

They must comply with all of the numerous requirements of the UPOV Act to which they are a party. However, they have no obligation to provide effective enforcement measures to breeders, their national treatment obligations are limited to those states that are members of the same UPOV Act and, in the case of those states bound by the 1978 Act, they need not protect all plant varieties and may impose certain reciprocity requirements on those varieties they do protect.

Options and Implementation Issues

Provided that their laws are in compliance with the relevant UPOV Act and unless they have ratified other international IPR agreements, these states need make no additional changes to their national plant variety protection laws.

Member of WTO Only

States that are parties to the TRIPs Agreement but are not members of either UPOV Act enjoy considerably greater discretion than those that are members of either UPOV Act alone. By virtue of their adherence to TRIPs, these states are required only to comply with the four core obligations of article 27.3(b). However, those countries that have ratified "TRIPs plus" treaties that require them to comply with one of the UPOV Acts).

Given the large number of states in this category (and the additional number of states likely to enter this category if their efforts to join the WTO succeed), a more detailed discussion of the numerous options available to them in balancing protection of plant breeders' rights against the competing policy objectives is provided in section 3.4 below.

Not a Member of WTO, UPOV or other IPR Agreements

Nations in this category have no obligation to protect plant varieties or breeders' rights in any form. Nevertheless, the mandatory requirements and discretionary choices facing other governments are likely to become matters of significant concern to these states as the world trading and intellectual property protection systems expand.

Attention to IPR issues will be of particular importance when these states implement the obligations of other international treaties to which they may be parties, such as the biodiversity goals of the CBD or the access requirements of the ITPGR. States in this category should consider whether it is sound policy to enact laws that are incompatible with TRIPs and the UPOV, since such an action will preclude their membership in these treaty systems unless they undertake a further revision of their national laws.

In addition, unilateral actions by or bilateral agreements with industrialized nations may pressure these states to comply with international IPR standards even if they remain outside of the WTO or UPOV.

For these reasons, states that are not yet Members of the WTO should give careful attention to the discretionary choices available to WTO Members analysed in the next section.

Bibliography

Ayala, F. J.: *The Genetic Structure of Populations*, W.H. Freeman & Co., San Francisco, California, 1977.

Beckner, M.: *The Biological Way of Thought*, University of California Press, Berkeley and Los Angeles, 1968.

Berg, Paul, and Singer, Maxine: *Dealing with Genes: The Language of Heredity. Mill Valley*, University Science Books, CA, 1992.

Brookes, Martin: *Get a Grip on Genetics*, *Time Life Books*, East Sussex, England, 1998.

Dando, M.: *Biological Warfare in the 21st Century: Biotechnology and the Proliferation of Biological Weapons*. London: Brassey's, 1994.

Darlington, C. D., and Mather, K.: *The Elements of Genetics*, London, Allen and Unwin, 1949.

Dawkins, R.: *The Selfish Gene*. Oxford: Oxford University Press, 1976.

DeVita, V.T.: *Cancer: Principles and Practice of Oncology*, Lippincott, Philadelphia, 2005.

Diaz, E.: *Microbial Biodegradation: Genomics and Molecular Biology,* Caister Academic Press, UK, 2008.

Fisher, R. A.: *The Genetical Theory of Natural Selection*, Clarendon Press, Oxford, 1930.

Fritz, A.: *International Classification of Diseases for Oncology*, World Health Organization, Geneva, 2000.

Gillis, Justin: *Drug Firms, Gene Labs to Map Genetic Code*, The Daily News, Longview, WA, 1999.

Glover, D. M., and Hames, B. D.: *Genes and Embryos*: New York, Oxford University Press, 1989.

Griffiths A.J.F.: *Introduction to Genetic Analysis*, W.H. Freeman and Company, New York, USA, 2005.

Hamerton, J. L.: *Human Cytogenetics*, Academic Press, New York, 1971.

Hartl D, Jones E: *Genetics: Analysis of Genes and Genomes,* Jones & Bartlett, 2005.

Hartl, Daniel L.: *Basic Genetics*, Jones and Bartlett Publishers, Boston, 1991.

Jonoska, N.: *Self-Assembling DNA Graphs, DNA-Based Computers VIII*, Springer-Verlag, Berlin, 2003.

Klotz, J. W.: *Genes, Genesis, and Evolution*: St. Louis, Mo., Concordia Publishing Co., 1970.

Lumsden C., and E. Wilson: *Genes, Mind and Culture: The Coevolutionary Process.* Cambridge, MA: Harvard University Press, 1981.

Mayr, E.: *Change of Genetic Environment and Evolution*, Allen and Unwin, London, 1954.

Nottingham, S.F.: *Eat Your Genes: How Genetically Modified Food is Entering Our Diet*, Zed Books, London, 1998.

Prakash, Chandra Prakash : *Cell Physiology*, Discovery, Delhi, 2010.

Robin Whyman: *Applied Organometallic Chemistry and Catalysis*, Oxford University Press, Delhi, 2005.

Roobeek, A: *The Biotechnology Revolution?,* Blackwell, Oxford, 1995.

Sturtevant, Alfred: *History of Genetics*, Harper and Row, New York, 1965.

Suzuki, D., and Knudtson, P.: *Genethics: The Clash Between the New Genetics and Human Values*, Harvard University Press, Cambridge, Mass, 1989.

Wilson, E. O., and Lumden, C.: *Genes, Mind, and Culture: The Evolutionary Process*, Harvard University Press, Cambridge, Mass, 1981.

Wilson, E. O., and Lumden, C.: *Genes, Mind, and Culture: The Evolutionary Process*: Cambridge, Mass., Harvard University Press, 1981.

Index

I

M

P

R

S

T

V

X

❑❑❑